THE PHYSICS OF ELECTRONIC AND ATOMIC COLLISIONS
XVIII INTERNATIONAL CONFERENCE

AIP CONFERENCE PROCEEDINGS 295

THE PHYSICS OF ELECTRONIC AND ATOMIC COLLISIONS

XVIII INTERNATIONAL CONFERENCE

AARHUS, DENMARK 1993

EDITORS:
TORKILD ANDERSEN
BENT FASTRUP
FINN FOLKMANN
HELGE KNUDSEN
UNIVERSITY OF AARHUS

N. ANDERSEN
UNIVERSITY OF COPENHAGEN

American Institute of Physics   New York

Authorization to photocopy items for internal or personal use, beyond the free copying permitted under the 1978 U.S. Copyright Law (see statement below), is granted by the American Institute of Physics for users registered with the Copyright Clearance Center (CCC) Transactional Reporting Service, provided that the base fee of $2.00 per copy is paid directly to CCC, 27 Congress St., Salem, MA 01970. For those organizations that have been granted a photocopy license by CCC, a separate system of payment has been arranged. The fee code for users of the Transactional Reporting Service is: 0094-243X/87 $2.00.

© 1993 American Institute of Physics.

Individual readers of this volume and nonprofit libraries, acting for them, are permitted to make fair use of the material in it, such as copying an article for use in teaching or research. Permission is granted to quote from this volume in scientific work with the customary acknowledgment of the source. To reprint a figure, table, or other excerpt requires the consent of one of the original authors and notification to AIP. Republication or systematic or multiple reproduction of any material in this volume is permitted only under license from AIP. Address inquiries to Series Editor, AIP Conference Proceedings, AIP, 500 Sunnyside Boulevard, Woodbury, NY 11797-2999.

L.C. Catalog Card No. 93-074103
ISBN 1-56396-290-X
DOE CONF-9309229

Printed in the United States of America.

CONTENTS

Preface xi
Sponsors and Exhibitors xii
ICPEAC Executive Committee 1991–93 xiii
ICPEAC General Committee 1991–93 xiv
XVIII ICPEAC Local Committee xvi

PLENARY

Atomic Physics Using Storage-Cooler Rings 3
    F. Bosch
Theory of Electron Collisions with Atoms,
Ions and Molecules 26
    P. G. Burke
Atmospheric Physics, Collision Physics
and Global Change 48
    K. P. Kirby

ATOMIC PHOTOIONIZATION

Threshold Double Photoionization
of Rare Gas Atoms 61
    R. I. Hall, L. Avaldi, G. Dawber, K. Ellis,
    G. C. King, A. G. McConkey, M. A. MacDonald,
    and M. Zubek
Angular Distribution and Spin Polarization
of Photoinduced Auger Electrons 73
    N. M. Kabachnik
Angle Resolved Photoelectron Spectroscopy
of Laser Excited and Aligned Atoms 83
    M. Pahler
Innershell Excitation of Atoms and Molecules
by Soft X-Rays 93
    K. Ueda

NEGATIVE IONS

Multiphoton Detachment of Negative Ions 105
    H. K. Haugen
Light-Induced Singly and Doubly Excited States
of the Negative Hydrogen Ion 115
    H. G. Muller and M. Gavrila

PHOTOMOLECULAR PROCESSES

Molecular Dynamics in Intense Laser Fields 127
    A. Giusti-Suzor, E. Charron, and F. H. Mies
Dissociation of Core Excited Polyatomic Molecules:
Sequential or Concerted Processes? 139
    P. Morin, M. Lavollée, M. Meyer, and M. Simon
Time Dependent Approach to Femtosecond Laser Chemistry 152
    R. Kosloff

RYDBERG ATOMS AND NEW TECHNOLOGIES

Radiative Collisions of Rydberg Atoms 173
    T. F. Gallagher, D. S. Thomson, and M. J. Renn
Collisions of Rydberg Atoms with Ions: A Review of
Recent Work on Charge Transfer
from Highly Excited Targets 183
    K. B. MacAdam
Rydberg and Continuum States of Atoms in External Fields 198
    F. Mota-Furtado, P. F. O'Mahony, and H. Marxer
X-UV Lasers Near 200Å 208
    P. Zeitoun, P. Jaeglé, A. Carillon, P. Dhez,
    P. Goedtkindt, G. Jamelot, A. Klisnick, and B. Rus

ELECTRON COLLISIONS

Fragmentation of Few-Body Coulomb Systems 221
    J. S. Briggs
Electron-Impact Ionization of Atoms, Molecules and Transient Species:
Current Statusand Perspectives for the Future 234
    V. Tarnovsky and K. Becker

ELECTRON-ATOM COLLISIONS

Recent Progress in Close-Coupling Electron-Atom Theory 251
    K. Bartschat
The Dirac R-Matrix Method for Scattering of
Slow Electrons from Alkali-Metal-Like Targets 263
    U. Thumm
Some New Developments in Polarized Electron
Science and Technology 276
    T. J. Gay, J. E. Furst, and W. M. K. P. Wijayaratna

Electron-Photon Correlation Studies Since Andersen *et al.* 286
    A. Crowe
Electron-Atom Collisions in Resonant Laser Fields 296
    W. R. MacGillivray and M. C. Standage
Electron Scattering from Laser-Excited Barium 306
    P. W. Zetner
L^2 Methods for Multichannel Continua 316
    F. Martín
Subshell and Innershell Excitation in Rare Gas
Atoms by Electron Impact 326
    T. Takayanagi

ELECTRON-MOLECULE COLLISIONS

Theory of Impulsive Rovibrational Excitation
of Molecules by Electrons 339
    H. J. Korsch
Molecular (e, 2e) Collisions—Present Status and Future Prospects
for Electron Momentum Spectroscopy 350
    C. E. Brion
Recent Theoretical Results on Electron-Polyatomic
Molecule Collisions 360
    C. W. McCurdy
Rydberg Atoms: A Nano-Laboratory for Collision Processes 371
    K. A. Smith and F. B. Dunning
Formation and Decay of Transient Anions Produced
by Electron Impact on Surface Molecules 381
    L. Sanche
Threshold and Resonant Features in the Vibrational
Excitation Functions of Hydrogen Halides 390
    S. Cvejanović

ELECTRON-ION COLLISIONS

Electron-Impact Excitation of Ions: Experimental Status
and Perspectives 405
    R. A. Phaneuf
Electron Scattering from Atomic Ions 415
    M. S. Pindzola, T. W. Gorczyca, D. C. Griffin, and N. R. Badnell
Experimental Studies of Dielectronic Recombination 432
    L. H. Andersen
Basic Problems in Dissociative Recombination
of Diatomic Molecules 442
    H. Takagi

POSITRON COLLISIONS AND EXOTIC SYSTEMS

Progress in Positron Scattering from Atoms and Molecules 455
    M. Charlton and G. Laricchia
Recent Development in the Theory of
Positron-Hydrogen Collisions 466
    S. J. Ward
The Theory of Muon Catalysed Fusion 478
    E. A. G. Armour, M. R. Harston, and M. P. Faifman
Electron-Positron Pair Production in Coulomb Collisions
at Ultrarelativistic Energies.. 491
    C. R. Vane, S. Datz, P. F. Dittner, H. F. Krause,
    C. Bottcher, M. Strayer, R. Schuch, H. Gao,
    and R. Hutton

HEAVY PARTICLE COLLISIONS

Collision Dynamics of Oriented States 505
    N. Andersen
Recent Advances and Challenges of
Heavy-Particle Collision Theories 520
    R. E. Olson, J. Wang, and J. Ullrich

COLLISIONS WITH MULTIPLY CHARGED IONS

Mechanisms for Radiative Decay of Multiply Excited
States Populated in Collision of Highly
Charged Ions with Rare Gases 537
    P. Roncin, M. N. Gaboriaud, Z. Szilagyi, and M. Barat
Theoretical Lifetime and Decay Modeof Multiply
Excited Ions Produced DuringIon-Atom Collisions 547
    H. Bachau
Excitation and Ionization Characteristics of Multicharged Ions 558
    V. P. Shevelko
Charge Changing Processes in Ion-Ion Collisions 574
    F. Melchert
One and Two Electron Processes in Collisions
of Highly Charged Ions with He
at Velocities Around 1 a.u. 585
    J. P. Giese, W. Wu, I. Ben-Itzhak, C. L. Cocke,
    R. Ali, P. Richard, M. Stöckli, and H. Schöne
Radiative Processes in Atomic Collisions at Storage Rings 595
    L. R. Andersson and J. Burgdörfer

LOW ENERGY HEAVY PARTICLE COLLISIONS

Guided Ion Beams, RF Ion Traps, and Merged Beams:
State Specific Ion-Molecule Reactions at meV Energies.................... 607
    D. Gerlich
Low-Energy Collisional Autoionization Processes 623
    B. Brunetti, S. Falcinelli, and F. Vecchiocattivi
Energy Sharing in Thermal Energy Collision Spectroscopy 635
    M. Allegrini and S. Milosevic
Associative and Penning Ionization in
$H^* + H^*$, $Na^* + Na^*$, and $K^* + K^*$ Collisions............................. 645
    X. Urbain
Reactive Scattering and Electron Detachment in
$H^- + H_2/D_2$ Collisions at Low eV Energies.............................. 654
    M. Zimmer and F. Linder
Photoabsorption Studies of Quasimolecules in
Thermal Collisions of Alkaline-Earth Atoms 665
    Y. Sato
Orbital Alignment and Vector Correlations in
Inelastic Atomic Collisions 675
    E. M. Spain, C. J. Smith, M. J. Dalberth, and S. R. Leone
Electron Capture by Low-Energy Proton Impact
on Atmospheric Gases 684
    B. Van Zyl

CLUSTERS

Collision Experiments with Fullerenes..................................... 697
    E. E. B. Campbell, R. Ehlich, M. Westerburg, and I. V. Hertel
Energetic Fullerene Ion-Atom Collisions.
Fragmentation, Charge Exchange, and Lifetimes 709
    P. Hvelplund
Diffraction and Vibrational Dynamics of
Large Clusters from He-Atom Scattering 719
    U. Buck

SURFACES

Hollow Atoms Below, Above, and at Surface 731
    J. P. Briand
Electron Spectroscopic Studies of the Neutralization of Slow Multicharged
Ions During Interactions with a Metal Surface 741
    F. W. Meyer, L. Folkerts, C. C. Havener,
    I. G. Hughes, S. H. Overbury, D. M. Zehner,

and P. A. Zeijlmans van Emmichoven
Ion-Surface Collisions at Grazing Incidence:
Atomic Collisions Under Special Conditions................................ 751
H. Winter
K-Auger Emission During Highly Charged
Ion-Surface Interaction 766
J. Das, H. Limburg, and R. Morgenstern

HOT TOPICS

Experimental Visualization of Photoelectron Angular Distributions........... 779
H. Helm, N. Bjerre, M. J. Dyer, M. Saeed, and D. L. Huestis
Classical-Quantum Correspondence for Ionization
of Atoms by Ion Impact and Electromagnetic Pulses 786
C. O. Reinhold and J. Burgdörfer
Radiative Stabilization in Double-Electron Capture 794
N. Vaeck, H. W. van der Hart, and J. E. Hansen
Dissociative Recombination of Stored and
Phase-Spaced Cooled Molecular Ions in Cryring.......................... 803
M. Larsson, G. Sundström, M. Carlson, H. Danared,
A. Källberg, K. G. Rensfelt, M. af Ugglas, L. Broström,
S. Mannervik, P. Sigray, A. Filevich, S. Datz, and J. R. Mowat
Absolute Cross Sections for Dissociation
of H_2O by Electron Impact... 811
M. Darrach and J. W. McConkey
Elastic Electron Ion Scattering... 820
B. A. Huber, C. Ristori, C. Guet, D. Jalabert,
M. Maurel, and J. C. Rocco
Electron Capture from Circular Rydberg Atoms 828
S. B. Hansen, T. Ehrenreich, E. Horsdal-Pedersen,
and K. B. MacAdam
Photoabsorption and Compton Scattering in
Ionization of Helium at High Photon Energies 836
L. R. Andersson and J. Burgdörfer
Measurement of Electron Capture from $e^+ - e^-$ Pair Production by
0.956 GeV/u U^{92+} on Au, Ag, Cu
and Mylar Targets 844
A. Belkacem, H. Gould, B. Feinberg, R. Bossingham,
and W. E. Meyerhof
Experimental and Theoretical Investigation of the Autoionization
Dynamics in the Excited Collision Complex $He^*(2\ ^3S) + H(1\ ^2S)$.............. 852
A. Merz, M.-W. Ruf, H. Hotop, M. Movre,
and W. Meyer
Author Index... 861

PREFACE

The Eighteenth International Conference on the Physics of Electronic and Atomic Collisions was held at the Concert Hall in Aarhus, Denmark, from July 21 to 27, 1993. The local organizing committee was from the Institute of Physics and Astronomy at Aarhus University.

The conference was attended by 717 delegates from 42 countries. Of these, 180 were students who had not yet completed their PhD thesis. In addition, 190 accompanying persons participated in the social programme. There were 866 contributed papers submitted to the conference. The 845 abstracts received before June 1, 1993 were printed in The Book of Abstracts. Over 90% of the papers were presented at the poster sessions in The Riding Hall where the commercial exhibitions were located.

This book contains written versions of the invited talks presented at the XVIII ICPEAC by 76 invited speakers, of which, however, three did not submit a manuscript. The talks included plenary lectures, review talks, progress reports, and ten Hot Topic papers selected among the abstracts by the International ICPEAC committee. The opening keynote lecture was given by Ben Mottelson from NORDITA, Copenhagen on "Shell Effects in Clusters", and a special lecture of public interest was presented Thursday evening by Willi Dansgaard, University of Copenhagen on "Ice and Climate."

The book of "Abstract of Contributed Papers for XVIII ICPEAC" contains 845 one-page abstracts, a Subject Index, and an alphabetic Author Index of 1884 co-authors. It is published in two A4-sized volumes (ISBN 87-8754834-8) available from the Institute of Physics and Astronomy, Aarhus University, Denmark at a price of 300 Danish Kroner (DKK).

We are grateful for financial support from the sponsors listed on the following page which made it possible to support participation of 250 delegates. Special emphasis was put on supporting young scientists and people from FSU. We appreciate all contributions from the exhibitors and the competent assistance of the Aarhus Convention Bureau and the Concert Hall in Aarhus.

Torkild Andersen
Bent Fastrup
Finn Folkmann
Helge Knudsen
N. Andersen

September 20, 1993

Sponsors

The Danish Science Research Council
The Danish Accelerator Committee
The Commission of European Communities
The International Union of Pure and Applied Physics
The International Science Foundation, New York
Thomas B. Thriges Fond
Otto Mønsteds Fond
Knud Højgaards Fond
Carlsbergfondet
Jysk Telefon A/S
Unibank
Scandinavian Airlines System (SAS)
Danfysik
Scandnordax AB
Caburn, MDC

Exhibitors

BBT Instrumenter ApS
Hewlett-Packard A/S
IOP Publishing
Nordiska Balzers AB
Optilas
Physica Scripta
Danfysik
Stakbogladen, Aarhus

International Conference on the Physics of Electronic and Atomic Collisions

Organization 1991–93

Executive Committee

Chairman
Sheldon Datz, USA

Vice-Chairman
Francois J. Wuilleumier, France

Secretary
J. Norman Bardsley, USA

Treasurer
Reinhard Morgenstern, The Netherlands

Members
T. Andersen, Denmark
A. Chetioui, France
L. J. Dubé, Canada
G. H. Dunn, Canada
B. Fastrup, Denmark
H. Hotop, Germany
F. A. Gianturco, Italy
J. H. Macek, USA
I. E. McCarthy, Australia
J. W. McConkey, Canada
J. B. A. Mitchell, Canada
M. C. Standage, Australia
G. Stefani, Italy
F. Torello, Italy
L. Roos, adm.secr., The Netherlands

General Committee

Argentina
R. D. Rivarola

Australia
M. C. Standage
I. E. McCarthy
J. F. Williams

Austria
O. Benka

Belgium
P. DeFrance

Brazil
M. A. P. Lima

Canada
L. J. Dubé
J. W. McConkey
J. B. A. Mitchell

Denmark
T. Andersen
B. Fastrup

Finland
T. Åberg

France
A. Chetioui
J. P. Gauyacq
J. Pascale
J. Remillieux
F. J. Wuilleumier

Germany
K. Bergmann
W. Domcke
B. Fricke
V. Heinzmann
H. Hotop
H. Loesch

India
D. P. Dewangan
S. Sathyamurty

Israel
M. Shapiro

Italy
F. A. Gianturco
G. Stefani
F. Torello

Japan
H. Nakamura
S. Othani

Latvia
E. G. Karule

Malaysia
W. C. Fon

Mexico
C. Cisneros

Romania
V. Zoran

SNG
L. A. Shmaenok
B. M. Smirnov

Spain
G. Delgado-Barrio

Sweden
A. Bárány

The Netherlands
R. Morgenstern
W. J. van der Zande

United Kingdom
A. Crowe
A. E. Kingston
A. C. H. Smith

USA
J. N. Bardsley
K. Bartschat
A. Chutjian
S. Datz
G. H. Dunn
P. S. Julienne
K. Kirby
P. M. Koch
C. D. Lin
J. H. Macek
A. S. Schlachter
A. F. Starace

XVIII ICPEAC Local Committee

Chairmen
T. Andersen
B. Fastrup

Secretary
H. Knudsen

Treasurer
F. Folkmann

Institute of Physics and Astronomy
Aarhus University, DK-8000 Aarhus C, Denmark

PLENARY

ATOMIC PHYSICS USING STORAGE-COOLER RINGS

Fritz Bosch
Gesellschaft für Schwerionenforschung, GSI, D-64220 Darmstadt, Germany

ABSTRACT

With the advent of the new generation of heavy ion storage-cooler rings brillant monochromatic beams of highly-charged ions are now available, allowing for new kinds of precision experiments in atomic physics. The basic principles of electron- and laser-cooler which make these rings powerful tools are outlined. Selected experiments conducted at cooler rings with a significant atomic physics program are briefly discussed. This overview covers a wealth of recent experiments, reaching from radiative and dielectronic recombination up to laser-stimulated electron capture, X-ray spectroscopy, test of relativity and atomic astrophysics.

INTRODUCTION

Already after a few years of operation, heavy ion storage-cooler rings present overwhelmingly rich and exciting results, in particular in the field of atomic physics. For extended periods of time brillant beams are available there with highly compressed phase-space density, i.e. with small size, divergence and with small momentum spread. These impressive features reflect the success of electron- and laser-cooling of fast, highly-charged ion beams. Multipass interaction of those bright and monochromatic beams with collinear electrons and/or photons opens the door to a wealth of hitherto not accessible experiments in the realm of fundamental physics, spectroscopy, and of ion-electron-photon interaction. Moreover, at heavy ion cooler rings for the first time the dependence of *nuclear* decays on the *atomic* charge state might be addressed — a topic of considerable impact for the nucleosynthesis in hot stellar plasmas.

Besides on cooling, the outstanding performance of the new rings is also based on very recent progress made in accelerator- and ion source-technology. It was not self-evident, however, that such a lot of different, sophisticated techniques could be combined ensuring a steady operation of the rings within an astonishingly short time. By now, odds on a bright future for storage-cooler rings would *not* bring much money back.

STORAGE-COOLER RINGS IN USE FOR ATOMIC PHYSICS

When S. van der Meer received the Nobel Prize for Physics in 1984 together with C. Rubbia, the physics community took for the first time notice of such exotic concepts as "beam cooling" or "stochastic cooling". It was very soon recognized, however, that stochastic cooling of antiprotons was the key for conducting the historical experiments in which the long-searched W vector boson was detected, the mediator of electroweak interaction. It became also obvious that these beam-cooling techniques would open a new horizon, in particular for atomic and nuclear physics.

Outside high-energy physics the "oldest" operational ion cooler ring is the Low Energy Antiproton Ring LEAR of CERN in Geneva [1]. There, electron cooling of antiprotons has been pioneered. Because this overview focuses on atomic physics at *heavy-ion* rings, the beautiful results of LEAR concerning antiproton physics must remain outside of our scope.

Among the "new" generation of ion-cooler rings about 50 % have major activities in atomic physics. These are, in alphabetical order: ASTRID (Aarhus Storage Ring Denmark) of the University of Aarhus, in operation since 1990 [2]; CRYRING (Cryogenic Ion Source Ring) of the Manne Siegbahn Institute of Physics in Stockholm, inaugurated in September, 1992 [3]; ESR (Experimental Storage Ring) at the Gesellschaft für Schwerionenforschung, Darmstadt, running since 1990 [4]; TARN II (Test Accumulation Ring for the Numatron Accelerator Facility) at the University of Tokyo which started in 1989 [5]; and, finally, TSR (Test Storage Ring) at the Max-Planck-Institut for nuclear physics in Heidelberg [6] which first succeeded in electron cooling of heavy ions (1988). Important contributions to atomic physics were also made at the cooler ring of the Indiana University Cyclotron Facility (IUCF) in Bloomington (1988), although this ring is mainly designed for nuclear physics experiments [7]. Here, the electron cooler was used in single-pass merged beam experiments on recombination of He^+ ions with electrons. Both ion-storage and ion-cooling were not employed.

Other rings as CELSIUS [8] in Uppsala or COSY [9] in Jülich focus on nuclear physics and presently do not have a significant program in atomic physics.

These five rings showing broad activities in atomic physics have some important things in common — besides obvious differences concerning energy-range or charge-state of the stored ions: They are all connected with existing facilities, in most cases as the final stage after linear or synchrotron-type accelerators; in one case (TARN II) the cooler ring serves as a pre-accumulation stage. All of them employ electron-cooling (at ASTRID the electron cooler was not installed in the first years of operation); all rings achieve ultra-high vacuum (typical residual gas pressure $\sim 10^{-10}$ to 10^{-11} mbar) which both ensure a long lifetime of the stored ions and a low background. Finally, at all rings similar beam-accumulation techniques have been developed in order to fill the total available phase-space with ions.

Those methods are e.g. multiturn injection (sweeping the beam energy during injection) or sophisticated combinations of RF-stacking and electron cooling. By the latter technique any new ion pulse is injected at an energy slightly above that of the ion beam already stored, and then ramped-down in a very short time to the stack energy. Because the circulating cooled beam fills a very small phase space (its phase-space *density* becomes highly compressed by cooling), many hundreds or even thousands of ion pulses can be accumulated in this way until the limits of space charge or ring acceptance are reached. Typical numbers of stored and cooled ions reach from 10^6 to 10^{10}.

The high intensity of stored beams is one result of electron cooling. Even more important is that electron cooling provides beams the quality and brightness of which are unique: Their longitudinal relative momentum spread $(\Delta p/p)_L$ is reduced to an order of 10^{-5}, i.e. the beams are monochromatic. Their emittance (i.e., the product of beam size and divergence) is reduced to less than 0.1 π mm mrad (typical emittances of tandem accelerators are 3 π mm mrad), i.e. the beams are small-sized and brillant. And, finally, the energy as well as the shape of the stored beams are kept constant because energy loss and small-angle scattering in reactions with the residual gas are compensated immediately in the electron cooler. Thus, the high quality of cooled and stored beams together with the typical large atomic cross sections ($\sim 10^{-18}$ to $\sim 10^{-21}$ cm^2) make cooler rings an ideal tool for precision experiments in the realm of atomic physics.

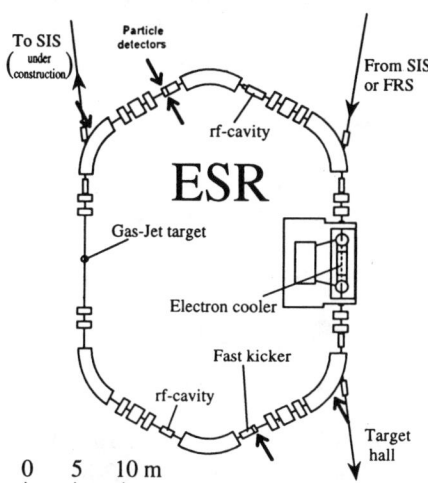

Figure 1: Sketch of the Experimental Storage Ring ESR at Darmstadt [15]. The position of detectors recording charge-changing processes is indicated by arrows.

Except the ESR, all rings are low-energy or medium-energy devices (in the order of a few MeV/u at most); they store, therefore, preferentially ions which are only partially dressed (except very light ones with $Z \leq 6$). Those ions have lifetimes in the order of seconds or minutes at most, whereas the lifetime of heavy bare or hydrogen-like ions reaches many hours — even bare uranium lives for ~ 1 hour in the ESR ! On the other hand, partially ionized atoms or molecules are especially well suited for laser-cooling and for the investigation of molecular dissociation. It is not surprising, therefore, that laser cooling of fast moving ions could be achieved first at the low-energy ring TSR [10] and, shortly later, at ASTRID [11].

Similarily, almost at the same time the race started at TSR, TARN II and CRYRING to study one "hot topic" of this conference, namely dissociative recombination of light, singly charged molecules [12–14].

The ESR cooler ring is, in some respect, a special case (see Fig. 1). It has been designed for storing and cooling all ions up to bare uranium at energies from ~ 500 MeV/u down to a few MeV/u [15]. For this reason RF cavities are installed (as at TSR or CRYRING) to accelerate or decelerate stored ion buckets. Furthermore, the ESR can be filled either with stable ions delivered by a synchrotron or with unstable ions provided by a fragment separator. For the nuclear physics and astrophysics program connected with the latter ion species an internal gas jet was installed which can serve, however, also for the purpose of atomic physics.

The five rings will have also in future some fields of interest in common: Laser interaction and dielectronic recombination studies will continue at almost all rings. Efforts to improve laser cooling will certainly not be weakened at TSR, ASTRID, and CRYRING. Studies of molecular dissociation will become most probably a "hot topic" at all five rings, whereas X-ray spectroscopy and "atomic astrophysics" will remain, in the foreseeable future, a special domain of the ESR.

BEAM-COOLING METHODS

Three beam-cooling methods have been successfully applied in the past: Stochastic-, electron-, and laser-cooling. Stochastic cooling which was invented to cool "very hot" beams like antiprotons or instable fragments, is installed at none of the five rings (its installation is scheduled for 1994 at the ESR) and will be, therefore, not discussed in our context.

Electron Cooling

The basic idea of electron cooling [16] was invented by G.I. Budker in 1966 [17] and developed lateron to a powerful tool for cooling of protons at the NAP-M ring in Novosibirsk. Electron cooling is the fast compressing of the phase-space density of "hot" ions by their repeated interaction (typically 10^6 times per second !) with "cold" electrons at matched velocity which move parallel to the ions over a certain distance (typically a few m). Coulomb collisions between electrons and ions lead to a steady momentum exchange and, finally, to a progressive assimilation of the properties of the two beams until thermal equilibrium is reached. Then the ion temperature T_i becomes equal to the electron temperature T_e which, in the transverse direction, is given by the cathode temperature of ~ 1200 K ($\hat{=}$ 0.1 eV). In the longitudinal direction it is significantly smaller due to the strong acceleration of the electrons. While the ions circulate continuously in the ring, getting colder and colder by the many times repeated interaction, the electrons are removed after one interaction cycle and replaced by cold ones from the gun. In this way, cooled down times τ_{cool} of ~ 1 s for protons can be achieved at electron currents of $\sim$ 1A and of $\sim$ 40 ms for uranium-ions, because τ_{cool} scales with A/Z^2, where Z and A are the nuclear charge and the atomic mass, respectively.

Electron cooling of protons has been demonstrated already in the seventies [18]. It was, however, by no means clear whether electron cooling could "work" also for heavy, highly-charged ions. There were mainly two reasons for these doubts: First, is electron cooling strong enough to compensate the heating due to intra-beam scattering (α Z^2) of the ions ? Even more uncertain was whether electron capture in the cooler would scale proportional to Z^2 as calculated; if this recombination would be much stronger (due to enhanced collective effects, for instance), the lifetime achievable for stored heavy ions would be severely restricted.

In 1988, electron cooling of ions heavier than protons could be achieved for the first time at the TSR in Heidelberg [19]. There, electron cooling of C^{6+} and O^{8+} ions was demonstrated with striking results. Most exciting was, however, that the lifetimes observed for those cooled beams were within the calculated values. Meanwhile, electron cooling has been successfully applied at many rings and for a plenty of heavy ion species even including bare uranium (ESR in 1991).

The initial relative longitudinal momentum spread $(\Delta p/p)_L$ of the stored ions of typically 10^{-3} could be reduced to about 10^{-5}. Transversal temperatures $T_\perp$ nearby the expected value of 0.1 eV were observed. Emittances as small as 0.05 π mm mrad after cooling were achieved [15].

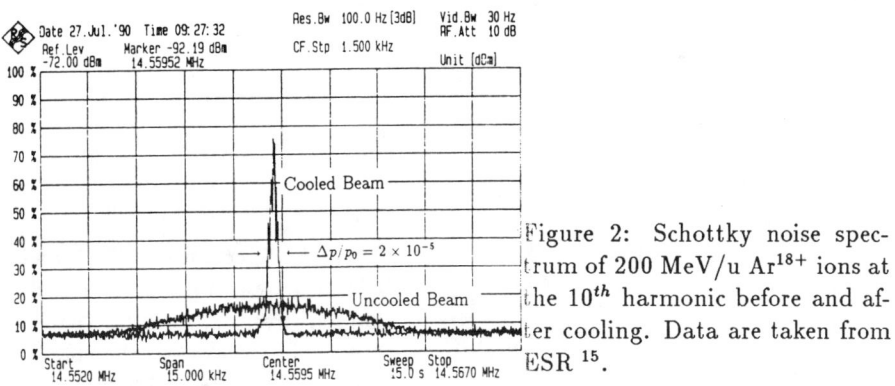

Figure 2: Schottky noise spectrum of 200 MeV/u Ar^{18+} ions at the 10th harmonic before and after cooling. Data are taken from ESR [15].

The most widely used diagnostics for cooled beams is the Schottky-noise spectrum of the circulating ions. It is basically the Fourier-transform in frequency space of the ion current fluctuations induced in pick-up plates. This noise shows a peak at any harmonic of the revolution frequency of the stored ions. The smaller their momentum spread is, the smaller is also the spread in frequency. Figure 2 shows the longitudinal Schottky noise intensity for $4 \cdot 10^6$ stored Ar^{18+} ions of

190 MeV/u before and after electron cooling was applied (data are taken from the ESR). The initial spread of the revolution frequency of 10^{-3} (broad distribution) was reduced within parts of a second to 1×10^{-5} by electron cooling with an electron current of 1 A. This final frequency spread corresponds to a longitudinal momentum spread $(\Delta p/p)_L$ of 2×10^{-5}.

Size and transverse temperature $T_\perp$ of cooled beams can be derived from the rate and profile of ions which are downcharged in the cooler. Because those ions have a changed magnetic rigidity with respect to the primary ones, they move on different orbits in the ring and can be recorded by (position-sensitive) detectors moved into the ring acceptance. Change of the atomic charge state q is in most atomic physics experiments a very specific signal for the physical process being investigated. Therefore, those detectors are installed in *all* cooler rings which focus on atomic physics. They are, for instance, indispensible in experiments addressing radiative (RR) or dielectronic recombination (DR), laser-induced recombination, molecular dissociation, X-ray spectroscopy and so on. Figure 3 shows a profile of 250 MeV/u U^{89+}-ions downcharged from primary U^{90+}-ions by recombination in the cooler section of the ESR. The beam size amounts, after unfolding the detector resolution, to 1.2 mm. From this width an emittance of 0.05 π mm mrad can be deduced.

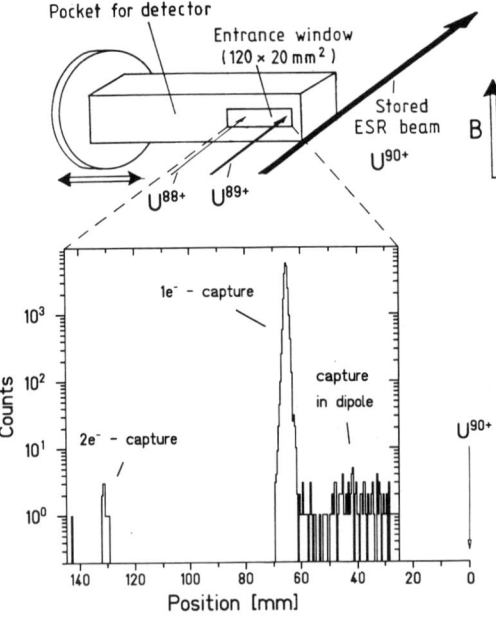

Figure 3: Spectrum of U^{89+} and U^{88+} ($2e^-$-capture) ions downcharged from stored U^{90+} ions be electron capture in the cooler. Data are taken from ESR.

Ions and electrons interacting in the electron cooler represent the exotic case of a "cold" plasma. Although the ions are in most cases highly-charged, the electrons — seen in the ion rest frame — are rather cold ($T_e \sim 0.1$ eV). Since the cooler electrons are kept together by a strong magnetic field of typically 0.1 Tesla in order to avoid blow-up, they exhibit in addition to their random motion circular orbits around the magnetic field lines at a cyclotron frequency of $\omega_c = 176\ \gamma^{-1}$ GHz $\cdot$ B[T], where γ is the Lorentz-factor of the electrons (and ions). Furthermore, the electrons will group themselves around the positive ionic charge shielding its Coulomb field. As a consequence, the electric field strength decreases exponentially over a typical distance, the Debye length, λ_D, which is related to the temperature T, the electron density n, and the ionic charge q by

$$\lambda_D^2 = \frac{kT}{n}(q\cdot e)^{-2}, \quad or \; \lambda_D^2 = kT/(m\cdot \omega_p^2) \tag{1}$$

where ω_p is the corresponding plasma frequency. Typical values for n are 10^6 to 10^8 cm^{-3} and, for λ_D, 10^2 to 10^3 μm.

The temperature of this cold plasma as well as the distribution of free or quasi-free electron states can be sensitively investigated by experiments on dielectronic recombination and/or laser-induced recombination.

Laser Cooling

The technique of resonance absorption of well-directed photons has been uitilized with striking success in the framework of ion-trapping [20]. Temperatures at the "Doppler limit" or even less could be achieved benefitting from intriguing effects of quantum-coupled microscopic systems. The same technique may also serve for reducing the momentum spread of fast-moving charged ions in a storage ring.

As for the case of electron cooling, the basic principle of laser cooling is extremely simple: Suppose there are moderately charged, circulating ions (like Li$^+$ or Be$^+$) which have a strong optical E1 transition. Their longitudinal momentum spread is, at first, rather broad. Now, the ion beam is in a straight section of the ring irradiated with co- and counterpropagating laser light. Ions the velocity of which corresponds to the resonance condition (i.e. in their rest frame the Doppler-shifted laser wavelength corresponds to the transition into the excited state) strongly absorb the laser light. In any absorption process, the photon momentum $\hbar k$ is transferred to the ion at 0^0 or 180^0 with respect to their motion, depending on whether the co- or counterpropagating light fulfills the resonance condition. The following spontaneous re-emission transfers, in the average, no momentum because it is isotropic in the ion rest frame.

Therefore, each absorption-reemission cycle transfers a net momentum $\hbar k$ parallel or antiparallel to the beam direction. Those ions which had absorbed a photon will be removed from their velocity class ("hole burning") and transferred to another class corresponding to the transferred momentum $\pm\hbar k$. By detuning both the co- and counterpropagating lasers in such a way that the wavelength of the copropagating laser gets smaller and that of the counterpropagating laser gets larger, the velocity distribution of the ions will be strongly compressed until both lasers show the same wavelength in the ion rest frame. In praxi, only one laser will be tuned, whereas the other one stays with fixed frequency [10,11].

Laser cooling of stored, fast moving ions was first demonstrated for Li^7 ions at the TSR [10] and, lateron, at ASTRID [11] (cf. Figs. 4,5). Since many absorption-reemission cycles (10^4 to 10^5) are needed to narrow the typical ion velocity distribution in a storage ring, optical pumping must be avoided and cyclic transitions have to be selected. It turns out that special ion species like Li$^+$, Be$^+$, or Mg$^+$ are most suited for laser cooling. Up to now, longitudinal beam temperatures down to a few mK could be achieved which corresponds to momentum spreads in the order of 10^{-7}.

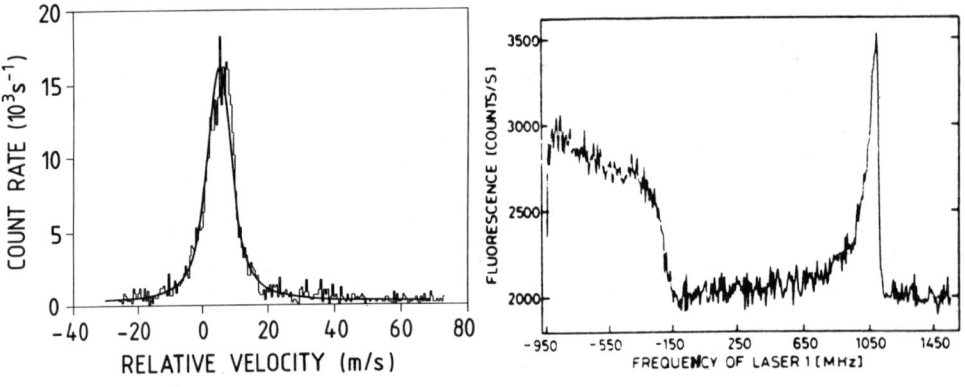

Figure 4: Longitudinal velocity distribution of laser cooled 7.3 MeV ^{9}Be$^+$ in the TSR [39].

Figure 5: Fluorescence yield vs frequency for laser cooled ^{7}Li$^+$ ions in the ASTRID ring [11].

Comparing laser and electron cooling it is obvious that, employing laser cooling, by far lower temperatures might be reached. However, laser cooling acts on the *longitudinal* velocity component only, heating at once the transverse velocity due to the isotropic re-emission. In contrast, electron cooling which works by momentum exchange (in *all* directions) in Coulomb collisions couples both, longitudinal and transverse temperature.

There are strong activities at TSR, ASTRID, and CRYRING to invent "better" laser cooling methods by incorporating also induced emission and by employing cooling techniques that act on both the longitudinal *and* transverse degrees of freedom. The goal of these efforts is to achieve one day the very end of cooling — namely "crystalline" ions circulating with "frozen" relative distances as a new kind of macroscopic, travelling Coulomb crystals [21].

SELECTED ATOMIC PHYSICS EXPERIMENTS

Lifetime Measurements

Like ion traps, storage rings are capable to trap charged particles for extended periods of time. This basic property allows for observing phenomena of relaxation and decay. In particular, the lifetime of a small (metastable) fraction of the stored ions can be measured rather easily.

Lifetimes (in the following lifetime is always understood as the 1/e mean lifetime) in a storage-cooler ring cover a broad time interval, from parts of a millisecond to many hours, depending on energy E, nuclear charge Z, and charge q of the stored ions on the one hand and on residual gas pressure and electron cooler current on the other hand. If, in addition, an internal gas jet is turned on also the e^--capture and e^--stripping losses in this target influence the lifetime.

First, we consider the operational mode where only the electron cooler is turned on. The main processes of ion loss (we consider an ion as "lost" when either its charge state q is changed or when it is scattered out of the ring acceptance) are in this case

1. Radiative recombination (RR) of electrons in the cooler

2. e^--capture or e^--loss in the residual gas

3. intrabeam scattering

Since all storage rings are baked out to guarantee the necessary ultra-high vacuum ($\sim 10^{-11}$ mbar), the residual gas consists of $\sim$ 90 % H_2 and of $\sim$ 10 % (C, N, O)-molecules, whereas the amount of heavy atoms (like argon) is negligible in most cases.

The observed lifetimes for partially dressed ions is in accordance with calculations based on the classical theory of N. Bohr for ionization on the one hand and on the Schlachter model for e^- capture on the other hand. A very important feature of cooler rings is that small-angle Coulomb scattering — either intra-beam or with residual gas atoms — will be counterbalanced by the cooling force of the electron cooler.

The dominant process which restricts the lifetime of *highly charged* atoms (bare, H- or He-like) is radiative recombination (RR) in the cooler. The RR-probability λ_{RR} should scale as

$$\lambda_{RR} = <\sigma_{RR} \cdot v_e> \cdot n_e \propto \frac{Z^2 n_e}{\tilde{n}\sqrt{kT_\perp^e}} \qquad (2)$$

where v_e, n_e, and $T_\perp^e$ are the electron velocity density and transversal temperature, respectively, and where $\tilde{n}$ is the mean electron shell into which the recombination occurs. $< \sigma_{RR} \cdot v_e >$ is the RR cross section averaged over the velocity distribution of the electrons. For the calculation of σ_{RR} the classical Stobbe theory [22] of the photoeffect can be employed, as RR is the time mirrored process of the latter one.

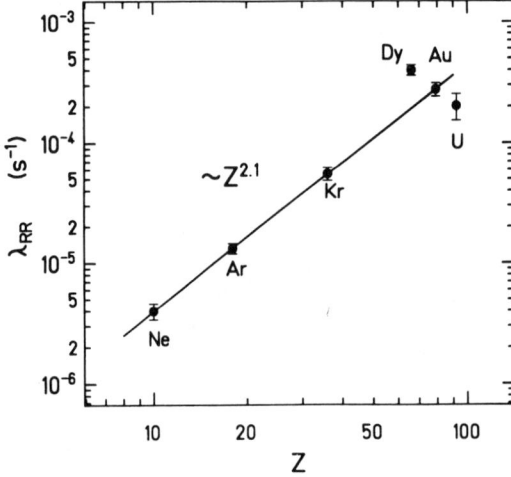

Figure 6: Radiative recombination probability of cooling electrons vs nuclear charge Z of stored ions. Data are taken from ESR [15].

This scaling was confirmed [15] for bare and H-like ions from Ne^{10+} to U^{92+} by measurements conducted at the ESR (cf. Fig. 6). There, lifetimes of bare ions between 70 h (Ne^{10+}) and $\sim$ 1 h (U^{92+}) have been observed at cooler curents of 0.1 A.

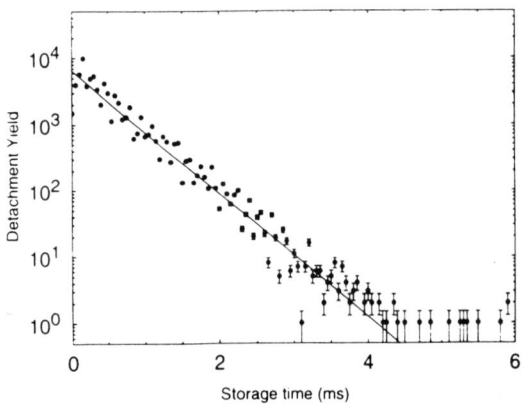

Figure 7: Detachment yield vs storage time of 100 keV Ca^- ions stored in the ASTRID ring [23].

The lifetime of moderately charged ions is, however, by many orders of magnitude shorter, because here ionization of the loosely bound electrons in the residual gas becomes the most dominant loss mechanism. For instance, bare Ni^{28+} at an energy of 150 MeV/u lives for 8 h in the ESR, whereas the corresponding lifetime of Ni^{15+} is only 12 s !

Storage rings are also ideal tools to measure very short lifetimes of special kinds of ions (even without cooling). A striking example is the observation of auto- and photodetachment of negative ions in the lifetime regime between a few microseconds and some tens of milliseconds. At ASTRID even the influence of black body radiation on the lifetime of Ca^- ions could be observed [23] (cf. Fig. 7).

Electron-Ion Interaction

Radiative and dielectronic recombination as well as photon-induced recombination and ionization are fundamental processes which occur in stellar and fusion plasmas. The electron cooler section of storage rings is, probably, the very best tool to study those interactions in detail and with unprecedented precision: In the ion rest frame the cooler electrons are "cold". They have only a small kinetic energy in the order of 0.1 eV. By observing the X-rays associated with RR into inner shells of bare or H-like ions, one gets directly the K and L binding energies of these ions. By employing fast detuning of the electron energy with respect to the ions, resonant dielectronic recombination (DR) can be enforced from which detailed and precise spectroscopic information of doubly excited electron states may be extracted. Finally, the interaction of ions and electrons with monochromatic photons can lead to a very selective population of Rydberg states.

Dielectronic Recombination

Dielectronic recombination (DR) denotes the resonant capture of a free electron with simultaneous excitation of a bound electron, followed by radiative stabilization of the doubly excited state. The first step of DR is the time-mirrored process of autoionization: A free electron with energy ΔE_e^* (with respect to the ion rest frame) is captured into a bound state with binding energy $E_{n\ell}$ and, simultaneously, a bound electron energy in the state $E_{n_1 \ell_1}$ is excited to a bound state $E_{n_2 \ell_2}$. For this process the resonance condition holds:

$$|E_{n_2 \ell_2}| - |E_{n_1 \ell_1}| = \Delta E_e^* + |E_{n\ell}| \tag{3}$$

In the second step radiative deexcitation of the doubly excited state might occur (or, of course, re-autoionization) by the emission of photons. Both processes can be summarized by

$$A^q + e^- \rightarrow A^{(q-1)**} \rightarrow A^{(q-1)} + h\nu + h\nu' \tag{4}$$

14 Atomic Physics Using Storage-Cooler Rings

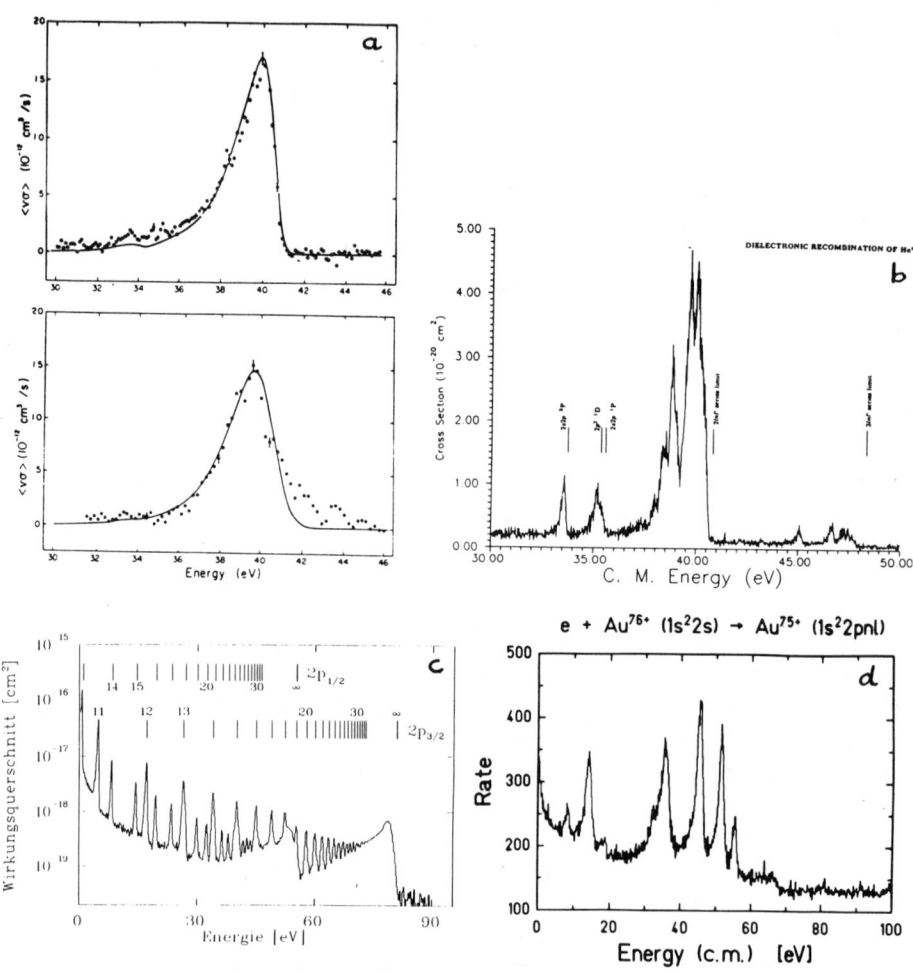

Figure 8: Dielectronic recombination data taken at TARN II [25] (He$^+$, a), CRYRING [26] (He$^+$, b), TSR [24] (Cu^{26+}, c), and ESR [27] (Au^{76+}, d).

Therefore, the net effect of DR (besides photon emission) is a change of the atomic charge q by one unit. The downcharged ions which follow another trajectory in the ring than the primary ions may easily be detected by a detector moved in the ring acceptance at an appropriate place. The signature of DR is, hence, a strong enhancement of the number of downcharged ions at the corresponding detuning energy ΔE_e^* of the cooler. Those resonant structures are superimposed onto a smooth distribution of nonresonant radiative recombination (RR).

A very important point is that the detuning energy ΔE_e^* in the ion rest frame is, for high energies of the stored ions, much smaller than the corresponding energy ΔE_e^{Lab} in the laboratory system, due to the kinematical transformation:

$$\Delta E_e^* \approx \frac{\left(\Delta E_e^{Lab}\right)^2}{4 E_e} \qquad (5)$$

where E_e is the laboratory energy of the cooler electrons, $E_e = 5.5 \cdot 10^{-4} \cdot E_{ion}$ (MeV/u). Hence, all structures of the DR spectra are stretched out over a broad energy band in the laboratory system. Therewith, DR experiments in cooler rings could reach a very high precision [24-27]. Spectroscopy with a resolution below the eV level could be achieved (cf. Fig. 8). In particular, for heavy, highly-charged ions DR experiments already showed data on QED (Lambshift)-corrections with an accuracy below 0.5 eV [27].

Storage-cooler rings deliver, therefore, the by far best data on dielectronic recombination. If it will become possible in the near future with advanced tuning techniques, to excite e.g. KLL resonances, data of unprecedented accuracy for the 1s – 2s energy difference in heavy He-like atoms can be expected almost certainly.

X-Ray Spectroscopy

X-ray studies on highly charged, stored and cooled ions are particularly interesting. The only ring where this kind of spectroscopy is being addressed is the ESR, where even bare or H-like uranium ions can be stored for times of about 1 h. X-rays associated with electron capture into bare or H-like heavy ions allow sensitive studies of relativistic and QED-effects on the atomic structure. For instance, a precise measurement of the 1s Lambshift of H-like U^{91+} (458 eV [28]) provides an information on QED in *strong fields* which can *not* be obtained by corresponding experiments on light ions.

In the ESR, at two places the X-rays associated with e^- capture are being observed: One place is the electron cooler where RR of *free* electrons into the K or L shell together with characteristic X-rays (if e^- capture populates higher shells) can be recorded. The X-rays are detected by a Ge(i) detector in coincidence with downcharged ions. In this way both the initial and final charge states of the ion are known and the long-standing problem of spectactor-electrons is completely avoided. These X-ray spectra are virtually free of any background and deliver directly the 1s and 2s binding energies besides many of characteristic lines. For H-like Au^{78+} the 1s Lambshift could be determined with an uncertainty of 15 eV only [29], where the main contribution to this error stems from the beam velocity β which is unknown within $\sim \Delta\beta/\beta \approx 10^{-4}$. Further improvement that would lead to a "magic" value of $\pm$ 1 eV uncertainty requires decelerated beams (in order to reduce $\Delta\beta/\beta$), absolute determination of β (e.g. by laser spectroscopy) and, finally, the use of crystal spectrometers. This is, obviously, still a long way to go.

The second place where X-rays are recorded is the internal gas jet of the ESR. Here, *bound* electrons are being captured. Therefore, the X-rays associated with the e$^-$ capture itself, show a broad line shape due to the kinematical broadening of the bound electrons (Compton profile). The characteristic X-rays, however, following onto e$^-$ capture into higher shells, show only the "usual" Doppler shift and Doppler broadening. Utilizing forward and backward Ge(i) detectors divided into small stripes, a Doppler correction as well as a precise determination of the X-ray emission angle could be obtained (cf. Fig. 9). In this way the 1s Lambshift of U^{91+} was determined with an uncertainty of $\sim$ 60 eV [30]. Also in these experiments the X-rays are taken in coincidence with downcharged ions which provides background-free spectra.

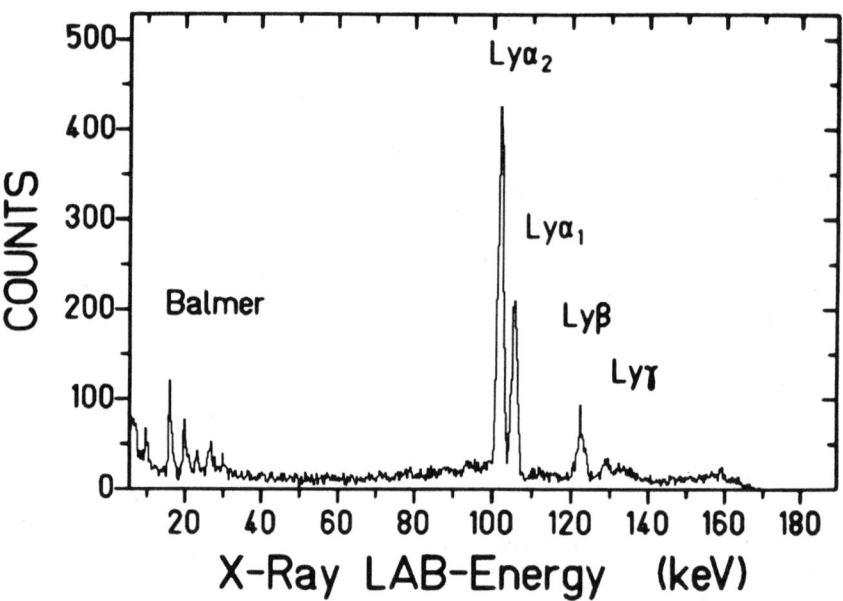

Figure 9: X-ray spectrum associated with electron capture from N$_2$ gas target to Pb^{82+}. Data are taken from ESR [30].

Electron-Ion-Photon Interaction

Laser-based investigation of ions contained in Penning- or Paul-type ion traps is a stormly developing field which will set very soon new landmarks in spectroscopy, in particular if the trapped ions can be cooled down. A similar bright future can be predicted for the corresponding spectroscopy on fast, cooled ions or molecules in a storage ring. The drawbacks of larger Doppler broadening and smaller efficiency in that case (only a part of the stored ions can interact at the same time with the laser beam) are nearly compensated by the unique possibility to tune the laser frequency in the ion rest frame over a broad band by varying the ion velocity β.

Laser-Stimulated Electron Recombination
A special kind of laser spectroscopy has ben addressed first in storage cooler rings: Stimulated transitions of free or quasi-free cooling electrons into Rydberg states by merged laser beams in the electron cooler section. Those resonant free-bound transitions are the counterparts of the well known stimulated bound-bound transitions. They require, however, larger photon intensities, typically a few MW/cm^2 for E1 transitions. Laser stimulated electron recombination is a highly specific tool for two purposes: First, Rydberg states might be populated very selectively. Secondly, from the observed line-shape of the resonance curve both the longitudinal and transverse temperature of the cooler electrons can be determined and also the distortion of the Coulomb potential around the ionization limit is probed with a sub-meV level resolution. Ions and cooling electrons form an inverted, cold plasma, where effects like Debye-screening of the Coulomb potential and sub-threshold quasi-bound electron states are expected to become increasingly pronounced as a function of the nuclear charge Z.

In order to get a clear signature for laser-stimulated electron capture, the rate of stimulated transitions into a well defined bound state must be comparable to the total rate of spontaneous transitions. The ratio of both numbers is called, in this context, the "gain factor" $\mathcal{G}$. If $\mathcal{G}$ is in the order of 1 or larger, a strong enhancement of downcharged ions at the corresponding resonant laser frequency occurs during the interaction time of (pulsed) laser-, electron- and ion-beam.

Laser stimulated recombination has been observed first (cf. Fig. 10) at the TSR for protons and C^{6+} ions [31]. The "Rydberg states" n = 2 for p and n = 14 for C^{6+} were resonantly populated. Surprisingly, also electrons up to 10 meV below the free ionization threshold were observed in the case of C^{6+} which cannot be explained solely by the effect of Stark- and motional fields in the electron cooler. Because this data strongly suggests the emergence of collective, Z-dependent effects — like the formation of Debye shells — corresponding experiments using high Z ions are highly desirable.

Those investigations already started at the ESR with Ar^{18+} ions. Also for these experiments strong and large-wavelength lasers are needed due to the λ^3 dependence of the gain factor $\mathcal{G}$. In the case of Ar^{18+} however, the stimulated e^- capture would lead to a bound electron state n_1 *above* the critical quantum number n_{cr} where field ionization occurs by the motional electric field in the dipole downstream of the electron cooler.

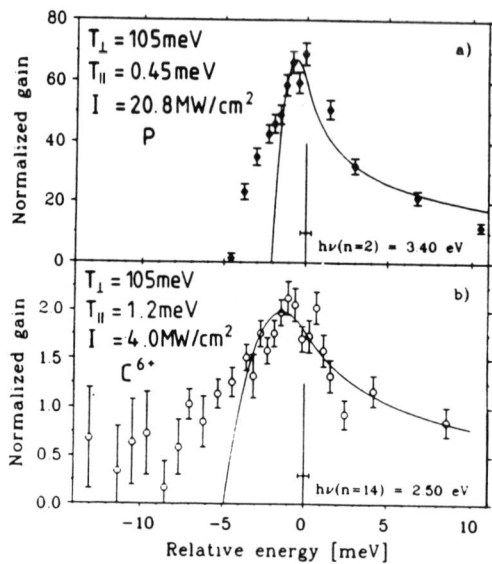

Figure 10: Laser stimulated electron recombination for protons (a) and C^{6+} vs the energy of cooling electrons in the ion rest frame. Data are taken from TSR [40].

For this reason *two* lasers have been employed for Ar^{18+}. The first one, a pulsed Nd:YAG laser was directed parallel to the ion beam in the electron cooler. Owing to the large Doppler effect at $\beta = 0.5$, the laser wavelength of $\lambda_1 = 1064$ nm was shifted into resonance with a transition from the electron continuum threshold to the $n_1 = 81$ state of H-like Ar^{17+}. The second laser, a tunable Ti:Sapphire laser was merged with the other beams but propagating in opposite direction. It stimulated a strong bound-bound transition from $n_1 = 81$ to the $n_2 = 36$ or 37, states, well below the critical field-ionization limit. This resonant two-step recombination has been realized for the first time very recently at the ESR [32].

Groundstate Hyperfine-Splitting of H-like Heavy Ions

Groundstate (1s) hyperfine-splitting of H-like heavy ions occurs when the magnetic moment of the nucleus is different from zero. The energy difference of the two hyperfine levels is sensitive on both the distribution of nuclear magnetization and on quantum electrodynamic (QED) effects. In storage rings, hyperfine transitions might be induced by co- or counterpropagating photons. By tuning the ion-beam velocity β, resonance between the laser frequency seen in the ion rest frame and the transition energy can be effected.

For Bi^{82+} the (1s)-hyperfine splitting corresponds to a wavelength of ~ 240 nm, far away from available wavelengths of powerful optical lasers. Photons of a standard dye laser with $\lambda = 480$ nm, counterpropagating to an ion beam at the velocity $\beta = 0.6$ "appear", however, exactly at the resonance wavelength in the ion rest frame. An experiment aiming to observe this induced transition (by recording the resonance fluorescence) recently started at the ESR. One hopes to reach, finally, with this technique a relative accuracy for the hyperfine splitting of better than 10^{-6}.

Test of Relativity

Whereas ions stored and cooled in an *ion trap* are expected to provide in the near future the new standards of time and mass, the cooled fast-moving ions in a storage-cooler ring can be used as fast-moving clocks which could give hints to the (possible) existence of a preferred reference-frame in the universe.

Ions moving at a velocity $v = (v_x^2 + v_y^2 + v_z^2)^{1/2}$ are "seeing" in their rest frame the (laboratory) frequency ν_0 of photons, travelling in $\pm$ z direction, at a shifted frequency, $\nu\pm$, given by:

$$\nu\pm = \gamma\nu_0 (1 \pm v_z/c) \cdot f \tag{6}$$

where γ is the usual Lorentz factor and where f is a "correction" factor which should be strictly equal to one in the framework of special relativity. If there would be a preferred frame of reference, for instance the direction of the overall velocity v' ($\approx$ 300 km/s) of our galaxy pointing towards the Virgo cluster of galaxies (in that direction the temperature of the 2.7 K blackbody radition of the Big Bang is slightly enhanced due to the Doppler effect), the factor f should be, according to the general "Mansouri-Sexl transformation" [33]:

$$f = 1 + \alpha \left(\left(\frac{v}{c}\right)^2 - \left(\frac{v'}{c}\right)^2 \sin^2\theta \right) \tag{7}$$

Here, α is a small, model-dependent factor [33], v' is the velocity of the preferred frame and θ is the angle between v and v'.

The famous experiment of Michelson has addressed first this topic. However, his search has been restricted on the question whether or not the "ether" is isotropic with respect to the direction of the rotation of the *earth*.

"General" Michelson-experiments, searching for a deviation of α from zero with respect to "cosmic" frames started soon after the detection of the Mössbauer effect. The topic has also been investigated by employing ions moving in linear accelerators, at rather moderate velocities of $\beta \approx 0.02$. In this context, in particular the pioneering experiments of O. Poulsen, conducted at Aarhus, have to be mentioned [34]. As the sensitivity for α grows $\propto v^2$, and as storage-cooler rings can provide both a high velocity v and a long interaction (observation) time, they are ideal tools for addressing general Michelson-experiments.

A first storage ring experiment in this field was conducted at the TSR with electron-cooled ^{7}Li$^+$ ions in the metastable triplet state [35]. The basic idea was to pump and repump a so-called Λ-system, to measure the resonance frequencies in the ion rest frame (i.e. $\nu^{\pm}$) and to compare them, finally, with the corresponding laboratory frequencies.

The system chosen consists of three hyperfine levels associated with the $^3S_1$ and $^3P_2$ states, namely $^3S_1$ (F = 5/2) – $^3P_2$ (5/2) – $^3S_1$ (F = 3/2). A first, parallel Ar$^+$

laser induces the 5/2 → 5/2 transition which results in optical pumping to the F = 3/2 level. (The 5/2 → 5/2 transition "burns a hole" in the velocity distribution of the 5/2 state, cf. Fig. 11). If at the same time the 3/2 → 5/2 transition is induced with an antiparallel dye laser, repumping occurs and a strong fluorescence signal is observed (Fig. 11).

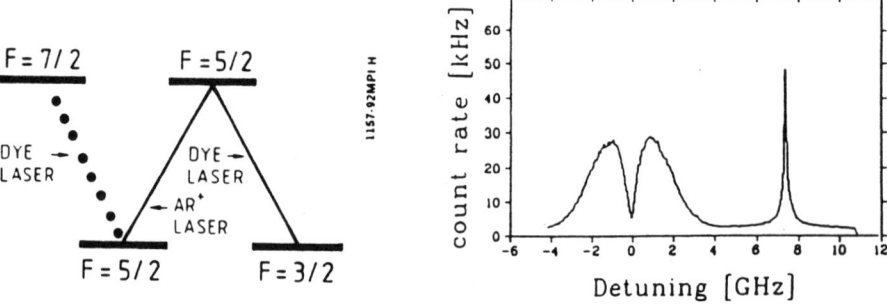

Figure 11: Partial level scheme of ^{7}Li$^+$ (left side) and induced laser fluorescence signal of the "Λ-system". Data are taken from TSR [40].

This experiment yielded an upper limit of $\alpha = \pm\, 1.5 \cdot 10^{-5}$; it is hoped [36] to reach, in a presently running experiment, a value nearby to 10^{-7} which would be comparable to the best present limits for α. An even better limit should be obtainable by benefitting, e.g. from the much higher velocities available at the ESR.

Atomic Astrophysics

The number of bound electrons in an atom influences its weak interaction decay, in some cases even drastically. In a stellar plasma during nucleosynthesis the temperature is high (typical s-process temperatures are about $4 \cdot 10^8$ K $\hat{=}$ 30 keV) and, therewith, also the mean atomic charge state is high. Hence, weak interaction lifetimes in a stellar plasma are *not* a priori equal to those measured in a low-temperature regime as, for instance in the earth.

Heavy-ion storage rings are presently — together with ion traps — the only tools by which the influence of the *atomic* charge state on weak interaction decay might be investigated, because there highly-charged ions can be stored for extended periods of time. A careful consideration is needed, however, to choose from a wealth of experiments the few ones that can shed light on fundamental questions of astrophysics.

First Observation of Bound-State β^--Decay (β_b-Decay)
The feasibility of the ESR to open this new field of research was convincingly demonstrated by the first observation of β_b-decay [37] at the example of the decay of bare ^{163}Dy^{66+} into hydrogen-like ^{163}Ho^{66+}. (It should be noted that *neutral* ^{163}Dy^0 is absolutely *stable*.) It turned out that this special kind of radioactive decay — predicted already 45 years ago — could be analyzed in the ESR by utilizing a rather simple technique.

Bare ^{163}Dy^{66+} atoms provided at 300 MeV/u by SIS are injected, accumulated, cooled and stored in the ESR for variable storage times t_s. Since t_s is always small as compared to the β_b halflife $T_{1/2}^{\beta_b}$, the number N_d of decay daughters, hydrogen-like ^{163}Ho^{66+} which bear one bound electron, is proportional to $N_d \sim \lambda_{\beta_b} \cdot t_s \cdot N_m$, where λ_{β_b} is the β_b-decay constant and N_m the actual number of mother atoms. Because the negative charge of the bound electron outweights the positive charge of the newly created proton in the daughter-nucleus, the daughters move along nearly the same trajectories as the mother atoms. After a variable storage time t_s, a gas jet will be turned on by which the bound electron of the daughter atom may be stripped off. Therewith, the magnetic rigidtiy and, hence, the trajectory of the daughter atom changes and it can be picked up by a position-sensitive detector moved into the acceptance of the ESR.

This number of bare daughter atoms, normalized onto the number of mother atoms (and upscaled for the number of daughter atoms which *captured* an electron in the gas jet) directly delivers, for each storage time t_s, the β_b-decay constant λ_{β_b} (in the lab. frame) searched for.

In a complementary experiment with hydrogen-like primary ions the ratio of (electron capture)/(electron loss) in the gas jet — needed for a proper normalization — will be determined. Finally, an experiment with bare atoms of a neighbouring isotope (e.g. ^{161}Dy) will be conducted from which an upper limit for daughter atoms created by (n,p)-reactions in the gas jet is delivered. In this way β_b-decay could be observed for the very first time [37] yielding a halflife of 48 ± 3 d, in excellent agreement with theoretial predictions.

A further, independent confirmation of these results came from the frequency analysis of the Schottky noise which was performed on the stored ions during and after their interaction with the gas jet. In addition to the intense primary beam, a significant signal from stored ^{163}Ho^{67+} ions has been observed after storage and turning on the gas jet (the particle detectors were removed in that case). This signal increased in proportion to t_s, in contrast to the much smaller signals from nuclear reaction products being *not* proportional to t_s [38].

This spectrum is shwon in Fig. 12. It demonstrates — besides the detection of β_b-decay — the excellent performance of a storage-cooler ring as a high-resolution mass spectrometer. Indeed, mass determinations of $\Delta m/m \leq 10^{-6}$ can be reached in storage-cooler rings like the ESR.

22 Atomic Physics Using Storage-Cooler Rings

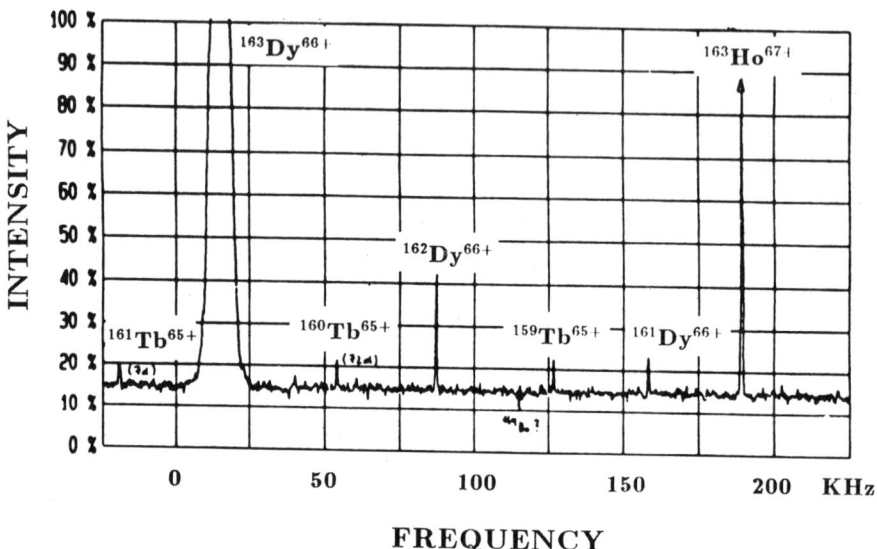

Figure 12: Schottky noise spectrum of stored and cooled ^{163}Dy^{66+} ions (strong line, l.h.s.) together with neighbouring nuclides (small lines) made by nuclear reactions, and with ^{163}Ho^{67+} created by bound-state β-decay (r.h.s.). Data are taken from ESR [38].

Further experiments will address the β_b-decay of bare ^{205}Tl^{81+} to ^{205}Pb^{81+} and of ^{187}Re^{75+} to ^{187}Os^{75+}. The ratio of ^{205}Tl/^{205}Pb can give information on the long-time solar neutrino flux, whereas the ^{187}Re/^{187}Os ratio is, besides ^{232}Th and ^{238}U, the most important "clock" for the age of the universe. For both cases the β_b-decay probability provides decisive information which *cannot* be obtained in any other way.

SUMMARY AND OUTLOOK

The averaged operation time of the five cooler rings under discussion amounts to three years exactly. Within this short time a wealth of impressive results has been obtained, far above all realistic expectations.

Only a few milestones can be enumerated here: First electron cooling of heavy ions; first laser cooling of fast ions; first laser stimulated recombination of electrons; first observation of bound-state β^--decay. The main reason for this striking success is the problem-less, effective operation of electron cooling which is the necessary condition for doing precision experiments. It turned out, furthermore, that the interplay of cooled, monochromatic, highly-charged ions on the one hand and of a tunable electron target, collinear laser beams and a thin internal target on the other hand has provided a fantastically rich "landscape" of precision experiments like radiative recombination, dielectronic recombination, X-ray spectroscopy, laser spectroscopy, molecular physics, atomic astrophysics and so on.

In this short overview not even all activities within atomic physics could be outlined. In particular, the fascinating aspects of dissociative molecular recombination had to remain outside the scope. For this specific case, another feature of cooler rings opens exciting perspectives: They are able to cool besides the translational also the *internal* degrees of freedom, i.e. they provide *completely cold* molecules for spectroscopic purposes.

The potential of cooler rings is not exhausted with the types of experiments described in this report. In contrary, it is almost certain that new topics will be detected for which cooler rings are the ideal tools. At the first place, therefore, "brainstorming" is needed and any (good) idea for new kinds of experiments is highly appreciated.

What progress will be made in the next future ? Certainly, the cooling methods in particular laser cooling will be improved. Whether or not we will see within the next few years crystalline, travelling beams is still an open question. On the other hand, it is urgently needed to improve the injection technique, deceleration and acceleration of stored beams, in order to provide the best possible conditions for spectroscopy.

Like ion traps, storage cooler rings have now been proven as real quantum jumps in experimental atomic physics. In some fields, like X-ray, laser, or molecular spectroscopy, there will be a competition of both. For other fields either the trap or the cooler ring will be a preferential place. For both of them, however, the future looks very bright.

REFERENCES

1. S. Baird et al., Part. Accel. $\underline{26}$, 223 (1990)

2. S. Möller, in: Proc. First Eur. Part. Accel. Conf., ed. S. Tazzari (World Scientific, Singapore), p. 328

3. C.J. Herlander, in: Cooler Rings and their Applications, ed. T. Katayama and A. Noda (World Scientific, Singapore 1991), p. 15 – 19

4. H. Eickhoff et al., in: Cooler Rings and their Applications, ed. T. Katayama and A. Noda (World Scientific, Singapore 1991), p. 11 – 14

5. T. Tanabe et al., Nucl. Instr. Meth., Phys. Res. $\underline{A307}$, 7 (1992)

6. E. Jaeschke et al., Part. Accel. $\underline{32}$, 97 (1990)

7. R.E. Pollock, in: Cooler Rings and their Applications, ed. T. Katayama and A. Noda (World Scientific, Singapore 1991), p. 4 – 10

8. A. Johansson and D. Reistad, in: Cooler Rings and their Applications, ed. T. Katayama and A. Noda (World Scientific, Singapore 1991), p. 20 – 25

9. R. Maier et al., in: Cooler Rings and their Applications, ed. T. Katayama and A. Noda (World Scientific, Singapore 1991), p. 52 – 58

10. S. Schröder et al., Phys. Rev. Lett. 64, 2901 (1991)

11. J.S. Hangst et al., Phys. Rev. Lett. 67, 1238 (1991)

12. P. Forck et al., Phys. Rev. Lett. 70, 426 (1993)

13. M. Larsson et al., Phys. Rev. Lett. 70 430 (1993)

14. T. Tanabe et al., Phys. Rev. Lett. 70, 422 (1993)

15. F. Bosch, Nucl. Instr. Meth. Phys. Res. 52, 945 (1989)

16. H. Poth, Phys. Rep. 196 135 (1990)

17. G.I. Budker et al., Part. Accel. 7, 197 (1967)

18. M. Bell et al., Nucl. Instr. Meth. 190, 237 (1981)

19. D. Habs et al., Nucl. Instr. Meth. B43, 390 (1989)

20. G. Birkl et al., Phys. Bl. 48, 359 (1992)

21. J. Schiffer and A. Rahman, Z. Physik A331, 71 (1988)

22. M. Stobbe, Ann. Phys. (Leipzig) 7, 661 (1930)

23. K.H. Haugen et al., Phys. Rev. A46, R1 (1992)

24. G. Kilgus et al., Phys. Rev. Lett. 64, 737 (1990)

25. T. Tanabe et al., Phys. Rev. A45, 276 (1992)

26. R. Schuch et al., to be published

27. W. Spies et al., Phys. Rev. Lett. 69, 2768 (1992)

28. G. Soff, Z. Physik D21, 7 (1991)

29. H.F. Beyer, private communication

30. Th. Stöhlker et al., private communication and submitted to Phys. Rev. Lett.

31. U. Schramm et al., Phys. Rev. Lett. 64, 2901 (1990)

32. R. Neumann, private communication

33. C. Will et al., Phys. Rev. D45, 403 (1992)

34. E. Riies et al., Phys. Rev. Lett. 62, 842 (1989)

35. R. Klein et al., Z. Physik A342, 455 (1992)

36. D. Habs, private communication

37. M. Jung et al., Phys. Rev. Lett. 69, 2064 (1992)

38. B. Franzke, private communication

39. H.J. Miesner, private communication

40. D. Habs, Nucl. Phys. A553 337c (1993)

THEORY OF ELECTRON COLLISIONS WITH ATOMS, IONS AND MOLECULES

Philip G Burke

The Queen's University of Belfast, Belfast BT7 1NN, Northern Ireland.

ABSTRACT

A review is given of recent progress in the theory of electron collisions with atoms, ions and molecules. In particular, new theoretical methods are described, stimulated by the rapidly increasing power and availability of high performance computers, which are enabling reliable calculations to be carried out for the first time for complex targets of importance in many applications. Emphasis is given to comparing theoretical predictions with experiments wherever possible.

INTRODUCTION

In recent years considerable progress has been made in this field stimulated by many applications particularly in astronomy, laser physics, plasma physics and upper atmosphere physics which require accurate scattering data for their interpretation. In addition, recent experimental advances have provided an important stimulus to theory. These advances include the use of coincidence techniques, polarized beams and targets and very high energy-resolution electron beams. On the theoretical side many of the new developments have been stimulated by the increasing availability of powerful computers which are allowing detailed calculations to be carried out for the first time for complex atoms, ions and molecules where electron correlation and relativistic effects are important.

This review commences by discussing scattering at low energies. In this case the theoretical and computational methods are well established and a number of major computer program packages are allowing accurate calculations to be carried out routinely for light atoms and ions. The following section then addresses the much more difficult problem of how to describe theoretically collisions at intermediate energies. A brief review is given of a number of approaches which have been developed to describe this situation where an infinite number of channels are open and ionizing collisions as well as elastic scattering and excitation are possible. Finally the review turns to a discussion of electron molecule collisions where the additional degrees of freedom associated with the nuclear motion and the non-central nature of the interaction must be considered.

COLLISIONS WITH ATOMS AND IONS AT LOW ENERGIES

We consider first the elastic and inelastic collision process

$$e^- + A_i \rightarrow e^- + A_j \qquad (1)$$

where A_i and A_j are the initial and final bound states of the target atom or ion and we assume that the velocities of the incident and scattered electrons are of the same order or less than that of the target electrons playing an active role in the collision. We also assume that all relativistic effects can be neglected which restricts our treatment to light atoms and ions. These restrictions will be relaxed later.

The Schrödinger equation describing the collision of an electron with an N-electron target with nuclear charge Z is

$$H_{N+1}\Psi = E\Psi \qquad (2)$$

where E is the total energy of the $(N+1)$-electron system and the Hamiltonian H_{N+1} is given in atomic units by

$$H_{N+1} = \sum_{i=1}^{N+1}\left(-\frac{1}{2}\nabla_i^2 - \frac{Z}{r_i}\right) + \sum_{i>j=1}^{N+1}\frac{1}{r_{ij}}. \qquad (3)$$

Here $r_{ij} = |\underline{r}_i - \underline{r}_j|$, $\underline{r}_i$ and $\underline{r}_j$ being the vector coordinates of electrons i and j respectively and we have taken the origin of coordinates to be the target nucleus which is assumed to have infinite mass.

We also introduce the target eigenstates and possibly pseudostates Φ_i and corresponding energies w_i by the equation

$$<\Phi_i \mid H_N \mid \Phi_j> = w_i \delta_{ij} \qquad (4)$$

where H_N is the target Hamiltonian defined by eq (3) with $N+1$ replaced by N. We will assume that sufficiently accurate target states can be calculated this being an essential criterion to obtain accurate collision cross sections.

In order to solve eq (2) most approximation methods describing low energy collisions start from an expansion of the total wave function of the form

$$\Psi = A \sum_i \Phi_i(\underline{x}_1,...,\underline{x}_N) F_i(\underline{x}_{N+1}) + \sum_i \chi_i(\underline{x}_1....\underline{x}_{N+1}) a_i \qquad (5)$$

where $\underline{x}_i \equiv \underline{r}_i \sigma_i$ represents the space and spin coordinates of the ith electron and the operator A antisymmetrizes the first summation with respect to exchange of any pair of electrons in accordance with the Pauli exclusion principle. The first summation in eq (5) goes over a finite number of target eigenstates and possibly pseudostates and the second expansion goes over a set of square integrable (L^2) correlation functions χ_i which allow for electron-electron correlation effects not adequately represented by the limited number of terms which can be retained in the first expansion. If the expansion over the correlation functions is omitted, eq (5) is referred to as the close coupling expansion which was first introduced by Massey and Mohr[1] and developed in its modern form by Seaton[2], who generalized the Hartree-Fock equations for continuum states and applied them to calculate transitions between the ^3P, ^1D and ^1S ground state terms of atomic oxygen.

Coupled equations for the radial components of the functions $F_i(\underline{x})$ representing the scattered electron and the coefficients a_i can be obtained by substituting eq (5) into eq (2) and projecting onto the target functions Φ_i and onto the L^2 functions χ_i. After separating

28 Theory of Electron Collisions

out the spin and angular variables and eliminating the coefficients a_i we obtain the following set of coupled integrodifferential equations

$$\left(\frac{d^2}{dr^2} - \frac{\ell_i(\ell_i+1)}{r^2} + \frac{2Z}{r} + k_i^2\right)\mathcal{F}_i(r) = 2\sum_j (V_{ij} + W_{ij} + X_{ij})\mathcal{F}_j(r) \qquad (6)$$

satisfied by the radial components $\mathcal{F}_i(r)$. In this equation, ℓ_i is the orbital angular momentum of the scattered electron and V_{ij}, W_{ij} and X_{ij} are the partial wave decompositions of the local direct, non-local exchange and non-local correlation potentials. These potentials are too complicated to write down explicitly except for the simplest atom, hydrogen, first discussed by Percival and Seaton[3]. Instead they are constructed by the general computer programs mentioned below.

The development of numerical and computational methods to solve eqs (6) to yield the K-matrix, S-matrix and scattering amplitudes, which can be compared with experiment, has occupied many workers over the last 40 years. These methods now form the basis of a number of computer program packages, some of which have been reviewed by Burke and Eissner[4], which are widely used. These methods include:
(i) linear algebraic equations method[5,6],
(ii) non-iterative integral equations method[7],
(iii) Kohn variational method[8,9],
(iv) Schwinger variational method[10] and
(v) R-matrix method[11,12].
Some of these methods have also been used to solve the corresponding equations arising in electron molecule collisions discussed below.

As an example of the application of this low energy theory we show in figure 1 the calculated and observed elastic differential cross section for $e^- - H$ scattering at an incident electron energy of 4.889 eV. The absolute measurements of Williams[13] are compared with calculations by Schwartz[14] for $\ell_i = 0$, Armstead[15] for $\ell_i = 1$ and Gailitis[16] for $\ell_i = 2$. The results of Schwartz and Armstead were obtained using the Kohn variational method where Hylleraas-type square integrable functions were retained in expansion (5). The phase shifts, accurate to about 4 figures, are seen to give cross sections in excellent agreement with experiment.

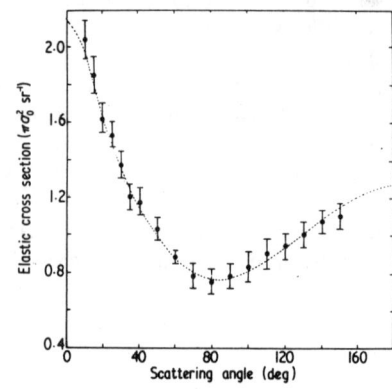

Figure 1.

e^- - H elastic differential cross section at 4.889 eV.
Theory: dashed line.
Experiment: points,
Williams[13].

In the case of low energy electron ion collisions an accurate knowledge of the corresponding collision rates is required in our understanding of atomic processes in hot gaseous plasmas. This data is usually provided by theoretical calculations and since the 1970s many hundreds of calculations of electron ion excitation cross sections have been carried out using the program packages mentioned above[12]. In a few very important cases experimental measurements have also been carried out. When such measurements are available it is imperative that they are used to test theory.

We show in figure 2 the collision strength for the $1^1S - 2^3P^o$ transition in $e^- - Li^+$ collisions, where the collision strength Ω_{ij} is defined in terms of the cross section $\sigma(i \rightarrow j)$ by

$$\Omega_{ij} = \sigma(i \rightarrow j) k_i^2 (2L_i + 1)(2S_i + 1) \qquad (7)$$

where k_i^2 is the incident electron energy in Rydbergs and L_i and S_i are the orbital and spin angular momenta of the target. The collision strength has the advantage of being symmetric (i.e. $\Omega_{ij} = \Omega_{ji}$) and dimensionless. A 19-state calculation, including all target states with $n \leq 4$ in the first expansion in eq (5), was carried out by Berrington and Nakazaki[17] and a 5-state calculation was performed by Christensen and Norcross[18]. The experimental measurements were made by Rogers et al[19] using a crossed charged beam technique. The 19-state results including cascade from the 3^3S and 3^3D states convoluted over a 1 eV energy resolution lie very close to the error bars of the experiment up to 6 Ryd. Important resonance effects below the n=3 and n=4 thresholds are also seen where the incident electron is captured into intermediate Li autoionizing states corresponding to the configurations $1s3\ell n'\ell'$ and $1s4\ell n'\ell'$. Rydberg resonance series of this type play an important role in most low-energy electron-ion collision cross sections and must be properly represented in the calculation if accurate collision rates are to be obtained. The application of Multichannel Quantum Defect Theory techniques developed by Seaton and collaborators is playing a crucial role in representing and understanding these resonance effects[20].

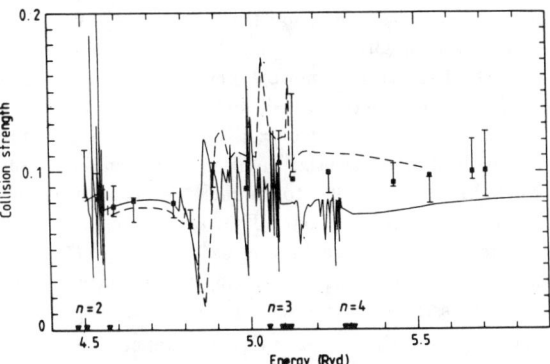

Figure 2.

$e^- - Li^+$ collision strength for $1^1S-2^3P^0$ excitation. Full curve, 19-state calculation[17]; broken curve, 5-state calculation[18]; full squares, 19-state plus cascade[17]; error bars, experiment[19].

INCLUSION OF RELATIVISTIC EFFECTS FOR HEAVY ATOMS AND IONS

As the nuclear charge Z of the target increases, relativistic effects become important even for low energy electron collisions. There are two ways in which these effects play a role. First, there is a direct effect corresponding to the relativistic distortion of the wave function describing the scattered electron by the strong nuclear Coulomb potential. Second, there is an indirect effect, caused by the change in the charge distribution of the target due to relativistic distortion. Both of these effects must be included to obtain accurate results.

For atoms and ions with relatively small and intermediate Z values, relativistic effects are small, so that the scattering calculation can first be performed in LS coupling using the nonrelativistic Hamiltonian defined by eq (3); the corresponding K-matrices are subsequently recoupled to yield transitions between fine-structure levels[21]. For atoms and ions with higher Z values, relativistic terms must be retained in the Hamiltonian. One method of achieving this is to use the Breit-Pauli Hamiltonian[22,23]

$$H_{N+1}^{BP} = H_{N+1}^{NR} + H_{N+1}^{REL} \tag{8}$$

where H_{N+1}^{NR} is the nonrelativistic Hamiltonian given by eq (3) and H_{N+1}^{REL} contains the relativistic terms resulting from the reduction of the Dirac equation and the Breit interaction to Pauli form. This approach has been implemented within the R-matrix framework in a computer program package by Scott and Taylor[24] retaining only the one-body relativistic terms.

Finally for high Z atoms and ions the most satisfactory approach is to adopt the Dirac Hamiltonian defined by[25,26]

$$H_{N+1}^{D} = \sum_{i=1}^{N+1} \left(c\underline{\alpha}_i \cdot \underline{p}_i + \beta c^2 - \frac{Z}{r_i} \right) + \sum_{i>j=1}^{N+1} \frac{1}{r_{ij}} \tag{9}$$

where $\underline{\alpha}_i$ and β are the usual Dirac matrices and c is the velocity of light. This approach has been implemented in a general computer program package by Norrington and Grant[27] within the R-matrix framework and Thumm and Norcross[28] have developed an R-matrix program for one-electron alkali-metal atoms which they have applied to $e^- - Cs$ scattering.

As an example of electron collisions with heavy atoms, we show in figure 3 R-matrix calculations by Bartschat et al[29] and measurements by Wolcke et al[30] of the Stokes' parameter η_1 for light emitted at 90° to the incident beam direction in excitation of the $6s6p$ $^3P_1^0$ state of Hg by polarized electrons. The calculations, carried out using the Breit-Pauli Hamiltonian, included the 5 target states $6s^2$ $^1S^0$, $6s6p$ $^3P_{0,1,2}^0$ and $6s6p$ $^1P_1^0$ in the expansion of the total wave function. The scattering is dominated by resonances close to threshold which have been assigned by Bartschat et al to be a $6s6p^2$ $^4P_{5/2}^0$ resonance at 4.9 eV, a $6s6p^2$ $^4P_{3/2}^0$ resonance at 5.0 eV, a $6s6p^2$ $^2D_{3/2}^0$ resonance at 5.5 eV and a $6s6p^2$ $^2D_{5/2}^0$ resonance also at 5.5 eV. Recent Dirac R-matrix calculations by Wijesundera et al[31] predict some changes in the positions of these resonances and further theoretical and experimental work is necessary to completely resolve this very complex situation.

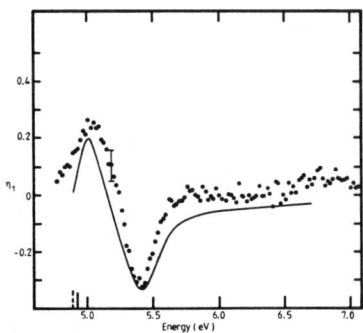

Figure 3.

e^- - Hg Stokes parameter η_1 for the $6s6p\ ^3P_1^0 - 6s^2\ ^1S_0$ transition in Hg excited by polarized electrons. Full curve, R-matrix calculations[29]; ●, experiment[30].

COLLISIONS WITH ATOMS AND IONS AT INTERMEDIATE AND HIGH ENERGIES

We now consider elastic and inelastic electron-atom and electron-ion collisions at electron impact energies greater than the ionization threshold of the target. We will consider ionizing collisions below. At these energies an infinite number of channels are open and we shall discuss a number of theoretical methods which have been used to obtain accurate results in this energy region. These include extensions of low energy methods based on eq (5), the use of optical potentials which allow for the loss of flux into the infinity of open channels and extensions of the Born approximation obtained by including higher order terms. For simplicity we will restrict our attention to light atoms and ions where relativistic effects can be neglected.

The first approach which we consider is to extend the first expansion in eq (5) by including a set of pseudostates as well as the target eigenstates of interest. These pseudostates are chosen to represent in some average way the infinity of high lying Rydberg states and continuum states of the target. An approach of this type was first discussed by Burke and Schey[32], who suggested that a pseudostate basis could be chosen to represent both short and long-range correlation effects, and by Rotenberg[33], who suggested that an expansion in Sturmian functions could be used. Later Damburg and Karule[34] introduced a $\overline{2p}$ polarized pseudostate that represented the full polarizability of the target in $e^- - H$ scattering. Polarized pseudostates have had wide use in low-energy electron collisions with complex atoms and molecules. Finally, an extension of the R-matrix method, called the Intermediate Energy R-matrix (IERM) method has been introduced[35] which enables a complete set of pseudostates to be defined in a finite region enclosing the target.

As an illustration of the effectiveness of this approach we show in figure 4 the integrated $1s - 2s$ excitation cross sections for $e^- - H$ scattering from threshold to 4 Ryd. We see that the $1s - 2s - 2p$ close coupling calculation, in which just the first 3 target states are retained in eq (5), overestimates the cross section by more than a factor of two close to threshold basically because this calculation does not allow for loss of flux into the ionizing channels. On the other hand, the IERM and pseudostate calculations, which do make allowance for this loss of flux, are in good agreement with experiment over the whole energy range. A similar result is obtained for the integrated $1s - 2p$ excitation cross section.

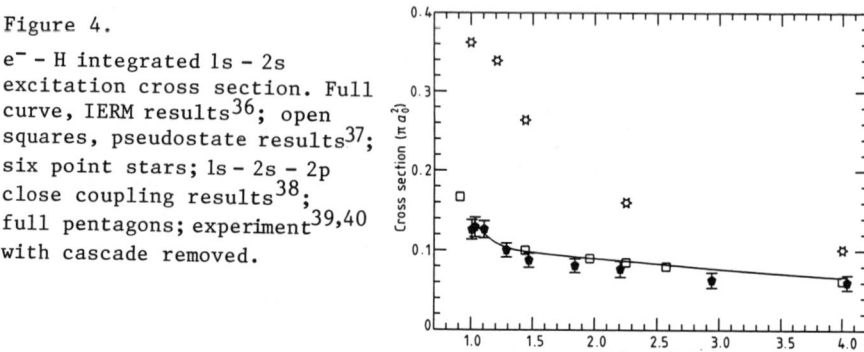

Figure 4.
e^- - H integrated 1s - 2s excitation cross section. Full curve, IERM results[36]; open squares, pseudostate results[37]; six point stars; 1s - 2s - 2p close coupling results[38]; full pentagons; experiment[39,40] with cascade removed.

A more stringent test of theory is to look at the angular distribution of the λ, R and I parameters for $e^- - H$ $1s - 2p$ excitation. These parameters are determined by measuring the scattered electron in coincidence with the decay Lyman-α photon. They are defined in terms of the magnitude and relative phases of the excitation amplitudes a_{m_ℓ} of the $2pm_\ell$ substate by

$$\lambda \equiv \frac{<a_0 a_0>}{\sigma_{2p}} \quad , \quad R \equiv \frac{Re <a_0 a_1>}{\sigma_{2p}} \quad , \quad I \equiv \frac{Im <a_0 a_1>}{\sigma_{2p}} = -\frac{L_\perp}{\sqrt{8}} \tag{10}$$

where σ_{2p} is the differential cross section, $<\ >$ indicates a spin weighted sum and $L_\perp$ is the angular momentum transfered perpendicular to the scattering plane (see review by Andersen et al[41]). We compare in figure 5, several theoretical predications for the angular distributions of these parameters with experiment at 54.4 eV. The theoretical results have been obtained by Bray and Stelbovics[42] who included 70 Sturmian functions in eq (5), Scholz et al[43] who used an IERM expansion, Bray et al[44] who included 6 states in a coupled-channel-optical calculation (see below), Madison et al[45] who used a distorted-wave second-order theory and van Wyngaarden and Walters[46] who used a pseudostate expansion. The theories tend to support each other and give good agreement with experiment in the forward direction, but poor agreement at intermediate angles for λ and backward angles for R. Bray and Stelbovics looked at the convergence of this result with increasing numbers of Sturmian functions and were forced to conclude that current theory is unable to explain the existing measurements. Clearly more work is needed in order to understand this discrepancy.

An approach which is capable in principle of including the effect of all excited and continuum states for complex atoms is the optical potential method introduced into electron atom collisions by Mittleman and Watson[50]. In recent years it has had wide use particularly by Bransden et al[51,52], Byron and Joachain[53] and McCarthy et al[54,55]. This approach can best be understood by rewriting eq (2) using Feshbach[56] projection operators P and Q. The operator P projects onto a few target states of interest, essentially the terms in the first close coupling expansion in eq (5), and $Q = 1 - P$. Eq (2) then becomes

$$(H_{PP} + V_{opt} - E) P\Psi = 0 \tag{11}$$

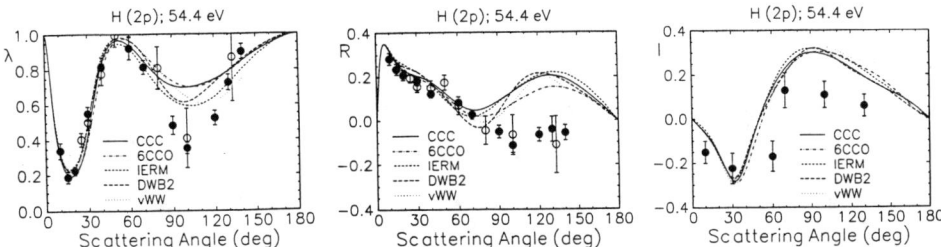

Figure 5. $e^- - H$ angular distribution of the λ, R and I correlation parameters for 1s − 2p excitation at 54.4 eV. Theory: see text for description of approximations. Experiment: ●, Williams[47,48]; ○, Weigold et al[49].

where $H_{PP} = PH_{N+1}P$ corresponds to the close coupling expansion and the optical potential V_{opt} defined by

$$V_{opt} = PH_{N+1}Q\frac{1}{Q(E-H_{N+1})Q}QH_{N+1}P \tag{12}$$

allows for virtual transition out of P-space into Q-space. In the work of Bransden et al the optical potential is treated in second-order while McCarthy et al have studied an approximation which goes beyond second order, called the coupled-channels-optical (CCO) method, which yields a set of coupled momentum-space integral equations.

As a further example of the power of the above methods we show in figure 6 calculated and measured results for the angular momentum transfer $L_\perp$ for $3s - 3p$ excitation in $e^- - Na$ collisions at 10 eV. The singlet and triplet results $L_\perp^s$ and $L_\perp^t$ have been measured separately by the NIST group[57,58] using spin-polarized electron and atom beams making this one of the most sophisticated experiments to date. The Sturmian function calculations of Bray[59] (CCC) are in almost perfect agreement with the experimental data while the 15 state CCO results[60] are in reasonable accord except in the neighbourhood of 60° for $L_\perp^s$. On the other hand the 15-state close coupling results (15CC) show significant deviations from experiment arising from omission of the continuum channels.

As we move to higher energies approximations based on the Born series for the scattering amplitude converge sufficiently fast to become appropriate. An important development in recent years has been the recognition of the need to retain consistently all terms in the Born series with similar energy and momentum transfer $\Delta = | \underline{k}_i - \underline{k}_f |$ dependence in order to obtain reliable results. For example, Byron and Joachain[61] showed that it is necessary to include Ref_{B3} as well as the first and second Born terms f_{B1} and f_{B2} in order to obtain the high energy limit correct through order k_i^{-2}. In view of the difficulty in calculating the third Born term they then showed that it could be accurately approximated by the third term in the expansion of the Glauber amplitude[62]. This led to the development of the Eikonal-Born Series (EBS) method which has had wide use.

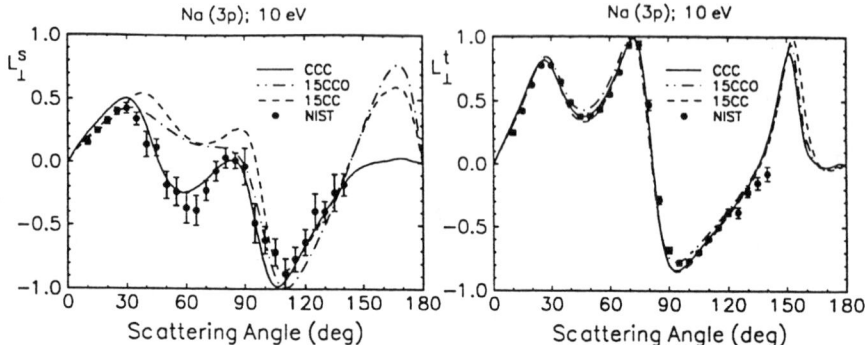

Figure 6. $e^- - Na$, Angular momentum transfer $L_\perp^s$ and $L_\perp^t$ for 3s–3p excitation at 10 eV. Theory: see text for description of approximations. Experiment: NIST[57,58].

In discussing electron impact excitation processes at intermediate and high energies it is also important to mention distorted wave methods[63,64]. The first-order distorted-wave approach has been very successful in understanding and interpreting electron atom differential cross section measurements. It has also proved useful in studies of electron collisions with highly ionized ions where the dominant diagonal nuclear Coulomb potential is treated exactly and the coupling between channels are treated in first-order. A review of perturbation theory methods in electron atom collisions which includes a discussion of distorted-wave Born series approaches has been given by Walters[65].

IONIZATION OF ATOMS AND IONS: THRESHOLD BEHAVIOUR

Wannier[66], in a seminal paper on the threshold behaviour of the ionization cross section in 1953, extended Wigner's general theory of threshold behaviour[67] to the single ionization of atoms and ions by electron impact. If we consider electron impact ionization by hydrogen-like ions, it is convenient following Fano[68], to adopt hyperspherical coordinates defined by

$$R^2 = r_1^2 + r_2^2 \quad , \quad \tan\alpha = r_1/r_2 \quad , \quad \cos\theta_{12} = \hat{r}_1 \cdot \hat{r}_2. \tag{13}$$

The potential energy in eq (2), written in terms of these coordinates, has a saddle point at $\alpha = \pi/4$ and $\theta_{12} = \pi$. Wannier showed using classical arguments that the ionization cross section is dominated by electron configurations in the neighbourhood of this saddle point where the electrons in the final state move away from the nucleus in opposite directions with the same velocity. By expanding the potential about this saddle point he showed that for a system with total angular momentum zero, the total ionization cross section is given by

$$\sigma_{ion} \sim E^m \quad , \quad m = -\frac{1}{4} + \frac{1}{4}\left(\frac{100Z-9}{4Z-1}\right)^{\frac{1}{2}} \tag{14}$$

where E is the excess energy above threshold. When $Z = 1$, then $m = 1.127..$ and as $Z \to \infty$ then $n \to 1$. A further important result, obtained by Vinkalns and Gailitis[69], is that the width of the angular distribution in θ_{12} given by $\triangle\theta_{12} \sim E^{\frac{1}{2}}$ so that the angular distribution becomes more concentrated about $\theta_{12} = \pi$ as the energy $E \to 0$. More recently Greene and Rau[70] have extended Wannier's result to other total angular momenta, spin and parities. They showed that in all states except the $^3S^e$ and $^1P^e$ states the wave function can remain finite at the saddle point and hence the Wannier threshold law will pertain. In the $^3S^e$ and $^1P^e$ states, the symmetry of the wave function forces a node to occur at the saddle point giving rise to to an exponent 3m rather than m given by eq (14). A review of this subject has been written by Rau[71].

Integral cross section studies have concentrated on verification of the energy dependence of the cross section but do not give a complete test of theory. However, in an important experiment, Cvejanovic and Read[72] measured the yield of low-energy electrons from ionization of helium as a function of energy and found the energy dependence to be consistent with the predicted behaviour, $E^{0.127}$.

Recently, triple differential cross sections (TDCS) have been measured by Selles et al[73], Rösel et al[74] and others (see review by Lahmam-Bennani[75]) to study this threshold behaviour. As an example, we show in figure 7 the absolute TDCS for $e^- - He$ scattering measured at an energy 2 eV above the ionization threshold[74] compared with various theoretical predications. In this figure the experiment was carried out in coplanar geometry with the electrons emitted in opposite directions ($\theta_{12} = 180°$). In the theory, Crothers[76] adopted a semi-classical approach which yields the Wannier threshold behaviour. The remaining calculations, JMS by Jones et al[77], JM by Rösel et al[74] and by Pan and Starace[78], use variants of the distorted wave Born approximation (see below) where the final state electron-electron interaction is not fully represented. The good agreement with experiment of both Crothers and Pan and Starace's calculations, which make substantially different assumptions, indicates that the physics governing this angular distribution is still not fully understood.

Figure 7.

$e^- - He$ TDCS at 26.6 eV for coplanar scattering with $\theta_{12} = 180°$. Theory: solid line, JM[74]; dashed-dotted line, JMS[77]; dotted line, Crothers[76]; long-dashed line, Pan and Starace[78]. Experiment: Rösel et al[74].

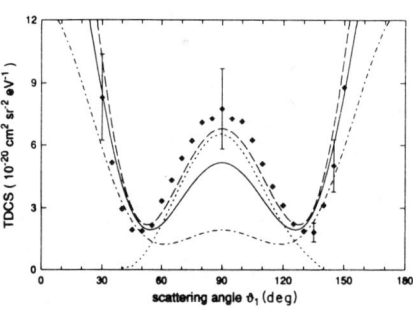

We conclude this section by noting that the correlation of two electrons near the saddle point leading to the Wannier threshold behaviour has also been shown to be responsible for the correlation of electrons in doubly-excited resonance states excited in electron and photon collisions, eg

$$e^- + He^+ \to He^{**}(2s2p\,^1P^0) \to e^- + He^+. \qquad (15)$$

This has been reviewed by Fano[68] and has been the subject of numerous recent studies (eg the recent topical review by Rost and Briggs[79]).

IONIZATION OF ATOMS AND IONS: INTERMEDIATE AND HIGH ENERGIES

The scattering amplitudes for electron impact ionization of atoms and ions at intermediate and high energies may be written in many different but equivalent forms[80]. Here we shall just consider the final-state form where the direct and exchange amplitudes in the case of $e^- - H$ scattering can be written as

$$f(\underline{k}_1, \underline{k}_2) = -\frac{1}{(2\pi)^{5/2}} < \Psi_f^-(\underline{r}_1, \underline{r}_2) | V_i(\underline{r}_1, \underline{r}_2) | e^{i\underline{k}_o \cdot \underline{r}_1} \psi_0(\underline{r}_2) > \tag{16}$$

and

$$g(\underline{k}_1, \underline{k}_2) = -\frac{1}{(2\pi)^{5/2}} < \Psi_f^-(\underline{r}_2, \underline{r}_1) | V_i(\underline{r}_1, \underline{r}_2) | e^{i\underline{k}_o \cdot \underline{r}_1} \psi_0(\underline{r}_2) > \tag{17}$$

where Ψ_f^- is the full scattering wave function satisfying ingoing wave boundary conditions and V_i is the initial state interaction potential

$$V_i(\underline{r}_1, \underline{r}_2) = -\frac{1}{r_1} + \frac{1}{r_{12}} \tag{18}$$

In an important paper Brauner, Briggs and Klar[81], to be referred to as BBK, argued that it is necessary to include exactly in Ψ_f^- the Coulomb interaction between all pairs of particles. For ionization in asymmetric (Ehrhardt) geometry at high energies the BBK approximation works well, as illustrated in figure 8 for $e^- - H$ scattering where the calculations at 250 eV are compared with measurements of Ehrhardt et al[82]. However, with decreasing incident energy, normalisation discrepancies with experiment appear although the shape of the data is still well described. This situation is not improved by including higher-order corrections to the BBK approximation as shown by Tong and Altick[83].

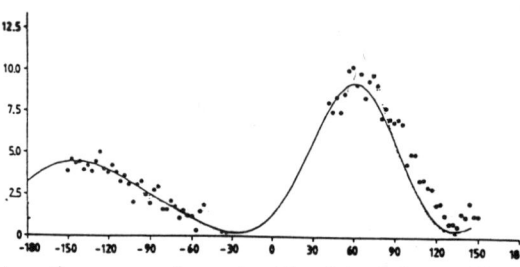

Figure 8.
$e^- - H$ TDCS at 250 eV for coplanar asymmetric scattering where $E_2 = 5$ eV. The TDCS is given as a function of the angle of the slow ejected electron where the direction of the fast electron is $-3°$. Theory: solid line, BBK[81]. Experiment: points, Ehrhardt et al[82].

Recently, there have been some interesting approaches using the distorted-wave Born approximation (DWBA) where effective charges are chosen to represent the final-state interactions. In these approaches the outgoing electrons are assumed to move in potentials $V_1(\underline{r})$ and $V_2(\underline{r})$ satisfying

$$V_{1,2}(\underline{r}) \underset{r \to 0}{\to} -\frac{Z}{r} \quad , \quad V_{1,2}(r) \underset{r \to \infty}{\to} -\frac{Z_{1,2}}{r}. \tag{19}$$

An obvious choice for the effective charges Z_1 and Z_2 is that they should satisfy the Peterkop-Rudge-Seaton[84,85] relation

$$\frac{Z_1}{k_1} + \frac{Z_2}{k_2} = \frac{1}{k_1} + \frac{1}{k_2} - \frac{1}{|\underline{k}_1 - \underline{k}_2|} \qquad (20)$$

but clearly a further condition must be imposed to completely determine Z_1 and Z_2. The choice of the effective charges and the potentials $V_1(r)$ and $V_2(r)$ for intermediate values of r are thus somewhat arbitrary and different choices will yield different results, as already seen in figure 7. A fuller discussion of these approaches has been given by Walters et al[80].

EXCITATION-AUTOIONIZATION OF ATOMS AND IONS

A major contribution to the ionization cross section, particularly for complex ions, often arises from the process known as excitation-autoionization (EA) first observed in the ionization of Ba$^+$ by Peart and Dolder[86]. In this process, the target is first excited to an autoionizing or resonant state lying in the continuum close to the ionization threshold and the resonant state subsequently autoionizes:

$$\begin{aligned} e^- + A_i &\rightarrow e^- + A_j^* \\ A_j^* &\rightarrow A_k^+ + e^- \end{aligned} \qquad (21)$$

The total effect is to ionize the original atom or ion. In addition the incident electron may sometimes be captured into a resonant state that decays by the emission of two (or occasionally more) electrons:

$$e^- + A_i \rightarrow A_\ell^{-*} \rightarrow A_j^* + e^- \rightarrow A_k^+ + 2e^- \qquad (22)$$

This process, known as 'resonant excitation double autoionization' or REDA[87], interferes with the direct process given by eq (21).

If the excitation and the autoionizing processes are weakly coupled then both EA and REDA can be calculated using eq (5), where the relevant autoionizing states as well as bound states are included in the first expansion. This approach has recently been used by Tayal and Henry[88] to describe REDA in electron impact ionization of C^{3+} ions by including 12 autoionizing states arising from the configurations $1s2s^2$, $1s2s2p$, $1s2p^2$, $1s2s3s$ and $1s2s3p$ together with the ground $1s^22s\ ^2S$ and excited $1s^22p\ ^2P^0$, $1s^23s\ ^2S$, $1s^23p\ ^2P^0$ and $1s^23d\ ^2D$ states. Their results were found to be in good agreement with the crossed-beam measurements of Müller et al[89].

Occasionally, Rydberg series of resonances, rather than isolated resonances as in eqs (21) and (22), can contribute to the EA process. If the excitation process is sufficiently weak this process can be described by a distorted-wave approach introduced by Jakubowicz and Moores[90]. However if both the excitation process and the decay process in eq (21) are strongly coupled, giving rise to 'post-collision-interaction' (PCI) effects first discussed by Smith et al[91], there is as yet no completely satisfactory theory.

COLLISIONS WITH MOLECULES

Processes that can occur in electron molecule collisions are more varied than those occuring in electron collisions with atoms and ions. As well as electronic excitation and ionization we now have the possibility of exciting degrees of freedom associated with the motion of the nuclei. In addition, the multicentered and nonspherical nature of the electron molecule interaction considerably complicates the solution of the collision problem by reducing its symmetry and introducing complicated multicentred integrals.

There are two other distinctive features of electron molecule collisions. First, there is the crucial role that resonances play in vibrational excitation and dissociative attachment[92]. Only when the incident electron energy is close to a resonance (or close to a threshold) does the electron spend sufficient time in the neighbourhood of the molecule to excite the nuclear motion with significant probability. Second, the long-range potential interaction between an electron and a molecule is more complicated than that between an electron and an atom. In the case of a linear molecule in a $\sum$ state this potential has the asymptotic form

$$V(r,\theta) \to -\frac{\mu}{r^2}P_1(\cos\theta) - \frac{Q}{r^3}P_2(\cos\theta) - \frac{\alpha_0}{2r^4} - \frac{\alpha_2}{2r^4}P_2(\cos\theta) \tag{23}$$

where r is the distance of the colliding electron from the centre-of-gravity of the molecule, which is taken as the origin of coordinates, and θ is the angle which its vector coordinate $\underline{r}$ makes with the internuclear axis. For polar molecules, the dipole moment μ is non-zero and the corresponding term in eq (23) causes the angular distribution to peak in the forward direction and often gives rise to bound and virtual states that lead to threshold peaks in the vibrational excitation cross sections first seen by Rohr and Linder[93,94]. Further, the long-range quadrupole (Q) and non-symmetric polarization (α_2) terms in this potential lead to enhanced rotational excitation cross sections.

LABORATORY FRAME REPRESENTATION

Most early theoretical and computational work on electron molecule collisions described the total wave function in the laboratory frame of reference[95,96,97]. The total collision wave function in this frame can be expanded, in analogy with eq (5), as

$$\Psi = \mathcal{A}\sum_i \Phi_i(\underline{x}_1,...\underline{x}_N;\underline{R})F_i(\underline{x}_{N+1}) + \sum_i \chi_i(\underline{x}_1...\underline{x}_{N+1};\underline{R})a_i \tag{24}$$

where the target states and possibly pseudostates Φ_i and the correlation functions χ_i now depend on the nuclear coordinates, denoted collectively by $\underline{R}$, as well as on the space and spin coordinates of the electrons. Further, in the first summation in eq (24) the index i goes over the vibrational and rotational states of the target as well as over the electronic states.

In principle a set of coupled integrodifferential equations for the radial components of the functions $F_i(\underline{x})$ similar to eqs (6) can be derived by substituting eq (24) into the Schrödinger equation describing the electron molecule system and projecting onto the rovibrational and electronic states of the target. However a major difficulty arises because of the very large number of ro-vibrational and electronic channels that need to be included in expansion (24) for all but the simplest electron molecule collision process. This difficulty can to a large extent be overcome by adopting the fixed-nuclei approximation in the molecular frame representation which is now the basis of most ab-initio electron molecule collision calculations.

MOLECULAR FRAME REPRESENTATION

In this representation a Born-Oppenheimer separation of the electronic and nuclear motion is adopted for the collision in a similar way to that used in molecular bound-state studies. The electronic motion is first determined with the nuclei held fixed. This is referred to as the fixed-nuclei approximation. The molecular rotational and vibrational motion is then included in a second step in the calculation. The whole procedure is called the adiabatic-nuclei approximation. It owes its validity to the large ratio of the nuclear mass to the electronic mass and can be adopted for electron molecule collision processes when the collision time is much shorter than the period of molecular rotation and/or molecular vibration. This approximation is thus expected to be valid when the scattered electron energy is not close to threshold or when the energy does not coincide with that of a narrow resonance. In these cases, further developments described below are needed to obtain reliable cross sections.

The fixed-nuclei approximation starts from the Schrödinger equation

$$(H_{el} + T + V)\Psi = E\Psi \tag{25}$$

where H_{el} is the electronic part of the target Hamiltonian defined by assuming that the target nuclei have fixed coordinates denoted by $\underline{R}$, T is the kinetic energy operator of the colliding electron and V is the electron molecule interaction potential. The solution of eq (25) for low energy electron molecule collisions proceeds in an onologous way to the solution of eq (2) describing electron atom collisions. We adopt an expansion for the total wave function Ψ which has the form eq (24), but where now the nuclear coordinates, denoted by $\underline{R}$, are held fixed and the summation index i only goes over the electronic states corresponding to the fixed value of $\underline{R}$. Substituting this expansion into eq (25) and projecting into the functions Φ_i and χ_i gives, after separating out the spin and angular variables of the scattered electron, coupled integrodifferential equations analogous to eq (6) for the radial components of the functions $F_i(\underline{x})$.

The final step is to solve these integrodifferential equations for the range of nuclear coordinates $\underline{R}$ of interest to yield the K-matrix, S-matrix and cross sections in the molecular frame. As in the case of electron atom collisions several general computer program packages have been developed for this purpose. Of the methods listed following eq (6) the last three, viz the Kohn variational method, the Schwinger variational method and the R-matrix method, based on standard quantum chemistry structure packages which have been modified to treat electron collisions, have received considerable attention in the last few years. Also, in the case of electron collisions with polyatomic molecules, considerable emphasis has been given over many years to single-centre approaches where the non-local exchange and correlation potentials are replaced by local potentials which make the resultant equations easier to solve[98]. These single-centre approaches have been applied mainly to studies of electron molecule collisions where the target molecule remains in its electronic ground state.

As an example of recent theoretical studies we first mention work on electron diatomic molecule collisions. Recent calculations for $e^- - H_2$ [99], $e^- - N_2$ [100], $e^- - O_2$ [101,102] and $e^- - CO$ [103,104] collisions have been concerned with electronic excitation where several electronic target states, usually represented by CI wave functions, have been retained in expansion (24). We show in figure 9 3-state and 9-state R-matrix calculations by Noble and Burke[102] and effective range theory (ERT) calculations by Gauyacq et al[105] for the $e^- - O_2 X^3 \Sigma_g^- \rightarrow a\ ^1\Delta_g$ integral cross section compared with experiments of Middleton et al[106] and Trajmar et al[107]. This cross section is strongly affected by two O_2^- resonance states. The low energy peak at $\sim$4 eV is due to the tail of the well known $1\pi_u^4 1\pi_g^3\ ^2\Pi_g$ resonance below the threshold at $\sim$1 eV while the peak at 8-10 eV is due to the $1\pi_u^3 1\pi_g^4\ ^2\Pi_u$ resonance. The 3-state R-matrix and ERT theories do not represent the $^2\Pi_u$ resonance while the 9-state R-matrix theory gives a strong peak due to this resonance but at an energy $\sim$2 eV below experiment. This discrepancy may be partly due to the nuclear motion, not yet included in the theory, and additional correlation effects omitted in the target and collision wave functions. These resonances play an important role in dissociation attachment in O_2 and the theoretical study of this process will be of particular importance.

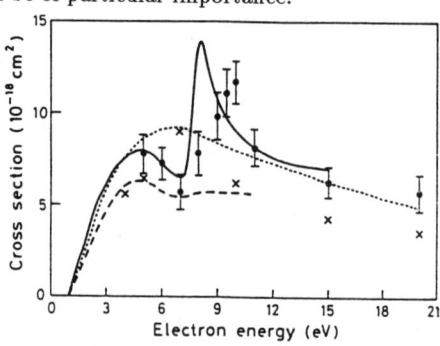

Figure 9.

$e^- - O_2$ integrated cross section for $X^3 \Sigma_g^- - a^1 \Delta g$ excitation. Theory: dotted line, 3-state R-matrix calculation[102]; solid line, 9-state R-matrix calculation[102]; dashed line, ERT[105]. Experiment: points, Middleton et al[106]; crosses, Trajmar et al[107].

A major development in recent years has been the application of the complex Kohn variational method to low energy electron molecule collisions[9]. It has been pointed out by Miller and Jansen op de Haar[108,109] that the S-matrix form of the Kohn variational principle avoids unphysical singularities which caused earlier workers difficulties with the K-matrix form of this method[8,14]. As a result, a major programme of work has been initiated to use this approach to calculated ab-initio electron polyatomic molecule collision cross sections. As one example of this work we show in figure 10 complex Kohn results for the electron-silane integrated cross section obtained by Sun et al[110] compared with experimental measurements of Wan et al[111]. Also shown in this figure are model potential calculations of Jain et al[112] and of Yuan[113]. The wave function used in the complex Kohn calculation is based on eq (24) where a Hartree-Fock description of the target ground state is included in the first expansion and single electron excitations from the Hartree-Fock target into a small set of virtual orbitals are included in the second expansion to allow for target polarization. We see that both the Ramsauer-Townsend minimum at ~ 0.3 eV and the broad d-wave shape resonance at ~ 3 eV are in good agreement with experiment indicating that this calculation gives a consistent description of the N-electron and $(N+1)$-electron correlation effects. This conclusion is confirmed by the corresponding differential cross sections which are also in good accord with the experiments of Tanaka et al[114].

Figure 10.
e^- – SiH_4 integrated cross section. Theory: solid line, Complex Kohn calculations[110]; dotted line, SEP model calculations[112]; dashed line, exact-exchange-model polarization calculations[113]. Experiment: open circles[111].

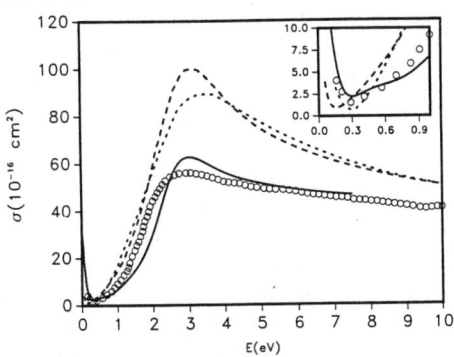

INCLUSION OF THE NUCLEAR MOTION

We conclude this review by discussing how processes involving the nuclear motion such as rotational and vibrational excitation and dissociative attachment can be described in terms of the solutions of the fixed-nuclei equations.

The most widely used approach is the adiabatic-nuclei approximation introduced by Drozdov[115] and Chase[116] in studies of neutron scattering by nuclei. In this approximation the scattering amplitude for an electronic, vibrational and rotational transition of a diatomic molecule is

$$A_{ivjm_j,i'v'j'm_j'}(\hat{\underline{k}} \cdot \hat{\underline{r}}) = \langle \chi_{iv}(R) Y_{jm_j}(\hat{\underline{R}}) \mid A_{ii'}(\hat{\underline{k}} \cdot \hat{\underline{r}}; \underline{R}) \mid \chi_{i'v'}(R) Y_{j'm_j'}(\hat{\underline{R}}) \rangle \qquad (26)$$

where $A_{ii'}(\hat{\underline{k}} \cdot \hat{\underline{r}}; \underline{R})$ is the scattering amplitude for an electronic transition $i \to i'$, obtained by solving the fixed-nuclei equations for a range of values of $\underline{R}$, and χ_{iv} and Y_{jm_j} are the appropriate vibrational and rotational target eigenfunctions. This approximation is valid provided that the collision time is short compared with the vibration and/or rotation time and is used for non-resonant collisions not close to threshold or for collisions in the neighbourhood of very broad resonances.

The adiabatic-nuclei approximation breaks down close to threshold or in the neighbourhood of narrow resonances as discussed by Morrison[97]. The most straightforward way of including non-adiabatic effects which arise in vibrational excitation is to retain the vibrational kinetic energy operator in the Hamiltonian in eq (25) but still treat the rotational motion adiabatically. We can then set up and solve a set of close coupling equations coupling the target electronic and vibrational states. This approach has been used with success by several authors[117] but is computationally very demanding and the number of coupled channels can become prohibitively large for complex molecules.

We have already remarked that vibrational excitation and dissociative attachment proceeds mainly through the formation and decay of intermediate resonance states. As a result a number of approaches have been developed based on electron molecule resonance theories (eg Bardsley[118], Herzenberg[119], O'Malley[120] and Domcke[121]). In these theories the resonance state Ψ_r is defined for a range of fixed-nuclei configurations $\underline{R}$ either by imposing Seigert outgoing wave boundary conditions on the wave function[122] or by introducing appropriate projection operators[56,121]. The amplitude for a transition from an initial electronic-vibrational state iv to a final state $i'v'$ is then given by

$$A_{iv,i'v'} = \langle \chi_{iv}(R)\zeta_i(R) \mid G_r(R,R') \mid \zeta_{i'}(R')\chi_{i'v'}(R') \rangle \tag{27}$$

where ζ_i are the 'entry amplitudes' from the resonance state Ψ_r into the initial and final electronic states and $G_r(R,R')$ is the Green's function that describes the propagation in the intermediate resonance state. Dissociative attachment can be described by an extension of this theory. The fixed-nuclei R-matrix method has also been extended to treat vibrational excitation and dissociative attachment[123] while a conceptually closely related vibrational frame transformation theory has also been developed[124,125]. Finally, an energy modified adiabatic approximation based on eq (26) has been developed by Nesbet[126] which includes non-adiabatic effects.

As an example of the above theory we show in fig. 11 the $e^- - N_2$ $v = 0 \to 1$ and $v = 0 \to 2$ vibrational excitation cross sections calculated by Hazi et al[127]. They determined an ab-initio fixed-nuclei N_2^- $^2\Pi_g$ resonance potential energy curve using a Stieltjes-moment technique, where the resonance wave function was represented by a linear combination of 193 configurations, and incorporated this in the 'boomerang' model of Herzenberg[119]. The results are seen to be in excellent agreement with the experimental data of Wong et al[128]. Similar results have been obtained by Morgan[129] using the non-adiabatic R-matrix theory[123] showing that the basic mechanism for resonant vibrational excitation in this important molecule is now well understood. More recently, considerable effort has been devoted to understanding the threshold peaks observed in the vibrational excitation cross sections of the hydrogen halides[130,131,132]. This work has confirmed the proposal by Gauyacq and Herzenberg[133] that a virtual state in the fixed-nuclei approximation gives rise to a 'nuclear excited Feshbach resonance' when the molecular vibrational motion is included. However much more work needs to be done to enable quantitative predications to be made of vibrational excitation and dissociative attachment cross sections in these and other molecules.

CONCLUSIONS

This field of physics has seen major advances made in the last few years with a secure understanding now emerging of low energy electron collisions with light atoms and ions and a rapidly increasing understanding of excitation processes at intermediate energies.

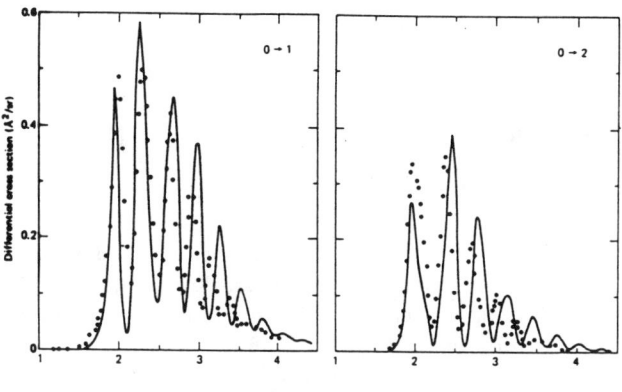

Figure 11.
$e^- - N_2$ $v = 0 - 1$ and $v = 0 - 2$ vibrational excitation cross sections. Theory: solid line, Hazi et al[127]. Experiment: points, Wong et al[128].

However much work remains to be done before reliable theories for all processes considered in this review emerge. There is still an incomplete understanding of electron collisions involving highly excited states and with heavy atoms and ions where relativistic effects are important. In addition, a full theoretical description of ionizing collisions close to threshold is still awaited, a particularly interesting and important example being excitation-autoionization close to threshold.

Turning to electron molecule collisions, in addition to the problems mentioned above for atoms and ions there is a need to account fully for the nuclear motion, with particular emphasis on the role it plays close to thresholds. Also, although there have recently been major developments in the study of electron-polyatomic molecule collisions in the fixed-nuclei approximation, much remains to be done particularly in developing reliable methods for describing vibrational excitation and dissociative attachment.

Finally, in all of the above work, a major incentive is the continuing and indeed increasing need for accurate data from neighbouring areas of science, in particular from astronomy, laser physics, plasma physics and upper atmosphere physics. This will continue to provide a motivation and stimulus for work in this field for many years to come.

REFERENCES

1. H S W Massey and C B O Mohr, Proc.Roy.Soc (London) A136 289 (1932).
2. M J Seaton, Phil.Trans.Roy.Soc. (London) A245 469 (1953).
3. I C Percival and M J Seaton, Proc.Camb.Phil.Soc. 53 654 (1957).
4. P G Burke and W B Eissner in Atoms in Astrophysics, eds P G Burke, W B Eissner, D G Hummer and I C Percival (Plenum Press, New York, 1983) 1.
5. M A Crees, M J Seaton and P M H Wilson, Comput.Phys.Commun. 15 23 (1978).
6. B I Schneider and L A Collins, Comput.Phys.Rep. 10 49 (1989).
7. E R Smith and R J W Henry, Phys.Rev. A7 1585 (1973), ibid A8 572 (1973).
8. R K Nesbet, Variational Methods in Electron-Atom Scattering Theory (Plenum Press, New York, 1980).
9. B I Schneider and T N Rescigno, Phys.Rev.A37 3749 (1988).

10. K Takatsuka and V McKoy, Phys.Rev. A24 2743 (1981), ibid A30 1734 (1984).
11. P G Burke and W D Robb, Adv.Atom.Molec.Phys. 11 143 (1975).
12. P G Burke and K A Berrington, Atomic and Molecular Processes: An R-matrix Approach (Inst.Phys.Publ., Bristol, 1993).
13. J F Williams, J.Phys.B.(At.Mol.Phys) 8 1683 (1975).
14. C Schwartz, Phys.Rev. 124 1468 (1961).
15. R L Armstead, Phys.Rev. 171 91 (1968).
16. M K Gailitis, Sov.Phys.JETP 20 107 (1965).
17. K A Berrington and S Nakazaki, J.Phys.B.(At.Mol.Opt.Phys.) 24 1411 (1991).
18. R B Christensen and D W Norcross, Phys.Rev. A31 142 (1985).
19. W T Rogers, J Olsen and G H Dunn, Phys.Rev. A18 1353 (1978).
20. M J Seaton, Rep.Prog.Phys. 46 167 (1983).
21. H E Saraph, Comput.Phys.Commun. 3 256 (1972), ibid 15 247 (1978).
22. M Jones, Phil.Trans.Roy.Soc. A277 587 (1975).
23. N S Scott and P G Burke, J.Phys.B.(At.Mol.Phys) 13 4299 (1980).
24. N S Scott and K T Taylor, Comput.Phys.Commun. 22 467 (1982).
25. D W Walker, J.Phys.B.(At.Mol.Phys.) 7 97 (1974).
26. J J Chang, J.Phys.B.(At.Mol.Phys.) 8 2327 (1975).
27. P H Norrington and I P Grant, J.Phys.B.(At.Mol.Phys.) 14 L261 (1981).
28. U Thumm and D W Norcross, Phys.Rev. A45 6349 (1992).
29. K Bartschat, N S Scott and P G Burke, J.Phys.B.(At.Mol.Phys.) 17 269 (1984).
30. A Wolcke, K Bartschat, K Blum, H Borgmann, G F Hanne and J Kessler, J.Phys.B.(At.Mol.Phys.) 16 639 (1983).
31. W P Wijesundera, I P Grant and P H Norrington, J.Phys.B.(At.Mol.Opt.Phys.) 25 2143 (1992).
32. P G Burke and H M Schey, Phys.Rev. 126 147 (1962).
33. M Rotenberg, Ann.Phys.(N.Y.) 19 262 (1962).
34. R Damburg and E Karule, Proc.Phys.Soc. 90 637 (1967).
35. P G Burke, C J Noble and M P Scott, Proc.Roy.Soc. (London) A410 289 (1987).
36. M P Scott, T T Scholz, H R J Walters and P G Burke, J.Phys.B. (At.Mol .Opt. Phys.) 22 3055 (1989).
37. J Callaway, K Unnikrishnan and D H Oza, Phys.Rev. A36 2576 (1987).
38. A E Kingston, W C Fon and P G Burke, J.Phys.B.(At.Mol.Phys.) 9 605 (1976).
39. W E Kauppila, W R Ott and W L Fite, Phys.Rev. A1 1099 (1970).
40. B L Long Jr., D M Cox and S J Smith, J.Rev.NBS A72 521 (1968).
41. N Andersen, J W Gallagher and I V Hertel, Phys.Rep. 165 1 (1988).
42. I Bray and A T Stelbovics, Phys.Rev. A46 6995 (1992).
43. T T Scholz, H R J Walters, P G Burke and M P Scott, J.Phys.B. (At. Mol. Opt. Phys.) 24 2097 (1991).
44. I Bray, D A Konovalov and I E McCarthy, Phys.Rev. A44 5586 (1991).
45. D H Madison, I Bray and I E McCarthy, J.Phys.B.(At.Mol.Opt.Phys.) 24 3861 (1991).

46. W L van Wyngaarden and H R J Walters, J.Phys.B.(At.Mol.Phys.) $\underline{19}$ 929 (1986).
47. J F Williams, J.Phys.B.(At.Mol.Phys.) $\underline{14}$ 1197 (1981).
48. J F Williams, Aus.J.Phys. $\underline{39}$ 621 (1986).
49. E Weigold, L Frost and K J Nygaard, Phys.Rev. $\underline{A21}$ 1950 (1979).
50. M H Mittleman and K M Watson, Phys.Rev. $\underline{113}$ 198 (1959).
51. B H Bransden and J P Coleman, J.Phys.B.(At.Mol.Phys.) $\underline{5}$ 537 (1972).
52. B H Bransden, T Scott, R Shingal and R K Raychoudhury, J. Phys. B.(At.M ol. Phys.) $\underline{15}$ 4605 (1982).
53. F W Byron and C J Joachain, Phys.Lett $\underline{A49}$ 306 (1974), Phys.Rev. $\underline{A15}$ 128 (1977).
54. I E McCarthy and A T Stelbovics, Phys.Rev. $\underline{A22}$ 502 (1980), J.Phys.B.(At.Mol. Phys.) $\underline{16}$ 1233 (1983).
55. I E McCarthy, B C Sola and A T Stelbovics, J.Phys.B.(At.Mol.Phys.) $\underline{14}$ 2871 (1981), $\underline{15}$ 2401 (1982).
56. H Feshbach, Ann.Phys.(N.Y.) $\underline{5}$ 357 (1958), ibid $\underline{19}$ 287 (1962).
57. M H Kelley, J J McClelland, S R Lorentz, R E Scholten and R J Celotta in Correlations and Polarization in Electronic and Atomic Collisions and (e,2e) Reactions, eds P J O Teubner and E Weigold (IOP Conference Series Bristol, Vol 122, 1992) 23.
58. S R Lorentz, R E Scholten, J J McClelland, M H Kelley and R J Celotta, Phys.Rev. $\underline{A47}$ 3000 (1993).
59. I Bray, submitted to Phys.Rev.Lett. (1993).
60. I Bray and I E McCarthy, Phys.Rev. $\underline{A47}$ 317 (1993).
61. F W Byron and C J Joachain, J.Phys.B.(At.Mol.Phys.) $\underline{7}$ L212 (1974), ibid $\underline{8}$ L284 (1975).
62. R J Glauber, in Lectures in Physics Vol 1 ed W E Brittin (Interscience, New York, 1959).
63. D H Madison and W N Shelton, Phys.Rev. $\underline{A7}$ 499 (1973).
64. D H Madison, K Bartschat and R P McEachran, J.Phys.B.(At.Mol.Opt.Phys.) $\underline{25}$ 5199 (1992).
65. H R J Walters, Phys.Rep. $\underline{116}$ 1 (1984).
66. G H Wannier, Phys.Rev. $\underline{90}$ 817 (1953).
67. E P Wigner, Phys.Rev. $\underline{73}$ 1002 (1948).
68. U Fano, Rep.Prog.Phys. $\underline{46}$ 97 (1983).
69. I Vinkalns and M Gailitis. Abstracts of the V ICPEAC (Nauka, Leningrad, 1967) 648.
70. C H Greene and A R P Rau, Phys.Rev.Lett. $\underline{48}$ 533 (1982), J. Phys. B. (At. Mol. Phys.) $\underline{16}$ 99 (1983).
71. A R P Rau, Phys.Rep. $\underline{110}$ 369 (1984).
72. S Cvejanovic and F H Read, J.Phys.B.(At.Mol.Phys.) $\underline{7}$ 1841 (1974).
73. P Selles, J Mazeau and A Huetz, J.Phys.B.(At.Mol.Opt.Phys.) $\underline{20}$ 5183 (1987), ibid $\underline{20}$ 5195 (1987).

74. T Rösel, J Röder, L Frost, K Jung, H Ehrhardt, S Jones and D H Madison, Phys.Rev. $\underline{A46}$ 2539 (1992).
75. A Lahmam-Bennani, J.Phys.B.(At.Mol.Opt.Phys.) $\underline{24}$ 2401 (1991).
76. D S F Crothers, J.Phys.B.(At.Mol.Phys.) $\underline{19}$ 463 (1986).
77. S Jones, D H Madison and M K Srivastava, J.Phys.B.(At.Mol.Opt.Phys.) $\underline{25}$ 1899 (1992).
78. C Pan and A F Starace, Phys.Rev. Lett. $\underline{67}$ 185 (1991).
79. J M Rost and J S Briggs, J.Phys.B.(At.Mol.Opt.Phys.) $\underline{24}$ 4293 (1991).
80. J R J Walters, X Zhang and C T Whelan, preprint (1993).
81. M Brauner, J S Briggs and H Klar, J.Phys.B.(At.Mol.Opt.Phys.) $\underline{22}$ 2265 (1989).
82. H Ehrhardt, G Knoth, P Schlemmer and K Jung, Phys.Lett. $\underline{110A}$ 92 (1985).
83. M Tong and P L Altick, J.Phys.B.(At.Mol.Opt.Phys.) $\underline{25}$ 741 (1992).
84. R K Peterhop, Theory of Ionization of Atoms (Colorado Associated University Press, 1977).
85. M R H Rudge and M J Seaton, Proc.Roy.Soc. $\underline{A283}$ 262 (1965).
86. B Peart and K T Dolder, J.Phys.B.(At.Mol.Phys.) $\underline{1}$ 872 (1968).
87. K T La Gattata and Y Hahn, Phys.Rev. $\underline{A24}$ 2273 (1981).
88. S S Tayal and R J W Henry, Phys.Rev. $\underline{A42}$ 1831 (1990).
89. A Müller, G Hofmann, K Tinschert and E Salzborn, Phys.Rev.Lett. $\underline{61}$ 1352 (1988).
90. H Jakobowicz and D L Moores, J.Phys.B.(At.Mol.Phys.) $\underline{14}$ 3733 (1981).
91. A J Smith, P J Hicks, F H Read, S Cvejanovic, G C King, J Comer and J M Sharp, J.Phys.B.(At.Mol.Phys.) $\underline{7}$ L496 (1974).
92. G J Schulz, Rev.Mod.Phys. $\underline{45}$ 423 (1973).
93. K Rohr and F Linder, J.Phys.B.(At.Mol.Phys.) $\underline{9}$ 2521 (1976).
94. K Rohr, J.Phys.B.(At.Mol.Phys.) $\underline{10}$ L389 (1977), ibid $\underline{11}$ 1849 (1978).
95. A M Arthurs and A Dalgarno, Proc.Roy.Soc. $\underline{A304}$ 465 (1960).
96. N F Lane, Rev.Mod.Phys. $\underline{52}$ 29 (1980).
97. M A Morrison, Adv.Atom.Molec.Phys. $\underline{24}$ 51 (1988).
98. F A Gianturco and A Jain, Phys.Rep. $\underline{143}$ 347 (1986).
99. S E Branchett, J Tennyson and L A Morgan, J.Phys.B.(At.Mol.Opt.Phys.) $\underline{23}$ 4625 (1990) ibid $\underline{24}$ 3479 (1991).
100. C J Gillan, O Nagy, P G Burke, L A Morgan and C J Noble, J. Phys. B. (At. Mol.Phys.) $\underline{20}$ 4585 (1987).
101. F J de Paixao, M A P Lima and V McKoy, Phys.Rev.Lett. $\underline{68}$ 1698 (1992).
102. C J Noble and P G Burke, J.Phys.B.(At.Mol.Phys.) $\underline{19}$ L35 (1986) ibid Phys.Rev.Lett. $\underline{68}$ 2011 (1992).
103. Q Sun, C Winstead and V McKoy, Phys.Rev. $\underline{A46}$ 6987 (1992).
104. L A Morgan and J Tennyson, J.Phys.B.(At.Mol.Opt.Phys.) to be published (1993).
105. J P Gauyacq, D Teillet-Billy, L Malegat, R Abouaf and C Benoit, Abstracts of the XV ICPEAC, eds J Geddes, H B Gilbody, A E Kingston, C J Latimer and

H R J Walters (Brighton, UK, 1987) 313.
106. A G Middleton, P J O Teubner and M J Brunger, Phys.Rev.Lett. $\underline{69}$ 2495 (1992).
107. S Trajmar, D C Cartwright and W Williams, Phys.Rev. $\underline{A4}$ 1482 (1971).
108. W H Miller and B M D D Jansen op de Haar, J.Chem.Phys. $\underline{86}$ 6213 (1987).
109. W H Miller, Comm.At.Mol.Phys. $\underline{22}$ 115 (1988).
110. W Sun, C W McCurdy and B H Lengsfield, Phys.Rev. $\underline{A45}$ 6323 (1992).
111. H X Wan, J H Moore and J A Tossell, J.Chem.Phys. $\underline{91}$ 7340 (1989).
112. A K Jain, A N Tripathi and A Jain, J.Phys.B.(At.Mol.Phys.) $\underline{20}$ L389 (1987).
113. J M Yuan, J.Phys.B.(At.Mol.Opt.Phys.) $\underline{22}$ 2589 (1989).
114. H Tanaka, L Boestan, H Sato, M Kimura, M A Dillon and D Spence, J.Phys.B.(At.Mol.Opt.Phys.) $\underline{23}$ 577 (1990).
115. S I Drozdov, Sov.Phys. JETP $\underline{1}$ 591 (1955), ibid $\underline{3}$ 759 (1956).
116. D M Chase, Phys.Rev. $\underline{104}$ 838 (1956).
117. N Chandra and A Temkin, Phys.Rev. $\underline{A13}$ 188 (1976), ibid $\underline{A14}$ 507 (1976).
118. J N Bardsley, J.Phys.B.(At.Mol.Phys.) $\underline{1}$ 349 (1968), $\underline{1}$ 365 (1968).
119. A Herzenberg, J.Phys.B.(At.Mol.Phys.) $\underline{1}$ 548 (1968).
120. T F O'Malley, Adv.At.Mol.Phys. $\underline{7}$ 223 (1971).
121. W Domcke, Phys.Rep. $\underline{208}$ 97 (1991).
122. A J F Siegert, Phys.Rev. $\underline{56}$ 750 (1939).
123. B I Schneider, M LeDourneuf and P G Burke, J.Phys.B.(At.Mol.Phys.) $\underline{12}$ L365 (1979).
124. E S Chang and U Fano, Phys.Rev. $\underline{A6}$ 173 (1972).
125. C H Greene and C Jungen, Phys.Rev.Lett. $\underline{55}$ 1066 (1985).
126. R K Nesbet, Phys.Rev. $\underline{A19}$ 551 (1979).
127. A U Hazi, T N Rescigno and M Kurilla, Phys.Rev. $\underline{A23}$ 1089 (1981).
128. S F Wong, J A Michejda and A Stamatovich, cited in ref 127.
129. L A Morgan, J.Phys.B.(At.Mol.Phys.) $\underline{19}$ L439 (1986).
130. L A Morgan and P G Burke, J.Phys.B.(At.Mol.Opt.Phys.) $\underline{21}$ 2091 (1988).
131. L A Morgan, P G Burke and C J Gillan, J.Phys.B.(At.Mol.Opt.Phys.) $\underline{23}$ 99 (1990).
132. H T Thümmel, R K Nesbet and S D Peyerimhoff, J.Phys.B.(At.Mol.Opt.Phys.) $\underline{25}$ 4553 (1992) ibid $\underline{26}$ 1233 (1993).
133. J P Gauyacq and A Herzenberg, Phys.Rev. $\underline{A25}$ 2959 (1982).

ATMOSPHERIC PHYSICS, COLLISION PHYSICS AND GLOBAL CHANGE

Kate P. Kirby
Harvard-Smithsonian Center for Astrophysics
60 Garden Street
Cambridge, Massachusetts 02138

ABSTRACT

A scientific discussion of current global change issues in the earth's atmosphere is presented. Predictions are that warming will occur in the troposphere (lower atmosphere) by 1.5 - 4.5 °C over the next century due to the build-up of greenhouse gases. The upper atmosphere is expected to cool by a significantly greater amount, as much as 40 - 50°C in the thermosphere. The amount of cooling is critically dependent on the rate coefficient for excitation of the ν_2 vibrational bending mode in CO_2 by collisions with atomic oxygen. Ozone formation and destruction in the stratosphere is discussed. The mechanism for ozone depletion in the Antarctic ozone hole has been shown to be a heterogeneous reaction occurring on the surfaces of polar stratospheric clouds. Gas phase catalytic cycles, alone, are not able to explain the current observational data regarding global ozone depletion. Attention is focussed on the "ozone deficit"--the inability of theoretical models to predict enough ozone at high altitudes in the stratosphere and mesosphere, compared to observations. Schemes for increasing odd-oxygen (O and O_3) production are discussed.

INTRODUCTION

The physics of electronic and atomic collisions, including, also, collisions of ions, molecules, and photons, has broad application in the study of planetary atmospheres. In this article I have chosen to focus entirely on the Earth's atmosphere and, in particular, on Global Change.

The topic of Global Change is far broader than just what is happening in the atmosphere. It goes beyond the issue of the harm that the human population has wreaked on the environment to look at the Global Earth System--including the atmosphere, oceans, land masses, and plant and animal populations--and the change to this system, both natural and human-induced. Within the vast array of unknowns the atmospheric issues most closely overlap with our area of expertise in atomic and molecular physics.

The major topics of discussion here are well-known: global warming, due to the build-up of infrared active gases; and stratospheric ozone depletion due to destructive catalytic cycles. I will also highlight two less-publicized, companion problems which pertain strongly to collision physics: substantial cooling in the upper atmosphere due to build-up of CO_2; and "the missing ozone"--the inadequacy of atmospheric models in predicting enough ozone to agree with observations in the upper stratosphere and mesosphere.

SOME DEFINITIONS

The overall temperature of a planet is determined by a balance between incoming and outgoing energy fluxes. In steady-state, the planet must radiate as much energy as it absorbs from the sun. If the earth radiates as a black-body at an effective temperature, T_e, it obeys the Stefan-Boltzmann law, and an equation can be written in which the flux absorbed equals the flux emitted:[1]

$$F_s \pi R_E^2 (1-A) = s T_e^4 4\pi R_E^2 \qquad (1)$$

where R_E is the radius of the earth, A is the albedo of the earth, s is the Stefan-Boltzmann constant, and F_s is the solar flux at the edge of the earth's atmosphere. Solving this equation, one obtains $T_e \simeq 255°K$ (-18°C). The mean surface temperature of the Earth, however, is approximately 288°K (15°C). The excess temperature of 33°C is due to the "Greenhouse Effect" of the Earth's atmosphere. The Greenhouse Effect is what makes the Earth habitable for life as we know it.

The vertical temperature structure of the earth's atmosphere, as shown in Figure 1, provides an important nomenclature that is widely used.[2] The atmosphere is divided into regions called "-spheres," in which the sign of the temperature gradient with respect to altitude, dT/dz, is constant. The regions in which the temperature gradient changes sign are called "-pauses."

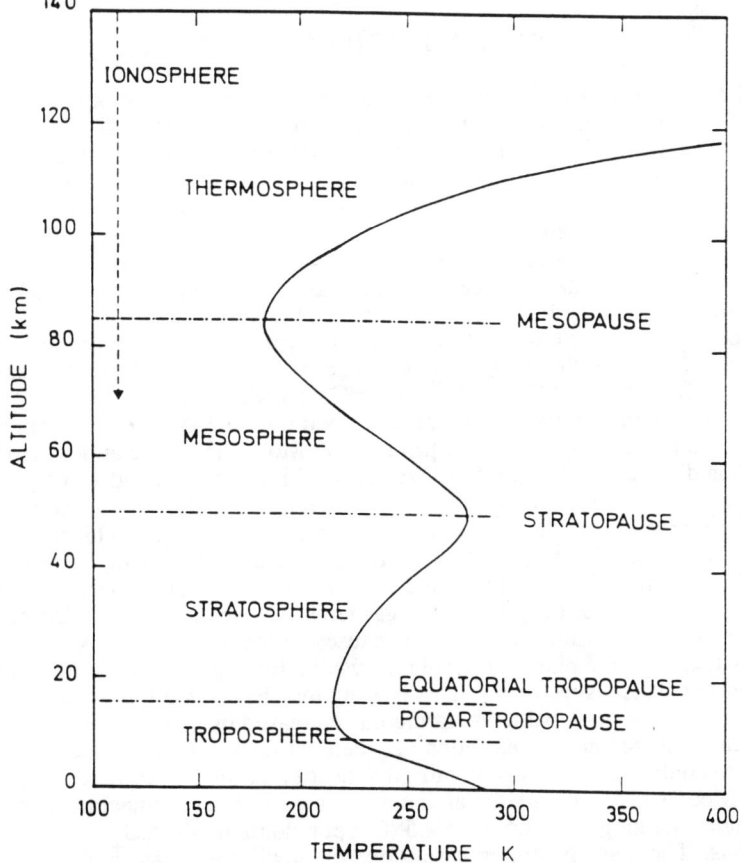

Figure 1: Vertical temperature profile of the atmosphere (courtesy of R. Roble).

In the troposphere, covering the range 0 to ~15 km, the temperature steadily decreases with increasing altitude. This is the region in which weather occurs. The change in sign of the temperature gradient at the tropopause to a positive dT/dz in the stratosphere is due to heating by absorption of solar ultraviolet radiation which photodissociates O_2 and O_3. The stratosphere extends from approximately 15 to 50 km. In the mesosphere, approximately 50-85 km, the heating processes have become too weak to compete with the cooling processes and the temperature decreases again with altitude. At the mesopause, where the temperature gradient once again becomes positive, the atmosphere has become so thin that collisions are very infrequent and the high translational temperatures of the atoms and ions are not efficiently equilibrated with the molecular vibrational modes which can radiate to space, causing cooling. Increased heating occurs due to this "bottleneck" for energy loss. In the thermosphere, which extends from approximately 85 km on up, the thermal inertia is very small and there are huge temperature variations, both diurnally, and with respect to solar activity. In this same region, from ~70 km on up, a very diffuse plasma called the ionosphere exists due to the photoionization of atoms and molecules in the thermosphere by short wavelength solar radiation.

GLOBAL WARMING IN THE TROPOSPHERE

In the ionosphere and thermosphere one finds the greatest number of applications of some of the collision physics topics covered at ICPEAC. These regions are at the outer edge of our atmosphere, farthest away from the biosphere, and only remotely involved in climate issues. Thus it is intriguing to consider that the global warming process occurring in the troposphere may, in fact, be more easily detectable in the thermosphere.

From air bubbles trapped at different depths in polar ice, it is possible to show that carbon dioxide levels have increased by 20% and methane concentrations have more than doubled, over the last two hundred years.[3] The manmade chlorofluorocarbons (CFC's), which have been widely used as refrigerants and in industry, have been increasing in the atmosphere at a rate of over 5% per year since the 1970's.[3] This build-up of CO_2, CH_4 and CFC's causes a problem.

The earth, with an effective radiating temperature of $255^\circ K$, emits mainly long wavelength radiation in the 4 -100 μm region. Molecules in the atmosphere, such as carbon dioxide, methane, water, ozone, and CFC's (known collectively as greenhouse gases) absorb much of this radiation, reradiating it back to the surface and thermalizing the energy through collisions with the ambient gas. This produces an additional warming of $33^\circ C$, mentioned above. There is a region of rather weak absorption by the atmosphere, from ~7 to 15 μm, known as the atmospheric window. Increased concentrations of the greenhouse gas tend to close this window, increasing the infrared opacity of the atmosphere. This causes an immediate decrease in thermal radiation from the planet-atmosphere system, forcing the temperature to rise until the energy balance is restored.[5] This heating may have already started. Over the last century the mean global temperature has increased by approximately $0.5^\circ C$.

Given the increased concentration of greenhouse gases that has occurred and is predicted to continue, the change in radiative heating of the troposphere can be calculated. Models generally predict an increase in tropospheric temperatures over the next century, ranging from $1.5^\circ C$ to $4.5^\circ C$, upon doubling the CO_2 concentrations. The range in temperature is due to the different ways in which climate feedbacks are incorporated into the models.[4] Feedbacks include water vapor, snow and sea ice, and clouds. The modelling of clouds and their radiative properties is at present fairly simplistic and is one of the biggest sources of uncertainty in climate models.

UPPER ATMOSPHERE COOLING

It turns out that the build-up of CO_2 has an even greater effect on the temperature in the upper atmosphere than in the troposphere.[6] In the stratosphere mesosphere, and thermosphere <u>cooling</u> is predicted to be considerably greater than the 1.5 - 4.5°C warming in the troposphere--as much as 40 - 50°C in the thermosphere. Carbon dioxide is both a heat source in the troposphere and a coolant in the stratosphere, mesosphere and thermosphere. The explanation revolves around the collision physics issue of quenching versus radiating. This is illustrated diagrammatically in Figure 2.

In the troposphere CO_2 absorbs infrared radiation coming from the earth, exciting its ν_2 vibrational bending mode. The excited molecule can either radiate or collisionally de-excite. In the lower atmosphere, where densities are large, the lifetime against collisions is very short and the excited molecule is rapidly quenched. Such collisions increase the kinetic energies of the colliding partners, causing a temperature rise. This net transfer of energy from radiation, through collisions, into kinetic energy results in a net heating.

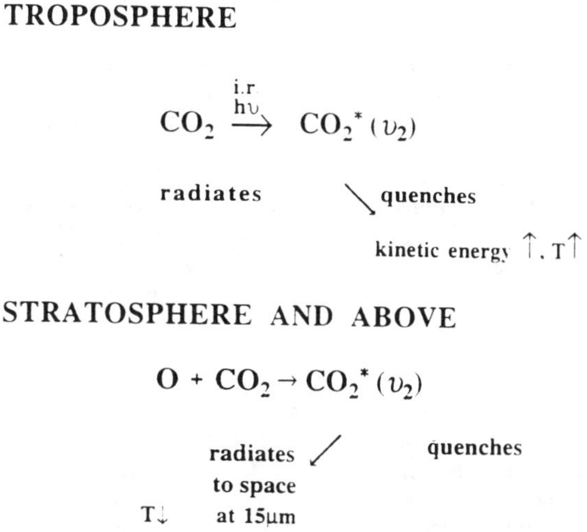

In the stratosphere and above, this same bending mode of CO_2 can be excited by collisions with atomic oxygen as well. But at these higher altitudes, densities are lower and quenching is greatly reduced. The excited molecule then can radiate, and because the opacity is low at these altitudes the radiation escapes to space. Thus, translational kinetic energy has been transferred, through collisional excitation of a bending mode, into radiation at 15 μm which escapes--resulting in a net cooling.

TROPOSPHERE

$$CO_2 \xrightarrow[h\nu]{i.r} CO_2^*(\nu_2)$$

radiates \ quenches

kinetic energy ↑, T↑

STRATOSPHERE AND ABOVE

$$O + CO_2 \rightarrow CO_2^*(\nu_2)$$

radiates / quenches
to space
T↓   at 15μm

Figure 2: Schematic illustration of the role of of CO_2 as heat source in the troposphere and coolant in the stratosphere and above.

Roble and coworkers[7,8] have investigated the doubling of CO_2 and CH_4 concentrations in the mesosphere and at the lower boundary of the thermosphere. Using sophisticated thermospheric/ionospheric general circulation models, they calculate that the thermosphere will cool by as much as 40 - 50°C.

The extent of this cooling very much depends on the rate coefficient for the O + CO_2 excitation of the ν_2 bending mode. Rishbeth and Roble[7] assumed a value for this rate coefficient of $1 \times 10^{-12} cm^3 s^{-1}$, intermediate between the value of Sharma and Wintersteiner[9] ($6 \times 10^{-12} cm^3 s^{-1}$) and an earlier value of $2 \times 10^{-13} cm^3 s^{-1}$ used by Dickenson.[10] The Sharma and Wintersteiner value, based on observations of 15 μm emission around 100 - 150 km was recently confirmed by Rodgers et al.[11] Using the larger rate coefficient would result in an even greater cooling. However, in recent laboratory measurements, Pollock et al.[12] obtain $1.2 \times 10^{-12} cm^3 s^{-1}$, very close to the value used by Rishbeth and Roble.[7]

The overall consequences of such a large temperature decrease in the thermosphere have not been fully explored--particularly the question as to how the dynamics of the atmosphere will be affected. However, as the atmosphere cools it contracts, and densities in the upper atmosphere should decrease by as much as 30-40%. Satellite drag will be reduced, thereby increasing the orbital lifetime of satellites.[7] Since many chemical reactions depend on temperature there may be considerable readjustments in the vertical distribution of minor species in the atmosphere. There may also be an increase in the occurrence of polar stratospheric clouds.

Most significantly, global warming and upper atmosphere cooling both result from the same general phenomenon--a build-up of greenhouse gases, CO_2 in particular. Thus, the large magnitude of the predicted temperature decrease in the thermosphere, approximately an order of magnitude greater than the temperature increases predicted for the troposphere, may allow for the possibility of detecting these trends sooner.

There is evidence that this cooling has already begun--particularly in the mesophere where temperatures appear to have decreased by 3-4°C over the last decade.[13, 14] Gadsden[15] has also found that the frequency of occurrence of noctilucent clouds, the highest lying clouds in the atmosphere, has more than doubled over the last twenty-five years. This change would result from a decrease in the mean temperature at the mesopause of 6.4°C during this time-period.

STRATOSPHERIC OZONE

In the stratosphere the layer of ozone, shielding the Earth's surface from harmful uv radiation in the 280-320 nm range, is remarkably thin. Column ozone abundances are often measured in Dobson Units (D.U.) which correspond to the thickness in units of a hundredth of a millimeter that the ozone column would occupy at standard temperature and pressure. A typical value for the atmosphere would be ~350 D.U., or 3.5 mm at STP.

Ozone production takes place continually in the stratosphere during daylight hours, as molecular oxygen is photodissociated and the resulting oxygen atoms undergo three-body recombination with O_2:

$$O_2 + h\nu \rightarrow 2O \qquad (2)$$
$$O + O_2 + M \rightarrow O_3 + M \qquad (3)$$

Ozone can be destroyed through photodissociation:

$$O_3 + h\nu \rightarrow O_2 + O \qquad (4)$$

but because an oxygen atom is produced which immediately recombines with another O_2 to form O_3, no net loss of O_3 results. The photodissociation of O_2 and O_3 are important heating processes in the stratosphere.

The amount of ozone in the stratosphere is quite variable, changing significantly with the seasons and with latitude. In the lower stratosphere, much of the ozone is created over the equatorial regions and then transported toward the poles.

Besides being photodissociated, O_3 is destroyed by reactions with radicals that are involved in catalytic cycles. These are given below, with the ozone-destroying step listed first, and the rate-limiting step closing the catalytic cycle and regenerating the ozone-destroying radical, listed second:

NO_x

$$NO + O_3 \longrightarrow NO_2 + O_2 \qquad (5)$$
$$NO_2 + O \longrightarrow NO + O_2 \qquad (6)$$
$$\text{NET: } O_3 + O \longrightarrow 2O_2$$

HO_x

$$OH + O_3 \longrightarrow HO_2 + O_2 \qquad (7)$$
$$HO_2 + O_3 \longrightarrow OH + 2O_2 \qquad (8)$$
$$\text{NET: } 2O_3 \longrightarrow 3O_2$$

ClO_x

$$Cl + O_3 \longrightarrow ClO + O_2 \qquad (9)$$
$$ClO + O \longrightarrow Cl + O_2 \qquad (10)$$
$$\text{NET: } O_3 + O \longrightarrow 2O_2$$

Bromine has an identical cycle to chlorine, but is 50 to 100 times more destructive than Cl. The net effect in each of these cases is to convert ozone and atomic oxygen (otherwise known as odd-oxygen) into molecular oxygen. The short-hand notation for the cycles, NO_x, HO_x and ClO_x refers to the catalytically active forms involved in the cycles. for instance, NO_x species are NO and NO_2. The amplification that takes place through a catalytic cycle is the reason that chemicals which are present at the parts per trillion level can have such a destructive effect.

The NO_x and HO_x cycles are naturally occurring, whereas the ClO_x and BrO_x cycles are due to man-made chemicals--CFC's and halons. All the cycles except one require atomic oxygen to complete the cycle. This is a problem in the lower stratosphere, where the ozone concentration peaks (~22 km) and the abundance of O atoms is low. There is an emerging consensus in the last year or so that gas-phase photochemical cycles, as listed above, are not the whole story with respect to ozone depletion.

It is useful to think in terms of a total chemical budget for a radical such as Cl which enters into a cycle. Chlorine is put into the stratosphere when chemicals such as CF_2Cl_2 are released into the atmosphere. Such compounds are chemically inert and insoluble in water and therefore are not easily cleansed out of the lower atmosphere. In the stratosphere, however, the CF_2Cl_2 is photodissociated, producing the Cl radical. Chlorine exists in the upper atmosphere in catalytically active forms, Cl and ClO, as well as in stable, reservoir species, HCl and $ClONO_2$. The total chlorine budget consists of both the catalytically active plus reservoir species. Reactions which reduce the formation of reservoir species, or convert reservoir species to catalytically active forms, contribute to the ozone destruction.

THE ANTARCTIC OZONE HOLE

The ozone depletion problem was largely theoretical until the discovery of the ozone hole over Antarctica. Following the 1985 announcement by Farman et al.[16] of ground-based observations of significant decline in O_3 concentrations during springtime in the Southern Hemisphere, it was possible to map this event using archived satellite data beginning in 1979. The data depict a worsening event throughout the early 1980's. In 1987, 70% of the total O_3 column over Antarctica was lost during the month of September and early October, and the areal extent of the hole was ~10% of the Southern Hemisphere.

The causal link between the release and build-up of man-made CFC's and the ozone hole over Antarctica has been quite convincingly established by Anderson et al.[17] through in situ observations from high altitude aircraft flights into the polar vortex during the end of polar night and the beginning of Antarctic spring in 1987.

The polar vortex is a stream of air circling Antarctica in the winter, creating an isolated region which becomes very cold during the polar night. Flights into the vortex were able to document a heightened, increasing level of ClO and a monotonically decreasing O_3 concentration over a 3-4 week time period during late September and early October.

The mechanism which appears to be repartitioning the chlorine from its reservoir form into its catalytically active form is a heterogeneous process occurring on the surfaces of polar stratospheric clouds. At the cold temperatures during the polar night, polar stratospheric clouds form, consisting of ice and nitric acid trihydrate. Gaseous $ClONO_2$ collides with HCl that has been absorbed into the surface of the cloud crystals. Chlorine gas is liberated and the nitric acid formed in the reaction remains in the ice:[17]

$$HCl + ClONO_2 \longrightarrow Cl_2(g) + HNO_3 \qquad (11)$$

As solar radiation starts to penetrate the region at the beginning of spring, the Cl_2 molecules are rapidly photodissociated, producing Cl atoms which initiate the catalytic destruction of O_3.

As there are no oxygen atoms around to complete the catalytic cycles, several mechanisms to regenerate the Cl and Br radicals have been proposed which involve only the ClO and BrO molecules themselves:

Mechanism I[18]
$$ClO + ClO + M \longrightarrow (ClO)_2 + M$$
$$(ClO)_2 + h\nu \longrightarrow Cl + ClOO$$
$$ClOO + M \longrightarrow Cl + O_2 + M$$
$$2 \times (Cl + O_3 \longrightarrow ClO + O_2)$$

NET: $\quad 2O_3 \longrightarrow 3O_2$

and

Mechanism II[19]
$$ClO + BrO \longrightarrow Cl + Br + O_2$$
$$Cl + O_3 \longrightarrow ClO + O_2$$
$$Br + O_3 \longrightarrow BrO + O_2$$

NET: $\quad 2O_3 \longrightarrow 3O_2$

While Mechanism I accounts for 75% of the observed ozone loss, the sum of I and II yield a destruction rate in harmony with the observed O_3 loss rates.[17]

GLOBAL OZONE DEPLETION

The Total Ozone Monitoring Spectrometer (TOMS) aboard the NIMBUS-7 satellite has been measuring the total ozone column globally since October, 1978 until May, 1993. Trends have been deduced from a statistical fit of the data through 1991 by Stolarski et al.[20] There appears to be no decrease of total ozone at the equator, but definite decreases towards the poles. At $50^\circ N$, the annually averaged trend is -0.5%/year. The trend in the southern hemisphere is larger, consistent with dilution effects from the Antarctic ozone hole.[20] The trend in the northern hemisphere appears to be larger than what can be attributed to homogeneous gas phase processes. Focus has shifted to consideration of heterogeneous reactions on aerosol surfaces.

In 1992 record low global ozone measurements, 2 to 3% lower than any previous year observed, were reported.[21] The trend has continued well into 1993. The increase in naturally occurring aerosols due to eruption of Mount Pinatubo in June, 1991 may explain the recent severe decline in global ozone.

A particularly important reaction appears to be the hydrolysis of N_2O_5 on sulfate aerosols.[22] This occurs very rapidly, converting reactive nitrogen, NO_2, into its reservoir species HNO_3:

$$N_2O_5 + H_2O \xrightarrow{\text{sulfate aerosol}} 2HNO_3 \qquad (12)$$

The N_2O_5 is formed at night by reaction of NO_2 and NO_3. Following the hydrolysis of N_2O_5, there is less reactive NO_2 around to convert ClO into its reservoir species $ClONO_2$ and less NO_2 around to convert OH into HNO_3. A heightened sensitivity of the ozone to increasing levels of CFC's develops.[22] Fahey et al.[22] have shown that measurements of the ratio of catalytically active nitrogen to total nitrogen can be reproduced using the above heterogeneous reaction, but not by using only gas phase processes. Study of further mechanisms at higher altitudes and varying latitudes is an active area of research.

Because ozone is the dominant heat source in the lower part of the stratosphere, declining ozone levels mean less heating and therefore declining temperatures. This effect, in combination with the cooling in the stratosphere due to increased amounts of CO_2, may cause temperatures to decline enough to increase the frequency of formation of polar stratospheric clouds--further accelerating ozone depletion.

THE OZONE DEFICIT

In the upper reaches of the stratosphere and into the mesophere, between 35 to 60 km, measurements of ozone are of particular interest in order to compare with theoretical models.[23] At these altitudes photochemical models of ozone concentrations should be most accurate because local production is considerably faster than transport processes.

There appears to be a significant problem, which has become known as "the ozone deficit" or "the missing ozone." Current theoretical models, using recommended rate coefficients,[24] systematically predict 15 - 50% less ozone at these altitudes than what is observed.[23,25] This problem is one which may lie solidly in the realm of collision physics and photophysics.

The ozone deficit in the models can be corrected by: (1) decreasing the ozone loss rate; and (2) increasing the ozone production. These issues have been examined in some detail by Eluskiewicz and Allen.[26] Recent research has tended to focus on (2), and, in particular, increasing the production of atomic oxygen which results from the photodissociation of molecular oxygen. The relevant potential courves of O_2 are shown in Figure 3. Below 65 km, the dominant photodissociation channel for O_2 is through the Herzberg continua (from 190 - 242 nm). From $v" = 0$ of the ground state of O_2, the Schumann-Runge bands (175 - 200 nm) and continuum (130 - 175 nm) dominate in the 65 -95 km range, and above 95 km, respectively. What is needed are new channels and new wavelengths (preferably long wavelengths) by which to photodissociate molecular oxygen.

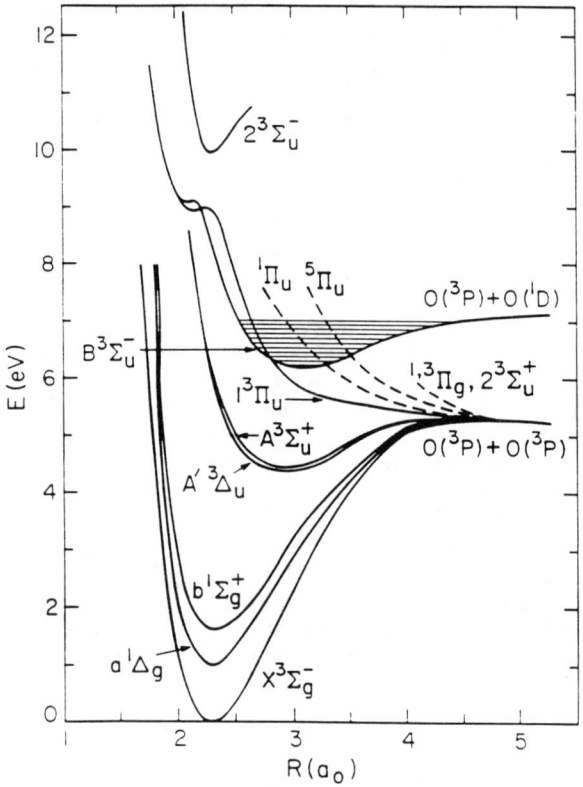

Figure 3. Low-lying potential curves of molecular oxygen. The photodissociation channels include: the Schumann-Runge continuum ($X^3\Sigma_g^-$ - $B^3\Sigma_u^-$), predissociation through the Schumann-Runge bands ($X^3\Sigma_g^-$ - $B^3\Sigma_u^-$), and the Herzberg continua ($X^3\Sigma_g^-$ - $A^3\Sigma^+$, $X^3\Sigma_g^-$ - $A'^3\Delta_u$).

In 1986, Dalgarno and McElroy[27] studied the possibility of photodissociating metastable O_2 ($a^1\Delta g$) created primarily through photodissociation of ozone. The proposed mechanism is a dipole forbidden transition into the upper state of the Herzberg system:

$$O_2(a^1\Delta g) \xrightarrow{h\nu} O_2 (A'^3\Delta_u) \longrightarrow O(^3P) + O(^3P)$$

The transition takes place through a spin-orbit coupling of the $1^3\Pi_g$ with the $a^1\Delta g$. Calculated cross sections are approximately four orders of magnitude too small for this mechanism to be a significant source of additional O atoms.

Slanger and coworkers,[28] upon observing autocatalytic production of ozone at 248 nm in the laboratory, proposed that this mechanism might be operative in the atmosphere. A small amount (~10%) of vibrationally excited ground state $O_2(X^3\Sigma_g^-)$ is produced in the photodissociation of ozone. The vibrationally hot O_2 photodissociates readily at wavelengths longer than for v" = 0, and the atomic oxygen produced rapidly recombines with molecular oxygen giving a net increase in O_3. Saxon and Slanger,[29] as well as Babb et al.[30], have explored the relative contributions of discrete and continuum absorption in the Schumann-Runge system for vibrationally excited O_2.

Toumi[31] has explored the issue of autocatalytic ozone production from vibrationally excited molecular oxygen using a one-dimensional photochemical model, and including a theoretical vibrational distribution of the O_2 as a function of ozone photodissociating wavelength. He included estimates of several quenching processes and concluded that little enhancement in atomic oxygen production could be obtained via this route because the lifetime of O_2 in the stratosphere appeared to be dominated by quenching.

Billing and Kolesnick[32] have reported using a semiclassical collision model to calculate vibrational-vibrational (V-V) and vibration-translational (V-T) rate constants for relaxation of vibrationally excited O_2. Their calculations, in which the vibrational motion is quantized and the translation is treated classically, have been shown to be in good agreement with recent measurements of Park and Slanger.[33] These latter authors[33] find that collisions with N_2 appear to be even more effective than O_2 in quenching vibrationally excited O_2. These measurements and calculations show definitively that O_2 cannot be a source of O atoms for increased production of O_3 in the upper stratosphere.

Processes currently under study include: (1) the creation of ozone directly in collisions of O_2 with an electronically excited O_2, and (2) the reaction of highly vibrationally excited O_2 with O_2 to produce O_3. As of now, the "ozone deficit" is still unsolved.

ACKNOWLEDGMENTS

I wish to thank a number of people whom I consulted in the preparation of this paper. They were most generous with their time, their expertise, and their viewgraphs: James Anderson, Kelly Chance, Alex Dalgarno, James Gleason, Michael Oppenheimer, Ray Roble, Ramesh Sharma and George Victor.

REFERENCES

1. R. P. Wayne, <u>Chemistry of Atmospheres</u> (Oxford Univ. Press, New York, 1991), p. 41.
2. R. G. Roble, "Atmospheric Structure" in Encyclopedia of Applied Physics, Volume 2, pp. 201-224 (VCH Publishers, 1991).
3. J. Firor, <u>The Changing Atmosphere</u> (Yale, New Haven, 1990).
4. J.F.B. Mitchell, Reviews of Geophysics <u>27</u>, 1 (1989).
5. J. Hansen, D. Johnson, A. Lacis, S. Lebedeff, P. Lee, D. Rind, and G. Russell, Science <u>213</u>, 957 (1981).
6. R. J. Cicerone, Nature <u>344</u>, 104 (1990).
7. H. Rishbeth and R.G. Roble, Planet Space Sci. <u>40</u>, 1011 (1992).

8. R.G. Roble and R.E. Dickinson, Geophys. Res. Lett. 16, 1441 (1989).
9. R.D. Sharma and P.P. Wintersteiner, Geophys. Res. Lett. 17, 2201 (1990).
10. R.E. Dickinson, J. Atmos. Terr. Phys. 46, 995 (1984).
11. C.D. Rodgers, F.W. Taylor, A.H. Muggeridge, M. Lopez-Puertas and M.A. Lopez-Valverde, Geophys. Res. Lett. 19, 589 (1992).
12. D.S. Pollock, G.B.I. Scott, and L.F. Phillips, Geophys. Res. Lett. 20, 727 (1993).
13. A.C. Aikin, M.L. Chanin, J. Nash, and D.J. Kendig, Geophys. Res. Lett 18, 416 (1991).
14. A. Hauchecorne, M.L. Chanin, and P. Keckhut, J. Geophys. Res. 96, No. D8, 15, 297 (1991).
15. M. Gadsden, J. Atm. Terr. Phys. 52, 247 (1990).
16. J.C. Farman, B.G. Gardiner, and J.D. Shankin, Nature 315, 207 (1985).
17. J.G. Anderson, D.W. Toohey, and W.H. Brune, Science 251, 39 (1991).
18. L.T. Molina and M.J. Molina, J. Phys. Chem. 91, 433 (1987).
19. M.B. McElroy, R.J. Salawitch, S.C. Wofsy and J.A. Logan, Nature 321, 759 (1986).
20. R.S. Stolarski, P. Bloomfield, R.D. McPeters and J.R. Herman, Geophys. Res. Lett. 18, 1015 (1991).
21. J.F. Gleason, P.K. Bhartia, J.R. Herman, R. McPeters, P. Newman, R.S. Stolarski, L. Flynn, G. Labow, D. Larko, C. Seftor, C. Wellemeyer, W.D. Komhyr, A.J. Miller and W. Planet, Science 260, 523 (1993).
22. D.W. Fahey, S.R. Kawa, E.L. Woodbridge, P. Tin, J.C. Wilson, H.H. Jonsson, J.E. Dye, D. Baumgardner, S. Borrmann, D.W. Toohey, L.M. Avallone, M.H. Proffitt, J. Margitan, M. Loewenstein, J.R. Podolske, R.J. Salawitch, S.C. Wofsy, M.K.W. Ko, D.E. Anderson, M.R. Schoeberl, and K.R. Chan, Nature 363, 509 (1993).
23. D. Robbins, J. Waters, P. Zimmerman, R. Jarnot, J. Hardy, H. Pickett, S. Pollitt, W. Traub, K. Chance, N. Louisnard, W. Evans and J. Kerr, J. Atmos. Chem. 10, 181 (1990).
24. Jet Propulsion Laboratory, Chemical kinetics and photochemical data for use in stratospheric modeling, evaluation number 10, Publ. 92-20, Pasadena, Calif., 1992.
25. M. Allen and M.L. Delitsky, J. Geophys. Res. 96, No. D7, 12883 (1991).
26. J. Eluskiewicz and M. Allen, J. Geophys. Res. 98, No. D1, 1069 (1993).
27. A. Dalgarno and M.B. McElroy, Geophys. Res. Lett. 13, 660 (1986).
28. T.G. Slanger, L.E. Jusinski, G. Black, and G.E. Gadd, Science 241, 945 (1988).
29. R.P. Saxon and T.G. Slanger, J. Geophys. Res. 96, No. D9, 17, 291 (1991).
30. J. Babb, Y. Sun, M.J. Jamieson, A. Dalgarno, and A. Allison, (to be published).
31. R. Toumi, J. Atmos. Chem. 15, 69 (1992).
32. G.D. Billing and R.E. Kolesnick, Chem. Phys. Lett. 200, 382 (1992).
33. H. Park and T.G. Slanger, to be published, 1993.

ATOMIC PHOTOIONIZATION

THRESHOLD DOUBLE PHOTOIONIZATION OF RARE GAS ATOMS

R. I. Hall
CNRS and Université Pierre et Marie Curie, 4 place Jussieu B75, 75252 Paris Cedex 05, France

L. Avaldi
IMAI del CNR, Area della Ricerca di Roma, 00016 Monterotondo, Italy

G. Dawber, K. Ellis, G. C. King, A. G. McConkey
Department of Physics, University of Manchester, Manchester M13 9PL, UK

M. A. MacDonald
SERC Daresbury Laboratory, Warrington WA4 4AD, UK

M Zubek
Department of Molecular Physics, Technical University, 80-952 Gdansk, Poland

ABSTRACT

Some predictions of Wannier theory and its extensions for double photoionization have been examined for the rare gases using different experimental methods. These methods are based on the detection of photoelectrons, photoions and coincidence techniques. While the cross section threshold law appears to be followed the angular behaviour of the electrons points to serious weaknesses in the theory.

INTRODUCTION

Double photoionization is one of the fundamental processes in atomic physics and is of primary interest since it only occurs because of the coulomb interaction between the target electrons. In addition, it is expected that, in the threshold region, this electron correlation will also control the dynamics of the two escaping electrons and thus determine the observable physical phenomena. The physical concepts governing this process were first discussed in a pioneering paper by Wannier[1]. Since then considerable effort has been put into both the extention of this theory or the development of quantal analogues and the experiment verification of the predictions and the determination of their range of validity (see the review by Read[2]). The Wannier model is based on the dynamics, of two electrons in an initial configuration wherein they are mutually shielded by the ionic core by being on opposite sides of the core at a mutual angle of 180° and equidistant from it.

For the $^1S^e$ final state symmetry that Wannier considered the predictions are :

(i) The cross section dependence is $\sigma(E) \propto E^\alpha$, where E is the excess energy above threshold and $\alpha=1.056$ for double photoionization. Note that, at threshold, the cross section is zero. This is a manifestation of the electron correlation that inhibits the double escape process. This is in strong contrast to single ionization, where the cross sections can be finite at threshold.

(ii) The partial cross section $\sigma(E_1)$=constant, where E_1 the energy of one of the electrons is given by $E_1 = E - E_2$ where E_2 is the energy of the other electron, i.e. the excess energy is randomly partioned between the two electrons and their energy distribution is uniform.

(iii) The distribution of the mutual angle between the two electrons is a Gaussian peaked at 180°, with a FWHM $\theta_{1/2} = \theta_0 E^{1/4}$. θ_0 is the angular correlation coefficient.

When the restriction to $^1S^e$ states is removed the predictions of Wannier theory can be different and α can have different values depending on the symmetry. Similarly, the distribution of the mutual angle between the two electrons is now determined by the symmetry of the system modified by the angular correlation.

Wannier theory has also led to the notion of final state selectivity in the double photoionization process[3,4]. The final state channels characterized by the symmetry of the two-electron wave functions can be classified "favoured" or "unfavoured" implying "large" and "small" cross sections, respectively. Double photoionization of helium is the prototype system for probing Wannier predictions. This system has "unfavoured" $^1P^o$ symmetry the consequence of which is a low cross section ($\sim 10^{-21} cm^2$). This makes its study difficult and would explain the rarity of observations.

Here we present observations of double photoionization of rare gas atoms in the threshold region which furnish information on predictions (i) and (iii) as well as on the final state selectivity of this process.

EXPERIMENTAL TECHNIQUES

The experiments reported here were performed at the Synchrotron Radiation Source (SRS) of the SERC Daresbury Laboratory (UK) where a toroidal grating monochromator provides a linearly polarized photon beam. The electron spectrometer consists of a gas beam and a 127° electrostatic filter and associated optics which analyses the energy of the photoelectrons. This instrument is used in two modes. In the threshold mode near-zero energy electrons are trapped in a potential well formed by the penetration into the scattering region of the field from an extractor electrode[5]. A threshold spectrum is then obtained by recording the electron intensity while scanning the photon energy. Here the energy resolution of the analyzer is determined by the well depth and can be as low as a few meV. The great advantage of this mode is that the detection efficiency is very high, since no angular discrimination is present and the electrons are collected over 4π steradians. In the other, more common mode of operation the analyzer detects electrons of finite energy produced above threshold which are emitted into the solid angle subtended by the electrostatic lens system. Here the energy resolution is that of the 127° filter. The spectrometer can rotate in a plane perpendicular to the photon beam direction.

This spectrometer was subsequently developed to detect ions by adding a time-of-flight tube opposite the electron analyzer on the other side of the gas beam. The ion detector also employed the penetrating field method to ensure high detection efficiency for the ions while negligibly perturbing the scattering centre. This system then allowed the photoelectrons to be detected in coincidence with their product ions (PEPICO technique), thus improving considerably the evaluation of the signal intensities compared to the background. This instrument and the different modes of operation have been described in detail elsewhere[6].

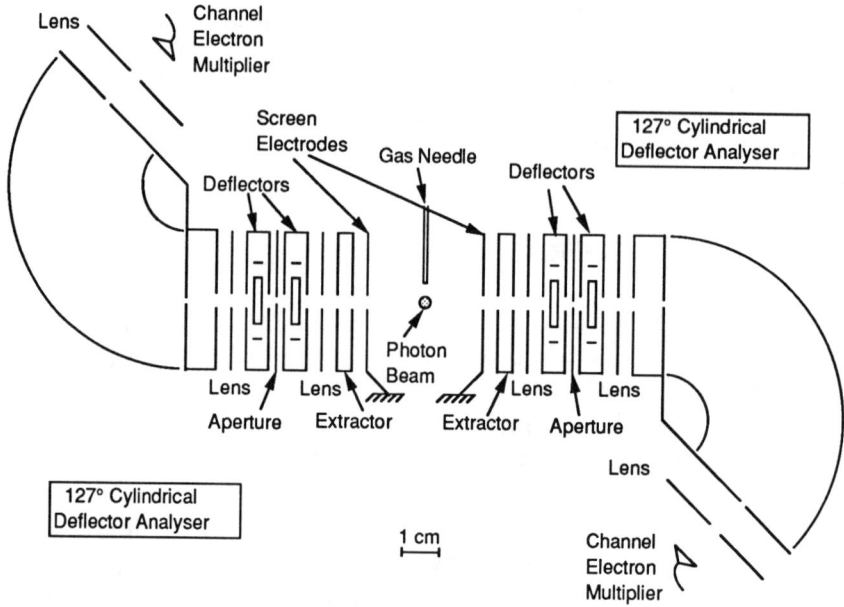

Fig. 1. Coincidence photoelectron spectrometer

More recently, the instrument has been modified so that two identical electron analyzers face each other on either side of the scattering centre[7], as is shown in fig. 1. Two photoelectrons produced by double ionization can now be detected in coincidence. The analyzers can be operated simultaneously in the threshold mode, sharing the signal of near-zero energy electrons between them. One or both of the analyzers can also operate in the conventional mode by detecting finite energy electrons.

THRESHOLD LAW FOR DOUBLE PHOTOIONIZATION

The threshold photoelectron spectrum of helium[8] is shown in fig. 2. This spectrum reveals the He^{+*} satellite manifold from n=8 to n=12. Higher levels merge into an almost flat shoulder which falls off rapidly reaching a sharp minimum at the double-ionization potential. This singularity, which has the form of a strongly asymmetric cusp, is a direct consequence of electron correlation.

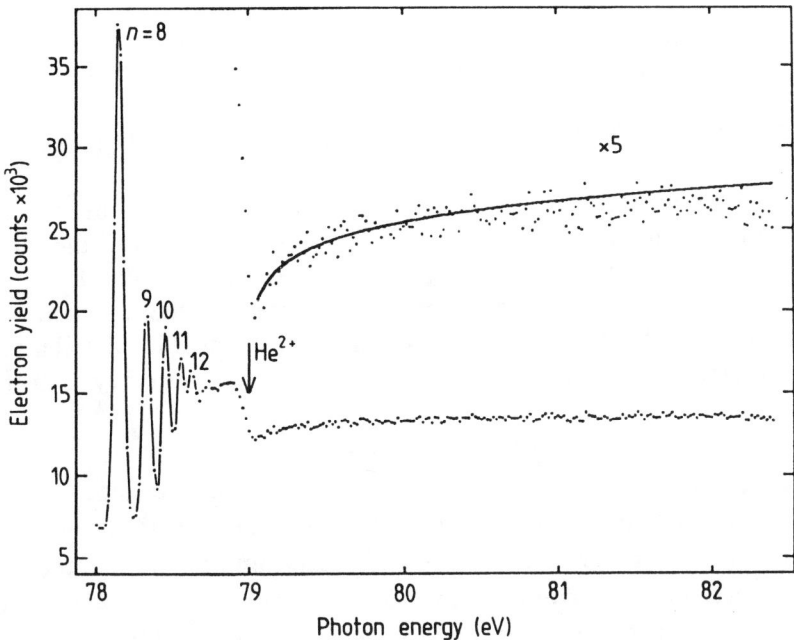

Fig. 2. Threshold ($E_e=0$) photoelectron spectrum of helium

The energy differential form of the threshold law when only one electron with energy E_e is detected can be written:

$$\sigma(E,E_e)=E^{\alpha-1}f(E_e/E)$$

where $f(E_e/E)$ is the energy partitioning function between the electrons and, in the Wannier model is independent of E. In principle, as α is near unity, measurement of the partial cross section corresponding to the measurement of near-zero energy ($E_e=0$) electrons thus provides a very sensitive method for determining small departures of α from unity.

The cusp structure was fitted with the above expression and the exponent was determined to be $\alpha=1.060\pm0.007$. This value is in excellent agreement with that of Wannier theory ($\alpha=1.056$). The continuous line shown in the inset of fig. 2 is a fit of the data up to $E=0.48$eV for the Wannier exponent $\alpha=1.056$ extrapolated to higher energies. The range of validity of the threshold law was evaluated to be 2.0 ± 0.5eV, which is in good accord with the observations of Kossmann et al [9]. A further study of the threshold cross section behaviour using the above mentioned PEPICO technique confirmed these findings[10].

ELECTRON ANGULAR CORRELATION IN DOUBLE PHOTOIONIZATION

Although the mutual angular distribution between the two ionized electrons was not studied, information on the angular correlation between them can, nevertheless, be obtained from the angular distribution of the electrons relative to the electric vector of the photon beam. In photoionization, the angular distribution of photoelectrons can be represented by the asymmetry parameter β using the relationship:

$$I(\theta) \propto (1+\beta/4(1+3P\cos2\theta))$$

where I is the photoelectron yield, P the degree of linear polarization of the photon beam and θ, the ejection angle with respect to the polarization axis. The asymmetry parameter β can take values ranging from -1 to +2. A β value of 2, with ejection along the electric vector axis, occurs when the electron is initially in an s state and is uncorrelated with other electrons, whereas $\beta=-1$ implies strong electron correlations.

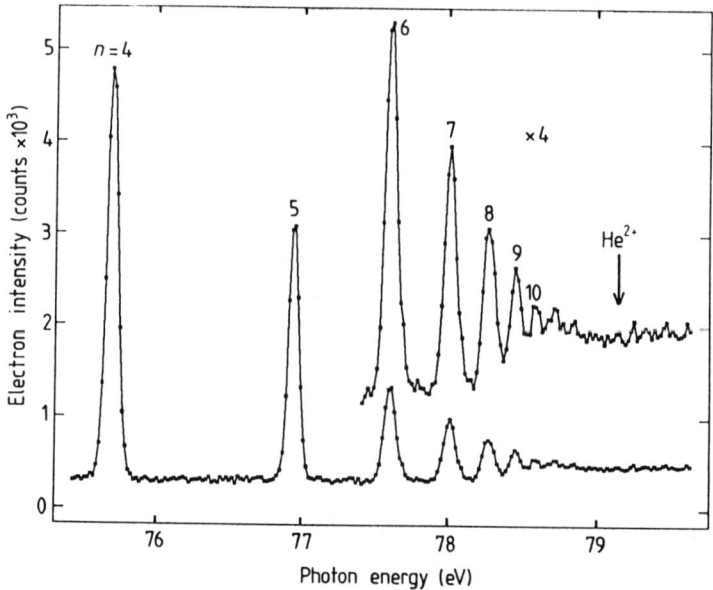

Fig. 3. Constant photoelectron energy spectrum for helium at 90° with $E_e=0.14eV$

Information on the angular behaviour of the photoelectrons was obtained from spectra of the type shown in fig. 3. This constant photoelectron energy spectrum was obtained at 90° by maintaining the collection energy constant at $E_e=0.14\,\mathrm{eV}$ and scanning the photon energy. The intensity of the double-ionization continuum corresponds to the yield of photoelectrons of energy E_e as a function of photon energy. The angular behaviour of this yield gives the β parameter for electrons with this particular value of E_e as a function of E.

The β parameters determined up to E=2eV are shown in fig. 4 for $E_e=0.25\,\mathrm{eV}$[8]. Also shown are the results of a separate study using the PEPICO technique[10] which, as can be seen, are in good agreement with those of the photoelectron study. The smooth curve is the result of calculations by Huetz et al[11] using Wannier theory and $\theta_0=91°$. At threshold β=-1 but rises as E increases reflecting the increasing effect of the photons which push the electrons in the direction of the electric field with respect to the torque exerted on the two electrons by the angular correlation which propels them in directions perpendicular to the polarization axis.

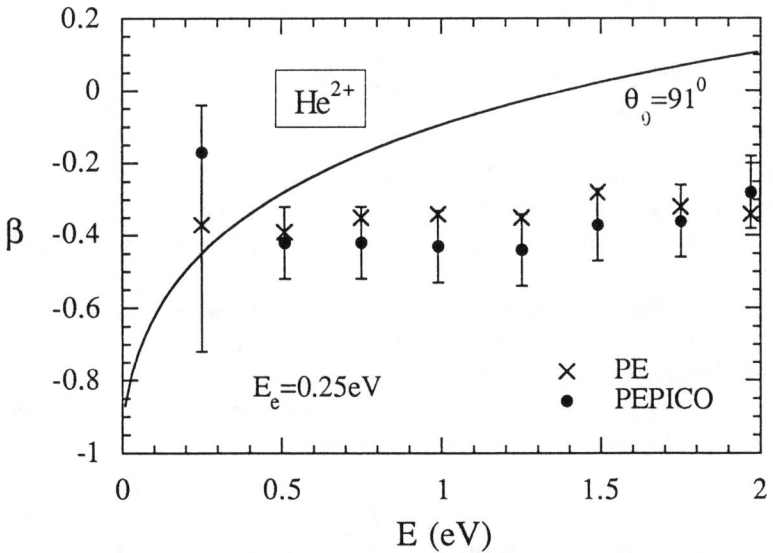

Fig. 4. Asymmetry parameters β for double photoionization of helium at energies E from threshold and for $E_e=0.25\,\mathrm{eV}$

The β values obtained above the double ionization threshold shown in fig. 4 do not follow the trends predicted by the Wannier model. These predictions are strongly dependent on the value of θ_0 used in the calculation. A rapid rise in β with excess energy implies weak angular correlation, whereas a slow rise in β indicates strong angular correlation or, equivalently, a more angularly rigid system. In this picture these measurements would imply that Wannier theory underestimates the degree of angular correlation between the two outgoing electrons. This was the first time that Wannier theory was found to fail in the many tests to which it had been put in recent years. Recently, Schwarzkopf et al[12] performed the very first study of the mutual angular distributions between the two electrons in helium. These observations were made 10eV from threshold. Nevertheless, Wannier theory could be made to agree with experiment, but only on condition that a value $\theta_0=43°$ was used.

The first *ab-initio* calculations of near threshold double photoionization in helium were made recently by Maulbetsch and Briggs[13,14], and these confirmed the dominant role of electron correlation in the final state. It was also shown that, contrary to the Wannier situation, the behaviour of β is strongly dependent on the energy partitioning between the two electrons. These calculations were in agreement with the values of β shown in fig. 4, which were obtained under the special conditions wherein only electrons with 0.25eV energy were registered while the photon energy was scanned. During such a scan the energy of the unobserved electron varies in the range $0<E_e<1.75eV$.

The asymmetry parameters for the extreme case of unequal energy partitioning between the two electrons, where one electron has near-zero energy and the other carries off the entire excess energy E have also been measured. This is the situation where electron correlation is now expected to have the least effect. These observations have been made for the $2p^{-2}$ ^1S state of neon[15] which has the same attributes as those of helium, i.e. only one continuum state ($^1P^o$), and is also free of indirect processes, but has the advantage of a larger cross section. The measurements were made using the instrument shown in fig. 1 by operating one analyzer in the threshold mode and setting the other to measure electrons with energy E and recording the coincidence yield at different angles.

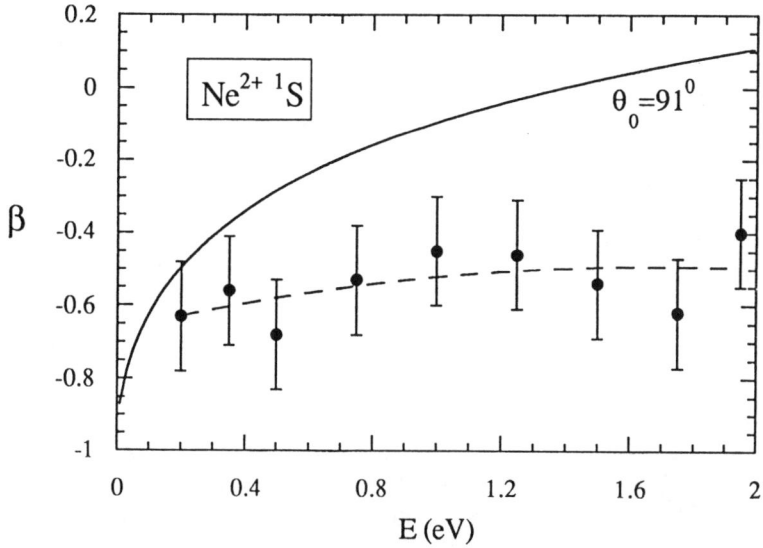

Fig. 5. Asymmetry parameters β for double photoionization of neon to the ^1S state for $E_e = E$

The measured values of β are displayed in fig. 5. As can be seen, the predictions of Wannier theory[12] which are independent of energy partitioning do not, as is the case for helium, give a good account of the experimental observations. As mentioned above, the *ab-initio* calculations[14,15] indicate that for unequal energy sharing β would be roughly flat over the present energy range as is observed. Only in the case of equal energy sharing is β expected to have considerable slope. In any case, the negative values of β, which are still near -0.5 at E=2eV, indicate that even in this extreme situation of unequal energy sharing the angular electron correlation still plays a dominant role.

STATE SELECTIVITY OF DOUBLE PHOTOIONIZATION

Analysis of the electrons in the Wannier geometry has shown that the wave function of the electron pair has special properties[3,4]. Particular symmetries are considered to be "favoured" or "unfavoured" depending whether the wave function has an anti-node or a node at this geometry, respectively. For example, in the case of the heavier rare gases, removal of two

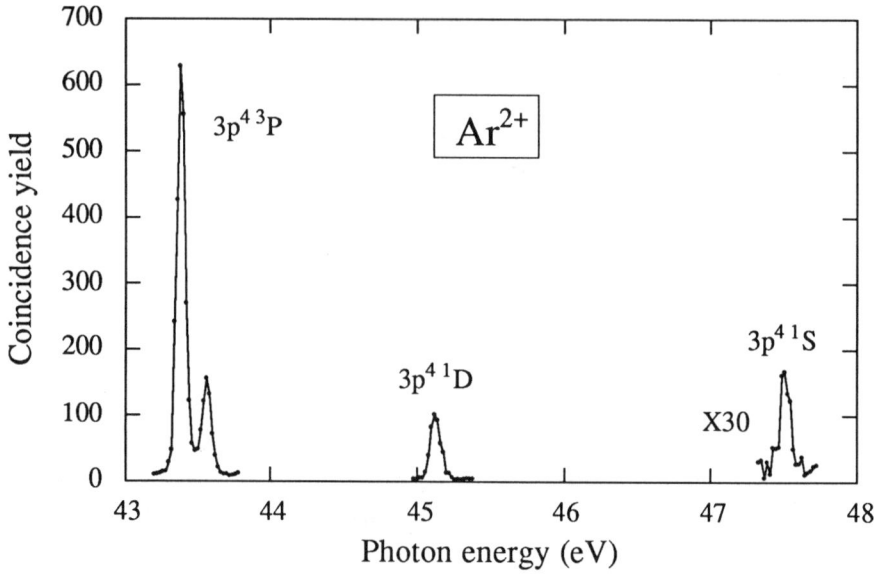

Fig. 6. Threshold photoelectrons coincidence (TPEsCO) spectrum for argon

valence p electrons leads to ^3P, ^1D and ^1S states, of which the ^3P state would be "favoured" and the ^1D and ^1S "unfavoured". This symmetry selectivity has been extended by Huetz et al[11] who determined that all the above states should be formed according to a weak propensity rule following a hierarchy of "favouredness" decreasing from ^3P to ^1D to ^1S.

Observations of the partial cross sections for near-threshold double photoionization of the heavier rare gases[7] have been made using a novel technique wherein coincidences between two essentially zero energy electrons are recorded. This is achieved by operating both analyzers of fig. 1 in the threshold mode and balancing the potentials on the extractor electrodes so that the yield of near-zero energy electrons is shared between them. A coincidence signal is then obtained each time the photon energy scans through a double ionization threshold. A spectrum obtained in this way is then a "threshold photoelectrons coincidence spectrum" (TPEsCO) and is the double ionization analogue of the threshold electron spectrum presented above in fig. 2.

A TPEsCO spectrum for argon is presented in fig. 6 showing the first three double ion states. As can be seen, the observed intensities appear to follow the predictions of Wannier theory. However, it must be kept in mind that near threshold double photoionization in the heavier rare gases is dominated by indirect autoionization processes[16] and that this is particularly true for the two lowest states.

The extension of the TEPEsCO technique to the study of molecules is proving to be a particularly interesting area of research and, for example, for oxygen has allowed, for the first time, the ground state of the doubly charged ion to be probed, revealing a vibrational progression containing 18 levels[17].

CONCLUSION

It is now some 40 years since Wannier proposed his classic theory which predicted that the dynamics of two escaping electrons would be governed by the correlation between them. However, it is only in recent times that experiments have been able to test the predictions proposed by this theory and its extensions. In two separate studies using threshold photoelectron and PEPICO techniques we have observed the near-threshold behaviour of the double photoionization cross section of helium. This behavior is compatible with the predictions of theory and would have a validity range of about 2eV. The heavier rare gases have also received attention[16] but the study of phenomena related to Wannier theory is hampered by the presence of indirect processes.

The angular behaviour of photoelectrons resulting from double ionization of helium and neon have been studied. These observations have revealed weaknesses in the Wannier theory which underestimates the angular correlation between the two electrons. Recent *ab-initio* theory confirms the important role of electron correlation but indicated that it depended on the relative energies of the two electrons.

Finally, the state selectivity predictions of Wannier theory have been examined using a new technique which involves the detection in coincidence of two near-zero energy electrons. The cross sections of the first three double ion states of the heavier rare gases formed by photoionization have been observed at threshold and would appear to follow the expectations of theory.

ACKNOWLEDGEMENTS

We are most grateful to the Daresbury SRS facility and staff for the excellent working conditions. We are particularly indebted to Dr Marion Hayes for providing the data acquisition and instrumental control program

REFERENCES

1. G. H. Wannier, Phys.Rev. 90, 817 (1953)
2. F. H. Read, Electron Impact Ionization. Eds. T. D. Märk and G. H. Dunn (Vienna, Springer-Verlag, 1985) p. 42.
3. A. D. Stauffer, Phys. Lett. 91A 114 (1982)
4. C. H. Greene and A. R. P. Rau, J. Phys. B. 16, 99 (1983)
5. S. Cvejanović and F. H. Read, J. Phys. B:. 7, 1180 (1974)
6. R. I. Hall, A. G. McConkey, K. Ellis, G. Dawber, L. Avaldi, M. A. MacDonald and G. C. King, Meas. Sci. Tech. 3, 316 (1992)
7. R. I. Hall, G. Dawber, A. G. McConkey, M. A. MacDonald and G. C. King, Z. Phys. D, 23, 377 (1992)
8. R. I. Hall, L. Avaldi, G. Dawber, M. Zubek, K. Ellis and G. C. King, J. Phys. B. 24, 115 (1991)
9. H. Kossmann, V. Schmidt and T. Andersen, Phys. Rev. Lett. 60, 1266 (1988)
10. R. I. Hall, A.G. McConkey, L. Avaldi, K. Ellis, M.A. MacDonald, G. Dawber and G. C. King, J. Phys. B. 25, 1195 (1992)
11. A. Huetz, P. Selles, D. Waymel and J. Mazeau, J. Phys. B. 24, 1917 (1991)
12. O. Schwarzkopf, B. Krässig, J. Elmiger and V. Schmidt, Phys. Rev. Lett. 70, 3008 (1993)
13. F. Maulbetsch and J. S. Briggs, Phys. Rev. Lett. 68, 2004 (1992)
14. F. Maulbetsch and J. S. Briggs, J. Phys. B. 26, 1679 (1993)
15. R. I. Hall, G. Dawber, A. G. McConkey, M. A. MacDonald and G. C. King, J. Phys. B. to be published
16. R. I. Hall, K. Ellis, A. G. McConkey, G. Dawber, L. Avaldi, M. A. MacDonald and G. C. King, J. Phys. B. 25, 377 (1992)
17. R. I. Hall, G. Dawber, A. G. McConkey, M. A. MacDonald and G. C. King, Phys. Rev. Lett. 68, 2751 (1992)

ANGULAR DISTRIBUTION AND SPIN POLARIZATION OF PHOTOINDUCED AUGER ELECTRONS

N. M. Kabachnik
Institute of Nuclear Physics, Moscow State University,
Moscow 119899, Russia

ABSTRACT

Anisotropy of the angular distribution of Auger electrons induced by photoionization or photoexcitation of inner atomic shells is discussed. Recent calculations of normal and resonant Auger transitions are compared with experimental data. Resonance photoionization of laser excited and aligned atoms is considered. Coincidence measurements of the angular correlation between photo and Auger electrons are discussed as well as spin polarization of Auger electrons produced in photoionization by circularly polarized light.

INTRODUCTION

When a highly excited state of an ion or atom with a vacancy in an inner shell decays via an Auger transition the angular distribution of the emitted Auger electrons may be nonisotropic[1]. The anisotropy is due to the nonstatistical population of the magnetic substates in the process of photoionization or photoexcitation[1,2] which reflects asymmetry of the photon-atom interaction. Study of the Auger electron angular distribution provides additional and more refined information as compared to the conventional measurements of intensities of the Auger electron spectra. This information concerns both the Auger decay process and the primary photoexcitation or photoionization processes which created the vacancy. Until recently, experimental investigations of the anisotropy of Auger emission produced by photoionization were rather rare[3]. However, the progress in synchrotron radiation sources, monochromators and energy analyzers has led to a rapid increase of the number of such experiments[4-13]. The results of these measurements and the problems connected with their interpretation as well as theoretical approaches to the Auger emission anisotropy have been discussed in several recent reviews[14-16].

Within the framework of a conventional model which considers Auger emission as a two-step process the angular distribution of Auger electrons produced in photoabsorption of unpolarized or linearly polarized light by an unpolarized target atom may be written in the following form[17,18]

$$W(\Theta) = \frac{W_0}{4\pi}\left(1 + \alpha_2 A_{20} P_2(\cos\Theta)\right) \tag{1}$$

Here Θ is the angle between the Z-axis and the direction of Auger electron emission, $P_2(\cos\Theta)$ is the second Legendre polynomial, and W_0 is the total probability of Auger emission per unit time. The Z-axis is chosen along the direction of the photon beam (for unpolarized radiation) or along the polarization vector of photons (for linearly polarized radiation). The angular distribution (1) is similar, naturally, to the angular distribution of photoelectrons in the dipole photoionization. However, for Auger electrons, the anisotropy of the angular distribution is determined by the product of two factors $\beta = \alpha_2 A_{20}$. One of them, A_{20}, is an alignment parameter which describes the anisotropy of the decaying state. The alignment parameter A_{20} can be expressed in terms of the substate cross sections $\sigma(J,M)$. For example, for $J = 3/2$ ionization one has

$$A_{20} = \frac{\sigma(3/2,3/2) - \sigma(3/2,1/2)}{\sigma(3/2,3/2) + \sigma(3/2,1/2)} \qquad (2)$$

The coefficient α_2 is the Auger decay anisotropy parameter, which describes the intrinsic anisotropy of the decay and depends on the contributing Auger amplitudes. The factorization of the anisotropy coefficient reflects the two-step character of the Auger emission.

ANGULAR DISTRIBUTION OF AUGER ELECTRONS IN NON-COINCIDENCE EXPERIMENTS

If one knows the alignment of an excited state from experiment (see below) or from theoretical considerations, then measurements of the angular distribution of Auger electrons for different transitions from this state give the values of the α_2 parameters which provide information on the Auger decay process. In this section we discuss the results of recent measurements of α_2 for Auger transitions in noble gases and their theoretical description.

In general the anisotropy parameter α_2 can be expressed in terms of Auger amplitudes as follows[19].

$$\alpha_2 = \sum_{lj,l'j'} \alpha_{lj,l'j'} \operatorname{Re}(M_{lj} M_{l'j'}^*) \Big/ \sum_{lj} |M_{lj}|^2 \qquad (3)$$

Here $M_{lj} = \langle \Psi_f(J')\varepsilon lj; J|V|\Psi_i(J)\rangle$ is an Auger decay amplitude corresponding to a definite channel characterized by the final ionic state and the total and orbital angular momenta of the emitted electron.

From the expression (3) it is clear that if, according to selection rules, only one partial wave (one channel) is allowed (for example, if $J' = 0$) then α_2 is only a geometrical quantity independent of the Auger amplitudes[19,20]. Such transitions have been used in experiments to determine the alignment A_{20} of the ionic state.

For multichannel transitions α_2 is, in principle, sensitive to the Auger decay dynamics and may serve as a good testing ground for theories. But even in this case, if one of the partial amplitudes is strongly dominating, the anisotropy parameter α_2 is weakly sensitive to the details of the theoretical description. Most interesting are those transitions with several more or less equally contributing channels.

A. Anisotropy of Normal Auger Transitions

As an example in Table I the experimental and some theoretical data for the N_5-$O_{2,3}$ $O_{2,3}$ Auger transition in Xe are compared. The transition to the $^3P_0$ state is a typical example of the transitions where only one partial wave is possible in any coupling scheme. The value of α_2 is identical in all calculations. On the contrary, the anisotropy of the transition to the $^1D_2$ state is especially sensitive to the models. Our early calculation was done in LS coupling with Hartree-Slater wave functions[21]. For the $^1D_2$ state the result contradicts experiment. It changes dramatically in calculations with intermediate coupling and Dirac-Fock wave functions[22]. Similar results were obtained by Chen[23] in the multiconfiguration Dirac-Fock approach (MCDF). Recently[24], an even more sophisticated model has been used for calculating α_2, which takes into account the continuum channel coupling, exchange and relaxation effect (Multichannel MCDF). The overall agreement with experiment is very good. Note that experimental data for some transitions are inconsistent as, for example, for the $^3P_2$ state (see Table I). Our calculations confirm the data of Kämmerling et al.[9].

Table I Angular anisotropy parameter α_2 for the Xe N_5-$O_{2,3}$ $O_{2,3}$ Auger transitions

Final state	Experiment			Theory		
	KKS[9]	Becker et al.[25]	Whitfield et al.[13]	KS[21] (LS + HS)	KLM[22] (IC + DF)	TKA[24] (MMCDF)
$^3P_2$	-0.47 (13)	-	-0.09 (10)	-0.382	-0.227	-0.385
$^3P_1$	-0.77 (17)	-1.0 (2)	-0.69 (10)	-0.748	-0.734	-0.743
$^3P_0$	-1.07 (10)	-1.2 (2)	-	-1.069	-1.069	-1.069
$^1D_2$	0.24 (10)	0.31	0.30 (12)	-0.592	0.139	0.094

B. Anisotropy of Resonant Auger Decay

In the Auger decay of resonantly excited atomic states an unusually large degree of angular anisotropy has been revealed by Carlson et al.[5] and confirmed in later experiments[6,8,9,25]. Partly, the large anisotropy of resonant Auger spectra is explained by a large value of the alignment produced in photoexcitation as compared with photoionization. In fact, if a closed shell atom is excited by a linearly polarized photon, only an $m = 0$ substate can be excited. Therefore, the alignment of the excited $J = 1$ state is $A_{20} = -\sqrt{2}$.

However, for a detailed explanation of the experimental data including the strong variation of the asymmetry parameter β_2 from line to line, one needs a systematic calculation of the anisotropy coefficient α_2. In this connection a spectator model of the resonant Auger decay was exploited[9,26,27]. Within a spectator model the excited electron is treated as a spectator which does not take part in the decay. Using this model one can calculate the corresponding Auger amplitudes and the asymmetry parameter with various degrees of sophistication In the simplest case of the so-called strict spectator model (SSM) [26], the interaction of the spectator electron with core electrons is ignored. In this approximation one can easily relate the amplitude of a resonant Auger transition with that of a corresponding normal transition[9,26]. Calculations within the SSM for resonant Auger transitions in Ar, Kr and Xe[26] give satisfactory agreement with experiments. However, a strong anisotropy of the most energetic lines of the spectra was not reproduced by the model (see Table II, lines 1 a, 1 b).

Table II   Asymmetry parameter $\beta = \alpha_2 A_{20}$ for some of the Xe $4d_{5/2}^{-1} 6p \to 5p^{-2} 6p$ transitions

Line	Final state	Experiment		Theory		
		KKS[9]	Becker et al.[25]	HKL[26] (SSM)	HKLB[27] (ICSM)	TAK[28] (MCDF)
1 a	$5p^4$ ($^3P_2$) $6p$ $[2]_{3/2, 5/2}$	-0.60 (3)	-0.67 (5)	0.002	-0.976	-0.647
1 b	$5p^4$ ($^3P_2$) $6p$ $[3]_{5/2, 7/2}$ $[1]_{1/2}$	-0.90 (2)	-0.93 (3)	-0.478	-0.726	-0.817
1 c	$5p^4$ ($^3P_2$) $6p$ $[1]_{3/2}$	1.31 (2)	1.35 (6)	1.014	0.972	1.025
3 a	$5p^4$ ($^1D_2$) $6p$ $[1]_{3/2}$, $[3]_{5/2,\,7/2}$	0.23 (2)	0.55 (5)	-0.132	-0.048	0.198
3 b	$5p^4$ ($^1D_2$) $6p$ $[1]_{1/2}$, $[2]_{3/2,\,5/2}$	0.33 (5)	0.46 (5)	-0.081	0.007	0.121

In a more elaborate model the interaction of the spectator electron with the core was taken into account which leads to the mixing of the core states[27]. However, the radial integrals for the resonant Auger transitions were supposed to be the same as for normal ones. This model yields much better results confirming the importance of the electron-core interaction. Finally, a complete MCDF calculation has recently been made[28,29] which gives very good agreement with experiment.

C. Core Resonances Excited in Laser Aligned Atoms

Recently, much attention has been paid to the photoionization of the laser excited and aligned atoms. Angular resolved photoelectron spectroscopy of excited atoms provides important information on the dynamics of photoionization. With the usage of synchrotron radiation it becomes possible to study the core-excited resonances by means of double photon resonant photoionization. In this way resonances could be studied which are inaccessible by single photon ionization. The spectrum of such resonances is very complicated and measurements of the angular distributions of autoionizing (Auger) electrons in this case are important for line identification.

Angular distribution of photoelectrons in photoionization of polarized atoms has been considered in the general case by Klar and Kleinpoppen[30]. However, in the region of resonances, the angular distribution has several peculiarities. Theoretically, resonance photoionization of polarized atoms was studied in refs.[31-33].

As an example we consider here a particular experiment[32] on Li with contrapropagating laser and synchrotron radiation beams. Both beams were linearly polarized and η is the angle between two polarization vectors (see Fig. 1). The angular distribution of autoionizing electrons was measured in the plane perpendicular to the photon beams. For this particular geometry the angular distribution may be presented as

$$W(\Theta) = \sum_{k=0,2,4} \sum_{q=0,1,2} a_{kq} \rho_{kq}(\eta) P_k^q(\cos\Theta) \qquad (4)$$

where P_k^q is the associated Legendre polynomial. The dependence on η is described by statistical tensors of the decaying state $\rho_{kq}(\eta)$. Note that due to the two-photon character of excitation the 4th order Legendre polynomials contribute making the angular distribution patterns more complicated than in the one-photon case. These patterns are very characteristic for every particular angular moment of the resonant state. Fig. 2 illustrates the angular distribution for the transition $1s^2 2p\ ^2P_{3/2} \to 1s2p^2\ ^2D \to 1s^2\ ^1S + \varepsilon l$ in Li. Peculiarities of the angular distributions have been used to advantage for line identification in ref.[32].

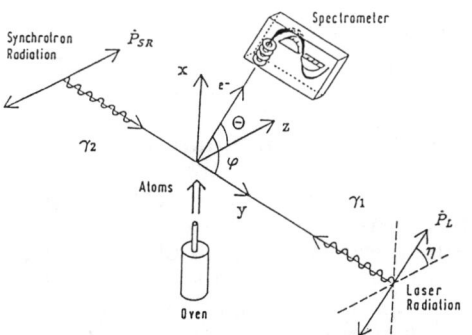

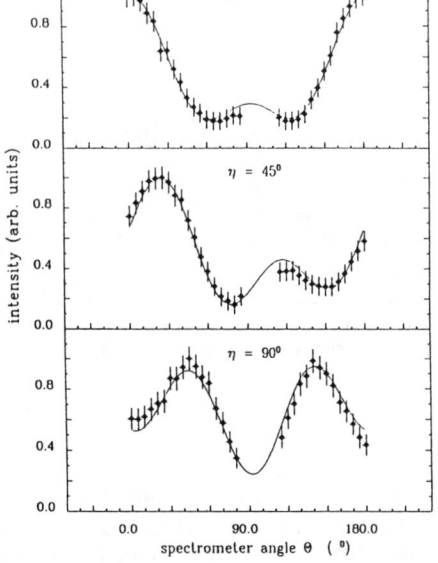

Fig. 1. Scheme of the experiment.

Fig. 2. Intensity of the electrons, emitted in the decay of 1s2p2 ^2D state of Li vs the spectrometer angle for three different angles between the two polarization axes.

ANGULAR CORRELATION BETWEEN AN AUGER ELECTRON AND A PHOTOELECTRON

In the previous section we discussed inner-shell photoionization experiments with only one electron (namely the Auger electron) detected. In earlier papers[34,35] coincidence measurements were suggested for inner-shell study with two electrons in the final state (i. e. photoelectron and Auger electron) detected in coincidence. It was shown that experimental investigation of the angular correlation between two escaping electrons can provide more detailed information on both stages of the process, ionization and decay, as compared with separate non-coincidence measurements of angular distributions of photoelectron and Auger electron. Difficulties and advantages of the coincidence experiments exploring the photon-induced double ionization processes were discussed recently in a review paper by V. Schmidt[36].

Consider the angular correlation between two escaping electrons in a conventional two-step model. At the first step, photoionization of an inner shell occurs and a photoelectron is detected in a direction $\bar{n}_1$. In general, the ionic state polarization can be described by a set of statistical tensors $A_{kq}(J,\bar{n}_1)$, $k \leq 2J$, $-k \leq q \leq k$ which are functions of the photoelectron emission angles. The values of the tensors are determined by the photoionization amplitudes and their relative phases[35,37]. If Auger decay of the excited state occurs, the angular distribution of emitted electrons is completeley determined by the tensors $A_{kq}(J,\bar{n}_1)$[35]:

$$W(\Theta_2,\varphi_2) = \frac{W_0}{4\pi}\left[1 + \sum_{k=2,4,\ldots} \alpha_k \sum_q A_{kq}(J,\bar{n}_1)\left(\frac{4\pi}{2k+1}\right)^{1/2} Y_{kq}(\Theta_2,\varphi_2)\right] \quad (5)$$

where α_k are the same anisotropy coefficients, determined by exp. (3). The symmetry conditions of a particular experiment impose some restrictions on the number of independent non-zero tensor components.

Recently, the first measurement of the angular correlation between two escaping electrons in two-step double photoionization of a Xe atom was reported[11]. The yield of the $4d_{5/2}$ photoelectrons produced by a linearly polarised photon beam was measured in coincidence with the N_5-$O_{2,3}$ $O_{2,3}$ ($^1S_0$) Auger electrons. Both electron analyzers were placed in a plane perpendicular to the photon beam. The position of the second analyzer which scans the Auger lines was varied. For this particular geometry the theoretical expression (5) can be reduced to the following form[11]:

$$W(\Theta_2) = A_0 + A_2 \cos 2\Theta_2 + B_2 \sin 2\Theta_2 + A_4 \cos 4\Theta_2 + B_4 \sin 4\Theta_2 \quad (6)$$

The results of the experiment are shown in Fig. 3 as points with error bars, and the solid line represents a least-square fit according to expression (6). As could be expected from (6) the experimental angular distribution exhibits the presence of at least two loops which confirm the existence of higher terms in the angular distribution expansion. Combining the experimental values of the ratios A_i/A_0 and B_i/A_0 with the experimental data for the photoionization cross section, the angular asymmetry parameter of the photoelectron and the alignment parameter in non-coincident measurements, one can derive the values of the dipole matrix elements and their relative phases for all ionization channnnels[11], which can serve as a stringent test for theories.

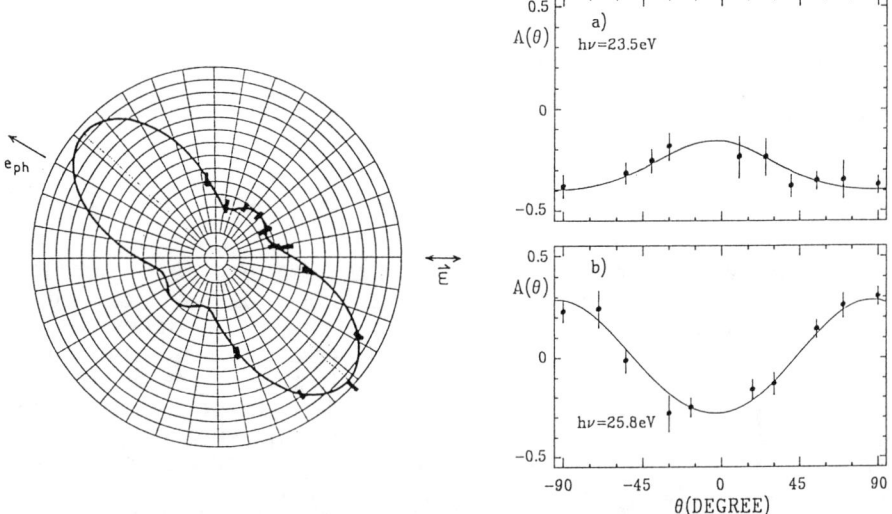

Fig. 3. Polar plot of coincidence intesities (bars) between $4d_{5/2}$ photo and N_5-$O_{23}O_{23}$ Auger electrons in Xe.

Fig. 4. Angular dependence of the Auger electron spin polarization for the Auger lines (a) O_3-$P_1P_1(^1S_0)$ and (b) O_2-$P_1P_1(^1S_0)$ in Ba.

SPIN POLARIZATION OF AUGER ELECTRONS

A new field of research including many interesting applications in surface and solid state physics is opening now with the use of spin-sensitive detectors in Auger electron spectroscopy. The spin polarization of Auger electrons is caused mainly by the spin-orbit interaction. It is convenient to discriminate two different mechanisms for the production of an electron spin polarization: dynamic polarization and polarization transfer[38]. The former arises even when the photon beam and the target are non-polarized[39]. If the intermediate state formed in photoabsorption is aligned, the outgoing Auger electron can be spin polarized by a final state interaction with the core charge distribution. On the other hand, in the case of ionization by circularly polarized light, an orientation of intermediate state can be transferred partially to the Auger electron spin. This so-called polarization transfer can be viewed as a consequence of the conservation of the angular momentum.

The few experiments performed[40,41] (however with particle beams) and theoretical calculations[21,22,24] show that the dynamic polarization is in general small. On the contrary, the polarization transfer has been predicted to be a very effective method for creating polarized Auger electrons[42,43].

Within the framework of the standard two-step model, the components of the Auger-electron polarization vector can be written in the form similar to the photoelectron polarization[42] (we use the reference frame with the Z axis along the photon beam, XOZ is the reaction plane, Θ is the angle between the direction of emission of an Auger electron and the quantization axis); for example the Z-component is:

$$P_z = \frac{\beta_1 A_{10} + \gamma_1 A_{10} P_2(\cos\Theta)}{1 + \alpha_2 A_{20} P_2(\cos\Theta)} \quad (7)$$

The denominator in expression (7) determines, as earlier, the angular distribution of the Auger electron. The tensor A_{10} describes the orientation of the photoion, and the tensor A_{20} describes its alignment. Both tensors depend on photoionization amplitudes. The orientation parameter A_{10} is proportional to the degree of circular polarization. The calculations[42] show that, in contrast to the alignment, A_{20}, which is large only in the near threshold region and in the vicinity of the Cooper minimum, the orientation parameter A_{10} is large (~ 0.6, 0.7) in a wide range of photon energies.

The parameters β_1, γ_1 depend only on the amplitudes of Auger decay. The parameter β_1 determines the net polarization (integral with respect to the angle Θ) arising due only to the polarization transfer. The parameter γ_1 determines the differential polarization. It is interesting that the parameters β_1 and γ_1 are non-zero even for a one channel Auger transition.

Recently, the first experimental study of the polarization transfer was reported. The spin polarization of Auger electrons from a Ba 5p hole state oriented by circularly polarized synchrotron radiation has been measured[44]. The experiment provides all parameters characterizing the spin polarization components. Fig. 4 shows the Z-component of the spin polarization vector, $P_z \equiv A(\Theta)$, as a function of the electron emission angle for two Auger lines $5p_{3/2}^{-1}\text{-}6s^{-2}$ and $5p_{1/2}^{-1}\text{-}6s^{-2}$ of Ba. Note that in both cases the polarization is large, but the angle dependence is quite different: whereas for the $5p_{3/2}$ hole the net polarization is large, and the variation is small, for the $5p_{1/2}$ hole the polarization is strongly angular dependent, but the net polarization is small.

A study of polarization transfer opens the possibility to explore a new physical characteristic of the vacancy – its orientation proportional to the mean value of the vacancy angular moment. Its measurements provide new independent information about the photoionization amplitudes. Theoretical calculations predict that the vacancy orientation is rather sensitive to the dynamics of the photoprocess.

Support from the Max-Planck-Gesellschaft and the hospitality of the Fritz-Haber-Insitute (Berlin) is gratefully acknowledged. I am grateful to N. Böwering, R. Kunze, and B. Langer for helping in preparation of this report.

REFERENCES

1. W. Mehlhorn, Phys. Lett. A **26**, 166 (1968).
2. S. Flügge, W. Mehlhorn, and V. Schmidt, Phys. Rev. Lett. **29**, 7 (1972).
3. S. H. Southworth, U. Becker, C. M. Truesdale, P. H. Kobrin, D. W. Lindle, S. Owaki, and D. A. Shirley, Phys. Rev. A **28**, 261 (1983).
4. A. Hausmann, B. Kämmerling, H. Kossmann, and V. Schmidt, Phys. Rev. Lett. **61**, 2669 (1988).
5. T. A. Carlson, D. R. Mullins, C. E. Beall, B. W. Yates, J. W. Taylor, D. W. Lindle, B. P. Pullen, and F. A. Grimm, Phys. Rev. Lett. **60**, 1382 (1988).
6. T. A. Carlson, D. R. Mullins, C. E. Beall, B. W. Yates, J. W. Taylor, D.W. Lindle, and F. A. Grimm, Phys. Rev. A **39**,1170 (1989).
7. B. Kämmerling, V. Schmidt, W. Mehlhorn, W. B. Peatman, F. Schaefers, and T. Schroeter, J. Phys. B **22**, L597 (1989).
8. U. Becker, in *The Physics of Electronic and Atomic Collisions (AIP Conf. Proc. 205)*, edited by A. Dalgarno, R. S. Freund, P. M. Koch, M. S. Lubell, and T. B. Lucatorto (AIP, New York, 1989) p. 162.
9. B. Kämmerling, B. Krässig, and V. Schmidt, J. Phys. B **23**, 4487 (1990).
10. C. D. Caldwell, in *X-ray and Inner-Shell Processes (AIP Conf. Proc. 215)*, edited by T. A. Carlson, M. O. Krause, and S. T. Manson, (AIP, New York, 1990) p. 685.
11. B. Kämmerling and V. Schmidt, Phys. Rev. Lett. **67**, 1848 (1991).
12. B. Kämmerling, B. Krässig, O. Schwarzkopf, J. P. Ribeiro, and V. Schmidt, J. Phys. B **25**, L5 (1992).
13. S. B. Whitfield, C. D. Caldwell, D. X. Huang, and M. O. Krause, J. Phys. B **25**, 4755 (1992).
14. W. Mehlhorn, in *X-ray and Inner-Shell Processes*, edited by T. A. Carlson, M. O. Krause, and S. T. Manson (AIP, New York, 1990) p. 465.
15. W. Mehlhorn, in *Proceedings of the International Workshop on Photoionization 1992*, edited by U. Becker and U. Heinzmann (AMS Press, New York, 1993) p. 13.
16. N. M. Kabachnik, ibid p. 20
17. E. G. Berezhko and N. M. Kabachnik, J. Phys. B **10**, 2467 (1977).
18. E. G. Berezhko, N. M. Kabachnik, and V. Rostovsky, J. Phys. B **11**, 1749 (1978).
19. N. M. Kabachnik and I. P. Sazhina, J. Phys. B **17**, 1335 (1984).
20. B. Cleff and W. Mehlhorn, J. Phys. B **7**, 605 (1974).
21. N. M. Kabachnik and I. P. Sazhina, J. Phys. B **21**, 267 (1988).

22. N. M. Kabachnik, B. Lohmann, and W. Mehlhorn, J. Phys. B **24**, 2249 (1991).
23. M. H. Chen, Phys. Rev. A **45**, 1684 (1992).
24. J. Tulkki, N. M. Kabachnik, and H. Aksela, Phys. Rev. A, in press.
25. U. Becker *et al.* The results are cited in ref. 22, 26.
26. U. Hergenhahn, N. M. Kabachnik, and B. Lohmann, J. Phys. B **24**, 4759 (1991).
27. U. Hergenhahn, B. Lohmann, N. M. Kabachnik, and U. Becker, J. Phys. B **26**, L117 (1993).
28. J. Tulkki, H. Aksela, and N. M. Kabachnik, Phys. Rev. A, in press.
29. M. H. Chen, Phys. Rev. A **47**, 3733 (1993)
30. H. Klar and H. Kleinpoppen, J. Phys. B **15**, 933 (1982).
31. V. V Balashov, A. N. Grum-Grzhimailo, and V. Zhadamba, Opt. Spectrosk. (USSR), **65**, 529 (1988).
32. M. Paler, C. Lorenz, E. v. Raven, J. Rüder, B. Sonntag, S. Baier, B. R. Müller, M. Schulze, H. Staiger, P. Zimmermann, and N. M. Kabachnik, Phys. Rev. Lett. **68**, 2285 (1992).
33. S. Baier, A. N. Grum-Grzhimailo, and N. M. Kabachnik, to be published.
34. S. C. McFarlane, J. Phys. B **8**, 895 (1975).
35. E. G. Berezhko and N. M. Kabachnik, J. Phys. B **12**, 2993 (1979).
36. V. Schmidt, in *X-ray and Inner-Shell Processes (AIP Conf. Proc. 215)*, edited by T. A. Carlson, M. O. Krause, and S. T. Manson (AIP, New York, 1990) p. 559.
37. N. M. Kabachnik, J. Phys. B **25**, L389, (1992).
38. H. Klar, J. Phys. B **13**, 4741 (1980).
39. N. M. Kabachnik, J. Phys. B **14**, L377 (1981).
40. U. Hahn, J. Semke, H. Merz, and J. Kessler, J. Phys. **18**, L417 (1985).
41. H. Merz and J. Semke, in *X-ray and Inner-Shell Processes (AIP Conf. Proc. 215)*, edited by T. A. Carlson, M. O. Krause, and S. T. Manson, (AIP, New York, 1990) p. 719.
42. N. M. Kabachnik and O. V. Lee, J. Phys. B **22**, 2705 (1989).
43. B. Lohmann, U. Hergenhahn, and N.M. Kabachnik, J. Phys. B, to be published.
44. R. Kuntze, M. Salzmann, N. Böwering, and U. Heinzmann, Phys. Rev. Lett, **70**,3716 (1993).

ANGLE RESOLVED PHOTOELECTRON SPECTROSCOPY OF LASER EXCITED AND ALIGNED ATOMS

M. Pahler
L.S.A.I., Université Paris Sud, Bât. 350, 91405 Orsay Cedex, France

ABSTRACT

Recent progress in angle dependent photoelectron spectroscopy of laser and synchrotron radiation excited and of laser aligned atoms is reported. The experimental results are discussed in comparison with the predictions of different theoretical approaches.

INTRODUCTION

The combination of dye lasers and vacuum ultraviolet (VUV) sources has opened up the possibility to investigate the relative innershell photoionisation cross sections of laser excited atoms[1-3]. These experiments allow for a stringent test of theoretical calculations. The experimental techniques applied and the results obtained in laser/synchrotron radiation combination experiments have been reviewed by Wuilleumier et al.[4]. Recent progress in the photoelectron detection efficiency and the considerable development of undulator sources allowed for measurements of the angular distribution of the outgoing VUV photoelectrons of laser excited atoms[5]. The determination of the angular distribution asymmetry parameter β leads not only to information on the photoionisation matrix elements like in measurements of the cross section, but also on the relative phases of the outgoing partial electron waves. Additionally to these attempts effort was concentrated on the extention of angular dependend VUV photoelectron spectroscopy by preparing the atoms in an aligned state[7,8]. These experiments can be used as a very sensitive probe for the symmetry character of the two photon excited core resonances.

LASER EXCITED ATOMS

Experimental arrangement

For relative cross section experiments it is very suitable to use the advantage of a partial angular integration of the electrons. This leads to a considerable increase in the electron counting rates, which enables even measurements of correlation satellite intensities emitted from laser excited atoms. Figure 1 shows the experimental arrangement of the pioneer-

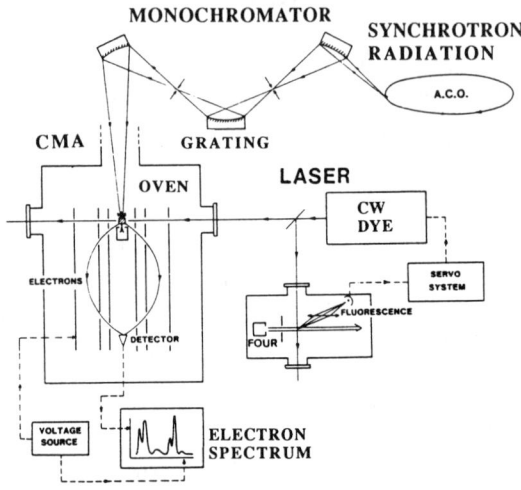

Figure 1. *Schematic diagram of the experimental set up used to study laser excited atoms (from Wuilleumier[4]).*

ing experiment combinding cw dye laser radiation with bending magnet synchrotron radiation used by Bizau and coworkers[9]. Synchrotron radiation from the storage ring ACO was monochromatized by a toroidal grating monochromator and focused onto the source volume of a 360° cylindrical mirror analyzer (CMA) colinear mounted to the VUV photon beam. The electron energy resolution was $\Delta E/E = 0.8\%$. Approximatlly 10^{12} photons s^{-1} in a 1% band pass were reached. A cw dye laser pumped the atoms emanating from a resistively heated furnace. No angular dependence is seen in the spectra because of the detection of the electrons under the magic angle $\Theta_m = 54.7°$ with respect to the propagation vector of the VUV radiation[9]. Alignment effects could be avoided due to the radiation trapping[10] caused by the high atomic density ($>10^{13}/cm^3$).

Relative Cross Sections

The branching ratio between satellite and photoline emission yields important information on the electron correlations. In figure 2 is compared the measured[1] ratio of the correlation satellite intensity to the 2p photoline intensity for ground state Na atoms (full circles) and for laser excited Na* atoms (empty circles). The resonant transition Na $2p^63s\ ^2S_{1/2} \rightarrow$ Na* $2p^63p\ ^2P_{3/2}$ was pumped with the cw laser radiation. The average branching ratio is about 50% in the excited state as opposed to 30% in the ground state. This ratio varies very slowly

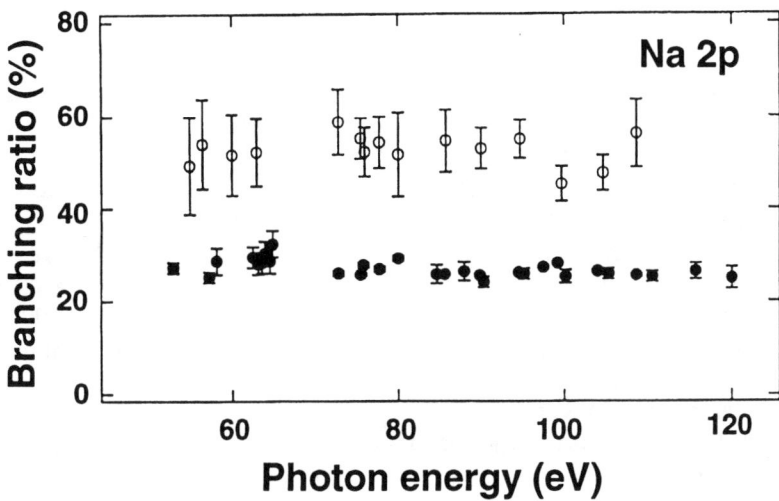

Figure 2. *Branching ratio of the correlation satellite intensity to the 2p photoline intensity for groundstate (full circles) and 3p laser excited (empty circles) Na (from Cubaynes[4]).*

with the photon energy. Further measurements[3] of ground state and 4p excited K atoms revealed also a strong enhancement of the correlation satellite intensity compared to the 3p photoline intensity of a factor of 2.
The most striking effect by exciting the valence electron for the transfer of oscillator strenght from the photoline towards the correlation satellites is expected by simple arguments[11] and theoretical calculations[13] for valence electron excited Li atoms. Very recent measurements[14] of ground state and 2p laser excited Li atoms at a photon energy of 92.2 eV have determined the branching ratio between the 1s photoline intensity and the corrosponding correlation satellite intensity. The result obtained for the branching ratio (1s3l ^{1,3}L)/(1s2p ^{1,3}P) for excited Li is 74% as compared to 25% for the (1s3l ^{1,3}L)/(1s2s ^{1,3}S) branching ratio in ground state Li atoms. Very recent R-Matrix calculations[15] are in excellent agreement with the measured value.

The shake up satellite enhancement by valence electron excitation is attributed to an ionic core relaxation[2,3] which leads to a higher overlap in the wave function of the final ionic and the initial state, which is higher for valence electron excited atoms. This behavior seems to be a general phenomenum in innershell photoionisation of excited alkaline atoms[4].

Angular Dependent Measurements

The measurement of the angular distribution asymmetry parameter β is complementary to the relative cross section experiments, because the determined quantity contains aditionally to the ionisation matrix elements the relative phases of the outgoing electron partial waves. The first angular dependent experiment of laser and synchrotron radiation excited direct photoionisation was performed by Cubaynes et al.[5] on Na. It was made by using six channeltrons placed in the focal plane of the CMA described above. Six electron spectra are taken simultaneously with the advantage to the classical experiments, where the electron spectrometer is rotated, that it minimizes data aquisition time and necessary calibrations. This measurement is a very sensitive probe of the theory because the relative phase of the outgoing electron waves can be affected by the change in the screening of the atomic core. Recent MBPT[17] calculations showed poor agreement with the measured 2p photoline cross sections[19] for laser excited Na and therefore is a comparison between the measured and calculated relative phases of the outgoing electron waves highly desirable. The results of the angle dependent 2p photoionisation of groundstate and 3p excited Na atoms are displayed in figure 3. The two asymmetry parameters cannot be distinguished experimentally beside a shift of the threshold by screening effects. This can be explained in a simple mo-

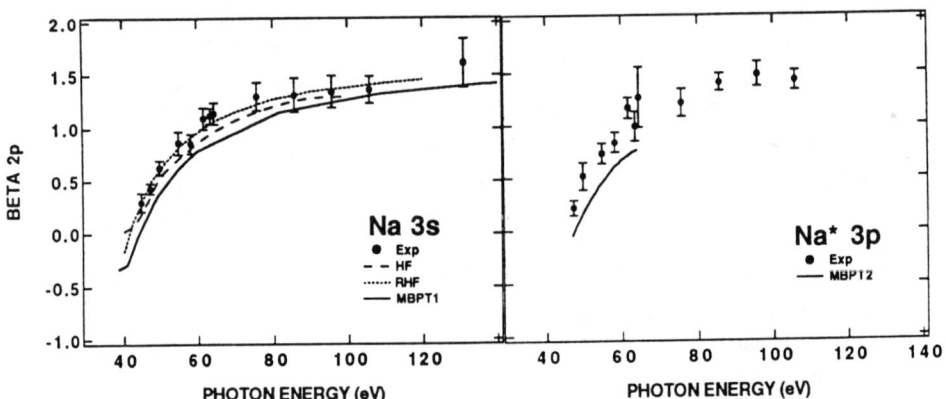

Figure 3. *Asymmetry parameter β of the 2p photoionisation from ground state (left spectrum) and 3p laser excited (right spectrum) Na atoms (from Rouvellou[12]).*

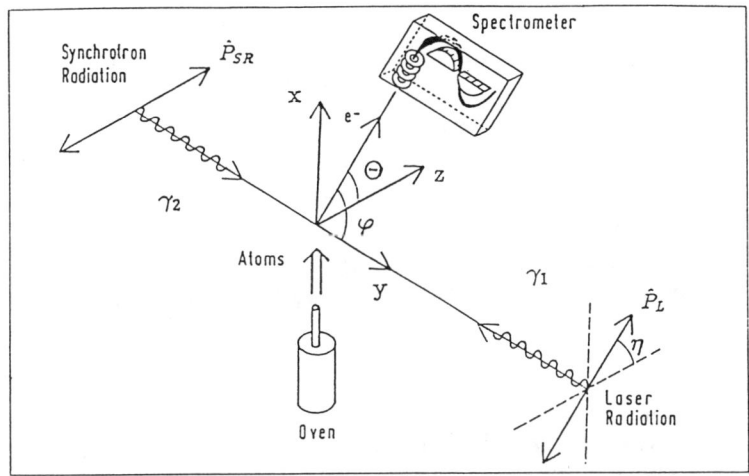

Figure 4. *Schematic representation of the experimental arrangement for two photon experiments on aligned atoms (from Pahler[9]).*

del by the fact that the two matrix elements of the outgoing partial electron waves are equally affected by the excitation of the valence electron and that the change in the relative phase is negligible[16]. Also this inter-pretation is supported by the MBPT[17,18] predictions which agree very well with the presented data for the β parameter.

LASER ALIGNED ATOMS

Alignment

Energy and angle resolved measurements allow for a detailed probing of the dynamics of the photoionisation process, but still they do not provide all the information for a full quantum mechanical characterisation. This information can be obtained by detecting the polarisation of the electrons[20], by measuring the angular distribution of the Auger electrons and photoelectrons[21], or by determinig the polarisation or the angular distribution of the fluorescence radiation[22]. The pioneering experiments on aligned atoms were made by two photon laser excitation to determine relative phases of the outgoing electron waves[23].

Alignment of an atom is a nonstatistical population probability of the states with different magnetic quantum numbers. In order to express the angular distribution of the outgoing electrons of an ensemble of aligned atoms in the most general form it is very suitable to use the density matrix formalism[24]. The angular distribution can be written as:

$$I(n_e, \alpha) = \mathrm{Tr}(\rho^T(n_e)) \quad (1)$$

n_e is the direction of emission of the electrons, α represents a set of quantum numbers and $\rho^T(n_e)$ the density matrix of the atoms. A very general application of the density matrix formalism on angular distributions was developed by Baier[25]. Within the LS coupling scheme it follows from formular 1 for the angular distribution of the outgoing electrons[26]:

$$I(n_e, L) = \sum_{KL} 2 \cdot \Pi \cdot \alpha_{KL} \cdot \sum_{QL} \rho_{KLQL}(\eta) \cdot Y_{KLQL}(n_e) \quad (2)$$

where α_{KL} is a kinematic parameter including the Auger matrix elements, ρ_{KLQL} are the statistical tensor components which depend on η the angle between the polarisation axes of the exciting laser- and VUV-radiation and Y_{KLQL} the spherical harmonics. The tensor rank K_L is restricted to $0 \leq K_L \leq 2L$. L is the total orbital angular momentum of the decaying state. The measurement of the angular distribution of the outgoing electrons from aligned atoms can be used as a very sensitive test for the symmetry of the decaying state, which is an important help for the assignment of the lines.

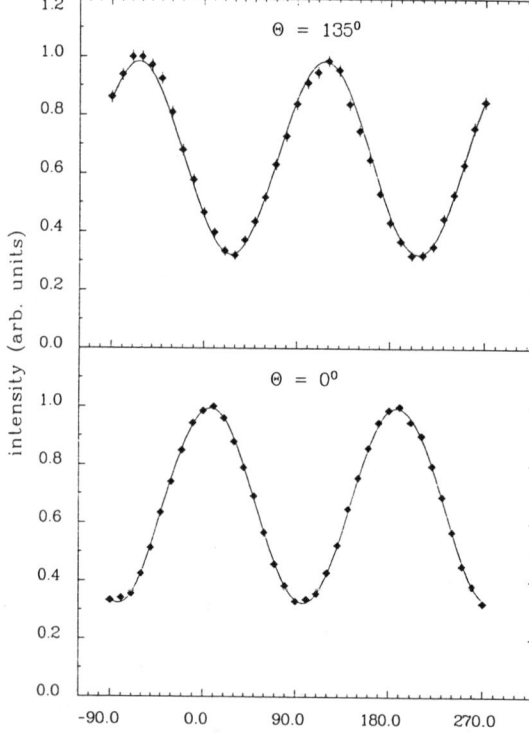

Figure 5. *Angle dependent spectrum taken in mode i) (see text) of laser and synchrotron radiation excited and aligned Li $1s2p^2\ ^2D$ autoionisation resonance (from Pahler[26]).*

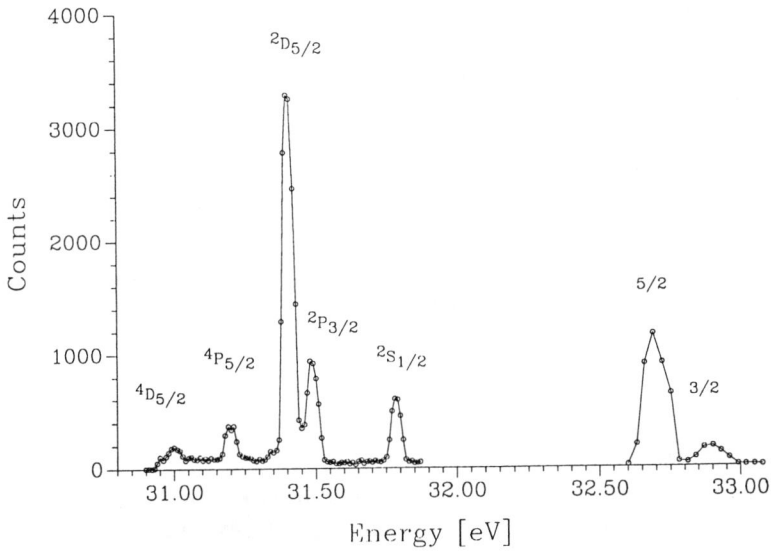

Figure 6. *Constant ionic state spectrum of the 2p subshell of 3p laser excited Na atoms (from Baier[25]).*

Angular Dependent Measurements

The experimental arrangement of the experiment used at BESSY[27] is schematically depicted in figure 4. The monochromatised high flux undulator radiation (bandpass 0.3eV at hν=60eV) and the laser beam (intensity ~100mW/mm^2) are propagating in opposite directions. The photons intersect an atom beam emanating from a resistively heated oven with a density of approximatly 10^{10} cm^{-3}. The electrons were analysed by a simulated hemispherical spectrometer (angular acceptance ±3°) with a four element electrostatic lens. The system can be operated in two modes: i) setting the spectrometer at a fixed angle and recording the electron intensity as a function of η by rotating the polarisation axis of the laser beam, and ii) by keeping η fixed and registering the electrons as a function of θ (n_e in formular 2) by rotating the electron analyser.

Figure 5 shows the results of 2p excited Li taken in mode i) at an energy of 61.06 eV with respect to the ground state. At this energy the process 1s^{2}2p ^{2}P$_{3/2}$ → 1s2p^2 ^{2}D$_{3/2,5/2}$ → 1s^2 ^{2}S$_{1/2}$ + εl is expected. The solid line represent a fit of formular 2 to the data and proves unambigously the

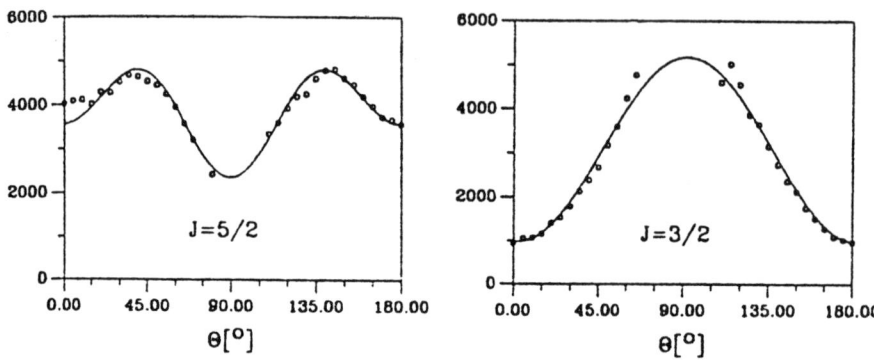

Figure 7. *Angle dependent photoionisation of the $2p^5(3s3p\ ^3P)$ $^2D_{5/2}$ autoionisation resonance of laser aligned Na in dependence of the laser polarisation axis (left) and of the electron spectrometer axis (right). (from Baier[25]).*

L=2 symmetry of this autoionisation resonance. The assignment and the resonance energies of several different resonances in this energy region was summarized by Pahler et al.[8].

Na is with its 11 electrons a more complicated atom than Li. Quartett terms and P autoionisation resonances are showing up in the spectra[26] due to violations of the LS coupling. Figure 6 shows the measured constant ionic state spectrum[26] for 3p excited Na[25]. This spectrum was recorded under the double magic angle[25] with the experimental set up described above, so it does not contain a dependency on the alignment of the intermediate state or the angular distribution. It consists of two groups of lines. First the $2p^5(3s3p\ ^3P)$ $^{2,4}L_J$ resonances with L=S,P,D and J=1/2, 3/2, 5/2 between 30.9 eV and 31.8 eV. The second group between 32.6 eV and 33.0 eV consists of the excitations $2p^5(3s3p\ ^1P)$ ^2L_J with L=S,P,D and J=1/2, 3/2, 5/2. As an example is displayed in figure 7 the angular distributions of two autoionisation resonances with J=5/2 (left) and J=3/2 (right) as measured by Baier[25]. The solid line represents a fit of formular (1) to the data considering the broken LS coupling. The results of the assignment and excitation energy are summarized in table 1.

The author thanks D.Cubaynes, B.F.Sonntag and F.J.Wuilleumier for discussions and a critical reading of the manuscript.

Table 1. *Assignment and VUV photon energy of the Na $2p^5(3s3p\ {}^{1,3}P)\ {}^{2,4}L_J$ autoionisation resonances (from Baier[25]).*

Assignment	hν [eV][25]	hν [eV][29]
$2p^5(3s3p\ {}^3P)\ {}^4D_{5/2}$	30.995	-
${}^4D_{3/2}$	31.072	-
${}^4P_{5/2}$	31.190	31.192
${}^4P_{3/2}$	31.239	31.242
${}^2D_{3/2}$	31.340	31.341
${}^2D_{5/2}$	31.397	31.398
${}^2P_{1/2}$	31.444	31.447
${}^2P_{3/2}$	31.481	31.483
${}^2S_{3/2}$	31.774	31.776
$2p^5(3s3p\ {}^1P)\ {}^2D_{5/2}$	32.686	32.687
${}^2S_{1/2}$	-	31.711
${}^2P_{3/2}$	32.738	32.741
${}^2D_{3/2}$	32.869	32.871
${}^2P_{1/2}$	32.902	32.904

REFERENCES

1. D.Cubaynes, J.M.Bizau, F.J.Wuilleumier, B.Carre and F.Gounand, Phys. Rev. Lett., 63, 2460 (1989).
2. M.Richter, J.M.Bizau, D.Cubaynes, T.Menzel, F.J.Wuilleumier and B.Carre, Europhys. Lett., 12, 35 (1990).
3. F.J.Wuilleumier, D.Cubaynes and J.M.Bizau, *Atomic and Molecular Physics*, eds C.Cisneros, I.Alvarez and T.J.Morgan (World Scientific: Singapore), 474 (1991).
4. F.J.Wuilleumier, D.L.Ederer and J.L.Picque, Adv. At. Mol. Phys. 23, 197 (1987).
5. D.Cubaynes, J.M.Bizau, C.Marinelli, T.J.Morgan, J.Novak, M.Pahler, B.Rouvellou, F.Schlachter, N.Berrah Mansour and F.J.Wuilleumier,Tenth Int. Conference on Vacuum Ultraviolet Radiation Phys., TU80 (1992).
6. D.Cubaynes, J.M.Bizau, T.J.Morgan, B.Carre and F.J.Wuilleumier,Tenth Int. Conference on Vacuum Ultraviolet Radiation Phys., TU79 (1992).
7. M.Meyer, M.Pahler, T.Prescher, E.v.Raven, M.Richter, B.Sonntag, S.Baier, W.Fiedler, B.R.Müller, M.Schulze and P.Zimmermann, J. Phys. B, T31, 28 (1990).
8. M.Pahler, C.Lorenz, E.v.Raven, J.Rüder, B.Sonntag, S.Baier, B.R.Müller, M.Schulze, H.Staiger and P.Zimmermann, Phys. Rev. Lett., 68, 2285 (1992).

9. J.MBizau, F.J.Wuilleumier, P.Dhez, D.L.Ederer, J.L.Picquet, J.L.Gouet, and P.Koch, *Laser Techniques for Extreme Ultraviolet Spectroscopy*, eds. T.J.McIlrath and R.J.Freeman, American Institute of Physics AIP Conf. Proc., 90, 331 (1982).
10. A.Fischer and I.V.Hertel, Z. für Phys. A304, 103 (1982).
11. L,Journel, B.Rouvellou, D.Cubaynes, J.M.Bizau, F.J.Wuilleumier and M.Ya.Amusia, X-93 conference, Debrecen, Hungary (1993).
12. B.Rouvellou, D.Cubaynes, J.M.Bizau, L.Journel, C.Marinelli, J.Novak, M.Pahler and F.J.Wuilleumier, see abstract presented at this conference.
13. Z.Felfi and S.T.Manson, Phys. Rev. Lett., 68, 1687 (1992).
14. L,Journel, D.Cubaynes, B.Rouvellou, J.M.Bizau and F.J.Wuilleumier, see abstract presented at this conference.
15. L,VoKy, A.Hibbert, L,Journel, B.Rouvellou, D.Cubaynes, J.M.Bizau and F.J.Wuilleumier, X-93 conference, Debrecen, Hungary (1993).
16. M.Y.Amusia private comunication.
17. T.N.Chang and Y.S.Kim, J. Phys. B, 15, L835 (1982).
18. M.Isenberg, S.Charter, H.P.Kelly and S.Salomonson, Phys. Rev. A, 32, 1472 (1985).
19. D.Cubaynes, J.M.Bizau, M.Richter and F.Wuilleumier, Tenth Int. Conference on Vacuum Ultraviolet Radiation Phys., Mo71 (1992).
20. A.Svensson, M.Müller, N.Böwering, U.Heinzmann, V.Radojevic and W.Wijesundra, J. Phys. B, 21, L179 (1988).
21. A.Haussmann, B.Kämmerling, H.Kossmann and V.Schmidt, Phys. Rev. Lett., 61, 2669 (1988).
22. J.Jimemenez-Mier, C.D.Caldwell and D.L.Ederer, Phys. Rev. Lett., 57, 2260 (1986).
23. J.C.Hansen, J.A.Duncanson R.L.Chien and R.S.Berry Phys. Rev. A, 21, 222 (1980):
24. K.Blum, *Density Matrix Theory and Application*, Plenum Press, New York (1981):
25. S.Baier, Thesis, Technische Universität Berlin (1993).
26. M.Pahler, Thesis, Universität Hamburg (1991).
27. M.Pahler, C.Lorenz, E.v.Raven, J.Rüder, B.Sonntag, S.Baier, B.R.Müller, M.Schulze, H.Staiger and P.Zimmermann, Phys. Rev. Lett., 68, 2285 (1992).
28. S.Wakid, A.K.Bhatia and A.Temkin, Phys. Rev. A, 21, 496 (1980).
29. J.Sugar, T.B.Lucatorto, T.J.McIlrath and A.W.Weiss, Opt. Lett., 4, 109 (1979).
30. T.Dohrmann, J.Rüder, B.Sonntag, S.Baier, M.Martins, M.Schulze, H.Staiger, M.Wedowsky and P.Zimmermann, BESSY Jahresbericht, 108 (1992).

INNERSHELL EXCITATION OF ATOMS AND MOLECULES BY SOFT X-RAYS

Kiyoshi Ueda

Research Institute for Scientific Measurements, Tohoku University, Sendai 980, Japan
Daresbury Laboratory, Daresbury, Warrington WA4 4AD, United Kingdom

ABSTRACT

Three case studies are reviewed on electronic decay processes following resonance innershell photoexcitation of atoms and molecules. In the case of B 1s → $2a_2$" excitation in BF_3, a spectator model is appropriate and spectator Auger decay is a dominant decay channel. In the case of 3p → 3d excitation in Ca, the spectator model breaks down and collapse of the 3d wave function and many body effects play important roles in the decay dynamics. In the case of 4p → 4d excitation in Sr, the phase difference between the s and d waves which leave the Sr^+ ion in the 5p $^2P_{3/2}$ level has been determined from the measured photoelectron asymmetry parameter β and the published alignment tensor A_{20} for the Sr^+ 5p $^2P_{3/2}$ level.

INTRODUCTION

Innershell photoexcitation of atoms and molecules has been studied extensively in the past decade, stimulated by the development of experimental techniques related to synchrotron radiation and the development of theoretical techniques. In this progress report three typical case studies on innershell photoexcitation will be reviewed; B 1s excitation in BF_3,[1] 3p excitation in Ca,[2] and 4p excitation in Sr.[3] The emphasis will be on electronic decay processes following resonant photoexcitation.

The experimental technique used for investigating electronic decay processes is electron spectroscopy using synchrotron radiation. The differential photoionization cross section is given by the well-known expression:

$$\frac{d\sigma}{d\theta} = \frac{\sigma}{4\pi} \left[1 + \frac{\beta}{4} (3p \cos 2\theta + 1) \right] \quad (1)$$

where p is the degree of linear polarization of incident light, θ is the angle of the photoelectron ejection relative to the principal axis of polarization, and σ and β are the partial cross section and photoelectron asymmetry parameter, respectively; σ and β can be determined by means of electron spectroscopy using a tunable light source such as synchrotron radiation.

In the case of the B 1s excitation in BF_3, we will mainly discuss a spectator model in which the electron excited in an unoccupied orbital acts as a spectator during Auger-like electronic decay. This spectator process is a dominant decay process in core excited atoms and molecules in many cases. The Ca 3p excitation is, however, a typical example in which the spectator model breaks down. In this example, we will discuss how collapse of the 3d wave function in the innershell excited configuration

$3p^54s^23d$ and electron correlation (many-body effect) play roles in the electronic decay processes. In the case of the Sr 4p excitation, we will mainly discuss the determination of the phase difference between s and d waves which leave the Sr^+ ion in the $5p\ ^2P_{3/2}$ level. The way we determine the phase difference is based on two separate measurements; the angular-distribution measurement of photoelectrons[3] and the polarization measurement of fluorescence decay by Hamdy et al.[4] A complete determination of the phase differences as well as the amplitudes of the dipole matrix elements involved in the photoionization process has been discussed by Heinzmann[5] who used angle-resolved spin analysis of the ejected photoelectrons from Xe. Other 'complete' experiments have been carried out by Jiménez-Mier et al.,[6] who determined the phase difference for He photoionization by measuring the angular distribution of the fluorescence and using the published photoelectron asymmetry parameter β, and by Hausmann et al.[7] for Mg on the basis of angle-resolved Auger-electron and photoelectron spectroscopy using synchrotron radiation.

B 1s EXCITATION OF BF_3

The ground state configuration of BF_3 is

$F\ 1s^2\ B\ 1s^2\ 1a_1'^2\ 1e'^4\ 2a_1'^2\ 2e'^4\ 1a_2''^2\ 3e'^4\ 1e''^4\ 1a_2'^2$.

The outer six orbitals at binding energies 15 – 22 eV are often called outer valence orbitals while the $1a_1'$ and $1e'$ orbitals at binding energies ~ 40 eV are often called inner valence orbitals. The binding energy of B 1s is 202.8 eV. The absorption spectrum of BF_3 near the B 1s threshold[8] shows two resonant features, B1s → $2a_2''$ ($2p\pi_B$) below the B 1s threshold and B 1s → 4e' ($2p\sigma*$) above the threshold. Kanamori et al.[9] observed B KVV normal Auger electron emission at the B 1s → 4e' excitation and resonance Auger electron emission at the B 1s → $2a_2''$ excitation. Ionic fragmentation following the normal and resonance Auger electron emission was also investigated using two types of coincidence techniques; energy-selected normal and resonance Auger electron – photoion coincidence[10] and energy-unselected electron – photoion – photoion coincidence.[11]

We have reinvestigated the electronic decay processes at the resonance excitation by scanning the photon energy across the two resonances B 1s → 4e' and B 1s → $2a_2''$.[1] The experiment has been carried out on the 24-m SGM monochromator[12] at the Photon Factory in Japan.

When we scanned the photon energy across the B 1s → 4e' resonance at 206 eV, we observed the B KVV normal Auger bands and valence photoelectron bands. The intensity of the normal Auger bands was enhanced by the resonance whereas the intensity of the photoelectron bands was not. When we scanned the photon energy across the B 1s → $2a_2''$ resonance at 195.5 eV, we found that some valence photoelectron main bands and satellite bands were strongly enhanced by the resonance. We have obtained the difference spectrum between the on-resonance spectrum taken at 195.5 eV and the off-resonance spectrum taken at 194 eV and compared it with the normal Auger electron spectrum in Fig. 1. The normal Auger spectrum was taken at the photon energy tuned to the B 1s → 4e' resonance (206 eV). The difference spectrum is plotted against the binding energy (*BE*); *BE* can be calculated as $BE = h\nu - KE$, where $h\nu$ and KE are photon energy and electron kinetic energy. The normal Auger spectrum is plotted against the double ionization potential *DIP* of the Auger final two-hole states; *DIP* can be calculated as $DIP = I(B\ 1s) - KE$, where $I(B\ 1s)$ is the ionization energy of the B 1s electron. The *DIP* scale is shifted by 16 eV with respective to the *BE* scale. Then a remarkable correspondence is clearly seen between the normal Auger bands A1 - A6 and the satellite bands R1 - R6. These resonance-enhanced satellite bands can be interpreted to result from resonance spectator Auger processes; the final states of the spectator Auger processes R1 - R6 correspond to the states which have two holes in the same configurations with the final states of the normal Auger processes A1 - A6 and one excited electron which has acted as a spectator during the Auger-like decay. In this spectator model, the energy difference, 16 eV, between the *DIP* scale and the the *BE* scale corresponds to the ionization energy of the excited electron leaving the two holes in the valence orbitals.

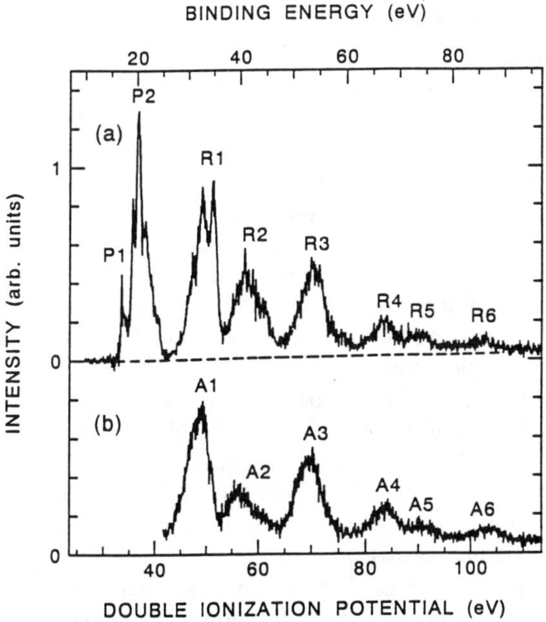

Fig. 1. (a) Difference spectrum between the on-resonance spectrum at 195.5 eV and the off-resonance spectrum at 194 eV. (b) Normal Auger spectrum excited by 206-eV photons.

The assignment of the Auger bands A1 - A6 requires some discussions. In Fig. 1 each broad peak includes many unresolved lines. Based on *ab initio* calculations[9,13] the A1 and A2 bands may be assigned to the states which have two outer valence holes, the A3 and A4 bands to the states which have one inner valence hole and one outer valence hole, and the A5 and A6 bands to the states which have two inner valence holes. The lower energy components in the pairs A1, A3, and A5 may be assigned to the states which have two holes in different F atomic sites and the higher energy components A2, A4, and A6 may be assigned to the states which have two holes in the same F atomic site.[9,13] In this model, the energy difference ~10 eV between the two peaks in each pair may be interpreted as the difference of the repulsive energy between the localized and delocalized two-hole configurations. It should be noted that two-hole states are highly correlated (configuration-mixed) and thus shake-up contributions might not be negligible in the higher energy components A2, A4, and A6.

I turn now to discuss the resonance-enhanced components P1 and P2 of the main bands (Fig. 1), where P1 is the 3e' band and P2 is a superposition of the $1a_2"$, 2e', and $2a_1$' bands. This resonance enhancement can be interpreted to result from resonance participant Auger processes; in these processes, the excited electron participates in the electronic decay and thus the final states are the same as those resulting from direct valence photoionization. The enhancement of the P2 band is far more significant that that of the P1 band, illustrating that the B atomic population in the valence orbital plays a role in the participant Auger decay following the B 1s excitation. Detailed discussions on the participant processes are given in Ref. 1 together with higher resolution photoelectron data. It should be emphasized here that the spectator processes are more dominant (~ 3 times) than the participant processes.

Ca 3p - 3d EXCITATION

The absorption spectrum of Ca[14] shows two remarkable features in the 3p innershell excitation region. First, the $3p^64s^2\ ^1S_0 \rightarrow 3p^54s^23d\ ^1P_1$ (abbreviated as $3p \rightarrow 3d$ in the following) excitation, often called a giant resonance, is much stronger than the excitations to higher n-members. This giant resonance is due to collapse of the 3d wave function in the Ca $3p^54s^23d$ configuration; because the central field has a significantly deeper well in the Ca $3p^54s^23d$ configuration than in the Ca $3p^54s^2nd$ configurations for n≥4 and draws the 3d wave function closer to the nucleus, the $3p \rightarrow 3d$ excitation will take a large share of the oscillator strength due to the large overlap between the 3d wave function and the innershell 3p wave function. Second, the absorption spectrum shows very complex structure in the vicinity of the giant resonance. This complexity can be attributed to strong configuration mixing in the innershell excited states such as $3p^54s^23d$, $3p^54s4p^2$, $3p^54s3d^2$, $3p^53d^3$, etc. The decay of the 3p innershell excited states following resonance excitation has also been investigated using synchrotron radiation; Sato et al.[15] measured the Ca^{2+}/Ca^+ ratio using photoion mass spectrometry and Bizau et al.[16] measured partial cross sections for many Ca^+ ionic levels using photoelectron spectroscopy.

We have investigated the electronic decay of the Ca $3p \rightarrow 3d$ giant resonance by means of angle-resolved photoelectron spectroscopy using synchrotron radiation.[2] The experiment has been carried out on the toroidal grating monochromator installed at the Atomic and Molecular Science beamline[17] at the Daresbury SRS.

Figure 2 show the resulting relative partial cross sections σ and asymmetry parameters β for the Ca^+ 4s and 4p levels in the Ca $3p \rightarrow 3d$ giant resonance region. The positions where the absorption lines were observed by Mansfield and Newsom[14] are marked in the figure by numbered vertical bars. The main peak at ~ 31.4 eV in the 4s partial cross section is the $3p \rightarrow 3d$ giant resonance (No. 3 in Fig. 2) and thus the assignment of the innershell excited level is $3p^54s^23d\ ^1P_1$. This peak clearly shows an asymmetric line shape with a tail towards lower energies. The second small peak at ~

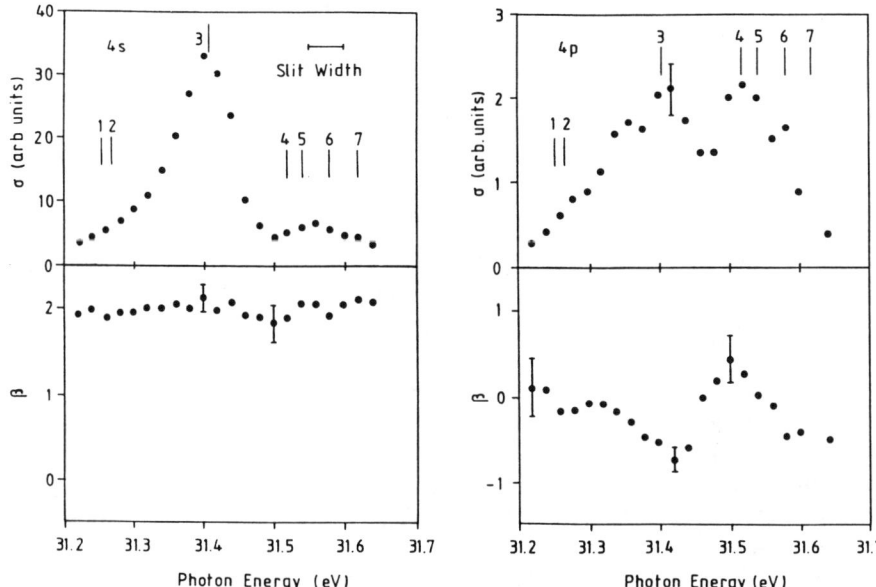

Fig. 2. Relative partial cross sections σ and photoelectron asymmetry parameters β for the Ca⁺ 4s and 4p levels in the Ca 3p → 3d giant resonance region. The bars (Nos. 1–7) indicate the positions of the absorption lines observed by Mansfield and Newsom.[14]

31.56 eV may correspond to the unresolved group of lines Nos. 4 - 7 assigned by Mansfield and Newsom[14] as $3p^53d^3$ $^1P_1$, $3p^54p^23d$ $^3D_1$, $3p^53d^24s$ $^1P_1$, and $3p^54s^24d$ $^3P_1$, respectively.

The results of our branching ratio measurement for the Ca⁺ levels, observed at 31.41 eV photon energy, can be summarized as follows; the data is presented in the form of branching ratios, as percentages of total intensity:

4s	3d	4p	5s	4d	5p	nl	$3p^54s3d$
51	4	4	2	7	3	13	16

The main decay channel is to the Ca⁺ 4s ground level, and its branching ratio is about a half of the entire decay channels. If the electron excited in the 3d orbital acted as a spectator, the final state of the decay would be the Ca⁺ 3d level. However, this spectator process is much weaker than the participant process which populates Ca⁺ 4s. This suppression of the spectator process results from the collapse of the 3d wave function in the Ca $3p^54s^23d$ configuration. Because of this collapse, the Ca innershell excited level $3p^54s^23d$ possesses an anomalously small effective principal quantum number n*=2.05, which is even smaller than the value of n*=2.14 for the Ca⁺ 4s ground level. Consequently, the 3d wave function in the Ca $3p^54s^23d$ configuration does not overlap the 3d wave function in the Ca⁺ $3p^63d$ configuration substantially. The branching ratio to the Ca⁺ 4p level is about the same as that to the 3d level. If the innershell excited level had the pure configuration $3p^54s^23d$, it would not autoionize to the Ca^+ 4p level because three electrons would be involved in this process. However, autoionization is made possible by mixing with other configurations of the same symmetry such as $3p^54p^24s$ and $3p^54p^23d$.

An elaborate theoretical analysis of the two decay channels to the Ca^+ 3d and 4p levels has been performed by Altun and Kelly[18] using many-body perturbation theory; they calculated the partial cross sections for populating the Ca^+ 3d and 4p levels through the Ca $3p \rightarrow 3d$ excitation and predicted the ratios of the 3d and 4p cross sections to the 4s cross section to be 0.077 and 0.354, respectively. The present measurement gives the ratio 0.073 for σ_{3d}/σ_{4s} and 0.075 for σ_{4p}/σ_{4s}. Though the agreement for σ_{3d}/σ_{4s} is excellent, σ_{4p}/σ_{4s} is clearly overestimated by the theory.

The above branching ratio data show that the final Ca^+ levels are very widely distributed among the high Rydberg members of Ca^+, as Bizau et al.[16] have already pointed out. Photoelectrons arising from the decay to the Ca^+ $3p^54s3d$ innershell hole level have almost zero kinetic energy. Thus, we have used the intensity of the Auger line from Ca^+ $3p^54s3d$ to Ca^{2+} $3p^6$, instead of attempting to detect near zero-energy photoelectrons, in order to measure this branching ratio. Note that the decay of this Ca^+ level is to the Ca^{2+} ground level. The measured branching ratio, 16 %, is in good agreement with the value 17 % expected from the intensity ratio of ~ 0.2 for Ca^{2+}/Ca^+ as measured by Sato et al.[15] at this photon energy.

We turn now to discuss the asymmetry parameter β. If LS-coupling is assumed, the two outgoing waves $p_{1/2}$ and $p_{3/2}$ which leave the Ca^+ ion in the 4s level are not separable and thus β should be equal to 2. As can be seen in Fig. 2, the β value for 4s is 2 within the experimental error over the whole range of photon energies, suggesting that LS-couping is still a good approximation despite the complex configurations of these resonances. The β value for Ca^+ 4p changes dramatically through the resonances, as can be seen in Fig. 2. If LS-coupling is assumed, there are two waves, s and d, which leave the Ca^+ ion in the 4p level. β can then be expressed as[19]

$$\beta = \frac{|D_d|^2 - 2\sqrt{2}\,|D_s||D_d|\cos\Delta}{|D_s|^2 + |D_d|^2} \qquad (2)$$

where D_s and D_d are the amplitudes of the two dipole matrix elements and $\Delta = \delta_s - \delta_d$ is their phase difference. For a single s wave or d wave, β would be 0 or 1, respectively, as can be seen from Eq. (2). However, the observed values of β go below zero at the $3p \rightarrow 3d$ resonance energy, indicating that both s and d waves are involved and the phase-difference term $\cos \Delta$ is positive.

Sr 4p -4d EXCITATION

The absorption spectrum of Sr in the 4p innershell excitation region has been measured by Mansfield and Newsom.[20] Complex structure was revealed, largely due to configuration-mixed excited states such as $4p^55s^24d$, $4p^55s5p^2$, $4p^55s4d^2$, $4p^54d^3$, etc. The decay of the 4p innershell excited levels has also been investigated extensively using synchrotron radiation. Nagata et al.[21] measured the Sr^{2+}/Sr^+ ratio using photoion mass spectrometry and Yagishita et al.[22] measured branching ratios by means of photoelectron spectroscopy. Recently, Jiménez-Mier et al.[23] made photoelectron spectroscopy measurements with a photon resolution of ~ 6 meV, and reinvestigated these branching ratios by resolving all the resonance features observed by Mansfield and Newsom[20].

We have investigated the electronic decay of the Sr 4p → 4d giant resonance by means of angle-resolved photoelectron spectroscopy using synchrotron radiation.[3] The experiment has been carried out on the 5-m normal incidence monochromator[24] at the Daresbury SRS.

Figure 3 shows the resulting relative partial cross sections σ and asymmetry parameters β for the Sr^+ 5s $^2S_{1/2}$ level and the sum of the 5p $^2P_{1/2}$ and 5p $^2P_{3/2}$ levels in the Sr 4p → 4d giant resonance region. The bars with numbers 1 - 7 given in Fig. 3 indicate the positions of the absorption peaks observed by Mansfield and Newsom.[20] It can be seen that most of the resonances are resolved in the partial cross sections with the present photon resolution (25 meV). It should be noted that the 25.26-eV main resonance (No. 4 in Fig. 3), assigned to the transition $4p^6 5s^2\ ^1S_0$ → $4p^5 5s^2 4d\ ^1P_1$ by Mansfield and Newsom,[20] has a larger width (~50 meV) than the present photon band pass. The results of our branching ratio measurement for the Sr^+ levels at the 4p → 4d resonance (25.26 eV) are 66 : 23 : 8 : 3 for 5s : 4d : 5p : nl. The dominant decay channel is Sr^+ 5s, reflecting collapse of the Sr 4d wave function. The branching ratios to higher-n members are considerably smaller than those in the Ca case.

In the case of Sr, the two spin-orbit components 5p $^2P_{1/2}$ and 5p $^2P_{3/2}$ have an energy separation of ~ 100 meV and we were able to resolve these. As can be seen in Fig. 3, the ratio between the 5p $^2P_{1/2}$ and 5p $^2P_{3/2}$ cross sections changes dramatically across some resonances. If we neglect the influence of spin-orbit interaction, the ratio is expected to be 1 : 2. This ratio is observed for the 4p → 4d resonance at 25.26 eV, indicating that the $4p^5 5s^2 4d$ level at 25.26 eV has strong LS-coupled $^1P_1$ character. For the resonance at 25.21 eV (No. 3 in Fig. 3), the ratio becomes 1 : 1, implying breakdown of LS-coupling.

Turning our attention now to the asymmetry parameter β, we see that for the Sr^+ 5s $^2S_{1/2}$ level β drops from 2 to ~1.6 at some resonance energies. If LS-couping were valid, the two p waves $p_{1/2}$ and $p_{3/2}$ would not be separable and β would be 2; thus the LS-coupling scheme is showing signs of breaking down here. It can be seen that β for the sum of the Sr^+ 5p $^2P_{1/2}$ and 5p $^2P_{3/2}$ levels changes dramatically across the resonances, reflecting variations in the amplitudes of the s and d waves and their phase difference. It should be noted also that the 25.32-eV resonance (No. 5 in Fig. 3), which is not well resolved in the σ curve, can be clearly identified in the β curve.

Hamdy et al.[4] reported measurements of the fluorescence intensity and polarization for the Sr^+ 5s $^2S_{1/2}$ ← 5p $^2P_{3/2}$ transition following the 4p innershell excitation of Sr, and calculated the alignment tensor A_{20} from these data. The branching ratio measurements confirm that the Sr 4p → 4d resonance at 25.26 eV decays predominantly to the Sr^+ 5s, 4d, and 5p levels. Therefore, in considering population and alignment of the Sr^+ 5p $^2P_{3/2}$ level arising from the Sr 4p → 4d resonance excitation, we can neglect cascades from higher Rydberg levels. Further we can assume LS-coupling; this is not

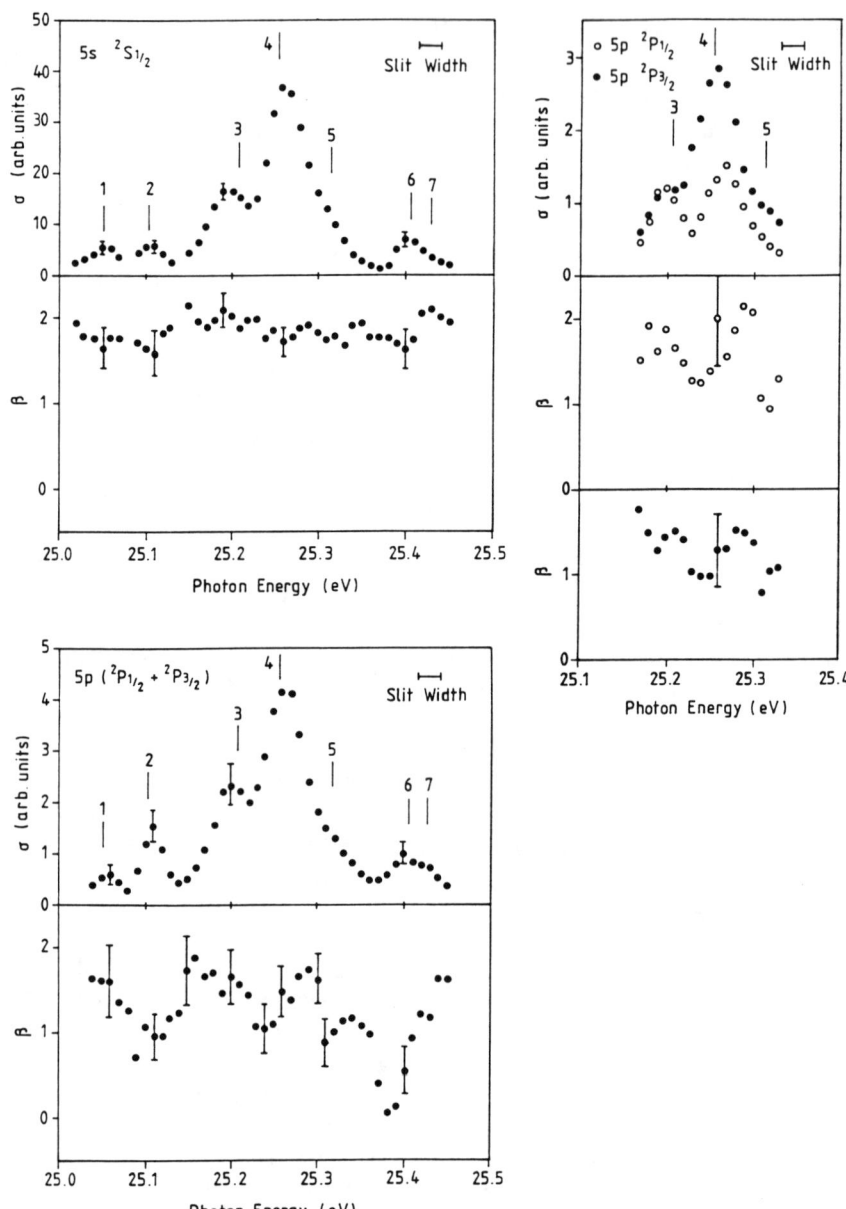

Fig. 3. Relative partial cross sections σ and photoelectron asymmetry parameters β for Sr+ 5s $^2S_{1/2}$, 5p $^2P_{1/2}$, 5p $^2P_{3/2}$, 5p $^2P_{1/2}$ + 5p $^2P_{3/2}$ in the Sr 4p → 4d giant resonance region. The bars (Nos. 1–7) indicate the positions of the absorption lines observed by Mansfield and Newsom.[20]

completely valid but from the above discussions is a reasonable approximation for the 4p → 4d resonance. Under these assumptions, the alignment tensor A_{20} of the Sr$^+$ 5p $^2P_{3/2}$ level reported by Hamdy et al.[4] can be expressed as

$$A_{20} = -\frac{|D_s|^2 + |D_d|^2/10}{|D_s|^2 + |D_d|^2} \quad (3)$$

and the asymmetry parameter β of the Sr$^+$ 5p $^2P_{3/2}$ level determined from the photoelectron measurement can be expressed by Eq. (2). Inserting the reported A_{20} value of –0.222 ± 0.016 in Eq. (3), we obtain a value for the ratio $|D_s|^2/|D_d|^2$ of 0.157 ± 0.024. Inserting this value and the measured β = 1.26 ± 0.43 into Eq. (2), we obtain a phase difference of $|\Delta|$ = 114 ± 30 degrees. It should be noted again that the error estimate includes only the uncertainties of the two values A_{20} and β and does not include possible systematic errors due to the influence of cascades on the A_{20} measurement or due to breakdown of LS-coupling.

CONCLUDING REMARKS

Three case studies on innershell photoexcitation of atoms and molecules have been reviewed with emphasis on electronic decay processes studied by electron spectroscopy using synchrotron radiation. The B 1s excitation of BF$_3$ is a typical case in which a spectator model is appropriate; the spectator Auger decay is dominant following the B 1s → 2a$_2$" excitation. In the case of the Ca 3p → 3d giant resonance, the spectator model breaks down and the collapse of the 3d wave function in the Ca 3p^{5}4s^{2}3d configuration and many-body effects play important roles in the decay dynamics of the giant resonance. In the case of the Sr 4p → 4d giant resonance, the phase difference between the s and d waves which leave the Sr$^+$ ion in the 5p $^2P_{3/2}$ level has been determined, using our photoelectron measurement and the flurorescence measurement by Hamdy et al.[4] This approach illustrates an alternative approach towards the 'complete' experiment for photoionization. However, in order to apply this approach to other cases such as the Ca 3p → 3d giant resonance, in which cascades are not negligible, it will be necessary to carry out a coincidence experiment between the photoelectrons and fluorescent photons.

ACKNOWLEDGMENTS

The experiment for BF$_3$ was carried out with the approval of the Photon Factory advisory committee in collaboration with Y. Sato, H. Chiba, T. Hayaishi, E. Shigemasa, and A. Yagishita. The experiments for Ca and Sr were carried out in collaboration with J. B. West, K. J. Ross, H. Hamdy, H. J. Beyer, and H. Kleinpoppen, supported by the Science and Engineering Research council. I am grateful to my colleagues listed above for a fruitful collaboration, to N. M. Kabachnik, F. P. Larkins, and J. P. Connerade for helpful discussions, and to J. B. West and K. J. Ross for a critical reading of this manuscript.

REFERENCES

1. K. Ueda, H. Chiba, Y. Sato, T. Hayaishi, E. Shigemasa, and A. Yagishita, to be published.
2. K. Ueda, J. B. West, K. J. Ross, H. Hamdy, H. J. Beyer, and H. Kleinpoppen, Phys. Rev. A48, Rxxxx (1993).
3. K. Ueda, J. B. West, K. J. Ross, H. J. Beyer, H. Hamdy, and H. Kleinpoppen, J. Phys. B26, Lxxxx (1993).
4. H. Hamdy, H. J. Beyer, J. B. West, and H. Kleinpoppen, J. Phys. B24, 4957 (1991).
5. U. Heinzmann, J. Phys. B13, 4367 (1980).
6. J. Jiménez-Mier, C. D. Caldwell, and D. L. Ederer, Phys. Rev. Lett. 57, 2260 (1986).
7. A. Hausmann, B. Kämmerling, H. Kossman, and V. Schmidt, Phys. Rev. Lett. 61, 2669 (1988).
8. E. Ishiguro, S. Iwata, Y. Suzuki, A. Mikuni, and T. Sasaki, J. Phys. B15, 1841 (1982).
9. H. Kanamori, S. Iwata, A. Mikuni, and T. Sasaki, J. Phys. B17, 3887 (1984).
10. K. Ueda, H. Chiba, Y. Sato, T. Hayaishi, E. Shigemasa, and A. Yagishita, Phys. Rev. A46, R5 (1992).
11. M. Simon, P. Lablanquie, M. Lavollee, and K. Ueda, to be published.
12. A. Yagishita, T. Hayaishi, T. Kikuchi, and E. Shigemasa, Nucl. Instrum. Methods A306, 528 (1991).
13. F. Tarantelli, A. Sgamellotti, and L. S. Cederbaum, J. Chem. Phys. 94, 523 (1991).
14. M. W. D. Mansfield and G. H. Newsom, Proc. R. Soc. A357, 77 (1977).
15. Y. Sato, T. Hayaishi, Y. Itikawa, Y. Itoh, T. Koizumi, J. Murakami, T. Nagata, T. Sasaki, B. Sonntag, A. Yagishita, and M. Yoshino, J. Phys. B18, 225 (1985).
16. J. M. Bizau, P. Gérard, F. J. Wuilleumier, and G. Wendin, Phys. Rev. A36, 1220 (1987).
17. J. B. West and H. A. Padmore, *Handbook on Synchrotron Radiation,* edited by G. V. Marr (North Holland, Amsterdam, 1987) Vol. II, p. 21.
18. Z. Altun and H. P. Kelly, Phys. Rev. A31, 3711 (1985).
19. See for example, V. Schmidt, Rep. Prog. Phys. 55, 1483 (1992).
20. M. W. D. Mansfield and G. II. Newsom, Proc. R. Soc. A377, 431(1981).
21. T. Nagata, J. B. West, T. Hayaishi, Y. Itikawa, Y. Itoh, T. Koizumi, J. Murakami, Y. Sato, H. Shibata, A. Yagishita, and M. Yoshino, J. Phys. B19, 1281 (1986).
22. A. Yagishita, S. Aksela, Th. Prescher, M. Meyer, E. von Raven, and B. Sonntag, J. Phys. B21, 945 (1988).
23. J. Jiménez-Mier, C. D. Caldwell, M. G. Flemming, S. B.Whitfield, and P. van der Meulen, Phys. Rev. A48, xxxx (1993).
24. D. M. P. Holland, J. B.West, A. A. MacDowell, I. H. Munro, and A. G. Beckett, Nucl. Instrum. Methods B44, 233 (1989).

NEGATIVE IONS

MULTIPHOTON DETACHMENT OF NEGATIVE IONS

H.K. HAUGEN
Departments of Physics and Astronomy and Engineering Physics
McMaster University, Hamilton, Ontario L8S 4M1, Canada

ABSTRACT

Multiphoton detachment of negative ions has been an area of intense experimental investigation in recent years. A revival of activity was stimulated in part by the prospect of exploring excess photon absorption in a short range potential. Selected experiments on excess photon absorption are reviewed. In addition, recent resonant multiphoton experiments are discussed. The resonant interaction work encompasses experiments which access both doubly excited states above the detachment limit, and which involve coupling of the fine structure of the electronic ground state. Finally, future experimental perspectives, both in terms of strong laser field phenomena and spectroscopic measurements, are briefly discussed.

1. INTRODUCTION

Multiphoton detachment of a negative ion was observed almost thirty years ago by Hall et al[1], and early theoretical calculations were performed by Robinson and Geltman[2]. However, a renewed interest, both experimental and theoretical has appeared only very recently. (See, e.g., Refs. 3-21). The current studies are stimulated by the qualitatively different features of these species versus analogous studies with neutral atoms or positive ions. Negative ions exhibit a short range potential ($\sim r^{-4}$) as opposed to the infinite-range Coulomb potential encountered in the ionization of neutral atoms. Negative atomic ions are bound by up to 3.6 eV[22], and thus they typically exhibit only a single bound electronic state. There is no infinity of bound states converging to the ionization limit as in a neutral atom, and the final state of a photodetached electron can to a good approximation be described as a free electron state[4,9]. In addition, the general absence of excited state allows the study of strong field phenomena without complications from Stark shifts of excited electronic states. The perspectives for future work fall into two categories: strong laser fields and spectroscopic studies of negative ion structure. The short range potential has led to important questions regarding the efficiency of excess photon absorption (analogous to above threshold ionization, ATI, observed in neutral atoms) in the short range potential. The absence of excited states, for example, should also simplify studies of laser-induced-continuum structure (LICS)[23], which is a topic of current interest in studies with neutral atoms. Resonant multiphoton studies will provide unique probes of atomic structure with high resolution. Technologically, the ion beam approach also demonstrates some unique possibilities for multiphoton studies. The present review represents largely a progress report on the experiments in which the author was directly involved.

2. EXPERIMENTAL TECHNIQUE

A schematic of a typical experimental setup is shown in Fig. 1. The experimental approach is based on keV ion beam technology[16,17,21]. A "universal" sputter ion source is employed which provides negative ion beams of most of the chemical elements. After acceleration to 4-25 keV, the ion beam is mass and charge

© 1993 American Institute of Physics

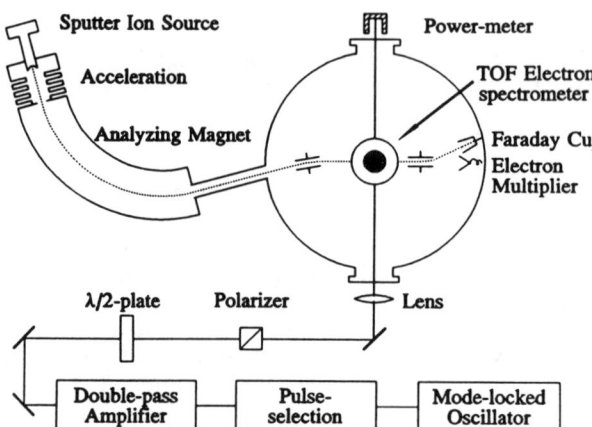

Fig. 1 Schematic of experimental apparatus. The specific setup shown was employed in the nonres-excess photon detachment work on the negative gold ion.

state analyzed, and is sent into an ultra-high vacuum (UHV) chamber. Typical ion currents in the UHV chamber are in the 10-100 nA range. The vacuum in the UHV chamber is typically 10^{-8}-10^{-9} mbar. Electric field deflection through 15 degrees removes collisionally detached neutral atoms. The ion beam is then crossed at right angles by the focussed laser beam. Normally a 35-cm focal length lens is utilized. Q-switched and mode-locked Nd:YAG lasers have been employed in the work, as well as a nanosecond dye laser and standard nonlinear optical techniques. After subsequent electric field analysis, the detached neutral atoms and positive ions are detected via discrete dynode electron multipliers. The electron multipliers exhibit a wide dynamic range such that a linear proportionality exists between the analog signal from the detector and the number of atoms/ions created by the laser pulse. In electron spectroscopic studies, the ejected electrons from the laser-ion interaction are analyzed by a 17-cm-long magnetically shielded time-of-flight (TOF) spectrometer equipped with a 2.5-cm diameter tandem-channel-plate detector. The lasers which are employed operate at a repetition rate of 10 Hz. Although the effective duty factor is very low, high signal-to-noise can be obtained in the experiments due to the time-gated nature of the signal collection. Furthermore, with the ion beam approach, even classic (long pulse) laser technology is afforded new advantages, and the kinematics of the particles can also be exploited in selective detection schemes. In the case of Q-switched laser sources, the pulse length 'seen' by the ions is dictated by the flight time through the laser focus. In this way, relatively short picosecond optical pulses can be obtained while employing only nanosecond laser technology. In the studies described here this possibility was not greatly exploited, but with a tighter laser focus and higher keV ion energies, this technical feature could prove very useful in future work.

3. EXPERIMENTAL RESULTS

3.1. EXCESS PHOTON DETACHMENT

In the late 1980's the revival of interest in multiphoton studies of negative ions resulted in experimental and theoretical work on multiphoton detachment cross sections. (See, e.g., refs. 4-8,17). However, one of the most intriguing experimental possibilities was the prospect of observation of excess photon detachment in a negative ion. Over one decade ago Agostini et al[24] discovered that an atom which is exposed to an intense laser field can absorb more than the minimum number of photons required for ionization. This process, which became known as ATI, has since been the subject of intense investigation. The absorption of excess photons must occur as an integral part of the ionization process, where the electron is still able to exchange energy and momentum with the residual ionic core. In photodetachment of negative ions this is a much more stringent requirement than in the case of photoionization, where in the latter situation the long-range Coulomb potential makes photon-electron interactions possible at rather large distances from the ion. However, about two years ago, three experimental groups observed excess photon detachment (EPD) in virtually simultaneous experiments. The experiments were very complementary. Blondel et al[12] utilized the fluorine negative ion; Davidson et al[15], the chlorine negative ion; and Stapelfeldt et al[16], the gold negative ion. All three groups employed Nd:YAG lasers; the first a nanosecond Q-switched system; and the latter two efforts, a picosecond mode-locked system. The works of references 12 and 16 used ion beam technology, while an ion trap approach was employed in reference 15. All groups arrived at the same qualitative conclusion: that nonresonant EPD is a significant process in a negative ion at intensities ~TW/cm^2. A maximum of two excess photon absorption was observed in the experiments. At comparable laser intensities, ATI is observed in experiments with neutral atoms.

Our EPD experiment will be described in some detail. The energy level diagram of Au$^-$ is shown in Figure 2. The absorption of two 1.165 eV photons exceeds the detachment threshold by only ~20 meV. Photodetachment is observed for absorption of a minimum of two and a maximum of four photons[16]. According to the Wigner threshold law, the two photon detachment cross section scales as $\varepsilon^{\ell+1/2}$, where

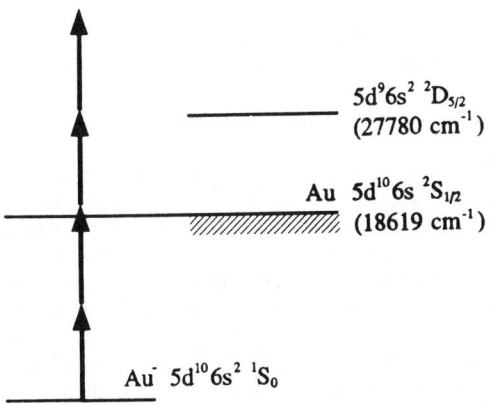

Fig. 2 Energy level diagram of the negative gold ion. A mode-locked (40 ps) Nd:YAG laser was employed, such that 2-photon absorption barely exceeds threshold. As a result, the Wigner threshold law plays a key role in the experiment.

ε is the energy of the ejected electron and ℓ is its angular momentum. Hence, in sharp contrast to the situation in neutral atoms, the depletion of target ions in the Au$^-$ near threshold situation is suppressed, which is advantageous in the search for higher-order EPD processes. A time-of-flight spectrum obtained for a 40 ps optical pulses and an irradiation intensity of ~3x10^{12} W/cm^2 is shown in Figure 3. It should be noted that an extraction field has been applied so that the flight times are not those expected from consideration of the energy level diagram alone. A blowup[16] of Figure 3 also reveals a second order peak. The second order peak was observed in several independent spectra but only at the highest laser intensities. We estimate

108 Multiphoton Detachment of Negative Ions

that the first order EPD peak represents ~5% of the total while the second order EPD peak is ~0.6% of the total electron yield. Several checks were performed on the integrity of the experimental data. Stepping the flight tube potential by 1.165 eV, for example, moved the position of the zeroth order peak exactly to the position of the first order peak. The intensity dependences of the 2-photon and 3-photon signals were also, within experimental error, given respectively by the square and cube of the laser intensity. Finally, measurements were performed on Ag^- where the lowest order peak corresponds to approximately a 1 eV electron. In this latter far-from-threshold situation, no excess photon absorption was observed, as anticipated.

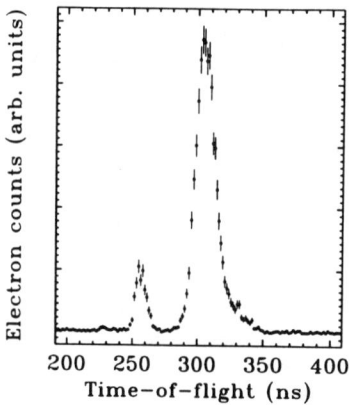 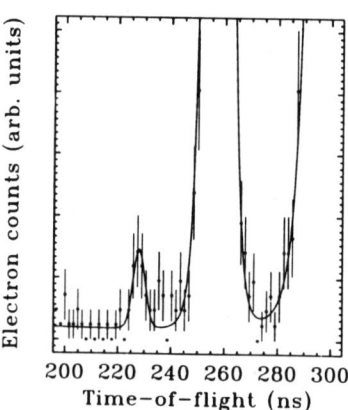

Fig. 3 Time-of-flight spectrum (left) and a blowup (right) for the irradiation of Au^- by 40 ps Nd:YAG laser pulses at an intensity of 3 TW/cm^2.

3.2. RESONANT MULTIPHOTON DETACHMENT

It was argued earlier that negative ions are often selected for strong field studies because of the absence of resonances and other specific features due to their short range potential. However, resonant phenomena can occur in negative ions in at least two different ways. One mechanism by which resonances can always occur is via doubly excited states above the detachment limit. We selected Cs^- to demonstrate this effect[21]. EPD from the ground state of Cs^- is strongly enhanced when the first photon reaches an energy that corresponds with the $Cs^-(6p_{1/2}7s)$ Feshbach state. The situation is illustrated in Figure 4(a) where a simplified energy level diagram is shown. At the energy of the Feshbach state, the cross section for single photon detachment exhibits a very deep minimum which is caused by the interference between the excitation of the $Cs^-(6p_{1/2}7s)$ state and the $6s\epsilon\ell$ continuum. This remarkable feature was originally observed by Patterson et al [25]. The resulting Fano profile exhibits a single photon detachment minimum which dips by two orders of magnitude over an energy range of 20 cm^{-1}. The minimum facilitates absorption of additional photons by suppressing the depletion of the ground state during the rise time of the laser pulse. In addition, the presence of the doubly excited state is expected to favour the absorption of excess photons. If one extra photon is absorbed, the ion will decay with high probability to an excited state of the cesium atom which can be ionized by absorption of one additional photon. Thus, where there is a minimum in the Cs atom signal, EPD leads to a maximum in the Cs^+ signal. The result[21] of a laser scan is shown in Figure 4(b). The irradiation intensity is ~1 GW/cm² and the pulse length, determined by the flight time through the laser focus, is ~0.7 ns. The broad Cs⁺ peak indicates resonant EPD while the sharp Cs⁺ peaks 'to the red' of this position derive from single photon detachment of Cs⁻(6s²) to the Cs(6s) ground state followed by (2+1) ionization of Cs(6s) via the excited Cs(11d) state. The absorption of two excess photons, as indicated by Fig. 4 (a), would represent double direct ionization.

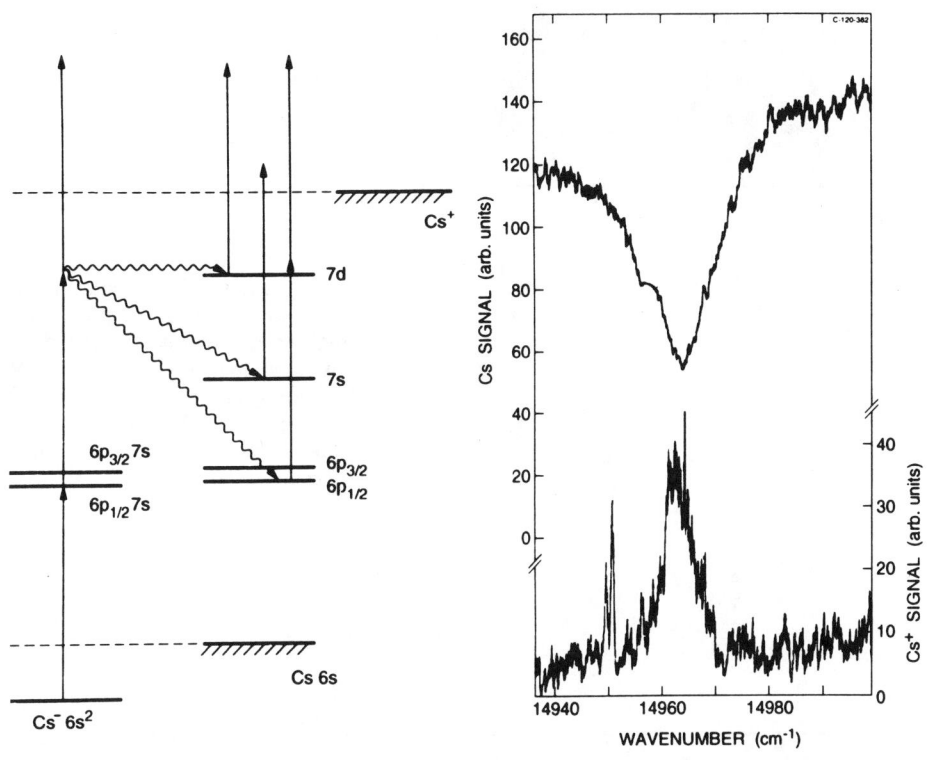

Figure 4(a)   Simplified energy level diagram Cs⁻ and Cs

Figure 4(b)   Cs signal (upper) and Cs⁺ (lower)

Resonant multiphoton detachment can occur in another way. Several negative ions exhibit fine structure in their ground electronic state. If two laser fields are applied such that the difference in the photon energies is equal to the energy splitting of the states (Raman condition), a modification of the multiphoton detachment probability can generally be expected. We selected the tellurium negative ion for a first demonstration of resonant multiphoton detachment involving only bound states[26]. An energy level diagram showing the relevant states in Te⁻ is shown in Fig. 5. The negative ion beam was irradiated with the fundamental laser wavelength of the Nd:YAG laser as well as a red dye laser beam. An enhancement in the detachment probability was obtained as the dye laser was scanned in the wavelength region of the Raman condition. Fig. 6 shows a typical signal. Some of the background can be attributed to nonresonant 2-photon detachment, but a significant portion stems from single photon detachment of the small population (< 2%) in the J = 1/2 state by the dye laser photons. Even though the population is small, single photon detachment is saturated well out into the wings of the focussed Gaussian beam. This background could be eliminated by an appropriate prepulse, if necessary. Via an optogalvanic calibration the Te$^-$(^{2}P; J=1/2, 3/2) fine structure splitting is determined to be 5004.7±0.2 cm^{-1}, which represents a 25-fold improvement over earlier measurements[22].

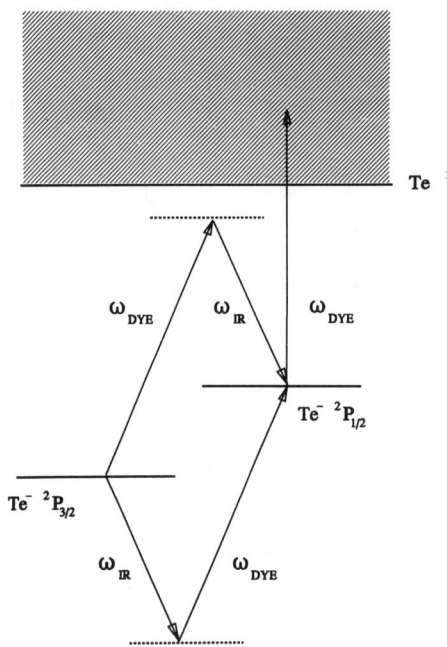

Fig. 5 Energy level diagram of Te$^-$. Three photon detachment, which is resonantly enhanced, is illustrated in both a Λ and a V configuration. The J=1/2-3/2 splitting is ~ 5000 cm^{-1}. A YAG fundamental and red dye laser were utilized.

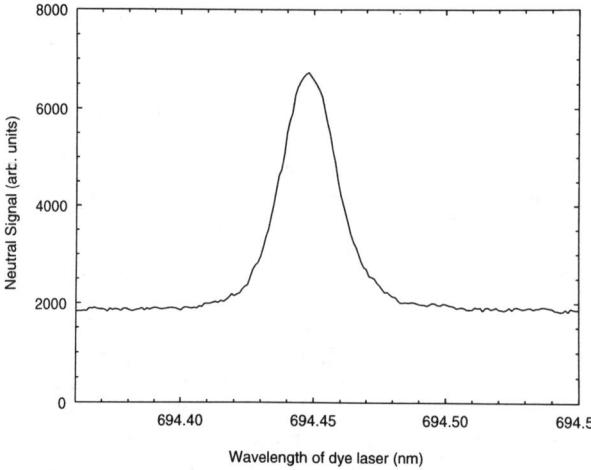

Fig. 6 Neutral atom signal versus dye laser wavelength for the two-colour laser irradiation of Te$^-$. The strong peak marks the resonant coupling of the J=1/2 and J=3/2 fine structure states.

4. DISCUSSION OF RESULTS AND FUTURE PERSPECTIVES

The results on excess photon detachment have established that this process is not significantly suppressed in negative ions. Since the absorption processes occur most effectively at short range, where the electron is close to the atomic core and energy and momentum are readily exchanged, the order of magnitude of the effect is not altered. The results for the halogen negative ions[12,15] are in good agreement with the calculations of Crance[4]. The experiments of Davidson et al[15] and Stapelfeldt et al[16] were the most contrasting. In the former case, four photons were required to detach Cl⁻, but the last photon exceeded threshold by approximately an eV. In the Au⁻ experiment, however, the lowest order (two-photon) process barely exceeded the detachment threshold, and threshold shift effects in the laser field played an important role. In fact, simple computer simulations which account for the spatial and temporal behaviour of the pulse[27], indicate that a ponderomotive shift of the order of the theoretically-expected shift is required to explain the data. The Au⁻ experiment was in stark contrast to typical ATI experiments in neutral atoms where the lowest nonvanishing order is much higher. Even though EPD turned out to be readily observed in the same intensity range as for neutral atoms, the details are different. These differences have been investigated in multiphoton detachment, for example, by the group of Blondel and coworkers, who have studied angular distributions for ejected electrons[10]. Comparisons with theory are particularly interesting since the outgoing electron is a good approximation to a plane wave. Other specific features of MPD have been explored by Davidson et al[20], who studied the saturation behaviour of the multiphoton detachment of Cl⁻ versus laser wavelength in the near-infrared, and discovered a sharply contrasting behaviour to that observed in neutral atoms.

Several new experiments are possible in the resonant interaction regime. Doubly excited states, which can be manifested as either a peak or window in the single photon detachment yield, can be exploited in studies of multiphoton phenomena and negative ion structure. A notable case of fundamental interest is represented by the hydrogen negative ion. Crance[28] has suggested two-photon spectroscopy of H⁻ to access states below H(n=2), with detection of the ~ 10 eV electrons which are formed in the rest frame of the ion. Single photon spectroscopy of doubly excited states of ¹P character has been performed earlier at Los Alamos via exploitation of the very large relativistic Doppler shift of an 800 MeV H⁻ beam[29], but the two-photon spectroscopy will access states of S and D character. Unfortunately, it appears that the considerations of Crance may have overestimated the signal by about two orders of magnitude[30]. Nevertheless, short picosecond pulse sources with high repetition rate would be optimal in such experiments, and for doubly excited states above H(n=2), a detection scheme analogous to that utilized in Cs⁻ could be employed with a strong enhancement in the signal. Further studies on Cs⁻ could employ a two-colour setup. If one laser colour is tuned to the window, a second laser could probe structure higher in the continuum. An even more interesting possibility than the cesium negative ion is afforded by the Rb⁻ case where an even narrower and (apparently) deeper window exists[31]. Such resonances can perhaps be utilized in studies of simultaneous double electron ejection in multiphoton absorption. Perhaps even more favourable cases in this latter context might be offered by systems where two electron ejection occurs to lowest nonvanishing order in multiphoton absorption. Two-photon absorption experiments utilizing a "window" in Si⁻ have been discussed by Balling et al[32]. The result of a laser scan representing single photon absorption is shown in Fig. 7. Absorption of two 5.29 eV photons by Si⁻ in the region of the $3s3p^4$ autoionizing resonance could lead to direct formation of Si⁺, but observation of a Si⁺ signal could in principle indicate stepwise double ionization or double direct ionization.

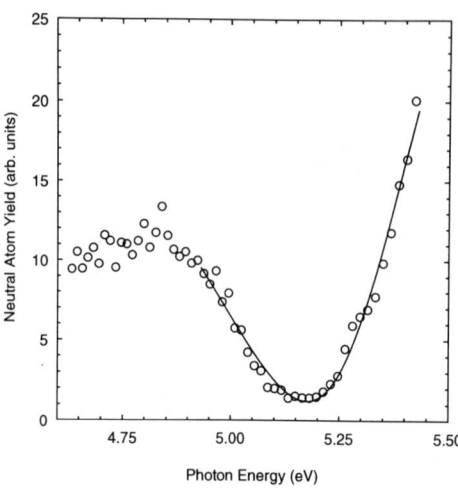

Fig. 7 UV laser scan of the photodetachment yield of Si^-. In contrast to the window in the cesium negative ion, the cross section here dips by only one order of magnitude, and the lifetime of the doubly excited state is much shorter than in the Cs^- case. The residual cross section at the window position may be partly due to detachment of excited Si^- states, but this is believed to be only a minor contribution.

However, no Si^+ signal was observed in a first experiment with long UV optical pulses (~ 0.5 ns) and intensities up to ~ 5 GW/cm² [32]. Long-lived metastable negative ions (lifetimes of 10's or 100's of microseconds) such as He^- [33] and Be^- [34] may also provide rather unique systems for multiphoton experiments. In these cases, the initial state lies above the ground state of the neutral atom. The beryllium negative ion is particularly interesting in that an optical transition is possible via the $1s^2 2p^3$ ^4S - $1s^2 2s 2p^2$ ^4P transition in Be^-. (See also review of Ref. 35). Resonant multiphoton pumping can also in this case explore the possible appearance of a component of direct double ionization.

The Raman studies which have been described for Te^- can be extended to several other negative ions. In fact, the present approach could readily improve the knowledge of the fine structure in a large number of negative ions by a factor which is typically between one and four orders of magnitude, and via a technique which does not require detection of photodetached electrons. The work can also be extended to cover regimes different from that explored in Te^-. A case could be selected where two-photon detachment is required from both fine structure states for both laser fields. An even more intriguing case is where the wavelengths are chosen such that the "blue" laser would be close to the Wigner threshold. A strong change in the behaviour of the total detachment signal might be expected when the two photon energies are varied in this region. Finally, multiphoton detachment can also be employed in the search for true bound states of a negative ion (i.e. a different electronic configuration which lies below the detachment limit). Greene[36] has proposed the existence of $6s6p$ ^3P_J (J=0,1,2) states in the cesium negative ion which are bound by ~10-30 meV, but Thumm and Norcross[37] have indicated that the actual position of the states may be just above the detachment limit.

5. CONCLUSIONS

In conclusion, the rapidly increasing activity in multiphoton studies of negative ions is presenting numerous possibilities for future work. Several unique qualitative features have been outlined in this brief overview. Beyond the typical features which have already been emphasized, it can also be noted that in contrast to neutral atoms, the interaction regime at a given laser wavelength can be widely varied through selection of the negative ion electron affinity. This is illustrated by the examples of the calcium and chlorine negative ions which exhibit binding energies of respectively 18 meV[38] and 3.6 eV[22]. The present description of negative ion studies, which is intended as a progress report, has focussed rather strongly on the work with which the author was associated: hence we have largely discussed excess photon absorption and in particular, perspectives for resonant multiphoton experiments. Future studies should be extended to much higher intensities and shorter optical pulses. There is generally a need for more rigorous comparisons of theory and experiment in the work on nonresonant strong field phenomena. Exciting theoretical predictions of new phenomena are appearing, including the new threshold structures for H$^-$ in a strong field, as described by Faisal and Scanzano [39], and the possibility of observation of Gailitis-Damburg oscillations in two-photon detachment of H$^-$ as predicted by Liu et al [40]. There will also be several spinoffs for work of a unique spectroscopic character at lower intensities and pulses of long or intermediate temporal duration. The present discussion has been restricted to atomic negative ions, but unique possibilities may also exist for multiphoton studies of molecular negative ions and clusters. (For a recent overview of negative ion structure see Andersen[41] and Bates[35].) Finally, from a technological viewpoint, it has been recently demonstrated[42] that negative ions can also provide "clean" sources of neutral atomic beams for multiphoton ionization studies, which would otherwise be difficult to obtain.

6. ACKNOWLEDGEMENTS

I would like to point out the very important contribution which was made to the work described here by several students and colleagues at the University of Aarhus in Denmark. Most central to the multiphoton experiments on negative ions were T. Andersen, P. Balling, C. Brink, P. Kristensen and H. Stapelfeldt. Finally, the generous financial support of the Danish National Research Council (SNF), the Carlsberg Foundation, and the University of Aarhus are gratefully acknowledged.

7. REFERENCES

1. J.L. Hall, E.J. Robinson and L.M. Branscomb, *Phys. Rev. Lett.* **14**, 1013 (1965).
2. E.J. Robinson and S. Geltman, *Phys. Rev.* **153**, 4 (1967).
3. H.R. Reiss, *Phys. Rev.* **A22**, 1786 (1980)
4. M. Crance, *J. Phys.* **B20**, 6553 (1987); and *J. Phys.* **B21**, 3559 (1988).
5. A. L'Huillier and G. Wendin, *J. Phys.* **B21**, L247 (1988).
6. C. Blondel, R.J. Champeau, M. Crance, A. Crubellier, C. Delsart and D. Marinescu, *J. Phys.* **B22**, 1335 (1989).
7. N. Kwon, P.S. Armstrong, T. Olsson, R. Trainham and D.J. Larson, *Phys. Rev.* **A40**, 676 (1989).
8. C. Pan, B. Gao and A.F. Starace, *Phys. Rev.* **A38**, 2347 (1990)
9. J.H. Eberly, *J. Phys.* **B23**, L619 (1990).
10. C. Blondel, M. Crance, C. Delsart, A. Giraud, and R. Trainham, *J. Phys.* **B23**, L685 (1990).
11. F.H.M. Faisal, P. Filipowicz, C. Rzazewski, *Phys. Rev.* **A41**, 6176 (1990).
12. C. Blondel, M. Crance, C. Delsart and A. Giraud, *J. Phys.* **B24**, 3575 (1991).
13. T. Mercouris and C.A. Nicolaides, *J. Phys.* **B24**, L57 (1991)
14. W.G. Greenwood and J.H. Eberly, *Phys. Rev.* **A43**, 525 (1991).
15. M.D. Davidson, H.G. Müller and H.B. van Linden van den Heuvell, *Phys.*

Rev. Lett. **67**, 1712 (1991).
16. H. Stapelfeldt, P. Balling, C. Brink and H.K. Haugen, *Phys. Rev. Lett.* **67**, 1731 (1991).
17. H. Stapelfeldt, C. Brink and H.K. Haugen, *J. Phys.* **B24**, L437 (1991).
18. C.Y. Tang et al, *Phys. Rev. Lett.* **66**, 3124 (1991).
19. W.W. Smith et al, *J. Opt. Soc. Am.* **B8**, 17 (1991).
20. M.D. Davidson, D. Schumacher, P.H. Bucksbaum, H.G. Muller and H.B. van Linden van den Heuvell, *Phys. Rev. Lett.* **69**, 3459 (1992).
21. H. Stapelfeldt and H.K. Haugen, *Phys. Rev. Lett.* **69**, 3459 (1992).
22. H. Hotop and W.C. Lineberger, *J. Phys. Chem. Ref. Data* **14**, 731 (1985).
23. See, e.g., O. Faucher et al, *Phys. Rev. Lett.* **70**, 3004 (1993).
24. P. Agostini, F. Fabre, G. Mainfray, G. Petite and N.K. Rahman, *Phys. Rev. Lett.* **42**, 1127 (1979).
25. T.A. Patterson et al, *Phys. Rev. Lett.* **32**, 189 (1974).
26. P. Kristensen, T. Andersen, P. Balling, H. Stapelfeldt and H.K. Haugen, submitted to *Physical Review Letters*.
27. H. Stapelfeldt, *PhD thesis*, University of Aarhus, Denmark (1993).
28. M. Crance, *J. Phys.* **B21**, L557 (1988).
29. H.C. Bryant et al, *Phys. Rev. Lett.* **38**, 228 (1977); P.G. Harris et al, *Phys. Rev. Lett.* **65**, 309 (1990).
30. D. Proulx and R. Shakeshaft, Phys. Rev. 46, R2221 (1992).
31. P. Frey et al, *Z. Phys.* **A304**, 155 (1982).
32. P. Balling, P. Kristensen, H. Stapelfeldt, T. Andersen and H.K. Haugen, submitted to *Journal of Physics* **B**.
33. J.W. Hiby, *Ann. Phys. (Leipzig)* **34**, 473 (1939); an update on the negative ion properties can be found in T. Andersen et al, *Phys. Rev.* **A47**, 890 (1993).
34. J.O. Gaardsted and T. Andersen, *J. Phys.* **B22**, L51 (1989); P. Balling et al, *Phys. Rev. Lett.* **69**, 1042 (1992).
35. D.R. Bates, *Adv. Atomic Molec. Phys.* **27**, 1 (1991).
36. C.H. Greene, *Phys. Rev.* **A42**, 1405 (1990).
37. U. Thumm and D.W. Norcross, *Phys. Rev. Lett.* **67**, 3495 (1991).
38. C.W. Walter and J.R. Peterson, *Phys. Rev. Lett.* **68**, 2281 (1992); M.J. Nadeau, X.L. Zhao, M.A. Garwan and A.E. Litherland, *Phys. Rev.* **A46**, R3588 (1992); H.K. Haugen et al, *Phys. Rev.* **A46**, R1 (1992).
39. F.H.M. Faisal and P. Scanzano, *Phys. Rev. Lett.* **68**, 2909 (1992).
40. R. Liu, N.Y. Du, and A.F. Starace, *Phys. Rev.* **A43**, 5891 (1991).
41. T. Andersen, *Physica Scripta*, **T34**, 23 (1991).
42. H. Stapelfeldt, P. Balling and H.K. Haugen, *Phys. Rev.* **A46**, R1177 (1992).

LIGHT-INDUCED SINGLY AND DOUBLY EXCITED STATES OF THE NEGATIVE HYDROGEN ION

H.G. Muller and M. Gavrila*
FOM-Institute for Atomic and Molecular Physics
Kruislaan 407, 1098 SJ Amsterdam, Netherlands

ABSTRACT

We present a fully correlated ab-initio calculation of electronic states of H⁻ in a strong electromagnetic field, in the high-frequency limit. We find a nonmonotonic increase of the detachment energy with increasing field intensity, to a value larger than that of the unperturbed H⁻ ion. Several light-induced bound states do appear, among which doubly excited states. Curves giving the binding energies of these states as a function of intensity are presented.

INTRODUCTION

The hydrogen negative ion is of considerable theoretical interest because of the large amount of correlation it exhibits. Indeed, if it were not for electron correlation, a hydrogen atom would not be able to bind a second electron at all[1]. Even with correlation the H⁻ system is only marginally bound, its detachment energy being a mere 0.75 eV. Singly excited states do not exist, since almost perfect screening by the compact 1s electron prevents the nuclear Coulomb potential from binding a second electron in an excited state. Doubly excited states all have energies much larger than the detachment energy, so they are embedded high in the continuum, where most of them form auto-detaching resonances.

The existence of only a finite number (namely one) of discrete states in H⁻ offers the interesting possibility that this number may change as a result of perturbations. An intense laser field can act as a strong perturbation, and additional bound states in this case would be called light-induced states. Such states have been found in calculations on simple model atoms[2,] and in this paper we show that they do also appear in real atomic systems. Shakeshaft et al[3] have reported the appearance of light-induced states in atomic hydrogen at finite frequency. At the moment it is not clear if the mechanism through which these states come in existence (adiabatic evolution from a 'shadow pole') is related to that in the two-electron case presented here.

It has been known for a long time that the ionization rate of an atom in an electromagnetic wave tends to zero when the light frequency ω tends to infinity. This fact in itself is not very interesting, since the straightforward high-frequency limit is simply the unperturbed atom. Recently Gavrila and Kaminsky[4] pointed out the possibility of taking a limit in which the frequency and intensity I are taken to infinity *simultaneously*. If this is done in such a way that the amplitude of oscillation α_0 of a free electron in the field is kept constant during the process, the limit obtained *does* show effects of the field, but *retains* the property of vanishing ionization rate. This

---

* Current address: Institute for Theoretical Atomic and Molecular Physics, Cambridge, MA

makes it possible to study this system with time-independent techniques, a fact that no doubt has greatly contributed to the recent popularity of this high-frequency limit [5].

The most straightforward description of an atomic system in a high-frequency field is obtained in a coordinate system that oscillates in the same fashion as a free electron. In these so-called Kramers-Henneberger (K-H) coordinates, the electric force

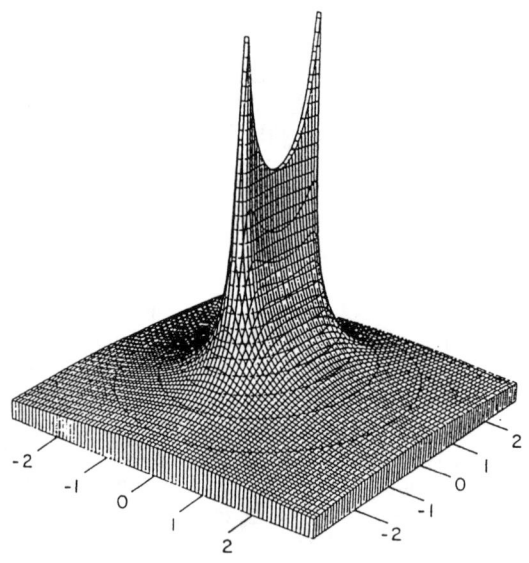

Fig. 1 - The 'dressed potential' V_0, generated by the time-average charge distribution of a harmonically quivering proton. (Taken from ref 4.)

on the electrons due to the radiation is exactly cancelled by the inertial forces due to the acceleration of the coordinate system. This situation is somewhat reminiscent of general relativity, where the effect of a homogeneous gravity field is equivalent to being in an accelerated frame. This equivalence exists because in a gravitational field all masses have the same acceleration, and in the case where one only has to describe electrons in a homogeneous electric field, the same holds true. In the K-H frame it is the nucleus that now exerts an apparent quiver motion, with amplitude $\alpha_0 = I^{1/2}\omega^{-2}$. As the frequency is taken to infinity while α_0 is kept constant, the acceleration of the nucleus becomes so high that the electron cannot possibly follow it, and in this limit the electron motion is determined by the time-average potential generated by the quivering nucleus, which we will call the dressed potential $V_0(r,\alpha_0)$. For linear polarization this potential is shown (in K-H coordinates) in fig. 1. It has a line-like logarithmic singularity along the quiver path of the nucleus, and slightly stronger $r^{-1/2}$ singularities at the end points, where the nucleus has its turning points.

Gavrila and Pont [6,7] have applied high-frequency theory to the three-dimensional hydrogen atom, leading to a number of surprising results. Among those is the contraction of the wavefunction into two spatially separated lobes ('dichotomy', fig.

2), which during the course of a field period alternately occupy the potential well created by the Coulomb field of the proton. By doing perturbation theory in the inverse frequency it was furthermore shown[8,9] that the ionization rate at fixed frequency becomes a decreasing function of intensity ('stabilization').

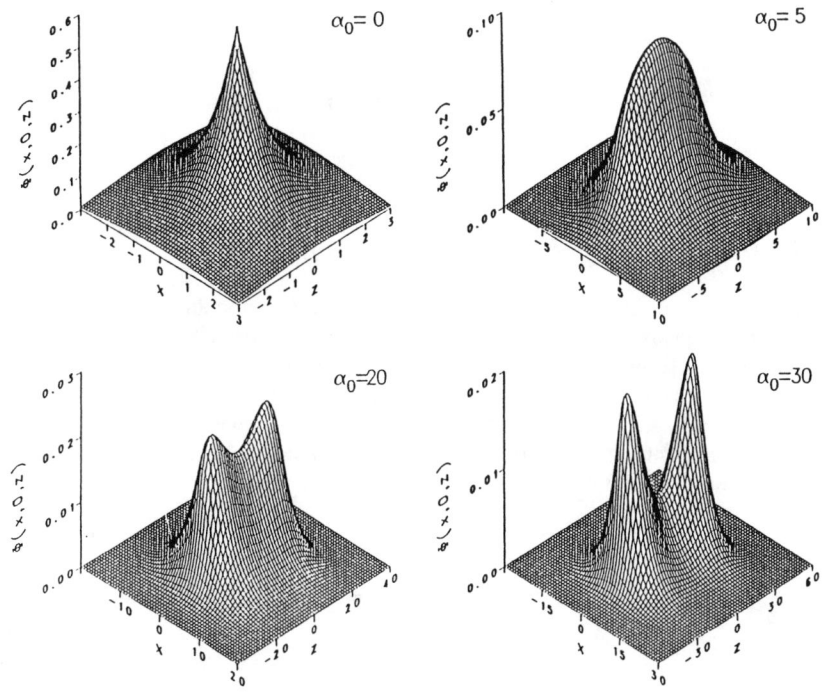

Fig. 2 - Wavefunctions of the one-electron system for various values α_0 of the stretching of the nuclear charge by the quiver motion. Note the separation into two different lobes ('dichotomy'), that sets in at $\alpha_0 = 20$. (Taken from ref 7)

Others have applied high-frequency theory to a number of other systems, often presented as model atoms. We mention the one-dimensional 'soft-Coulomb potential' [10,11], as well as a number of short-range potentials. It is often suggested that the latter provide good models for negative ions. Our present calculations, however, do not seem to bear this out, and the behaviour of a many-body Coulomb system seems quite different from that of a short-range model potential.

In all cases the effect of the field is the smearing of the potential well in the direction of the light polarization (which, for convenience, is assumed to be linear). In the short-range models, as in any one-electron system, this leads to a more diffuse binding of the electron in the now extended potential well. For the consequences the dimensionality of the system is of critical importance. In one dimension the loss of depth of the well due to smearing is more than compensated for by the lowering of kinetic energy of bound electrons in the larger well, and the number of states asymptotically increases without bound. In three dimensions, however, stretching will always cause sufficient shallowing to make the well lose its binding power *perpendicular* to the polarization vector, so all states must eventually disappear. Two-electron Coulomb systems behave quite differently form either of these cases. As we will show in this paper inter-electronic repulsion helps to concentrate the electrons in seperate regions, leading to a dramatic decrease of electron screening resulting in increased binding power.

CALCULATIONAL METHOD AND ACCURACY

We performed an ab-initio calculation of the wavefunctions and energies of the H$^-$ ion in the high-frequency limit for linear polarization, by putting the Hamiltonian

$$H = 1/2\, p_1^2 + 1/2\, p_2^2 - V_0(r_1;\alpha_0) - V_0(r_2;\alpha_0) + 1/r_{12} \quad (1)$$

(where V_0 is the dressed potential[4] and p_i and r_i the momenta and positions of the electrons) on a basis set of products of one-electron orbitals. These orbitals were given as simple functions (exponentials and powers) of prolate spheroidal coordinates[12] centered on the end-points of the nuclear charge line. This basis set is essentially a generalization of Slater type orbitals to the two-center case. In this simple basis all matrix elements can be evaluated analytically. The resulting matrix was then diagonalized. Details of the method will be published elsewhere.[13]

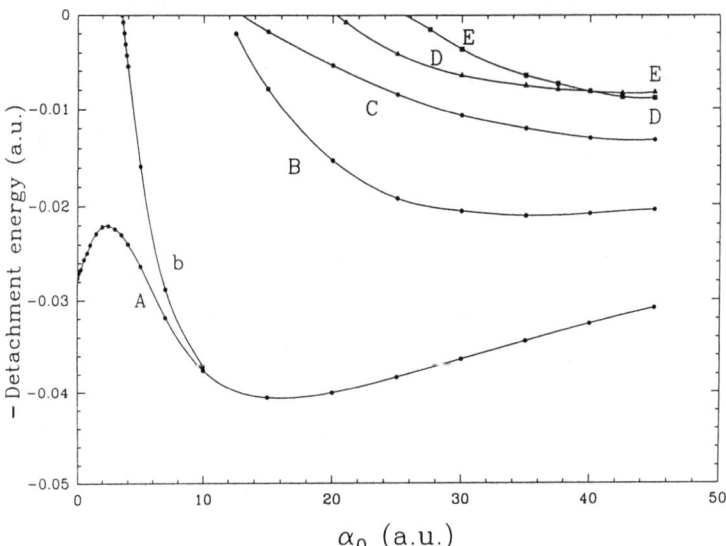

Fig. 3 - Energy of the five lowest-lying bound states of $^1\Sigma_g$ symmetry of H$^-$ in a laser field, as a function of the quiver amplitude α_0. The lowest $^3\Sigma_u$ state is also shown (curve b). (Taken from ref 17)

The symmetry and system of labeling of states in dressed H⁻ is the same as that of the H_2 molecule. We only incorporated basis functions belonging to the same irreducible representation as the desired result, by fixing spin multiplicity, total angular momentum around the symmetry axis and parity. The results for $^1\Sigma_g$ states can be seen in fig. 3.

The true two-electron wave function has a cusp where the two electrons meet. The basis functions we use are pure configurations that lack this cusp. Thus the description of the cusp must come from configuration interaction (CI), which makes the convergence of the correlated wavefunction in this basis set slow when the cusp is pronounced, i.e. at small α_0. Including basis functions that have explicit r_{12} dependence, as was done by Kolos[14] for the H_2 molecule, greatly speeds up convergence. In our case it was difficult to calculate the matrix elements of V_0 for r_{12}-dependent basis functions, however. We estimated the accuracy of our small-α_0 results by comparing them to the very similar problem of H_2 with the proton charge set to 1/2. For this problem we could use the Kolos basis, to get rapid convergence of the energy, downto six digits. We then checked the results for this model problem from the CI basis against these accurate values. Our conclusion was that the total energy obtained from the CI calculation for the number of basis functions used in dressed H⁻ is accurate to at least one part in 2000.

RESULTS AND DISCUSSION

All total energies were converted to detachment energies by subtracting the ground-state energy of the corresponding one-electron system, dressed H. The latter energy could be obtained by a diagonalization with about 40 one-electron functions of our basis set to better than 6 digits. The behaviour of the ground-state detachment energy as obtained through this calculation is quite surprising (fig. 3). At first the detachment energy decreases from its H⁻ value of 0.75eV, to reach a minimum around α_0=2.5 of 0.6eV. If α_0 is increased further the detachment energy increases (and even surpasses its unperturbed value) until it reaches a maximum of 1.1eV around α_0=17. From then on the detachment energy decreases again. Eventually it approaches zero, as it must, since the total energy is bounded from below by twice that of the one-electron system, which asymptotically scales[7] as $\alpha_0^{-2/3}$. The lowest expected excited state is $^3\Sigma_u$, just like in H_2. When α_0 is large enough for dichotomy to set in, each electron retreats in its own potential well. As a result the exchange interaction vanishes and singlets and triplets become degenerate. The energy of lowest state in the $^3\Sigma_u$ manifold is also shown in fig. 3. This state becomes stable for α_0 larger then 3.5. At α_0=3.55 its binding energy was at least 10 meV, and this approaches the $^1\Sigma_g$ ground state quite rapidly when α_0 is increased. This behaviour reflects the rapid onset of dichotomy under influence of the inter-electron repulsion. In the one electron system (fig. 2) dichotomy does not set in before α_0 = 20, but in H⁻ the electron repulsion forces complete separation of the two electron clouds at α_0 values as small as 10.

THE NATURE OF THE LIGHT-INDUCED STATES

Thus light-induced excited states do indeed exist in H⁻. In looking for other such states, specifically the predicted doubly excited states, we limited our search to the $^1\Sigma_g$ manifold, since in the one-electron system there are several excited σ-states below the lowest π-state. In the limit of complete dichotomy there is degeneracy between singlet and triplet states as well as between gerade and ungerade states, so for each light-induced $^1\Sigma_g$ excited state in figure 1 there will be nearby $^1\Sigma_u$, $^3\Sigma_u$ and $^3\Sigma_g$ states, in so far the Pauli exclusion principle allows them. Based on calculations of single-electron states in the end-point potential[15], the lowest states in the $^1\Sigma_g$ manifold (in the independent-particle approximation) should be the $(1\sigma)^2$, $(1\sigma)(2\sigma)$, $(1\sigma)(3\sigma)$, $(1\sigma)(4\sigma)$ and $(2\sigma)^2$, where nσ denotes the nth σ-state in the end-point potential.

In order to assign configurations to the states obtained though the *correlated* calculations we plotted several cuts through the corresponding six-dimensional wavefunctions and identified their nodal structure. The plots are best made in terms of cylindrical coordinates (ρ, z, ϕ) of individual electrons. An example of a useful cut, throught the z_2-O-ρ_2 plane, is presented in fig. 4. This cut corresponds to fixing one electron in one of the endpoints, and then plotting the (now ϕ-independent) charge distribution of the other electron.

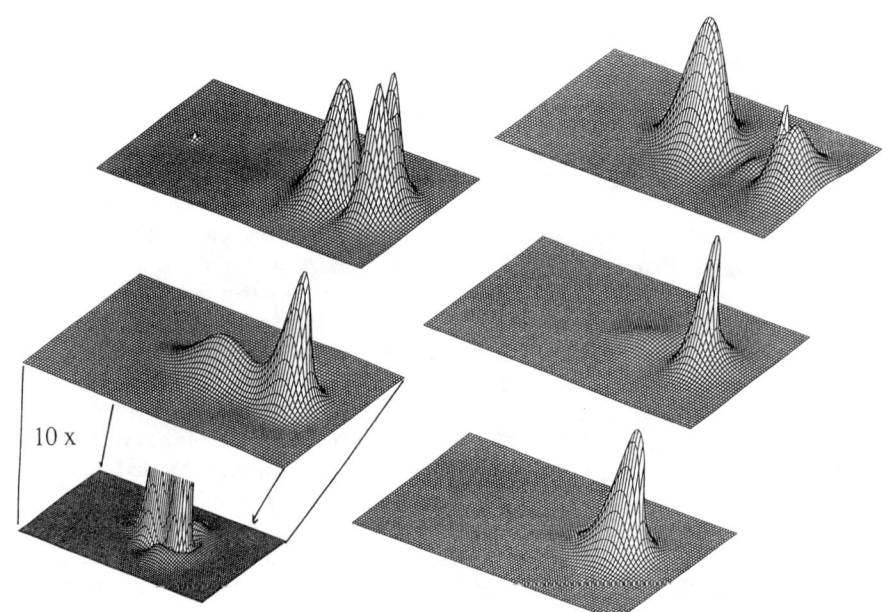

Fig. 4 - Wavefunctions of the lowest states of H$^-$ in a laser field. Plotted is Ψ^2 as a function of the ρ and z coordinate of electron 2, when electron 1 is positioned on the left-most end of the stretched nuclear charge ($\alpha_0 = 30$). On the lower right is the ground state (curve A in fig. 3), and above it the lowest axially excited states (B and C). On the left are the lowest doubly excited state (E) and a radially excited state (D). The radial nodal plain of the latter shows up more clearly in the magnified plot.

With the exception of $D^1\Sigma_g$ the wavefunctions have little structure perpendicular to the axis (i.e. only the axial motion is excited), and a cut through the z_1-O-z_2 plane completely describes the wavefunction. From plotting along this cut it turned out that for all excited states seen in fig. 3 it was possible to assign quantum numbers for each of the six degrees of freedom. For instance at $\alpha_0=30$ the lowest four excited $^1\Sigma_g$ states (B to E in fig. 3) correspond to single and double excitation of the axial motion of one electron, single excitation of the radial motion of one electron, and single excitation of the axial motion of both electrons, respectively. Thus this last state is the predicted[16] *doubly-excited* state.

Even after dichotomy has set in, some of the excited states show a fair degree of correlation (apart from the trivial left-right or 'valence-bond type' correlation which keeps the electrons localized on separate ends of the potential). Fig. 5 shows a number of cuts through the wavefunction of the $(1\sigma)(3\sigma)$ state, corresponding to situations where one of the electrons is fixed at various positions on the z-axis. As can be seen,

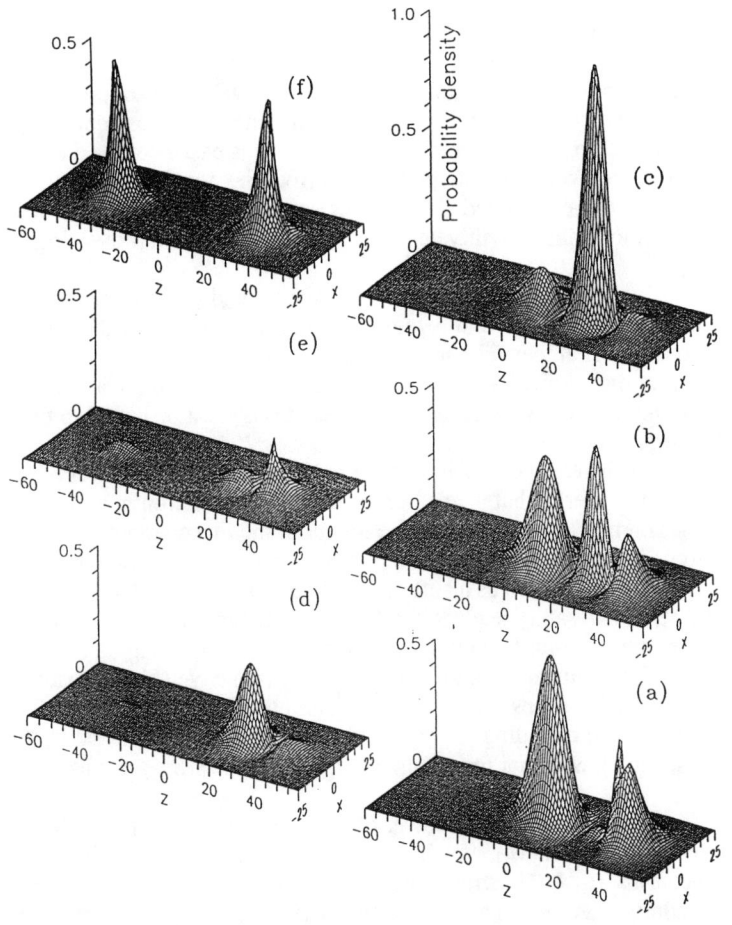

Fig. 5 - Correlation in the second excited state for $\alpha_0 = 30$. In the successive plots electron 1 is positioned at various positions on the nuclear-charge line, with a dramatic effect on the plotted charge distribution of electron 2. Taken from ref 17.

the charge distribution of the other electron is quite dependent on this, even when that electron is far away near its own end point.

Another point of interest is the behaviour of the $(1\sigma)(4\sigma)$ and $(2\sigma)^2$ states. The curves belonging to these states seem to cross around $\alpha_0=39$. In principle this crossing should be of the avoided type, since these states belong to the same symmetry species ($^1\Sigma_g$). The fact that this avoidance is not obvious from our calculation is another indication of the high degree to which the several degrees of freedom are uncoupled. Converting the axial excitation of two electrons into the radial excitation of a single electron violates two approximate symmetries, namely the independent particle and the axial-radial energy conservation. Therefore this 'doubly forbidden' coupling is especially weak. Since the curves are so nearly parallel it is hard to calculate the exact point of crossing with much confidence.

MULTIPLY CHARGED NEGATIVE ATOMIC IONS

The proliferation of the light-induced states at high values of α_0 points to an increased binding power of an extended charge configuration over a point charge, at least where several electons are concerned. Apparently it is easier for the electrons to minimize their repulsion without straying to much from the positive charge that binds them if this charge is more extended. This makes one wonder if it would be possible to bind three electrons to a single positive charge, a feat that has been proven impossible for point charges[18].

In the linearly polarized case, the obvious place to bind the third electron is in the center of the nuclear charge line. In this region there is an exposed positive line charge. The two electrons bound at the end points of the charge line do not offer much screening here after dichotomy has set in. In the case of circular polarization the apparent nuclear motion follows a circle, and all electrons must be bound to this one-dimensional charge structure.

To judge if this is feasible, it is important to know the binding power of a piece of line charge, i.e. the energy that is released on binding an electron in a two-dimensional logarithmic potential well. If the length of the line charge is not infinite, there will be some kinetic energy due to the motion along the line (which is essentially free), but this will approach zero as a sufficiently high power of the length (which scales as α_0) that we can safely neglect it. Obviously the binding power will depend on the amount of charge ρ per unit length (which scales as α_0^{-1}). Plugging this information in the Schrödinger equation with the logarithmic well shows that the extension of the wave functions scales as $\alpha_0^{-1/2}$, and the binding power as $(E_0 + 1/2 \ln \alpha_0)/\alpha_0$, where E_0 is the binding energy in a logarithmic well of unit strength. (Compare this scaling law to that for a point charge Z, where sizes scale as Z^{-1}, and the energy as Z^2.)

The leading term of the binding power of the exposed pieces of line charge for large α_0 thus becomes $\alpha_0^{-1}\ln \alpha_0$, which decreases slower than the Coulomb repulsion between the electrons, α_0^{-1}. Thus for a large enough value of α_0, the energy released on binding the additional electron to the nucleus, will always exceed the repulsion of that electron by the electrons already present, no matter how large the pre-factor (and thus no matter how many electrons were present already). In the high-frequency laser field negative ions of arbitrarily high charge exist!

If one plugs in the numbers it turns out that for the linearly polarized case a rather large value ($\alpha_0=95$) of the quiver amplitude is required to bind a third electron. For circular polarization, where the proton passes only one time at each point of its trajectory during a light cycle, the binding is even weaker, and a quiver radius around 200 is required. These numbers are already pretty steep, but for four and more

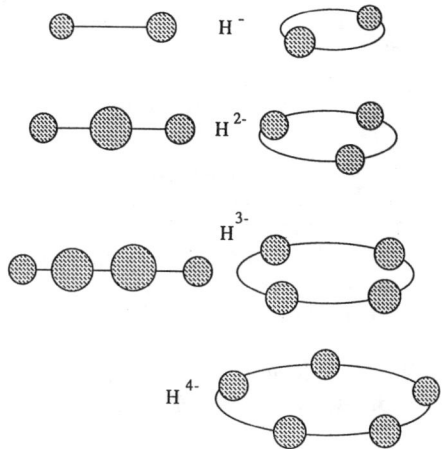

Fig. 6 - Configuration of various multiply charged negative ions for linear polarization (left) and circular polarization (right). In the latter case the nucleus moves along a circle with radius α_0.

electrons, they rapidly become ridiculous. The easiest case is circular polarization, where it is immediately obvious how to place the electrons on the circle to minimize their repulsion, namely at the corners of a regular polygon (fig. 6). The quiver radii required to bind 4, 5 or 6 electrons then are predicted to be $4 \cdot 10^4$, $16 \cdot 10^6$ and 10^{10} atomic units, respectively. The latter value would correspond to a quiver radius of 53 cm, and is obviously beyond the range of validity of the theory presented! Nevertheless, the H^{2-} values are still within the range of quiver amplitudes that can be achieved with present day lasers. For photons of 6eV, an intensity of $3.5 \cdot 10^{18} W/cm^2$ would suffice to cause an α_0 of 200, for example.

CONCLUSIONS

In conclusion, we have shown that light-induced excited states in H^- do exist at moderate, possibly even experimentally accessible values of the quiver amplitude α_0. The problem of experimental detection, however, will be even greater than that of detecting stabilized atomic states: In this latter case the states involved might have too short a lifetime at low intensity to survive the leading edge of the laser pulse. In the H^- case, however, the states do not even exist at low intensity, and would have to be made *in situ*, for instance by excitation with a second laser. The situation is even more bleak for the multiply charged negative ions, the production of which would require the attachment of an electron somewhere in the laser pulse. Nevertheless it seems that super-intense laser fields have a profound effect on the qualitative behaviour of negative ions, and their interaction might provide an interesting subject for experimental physics.

Acknowledgements - This research was supported by Fundamental Research on Matter (FOM) and the Netherlands Organization for the Advancement of Research (NWO). We thank the Institute for Theoretical Physics at the University of California, Santa Barbara for its hospitality and the computer time that lead to these results. This research was partly funded under European-Community Stimulation Act SCI-0103-C, and by a NATO travel grant.

REFERENCES

1. H.A. Bethe and E.E. Salpeter, *Quantum Mechanics of One- and Two-Electron Atoms* (Springer, Berlin, 1957), p.154
2. R. Bhatt, B. Piraux and K. Burnett, Phys. Rev. A **37** (1988) 98
3. R.M. Potvliege and R. Shakeshaft, Phys. Rev. A **38** (1988) 6190
4. M. Gavrila and J.Z. Kaminski, Phys. Rev. Lett. **52**, 614 (1984)
5. M. Gavrila in *Atoms in Intense Laser Fields*, Ed. M. Gavrila (Academic, 1992), p.435
6. M. Pont and M. Gavrila, Phys. Rev. Lett. **65**, 2362 (1990)
7. M. Pont, N, Walet, M. Gavrila and C.W. McCurdy, Phys. Rev. Lett. **61**, 939 (1988)
8. M. Pont, N. Walet and M. Gavrila, Phys. Rev. A **41**, 477 (1990)
9. R.J. Vos and M. Gavrila, Phys. Rev. Lett. **68**, 170 (1992)
10. R. Grobe and J.H. Eberly, Phys. Rev. Lett. **68**, 2905 (1990)
11. R. Grobe and J.H. Eberly, Phys. Rev. A **47**, R1605 (1993)
12. M. Abramowitz and I. Stegun, *Handbook of Mathematical Functions*, (Dover, N.Y., 1965), §21.
13. H.G. Muller and M. Gavrila, Phys. Rev. A., to be published
14. W. Kolos and C.C.J. Roothaan, Rev. Modern Phys. **32**, 219 (1960)
15. M. Gavrila and J. Shertzer, to be published
16. M.H. Mittleman, Phys. Rev. A **42**, 5645 (1990)
17. H.G. Muller and M. Gavrila, Phys. Rev. Lett. **71** (1993) accepted for publication
18. E.H. Lieb, Phys. Rev. A **29** (1984) 3018

PHOTOMOLECULAR PROCESSES

MOLECULAR DYNAMICS IN INTENSE LASER FIELDS

A. Giusti-Suzor [1] and E. Charron
Laboratoire de Photophysique Moléculaire, Université Paris-Sud, 91405 Orsay France
(1)*also* Laboratoire de Chimie Physique, 11 rue Pierre et Marie Curie, 75231 Paris

F.H. Mies
National Institute of Standards and Technology, Gaithersburg Md 20899, USA

ABSTRACT

The fragmentation dynamics of small molecules submitted to intense laser fields can be described in the framework of *laser-assisted multichannel half-collisions*. A number of unexpected features have been observed or predicted in the past few years. We review these processes for the case of the molecular ion H_2^+. The theoretical results, obtained mostly from *time-dependent wave-packet calculations*, are analyzed in terms of "dressed" potential curves, and compared to available experimental results.

INTRODUCTION

The production of short, intense light pulses in the optical range has opened an exciting new chapter in the study of molecular dynamics. Under these exotic conditions recent experiments and theories have underscored the combined action of interparticle couplings and particle-field interactions in the time-development of *half-collisions*, namely photoionization and photodissociation. Measurements of the energy and angular distribution of the ejected electrons and/or of the ionic fragments provide an useful insight into the fragmentation dynamics, helping in particular to disentangle multistep processes from the so-called "Coulomb explosion" where dissociative multielectron ionization is expected to happen abruptly through a "vertical" multiphoton absorption. Specific studies of the *sequential* versus *direct* processes in a variety of molecules, mostly diatomic (HI, I_2, N_2, CO...) but also polyatomic (alkanes), have been reviewed recently by Codling and Frasinski[1]. They show how the gross features of the experimental data can be discussed in terms of a simple classical field-ionization model.

Here we concentrate instead on the simplest molecule, H_2, which has been also the most extensively studied in recent years. Even more, we restrict our review to a specific step of the fragmentation dynamics (clearly sequential in this case [2]) observed in strong field, namely the *photodissociation of the molecular ion* H_2^+,

$$H_2^+(1s\sigma_g, J, M_J, v) + n\hbar\omega \longrightarrow H(1s) + H^+ \tag{1}$$

prepared in a given rovibrational level of the ground state by multiphoton ionization of the neutral. Experiments[3-5] as well as theoretical studies[6-13] have revealed a number of interesting features which can only be understood in terms of complex multiphoton interactions, involving absorption as well as stimulated emission. They are best analyzed within the framework of the "dressed" molecular potential curves, although most of the calculations are now done by directly solving the time-dependent Schrödinger equation. We stress that in the intensity range of these studies (10^{12} to

10^{15} W/cm^2) perturbative treatments are excluded. We will see that the radiative field plays an active role not only in the excitation step, as in the case of weaker intensity (REMPI), but all along the decay process, and strongly acts on the subsequent molecular dynamics.

THEORETICAL TREATMENTS

General framework

Both time-independent[6-9] and time-dependent[10-13] treatments have been used lately to study the fragmentation dynamics of H_2^+ in a strong laser field. All these calculations only include radiative coupling between the attractive ground state $1s\sigma_g$ and the first repulsive $2p\sigma_u$ state which both asymptotically dissociate to $H(1s) + H^+$. The radiative dipolar interaction is expressed either in the electric field gauge representation (*length* form) or in the Coulomb gauge (*velocity* form). Actually, recent field-dressed variational calculations by Muller[14] have demonstrated that this two-state approximation is justified only for the length form of the dipolar transition moment, which varies as R/2 (R being the internuclear distance). The much larger values of the velocity dipolar moment at short internuclear distance, where most of the dynamics occurs, makes it invalid to neglect higher electronic states in the Coulomb gauge representation. On the other hand, the asymptotic divergence (R --> ∞) of the transition moment in the length form complicates the scattering formulation of the fragmentation process when the laser intensity is assumed to stay constant (long pulse, time-independent treatment). Various remedies have been proposed to overcome this difficulty, which is avoided anyway for short laser pulses treated in a time-dependent approach since then the radiative interaction vanishes before the particles reach large internuclear distances.

Another approximation used in most of the existing calculations (all performed for linear laser polarization, as in the experiments) is to restrict the nuclear motion to one dimension, assuming the molecular ion to be strongly aligned along the laser polarization axis : this is based on experimental evidence[3,15] and results from rapid "optical pumping" to high J, low M_J rotational levels by multiphoton interactions. More rigorous calculations including an expansion on a reasonably complete set of rotational states have recently been performed[16,17] and confirm that the proton angular distribution mainly consists of two peaks centered around 0° (forwards ions) and 180° (backwards ions) and becoming narrower with increasing intensity.

Time-independent approaches

The non-perturbative time-independent theory of strong field photodissociation can be derived in two ways, leading to practically identical formulations. One can start from a classical representation of the laser oscillating electric field and perform a Fourier expansion of the radiative interaction and of the resulting molecular wavefunctions (Floquet theory)[6]. Alternatively, one can use the quantal formulation of the radiative field with creation and annihilation operators associated with photon emission and absorption. In both cases, one defines a set of photon number states |N-n> centered around a large N photon number proportional to the field intensity. The molecule-field channel state |g>|N-n> with n=0 defines the initial state of the system before any absorption (n>0) or stimulated emission (n<0) has occurred. In Fig.1 the n=0 asymptote is scaled to zero and joins to the $V_g(R)$ attractive ground state potential

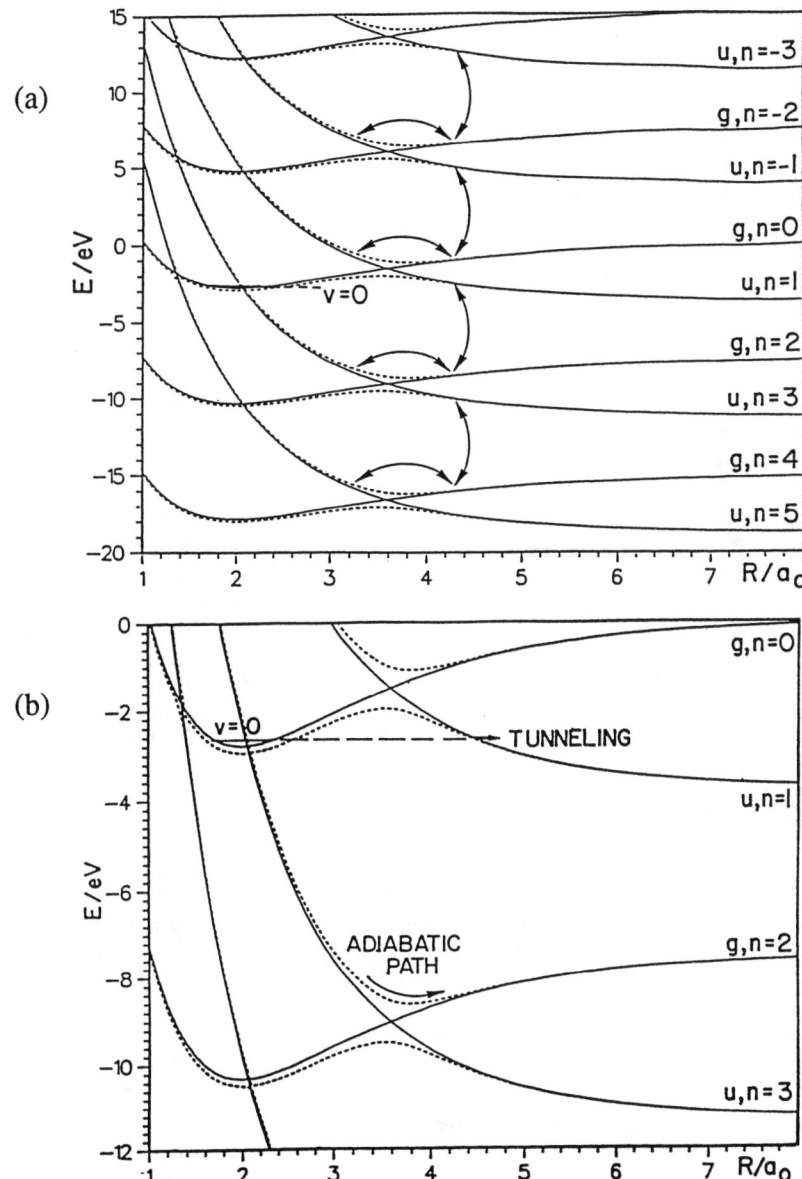

<u>Figure 1</u>: (a) Potential energy curves for the $1s\sigma_g$ and $2p\sigma_u$ states of H_2^+, dressed with ±5 photons of wavelength 330nm. The full curves correspond to the 10 interacting "diabatic" channels, with arrows indicating the direct radiative couplings. The dotted curves represent the field-dressed "adiabatic" potentials obtained by diagonalizing the radiative interaction for $I=1.4 \cdot 10^{13}$ W/cm^2.

(b) Enlargement of (a) in the region of the avoided crossing points which dominate the dissociation mechanisms. From ref.(9)

which supports the initial vibrational state. The n=1 asymptote correlates with the |u>|N-1> channel and joins to the $V_u(R)$ repulsive potential, shifted down in energy by $\hbar\omega$ due to the loss of one absorbed photon. These two channels, augmented by similar ones (g,n=odd and u,n=even) corresponding to additional photons absorbed (n>0) or emitted (n<0) can be viewed as *diabatic field-dressed states* for the combined molecule-field system. In Fig.1 are also shown, by dotted lines, the *adiabatic* counterpart of these dressed molecular curves[7], obtained by full diagonalization of the (tridiagonal) radiative couplings between the diabatic dressed curves. Although most calculations are performed within the diabatic basis set, the adiabatic set of curves has proven to be very helpful in analyzing the numerical results and yield a good insight into the fragmentation dynamics even when time-dependent treatments are used, as we will see later.

Starting from a set of dressed channels coupled by time-independent radiative couplings, any of the numerical techniques used in close-coupling scattering theory can be applied for solving the corresponding system of coupled equations (complex scaling of the R-coordinate[6,8], Fox-Goodwin propagation with adjonction of an artificial channel[7], renormalized Numerov[9]...). The initial vibrational level is viewed as a *resonant state* in a specific closed channel, and its laser broadening (inverse of its dissociation rate) and AC-Stark shift are easily extracted from the numerical results. The methods based on a full-collision formalism[18] with outgoing waves in each open dressed channel are particularly fruitful: from the half-collision amplitude associated with the initial level (or any prescribed vibrational level in a given closed channel), one can easily get the *branching ratios* for dissociation into specific open channels corresponding to varying amounts of absorbed photons. These quantities are the most clear signature of the dynamical events induced by the strong radiative field in the dressed molecule, and are directly related to the photofragment energy spectra recorded in the experiments.

Time-dependent approaches

For short laser pulses, in the subpicosecond domain, the field intensity varies on the time scale of the fragmentation process and the pulse shape has to be included in the calculations. Time-dependent treatments then become more convenient and have been adopted by most theoretical groups in the last two years, due to the rapid development of the short pulse laser technology.

A molecule in the initial stable state v is subjected to a gaussian laser pulse of width T_p, corresponding to the classical electric field, linearly polarized along $\hat{e}$:

$$E(t) = f(t) \, E_0 \cos(\omega t) \, \hat{e} \qquad (2)$$

with ω the laser frequency and E_0 the field amplitude. The pulse shape is often approximated by the gaussian-like expression $f(t) = \sin^2(\pi t / 2T_p)$ with total pulse duration $2T_p$, to avoid the extended wings of the true gaussian shape which lengthen the calculations without any noticeable contribution to the physical processes. The time-dependent Schrödinger equation is directly solved by any of the available methods for wave-packet propagation[19]. The most commonly used in this domain is the splitting technique of the short-time propagator[20], which transforms the wave function back and forth between the coordinate (R) and the momentum (k) representations, using fast Fourier transformations (FFT).

These time-dependent calculations provide the time evolution of the nuclear wave packet, in particular the decay of the initial bound state population and the dissociation probability at the end of the laser pulse. The fragment energy spectrum is also obtained by projecting the final wave packet onto the complete set of dissociative continuum wave functions of the field-free molecular ion. Thus the two types of treatments, time-independent and time-dependent, lead to the same "observables", namely the branching ratios between alternative decay channels, with a more direct insight into the dynamics of the fragmentation in the case of time-dependent calculations[21]. The choice between the two approaches is mainly dictated by the length of the laser pulse one is trying to simulate.

FRAGMENTATION MECHANISMS

We present now a summary of the theoretical results obtained by various groups in the last three years. Calculations have been performed for a wide range of initial vibrational levels (v=0-14) and for several wavelengths used in recent experiments (248 nm, 330 nm; 1064 nm, 532 nm and 355 nm-the three first harmonics of the Nd:YAG laser; 769 and 780 nm, in the wavelength domain of the Ti:Sapphire subpicosecond laser). The intensity ranged from 10^{13} to 5.10^{14} W/cm^2. The first observation to be pointed out is that the results change drastically from the lowest vibrational levels to the highest ones, and they will be discussed in turn in this section. The limiting values of v for different behaviors depend of course on the laser wavelength and intensity, but the discussion qualitatively applies to all the cases studied.

Low vibrational levels: above threshold dissociation (ATD)

Except for the shortest wavelengths, the lowest vibrational levels, deep in the $1s\sigma_g$ potential well, either cannot dissociate with a single photon absorption (e.g. v=0,1 for 532 nm) or dissociate only via tunneling below the adiabatic potential barrier (e.g. v=2 at 532 nm or v=0 at 330 nm, see Fig.1b). With such a weak one-photon dissociation, three-photon dissociation, although resulting from an indirect third order coupling, is dominant due to the more favorable crossing between the n=3 and n=0 field-dressed curves (see Fig.1b). As a result, *above threshold dissociation* (named after the well-known above threshold ionisation-ATI-process[22]) is observed, corresponding to dissociation with an "excess" of photons absorbed in the dissociation continuum. Actually, rather than the expected three-photon dissociation, the energy spectrum for v=2, 532 nm is dominated by the peak corresponding to two-photon dissociation (Fig.2). This result indicates that most of the dissociating flux follows the adiabatic path in Fig. 1b which connects the n=3 dressed curve at short distance to the n=2 one at long distance due to the avoided crossing induced by the radiative interaction. We stress that this global two-photon dissociation process actually involves four photons (three absorbed at short internuclear distance, one emitted farther away), a somewhat surprising feature which illustrates our early remark that the photon field plays an active role all along the fragmentation process. The ATD process has been experimentally demonstrated[3,4] at 532 nm, as we discusss later.

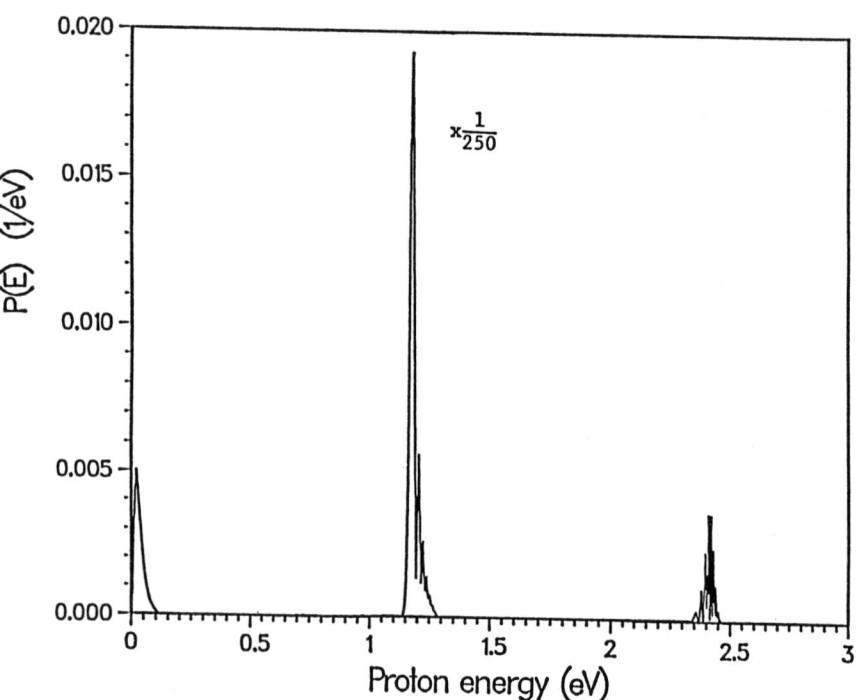

Figure 2: Proton energy distribution in the dissociation process of H_2^+ with λ=532nm, I=10^{14} W/cm^2 and initial vibrational level v=2. The succesive peaks are separated by half a photon energy (1.15 ev), the second and third peaks corresponding to ATD with excess-photon energy equally shared between H and H^+.

532 nm PULSE 150 fs I=10(14) W/cm2 v=3

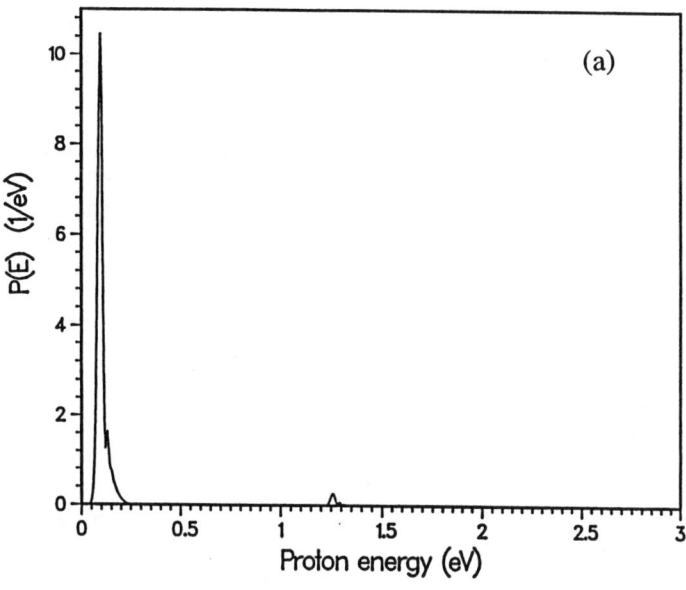

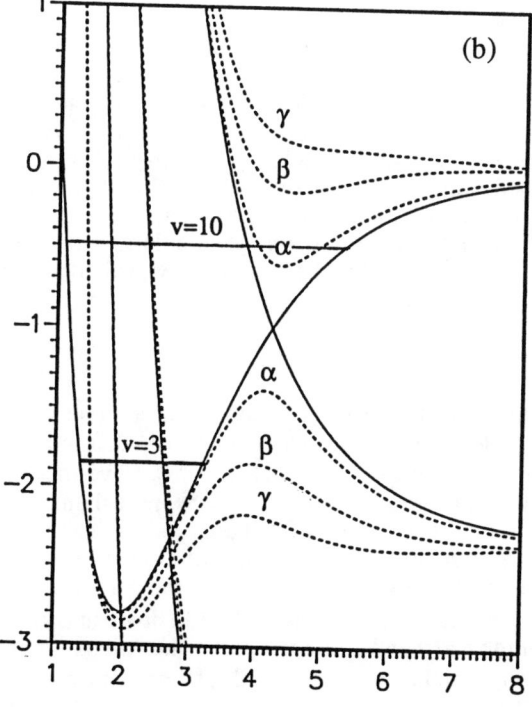

Figure 3: (a) same as figure 2, for initial vibrational level v=3

(b) dressed potential curves (——) diabatic (...) adiabatic for λ=532nm and 3 intensities:

(α) 10^{13} W/cm²
(β) 5.10^{13} W/cm²
(γ) 10^{14} W/cm²

v=3: "bond-softening" mechanism
v=10: "trapping" mechanism

Intermediate vibrational levels: bond-softening

For a given vibrational level and a given wavelength, there is an intensity threshold beyond which the adiabatic potential barrier is pushed below the initial vibrational level (including its AC-Stark shift), and the molecule undergoes fast one-photon dissociation. As soon as the peak intensity of the laser field exceeds this threshold value (e.g. $I=5.10^{13}$ W/cm^2 for v=3, λ=532 nm, see Fig.3b), this one-photon process will dominate the dynamics of the fragmenttion at the expense of ATD. It has been termed *bond-softening* [3] since it facilitates the molecule dissociation due to the strong laser induced avoided crossing in the dressed curve potential curves. Indeed, we still get some n=2 dissociation for initial level v=3 at 10^{13} W/cm^2, as shown on Fig.(3a), but it becomes negligible for v≥4 which are 100% dissociated with one-photon absorption, even for a pulse duration less than 100 fs. For the shortest wavelengths like 248 nm, even the lowest levels (v=0-3) are dominated by the bond-softening process and dissociate completely within the first half of a 100 fs pulse, as shown in Fig.(4a).

High vibrational levels: wave-packet trapping

Further inspection of Fig.4 reveals a new behavior for higher vibrational levels. For λ=248 nm, the levels v≥4 do not totally dissociate any more, and part of the population is "trapped" in high vibrational states at the end of the pulse. This new behavior, which has been found for other wavelengths as well (beginning at higher v for longer λ)[11-13], may be understood again from the adiabatic dressed potential curves in Figs.(1b) or (3b). As the pulse rises, the portion of the wave packet that lies above the diabatic crossing between the (g, n=0) and (u, n=1) dressed curves, to the right of their intersection, is trapped and pushed up by the rising adiabatic potential, thus escaping from dissociation. This stabilization mechanism, named "bond-hardening" by Yao and Chu[12] since it is opposite (or rather complementary) to the bond-softening process, may also result from trapping in other laser induced potential wells (e.g. the well formed by the avoided crossing between the n=0 and n=3 dressed curves, at 769nm). Note however that the relative amount of population trapped, starting from a given initial vibrational level, has been found by Aubanel and Bandrauk[17] to be very sensitive to rotational effects.

COMPARISON TO EXPERIMENTS

Except for the most recent experiment[5] performed with a Ti:Sapphire laser (T_p=150 fs, $I=10^{15}$W/cm^2), the available experimental data on H_2^+ photodissociation in strong field have been obtained with a relatively long pulse (T_p>1 ps). Nevertheless, we think that they can be qualitatively understood in terms of the theoretical analysis presented above, based on calculations performed mostly for shorter pulses. The main question raised by a first glance to the proton energy distribution at various wavelengths[3-5] is: why are the ATD peaks so weak, compared to the ATI peaks in atomic ionization[21] ? Indeed, ATD peaks are clearly visible in the 532 nm results only (see Fig. 5b), and much weaker than the n=1 bond-softening peak. The "preparation" of H_2^+ in these experiments, via resonant and non-resonant multiphoton ionization of ground state H_2 molecules, may explain this observation. Generally a distribution of ionic vibrational levels is formed, each of which yields its own proton energy

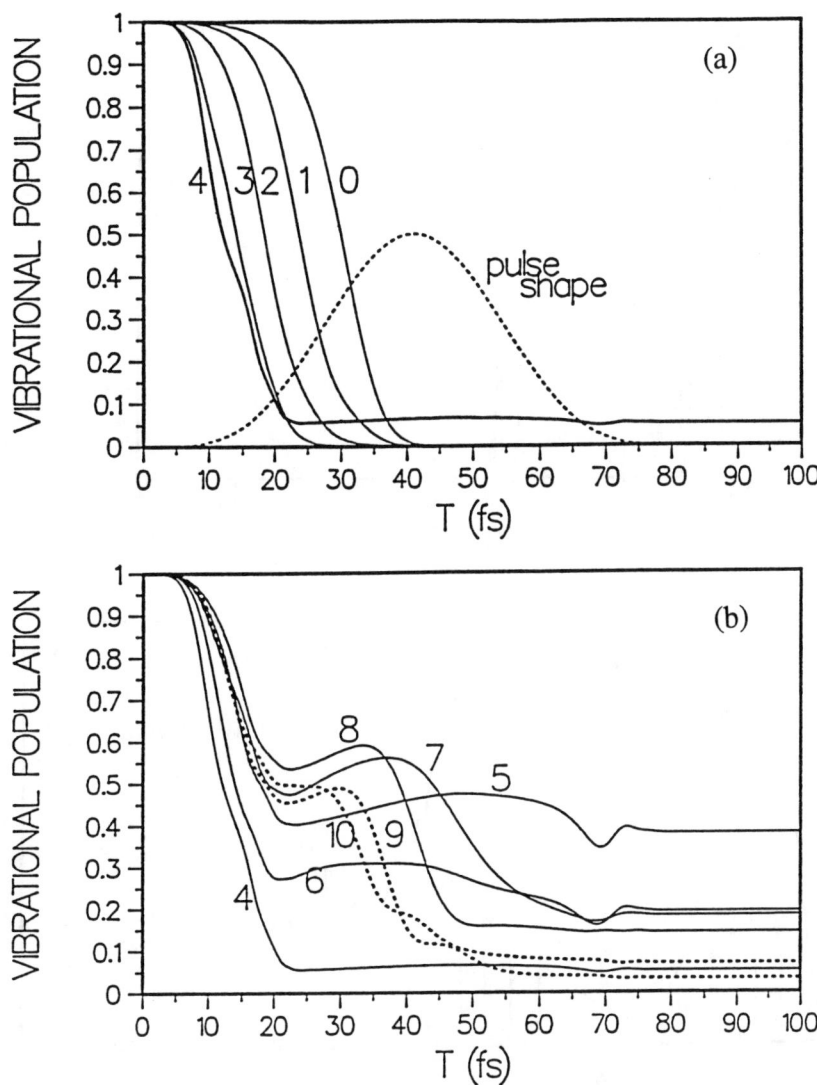

Figure 4: Decay of the total bound state population (v=1-18) of H_2^+ ground state for a 100 optical cycles (83 fs) laser pulse (dotted line in fig.4a) with λ=284 nm and peak intensity 5.10^{14} W/cm^2. The numbers denote the initial vibrational level for each calculation. Trapping occurs for v≥4 (see fig.4b). From ref.(11)

136 Molecular Dynamics in Intense Laser Fields

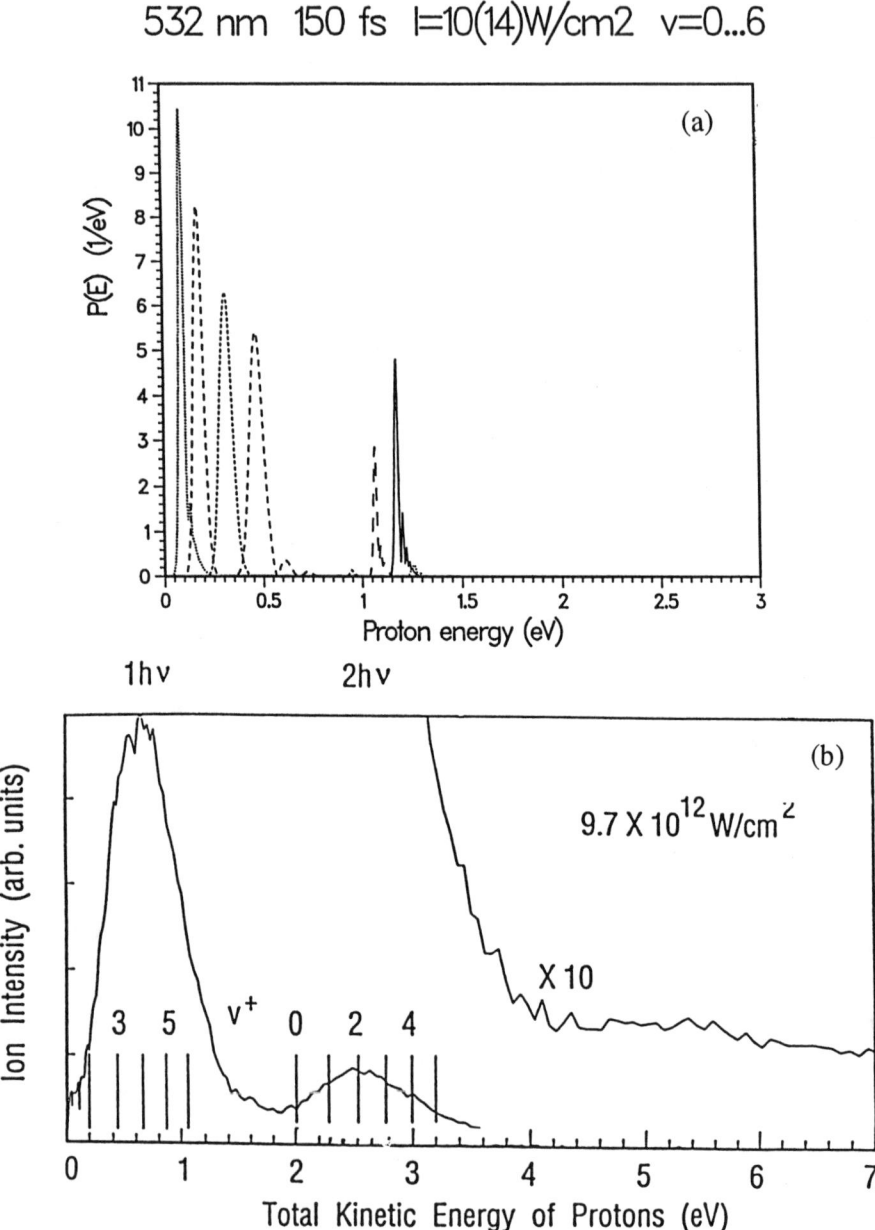

Figure 5: (a) Proton energy distribution in the dissociation of H_2^+ with λ=532 nm, I= 10^{14} W/cm^2 and equal contribution of initial v=0 to 6.
(b) Experimental results from ref.(4)

distribution. Thus one measures an ill-defined superposition of proton energy spectra, dominated by the fast one-photon dissociation of the intermediate levels through bond-softening. To sustain this explanation, we have superimposed our theoretical results obtained for several v's, assumed to be equally populated (which is obviously not true in resonant MPI). The result for $\lambda=532$ nm (Fig.5a) is consistent with the experimental observations (Fig. 5b), showing a weak ATD peak mostly due to the v=2 contribution (cf Fig.2). For other wavelengths (355 or 1064 nm), the ATD peaks are either too weak, or too close in energy (especially at long wavelength like 1064 nm) to the n=1 dissociation peaks associated with higher v's, thus being "hidden" in the wings of the dominant one-photon contribution to the energy spectrum as observed in the experiments. Only a vibrational-state selective preparation of H_2^+ would allow a precise comparison with theoretical results, and demonstration of intensity-dependent stimulated emission processes due to avoided crossings in the dressed curves[7, 9-10], as shown in Fig. (2b).

Finally, some indirect evidence of the "trapping" process predicted for high vibrational levels[11-13] might be inferred from two sets of experiments. First, the REMPI results ot Allendorf et al[23] around 304 nm suggest that temporary bound states in the upper field-induced potential well (see e.g. Fig. 3b) might act as final "dressed" states in the photoionization of the neutral molecule. Second, the fast protons collected in the Ti:Sapphire (769 nm) experiment of Zavriyev et al.[5] could come from further ionization to H_2^{++} repulsion curves of high v H_2^+ population, trapped long enough to survive until the laser peak intensity (10^{15} W/cm^2) is reached.

CONCLUDING REMARKS AND FUTURE

The present review has been limited, for the sake of clarity, to one-color, strong field photodissociation of H_2^+ molecular ions. Recent studies have already broaden the domain and indicate future trends in theoretical as well as experimental work. First, *ionization rates* for H_2^+ have been calculated[24] for various wavelengths and internuclear distances (in the Born-Oppenheimer approximation). The results suggest that in the case of short wavelengths (248 nm, 355 nm), ionization becomes operative around 10^{14} W/cm^2 at intermediate R-values, thus making questionnable the previous two-states dissociation calculations where not only ionization but also the high Rydberg states were neglected. But for the longer wavelengths (532 nm, 780 nm, 1064 nm) which are commonly used in the short pulse experiments, ionization does not efficiently compete with dissociation in the regions important for the nuclear dynamics. Further progress in this area would require simultaneous treatment of competing ionization and dissociation, beyond the B.O. approximation. Current computer technology gives us hope that the double continuum problem can be treated, and H_2^+ would be an ideal candidate for such a study, both experimentally and theoretically.

Another promising direction of research is to study *two-color photodissociation* by a *coherent* superposition (phase-locked) of an intense laser radiation and one of its harmonics. The first results[25] indicate that the total dissociation probability, the energy distribution and the direction of ejection of the protons are very sensitive to the relative phase of the two radiations. A high degree of "coherent control" [26] may thus be achieved, and this process should allow interesting experimental manipulation of dissociation dynamics.

This work was supported in part by a NATO grant for International Collaborative Research.

REFERENCES

1. K. Codling and L. J. Frasinski, Topical Review in J. Phys.B **26**, 783 (1993)
2. D. Normand. C. Cornaggia and J. Morellec, J. Phys. B **19**, 2881 (1986)
3. A.Zavriyev, P.H.Bucksbaum, H.G.Muller, D.W.Schumacher, Phys.Rev.Lett. **64**, 1883 (1990); Phys.Rev.A **42**, 5500 (1990)
4. B. Yang, M. Saeed, L.F.DiMauro, A. Zavriyev and P.H. Bucksbaum, Phys. Rev.A **44**, R1458 (1991)
5. A. Zavriyev, P.H.Bucksbaum, J. Squier and F. Saline, Phys. Rev. Lett. **70**, 1077 (1993)
6. S.-I Chu, J. Chem. Phys. **75**, 2215 (1981); J. Chem. Phys, **94**, 7901 (1991)
7. A.D.Bandrauk, M.L.Sink, J.Chem.Phys. **74**, 1110 (1981); A. D. Bandrauk, E. Constant and J.M. Gauthier, J. Phys. II France **1**,1033 (1991), and ref. therein
8. X. He, O. Atabek, A.Giusti-Suzor, Phys.Rev. A **38**, 5586 (1988);**42**, 1585 (1990)
9. A. Giusti-Suzor, X.He, O.Atabek, F.H. Mies, Phys.Rev.Lett. **64**, 515 (1990)
10. G.Jolicard, O.Atabek, Phys.Rev.A **46**, 5845 (1992)
11. A.Giusti-Suzor, F.H.Mies, Phys.Rev.Lett. **68**, 3869 (1992)
12. G.Yao and S.-I Chu,Chem.Phys.Lett.**197**,413 (1992); Phys.Rev.A**48**,485(1993)
13. A.D.Bandrauk, J.M. Gauthier, J.F.McCann, Chem. Phys. Lett. **200**, 399(1992)
14. H. G. Muller, in "Coherence Phenomena in Atoms and Molecules in Laser fields", ed. A.D. Bandrauk and S.C. Wallace, Plenum Press NY(1992), p.89
15. D.Normand, L.A.Lompré, C.Cornaggia, J.Phys.B **25**, L497 (1992) ; D.T. Strickland, Y.Beaudoin, P.Dietrich, P.B.Corkum, Phys.Rev.Lett. **68**, 2755 (1992)
16. E. Charron, A. Giusti-Suzor and F. H. Mies, to be published
17. E. Aubanel and A. D. Bandrauk, to be published
18. F. H. Mies and A. Giusti-Suzor, Phys. Rev. A **44**, 7547 (1991)
19. K. Kulander, K. J. Schafer, J. L. Krause, "Atoms in intense fields", Adv. in At.Mol. Opt.Phys. (Supplement 1), Ed. M. Gavrila, Academic Press (1992), p.247
20. M. J. Feit, J. A. Fleck, A. Steiger, J. Comput. Phys. **47**, 412 (1982); R. W. Heather, Comput. Phys. Com.**63**, 446 (1991)
21. R. Heather and F. H. Mies, Phys. Rev. A **44**, 7560 (1991)
22. P. Agostini, F. Fabre, G. Mainfray, G. Petite and N. K. Rahman, Phys. Rev. Lett. **42**, 1127 (1979); see various papers in"Atoms in intense fields", Adv. in At.Mol. Opt.Phys. (Supplement 1), Ed. M. Gavrila, Academic Press (1992)
23. S. W. Allendorf and A. Szöke, Phys. Rev. A **44**, 518 (1991)
24. S. Chelkowski, T. Zuo. A. D. Bandrauk, Phys. Rev. A **46**, R5342 (1992); F.H.Mies, A.Giusti-Suzor, K.J.Schafer, K.C. Kulander, proceedings of SILAP III, NATO ASI Series (in press, 1993)
25. E. Charron, A. Giusti-Suzor and F.H.Mies, Phys.Rev.Lett. **71**, 682 (1993)
26. M.Shapiro, J.W.Hepburn, P.Brumer, Chem.Phys.Lett.**149**, 451 (1988)

DISSOCIATION OF CORE EXCITED POLYATOMIC MOLECULES: SEQUENTIAL OR CONCERTED PROCESSES?

P.Morin[a,b], M.Lavollée[a], M.Meyer[a] and M.Simon[a,b]

a) LURE, Bat 209D, Université PARIS-SUD, 91405 ORSAY France
b) DRECAM/SPAM, Bat 522 CENS/CEA, 91191 SACLAY France

ABSTRACT

We show that molecular dissociation dynamics of free molecules, as induced by core excitation using synchrotron radiation and probed with 2D mass spectrometry, may be described either as sequential or instantaneous processes, depending on the size of the excited target. The studied molecules are the prototype molecules $FeCO_2NO_2$, N_2O and CO_2 excited around the carbon and nitrogen K shell. Complex polyatomic molecules are shown to undergo fast internal energy conversion, followed by slower stepwise dissociation. In case of triatomics, on the contrary, fragmentation is very fast and we propose an impulsive model to describe the process, pointing out the breakdown of the Coulomb explosion picture.

INTRODUCTION

Ionisation and/or excitation of core shells of molecules in the soft X-Ray regime, allows production of multiply charged ions, without a too strong radiation damage, which may occur when very deep core holes are created. Thus monochromatised synchrotron radiation in the 100-1000 eV photon energy range is an ideal tool to study dissociation of free molecules, especially into three body processes which are of highest interest as far as dynamics is concerned.

In the last ten years numerous studies have been devoted to photofragmentation of such excited molecules[1-9], trying to reveal, by use of various mass spectrometry techniques, any "memory effects" induced by the very localised character of the core hole. In the present paper we will essentially focus on three body (or more) dissociation of polyatomic systems and try to determine if such a process is slow as compared to rotation of molecular fragments giving rise to stepwise decays, or if it is on the contrary a very fast process more likely described as an instantaneous "explosion". Indeed, in the case of hydrides for instance, it has been shown[10-15], using photoelectron spectroscopy, that core to empty orbitals transition gives rise to a very fast dissociation into neutral fragments, followed by further Auger decay of the core excited fragment. Such an Auger decay, used as an "internal clock" of the system, shows that dissociation occurs in the 10 femtosecond time scale, which is very fast as compared to rotation.

After photoabsorption in the ionisation continuum of a core shell, the photoelectron is quickly (10^{-17}s) ejected from the molecule, leaving a highly excited ion. The core hole i then undergoes an Auger decay where one electron from shell j fills the hole and another one k is ejected in the continuum. This normal Auger decay takes place in the 10^{-14} s time scale (for the considered photon energy range) and justify in most of the cases this two step picture. During this process, due to the rapid change of the charge experienced by the electrons, additional excitation may be transferred to the outer electrons, jumping from orbital j (or k) to an empty one l (shake up) or to the continuum (shake off). The resulting ion can thus be found not only in a doubly charged state but also in an excited doubly charged, triply charged state (even more if any cascade process is open). Because of the core hole localisation, the Auger process is basically an intra-atomic process, which means that j and k orbitals must have a good overlap with k to give rise to a significant Auger rate.

In case of discrete excitation, the situation is similar if we consider the excited electron as a spectator temporarily trapped in the empty orbital. In addition to this process, further excitation may be transferred to the spectator electron by shake up and shake off processes, leading to highly excited monocations (that may undergo second step autoionisation) as well as doubly charged ions. Participator decay, including directly the excited electron in the relaxation, is of minor importance.

The resulting molecular ions are usually unstable, as the chemical bonds are weakened, when bonding electrons are missing. They are likely to dissociate into several fragments and appropriate coincidence technique is needed for proper analysis.

2D MASS SPECTROMETRY

In order to study fragmentation of multiply charged ions, one has to detect several particles in coincidence for each ionisation event. Time of flight spectrometry is ideally suited for this kind of study because it naturally allows coincidence measurements and also because ions are extracted through strong electric fields, which reduces angular discrimination among energetic fragments. Schematics of the apparatus is given in figure 1. A detailed presentation of the experimental set up as well as the basic features of the 2D mass spectrometry analysis is given in reference [16] and in the pioneering works of Frasinsky et al[17] and Eland[18].

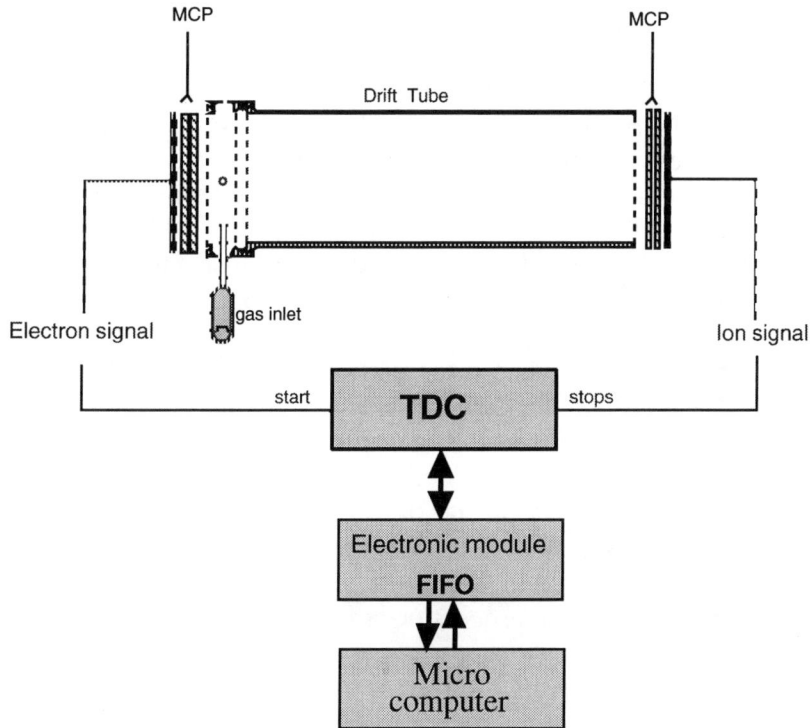

Figure 1: Schematics of the PEPIPICO apparatus

Monochromatized synchrotron radiation crosses at right angle an effusive jet of gas, at the centre of the ionisation chamber where a strong extraction field (2kV/cm) is applied. Electrons are immediately detected and give the time origin (start) of the ionisation event. Ions after flying into the drift tube give successive stops, related to their m/q ratio. Each set of corresponding time of flight data is stored in a buffer memory before further analysis. At the end of the experiment, the data are sorted as a one dimension histogram mass spectrum (PEPICO: photoelectron-ion coincidence) corresponding to the detection of only one ion, a set of doublet of times (t_1, t_2) and their occurrence z (PEPIPICO: photoelectron-photoion-photoion coincidence) corresponding to the detection of two ions and finally a set of triplet of times (t_1,t_2,t_3) when three ions have been detected.

The advantage of the time of flight (TOF) technique is that it also gives informations on dynamics. Indeed, the TOF of a particle A^{q+}, ejected with a kinetic momentum **P** is given by:

$$t = t^0 + \frac{\alpha}{q} P \cdot \cos(\theta) \qquad (1)$$

where t^0 is the TOF value of the particle at rest, α a normalisation constant and θ measures the ejection angle with respect to the detection axis. From this simple formula, one deduces that when only two particles are ejected (giving $\mathbf{P_2}=-\mathbf{P_1}$), the corresponding TOF data are related through the formula: $t_2=-t_1$ (if we neglect the constant). The plot of t_2 vs t_1 (PEPIPICO pattern) represents a "stick" of slope -1, which length reflects the total kinetic energy released in the dissociation. More interesting is the situation when three particles are ejected. In this case, one has $\mathbf{P_1}+\mathbf{P_2}+\mathbf{P_3}=0$. In the general case, one can conclude that the knowledge of $\mathbf{P_1}$ and $\mathbf{P_2}$ allows the determination of $\mathbf{P_3}$. For instance, if particle 1 and 2 are detected, one deduces that t_1+t_2 is proportional to $-\mathbf{P_3}$: we can get information on the third undetected particle (neutral). The effect of the third particle is to transform the previous "stick" peak into an oviform one. In some situations, one can make more precise predictions. When a stepwise process occurs, for instance like

$$ABC^{2+} \rightarrow A^+ + BC^+ \text{ followed by } BC^+ \rightarrow B^+ + C$$

one can write the momentum conservation at each step. The final peak is a parallelogram whith dimensions reflecting the kinetic energy released at each step, with a slope given in this case by $\alpha = -m_B/m_{BC}$. A more detailed analysis of the PEPIPICO peak shape is given in references [19-22].

DISSOCIATION OF $FeCO_2NO_2$: sequential evaporation of ligands

The $FeCO_2NO_2$ molecule is of special interest concerning photochemistry as the NO and CO ligands can be core-excited separately into a Π^* orbital at 287.8 and 400.3 eV respectively, resulting in intense discrete resonances. The study of its fragmentation is presented in more detail in reference [21] and is only summarised here. Various spectra

(PEPICO and PEPIPICO) have been measured at the relevant photon energies, from which have been deduced mass spectra corresponding to single and double ionisation channels. The dominant singly charged ion is Fe+, and one stable doubly charged fragment is found: Fe(CO)$^{2+}$. The comparison of the spectra reveals that the fragmentation pattern is mostly photon energy independent, except the single to double ionisation ratio. This shows clearly that no "memory effects" of the initial core hole localisation are observed in this molecule. This fact has been interpreted in terms of very fast internal energy conversion after Auger decay (which creates still localised electronic states). The molecule thus undergoes energy redistribution of the available energy into all its vibration modes, with a statistical partitioning, smearing out any energy localisation.

Selected peaks of the PEPIPICO spectrum are shown in figure 2, corresponding to the FeCO+/NO+, Fe+/NO+ and Fe+/CO+ ion pair formation.

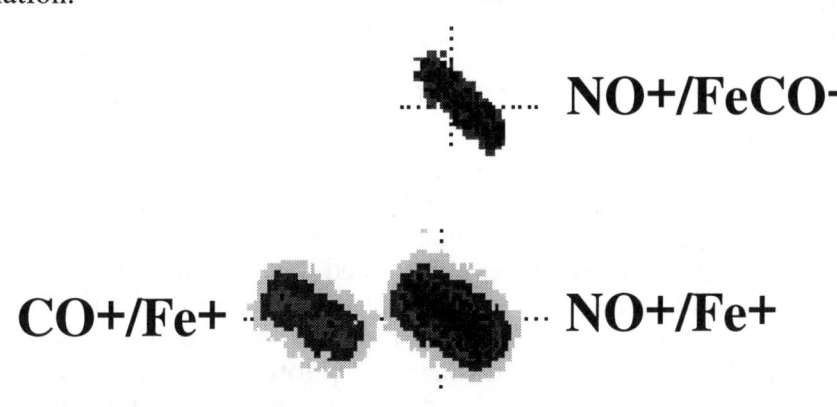

Figure 2: Selected peaks of the FeCO$_2$NO$_2$ PEPIPICO spectrum (resulting from both C1s and N1s continuum).

They clearly show parallelogram shapes, with specific slopes, as summarised in the following table:

ion pair	measured slope*	calculated slope
FeCO+/NO+	-0.80	-0.75
Fe+/NO+	-0.65	-0.66
Fe+/CO+	-0.67	-.066

* as determined from a mean square fitting procedure.

Moreover, these peaks do not show any angular distribution structure, indicating a slow dissociation process with respect to rotation.

The following stepwise processes are proposed to explain the observed peaks:

$FeL_4^{2+} \Rightarrow FeL_3^{2+}+L$ $FeL_4^{2+} \Rightarrow FeL_2^{2+}+2L$ $FeL_4^{2+} \Rightarrow FeL_2^{2+}+2L$
⇓   ⇓   ⇓
$FeL_2^+ + NO^+$ **$FeL^+ + NO^+$**   **$FeL^+ + CO^+$**
⇓   ⇓   ⇓
$FeCO^+ + L$ **$Fe^+ + L$**   **$Fe^+ + L$**

where L states for CO or NO. The corresponding calculated slopes, as deduced from the mass ratio, are given in last column of the table. Note that the proposed mechanism is the same in the three reactions: elimination of a neutral ligand, followed by charge separation including a charged ligand, followed by further elimination of a neutral ligand. This successive evaporation of ligands is similar to what was observed by Waller and Hepburn[23], after irradiation of $FeCO_2NO_2$ with a visible laser.

DISSOCIATION OF N_2O AND CO_2: a concerted process

The N_2O molecule is of particular interest for core excitation study, because of the presence of two unequivalent nitrogen atoms: a terminal one (Nt) and a central one (Nc). Due to the electronegativity of the oxygen atom, the Nc 1s level is shifted down about 3eV in energy as compared to Nt, allowing a selective photon excitation of Nt or Nc. The excitation spectrum of this linear molecule around the N K-shell is dominated by two sharp resonances below the ionisation onset - Nt 1s->π* (401.1 eV) and Nc 1s->π* (404.6 eV) - and a broad one - N 1s->σ* (max at 422eV) - located in the ionisation continuum. A full discussion including relaxation of all resonances and fragmentation channels in presented in reference [9,24]. We focus here only on the comparison of the atomisation processes (leading to three fragments) along the Nt 1s->π* and N 1s->σ* resonances.

PEPIPICO spectra recorded along those resonances are shown in figure 3. In order to distinguish Nt from Nc in the time of flight measurements, we have used the isotopically labelled $^{14}N^{15}NO$ molecule. When ionised in the N1s continuum, the PEPIPICO pattern gives rise to elongated "cigar like" peaks, with different orientations: ≈horizontal, diagonal and ≈vertical for Nt^+/Nc^+, Nt^+/O^+ and Nc^+/O^+ peaks respectively. This clearly indicates that the central particle is ejected in each case with very little kinetic energy (≈1 eV), whatever its charge state is (0 or 1). Similar observations (not reported here) are made with Nt^{2+}, Nc^{2+} and O^{2+}. On the contrary, when the Nt 1s->π^* resonance is excited, the PEPIPICO pattern is dramatically changed: the three peaks are enlarged and become more or less parallel. This indicates now that the central particle is ejected with a significant kinetic energy (≈ 4 eV), independent of its charge state. The kinetic energy values one can deduce from the various projections of the peaks are not compatible with any sequential process.

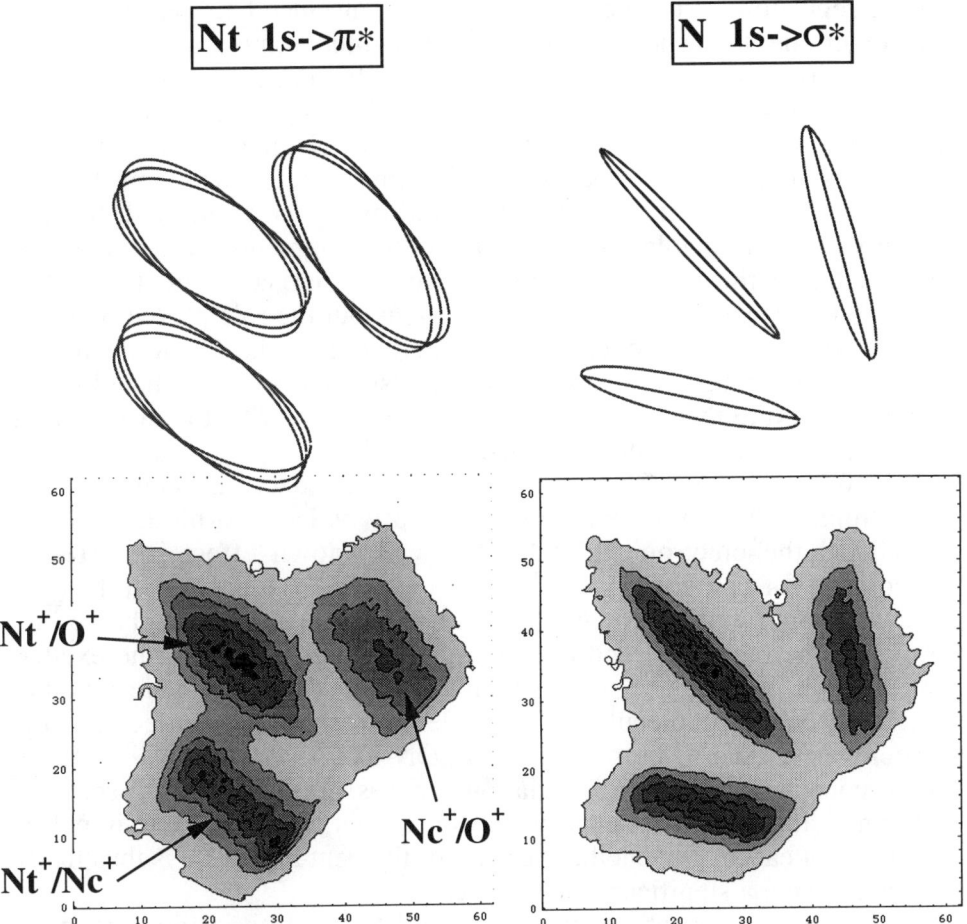

Figure 3: PEPIPICO spectra of N_2O recorded in the N1s region. Only selected peaks corresponding to three body dissociation are reported. Above each peak is reported a simulation using the impulsive model (see text)

Our interpretation is the following: ionisation of a core N1s electron does not change significantly the geometry of the molecule, because this electron doesn't contribute directly to the chemical bond. It is thus likely to consider that the molecule is still in a linear geometry when the Auger process (and consequently the dissociation) occurs. In the created doubly charged ion, removal of two bonding electrons may result in strong repulsion between the atoms, even before Coulomb forces act (when charges are fully localised on the atoms). Note that even if the repulsion is stronger on one side, for instance between N-N, than the other one, the central particle will then collide onto the third one and transfer its kinetic momentum (head on collision). Assuming thus that the NN and NO bonds are simultaneously or quasi simultaneously broken in a linear geometry, we conclude that the Nc particle receive two impulses in opposite directions and is thus ejected with low kinetic energy. The rapidity of the dissociation is corroborated by the observation of angular distribution of the fragments (as induced by symmetry of the resonances and the linear polarisation of the photon beam) not discussed here. Our model points out the role played by short range forces due to bonds breaking, rather than the Coulomb forces acting at long range. We thus call it "impulsive model", as detailed in the following. Note that such a model was used to describe the dissociation of ICN into I and CN after laser irradiation[25].

When the Nt 1s->π* resonance is excited on the contrary, the additional electron trapped in the empty valence orbital, strongly modified the electronic cloud. This well known effect is correctly accounted for using the equivalent core model[26]: it states that the core excited atom (atomic number Z), with one additional positive charge (the hole) in the core shell and one additional negative charge (the excited electron) in the valence shell is equivalent the Z+1 atom. It results that the core excited molecule is "equivalent" to the ONO molecule. This species is known to be bent, with an ON-NO angle of 134°[27]. If we now consider the same dissociation process previously discussed, but starting with a bent geometry, we conclude that Nc receive two impulses that do not cancel one another because of the bent angle. Nc is thus likely to carry away a significant kinetic energy.

In order to rationalise this qualitative description, we have modelled the potential between the various particles and carried out classical trajectory calculations. The system of differential equations, as resulting from classical mechanics law, was solved numerically.

The following potential was used:

$$V = D_1 e^{-\alpha_1(R_1-Re_1)} + D_2 e^{-\alpha_2(R_2-Re_2)} + \frac{e^2}{R}$$

were R_1 and R_2 are the NN and NO distances, R represents either R_1, R_2 or R_3 (the NtO distance), depending on the various possible location of the charge of the detected ions. This potential accounts for repulsion at short range between NN and NO (with characteristic values D_1 and D_2 at equilibrium distances Re_1 and Re_2), as well as the standard Coulomb repulsion. A schematics of such a potential is given in figure 4 in the case of Nt/Nc+/O+ dissociation.

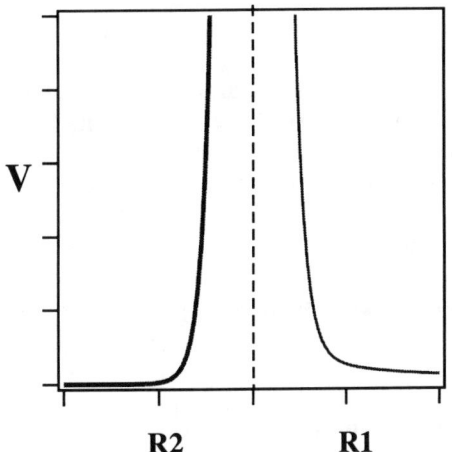

figure 4: schematics of the repulsive potential between NN(R_1) and NcO(R_2)

Starting with initial conditions given by physical assumptions (i.e. equivalent core model), trajectories are calculated up to 25Å distances, for which the residual potential is negligible. The corresponding kinetic momenta P_1, P_2 and P_3 of the three departing particles are thus used to simulate a PEPIPICO spectrum. Indeed, the dissociation plan (defined by $P_1+P_2+P_3=0$) may have any orientation with respect to the detection axis, determined by the two angles θ and δ.

The time of flight values t_1 and t_2 of the detected particle (neglecting the constant part) is simply given by:

$t_1 = P_1 \cos\delta \sin\theta$ and $t_2 = P_2 \cos(\delta+\alpha_3) \sin\theta$

where α_3 is the angle between P_1 and P_2. The resulting PEPIPICO pattern is a filled ellipse ($\sin\theta$ acting as a scaling factor). In the following only the external ellipse is drawn. The resulting simulations are reported in figure 3, on top of each PEPIPICO peak, using the following conditions:

- equal dissociation energies D_1 and D_2
- initial angle of $180°\pm7°$ (σ^*) and $134°\pm7°$ (π^*)

The agreement between experiment and calculation is surprisingly good, despite the simplicity of the model. It shows that the impulsive model is quite valid to describe such very fast dissociation processes, and that dissociation dynamics in case of N_2O, is primarily governed by the geometry (and also initial velocities) of the intermediate excited state.

In order to confirm this description, we performed similar experiment on CO_2 around the C1s edge. Indeed, this molecule can be core ionised in the C1s shell and core excited into the first π^* empty orbital in the 300eV energy range. The interest of the comparison with N_2O is that in both cases the equivalent core molecule after π^* excitation is the same: $OC^*O<=>ONO<=>N^*NO$. The results are shown in figure 5, were we have reported the C^+/O^+ and O^+/O^+ pairs. In order to detect in coincidence the two oxygen ions, we had to use the isotopically labelled $^{16}OC^{18}O$ molecule.

The observations are very similar to those obtained for N_2O: the carbon is ejected with very little kinetic energy in the continuum (σ^* resonance), producing elongated peaks. On the contrary, along the π^* resonance, the carbon atom carries away some significant kinetic energy and the PEPIPICO peaks have an oviform shape. Simulations are reported on top of each peak, using the impulsive model with the same initial conditions than in the N_2O case (using of course appropriate masses and nuclear length). The agreement is again very good, showing definitively the adequacy of this model to describe fast concerted dissociation of light molecules.

CONCLUSION

We have shown, using $FeCO_2NO_2$, N_2O and CO_2 as prototype molecules, that the dissociation of polyatomic molecules undergoes quite different path, depending of the size of the system. In case of large molecules, memory effects induced by the photoexcited intermediate state are difficult to established, at least from a quantitative point of view. This is due to the fast energy conversion following the Auger effect. The dissociation, governed by the energy redistribution over the large amount of vibration modes, exhibits stepwise processes, as far as molecular fragments are concerned. Similar findings were found for a series of silicon containing molecules[28]. On the contary, core induced fragmentation of light molecules is more likely to point out the role of the intermediate state geometry. Instantaneous dissociation of the chemical bonds

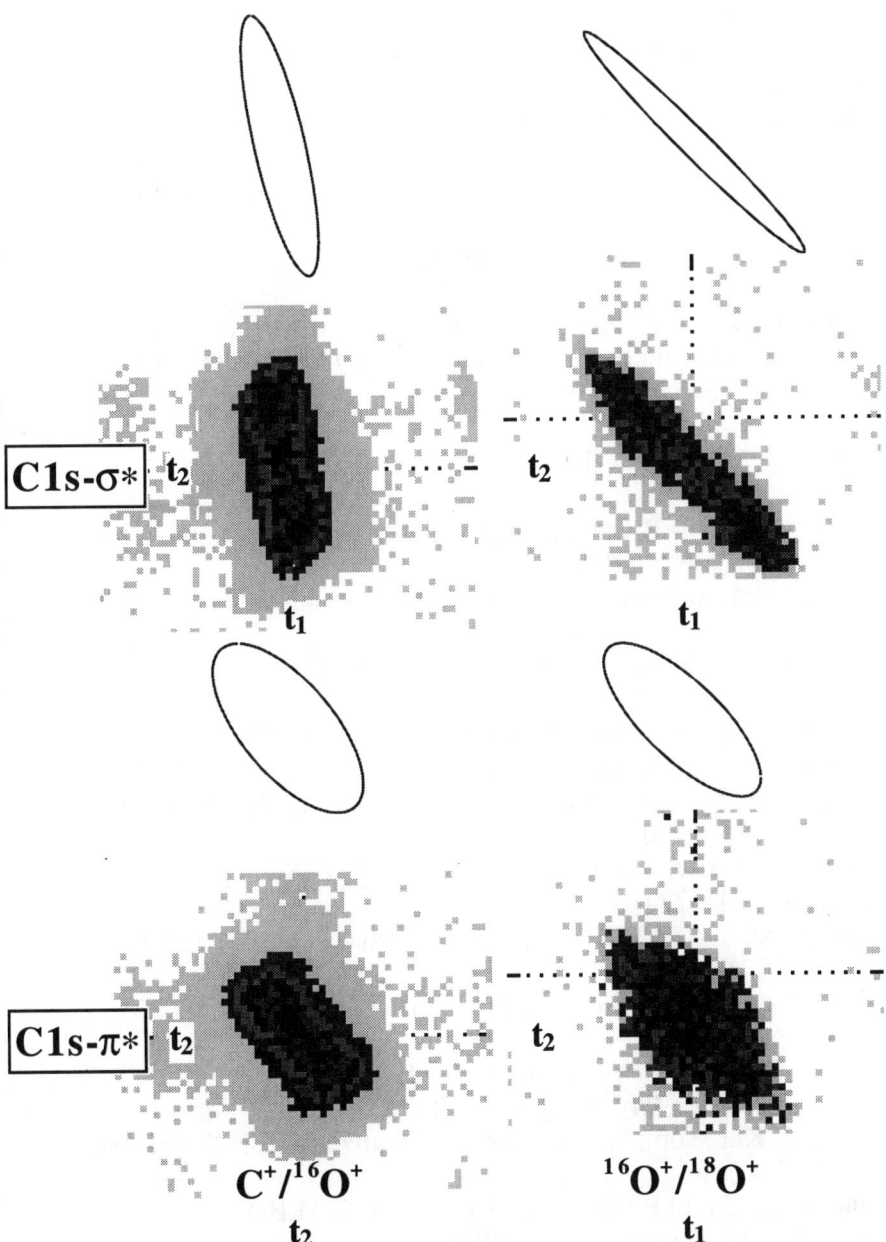

Figure 5: PEPIPICO spectra of CO_2 recorded in the C1s region. Only selected peaks corresponding to three body dissociation are reported. Above each peak is reported a simulation using the impulsive model (see text)

is well reproduced by an impulsive picture, which rules out the Coulomb explosion model where the dissociation dynamics is only driven by electrostatic charges.

Further systematic work is needed to confirm this overall picture, with additional information like the internal energy of the dissociating states, as well as a deeper understanding of angular distribution of fragments.

ACKNOWLEDGEMENTS

We wish to warmly thank A.Beswick and M.Miranda for very fruitful discussions, especially on N_2O and for providing us the classical trajectory calculation code used in this work.

REFERENCES

1: W.Eberhardt, T.Sham, R.Carr, S.Krummacher,M.Strongin, S.Weng, D.Wesner *Phys.Rev.Lett.* 50(1984)1038
2: K.Muller-Dethslef, M.Sander, L.A.Chewter, E.Schlag *J.Chem.Phys.* 85(1986)5755
3: J.Murakami, M.C.Nelson, S.L.Anderson and D.M.Hanson *J.Chem.Phys.* 85(1986)5755
4: R.Murphy and W.Eberhardt *J.Chem.Phys.* 89(1988)4054
5: G.G.B.de Souza, P.Morin and I.Nenner *J.Chem.Phys.* 90(1989)7071
6: D.M.Hanson, C.I.Ma, K.Lee, D.Lapiano-Smith and D.YKim *J.Chem.Phys.* 93(1990)9200
7: W.Habenicht, H.Barter, K.Muller-Dethlefs and E.W.Schlag *J.Chem.Phys.* 95(1991)6774
8: K.Ueda, H.Chiba, Y.Sato, T.Hayaishi, E.Shigemasa and A.Yagishita *Phys.Rev.A* 46(1992)R5
9: T.LeBrun, M.Lavollée, M.Simon and P.Morin*J.Chem.Phys.* 98(1993)2534
10: P.Morin and I.Nenner *Phys.Rev.Lett.* 56(1986)1913
11: P.Morin and I.Nenner *Phys.Scripta* T17(1987)171
12: S.Svensson, L.Karlsson,, N.Martensson,, P.Baltzer and B.Wannberg *J.Elec.Spec.* 50(1990)c1
13: H.Aksela, S.Aksela, O.P.Sairanen, M.Hotokka, G.M.Bancroft, K.H.Tan and J.Tulkki *Phys.Rev A* 41(1990)6000
14: H.Aksela, S.Aksela, M.Hotokka, A.Yagishita and E.Shigemasa *J.Phys.B* 25(1992)3357
15: H.Aksela, S.Aksela, A.Naves de Brito, G.M.Bancroft and K.H.Tan

Phys.Rev.A 45(1992)7948
16: M.Simon, T.LeBrun, P.Morin, M.Lavollée and J.L.Maréchal *Nucl.Instr.&Meth.* B62(1991) 167
17: L.J.Frasinsky, M.Stankiewicz, K.J.Randall, P.A.Haterley, K.Codling *J.Phys.B* L819(1986)
18: J.H.D.Eland, F.S.Wort and R.N.Royds, *J.Electr.Spectr.* 41(1986)297
19: J.Eland *Accounts of chemical research* 22(1989)381
20: T.LeBrun, *Thèse d'Université*, Orsay (1991)
21: M.Simon, *Thèse d'Université*, Orsay (1992)
22: M.Lavollée and H.Bergeron, *J.Phys.B* 25(1992)3101
23: I.Waller and J.Hepburn, *J.Chem.Phys.* 10(1988)6658
24: T.LeBrun, M.Lavollée, P.Morin *AIP Conference Proceedings 215, X-Ray and inner shell processes, Knoxville, TN, 1990* T.A.Carlson, M.O.Krause, S.T.Manson eds, p846
25: K.Holdy, L.C.Klotz and K.R.Wilson *J.Chem.Phys.* 52(1970)4588
26: M.E.Schwartz and J.D.Switalski *J.Am.Chem.Soc.* 94(1972)6298
27: G.Hertzberg, *Molecular Spectra and Molecular Structure* (V.Nostrand Reinhold, NY,1950)
28: see for instance M.Simon, T.LeBrun, R.Martins, G.G.B.de Souza, I.Nenner, M.Lavollée and P.Morin *J.Phys.Chem.* 97(1993)5228

TIME DEPENDENT APPROACH TO FEMTOSECOND LASER CHEMISTRY

Ronnie Kosloff
*Department of Physical Chemistry and the Fritz Haber Research Center,
the Hebrew University, Jerusalem 91904, Israel*

Active control of molecular motion is perused using shaped ultrafast laser pulses. The theoretical framework emphasizing the role of interference is presented for a molecule with two electronic surfaces coupled by radiation. The flow of energy and population from one surface to the other is analyzed and compared to the power consumption from the radiation field. The analysis of the flows shows that the phase of the radiation becomes the active control parameter which promotes the transfer of one quantity and stops the transfer of another. The phase relations which are responsible for laser catalysis impulsive excitation and stimulated emission are worked out. An interesting possibility of laser cooling emerges by a process that promotes transfer of energy simultaneously with stopping population transfer.

I. INTRODUCTION

Finding ways to control the temporal evolution of complex molecular systems has recently been motivated by advances in laser technology, molecular spectroscopy and in theoretical insight of molecular dynamics. An important example is control of selection of products of a chemical reaction. Typically, a polyatomic molecule will, for defined initial conditions, react to form several products in different amounts. In most instances one of these products is wanted much more than the others. What is sought is a method for controlling the molecular dynamics actively, in real time, so as to guide the molecule to form the desired product.

The developments in laser technology that have stimulated interest in actively controlling selectivity of product formation in a chemical reaction include methods for the generation of very short pulses, of shaped pulses, of pulses with well-defined phase relationship, of light fields with extraordinary pure frequency, and of light fields with very high intensity. Application of laser technology to molecular spectroscopy has yielded a wealth of information concerning molecular potential energy surfaces. It has also led to an increased awareness that exploitation of interference effects inherent to the quantum mechanical description of a system can be used to guide system evolution; to recognition that the dynamics of a strongly coupled light-matter system can be influenced by alteration of the temporal and spectral distributions of the light; and to the realization that a polyatomic molecule may have sufficiently few degrees of freedom that it is not impossible to control all of them. The underlying principle of the new approach to controlling product selectivity in a reaction is different from that used in earlier attempts to achieve "bond selective chemistry". The new approach is based on exploitation of quantum interference effects whereas the old approaches are, typically, based on use of a very intense laser field to generate a very high level of local bond excitation and the hope that the rate of bond breakage will then greatly exceed the rate of transfer of energy from the excited bond to the rest of the molecule. Two different ways of using quantum mechanical interference to control product selectivity have been proposed [1-9].

Suppose there are two independent excitation pathways between a specified initial state of a molecule and a specified final state of the products; these might be transitions involving absorption of one and three photons, respectively. Quantum mechanics tells us that the probability of forming the specified product is proportional to the square of a sum of the transition amplitudes for the two pathways; because the amplitudes can have different signs, the magnitude of that probability is determined by the extent of their interference. For example, when one- and three-photon transitions generate the independent pathways between the initial and final states, the extent of interference can be controlled by altering the relative phase of the two excitation sources. The situation is analogous to the formation of a diffraction pattern in a two-slit experiment in that the excited state amplitude in each molecule is the sum of the excitation amplitudes generated by two routes which are not distinguished from each other by measurement. An example of control of the population of a level in HCl with this method has been done [10].

An alternative method of influencing the selectivity of product formation in a reaction is to modulate the product yield via interference between two excitation pulses with a variable time delay between them [6,4]. In the simplest case, when only two electronic potential energy surfaces are involved, the first pulse transfers probability amplitude from the electronic ground state, forming a "replica" of the ground state amplitude on the excited state potential energy surface; that replica then evolves on the excited state potential energy surface during the time interval between pulses. The second pulse of the sequence, whose phase is locked to that of the first one, also creates amplitude in the excited electronic state, which is in superposition with the initial, propagated, amplitude. Such an intramolecular superposition of amplitudes can lead to interference. Whether the interference is constructive or destructive, giving rise to larger or smaller excited state population for a given inter-pulse delay, depends on the optical phase difference between the two pulses and on the detailed nature of the evolution of the initial amplitude. This situation is also analogous to a two-slit experiment. The method described has been used to control the population of a level of I_2 [5].

In principle, the methods available for guiding the evolution of a quantum system by coupling it to an external field are not restricted to the use of a time-independent field or a simple pulse sequence. If the goal to be achieved is, say, maximization of the amount of a product in a reaction, the design of the external field which accomplishes the goal is an inverse problem: given the goal and the quantum mechanical equations of motion, calculate the guiding field which is required. The solution to this inverse problem is very likely not unique, which for the case under consideration is a strength since it is then plausible that one of the possible guide fields is more easily generated than others.

The methodology used in these calculations is optimal control theory [7]. It is usually found that the optimal guiding field has a complicated spectral and temporal structure. The underlying physical principle that determines the efficiency of the guiding field is, again, interference between the amplitudes associated with its different spectral and temporal components. In the model problems studied to date it is predicted that the use of an optimal guide field can increase the product yield by many orders of magnitude relative to the yield from a two-pulse control field.

To facilitate the understanding of the control strategies of molecular motion it is important to study the control of basic quantities. For example if the object to control is the yield of a chemical species, one has to understand the elementary control of population transfer within an electronic surface and from one surface to the other. If the final objective is to coherently excite or on the contrary cool the motion of the molecule on the ground electronic surface. The elementary step involves the control of energy flow within the molecule. The combination of these elementary controls enables achieving quite elaborate goals including constraints on the dynamics the most important of which is radiation damage control [9,11]. In this overview a time dependent approach to active control is formulated with the purpose of shedding light on the role of interference processes.

II. THE MATTER RADIATION QUANTUM MECHANICAL FORMULATION

To explore the basic possibilities of active control of molecular motion the basic formalism of the interaction of radiation with a molecule has to be revealed. Consider two electronic states of a specific molecule coupled by radiation. The radiation which couples the two surfaces is the means of control. As an explicit example the electronic potential surfaces of a diatomic molecule is shown in figure 1.

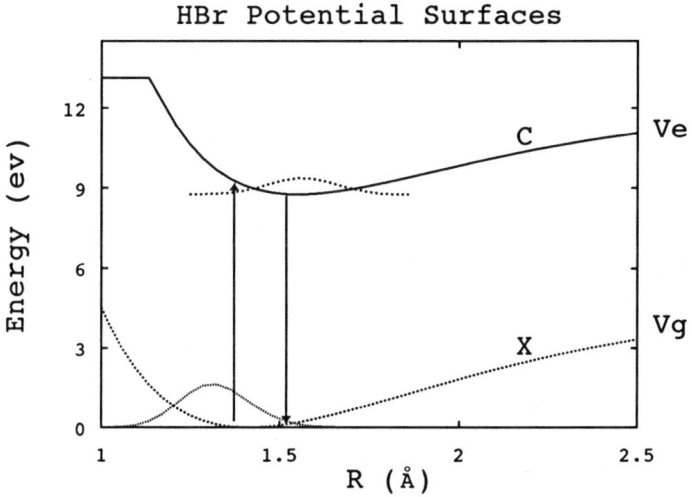

FIG. 1. Ground ($X\ ^1\Sigma^+$) and the ($C\ ^1\Pi$) excited potential energy surfaces of HBr.

The internal state of the molecule is defined by the density operators ρ_l, $l \in g,\ e$ which are the indexes of the ground and excited surfaces respectively [12]. The combined density operator describing the state of the combined ensemble can be written as:

$$\hat{\rho} = \hat{\rho}_g \otimes \hat{P}_g + \hat{\rho}_e \otimes \hat{P}_e + \hat{\rho}_i \otimes \hat{S}_+ + \hat{\rho}_i^\dagger \otimes \hat{S}_- \tag{1}$$

where the tensor product of the internal space and the surface designation is used. $\hat{P}_l$ are the projection operators on the surface, l and $\hat{S}_\pm$ are the raising and lowering operators from one surface to the other. The first two terms in equation (1) represent the state of the molecules residing on the ground and excited surface. The last two terms represent the electronic coherence induced by the radiation.

The Hamiltonian of the system consists of the sum of internal Hamiltonians and an interaction term

$$\hat{H} = \hat{H}_0 + \hat{V}_t \tag{2}$$

where the zero order Hamiltonian is the sum of the separate parts,

$$\hat{H}_0 = \hat{H}_g \otimes \hat{P}_g + \hat{H}_e \otimes \hat{P}_e = \frac{\hat{P}^2}{2m} \otimes \hat{I} + \hat{V}_g \otimes \hat{P}_g + \hat{V}_e \otimes \hat{P}_e \tag{3}$$

Each surface Hamiltonians $\hat{H}_l$ or surface potential $\hat{V}_l$ is a functions of the internal coordinates. The interaction part controls the transport properties of the between the two electronic manifolds. In this study, it contains only radiative coupling terms:

$$\hat{V}_t = -\hat{\mu} \otimes \left(\hat{S}_+ \epsilon + \hat{S}_- \epsilon^*\right) \qquad (4)$$

were $\hat{\mu}$ is the transition dipole operator and $\epsilon(t)$ represents a semiclassical time dependent radiation field. The field amplitude $\epsilon(t)$ represents the main control parameter. Experimental methods to shape the field amplitude as well as its phase enable this control.

The evolution of the molecule is generated by the Liouville von Neumann equation [12]:

$$\frac{\partial \hat{\rho}}{\partial t} = -\frac{i}{\hbar}[\hat{H}, \hat{\rho}] + \mathcal{L}_D(\hat{\rho}) \qquad (5)$$

where the first term represents unitary dynamics generated by the Hamiltonian, and the second term represents the dynamics generated by the dissipation. This equation represents the dynamics of an open quantum mechanical system under the restriction of a dynamical semigroup evolution [13–16]. The source of the dissipative term is the reduction of the combined system and bath dynamics to the dynamics of the system only. Within the semigroup approach an explicit form of the dissipative operator exists.

Since the objective is to control observable it is convenient to use the Heisenberg equations of motion to calculate the change in time of an explicitly time dependent operator:

$$\frac{d\hat{A}}{dt} = \frac{\partial \hat{A}}{\partial t} + \frac{i}{\hbar}[\hat{H}, \hat{A}] + \mathcal{L}_D^*(\hat{A}) \qquad (6)$$

where the first term represents the explicit time dependence of the operator, the second term the Hamiltonian evolution, and the third term the dissipation which is the Heisenberg version of the dissipative superoperator in equation (5).

III. THE TRANSPORT EQUATIONS.

The objective of the control is to change a specific observable. The transport of quantities is the agent of change. Thermodynamic insight can be gained by studying the total change in energy representing the first law of Thermodynamics. Additional insight is gained by considering the population and energy currents from one system to the other. These currents, represent the balance equations. Within the setup of section II, the total energy change becomes:

$$\frac{d<E>}{dt} = <\frac{\partial \hat{H}}{\partial t}> + <\mathcal{L}_D^*(\hat{H})> \tag{7}$$

where the second term of equation (6) is zero, since the Hamiltonian commutes with itself. This equation can be interpreted as the time derivative of the first law of thermodynamics [17–19] with the identification of power as:

$$\mathcal{P} = <\frac{\partial \hat{H}}{\partial t}> \tag{8}$$

which is the time derivative of the work. The heat flow is:

$$\dot{\mathcal{Q}} = <\mathcal{L}_D^*(\hat{H})>. \tag{9}$$

With these definitions, the power absorbed from the field into the system becomes:

$$\mathcal{P} = -<\hat{\mu} \otimes \left(\hat{S}_+ \frac{\partial \epsilon}{\partial t} + \hat{S}_- \frac{\partial \epsilon^*}{\partial t}\right)> = -2Real\left(<\hat{\mu} \otimes \hat{S}_+> \cdot \frac{\partial \epsilon}{\partial t}\right) \tag{10}$$

The term $<\hat{\mu} \otimes \hat{S}_+>$ is the main control handle and therefore will reappear in many of the derivations and will be called the instantaneous dipole expectation.

Considering the population balance conditions, the total population is conserved since with only two electronic surfaces the molecule has to reside on one of them therefore:

$$dN_g + dN_e = 0. \tag{11}$$

The flow of population from one electronic surface to the other can be calculated using equation (6):

$$\frac{dN_g}{dt} = \frac{d<\hat{P}_g>}{dt} = \frac{i}{\hbar}<[\hat{V}_t, \hat{P}_g]> + <\mathcal{L}_D^*(\hat{P}_g)> \tag{12}$$

Ignoring nonradiative couplings between the ground and excited surfaces i.e. $\mathcal{L}_D^*(\hat{P}_g) = 0$, the ground surface population change becomes:

$$\frac{dN_g}{dt} = -\frac{i}{\hbar}<\hat{\mu} \otimes (\hat{S}_+\epsilon - \hat{S}_-\epsilon^*)> = \frac{2}{\hbar}Imag\left(<\hat{\mu} \otimes \hat{S}_+> \epsilon\right) \tag{13}$$

The flow of energy from the ground state can be calculated under the assumptions of small electronic dephasing i.e. $\mathcal{L}_D^*(\hat{P}_g) = 0$ and also pure vibrational dephasing i.e. $\mathcal{L}_D^*(\hat{H}_g) = 0$. Physically these conditions apply when the rate of relaxation to equilibrium is slow compared to the loss of phase ($\tau_1 \gg \tau_2$). Under these conditions:

$$\frac{dE_g}{dt} = \frac{2}{\hbar} Imag \left(< \hat{\mu} \hat{H}_g \otimes \hat{S}_+ > \epsilon \right) \tag{14}$$

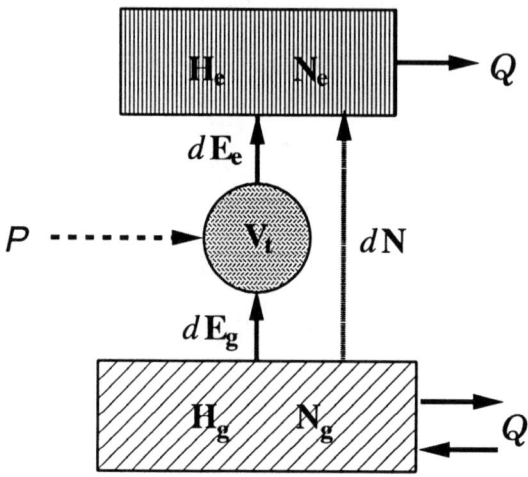

FIG. 2. Schematic diagram of the thermodynamic flows in the radiative quantum mechanical model.

When the total energy balance is considered, the power uptake can be distributed into the "bulk" of ground and excited surfaces as well as into energy stored by the interaction. Since the interaction term is proportional to the field ϵ, for a pulsed field for which $\epsilon = 0$ at $t = \pm\infty$, one can consider the energy flow into the bulk by integrating equation (10) by parts, obtaining:

$$\int_{-\infty}^{\infty} P dt = \int_{-\infty}^{\infty} \frac{\partial <\hat{\mu}>}{\partial t} \epsilon dt = \int_{-\infty}^{\infty} \frac{dE_g}{dt} dt + \int_{-\infty}^{\infty} \frac{dE_e}{dt} dt \tag{15}$$

This equation can be interpreted as the energy balance equation. The boundary term $-2Real\left(<\hat{\mu} \otimes \hat{S}_+ > \epsilon(t)\right)$ which goes to zero when deriving equation (15), represents the transient loading of the system. If the loading term becomes small in comparison to the volume terms E_g and E_e then one can omit the integral in equation (15) and obtain:

$$dE = dE_g + dE_e \tag{16}$$

Equation (16) can also be obtained within the limit of weak fields. Such balance equations represent one of the main tools of thermodynamic investigation. A schematic view of the thermodynamic currents can be seen in figure 2.

The spatial derivative of the loading term represents the internal force which the electromagnetic field exerts on the molecule:

$$\vec{F} = -2 \text{Real} \left(< \nabla_{\vec{r}} \hat{\mu} \otimes \hat{S}_+ > \epsilon \right) \tag{17}$$

where $\nabla_{\vec{r}}$ is the gradient with respect to the internal co ordinates. Combined with the conditions imposed in the next section in equation (27); one can apply a directional force on the molecule.

Equations (10), (13), (14) and (17) constitute the transport equations from one surface to the other. It is the objective of control to manipulate these quantities. The next section, describes the control of the transport by manipulating the phase of the transient field ϵ.

IV. TIME DEPENDENT CONTROL OF TRANSPORT.

Molecular transport processes can be promoted either by controlling the field ϵ or its time derivative $\frac{\partial \epsilon}{\partial t}$. This possibility stems from the fact that the transport equations Eq. (13) and Eq. (14) have a similar structure consisting of the real part of a product of a molecular expectation value $< \hat{X} >$ by the field ϵ. A similar structure is found in equation (10) where the transport is controlled by the real part of the product of a molecular expectation and the time derivative of the field. This structure can be expressed for equation (13) as:

$$\frac{dN_g}{dt} = \frac{2}{\hbar} | < \hat{\mu} \otimes \hat{S}_+ > | |\epsilon| \sin(\phi_\mu + \phi_\epsilon) \tag{18}$$

where ϕ_μ is the phase angle of the instantaneous dipole and ϕ_ϵ is the phase angle of the radiation field. The physical interpretation of this angle is the sum of the angle of the induced polarization of the molecule and the angle of the polarization of the light. In a similar way:

$$\mathcal{P} = -2| < \hat{\mu} \otimes \hat{S}_+ > | |\frac{\partial \epsilon}{\partial t}| \cos(\phi_\mu + \phi_\epsilon) \tag{19}$$

and

$$\frac{dE_g}{dt} = \frac{2}{\hbar} | < \hat{\mu}\hat{H}_g \otimes \hat{S}_+ > | |\epsilon| \sin(\phi_{\mu H} + \phi_\epsilon) \tag{20}$$

where $\phi_{\mu H}$ is the phase angle of $<\hat{\mu}\hat{H}_g \otimes \hat{S}_+>$. and

$$\vec{F} = -2\left|<\nabla_{\vec{r}}\hat{\mu} \otimes \hat{S}_+>\right| |\epsilon| \sin(\phi_{\nabla\mu} + \phi_\epsilon) \qquad (21)$$

where $\phi_{\nabla\mu}$ is the phase angle of $<\nabla\hat{\mu} \otimes \hat{S}_+>$. From this description, it is clear that the control parameter of the transport quantities is the phase angle $\phi = \phi_i + \phi_j$. The most simple example is a monochromatic circularly polarized pulse for which $\epsilon(t) = Ae^{-i\omega t}$ where A is a slowly varying envelope function. For this pulse the phase angle of $\frac{\partial \epsilon}{\partial t}$ is $90°$ rotated from the direction of ϵ, $\phi_{\dot\epsilon} = \phi_\epsilon - \pi/2$. Examining equations (18) and (19) one obtains the relation

$$\mathcal{P} = -\hbar\omega \frac{dN_g}{dt} \qquad (22)$$

This means that the system behaves as an effective two-level system where the power absorbed is linearly proportional to the population transfer.

For more general pulse shapes, using the above analysis, the control of the transport can be classified as being active or passive. For active control conditions the angle is:

$$\phi_\mu + \phi_{\dot\epsilon} = \begin{matrix} 0 \\ \pi \end{matrix} \Rightarrow \begin{matrix} maximum\ energy\ absorption \\ maximum\ energy\ emission \end{matrix} \qquad (23)$$

$$\phi_\mu + \phi_\epsilon = \begin{matrix} \frac{1}{2}\pi \\ -\frac{1}{2}\pi \end{matrix} \Rightarrow \begin{matrix} maximum\ positive\ mass\ transport \\ maximum\ negative\ mass\ transport \end{matrix} \qquad (24)$$

$$\phi_{\mu H} + \phi_\epsilon = \begin{matrix} \frac{1}{2}\pi \\ -\frac{1}{2}\pi \end{matrix} \Rightarrow \begin{matrix} maximum\ energy\ intake\ to\ ground\ surface \\ maximum\ energy\ disposal\ from\ ground\ surface \end{matrix} \qquad (25)$$

$$\phi_{\nabla\mu} + \phi_\epsilon = \begin{matrix} \frac{1}{2}\pi \\ -\frac{1}{2}\pi \end{matrix} \Rightarrow \begin{matrix} maximum\ positive\ force \\ maximum\ negative\ force \end{matrix} \qquad (26)$$

The active control of mass transport using the control relation of equation (24) has been demonstrated experimentally by Sherer et. al. [5]. In this experiment in I_2 the first pulse transfers population from the ground to the excited surface thus creating an instantaneous transition dipole. This dipole function is modulated by the excited surface vibration. At the proper instant when the dipole expectation amplitude is maximized a second pulse is applied. The direction of mass transport was then controlled by changing the relative phase of the second pulse.

Another interesting possibility is the induction of stimulated emission which is obtained when the combined phase angle is π in equation (23). This means that a pulsed laser can be obtained from an ensemble of molecules without the usual condition of population inversion provided the phase relation is right.

Even more important are passive controls where the transport of a certain quantity is blocked. If the phase angle is

$$\phi_\mu + \phi_\epsilon = \pm\frac{1}{2}\pi \Rightarrow \; zero\; total\; energy\; change \tag{27}$$

$$\phi_\mu + \phi_\epsilon = 0, \pi \Rightarrow \; zero\; mass\; transport \tag{28}$$

$$\phi_{\mu H} + \phi_\epsilon = 0, \pi \Rightarrow \; zero\; change\; in\; ground\; surface\; energy \tag{29}$$

$$\phi_{\nabla\mu} + \phi_\epsilon = 0, \pi \Rightarrow \; zero\; force \tag{30}$$

On examining the passive control conditions in equation (27),(28),(29),(30) it can be noticed that there are two values of $\phi_i + \phi_j$ for which zero transport conditions apply. This fact can be utilized to simultaneously block the transport of one quantity and to select the direction of the transport of another quantity.

A note of caution is needed at this point because naively it would seem that there is a meaning to the absolute phase of the field and to the phase of the molecular expectation values. But it should be remembered that the phase of the molecular quantity is induced by the radiation in a former time. Therefore all phases have to be related to a previous synchronization pulse which synchronized the molecular clock with the field clock. This is the meaning of quantum mechanical interference between events which are induced in the past propagate and then are used to control transport at a later time.

Figure 3 shows the phase angle conditions leading to different transport quantities.

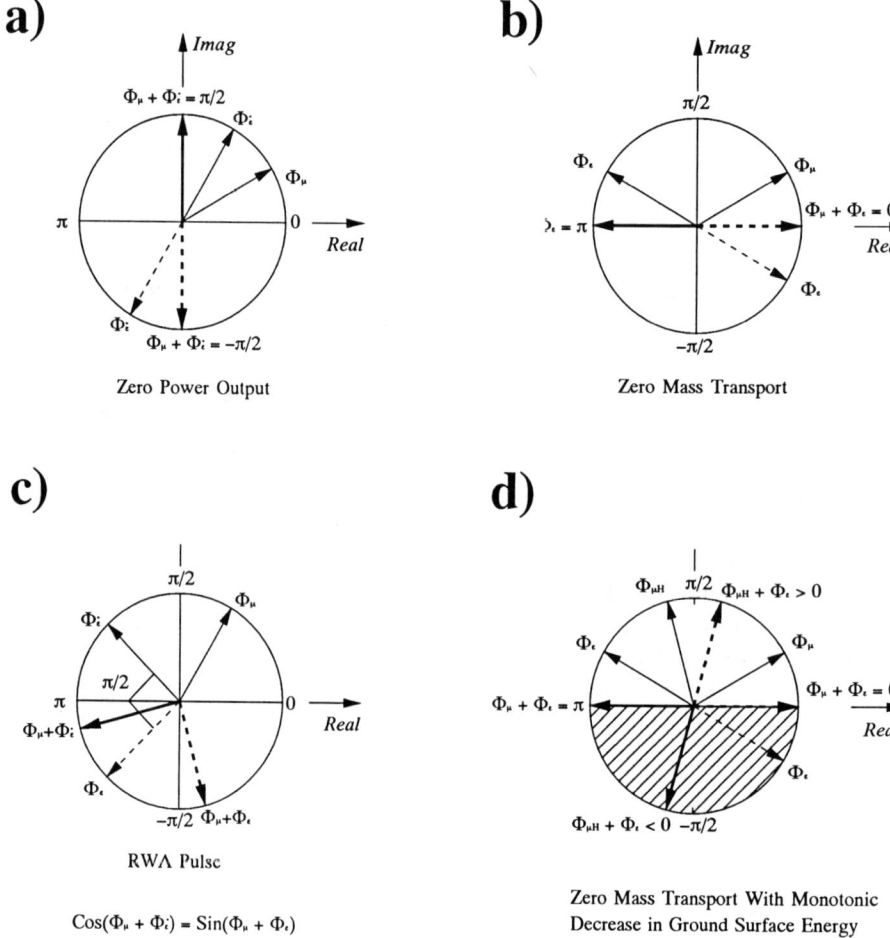

FIG. 3. a) Phase angle diagram for zero power output. b) Phase angle diagram for zero mass transport. c) Phase angle diagram for a RWA pulse. d) phase angle diagram for zero mass transport and for a monotonic decrease in energy. The shaded area represents the region of phase angle of energy decrease in the ground surface.

Figure 3a shows the phase angles for which the total power absorption is zero. Since no energy is absorbed or emitted from the field these conditions define laser catalysis [20]. Figure 2b shows the phase angle relations leading to zero mass transport. Figure 3c is of a RWA pulse corresponding to equation (22), where the mass transport is directed from the ground to the excited surface. Figure 3d shows the phase conditions which stop mass transport while monotonically directing energy out of the ground surface.

V. ENERGY TRANSFER WITH ZERO MASS TRANSPORT.

As an explicit example the local control of energy transport under the constraint of zero mass transport is analysed. The phase relation with null population transfer is obtained when the radiation field has the form:

$$\epsilon = C(t)\Phi\left(<\hat{\mu}\otimes\hat{S}_-> \right) \quad (31)$$

where $C(t)$ is a real function of time and $\Phi(X) = X/|X| = e^{i\phi}$ is the phase factor. Equation (31) is another formulation of the conditions in equation (28). Active control requires the expectation value of the instantaneous dipole $<\hat{\mu}\otimes\hat{S}_+>$ to be non zero. This means that the present control is related to the instantaneous dipole created in the past.

Under the conditions of zero mass transport, the change of the energy on the ground state becomes:

$$\frac{d<E_g>}{dt} = C(t)\frac{2}{\hbar} Imag\left(\Phi(<\hat{\mu}\otimes\hat{S}_->)<\hat{\mu}\hat{H}_g\otimes\hat{S}_+>\right) \quad (32)$$

This means that the sign of the function $C(t)$ determines the direction of the energy transport.

$$C(t) \propto \pm\frac{2}{\hbar} Imag\left(\Phi(<\hat{\mu}\otimes\hat{S}_->)<\hat{\mu}\hat{H}_g\otimes\hat{S}_+>\right) \quad (33)$$

A positive sign will excite the ground surface internal energy. This is the base of an excitation without demolition strategy [9]. Figure 4 shows a monotonic increase in the ground surface energy while maintaining a constant population on the excited surface.

A different excitation strategy has been proposed by Cina and Smith [22] where the excitation is achieved by a double pulse train where the first pulse transfers population to the excited surface and creates an instantaneous dipole. The second pulse is chosen with the inverse phase relation of the first one Cf. (24) transferes all the population back to the ground surface the net result is an impulsive vibrational excitation.

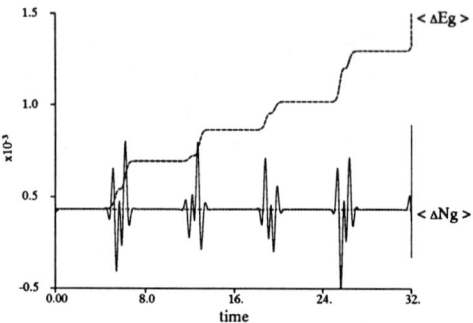

FIG. 4. The change in ground surface energy $\Delta \bar{E}_g$ change in norm $< \Delta \hat{N}_g >$ and field amplitude ϵ as a function of time for an excitation pulse. Notice that the field amplitude comes with bursts corresponding to the vibrational period ($\omega = 1$).

A negative sign of $C(t)$ causes a decrease of the ground surface energy. Figure 5, shows a monotonic decrease in energy as a result of a pulse shape calculated according to equation (33).

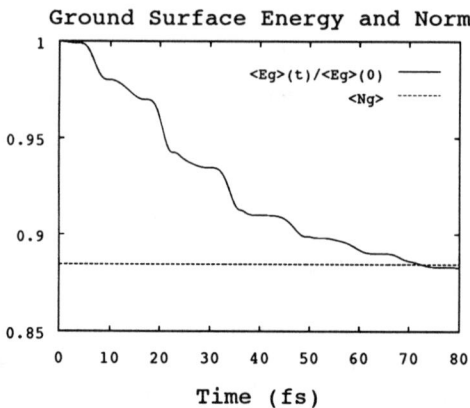

FIG. 5. The ground surface energy $\bar{E}_g$ and norm $< \hat{N}_g >$ as a function of time for a cooling pulse, for the HBr molecule starting at a thermal state $T = 5000K$. The energy is normalized to its initial value.

The resulting cooling pulse shape is shown in figures 6. Notice that the pulse amplitude is modulated by the vibrational period of HBr. The reduction of energy on the ground surface is accompanied by a reduction in entropy while the total entropy of the system increses. Once the system approaches the ground state the it decouples from the radiation in complience with the third law of thermodynamics.

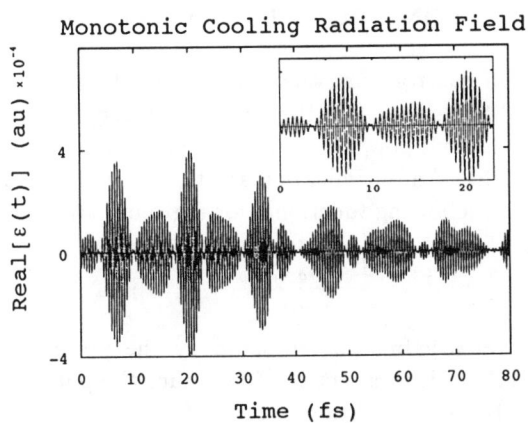

FIG. 6. Real part of the pulse as a function of time for a monotonic decrease in the ground surface energy.

The spectrum of the pulse is shown in figure 7.

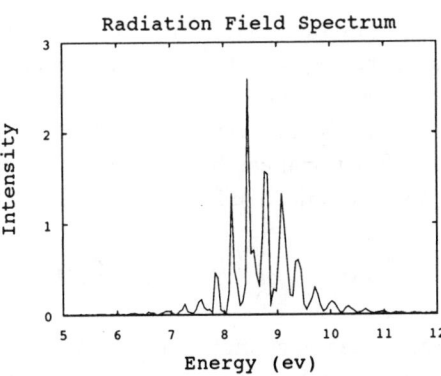

FIG. 7. Spectrum of a cooling pulse for the HBr molecule.

A close analysis shows that the spectrum has its maximum exactly in between the Raman lines of HBr This decrease in energy can be interpreted as laser cooling [19].

In a similar fashion a force can be applied to the ground surface creating a "squeezed state" with a minimum amount of radiation damage [21].

VI. FROM LOCAL TO GLOBAL CONTROL

The strategy of control using the pulse shape of equation (33), is only one of many possibilities. Therefore the optimal strategy for transporting an observable with minimum power consumption under the restriction of zero mass transport is sought. An optimization solution local in time can be obtained by optimizing the following functional for a ground state quantity $<\hat{A}_g>$

$$J = \frac{d<\hat{A}_g \otimes \hat{P}_g>}{dt} + W|\epsilon|^2 \quad (34)$$

where W is a penalty function imposed by the power consumption. The variation of J, δJ is with respect to $\delta C(t)$, since the phase is restricted by the zero population transfer condition equation (31). Carrying out the calculation leads to equation (33) with the proportionality constant equal to $1/2W$. This local solution can be replaced with a global one, optimizing the ground surface quantity $J = <\hat{A}_g>$ at a specific final time t_f with the minimal cost of external pumping energy $\int_0^{t_f} \mathcal{P} dt$. This problem can be cast into optimal control theory.

The objective is to minimize the ground surface observable $A_g(t_f)$. In the case of cooling the observable is the ground surface energy $E_g(t_f) = <\hat{H}_g>$.

The constraints are:

a) evolution governed by the Liouville von Neumann equation $\frac{\partial \hat{\rho}}{\partial t} = \mathcal{L}(\hat{\rho})$

b) zero mass transport $dN_g = 0$ or other constraints.

c) restricted total power consumption $\bar{E} = \int_0^{t_f} |\epsilon|^2 dt$.

The initial state of the system is set as $\hat{\rho}(0)$. By defining the Lagrange multipliers, the objective functional can be minimized subject to the constraints. This leads to the modified objective functional $\bar{J}$ [4]:

$$\bar{J} = tr\left(\hat{A}_g \otimes \hat{P}_g \hat{\rho}(t_f)\right) + \int_0^{t_f} tr\left((\frac{\partial \hat{\rho}}{\partial t} - \mathcal{L}(\hat{\rho}))\hat{B} + \lambda|\epsilon|^2\right) dt \quad (35)$$

where $\hat{B}$, is an operator Lagrange multiplier and λ is a scalar Lagrange multiplier. The variation of $\bar{J}$ is with respect to $\delta\hat{\rho}$ and $\delta|\epsilon|$. The condition $\frac{dN_g}{dt} = 0$ determines the phase of ϵ through equation (28) or (31). It therefore is omitted from the variation. Calculating the variation of equation (35) and integrating by parts leads to the following equations:

a) a forward equation for the density operator $\hat{\rho}$

$$\frac{\partial \hat{\rho}}{\partial t} = \mathcal{L}(\hat{\rho}) \tag{36}$$

subject to the initial condition $\hat{\rho}(0)$.
b) a backward equation for the Lagrange operator $\hat{B}$:

$$-\frac{\partial \hat{B}}{\partial t} = \mathcal{L}^*(\hat{B}) \tag{37}$$

subject to the final condition $\hat{B}(t_f) = \hat{A}_g \otimes \hat{P}_g$. The dissipative part of equation (37) is symmetric in time, meaning that dissipation takes place in the forward as well as backward evolution.
c) and the condition on the field:

$$|\epsilon(t)| = \frac{1}{2\lambda} tr\left(\frac{\partial \mathcal{L}(\hat{\rho}(t))}{\partial |\epsilon|}\hat{B}(t)\right) = -\frac{1}{2\lambda} < \frac{\partial \mathcal{L}^*(\hat{B}(t))}{\partial |\epsilon|} > \tag{38}$$

Equation (38) can be interpreted as the scalar product of a forward moving density, and of a backward moving time-dependent operator. The optimal field at time t is determined by a time-dependent objective function propagated from the target time t_f backwards to time t. A first order perturbation approach to obtain a similar equation for optimal chemical control in Liouville space has been derived by a different method by Yan et. al. [23].

In dissipative dynamics, the backwards propagating target operator decays into a stationary operator and therefore, its change in time becomes zero $\mathcal{L}^*(B(-\infty)) = 0$. This leads to loss of control, as can be seen in equation (38).

In the cooling process, the target operator is the ground surface operator $\hat{H}_g(t_f)$. In general for weak field excitation, the target operator $\hat{B}(t)$ can be expanded in powers of the field ϵ^k. For the cooling scheme under these conditions the target function becomes time independent and is just the Schrödinger picture operator $\hat{H}_g$. Inserting the target function into equation (38) one obtains the result of equation (33) with the proportionality constant becoming $1/2\lambda$. As a conclusion the phase and amplitude conditions for cooling suggested in equation (33) are the optimal control solutions for weak fields. They can be used also as a first guess for the optimal field in an iterative solution for strong fields [24].

VII. DISCUSSION

Considering the control processes studied they are all based on the quantum mechanical interference phenomena. Interference is a global phenomena which means that in the context of a time dependent description events created in the past interfere with events created later in time. To unravel the control mechanism, a local in time description is more easy to interpret since one can optimize the transport of the desired quantity which will accumulate to produce the final outcome. The link between the local and global picture is the phase of the molecular expectation the most important of which is the instantaneous dipole $< \hat{\mu} \otimes \hat{S}_+ >$ Although the phase is not measurable it links the past events to the present ability to control transport.

The interaction of light with a two electronic surface system, bears close analogies to the thermodynamics of two coupled systems which can transfer mass and energy via external work (figure 2). Versions of the first, second and third law of thermodynamics have been developed [19]. Various processes which transfer energy without mass and mass without energy are identified. An interesting consequence of the analysis is the possibility of a pulsed laser with an active medium without population inversion.

The realization of control scenarios is a matter for the near future. In recent years the techniques of pulse shaping have advanced considerably [32–34]. Methods to shape both the phase and the amplitude of a ultrashort pulse are becoming available. Advances in simulation of quantum dynamics including the solution of the Liouville von Neumann equation methods [25–28] enable the study of molecules under the influence of strong fields and dissipative forces. The third element is the progress in optimal control strategies [24,35].

Finally one has to consider the practicality of using light as a chemical substance. Since the price of photons is high, optical control of chemical reactions is justified only for very expensive materials. Nevertheless the control methods developed have already found their use in new spectroscopic techniques which are based on impulsive excitation of coherent motion [29–31]. It is well known that measurements based on interference effects are of high quality. The control mechanisms can be used to optimize the information content of the experiment. Since in the modern era information is more valuable than materials the effort to seek control is justified.

ACKNOWLEDGMENTS

It is pleasure to thank my collaborators Allon Bartana, David Tannor and Stuart Rice for active participation in this work. This research was supported by the Binational United States - Israel Science Foundation. The Fritz Haber Research Center is supported by the Minerva Gesellschaft für die Forschung, GmbH München, FRG.

---

[1] S. Rice, Science **258**, 412 (1992).
[2] G. W. Coulston and K. Bergmann, J. Chem. Phys. **96**, 3467 (1992).
[3] M. Shapiro and P. Brumer, J. Chem. Phys. **84**, 4103 (1986); Acc. Chem. Res. **22**, 407 (1989).
[4] R. Kosloff, S. A. Rice, P. Gaspard, S. Tersigni and D. J. Tannor, Chem. Phys. **139**, 201 (1989).
[5] N. F. Scherer, R.J. Carlson, A. Matro, M. Du, A.J. Ruggiero, V. Romero-Rochin J.A. China, G. R. Fleming and S. A. Rice, J. Chem. Phys. **95**, 1487 (1991).
[6] D. Tannor and S. A. Rice, J. Chem. Phys. **83**, 5013 (1985); D. Tannor, R. Kosloff and S. A. Rice, J. Chem. Phys. **85**, 5805 (1986).
[7] S. Shi, A. Woody and H. Rabitz, J. Chem. Phys. **88**, 6870 (1988); S. She and H. Rabitz, Comp. Phys. Comm. **63**, 71 (1991).
[8] P. Gross, D. Neuhauser, and H. Rabitz, J. Chem. Phys. (1992).
[9] R. Kosloff, A. Hammerich and D. Tannor, Phys. Rev. Lett. **69**, 2172 (1992).
[10] S. M. Park, S-P. Lu, J. Gordon, J. Chem. Phys. **94**,8622 (1991);S-P. Lu S. M. Park Y. Xie and R. J. Gordon, ibid **96**, 6613 (1992).
[11] K. Nelson, in J. Jortner (eds.) Mode Selective Chemistry, Jerusalem Symposium on Quantum Chemistry and Biochemistry 24 527 (Kluwer Academic 1991).
[12] J. von Neumann, "Mathematical Foundations of Quantum Mechanics" (Princeton University Press, Princeton NJ 1955)
[13] G. Lindblad, Commun. Math. Phys. **48** 119 (1976).
[14] V. Gorini, A. Kossakowski and E.C.G. Sudarshan J. Math. Phys. **17** 821 (1976).
[15] R. Alicki, K. Landi, "Quantum Dynamical Semigroups and Applications" Springer-Verlag (1987).
[16] E. B. Davis, "Quantum Theory of Open Systems", (Academic Press, London 1976).
[17] R. Kosloff, J. Chem. Phys. **80**, 1625 (1984).
[18] E. Geva and R. Kosloff, J. Chem. Phys. **97**, 4398 (1992).
[19] A. Bartana, R. Kosloff and D. Tannor, J. Chem. Phys. **99** 196 (1993).

[20] T. Seideman and M. Shapiro, J. Chem. Phys. **88**,5525 (1988).
[21] T. Szakas, J. Somoloi and A. Lorincz, Chem. Phys. Lett. **172**, 1 (1993).
[22] J. A. Cina and T. J. Smith, J. Chem . Phys. **98**, 9211 (1993).
[23] Y. Yan, R. E. Gillian, R. W. Whitnell, K. Wilson and S. Mukamel, J. Phys. Chem. (1992).
[24] D. Tannor and V. Orlov, Proceedings of the Snowbird NATO workshop on "time dependent quantum mechanical methods" (1992).
[25] M. Berman and R. Kosloff, Compu. Phys. Commun. **63**, 1 (1991).
[26] M. Berman, R. Kosloff and H. Tal-Ezer, J. Phys. A **25**, 1283 (1992).
[27] C. Leforestier, R. Bisseling, C. Cerjan, M. Feit, R. Friesner, A. Guldberg,A. D. Hammerich, G. Julicard, W. Karrlein, H. Dieter Meyer, N. Lipkin, O. Roncero and R. Kosloff, J. Comp. Phys. **94**, 59 (1991).
[28] H. Tal Ezer, R. Kosloff, and C. Cerjan, J. Comp. Phys. **100**, 179 (1992).
[29] B. Hartke, R. Kosloff and S. Ruhman, Chem. Phys. Lett. **158**, 238 (1989).
[30] T. Baumert, U. Engel, C. Meier and G. Gerber, Chem. Phys. Lett. (1992).
[31] U. Banin A. Bartana S. Ruhman and R. Kosloff J. Chem. Phys. (1993).
[32] J. Chesnoy and A. Mokhtari, Phys. Rev. A **38** , 3566 (1988).
[33] A. M. Weiner, D. E. Leaird, G. P. Weiderrecht and K. Nelson, J. Opt. Soc. Am. B **8**, 1264 (1991).
[34] A.M. Weiner, J.P. Heritage and E.M. Kirshner, J. Opt. Soc. Am. B **5**, 1563 (1988).
[35] R. S. Judson and H. Rabitz, Phys. Rev. Lett **68**, 1500 (1992).

RYDBERG ATOMS AND NEW TECHNOLOGIES

RADIATIVE COLLISIONS OF RYDBERG ATOMS

T.F. Gallagher, D.S. Thomson, and M.J. Renn
Department of Physics, University of Virginia, Charlottesville, VA 22901

ABSTRACT

We describe the radiative collisions between Rydberg atoms in which several radio frequency photons are absorbed during the collisions. Specifically, we show the evolution of the process as the photon frequency is raised from much less than the collision time to much greater than the collision time. In addition we show how the theoretical descriptions in these frequency regimes are connected.

INTRODUCTION

Radiatively assisted collisions, collisions in which two atoms or molecules absorb a photon during the collision, were first suggested by Gudzencko and Yakovlenko.[1] In Fig. 1a we show an example of a radiatively assisted collision. If an excited atom A* and a ground state B atom absorb a photon of energy equal to the difference between the energies of A* and B* while colliding, the B atom can be excited to the B* state with the A* atom returned to the ground state. As pointed out by Gallagher and Holstein,[2] this process is intimately related to line broadening or collision induced absorption, as shown by Fig. 1b, a plot of the energy levels of the molecule AB as a function of internuclear separation R. The photon simply drives the transition A*B → AB*. The transition moment for this transition is zero at large R and is only significant for internuclear separations of a few Angstroms. At small R the molecular potential curves are R dependent, and as a result tuning the frequency of the light often leads to asymmetric resonances for the A*B → AB* transition. The A*B → AB* transition can be studied with high atomic densities and low-light intensities, as line broadening or collision induced absorption, or with low-atomic densities and high light intensities, as radiatively assisted collisions.

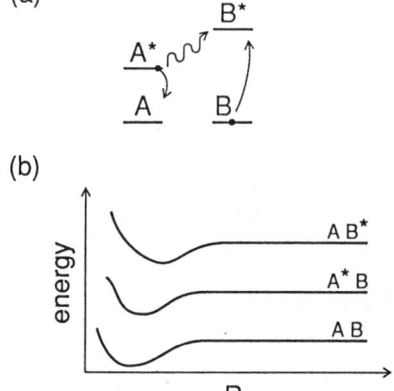

Fig. 1. (a) Schematic diagram of a collision in which the excited atom A* and ground state atom B collide while absorbing a photon to produce A and B*.

Radiatively assisted collisions were first studied experimentally by Harris and collaborators and have since been explored extensively.[3-9] Due to the relatively short collision times, ~ 10^{-12} s, and small dipole moments, ~ 1 ea_0, optical powers of 1 Mw/cm^2 are required to observe radiatively assisted collisions involving ground state atoms, and observing multiphoton assisted collisions is not really very likely. On the other hand, using Rydberg atoms and microwave fields Pillet et al.[10,11] showed that it was straightforward to observe such collisions. They used a variant of the resonant Rydberg atom collisions which have very long collision times, $>10^{-8}$ s. In Fig. 2a we show the initial and final levels of the resonant dipole-dipole collision process involving atoms A and B.

In atom A the initial and final states are connected by the dipole moment μ_A and in atom B they are connected by μ_B as shown. The two atoms are coupled by the dipole-dipole interaction $\mu_A \mu_B / R^3$. If the two atoms pass by each other with collision velocity v and impact parameter b, to a reasonable approximation they have an interaction strength $\mu_A \mu_B / b^3$ for a time b/v. They undergo the transition from the initial to the final state if the product of the interaction and the time is unity, i.e. if

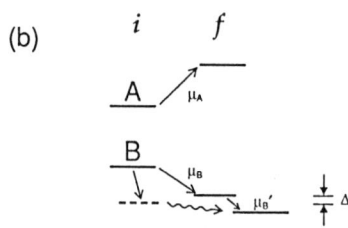

$$\frac{\mu_A \mu_B}{b^3} \cdot \frac{b}{v} = 1, \qquad (1)$$

leading to

$$\sigma \approx b^2 = \frac{\mu_A \mu_B}{v}. \qquad (2)$$

For a Rydberg atom $\mu_A \approx \mu_B = n^2$ and $\sigma \sim n^4/v$, which is 10^9 Å^2 at n = 20 and thermal velocities. In Fig. 2b we show the levels of a prototype radiative collision. In this case the interaction strength is $\mu_A \mu_B \mu_B' \cdot E/\Delta R^3$, differing from the interaction strength of the

Fig. 2. (a) Energy levels of atoms A and B involved in a dipole-dipole collision. The levels are connected by dipole matrix elements μ_A and μ_B. The initial and final states, i and f, are on the left and right.

(b) Energy levels of atoms A and B for a radiatively assisted collision. The dipole matrix elements are shown as well as the detuning Δ between the real and virtual intermediate states.

process of Fig. 2a by the factor $\mu_B' E/\Delta$. Here E is the amplitude of the applied radiation field, and μ_B' and Δ are the dipole moments and detunings indicated in Fig. 2b. Applying the same reasoning we used in Eq. (1), we can see that the radiatively assisted cross section for the process shown in Fig. 2b is comparable to the resonant collision process of Fig. 2a when

$$\mu_B' E/\Delta \approx 1. \qquad (3)$$

For $\mu_B' \approx n^2$ and $\Delta = 0.1/n^3$, 10% of the spacing between n levels, we arrive at a field of $E \approx 0.1 n^{-5}$, which at n=20, is a field of 161 V/cm, corresponding to an intensity of 68 W/cm^2. Since it is straightforward to generate radio frequency and microwave power densities far in excess of this value, it is in fact possible to observe multiphoton radiatively assisted collisions. Here we shall describe such collisions using radiation fields at frequencies covering the range from much higher than to much lower than the inverse of the collision time.

RESONANT COLLISION OF K RYDBERG ATOMS

The experiments described here are all based on the resonant collision process[12]

$$K(n+2)s + K\,nd \to K(n+2)p + K(n+1)p, \quad (4)$$

which can be tuned into resonance with small static fields, as shown by the energy level diagram of Fig. 3 for n = 27. As shown by Fig. 3, most of the tuning comes from the 27d state, and at a field of 6.5 V/cm the 29p state is the same energy above the 29s state as the 28p state is below the 27d state.

A useful way of describing these collision processes is to construct the molecular states

$$\Psi_i = \Psi_{29s} \otimes \Psi_{27d}$$

and (5)

$$\Psi_f = \Psi_{2np} \otimes \Psi_{28p}$$

which have energies W_i and W_f given by

$$W_i = W_{29s} + W_{27d}$$

and (6)

$$W_f = W_{29p} + W_{28p}.$$

In Fig. 4 we show by the solid lines the energies W_i and W_f as a function of static field near the resonance field E_R = 6.5 V/cm. As shown by Fig. 4 the levels are degenerate at E_R. In fact, they do not cross but have an avoided crossing given by $\mu_A\mu_B/R^3$, the dipole-dipole interaction where μ_A and μ_B are the 29s-29p and 27d-28p dipole matrix elements. The avoided crossing size

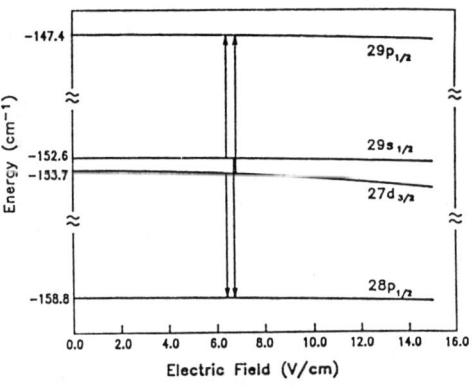

Fig. 3 Energy level diagram for the Rydberg states of K involved in the resonant energy-transfer collision K 29S + K 27d → K 29p + K 28p. The two resonances shown have been separated for clarity; they are degenerate in electric field.

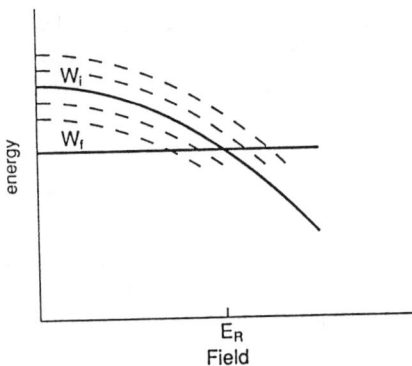

Fig. 4. Energy levels W_i and W_f of the initial and final molecular energy levels, as a function of field with no rf field (———). The initial molecular state is composed of the (n+2)s and nd states. In the absence of dipole-dipole coupling the two molecular states are degenerate at the resonance field E_R. Adding the rf field produces sidebands on the initial state (- - -) but does not alter the final state.

depends on R, so Fig. 4 is only strictly correct for $R = \infty$. Consider now adding a radio frequency (rf) field of the frequency $\nu = \omega/2\pi >> 1/\tau$ where τ is the collision time. The rf field modulates W_i about its static field value. Just as frequency modulating a radio wave produces sidebands, adding the high frequency field $E_{rf} \cos\omega t$ breaks Ψ_i into sideband states, i.e. in the field[11]

$$\psi_i(n) \rightarrow \psi_i(r) e^{-iW_i t} \sum_m J_m(kE_{rf}/w) e^{-im\omega t} \qquad (7)$$

where J_m is the mth Bessel function and k is dW_i/dE evaluated at E_R. In Fig. 4 the broken lines are the sideband states of W_i. As shown by Fig. 4 they are displaced from W_i by $m\omega$ and cross W_f at fields displaced from E_R by

$$\Delta E = m\omega/k. \qquad (8)$$

Consequently, the radiatively assisted collisions are displaced from the resonant collision by the amounts given by Eq. (8). The strengths of the m photon assisted resonances relative to the resonant collision process can be determined from the avoided crossings of the sideband states of Ψ_i with Ψ_f. Using Eq. (8) the avoided crossing of the mth sideband of Ψ_i with Ψ_f is given by $(\mu_A \mu_B/R^3) J_m(kE_{rf}/\omega)$, and since the cross section is proportional to the interaction strength, as shown by Eq. (2), the m photon assisted cross section is given by

$$\sigma_m = \sigma_0 |J_m(kE_{rf}/\omega)|. \qquad (9)$$

The sign of the interaction does not matter, hence the absolute value.

If the field has a frequency $\nu << 1/\tau$ it appears as a quasistatic field during the collision, and the position of the resonance is shifted. If the collision occurs in a small time interval about $t=0$ and the added rf field is $E_{rf}(\cos\omega t + \phi)$ the collisional resonance will be observed at the static field for which the combined static and rf fields at $t=0$ equal E_R, i.e. for

$$E_s = E_R - E_{rf}\cos\phi \qquad (10)$$

EXPERIMENTAL APPROACH

In our experiments we excite K atoms in a velocity selected beam to the 29s and 27d Rydberg states by two pulsed lasers via the route 4s 4p → 29s, 27d. The excited atoms collide with each other for 1μs after which they are field ionized and the ionization signal from the 29p state detected. An rf field of known amplitude and phase is present as well as a static field, which is slowly scanned through the collisional resonances over many shots of the laser.

The apparatus is shown schematically in Fig. 5. The atomic beam from a heated oven passes through a chopper wheel which has a small 1-4 mm slit allowing bursts of atoms to pass through it. About 20 cm downstream and 200 μs later the atoms in a burst are crossed by the laser beams, providing the velocity selection as well as the excitation. There is a hexapole electromagnet to focus atoms of the selected velocity, increasing the density almost tenfold. The excitation is between a pair of field plates 1.5 cm apart. The rf voltage, if any, is applied to the upper plate and the tuning static field voltage and the ionizing pulse are applied to the bottom plate. Ions resulting from field ionization leave the interaction region through small holes in the upper plate and impinge on a microchannel plate detector. The interaction region is surrounded by a magnetic shield which is thermally connected to a liquid nitrogen reservoir to condense any K atoms not in the velocity selected beam.

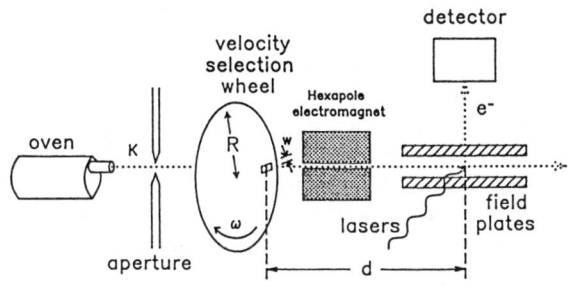

Fig. 5. Experimental Apparatus.

OBSERVATIONS

Using the velocity selected beam we can observe collisional resonances of ~ 1 MHz width, so a frequency as low as 4 MHz is in fact in the high frequency regime. In Fig. 6 we show recordings of the collisional resonances versus the static tuning field for 4 MHz fields from amplitude zero to 0.42 V/cm.[12] Inspecting Fig. 6 we can see resonances evenly spaced by 0.125 V/cm in agreement with Eq. (8). Furthermore, it is apparent that at higher rf fields collisions requiring the absorption or emission of more photons can be observed, as expected. However, it is difficult to see if the data of Fig. 6 exhibit the Bessel function dependence of Eq. (9). To show this dependence explicitly we show in Fig. 7 the collision signals as a function of rf field amplitude with the static field set to the resonances corresponding to the emission (absorption) of 0, 1, 2, and 3 photons. Also shown are the predictions of Eq. (9) and a numerical solution of the Schroedinger equation, and the excellent agreement is quite apparent.

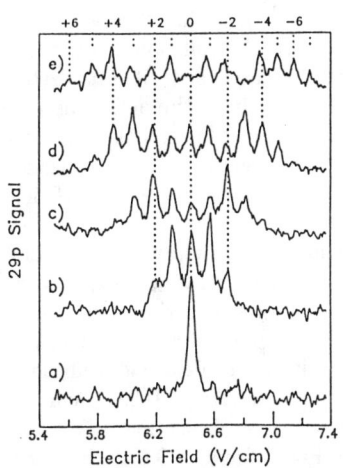

Fig. 6. Radiative collision signal for the K29s + 27d collisions in the presence of a 4 MHz rf field of various amplitudes. (a) $E_{rf} = 0$ (b) $E_{rf} = 0.19$ V/cm (c) $E_{rf} = 0.38$ V/cm (d) $E_{rf} = 0.57$ V/cm, and (e) E_{rf} 0.76 V/cm. The numbers at the top of the figure indicate how many rf photons are absorbed.

The Bessel function dependence of σ_m leads, since $\Sigma_m J_m^2(x) = 1$, to the conclusion[12]

$$\Sigma_m \sigma_m > \sigma_0 \qquad (11)$$

for $E_{rf} \neq 0$. When the resonant collision cross section is spread over many sidebands by the rf field it is larger than the resonant collision cross section with no rf field. Eq. (11) has been verified experimentally and reflects the coherent nature of the "broadening", in contrast to pressure broadening, for example.

The phase sensitive low frequency prediction of Eq. (10) is verified using frequencies much less than 1 MHz.[13] However, as the frequency approaches 1 MHz, it is only possible to specify the phase of the rf field if we know when individual collisions start and end relative to the laser pulse. This requirement is equivalent to the collisional resonances' being transform limited. When the atoms are velocity selected and allowed to collide for 0.7 μs, before the ionization pulse, the observed resonances are 1.4 MHz wide, i.e. they are transform limited and must have lasted 0.7 μs. In other words we know when each collision started and ended, at the laser and field ionization pulses, respectively.[14] We define the time t=0 as the center of the collision, halfway between the laser and field ionization pulses, and we specify the applied rf field as $E_{rf} \cos(\omega t + \phi)$.

Knowing when individual collisions start and end we can explore the $\omega/2\pi \approx 1/\tau$ regime experimentally. In Fig. 8 we show the collisional resonances observed with no rf field and with an rf field of amplitude $E_{rf} = 0.21$ V/cm and three phases $\phi = 0$, π, and $\pi/2$.[15] As shown by Fig. 8 for $\phi = 0$ there is a large

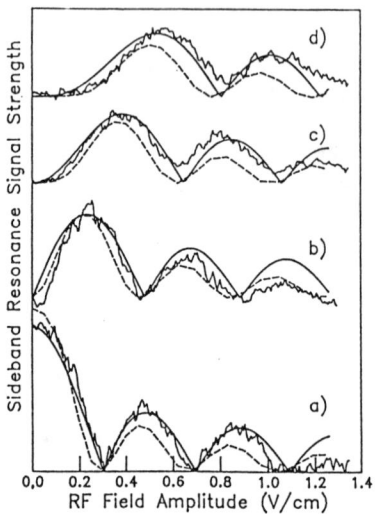

Fig. 7. Cross sections as a function of rf field strength for the first four orders of m photon resonances of the K29s + 27d radiative collisions in a 4 MHz rf field. (a) The zero-photon resonant collision cross section, (b) the +1 photon resonance, (c) the -2 photon resonance, and (d) the +3 photon resonance. The solid line shows the experimental data, the bold line indicates the prediction of the Floquet theory, and the dashed line is the result of numerical integration of the Schroedinger equation.

resonance at $E_S = 6.30$ V/cm, as expected from Eq. (10), and subsidiary resonances at 6.38, and 6.44 V/cm. For $\phi = \pi$, the observed signal is simply reflected about $E_R = 6.5$ V/cm. The puzzling point is the origin of the subsidiary peaks. They are certainly not spaced by the field corresponding to 0.75 MHz; they are too far apart. In marked contrast to $\phi = 0$ and π, $\phi = \pi/2$ (and $\phi = \pi/2$, not shown) leads to one broad resonance extending from $E_R - E_{rf}$ to $E_R + E_{rf}$.

When the frequency is raised to 1.48 MHz, the results shown in Fig. 9 are obtained. For $\phi=0$ there is a large peak at $E_S = E_R - E_{rf}$ and subsidiary peaks extending up to $E_S = E_R + E_{rf}$. The choice $\phi = \pi$ leads to a spectrum which is that of Fig. 9a reflected about E_R. Unlike Fig. 9 $\phi = \pi/2$ leads to subsidiary peaks as shown in Fig. 10. In both cases shown in Fig. 10 the spacing between the subsidiary peaks is too large to correspond to 1.48 MHz. As the frequency is raised to 4 MHz the observed spectra evolve smoothly to the high frequency regime.

THE INTERMEDIATE FREQUENCY AS THE CONNECTION BETWEEN HIGH AND LOW FREQUENCIES

We begin by considering the subsidiary peaks of Fig. 8, which seem to be a clear evolution from the low frequency regime. Accordingly we use a low frequency picture. In Fig. 10 we show the energy levels W_i and W_f as a function of time during the collision, for several values of the static field. For $E_s < E_R - E_{rf}$ the two levels never come into resonance and no collisional signal is observed. When $E_s \approx E_R - E_{rf}$ the two levels are brought into resonance at the peak of the rf field cycle, at $t \approx 0$, leading to the large peak at $E_s = 6.3$ V/cm in Fig. 8c. For $E_R - E_{rf} < E_s < E_R + E_{rf}$ the two levels come into resonance twice, at $t = \pm t'$ or $t = \pm t''$, during the collision as shown in Fig. 10. When they first come into resonance, at $t = -t'$, the initial state is split into a coherent superposition, which is recombined at $t = t'$. The two paths are shown in Fig. 10. Depending on the phase difference Φ accumulated,[14] where

$$\Phi = \int_{-t'}^{t'} (W_f - W_i) \, dt , \qquad (12)$$

the i→f transition amplitudes at the two resonances can add constructively or destructively. Constructive interference occurs for

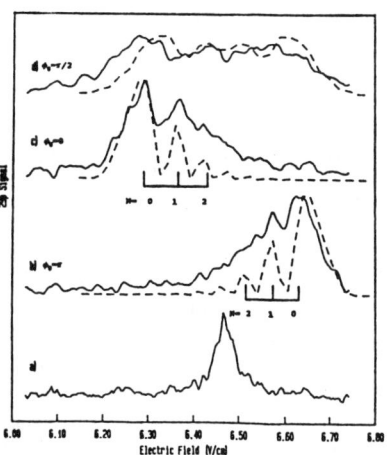

Fig. 8. The experimental lineshapes of the n = 29 radiatively assisted resonance of K as functions of static field E_s. (a) The unperturbed resonance with no rf field present, FWHM = 1.4 MHz. (b) - (c) The resonance in the presence of a 0.75 MHz, 0.21 - V/cm rf electric field and a phase of (b) $\phi = 0$, (c) $\phi = \pi$ and (d) $\phi = \pi/2$. Indicated below the spectra are the calculated positions of the peaks using the $\Phi = 2\pi N$ condition and Eq. (13).

$$\Phi = 2\pi N \qquad (13)$$

and destructive interference occurs for

$$\Phi = 2\pi(N + 1/2) \qquad (14)$$

where N is an integer. In other words, the subsidiary peaks of Fig. 8 for $\phi = 0$ and π are interference fringes which are analogous to those observed in Young's two slit experiment, Ramsey's separated oscillatory field method,[16] or Stuckelberg oscillations.[17] In Figs. 8 and 9 we show the predicted locations of the interference fringes based on Eqs. (12) and (13).

This picture also explains why no interference fringes are observed for $\phi = \pi/2$ in Fig. 8. The two levels only come into resonance once, and no interference is possible.

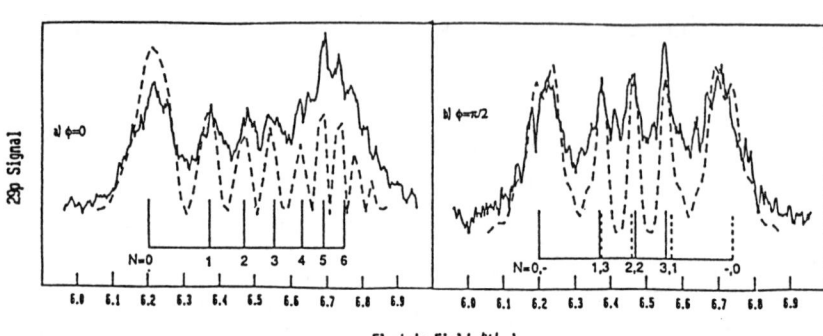

Fig. 9. The experimental (—) and calculated (- - -) lineshapes as functions of the static field E_s in a 0.31 V/cm, 1.48 - MHz rf field of phases (a) $\phi = 0$ and (b) $\phi = \pi/2$. The allowed collision time is 0.8 μs. The calculated fringe positions using $\Phi = 2\pi N$ are indicated below the spectra. In (b) the fringe positions due to interference between the first and second (solid vertical marker) and the second and third (dashed marker) interactions are shown.

However, they do come into resonance at some time during the collision for $E_R - E_{rf} < E_s < E_R + E_{rf}$. The above picture explains why for the higher frequency of 1.48 MHz interference fringes are seen for $\phi = \pi/2$. At this frequency the two levels come into resonance at least twice during the allowed collision time.

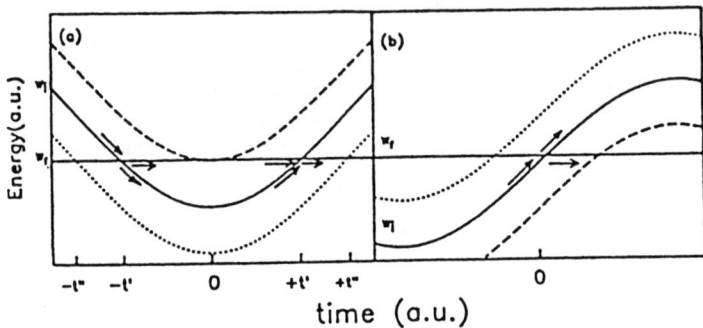

Fig. 10. The initial-and final-state energies, W_i and W_f, in a 0.75 MHz rf field with phases (a) $\phi = 0$ and (b) $\phi = \pi/2$. In both cases, the allowed collision time is 0.7 μs. As the static electric field is scanned, the single interaction period at t=0, when $E_s = E_R - E_{rf}$ (- - -), evolves into two interaction periods at $t = \pm\, t'$ (—) and $\pm\, t''$ (···) when $E_s > E_R - E_{rf}$, as shown in (a). The transition amplitudes interfere constructively when the total phase shift of ψ_i between -t' and t' is $2\pi N$. In (b) only one interaction period occurs irrespective of E_s.

Finally, it is possible to connect the above picture, derived from the low frequency limit, to the high frequency large amplitude rf field limit. The connection can be made more rigorously,[14] but here we present an intuitive picture. In Fig. 11 we show the energy levels W_i and W_f over one and a half cycles of the field. For constructive interference in the transition amplitude Φ_0, the accumulated phase over one field cycle must satisfy

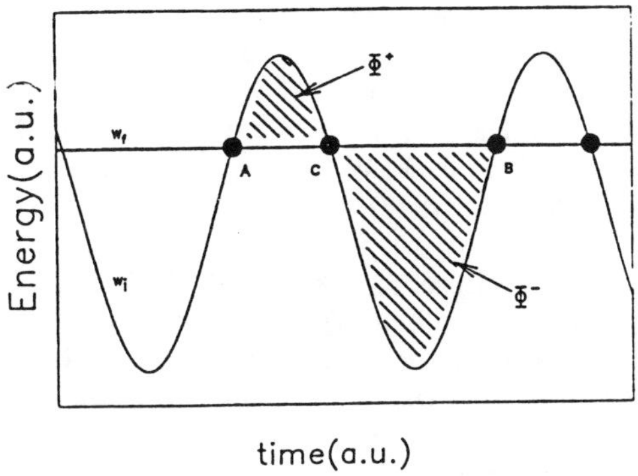

$$\Psi = \int_{-\pi/\omega}^{\pi/\omega} (W_f - W_i)\, dt = 2\pi N \quad (14)$$

where N is an integer. Since

$$W_f - W_i = k(E) + E_{rf}\cos\omega t - E_R), \quad (15)$$

Eq. (14) is satisfied for

$$\Delta E = E_s - E_R = N\omega/k. \quad (16)$$

Eq. (16) is identical to Eq. (8) the high frequency location of the radiatively assisted collisional resonances. Eq. (16) tells us where the collisional resonances are but not how strong they are.

Fig. 11. The initial- and final- state energies, W_i and W_f, in a high frequency rf field. Only two cycles of the rf field are shown. Each time $W_i = W_f$ a collisional interaction occurs and the relative amplitudes of ψ_i and ψ_f change. If $\Phi^+ + \Phi^- = 2\pi N$, then the transition amplitudes over successive cycles add constructively and a resonance in the cross section is observed.

Requiring that $\Phi = 2\pi N$ over an entire field cycle implies that the amplitudes of each cycle interfere constructively with the amplitudes of successive cycles. More specifically, it ensures that the amplitudes on the rising field traversals interfere constructively with themselves and that the amplitudes from the falling field traversals interfere constructively with themselves. To ensure that they interfere constructively with each other we require that the phases Φ_+ and Φ_- between these level crossings both be equal to $2\pi N$. Requiring that the area Φ_+, or Φ_-, be equal to $2\pi N$ leads to the same locations of the maxima of the collisional resonances as does Eq. (9).

CONCLUSION

This work demonstrates the connection between radiatively assisted collisions in high and low frequency fields. Many details such as the shapes of the collisional resonances, and the velocity dependences need to be explored, but many of the essential ideas are understood.

ACKNOWLEDGEMENTS

This work has been supported by the Air Force Office of Scientific Research.

REFERENCES

1. L.I. Gudzenko and S.S. Yakovlenko, Zh. Eksp. Teor. Fiz. 62, 1686 (1972) (Sov. Phys. - JETP 35, 877 (1972)).
2. A. Gallagher and T. Holstein, Phys. Rev. A16, 2413 (1977).
3. S.E. Harris and D.B. Lidow, Phys. Rev. Lett. 33, 674 (1974).
4. R.W. Falcone, W.R. Green, J.C. White, J.F. Young, and S.E. Harris, Phys. Rev. 15, 1333 (1977).
5. W.R. Green, J. Lukasik, J.R. Wilson, M.D. Wright, J.F. Young, and S.E. Harris, Phys. Rev. Lett. 42, 970 (1979).
6. P. Cahuzac and P.E. Toschek, Phys. Rev. Lett. 40, 1087 (1978).
7. A.V. Hellfield, J. Caddick, and J. Weiner, Phys. Rev. Lett. 40, 1369 (1978).
8. W.R. Green, M.D. Wright, J. Lukasik, J.F. Young, and S.E. Harris, Opt. Lett. 4, 265 (1979).
9. J.C. White, Opt. Lett. 6, 242 (1981).
10. P. Pillet, R. Kachru, N.H. Tran, W.W. Smith, and T.F. Gallagher, Phys. Rev. Lett. 50, 1763 (1983).
11. P. Pillet, R. Kachru, N.H. Tran, W.W. Smith, and T.F. Gallagher, Phys. Rev. A36, 1132 (1987).
12. D.S. Thomson, M.J. Renn, and T.F. Gallagher, Phys. Rev. A45, 358 (1992).
13. M.J. Renn, D.S. Thomson, and T.F. Gallagher, (unpublished).
14. D.S. Thomson, M.J. Renn, and T.F. Gallagher, Phys. Rev. Lett. 65, 3273 (1990).
15. M.J. Renn and T.F. Gallagher, Phys. Rev. Lett., 67, 2287 (1991).
16. N.F. Ramsey, Molecular Beams, (Oxford Univ. Press, New York, 1956).
17. D. Coffey, Jr., D.L. Lorents, and F.T. Smith, Phys. Rev. 187, 201 (1969).

COLLISIONS OF RYDBERG ATOMS WITH IONS: A REVIEW OF RECENT WORK ON CHARGE TRANSFER FROM HIGHLY EXCITED TARGETS

Keith B. MacAdam
University of Kentucky, Lexington, KY 40506 USA

ABSTRACT

Charge transfer from high-n atomic targets populates comparably excited states of ionic projectiles. This paper reviews a decade of accelerating experimental and theoretical progress in the field, which offers an unusual variety of challenges in the preparation of initial states and in the analysis of final states. Interesting dynamical questions at the interface of classical and quantum mechanics are posed by new experimental results. Rydberg atom phenomena are increasingly found to be important for other areas of atomic collision physics in all energy regimes.

INTRODUCTION

Many physicists have been challenged by Rydberg atoms over the past twenty years, the period marked mainly by the technology of selective level excitation by tunable lasers.[1,2] Although the impact of Rydberg atom physics has been felt also in precision spectroscopy, quantum optics, coherent-state phenomena, buffer-gas collisions, beam/solid interactions and other areas, this review will focus on the collisions of Rydberg atoms with ions that result in charge transfer,[3] in order to highlight the continuum Coulomb three-body problem in the regime of high quantum numbers. The principal issues are: (1) At intermediate velocities (where a charged particle has a speed comparable to the speeds of internal motions in the target) what are the basic phenomena and their relative strengths in ion-Rydberg collisions? (2) How can one deal experimentally and theoretically with the plethora of nearly degenerate initial and final states of the system, which are strongly coupled by the relative motion of the heavy constituents? and (3) To what extent may classical physics be applied to these problems, and what is the correct classical limit of quantum theory for this Three-Body Problem?

The past decade has seen the development of several new experimental and theoretical techniques in this field. This review is intended to summarize and synthesize them and point to directions of likely future progress from the author's point of view.

MEASUREMENT OF FINAL n DISTRIBUTIONS IN CHARGE TRANSFER FROM RYDBERG-ATOM TARGETS

Positive ions X^+ incident at near unit reduced velocity on Rydberg atoms $A(n\ell)$ with principal quantum number n and orbital angular momentum ℓ capture electrons preferentially into Rydberg states n' of the projectile with $n' \approx n$. (Reduced velocity $\tilde{v} \equiv v_{\text{ion}}/v_{\text{Bohr}} = v_{\text{ion}}(n/v_0)$, where $v_0 = 2.19 \times 10^6$ m/s is the electronic speed in the first Bohr orbit of hydrogen and is defined as the atomic unit of velocity. In practical units, $\tilde{v} = n\sqrt{(E_X/24.8\,\text{keV})/(M_X/1\,\text{amu})}$. For states having a significant quantum defect δ, n is replaced here by $n - \delta$.[4]) This fact can be anticipated by simple energetic arguments: The change of n corresponds to a change of orbital energy, which must be gained or lost by transfer from the heavy projectile motion. This energy transfer is inefficient between particles of such unequal masses $M_X/m_e \gtrsim 10^4$, and thus minimal changes of n are the most likely. The charge-transfer cross sections near $\tilde{v} = 1$ are of order $\sigma_{\text{CT}} = \pi n^4$ a.u., which can be as great as 10^6Å^2 for $n \approx 30$. It is this enormous magnitude that allows investigation by crossed or merged beam techniques or by pulsed laser experiments with very small duty factors.

In a lengthy series of experiments at the University of Kentucky we have measured the final n' state distributions of capture,

$$X^+ + \text{Na}(n\ell) \rightarrow X(n') + \text{Na}^+ \tag{1}$$

for projectiles Na^+ and Ar^+ over a range $\tilde{v} = 0.19$ to 1.95 for initial ns ($n = 25$, 29 and 34) and nd ($n = 24$, 28 and 33) states using 60–2100 eV ions bombarding a laser-excited target.[5] Essential for this work was an inhomogeneous-electric-field stripper and energy analyzer (Fig. 1a).[6] The beam of neutralized projectiles $X(n')$ in Rydberg states entered the region between two coaxial half-cylinder electrodes along a radial path. The magnitude of the electric field between the cylinders, established by the negative potential applied to the inner one, varied as $1/r$, covering a range of about 5:1 along the flight path of the projectiles. When the electric field at a point on this radial path was sufficient to field-ionize the projectile, the fragment ion X^+ was accelerated forward through the potential difference remaining before X^+ exited through a grid in the face of the inner cylinder. Its energy gain relative to the known initial kinetic energy of the X^+ beam was measured by a conventional electrostatic analyzer, and, based on careful calibrations with known n' values,[5] the resulting distribution of projectile energies could be transformed into an inferred distribution of final n' states. With this detector, projectiles that had gained or lost only *milli*electron-volts of kinetic energy through capture and whose electrons occupied states differing by merely meV in binding energy could be separated and analyzed out of a projectile beam at keV energy.

The inhomogeneous-field analyzer used for this work has since been implemented in other laboratories. At Kansas State University, DePaola has adapted

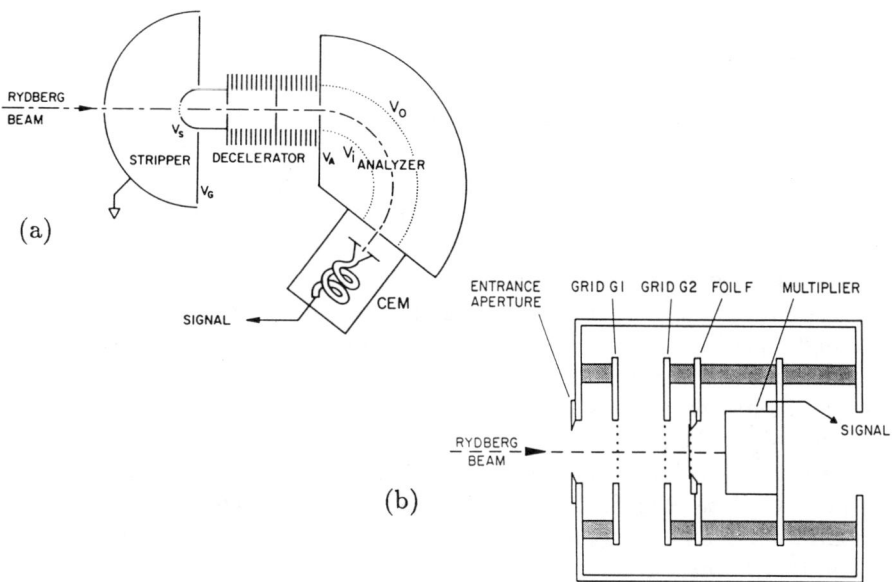

Fig. 1. (a) Inhomogeneous-field cylindrical stripper and electrostatic analyzer for determining n distributions of Rydberg atoms or ions in fast beams (Ref. 6). (b) Detector for measuring velocity dependences of total charge transfer that has a nearly flat response over a wide range of n states and for beam energies $0 < E \lesssim 2$ keV (Ref. 7).

the design along lines suggested by MacAdam and Rolfes[6] for use with MeV projectile beams by detecting the electrons released in field ionization instead of the ions and by using concentric spheres to gain an advantage of additional field variation for a given applied potential.[8,9] DePaola and coworkers have observed very high Rydberg states produced through electron capture by 9.5-MeV F^{q+} ($q = 3, 6$ and 7) in Ne. This work did not involve Rydberg-state targets, but capture into final Rydberg states is a consequence of the high projectile charge in this case. A similar detector has been used at the National Tsing Hua University in Taiwan to observe capture to high n states by protons in Ar up to 100 keV.[10] Very recently, DePaola has reported use of a stripper of this type with a position-sensitive detector (PSD) to observe capture by Xe^{35+} from laser-excited Na Rydberg states around $n = 25$.[11] Eventually, plans call for a study of resonant transfer and excitation (RTE) from a Rydberg-state target at MeV energies.[12,13]

Underwood et al.[14] applied the inhomogeneous-field stripper principle in a study of Rydberg capture by 0.1-MeV/amu Ag^{4+} ions in gas targets in which capture to a bound state is accompanied by cusp electrons. The advantage, however, of using inhomogeneous fields to tag the location of individual field-ionization events by their differential energy gain was lost by selecting spheres whose radii differed by only about 25% rather than a factor of about 5.

An alternative method for measuring the n distribution of Rydberg states populated in a fast beam by a collision process was described by Müller et al.[15,16] in their study of dielectronic recombination, Mg$^{+}(3s) + e^{-} \rightarrow$ Mg$(3s\,n\ell) + h\nu$. The post-collisional beam entered the region between two metal plates set at a wedge angle of 30° and with equal and opposite potentials applied between them. The transverse electric field, increasing in intensity as projectiles approached the apex of the wedge, caused field ionization of high-n states. Positive ions were deflected by the field into a PSD behind the negative plate, or alternately, electrons could be collected transversely and with spatial resolution on the side of the positive plate. Careful trajectory calculations and modeling of hydrogenic field ionization determined the relation between the position of impact on the PSD and the Rydberg state n.

Aseyev et al.[17] have recently compared the properties of longitudinal and transverse electric-field ionizers for fast Rydberg atoms both experimentally and theoretically and have determined that the energy spread of ions formed by field ionization of Rydberg atoms in the longitudinal scheme can be smaller than in the transverse one, offering some advantages of resolution for subsequent energy or mass analysis.

Another technique based on field ionization in an *einzel* lens and electrostatic spatial separation of products has been described recently by Harder et al.[18] and by Kuiper et al.[19] This RYDFISS method has been applied to the Rydberg projectile states formed by 13–24 MeV O^{7+} beams penetrating carbon foils.

The results of our Rydberg charge-transfer experiments[5] were most easily parametrized in terms of a final-state distribution based on the maximum-entropy principle. Åberg et al.[20] determined the distribution having maximum entropy, i.e., maximum thermodynamic randomness among final states, consistent with constraints of conservation of energy and momentum in the Rydberg capture process. The results can be expressed as

$$P(n') = Cf(n')\exp\left[-\mu\left(\frac{n^*}{n'} - 1\right)^2\right], \qquad (2)$$

where $P(n')$ is the number of atoms having principal quantum number n', C is a normalization constant, $f(n')$ is a "prior" distribution or statistical weight, μ controls the width of the peaked distribution, and n^* controls its center. MacAdam et al.[5] have argued that $f(n') = n'$ best represents the weight of sublevels populated by capture and that these n' states are then distributed

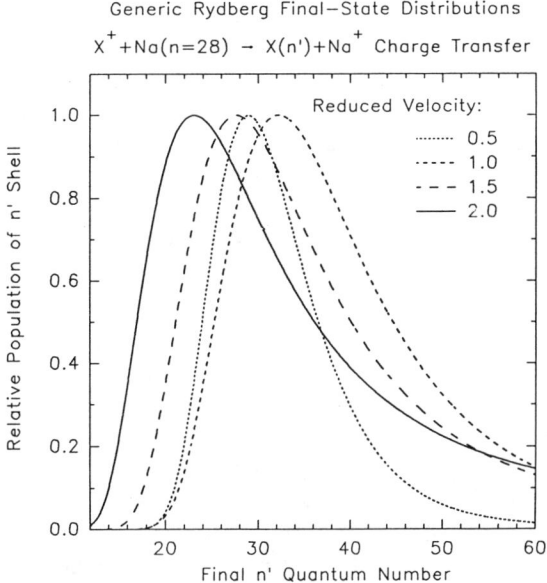

Fig. 2. Generic $P(n')$ final-state distributions for an $n = 28$ Rydberg target, based on fits of Eq.(2) to a large number of experimental results in the velocity-matching region (Ref. 25).

among the n'^2 states of the manifold by Stark and rotational coupling. Recent theory[21-23] has provided more detailed mechanisms that lead to a similar conclusion. For this choice of $f(n')$ the maximum of $P(n')$ occurs slightly above n^* at $n' = n_{\max}$, where

$$n_{\max} = \mu n^* \left(1 - \sqrt{1 - 2/\mu}\right). \qquad (3)$$

By using the Åberg distribution, Eq.(2), empirical final-state distributions could be represented by the two fitted parameters μ and n^* over a wide range of initial $n\ell$, reduced velocity and projectile species X. In the matching velocity region $\tilde{v} \approx 1$, where currently available quantal methods of collision theory are intractable,[24] it is useful to note some features of the results that may be generic. MacAdam et al.[5] described two ratios $R_{\rm KE}$ and $R_{\rm Bohr}$ that appear to place the existing Rydberg charge-transfer data on a scale that is independent of n and ℓ. The results of over 150 combinations of projectile energy and initial state were combined to allow simple predictions for typical $X^+ + A(n\ell)$ capture experiments (Fig. 2) based on the Åberg distribution.[25]

An important advance in the Rydberg charge-transfer experiments was the observation that a wide range of magnetic sublevels m_ℓ (referred to the projectile beam as the quantization axis) is populated in intermediate-velocity capture.[5,26,27] This has implications for detection of Rydberg states by field ionization because alkalis (and presumably other nonhydrogenic atoms) field ionize at widely different electric-field strengths by adiabatic and diabatic selective field ionization (SFI) at low and high m_ℓ, respectively.[28]

VELOCITY DEPENDENCE

The detector[6] that was so useful for determining final-state distributions was not at all suited for measuring the velocity dependence of charge transfer. Its sensitivity for a given final n' had to be determined as a function of projectile energy, and the total charge transfer had to be found by summing over all n' of the properly inferred n' distribution.[29] Nonetheless, velocity dependences of charge transfer at $\tilde{v} = 0.567$ to 1.40 obtained in this way were combined[30] with electron loss (target ionization plus charge transfer) at $\tilde{v} = 0.8$ to 3.7 to infer the velocity dependence of charge transfer over a wide range of $\tilde{v}$ and its scaling with n.[31] Unfortunately, the low-velocity dependences have not been borne out by subsequent work,[7] and low-velocity electron-loss data by Koch and Bayfield[32] were misplotted by MacAdam[31] where they appeared to support our work. This unsatisfactory situation motivated the design of a new Rydberg detector intended specifically for reliable velocity-dependence studies of charge transfer from Rydberg states at energies 0 to 2 keV.

The new detector[7] (Fig. 1b) is cylindrically symmetric about the projectile beam axis and consists of a grounded high-transmission grid, a 2 μgram/cm^2 carbon foil elevated to about $\Delta V = -14$ kV, and a charged-particle detector (MCP, CEM or discrete-dynode device). Rydberg atoms formed from a Rydberg target by charge transfer pass through the grid and are field ionized (if $n' > 16$, which includes all but a small percentage of final states in our work). The positive ions are accelerated through ΔV, pass with transmission probability $> 50\%$ through the foil, and are registered by the particle detector as individual counts. Unneutralized projectiles are gently swept aside before they reach the detector without field-ionizing any Rydberg projectiles that have $n' < 60$. The very numerous fast ground-state projectiles, neutralized in the beam line off the background gas, are not field ionized and are thus not accelerated, and they are stopped in the foil. The foil detector has a sensitivity of about 50% for Rydberg products in the fast beam over an area at least 1 cm^2, and this sensitivity is nearly flat for n' between 16 and 60 and for projectile energies much less than 14 keV.

Using the foil detector, Hansen et al.[7] determined the relative velocity dependence of capture by Na$^+$ from a wide range of selectively excited Na ns and nd targets, $n = 22$ to 41. The velocity dependences were remarkably similar in shape as a function of $\tilde{v}$ over the range $\tilde{v} = 0.3$ to 2, with initial ns states

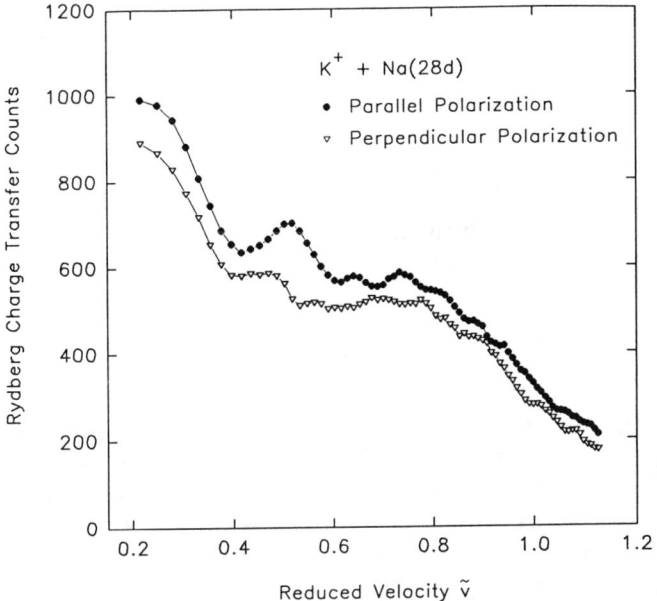

Fig. 3. Relative total cross sections for charge transfer in $K^+ + Na(28d)$ collisions with two different exciting-laser polarizations near $\tilde{v} = 0.5$ (Ref. 35).

and initial nd states forming two distinct classes when the curves were overlaid. Furthermore, reproducible undulations (peaks and/or dips) were observed in the curves, which had not been anticipated on the basis of benchmark p + $H(1s)$ total capture cross sections.[33] Each cross section exhibited a broad peak or shoulder at $\tilde{v} = 0.8$ before the rapid fall of σ_{CT} toward higher velocities, and each showed a small bump at $\tilde{v} = 0.5$. Most remarkably the bump at $\tilde{v} = 0.5$ in nd-state targets, about 10% above a smooth background, was sensitive to the plane of linear polarization of the laser (410 nm) used to excite Na(nd) from Na($3p\ ^2P_{3/2}$). The undulations did not look like those noted long ago in differential and total cross sections in near-resonant charge transfer of ground-state alkalis,[34] which result from interferences between g and u scattering potentials. Day et al.[35] (Fig. 3) determined that the bump at $\tilde{v} = 0.5$ was equally apparent for the unsymmetrical case

$$K^+ + Na(28d) \rightarrow K(n') + Na^+ \tag{4}$$

and confirmed that the capture cross section was enhanced selectively at that velocity, but not nearby, by a laser polarized *parallel* to the projectile beam.

Berkowitz and Zorn[36] had observed some undefined structure in near-resonance charge transfer of protons in K and Na leading to H(2s) near the cross section maxima. Avakov et al.[37] have recently obtained reasonable agreement with those experiments in calculations based on the Fadeev three-body equations with a purely Coulomb interaction and hydrogenic wave functions in the impact parameter representation. These results suggest that the structure in the cross sections for those ground-state targets and also for our Rydberg-state targets have little to do with the many-electron nature of the problem but may be a generic feature of the three-body Coulomb problem.

ALIGNMENT EFFECTS

A flood of results on the capture from aligned and oriented Na(3p) targets has recently appeared from Orsay[38,39] with related theoretical work from Copenhagen, Bergen, Paris and elsewhere.[40] In Freiburg, alignment and orientation effects in K$^+$ + Na(4d) capture are under study.[41] At Kansas State University capture by a 63-keV Ar^{7+} beam in Na(3p) has been observed as a prelude to studies with Rydberg-state targets and other highly charged ions.[42]

The first substantial report of the velocity dependence of alignment effects in charge transfer from Rydberg states was the recent experiment of Wörmann et al.[43] in which laser-excited Na($n\ell m$) for $\ell = 2$ states, $n = 20$ to 28, were bombarded by Ar$^+$ at 0.5 to 5 keV. Individual m_ℓ sublevels of the nd states could be excited by selecting appropriate laser frequencies and polarizations, pumping Na out of the $3p\ ^2P_{3/2}$, $F = 3$, $M_F = 3$ hyperfine level in a 75-G magnetic field. The authors obtained a significant variation with velocity of the ratio σ_2/σ_0 of cross section from $|m_\ell| = 2$ and 0 target states near the matching velocity. The cross sections seem to become independent of initial m_ℓ at $\tilde{v} < 0.5$.[44]

These many new results underscore the richness in the directional properties of these relatively simple collision systems. An orbitally *aligned* target is one whose distribution among (degenerate) sublevels m_ℓ varies with $|m_\ell|$ but not necessarily between $+|m_\ell|$ and $-|m_\ell|$.[45] If the distribution is unbalanced between corresponding positive and negative magnetic sublevels then the electron cloud may have a net direction of circulation, and the state is then said to be *oriented*. Orbital alignment and orientation of a target (which is possible only for $L > 0$) can affect the collisional outcomes if the partial cross sections (e.g., for capture) differ among m_ℓ sublevels. Different physical mechanisms of capture couple differently to m_ℓ sublevels, and thus the alignment or orientation effects probe deeper into the physics of the collision process than the more conventional total and differential cross sections.

Unfortunately, the deflection of an ion in a typical Rydberg capture process is only about 50 μrad—5 mm deflection in 100 meters of flight! In inelastic processes mediated by larger impact parameters it is even smaller. If the direction of scattering is too small to be determined, then an experimental setup effectively averages cylindrically about the incident beam, and all effects of target orientation are averaged to zero. Thus the chance of measuring orientation effects in ion-Rydberg collisions seems remote.

However, alignment effects such as those reported by Hansen et al.,[7] Day et al.,[35] and Wörmann et al.[43] represent an important new area of work in Rydberg collisions. MacAdam and Morrison[46] have completed a thorough theoretical study of the alignment produced in Na(nd) Rydberg states by two-step pulsed-laser excitation in a mutually perpendicular geometry of laser and particle beams as a prelude to inferring partial cross sections from laser polarization studies. Through such studies it is hoped to correlate the polarization-sensitive contribution to the charge transfer cross sections with a particular sublevel $|m_\ell|$ (or definite combination) and thus to single out dynamical mechanisms that act preferably for such $|m_\ell|$'s.

A simple "synchrony" model that is consistent with the observed alignment sensitivity of charge transfer from Na(nd) states at certain velocities can be stated as follows. An ion passes a Rydberg atom on a straight-line trajectory at an impact parameter b equal to an undetermined multiple of the orbital radius, $b = \lambda n^2$ a.u. Suppose that charge transfer is *enhanced* slightly when the projectile passes the target atom in a time that is approximately an integer multiple of the electronic orbit time. (Parents who push their children on swings know that the pushing is most effective if it comes at a subharmonic of the frequency of the swing!) The orbit time is $T_n = 2\pi n^3$ a.u., and we define the time of passage to be $T_\text{pass} = 2b/v = 2\lambda n^2/(\tilde{v}/n) = 2\lambda n^3/\tilde{v}$. If we put $T_\text{pass} = pT_n$ for integer $p = 1, 2, 3, \ldots$, we find that $\tilde{v} = \lambda/\pi p$. Enhancements of the charge transfer cross section at $\tilde{v} = 1, \frac{1}{2}, \frac{1}{3}, \ldots$ would occur if a portion of capture that is sensitive to synchrony effects were dominated by impact parameters corresponding to $\lambda \approx 3$. Such a crude model should not be expected to distinguish between $\tilde{v} = 1.0$ and $\tilde{v} = 0.8$ (where a peak or shoulder that we propose to associate with $p = 1$ appears in Hansen's results), and $p = 3, 4$, etc. correspond to high-order subharmonics for which adjacent orders would blend. But $p = 2$ ($\tilde{v} = \frac{1}{2}$) may be the right balance of big and small, and CTMC calculations[47,48] indicate that the total cross section for capture is dominated by impact parameters 1.5 to 3 times the orbital radius in the region $\tilde{v} = 0.2$ to 1.

This synchrony model furthermore predicts that orbits having low $|m_\ell|$, which are the ones preferentially excited by light polarized along the ion-beam direction, should exhibit enhanced charge transfer as compared with those having "high" m_ℓ, ($|m_\ell| = 2$). The $m_\ell = 0$ orbits tend to be approached edgewise by projectiles while $|m_\ell| = 2$ orbits are struck broadside, for which the synchrony concept does not apply. The model's predicted effects scale with n so that bumps or other characteristic features should occur at the same reduced velocity $\tilde{v} = v_\text{ion}/n$ for different n, in agreement with our experiments.

NEW THEORY (MOSTLY CTMC)

Several relevant new theoretical results have appeared that are based on the classical-trajectory Monte Carlo (CTMC) method. Meng, Reinhold and Olson[49] have analyzed capture to high n states that occurs with multicharged projectiles at 20–200 keV/amu and have found that for capture to $n' < n_{max}$ the highest ℓ's receive the most population, but that capture to states n' above the maximum of the distribution favors $\ell_{max} \approx n_{max}$, which is thus independent of n'. Burgdörfer and Dubé,[50] using the continuum-distorted-wave (CDW) approximation modified to account for Stark mixing in the exit channel, find that the residual electric field of the target induces a shift of capture probability to lower ℓ values. By comparing CDW and Oppenheimer-Brinkmann-Kramers (OBK) approximations they also find that the capture into large ℓ values is mediated by the higher-order terms in the interaction. It is not clear to what extent CTMC represents the physics of these higher order effects at intermediate velocities, but the CDW and OBK results may help to interpret the results obtained by CTMC.

In a detailed comparison with our previous experimental work, Pascale, Olson and Reinhold[27] used CTMC to calculate n', ℓ', and m' distributions for $Na^+ + Na(28d)$ capture at $\tilde{v} = 1.0$ and 1.658. They found that the contribution of final states having $m'_\ell \geq 2$ to the total capture cross section near the maximum of the n' distribution is greater than 60% and that the final ℓ' distribution is very broad, extending to the maximum allowed $\ell' = n' - 1$ when $n' \lesssim n_{max}$, although not reaching so high a value for $n' \gtrsim n_{max}$. These observations agree with the CDW results, which are presumably valid at a higher range of velocity.[50] A reanalysis of the experimental data by MacAdam that properly accounts for the broad distribution of m'_ℓ's (and consequently the mixture of SFI types) shows that for the case $X^+ + Na(n = 28)$ at $\tilde{v} = 1$ the interpolated average of experimental data[5,25] gives $n_{max} = 32.2$ (Eq. (3)) and half-maximum points at $n' = 25.3$ and 44.6, in very good agreement with Pascale et al.[27]

Lundsgaard and Lin[22] and Toshima and Lin[23] have carried out close-coupling calculations, confirming their results with CTMC, that indicate a propensity rule[51] for electron capture concerning m_ℓ: If the quantization axis is chosen to be perpendicular to the scattering plane and the geometry is defined so that the internuclear axis rotates clockwise in the collision plane during the collision, then capture favors initial $m_\ell = -\ell$ substates and tends to populate final $m'_\ell = -\ell'$ substates for any ℓ'. Expressed with an axis of quantization along the initial projectile momentum direction, this propensity would hardly be apparent because of the scrambling of m_ℓ that occurs upon rotation of coordinates.[45] The propensity rule gives a very useful pointer to future theoretical (and experimental!) work because it would allow a drastic reduction in the number of magnetic substates that need to be included to describe capture in the velocity-matching region.

MacKellar et al.[48] have applied the CTMC method to calculate charge transfer from an aligned Na(28d) target in the range $\tilde{v} = 0.2$ to 1.2. The object of these calculations was to look for structure in the low-velocity cross section and in particular to see whether the polarization-sensitive bump in the capture cross section[7] at $\tilde{v} = 0.5$ could be predicted classically. An open question is the validity of CTMC in this velocity range, and it has often been said that classical methods would not apply below $\tilde{v} = 1$. However, the relative contribution of tunneling at low velocity *for highly excited states* is not completely understood,[52,53] and a velocity selective effect of target alignment might equally well appear in the classically accessible part of phase space even if tunneling were significant. The result of these calculations,[48] which had to be very carefully done because of problems of numerical accuracy and complicated orbiting at low velocity, was that, although the CTMC capture cross section falls off more gradually than experiment for $\tilde{v}$ increasing from 0.2, a distinct bump is visible at $\tilde{v} = 0.5$ for $m_\ell = 0$ initial states that is absent for $m_\ell = 2$. Furthermore a broader peak or shoulder is present at about $\tilde{v} = 0.85$ for both m_ℓ states, and there is even a suggestion of another peak around $\tilde{v} = 0.3$. These results are in remarkably fine qualitative agreement with experiment (and with the simple synchrony model offered above) and should be a strong incentive to examine more closely the classical and semiclassical models for ion-Rydberg charge transfer in the velocity-matching region.

OTHER NEW EXPERIMENTAL DEVELOPMENTS

Another recent advance in the techniques of electron capture in ion-Rydberg collisions has been made by Deck, Hessels and Lundeen[54] at U. Notre Dame. For some years Lundeen and coworkers[55] have been developing radiofrequency and microwave techniques of extremely high precision for studies of fine structure in the high angular momentum states of Rydberg He and H_2. Rydberg levels were formerly populated nonselectively by electron capture by the passage of a fast ion through a gas cell, and very few atoms would thus be found in any one desired sublevel around $n' = 10$. In the latest work,[54] a Rydberg target of Rb $8f$ or $10f$ states has been populated by a *cw* three-laser stepwise excitation, and an ion beam passing through the target, for instance 11-keV S^+, can efficiently capture electrons into Rydberg states near $n' = 10$. Even though at the presently achieved Rb target thickness only about 10^{-3} of the ion beam is neutralized by capture, more than a tenfold increase of population in $n' = 10$ sublevels of the projectile has been obtained as compared with the gas-cell neutralization. The same increase, and probably a considerable reduction in background at the same time, should be obtainable for any ion that can be formed into an ion beam around 10 keV. This new omnibus quasi-selective Rydberg-beam preparation method offers great possibilities for future experiments with Rydberg atoms.

Havener et al.[56] and Haque et al.[57] have used an electron-cyclotron resonance (ECR) ion source and merged beams at the Oak Ridge National Laboratory to measure electron capture from hydrogen Rydberg states by multicharged ions. Initial states $n = 11$ to 24 are populated non-selectively by collisional detachment of H$^-$ and D$^-$. A stripper is used to differentially truncate the $1/n^3$ distribution of initial n's and to infer the n dependence of the ion-Rydberg capture by O^{q+} ions ($q = 2, 3$ and 5) in the region $\tilde{v} \approx 0.05$ to 3. Macek and Ovchinnikov[21] have predicted an anomalous n dependence for slow capture from H(n) by multicharged ions, $\sigma_{\mathrm{CT}} \propto n^\beta$, where $\beta = 5$ to 7. So far, the experiments are consistent with $\beta = 4$, which is the geometrical scaling expected on classical grounds, but the predictions of a strong dependence on the dipole moment of the initial state[21] have not yet been tested. Experiments in Aarhus by Ehrenreich et al.,[58] although they involve collisions of eccentric oriented Li Rydberg states with *singly* charged ions, may provide such a test.

CAPTURE FROM CIRCULAR STATES

Perhaps the most exciting new experimental developments in ion-Rydberg charge transfer are underway here in Aarhus. Horsdal-Pedersen and coworkers[59] have built an apparatus that uses the crossed-field adiabatic method[60,61] for preparing a target of Li atoms in the oriented "circular" Rydberg state $n = 25$, $\ell = 24$, $m_\ell = +24$. These orbits are stabilized by a weak (<100 G) magnetic field, which determines the direction of ℓ and thus the plane of the orbit. This plane can be readjusted at will by setting the magnetic field direction, and the oriented target is bombarded by a beam of 2.5-keV Na$^+$ ions ($\tilde{v} = 1.65$). The total capture cross section has been determined in a foil detector[7] as a function of the angle φ between the normal to the orbit and the incident beam direction. By a ratio of 4:1, capture is more probable for edgewise impact ($\varphi = 90°$) than for face-on impact ($\varphi = 0°$), and the φ dependence departs from sinusoidal. Just as in the case of capture from aligned nd-state Rydberg targets discussed above,[7,35,48] this result strongly suggests that the classical velocity matching discussed by Kohring, Wetmore and Olson[62] is a critical element of the capture process at intermediate velocities. The Aarhus circular-state capture experiment will be described in a "Hot Topic" presentation at this meeting by Søren Hansen.[63]

SUMMARY AND ACKNOWLEDGEMENTS

This paper has reviewed the currently active areas of ion-Rydberg collisions leading to charge transfer. Important advances have occurred in ion-Rydberg inelastic collisions, non-adiabatic effects, multicharged-ion collisions producing Rydberg states and other areas that might also have been described but will have to fit elsewhere. It is enough to say that the experimental and theoretical work on collisions of ions with Rydberg atoms have in the last ten years helped to pose many interesting questions and motivate much progress in collision theory, the relation of quantal and classical methods, and the venerable Three-Body Problem.

The author is especially grateful to his coworkers Drs. Lewis Gray and Richard Rolfes for sustained concentration in the laboratory over many years. This work was supported in part by the National Science Foundation under Grants PHY-9122377 and INT-9117374.

REFERENCES

1. *Rydberg States of Atoms and Molecules*, edited by R. F. Stebbings and F. B. Dunning (Cambridge U.P., New York, 1983).
2. T. F. Gallagher, Rep. Prog. Phys. **51**, 143 (1988).
3. K. B. MacAdam, in *Atomic Physics 12*, edited by J. C. Zorn and R. R. Lewis (Am. Inst. Phys., New York, 1991), p. 310.
4. K. B. MacAdam, in *Electronic and Atomic Collisions*, edited by J. Eichler, I. V. Hertel and N. Stolterfoht (Elsevier, New York, 1984), p. 265.
5. K. B. MacAdam, L. G. Gray and R. G. Rolfes, Phys. Rev. A **42**, 5269 (1990).
6. K. B. MacAdam and R. G. Rolfes, Rev. Sci. Instrum. **53**, 592 (1982).
7. S. B. Hansen, L. G. Gray, E. Horsdal-Pedersen and K. B. MacAdam, J. Phys. B **24**, L315 (1991).
8. B. D. DePaola *et al.*, Nucl. Instrum. Methods **B40/41**, 187 (1989).
9. D. H. Lee *et al.*, Nucl. Instrum. Methods **B40/41**, 1229 (1989).
10. B. D. DePaola, C. C. Hsu and Y. C. Chang (private communication, 1991).
11. B. D. DePaola *et al.*, Bull. Am. Phys. Soc. **38**, 1110 (1993).
12. R. Parameswaran *et al.*, Nucl. Instrum. Methods **B40/41**, 158 (1989).
13. B. D. DePaola, Nucl. Instrum. Methods **B56/57**, 154 (1991).
14. T. A. Underwood, M. Breinig and C. C. Gaither, Phys. Rev. A **44**, 1668 (1991).
15. A. Müller *et al.*, Phys. Rev. A **36**, 599 (1987).
16. A. Müller *et al.*, Nucl. Instrum. Methods **B23**, 254 (1987).
17. S. A. Aseyev, Yu. A. Kudryavtsev and V. V. Petrunin (1992, unpublished).
18. R. Harder *et al.*, Nucl. Instrum. Methods **B48**, 111 (1990).
19. H. Kuiper *et al.*, J. Phys. B **25**, 687 (1992).
20. T. Åberg, A. Blomberg and K. B. MacAdam, J. Phys. B **20**, 4795 (1987).
21. J. Macek and S. Y. Ovchinnikov, Phys. Rev. Lett. **69**, 2357 (1992).
22. M. F. V. Lundsgaard and C. D. Lin, J. Phys. B **25**, L429 (1992).
23. N. Toshima and C. D. Lin, Phys. Rev. A **47**, 4831 (1993); and private communication, 1993.
24. J. S. Briggs, in *Semiclassical Description of Atomic and Nuclear Collisions*, edited by J. Bang and J. De Boer (Elsevier, Amsterdam, 1985), p. 183.
25. K. B. MacAdam, Nucl. Instrum. Methods **B56/57**, 253 (1991).

26. A. D. MacKellar and R. L. Becker, in *Electronic and Atomic Collisions. XIII ICPEAC, Abstracts of Contributed Papers*, edited by J. Eichler et al. (Elsevier, New York, 1983), p. 660.
27. J. Pascale, R. E. Olson and C. O. Reinhold, Phys. Rev. A **42**, 5305 (1990).
28. T. H. Jeys et al., Phys. Rev. Lett. **44**, 390 (1980).
29. K. B. MacAdam and R. G. Rolfes, J. Phys. B **16**, 3251 (1983).
30. K. B. MacAdam et al., Phys. Rev. A **34**, 4661 (1986).
31. K. B. MacAdam, Phys. Rev. A **34**, 2767 (1986).
32. P. M. Koch and J. E. Bayfield, Phys. Rev. Lett. **34**, 448 (1975).
33. C. F. Barnett, *Atomic Data for Fusion, Vol.1: Collisions of H, H_2, He and Li Atoms with Atoms and Molecules* (Oak Ridge National Laboratory Report ORNL-6086/V1, 1990).
34. J. Perel, R. H. Vernon and H. L. Daley, Phys. Rev. **138**, A937 (1965).
35. J. C. Day, S. B. Hansen, L. G. Gray and K. B. MacAdam, Bull. Am. Phys. Soc. **36**, 1257 (1991).
36. J. K. Berkowitz and J. C. Zorn, Phys. Rev. A **29**, 611 (1984).
37. G. V. Avakov, L. D. Blokhintsev, A. S. Kadyrov and A. M. Mukhamedzhanov, J. Phys. B **25**, 213 (1992).
38. D. Dowek, J. C. Houver and C. Richter, in *Electronic and Atomic Collisions*, edited by W. R. MacGillivray, I. E. McCarthy and M. C. Standage (IOP Publishing, Bristol, 1992), p. 537.
39. C. Richter et al., J. Phys. B **26**, 723 (1993) and references therein.
40. A. Dubois, S. E. Nielsen and J. P. Hansen, J. Phys. B **26**, 705 (1993), and references therein.
41. E. E. B. Campbell, H. Hülser, R. Witte and I. V. Hertel, Z. Phys. D **16**, 21 (1990).
42. B. P. Walch et al., Phys. Rev. A **47**, R3499 (1993).
43. T. Wörmann, Z. Roller-Lutz and H. O. Lutz, Phys. Rev. A **47**, R1594 (1993).
44. H. O. Lutz (private communication, 1993).
45. R. N. Zare, *Angular Momentum: Understanding Spatial Aspects in Chemistry and Physics* (Wiley-Interscience, New York, 1988).
46. K. B. MacAdam and M. A. Morrison, Phys. Rev. A (to be published).
47. R. L. Becker and A. D. MacKellar, J. Phys. B **17**, 3923 (1984).
48. A. D. MacKellar, G. L. Li, M. J. Cavagnero, J. C. Day and K. B. MacAdam (unpublished, 1993).
49. L. Meng, C. O. Reinhold and R. E. Olson, Phys. Rev. A **42**, 5286 (1990).
50. J. Burgdörfer and L. J. Dubé, Phys. Rev. A **31**, 634 (1985).
51. S. E. Nielsen, J. P. Hansen and A. Dubois, J. Phys. B **23**, 2595 (1990).
52. D. Banks, K. S. Barnes and J. M. Wilson, J. Phys. B **9**, L141 (1976).
53. M. J. Cavagnero, "A tunneling-shell hypothesis for charge transfer from Rydberg atom targets," APS-DAMOP Meeting, May 1993, Reno NV, Postdeadline Paper TE-116.

54. F. J. Deck, E. A. Hessels and S. R. Lundeen, Bull. Am. Phys. Soc. **38**, 1110 (1993) and private communication.
55. E. A. Hessels, P. W. Arcuni, F. J. Deck and S. R. Lundeen, Phys. Rev. A **46**, 2622 (1992).
56. C. C. Havener et al., in Proceedings of the VIth International Conference on the Physics of Highly Charged Ions, edited by P. Richard, et al. (AIP, New York, 1993), p. 43; and private communication.
57. M. A. Haque et al., Bull. Am. Phys. Soc. **38**, 1160 (1993).
58. T. Ehrenreich, J. C. Day, S. B. Hansen, E. Horsdal-Pedersen, K. S. Mogensen and E. Wolfrum, in Electronic and Atomic Collisions. XVIII ICPEAC, Abstracts of Contributed Papers (Aarhus, 1993).
59. S. B. Hansen, T. Ehrenreich, E. Horsdal-Pedersen, K. B. MacAdam and L. J. Dubé (unpublished, 1993).
60. D. Delande and J. C. Gay, Europhys. Lett. **5**, 303 (1988).
61. J. Hare, M. Gross and P. Goy, Phys. Rev. Lett. **61**, 1938 (1988).
62. G. A. Kohring, A. E. Wetmore and R. E. Olson, Phys. Rev. A **28**, 2526 (1983).
63. S. B. Hansen, T. Ehrenreich, E. Horsdal-Pedersen and K. B. MacAdam, in Electronic and Atomic Collisions. XVIII ICPEAC, Abstracts of Contributed Papers (Aarhus, 1993).

RYDBERG AND CONTINUUM STATES OF ATOMS IN EXTERNAL FIELDS

F. Mota-Furtado
Departamento de Física, Faculdade de Ciências e Tecnologia,
Universidade Nova de Lisboa, 2825 Monte de Caparica, Portugal

P.F. O'Mahony and H. Marxer
Department of Mathematics, Royal Holloway, University of London,
Egham TW20 OEX, United-Kingdom

ABSTRACT

A general theory is presented to calculate the Rydberg and continuum states of an atom or molecule in external magnetic and parallel or crossed magnetic and electric fields of arbitrary strength. We show how the photoabsorbtion and photoionisation spectra can be obtained for an atom or molecule at any excitation energy and we demonstrate how the complex resonance structure present in the classically chaotic energy region of the spectrum can be analysed and understood. Results are presented for lithium in magnetic fields of 6.1 T and for hydrogen and rubidium in crossed magnetic and electric fields of intensity 6 T and 2975 Vcm^{-1}, respectively. The theory combines quantum-defect theory, R-matrix propagation and two and three-dimensional frame transformations. Much of the observed experimental resonance structure in the continuum can be understood in terms of complex perturbed Rydberg series.

INTRODUCTION

Rydberg and continuum states of atoms or molecules in external fields are simple examples of Hamiltonian systems which classically lead to solutions exhibiting chaotic motion and whose quantum spectra are extremely complex[1]. Its interest is enhanced by the fact that one can investigate such systems experimentally in great detail, by high resolution laser spectroscopy. At present there are a considerable amount of experimental spectra for both bound and continuum states of many atoms in laboratory strength fields of about 6 T [2] and in crossed magnetic and electric fields strength of about 0.7T to 6T and 0 to 4 kV.cm^{-1}, respectively[3, 4]. Up to recently comparisons between theory and experiment were confined to the bound state spectrum and the theoretical methods developed to understand the behavior of an atom or molecule under the influence of external fields have been mainly confined to these states and only in the presence of an external magnetic field[1]. Classically this was achieved through trajectory calculations of the shortest periodic orbits of the systems and, quantum-mechanically, by expansion of the wavefunction in a basis set, using a very large number of basis functions.

We present a general *ab initio* quantum theory to calculate the bound and continuum states of any atom or molecule in an arbitrary strength magnetic field and in crossed magnetic and electric fields also of arbitrary strength. The basis of the theory is to identify different regions of configuration space where particular forces and symmetries dominate and to solve the Schrödinger equation in these regions separately in the appropriate coordinate system. The solution over all space is then obtained by matching the solutions on the boundary surfaces between these regions.

Firstly we consider an atom in an external magnetic field. As the theory is directly applicable to any non-hydrogenic atom or molecule, direct comparison between theory and experiment for lithium at laboratory energy fields is possible and we calculated[5] the experimental continuum spectra of the hydrogen and lithium atoms, at a field of 6.1 T. In an inner spherical region surrounding the origin where the Coulomb potential dominates, the wavefunction is expanded in spherical coordinates. In the outer region, extending to infinity, the wavefunction is represented in cylindrical coordinates. The matching of the solutions between these two regions on a spherical surface gives the reaction matrix or scattering matrix and hence, the photoionisation cross section, through evaluation of the dipole matrix elements.

The bound state and the related complex scaling methods[6], together with all previous photoionisation calculations[7] were performed by representing the wavefunction either purely in cylindrical or in spherical coordinates. The number of eigenfunctions needed in such expansions is vast as one ends up, for example, using a spherical basis to try to represent asymptotic wavefunctions which have essentially cylindrical symmetry.

Secondly we apply the above theory to the hydrogen and rubidium atoms in crossed magnetic and electric fields showing preliminary results for a magnetic field of 6 T and an electric field of 2975 Vcm^{-1}. We analyse the main modifications needed for the crossed field case. In particular the wavefunction in the outer region will be represented in cartesian coordinates as this will be the dominating symmetry with the excited electron moving along the $\underline{E} \times \underline{B}$ drift direction.

Two other theoretical quantum approaches were also successful on handling the problem of an atom in a magnetic field. One is based on the complex coordinate method using a spherical basis set to proceed to a L^2 discretization of the continuum[6]. Due to the finite L^2 basis set one of the limitations of this method is its present inability to reproduce the complex structure of many resonance lines corresponding to Rydberg series converging to

Landau thresholds. Another approach[8] is similar to the theory presented here but uses different numerical procedures.

THEORY

A — Atom in an external magnetic field.

The difficulty associated with the motion of a Rydberg electron in a magnetic field arises from the competition between two potentials of different symmetry and the subsequent nonseparability of the Schrödinger equation. The Hamiltonian for the hydrogen atom in an external magnetic field which is taken to be pointing along the z axis is, in the symmetric gauge, using atomic units,

$$H = -\frac{1}{2}\nabla^2 - \frac{1}{r} + \frac{\gamma}{2} L_z + \frac{1}{8}\gamma^2 r^2 \sin^2\theta \qquad (1)$$

where $\gamma = \omega_c$ and ω_c is the cyclotron frequency. The z component of the orbital angular momentum L_z is a conserved quantity since the potential does not depend on the coordinate φ. Quantum mechanically the linear Zeeman term, $\gamma L_z / 2$, has eigenvalues $m\gamma/2$ and the corresponding eigenstates of (1) can thus be labeled with a fixed value of the magnetic quantum number m. In addition, the potential is symmetric under the transformation $z \to -z$ so that the eigenstates can be further classified according to their z-parity, π_z, which can be even or odd. Eigenstates of the Hamiltonian in (1) can therefore be studied within a fixed subspace defined by definite values of m and parity π_z.

There are two readily identifiable asymptotic limits in which the above Hamiltonian is separable. As $r \to 0$ the equation is separable in spherical coordinates since the Coulomb potential dominates over the external potential. In the other limit, as $r \to \infty$, the equation is again separable but now in cylindrical coordinates

$$V(\vec{r}) = -\frac{1}{\sqrt{\rho^2+z^2}} + \frac{1}{8}\gamma^2\rho^2 \approx -\frac{1}{z} + \frac{1}{8}\gamma^2\rho^2 + \frac{\rho^2}{2z^3} + O\left(\frac{1}{z^5}\right) \qquad (2)$$

since the motion in ρ is bounded due to the magnetic field. At intermediate values of r the Hamiltonian is non separable. This region marks the transition from spherical to cylindrical symmetry.

The overall effect of the diamagnetic potential in (1) on the motion of an electron depends **both** on the strength of the external field and on the radial extent of the electron from the nucleus or equivalently the excitation energy given to the electron. Thus, if we take $\gamma = 10^{-5}$, the initial state in the photoionisation process, namely the ground or a low lying excited state is unaffected by the presence of the field since $<r> \sim 1\text{--}10$ a.u. and so the magnetic potential, $\gamma^2\rho^2/8$, is of the order 10^{-9} which is a negligible perturbation to these states. However the continuum states have an infinite radial extent and thus the electron moves in all of the regions of configuration space described above. Hence for these states one has to pass from solutions expressed in spherical coordinates at short radial distances to solutions in cylindrical coordinates at large distances. Letting $\psi_\varepsilon(\vec{r})$ represent the wavefunction of the excited electron with an energy ε, we divide configuration space into different regions. The first of these is a spherical region, $0 \le r \le a$, where a is chosen large enough to contain the atomic or molecular core but small enough that the quadratic Zeeman potential in (1) is negligible. Once the electron is outside of the multielectronic interacting region of the core, typically for $r \sim 10$ a.u., it moves in the Coulomb field of the ion. So for $r > 10$ a.u. one can write the ℓth partial wavefunction as

$$\psi_{\varepsilon\ell}(\vec{r}) = (s_{\varepsilon\ell}(r) + c_{\varepsilon\ell}(r) \tan \pi\mu_\ell) Y_{\ell m}(\theta, \varphi) \tag{3}$$

where s and c are the regular and irregular Coulomb functions respectively[9] and $Y_{\ell m}$ are the spherical harmonics. The effect of the core on the excited electron's wavefunction is represented by the quantum defects μ_ℓ. They can be calculated *ab initio* or deduced from experimental energy levels. The total wavefunction at r=a can then in general be written as

$$\psi_\varepsilon(\vec{a}) = \sum_\ell A_{\varepsilon\ell} \psi_{\varepsilon\ell}(\vec{a}) \tag{4}$$

where the A's are constants to be determined through matching to the asymptotic solutions.

When the electron moves beyond r=a the magnetic potential in (1) is no longer negligible and the full potential is no longer Coulombic. We therefore consider a second spherical region $a \le r \le b$ where we expand the wavefunction ψ_ε for the excited electron in terms of the eigensolutions ψ_k

$$\psi_\varepsilon(\vec{r}) = \sum_k B_{\varepsilon k} \psi_k(\vec{r}) \tag{5}$$

which are obtained from a spherical expansion

$$\psi_k = \sum_{n\ell} c_{n\ell}^k \frac{f_{n\ell}}{r} Y_{\ell m}(\theta, \varphi) \tag{6}$$

where the $f_{n\ell}(r)$ are a radial basis set and the eigenvectors, $c_{n\ell}$, are determined by diagonalising the Hamiltonian in (1) (plus the Bloch operator) over the finite region $a \le r \le b$. Again the correct superposition coefficients B are determined by matching to the solutions at the boundary surfaces. Beyond r=b the electron is moving in the asymptotic zone and the symmetry is essentially cylindrical.

We therefore choose the third region to be cylindrical with $c \leq z \leq d$ and $0 \leq \rho \leq \infty$ (This region overlaps the spherical region $a \leq r \leq b$ since we will match the cylindrical solutions with the spherical solutions at r=b.). We take the boundary at z=d to be large enough that the potential can be approximated by its separable limit in (2). Then one can write down the following analytic expression for the jth linearly independent solution in cylindrical coordinates in the region $d \leq z \leq \infty$,

$$\psi_{\varepsilon j} = \sum_i \Phi_i (\rho, \varphi) \left(s_i(z) \delta_{ij} + c_i(z) K_{ij} \right) \qquad (7)$$

where Φ_i denotes the ith Landau or harmonic oscillator level, s_i and c_i are the regular and irregular Coulomb functions and K_{ij} is the reactance matrix. In the region $c \leq z \leq d$ we use a basis set expansion or the R-matrix method as in (6) but this time using a set consisting of a product of Landau states Φ_i and a radial basis set in z. Thus knowing the regular and irregular solutions at z=d from (7) we can obtain the corresponding coupled channel solutions over the whole region $c \leq z \leq d$, namely,

$$\psi_{\varepsilon j} = \sum_{ik} \Phi_i (\rho, \varphi) \left(S_{ik}(z) \delta_{kj} + C_{ik}(z) K_{kj} \right) \qquad (8)$$

We now have a set of linearly independent solutions over all space but expressed in different regions and in different coordinate systems. The final step is to match these solutions and to calculate the K-matrix.

Since we know the logarithmic derivative or R-matrix (the inverse of the logarithmic derivative), $R_{\ell\ell'}$, at r=a from (3), we can calculate the R-matrix at r=b from the solutions in (5), i.e. we can propagate the R-matrix from r=a to r=b [10]. The calculated R-matrix at r=b contains many unwanted channels. A key point is to eliminate these channels. We first calculate the local adiabatic solutions by diagonalising the Hamiltonian in (1) at fixed r=b in a basis set of spherical harmonics. The corresponding solutions are $\phi_\lambda(\theta, \varphi)$. We then perform a frame transformation from the basis set of spherical harmonics to the new local set at r=b. The R-matrix $R_{\ell\ell'}$ is thus transformed to $R_{\lambda\lambda'}$ and can be contracted to the number of channels of physical interest, i.e. the number of Landau channels. The ϕ_λ are bounded in ρ and are similar to the Landau states. We then evaluate the regular and irregular solutions and their radial derivatives on the arc r=b by projection onto the ϕ_λ, i.e. we integrate over θ and φ, giving, for example, for the regular solution

$$F_{\lambda j}(b) = \left(\Phi_\lambda(\theta, \varphi) \mid \sum_i \Phi_i(\rho, \varphi) \, S_{ij}(z) \right)_{r=b} \tag{9}$$

This procedure is a two-dimensional frame transformation. Knowing these four functions from outside r=b and the R-matrix $R_{\lambda\lambda'}$ from inside we can determine the K-matrix in (8) and hence the S-matrix. Knowing the S matrix and the wavefunction over all space one can calculate the dipole matrix elements for a given transition and hence the photoionisation cross section.

B — Atom in crossed magnetic and electrical fields.

The additional electric field adds one extra potential with different symmetry which competes with the Coulomb and magnetic potentials. The corresponding Schrödinger equation leads us to a fully 3-dimensional non-separable problem. If we take the magnetic field along the z-axis and the electric field along the negative x-axis, the Hamiltonian for the hydrogen atom is, assuming the symmetric gauge,

$$H = -\frac{1}{2}\nabla^2 - \frac{1}{r} + \frac{\gamma}{2}L_z + \frac{1}{8}\gamma^2 r^2 \sin^2\theta - Fx \tag{10}$$

The external electric field adds an extra term in the cartesian coordinate x. F is the electrical field strength and has typical laboratory values in the range 0 to 3000 V.cm^{-1}. Only the z-parity parity is maintained as a good quantum number since L_z is no longer conserved and so we can label the eigenstates with a fixed value of π_z.

In the limit r→0 the Schrödinger equation maintains its separability in spherical coordinates as the Coulomb field remains dominant in comparison with both magnetic and electric fields. The first spherical symmetric region, $0 \leq r \leq a$, is then treated as before. In the next region, $a \leq r \leq b$, the competition between the fields will be felt as the Coulomb field decreases in strength. This region is treated in a similar way to the magnetic field case alone but with the inclusion of the extra terms in the Hamiltonian. In the limit $r \to \infty$, the Schrödinger equation becomes separable in cartesian coordinates if we neglect the Coulomb interaction. The electron is free to move in the z direction and the y direction in the E×B drift but is bound in the x direction. This corresponds to the third region, $c \leq r \leq d$, where now the dominant symmetry is cartesian.

The evaluation of the K-matrix now requires the development of a three-dimensional matching procedure as the Rydberg electron can escape in two different directions of space, namely the z and y. This is achieved much in the same way as previously by performing the appropriate frame transformation but it becomes essential to propagate the R-matrix from r=a to r=b in order to keep the calculation within an acceptable size. When studying the bound state spectrum this matching procedure is simplified as we can match to exponentially decreasing eigenfunctions represented in spherical coordinates. Non-hydrogenic atoms are included in the theory via the non-zero quantum defects of equation (3).

RESULTS

In Fig. 1 we show a part of the calculated photoionisation cross-section for lithium in a field B=6.1143T, to illustrate our results.[5] We

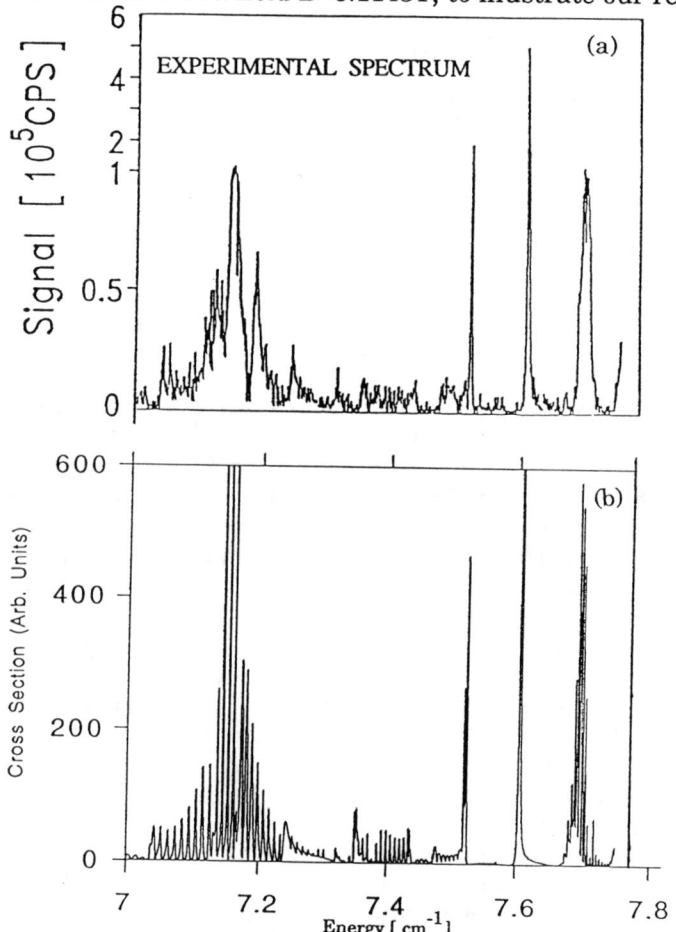

Figure 1 — *The photoionisation cross section from the 3s state of lithium to the m=0, odd parity final states in a field of 6.1143T as a function of energy in cm^{-1} between the ionisation threshold and the first excited Landau level. The energy is measured relative to the field free ionisation threshold. (a) Experimental cross section from Ref. [2]; (b) present theory convoluted with a $0.001 cm^{-1}$ experimental linewidth.*

compare it with a recent experiment[2] in this energy region, which is just above the ionisation threshold but below the first excited Landau threshold.

The radii used in the calculation had to be scaled accordingly so that a=200, the matching radius b=7500, c=7350 and the asymptotic radius d=12,000 a.u.. 125 channels or ℓ values were used in (6) but no large matrix diagonalisations were required since the R-matrix was propagated from a to b.[10] One major advantage of the R-matrix method used here is that we only need to calculate the eigensolutions in (6) once and then the spectrum can be calculated for *any* atom or molecule with little further computation. One just uses the appropriate quantum defects at r=a for the system in question.

The rapidly varying resonance structure seen in Fig. 1(b) was not seen in a calculation of the cross section using the complex scaling method[2,6]. This is due to the limitations of such finite basis set methods which have inherent difficulties near thresholds. The experiment, Fig. 1(a), shows evidence of this resonance structure even if some of these very weakly bound states are not present due perhaps to field ionisation by stray electric fields. The overall agreement however between theory and experiment is excellent. Much of the resonance structure in Fig. 1(b) is simply due to perturbed Coulomb Rydberg states converging to one of the first few Landau thresholds. To see this more clearly we can calculate the Gailitis average of the spectrum converging to the nearest Landau threshold (the first excited level), i.e. we average the cross section over Rydberg resonances converging to this threshold.[9]

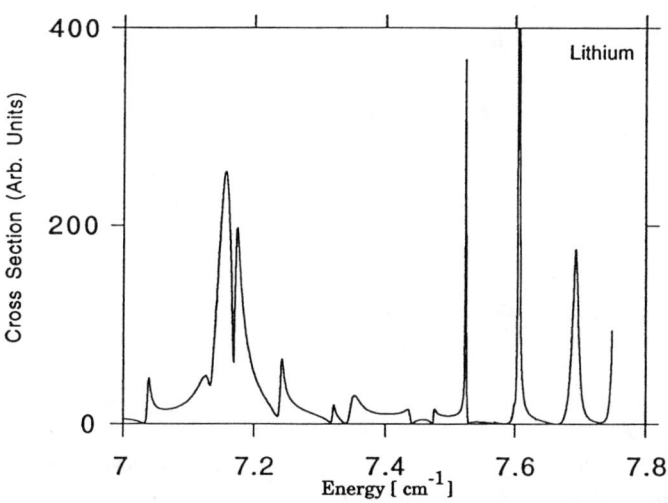

Figure 2 *The spectrum in Fig. 1(b) averaged over resonances converging to the first excited Landau level. (Gailitis average).*

The result is shown in Fig. 2. Most of the resonance structure disappears and the averaged spectrum now agrees very well with the complex scaling calculation.[2] Clearly the spectrum calculated by this complex coordinate method gives an approximate Gailitis averaged cross section. Many other resonances in Fig. 2 can be identified with perturbed Rydberg resonances converging to higher Landau thresholds by using similar averaging techniques and hence can be labelled by approximate quantum numbers.

In figure 3 we show some preliminary results for the bound state oscillator strength of H and Rb in external applied crossed magnetic and electric fields with strength, in a range of energy corresponding to the manifold n=9. The levels obtained for hydrogen, fig3(a), are almost degenerate and agree very well with the second order perturbative calculations of Soloviev[11] which are accurate for

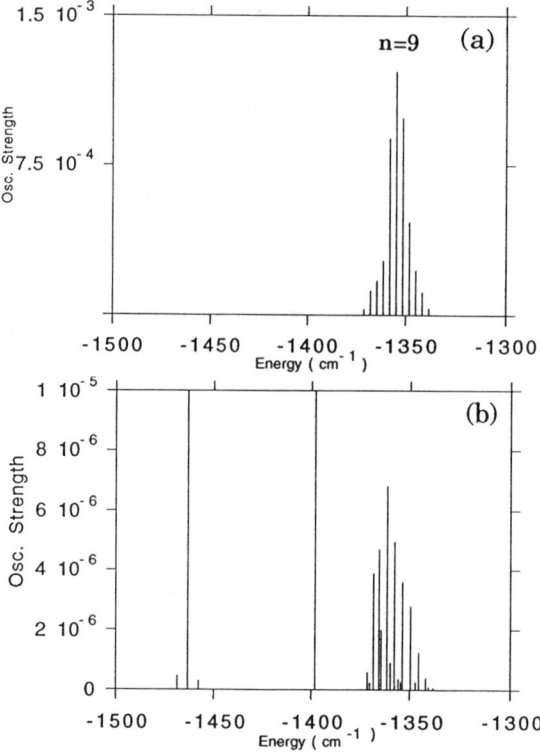

Figure 3 — *The oscillator strength in crossed magnetic and electric fields of strength 6T and 2975V.cm^{-1} respectively, from (a) the m=0, 2p state of hydrogen to the even z-parity final states (b) the m=0, 5p state of Rubidium to the even z-parity final states, as a function of energy in cm^{-1}.*

these energies. Fig. 3(b) shows the interesting effect observed in the spectrum for a non-hydrogenic atom, rubidium, where the near degeneracy in the spectrum is clearly broken due to the core. This shows how drastically the spectrum of non-hydrogenic atoms can be modified for crossed magnetic and electric fields.

CONCLUSIONS

The above theory shows that it is now possible to calculate *ab-initio* the Rydberg and continuum energy spectrum of an atom or molecule in a magnetic field or crossed magnetic and electric fields of arbitrary strength. This is achieved by identifying asymptotic regions of separability, construction of local solutions in the appropriate coordinate system and matching of solutions across common boundaries using two or three dimensional frame transformations. In addition, one can use the power of multi-channel quantum defect theory to analyse and classify the resonance structure in the spectrum. The approach is very general and can be applied at any excitation energy. The case of atoms in parallel electric and magnetic fields is just an extension of the magnetic field alone and is being studied at the moment as well as the crossed fields problem.

ACKNOWLEDGEMENTS

This research work has been supported in part by the European Community under contract number SC1*-CT91-0742 (TSTS) by NATO under grant no. CRG 890543 and by J.N.I.C.T. (Junta Nacional de Investigação Científica e Tecnológica). Computer time was provided through a grant by S.E.R.C., U.K.

REFERENCES

1. 'Irregular Atomic Systems and Quantum Chaos', Comments At.Mol.Phys. 25, Nos. 1-6, ed. J.C. Gay (1991) and references therein.
2. C. Iu, G.R. Welch, M.M. Kash, D. Kleppner, D. Delande and J.C.Gay, Phys. Rev. Letts. 66, 145 (1991) and references therein.
3. G. Wiebush, J. Main, K. Krüger, H. Rottke, A. Holle and K. Welge, Phys. Rev. Letts. 62, 2821 (1989)
4. G.Raithel, M. Fauth and H. Walther, Phys. Rev.A 44, 1898 (1991).
5. P.F. O'Mahony and F. Mota-Furtado, Comments At.Mol.Phys. 25, Nos. 4-6, 309 (1991); P.F. O'Mahony and F. Mota-Furtado, Phys. Rev. Letts. 67, 2283 (1991).
6. D. Delande, A. Bommier and J.C. Gay Phys. Rev. Letts. 66, 141 (1991); M. Halley, K. Taylor and D. Delande, J. Phys. B 25, L525 (1992)
7. A. Alijah, J. Hinze and J.T. Broad, J. Phys. B 23, 45 (1990) and references therein.
8. S. Watanabe and H. Komine, Phys. Rev. Letts. 67, 3327 (1991).
9. M.J. Seaton, Rep. Prog. Phys. 46, 167 (1983).
10. K.L. Baluja, P.G. Burke and L.A. Morgan, Comput. Phys. Commun. 27, 299 (1982); J.C. Light and R.B. Walker, J.Chem. Phys 65, 4272 (1976).
11. E.A.Soloviev, Sov. Phys. JETP 59, 38 (1984).

X-UV LASERS NEAR 200 Å

P. Zeitoun, P. Jaeglé, A. Carillon, P. Dhez, P. Goedtkindt, G. Jamelot, A. Klisnick, B. Rus

Laboratoire de Spectroscopie Atomique et Ionique, Université Paris-Sud, Bât. 350, 91405 Orsay, France

ABSTRACT

Recent experiments, performed with pumping laser of relatively moderate energy (E ≤ 1KJ per pulse), have considerably increased the gain factor of X-UV lasers in the wavelength region of 200 Å. In this paper we recall the pumping mechanisms operating in laser-produced plasmas and we present the large gain experiments performed with neonlike germanium, in using the laser VULCAN (E ≈ 1KJ), as with neonlike zinc and lithiumlike sulphur, in using the laser of LULI (E ≈ 0.5KJ).

INTRODUCTION

Since hot plasmas produced by lasers have proved to amplify X-UV radiation, several tens of lasing lines have been discovered between 40 Å and 250 Å [1]. Recent experiments have considerably increased the gain factor of X-UV lasers - the product of the gain coefficient, G, and the laser length, L. This quantity well characterizes the level of amplification at which experiments arrive and it is closely related to the brightness, the directivity and the coherence of the X-UV laser. Another important index of efficiency is the ratio GL/J where J is the energy contained in the pulse of the pumping laser. The enhancement of this ratio is necessary in order to decrease the economical cost of the laser. The experiments, which are presented hereafter, illustrate the recent important improvement of these two indexes for lasers emitting in the 200 Å-wavelength region.

To start with, let us recall the principle for the creation of population inversion in hot plasmas produced by lasers. For ions of neonlike and nickellike iso-electronic series, generally there exists a range of plasma temperatures and densities in which free-electron collisions with ions in their ground state levels give rise to a large population of specific excited levels while the lower levels remain weakly populated. In the case of neonlike ions this produces 3s-3p population inversions. On the other hand, after the end of the pumping laser pulse, for ions such as lithiumlike and hydrogenic ions, the electron density can remain large enough in the expanding plasma to allow three-body recombination to populate the highest lying levels via electron capture on helium-like ions or bare nuclei. In the case of lithiumlike ions, this process, together with the resulting radiative and collisional cascades to intermediate levels, leads to 3-4, 3-5 and 4-5 population inversions. In order to exploit these properties in an X-UV laser, the plasma must be produced under the form of a column, along the axis of which the X-UV amplification can take place.

SATURATED EMISSION AT 236 Å IN GERMANIUM LASER

The first successful experiments on collisional pumping of neonlike ions, namely neonlike selenium [2], required a huge pumping laser which yielded several KJ per pulse. Later on, germanium slab targets proved to produce lasing lines at 232 Å and 236 Å with pumping laser pulses of energy less than 1 KJ [3]. An important gain optimization for these lasing lines was then achieved at the Rutherford Laboratory with the 1-KJ -laser Vulcan, using a doubled target system which provides a partial compensation of refractive deviation [4].

However emission saturation was still not achieved. Saturation occurs when the laser intensity becomes so large that the rate of new photon creation prevails over the rate of emitter creation, i.e. the rate of population inversion recovery. Then the density of population inversion decreases, which leads to the gain coefficient being lowered. Thus the X-ray laser intensity stops increasing exponentially as a function of the plasma length . In the common experiment of the Rutherford Appleton Laboratory (G.B.) and the Laboratoire de Spectroscopie Atomique et Ionique (France) , this crucial step in X-UV laser development has been passed in providing the optimized target system of R.A.L. with a multilayer mirror acting as a 'half-cavity' [5,6]. The principle of this experimental device is shown in Figure1. The total length of the targets was 36 mm. With this arrangement, the double-pass of the X-UV beam through the plasma brought the gain factor to GL = 22 instead of 14 without mirror. On the other hand, varying the plasma length showed saturation to start at GL= 17 - 18.

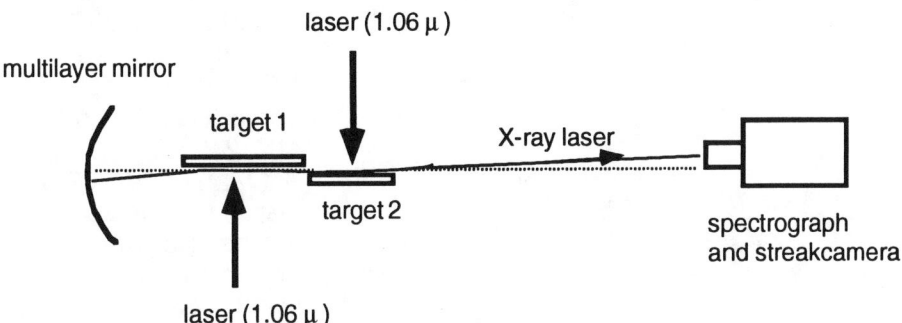

Fig. 1: Sketch of the germanium laser. The laser-irradiated target surfaces are opposite each other [4]. The concave mirror, of 130 mm curvature radius, stands at 22 mm from the edge of the first target. The mirror reflexion coefficient measured with synchrotron radiation was of 28% at 236 Å.

The brightness of this laser is such that a shot supplies $2-8 \times 10^{12}$ X-UV photons in about 300 ps and 10^{-5} steradian. This is larger by several order of magnitude than emission from any other laboratory source, including undulators. An experimental technique developed at L.L.N.L. (U.S.A.) [7] enabled us to make an investigation of the transverse coherence of the X-UV laser. The transverse coherence length has been found to be about 15 µm at the edge of the plasma (on the detection side) and of 390 µm at 150 cm from the edge. We also observed that the coherence length was larger by a factor three when saturation was achieved. Thus this experiment has been the first complete demonstration of an X-UV laser.

LARGE GAIN OBSERVED FOR THE $J_{0 \rightarrow 1}$ LINE OF COLLISIONALLY-PUMPED Zn^{20+}.

The LULI (Laboratoire pour l'Utilisation des Lasers Intenses, Palaiseau, F.) has recently developed a new X-ray laser facility [8]. A permanent experimental set-up can now make use of the six beams of the laser, allowing to irradiate targets with near 0.5 KJ light pulses of 600 ps duration at 1.06 µm wavelength. All the six line focal spots (~100 µm wide and ~2.6 cm long) are superimposed on each other with precision being better than 30 µm. The resulting spot is thus about ~150 µm. The maximal attainable intensity on the target is accordingly $\sim 1.4 \times 10^{13}$ Wcm^{-2}. Considering this driving power density, an element with Z=30 is the first one in the Z-scaling "chart" where concomitant plasma characteristics, analogous to those for Se (7×10^{13} Wcm^{-2} - e.g.[9]) or Ge (2.4×10^{13} Wcm^{-2} - e.g.[10]), can be achieved. In conjunction with the relatively few studies of lower-Z elements performed yet, this made zinc an interesting subject of our experimental focus.

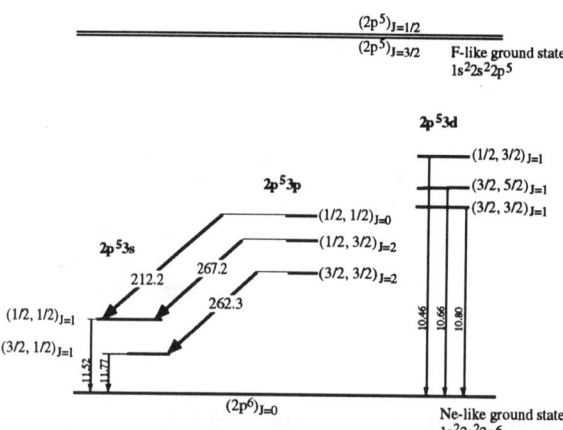

Fig.2: Simplified Grotrian diagram of Ne-like zinc; 3s-3p transitions give rise to lasing lines at 212 Å, 267 Å and 262 Å.

The simplified Grotrian diagram for neonlike Zn is in Figure 2. The levels are described in $(j,j)_J$ notation. Three 3s-3p lasing lines at 212.2 Å ($J_{0 \rightarrow 1}$), 267.2 Å ($J_{2 \rightarrow 1}$) and 262,3 Å ($J_{2 \rightarrow 1}$) [11] are specified in the diagram. It is important to have in mind that, whereas numerical simulation of lasing in neonlike ions predicts a larger gain for the $J_{0 \rightarrow 1}$ than for the

$J_{2\to 1}$ lines, the $J_{0\to 1}$ gain observed till now has been systematically lower than predicted and generally lower than the $J_{2\to 1}$ gain. In the field of collisional-excitation Ne-like laser, this an issue which still needs to be elucidated as the experimental results are not entirely consistent with the modelling.

The targets used in our experiment were 1.25 mm thick Zn slabs 2 cm wide. They were positioned at the centre of the 2.6 cm line-focus length which thus extended beyond the target ends providing a good illumination. A multipinhole camera with a CCD readout was used to monitor the plasma column via bremsstrahlung emission in the kev spectral region [12]. This system provides information about the geometrical dimensions of the plasma as well as about its uniformity. The diagnostic used to analyse the soft X-ray plasma emission and measure the gain is the 'Focal' grazing incidence spectrometer which incorporates Wadsworth geometry [13]. To use it in collisional-excitation laser experiments, where refraction causes X-ray emission deflection outwards of the target surface and the deflection angle undergoes time evolution, an additional collecting X-ray optics has been implemented to enhance its angular coverage necessary to intercept X-ray laser emission from plasmas of different lengths. This collecting optics consists of two elliptical grazing-incidence mirrors whose two focal points are matched while the second two ones correspond respectively to the target centre and spectrometer entrance slit. The actual system used in this work was capable of collecting X-ray emission from an angular range of 2-11.3 mrad [14]. A streak-camera coupled to the spectrometer provided time resolution of the X-UV laser signal. The gain of all three lasing transitions has been measured by systematically varying the plasma length with the help of beam aperture.

Figures 3 and 4 summarize the results of this experiment. Figure 3 presents the measured intensity as a function of target length from 1 cm to 2 cm for the three lines at 212, 262 and 267 Å. The intensity values are taken at the time of the peak maximum. Curves have been fitted to the experimental data according to a model function which is a modified Lindford formula evaluating the spectrally integrated intensity, I, of non-saturated emission [15]:

$$I(l) = \sqrt{\pi} \, \alpha \, \Delta v \, \frac{j}{g} \, \frac{(e^{gl} - 1)^{3/2}}{(gl \, e^{gl})^{1/2}} \qquad (1)$$

where l is the target length, j the spectral peak emissivity, g the spectral peak gain coefficient, Δv the intrinsic frequency line width and α a numerical factor ranging from 0.5 (Lorentzian profile) to 0.6 (Gaussian profile).

For the $J_{0\to 1}$ line at 212 Å the best fit of the Linford function to the experimental points results in the large gain coefficient 4.9($\pm$0.2) cm^{-1}. At the same time the best fit corresponds to a gain of 2.3($\pm$0.6) cm^{-1} for the $J_{2\to 1}$ line at 262 Å and to 2.6($\pm$0.4) cm^{-1} for 267 Å. These results exhibit one

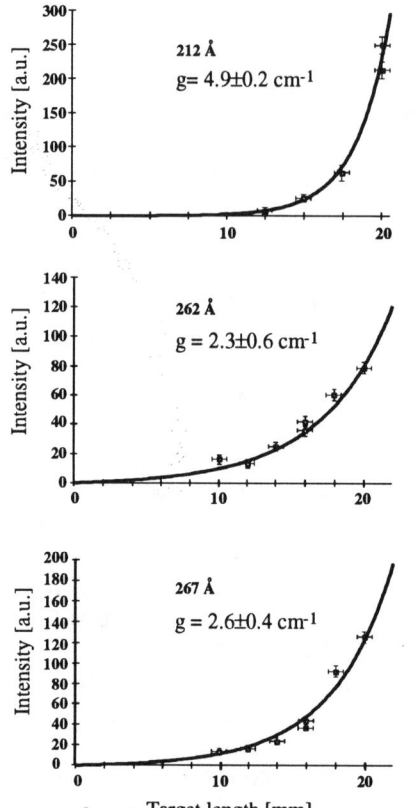

Fig. 3: Recorded intensity of the 212.2 Å ($J_{0\to1}$), 267.2 Å ($J_{2\to1}$) and 262,3 Å ($J_{2\to1}$) lasing lines vs plasma length. The gain coefficients are respectively 4.9 cm^{-1}, 2.3 cm^{-1} and 2.6 cm^{-1}.

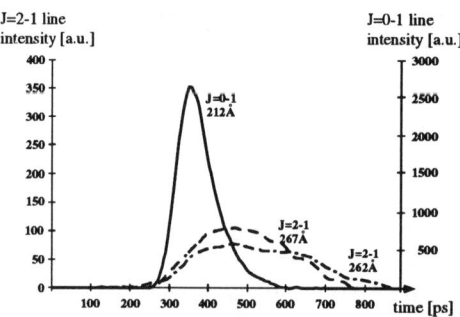

Fig.4: Temporal evolution of all three $J_{0\to1}$ and $J_{2\to1}$ lines. The left intensity scale applies to the $J_{2\to1}$ line, the right one to the $J_{0\to1}$.

apparent central point: while the gains and other characteristics of the $J_{2\to1}$ lasing lines are quite consistent with those so far observed in other elements, the $J_{0\to1}$ line clearly diverges from the conventional picture. Although its gain was initially supposed to be close or perhaps slightly higher than for $J_{2\to1}$ transitions, the measured value proved to be surprisingly large.

The temporal evolution of the three lasing line intensities, displayed in figure 4, also exhibits an unexpected behaviour of the $J_{0\to1}$ line at 212 Å. Its brightness is about 25 times greater than the brightness of the $J_{2\to1}$ lines while its duration (≈ 100 ps) is considerably shorter than the duration of the $J_{2\to1}$ lines at 262 Å and 267 Å (≈ 360 ps and ≈ 300 ps respectively). Moreover one sees that the 212 Å line goes through its maximum 100 ps before the two other lines.

Without undertaking a detailed discussion of these features, it is noteworthy that the study of the time-dependent emissivity ,$j(t)$, and the gain coefficient, $g(t)$, appearing in expression (1) (now expressed as a function of time), gives valuable information on the evolution of the J=2, J=0 and J=1 populations. As the 212 Å and the 267 Å lines have the same lower J=1 level (see Figure 2) one expects the same behaviour of this J=1 population to be derived from the time dependence of each of the two lines. In fact this is not what is observed. The J=1 population varies much faster for the 212 Å line than for the 267Å one. Thus we come to the conclusion that these evolutions are strongly incompatible even though the lines are emitted from the same plasma element. The necessary conclusion is that the $J_{0 \to 1}$ and $J_{2 \to 1}$ lasing lines don't originate from the same plasma region. Further, as regards the unexpectedly large gain coefficient observed for the $J_{0 \to 1}$ line, interesting remarks have recently been made about the disappearance of one of the line broadening factors in atoms of even atomic number [16], as Zn is. Since the nuclear spins are equal to zero, there is no hyperfine splitting of the levels of these atoms. Now it is true that the narrower the lasing line is, the larger the gain coefficient must be.

GAIN FACTOR INCREASING IN RECOMBINATION PUMPING FOR THE 4-5 TRANSITIONS OF LITHIUMLIKE IONS.

For the recombination pumping, it has the well-known advantage to require a moderate pumping energy but, until now, it has had the disadvantage to offer gain factor generally no larger than 3 or 4. We have shown that one reason of the gain limitation is hot points in plasma causing reabsorption of lasing emission [17,18]. In the case of lithiumlike ions, the transitions for which a gain has been measured are the 4 (or5) f- to 3d-transitions [19,20]. Now two peculiarities of the lithiumlike electronic structure are the very large transition matrix-element between the 3d- and the 2p-levels and the fact that the 2p-level is a quasi-ground level in the sense that, lying close to the 2s-ground level, the 2p-level is approximately populated as much as the ground level itself. During plasma recombination, the large 3d-2p decay probability does foster population inversions between the 3d- and some upper levels but the large 2p-population limits the efficiency of the mechanism and may even quench it completely. The quenching depends strongly on the plasma parameters, N_e and T_e. In non-uniform plasmas, spots displaying different temperature history are distributed along the plasma column axis. At a given instant, some of them contribute to X-ray amplification while their neighbours may behave as purely absorbing buffers.

Therefore we have been led to seek after transitions which would be less subject to the effect of radiation trapping or similar processes than the 5-3 and 4-3 transitions are. Since the levels of main quantum number 4 are widely separated from the ground level, one can assume that radiation trapping and collisional excitation of the ground level must have less effect on 5-4 than on 5(or 4)-3 population inversions [21]. Computer simulations performed in the case of lithiumlike sulphur allow a more quantitative picture of the differences between 5-4 and 4-3 transitions. The calculated wavelengths of 3d-4f and 4f-5g transitions are 95.6 Å and 206,5 Å respectively. Figure 5 shows the spatial distribution of the gain coefficient calculated in the direction perpendicular to the surface of a massive target.

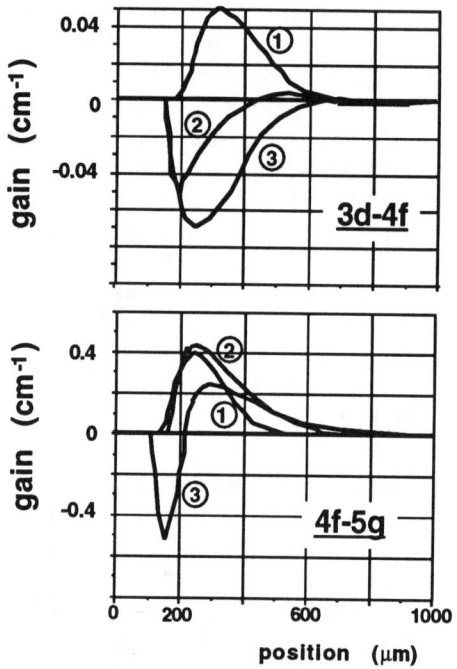

Fig. 5 - Numerical simulation of the gain radial distribution for the 3d-4f and 4f-5g lines of lithiumlike sulphur and for three driving laser intensities, 2×10^{12} W/cm^2 (curves 1), 3.7×10^{12} W/cm^2 (curves 2), 5×10^{12} W/cm^2 (curves 3); for the 3d-4f transition, one sees that 'hot plasma spots' reduce the gain.

The curves displayed in this figure can be considered as giving the gain radial distributions over plasma column cross-sections. The gain is calculated for the two 3d-4f and 4f-5g lines and for three pump intensities. The lowest pump intensity is the one which gives the gain maximum for the corresponding lasing transition. Thus the curves obtained for higher intensities in a way simulate what occurs with "hot spots".

For the 3d-4f line one can see that hot plasma spots will contribute negatively to the gain in almost all the space. This comes from the large population transfer from the 2p- to the 3d-level in conditions prevailing in hot spots. But for the 4f-5g line one sees that reabsorption in hot spots takes place only in a small part of the space while, at larger distance from the target, all the plasma contributes positively to the gain. This illustrates that radiation trapping and collisional excitation of the ground level must have less effect on 5-4 than on 4-3 population inversions.

For the experimental study the target was a 7500 Å sulphur layer coating a massive aluminium support. We used an irradiance of 7.5 10^{12} W/cm^2. A 300 µm-wide vertical slit limited the plasma region viewed by the detection system. The result presented hereafter in figure 6 proceeds from plasma lying approximately between 300 µm and 600 µm from the target. Densitometries of plasma pinhole images show that the plasma displayed non-uniformities: in addition to a slow intensity decrease near the ends it shows small hot spots with typical scale length of 1 mm.

At the predicted wavelength of 206,5 Å, the 2 cm-long plasma exhibited a strong line which disappeared almost completely when the plasma length was reduced to 1 cm. We measured the time-resolved intensity of this line for lengths of 0.5, 1, 1.5 and 2 cm. The values corresponding to the instant of strongest length-intensity dependence are plotted in fig. 6. The best fit of the function (1) to the experimental points

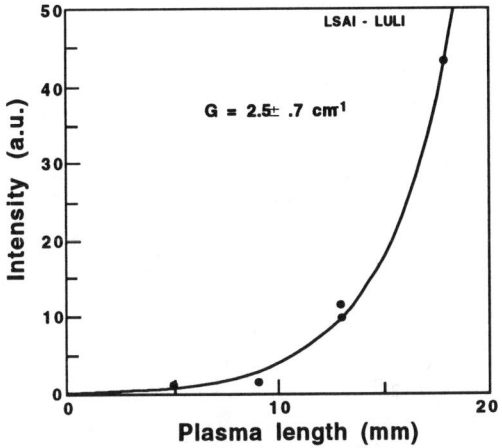

Fig. 6 - Recorded intensity of the 4f-5g line of Li-like sulphur vs. plasma length; the gain coefficient is 2.5 cm^{-1}.

results in the gain coefficient of 2.5($\pm$0.7) cm^{-1}. Thus the gain factor is of the order of 5.

We must first make a note of the fact that 5 is the largest gain factor value we ever obtained for a lithiumlike ion laser. The previous largest value had been 4 for the 5-3 line at 105.6 Å in an experiment making use of a multilayer mirror double-pass system [22]. In single pass experiments most of the values reported in literature for the 5(or4)-3 lines are lying between 1 and 3. On the other hand, for the first time with recombination pumping, we did not observe any reduction of the gain coefficient when the plasma length was raised from 1 to 2 cm. As a last remark, preliminary results obtained about the 5g-4f line of Li-like aluminium near 334 Å give rise to similar observations, the gain coefficient being of 1.5 cm^{-1} with a plasma length of 3 cm. We can thus conclude that using the 5-4 lines may bring about a significant improvement to the prospect of the recombination pumping scheme.

CONCLUSION

The results presented in this paper show strong X-UV lasing in the wavelength range of 200 Å, the pumping being obtained by collisional excitation, on the one hand, and by plasma recombination, on the other. The J=0-1 transition of neonlike zinc, at 212 Å, proved an originally unexpected gain of 4.9 cm^{-1}. This result gives valuable information about the discrepancies generally observed between expected and measured gain coefficients for the various lasing lines of neonlike ions.

The fact that a gain factor of 10 has been obtained with a driving laser of 0.5 KJ, at 212 Å, in a single-pass experiment, is an important step in the reduction of size and cost of equipment required for building X-ray lasers. A previous step had been the achievement of gain-length of 22 in a-double pass device with the 1 KJ-Vulcan laser as driver. It is now expected that double-passing will carry the gain-length of the 212 Å line of zinc at 16-17 with the half-kilojoule laser. Other progress could come from the 5-4 lines of lithiumlike ions. Single-passing in sulphur plasma column of 4 cm length should also provide gain-length about 10 at 206 Å, making feasible a large gain recombination X-ray laser.

This work has been supported by the DRET under contract n° 91/109.

REFERENCES

1 - X-Ray Lasers 1992, Proceedings of the third International Colloquium on X-Ray Lasers,. edited by E Fill, Inst. Phys. Conf. Ser. N° 125, Section 1 (1992)

2 - D.L.Matthews, P.L.Hagelstein, M.D.Rosen, M.J.Eckart, N.M.Ceglio, A.U.Hazi, H.Medecki, B.J.MacGowan, J.E.Trebes, B.L.Whitten, E.M.Campbell, C.W.Hatcher, A.M.Hawryluk, R.L.Kauffman, L.D.Pleasance, G.Rambach, J.H.Scofield, G.Stone and T.A.Weaver: Phys.Rev.Lett., **54**, No.2 (1985), p.110

3 - T. N. Lee ,E .A . McLean , R. C.Elton , Phys Rev Lett. **59** (1987) p. 1185

4 - D. Neely , C. L. S. Lewis , D. M. O'Neill, J. Uhomoibhi , M. H.Key , S. J.Rose , G. J. Tallents , S. A. Ramsden , Opt. Commun. **87** (1992)p.231

5 - A. Carillon, H.Z. Chen, P. Dhez, L. Dwidedi, J. Jacoby, P Jaeglé , G Jamelot, J. Zhang, M.H. Key, A. Kidd, A. Klisnick, R. Kodama, J. Krishnan, C.L.S. Lewis, D. Neely, P. Norreys, D.M. O'Neill, G.J. Pert, S.A. Ramsden, J.P. Raucourt, G.J. Tallents, J.O. Uhomoibhi, Phys. Rev. Letters 68 (1992) 2917-20

6 - P. Jaeglé , A Carillon, H Z Chen, P Dhez, L. Dwivedi, J Jacoby, G Jamelot, Jie Zhang, M H Key, A Kidd, A Klisnick, R Kodama, J Krishnan, C L S Lewis, D Neely, P Norreys, D O Neill, G J Pert, S A Ramsden, J P Raucourt, G J Tallents, J Uhomoibhi' in vol. ref.n°1, p 1

7 - C L S Lewis, D Neely, D O Neill,J Uhomoibhi' G. Cairns, A. MacPhee, G J Tallents, J. Krishnan, L. Dwidedi, M H Key, P. Norreys, R. Kodama, R.E. Burge, G. Slark, M. Brown, G.J. Pert, P. Holden, M. Lightbody, S A Ramsden, J. Zhang, C.S.Smith, P Jaeglé , G Jamelot, A. Carillon, A. Klisnick, J.P. Raucourt, J. M.Trebes, M.R. Carter, S. Mrowka, K.A. Nugent, in vol. ref.n°1, p 23

8 - B Rus, P Zeitoun, A Carillon, B Gauthé, P Jaeglé, G Jamelot, A Klisnick, JC Lagron, C Vinsot, X-Ray Lasers 1992, edited by E Fill, Inst. Phys. Conf. Ser. N° 125: Section 7 (1992) 361-6

9 - C.J.Keane, N.M.Ceglio, B.J.MacGowan, D.L.Matthews, D.G.Nilson, J.E.Trebes and D.A.Whelan: J.Phys.B: At.Mol.Opt.Phys.**22** (1989),.p.3343

10 - D.Neely, C.L.S.Lewis, D.M.O'Neill, J.H.Uhomoibhi, M.H.Key, S.J.Rose, G.J.Tallents, S.A.Ramsden: Optics Communications **87** (1992), p.231

11 - E.A.McLean, J.A.Stamper, C.A.Brown, C.K.Manka, H.R.Griem, J.C.Rife and B.H.Ripin: X-Ray Lasers 1990, 2nd Int.Colloquium on X-ray Lasers, p.339, Int.Phys.Conf.Ser.No 116, ed by G.Tallents. Also R.C. Elton: X-Ray Lasers, Academic Press, 1990, pp.115-117.

12 - B Rus, P Zeitoun, A Carillon, B Gauthé, P Jaeglé, G Jamelot, A Klisnick, JC Lagron, C Vinsot, in vol. ref.n°1, p 361

13 - A.Carillon, J.C.Lagron, P.Jaeglé, G.Jamelot: LULI Annual Rapport 1988, p.233

14 - A.Carillon: LULI Annual Rapport 1992
15 - G.J.Linford, E.R.Peressini, W.R.Sooy and M.L.Spaeth, Appl.Optics, **13**, No.2 (1974), p.379
16 - J. Nielsen, J.A. Koch, J.H. Scofield, B.J. MacGowan, J.C. Moreno, L.B. Da Silva, Phys. Rev. Letters, **70** (1993) 3713
17 - P. Jaeglé, P. Dhez, A. Klisnick, A. Carillon, G. Jamelot, B. Gauthé, J.P. Raucourt, P. Goedtkindt and J.C. Kieffer, Proc. of Short-Wavelength Coherent Radiation OSA, Vol. 11, P.H. Bucksbaum and N.M. Ceglio eds,.pp. 123 (1991)
18 - G. Jamelot, A Carillon, P Dhez, B Gauthé, P Jaeglé, A Klisnick, JP Raucourt, X-Ray Lasers 1992, edited by E Fill, Inst. Phys. Conf. Ser. N° 125: Section 2 (1992) 89-95
19 - A. Carillon, J. Edwards, M. Grande, M. Henshaw, P. Jaeglé, G. Jamelot, M.H. Key, G.P. Kiehn, A. Klisnick, C.L.S. Lewis, D. O'Neill, G.J. Pert, S.A. Ramsden, C. Regan, S.J. Rose, R. Smith, O. Willi, J. Phys. B 23 , p. 147-163 (1990)
20 - Moreno JC , Griem HR , Goldsmith S , Knauer J , Phys. Rev. A , 39 (1989) 6033
21 - H. Guennou, A. Sureau, C. Möller, International Colloquium on X-ray Lasers, ed.P.Jaeglé and A.Sureau, in Journal de Physique **47** (1986) C6-351.
22 - P. Jaeglé, A. Carillon, P. Dhez, B. Gauthé, F. Gadi, G. Jamelot and A. Klisnick, Europhysics Letters, 7 , 337, (1988)

ELECTRON COLLISIONS

FRAGMENTATION OF FEW-BODY COULOMB SYSTEMS

J.S.Briggs

Fakultät für Physik, Hermann-Herder-Str. 3, 79104 Freiburg i. Br.

ABSTRACT

The problem of the fragmentation of few-body Coulomb systems remains a difficult outstanding problem of atomic and molecular physics. Recent progress in the resolution of this problem, both theoretical and experimental, is described. Particular attention is given to the fundamental three-body Coulomb problems of the photo-ionisation of helium and the electron impact ionisation of atomic hydrogen.

1 INTRODUCTION

In recent years, significant progress has been made in our understanding of the structure of multiply-differential cross-sections (MDCS) for the fragmentation of few-body Coulomb systems. As is usual in a rapidly developing subject, experiment and theory have advanced 'hand-in-hand', the one providing the impetus for the other. Here attention will be focussed on particular atomic fragmentation processes leading to two free electrons and a doubly or singly charged residual ion in the final state. The processes of main interest are a) the double photoionisation of a neutral atom and b) the electron-impact single ionisation of a neutral atom. These two processes have been much studied experimentally and recently even detailed triply-differential cross-sections (TDCS) have been measured in some cases. The main aim of this review is to describe the theoretical advances made in the description of these processes. The pervading theme will be that correlation between all three charged particles is always significant in deciding the final relative momentum distribution of the three-body continuum. Whilst one might expect this feature energetically close to the three-body break-up threshold it will be shown that surprisingly large electron-electron correlation effects persist even in high-energy (with respect to the ionisation energy) collisions.

Essentially the problem separates into two parts. The first is the preparation of the collision complex leading to two electrons emanating from the positive ion core. For double photoionisation this is relatively simple; the initial state is multiplied by the sum of the electron-dipole operators corresponding to the sudden (first-order perturbation theory) absorption of a photon. For electron-impact ionisation it is more complicated since the incident electron interacts both with the target electron and the positive ion core and the preparation of the collision complex cannot be viewed as a perturbative process. The second part of the problem is the description of the correlated three-body motion describing two electrons breaking off from the collision complex. It is here that the long-range correlation due to Coulomb forces come into play.

The problem of two-electron ionisation at threshold has mostly been treated separately from the higher-energy region, although quite what constitutes 'the threshold region' has never been defined clearly. Such threshold treatments go under the general heading of 'Wannier theory', following Wannier's 1953 analysis[1] of the problem, using classical mechanics. Subsequent semi-classical treatments of the problem[2,3,4] provide the energy dependence of the threshold cross-section and make predictions as to the form of the angular distribution[5,6]. Such treatments are generally quite involved and require several approximations whose validity is difficult to assess. Here it will be shown that a much simpler approach to the electron correlation problem provides a good description of most features of the relative momentum distribution of the two continuum electrons, even down to total energies less than 1eV above threshold.

In section 2 the general parametrisation of the TDCS is discussed and the importance of bipolar harmonics in the expansion of the continuum 3-body wavefunction is emphasised. In section 3 some previous analyses of the 3-body wavefunction in the Wannier configuration are reviewed. The process of double photoionisation is discussed in detail in section 4. The more difficult e-2e ionisation process is the subject of section 5 and section 6 contains some remarks on the relationship of the three-body continuum states to the doubly-excited resonant states existing just below the three-body break-up threshold. Atomic units will be used throughout.

2 THE MULTIPLY-DIFFERENTIAL CROSS-SECTION

Klar and Fehr[7] have given a general parametrisation of MDCS for both double photoionisation and electron impact ionisation. The simpler case of photoionisation is considered first. The idea is that if the initial state of the neutral atom is randomly orientated and if the orientation of the residual atom is not detected, then the MDCS is a scalar quantity. Then it depends only on certain scalar invariants formed from the unit vectors $\hat{\mathbf{k}}_a, \hat{\mathbf{k}}_b$ (defining the emission directions of electrons 'a' and 'b') and the unit polarisation vector $\hat{\varepsilon}$ of the photon. In terms of Racah tensors $\underline{C}_l$, the invariants are

$$I_{l_a l_b L} = [\underline{C}_{l_a}(\hat{\mathbf{k}}_a) \wedge \underline{C}_{l_b}(\hat{\mathbf{k}}_b)]_L \cdot \underline{C}_L(\hat{\varepsilon}) \qquad (1)$$

$$= \sum_M \mathcal{Y}_{L,M}^{l_a l_b}(\hat{\mathbf{k}}_a, \hat{\mathbf{k}}_b) C_{LM}(\hat{\varepsilon}) \qquad (2)$$

where $\mathcal{Y}_{LM}^{l_a l_b}$ is a bipolar spherical harmonic

$$\mathcal{Y}_{L,M}^{l_a l_b}(\hat{\mathbf{k}}_a, \hat{\mathbf{k}}_b) = \sum_{m_a m_b} <l_a m_a l_b m_b \mid LM> C_{l_a m_a}(\hat{\mathbf{k}}_a) C_{l_b m_b}(\hat{\mathbf{k}}_b) \qquad (3)$$

and the C_{lm} are proportional to spherical harmonics. The double photoionisation cross-section for fixed energies of the two electrons is the TDCS

$$\frac{d\sigma}{d\hat{\mathbf{k}}_a d\hat{\mathbf{k}}_b dE_a} = 4\pi^2 \alpha \frac{k_a k_b}{\omega} \sum_{M_f} \frac{1}{2J_i + 1} \sum_{M_i} |<\psi_f^-(\hat{\mathbf{k}}_a, \hat{\mathbf{k}}_b) \mid \hat{\varepsilon} \cdot \sum_i \nabla_i \mid \phi_i >|^2 \qquad (4)$$

where α is the fine-structure constant, ω the photon frequency, J_i the angular momentum of the initial state and $\psi_f^-(\hat{\mathbf{k}}_a, \hat{\mathbf{k}}_b)$ the exact scattering state of a pair of electrons moving in the doubly-charged ion core with quantum numbers J_f, M_f. In the particularly simple case of double ionisation of helium the TDCS reduces to

$$\frac{d\sigma}{d\hat{\mathbf{k}}_a d\hat{\mathbf{k}}_b dE_a} = 4\pi^2 \alpha \frac{k_a k_b}{\omega} |<\psi_f^-(\hat{\mathbf{k}}_a, \hat{\mathbf{k}}_b) | \hat{\varepsilon} \cdot (\nabla_a + \nabla_b) | \phi_i >|^2 \tag{5}$$

where ψ_f^- is an exact scattering state describing the motion of two free electrons in the field of the bare nucleus of charge two. Since the cross-section (4) is bi-linear in $\hat{\varepsilon}$, only the L values 0 and 2 are allowed, so that in terms of the scalar invariants (2), with $L = 0$ or 2, the TDCS may be written

$$\frac{d\sigma}{d\hat{\mathbf{k}}_a d\hat{\mathbf{k}}_b dE_a} = \sum_l A_l P_l(\hat{\mathbf{k}}_a \cdot \hat{\mathbf{k}}_b) + \sum_{l_a l_b} B_{l_a l_b} \sum_M \mathcal{Y}_{2M}^{l_a l_b}(\hat{\mathbf{k}}_a, \hat{\mathbf{k}}_b) C_{2M}(\hat{\varepsilon}) \tag{6}$$

If one restricts to linear polarisation and chooses the z-axis along $\hat{\varepsilon}$ this parametrisation assumes the simpler form

$$\frac{d\sigma}{d\hat{\mathbf{k}}_a d\hat{\mathbf{k}}_b dE_a} = \sum_l A_l P_l(\hat{\mathbf{k}}_a \cdot \hat{\mathbf{k}}_b) + \sum_{l_a l_b} B_{l_a l_b} \mathcal{Y}_{20}^{l_a l_b}(\hat{\mathbf{k}}_a, \hat{\mathbf{k}}_b) \tag{7}$$

where the coefficients A_l and $B_{l_a l_b}$ are functions of the energies E_a and E_b. In ref. 7 a similar parametrisation is given in the case of circularly-polarised light. It is also shown how, by integrating over the angles $\hat{\mathbf{k}}_b$, the doubly-differential cross-section DDCS may be written

$$\frac{d\sigma}{d\hat{\mathbf{k}}_a dE_a} = \frac{1}{4\pi} \frac{d\sigma}{dE_a} [1 + \beta_{20} P_2(\hat{\mathbf{k}}_a \cdot \hat{\varepsilon})] \tag{8}$$

where $\beta_{20} \equiv B_{20}/A_0$ and $d\sigma/dE_a$ is the singly-differential cross-section (SDCS), equal to $16\pi^2 A_0$. Finally, in the case of ionisation by a beam of electrons with initial momentum $\mathbf{k}_i$, the TDCS for the e-2e process can be parametrised as

$$\frac{d\sigma}{d\hat{\mathbf{k}}_a d\hat{\mathbf{k}}_b dE_a} = \sum_{l_a l_b L} B_{l_a l_b L} I_{l_a l_b L}$$
$$= \sum_{l_a l_b LM} B_{l_a l_b L} \mathcal{Y}_{LM}^{l_a l_b}(\hat{\mathbf{k}}_a, \hat{\mathbf{k}}_b) C_{LM}(\hat{\mathbf{k}}_i) \tag{9}$$

or, with $\hat{\mathbf{k}}_i$ as the z-direction, in the simpler form

$$\frac{d\sigma}{d\hat{\mathbf{k}}_a d\hat{\mathbf{k}}_b dE_a} = \sum_{l_a l_b L} B_{l_a l_b L} \mathcal{Y}_{L0}^{l_a l_b}(\hat{\mathbf{k}}_a, \hat{\mathbf{k}}_b) \tag{10}$$

and again the coefficients $B_{l_a l_b L}$ are functions of the energies E_a, E_b and E_i.

For later use it is also useful to note that the final-state two-electron wavefunction itself can be expanded in bipolar harmonics

$$< \mathbf{r}_a, \mathbf{r}_b \mid \psi_f^-(\mathbf{k}_a, \mathbf{k}_b) > = \sum_{l_a, l_b, L, M} \mathcal{Y}_{L,M}^{l_a, l_b *}(\hat{\mathbf{k}}_a, \hat{\mathbf{k}}_b) \Big\{ \mathcal{Y}_{L,M}^{l_a, l_b}(\hat{\mathbf{r}}_a, \hat{\mathbf{r}}_b) R_{L,M}^{l_a, l_b, k_a, k_b}(r_a, r_b) \quad (11)$$

$$\pm \mathcal{Y}_{L,M}^{l_a, l_b}(\hat{\mathbf{r}}_b, \hat{\mathbf{r}}_a) R_{L,M}^{l_a, l_b, k_a, k_b}(r_b, r_a) \Big\}$$

where the + sign denotes the singlet and the − sign the triplet state. Such an expansion in terms of states of well-defined two-electron angular momentum L is the generalisation of the usual expansion of a one-electron wavefunction of given momentum **k** in terms of partial waves.

3 ANALYSIS OF SYMMETRIES IN THE WANNIER CONFIGURATION

Several authors[8,9,10,11,12] have examined the properties of the two-electron wavefunction, as functions of $\mathbf{r}_a, \mathbf{r}_b$ in the configuration $\mathbf{r}_a \approx -\mathbf{r}_b$ considered by Wannier to be important in the region of the double-ionisation threshold. Note that strictly speaking the spatial wavefunction itself is not the object of interest (since its coordinates are integrated over in the transition matrix element). Rather it is the symmetries of particular $\mathcal{Y}_{LM}^{l_a l_b}(\hat{\mathbf{k}}_a, \hat{\mathbf{k}}_b)$ which decide the detected angular distribution, However from (11) one sees that the spatial and momentum bipolar harmonics appear symmetrically so that the properties in the region $\hat{\mathbf{k}}_a \approx -\hat{\mathbf{k}}_b$ are the same as those in the classical Wannier configuration $\hat{\mathbf{r}}_a \approx -\hat{\mathbf{r}}_b$. Alternatively one assumes that the directions $\hat{\mathbf{k}}_a, \hat{\mathbf{k}}_b$ are identical with the directions $\hat{\mathbf{r}}_a, \hat{\mathbf{r}}_b$ *asymptotically* and this is the region deciding the form of the cross-section.

The Wannier configuration wavefunction has been examined in a variety of coordinate systems. Here a form will be used that, at least in my opinion, most directly brings out the essential symmetries and has been used extensively to describe resonant states[13]. The key is to recognise that $\mathbf{r}_a = -\mathbf{r}_b$ implies

$$\mathbf{r} = \frac{1}{2}(\mathbf{r}_a + \mathbf{r}_b) = 0 \quad (12)$$

i. e. the centre-of-mass (and charge) of the two electrons is at the origin. The second orthogonal Jacobi coordinate is then naturally

$$\mathbf{R} = \mathbf{r}_a - \mathbf{r}_b \quad (13)$$

the interelectronic distance. These Jacobi coordinates are then expressed in terms of body-fixed coordinates R, $\eta = (r_a + r_b)/R$, $\gamma = (r_a - r_b)/R$ and Euler angles ψ, θ, ϕ, the latter angle ϕ corresponding to the projection m of the total angular momentum L on the body-fixed axis $\hat{\mathbf{R}}$. The projection on a space-fixed axis is M. Expressing a wavefunction of given L, S and total parity π in these coordinates one can show[12] (see also refs. 8–11)

a) Only states with $(-1)^t = \pi(-1)^S = 1$ can be finite at the Wannier saddle $\hat{\mathbf{r}}_a = -\hat{\mathbf{r}}_b$.

b) Additionally only states with *body-fixed* $m = 0$ are finite at $\hat{\mathbf{r}}_a = -\hat{\mathbf{r}}_b$. All others have a zero at this point. Note that for this rule to hold the space and body-fixed frames must coincide. As a corollary of a) and b) one can show that only states with $\pi(-1)^L = 1$ can be finite at $\hat{\mathbf{r}}_a = -\hat{\mathbf{r}}_b$. One also has the result

c) Only $^3S^e$ and $^1P^e$ states have a node at $r_a = r_b$.

As an example of the application of these rules one can consider the pure $^1P^o$ wavefunction arising from double photoionisation of the $^1S^e$ ground state of helium. From a) this state is zero along $\hat{\mathbf{r}}_a = -\hat{\mathbf{r}}_b$, or relative angle $\theta_{ab} = 180°$. The $^1P^o$ states have $|M| = 0$ or 1 and therefore $|m| = 0$ or 1 also. As the double-ionisation threshold is approached (infinitely slow electrons) one can expect the space-fixed and body-fixed frames to coincide (no rotation of the interelectronic axis) so that according to rule b) only $^1P^o$ states with $|M| = |m| = 1$ will be populated. This leads immediately to the result that if only one electron is detected its angular distribution will be of $|m| = 1$ i. e. $\sin^2\theta$ form. Hence in the DDCS of equation (8) one has $\beta_{20} = -1$. This result was arrived at by Greene[11] from a rather involved argument based on Herrick's[15] SO_4 classification of doubly-excited states. It has been shown[13] that Herrick's T quantum number is identical with the body-fixed m quantum number.

4 DOUBLE PHOTOIONISATION

Attention will be focussed here on the 'pure' three-body Coulomb problem represented by the double photoionisation of helium (H^- would also satisfy this condition). Measurements exist of the DDCS (8)[16,17] and very recently of the TDCS (5)[18]. The measurements of the DDCS near to threshold will be described first.

The main parameter of interest in the DDCS is the asymmetry parameter β_{20} of equation (8). In this section the notation will be simplified and this parameter called β. Greene[11] was the first to suggest that at threshold this parameter should take the value -1, as explained in section 3. Experiment[16] appeared not to agree with this prediction, rather giving values $\beta \approx -0.5$. A possible explanation was provided on the basis of calculations[19] of the photoionisation cross-section using an explicit 3-body correlated (3C) wavefunction for the final state. Since the use of this wavefunction will figure prominently in the results to be presented on angular distributions, both in photoionisation and in impact ionisation, its form will be given in detail.

The correlated 3-body wavefunction has the symmetric form of a product of three two-body on-shell Coulomb wavefunctions, one for each pair of interacting particles i. e.

$$\begin{aligned}\Psi_{3C} &= (2\pi)^{-\frac{3}{2}}\exp(-\pi\alpha_a/2)\Gamma(1-i\alpha_a)e^{i\mathbf{k}_a\cdot\mathbf{r}_a}{}_1F_1(i\alpha_a;1;-i[k_ar_a+\mathbf{k}_a\cdot\mathbf{r}_a])\\ &\times (2\pi)^{-\frac{3}{2}}\exp(-\pi\alpha_b/2)\Gamma(1-i\alpha_b)e^{i\mathbf{k}_b\cdot\mathbf{r}_b}{}_1F_1(i\alpha_b;1;-i[k_br_b+\mathbf{k}_b\cdot\mathbf{r}_b])\\ &\times \exp(-\pi\alpha_{ba}/2)\Gamma(1-i\alpha_{ba}){}_1F_1(i\alpha_{ba};1;-i[k_{ba}r_{ba}+\mathbf{k}_{ba}\cdot\mathbf{r}_{ba}])\end{aligned} \quad (14)$$

with

$$\alpha_a = \frac{-Z}{k_a} \quad \alpha_b = \frac{-Z}{k_b} \quad \alpha_{ba} = \frac{1}{2k_{ba}} \quad \text{and} \quad k_{ba} = \frac{|\mathbf{k}_b-\mathbf{k}_a|}{2} \quad (15)$$

This wavefunction has been used in expression (5) for the TDCS in helium and, after integration over the emission angles of electron 'b', in (8) to calculate the energy dependence of the β parameter. The initial-state wavefunction was a relatively simple Hylleraas $^1S^e$ correlated wavefuntion. The results for the TDCS

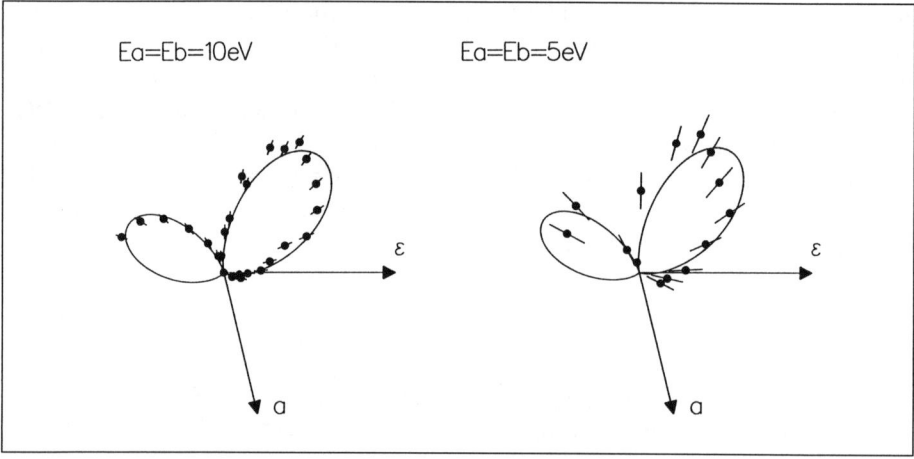

Fig. 1: The angular distribution of electron 'b' following double photoionisation of helium. Electron 'a' is emitted with equal energy in a direction fixed with respect to the polarisation vector $\hat{\mathbf{e}}$. Data from ref. 18. The continuous curve is theory from ref. 20.

for equal final energies are shown in figure 1 where excellent agreement in both length and velocity forms is obtained with the experimental results[18]. In fact, it has been shown[20] that even though the total energy above threshold is as low as 10eV, the angular distribution is essentially explained by a form

$$\frac{d\sigma}{d\hat{\mathbf{k}}_a d\hat{\mathbf{k}}_b dE_a} \sim (\cos\theta_a+\cos\theta_b)^2 C(\theta_{ab}) \quad (16)$$

where $C(\theta_{ab}) = |N(k_{ab})|^2$ is the Coulomb density of states (CDS) factor for the repulsive electron-electron interaction. It is this fundamental form that explains the oriented double-lobe structure of the TDCS shown in figure 1. The β values of the DDCS calculated with wavefunction (14) do indeed approach the limit -1 at threshold[19], but only if threshold is approached in such a way that the ratio $R = E_a/E_b$ is fixed. In addition the approach to $\beta = -1$ is very slow, being most quickly approached in the $R \approx 1$ equal energy case. Calculated β values agree with experiment over a range of total energies above threshold from less than 0.1eV to of the order of 80eV as shown in figure 2. The major features of the shape of β in this region arise from the effects of the electron-electron interaction in the final state. Proulx and Shakeshaft[22] have used a product of only two electron-nucleus wavefunctions, but with momentum-dependent effective charges designed to mimic this repulsion effect, and obtained almost the same quality of agreement with experiment.

In the case of absorption of one photon from a $^1S^e$ state, the TDCS can be written in the form[22]

$$\frac{d\sigma}{d\hat{\mathbf{k}}_a d\hat{\mathbf{k}}_b dE_a} \sim | g(k_a, k_b, \cos\theta_{ab}) \cos\theta_a + g(k_b, k_a, \cos\theta_{ab}) \cos\theta_b |^2 \qquad (17)$$

which clearly reduces to (16) for $k_a = k_b$. On the basis of Wannier theory, Selles et al[10] have argued that g is a symmetric function of k_a and k_b for all energy-sharing ratios in the threshold region. On this basis they derive a cross-section form identical to (16) but with the factor $C(\theta_{ab})$ of Gaussian form. Proposed forms of this Gaussian[5,6] do not give good agreement with experiment however[18].

The particular case of double photoionisation with left or right circularly polarised light has been examined in detail by Berakdar et al[23]. They point out the interesting result that the TDCS exhibits circular dichroism in that the momentum distribution is sensitive to the circular polarisation.

5 e-2e CROSS-SECTIONS

The major features of the TDCS for electron-impact ionisation at higher energies and for very asymmetric energy sharing have been understood for some time. For fixed scattering angle of the projectile electron, a slow ionised electron angular distribution shows a double-peak structure; the binary peak at forward angles and the recoil peak at backward angles. It is also well-appreciated that the first Born cross-section fails to reproduce the fine details of this distribution, even at impact energies $\approx 500eV$ on hydrogen. Higher-order processes must be taken into account in order to predict accurately the shape and position of the binary and recoil peaks.

In addition to the use of the 3C correlated wavefunction (14) for the final state, various types of distorted-wave theory have been developed which more- or less-successfully describe the high-energy TDCS[24,25].

Here only two new features of higher-energy e-2e processes that have emerged recently will be discussed. The first is the recognition of the importance of explicit double-binary[26] collisions giving rise to peaks in TDCS. A secondary peak, arising

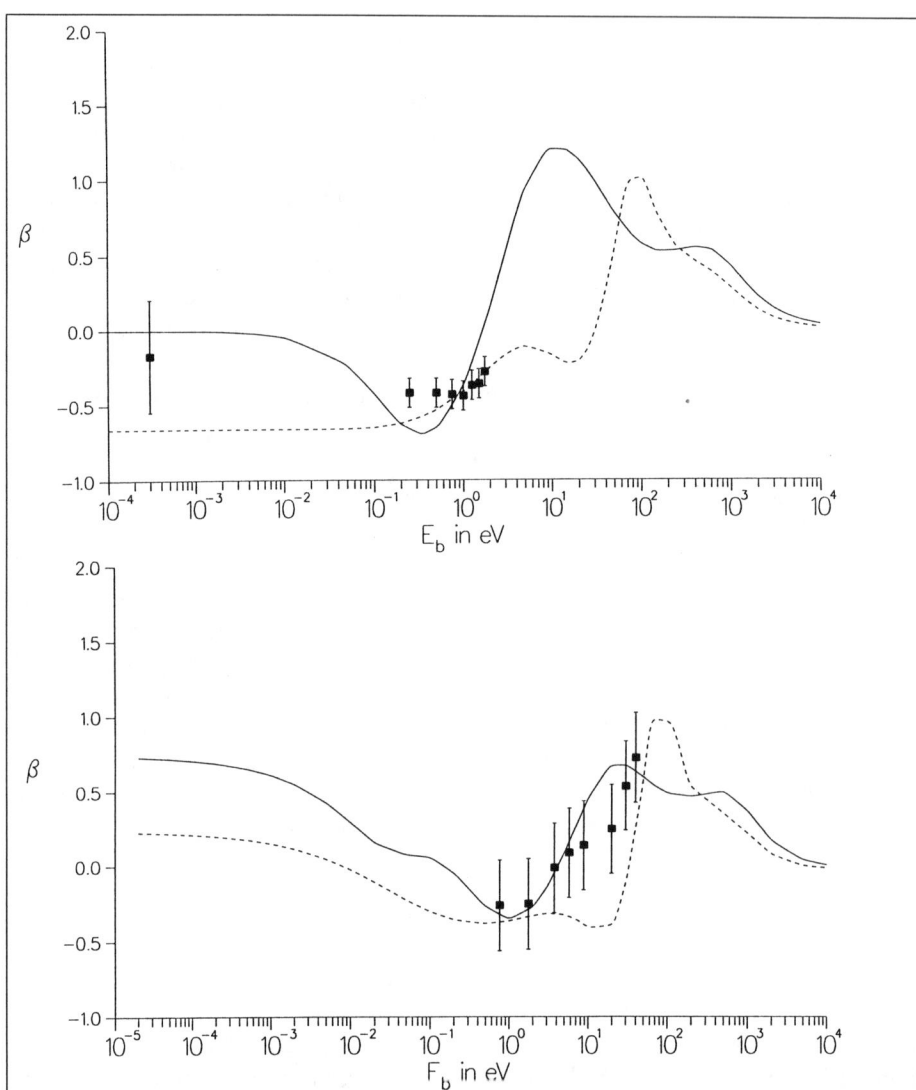

Fig. 2: The asymmetry parameter of equation 8. Upper curve: data from ref. 16 with fixed $E_a = 0.25 eV$. Lower curve: data from ref. 17 with fixed $E_a = 1.24 eV$. Theory is from ref. 21; continuous curve, velocity form; dashed curve, orthogonalised form.

only in second Born calculations, was identified ten years ago in large-angle coplanar symmetric (equal energy, emission angles equal and opposite in sign) e-2e processes[27]. Later the appearance of this peak close to an angle of 135° was explained[26] as arising from a double-binary collision. The fast incident electron

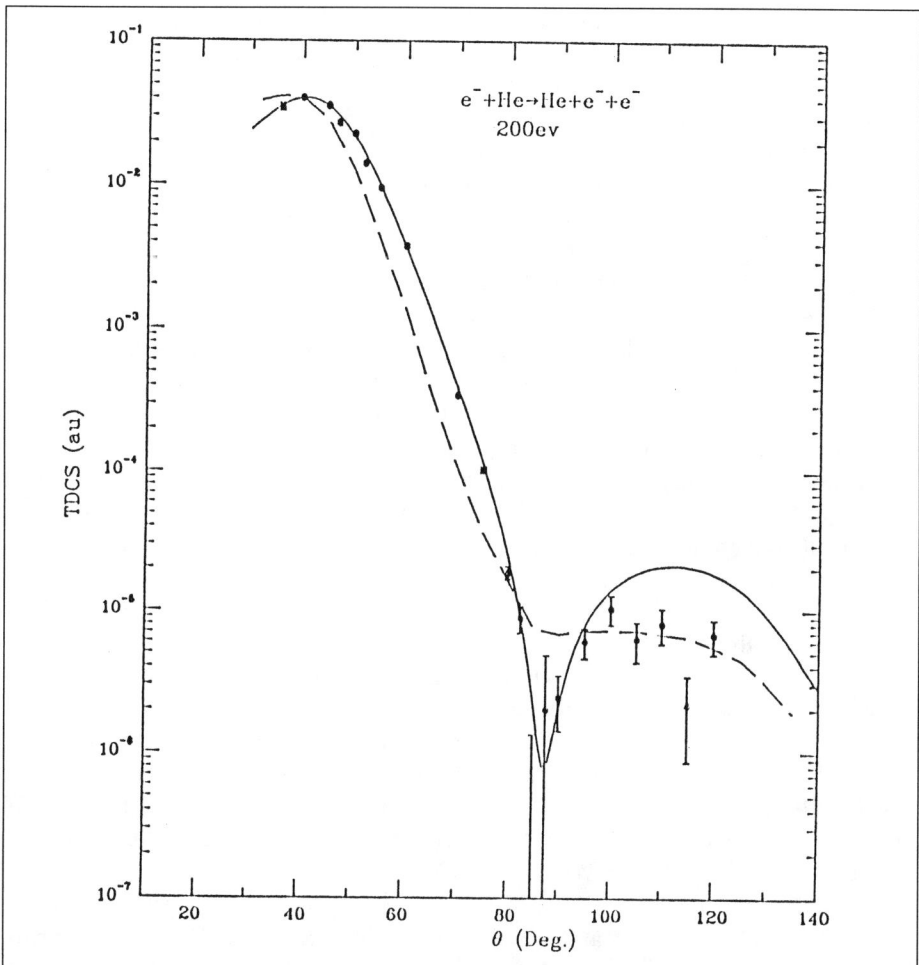

Fig. 3: TDCS for the electron impact ionisation of helium in coplanar symmetric geometry at incident energy 200eV (from ref. 32). Theory: continuous curve, DWBA from ref. 32; dashed curve, second Born theory from ref. 31.

scatters first elastically through 180° off the nucleus. As it emerges, it makes a second binary collision with the target electron so that the electrons with equal energy appear at 90° to each other and therefore at 135° to the beam direction. The symmetric co-planar e-2e measurement is shown in figure 3.

The major peak is around 45° and is due essentially to a single electron-electron binary collision. The single and double binary processes interfere and produce a remarkably sharp dip. Again distorted-wave theories are successful in explaining the shape[28,32].

The double-collision process has been shown to be important in two other situations. The first is in asymmetric geometry but where the angular distribution of the fast projectile (exchange can be neglected) electron is measured for fixed momentum of the slow electron. Then it has been shown[29] that peaks due to double-binary collisions should be visible in this angular distribution. The second situation is where the initial beam direction and the final electron momenta are not co-planar. For example, in 'equatorial' geometry the two electrons are detected at $90°$ to the beam direction. Clearly in this case a close collision with the nucleus is necessary to steer the fast incident electron through a large angle. A subsequent collision with the target electron leads to ionisation. Peaks due to such double-binary events have been observed[30] and calculated[31,32].

One feature is not well-explained by distorted-wave or second Born theories[32]. This is the sharp fall in the cross-section for small emission angles i. e. when two electrons of the same energy are emitted close to the forward direction. Then it is clear that it is the electron-electron repulsion that is causing the dip, the analogue of the well-known cusp electrons seen in impact ionisation by positively-charged particles. This effect has been studied in detail by Pan et al[33] where it was shown that the 3C wavefunction, since it includes electron correlation explicitly, describes the dip in the cross-section well.

More challenging for the theorist has been the explanation of new data[34-36] in the intermediate to low-energy regime. It is fair to say that the intermediate energy data (impact energy $\approx 50 eV$) for asymmetric energy sharing is not well explained by theory. Surprisingly perhaps, in certain geometries near threshold (a few eV above) TDCS are quite well-described by both the 3C and distorted-wave theories[34,28,37]. This is important since the geometries are those in which the two electrons have equal energies and emerge at fixed angle to each other. For relative angles around $180°$ this is precisely the configuration that should be important near threshold according to Wannier's theory.

The TDCS for a hydrogen target with electrons emerging with 1eV each at fixed relative angle of $180°$ and variable angle with respect to the incident beam direction, is shown in figure 4 in comparison with the calculated TDCS from the 3C wavefunction. The cross-section is smooth, with a minimum at $90°$ and maxima in the forward and backward directions. This tendency of the *relative* momentum to stay aligned along the beam direction can be explained simply by a plane-wave as final state[34]. The same tendency is also seen when the relative angle is fixed at $150°$. The TDCS exhibits a maximum when the relative momentum is along the beam direction and a minimum when at $90°$.

The shape of the TDCS for fixed angle of $180°$ in hydrogen indicates predominantly P-wave in the final state. The TDCS for a helium target in the same geometry shows the same overall tendency of maximum cross-section in the forward and backward directions. However precisely at $90°$ there is a subsidiary maximum indicating an admixture of even-L waves. Distorted-wave calculations [28,37] allowing for scattering of the two electrons in the He^+ core reproduce this maximum.

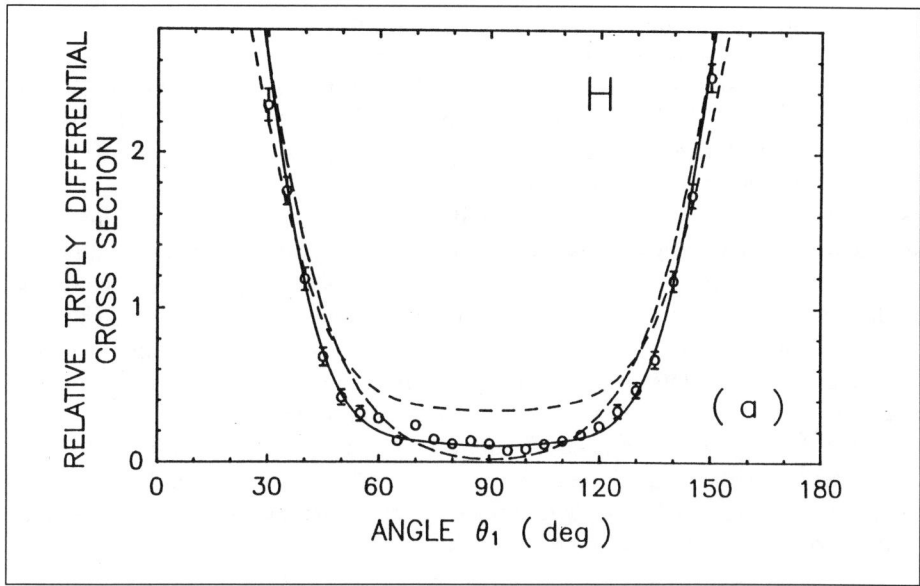

Fig. 4: Relative TDCS for $E_a = E_b = 2eV$ and relative angle fixed at 180°. The angle θ_1 is that of the interelectronic axis relative to the beam direction (from ref. 28). Data and theory (long dashed curve) from ref. 34. Theory: continuous curve, ref. 28; dashed curve ref. 37.

6 THE THRESHOLD REGION

The traditional approach to the threshold region is to employ the semi-classical version of Wannier's classical theory[2,4]. Here it is assumed that the dominant configuration at threshold is one in which the electrons appear predominantly at 180° to each other. Furthermore, for a given total energy, all combinations of individual electron energies are assumed to be equally probable. The Wannier theory provides threshold laws[38], angular distributions[3,4,6] and has also been used to calculate TDCS[39]. The theory is quite different from the approaches described in section 5, which however give accurate results for the shape of the TDCS near threshold. There appears to be an urgent need for an analysis indicating precisely the conditions upon which Wannier theory is valid and whether these conditions are experimentally realisable. Friedman et al[40] have thrown doubt upon experimental results which appear to establish the Wannier threshold law. Hence the situation close to threshold is still unclear. Although the theory using the 3C wavefunction predicts angular distributions very well, it appears not to give absolute cross-section values accurately, or to describe the relative probability of energy sharing correctly.

One aspect of the threshold region has been clarified recently however and that is the connection of the energy region just above threshold with the infinity of doubly-excited resonant states residing just below threshold. In the two-body Coulomb problem the bound state Rydberg spectrum extrapolates smoothly through the continuum and quantum defect theory makes the formal smooth connection between states just above and just below the two-particle break-up threshold. Although the doubly-excited states in helium or H$^-$ are unbound they can have very long lifetimes and it appears to have been assumed generally that the character of these states also connects smoothly with the Wannier mode $\mathbf{r}_a \approx -\mathbf{r}_b$, assumed to be the dominant mode leading to three-body breakup. Recently, very accurate calculations have become available for rather high-lying states of helium[41]. An analysis of the structure of these states, together with a knowledge of the nature of the underlying classical mechanics has shown, however, that the symmetric doubly-excited states are essentially *orthogonal* to the Wannier mode. Whilst it is true that there is a large probability for the electrons to be located near the Wannier ridge i. e. $\mathbf{r} = -\mathbf{r}_b$, the predominant character of the motion is not a symmetric stretch as in the Wannier mode, but an asymmetric stretch[42,43]. The consequence that in this mode the electrons avoid being simultaneously in the presence of the nucleus (a highly unstable configuration) explains the relative stability of the doubly-excited states.

REFERENCES

1. G.H.Wannier, Phys.Rev. 90, 817, (1953)
2. R.Peterkop, J.Phys.B 4, 513 (1971)
3. A.R.P.Rau, Phys.Rev.A 4, 207 (1971)
4. J.M.Feagin, J.Phys.B 17, 2433 (1984)
5. A.R.P.Rau, J.Phys.B 9, L283 (1976)
6. A.Huetz, P.Selles, D.Waymel and J.Mazeau, J.Phys.B 24, 1917 (1991)
7. H.Klar and M.Fehr, Z.Phys.D 23, 295 (1992)
8. C.H.Greene and A.R.P.Rau, Phys.Rev.Letts. 48, 533 (1982)
9. A.D.Stauffer, Physics Letters 91A, 114 (1982)
10. P.Selles, J.Mazeau and A.Huetz, J.Phys.B 20, 5183 (1987)
11. C.H.Greene, J.Phys.B 20, L357 (1987)
12. J.M.Rost and J.S.Briggs, J.Phys.B 24, L393 (1991)
13. J.M.Feagin and J.S.Briggs, Phys.Rev.A 37, 4599 (1988)
14. J.M.Rost and J.S.Briggs, J.Phys.B 24, 4293 (1991)
15. D.R.Herrick, Adv.Chem.Phys. 52, 1 (1983)
16. R.I.Hall, L.Avaldi, G.Dawber, M.Zubek, K.Ellis and G.C.King, J.Phys.B 24, 115 (1991)
17. R.Wehlitz, F.Heiser, O.Hemmers, B.Langer, A.Menzel and U.Becker, Phys.Rev.Letts. 67, 3764 (1991)
18. O.Schwarzkopf, B.Krässig, J.Elmiger and V.Schmidt, Phys.Rev.Letters 70, 3008 (1993)
19. F.Maulbetsch and J.S.Briggs, Phys.Rev.Letts. 68, 2004 (1992)
20. F.Maulbetsch and J.S.Briggs, submitted to J.Phys.B

21. F.Maulbetsch and J.S.Briggs, J.Phys.B 26, 1679 (1993)
22. D.Proulx and R.Shakeshaft, Phys.Rev.A, to be published
23. J.Berakdar, H.Klar, A.Huetz and P.Selles, J.Phys.B 26, 1463 (1993)
24. I.E.McCarthy Zeit.für Phys. 23, 287 (1992)
25. X.Zhang, C.T.Whelan and H.R.J.Walters, Zeit. für Phys. 23, 301 (1992)
26. J.S.Briggs, J.Phys.B 19, 703 (1986)
27. Byron et al , J.Phys.B 16, L769 (1983)
28. C.Pan and A.F.Starace, Phys.Rev.Letts. 67, 185 (1992)
    —, Phys.Rev.A 45, 4588 (1992)
29. M.Brauner and J.S.Briggs, J.Phys.B, to be published
30. A.J.Murray and F.H.Read, Phys.Rev.Letts. 68, 2912 (1992)
    —, J.Phys.B 25, 3021 (1992)
31. F.Mota Furtado and P.F.O'Mahony J.PhysB 22, 3925 (1989)
32. X.Zhang, C.T.Whelan and H.R.J.Walters, J.Phys.B 23, L509 (1990)
33. G.Pan, P.Hvelplund, H.Knudsen,Y.Yamazaki,M.Brauner and J.S.Briggs, Phys.Rev.A 47, 1531 (1993)
34. M.Brauner, J.S.Briggs, H.Klar, J.T.Broad, T.Rösel, K.Jung and H.Ehrhardt, J.Phys.B 24, 657 (1991)
35. M.Cherid, F.Gelebart, A.Pochat, R.J.Tweed, X.Zhang, C.T.Whelan and H.R.J.Walters, Zeit.für Phys.D 23, 347 (1992)
36. T.Rösel, P.Schlemmer, J.Röder, L.Frost, K.Jung and H.Ehrhardt, Zeit.für Phys.D 23, 359 (1992)
37. S.Jones, D.H.Madison and M.K.Srivastava, J.Phys.B 25, 1899 (1992)
    T.Rösel, J.Röder, L.Frost, K.Jung, H.Ehrhardt, S.Jones and D.H.Madison, Phys.Rev.A 46, 2539 (1992)
38. H.Klar, Zeit.für Phys.A 307, 75 (1982)
39. D.R.J.Carruthers and D.S.F.Crothers, Zeit.für Phys.D 23, 365 (1992)
40. J.R.Friedman, X.Q.Guo, M.S.Lubell and M.R.Frankel, Phys.Rev.Letts. (1991)
41. J.M.Rost, R.Gersbacher, K.Richter, J.S.Briggs and D.Wintgen, J.Phys.B 24, 2455 (1991)
42. G.S.Ezra, K.Richter, G.Tanner and D.Wintgen, J.Phys.B 24, L431 (1991)
43. D.Wintgen, K.Richter and G.Tanner, CHAOS 2, 19 (1992)

ELECTRON-IMPACT IONIZATION OF ATOMS, MOLECULES AND TRANSIENT SPECIES: CURRENT STATUS AND PERSPECTIVES FOR THE FUTURE

V. Tarnovsky and K. Becker
Physics Department, City College of C.U.N.Y., New York, NY 10031 USA

ABSTRACT

The ionization of an atom or molecule is one of the most fundamental collision processes in atomic and molecular physics both from a basic as well as from a more applied view point. The final state of an ionizing collision consists of a positively charged ion and at least two outgoing electrons. This is a formidable challenge to theory and explains why cross section calculations have overall not been as successful in predicting the shapes and absolute values of ionization cross sections as they have been in predicting cross sections for other inelastic electron collisions e.g. differential excitation cross sections. Ionization cross sections play an important role in many areas of application ranging from discharges and plasmas to excimer lasers to planetary and stellar atmospheres to radiation chemistry. This article reviews the current status of the experimental determination of absolute electron-impact ionization cross sections for atoms, molecules and transient species and attempts to highlight the most promising avenues for future research in this field. Improved semi-classical and semi-empirical methods which are widely used in efforts to predict electron-impact ionization cross sections will also be discussed.

INTRODUCTION

Electron impact ionization processes have been studied extensively since the 1930s by experimentalists and theorists alike. Ionization processes play an important role in many areas of application ranging from discharges and plasmas to excimer lasers to planetary, cometary and stellar atmospheres to radiation chemistry[1-3]. To the theorist, the final state of even the simplest electron impact ionization process, a positively charged ion and two outgoing electrons, presents a system with an infinite number of final states - unless further restrictions are imposed on the kinematics of the process, as e.g. in (e, 2e) experiments. This is a formidable challenge and explains why theoretical calculations have not been as successful in predicting the absolute values and shapes of ionization cross sections as they have been in predicting e.g. excitation cross sections[4].

Advances in experimental instrumentation and techniques such as the advent of high-resolution electron spectrometers, the use of spin-polarized electrons and coincidence techniques make possible studies of electron impact ionization processes (e.g. absolute or relative double and triple differential ionization cross section measurements - in some cases spin-resolved) which probe the ionization process at a very fundamental level and allow stringent tests of theoretical predictions. At the same time, it remains an equally challenging task to obtain a satisfactory answer from either experiment or theory to the seemingly much simpler question "What is the absolute, angle-integrated single ionization cross section for a particular atom as a function of impact energy?" In the absence of reliable rigorous theoretical calculations of ionization cross sections, which are complicated by the inherent complexity of the process (viz. the multitude of possible final states), various classical, semi-classical and semi-empirical formulae have been widely used in the past in efforts to predict ionization cross sections[5].

To date, absolute single ionization cross sections have been measured for less than 40 atoms, cross sections for multiple ionization for even fewer atoms. Many of these cross sections have only been measured once using one particular experimental technique. This is a highly unsatisfactory situation in view of the fact that measurements of absolute cross sections are tedious, cumbersome, difficult and prone to systematic errors whose presence and importance sometimes becomes evident only after the same absolute cross section is re-measured using a different experimental technique. The status of ionization cross section calculations is equally unsatisfactory in many cases. In instances where experimental results and theoretical predictions can be compared, it is not uncommon to find discrepancies of 50% and more. The discrepancies are particulary serious in the low energy regime below about 100 eV, which is, on the other hand, the energy regime that is most relevant for many practical applications of ionization cross section data. Fig. 1 highlights as an example the poor agreement between the measured absolute cross section for the process Ne + e⁻ → Ne⁺ + 2e⁻ and various predictions from theoretical models as well as from semi-emirical and semi-classical approaches. The situation gets progressively worse as one goes from simple to complex atoms to molecules to free radicals and other transient species.

This article summarizes the status of absolute partial, angle - integrated

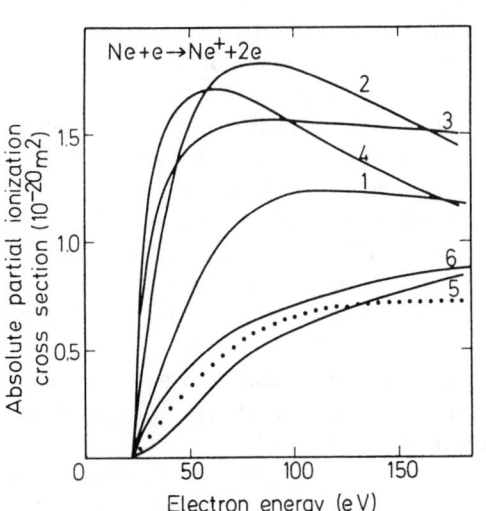

Fig.1: Absolute partial cross section for the single ionization of Ne after Märk[44]. The experimental data (dots) of Stephan et al.[45] are compared to the preictions from various theories and semi-empirical formulae. (For further details, see Märk[44].)

electron-impact ionization cross section measurements for various targets ranging from simple atoms and molecules to complex polyatomic molecules, free radicals and transients. Results from different groups using various experimental techniques are reviewed and perspectives and opportunities for future work are discussed. We note that the wide field of electron-impact ionization experiments and theories that are differential in angle and/or in energy are outside the scope of this review.

EXPERIMENTAL DETAILS

The subject of electron-impact ionization cross section measurements was reviewed by Kieffer and Dunn[6] and by Märk and Dunn[2]. Special emphasis was devoted in both reviews to a detailed analysis and a critical evaluation of the various experimental techniques that are used for the measurement of electron-impact ionization cross sections. It is not the purpose of this article to reiterate the detailed findings outlined in these reviews to which we refer the reader for further details. We mention, however, the article by Märk[7] which discusses the salient features of the various experimental techniques that are currently being used by experimentalists. We will briefly review the fast-neutral-beam apparatus that is used in our group. This apparatus is similar to the experimental set-up used by Peterson[8] and essentially identical to the apparatus of Freund and collaborators[9].

A schematic diagram of our fast-beam apparatus is shown in figure 2. A fast (3-3.5 kV) neutral target beam is prepared by resonant or near-resonant charge exchange of a primary mass selected ion beam with an appropriate charge transfer gas. The primary ion beam is extracted from a dc Colutron discharge and a single mass is subsequently selected by passing the primary ion beam through a Wien filter. After the charge exchange cell the residual ions are removed from the beam by electrostatic deflection and most target species formed in Rydberg states are ionized and removed from the beam in a region of high electric field (typically 5 kV/cm). The remaining neutral beam passes through a beam-defining aperture before it is crossed with a well-characterized electron beam of variable energy (5-200 eV). The neutral beam is monitored by a secondary emission detector (Nichrome surface) and its absolute flux can be determined with a calibrated pyroelectric detector. The product ions produced by electron - impact ionization in the interaction region largely retain their tight initial collimation and narrow energy spread because of the negligible momentum transfer to the ion. The product ions are focused in the entrance plane of an electrostatic hemispherical analyzer which separates ions of different energy (i.e. charge state or mass) and, of course, the product ions from the neutrals. After the analyzer, the ions are detected by a channel electron multiplier (CEM) operated in the pulse counting mode. Further details of the apparatus and of the experimental procedure and a comprehensive analysis of potential sources of systematic uncertainties along with an estimate of the overall accuracy can be found in previous publications[9-11].

The various experimental techniques all have strengths and weaknesses. The main advantages of the fast-beam technique are a product ion collection efficiency of 100% (except in certain cases of dissociative ionization processes which will be discussed later) and the capability to study the ionization of target species which cannot be produced readily by conventional methods such as free radicals and transients such as metastables, the halogen atoms, and atomic oxygen, nitrogen and

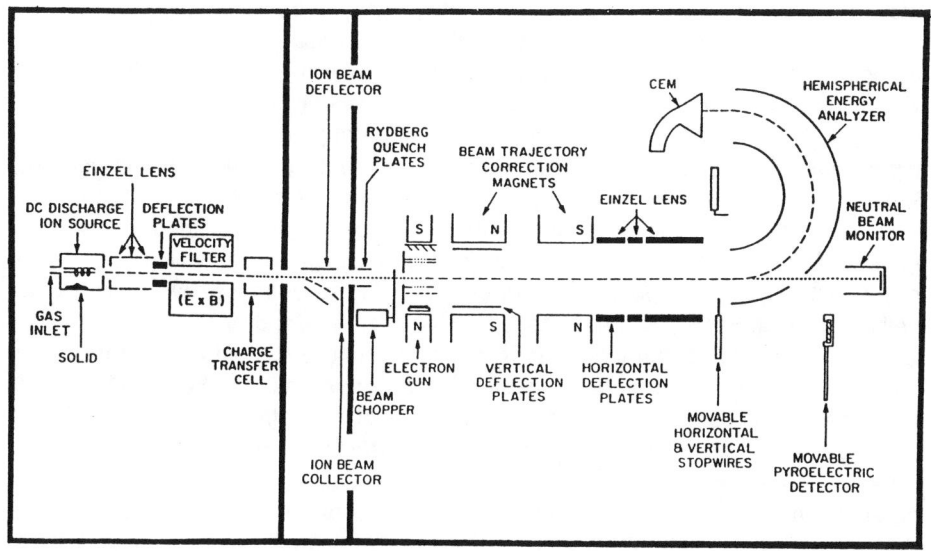

Fig. 2: Schematic diagram of our fast-beam apparatus.

hydrogen. Complications can arise from the preparation of the fast-beam through charge exchange. This can cause contamination of the target beam by neutrals which are not in the ground state, but in a metastable state or in a Rydberg state and survive long enough to be ionized in the interaction region. Along the same lines, vibrational excitation of molecular targets formed through charge exchange is not uncommon. This requires a careful and in-situ monitoring of the neutral beam composition and/or corrections to the measured data to remove contributions to the recorded ion signal due to the ionization of target species which are not in the electronic and/or vibrational ground state.

SINGLE AND MULTIPLE IONIZATION OF ATOMS

Absolute partial electron-impact cross sections for the single ionization of atoms have been measured to date for 38 atomic targets, reliable cross sections for multiple ionization for far fewer targets. It is interesting to note that at the time of the 1985 review of Märk[7] cross section were available for only 23 atoms. We furthermore point out that more atoms, viz. 27, have been studied using the fast-beam technique than by any other experimental method. Targets measured by more than one group using different experimental techniques include the noble gases He,

Ne, Ar, Kr, Xe as well as the alkalis and alkaline earths. The agreement among the reported values from different sources is best for the noble gases which have been studied most thoroughly. In many other instances, however, the ionization cross section has only been measured once using one particular experimental technique.

The experimental situation is less than satisfactory for many simple atoms such as H, O, and N, which are interesting from a basic point of view (e.g. interesting electronic configuration and/or unique physical or chemical properties) in addition to their important role in many areas of application. The major problem with these atoms lies in the determination of the neutral beam density. In many cases, relative measurements were subsequently normalized to a Born approximation at sufficiently high energies. An interesting alternative was used by Shah et al.[12] who normalized their e⁻ + H ionization cross section to absolute cross sections for the H⁺ + H charge transfer at low impact energies obtained in the same apparatus. Much activity in recent years focused on target atoms which are difficult to produce for the purpose of electron collision experiments by conventional methods such as Fe, Si, Ge, Sn, P, As, Sb, Bi, Se, Te, V, Ti, and Ni.

While the fast-beam technique is the most versatile method to generate target beams of many difficult-to-produce species, there were cases, e.g. Fe and Cu, where the target beam contained a significant fraction of metastable atoms. Recently, ionization cross sections for Fe and Cu were measured using techniques where the target beam was not affected by metastable contamination. Kopnarski et al.[13] used ion sputtering in a low-pressure, high frequency plasma as a source of Cu and Fe atoms and Gilbody and collaborators[14,15] produced thermal beams of these atoms from a specially designed oven. The Fe^+ and Cu^+ ionization cross sections of Gilbody and collaborators were systematically smaller than those of Freund et al.[10] by about 30% (Fe) and 15% (Cu) at 200 eV with the discrepancy increasing towards lower energies. Kopnarski et al., on the other hand, reported Fe^+ and Cu^+ cross sections in good agreement with the fast-beam results of Freund et al.[10] for energies from 10 - 200 eV and with a discrepancy from the fast-beam results only at energies below 10 eV. There is a need for further experimental work involving difficult-to-produce target beams such as Cu, Fe, V, Ni, Ti using different experimental techniques.

Even in the case of the thoroughly researched and presumably well understood ionization cross sections for the noble gases, one troublesome aspect emerged regarding the cross sections for multiple ionization of these species, most notably in Ar. While there was excellent agreement (to within 8%) in the Ar single ionization cross sections measured by several groups since 1980, there was a spread of almost 30% in the Ar double ionization cross sections reported by these groups. This was particularly troubleome, since all groups determined the absolute double ionization cross sections by normalizing a relative Ar^{2+} ionization cross section to the previously measured absolute Ar^+ cross section in their apparatus and the measured double-to-single cross section ratio at a fixed impact energy. These ratios were typically reported with statistical errors of not more than ±3% assuming that the singly and doubly charged product ions were detected under identical experimental conditions, so that all systematic uncertainties cancel. Despite the quoted ±3% errors in the individual cross section ratios, the reported

Ar^{2+}/Ar^+ ratios at 100 eV impact energy ranged from 0.091 to 0.060, a spread of more than 30%. Such an inconsistency for a target as fundamental and simple as Ar which is experimentally easy to handle and which has been studied by many groups using different experimental techniques suggested a serious problem with the measurement of ionization cross section ratios.

Tarnovsky and Becker[11] and independently Bruce and Bonham[16,17] and Syage[18] carried out a comprehensive analysis of potential sources of systematic uncertainties and concluded that the previously reported error bars of ±3% had been much too optimistic and that the error limits on measured multiple-to-single ionization cross section ratios for the same target atom had to be raised to at least about ±8% due to systematic uncertainties which did not cancel when cross section

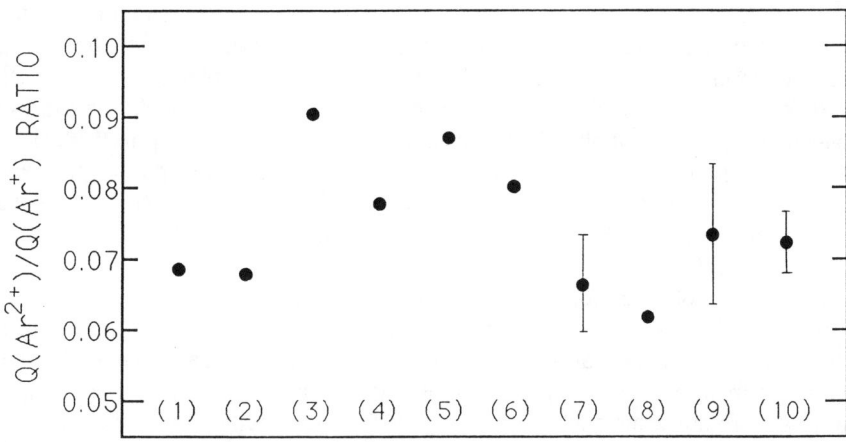

Fig. 3: Summary of double-to-single ionization cross section measurements in Ar at 100 eV since 1980. The numbers in parentheses refer to (1): ref. 44, (2): ref. 9, (3): ref. 46, (4): ref. 47, (5): ref. 16, (6): ref. 48, (7): ref. 17, (8): ref. 18, (9): ref.11, and (10): ref. 52.

ratios were measured under similar, but not identical experimental conditions. Figure 3 shows a summary of the most recently (since 1980) reported double-to-single ionization cross section ratios in Ar at 100 eV. With two exceptions, the cross section ratios appear to converge towards a value of 0.068±0.07

IONIZATION OF STABLE MOLECULES

The 1985 review by Märk[7] lists close to 40 molecules for which absolute cross sections for the ionization and dissociative ionization had been obtained. The list ranged from the simple diatomic molecules H_2, N_2, O_2, CO, NO, and HCl to triatomic molecules such as H_2O, CO_2, N_2O, NO_2 and SO_2 to complex polyatomic molecules such as C_2H_2, NH_3, CH_4, CCl_4, CF_4, SF_6 and Si_2H_6. As pointed out by

Märk[7] in many cases reliable absolute cross sections were available only for the parent ionization of these molecules. Fragment ions resulting from the dissociative ionization are often produced with excess kinetic energies of up to several electronvolts which makes it difficult to collect all the fragment ions.

Much progress has been made in the experimental determination of molecular ionization cross sections in the past few years. The list of molecules for which absolute cross sections are available has roughly doubled since 1985. More importantly, much attention has been devoted by several groups to the question of how to improve the reliability of cross sections obtained for energetic fragment ions resulting from dissociative ionization. In many cases, careful experimental studies combined with extensive and detailed ion trajectory simulation provided a quantitative understanding of the collection efficiency of a particular apparatus for fragment ions formed with a given excess kinetic energy. A drastic example are the partial ionization cross sections for the formation of F^+ fragment ions from CF_4 reported by Bonham and collaborators[19] and by Märk and collaborators[20] which initially differed by a factor of 8. After both groups carried out a careful analysis of the collection efficiency of their respective apparatus for the comparatively light fragment ion F^+ produced by dissociative ionization of a heavy CF_4 parent molecule and applied the appropriate corrections to their original data, both cross sections agreed to within 20% which was well within the combined quoted uncertainties of the two experiments[19,20].

The list of molecules for which ionization cross sections have been measured since 1985 includes several diatomic molecules (GaCl, GeCl, SnCl, N_2, CO) which were measured using the fast-beam technique[21,22] as well as larger molecules ranging from simple polyatomic species to more complex hydrocarbons, e.g. C_3H_8 to vary large and complex molecules such as HMDSO, hexamethyl-disiloxane, $(CH_3)_3$-Si-O-Si-$(CH_3)_3$, TEOS, tetraethoxysilane, $Si(O-C_2H_5)_4$, and the metal-organic compounds $(CH_3C_5H_5)_2Ru$, (C_5H_5)-Pt-$(CH_3)_3$ and $(CH_3C_5H_5)_2Fe$[23-26]. In many, if not all, cases the primary motivation for the experimental studies involving the more complex molecules was the need for ionization data in various areas of applications rather than the desire to investigate molecular ionization systematically from a fundamental point of view.

Recently, Bonham and collaborators[49] implemented the technique of covariant mapping mass spectroscopy[50] to their pulsed electron-beam time-of-flight ionization apparatus and applied the technique to the double positive ion pair formation following electron impact ionization of the CF_4 molecule. These authors observed several previously unreported break-up channels of this molecule and determined their absolute cross sections. With very few exceptions, these cross sections were found to be relatively small (1×10^{-18} cm^2 or less at 100 eV). It is interesting to note that the by far largest cross section (1×10^{-17} cm^2 at 100 eV) was found for the $CF^+ + F^+$ double positive ion pair formation from CF_4. Very recently, Tarnovsky et al.[30] found experimental evidence of an even larger cross section for the double positive ion formation $CF^+ + F^+$ in the dissociative ionization of the CF_2 free radical (4×10^{-17} cm^2 at 70 eV).

IONIZATION AND DISSOCIATIVE IONIZATION OF FREE RADICALS

Cross sections for the electron-impact ionization of free radicals have received an increased level of attention in recent years. This interest was stimulated by the important role of some radicals in the modelling of planetary atmospheres (e.g. the plasma torus around Jupiter and its satellite Io and the magnetosphere of Saturn) and in many plasma-assisted materials processing applications. Our fast-beam apparatus provides a unique experimental tool to investigate the ionization and the dissociative ionization of free radicals. These species cannot be generated using conventional methods (e.g. effusive gas beam or static gas target). The Colutron discharge source, on the other hand, in conjunction with the Wien filter (see figure 1) enables us to produce mass selected beams of just about any stable primary ion that can be produced in the dc discharge. Assuming that an appropriate charge transfer gas can be identified, fast neutral beams of a variety of free radicals can be generated in our apparatus. Free radicals which were investigated previously using the fast-beam method are CD_3, CD_2, and the SiF_x (x=1-3) radicals[27,28]. Recently, in the context of our systematic investigations of electron-molecule collisions with the commonly employed constituents of low-temperature processing plasmas such as SF_6, CF_4, NF_3, CCl_2F_2 and BCl_3 we investigated the ionization and dissociative ionization of the CF_x (x=1-3) radicals[29,30]. Preliminary data have also been obtained for NF_2 and NF radicals and for the stable NF_3 molecule[31].

Figure 4 shows the measured cross sections for the parent ionization of the CF_x (x=1-3) radicals and in figure 5 we show the results of the dissociative ionization of the CF_3 radical. Measurements of ionization cross sections for radicals (and molecules) using the fast-beam technique can be affected by the presence of vibrationally excited molecules in the incident neutral beam, a problem which is not a factor in experiments which use thermal atom beams. The charge exchange process can produce molecules not just in the vibrational ground state, but also in higher vibrational states, since the vibrational energy spacing is typically very small. As a consequence, the incident molecular beam may contain molecules in various vibrational states and the measured cross section does not represent the ionization cross section of a molecules in a well-defined (and preferably the lowest) vibrational level. The presence of vibrationally excited molecules in the incident neutral beam results in an enhanced curvature of the measured cross section in the near-threshold region extending to energies slightly below the spectroscopic ionization threshold for molecules in the vibrational ground-state. Figure 6 shows the near-threshold region of the SiF_3^+ cross section in comparison with the Xe^+ cross section. Note the extended curvature in the SiF_3^+ cross section over a region of 2 eV, while the xenon data show curvature over only 0.6 eV which reflects the energy spread in the ionizing electron beam. There is no reliable way of predicting the possible presence and the amount of vibrational excitation in any particular case. The only way to check for the presence or lack of vibrational excitation in a specific case is through a tedious experimental trial-and-error procedure, i.e. through a careful mapping of the ionization cross section in the near-threshold

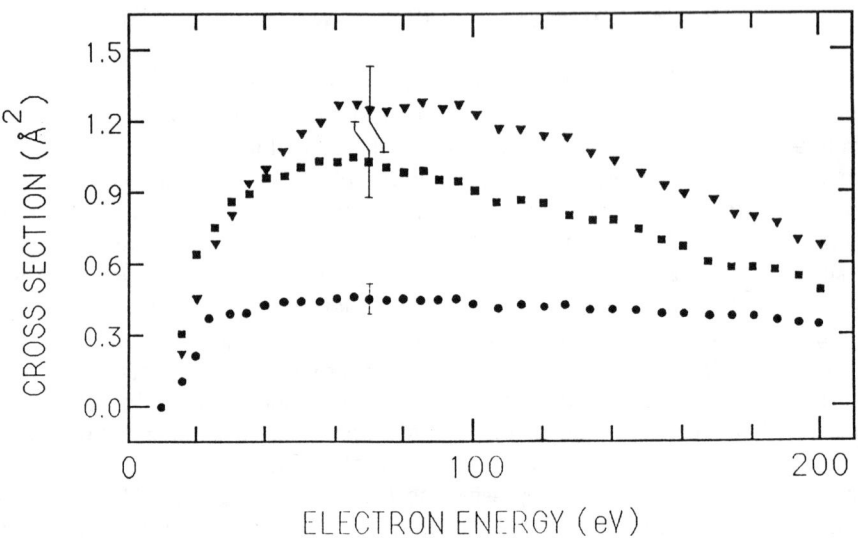

Fig. 4: Absolute parent ionization cross sections of the CF_3 (●), the CF_2 (■), and the CF (▼) free radicals as a function of the electron energy. (For further details, see Tarnovsky and Becker.[29])

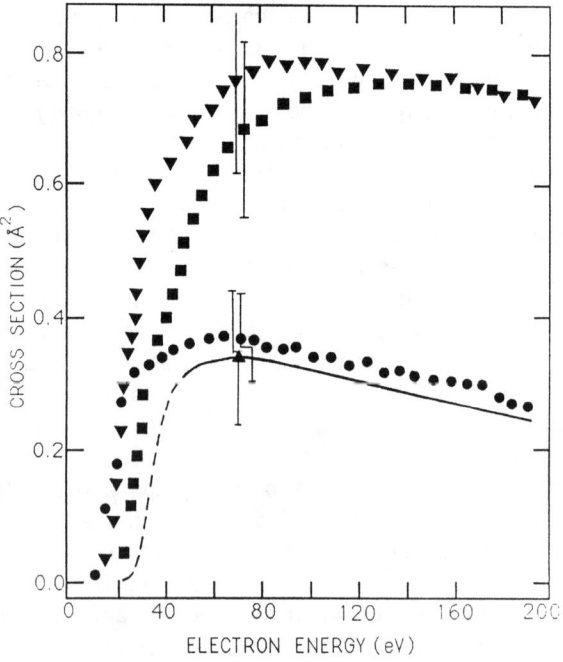

Fig. 5: Absolute ionization cross sections for the formation of the CF_3^+ parent ion (●) and the CF_2^+ (▼), CF^+ (■) fragment ions from CF_3 as a function of the electron energy. Also shown (▲ and solid line) is the cross section for F^+ formation. (For further details, see Tarnovsky et al.[30]).

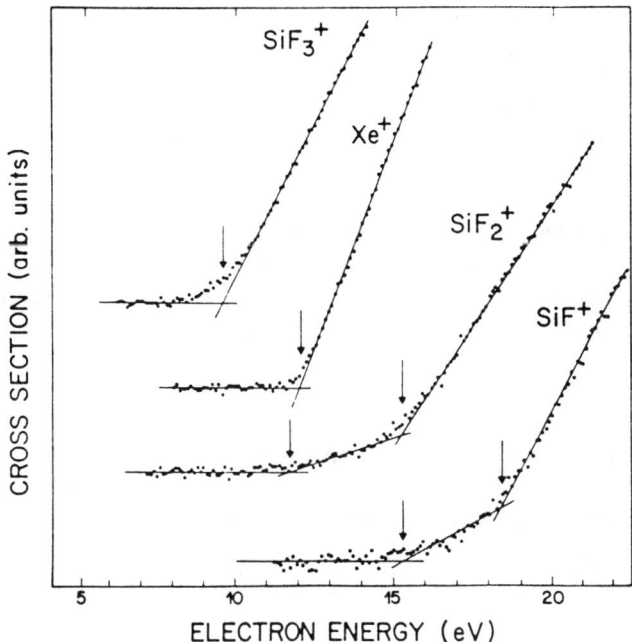

Fig.6: Thresholds for the ionization of SiF$_3$ to its parent and fragment ions, together with the Xe threshold. (For further details, see Hayes et al.[28]).

region for different charge transfer gases, until a gas is found where little or no extended curvature is found. We note that there was little, if any, evidence that our recent cross section measurement for the CF$_x$ radicals were seriously affected by the presence of vibrationally excited neutral radicals in the incident fast beam.

The dissociative ionization of free radicals is also an important process - just as it is for stable molecules (see previous discussion). The reliable measurement of the dissociative ionization cross sections for the CF$_3$ radical shown in fig. 5 (and, of course, also for other free radicals) required a detailed analysis of the capability of our fast-beam apparatus to collect energetic fragment ions with essentially 100% efficiency. Careful experimental studies in conjunction with extensive ion trajectory modelling demonstrated that no significant ions losses occur for fragment kinetic energies which are lower than about 4 eV per fragment ion. In most cases, characteristic excess kinetic energies in dissociative ionization processes are significantly smaller[19,22,27-30].

IONIZATION CROSS SECTION CALCULATIONS

As discussed in detail by Younger[4] and by Younger and Märk[5] there are no established rigorous theoretical approaches to calculate absolute electron-impact ionization cross sections for targets more complex than the simplest atoms with the same precision and level of confidence with which other electron-impact cross sections, e.g. excitation cross sections can be calculated. Various classical, semiclassical and semi-empirical formulae have been used to determine ionization cross sections for atoms, albeit with limited success. While the agreement between measured cross sections and predictions from these formulae was quite good for many low-Z atoms (Z=nuclear charge), these formulae failed for a few simple atoms such as Ne, and F (see fig. 1) and for most high-Z atoms[5,7,32]. Reliable calculations of ionization cross sections for molecules are even further beyond the capabilities of rigorous theoretical approaches. Molecular ionization cross section predictions have instead relied largely on rather simplistic additivity rules whose success has been very limited[33,34]. Recently, a new additivity rule has been introduced[35,36] which incorporates appropriate weighting factors for the ionization cross sections of the constitutents of a molecule or radical. In this approach, the ionization cross section $Q(AB_n)$ of a molecule AB_n is given in terms of the individual ionization cross sections $Q(A)$ and $Q(B)$ by an expression of the form

$$Q(AB_n) = f_A(r_A,\xi_A) \cdot Q(A) + f_B(r_B,\xi_B) \cdot n \cdot Q(B)$$

where the weighting factors f_A and f_B are given by

$$f_A = [r_A/(n \cdot r_B)]^{1/2} \cdot [\xi_A/(\xi_A + n \cdot \xi_B)]; \quad f_B = [n \cdot r_B/(r_A)]^{1/2} \cdot [n \cdot \xi_B/(\xi_A + n \cdot \xi_B)]$$

Here r_A, r_B refer to the radii of atoms A and B and ξ_A and ξ_B denote the effective number of electrons. The first success of the new additivity rule was the prediction of the ordering of the total single ionization cross sections for the SiF_x (x=1-3) radicals[37] $Q[(SiF)^+] > Q[(SiF_2)^+] > Q[(SiF_3)^+]$ which was consistent with experimental results[28] and reversed the cross section ordering predicted by previous additivity rules[34] which disagreed with the measured cross sections[28]. This new additivity rule has also been rather successful in the case of a few other molecules such as the H_2, N_2, O_2, CO_2, C_2H_2, SF_6, CF_4, CH_3OH, C_6H_6 and C_2H_6 molecules[38] as well as with the CF_x (x=1-3) free radicals[39] and the NF_3 molecule[39]. We note, however, that the data base for this additivity rule is still limited. A meaningful comparison with experimental data requires that cross section for all single ionization channels of a molecule be measured, i.e. the parent (single) ionization cross section and all partial (single) ionization cross sections. To date, reliable data of that kind exist only for a few molecules and radicals.

In a different approach, the so-called DM-formalism, Deutsch and Märk introduced a modified semi-empirical approach[32,40] which combines a Gryzinski-type energy dependence[41] with detailed atomic structure information

$$Q = \sum_{n,l} g_{nl} \cdot \pi \cdot (r_{nl})^2 \cdot \xi_{nl} \cdot f(u)$$

where ξ_{nl} is the number of electrons in the shell with quantum numbers n and l, $(r_{nl})^2$ is the mean square radius of that shell and g_{nl} are weighting factors related to the "ionization factors" calculated by Bethe[42]. The function f(u) is the well-known energy dependence given by the Gryzinski-model[41] in terms of the normalized energy $u = E/E_{i,n}$ where E and $E_{i,n}$ are respectively the incident electron energy and the ionization energy of the nth shell. When applied to molecules and radicals, the application of the above cross section formula requires an analysis of the respective molecular/radical orbitals in terms of the atomic orbitals of its constituents, e.g. a Mulliken population analysis[38,43].

PERSPECTIVES FOR THE FUTURE

During the past 60 years the level of activity devoted to the study of electron impact ionization cross sections has undergone several oscillations in the sense that periods of increased activity were followed by periods of relatively low activity and vice versa. It appears that we are now at a point where ionization cross section measurements have become fashionable again. It is also obvious that the major driving force and motivation for these activities comes from a need to know cross section data in various areas of application and modelling (low and high tempertaure plasmas, planetary and cometary atmospheres, astrophysics, radiation chemistry, excimer lasers - to name just a few). Experimental studies which are aimed primarily at improving our fundamental understanding of the ionization process and at elucidating its finer details have to be investigations which are differential in energy and/or angle (such as (e,2e) and (e,3e) experiments) and experiments involving spin-polarized projectiles and targets.

The obvious opportunities for future ionization cross section measurements of atomic targets involve (1) obtaining cross sections for the remaining roughly 50 atoms in the periodic table for which no data exist at this point in time - preferably using more than just one experimental technique and (2) verifying cross sections which have only been measured once using one particular experimental technique. Both tasks are rather tedious and time-consuming and might be conceived as being of limited intellectual stimulus. One should keep in mind, however, that (1) there are many experimental challenges in the preparation of well-characterized target beams of many of the remaining atoms, which could prove advantageous to other fields of atomic physics and that (2) most of these atoms have rather complex electron configurations and the study of their ionization cross section might reveal unexpected and interesting features. A perhaps more challenging research avenue lies in the direction of new methods for the experimental determination of partial atomic ionization cross section, i.e. ionization cross sections for a specific shell or subshell of an atom with particular emphasis on innershell ionization processes.

Given the large number of molecules, free radicals and transient species, there are vast opportunites for further measurements of ionization cross sections. However, the field of ionization cross section measurements for molecules, radicals and transients provides many more challenges besides the obvious one of investigation targets that have not been studied before. Dissociative ionization and the ability to measure reliable cross sections for the formation of the various fragments, in particular cross sections for fragments which are formed with a certain amount of excess kinetic energy, require serious experimental considerations as to the efficiency of the ion extraction and the ion transport to the detector. The significance of these effects for the reliability of the measured cross sections has only been recognized recently. Experimental efforts are now under way by many groups to quantify the capability of their apparatus to detect energetic fragment ions. This is done in almost all cases by combining experimental studies with ion trajectory modelling. The fast-beam technique has proved to be rather insensitive to the deleterious effects of excess kinetic energies on measured ionization cross sections as long as these energies are not unreasonably large (i.e. as long as they are less that 4-5 eV per fragment ion). Time-of-flight techniques, particularly when combined with position-sensitive detectors represent another approach to the reliable measurement of fragment ionization cross sections.

Free radicals and transients which are difficult, if not impossible, to produce for electron collision experiments by conventional techniques can be studied using the fast-beam technique in conjunction with a versatile discharge souce. Such a source is capable of generating a large variety of primary ions which, in turn, can be neutralized to produce well-characterized beams of radicals and other transient species. Many of these chemically reactive species play an important role in various areas of application and modelling and there is an urgent need for ionization and dissociative ionization cross section data of these species.

Progress has been made recently in using improved and novel semi-empirical approaches to the calculation/prediction of single ionization cross sections. While the initial success of these new methods has been quite impressive, the data base for a meaningful comparison with experimental data is still too limited to feel comfortable in extending these methods with any level of confidence to species for which there are no experimental data at this point in time. Future directions in this area should focus in addition to efforts of building a broader data base for comparison with experimental data on attempts to provide a more scientific basis for these semi-empirical approaches and, if at all possible, establish a connection to more rigorous theories. Activities along these lines are currently being pursued[51].

While improvements in these semi-empirical methods are commendable and should be encouraged, there is a serious lack of advances in rigorous theoretical methods for the calculation of ionization cross sections. It appears that novel theoretical concepts are urgently needed and highly desirable, perhaps in conjunction with the use of ever more powerful and faster computers, viz. highly parallel computers.

ACKNOWLEDGMENTS

We acknowledge the financial support of this work by the U.S. National Science Foundation through grants CTS-8902405, CTS-9017211, PHY-8910360, by the City University of New York through PSC-CUNY grant 663353 as well as through partial support from a NATO Collaborative Research Grant (CRG-920089). Acknowledgment is also made to the Donors of The Petroleum Research Fund, administered by the American Chemical Society, in partial support of this research through grant 26380-ACS. We are grateful to Dr. R.S. Freund, Mr. R.C. Wetzel, Dr.

M. Schmidt, Dr. R. Basner, Prof. R.A. Bonham, Prof. H. Deutsch and Prof. T.D. Märk for their continued interest in this work and for many helpful and stimulating discussions. We further acknowledge Prof. H.B. Gilbody, Prof. R.F. Stebbings and Dr. J.A. Syage for making some of their data available to us prior to publication.

REFERENCES

1. T.D. Märk, in "Electron-Molecule Interactions and their Applications", editor: L.G. Christophorou, Academic Press, London,1984
2. T.D. Märk and G.H. Dunn (editors), "Electron Impact Ionization", Springer Verlag, Wien, 1985
3. R.S. Freund, in "Swarm Studies and Inelastic Electron-Molecule Collisions", editors: L.C. Pitchford, V.B. McKoy, A. Chutjian and S. Trajmar, Springer Verlag New York, 1987
4. S.M. Younger, "Quantum Theoretical Methods for Calculating Ionization Cross Sections" in "Electron Impact Ionization", Springer Verlag, Wien, 1985, p. 1-23
5. S. Younger and T.D. Märk, "Semi-Empirical and Semi-classical Approximations for Electron Ionization" in "Electron Impact Ionization", Springer Verlag, Wien, 1985, p. 24-41
6. L.J. Kieffer and G.H. Dunn, Rev. Mod. Phys. 38, 1 (1966)
7. T.D. Märk, "Partial Ionization Cross Sections" in "Electron Impact Ionization", Springer Verlag, Wien, 1985, p. 137-97
8. J.R. Peterson, in "Proc. 3rd International Conference on the Physics of Electronic and Atomic Collision Processes (ICPEAC)", editor: M.R.C. McDowell, North Holland Publ. Com., Amsterdam, 1964, pp. 465
9. R.C. Wetzel, F.A. Biaocchi, T.R. Hayes and R.S. Freund, Phys. Rev. A 35, 559 (1987)
10. R.S. Freund, R.C. Wetzel, R.J. Shul and T.R. Hayes, Phys. Rev. A 41, 3575 (1990)
11. V. Tarnovsky and K. Becker, Z. Phys. D 22, 603 (1992)
12. M.B. Shah, D.S. Elliott and H.B. Gilbody, J. Phys. B 20, 3501 (1987)
13. M. Kopnarski, H. Oechsner and T. Halden, Contributed Paper, XVII. ICPEAC, Brisbane, Australia (1991), p. 176
14. M.A. Bolorizadeh, C.J. Patton, M.B. Shah and H.B. Gilbody, Contributed Paper, XVIII. ICPEAC, Aarhus, Denmark (1993)
15. M.B. Shah, P.McCallion, K. Okuno and H.B. Gilbody, Contributed Paper, XVIII. ICPEAC, Aarhus, Denmak (1993)
16. C. Ma, C.R. Sporleder and R.A. Bonham, Rev. Sci. Instr. 62, 909 (1991)
17. M.R. Bruce and R.A. Bonham, Z. Phys. D 24, 149 (1992)
18. J.A. Syage, J. Phys. B 24, L527 (1991)
19. C. Ma, M.R. Bruce and R.A. Bonham, Phys. Rev. A 44, 2921 (1991); M.R. Bruce and R.A. Bonham, Int. J. Mass Spectrom. Ion Proc. 123, 97 (1993)
20. K. Stephan, H. Deutsch and T.D. Märk, J. Chem. Phys. 83, 5722 (1985); H.U. Poll, C. Winkler, D. Margreiter, V. Grill and T.D. Märk, Int. J. Mass

Spectrom. Ion Proc. 112, 1 (1992)
21. R.J. Shul, R.S. Freund and R.C. Wetzel, Phys. Rev. A 41, 5856 (1990)
22. R.S. Freund, R.C. Wetzel and R.J. Shul, Phys. Rev. A 41, 5861 (1990)
23. V. Grill, G. Waldner, P. Scheier, M. Kurdel and T.D. Märk, Int. J. Mass Spectrom. Ion Proc. (1993), in press
24. J.A. Syage, J. Chem. Phys. 97, 6085 (1992)
25. R. Basner and M. Schmidt, Contributed Paper, XVIII. ICPEAC, Aarhus, Denmark (1993); R. Basner and M. Schmidt, private communication (1993)
26. J. Holtgrave, K. Riehl and P. Haarland, Bull. Am. Phys. Soc. 37, 2004 (1992)
27. F.A. Biaocchi, R.C. Wetzel and R.S. Freund, Phys. Rev. Lett. 53, 771 (1984)
28. T.R. Hayes, R.C. Wetzel, F.A. Biaocci and R.S. Freund, J. Chem. Phys. 88, 823 (1988); ibid. 89, 4035 (1989); ibid. 89, 4042 (1989)
29. V. Tarnovsky and K. Becker, J. Chem. Phys. 98, 7868 (1993)
30. V. Tarnovsky, P. Kurunczi, D. Rogozhnikov and K. Becker, Int. J. Mass Spectrom. Ion Proc. (1993), in press
31. V. Tarnovsky and K. Becker, 46th Gaseous Electronics Conference, Montreal (1993), Abstracts of Contributed Papers, in press
32. H. Deutsch and T.D. Märk, Int. J. Mass Spectrom. Ion Proc. 79, R1 (1987)
33. H. Deutsch, C. Cornelissen, L. Cespiva, V. Bonacic-Koutecky and T.D. Märk, Int. J. Mass Spectrom. Ion Proc. (1993), in press
34. H. Deutsch, D. Margreiter and T.D. Märk, Int. J. Mass Spectrom. Ion Proc. 93, 259 (1989)
35. H. Deutsch and T.D. Märk, private communications (1992, 1993)
36. V. Tarnovsky, K. Becker, H. Deutsch and T.D. Märk, Bull. Am. Phys. Soc. 37, 2009 (1992)
37. H. Deutsch, C. Cornelissen, L. Cespiva, V. Bonacic-Koutecky and T.D. Märk, Int. J. Mass Spectrom. Ion Proc. (1993), in press
38. D. Margreiter, H. Deutsch, M. Schmidt and T.D. Märk, Int. J. Mass Spectrom. Ion Proc. 100, 157 (1990)
39. H. Deutsch, private communications (1992, 1993)
40. T.D. Märk, Plasma Phys. Controlled Fusion 34, 2086 (1992)
41. M. Gryzinski, Phys. Rev. 115, 374 (1959); ibid. 138, A305 (1965); Phys. Rev. Lett. 14, 1059 (1965)
42. H. Bethe, Ann. Phys. 5, 325 (1930); Z. Phys. 76, 293 (1932)
43. R. Tang and J. Callaway, J. Chem. Phys. 84, 6858 (1986)
44. T.D. Märk, Beitr. Plasmaphys. 22, 257 (1982)
45. K. Stephan, H. Helm and T.D. Märk, J. Chem. Phys. 73, 3763 (1980)
46. E. Krishnakumar and S.K. Srivastava, J. Phys. B 21, 1055 (1988)
47. D. Vikor, M. Mimic, I. Cadez and M. Kurepa, Fizika 21, 345 (1989)
48. P. McCallion, M.B. Shah and H.B. Gilbody, J. Phys. B 25, 1061 (1992)
49. M.R. Bruce, C. Ma and R.A. Bonham, Chem. Phys. Lett. 190, 285 (1992)
50. K. Codling, L.J. Franzinski, P.A. Hatherly, M. Stankiewicz and F.P. Larkin, J. Phys. B 24, 951 (1991)
51. H. Deutsch and T.D. Märk, private communication (1993)
52. R.F. Stebbings, private communication (1993), unpublished

ELECTRON-ATOM COLLISIONS

Recent Progress in Close-Coupling Electron-Atom Theory

Klaus Bartschat†

Department of Physics and Astronomy, Drake University, Des Moines,
Iowa 50311, USA

ABSTRACT

This talk reviews recent theoretical work on electron-atom collisions in the framework of the close-coupling method. It is shown how the inclusion of target continuum states has lead to significant advances in the solution of the three-body Coulomb problem for elastic scattering and excitation of hydrogen and light quasi one-electron targets. Some results for heavier and more complex target systems are also presented, together with some examples to illustrate the need for further theoretical work.

INTRODUCTION

In recent years, much progress has been made in experimental and theoretical studies of atomic collision processes. The use of spin-polarized collision partners[1] has enabled experimentalists to perform very detailed tests of theoretical models, particularly with regard to the description of spin-dependent effects such as electron exchange or the spin-orbit interaction.[2] At the same time, many new developments have taken place in the numerical treatment of such processes. In particular, the nonperturbative close-coupling method has been applied very successfully to various collision systems.[3,4]

In this talk, recent progress towards solving the close-coupling equations is reviewed. After a short summary of the basic equations in a nonrelativistic framework, I will concentrate on one of the most important problems, namely inclusion of the target continuum states. Much progress in this area has been achieved through the "Close-Coupling plus Optical Potential" approach (CCO) of the Adelaide group[5] and the "Intermediate Energy R-Matrix" method (IERM) of the Belfast group.[6] Another important step towards solution of the three-body Coulomb problem has been achieved with the "Convergent Close-Coupling" approach developed by Bray and Stelbovics[7] which has been applied to electron scattering from atomic hydrogen[7,8] and light quasi one-electron systems such as sodium.[9]

The second part of my talk deals with heavier and more complex target systems where some progress has been made in using R-matrix approaches based on both the Breit-Pauli and the Dirac-Breit Hamiltonians. After discussing some selected results for electron scattering from cesium and mercury atoms, I will finish with some unresolved problems that clearly demonstrate the need for further theoretical work.

THE CLOSE-COUPLING EXPANSION

The close-coupling approximation[10] has been a standard method of treating low-energy scattering, both elastic and inelastic, for many years. This nonperturbative method is based on an expansion of the total wavefunction for a collision system in

---

† 1992-93 Visiting Fellow, Joint Institute for Laboratory Astrophysics, University of Colorado, Boulder, Colorado 80309-0440, USA

terms of a sum of products which are constructed from target states Φ_i that diagonalize the N-electron target Hamiltonian

$$H_T^N = \sum_{i=1}^{N} \left[-\frac{1}{2}\nabla_i^2 - \frac{Z}{r_i} + \frac{1}{2}\sum_{j \neq i}^{N} \frac{1}{|\mathbf{r}_i - \mathbf{r}_j|} \right] \qquad (1)$$

according to

$$\langle \Phi_{i'} | H_T^N | \Phi_i \rangle = E_i \delta_{i'i} \quad , \qquad (2)$$

and unknown functions F_i describing the motion of the projectile. If relativistic effects are neglected, the wavefunction for each total orbital angular momentum L, total spin S and parity π is expanded as

$$\Psi^{LS\pi}(\mathbf{r}_1,\ldots,\mathbf{r}_{N+1}) = \mathcal{A} \sum_i \Phi_i^{LS\pi}(\mathbf{r}_1,\ldots,\mathbf{r}_N,\hat{\mathbf{r}}) \frac{1}{r} F_i(r) \quad . \qquad (3)$$

In Eq. (3), $\sum$ denotes a sum over all discrete and an integral over all continuum states of the target, while $\mathcal{A}$ is the antisymmetrization operator that accounts for the indistinguishability of the projectile and the target electrons. Furthermore, the angular and spin coordinates of the projectile electron (collectively denoted by $\hat{\mathbf{r}}$) have been coupled with the target states to produce the "channel functions" $\Phi_i^{LS\pi}(\mathbf{r}_1,\ldots,\mathbf{r}_N,\hat{\mathbf{r}})$.

After some algebraic manipulations, the unknown radial wavefunctions F_i are determined from the solution of a system of coupled integro-differential equations given by

$$\left[\frac{d^2}{dr^2} - \frac{\ell_i(\ell_i+1)}{r^2} + k^2 \right] F_i(r) = 2 \sum_j V_{ij}(r) F_j(r) + 2 \sum_j W_{ij} F_j(r) \qquad (4)$$

with the direct coupling potentials

$$V_{ij}(r) = -\frac{Z}{r}\delta_{ij} + \sum_{k=1}^{N} \langle \Phi_i | \frac{1}{|\mathbf{r}_k - \mathbf{r}|} | \Phi_j \rangle \qquad (5)$$

and the exchange terms

$$W_{ij} F_j(r) = \sum_{k=1}^{N} \langle \Phi_i | \frac{1}{|\mathbf{r}_k - \mathbf{r}|} | (\mathcal{A}-1) \Phi_j F_j \rangle \quad . \qquad (6)$$

For each "i", several sets of independent solutions (labeled by a second subscript "j") must be found, subject to the appropriate boundary conditions. For scattering from and excitation of neutral targets, these are given by[11]

$$F_{ij}(r=0) = 0 \qquad (7a)$$

$$\lim_{r \to \infty} F_{ij} = \delta_{ij} \sin\left(k_i r - \tfrac{1}{2}\ell_i \pi\right) + \mathcal{K}_{ij} \cos\left(k_i r - \tfrac{1}{2}\ell_i \pi\right); \; i = 1, n_{open} \qquad (7b)$$

$$\lim_{r \to \infty} F_{ij} = C_{ij} \exp(-|k_i|r); \; i > n_{open} \quad . \qquad (7c)$$

In Eqs. (7), $k_i = \sqrt{E - E_i}$ is the linear momentum in channel i for the total energy E (in Rydberg) and the channel energy E_i while n_{open} is the number of "open" channels for which k_i is a real number; for the "closed" channels, k_i is purely imaginary.

The coupled integro-differential equations (4) can be generalized to a relativistic framework, and the collision problem essentially consists of finding the solution to this system for each total energy. This can be done by various iterative, noniterative or algebraic methods (for an introductory overview, see Burke and Seaton[11]). Without going into the details, we note here that the nature of the physical problem makes some simplifications possible. For example, no exchange effects need to be considered for electron scattering outside a sphere of radius a, approximately the "size" of the target. Furthermore, the "centrifugal barrier" associated with partial waves of large angular momenta ensures that these waves only "see" the long-range part of the interaction potential. Consequently, the analytic "effective range formula"[12] and simpler Born-type approximations[13] can be used to speed up the calculation.

The final step consists of the calculation of scattering amplitudes[14] from the scattering ($\mathcal{S}$), transition ($\mathcal{T}$) or reactance ($\mathcal{K}$) matrices. Their relationship is given by

$$\mathcal{S} = 1 + \mathcal{T} = [1 + i\mathcal{K}][1 - i\mathcal{K}]^{-1}. \tag{8}$$

Any experimental observable of interest can then be obtained from the scattering amplitudes.[3,15]

NUMERICAL METHODS

One of the most popular and general methods used to solve the close-coupling equations is the R-Matrix method of the Belfast group. Its nonrelativistic formulation has been reviewed by Burke and Robb;[16] it was later extended to include parts of the Breit-Pauli Hamiltonian[17] and has recently also been used in fully relativistic formulations based on the Dirac-Breit Hamiltonian.[18-20] A summary of its applications to various atomic collision processes will be given in the plenary lecture by P.G. Burke[21] at this conference.

A recent advance using this approach that was discussed by Scholtz[6] at the XVII ICPEAC is the "Intermediate Energy R-Matrix" method (IERM). In this approach, the wavefunction for a given total energy E is expanded inside the R-Matrix box of radius a as

$$\Psi_E^{LS\pi}(\mathbf{r}_1,\ldots,\mathbf{r}_{N+1}) = \sum_k A_{Ek} \Psi_k^{LS\pi}(\mathbf{r}_1,\ldots,\mathbf{r}_{N+1}) \tag{9}$$

with *energy dependent* coefficients A_{Ek} and *energy independent* basis functions given by

$$\Psi_k^{LS\pi}(\mathbf{r}_1,\ldots,\mathbf{r}_{N+1}) =$$

$$\mathcal{A}\Bigg\{\sum_{i,j,m}\varphi_i^{LS\pi}(\mathbf{r}_1,\ldots,\mathbf{r}_{N-1},\hat{\mathbf{r}}_N,\hat{\mathbf{r}}_{N+1})\,u_{ij}(r_N)\,u_{im}(r_{N+1})\,a_{ijmk}$$

$$+\sum_{i,j}\Phi_i^{LS\pi}(\mathbf{r}_1,\ldots,\mathbf{r}_N,\hat{\mathbf{r}}_{N+1})\,u_{ij}(r_{N+1})\,b_{ijk}$$

$$+\sum_i\Theta_i^{LS\pi}(\mathbf{r}_1,\ldots,\mathbf{r}_{N+1})\,c_{ik}\Bigg\}. \tag{10}$$

In addition to the usual products of channel functions $\Phi_i^{LS\pi}(\mathbf{r}_1,\ldots,\mathbf{r}_N,\hat{\mathbf{r}}_{N+1})$ times "continuum orbitals" $u_{ij}(r_{N+1})$ and the $(N+1)$-electron "correlation functions" $\Theta_i^{LS\pi}(\mathbf{r}_1,\ldots,\mathbf{r}_{N+1})$ in Eq. (10), the products of ionic channel functions

$\varphi_i^{LS\pi}(\mathbf{r}_1,\ldots,\mathbf{r}_{N-1},\hat{\mathbf{r}}_N,\hat{\mathbf{r}}_{N+1})$ with two continuum wavefunctions account for the inclusion of the target continuum states. The coefficients a_{ijmk}, b_{ijk} and c_{ik} are obtained by diagonalizing the scattering Hamiltonian inside the R-matrix box.

The IERM, which is a systematic extension of earlier "pseudo-state" work,[22,23] has so far only been applied to electron scattering from hydrogen atoms where the ionic target wavefunctions in the first part of the expansion (10) describe the naked proton and simply reduce to a factor of 1. Extensive calculations have recently been performed, for example, dealing with excitation of the $n = 3$ and $n = 4$ levels.[24,25] To make these calculations numerically tractable, detailed checks regarding the necessary size of the basis were required.[24] Some recent results for differential cross sections and angular correlation parameters obtained with this method will be shown below.

Another very successful approach to representing the target continuum states is the "n-state Close-Coupling plus Optical Potential" approach (nCCO) of the Adelaide group[26] which was also discussed at the ICPEAC XVII by Bray.[5] The main idea is to solve the Lippmann-Schwinger equation

$$< \mathbf{k}_f f|T^S|\mathbf{k}_i i > = < \mathbf{k}_f f|V_Q^S|\mathbf{k}_i i > + \sum_{n \in P}\int d\mathbf{k}^3 \frac{< \mathbf{k}_f f|V_Q^S|\mathbf{k}n >}{E^{(+)} - \varepsilon_n + k^2/2} < \mathbf{k}|T^S|\mathbf{k}_i i > \quad (11)$$

for the T-matrix in a partial wave formalism. In (11), $\mathbf{k}_i$ and $\mathbf{k}_f$ are the momenta of the projectile in the initial and final channels, respectively, while V_Q^S is an "optical potential". This potential contains a sum/integral over all target states in Q-space and generally depends on the total spin S of the scattering channel. Since ICPEAC XVII, this model has been extended to include an increasing number of discrete target states in P-space, and also to quasi one-electron systems such as sodium[27] where the closed ionic core of Na$^+$ was represented by the frozen-core Hartree-Fock potential together with a small phenomenological polarization potential. The main approximation in the nCCO method is that of "weak coupling" in Q-space, i.e., the method contains full coupling of all channels related to the n target states that make up P-space, as well as couplings between Q- and P-space. Only strong coupling within the Q-space is neglected.

An important step towards removing this latter approximation in the nCCO method was recently undertaken by Bray and Stelbovics[7] who developed the "Convergent Close-Coupling" (CCC) approach. The main idea behind the CCC-method is to represent both discrete and continuum target states by diagonalizing the target Hamiltonian in a large Laguerre basis of the form

$$\xi_{kl}(r) = \left(\frac{\lambda(k-1)!}{(2l+1+k)!}\right)^{1/2}(\lambda r)^{l+1}\exp(-\lambda r/2)L_{k-1}^{2l+2}(\lambda r) \quad , \quad (12)$$

where $L_{k-1}^{2l+2}(\lambda r)$ is a standard associate Laguerre polynomial with λ an arbitrary parameter and k ranging from 1 to the size of the basis. By increasing the basis size, Bray and Stelbovics showed that the lower discrete target states are obtained to sufficient accuracy, while a representation of the target continuum including full coupling of all channels can be achieved at the same time. Their results describing transitions between the lowest few target states of hydrogen were estimated to be converged to about the 2% level.[7] Note, however, that the convergence depends on both the collision energy and the actual physical observables. For example, while the angular momentum transfer for the (3s ⟶ 3p) transition in atomic sodium might be converged in a given calculation for the (total spin) triplet channel, this may not necessarily be the case for the singlet channel (see figure 3 below).

CLOSE-COUPLING RESULTS FOR ELECTRON SCATTERING FROM HYDROGEN AND SODIUM ATOMS

Figure 1 shows differential cross sections for elastic scattering from atomic hydrogen in the initial $(1s)^2S$ state and for excitation of the $(2s)^2S$ and $(2p)^2P$ states, respectively. For an incident electron energy of 54.4 eV, the agreement between the results obtained with the CCC method[7] (a close-coupling calculation with 70 (!) channels), the 6CCO,[7] the IERM,[30] the large (40 channel) pseudo-state method of van Wyngaarden and Walters,[23] (vWW) and the perturbative "exact second order" distorted wave approach of Madison et al.[31] (DWB2) with the experimental data of Williams[28,29] is very good, except perhaps for the 1s $\longrightarrow$ 2s transition where the experimental data lie systematically above the theoretical predictions.

This rather satisfactory situation changes dramatically, however, when one looks at the λ, R and I parameters shown in figure 2.* These observables, which can be measured in electron-photon coincidence experiments, are related to the scattering amplitudes $f_{M_L}^S$ for excitation of the magnetic sublevels with $M_L = 0$ and $M_L = 1$ in the 1s $\longrightarrow$ 2p transition ($\lambda \equiv <f_0 f_0^*>/\sigma$, $R \equiv \text{Re}\{<f_1 f_0^*>\}/\sigma$, $I \equiv \text{Im}\{<f_1 f_0^*>\}/\sigma$, where the brackets denote the average over the singlet ($S = 0$) and triplet ($S = 1$) total spin channels while σ is the differential cross section). For the same incident energy, 54.4 eV, the differences in the predictions of the individual theories become more pronounced, and there is significant disagreement at larger scattering angles between all theoretical results and the two sets of independent experimental data,[32,33] which do agree with each other within the quoted error bars. Based on this rather disappointing finding, Bray and Stelbovics conclude that "[h]aving established convergent close-coupling results, we are forced to claim that the close-coupling theory is unable to explain the existing measurements of the angular correlation parameters.[7]" On the other hand, the strength of the close-coupling formalism was emphasized by the application of the CCC method to the calculation of the total ionization cross section and the spin asymmetry.[8] This yielded remarkable agreement with experiment,[34-36] even though no incorporation of the three-body ionization boundary condition[37] was implemented. Despite the difficulty of the experiments, I hope that some day new experimental data with smaller error bars will become available, as well as data for other observables that can be measured with spin-polarized electron and target beams. Such data might help to clarify the ongoing mystery in one of the most basic atomic collision processes.

Another very impressive calculation involving the CCC method has recently been performed by Bray[9] for electron scattering from atomic sodium. Figure 3 presents his results for the angular momentum transfer $L_\perp$ in the (3s $\longrightarrow$ 3p) excitation process for the individual singlet and triplet spin channels. The parameters $L_\perp^s$ and $L_\perp^t$ can be determined via the circular light polarization in an electron-photon coincidence experiment or in superelastic scattering from laser-excited sodium atoms.[38] Since both spin-polarized electron and atom beams are required to separate the two spin channels, the comparison presented in figure 3 allows for one of the most detailed tests of electron-atom collision theory to date.[39]

As can be seen from figure 3, for an incident electron energy of 10.0 eV the CCC results are in almost perfect agreement with the experimental data of the NIST

---

* The IERM data in this figure have been smoothed by supplementing the original $\mathcal{T}$-matrices for total angular momenta up to $L = 4$ by CCC results for $L \geq 5$.

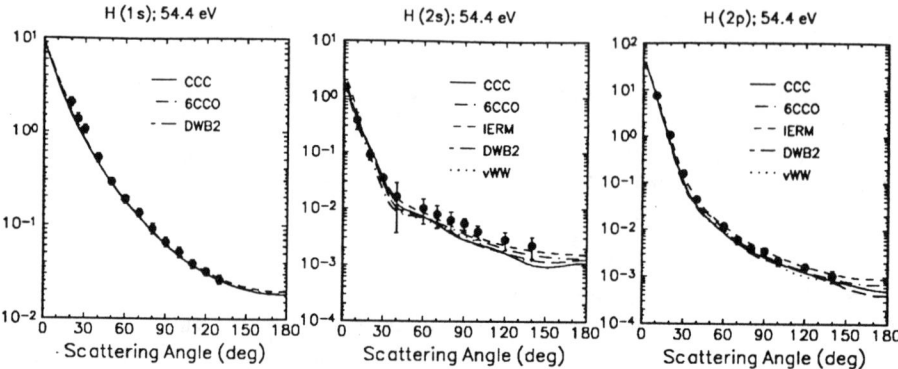

Figure 1: Differential cross sections for elastic electron scattering from the $(1s)^2S$ state and excitation of the $(2s)^2S$ and $(2p)^2P$ states of atomic hydrogen for an incident electron energy of 54.4 eV. The experimental data are from Williams.[28,29]

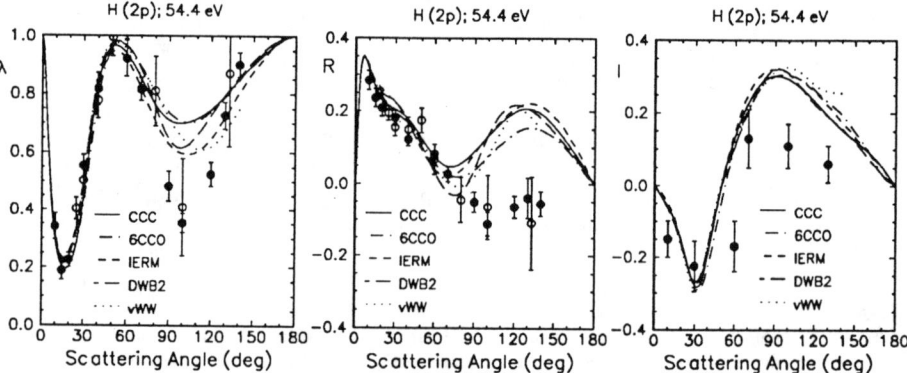

Figure 2: Angular correlation parameters for electron-impact excitation of the $(1s)^2S \longrightarrow (2p)^2P$ state of atomic hydrogen for an incident electron energy of 54.4 eV. The experimental data are from Weigold et al.[32] (o) and Williams.[33] (•)

group[40,41] and clearly represent a further improvement on the previous 15CCO results.[27] Furthermore, it is very clear that even a 15-state close-coupling expansion (15CC in figure 3) that includes only the discrete spectrum and is converged within this subspace[9] cannot reproduce the experimental data structure for $L^*_\perp$ at this energy in the ionization continuum. This is further illustrated in figure 4 where the spin asymmetry function A_{exch}, which can be measured by comparing cross sections for parallel and antiparallel projectile and target spin polarizations,[2] is shown for electron-sodium scattering. The importance of including the target continuum can be seen from the fact that the second-order distorted-wave results[42] (DBW2) are in better agreement with the experimental data[41] than the 15CC predictions.

As expected, simpler close-coupling expansions are sufficient at lower energies such as 4.1 eV. For this energy, independent results of the Adelaide[27] and the JILA[43] groups are in very good agreement with each other and with experiment.

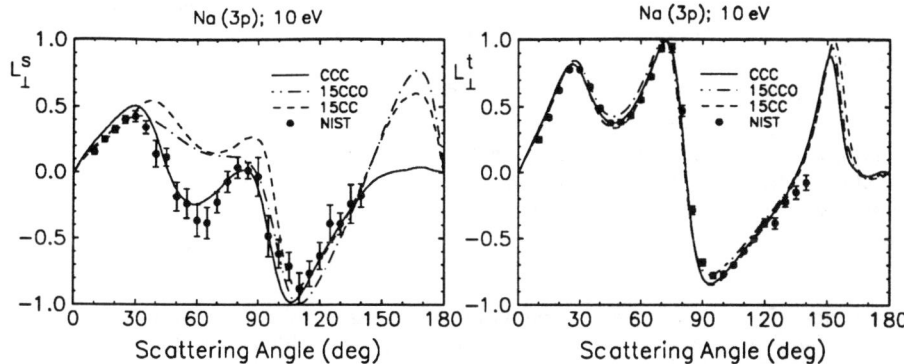

Figure 3: Angular momentum transfers $L_\perp^s$ (singlet channel) and $L_\perp^t$ (triplet channel) for electron-impact excitation of the $(3s)^2S \longrightarrow (3p)^2P$ state of atomic sodium for an incident electron energy of 10.0 eV.

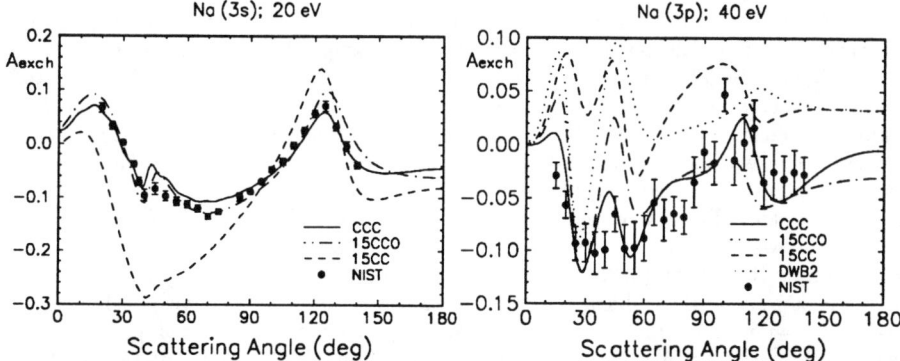

Figure 4: Spin asymmetry function A_{exch} for elastic electron-sodium scattering at 20 eV and for electron-impact excitation of the $(3s)^2S \longrightarrow (3p)^2P$ transition at an incident energy of 40.0 eV.

CLOSE-COUPLING RESULTS FOR ELECTRON SCATTERING FROM HEAVY ATOMS

I will now move on to a few selected cases of electron scattering from heavier and more complex target atoms. This topic will be further discussed in P.G. Burke's plenary lecture at this conference.[21]

A collision problem that has drawn a lot of interest in the past is low-energy electron scattering from atomic cesium. A fullrelativistic R-matrix approach to treat this problem has been developed by Thumm and Norcross[18] and will be discussed in detail by U. Thumm at this conference.[44] In their work, the Cs atom was treated as a quasi one-electron system, and a very sophisticated model potential was used to generate the target valence orbitals. While the first results indicated rather large discrepancies between the fullrelativistic results and early Breit-Pauli calculations of the Belfast group,[45,46] an improved Breit-Pauli calculation[47] with a model potential of similar quality[48] removed many of the discrepancies. As an example, figure 5 shows total cross section results just above the elastic threshold. Inclusion of the dielectronic core-polarization term[49] changes the $(6s6p)^3P_{0,1,2}$ configurations from bound states to

Figure 5: Total (= elastic) cross section for electron scattering from cesium atoms in their ground state $(6s)^2 S_{1/2}$ near the $(6s6p)^3 P^o_{0,1,2}$ resonances. Included for comparison are the results of a recent eight-state Breit-Pauli calculation[47] and the five-state Dirac approach of Thumm and Norcross.[51]

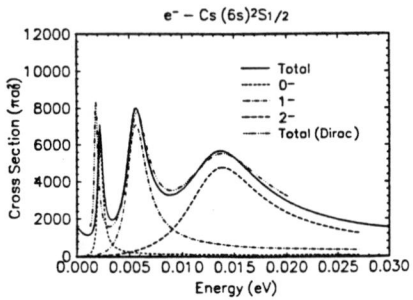

resonances, in agreement with some indirect experimental and theoretical evidence.[50] For a detailed discussion of the similarities and the differences found in the two calculations for differential cross sections[52] as well as spin[53] and light[54] polarization parameters, see the paper by U. Thumm.[44]

Another, more complex, collision system that has been investigated very thoroughly is electron-mercury scattering. Figure 6 shows recent total cross section results of the Belfast/Oxford group[19] for elastic scattering from the $(6s^2)^1 S_0$ ground state and for electron-impact excitation of the $(6s6p)^3 P_1$ and $(6s6p)^3 P_2$ states of mercury. These results have been obtained in a fullrelativistic all-electron (i.e., no model potentials) Dirac R-matrix calculation and they are compared with those from an earlier Breit-Pauli calculation of the Belfast group.[55] Perhaps surprisingly for such a heavy target system, the two theoretical models give rather similar results, although differences were found in the actual resonance positions. Due to the complexity of the target description and the lack of highly reliable experimental data for absolute total cross sections, however, it is not clear which of the calculations yields the better results. For example, the sharp $(6s6p^2)^4 P_{5/2}$ resonance, just above the $(6s6p)^3 P_1$ threshold at 4.89 eV, is very well established experimentally and is essentially responsible for the features seen in the Franck-Hertz experiment. Its position is indeed predicted very well by the Breit-Pauli calculation (at 4.90 eV) while the Dirac calculation finds this resonance at 4.81 eV – just below the excitation threshold, which changes the physical picture dramatically. It is also worth pointing out that the Breit-Pauli calculation gives very satisfactory agreement[3,4,56] with very detailed experimental data obtained in electron-photon coincidence experiments with spin-polarized electrons.[57] As an example, figure 7 shows two alignment angles that provide information about the charge cloud distribution of mercury atoms after excitation by spin-polarized electrons.[56,57] No such results from the Dirac calculation have been reported to date, but a similarly detailed test would be highly desirable.

I would like to conclude this talk by presenting two problem cases that clearly require more work in the future. These problems usually become very apparent in comparisons of theoretical results with experimental data that are sensitive to both the magnitudes and the relative phases of the scattering amplitudes. One of the more readily accessible observables is the asymmetry function S_A that determines the left-right asymmetry in the differential cross section for scattering of spin-polarized electrons.[1] For an electron polarization P perpendicular to the scattering plane, this asymmetry is given by

$$A \equiv \frac{\sigma_l(E,\theta) - \sigma_r(E,\theta)}{\sigma_l(E,\theta) + \sigma_r(E,\theta)} = S_A(E,\theta)\, P \quad , \qquad (13)$$

where θ is the scattering angle.

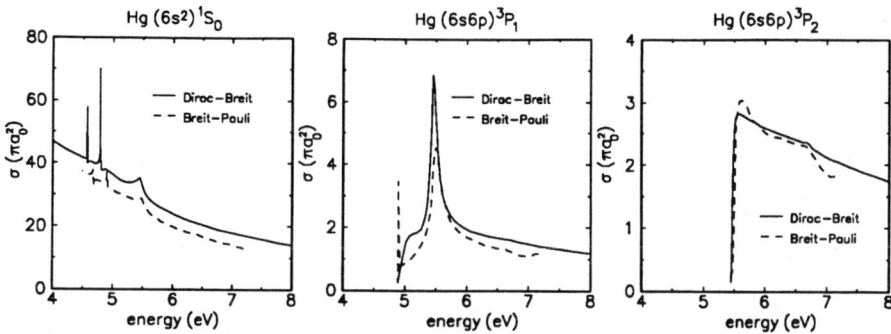

Figure 6: Total cross sections for electron scattering from the $(6s^2)^1S_0$ state and excitation of the $(6s6p)^3P_{1,2}$ states of mercury as a function of the incident projectile energy. Compared are the results of a semirelativistic Breit-Pauli[55] and a fullrelativistic Dirac[19] R-matrix calculation.

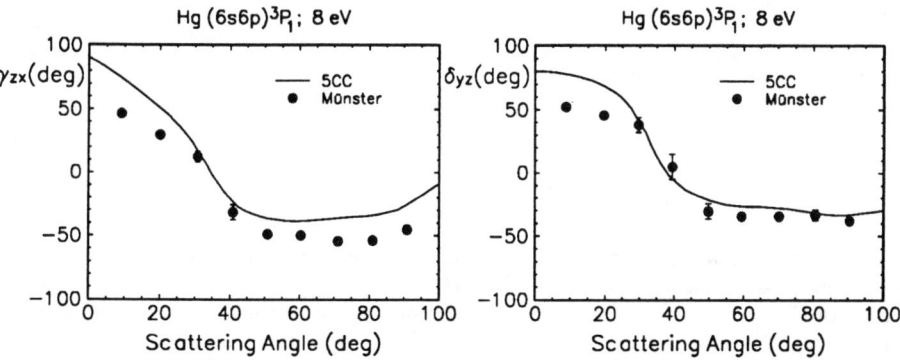

Figure 7: Alignment angles γ_{zx} and δ_{yz} for electron-impact excitation of the $(6s6p)^3P_1$ state of mercury for an incident projectile energy of 8 eV. The experimental data[57] are compared with results based on a five-state Breit-Pauli R-matrix calculation.[55]

Results for electron scattering from lead atoms at an incident electron energy of $E = 5$ eV are shown in figure 8 where the experimental data of the Münster group[58] are compared with the predictions of a five-state Breit-Pauli calculation[59,60] and (for elastic scattering only) with those from a fullrelativistic "generalized density functional" (GDF) method.[61] While there is at least qualitative agreement between theory and experiment for elastic scattering, the discrepancies are very pronounced for the inelastic transitions. It is not clear at the moment whether the Breit-Pauli approach is simply insufficient to describe this collision problem or whether the disagreement can be removed by a many-state close-coupling calculation, either by treating all electrons explicitly or by using improved model potentials and target wavefunctions.

As a final example, figure 9 shows another surprise that one might encounter in close-coupling calculations for electron scattering from complex target systems. While a very simple one-state (i.e., "static exchange") Breit-Pauli model potential calculation[62] yields near perfect agreement with experiment[63] for the asymmetry function S_A in

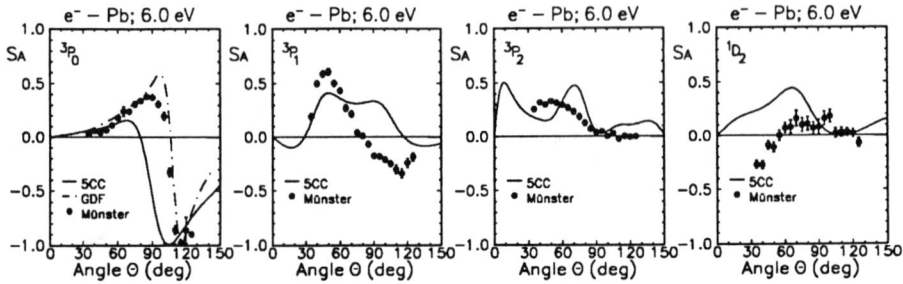

Figure 8: Asymmetry function S_A for elastic scattering from the $(6p^2)^3P_0$ ground state and for excitation of the $(6p^2)^3P_{1,2}$ and $^1D_2$ states of lead at an incident electron energy of 6 eV.

Figure 9: Asymmetry function S_A for elastic scattering from the $(6p)^2P_{1/2}$ ground state of atomic thallium at an incident electron energy of 1 eV.

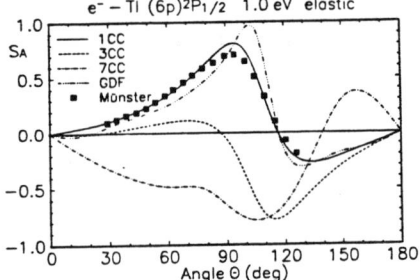

elastic electron scattering from atomic thallium, the agreement disappears completely when more target states are included in the close-coupling trial function. While the origin of the problem (which vanishes for collision energies above 2.5 eV) is not known precisely at the moment, it is most likely related to a large number of overlapping resonances in the very low-energy regime. While a recent GDF calculation[64] which optimized the nonlocal potential "seen" by the projectile in the elastic scattering channel alone also gives very satisfactory results for this channel, a fullrelativistic all-electron close-coupling calculation would be highly desirable to shed some light on this situation.

SUMMARY

I have tried to review some recent progress in the use of the close-coupling method to treat electron-atom scattering. Due to limited space and time, I have concentrated on neutral atoms only, with some emphasis on light quasi one-electron systems where the theory has advanced tremendously, and on some heavy target systems where – despite the progress achieved in recent years – much more work will need to be done. Given advances in the construction of electronic computers, however, the time may come when "many-electron multi-state close-coupling" calculations can be performed routinely on small workstations, but at the moment there is certainly still room for a lot of improvement.

ACKNOWLEDGEMENTS

I would like to thank my colleagues from the theoretical and experimental groups at Adelaide, Belfast, Boulder (JILA), Gaithersburg (NIST) and Münster for making their data readily available, sometimes even prior to publication. The author's work reported in this review was supported, in part, by the National Science Foundation under grant PHY-9114103, by the Research Corporation under grant C-2640, by the Deutsche Forschungsgemeinschaft in SFB 216, by NATO under grants CRG-900144 and CRG-930056, and by a JILA Visiting Fellowship.

REFERENCES

1. J. Kessler: *Polarized Electrons*, 2nd edition (Springer, Heidelberg 1985)
2. J. Kessler, Adv. At. Mol. Opt. Phys. **27**, 81 (1991)
3. K. Bartschat, Phys. Rep. **180**, 1 (1989)
4. K. Bartschat, Comm. At. Mol. Phys **27**, 239 (1992)
5. I. Bray in: *Electronic and Atomic Collisions, Invited Papers for XVII ICPEAC* (IOP Publishing 1992), 191 (and references therein)
6. T.T. Scholtz in: *Electronic and Atomic Collisions, Invited Papers for XVII ICPEAC* (IOP Publishing 1992), 181 (and references therein)
7. I. Bray and A.T. Stelbovics, Phys. Rev. A **46**, 6995 (1992)
8. I. Bray and A.T. Stelbovics, Phys. Rev. Lett. **70**, 746 (1993)
9. I. Bray, submitted to Phys. Rev. Lett. (1993) and private communication
10. N.F. Mott and H.S.W. Massey: *The Theory of Atomic Collisions* (Clarendon, Oxford 1965)
11. P.G. Burke and M.J. Seaton, Meth. Comput. Phys. **10**, 1 (1971)
12. L. Rosenberg, T.F. O'Malley and L. Spruch, J. Math. Phys **2**, 491 (1961)
13. M.J. Seaton, Proc. Phys. Soc. **77**, 174 (1961)
14. K. Bartschat and N.S. Scott, Comp. Phys. Commun. **30**, 369 (1983)
15. K. Bartschat, Comp. Phys. Commun. **30**, 383 (1983)
16. P.G. Burke and W.D. Robb, Adv. At. Mol. Phys. **11**, 143 (1975)
17. N.S. Scott and P.G. Burke, J. Phys. B **13**, 4299 (1980); for an overview of recent calculation, see refs. [3,4]
18. U. Thumm and D.W. Norcross, Phys. Rev. Lett. **67**, 3495 (1991)
19. W.P. Wijesundera, I.P. Grant and P.H. Norrington, J. Phys. B **25**, 2143 (1992)
20. W.P. Wijesundera, I.P. Grant and P.H. Norrington, J. Phys. B **25**, 2165 (1992)
21. P.G. Burke in: *Electronic and Atomic Collisions, Invited Papers for XVIII ICPEAC*, this conference
22. J.W. Callaway, Phys. Rev. A **32**, 775 (1985)
23. W.L. van Wyngaarden and H.R.J. Walters, J. Phys. B **199**, 929, 1817, and 1827 (1986)
24. M.P. Scott and P.G. Burke, J. Phys. B **26**, L 191 (1993)
25. M.P. Scott, P.G. Burke, B.R. Odgers and B.M. McLaughlin, *Electronic and Atomic Collisions, Contributed Papers for XVIII ICPEAC*, this conference
26. I. Bray, D.A. Konovalov and I.E. McCarthy, Phys. Rev. A **43**, 1301 (1991)
27. I. Bray and I.E. McCarthy, Phys. Rev. A **47**, 317 (1993)
28. J.F. Williams, J. Phys. B **8**, 2191 (1975)
29. J.F. Williams, J. Phys. B **14**, 1197 (1981)
30. T.T. Scholtz, H.R.J. Walters, P.G. Burke and M.P. Scott, J. Phys. B **24**, 2097 (1991)
31. D.H. Madison, I. Bray and I.E. McCarthy, J. Phys. B **24**, 3861 (1991)
32. E. Weigold, L. Frost and K.J. Nygaard, Phys. Rev. A **21**, 1950 (1979)

33. J.F. Williams, Aust. J. Phys. **39**, 621 (1986)
34. M.B. Shah, D.S. Elliott and H.B. Gilbody, J. Phys. B **20**, 3501 (1987)
35. G.D. Fletcher, M.J. Alguard, T.J. Gay, V.W. Hughes, P.F. Wainwright, M.S. Lubell and W. Raith, Phys. Rev. A **31**, 2854 (1985)
36. D.M. Crowe, X.Q. Guo, M.S. Lubell, J. Slevin and M. Eminyan, J. Phys. B **23**, L325 (1990)
37. M. Brauner, J.S. Briggs and H. Klar, J. Phys. B **22**, 2265 (1989)
38. J.J. McClelland, M.H. Kelley and R.J. Celotta, Phys. Rev. Lett. **56**, 1362 (1986)
39. N. Andersen and K. Bartschat, Comm. At. Mol. Phys., in press (1993)
40. J.J. McClelland, M.H. Kelley and R.J. Celotta, Phys. Rev. A **40**, 2321 (1989)
41. M.H. Kelley, J.J. McClelland, S.R. Lorentz, R.E. Scholten and R.J. Celotta in: *Correlations and Polarization in Electronic and Atomic Collisions and (e,2e) Reactions* (eds. P.J.O. Teubner and E. Weigold), IOP Conference Series 122, 23 (Bristol 1992)
42. D.H. Madison, K. Bartschat and R.P. McEachran, J. Phys. B **25**, 5199 (1992)
43. H.L. Zhou, W.K. Trail, B.L. Whitten, M.A. Morrison, T.L. Goforth, K.B. MacAdam, E.G. Layton, K. Bartschat and D.W. Norcross, *Electronic and Atomic Collisions, Contributed Papers for XVIII ICPEAC*, this conference
44. U. Thumm in: *Electronic and Atomic Collisions, Invited Papers for XVIII ICPEAC*, this conference
45. N.S. Scott, K. Bartschat, P.G. Burke, W.B. Eissner and O. Nagy, J. Phys. B **17**, L 191 (1984)
46. N.S. Scott, K. Bartschat, P.G. Burke, O. Nagy and W.B. Eissner, J. Phys. B **17**, 3775 (1984)
47. K. Bartschat, J. Phys. B **26**, in press (1993)
48. B.J. Albright, K. Bartschat and P.R. Flicek, J. Phys. B **26**, 337 (1993)
49. C.D.H. Chisholm and N. Öpik, Proc. Phys. Soc. London **83**, 541 (1964)
50. V.M. Borodin, I.I. Fabrikant and A.K. Kazansky, Phys. Rev. A **44**, 5725 (1991)
51. U. Thumm and D.W. Norcross, Phys. Rev. A **45**, 6349 (1992)
52. U. Thumm and D.W. Norcross, Phys. Rev. A **47**, 305 (1993)
53. U. Thumm, K. Bartschat and D.W. Norcross, J. Phys. B **26**, in press (1993)
54. K. Bartschat, U. Thumm and D.W. Norcross, J. Phys. B **25**, L 641 (1992)
55. N.S. Scott, P.G. Burke and K. Bartschat, J. Phys. B **16**, L 361 (1983)
56. A. Raeker, K. Blum and K. Bartschat, J. Phys. B **26**, 1491 (1993)
57. M. Sohn and G.F. Hanne, J. Phys. B **25**, 4627 (1992)
58. H. Geesmann, M. Bartsch, G.F. Hanne and J. Kessler, J. Phys. B **24**, 2817 (1991)
59. K. Bartschat, J. Phys. B **18**, 2519 (1985)
60. H.-J. Goerss, R.-P. Nordbeck and K. Bartschat, J. Phys. B **24**, 2833 (1991)
61. R. Haberland and L. Fritsche, J. Phys. B **20**, 121 (1987)
62. K. Bartschat, in preparation (1993)
63. M. Dümmler, M. Bartsch, H. Geesmann, G.F. Hanne and J. Kessler, J. Phys. B **25**, 4281 (1992)
64. L. Fritsche, C. Kroner and Th. Reinert, J. Phys. B **25**, 4287 (1992)

THE DIRAC R-MATRIX METHOD FOR SCATTERING OF SLOW ELECTRONS FROM ALKALI-METAL-LIKE TARGETS

Uwe Thumm
Department of Physics, Kansas State University, Manhattan, KS 66506-2601 USA

ABSTRACT

The Dirac R-matrix theory is a close-coupling method that includes a fully relativistic representation of the target. We outline this method for the interaction of slow incident electrons with (effectively) one-electron, alkali-metal-like targets. Our numerical results for atomic cesium targets include total, angle-differential, and momentum-transfer scattering cross sections, spin-polarization parameters, and light-polarization parameters for the radiative decay of impact-excited target states. We discuss the influence of induced core polarization and relativistic effects on the lowest bound and resonant states of Cs^-.

1. INTRODUCTION

Being rather simple one-electron-like systems, alkali-like targets are an attractive subject for the study of their interactions with slow incident electrons.Theoretically, these relatively uncomplicated systems are accessible to an effective-potential description which reduces the dynamics of the multi-electron targets to the motion of a valence electron in a suitably chosen model potential for the noble-gas-like core. Over the years, such models turned out to be very successful in describing low-energy electron-impact phenomena on the condition that the effective core potentials are carefully matched to the lower part of the target spectrum [1, 2].

Heavy alkali-like targets can be used to investigate relativistic effects. Cesium, in particular, appears to be a good candidate for a detailed study of spin-, fine-structure-, or hyperfine-structure dependent effects; it is relatively easily accessible in the laboratory, has a long record of experimental investigations spanning more than 60 years [1, 3], is of some technological interest [3] and has been presented in several calculations [1]–[15]. Nevertheless, the electron-Cs collision system is a subject of speculation, in particular about the existence of low-lying resonances in the elastic scattering cross section [1] that had previously been identified as bound excited states of Cs^- [2, 10, 11, 16].

In this contribution, we first outline our formulation [1] of the Dirac R-matrix method [7] for the effective three-body system formed by two active electrons (the incident and the valence electron) and a relatively inert, but polarizable, noble-gas-like core (section 2). This non-perturbative method provides access to three-body bound states, as well as to the complete description of the scattering process with possibly spin-polarized collision partners. We then discuss our numerical results [1, 3, 13, 15] for atomic Cs targets and incident electrons with less than 3 eV kinetic energy in comparison with available experimental and theoretical data (section 3). Unless otherwise stated, we use atomic units.

2. THE DIRAC R-MATRIX METHOD FOR ALKALI-LIKE TARGETS

The key feature of the R-matrix theory is a division of configuration space into two (or more) subspaces in which the many-particle problem is treated at different levels of approximation. Introducing a sphere of radius ρ, large enough to accommodate the target in its ground and lowest excited states, we shall first treat the inner and outer regions separately and then combine the two in formulating the matching condition for the two-electron wave function at the surface of the 'R-matrix sphere'. The solution of the matching equation contains the complete information of the scattering event, is easily related to measurable quantities, and allows us to determine the affinities of bound states of the effective three-body system.

In the inner region ($r < \rho$), we solve the two-electron problem at a rather high level of approximation and include all one-electron relativistic effects, exchange, and the dielectronic response of the core to the two active electrons. We split the full Hamiltonian

$$H = H_1 + H_2 + H_{12} \tag{1a}$$

into one-electron Dirac Hamiltonians

$$H_k = c \sum_{i=1}^{3} \alpha_i p_{ki} + (\beta - 1)c^2 + V_{core}(r_k), \tag{1b}$$

that describe the interaction of each outer electron with the core through the adjustable potential

$$V_{core}(r) = V_{TFDA}(\lambda, r) + V_{pol}(\alpha_d, \alpha_q, \bar{r}_c, r) \tag{2a}$$

$$V_{pol} = -\frac{\alpha_d}{r^4} W_6(\bar{r}_c, r) - \frac{\alpha_q}{r^6} W_{10}(\bar{r}_c, r) \tag{2b}$$

$$W_n = 1 - \exp\left[-(r/\bar{r}_c)^n\right], \tag{2c}$$

and the interaction

$$H_{12} = \frac{1}{r_{12}} + V_{diel}(r_c, \vec{r}_1, \vec{r}_2) \tag{3a}$$

$$V_{diel} = -\frac{\alpha_d}{r_1^2 r_2^2} \left[W_6(r_c, r_1) W_6(r_c, r_2)\right]^{1/2} P_1(\cos\theta_{12})$$

$$-\frac{\alpha_q}{r_1^3 r_2^3} \left[W_{10}(r_c, r_1) W_{10}(r_c, r_2)\right]^{1/2} P_2(\cos\theta_{12}). \tag{3b}$$

The scaling parameter λ in the Thomas-Fermi-Dirac-Amaldi potential V_{TFDA}, the static dipole and quadrupole polarizabilities α_d and α_q, and the cut-off radius $\bar{r}_c$ of the induced core-polarization potential V_{core} are determined by fitting the target spectrum such that the eigenvalues of H_k reproduce the lowest energy levels of the target.

Apart from the Coulomb repulsion, the interaction H_{12} between the two active electrons is determined by the 'dielectronic term' V_{diel}, which adds the influence of the core polarization caused by one active electron on the motion of the other active electron. P_l are Legendre polynomials and θ_{12} is the angle between the radius vectors $\vec{r}_1$ and $\vec{r}_2$ of both electrons. The cut-off radius r_c in V_{diel} is used to 'gauge' the two-electron spectrum. For atomic alkali targets, r_c is obtained by matching the lowest eigenvalue of H to the (well known) electron affinity of the negative ion in its stable ground state.

By diagonalizing H (inside the R-matrix sphere) we obtain eigenvalues E_K and coefficients of the eigenfunctions

$$\Psi_K(\vec{r}_1, \vec{r}_2) = \sum_{ij} c_{ijK} \left[\phi_i^b(\vec{r}_1)\, \phi_{ij}^c(\vec{r}_2)\right] + \sum_{ii'} d_{ii'K} \left[\phi_i^b(\vec{r}_1)\, \phi_{i'}^b(\vec{r}_2)\right], \quad r_{1,2} \leq \rho \quad (4)$$

with respect to a basis of antisymmetrized (symbol) products of bound and continuum orbitals, ϕ^b and ϕ^c. Typically (for sufficiently low impact energies), only a small number of bound orbitals is needed in order to represent the valence electron in the ground and lowest excited states, in accord with the basic idea of a close-coupling ansatz. The bound orbitals are eigenfunctions of (1b) with the usual boundary conditions (they practically vanish at $r = \rho$). The Dirac-R-matrix theory requires the large and small radial components, $p^c(r)$ and $q^c(r)$, of ϕ^c to satisfy unphysical boundary conditions at $r = \rho$,

$$\frac{q^c}{p^c}\Big|_{r=\rho} = \frac{b+k}{2\rho c}, \quad (5)$$

where $k = \pm(j + 1/2)$ for $j = \ell \mp 1/2$ depends on the orbital angular momentum ℓ and the coupling of electron spin and ℓ to j. In the nonrelativistic limit (velocity of light = c $\to \infty$) and for our choice $b = 0$, Eqn.(5) becomes identical to the logarithmic derivative of p^c.

The Dirac R-matrix is essentally a multi-channel and relativistic generalization of the logarithmic derivative. It contains in compact form information on the dynamics of the two active electrons inside the R-matrix sphere. In terms of the surface amplitudes W_{iK} with the large component p_{ij}^c of the jth continuum orbital in channel i, the R-matrix elements are

$$R_{ii'} = \frac{1}{2\rho} \sum_K \frac{W_{iK} W_{i'K}}{E_K - E}, \quad W_{iK} = \sum_j c_{ijK}\, p_{ij}^c(\rho), \quad (6)$$

where E is the total energy and where each channel i is given by a specific electronic state of the target, the quantum number k of the scattered electron, and the total angular momentum JM_J of the two-electron system [1, 7].

For all practical applications only a finite number of continuum states ϕ_{ij}^c can be retained. This incompleteness with respect to the representation of the scattered electron inside the R-matrix sphere needs to be corrected. This is conveniently and with sufficient precision accomplished by adding the so-called 'Buttle correction' to the diagonal elements of the R-matrix [1].

In the outer region ($r > \rho$), exchange can be neglected (for large enough ρ), and the Dirac equation can be simplified by neglecting terms of the order $1/c^2$ (for slow incident electrons). The large radial components of the scattered-electron wavefunction then obey the coupled radial equations

$$\left[\frac{d^2}{dr^2} + 2(\epsilon_i + V)\right] p_i^c + \sum_{i'} \sum_{\lambda=1} \frac{b_{ii'}^\lambda}{r^{\lambda+1}}\, p_{i'}^c = 0, \quad r > \rho \quad (7)$$

where ϵ_i is the energy of the scattered electron in channel i. The monopole term V of the target potential vanishes identically for neutral targets, and p_i^c in different channels are coupled by the long-range potential of the perturbed target with multipole coefficients $a_{ii'}^\lambda$. The $a_{ii'}^\lambda$ and the diagonal centrifugal potentials are included in the coupling matrices $b_{ii'}^\lambda = -2a_{ii'}^\lambda - \ell_i(\ell_i+1)\delta_{ii'}\delta_{\lambda 1}$, where ℓ_i is the angular momentum of the scattered electron in channel i. If N_o channels are open, each channel contains N_o independent solutions of (7), labeled by the index j. Their asymptotic behavior in open ($\epsilon_i > 0$) and closed channels ($\epsilon_i < 0$) is

$$p_{ij}^c \xrightarrow[r \to \infty]{} \begin{cases} \mathfrak{x}_i^{-1/2}(\delta_{ij}\sin\phi_i + K_{ij}\cos\phi_i) & \epsilon_i > 0 \\ \tilde{K}_{ij}\exp(-|\mathfrak{x}_i|r) & \epsilon_i < 0 \end{cases}, \quad (8)$$

where $\mathfrak{x}_i^2 = 2\epsilon_i$. K_{ij} and $\tilde{K}_{ij}$ are elements of the 'K matrix' and its closed-channel extension, and ϕ_i is the usual phase (e.g., $\phi_i = \mathfrak{x}_i r - \ell_i \pi/2$, for neutral targets).

The K matrix is related to the scattering matrix by $S = (1+iK)(1-iK)^{-1}$ as well as to the spin-dependent scattering amplitude (by a recoupling formation, see [3] for details). We determine K by solving (7) and (8) numerically and by matching inner and outer space solutions at $r = \rho$. Scattering cross sections (for not spin-polarized collision partners) are now conveniently expressed in terms of S. The partial angle-integrated cross-section matrix for transitions between channels i and i' is (in units of πa_o^2)

$$\sigma_{i \to i'}^{J\Pi} = \frac{2J+1}{4\epsilon_i(2\underline{j_i}+1)}|S_{ii'} - \delta_{ii'}|^2, \quad (9)$$

where $\underline{j_i}$ is the initial angular momentum of the valence electron. Transitions between specific target states are addressed by the partial cross section $\sigma_{\underline{n\ell j} \to \underline{n'\ell'j'}}^{J\Pi}$, which is the sum of $\sigma_{i \to i'}^{J\Pi}$ over all channels i and i' that include $\underline{n\ell j}$ and $\underline{n'\ell'j'}$ as initial and final target states, respectively. Finally, as a measurable quantity, the total, angle-integrated cross section (AICS) $\sigma_{\underline{n,\ell,j} \to \underline{n',\ell',j'}}$ for elastic, inelastic or superelastic scattering is obtained by summing $\sigma_{\underline{n\ell j} \to \underline{n'\ell'j'}}^{J\Pi}$ over sufficiently many J and both (total) parities Π.

If all channels are closed ($N_o = 0$), the R-matrix method can still be applied [1], although this case does not describe a scattering problem. Matching of inner and outer space solutions is now only (if at all) possible for discrete values of the total energy E that correspond to bound states of H.

Apart from AICSs, quantities that are more sensitive to details of the collision dynamics, such as momentum-transfer cross sections (MTCS), angle-differential cross sections (ADCS), and spin-polarization parameters, can be expressed as bilinear products of the scattering amplitudes [3, 15, 17]. The information available in the K-matrix (or scattering amplitude) allows for a complete description of the scattering event and parallels the maximum information accessible in a 'complete scattering experiment' [17, 18].

3. APPLICATION TO ELECTRON - Cs INTERACTION

Being an alkali atom, Cs has a small (first) ionization energy (3.9 eV) that lies well below the minimal energy necessary for core excitation (12.3 eV). Therefore, for the interaction with slow electrons of incident kinetic energy clearly below the first ionization threshold, Cs satisfies the basic requirements of section 2. In addition, with a nuclear charge of Z=55, it has noticeable fine-structure splittings (70 meV for its first excited state) that, as we will see in a moment, exceed narrow resonance structures in the low-energy scattering cross section, thereby justifying a fully relativistic treatment.

Our calculation is based on an ℓ- and j- dependent fit of the parameters in V_{core} (2) to the most reliable energy levels available in the literature [1]. We included close coupling between the five lowest target states ($6s_{1/2}$, $6p_{1/2,3/2}$, $5d_{3/2,5/2}$), 24 continuum orbitals in each channel, and Buttle-corrected for the incompleteness of the basis of continuum orbitals. We found that $\rho = 40$ a.u. is sufficiently large and adjusted the remaining undetermined parameter r_c in V_{diel} (3) to the measured electron affinity of Cs$^-$ (471.5 meV).

3.1 Low-lying ^{3}P Resonances

We start the discussion of our numerical results with two multiplets of narrow ^{3}P resonances in the elastic scattering domain [1]. Table I shows affinities E_{Aff} and widths Γ of these resonances for the three normally-ordered terms ($J = 0, 1, 2$) in each multiplet. The surprisingly dramatic influence of the dielectronic potential (3b) is easily seen by comparing the resonance positions and width with (E_{Aff}, Γ_{Aff}), and without ($E_{Aff}^{(-d)}, \Gamma^{(-d)}$) inclusion of V_{diel} in the interaction H_{12} (2). The effect of V_{diel} can be understood by the following qualitative consideration. One of the two active electrons induces a dipole in the Xe-like core, which has its negative end where the other active electron is most likely be found, due to the predominant $1/r_{12}$ repulsions in (3a). The dielectronic term therefore enhances the electronic $1/r_{12}$ repulsion and tends to shift the negative ion spectrum towards higher energies. This shift amounts to about 20-30 meV, such that the $^3P_J^o$ resonances appear in the bound spectrum of Cs$^-$ if we neglect V_{diel}, in agreement with earlier suggestions [2, 10, 11, 16] that Cs may have *bound* excited negative-ion states. At very low scattering energies, the $^3P_J^o$ resonances strongly enhance the elastic scattering cross section (Fig. 1a). So far, the only experimental evidence that the 6s6p $^3P^o$ state of Cs$^-$ is a resonance is indirect and based on the analysis of the perturbation of Cs Rydberg levels by ground-state Cs and on the transport properties of electrons in weakly ionized Cs vapor [19].

We find the even-parity $6p^2$ $^3P_J^e$ states to be bound with respect to the first excitation threshold of Cs, in agreement with an earlier [6] nonrelativistic calculation. We note that in a nonrelativistic approach these states are strictly uncoupled to the adjacent continuum, such that their finite autoionization widths (Table I, Fig. 1b) are entirely due to relativistic interactions.

3.2 Angle Integrated Cross Sections

For incident electrons with energies between ~ 0 and 2.8 eV, we obtained converged AICS for elastic scattering and target excitation to $6p_{1/2,3/2}$ and $5d_{3/2,5/2}$ final states by summing partial cross sections over the 20 lowest J^Π symmetries (Fig. 2). The $6p^2$ $^3P_J^e$ resonances in Fig. 2 are not fully resolved and appear as a spike below the first excitation threshold, whereas the high-energy tail of the 6s6p $^3P_2^o$ term explains the rise of the cross section at the lowest shown energies. We believe that the peaks in the inelastic cross sections for $6p_{1/2,3/2}$ excitation in Fig. 2a originate in a multiplet of 6p nd $^3F_J^o$ resonances [1, 3, 22].

For the displayed energies, the sum of the elastic and inelastic cross sections in Fig. 2a is practically identical with the total cross section for electron-Cs scattering (Fig. 2b), in which the final target state remains unresolved. The most striking feature in the

268 The Dirac R-Matrix Method

Table 1: Negative-ion energies and resonance positions E_{Aff} and widths Γ in meV for Cs$^-$ with respect to the $6s_{1/2}$ threshold (for 6s6p $^3P^o_J$) or the $6p_{1/2}$ threshold (for $6p^2$ $^3P^e_J$). $E^{(-d)}_{Aff}$, $\Gamma^{(-d)}$ and E_{Aff}, Γ are obtained without and with the dielectronic term. Negative values indicate states not bound relative to the given threshold.

	J^π	$E^{(-d)}_{Aff}$	$\Gamma^{(-d)}$	E_{Aff}	Γ
6s6p $^3P^o_J$	0°	28.35		−1.78	0.42
	1°	21.39		−5.56	2.43
	2°	8.60		−12.76	9.32
$6p^2$ $^3P^e_J$	0^e	159.68	0.005	141.24	0.009
	1^e	148.02	0.19	129.71	0.15
	2^e	126.13	1.12	108.11	1.33

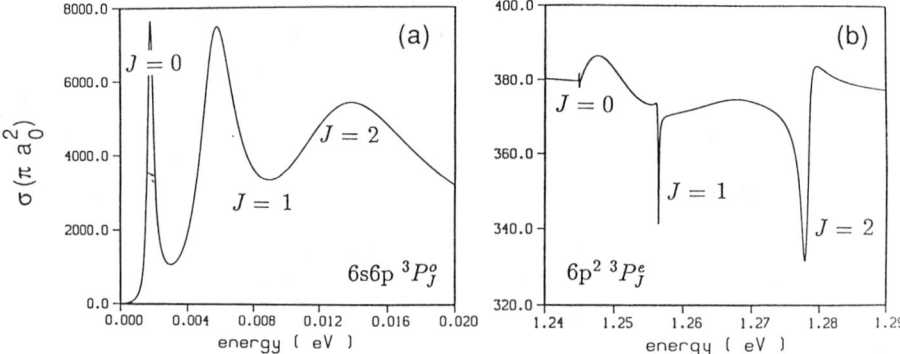

Figure 1: Angle-integrated cross sections near the $6s6p^3P^o_J$ (a) and $6p^2$ $^3P^e_J$ shape resonances.

comparison of our cross section with previous calculations is the strong $^3P^o$-resonance enhancement (Fig. 2b). The nonrelativistic two-state calculations in Ref. [4] and Ref. [5] predict bound $^3P^o_J$ states of Cs$^-$ which explains the lack of the resonance enhancement in the scattering cross sections (both previous calculations do not include core polarization through a dielectronic term). The more recent semirelativistic five-state calculation of Ref. [8] shows a large discrepancy with our results. This disagreement was recently [14] traced to the inappropriate use of orbitals that were derived from a trial core potential in Ref. [8]; recalculation of the orbitals in an adjusted core potential brings the results of the corrected calculation in close agreement with ours [14]. The absolute measurement of Ref. [20] has an experimental uncertainty of ± 20%, which is not quite enough to overlap our results. However, this 20-years-old investigation was a result of pioneering technology, and we therefore maintain the thought, that new absolute measurements are in order, even though a more recent experiment [21] may seem to provide evidence

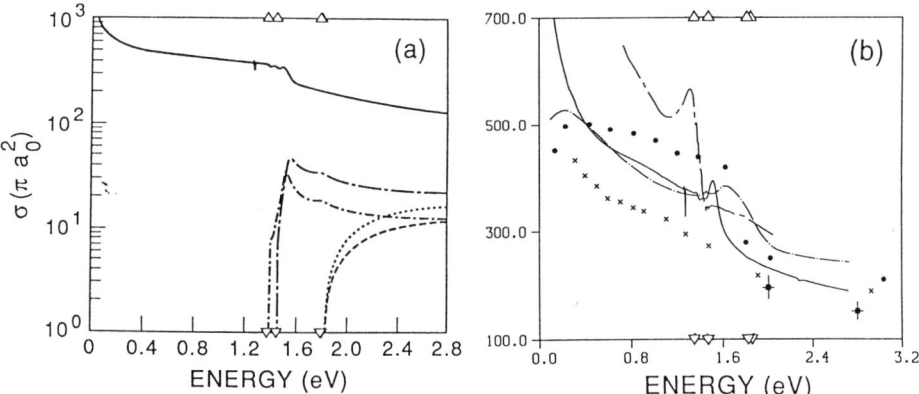

Figure 2: (a) Elastic and inelastic angle-integrated cross section. (b) Total angle-integrated cross section. Theory: present Dirac R-matrix calculation (Ref. [1]) (———), Ref. [5] (— · — · — ·), Ref. [8] (— - — -), Ref. [4] (●). Experiment: Ref. [20] (x), Ref. [21] (■). The $6p_{1/2,3/2}$ and $5d_{3/2,5/2}$ thresholds are marked by the small triangles.

that our results slightly overestimate total AICS.

3.3 Momentum-Transfer Cross Sections (MTCSs)

The MTCS is given by the integral over scattering angles θ of the ADCS weighted with the factor $(1-\cos\theta)$ that is proportional to the momentum transfer during the collision. Including a higher moment of the scattering distribution, the MTCS is more sensitive to the detailed collision dynamics than the AICS. For collisions with 0.1 to 1 eV impact energy, the MTCS is of interest for the development of thermionic energy converters [3] and as an input quantity for transport calculations that attempt to predict conductivities in swarm experiments [19, 23]. Our MTCSs [3] in Fig. 3 cover the energy range of primary interest for these plasma applications and also include the immediate vicinity of the elastic threshold. Being the result of a direct close-coupling calculation, we expect them to be of higher accuracy than rather indirect methods, shown for comparison, that refer to measured thermal and electrical conductivities.

The agreement with our prediction is best for the semiempirical effective-range extrapolation [19] (below 0.1 eV) of theoretical phase shifts calculated (above 0.1 eV) in Ref. [4]. It is interesting that this extrapolation per se leads to a bound $^3P^o$ state of Cs$^-$ and disagrees with experiment. However, the disagreement can be resolved if the polarizability of Cs as well as effective-range parameters are adjusted using experimental results for the shift and broadening of Cs Rydberg levels by Cs atoms [19]. This indirectly supports our predictions of an autoionizing $^3P^o_J$ state of Cs$^-$ and, indeed, as shown in Fig. 3, the adjusted extrapolation of Ref. [19] produces some of the $^3P^o_J$ resonance structure.

270 The Dirac R-Matrix Method

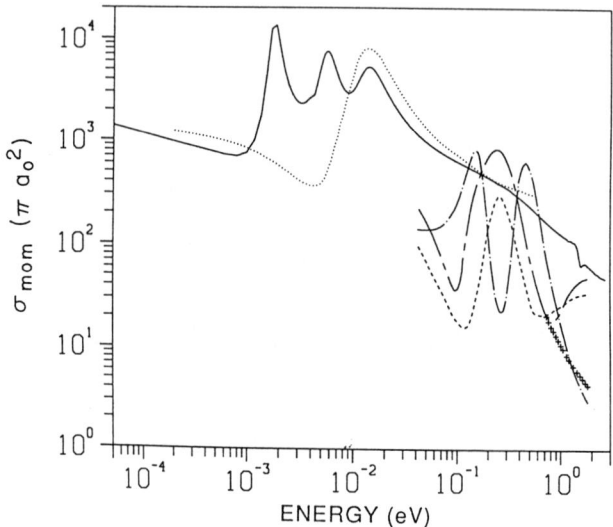

Figure 3: Momentum-transfer cross section for elastic scattering: ——— Dirac R-matrix calculation (Ref. [3]); ····· Fabrikant (Ref. [24]); —·—·— Stefanov (Ref. [23]); — — — — Nighan and Postma (taken from Ref. [23]); - - - - - - Crown and Russek (taken from Ref. [23]); ++++++ Postma (taken from Ref. [23]).

3.4 Angle-Differential Cross Sections

Regarded as a function of θ and for fixed scattering energy E, the very-low-energy ADCS [3] shows a pronounced $\cos^2 \theta$-like behavior due to dominant contributions from p-wave scattering (Fig. 4a). At E=100 meV this behavior still dominates (Fig. 4b), but loses its influence quickly as E is further increased due to strong admixture of higher partial waves. At 1 eV a local maximum appears at $\theta \approx 90°$ that can be assigned to increasingly important d-wave scattering. Regarded as a function of E and for fixed θ, we recover the three terms of the $^3P^o_J$ resonance (Fig. 4a) and a strong decline of elastic scattering at the onset of the first inelastic threshold (Fig. 4b). At energies above 1.4 eV, our elastic ADCSs are in fair agreement with the relative experimental results of Ref. [22] and also with the previous two-state calculation of Ref. [4] (Figs. 4c,d).

The inelastic ADCSs for 6s→6p$_{1/2}$ (Fig. 5a) and for 6s→6p$_{3/2}$ excitation (Fig. 5b) display a striking structure near the inelastic threshold that can be related to a $^3F^o$ resonance at the J-averaged energy E=1.62 eV [3, 22]. As for elastic scattering, the ADCS becomes strongly forward peaked as the scattering energy exceeds the close-to-threshold region. Recently, strong backward scattering close to the first inelastic threshold was also found in a merged-beam experiment for 3s→3p electron-impact excitation of Ar^{7+} [25]. Our results suggest that this phenomenon may be related to near-threshold resonances.

Comparison of the two subplots in Fig. 5 elucidates relativistic interactions through (i) the fine-structure split threshold for 6s→6p excitation and (ii) the deviation from

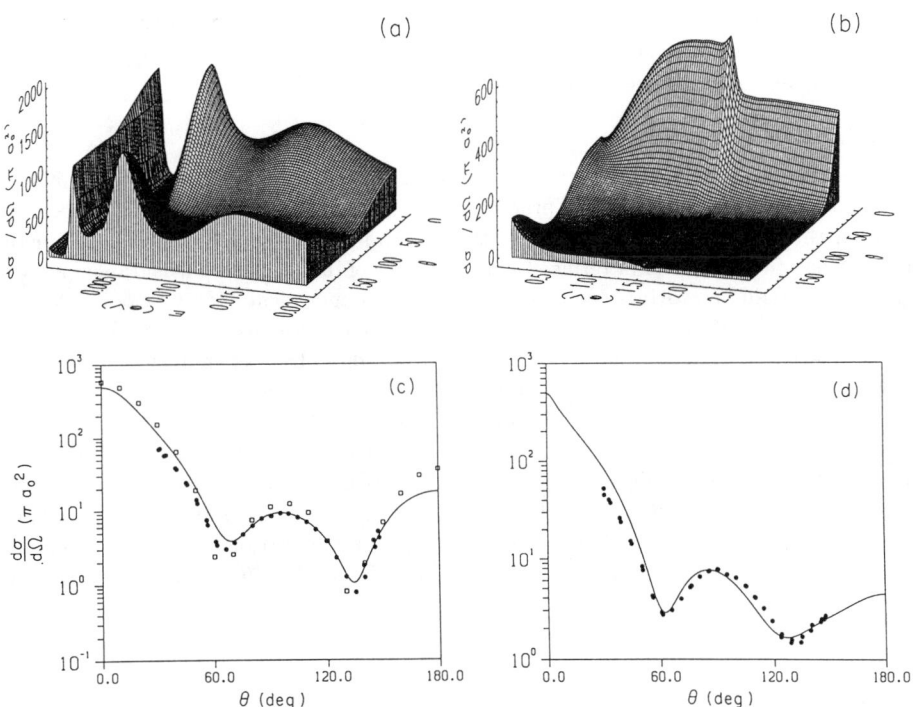

Figure 4: Angle-differential cross sections for elastic scattering. (a) E=0—20 meV; (b) E=0.1—2.7 eV; (c) E=1.4 eV; (d) E=1.6 eV. Experiment (relative): Ref. [22] (●). Theory: Dirac R-matrix calculation (Ref. [3]) (———); Ref. [4] (□).

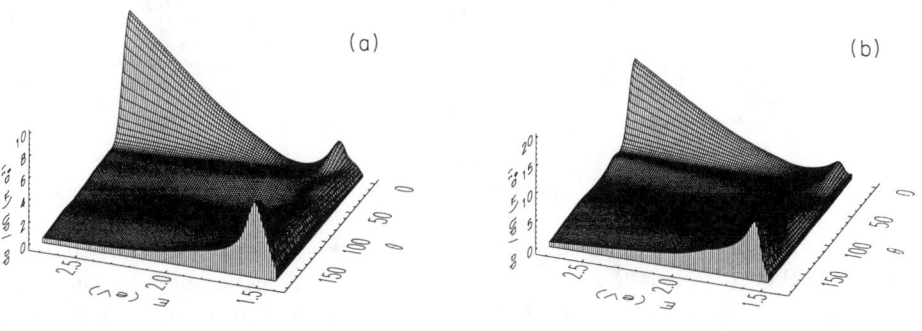

Figure 5: Angle-differential cross section (a) for 6s→$6p_{1/2}$ and (b) for 6s→$6p_{3/2}$ excitation.

the statistical branching ratio (=2), which is particularly striking for forward scattering at the highest shown energies and close to threshold. Note that without relativistic interactions, the two subplots would look alike.

3.5 Spin-Polarization Parameters

Cross sections for the scattering of unpolarized electrons extract only part of the information about the scattering process made available by our Dirac R-Matrix calculation through the spin-dependent scattering amplitudes [3]. In view of a complete description of the collision in terms of a maximal set of independent measurable quantities, which would be provided by a 'perfect scattering experiment' [18], we calculated the complete set of variables that parametrize the reduced spin density matrix for scattering of polarized electrons from unpolarized cesium atoms [15]. As an example, we show in Fig. 6 our results [15] for the seven 'generalized STU-parameters' [17] for $6s \rightarrow 6p_{1/2}$ excitation at $E=1.6$ eV that complement the ADCS in Fig. 4b. Comparison with the same parameters calculated for $6s \rightarrow 6p_{3/2}$ excitation (not shown here, cf. [15]) shows significant deviations for all seven variables. (This might not be surprising since we have already seen that the ADCSs strongly deviate from the statistical branching ratio.) Furthermore, relativistic interactions become apparent in comparison with the predictions $S_A[6s \rightarrow 6p_{1/2}] = -2S_A [6s \rightarrow 6p_{3/2}]$, $U_{xz} = U_{zx}$ and $T_x = T_y = T_z$ of a non-relativistic calculation [26].

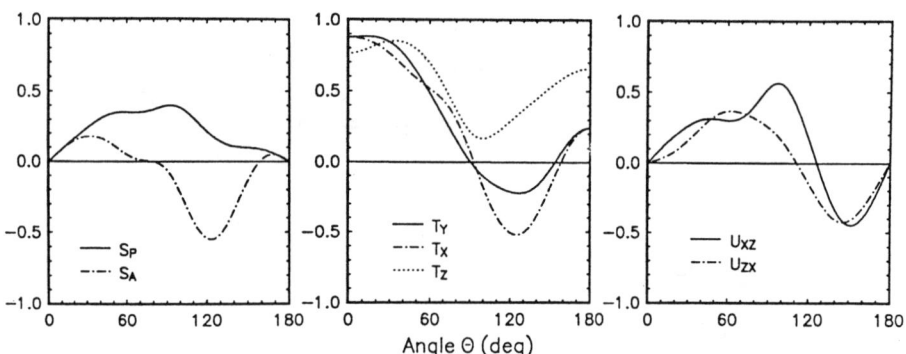

Figure 6: Generalized STU parameters for $6s \rightarrow 6p_{1/2}$ excitation at $E=1.6$ eV.

If both incident electrons and target are initially spin-polarized with parallel or antiparallel polarizations $\vec{P}_e$ and $\vec{P}_t$, parameters $A_\parallel$ and $A_\perp$ can be determined according to $(\sigma_{\uparrow\uparrow} - \sigma_{\uparrow\downarrow}/(\sigma_{\uparrow\uparrow} + \sigma_{\downarrow\uparrow}) = A_{\parallel,\perp} P_e P_t$, by measuring the ADCSs $\sigma_{\uparrow\uparrow}$ (for $\vec{P}_e \uparrow\uparrow \vec{P}_t$) and $\sigma_{\uparrow\downarrow}$ (for $\vec{P}_e \uparrow\downarrow \vec{P}_t$), where $\vec{P}_e$ and $\vec{P}_t$ either lie both in the scattering plane ($A_\parallel$) or are perpendicular to it ($A_\perp$). Without relativistic effects one finds $A_\parallel = A_\perp$, in sharp contrast to our numerical results (Fig. 7), that stress the dependence of sensitive spin-polarization parameters on relativistic interactions (most importantly on the spin-orbit coupling).

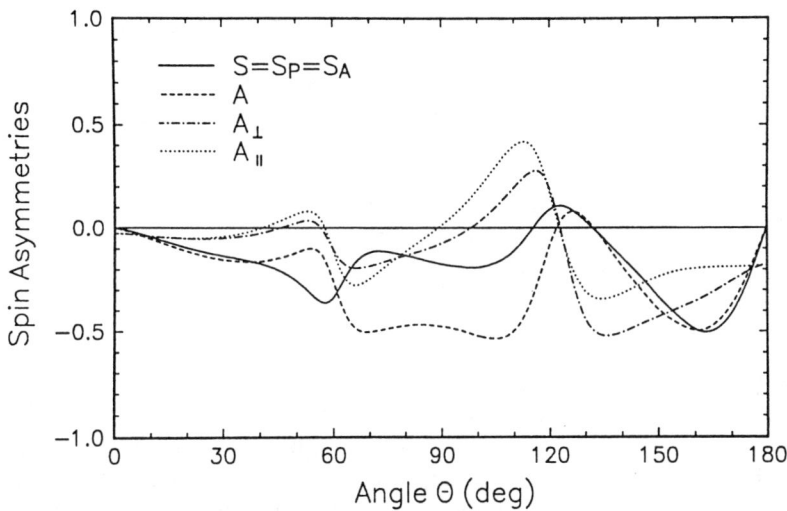

Figure 7: Sherman function S and spin-asymmetry parameters for elastic scattering of spin-polarized electrons from spin-polarized Cs atoms at E=1.6 eV.

3.6 Radiative Decay after Electron-Impact Excitation

The polarization characteristics of light emitted after electron-impact excitation of atoms can be expressed through 'integrated' (over the scattering angle of the unobserved projectile) 'Stokes parameters' (ISP). For electrons incident along the z-axis, for photons observed along the y-axis, and for emitted light intensities $I(\beta)$ (for linear polarization at angle β to the z-axis) and $I_{+,-}$ (for circular polarization), these parameters are given by $\eta_1^y = (I(45°) - I(135°))/(I(45°) + I(135°))$, $\eta_2^y = (I_+ - I_-)/(I_+ + I_-)$, for transversely polarized electrons with spin polarization $\vec{P}_e = P_y \hat{e}_y$, and by $\eta_3^y = (I(0°) - I(90°))/(I(0°) + I(90°))$ for unpolarized incident electrons. For longitudinally polarized electrons ($\vec{P}_e = P_z \hat{e}_z$) and photons observed along the z-axis, for symmetry reasons only η_2^z can be nonzero [13].

¿From the scattering amplitudes of Ref. [3], we calculated the ISPs for the optical decay of impact-excited $6p_{1/2,3/2}$ states of Cs atoms [13]. The comparison of our Dirac-R-matrix results with experiment [27, 28, 29] and a previous semirelativistic Breit-Pauli R-matrix calculation [9] is quite satisfactory for some ISPs, in particular in view of the high sensitivity of the parameters and experimental uncertainties [13], while severe discrepancies exist for others (Fig. 8). We note that a nonrelativistic approach (i.e., the neglect of fine-structure splitting and all spin-dependent forces) would predict $\eta_1^y = 0$, whereas we find small values of the order of 1%.

4. CONCLUSIONS

We have used the Dirac R-matrix method as a vehicle for an accurate description of bound and low-lying continuous states of the e^- + Cs system in terms of a variety of parameters such as resonance positions and widths, cross sections, and spin- and light-polarizations. We find that core-polarization and relativistic effects are essential for the

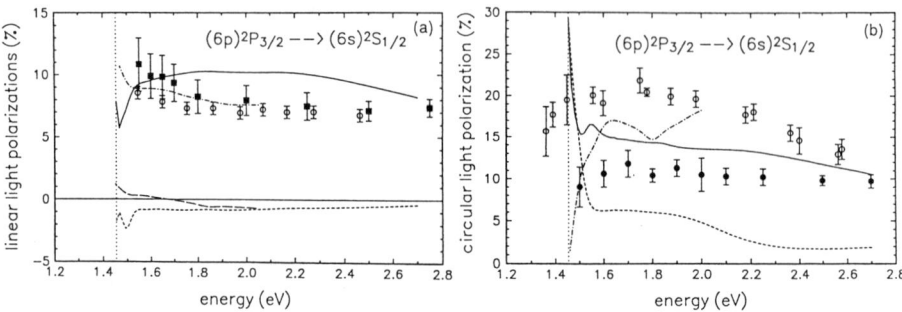

Figure 8: (a) Linear light polarizations for the optical transition $6p_{3/2} \to 6s_{1/2}$ in Cs after electron impact excitation: ———, present R-matrix calculation (Ref. [13]); for η_3^y; — — — — —, present result for η_1^y/P_y; — · — · — ·, Breit-Pauli result Ref. [9] for η_3^y; — — —, Breit-Pauli result for η_1^y/P_y. Experiment: (o), Ref. [27] for η_3^y; (■), Ref. [29]. (b) Circular light polarizations for the optical transition $6p_{3/2} \to 6s_{1/2}$ in Cs after impact excitation by spin polarized electrons. The individual curves are: ———, present result for η_2^y/P_y; — — — — —, present result for η_2^z/P_z; — · — · — ·, Breit-Pauli result for η_2^y/P_y. Experiment: (o), Ref. [27] for η_2^y/P_y; (●), Ref. [28] for η_2^z/P_z. The dotted line marks the excitation threshold of the $6p_{3/2}$ state.

detailed understanding of the scattering dynamics. In order to uniquely resolve existing discrepancies in the literature and to further test theoretical predictions, improved experimental investigations are highly desirable.

ACKNOWLEDGMENTS

This work was done in collaboration with David Norcross under the auspices of DOE. The generalized STU and Stokes parameters were calculated in collaboration with Klaus Bartschat. In the course of this work, I have benefited considerably from discussions with David, Klaus, Chris Greene, Michael Morrison, and many others I was lucky to meet at JILA.

References

[1] U. Thumm and D.W. Norcross, Phys. Rev. Lett. **67**, 3495 (1991); Phys. Rev. **A45**, 6349 (1992), and references therein.

[2] C.H. Greene, Phys. Rev. **A42**, 1405 (1990).

[3] U. Thumm and D.W. Norcross, Phys. Rev. **A47**, 305 (1993), and ref.s therein.

[4] E.M. Karule (1965); E.M. Karule and R.K. Peterkop (1965), as given in [1].

[5] P.G. Burke and J.F.B. Mitchell, J. Phys. B6, L161 (1973); 7, 214 (1974).

[6] D.W. Norcross, Phys. Rev. Lett. 32, 192 (1974).

[7] J.J. Chang, Phys. Rev. A12, 791 (1975); J. Phys. B8, 2327 (1975).

[8] N.S. Scott, K. Bartschat, P.G. Burke, O. Nagy, and W.B. Eissner, J. Phys. B17, 3775 (1984); 17, L191 (1984).

[9] O. Nagy, K. Bartschat, K. Blum, P.G. Burke, and N.S. Scott, J. Phys. B17, L527 (1984).

[10] J.L. Krause and R.S. Berry, Comments At. Mol. Phys. 18, 91 (1986).

[11] C.F. Fischer and D. Chen, J. Mol. Struct. 199, 61 (1989).

[12] W.P. Wijesundera, I.P Grant, P.H. Norrington, and F.A. Parpia, J. Phys B24, 1017 (1991).

[13] K. Bartschat, U. Thumm and D. Norcross, J. Phys. B25, L641 (1992), and ref.s therein.

[14] K. Bartschat, private communication, and to be published.

[15] U. Thumm, K. Bartschat and D.W. Norcross, J. Phys. B26, 1587 (1993).

[16] I.I. Fabrikant, Opt. Spectrosc. (USSR) 53, 131 (1982), and ref.s therein.

[17] K. Bartschat, Phys. Rep. 180 1 (1989).

[18] B. Bederson, Comments At. Mol. Phys. 1, 71 and 2, 65 (1969).

[19] V.M. Borodin, I.I. Fabrikant and A.K. Kazansky, Phys. Rev. A44, 5725 (1991).

[20] P.J. Visconti, J.A. Slevin, and K. Rubin, Phys. Rev. A3, 1310 (1971).

[21] B. Jaduszliwer and Y.C. Chan, Phys. Rev. A45, 197 (1992).

[22] W. Gehenn and E. Reichert, J. Phys. B10, 3105 (1977).

[23] B. Stefanov, Phys. Rev. A22, 427 (1980).

[24] I.I. Fabrikant, J. Phys. B19, 1527 (1986).

[25] X.Q. Guo, et al., Phys. Rev. A47, R9 (1993).

[26] G.H. Hanne, Phys. Rep. 95, 95 (1983).

[27] F. Eschen, G.F. Hanne, K. Jost, and J. Kessler, J. Phys. B22 L455 (1989).

[28] P. Nass, M. Eller, N. Ludwig, E. Reichert, and M. Webersinke, Z. Phys. D11 71 (1989).

[29] S.T. Chen and A.C. Gallagher, Phys. Rev. A17, 551 (1978).

SOME NEW DEVELOPMENTS IN POLARIZED ELECTRON SCIENCE AND TECHNOLOGY[o]

T.J. Gay, J.E. Furst[*], and W.M.K.P. Wijayaratna[§]
Laboratory for Atomic and Molecular
Research and Physics Department, University of
Missouri-Rolla, Rolla, Missouri 65401, U.S.A.

ABSTRACT

We discuss some recent developments in polarized electron physics and technology, emphasizing the current status of tests of relativistic inelastic electron-atom scattering theory. These tests involve the precision measurement of the polarization of light emitted following collisions between polarized electrons and heavy noble gas atoms. An ancillary benefit of such studies has been the development of high-efficiency optical electron polarimeters. Future directions this work might take are suggested.

INTRODUCTION

When electrons are scattered by heavy atomic targets, they generally experience magnetic forces as a result of their interaction with the highly-charged nuclei. These forces in turn produce a variety of physical phenomena, which are generally categorized under the heading "relativistic effects". An example of such an effect is the asymmetric scattering of polarized electrons in a plane perpendicular to their ensemble polarization vector. The asymmetry, which is proportional to the degree of the electron polarization, P_e, is a measure of the importance of the relativistic spin-orbit forces acting on the continuum electron, compared with the Coulomb interaction. Polarized electrons are indispensable as probes of spin-dependent relativistic effects of this type.[1]

The elastic scattering of polarized electrons by heavy nuclei or atoms is generically called "Mott scattering". More generally, the spin-orbit forces responsible for Mott scattering asymmetries can occur in inelastic scattering as well. Here we wish to consider what experimental tests might be made for the existence of these relativistic forces in inelastic scattering. This is a timely question, given the recent development of new, sophisticated atomic scattering theories which consider relativis-

---

o   Supported by National Science Foundation Grant PHY-9201289 and the University of Missouri

*   Present address: Physics Department, University of Western Australia, Nedlands, W.A.

§   Present address: Physics Department, University of Colombo, Colombo 03, Sri Lanka

tic effects in inelastic processes explicitly.[2]

An experimental search for spin-orbit forces on the continuum electron in inelastic collisions (we will refer to this as Mott scattering also) is complicated by the "fine-structure effect", discovered by Hanne in the mid-70's.[3] While Mott scattering asymmetries require significant magnetic forces to act on the continuum electron, the fine-structure effect can cause comparable asymmetries solely as a result of quantum interference between spin-flip and non-spin-flip channels in pure exchange scattering. As such, appreciable left-right scattering differences can be observed even with relatively light targets such as sodium.[4] In the case of inelastic collisions, all that is required for fine structure asymmetries to occur is that the final target state be created in an exchange collision with an incident polarized beam, that the initial and/or final target states be at least partially resolved within their respective fine-structure multiplets, and that orbital angular momentum perpendicular to the scattering plane (and parallel to the incident electron-spin direction) be transferred to or from the target. In elastic scattering (from S ground states), an unambiguous signature of relativistic spin-orbit forces on the continuum electron is the existence of a Mott asymmetry at a given polar scattering angle. In inelastic scattering, however, the existence of a left-right asymmetry might be due, at least in part, to the fine-structure effect. Thus clean tests of the relativistic effects predicted by the new theories must be made in some other way.

In 1982, Bartschat and Blum pointed out that for inelastic excitation processes involving polarized electrons in which the scattered electrons were not detected, certain restrictions on the Stokes polarization parameters of the light emitted by the excited states would hold, depending on the dynamics of the collision.[5] Specifically, they showed that if the combined orbital and spin angular momenta of the collision partners were separately conserved, the Stokes parameter η_1 would be identically zero. Here

$$\eta_1 \equiv P_2 \equiv \frac{C}{I} \equiv \frac{I(45°) - I(135°)}{I(45°) + I(135°)}, \qquad (1)$$

where $I(\theta)$ is the detected intensity for photons with linear polarization along an axis that makes an angle θ with the (electron beam) $\hat{z}$-axis in the xz-plane. The direction of photon emission and the electron polarization vector are taken to be along $\hat{y}$.

The value of η_1 can be non-zero either if the excited target state is not well-LS coupled (i.e. if L^2 and S^2 are time-dependent), or if Mott scattering occurs during the collision. The latter case means that the incident electron beam is, on average, deflected by the scattering to one side of the zy-plane. The symmetry of such collisions is illustrated in Fig. 1.

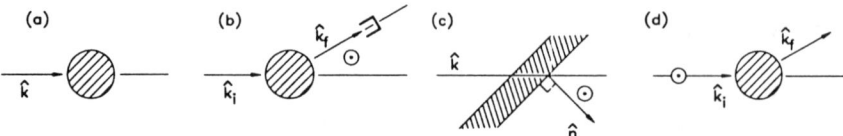

Fig. 1. Collision geometries (see text). Photons are emitted perpendicular to the plane of the diagram; axial vectors are indicated by a dot and concentric circle.

In scattering with unpolarized electrons with no electron detection, the only Stokes parameter that is non-zero is

$$\eta_3 \equiv P_1 \equiv \frac{M}{I} \equiv \frac{I(0°) - I(90°)}{I(0°) + (I(90°)} \quad (2)$$

(Fig. 1a). If the photons are detected in coincidence with electrons scattered to a given angle in a plane perpendicular to the direction of photon emission (Fig. 1b), an axial vector is now defined by the collision geometry and all three Stokes parameters, including the circular polarization fraction

$$\eta_2 \equiv P_3 \equiv \frac{S}{I} \equiv \frac{I(\sigma^+) - I(\sigma^-)}{I(\sigma^+) + I(\sigma^-)} \quad (3)$$

can be non-zero. (Here $\sigma^\pm$ designates photons with positive or negative helicity.) An equivalent symmetry reduction occurs in the case of the beam-foil interaction (Fig. 1c); when the foil normal is tilted off the incident beam axis it defines an axial vector in combination with the beam direction. In Mott scattering, which can also yield non-zero values of all three Stokes parameters, the symmetry is broken not by the initial collision geometry but by the axial vector of electron polarization. This vector in turn causes a "post collisional" axial vector associated with the final average beam momentum crossed with its initial value. While Mott scattering can produce a net atomic orbital angular momentum along $\hat{y}$ ($<L_y> \neq 0$; "orientation"), a non-zero value of η_2 does not constitute an unambiguous test for spin-orbit forces because orientation is also produced simply by exchange excitation with polarized electrons. (This is the basis of optical electron polarimetry.) Both exchange excitation and relativistic effects must occur for η_1 to be non-zero. It is amusing to note that in coincidence measurements, a deflected beam produces both non-zero η_2 and η_1 as well as η_3 (which does not require an axial vector), whereas the fine-structure effect, which can also yield beam deflections without relativistic effects, produces only circular polarization.

A non-zero η_1 value signifies Mott scattering unambiguously only if the excited state is well-LS coupled.[6] Other than the recent work in our laboratory which we report here,[6] the only other experiments done with well-LS coupled states of moderately heavy atoms in which final-electron-trajectory-integrated Stokes parameter measurements were made with polarized electrons were those of Jitschin et al.[7] using Na, and Eschen et al. with Cs.[8] The values of η_1 they reported were consistent with zero, although their measurements are not of great precision because they were not looking for relativistic effects. A recent calculation by Bartschat, Thumm, and Norcross[9] indicates that η_1 with Cs should be between 1% and 2% near threshold.

We decided to use a target of comparable nuclear charge to Cs, Xe, to make a precision optical test for relativistic effects. We chose to look at excitation of the $5p^56p$ $^3D_3$ state because it is well-LS coupled, decays with a convenient photon wavelength, and has 0.3 eV of "headroom" below the lowest excitation threshold of a state that can cascade into it. A relativistic first Born calculation of η_1 by Bartschat and Madison indicates that η_1 should have a very small value: 0.0001 with Pe = 1.[6] This result stands in curious contrast to the R-matrix Cs calculation,[9] and also to the fact that the final-trajectory-integrated value of the Xe elastic scattering Mott asymmetry is about 0.03 at 10 eV.[10]

EXPERIMENTS

The apparatus we used in these experiments has been described previously.[6,11] It consists of a GaAs polarized electron source, a concentric-electrode Mott polarimeter, and a target chamber illustrated in Fig. 2.

Just prior to entering the target chamber, the electron spins are rotated by a solenoidal spin rotator to point parallel or anti-parallel to the axis of the optical polarimeter. The beam is then decelerated from its transport energy of 2 keV and directed to intersect an effusive gas target before being collected in a Faraday cup. Light emitted by excited target atoms along the direction of electron polarization is collected by a 5 cm diameter lens which is an integral part of the vacuum wall. For linear polarization measurements it then passes through a dichroic film polarizer followed by a quarter-wave retarder, an interference filter, and a refocusing lens whose focal point is the GaAs photocathode of a photon-counting PM tube. For circular polarization measurements, the first optical element after the collection lens is the retardation plate, followed by the polarizer. The Stokes parameters η_1 and η_3 are measured by rotating the retarder and polarizer together, with the fast axis of the retarder held at 45° relative the polarizer transmission axis. This renders the light circularly polarized after these two elements, and significantly reduces instrumental polarization of the optical train's response to linearly polarized light. In η_2 measurements, the linear polarizer is kept fixed, which accomplishes the same purpose. Optical train instrumental asymmetries can be measured and corrected for in the η_1 and η_2 data simply by flipping the electron polarization.

In addition to our attempts to find optical evidence for Mott scattering in Xe by measuring η_1, we made general survey measurements of all the Stokes parameters

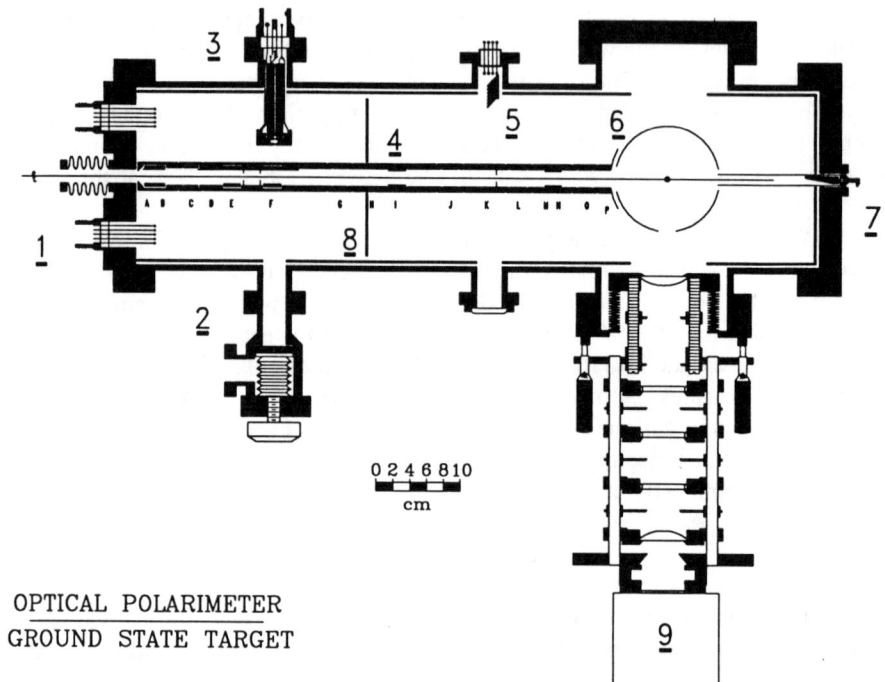

OPTICAL POLARIMETER
GROUND STATE TARGET

Fig. 2. Target chamber with numbered elements: (1) entrance aperture and electron gun electrical feedthroughs; (2) roughing valve; (3) quadrupole mass analyzer head; (4) electron gun with zoom lens and deflector elements; (5) nude ion gauge; (6) electrical shield for target region; (7) Faraday cup; (8) magnetic shield and differential-pumping baffle; (9) optical train ending in cooled photomultiplier tube housing. The effusive gas beam is directed into the plane of the diagram at the center of the target region.

for the heavy noble gases Ne-Xe for their $np^5(n+1)p\ ^3D_3 \rightarrow np^5(n+1)s\ ^3P_2$ transitions (n=2..5 in Ne..Xe, respectively). We also measured the Stokes parameters of light emitted by the non-well-LS coupled $4p^55p\ ^3D_2$ state in Kr.

RESULTS

Our measurements of η_1 for the $^3D_3$ states in the heavy noble gases yielded the following results: Ne, -0.0002(12); Ar, -0.0021(35); Kr, 0.0037(48); Xe, 0.0049(85).[12] These data were taken at energies just below the respective cascading thresholds of the various targets to maximize count rates without invalidating the requirement of pure LS coupling for the excited state. The Xe data, which thus had to be taken within 0.3 eV of the excitation threshold, required 50 h of running time (not including background runs and energy checks) to obtain statistical uncertainties of less than 0.01. Even with this precision the Xe result is consistent with zero, as are the other η_1 values, This is a somewhat surprising result, given that the Mott asymmetries for elastic and inelastic scattering at low energy from, e.g., Hg are of comparable magnitude at all scattering angles[1,13] and that, as mentioned before, the integrated elastic Mott asymmetry for Xe at 10 eV is about 3%.[10] Having said this, our result does agree with the relativistic Born result of -2.6×10^{-5} (for $P_e = 0.26$[11]). Why theory predicts a result more than two orders of magnitude larger for Cs is an open question.

A non-zero value of η_1 is allowed even in the absence of Mott scattering for a non Russell-Saunders target state. To test this case, we measured η_1 for the $4p^55p[5/2]_2 \rightarrow 4p^55s[3/2]_1$ transition, where the upper state is nominally designated as a $^3D_2$ level. In the intermediate coupling scheme it can be written as[6]

$$\psi(^3D_2) = 0.708\psi(^3D_2) + 0.684\psi(^1D_2) - 0.173\psi(^3P_2) , \qquad (4)$$

where the wavefunctions on the right-hand side of the equation represent pure Russell-Saunders states. Blum has noted a further restriction on η_1: it will be zero even for an intermediately coupled state unless L-values are mixed.[14] Thus if the coefficient of the last term in eq. (4) were nil, η_1 would still vanish.

However,, as can be seen in Fig. 3, we find η_1 to be non-zero (albeit small) for this transition. It is instructive to consider the mechanism whereby this occurs. A finite η_1 requires the existence of a $\mathfrak{I}_{21}(J)$ integrated state multipole moment of the excited $^3D_2$ ensemble of atoms. This corresponds in turn to a non-zero expectation value of the operator $(J_xJ_z+J_zJ_x)$. Physically, $\mathfrak{I}_{21}(J)$ represents "off-axis" alignment of the excited state or a tilt off the $\hat{z}$-axis in the zx plane of the oscillator distribution ensemble (which is correlated with the atomic charge distribution).

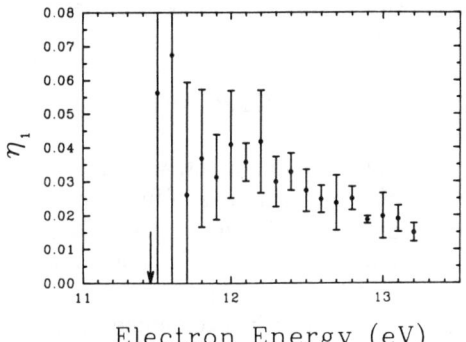

Fig. 3. Stokes parameter η_1 vs. electron energy for the Kr $^3D_2$ state. Arrow indicates excitation threshold.

In the absence of spin-orbit forces on the continuum electron, the only integrated multipole moments of L that can be non-zero immediately following the collision are $\mathfrak{I}_{20}(L)$ and $\mathfrak{I}_{00}(L)$. "Immediately" in this context means for times short compared with the fine-structure relaxation time. Similarly, only $T_{11}(S)$ and $T_{00}(S)$ are non-zero. Under LS-coupling, no combination of these orbital and spin multipole moments yields a $\mathfrak{I}_{21}(J)$. Thus immediately after the collision, the atoms have spin orientation along $\hat{y}$ and, separately, alignment along $\hat{z}$ only (left hand side of Fig. 4). For the $^3D_3$ state, the alignment and orientation remain "compartmentalized". But in the $^3D_2$ state, the strong magnetic forces responsible for the breakdown of LS coupling mix the spin alignment with the orbital orientation in a time of the order of the fine-structure relaxation time (Fig. 4). This mixing causes the atomic oscillators to tilt off $\hat{z}$, yielding a $\mathfrak{J}_{21}(J)$ component. It is important to note that during the collision, L and S of the collision system are separately conserved. The off-axis alignment, and hence η_1, is caused by breakdown of the atomic L and S values after the collision is over. The values of η_1 that we find for the $^3D_2$ state in Kr are presumably due entirely to these intra-atomic relativistic forces, given that the $^3D_3$ state in the same fine-structure manifold shows no evidence of tilting due to Mott scattering.

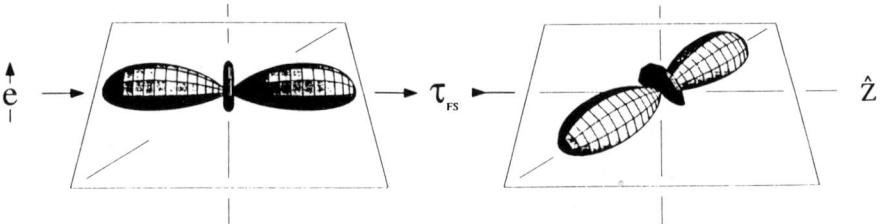

Fig. 4. Schematic representation of the rotation of an initially aligned, oriented charge cloud under the influence of strong intra-atomic magnetic forces. For simplicity, an n = 3, ℓ = 2, m = 0 hydrogenic wavefunction has been plotted. In our experiment, the charge cloud or oscillator density would have to be averaged over all the final (axially symmetric) electron trajectories.

RESONANCE EFFECTS AND ELECTRON POLARIMETRY

In making survey measurements of the heavy noble gases, we were struck by the weak energy dependence of η_2 and η_3 near threshold for all four gases. The excitation functions for these targets, both in our work and that of Kisker,[15] exhibit significant energy dependence due to negative-ion resonances. This is also true for He, specifically in the case of the $3^3P \rightarrow 2^3S$ transition. In He, however, the resonance features observed in the excitation function are mimicked in the η_3 energy dependence.[16] The values of η_2 do not depend on energy below the first cascade threshold. This is because the fine-structure relaxation time in He is much longer (~10^{-11}s) than typical resonance lifetimes (5×10^{-13}s).[17] This is not the case in the heavy noble gases, as can be seen in Fig. 5 for the $^3D_3 \rightarrow {}^3P_2$ transition in Kr. The lack of obvious resonance structure in the heavy noble gas $^3D_3$ η_2 curves is somewhat surprising, given that the fine-structure relaxation time of the $np^5(n+1)p$ manifolds is much shorter (< than 10^{-14}s) typical resonance lifetimes. Such resonance effects would also be likely to cause non-zero values of η_1, which compounds the puzzle.

The relatively large excitation cross sections for the heavy noble gases when compared with He led us to consider their use for optical electron polarimetry,

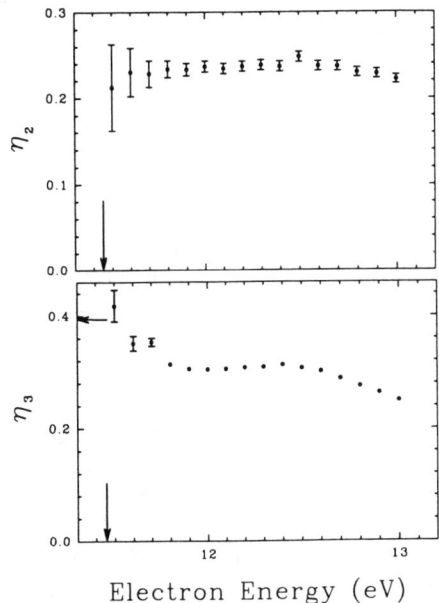

Fig. 5. Stokes parameters η_2 and η_3 vs. electron energy for the Kr^3D_3 state. Vertical arrows indicate the excitation threshold; the horizontal arrow is the kinematically required threshold value of η_3, assuming only $M_L = 0$ states are populated initially in the collision.

instead of the He ($3^3P \rightarrow 2^3S$) transition, which has been used previously.[16-18] In general, the electron polarization P_e can be related to fluorescence polarization following exchange excitation by the electrons whose polarization is to be measured:

$$\eta_2 = P_e \left[a + b\eta_3 \right], \tag{5}$$

where the constants a and b depend on the atomic transition in question and are determined solely by angular momentum algebra (kinematics) if the excited state is well-LS coupled. Table I lists the transition we have considered for electron polarization analysis.[6]

TABLE I. OPTICAL ELECTRON POLARIMETER TRANSITIONS

TRANSITION	WAVELENGTH	EXCITATION THRESHOLD	CASCADE THRESHOLD	a	b	THRESHOLD η_3	EFFICIENCY RELATIVE TO He
He($3^3P \rightarrow 2^3S$)	3889 Å	23.00 eV	23.59 eV	0.500	-0.167	0.366	1
Ne($3^3D_3 \rightarrow 2^3P_2$)	6402 Å	18.55 eV	19.66 eV	0.666	0.149	0.439	12
Ar($4^3D_3 \rightarrow 3^3P_2$)	8115 Å	13.07 eV	14.07 eV	0.667	0.148	0.439	10
Kr($5^3D_3 \rightarrow 4^3P_2$)	8112 Å	11.44 eV	12.11 eV	0.621	0.172	0.400	33
Xe($6^3D_3 \rightarrow 5^3P_2$)	8819 Å	9.72 eV	9.94 eV	0.632	0.196	0.387	2

The heavy noble gases have better analyzing power (a + bη_3) than does He. The only energy dependence of the analyzing power is due to the bη_3 term. Although this term is small in all cases compared to the constant a, the heavy noble gases minimize the energy dependence of the analyzing power even further because of the weak energy dependence of their η_3's, when compared with He, over the energy range below the first cascade threshold. In He η_3 falls rapidly from its kinematically required threshold value of 0.37 to less than 30% of this within several tenths of an electron volt. It then exhibits prominent energy-dependent resonance features. In Kr, which is typical of the heavy noble gases, the drop is much less dramatic, and no resonance effects are apparent. All these features, in combination with their larger excitation cross sections, make the heavy noble gases promising candidates to replace He as a polarimeter target.

The appropriate figure of merit to use when comparing various polarimeters is their "efficiency", $\varepsilon \equiv (I/I_0)^{1/2} A$, where I_0 in the incident electron particle current, I is the detected signal, and A is the device's analyzing power.[1] The efficiency is inversely proportional to the time required to make a measurement of P_e with a given statistical uncertainty. We measured this factor for all the transitions listed in Table I. While the excitation cross section maximum with energy increases monotonically with target Z (corresponding to I/I_0), one is in practice limited to running at or below the first cascade threshold. Thus xenon's efficiency, for example, is small because the cascade threshold is less than 0.3 eV above the threshold for excitation. The efficiencies listed in Table I are normalized to target density, but no attempt was made to provide constant detection efficiency of the optical train from transition to transition. Thus these values should be taken solely as qualitative guides to polarimeter efficiency.

We have recently demonstrated the efficiency of a neon in-line polarimeter to test a new polarized electron source developed in our lab.[19] We expect that the higher efficiency of, e.g., the Kr polarimeter will lead to calibration standards capable of statistical precisions significantly better than those available with a He-based device.

COINCIDENCE MEASUREMENTS

The null result of our Xe experiment raises the question: Can a more sensitive optical test for spin-orbit forces on the continuum electron be made? Certainly the trajectory averaging inherent in an integrated Stokes parameter measurement tends to wash out the relativistic effects prominent in angle-differential scattering.[10] We thus turn our attention to coincidence measurements. Such experiments with heavy noble gas targets have been reviewed recently.[20] The Münster group has succeeded in combining polarized electron and coincidence measurement technologies with Hg scattering.[21] The simplest example of a coincidence measurement that would provide a clean signature of Mott scattering would be the determination of η_1 in the geometry considered above, except with electron detection in the xz-plane. Any transverse $\hat{y}$ spin dependence of η_1 would constitute such evidence, providing that the excited target state were well-LS coupled. This is because $\mathfrak{I}_{21}(J)$ depends only on longitudinal electron spin when relativistic forces are negligible. No such experiment has been performed to date.

REFERENCES

1. J. Kessler, *Polarized Electrons*, 2nd ed. (Springer-Verlag, Berlin, 1985).
2. See, e.g., T. Zuo, R.P. McEachran, and A.D. Stauffer, J. Phys. B **24**, 2853 (1991); U. Thumm, K. Bartschat, and D.W. Norcross, J. Phys. B **26**, 1587 (1993); and references therein.
3. G.F. Hanne, J. Phys. B **9**, 805 (1976); G.F. Hanne, Phys. Rep. **95**, 97 (1983).
4. J.J. McClelland, S.J. Buckman, M.H. Kelley, and R.J. Celotta, J. Phys. B **23**, L21 (1990) and references therein.
5. K. Bartschat and K. Blum, Z. Phys. A **304**, 85 (1982).
6. J.E. Furst, W.M.K.P. Wijayaratna, D.H. Madison, and T.J. Gay, Phys. Rev. A **47**, 3775 (1993); J.E. Furst, T.J. Gay, and W.M.K.P. Wijayaratna, J. Phys. B. **25**, 1089 (1992).
7. W. Jitschin, S. Osimitsch, H. Reihl, H. Kleinpoppen, and H.O. Lutz, J. Phys. B. **17**, 1899 (1984).
8. F. Eschen, G.F. Hanne, K. Jost, and J. Kessler, J. Phys. B. **22**, L455 (1989).
9. K. Bartschat, U. Thumm, and D.W. Norcross, J. Phys. B **25**, L641 (1992).
10. R.P. McEachran and A.D. Stauffer, J. Phys. B **19**, 3523 (1986); M. Klewer, M.J.M. Beerlage, and M.J. van der Wiel, J. Phys. B **12**, 3935 (1979).
11. T.J. Gay, M.A. Khakoo, J.A. Brand, J.E. Furst, W.V. Meyer, W.M.K.P. Wijayaratna, and F.B. Dunning, Rev. Sci. Instrum. **63**, 114 (1992).
12. W.M.K.P. Wijayaratna, Ph.D. Thesis, U. of Missouri-Rolla (1993; unpublished).
13. D.H. Madison and W.N. Shelton, Phys. Rev. A **7**, 514 (1973).
14. K. Blum, *Fundamental Processes in Energetic Atomic Collisions*, edited by H.O. Lutz, J.S. Briggs, and H. Kleinpoppen, (Plenum, New York, 1983), p.621.
15. E. Kisker, Z. Phys. **256**, 121 (1972) and reference therein.
16. D.W.O. Heddle, R.G.W. Keesing, and A. Parkin, Proc. Roy. Soc. Lond. A. **352**, 419 (1977).
17. T.J. Gay, J. Phys. B **16**, (1983) L553; J. Goeke, J. Kessler, and G.F. Hanne, Phys. Rev. Lett. **59**, 1413 (1987).
18. See, e.g., I. Humphrey, C. Ranganathaiah, J.L. Robins, J.F. Williams, R.A. Anderson, and W.C. Macklin, Meas. Sci. Technol. **3**, 884 (1992).
19. K.W. Trantham, R.J. Vandiver, T.J. Gay, and L.D. Schearer, contributed paper, these proceedings.
20. K. Becker, A. Crowe, and J.W. McConkey, J. Phys. B **25**, 3885 (1992).
21. M. Sohn and G.F. Hanne, J. Phys. B **25**, 4627 (1992), and references therein.

ELECTRON-PHOTON CORRELATION STUDIES SINCE ANDERSEN ET AL

A. Crowe

Physics Department, University of Newcastle upon Tyne, NE1 7RU, United Kingdom

ABSTRACT

Some of the more recent developments in the detailed study of electron-atom excitation processes using electron-photon angular and polarisation correlation techniques are discussed. Particular attention is given to both experimental and theoretical studies carried out since the major review of Andersen, Gallagher and Hertel in 1988. The disagreement between experiment and recent sophisticated calculations, using both perturbative and non perturbative methods, for excitation of the 2p state of atomic hydrogen is highlighted. A brief discussion is given of a new generation of experiments in which the atom is excited by spin polarised electrons.

INTRODUCTION

The ability to observe only an identically prepared subset of excited states produced in electron-atom collisions was first realised by Eminyan et al.[1,2] They measured the angular correlation between electrons scattered through some angle and the photon emitted in the decay of the excited state, analysing the data in terms of the magnitude and phase of the excitation amplitudes. This same group then obtained similar information by measurement of the linear and circular polarisation of the radiation emitted in specific momentum transfer collisions.[3] An alternative approach in which electrons were superelastically scattered from laser excited atoms was used by Hertel and Stoll.[4] The theoretical foundations of the work were laid in the papers of Macek and Jaecks[5] and Fano and Macek.[6]

The comprehensive review of Andersen, Gallagher and Hertel[7] presents all experimental and theoretical results, together with extensive analysis, up to a cut off date in mid 1987. Since then only two partial reviews, Slevin and Chwirot[8] and Becker, Crowe and McConkey[9] have been published. The aim of this review is to highlight some of the more important developments in the field since 1987. The general plan of the review of Andersen et al is followed. After the definition of a co-ordinate system and a consistent set of parameters, consideration is given to various atomic systems in turn from the simplest case of the fully coherent excitation of $He(n^1P)$ states through to various examples of partial coherence. A brief discussion is given of a new generation of electron-photon angular correlation experiments using spin polarised electrons.

FRAMEWORK OF ANALYSIS

The data are analysed in the natural co-ordinate frame (Hermann and Hertel[10]) where the (x^n, y^n) collision plane is defined by the incident electron of momentum $\underline{k}_{in}$ and the scattered electron of momentum $\underline{k}_{out}$. A schematic illustration of an excited p-state charge cloud with respect to this co-ordinate frame is shown in figure 1.

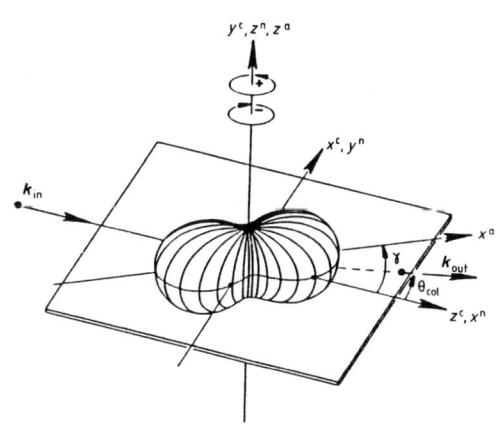

Fig. 1. Schematic illustration of excited p-state charge cloud shown with respect to natural coordinate frame, (x^n, y^n, z^n)

A number of different parameter sets have been used to describe the excitation process and the excited state charge cloud produced in the collision. Here the parameter set $\left(\gamma, P_\ell, L_\perp, \rho_{oo}\right)$ of Andersen et al[7] in terms of the shape and dynamics of the excited state charge cloud is used. γ defines the direction of the major axis of the charge cloud with respect to $\underline{k}_{in}$, P_ℓ the shape of the charge cloud (relative width to length) in the collision plane, $\langle L_\perp \rangle$ the angular momentum of the excited state, and ρ_{oo} the relative height of the charge cloud.

In an angular correlation experiment, the charge cloud alignment angle γ is given directly by the position of the electron-photon correlation minimum while the correlation amplitude gives the shape parameter P_ℓ. These measurements do not yield the angular momentum $L_\perp$ although its magnitude can be found in the fully coherent cases. Where appropriate, the ρ_{oo} parameter can also be determined from angular correlation measurements in different geometries.[11,12]

In polarisation correlation experiments the relative Stokes parameters for the radiation emitted perpendicular to the scattering plane (+z direction) are measured. These are:

$$IP_1 = I(0°) - I(90°)$$

$$IP_2 = I(45°) - I(135°)$$

$$IP_3 = I(RHC) - I(LHC)$$

where I is the total photon intensity, $I(\alpha)$ is the photon intensity transmitted by a linear polariser with transmission axis at an angle α relative to the x^n direction and R(L)HC refers to the circular polarisation of the radiation. Additionally, the ρ_{oo} parameter can be determined from a linear polarisation measurement in a direction having a component in the scattering plane. This is usually carried out in the scattering plane and defined as:

$$IP_4 = I(para) - I(perp)$$

where I(para) and I(perp) refer to the linear polariser axis parallel and perpendicular to the scattering plane. The excited state parameters are related to the Stokes parameters by:

$$\gamma = \tfrac{1}{2}\arg(P_1 + iP_2)$$
$$P_\ell = (P_1^2 + P_2^2)^{1/2}$$
$$L_\perp = -P_3$$

It is also useful to define the degree of coherence P^+ as:

$$P^+ = (P_\ell^2 + P_3^2)^{1/2}$$

REVIEW OF RECENT DEVELOPMENTS

1. The Fully Coherent Case

The most studied system until 1987 was the 2^1P state of helium. In contrast, the only published work since then has been theoretical - the 29-state R-matrix calculation of Fon et al[13] at energies close to threshold, FOMBT calculations[14,15] from near-threshold to 500 eV and DMET calculations[16] at 81.2 eV. Interest in helium has switched to studies of the n=3 states. For the 3^1P state of helium it is interesting to compare the recent 19-state R-matrix calculations of Fon et al[17] with the experiments of Neill et al[18] and Neill and Crowe[19] at 29.6 eV. In figure 2 it can be seen that the P_ℓ parameter is well reproduced by theory whereas there is disagreement with theory at scattering angles above 60° for γ. ^1P excitation studies have been extended to the 3^1P state of magnesium at small scattering angles by Brunger et al[20], supported by close coupling and FOMBT calculations, and to the heavier alkaline earth atoms (see conference abstracts).

Excitation of the ^3P states in helium provides an opportunity to study an electron exchange process without the need to carry out experiments involving spin polarised electrons. A number of both theoretical and experimental studies have been carried out over a range of electron energies. A comparison between the 19-state R-matrix calculations[17] and experiment[21,22] at 29.6 eV (figure 3) shows much better agreement for the γ parameter at all angles compared with the ^1P state.

Excitation of the $3^{1,3}D$ states of helium introduces a number of additional complexities relative to the P states. The M=0 magnetic substate has positive reflection symmetry with respect to the scattering phase, leading to excitation of a dipole lying along the z-axis and to the charge cloud having a finite height in this direction.[7] Full definition of the dipole radiation pattern therefore also requires a measurement of P_4 to determine ρ_{00}. Values of ρ_{00} have been presented by Crowe[23] for the 3^1D state at 29.6 eV and by Batelaan et al[24] at 40, 50, 60 eV and for the 3^3D state at 40 eV. As yet only DMET calculations[25] are available for this parameter.

Fig. 4 shows results for the Stokes parameters P_1-P_4 at 40eV for the 3^1D state of helium. Data from the Dutch group[24,26] over a restricted range of scattering angles are totally consistent with the large angle data of Donnelly and Crowe[27] for P_1-P_4 but

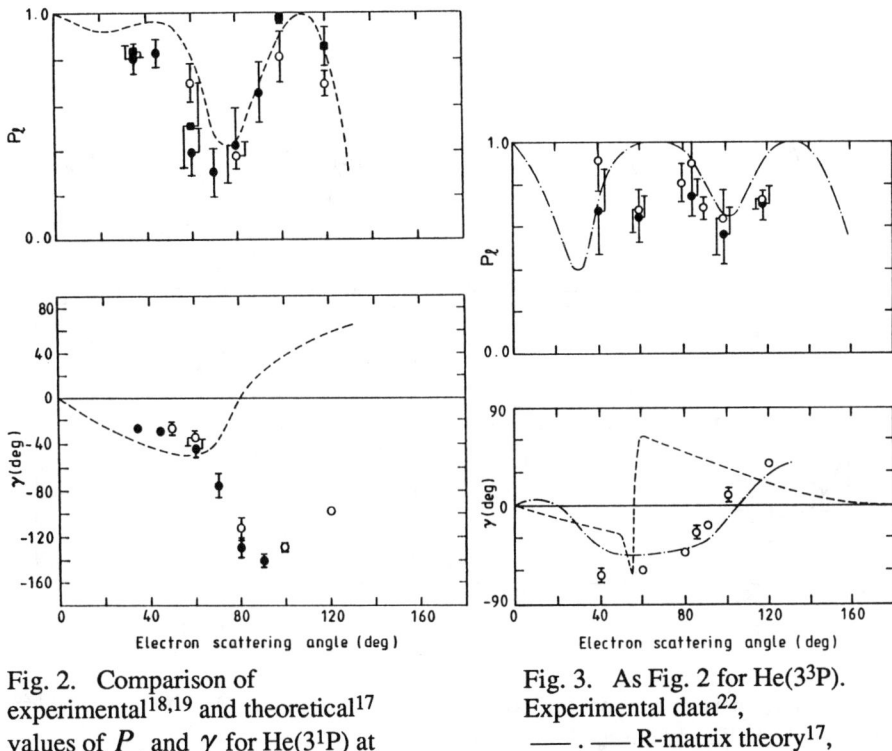

Fig. 2. Comparison of experimental[18,19] and theoretical[17] values of P_ℓ and γ for He(3^1P) at 29.6eV.

Fig. 3. As Fig. 2 for He(3^3P). Experimental data[22],
—·— R-matrix theory[17],
— — — FOMBT[62].

there is no agreement with the DWBA[28] or FOMBT[29] calculation. However, very recent small angle data by Wedding et al[30] are in disagreement with previous P_1 measurements but in agreement with the DWBA and FOMBT theories and with the previous P_2 and P_3 experimental data. As a result we have extended our P_1-P_3 measurements to lower angles[31] and find results totally consistent with our earlier values and those of Beijers et al[26]. Extensive checks have failed to indicate the source of this discrepancy although P_1 measurements which we have made at 60eV are in agreement with data of Perera and Burns[12] using an angular correlations method.

2. The Partially Coherent Case

Atomic hydrogen excitation processes provide the most important examples of partially coherent excitation. Relative to helium this system is complicated by the fact that electron scattering by a spin ½ atom leads to singlet and triplet scattering. Moreover, unlike helium, the direct and exchange scattering can be disentangled in experiments where the electron spin is defined. Hence the studies described here,

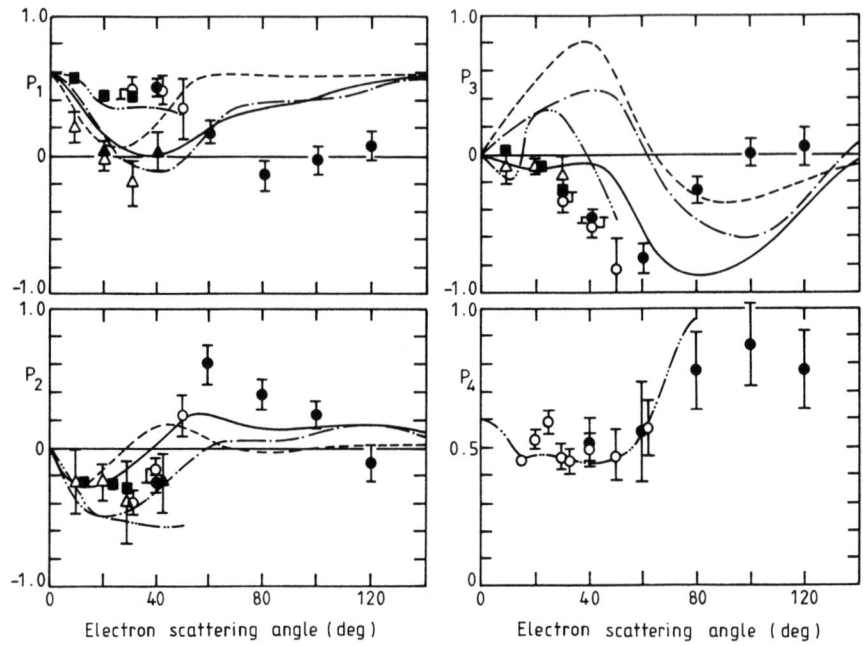

Fig. 4 Stokes parameters for He(3^1D) at 40 eV. Solid points[27,31], Open circles[24,26], Triangles[30], Solid and dashed lines[28], —.—[29], and —..—[25].

where spin states are not determined, average over different processes leading to a loss of coherence.

Results for the H(2p) excitation process at 54.4eV are presented in figure 5 in terms of the $\gamma, P_\ell, L_\perp$ and P^+ parameters. For all parameters the recent exact second-order distorted wave with second order exchange calculation of Madison et al[32], the intermediate energy R-matrix calculation of Scholz et al[33], and the convergent close coupling calculations of Bray and Stelbovics[34] are in excellent accord at all scattering angles. Similarly for P_ℓ and γ there is good accord between the only two independent data sets[35-8] extending over a large angular range. However, the serious discrepancies between the measured and calculated phases of the angular correlations at larger scattering angles is probably the major unresolved problem currently in the study of electron-atom collisions. It is unclear whether this is due to some subtle experimental artefact or the absence of some unidentified physics in intermediate energy scattering theories. A similar discrepancy is also noted at 35eV between theory and experiment[39].

Although the interpretation must be treated with care[40], the P^+ parameter has been used increasingly to indicate the importance of exchange scattering. Recent theories,

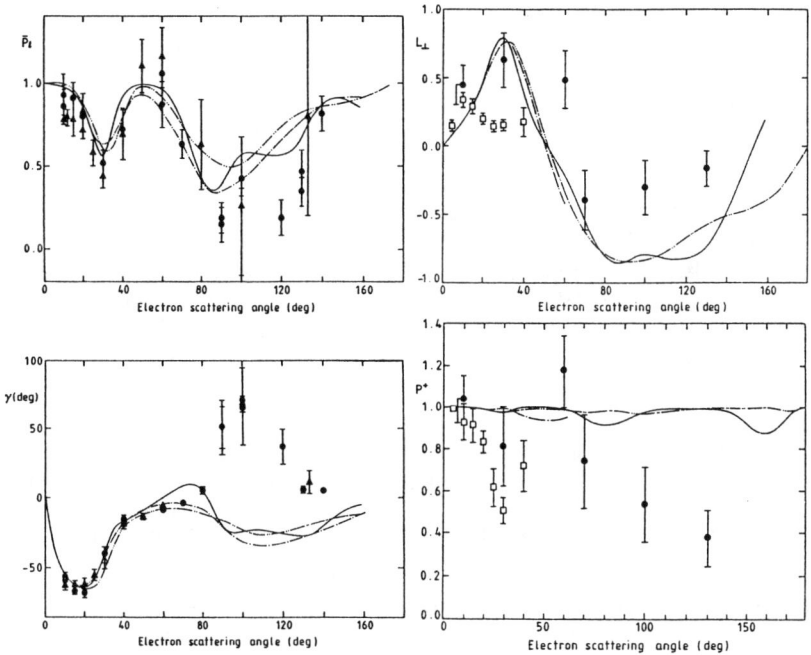

Fig. 5 H(2p) parameters at 54.4 eV. Solid points[35-38], open points[41], solid curve[33], —.—[32], —..—[34].

showing little variation of P^+ from unity, are not in agreement with the large angle data of Williams[38]. Even more disturbing is the large discrepancies with theory in $L_\perp$ and P^+ at small scattering angles in the very recent data of Nic Chormaic et al[41]. Although extensive experimental checks have been carried out, the data are clearly inconsistent with current understanding of low angle intermediate energy electron scattering. For example, for the equivalent 3^2P state of sodium at 12.1 and 22.1eV, the superelastic scattering data of Scholten et al[40] show P^+ close to unity in agreement with theory and in contrast to some earlier experimental data.

Despite these fairly fundamental outstanding problems for the 2p state of H, there is increasing interest in the H(n=3) states. Both Williams et al[42] and Chwirot and Slevin[43] have made small angle measurements for the 3^2P state which can be compared with convergent close coupling calculations[42]. A complete set of parameters has also been measured for the 3^2D state[44-46] by observation of the L_α cascade photon. However limited data has not the quality to evaluate DWA calculations[44].

The electron-photon correlation technique has also been used to study S-P coherence for the n=2 states of hydrogen[47,48], while recent measurements of angular correlations between cascading photons have provided new detailed information on the n=3 states[49].

4. Studies with spin polarised electron beams.

Except for a few special favourable cases, such as triplet state excitation in helium already discussed, the complete unravelling of spin dependent effects is possible only in experiments with definition of the electron spin.

Two types of experiments have been carried out. Kelly et al[57] have studied the superelastic scattering of polarised electrons from sodium optically pumped to the 3P state. In this case spin effects are due only to exchange. The technique enables isolation of the singlet and triplet scattering channels. Figure 7 shows their results for $L_\perp$ at 4.1 are 40 eV. The striking difference in the singlet and triplet data at the lowest energy indicates the importance of electron exchange. On the other hand, at 40 eV, the similarity of the data shows that exchange is no longer important. The data are well reproduced by a recent coupled channel optical calculation[58] at all energies and by a DWB2PE calculation[59] at the higher energies.

Sohn and Hanne[60] have used the electron-polarised photon correlation method to study excitation of the 6^3P_1 state of mercury excited by transversely polarised incident electrons. Now, in addition to exchange, explicitly spin-dependent forces such as the spin-orbit interaction between the continuum electron and the target may be important. In general the symmetry of the system about the scattering plane is broken in these experiments and the tilt of the charge cloud with respect to the scattering plane is defined by the additional alignment angles δ and ϵ[61]. Figure 8 shows data obtained for one of the angles δ defining the tilt of the charge cloud out of the scattering plane. Sohn and Hanne have shown that the non-zero values observed at all scattering angles is confirmation that the collision system cannot be described in LS coupling.

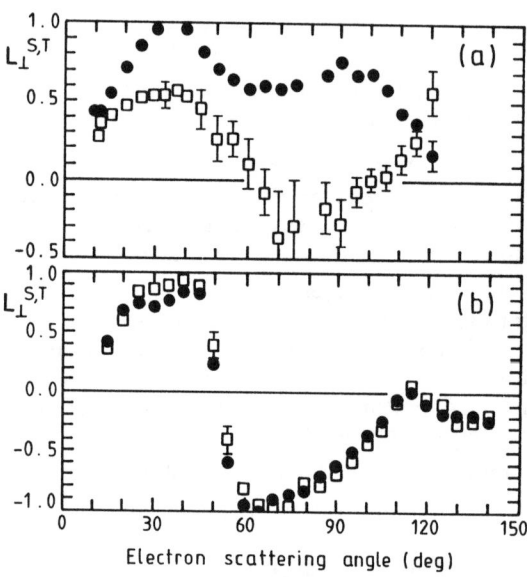

Fig. 7. $L_\perp$ for Na(3p)
(a) 4.1 eV (b) 40 eV
Full points - triplet scattering
Open points - singlet scattering

3. The Incoherent Case Without Conservation of Atomic Reflection Symmetry.

In all previous cases discussed it could be reasonably assumed that only systems with excited state wave functions having positive reflection symmetry about the scattering plane are excited. In general though this assumption cannot be made. For example for a J=1 state of a heavy atom the $M_J=0$ state with negative reflection symmetry may be excited due to spin flip in addition to the $M_J=\pm 1$ states with positive reflection symmetry. Excitation of $M_J=\pm 1$ states can be parameterised using $\gamma, P_\ell^+, L_\perp^+$ and P^+, where the "+" refers to the component of the J=1 oscillator density with positive reflection symmetry. The density matrix element ρ_{oo} is used to describe the component with negative reflection symmetry. Physically it can be regarded as the relative height of the charge cloud in the z'' direction and as a measure of the relative importance of spin flip transitions.

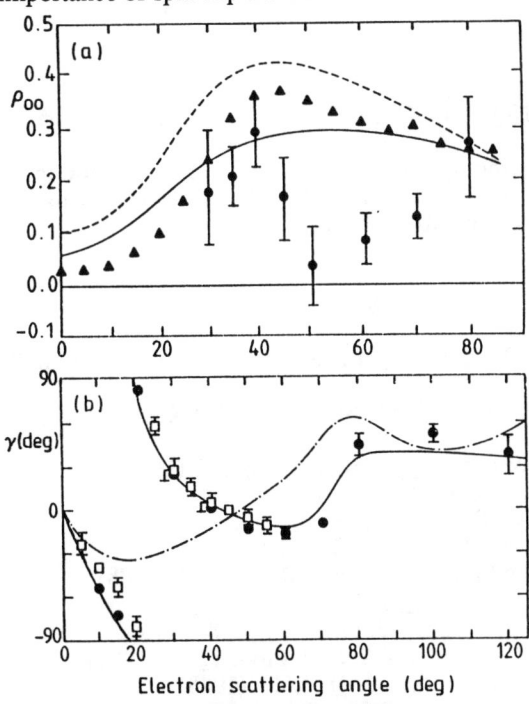

Calculations[50-52] indicate that ρ_{oo} is very small for the $(^2P_{1/2})ns'[1/2]_1^0(^1P_1)$ and $(^2P_{3/2})ns[3/2]_1^0(^3P_1)$ states of the heavier rare gases except at energies close to threshold. Early experiments showing significant values of ρ_{oo} at higher energies and small scattering angles, have been shown to be erroneous[9]. A recent example of data free from experimental artefacts[53] is shown in figure 6a for the $4s[3/2]_{J=1}$ state of argon excited by 20 eV electrons. Although there are some discrepancies with the angular variation predicted by theory the general size of ρ_{oo} is as predicted. Another common feature of the heavy rare gas data is the generally good agreement between the DWBA theory[50] and experiment[54,55] for the γ parameter over a wide range of energies and angles. A typical example is shown for the $5s[3/2]_{J=1}$ state of krypton at 30 eV in figure 6b.

Fig. 6(a) ρ_{oo} for $4s[3/2]_{J=1}$ state of argon. Points[53], triangles[51], full curve[50], dots[52].

(b) γ for Kr $5s[3/2]_{J=1}$ state. Closed points[54], open points[55], full curve[50], dashed curve[56].

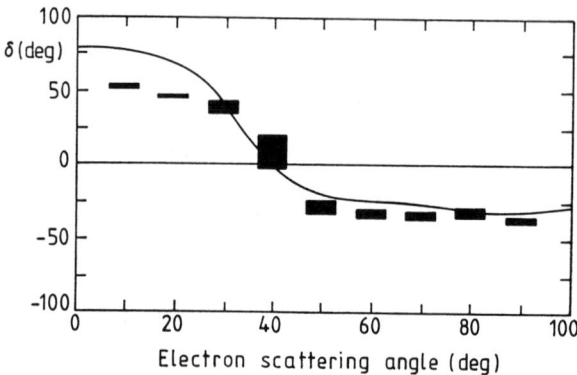

Fig. 8. Tilt of charge cloud axis out of scattering plane for Hg(6^3P_1) at 8 eV. Experiment[60]. Theory[61].

ACKNOWLEDGMENTS

The author acknowledges the continuing support of SERC for this work in his laboratory. I am grateful to I Bray, A Stelbovics and J F Williams for supplying data prior to publication and published work in tabular form.

REFERENCES

1. M Eminyan, K B MacAdam, J Slevin and H Kleinpoppen, Phys. Rev. Lett. 31, 576 (1973)
2. M Eminyan, K B MacAdam, J Slevin, M C Standage and H Kleinpoppen, J. Phys. B 7, 1519 (1974)
3. M C Standage and H Kleinpoppen, Phys. Rev. Lett. 36, 577 (1976)
4. I V Hertel and W Stoll, J. Phys. B 7, 570 (1974)
5. J Macek and D H Jaecks, Phys. Rev. A4, 2288 (1971)
6. U Fano and J H Macek, Rev. Mod. Phys. 45, 553 (1973)
7. N Andersen, J W Gallagher and I V Hertel, Phys. Rep. 165, 1 (1988)
8. J A Slevin and S Chwirot, J. Phys. B 23, 165 (1990)
9. K Becker, A Crowe and J W McConkey, J. Phys. B 25, 3885 (1992)
10. H W Hermann and I V Hertel, Comments A. Mol. Phys. 12, 61(1982)
11. S J King, P A Neill and A Crowe, J. Phys. B 18, L589 (1985)
12. N W P H Perera and D J Burns, J. Phys. B 23, 3007 (1990)
13. W C Fon, K P Lim and P M J Sawey, J. Phys. B 25 L599 (1992)
14. G Csanak and D C Cartwright, Phys. Rev. A 38, 2740 (1988)
15. G D Meneses and G Csanak, Z.Phys. D 8, 219 (1988)
16. E J Mansky and M R Flannery, J. Phys. B 23, 4573 (1990)
17. W C Fon, K A Berrington and A E Kingston, J. Phys. B 23, 4347 (1990)
18. P A Neill, B P Donnelly and A Crowe, J. Phys. B 22, 1417 (1989)
19. P A Neill and A Crowe, J. Phys. B 21, 1879 (1988)
20. M J Brunger, J L Riley, R E Scholten and P J O Teubner, J. Phys. B 22, 1431 (1989)

21. B P Donnelly and A Crowe, J. Phys. B 21, L321 (1988)
22. B P Donnelly and A Crowe, Z. Phys. D 14, 333 (1989)
23. A Crowe, Proc. Int. Sym. on Correlation and Polarisation in Electronic and Atomic Collisions, NIST Special Publication 789, p39 (1990)
24. H Batelaan, J. van Eck and H G M Heideman J. Phys. B 24, L397 (1991)
25. E J Mansky and M R Flannery, J. Phys. B 24, L551 (1991) and private communication
26. J P M Beijers, S J Doornenbal, J van Eck and H G M Heideman J. Phys. B 20 6617 (1987)
27. B P Donnelly and A Crowe, J. Phys. B 21, L637 (1988) and private communication
28. K Bartschat and D H Madison, J. Phys. B 21, 153 (1988)
29. D C Cartwright and G Csanak, J. Phys. B 20, L583 (1987)
30. A B Wedding, A G Mikoza and J F Williams, J. Phys. B 26, 321 (1993)
31. D T McLaughlin, D G McDonald and A Crowe (to be published)
32. D H Madison, I Bray and I E McCarthy, J. Phys. B 24, 3861 (1991)
33. T T Scholz, H R J Walters, P G Burke, M P Scott, J. Phys. B 24, 2097 (1991)
34. I Bray and A T Stelbovics, Phys. Rev. A 46, 6995 (1992)
35. S T Hood, E Weigold and A J Dixon, J. Phys. B 12, 631 (1979)
36. E Weigold, L Frost and K J Nygaard, Phys. Rev. A 21, 1950 (1980)
37. J F Williams, J. Phys. B 14, 1197 (1981)
38. J F Williams, Aust. J. Phys. 39, 621 (1986)
39. J Slevin, M Eminyan, J M Woolsey, G Vassilev, H Q Porter, C G Back and S Watkin, Phys. Rev. A 26, 1344 (1982)
40. R E Scholten, G F Shen and P J O Teubner, J. Phys. B 26, 987 (1993)
41. S Nic Chormaic, S Chwirot and J Slevin, J. Phys. B 26, 139 (1993)
42. J F Williams, A T Stelbovics and I Bray (private communication)
43. S Chwirot and J Slevin, J. Phys. B 20, 6139 (1987)
44. D Farrell, S Chwirot, R Srivastava and J Slevin, J. Phys. B 23 315 (1990)
45. J F Williams, Inst. Phys. Conf. Ser. No.122, p 71(1992)
46. A T Stelbovics, M Kumar and J F Williams, J. Phys. B 26, L237 (1993)
47. J F Williams and E L Heck, J. Phys. B 21, 1627 (1988)
48. S. Chwirot, S Legowski, J Slevin and J Zaremba, J. Phys. B 22, 1411 (1989)
49. J F Williams, M Kumar and A T Stelbovics, Phys. Rev.Lett. 70, 1240 (1993)
50. K Bartschat and D H Madison, J.Phys. B 20, 5839 (1987)
51. F J da Paixao, N T Padial and G Csanak, Phys. Rev. A 30, 1697 (1984)
52. T Zuo, R P McEachran and A D Stauffer, J. Phys. B 24, 2853 (1991)
53. J J Corr, S Wang and J W McConkey, J. Phys. B 25, 4929 (1992)
54. P B Murray, S F Gough, P A Neill and A Crowe, J. Phys. B 23, 2137 (1990)
55. S H Zheng and K Becker, J. Phys. B 26, 517 (1993)
56. G D Meneses, F J da Paixao and N T Padial, Phys. Rev. A 32 156 (1985)
57. M H Kelley, J J McClelland, S R Lorentz, R E Scholten and R J Celotta, Inst. Phys. Conf. Ser. No 122, 123 (1992)
58. I Bray and I E McCarthy, Phys. Rev. A 47, 317 (1993)
59. D H Madison, K Bartschat and R P McEachran, J. Phys. B 25, 5199 (1992)
60. M Sohn and G F Hanne, J. Phys. B 25, 4627 (1992)
61. A Raeker, K Blum and K Bartschat, J. Phys. B 26, 1491 (1993)
62. D C Cartwright and G Csanak, J. Phys. B 19, L485 (1986)

ELECTRON-ATOM COLLISIONS IN RESONANT LASER FIELDS

W. R. MacGillivray and M. C. Standage
Laser Atomic Physics Laboratory
School of Science
Griffith University
Nathan
Queensland 4111
Australia

ABSTRACT

In this paper, we describe two methods in which resonant laser excitation can be employed to obtain a complete description of electron-atom collisions. The electron superelastic scattering technique is illustrated by presenting data for the Na D line, while collision parameters for the 6^1S_0-6^1P_1 transition for the I=0 isotope of Hg are shown to demonstrate the feasibility of a stepwise electron/laser excitation method.

INTRODUCTION

Measurement of the complex excitation amplitudes a_m for inelastic collisions between electrons and atoms has been achieved through electron and photon correlation experiments where the photon is derived from the atomic excited state. In particular, experiments in which the electrons scattered at a given angle are detected in coincidence with photons emitted in a fixed direction (normally perpendicular to the scattering plane defined by the incident and scattered electron directions) have been used to deduce collision parameters by the analysis of the photons in terms of Stoke's parameters. This area has been extensively reviewed[1]. The resolution of coincidence experiments are effectively limited by the energy resolution of the incident electrons. In most atomic systems this method will be unable to resolve the contribution of the atomic fine or hyperfine structure. Another limitation is reached when the electron excited state is metastable so that no fluorescence channel is available.

In an attempt to overcome these and other problems that may arise in the conventional coincidence method, a number of techniques involving the resonant excitation of an atomic transition associated with the collisional event have been developed. Two of these methods, the superelastic scattering method and the stepwise electron/photon coincidence technique, are currently being employed in our laboratory for the measurement of collision parameters. This article will review these two methods and discuss data derived from them. We are also developing a third method for measuring electron collision parameters based on the laser deflection of an atomic beam. This new work will be discussed in detail in another article[2].

To introduce the necessary parameters, we first discuss the conventional coincidence experiment shown schematically in figure 1. The atom is excited from state |g> to state |e> by an inelastic electron collision. A time-resolved, fluorescence signal is obtained by detecting the photon emitted from the relaxation of the atom in coincidence with the scattered electron. The fluorescence is analysed to obtain the Stoke's parameters, P_i, which completely describe the light;

$$P_1 = (I_0-I_{90})/I, \quad P_2=(I_{45}-I_{135})/I, \quad P_3=(I_{RHC}-I_{LHC})/I \tag{1}$$

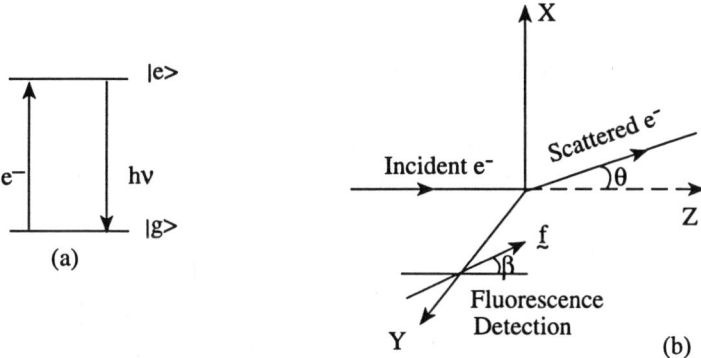

Figure 1. Schematic of (a) the process and (b) an experiment geometry for the coincidence method.

where the subscripts for the intensity denote either the angle β of the linear analyser to the incident electron direction (the collision frame quantisation axis) or the handedness of a $\lambda/4$ plate analyser, and I is the total intensity. The electron excited state of the atom ensemble can be represented by the density matrix, ρ, whose elements are $<a_n a_m^*>$. We will confine our discussion to the excitation of an S-P transition with no spin - orbit interactions (such as Na 3^2S-3^2P). Under these conditions, the atomic states exhibit positive reflection symmetry about the scattering plane. Sets of real parameters have been introduced which describe the collision. One such set is the so called "collision frame" parameters which are defined as[3],

$$\lambda = \rho_{00}/(\rho_{00}+2\rho_{11}), \quad \cos\chi = Re(\rho_{10})/(\rho_{00}\rho_{11})^{1/2} \quad \sin\phi = Im(\rho_{10})/(\rho_{00}\rho_{11})^{1/2} \quad (2)$$

It can be shown that the Stoke's parameters are related to the collision parameters by

$$P_1 = 2\lambda - 1, \quad P_2 = -2[\lambda(1-\lambda)]^{1/2}\cos\chi, \quad P_3 = 2[\lambda(1-\lambda)]^{1/2}\sin\phi \quad (3)$$

An alternative parameterisation is with reference to the "natural frame"[1];

$$P_{lin} = (P_1^2 + P_2^2)^{1/2}, \quad \gamma = \frac{1}{2} \arg(P_1 + iP_2), \quad L_\perp = -P_3 \quad (4)$$

which are more physically interpretable as the shape and spatial orientation of the atomic charge cloud excited in the collision process and the transfer of angular momentum. Finally, the coherence of the excitation process can be described by the degree of polarisation, P_{tot}, which is defined as

$$P_{TOT} = (P_1^2 + P_2^2 + P_3^2)^{1/2} \quad (5)$$

Thus, a measurement of the Stoke's parameters of the coincidence fluorescence intensity will yield, from equations (3) or (4), a complete description of the electron excitation process. If the target atom has fine and hyperfine interactions unresolved by the electron excitation, as is the case for the cited Na transition, the measured Stoke's parameters will be depolarised due to the precession of the atomic state in the time between the excitation and its relaxation. The depolarisation can be calculated from a knowledge of the fine and hyperfine structure.

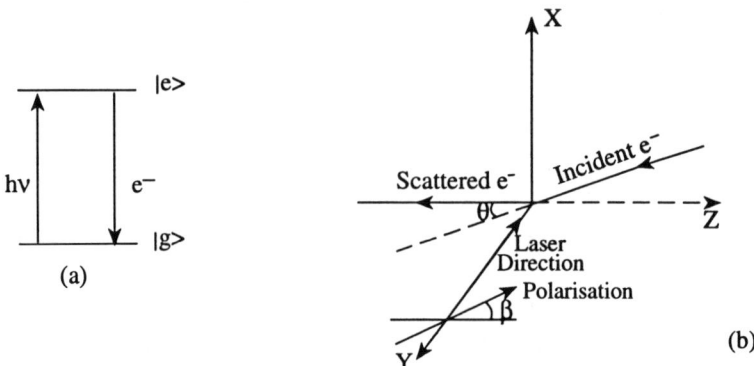

Figure 2. Schematic of (a) the process and (b) an experiment geometry for the superelastic scattering method.

Figure 2 illustrates the superelastic scattering of an electron from an atom that has been resonantly excited by laser radiation. The interaction of the electron with the excited atom causes the atom to undergo radiationless relaxation and the excess energy is transferred to the electron. This method was first demonstrated by Hertel and coworkers[3]. The geometry, as shown, suggests that this process is the time-reverse of the conventional coincidence experiment. The formal equivalence between the two methods has been examined in detail[4] and confirmed, provided that the principle of micro-reversibility holds for electron-atom collisions. Psuedo Stoke's parameters P_i^S are measured where;

$$P_1^S = (S_0 - S_{90})/S, \quad P_2^S = (S_{45} - S_{135})/S, \quad P_3^S = (S_{RHC} - S_{LHC})/S \tag{6}$$

where S_β is the electron superelastic differential cross section with the laser radiation linearly polarised at angle β to the quantisation axis (z axis) or right or left circularly polarised and S is the total differential cross section. The superelastic scattering process results from the overlap of the laser excitation and the electron deexcitation events. Therefore, to obtain the electron-atom collision parameters, the effect of the optical excitation contained in the psuedo Stoke's parameters must be deconvoluted.

The sodium D_2 line ($3^2S_{1/2} - 3^2P_{3/2}$) has been the most extensively studied system using electron superelastic scattering. Using narrow bandwidth, cw laser excitation allows the hyperfine structure of this transition to be resolved. It has been shown[4] that the psuedo Stoke's parameters are related to those associated with the electron excitation of a pure P state by;

$$P_1^S = KP_1, \quad P_2^S = KP_2, \quad P_3^S = K'P_3 \tag{7}$$

The evaluation of K and K' for this transition has been performed using a full quantum electrodynamic model[5]. The value of these parameters depends on such variables as the Doppler width of the transition and the laser intensity and detuning. K' does not deviate significantly from minus one and this value is adopted. However, K can vary considerably under certain conditions and so must be measured. K is identical to the line polarisation of the resonance fluorescence associated with this transition scattered at right angles to the plane defined by the laser polarisation vector and propagation direction, so a measurement of that parameter will yield the value of K. Care must be taken to eliminate radiation trapping effects in the fluorescence and to ensure that the fluorescence is collected from the superelastic scattering interaction region. Another method for determining K is to measure the r parameter[6] for laser radiation incident in the scattering plane in the direction of the quantisation axis. r is the ratio of superelastic differential scattering cross sections measured for the laser polarisation set respectively parallel and perpendicular to the scattering plane. It can be shown that[5];

$$r = (1-\lambda)\left(\frac{1+K}{1-K}\right) + \lambda \tag{8}$$

Substituting for λ from equation (3) and using equation (7) yields,

$$K = (r + P_1^S - 1)/r \tag{9}$$

In-plane measurements at low electron scattering angles are susceptible to the finite interaction volume[7], so procedures must be employed to eliminate this effect from the data.

The details of our experiment are given elsewhere[8]. Briefly, a 500nA electron beam is produced from a two-stage electrostatic lens design gun with a tungsten filament. The gun can be operated over the energy range 1-100eV with an energy resolution of 500meV. The scattered electrons are energy selected by a cylindrical analyser whose input lens images the interaction region with a solid angle of about 0.002sr. The electrons are detected by a channel electron multiplier. The atomic beam is produced from an effusive oven and collimator yielding a measured Doppler width of 300MHz with an estimated density of $10^{16} m^{-3}$. This instrumentation is contained in a vacuum chamber constructed from non-magnetic stainless steel, lined with μ metal and surrounded by three orthogonal pairs of Helmholtz coils to minimise residual magnetic fields in the interaction region. The electron gun and analyser are both mounted on turntables for selection of the scattering angle.

An actively stabilised Spectra Physics 380D single mode ring dye laser pumped by a Spectra Physics 2020 Ar+ laser provides the radiation to resonantly excite the atomic transition. Choice of appropriate optics produces light of the desired polarisation. The degree of polarisation of the light was carefully measured at the interaction region. In this region, the intensity of the laser light is typically in the range 80-100mW/mm^2 in a beam with a 1/e diameter of 1mm. Standard digital counting electronics interfaced to a personal computer collect and analyse the output from the channeltron. The computer also controls stepper motors which rotate the polarising optics.

At a laser intensity of 80mW/mm^2 under usual operating conditions for the collection of superelastic scattering data, the value of the line polarisation was measured to be 0.39±0.01 while the value of K obtained from measurements of r and P_1^S at 15° scattering angle and 20eV incident electrons was 0.39±0.04. Stoke's parameters have been measured for a range

300 Electron-Atom Collisions in Resonant Laser Fields

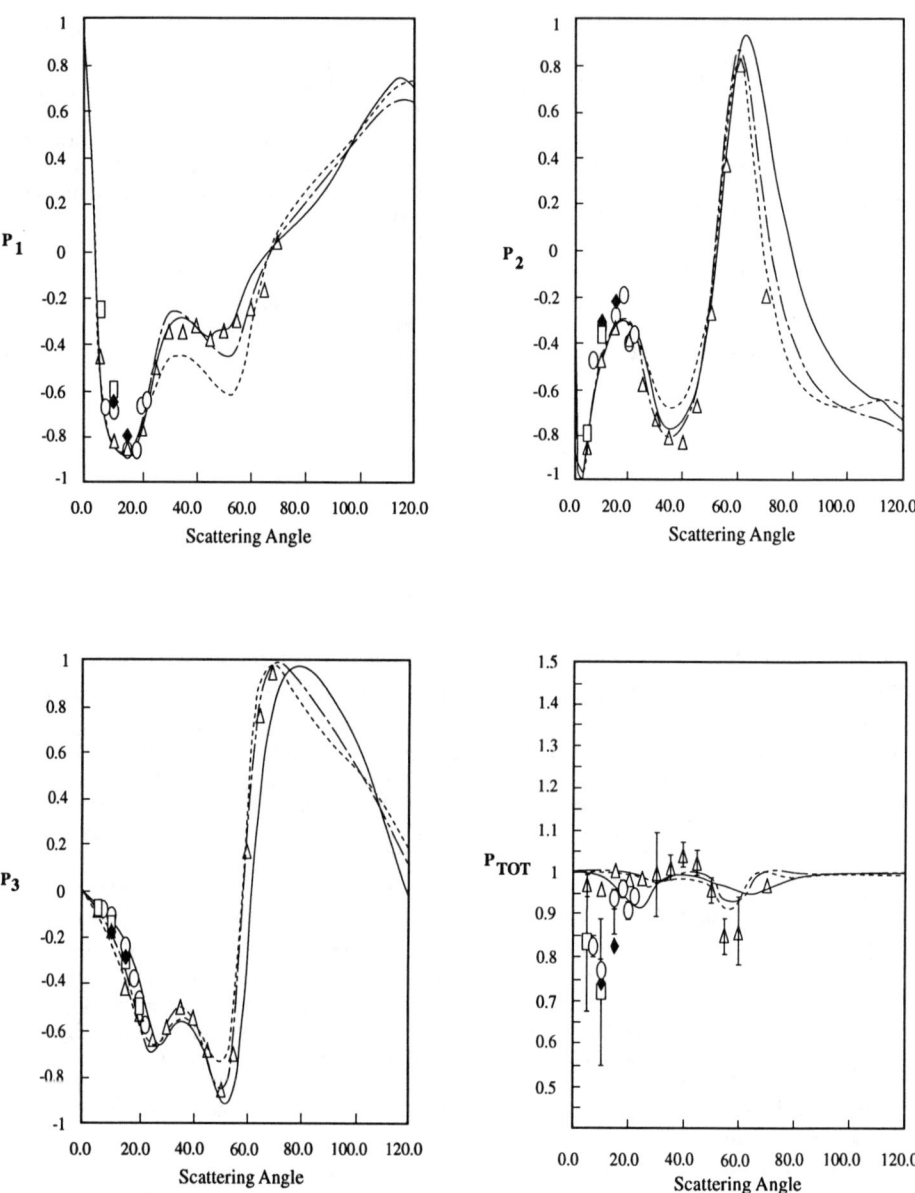

Figure 3. Measured and calculated Stoke's parameters and total polarisation for Na D_2 line for 20eV incident electron energy. ⎯⎯ (DWBA2[9]) ⎯·⎯ (CCO[10]) ⎯ ⎯ ⎯ (CC[10]) O (Sang et.al.[8]) ◆ (Farrell et al.[11]) △ (Teubner et al.[12]) □ (Hertel and coworkers [6,13])

of angles and incident electron energy by a number of groups. Most data has been collected for 20eV electrons and the Stoke's parameters and P_{TOT} at this energy are displayed in figure 3 where they a compared with calculations from a second order distorted wave model, DWBA2[9], a coupled channel optical method, CCO[10], and a 15 state close coupling method, CC[10]. While there are regions of good agreement between all data both measured and calculated, significant differences do occur. As far as the experimental data is concerned, the results from the Teubner group's work deviates from the other measurements around 10° for P_1 and P_2. All of the P_3 experimental data is in generally good agreement. The disagreement flows on to P_{TOT} where, except for the data from Teubner's group, the experiments indicate a sharp minimum in the parameter at 10°. The Teubner data does show a broader minimum at 60°. No theory predicts the magnitude of these features although the close coupling models do produce a minimum just below 60° while the DWBA2 exhibits two minima but at larger angles than the experimental features. No resolution of these differences has yet been achieved.

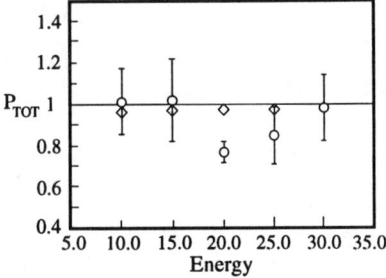

Figure 4 is a plot of P_{TOT} versus energy for 10° scattering angle. Our experimental results indicate that the loss of coherence in the excitation process observed for 20eV is confined to a small energy range. The DWBA2 calculation does not exhibit the same minimum in P_{TOT}.

Figure 4. P_{TOT} versus incident electron energy at 10^0 scattering angle.
o (Sang et. al.[8]) ◇ (DWBA2[9]).

STEPWISE ELECTRON/PHOTON COINCIDENCE

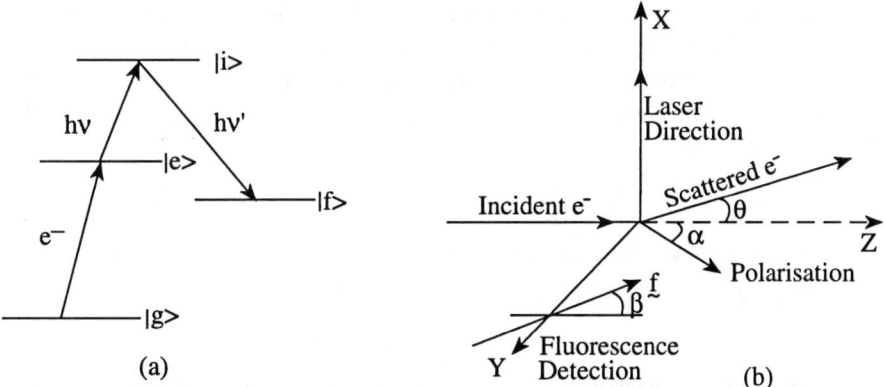

Figure 5. Schematic of (a) the process and (b) an experiment geometry for the stepwise coincidence method.

In this method, the electron excited atom is further excited by resonant laser radiation. Fluorescence photons from the laser excited state are then detected in coincidence with the inelastically scattered electrons. A schematic of the principle of this method is shown in figure 5 together with the experimental geometry currently employed. A Stoke's parameter analysis of the time integrated fluorescence coincidence signal is carried out in the same way as for the conventional coincidence technique. Because of the optical coherence of the laser excitation, the information carried in the electron excited state related to the complex scattering amplitudes is transferred to the higher excited state. Using a narrow band, cw laser permits the resolving of fine and hyperfine structure in many atomic species so that this technique offers a method of obtaining collision parameters for single isotopes. This feature of the method has been utilised in our work on electron collisions with the I=0 isotope of mercury[14,15]. The excitation scheme studied has been electron excitation of the 6^1S_0-6^1P_1 (185nm) transition followed by the laser excitation of the 6^1P_1-6^1D_2 transition (579nm). Fluorescence from the 6^1D_2-6^3P_1 (313nm) relaxation channel was detected. An added advantage of employing this scheme over the conventional coincidence method is that fluorescence measurements are transferred from the VUV to the UV where polarising and detection optics are standard.

The stepwise method offers a suitable technique for obtaining collision parameters for the electron excitation of a metastable state. Partial relative differential cross sections have been measured for the 6^3P_2 metastable level of mercury using this method[16,17].

In heavy atoms such as mercury, the effect of spin-orbit interaction between the incident electron and the target may be strong enough to flip the spin of the scattered electron. This spin-flip destroys the positive reflection symmetry about the scattering plane and results in the need for a fourth collision parameter to completely describe the excitation. In the collision frame, this parameter is designated as $\cos\delta$ and is defined as ρ_{1-1}/ρ_{11}, while, in the natural frame, the spin-flip is described by populating the ρ_{00} state in that frame and so is designated here as ρ_{00}^N. To determine the spin-flip parameter requires a fourth Stoke's parameter to be measured. In the conventional coincidence experiment, this measurement is made by detecting fluorescence in the scattering plane in addition to the perpendicular geometry of figure 1.

The collision parameters can be obtained directly from the Stoke's parameters associated with the stepwise coincidence signal provided the optical excitation can be considered as weak. A thorough investigation of the resonant excitation of the 6^1P_1-6^1D_2 transition confirms that this approximation is good for the range of laser intensities employed in our experiments[18]. The relationships between the Stoke's parameters and the collision frame parameters are given by;

$$IP_{1\alpha} = \lambda \left(\frac{27}{2} - 6\cos^2\alpha\right) + \frac{9}{2}\cos 2\alpha + \frac{9}{2}(1-\lambda)\cos\delta \qquad (10a)$$

$$IP_{2\alpha} = -3(5 - \cos 2\alpha)[\lambda(1-\lambda)]^{1/2}\cos\chi \qquad (10b)$$

$$IP_{3\alpha} = 9(3 + \cos 2\alpha)[\lambda(1-\lambda)]^{1/2}\sin\phi \qquad (10c)$$

$$I = \lambda\left(6\cos^2\alpha + \frac{1}{2}\right) + 5\left(\cos^2\alpha + \frac{17}{2}\right) + \frac{1}{2}(1-\lambda)(8\sin^2\alpha - 9)\cos\delta \qquad (10d)$$

where α is the angle that the laser polarisation makes with the quantisation axis (figure 5). α is an extra experimental variable which allows the full determination of the collision parameters from the Stoke's parameters measured perpendicular to the scattering plane. For example, λ and δ can be evaluated from equation (10a) by making measurements with α at 0° and 90°. χ and ϕ are then obtained from equations (10b) and (10c) respectively for α set at 0°.

The experimental arrangement is very similar to that used in the superelastic scattering experiment with the electron optics being identical and the atomic oven being suitably modified to produce a beam of mercury with a residual Doppler width of 300MHz. The atomic beam density at the interaction region is estimated in the range $2-5 \times 10^{16}$ m^{-3}. The vacuum chamber and magnetic field minimising procedures are also similar to the superelastic experiment. The radiation is again from a Spectra Physics 380D actively stabilised cw ring dye laser. The linearly polarised light is injected into the interaction region in the scattering plane (x-z plane) along the x axis with the polarisation vector either in the plane (α=0°) or at right angles (α=90°). The fluorescence is detected along the y axis where it is collected by a lens and focussed onto the cathode of a cooled photomultiplier tube through analysing optics and a filter. Standard electronic coincidence techniques are employed to obtain the signal. Details of the experiment have been published elsewhere[15].

The collision parameters are obtained from the measured Stoke's parameters. We will quote here just the results for the parameters in the natural frame. These are related to the measured Stoke's parameters by the relationships;

$$\rho^N_{00} = [9P_{10}(19+P_{190})-54(1+P_{190})]/[P_{10}(90+P_{190})+3(9-2P_{190})] \tag{11a}$$

$$P^+_{lin} = \left\{ [2P_{190}(71P_{10}+33)]^2 + [9P_{20}(17P_{190}+39)]^2 \right\}^{1/2}/2[81(1-P_{10})+8P_{190}(6-P_{10})] \tag{11b}$$

$$\tan 2\gamma = 9P_{20}(17P_{190}+39)/2P_{190}(71P_{10}+33) \tag{11c}$$

$$L^+_\perp = -3P_{30}(17P_{190}+39)/2[81(1-P_{10})+8P_{190}(6-P_{10})] \tag{11d}$$

where the superscript,+, indicates that the parameter relates to the states with positive reflection symmetry[1]. The natural frame parameters deduced from Stoke's parameters measured at 5°, 10°, 20° and 30° for 16eV incident electrons are shown in figure 6 where they are compared with calculations from a DWBA theory[19]. The experimental data have been corrected for the finite volume effect. Even though the uncertainties of the experimental data are large, the agreement with the theory is not good overall. In particular, ρ^N_{00} indicates evidence of a non-zero spin-flip at 30° and there is a substantial loss of coherence at all measured scattering angles which is not predicted by the theory. As well, the measured values of the angular momentum transfer, $L^+_\perp$, are an order of magnitude smaller than the predicted values.

These parameters are now being measured for the same atomic scheme but at 50eV incident electron energy. At this energy, it is expected that the DWBA model will describe the electron atom collision more accurately.

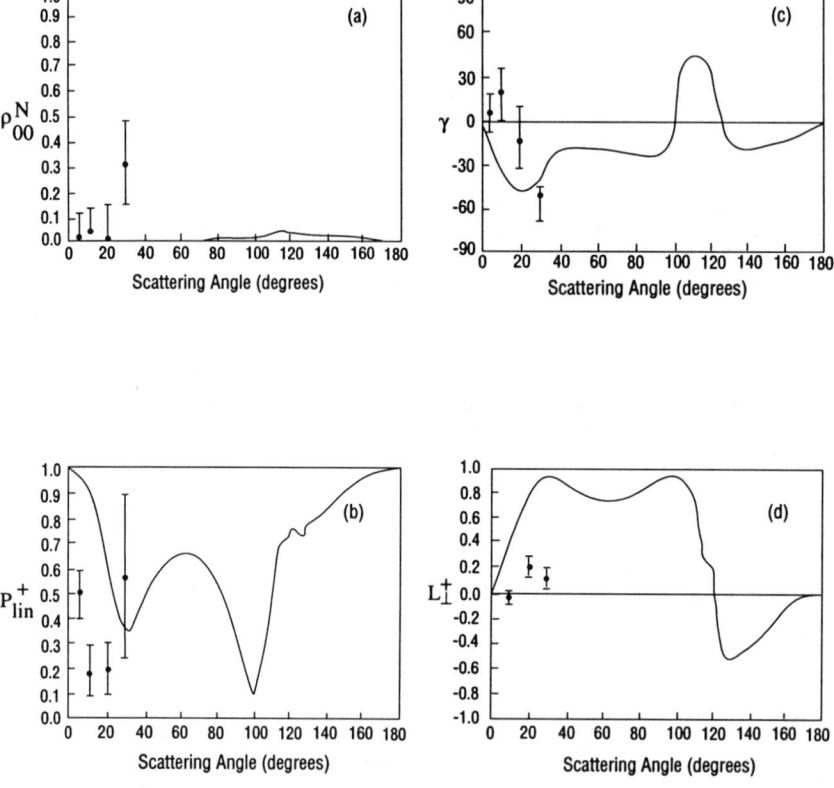

Figure 6. Experimental and theoritical (DWBA[19]) natural frame parameters for $6^1S_0 - 6^1P_1$, Hg transition excited by 16eV electrons.

CONCLUSION

In this progress report, we have reviewed two methods of measuring parameters for electron-atom collisions which involve the resonant excitation of the atom by laser radiation. The data produced from such experiments give a complete description of the collision process including the amplitude and phase of the scattering amplitudes. With such data, more stringent tests of theoretical models can be made. From the superelastic scattering data for the sodium D line, the agreement with the models is quite good except in regions where the measurements show a loss of coherence in the collision. Close agreement between 16eV data and a DWBA theory is not obtained for the $6^1S_0-6^1P_1$ of mercury for a stepwise electron/laser excitation experiment. Considerably more data for a variety of systems is required to further test the theoretical models.

ACKNOWLEDGEMENTS

The authors wish to thank P.M.Farrell, R.T.Sang, A.J.Murray, R.Pascual and D.H.Madison for their significant contributions to the work described here. The programme is funded by grants from the Australian Research Council and the Griffith University Research Grants Scheme. We also wish to acknowledge access to unpublished theoretical results provided by Professor I E McCarthy, Dr I.Bray and Professor D.H.Madison.

REFERENCES

1. N.Andersen, J.W.Gallagher and I.V.Hertel, Phys.Rep. 165, 1 (1988).
2. G.S.Summy, W.R.MacGillivray and M.C.Standage, Z.Phys.D submitted.
3. I.V.Hertel and W.Stoll, Adv.At.Mol.Phys. 13, 113 (1977).
4. W.R.MacGillivray and M.C.Standage, Com.At.Mol.Phys. 26, 179 (1991).
5. P.M.Farrell, W.R.MacGillivray and M.C.Standage, Phys.Rev. A44, 1828 (1991).
6. H.W.Hermann, I.V.Hertel, W.Reiland, A.Stamatovic and W.Stoll, J.Phys.B. 10, 251 (1977).
7. P.W.Zetner, S.Trajmar, G.Csanak and R.E.H.Clark, Phys.Rev. A39, 6022 (1989).
8. R.T.Sang, P.M.Farrell, D.H.Madison, W.R.MacGillivray and M.C.Standage, J.Phys.B. submitted.
9. D.H.Madison, K.Bartschat and J.McEachran, J.Phys.B. 25, 5199 (1992).
10. I.Bray and I.E.McCarthy, Phys. Rev. A submitted
11. P.M.Farrell, C.J.Webb, W.R.MacGillivray and M.C.Standage, J.Phys.B. 22, L527 (1989).
12. P.J.O.Teubner, R.E.Sholten and G.F.Shen, in "Proceedings of the International Symposium on Correlation and Polarization in Electronic and Atomic Collisions" eds. P.A.Neill, K.H.Becker and M.H.Kelley (NIST Special Publication 789, 1990) p.45.
13. H.W.Hermann, I.V.Hertel and M.H.Kelley, J.Phys.B. 13, 3465 (1980).
14. A.J.Murray, C.J.Webb, W.R.MacGillivray and M.C.Standage, Phys.Rev.Lett. 62, 411 (1989).
15. A.J.Murray, R.Pascual, W.R.MacGillivray and M.C.Standage, J.Phys.B. 25, 1915 (1992).
16. C.J.Webb, W.R.MacGillivray and M.C.Standage, J.Phys.B. 18, L259 (1985).
17. G.F.Hanne, V.Nickich and M.Sohn, J.Phys.B. 18 2037 (1985).
18. A.J.Murray, W.R.MacGillivray and M.C.Standage, J.Phys.B. 23, 3373 (1990).
19. D.H.Madison, private communication.

ELECTRON SCATTERING FROM LASER-EXCITED BARIUM

P. W. Zetner
Department of Physics, University of Manitoba
Winnipeg, Manitoba, Canada R3T 2N2

ABSTRACT

We report on progress in the investigation of superelastic and inelastic electron collisions involving the laser-excited ^{138}Ba (...6s6p $^1P_1$) state. Measurements are compared with available theory.

INTRODUCTION

The investigation of electron collisions with excited state atomic species is a subject of considerable current interest. As shown in two recent reviews[1,2], our understanding of such processes is quite limited in comparison with the substantial body of knowledge available for electron impact on ground state atoms. The major obstacle to progress in experimental studies of this type is the difficulty encountered in producing sufficiently high number densities of excited state target atoms. When it is applicable, optical preparation of the excited species by a continuous wave (cw) dye laser offers an attractive solution to this problem. Such a scheme can be used to generate steady-state concentrations of short-lived excited atoms large enough to allow, for example, the study of angle-resolved excited state to excited state (inelastic) and excited state to ground state (superelastic) scattering by conventional crossed-beam methods. The scope of the present report will be restricted to the discussion of inelastic and superelastic processes investigated in this way.

The theoretical framework for the interpretation of measurements of electron scattering from laser excited atoms was developed in the 1970's[3,4] and summarized in the review[5] of Hertel and Stoll who described pioneering experiments with sodium. It was shown that the scattered electron intensity depends, in general, on a complicated bilinear combination of scattering amplitudes which we will refer to as a partial differential cross section (PDCS). This arises as a result of the highly directional and highly polarized nature of laser light which leads to the preparation of a strongly anisotropic target (ie. a target in which the magnetic sublevels are non-uniformly populated). In conventional terminology, the anisotropic target is said to be aligned if, for a given magnetic quantum number, M, the $|J,M\rangle$ and $|J,-M\rangle$ magnetic sublevels are equally populated or oriented if these sublevels are unequally populated.

The PDCS is a function of laser beam incidence direction and polarization state. In our discussion of superelastic scattering we will show that this functional dependence can be exploited to give insight into fine details of the electron atom collision. On the other hand, if one is interested in measuring the conventional differential cross section (DCS) for excited state inelastic or superelastic processes, then some means of eliminating the dependence of the scattering signal on laser photon direction and polarization will be desirable. Experimental approaches to accomplish this are discussed below.

Studies of this type applied to barium are interesting both from theoretical and experimental standpoints. Experimental investigation is straightforward due to the relative ease with which vapour beams of this substance can be produced and also because of the availability of stable cw dye lasers that can transfer an appreciable fraction (~40%) of the ground state (...$6s^2$ $^1S_0$) population to the excited (...6s6p $^1P_1$) state. The most abundant isotope, ^{138}Ba (easily resolved by the laser when a collimated vapour beam is transversely illuminated), has zero nuclear spin which gives a laser excitation scheme uncomplicated by hyperfine structure (unlike the alkali elements, for example). Laser excitation of the (...6s6p $^1P_1$) state can also give rise, by means of radiative cascade, to significant populations of metastable (...6s5d $^1D_2$) and (...6s5d $^3D_2$) atoms. Inelastic and superelastic collisions involving these metastable states have been observed[6] but, in this report, we concentrate on scattering from the (...6s6p $^1P_1$) state.

Theoretical interest in barium is stimulated partly by the fact that it is one of the few atoms that can be efficiently pumped by currently available laser systems. It also has intrinsic theoretical appeal in that it consists essentially of a "simple" system of two valence electrons bound to a relatively inert core. By virtue of its large atomic number (Z = 56) barium may provide a useful test species for relativistic theories since the collisional interaction can be expected to exhibit spin-dependence.

COLLISIONAL DEEXCITATION OF ^{138}Ba(...6s6p $^1P_1$) TO THE GROUND STATE

The PDCS for deexcitation of the anisotropic, laser-prepared, J = 1, state to the ground state (J_0 = 0) by superelastic scattering is related to the PDCS for excitation of the same J = 1 anisotropic state from the J_0 = 0 ground state by inelastic scattering through time-reversal symmetry of the scattering amplitudes[4]. The sensitivity of the superelastic PDCS to laser photon direction and polarization provides a means by which to extract observables involving the relative phases and moduli of scattering amplitudes for this "time-inverse" inelastic collision process. Various sets of these observables, the so-called electron impact coherence parameters (EICP), have been defined by different authors and we refer the reader to the review by Andersen et al[7] for a full discussion. In addition to the DCS, four parameters are required to describe a J_0 = 0 to J = 1 collisional excitation (by unpolarized electrons and spin-insensitive detection) when the interaction Hamiltonian exhibits spin dependence. The set of EICP defined by Andersen et al[7] for this case have a particularly appealing physical significance. These parameters are; the spatial anisotropy, P_ℓ^+, and alignment angle, γ, of the positive reflection symmetry component (with respect to the scattering plane) of the excited charge cloud, the orbital angular momentum, $L_\perp$, transferred to the charge cloud and the "height parameter", ρ_{00}, which characterizes the negative reflection symmetry component of the charge cloud. In the absence of spin dependence in the interaction Hamiltonian the charge cloud possesses purely positive reflection symmetry and only three parameters (namely, DCS, P_ℓ^+ and γ) are necessary.

For this case (exemplified by the He 1 $^1S_0$ to 2 $^1P_1$ excitation) $\rho_\infty = 0$ and $(P_i^+)^2 + (L_\perp)^2 = 1$.

Direct measurements of these EICP using superelastic scattering were carried out by Zetner et al[8,9] and Li and Zetner[10] using the apparatus depicted schematically in Figure 1. The electron gun (impact energy resolution FWHM = 0.6 eV) is rotatable with respect to a fixed, hemispherical electron detector and can deliver currents on the order of 1 µA at the target. The barium vapour source consists of a stainless steel tube resistively heated to about 760°C and an uncooled stainless steel heat shield with collimating aperture. The laser beam is incident upon the target Ba vapour beam from below the scattering plane and strikes the scattering plane perpendicularly. The polarization state of the laser beam at the target is determined by a retardation plate with a phase retardation of δ whose fast axis makes an angle β with respect to the forward scattering direction. The Glan-Taylor prism is employed to ensure that pure linearly-polarized laser light is incident upon the retardation plate. The angle α gives the angular displacement of the Glan-Taylor prism transmission axis with respect to the forward scattering direction.

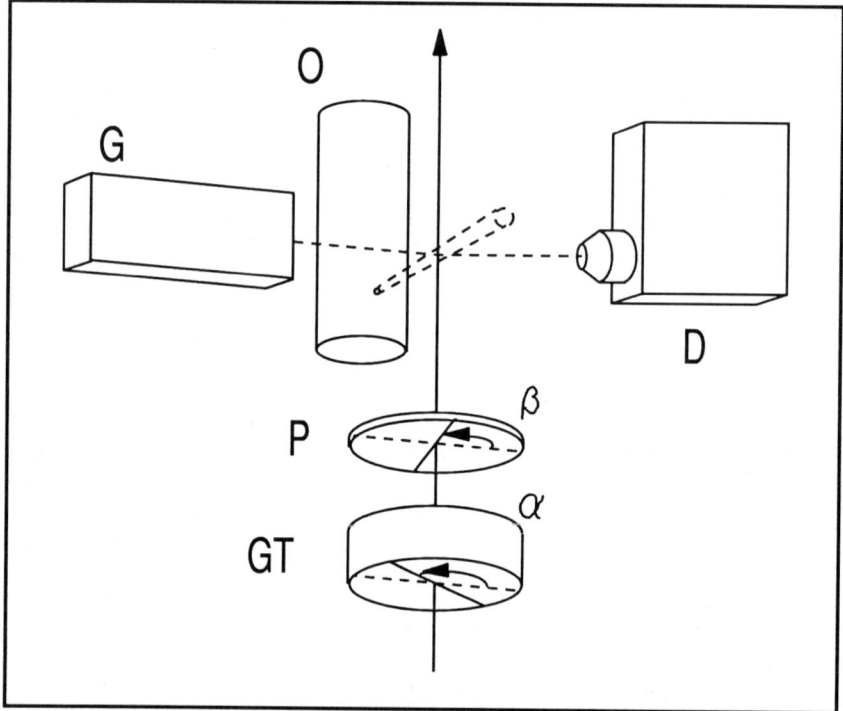

Figure 1. Schematic diagram of the experimental arrangement of Zetner et al[8,9] and Li and Zetner[10]. The electron gun (G), scattered electron detector (D), metal vapour beam source (O), retardation plate (P) and Glan-Taylor prism (GT) are indicated.

For this arrangement, the superelastic scattering intensity is

$$I^S(\beta-\alpha) = \overline{C} \left\{ \begin{array}{l} 1 + \dfrac{P_\ell^+}{2}(1+\cos\delta)\cos(2\alpha-2\gamma) \\[1ex] + \dfrac{P_\ell^+}{2}(1-\cos\delta)\cos(4\beta-2\gamma-2\alpha) \\[1ex] + L_\perp \sin\delta \sin(2\beta-2\alpha) \end{array} \right\} \quad (1)$$

where $\overline{C}$ is a constant consisting of multiplicative factors such as detection solid angle, detection efficiency, laser-excited atom population, incident electron flux and the superelastic DCS and $I^S(\beta-\alpha)$ indicates explicitly the functional dependence of I^S on $\beta-\alpha$.

Examination of equation (1) reveals that a half-wave retardation plate ($\cos\delta=-1$) will allow the straightforward extraction of P_ℓ^+ and γ through a measurement of I^S as a function of $\beta-\alpha$. In this case the laser light is linearly polarized at the target. A quarter-wave retardation plate ($\cos\delta=0$) gives circularly polarized light at the target for $\beta-\alpha = \pm \pi/4$ and allows the extraction of $L_\perp$.

Experimental conditions can sometimes lead to a significant reduction in the sensitivity of the superelastic scattering signal to the polarization state of the laser beam and, hence, adversely affect the measurement of EICP. This depolarizing influence can arise from the well known process of radiation trapping and from the presence of a scattering volume of finite spatial extent. With regard to superelastic scattering experiments the former problem has been discussed by Hertel and Stoll[5] and Fischer and Hertel[11] while the latter effect has been discussed in detail by Zetner et al[12]. Measured quantities presented here were estimated to be minimally affected by radiation trapping (<5%) while the effect of the finite interaction volume was demonstrated to be significant only for near-forward scattering directions.

Theoretical approaches based on first order perturbation methods have recently been applied to the calculation of EICP for the barium ($6s^2\,{}^1S_0$) to ($6s6p\,{}^1P_1$) excitation[13,14]. The unitarized distorted wave approximation (UDWA) has been utilized by Clark et al[13]. This approach employs a description of the atomic structure which is relativistic to the extent that the mass correction and Darwin terms are included in the calculation of radial wavefunctions. The continuum electron is treated non-relativistically. Spin dependence of the collision through exchange excitation can, therefore, be predicted with this theory while the amplitude for direct spin flip of the projectile electron is not calculable. The atomic structure calculations show that the barium ${}^1P_1$ state is predominantly LS coupled and, as a consequence, UDWA theory predicts $\rho_\infty=0$. Calculations of barium EICP with a fully relativistic distorted wave (RDW) theory have recently been reported (Srivastava et al[14]). This theory utilizes Dirac-Fock target states and solutions to the Dirac equation in the presence of a distortion potential. Although the atomic structure calculations used in this theory also

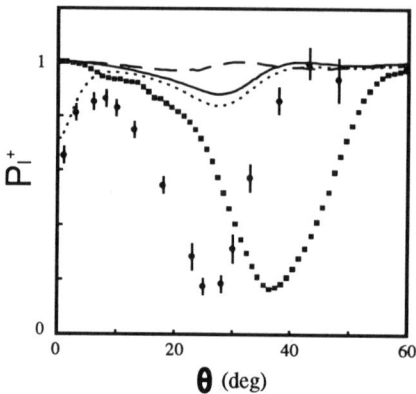

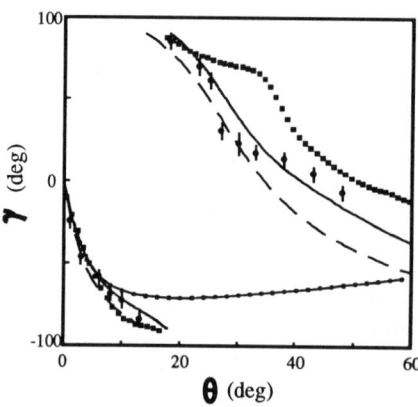

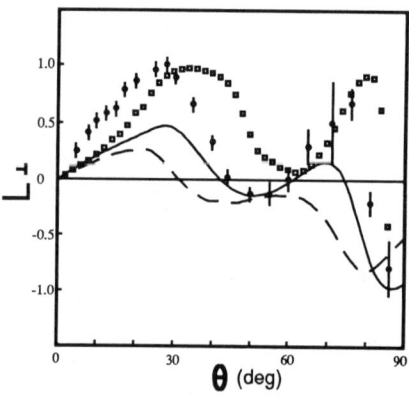

Figure 2. A comparison of measured EICP with available theory for $E_0 = 20$ eV.

give a (nearly) purely LS coupled $^1P_1$ state, some relativistic (spin-dependent) properties of the scattering were predicted, namely a non-zero spin-polarization function, S_p. The manifestation of these relativistic effects through the EICP was, however, again predicted to be negligible (ie. $\rho_\infty \leq 10^{-4}$). Other theoretical developments in the description of electron impact excitation of the barium ($6s^2\ ^1S_0$) to ($6s6p\ ^1P_1$) transition have been reported by Fabrikant[15]. Two state close-coupling calculations were carried out to predict EICP at 20 eV impact energy. These calculations involved a non-relativistic approach in which semi-empirical atomic wavefunctions were employed and exchange between the incident electron and core target electrons was ignored.

Figures 2, 3 and 4 compare the measurements of Zetner et al[9] and Li and Zetner[10] (circles with error bars) with UDWA calculations of Clark et al[13] (dashed curve), RDW calculations of Srivastava et al[14] (solid curve) and the close-coupling calculations of Fabrikant[15] (open squares). A calculation of the alignment angle based on the first Born approximation is also shown (lines joining squares). The dotted curve represents the results of a modelling calculation (which employs the RDW EICP) carried out to show the effect of an interaction volume which subtends an angle of 4° at the detector. EICP extracted from superelastic scattering experiments with incident electron energy E_o^s are compared to theoretical EICP calculated for the time-inverse inelastic collision involving incident electrons with energy $E_o = E_o^s + \Delta E$ where ΔE is

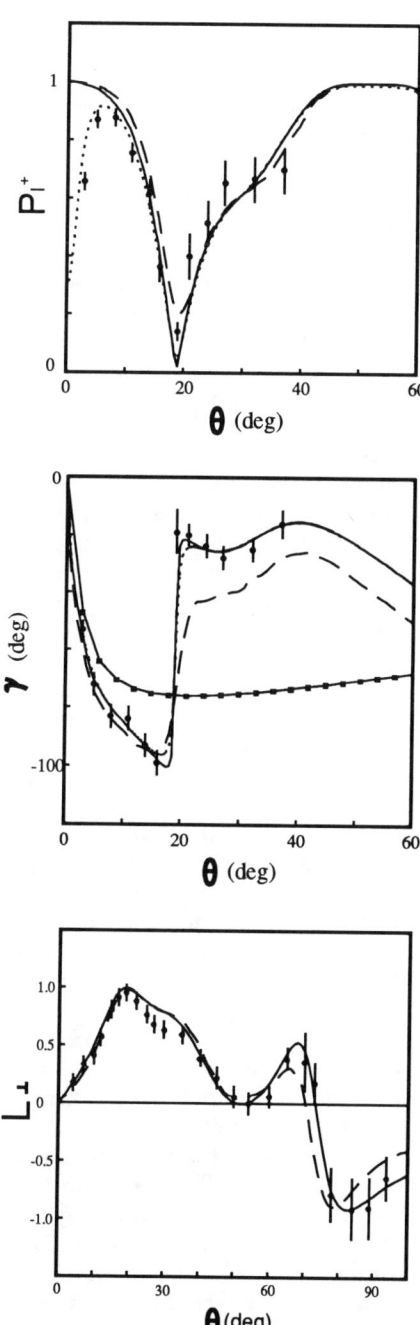

Figure 3. Same as Figure 2. except $E_0 = 36.7$ eV.

the excitation energy of the Ba(...6s6p $^1P_1$) state ($\Delta E = 2.24$ eV).

In a general sense, the experimental results exhibit behaviour similar to that which has been observed in electron-photon coincidence studies of 1^1S_0 to 2^1P_1 excitation in helium. At near-forward scattering angles $\gamma < 0$ and the highly anisotropic charge cloud is aligned roughly along the momentum transfer direction in agreement with predictions of the first Born approximation (FBA). The behaviour of γ with increasing scattering angle quickly diverges from FBA values. The transferred orbital angular momentum, $L_\perp$, takes on positive values for small scattering angles. It then attains a maximum value close to one accompanied by a deep minimum in P_l^+ and, at higher impact energies, a rapid excursion of γ to less negative values. All these behaviours are indicative of the excitation of a highly oriented (circular) $J = 1$ state. At $E_o = 20$ eV we note that the alignment angle undergoes a comparatively smooth decrease with increasing scattering angle suggestive of He $1\,^1S_o$ to $2\,^1P_1$ observations closer to the excitation threshold.

The $L_\perp$ measurements extend to relatively large scattering angles and exhibit pronounced structure. This is significantly different behaviour than observed in helium and can be understood within the framework of distorted wave theory. Scattering by a heavy atom, like barium, is characterized by non-negligible contributions from distorted partial waves of high angular momentum. It has been demonstrated by Madison et al[16]

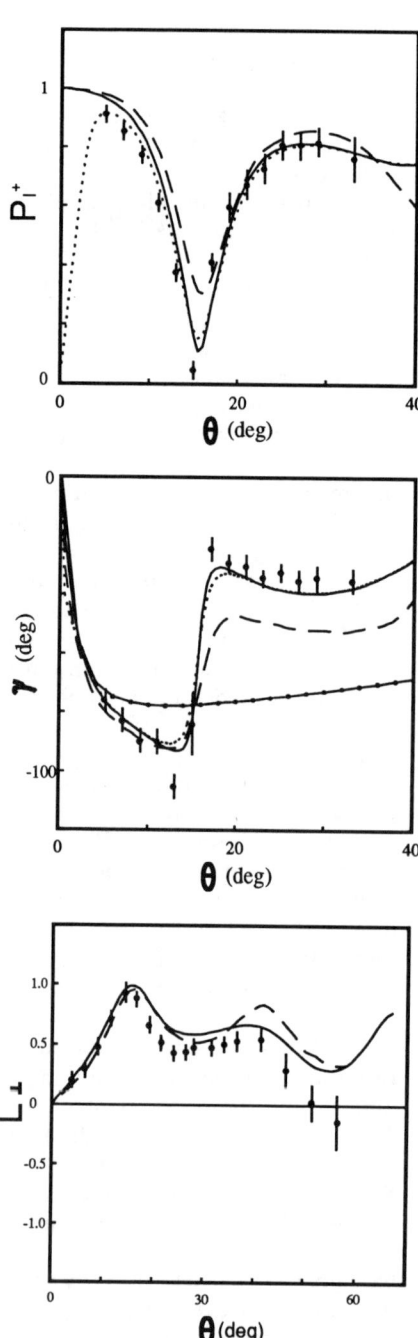

Figure 4. Same as Figure 2. except $E_0 = 50$ eV

that interference among these partial waves determines the angular behaviour of $L_\perp$. With many more contributing partial waves in the barium excitation we therefore expect a richer interference pattern. This invites comparison with excitation of the so-called "singlet" states of the heavier rare gases and in fact similarities can be seen with observed[7,17] and predicted[7,18,19] behaviours of $L_\perp$ for these targets.

Comparison between experiment and theory at impact energy $E_o = 20$ eV reveals that both distorted wave theories fail completely to reproduce the observed deep minimum in P_l^+ and underestimate the magnitude of $L_\perp$ at its maximum near 25 degrees scattering angle by about a factor of two, although an overall qualitative agreement between observed and predicted behaviour of $L_\perp$ is apparent. Both theories are successful, however, in predicting the observed alignment angle behaviour.

It is well understood that distorted wave theories should begin to fail at lower impact energies although, in the present case, it is somewhat surprising that 20 eV (approximately nine times the barium (6s6p $^1P_1$) excitation threshold) should be categorized as a low impact energy. Figure 2 includes a comparison of present experimental results with the close-coupling calculations of Fabrikant[15] for $E_o = 20$ eV. The deep minimum in P_l^+ is clearly predicted by this theory, however, the location of this minimum is shifted in scattering angle with respect to the experimentally observed minimum. Agreement with the observed behaviour of the alignment angle is

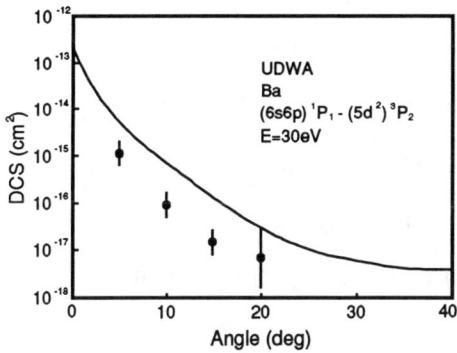

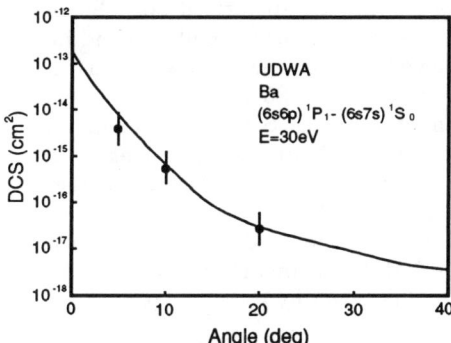

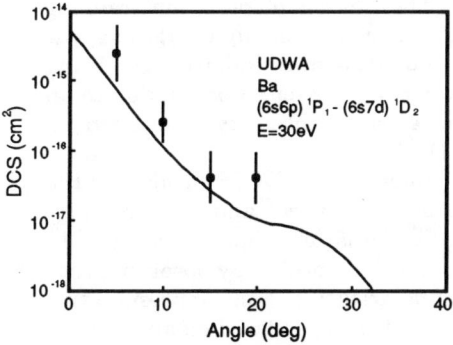

Figure 5. Comparison of the measurements of Register et al[6] with the UDWA calculations of Clark et al[20] (from Clark et al[20]).

excellent below about 20 degrees scattering angle, but significant discrepancies exist outside of this region.

Over the range of scattering angles studied, the RDW and UDWA theories provide an excellent description of observed EICP at E_o = 36.7 eV and 50 eV with the exception of a notable discrepancy for $L_\perp$ at E_o = 50 eV and larger scattering angles.

EXCITED-STATE DCS MEASUREMENTS

When the initial state is populated by laser excitation, the scattering signal becomes proportional to a PDCS which, as discussed above, depends on laser photon incidence direction and polarization state. Consider, for example, the case of laser excitation of a $J = 1$ superposition state involving magnetic substates $|JM>$ where the laser light is linearly polarized and its incidence direction is defined by spherical polar angles θ_v and ϕ_v measured with respect to the incident electron beam direction (which defines the quantization axis). Restricting the discussion to an experimental configuration in which the laser beam lies within the scattering plane ($\phi_v = 0°$ or $180°$), the spin-averaged PDCS for collisional excitation of a state labelled by angular momentum quantum number J' is given by

$$PDCS\,(\psi,\theta_v,\phi_v) = \frac{k'}{2k}\sum_{M'} \left\{ \begin{array}{l} (\sin^2\psi\cos^2\theta_v + \cos^2\psi)\,|f(|M'|,-1)|^2 \\[4pt] + 2\sin^2\theta_v\sin^2\psi\,|f(|M'|,0)|^2 \\[4pt] + (\sin^2\psi\cos^2\theta_v + \cos^2\psi)\,|f(|M'|,1)|^2 \\[4pt] -2(\sin^2\psi\cos^2\theta_v - \cos^2\psi)\,Re\,[f(|M'|,-1)f(|M'|,1)^*] \\[4pt] \mp 2\sqrt{2}\sin^2\psi\sin\theta_v\cos\theta_v\,Re\,[f(|M'|,-1)f(|M'|,0)^*] \\[4pt] \pm 2\sqrt{2}\sin^2\psi\sin\theta_v\cos\theta_v\,Re\,[f(|M'|,1)f(|M'|,0)^*] \end{array} \right\} \quad (2)$$

where f(M',M) is the scattering amplitude for collisional excitation of the |J'M'> magnetic sublevel from the |JM> magnetic sublevel (J = 1), $\vec{k}, \vec{k}'$ are the momentum vectors of the incident and scattered electrons respectively, the angle ψ defines the polarization direction of the linearly-polarised laser beam (measured with respect to an axis perpendicular to the scattering plane) and Re [ ] denotes the "real part of". The upper (lower) sign of the last two terms in equation (2) arises when the azimuthal angle, ϕ_v, is 0° (180°).

It is often desirable to know the conventional DCS which involves an average over initial magnetic quantum numbers, M, and a sum over final magnetic quantum numbers, M', of the quantity |f (M', M)|^2. Direct measurement of the DCS is achievable in a number of ways. For example, an appropriate combination of PDCS measurements can be used. It can be shown using equation (2) that, for excitation of a J = 1 state by linearly-polarized laser light incident in the scattering plane, the DCS is given by

$$DCS = \frac{1}{2}\,(\,PDCS(54.7°,\,45°,\,0°) + PDCS(54.7°,\,45°,\,180°)\,).$$

Alternatively some means of destroying the alignment and orientation of the laser-excited target can be employed. The process of radiation trapping offers a direct means for accomplishing this. For large enough target vapour beam densities the reabsorption of spontaneous resonance radiation can bring about an efficient conversion of the anisotropic, laser-excited population to an isotropically excited population. Scattered electron intensity from such an isotropic target is then proportional to the DCS.

Measurements of inelastic and superelastic DCS involving the ^{138}Ba(...6s6p $^1P_1$) excited state target have been reported[6] and were recently compared with calculations based on UDWA[20]. Radiation trapping was used to produce an isotropic target population and this was verified by observing that the scattering signal was independent of laser polarization. Some representative results are reproduced in Figure 5. Measured DCS were placed on an absolute scale by normalizing against the (...6s^2 $^1S_0$) to (...6s6p $^1P_1$) inelastic DCS, making use of the relative intensities of (...6s^2 $^1S_0$) to (...6s6p $^1P_1$) inelastic scattering and (...6s6p $^1P_1$) to (...6s^2 $^1S_0$) superelastic scattering to determine excited-state to ground-state relative populations by the principle of detailed balance.

Measurements so far comprise a rather sparse coverage of scattering angles and impact energies but agreement between the limited data set and UDWA theory is reasonable, especially in the case of dipole-allowed transitions.

ACKNOWLEDGEMENTS

The author wishes to thank Dr. S. Trajmar for many helpful discussions, and for a generous equipment loan. The important contributions of Mr. Y. Li to the success of this work are also gratefully acknowledged as are unpublished results provided by Drs. R.E.H. Clark and A. Stauffer. Financial support was provided by the Natural Sciences and Engineering Research Council of Canada.

REFERENCES

1. S. Trajmar and J.C. Nickel, Adv. At. Mol.Opt.Phys. 30 45 (1992).
2. C.C. Lin and L.W. Anderson, Adv.At.Mol.Opt.Phys. 29 1 (1991)
3. I.V. Hertel and W. Stoll, J.Phys. B: At.Mol.Phys. 7 583 (1974).
4. J. Macek and I.V. Hertel, J.Phys.B: At.Mol.Phys. 7 2173 (1974).
5. I.V. Hertel and W. Stoll, Adv.At.Mol.Phys. 13, 113 (1977).
6. D.F. Register, S. Trajmar, S.W. Jensen and R.T. Poe, Phys.Rev.Lett. 41 749 (1978).
7. N.Andersen, J.W. Gallagher and I.V. Hertel, Phys.Rep. 168 1 (1988).
8. P.W. Zetner, Y.Li and S. Trajmar, J.Phys.B.:At.Mol.Opt.Phys. 25 3187 (1992).
9. P.W. Zetner, Y. Li and S. Trajmar, Phys.Rev.A (in press, 1993).
10. Y. Li and P.W. Zetner (in preparation, 1993).
11. A. Fischer and I.V. Hertel, Z.Phys.A: At.Nucl. 304 103 (1982).
12. P.W. Zetner, S. Trajmar and G. Csanak, Phys.Rev.A 41 5980 (1990).
13. R.E.H. Clark, J. Abdallah Jr., G. Csanak and S.P. Kramer, Phys. Rev. A 40 2935 (1989).
14. R. Srivastava, T. Zuo, R.P. McEachran and A.D. Stauffer, J.Phys.B: At.Mol.Opt.Phys. 25 3709 (1992).
15. I.I. Fabrikant, Izvest.Akad.Nauk.Latv. SSR, Ser.Fiz.Tekhn., No.6, 11 (1985).
16. D.H. Madison, G. Csanak and D.C. Cartwright, J.Phys.B: At.Mol.Phys. 19 3361 (1986).
17. J.J. Corr, P. Plessis and J.W. McConkey, Phys. Rev. A 42 5240 (1990).
18. L.E. Machado, G.D. Meneses, G. Csanak and D.C. Cartwright, "Proc. 17th Int. Conf. on the Physics of Electronic and Atomic Collisions (Brisbane)", ed. I.E. McCarthy, W.R. MacGillivray and M.C. Standage, Griffith University Press, p.127 (1991).
19. T. Zuo, R.P. McEachran and A.D. Stauffer, J.Phys.B: At.Mol.Opt.Phys. 25 3393 (1992).
20. R.E.H. Clark, G. Csanak, J. Abdallah and S. Trajmar, J.Phys.B: At.Mol.Opt.Phys. 25 5245 (1992).

L^2 METHODS FOR MULTICHANNEL CONTINUA

F. Martín
Departamento de Química C-14.
Universidad Autónoma de Madrid. 28049-Madrid. Spain.

ABSTRACT

A completely L^2 method to evaluate continuum states in multichannel scattering problems such as electron-atom collisions, photoionization, etc, is presented. The continuum wave function is built from a basis of discretized uncoupled channels. The coupling between channels is introduced, using formal scattering theory, in a pure algebraic way by simply solving a system of linear equations. The advantage of the present approach is that it can be used irrespective of the interaction strength between the uncoupled states. The accuracy of the method is illustrated with some results for both total and partial cross sections in He and H^- photoionization.

INTRODUCTION

Representation of scattering states by L^2 integrable functions is a topic of continuous interest in atomic physics. Discretization of the continuum spectrum of a hamiltonian permits the use of algebraic techniques, thus avoiding the solution of (coupled) differential equations. Furthermore, the use of large basis sets in order to achieve good convergence of the results as well as a good description of the resonance structure is possible with moderate computational effort.

On the other hand, discretized wave functions do not satisfy the proper δ function normalization of continuum states. Therefore, a discretization technique must not only provide the wave functions, but also the correct renormalization factors.

Many different L^2 methods have been proposed in the literature for the case of a single open channel. On the other hand, when more than one channel is open, the use of L^2 methods is scarce[1-4]. The reason is that, besides the renormalization problem, one has to satisfy boundary conditions that couple all channels asymptotically. As direct diagonalization of a hamiltonian with L^2 functions yields couplings that depend on the basis set[5], such a simple procedure cannot be used for multichannel continua. In this paper I present an L^2 method that solves this problem by making use of the formal scattering theory to ensure the correct boundary conditions. The theoretical method will be presented in the general context of electron-atom collisions. Results will be shown for He photoionization and H^- photodetachment, for which there are accurate experimental data for both total and partial cross sections.

Atomic units are used throughout the paper unless otherwise stated.

ALGEBRAIC CLOSE-COUPLING APPROACH

In order to simplify the presentation I shall consider the two electron problem of one electron colliding with a hydrogenlike ion. Generalization to systems with more electrons offers no difficulty. In this work, a partial wave decomposition of the two-electron wave function will be used. For each partial wave, the solutions $\Psi_{\mu E}^{\Omega-}$ of the Schrödinger equation will be eigenfunctions of $\mathbf{L}^2$, $\mathbf{S}^2$, L_z, S_z, and the parity operator π. Then, one can use the following expansion:

$$\Psi_{\mu E}^{\Omega-}(\mathbf{r}_1, \mathbf{r}_2) = \Theta \sum_{\mu'=1}^{N_c} (\psi_{\mu'}^{\Omega}(\mathbf{r}_1, \hat{r}_2) F_{\mu\mu'}^{\Omega}(r_2)) + \sum_i c_i \Lambda_i^{\Omega}(\mathbf{r}_1, \mathbf{r}_2) \quad (1)$$

where Θ is the symmetrization (antisymmetrization) operator for singlet (triplet) states and the channel functions, $\psi_{\mu'}^{\Omega}$, are combinations of the target states with the angular functions of the scattered electron to form eigenfunctions of $\mathbf{L}^2$, $\mathbf{S}^2$, L_z, S_z, and π. The corresponding eigenvalues are $\Omega = L, S, M_L, M_S, \pi$. The first summation in eq. (1) runs over N_c target states and the second includes a set of L^2 integrable functions Λ_i^{Ω} which are needed to properly represent electron correlation, specially in the resonance regions. The radial wave functions $F_{\mu\mu'}^{\Omega}$ have the following asymptotic behaviour:

$$F_{\mu\mu'}^{\Omega}(r) \to k_\mu^{-1/2}(\delta_{\mu\mu'} e^{i\theta_\mu} - S_{\mu\mu'}^{\Omega\dagger} e^{-i\theta_\mu}) \quad k_\mu^2 \geq 0 \quad (2)$$

$$F_{\mu\mu'}^{\Omega}(r) \to 0 \quad k_\mu^2 < 0 \quad (3)$$

where

$$\theta_\mu = k_\mu r - \frac{1}{2} l_\mu \pi + \frac{Z-1}{k_\mu} \mathrm{Ln}(2k_\mu r) + \arg\Gamma(l_\mu + 1 - i(Z-1)/k_\mu) \quad (4)$$

The first summation in eq. (1) is the non L^2 integrable part of the wave function. Let us define a projection operator P associated to the space spanned by the N_c channels included in that expansion, and $Q = 1 - P$, its complementary orthogonal counterpart. Then, from substitution of $\Psi_{\mu E}^{\Omega-} = P\Psi_{\mu E}^{\Omega-} + Q\Psi_{\mu E}^{\Omega-}$ in the Schrödinger equation, one obtains[6]:

$$\Psi_{\mu E}^{\Omega-} = (1 + G_Q(E) Q \mathcal{H} P) P \Psi_{\mu E}^{\Omega-} \quad (5)$$

where $P\Psi_{\mu E}^{\Omega-}$ satisfies a P projected Schrödinger equation, including an optical potential:

$$[P\mathcal{H}P + P\mathcal{H}QG_Q(E)Q\mathcal{H}P - E]|P\Psi_{\mu E}^{\Omega-}> \equiv [\mathcal{H}' - E]|P\Psi_{\mu E}^{\Omega-}>= 0 \quad (6)$$

$G_Q(E)$ is the Green operator in Q subspace:

$$G_Q(E) = \sum_n \!\!\!\!\!\!\int \frac{|\phi_n><\phi_n|}{E - \mathcal{E}_n} \quad (7)$$

and ϕ_n are the eigenfunctions of $Q\mathcal{H}Q$:

$$(Q\mathcal{H}Q - \mathcal{E}_n)\phi_n = 0 \tag{8}$$

Now, one is led to solve the two projected Schrödinger equations (6) and (8). For the second one there exist standard codes which make use of L^2 integrable basis, such as the conventional[7] or the pseudopotential[8] Feshbach methods. Therefore, this will not be described here. On the other hand, the present paper will focus on the solution of eq. (6), whose eigenfunctions are not L^2 integrable. It must be noted that when $N_c \to \infty$, $Q \to 0$ and therefore $G_Q(E) \to 0$, so that one would not need to consider the last term in eq. (1). However, N_c cannot be very large in practical calculations and, therefore, such a term is indispensable in order to have a proper description of electron correlation.

Discretization of a **single** continuum μ can be performed easily. In this case, diagonalization of the hamiltonian in a basis of configurations $\Theta(\psi_\mu^\Omega(\mathbf{r}_1,\hat{r}_2)\varphi_{\mu k}(r_2))$, where $\{\varphi_{\mu k}\}$ is a set of L^2 radial functions, leads to discretized states:

$$\tilde{\chi}_{\mu n}^0 = \Theta(\psi_\mu^\Omega(\mathbf{r}_1,\hat{r}_2)\sum_i c_{in}^\mu \varphi_{\mu i}(r_2)) \equiv \Theta(\psi_\mu^\Omega(\mathbf{r}_1,\hat{r}_2)\tilde{\xi}_{\mu n}(r_2)) \tag{9}$$

which are similar to the static exchange wave functions:

$$\chi_{\mu E}^0 = \Theta(\psi_\mu^\Omega(\mathbf{r}_1,\hat{r}_2)\xi_{\mu E}(r_2)) \tag{10}$$

in a finite domain D of configuration space, except for a renormalization factor ρ_μ:

$$\chi_{\mu E_n}^0 \simeq \rho_\mu^{1/2} \tilde{\chi}_{\mu n}^0 \tag{11}$$

where ρ_μ is called density of states. The asymptotic form of the discretized states $\rho_\mu^{1/2}\tilde{\chi}_{\mu n}^0$ (before reaching their exponential decreasing behaviour) differs from that of eqs. (1) and (2) by a global phase factor, which is irrelevant, but can be evaluated if needed. In other words, the use of such a basis of configurations is coherent with the physical boundary conditions.

On the other hand, when the number of open channels is greater than one, a direct diagonalization of the hamiltonian cannot provide a correct discretization of the continuum. Indeed, one cannot use a single set of configurations $\Theta(\psi_\mu^\Omega(\mathbf{r}_1,\hat{r}_2)\varphi_{\mu k}(r_2))$ for a given channel μ, since different μ are coupled asymptotically (see eqs. (1) and (2)). Moreover, a direct diagonalization of $\mathcal{H}$ in a compound basis including all open channels μ does not reproduce the asymptotic behaviour of eq. (2) because, as a result of the L^2 normalization of the basis, interchannel coupling is not described properly[5]. In other words, there is no asymptotic proportionality between $P\Psi_{\mu E}^{\Omega-}$ (eqs. (1) and (2)) and the wave function resulting from such a diagonalization.

The present method[9] adopts a different point of view and takes advantage of the fact that discretization techniques are well adapted for single open channels. Then, I start by defining **orthogonal uncoupled continuum states** (UCS) which are written in the static exchange approximation (eq. (10)) and, therefore, can be discretized as described above. Each UCS is formally associated to a projection operator P_μ which satisfies:

$$P_\mu \chi^0_{\mu E} = \chi^0_{\mu E} \tag{12}$$

$$P_\mu P_{\mu'} = \delta_{\mu\mu'} P_\mu \tag{13}$$

$$P = \sum_\mu P_\mu \tag{14}$$

so that the UCS are eigenfunctions of a zeroth order uncoupled hamiltonian:

$$(\sum_{\mu'} P_{\mu'} \mathcal{H} P_{\mu'} - E) \chi^0_{\mu E} = 0 \tag{15}$$

The scattering states are related to the UCS through the well known Lippman-Schwinger equation:

$$P\Psi^{\Omega-}_{\mu E} = \chi^0_{\mu E} + G^-_P(E) V \chi^0_{\mu E} \tag{16}$$

where

$$V = \sum_{\substack{\mu\mu' \\ \mu \neq \mu'}} P_\mu \mathcal{H} P_{\mu'} + P\mathcal{H}Q G_Q(E) Q\mathcal{H} P \tag{17}$$

and

$$G^-_P(E) = \lim_{\eta \to 0} \frac{1}{E - \mathcal{H}' - i\eta} \tag{18}$$

The aim is to evaluate $G^-_P(E)$ using algebraic methods. For this purpose it is very useful to consider the following relation:

$$G^-_P(E) = G^-_0(E) + G^-_0(E) V G^-_P(E) \tag{19}$$

where G^-_0 is the Green operator associated to the zeroth order uncoupled hamiltonian:

$$G^-_0(E) = \lim_{\eta \to 0} \frac{1}{(E - \sum_\mu P_\mu \mathcal{H} P_\mu - i\eta)} \tag{20}$$

Now, I assume that the discretized UCS satisfy the closure relation in the domain D:

$$\sum_\mu \sum_n |\tilde{\chi}^0_{\mu n} ><\tilde{\chi}^0_{\mu n}| = P \tag{21}$$

In the basis of discretized zeroth order states the latter operator takes the form:

$$G^-_0(E_n) = \sum_{\mu'} \sum_{\substack{n' \\ E_{n'} \neq E_n}} \frac{|\tilde{\chi}^0_{\mu' n'} ><\tilde{\chi}^0_{\mu' n'}|}{(E_n - E_{n'})} + i\pi \sum_\mu \rho_\mu(E_n) |\tilde{\chi}^0_{\mu n} ><\tilde{\chi}^0_{\mu n}| \tag{22}$$

Using eq. (22) and the closure relation (21) in eq. (19), and projecting from the left and from the right with the $\tilde{\chi}^0_{\mu n}$ functions, one obtains:

$$\sum_{\mu'',n''} \mathcal{C}_{\mu'n'\mu''n''} < \tilde{\chi}^0_{\mu''n''} \mid G_P^-(E) \mid \tilde{\chi}^0_{\mu n} > = \mathcal{D}_{\mu'n'} \tag{23}$$

where

$$\mathcal{C}_{\mu'n'\mu''n''} = \delta_{\mu''\mu'}\delta_{n''n'} - \Xi_{\mu'}(E_{n'}) < \tilde{\chi}^0_{\mu'n'} \mid V \mid \tilde{\chi}^0_{\mu''n''} > \tag{24}$$

$$\mathcal{D}_{\mu'n'} = \delta_{\mu'\mu}\delta_{n'n}\Xi_{\mu'}(E_{n'}) \tag{25}$$

and

$$\Xi_{\mu'}(E_{n'}) = \begin{cases} i\pi\rho_{\mu'}(E_{n'}) & \text{for } E_{n'} = E_n \\ 1/(E_n - E_{n'}) & \text{for } E_{n'} \neq E_n \end{cases} \tag{26}$$

Eq. (23) represents a system of linear equations in the complex plane for each channel μ. The coefficient matrix **C** multiplying the unknowns is the same for all μ, so that each system of equations differs exclusively in right-hand side column vector **D**. Therefore, only one matrix inversion is required to solve eq. (23), which means that computer time is independent of the number of channels included in the calculations. Finally, for each open channel μ, the $P\Psi^{\Omega-}_{\mu E}$ wave function is written:

$$P\Psi^{\Omega-}_{\mu E_n} = \rho_\mu^{1/2}(E_n) \left(\tilde{\chi}^0_{\mu n} + \sum_{\mu'n'}\sum_{\mu''n''} < \tilde{\chi}^0_{\mu'n'}|G_P^-(E_n)|\tilde{\chi}^0_{\mu''n''} >< \tilde{\chi}^0_{\mu''n''}|V|\tilde{\chi}^0_{\mu n} > \tilde{\chi}^0_{\mu'n'} \right) \tag{27}$$

so that, from eqs. (1) and (9),

$$F^\Omega_{\mu\mu'} = \rho_\mu^{1/2}(E_n) \sum_{n'} \left[\delta_{\mu\mu'}\delta_{nn'} + \sum_{\mu''n''} < \tilde{\chi}^0_{\mu'n'}|G_P^-(E_n)|\tilde{\chi}^0_{\mu''n''} >< \tilde{\chi}^0_{\mu''n''}|V|\tilde{\chi}^0_{\mu n} > \right] \tilde{\xi}_{\mu'n'} \tag{28}$$

PHOTODETACHMENT OF H^-

The present method have been applied to evaluate photodetachment cross sections of H^- near the $N = 2$ and $N = 3$ thresholds in the dipole approximation[10]. Only some results near the $N = 2$ threshold are presented here. All $N = 1$, $N = 2$ and $N = 3$ states of hydrogen have been included in P subspace, i.e. in the first summation of eq. (1), since they are associated to the strongest interacting channels in that region. Therefore, channels with $N \geq 4$ are obtained in Q subspace and are included through the $G_Q(E)$ operator. The wave functions belonging to Q subspace are evaluated in the framework of the pseudopotential Feshbach method[8]. Specific details of these calculations can be found elsewhere[10]. The discretized UCS, $\tilde{\chi}^0_{\mu n}$, are calculated with an even tempered basis of STOs using the standard codes of Macías et al[11], which also provide the renormalization factor $\rho_\mu(E_n)$ for each μ and each energy E_n.

The total and $N = 2$ cross sections are shown in Figs. 1 and 2, respectively. Gauge invariance is very good. Just below the $N = 2$ threshold, one can see two narrow Feshbach resonances. The broad structure just above this threshold is the

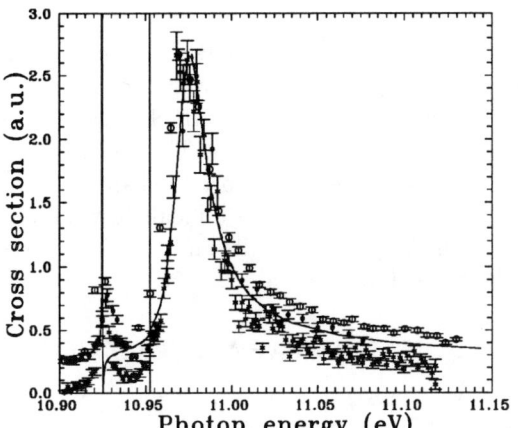

Figure 1: Calculated total photodetachment cross section of H^- in the vicinity of the $N = 2$ threshold at 10.9530 eV. Experimental values, obtained in arbitrary units, have been normalized to the theoretical amplitude at the maximum of the cross section: Bryant et al[12] (•), Halka et al[13] (×) and Butterfield reported in ref. 13 (o). (———): length gauge; (– – –): velocity gauge.

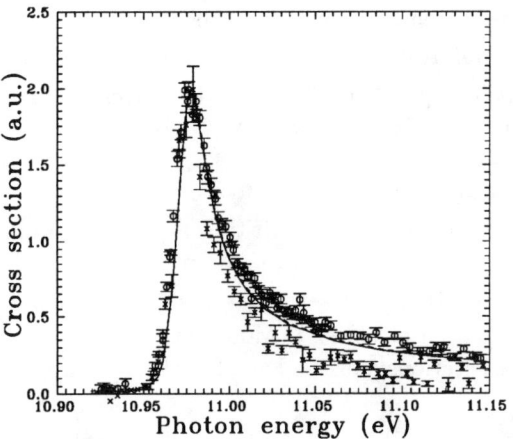

Figure 2: Calculated $\sigma_{N=2}$ photodetachment cross section of H^- in the vicinity of the $N = 2$ threshold at 10.9530 eV. Experimental values, obtained in arbitrary units, have been normalized to the theoretical amplitude at the maximum of the cross section: Halka et al[13] (×), Butterfield reported in ref. 13 (o). (———): length gauge; (– – –): velocity gauge.

well known $^1P^o$ shape resonance of H^-. In these figures, the present results are also compared with the available experimental data[12,13]. The general agreement is very good. The comparison with previous calculations reported in the literature[1,14] is also satisfactory. It is worth noticing that the UCS wave functions present **no** resonant behaviour in the shape resonance region. The appearance of such a resonance in the cross sections is then entirely due to interchannel coupling in P subspace. Since description of such a coupling is very difficult just above threshold, the good agreement found between the present results and the experimental ones constitutes a stringent test of the present L^2 method.

RESONANCE PARAMETRIZATION

When the number of channels N_c included in P is restricted to the physical open channels, *i.e.* when only energetically accesible target states are included in the first summation of eq. (1), the exact continuum wave function $\Psi_{\mu E}^{\Omega^-}$ can be written[15]:

$$|\Psi_{\mu E}^{\Omega^-}>= \frac{<\phi_s|Q\mathcal{H}P|P\Psi_{\mu E}^{0-}>}{E-\mathcal{E}_s-\Delta_s(E)-i(\Gamma_s(E)/2)}|\phi_s>+$$

$$(1+G_Q^{(s)}(E)Q\mathcal{H}P)\left[|P\Psi_{\mu E}^{0-}>+\frac{<\phi_s|Q\mathcal{H}P|P\Psi_{\mu E}^{0-}>}{E-\mathcal{E}_s-\Delta_s(E)-i(\Gamma_s(E)/2)}G_P^{(s)-}(E)P\mathcal{H}Q|\phi_s>\right]$$

(29)

where $G_Q^{(s)}(E)$ is the Green operator in Q subspace in which the s state has been excluded, $P\Psi_{\mu E}^{0-}$ is a non-resonant wave function of energy E which is the eigenstate of $P\mathcal{H}P + P\mathcal{H}QG_Q^{(s)}(E)Q\mathcal{H}P$, $G_P^{(s)-}(E)$ is the corresponding Green operator in P subspace, $\Gamma_s(E)$ is the "width", and $\Delta_s(E)$ is the "shift" of the ϕ_s Feshbach resonance at the energy E. Both $P\Psi_{\mu E}^{0-}$ and $G_P^{(s)-}(E)$ can be evaluated using the L^2 UCS as explained above.

Eq. (29) is exact, but the way of writing $\Psi_{\mu E}^{\Omega^-}$ is not unique. This flexibility allows one to single out any ϕ_s state to build $G_Q^{(s)}(E)$ and $G_P^{(s)-}(E)$. In particular, selection of the ϕ_s state that is closer to the Feshbach resonance s accelerates convergence in the calculation of $G_Q^{(s)}(E)$. Moreover, the use of eq. (29) leads to total photoionization cross sections[15] of the Fano form[16]:

$$\sigma(E) = \sigma^0\left[\rho_s^2\frac{(q_s+\epsilon)^2}{1+\epsilon^2}+1-\rho_s^2\right] \quad (30)$$

where $\epsilon = 2(E-\mathcal{E}_s-\Delta_s(\mathcal{E}_s))/\Gamma_s(\mathcal{E}_s)$, σ^0 is the background nonresonant cross section, q_s is the line profile parameter, and ρ_s^2 is the correlation parameter. For the partial cross section, it leads[15] to the Starace equations[17]:

$$\sigma_\mu(E) = \frac{\sigma_\mu^0}{1+\epsilon^2} \left\{ \epsilon^2 + 2\epsilon \left(q_s \text{Re}[\alpha_\mu^s] - \text{Im}[\alpha_\mu^s] \right) + 1 \right.$$
$$\left. -2q_s \text{Im}[\alpha_\mu^s] - 2\text{Re}[\alpha_\mu^s] + (q_s^2 + 1)|\alpha_\mu^s|^2 \right\} \quad (31)$$

where σ_μ^0 is the partial background nonresonant cross section and α_μ^s are the Starace parameters. Therefore, no fitting procedure is needed to evaluate the resonance parameters, which are readily obtained for $E = \mathcal{E}_s$. These parameters remain practically constant around $\mathcal{E}_s$, and take into account the effect of neighboring resonances through the terms including the $G_Q^{(s)}(E)$ operator.

HELIUM PHOTOIONIZATION

The wave function of eq. (29) has been used to calculate photoionization cross sections of He below the $N = 3$ threshold from the ground[15,18,19] and $2^{1,3}S^e$ metastable[20,21] states. Both total and partial cross sections (integrated and differential) have been evaluated and the results are in good agreement with the experimental data. Recently, the method has been applied to study photoionization of He in the $4lnl'$ resonance region which lies between the $N = 3$ and $N = 4$ thresholds[22]. In this region, there are nine open channels, $1s\varepsilon p$, $2s\varepsilon p$, $2p\varepsilon s$, $2p\varepsilon d$, $3s\varepsilon p$, $3p\varepsilon s$, $3p\varepsilon d$, $3d\varepsilon p$, and $3d\varepsilon f$, of which the last five ones are strongly coupled. The $N = 2$ and $N = 3$ cross sections are shown in Fig. 3, and the sum of both is presented in Fig. 4. The available experimental data have been also included[23,24]. Only one strong series of resonances is observed: the peaks at 73.71, 74.61 and 74.98 eV, which correspond to the first three $4lnl'$ doubly excited states of the (2,1) series[†]. The small features seen at 74.15 and 74.85 eV in Fig. 3 are the first two (0,1) resonances, and the small hump at 74.91 corresponds, probably, to the first resonance of the (-2,1) series. Results for each individual partial cross sections as well for resonance parameters can be found elsewhere[22].

FINAL REMARKS

In this paper, I have presented a fully algebraic close-coupling method to evaluate continuum wave functions in multichannel scattering. In this method, one builds a representation of the Green operator in a basis of discretized uncoupled continuum states, so that one is led to solve a system of linear equations in the complex plane. It can be applied irrespective of the interaction strength between the uncoupled channels.

Total and partial photoionization cross sections of H^- and He have been presented in order to illustrate the accuracy. Agreement with the existing experimental measurements is very good. Application to electron-atom scattering, which implies the evaluation of several partial waves, is in progress.

[†]Resonances are labeled with the (K, T) quantum numbers[25]

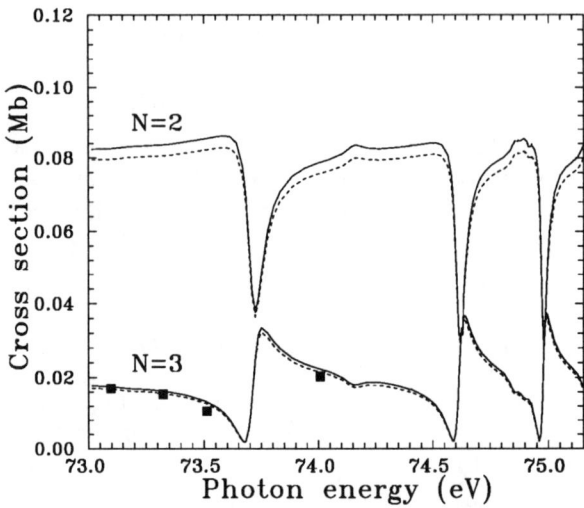

Figure 3: Partial $\sigma_{N=2}$ and $\sigma_{N=3}$ photoionization cross sections of helium above the $N = 3$ threshold. (——): length gauge; (- - -): velocity gauge; (■): experimental values of Heimann et al[23].

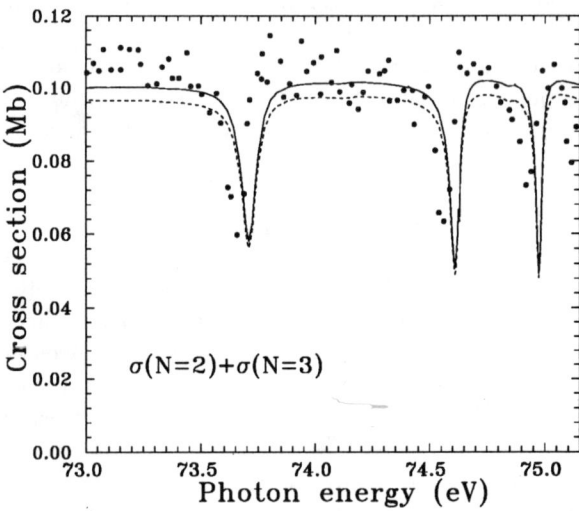

Figure 4: $\sigma_{N=2} + \sigma_{N=3}$ photoionization cross section of helium above the $N = 3$ threshold. (——): length gauge; (- - -): velocity gauge; (●): experimental values of Woodruff and Samson[24].

ACKNOWLEDGMENTS

Many interesting discussions with Drs. M. Cortés, A. Salin and I. Sánchez are acknowledged. I also want to thank Dr. M. Cortés for her kind help in the ellaboration of this manuscript. This work has been partially supported by the DGICYT project No. PB90-213 and the EEC twining program No. SCI*.0138.C(JR)

REFERENCES

1 J. T. Broad and W. P. Reinhardt, Phys. Rev. A **14** 2159 (1976)
2 B. H. Bransden and A. T. Stelbovics, J. Phys. B **17** 1877 (1984)
3 M. L. Du and A. Dalgarno, Phys. Rev. A **43** 3474 (1991)
4 R. Moccia and P. Spizzo, Phys. Rev. A **43** 2199 (1991)
5 F. Martín, A. Riera, and I. Sánchez, J. Chem. Phys. **94** 4275 (1991)
6 H. Feshbach, Ann. Phys. (N.Y.) **19** 287 (1962)
7 H. Bachau, J. Phys. B **17** 1771 (1984)
8 F. Martín, O. Mó, A. Riera, and M. Yáñez, Europhys. Lett. **4** 799 (1987)
9 F. Martín, Phys. Rev. A **48** July (1993)
10 M. Cortés and F. Martín, Phys. Rev. A **48** August (1993)
11 A. Macías, F. Martín, A. Riera, and M. Yáñez, Phys. Rev. A **36** 4179 (1987)
12 H. C. Bryant, B. D. Dieterle, J. Donahue, H. Sharifian, H. Tootoonchi, D. M. Wolfe, P. A. M. Gram, and M. A. Yates-Williams, Phys. Rev. Lett. **38** 228 (1977)
13 M. Halka, H. C. Bryant, C. Johnstone, B. Marchini, W. Miller, A. H. Mohagheghi, C. Y. Tang, K. B. Butterfield, D. A. Clark, S. Cohen, J. B. Donahue, P. A. M. Gram, R. W. Hamm, A. Hsu, D. W. MacArthur, E. P. MacKerrow, C. R. Quick, J. Tiee, and K. Rózsa, Phys. Rev. A **46** 6942 (1992)
14 H. R. Sadeghpour, C. H. Greene and M. Cavagnero, Phys. Rev. A **45** 1587 (1992)
15 I. Sánchez and F. Martín, Phys. Rev. A **44** 7318 (1991)
16 U. Fano and J. W. Cooper, Phys. Rev. **137** A1364 (1965)
17 A. F. Starace, Phys. Rev. A **16** 231 (1977)
18 I. Sánchez and F. Martín, Phys. Rev. A **44** 13 (1991)
19 I. Sánchez and F. Martín, Phys. Rev. A **45** 4468 (1992)
20 I. Sánchez and F. Martín, Phys. Rev. A **47** 1520 (1993)
21 I. Sánchez and F. Martín, Phys. Rev. A **47** 1878 (1993)
22 I. Sánchez and F. Martín, Phys. Rev. A **48** August (1993)
23 P. A. Heimann, U. Becker, H. G. Kerkhoff, B. Langer, D. Szostak, R. Wehlitz, D. W. Lindle, T. A. Ferrett, and D. A. Shirley, Phys. Rev. A **34** 3782 (1986)
24 P. R. Woodruff and J. A. R. Samson, Phys. Rev. A **25** 848 (1982)
25 D. R. Herrick and O. Sinanoglu, Phys. Rev. A **11** 97 (1975)

SUBSHELL AND INNERSHELL EXCITATION IN RARE GAS ATOMS BY ELECTRON IMPACT

T. Takayanagi
Department of Physics, Sophia University
Chiyoda-ku, Tokyo 102, Japan

ABSTRACT

Experimental studies on the electron impact excitation of atomic inner-shell electrons are discussed, laying an emphasis on an experimental technique using the electron-energy-loss spectroscopy, which has been developed recently in the atomic physics group of Sophia University.

A detail of the procedure is described on the measurements of cross sections for excitation and ionization of the 4d-electrons in Xe. The electron-energy loss spectra (EELS) for the region of the 4d-excitation and ionization in Xe are measured for impact energies 1, 1.5, and 2keV at scattering angles from 0° to 5°. The spectra reveal a giant resonance feature, which is well known from the photo-absorption experiments.

The generalized oscillator strengths (GOS) for the 4d-ionization deduced from the DCS are displayed as a function of the squared momentum transfer; this procedure is useful for the comparison of the electron impact cross section data with the optical oscillator strengths, which are determined by the photo-absorption experiments.

Results of cross sections for the 4d-ionization determined by the new procedure are presented as a function of the impact energy from the threshold to 4keV.

Finally, the comparison of the double and triple ionization cross sections among Ba^+, Cs^+, and Xe^+ are discussed in connection with the 4d-ionization processes in these species.

INTRODUCTION

Recently studies of excitation and ionization of atomic inner-shell have been greatly developed by photo-absorption experiments using synchrotron radiation sources.

Though a lot of works have been done on the valence electron excitation by the electron impact, very small number of works have been done on the inner-shell excitations. In the photo-absorption processes, the excitation energy is known decidedly from the incident photon energy, and usually, only dipole allowed transitions occur. So the analysis of experimental results are rather simple and clear.

In the case of the electron impact, excitation energy cannot be uniquely determined from the incident electron energy, because electrons give their energies to target atoms only partially. Besides many forbidden transitions may occur. Then, the analysis of experimental results is not so simple as the photo-absorption case. Electron impact excitation is, however, one of the fundamental processes in atomic collision physics, and many interesting phenomena must be caused by the correlation effects between incident or scattered electron with those in the atom.

A series of experimental studies, has been carried on in our

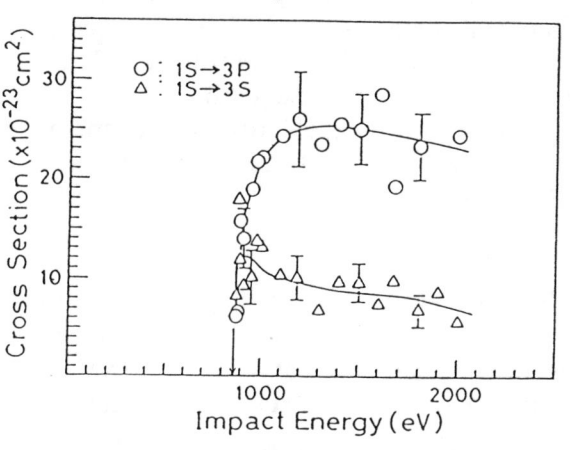

Fig.1.
Cross sections for the 1s→3p and the 1s→3s excitation in Ne.

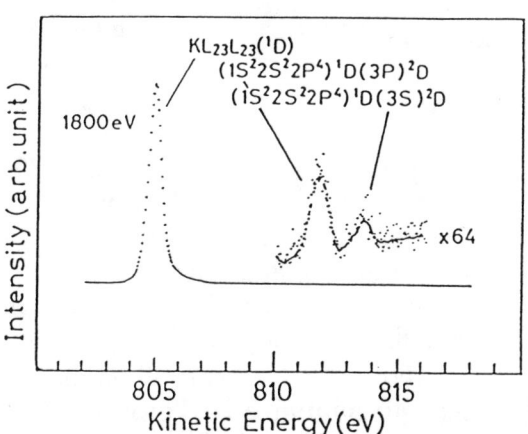

Fig.2.
A typical spectrum of the excitation-Auger-satellite lines.

laboratory, on subshell and inner-shell excitation (including ionization) processes by electron impact, using the emitted electron spectroscopy and the VUV spectroscopy.

As an example, Fig.1 shows cross sections for the 1s→3p and 1s→3s excitation in Ne as a function of the impact energy. They were determined from the intensities of the excitation-Auger-satellite lines $(1s^22s^22p^4)$ ^1D (3p) ^2D and $(1s^22s^22p^4)$ ^1D (3s) ^2D, which follow the 1s vacant states caused by the 1s→3p and 1s→3s excitation, relative to $KL_{2,3}L_{2,3}(^1D)$ normal Auger line (see Fig.2). It is interesting that the energy dependence of the cross sections for the 1s→3p and the 1s→3s excitation behaves in a typical way of the optical allowed and the forbidden transition, respectively, even in the inner-shell excitation.

Differential cross sections (DCS) and generalized oscillator strengths (GOS) are fundamental quantities to understand excitation processes by electron impact. But very few experimental data are available for such quantities for the inner-shell excitation.

We constructed a specially designed electron spectrometer to measure the DCS for inner shell excitation of atoms for high impact energies and at minute scattering angles including 0°. Measurements for rare gas atoms are in progress, and we get some interesting results for the Xe 4d excitation and ionization.

It is known by photo-absorption experiments that the Xe 4d-electron ionization is characterized by the behavior called the giant resonance[1], with a large broad peak at around 100 eV excitation energy, which is understood as the collective excitation of electrons in an atom[2]. We determined the DCS and the GOS for these processes utilizing the electron energy-loss spectra (EELS).

APPARATUS

Schematic diagram of the apparatus is shown in Fig.3. The apparatus consists of four parts ; a monoenergetic electron source (electron gun, energy selector and lenses), a collision region, a tandem-type energy analyzer with appropriate lenses, and a channel electron multiplier used as a detector. The energy selector and both component of the tandem analyzer are the simulated hemi-spherical type ones, which were first designed by Jost[4], whose mean radii are all 52-mm.

The tandem-analyzer is useful to minimize background noises especially at minute scattering angles including 0°. The conventional constant resolution mode was used to measure the EELS, supplying

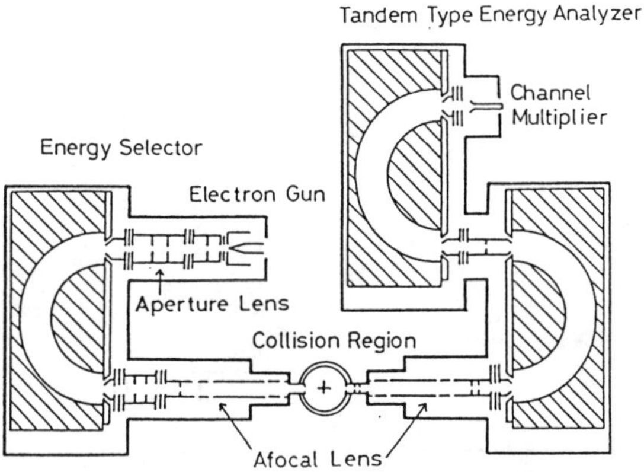

Fig.3. Schematic diagram of the high energy electron impact spectrometer.

typical resolution of 80 - 100 meV in full width at half maximum (FWHM).

The calibration of scattering angles and the angular resolution of the apparatus are important, because the DCS around zero degree decreases drastically as the scattering angle increases. In our apparatus, scattering angles can be set with an accuracy of 0.08° using a stepping motor controlled by a micro-computer system. We measured the angular distributions of the outer 5p-electron excitations at each impact energy and calibrated the zero scattering angle utilizing the symmetry nature of them around zero degree. The angular resolution of the spectrometer was examined by measuring the angular distributions of the incident electron beam which came without scattering. They were from 0.4° to 0.6° (FWHM) for 1 to 2 keV.

ENERGY LOSS SPECTRUM OF Xe 4d EXCITATION

A typical energy-loss-spectrum for the ionization region of the 4d-electron in xenon, at impact energy of 2 keV and scattering angle of 0°, is shown in Fig.4. It shows a well known giant resonance feature of the 4d-ionization. Two sharp peaks bellow 70 eV correspond to excitations of the 4d-electrons to the 6p orbital. An expanded spectrum of the excitation region is shown in Fig.5, where we can see the $4d^9(^2D_{5/2})6p,7p$ and $4d^9(^2D_{3/2})6p,7p$ structures evidently, although others are not clear. The energy-loss-spectrum shown in Fig.4 resembles the photo-absorption spectrum as a whole, but some interesting differences exist. In Fig.4 the maximum position of the 4d-ionization is about 88eV, but in the photo-absorption case, the maximum is located around 100eV. The relative intensities of the 4d→6p excitation peaks to that of the ionization are large compared

330 Subshell and Innershell Excitation

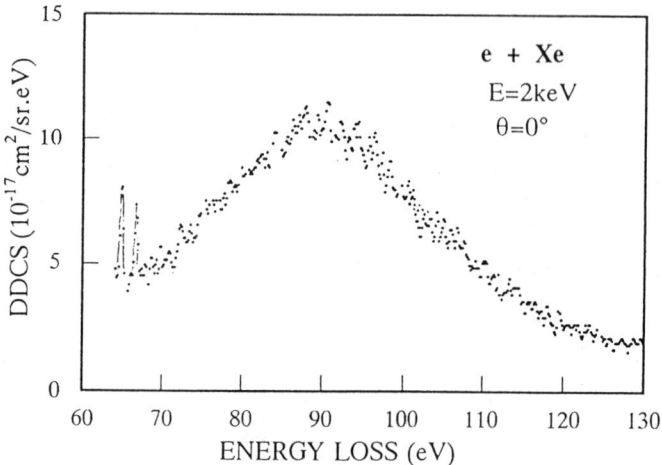

Fig.4. The energy-loss-spectrum for the 4d-ionization in Xe. The energy is 2 keV and the scattering angle is 0°.

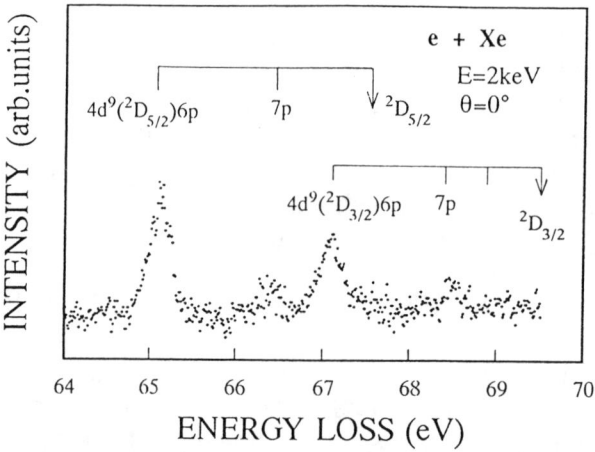

Fig.5. A typical energy-loss-spectrum for excitation of the 4d-electron in Xe. Impact energy is 2 keV, and the scattering angle is 0°. Two series of the 4d→np excitation(n≥6) are observed.

with photo-absorption spectrum. Figure 6 is the spectrum at the same impact energy but at the scattering angle of 2.6°. This spectrum shows some fine structures at about 80 eV, 95 eV, and 115 eV. The structures at 80 eV correspond to the combined excitation of the 4d and 5p-electrons, but we can not understand the origin of other structures yet. Besides these fine structures, the maximum position of the giant

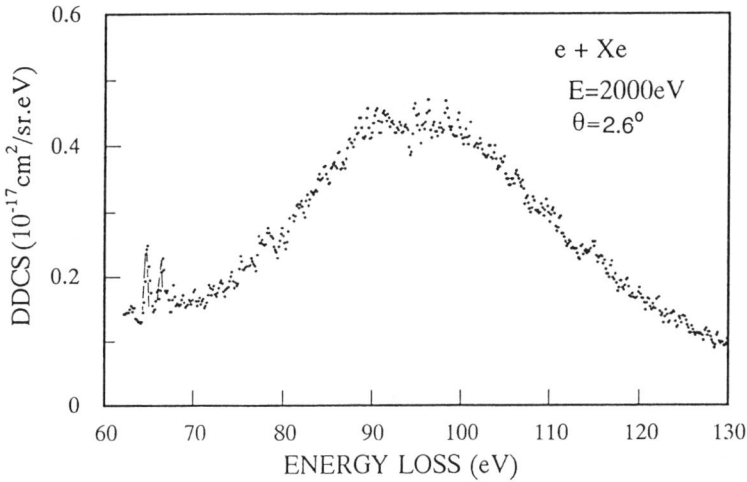

Fig.6. Same as Fig.4 except that the scattering angle is 2.6°.

resonance shifts from 88 eV (in Fig.4) to 97 eV.

Xe 4d IONIZATION

We determined the double differential cross section (DDCS) for the energy- loss region mentioned above, relative to the DCS for the excitation of the 5p-electron to the lowest excited state $5p^5(^2P_{3/2})6s$, which can be deduced from GOS values determined experimentally by T.Y. Suzuki et al.[5] The vertical scales in Fig.4 and FIg.6 are those determined by this procedure. Once we know the absolute values of the DCS at several scattering angles at sufficiently high impact energy, we can transform them to the GOS according to Bethe's formula.[6][7] We did it for the energy-loss range from 70 to 120 eV, which corresponds to almost all 4d-ionization. In Fig.7, we show some examples of them. They are for the 80, 100 and 120 eV energy-losses. It is well known that the GOS become the optical oscillator strength (OOS) at the limit of the momentum transfer $K^2 \to 0$. We show the OOS, deduced from photo-absorption experiment by Hänsl et al.[1], in each figure. Every GOS curve in Fig.7 doesn't converge smoothly to the OOS. The measured GOS contains rather large errors especially at small momentum transfer (12%), but the GOS behavior mentioned above can

332 Subshell and Innershell Excitation

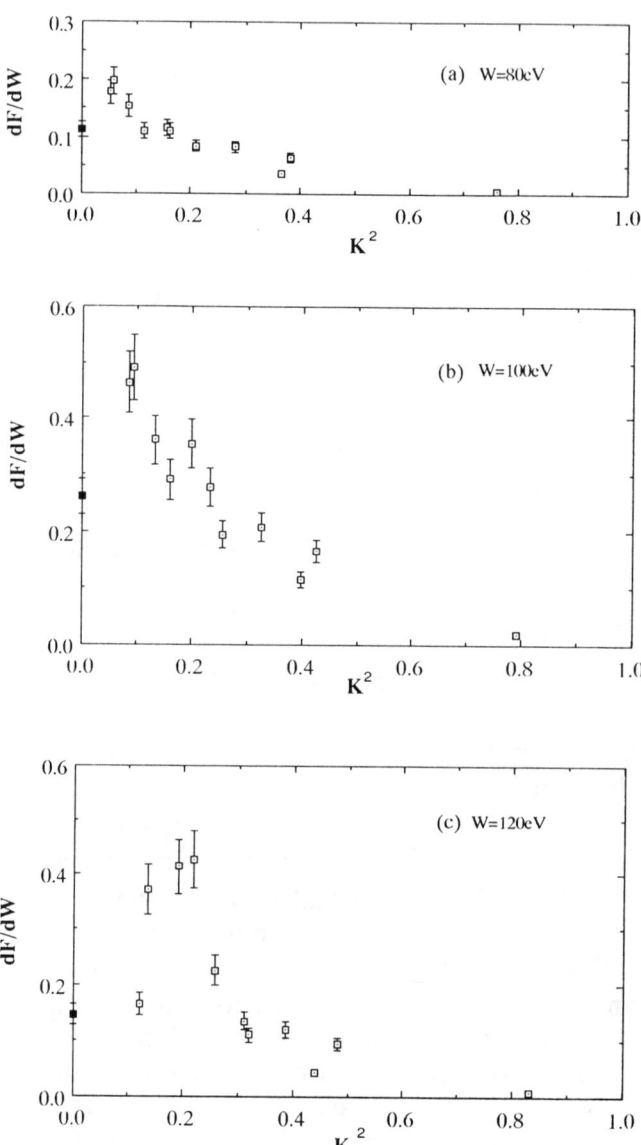

Fig.7. The differential generalized oscillator strength DGOS in the 4d-ionization in Xe as a function of squared momentum transfer K^2, at the excitation ecergies W: (a) W=80 eV, (b) W=100 eV, (c) W=120 eV. In the figures, data point on the vertical axis at zero K^2 is the optical oscillator strength OOS measured by Hänsel et. al.[1])

not be attributed to these errors. The GOS for the energy loss W=120eV (in Fig.7-(c)) clearly shows the peak at $K^2=0.2$, and seems to converge to the OOS. A manner how the GOS converge to the OOS depends on the type of excitation. The present curves of the DGOS against K^2 are the first examples for the 4d-ionization.

We get the 4d-ionization cross section by integrating the measured GOS for both momentum transfer and excitation energy. We see in Fig.7 that the measured GOS become very small in the large momentum transfer region ($K^2 >1$), therefore we can ignore the contribution from this region. We obtained the absolute value of the 4d-ionization cross section $(8.9\pm1.8)\times10^{-18}$ cm^2 for the impact energy of 2 keV.

The impact energy dependence of the 4d-ionization cross section can be known by measuring the intensity of Auger-electrons emitted after the 4d-ionization at each impact energy. We measured $N_5O_{2,3}O_{2,3}$ (^1S) Auger line intensity at the impact energy from 70 eV to 4 keV region. We have normalized this relative impact energy dependence at 2 keV using the value obtained by the EELS method. In Fig.8, the 4d-ionization cross sections obtained by the present procedure are shown as a function of the impact energy. In Fig.8, we also show the previous experimental data (Suzuki et al.[8]) and the theoretical

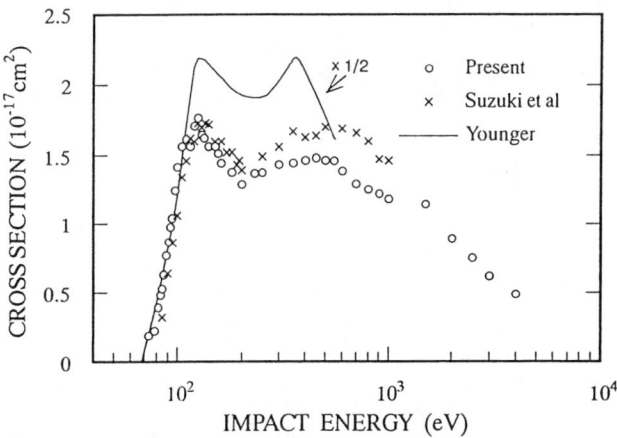

Fig.8. The impact energy dependence of the Xe 4d-ionization by electron impact. Present results are compared with the previous experimental results of Suzuki et al.[8] and the theoretical calculation of Younger[9].

calculation of Younger[9]. In our previous experiment, we determined the intensities of all Auger lines concerned with the 4d-ionization relative to the elastic scattering of 30 eV electrons. Using DCS of the elastic scattering, and taking account of double Auger processes, we obtained the absolute values of the 4d-ionization cross sections. The normalization methods are quite different in two experiments, but, as we see in Fig. 8 , the results agree well each other. Younger's theoretical calculation is too large by a factor of 2, and decreases so fast compared with our experiments, though it succeeds to reproduce the double maximum behavior of impact energy dependence. He explained theoretically such impact energy dependence of 4d ionization as a kind of shape resonance and called it giant resonance in electron impact ionization.

SOME REMARKS ON DOUBLE AND TRIPLE IONIZATION CROSS SECTIONS OF Ba^+ and Sr^+

The 4d-ionization processes have been often mentioned in the discussions of double ionization cross section of some ions by electron impact. But very few reliable cross section data are available for the 4d-ionization.

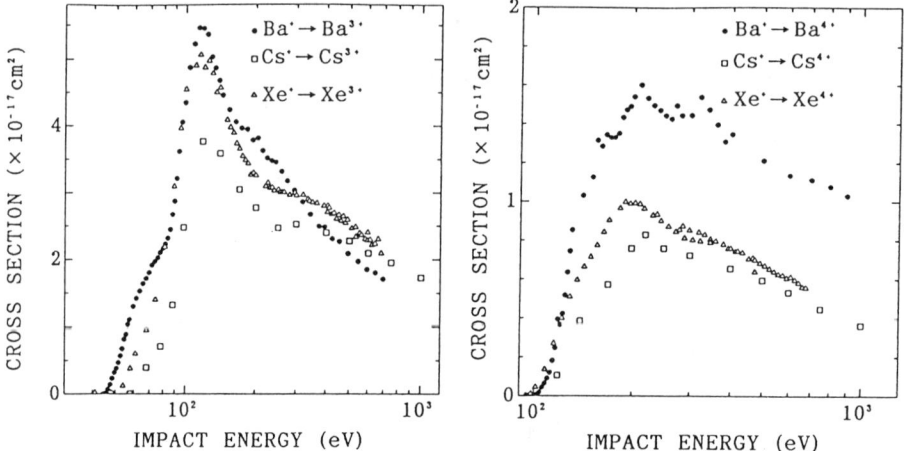

Fig. 9. (left) Comparison of the double ionization cross sections for Ba^+ with those for Cs^+ (D.R.Heartling et.al.[10])and Xe^+(A.Muller et.al.[11]).

Fig.10. (right) Comparision of the triple ionization cross sections among the same ions as Fig.9.

Atomic physics group in Sophia University have measured double and triple ionization cross sections of Ba^+ and Sr^+ by electron impact. In Fig.9 and Fig.10 we show those of Ba^+ in comparison with the data for Cs^+ and Xe^+.

Double ionization cross section of Ba^+ decreases very rapidly in the high impact energy region, and becomes smaller than those of Xe^+ and Cs^+. On the other hand, triple ionization cross section of Ba^+ is about a factor of 2 larger than those of Xe^+ and Cs^+. Binding energy of the Ba^+ 4d electron is larger than the threshold energy of outer three electron ionization, therefore the 4d ionization can contribute to the triple ionization of Ba^+ by auto-double ionization or double auto-ionization. If the impact energy dependence, shown in Fig.8, is the general behavior of the 4d ionization at high impact energy region, our results suggest that the 4d-ionization of Ba^+ mainly contributes to triple ionization, and the peak at about 100eV in double ionization mainly comes from the auto-double ionization of the 4d-excited states.

REFERENCES

1) R.Hänsel, G.Keitel, P.Schreider and C.Kunz: Phys.Rev. **188** (1969) 1375.
2) M.Ya.Amusia, V.K.Ivanov, N.A.Cherepkov and L.V.Chernysheva : Sov.Phys. J.E.T.P. **66** (1974)1537.
3) Y.Sakai, T.Y.Suzuki, B.S.Min, T.Takayanagi, K.Wakiya, H.Suzuki, S.Ohtani and H.Takuma: Phys. Rev. **A43** (1991) 1656.
4) K.Jost: J.Phys. E **4** (1971) 981.
5) T.Y.Suzuki, Y.Sakai, B.S. Min, T.Takayanagi, K.Wakiya, H.Suzuki, T.Inaba and H.Takuma: Phys.Rev. **A43** (1991) 5867.
6) H.Bethe: Ann. Phys. (Leipzig) **5** (1930) 325.
7) M.Inokuti: Rev.Mod.Phys. **50** (1978) 23.
8) H.Suzuki, T.Takayanagi, K.Morita, and Y.Iketaki: Measurements of Cross Sections for Inner Shell Ionization in Rare-Gas Atoms by Electron Impact, in "Electron-Molecule Collisions and Photoionization Processes" ed. V.McKoy, H.Suzuki, K.Takayanagi and S.Trajmar (Verlag Chemie International, 1983).
9) S.M.Younger: Phys.Rev. **A35** (1987) 2841.
10) D.R.Heartling, R.K.Feenery, D.W.Hughes, and E.Stayle: J.Appl. Phys. **53** (1982) 5427.
11) A.Muller, C.Achenbach, E.Salzborn and R.becker: J.Phys. B **17** (1984)1427.

ELECTRON-MOLECULE COLLISIONS

THEORY OF IMPULSIVE ROVIBRATIONAL EXCITATION OF MOLECULES BY ELECTRONS

H. J. Korsch
Fachbereich Physik, Universität Kaiserslautern,
D-67653 Kaiserslautern, Germany

ABSTRACT

Rotational–vibrational transitions in angular resolved state–to–state cross sections in electron–molecule scattering are discussed by means of the N–centre spectator model. Rotational rainbow structures in the cross sections for diatomic and highly symmetric polyatomic moolecules (symmetric tops) are discussed as well as combined rotational and vibrational rainbow features for highly vibrationally excited diatomics.

INTRODUCTION

Rotational transitions in scattering of anisotropic, nonspherical systems are of interest in many branches of physics, ranging from nuclear to astrophysics. Here we will discuss electron–molecule collisions. As in atom–molecule systems before it was realized recently that the folklore of 'small rotational transitions in single encounters' is not generally true and very large rotational momentum changes can occur with high probability. Under such strong coupling conditions

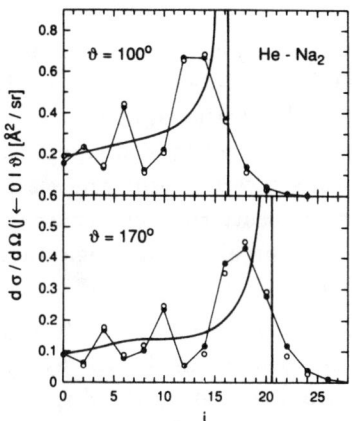 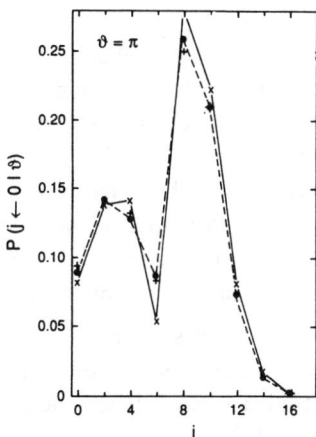

Fig. 1: Differential cross sections for rotational transitions in He–Na$_2$ collisions and nuclear Coulomb excitation. Also shown are semiclassical and classical results.

characteristic *'rotational rainbow structures'* are observed in the rotationally inelastic differential cross sections[1]. Examples from nuclear[2] and molecular[3] scattering in Fig. 1 (Coulomb excitation of ^{238}U colliding with ^{84}Kr at an energy of 300 MeV and He–Na$_2$ collisions at 100 meV). The distributions show oscillations (*'rotational rainbow oscillations'*) and a pronounced maximum (the *'rotational rainbow'*) at a slightly smaller value of the rotational quantum number than the rainbow singularity of the classical distributions.

Rotational rainbows have been observed experimentally for many atom–molecule systems in state–to–state differential cross sections. The theoretical description of rotational rainbow structures is largely based on the infinite–order sudden (IOS) approximation in combination with a semiclassical analysis[4].

More recently rotational rainbow structures have been observed in electron–molecule collisions at intermediate collision energies[5-8], which are even more pronounced than those found before. This is due to the extreme anisotropy of the interaction, which makes a treatment within the framework of the IOS approximation impossible. A very satisfactory description can be given by the impulsive spectator model discussed below.

For a general discussion of rotational rainbows in atom– and electron–molecule collisions see the recent review by Korsch and Ernest[1].

ROTATIONAL TRANSITIONS FOR DIATOMICS

The spectator model is based on a very simplistic picture of the collision dynamics: (i) The collision is sudden (impulsive), i.e. the interaction time between projectile and target is short compared to typical vibrational or rotational periods of the target. (ii) The interaction is approximated by a superposition of pair interactions, which are treated as potential scattering. (iii) The scattering amplitude is approximated by a superposition of scattering amplitudes of the target constituents. Multiple collisions and screening effects are neglected. (iv) The energy transfer is assumed to be negligible.

This leads to simple expressions for the transition operator: If $\tilde{T}$ is the transition operator in the centre of mass for scattering of an electron by a single target atom, the transition operator for the atom shifted by the displacement operators $D(\vec{d}) = \exp(-i\vec{d}\cdot\vec{k})$ and $D'(\vec{d}) = \exp(-i\vec{d}\cdot\vec{k'})$ to a position $\vec{d}$ displaced from centre of mass of the target molecule ($\vec{k}$ and $\vec{k'}$ are the initial and final wave numbers) is given by

$$T = D'(-\vec{d})\,\tilde{T}\,D(\vec{d}). \tag{1}$$

For homonuclear molecules the transition matrix elements between asymptotic

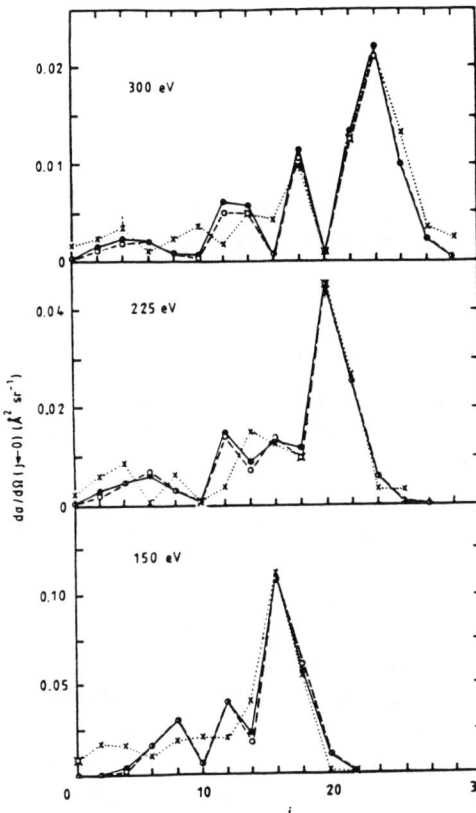

Fig. 2: Differential cross sections for rotational transitions $0 \to j$ for electron–Na$_2$ scattering at a scattering angle of 150^0. Experimental (×) close coupling (•) and spectator model (○) results are compared.

rotor states $|jm;\vec{k}\rangle$ factorize into a molecular form factor

$$F_{j'm'jm} = \langle j'm' | e^{i\Delta \cos\gamma} + e^{-i\Delta \cos\gamma} | jm \rangle \qquad (2)$$

and an elastic scattering amplitude $f(\vartheta) = \langle \vec{k'}|\hat{T}|\vec{k}\rangle$. In Eq. (2) the contributions from the two constituents of the diatomic molecule are added. The molecular anisotropy enters via the parameter

$$\Delta = d\Delta k = r\Delta k/2 \approx rk \sin(\vartheta/2) \qquad (3)$$

under energetically sudden conditions $|\vec{k'}| \approx |\vec{k}| = k$, where d is the distance of the two atoms from the centre of mass, i.e. half of the equilibrium distance r, and $\Delta\vec{k} = \vec{k'} - \vec{k}$ the momentum transfer. The angle between the molecular axis and $\Delta\vec{k}$ is denoted by γ.

Specializing to transitions from the rotational ground state and choosing the quantization axis in the direction of $\Delta\vec{k}$ (kinematic apse) this simplifies to

$$F_{jm00} = \delta_{m0} F_{j000} \quad \text{with} \quad F_{j000} = (1 + (-1)^j)\sqrt{2j+1}\, i^j\, j_j(\Delta), \qquad (4)$$

where j_j is the spherical Bessel function. The factor $(1 + (-1)^j)$ expresses the selection rule '$\Delta j = j = even$'. The differential cross section is given by

$$\frac{d\sigma}{d\Omega}(j \leftarrow 0|\vartheta) = |F_{j000}|^2 \left(\frac{d\sigma}{d\Omega}\right)_{atom}, \tag{5}$$

where $(d\sigma/d\Omega)_{atom} = |f(\vartheta)|^2$ is the (elastic) cross section for electron–atom scattering. Transition probabilities at fixed scattering angle are given by

$$P(j \leftarrow 0|\vartheta) = A(\vartheta)(2j+1)j_j^2(\Delta) \tag{6}$$

for even j, where $A(\vartheta)$ is the normalization constant. Transition cross sections from initially excited rotational states can be directly obtained from the $0 \to j$ data by means of the so-called *'factorization formula'*

$$\frac{d\sigma}{d\Omega}(j' \leftarrow j|\vartheta) = (2j'+1)\sum_{j''}\begin{pmatrix} j & j' & j'' \\ 0 & 0 & 0 \end{pmatrix}^2 \frac{d\sigma}{d\Omega}(j'' \leftarrow 0|\vartheta). \tag{7}$$

In Fig. 2 the results of spectator model (6) are compared with results of elaborate close coupling computations and experimental data for electron–Na_2 at about 10^2 eV [6,5,7]. The agreement is very good, in particular in view of the limited experimental resolution and the extreme simplicity of the spectator model. Improvements of the close coupling computations [9] could reduce the deviation from the the spectator model observable in Fig. 2 for 300 eV. Similar impressive agreement has been found for other angles (down to 30°) and energies down to 25 eV [5,9].

It should be stressed that the rotational transition probabilites at fixed scattering angle are independent of the interaction. As a check of the underlying assumption of simple potential scattering for the electron–atom collisions the electron–Na potential has been fitted by a simple superposition of two screened Coulomb potentials, which could reproduce [9] the theoretical electron–atom cross sections [10] very well in the considered energy region of about 10^2 eV.

ROTATIONAL TRANSITIONS FOR POLYATOMICS

An extension of the two–centre spectator model for rotational excitation to N–atomic molecules is straightforward. A number of closed form expressions of several quantities of interest have been derived for the special case of symmetric tops, in particular for regular planar and bipyramidal molecules. The molecular structure has been assumed to be rigid.

We use symmetric top wave functions $|jkm\rangle$, where j is the rotational quantum number and the quantum numbers k and m are the projections onto the body–fixed axis or a space–fixed axis, respectively. The following results are obtained for the highly symmetric case, where all the N atomic constituents are the same. In this case simple closed form results for differential rotational state–to–state cross sections or transition probabilities can be derived [11-14].

As a direct extension of Eq. (2) the molecular form factor is given by

$$F_{j'k'm'\,jkm} = \langle j'k'm'| \sum_{l=1}^{N} e^{i\vec{d}_l \cdot \Delta \vec{k}} |jkm\rangle, \qquad (8)$$

where $\vec{d}_l$ denotes the position vector of the lth scattering centre measured from the molecular centre of mass.

For transitions from the rotational ground state $|000\rangle$ the summation over the final m and k quantum numbers can be done in closed form. In the case of rotational excitation of planar molecules forming a regular polygon the distances of all N atoms from the centre of mass are equal ($d = |\vec{d}_j|$), all angles between the position vectors are equal to $2\pi/N$, and the transition probabilities are [11,12]

$$P(j \leftarrow 0|\vartheta) = A(\vartheta)\,(2j+1)\,j_j^2(\Delta) \sum_{l=1}^{N} P_j\left(\cos\left[\frac{2\pi}{N}(l-1)\right]\right), \qquad (9)$$

where the $P_j(x)$ are Legendre polynomials. Again $A(\vartheta)$ normalizes the probabilities to unit sum. For an even number N of atoms the selection rule '$j = even$' is valid.

For bipyramidal symmetric top targets (a regular polygon formed by N atoms and one additional atoms attached to each side) the rotational transition probabilities are found to be [13]

$$\begin{aligned}P(j \leftarrow 0|\vartheta) &= A(\vartheta)\,(2j+1)\Big\{2[1+(-1)^j]\,j_j^{\,2}(\Delta_t) + 4N\,j_j(\Delta_p)\,j_j(\Delta_t)\,P_j(0) \\ &\quad + N\,j_j^2(\Delta_p) \sum_{l=1}^{N} P_j(\cos[(l-1)\,2\pi/N])\Big\} \end{aligned} \qquad (10)$$

with

$$\Delta_i = 2d_i k \sin\vartheta/2 \qquad \text{for} \quad i = t, p, \qquad (11)$$

where d_p is the distance of the top atoms and d_p the distance of the ground plane atoms from the centre of mass. Also in this case we find the selection rule '$\Delta j = even$ for even N'.

As an example, Fig. 3 shows the rotational transition probabilities [13,9] for different values of N ($\Delta_p = 14.5$ and $\Delta_p = \Delta_p/2$). Also shown are the classical distributions. We observe a double rainbow structure, where the two rainbows are due to excitation by collisions with the ground plane atoms or the atoms at the tops, respectively, whose relative importance decreases for increasing N.

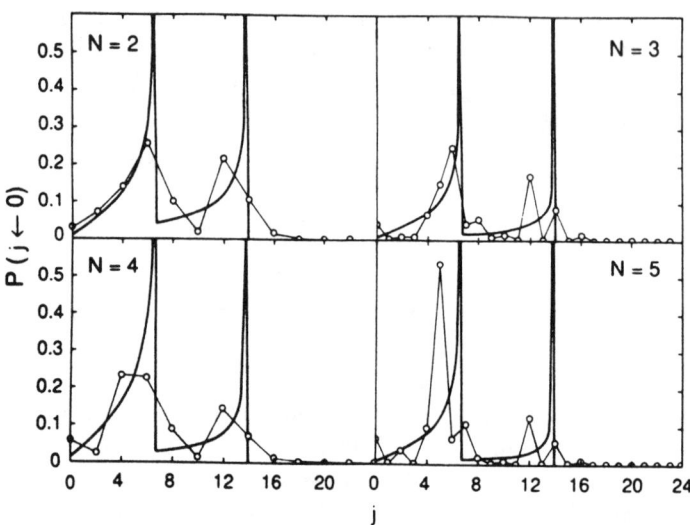

Fig. 3: Rotational rainbow pattern (o) of bipyramidal molecules compared with classical probabilities for an increasing number of N atoms in the ground plane.

Recently, a rotational factorization relation has been derived for symmetric top molecules[14], which is most simple for spherical tops:

$$\frac{d\sigma}{d\Omega}(j' \leftarrow j|\vartheta) = \sum_{j''=|j'-j|}^{j'+j} \frac{2j'+1}{(2j+1)(2j''+1)} \frac{d\sigma}{d\Omega}(j'' \leftarrow 0|\vartheta) . \quad (12)$$

This relation has been stated before in context with low energy electron–spherical top scattering[15].

The quantum mechanical transition probabilities for excitation from the ground state of an octahedron ($N = 4$) are given by

$$P(j \leftarrow 0|\vartheta) = A(\vartheta)\,(2j+1)\,j_j^2(\Delta)\left[1+(-1)^j+4P_j(0)\right] \quad (13)$$

and for a trigonal bipyramidal target ($N = 3$) with $r_p = r_t$ by

$$P(j \leftarrow 0|\vartheta) = \frac{A(\vartheta)}{5}(2j+1)\,j_j^2(\Delta)\left[2(1+(-1)^j)+15P_j(0)+(-1)^j 6P_j(1/2)\right], \quad (14)$$

where j_j are spherical Bessel functions and P_j are Legendre functions.

Quantum and classical distributions for transitions from the highly excited initial state $j = 30$ are shown in Fig. 4 for a spherical top with $N = 4$. Due to

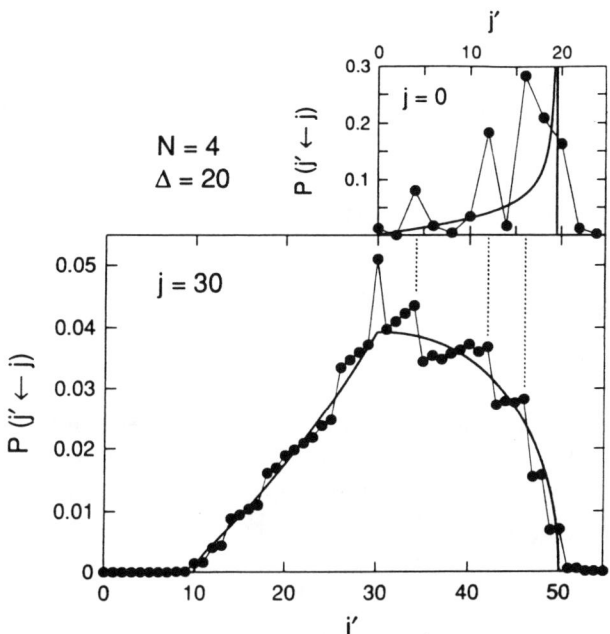

Fig. 4: Classical (—) and quantum (•) rotational transition probabilities for quadratic–bipyramidal spherical top molecules ($N = 4$) for initial states $j = 30$ and $j = 0$.

quenching by the summation in (12) the quantum probabilities and the classical distribution are in very close agreement, in contrast to excitation from the ground state (shown on the upper right for comparison), where the quantum probabilities oscillate strongly around the classical curves. The observed steps in the distributions are strongly correlated with the rainbow maxima for the excitation from the ground state.

VIBRATIONAL–ROTATIONAL TRANSITIONS

Rotational rainbows are well-understood features of rotationally inelastic differential cross sections. The related *vibrational rainbows* have been much less explored.

Recently, measurements of differential state–to–state cross sections for combined rotational and vibrational transitions have been reported for Na_2 in collisions with electrons at about 10^2 eV, where the molecules are initially in a well

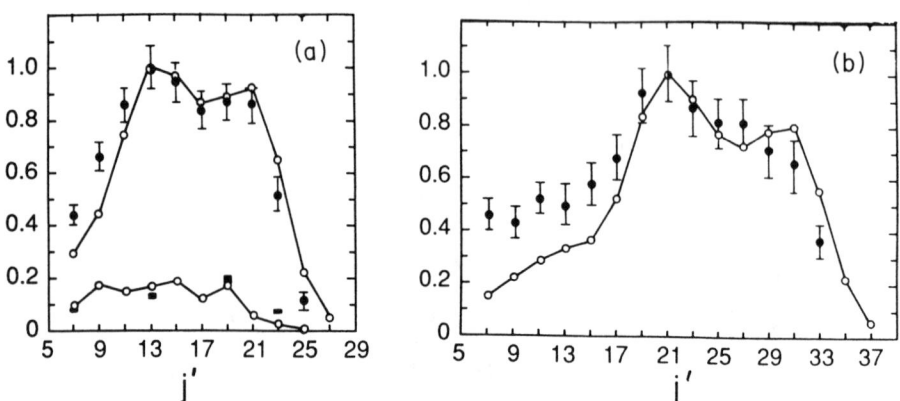

Fig. 5: Probabilties for rovibrational excitation from state $j = 5$ within or into the level $n = 2$ (a) and from states $j = 5$ or 7 within or into the level $n = 31$ (b) as a function of the final value of j'. Experimental results (•) are compared with the spectator model (o).

defined *high* vibrational state[8]. The experimental results are found to be in excellent agreement with an extension of the impulsive spectator model. For homonuclear diatomic targets the molecular form factor is

$$F_{n'j'm',njm} = \left\langle n'j'm' \left| e^{i\frac{1}{2}r\Delta k \cos\gamma} \right| njm \right\rangle + c.c. ,\qquad(15)$$

where n is the vibrational quantum number and r the intermolecular distance. Summing and averaging over m' and m can be done in closed form and the final result is simply a replacement of the Bessel function in Eq. (6) by an integral over radial vibrational wavefunctions[9]

$$j_j^2(\Delta) \to \langle n'j'm' | j_j^2(kr\sin(\vartheta/2)) | njm \rangle .\qquad(16)$$

The rotational factorization formula (7) is also valid for the individual vibrational transition probabilities.

Figure 5 shows the relative probabilties for rovibrational excitation from state $j = 5$ within or into the level $n = 2$ and from states $j = 5$ or 7 within or into the level $n = 31$ as a function of the final value of j'. All contributions into final n' states are summed. Also shown are vibrationally inelastic transitions $n = 3 \to n' = 2$ (■). The agreement with the spectator model (–o–) is amazingly good.

Recently the rainbow structure of the combined rotational–vibrational state–to–state differential cross sections has been analyzed[16] for high initial vibratio-

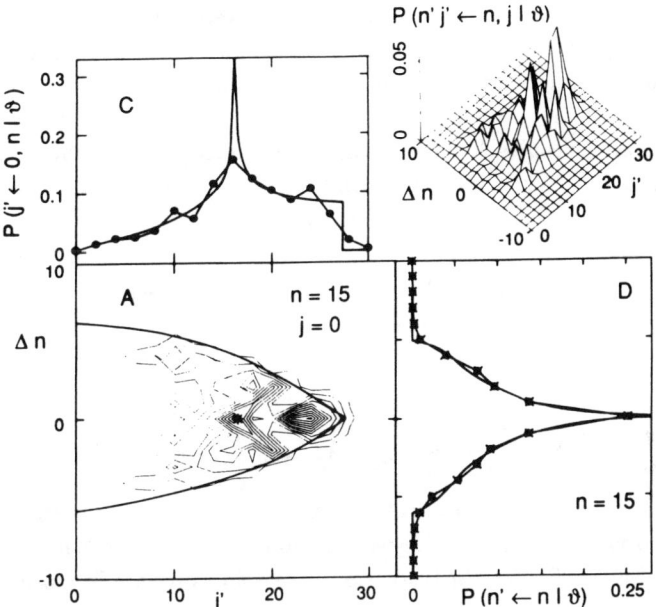

Fig. 6: Rotational–vibrational transition probabilites of Na_2 for an electron impact energy of 200 eV, scattering angle $\vartheta = 180°$, and initial state $n = 15$, $j = 0$.

nal ecitation within the spectator model. A typical state–to–state rotational–vibrational rainbow pattern of the transition probabilities $P(n'j' \leftarrow nj|\vartheta)$ at fixed scattering angle is shown in Fig. 6 as a function of $\Delta n = n' - n$ and j' ($n = 15$ and $j = 0$ initially) in comparison with classical results. The contour plot A demonstrates a confinement to the classically allowed region bounded by a classical rainbow curve, where the classical cross section shows the typical square–root rainbow singularity. A new and unexpected feature is an additional first–order singularity at $\Delta n = 0$, $j' = 16.4$ (marked by '∗' in Fig. 6A). This isolated rainbow singularity originates from a perpendicular orientation of the molecular axis with respect to the kinematic apse $\Delta \vec{p}$, while the vibrational motion is at its inner turning point. This nongeneric behaviour unfolds in a characteristic manner under generic perturbations[16].

The (rotationally or vibrationally summed) marginal distributions (C and D in Fig. 6) are very close to the classical results and clearly show the new vibrational–rotational rainbow singularity. The total rotational (vibrationally summed) transition probabilities oscillate around the classical distribution, which possesses a logarithmic singularity inside the allowed region. The total vibrational (rotationally summed) probabilities (D) show a classical singularity for elastic transitions. Quantum mechanically the singularities are softened into maxima.

More spectator results for (rotationally summed) vibrational transitions have been obtained[1,16-18]. Here we will only point out the appearence of a remarkable *vibrational factorization* formula

$$P(n' \leftarrow n|\vartheta) = \sum_{n''=|n'-n|}^{n'+n} C_{n''n'n} \, P(n'' \leftarrow 0|\vartheta) \quad , \qquad (17)$$

with coefficients

$$C_{n''n'n} = (-1)^s \sum_k \binom{n''}{k-n_>} \binom{n_<}{k-n''} \binom{n_>}{k-n_<} \quad , \qquad (18)$$

in almost complete analogy to the rotational factorization (7). It can be shown that (17) is also valid for general forced harmonic oscillator transitions.

The probabilities for $0 \to n$ transitions are also available in closed form[17]:

$$P(n' \leftarrow 0|\vartheta) = \frac{1}{n'! \, \Delta k \, d} \gamma\left(n' + 1/2, [\Delta k \, d/2]^2\right) \quad , \qquad (19)$$

where $\gamma(m,x)$ is the incomplete Gamma function. These results are derived for harmonic vibration. The factorization (17) is, however, also applicable to strongly anharmonic potentials. In Fig. 7 the total vibrational transition probabilities are shown for electron–Na_2 collisions in comparison with numerical results using anharmonic wave functions. The initial vibrational state is $n = 31$ (vibrational energy $\approx$ 70and the vibrational probabilities do *not* show the anharmonicity. They appear, however, clealy in the rotational excitation cross sections[18]. More details on vibrational factorization can be found in recent articles[1,16-19].

REFERENCES

1. H. J. Korsch and A. Ernesti, J. Phys. B 25 (1992) 3565

2. S. Levit, U. Smilanski, and D. Pelte, Phys. Lett. B 53 (1974) 39

3. H. J. Korsch and R. Schinke, J. Chem. Phys. 73 (1980) 1222

4. R. Schinke and J. M. Bowman, in: J. M. Bowman, Ed., *Molecular Collision Theory*, Springer, Berlin, 1983

5. H. J. Korsch, H. Kutz, and H. D. Meyer, J. Phys. B 20 (1987) L433

6. G. Ziegler, M. Rädle, H. Ehrhardt, K. Bergmann, and H. D. Meyer, Phys. Rev. Lett. 58 (1987) 2642

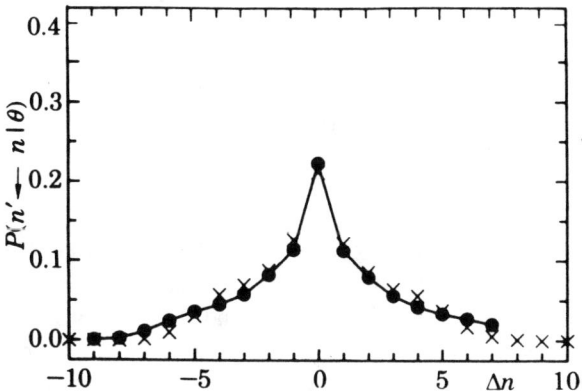

Fig. 7: Vibrational transition probabilities (×) for Na_2 ($\vartheta = 150°$, collision energy 150 eV) for initial state $n = 31$ as a function of $\Delta n = n' - n$ compared with numerically calculated values for an anharmonic oscillation (•).

7. G. Ziegler, S. K. Kumar, M. Rädle, K. Jung, H. Ehrhardt, K. Bergmann, and H. D. Meyer, Z. Phys. D 16 (1990) 207

8. S. V. K. Kumar, G. Ziegler, H. J. Korsch, K. Bergmann, and H.-D. Meyer, Phys. Rev. A 44 (1991) 268

9. H. J. Korsch, H. D. Meyer, and C. P. Shukla, Z. Phys. D 15 (1990) 227

10. J. Mitroy, J. E. McCarthy, and A. T. Stelbovics, J. Phys. B 20 (1987) 4827

11. A. Ernesti and H. J. Korsch, Z. Phys. D 16 (1990) 201

12. A. Ernesti and H. J. Korsch, J. Phys. B 23 (1990) L379

13. A. Ernesti and H. J. Korsch, J. Phys. B 24 (1991) 1877

14. A. Ernesti and H. J. Korsch, J. Phys. B (submitted for publication)

15. I. Shimamura, in: I. Shimamura and K. Takayanagi, Eds., *Electron–molecule collisions*, Plenum, New York, 1984, p. 89

16. A. Ernesti and H. J. Korsch, Z. Phys. D 27 (1993) 173

17. A. Ernesti and H. J. Korsch, Phys. Rev. A 44 (1991) 4095

18. A. Ernesti and H. J. Korsch, Europhys. Lett. 16 (1991) 433

19. H. J. Korsch, A. Ernesti, and J. A. Núñez, J. Phys. B 25 (1992) 773

MOLECULAR (e,2e) COLLISIONS—PRESENT STATUS AND FUTURE PROSPECTS FOR ELECTRON MOMENTUM SPECTROSCOPY

C.E. Brion
Department of Chemistry, The University of British Columbia,
Vancouver, BC, CANADA, V6T 1Z1

ABSTRACT

The theoretical background of electron momentum spectroscopy is summarized together with recent developments in the field. The present status of the technique is illustrated with examples from orbital symmetry and ordering, wavefunction evaluation and design, and reactivity studies. The limitations of single-channel instrumentation are discussed in the context of existing and new approaches to multichannel EMS techniques. Future possibilities for EMS are considered.

INTRODUCTION AND BACKGROUND

Over the past two decades electron momentum spectroscopy (EMS), formerly known as binary (e,2e) spectroscopy, has been developed into a unique and powerful technique for the investigation of molecular electronic structure, and in particular for studies in experimental quantum chemistry[1-4]. It is in the latter area and in the interplay with quantum mechanical theory[5,6] that some of the most significant achievements have been made, and where many future prospects exist. These possibilities are due to the fact that EMS provides very direct information on the behavior of electrons in the reactive outer-valence orbital parts that can be selected from the total electron distribution.

EMS is a coincidence technique involving the complete kinematic determination of high-energy electron impact ionization processes at large momentum transfer under binary encounter conditions. Although other geometries are possible, EMS is most conveniently carried out in the symmetric noncoplanar geometry where the variables are impact energy (binding-energy mode) or the relative azimuthal angle (ϕ) between the two outgoing electrons (momentum-distribution mode). Then all kinematic terms are effectively constant and, within the plane wave impulse approximation (PWIA), the cross section for randomly oriented molecules is proportional to the spherically averaged square of the ion-neutral overlap (electronic structure factor) as given by

$$\sigma(e,2e) = (\frac{4\pi^3 p_1 p_2 \sigma_{Mott}}{p_0}) \int d\Omega \left| \left\langle p\Psi_f^{N-1} | \Psi_o^N \right\rangle \right|^2 \quad (1)$$

where p refers to the momentum of the "knocked-out" electron and Ψ_f^{N-1} and Ψ_o^N are the many-electron wavefunctions for the (N-1) and N electron final and initial states, respectively. If the target Hartree-Fock approximation (THFA) is further applied by using antisymmetrized products of the same one-electron wavefunctions for both the neutral molecule and final ion state, then equation (1) reduces to

$$\sigma(e,2e) \propto S_{jo}^f \int d\Omega \left| \psi_j(p) \right|^2 \qquad (2)$$

where $\psi_j(p)$ is the one-electron momentum-space wavefunction of the removed electron and the spectroscopic factor S_{jo}^f is the probability that the fth ion state contains a hole in orbital ψ_j. To this extent EMS provides "orbital imaging" of the electron density. The quantity $\psi_j(p)$ is the Fourier transform of the position-space wavefunction $\psi_j(r)$. Using this reaction theory, EMS measurements may be used for the evaluation and design of molecular wavefunctions[1,5,6].

Summaries of the studies completed by groups in Italy, Australia, Canada, and the USA prior to 1989 have been published earlier[7,8]. More recently many further studies have been reported, including targets as complex as $(CH_3)_3NBF_3$[9], methylated siloxanes[10], $(\eta^4$-1,3-$C_4H_6)Fe(CO_3)$[11], $B_3N_3H_6$[12], $(CH_3)_4Si$[13], 1,2-propadiene[14], cyclopropane[15], formic acid[16], dimethyl ether[17,18], $(CH_3)_2CO$ and CH_3CHO[19], and NO_2[20]. Definitive studies of momentum-resolution effects on momentum profiles, essential for the quantitative comparison of theory and experiment, have also been reported[21-23]. New EMS instruments include a single-channel CMA-based system by Bawagan et al.[24], and a very high-sensitivity multichannel EMS spectrometer by Todd, Lermer and Brion[25].

THE PRESENT STATUS OF ELECTRON MOMENTUM SPECTROSCOPY

The present capabilities, limitations, and achievements of EMS are illustrated by the following examples that have been taken from some of the more important application areas of the technique.

(a) Determination of Orbital Symmetry and Ordering.

In this respect EMS provides a much more definitive diagnostic than photoelectron spectroscopy (PES), albeit at much lower energy resolution. For example, Fig. 2 shows binding-energy spectra at two azimuthal angles (momenta) for formaldehyde[26]. The peaks have been fitted using positions and widths from high-resolution PES. Previous theoretical and experimental studies[26] conflicted as to the relative ordering of the $5a_1$ and $1b_2$ ion states. The controversy is immediately solved by a consideration of the relative intensities at the two ϕ angles in Fig. 1, which shows clearly that peak 3 is "s-type" ($5a_1$) and that peak 4 is "p-type" ($1b_2$), as confirmed by the measured momentum distributions[26]. This type of information can also be used

352   Molecular (e, 2e) Collisions

to assign satellite parentage in inner-valence binding-energy spectra, as has been carried out, for example, in H_2S[27] and C_2H_2[28].

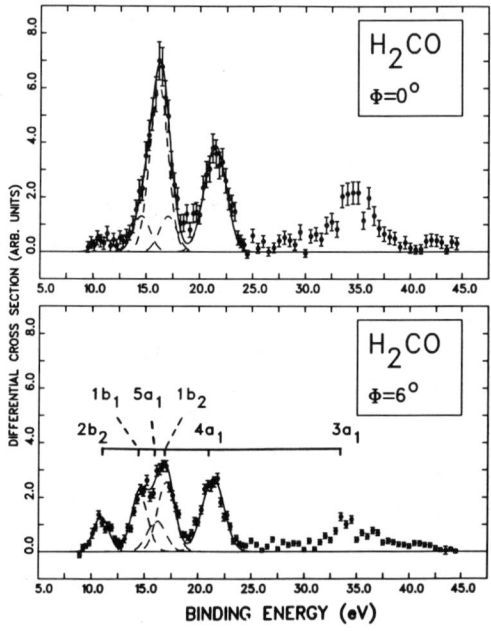

Fig. 1: Binding energy spectra of formaldehyde

(b) Wavefunction Evaluation and Design

EMS measurements of binding-energy-selected momentum profiles provide a unique and powerful tool for the evaluation and design of SCF and CI wavefunctions[1,5,6]. SCF Wavefunctions ranging from minimum basis set to Hartree-Fock limit can be tested using equation (2). Going beyond the independent-particle model, the full ion-neutral overlap in equation (1) can be evaluated with many-body treatments of electron-correlation and relaxation such as MR-SDCI methods. The most detailed and exhaustive studies have been carried out for the small hydrides CH_4, SiH_4, NH_3, PH_3, H_2O, H_2S, HF and HCl[1,5,6,27,29-31], for which results for the HOMO orbitals are shown in Fig. 2. The case of H_2O, a "bench-mark" system in molecular quantum mechanics, has been studied most exhaustively. EMS experiments[1,5,6] indicated that even the best existing (84-GTO) SCF basis sets[32] available at the time had to be considerably modified to further saturate the basis set and to include very diffuse functions. Only then did the SCF wavefunction give effectively converged results at the true Hartree-Fock limit, not only for total energy, but also for

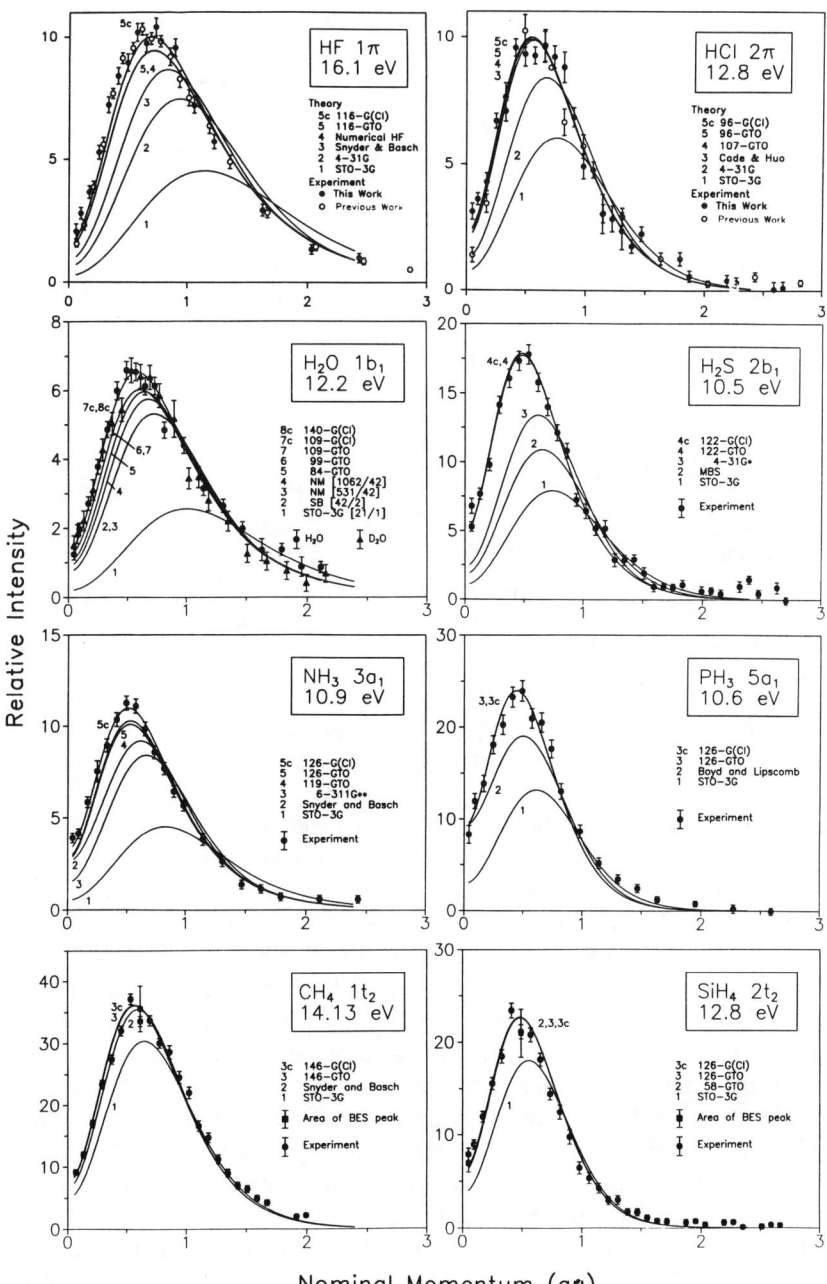

Fig. 2: The HOMO orbitals of the hydrides

Table 1. Wavefunctions and Properties for H_2O

Wavefunction (basis set)[5]	Total Energy (au)	Dipole Moment (D)	p_{max} ($1b_1$) (au)
STO-3G	-74.9629	1.7258	1.02
14-GTO	-76.0035	2.681	0.74
27-STO	-76.00468	2.035	0.67
36-GTO	-76.044	2.092	0.74
58-GTO	-76.059	1.996	0.70
39-STO	-76.0642	1.995	0.66
84-GTO	-76.06661	2.021	0.66
99-GTO	-76.06689	2.006	0.62
109-GTO	-76.0671	2.006	0.62
140-GTO	-76.0673	1.981	0.62
HF Limit	-76.0675	1.980	0.62
84-G(CI)	-76.3210	1.929	0.65
109-G(CI)	76.3761	1.895	0.58
140-G(CI)	-76.3963`	1.870	0.58
Experiment	-76.4376	1.8546	0.58

dipole-moment and momentum distribution (see table 1). The EMS studies and associated calculations[5] showed that these three properties exhibit very different convergence behavior since they emphasize small r (large p), medium r (medium p) and large r (small p) regions of phase space, respectively. These three properties (rather than just the total energy) have been found to provide effective criteria for developing very accurate "universal" wavefunctions, applicable in all regions of phase space. The remaining discrepancies between theory and experiment for H_2O, NH_3, and HF were reconciled by MR-SDCI calculations of the full ion-neutral overlap[5], together with proper quantitative treatment of momentum-resolution effects[1,21,22,29]. In the case of H_2O the CI calculations recovered up to 88% of the correlation energy[5,6]. These CI and SCF "universal" wavefunctions for H_2O, designed using the EMS experimental results, provide the most accurate calculated properties yet obtained[6] and are listed as the best available wavefunctions in the latest edition of Levine's *Quantum Chemistry*[33]. The excellent agreement in the case of these small hydrides confirms the reliability of the PWIA reaction model at impact energies of 1200 eV. It is also of interest to note that the measured results for all valence orbitals of the hydrides closely conform to the canonical Hartree-Fock molecular orbital picture and not, for example, in CH_4[31] to sp^3 hybrid orbitals or localized valence bond orbitals. This observation applies to all EMS studies made thus far, and even in the cases of H_2O, NH_3, and HF (Fig. 2), where correlation is important in predicting the XMP, the results are very similar to Hartree-Fock canonical orbitals[1,29]. Such detailed SCF and CI wavefunction evaluation and design studies have recently been applied to EMS studies of

larger systems such as CH_3NH_2[34,35], $(CH_3)_2CO$ and CH_3CHO[19] and C_2H_2[28].
(c) Reactivity and Electronic Structure

Theories of chemical reactivity such as the Frontier Orbital Theory of Fukui[36] suggest that the denser portions of the outermost valence molecular orbitals are the main reactive parts of the total electron distribution. EMS is most sensitive to the chemically important low-momentum (i.e., larger r) parts of the electron distribution and measurements for valence electrons therefore provide a simple and rather direct momentum-space probe of chemical effects in molecules. As such, EMS measurements and calculations of orbital-momentum profiles and density maps provide a new experimental means of investigating the relation between the potential for chemical reactivity and electron-density distributions. A clear example of this is given (see Fig. 3) by the EMS studies[35,37] of the HOMO orbitals of trimethylamine and NH_3 that indicate that CH_3 is electron-attracting compared to H when bonded to N (see density difference map). This finding, contrary to commonly used chemically

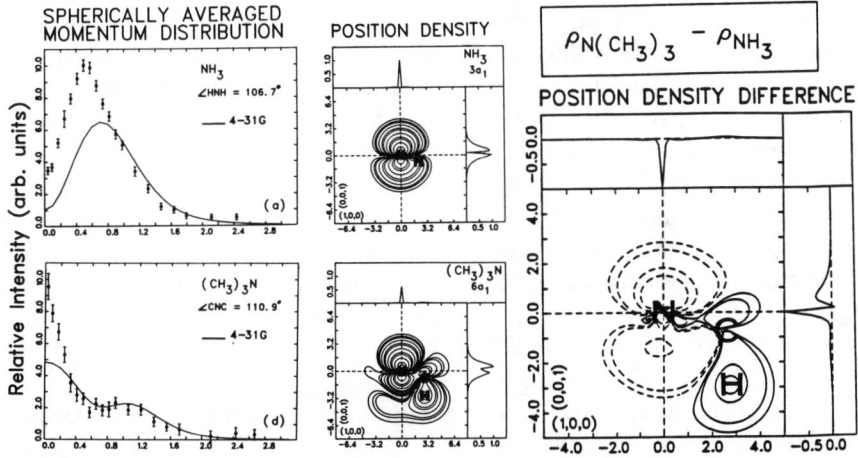

Fig. 3: EMS measurements and calculations for ammonia and trimethylamine

intuitive arguments, is supported by ^{14}N NMR and dipole moment measurements and also by MO calculations[35]. Recent EMS studies on the $(CH_3)_3N$-BF_3 electron donor-acceptor complex[9] and the methylated siloxanes[10] also indicate the ability to study the relation between reactivity and the HOMO electron distribution.

PRESENT LIMITATIONS OF EMS

Existing EMS studies indicate that further progress in the field is limited by several factors. Firstly, the requirements of high-impact energy (plane wave) and large scattering angle (binary encounter) result in very low count rates and therefore limited statistical precision, even with incident electron beam currents of 50 microamps of unmonochromated electrons and measuring times of several days or weeks per orbital The resulting coincidence energy resolution is typically at best ~1.5 eV fwhm. The required beam current precludes the use of monochromated electron sources to study vibrational effects or for the separation of closely spaced orbitals in larger molecules such as organics of biochemical interest. As such, the range of EMS studies has been restricted and most accessible targets have now been studied. If long-term operation and poisoning effects could be overcome, then use of GaAs low-temperature photoemission electron sources could provide some improvement in this area. Secondly, the low coincidence count rates are a limiting factor even in the case of smaller stable molecules and further preclude the study of chemical interesting reactive species such as free radicals, excited molecules, ions, cluster, and van der Waals molecules, which can only be prepared at much lower target densities. Thirdly, the random orientation of free molecules results in some loss of information that could be recovered if studies of oriented or aligned species were possible.

All of these limitations could be overcome if the EMS coincidence intensity could be significantly increased above that available from single-channel angle scanning instruments. While multichannel instruments sampling a range of angles[38] or binding energies[39] have been reported, these developments have various limitations and have, thus far, only increased the coincidence count rate by approximately an order of magnitude. In a notable experiment, the energy-dispersive approach has permitted EMS data to be obtained for laser-excited sodium atoms[40]. An order of magnitude improvement in coincidence count rates has also been obtained by using asymmetric kinematics[41]. However, it is clear that new-generation, multichannel EMS instruments with much larger coincidence count rates are required.

A NEW HIGH-SENSITIVITY MULTICHANNEL EMS SPECTROMETER

In this laboratory Todd, Lermer, and Brion[25] have recently developed a new momentum-dispersive multichannel EMS spectrometer (Fig. 4) based on a full cylindrical mirror analyzer (CMA) designed to record essentially all symmetric (e,2e) events ocurring around the full 2π azimuth of ϕ angles. In the first phase of development, coincidences have been recorded between a fixed single-channel electron multiplier (CEM) and a microchannel plate (MCP) with a resistive anode encoder covering a $\pm 30°$ range of azimuth that permits an orbital momentum distribution to be obtained in parallel. Specially developed pulse pile-up (PPU) coincidence circuitry using power combiners gives a time resolution less that two nanoseconds and has resulted in minimal dead times. The position computer has also been gated so as to compute only the positions of coincident events. The present arrangement provides approximately two orders of magnitude improvement in coincidence count rate. Later implementation of the full 2π azimuth of detection, with appropriate detectors and read-out circuitry, is estimated to provide a 10,000-fold overall increase in coincidence count rate. Typical results with the present arrangement are shown in Figs. 5-7. The 2D-EMS spectrum (Fig. 5) is obtained by varying the impact energy.

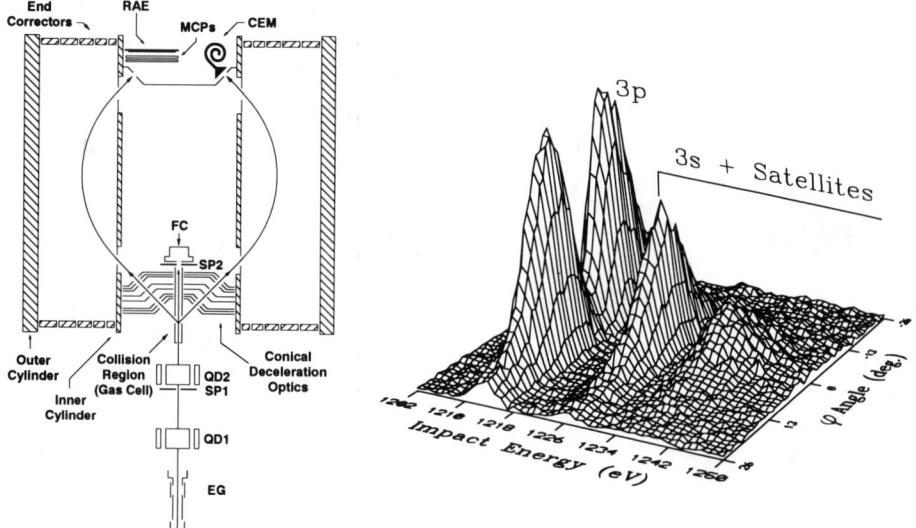

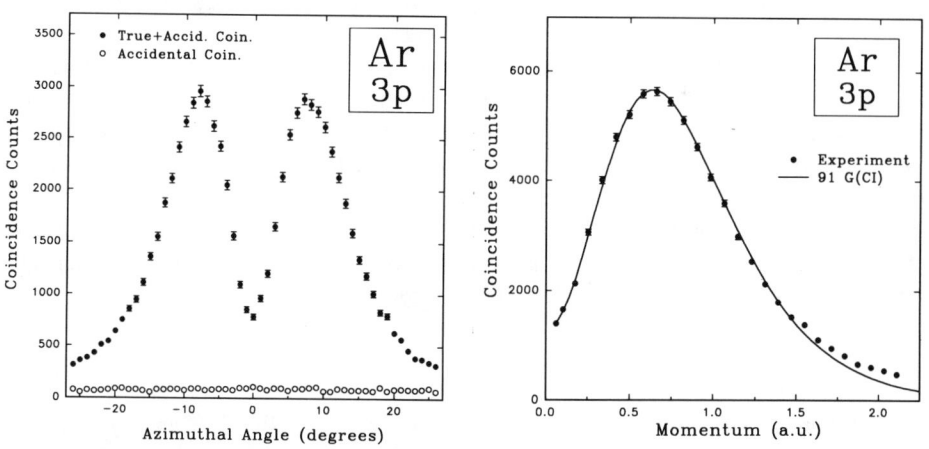

Fig. 4: Multichannel EMS spectrometer

Fig. 5: Two-dimensional EMS measurement for argon

Fig. 6: φ angular distribution for Ar 3p

Fig. 7: Measured and calculated momentum profiles for Ar 3p

FUTURE PROSPECTS

With further improvements in multichannel EMS spectrometers, the higher sensitivity and higher energy resolution capability will permit a wide range of new studies to be made on vibrational effects, large molecules, free radicals, ions, excited species, clusters and van der Waals molecules. The study of oriented species in the gas phase or on surfaces may also be possible. Much higher statistical precision and shorter measuring times will facilitate a rigorous investigation of the (e,2e) reaction model including the effects of distorted waves. Higher count rates will also afford the dynamic tuning and instrument optimization in the coincidence mode that is not possible with single-channel spectrometers.

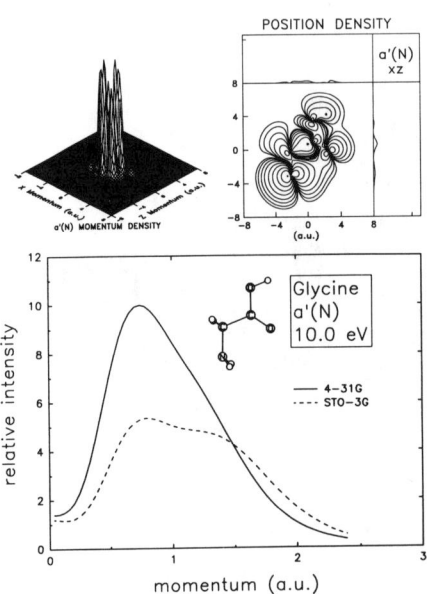

Fig. 8: Calculations for glycine

A fruitful application area is likely to be in larger molecules of biochemical and biomedical interest. Such EMS studies are now under way in this laboratory for the simplest amino acid, glycine, and Fig. 8 shows calculations in position and momentum space[42]. The successful use of EMS in wavefunction evaluation and design for small molecules and the preliminary investigations of reactivity and frontier orbitals suggest that similar studies for larger molecules could provide significant improvements into computer-aided molecular design and reactivity studies. At present such software packages rely on total charge distributions obtained from molecular potentials. The orbital selectivity of EMS should provide much more specific electron-density information on the reactive portions of the total charge distribution. Since high-level SCF and CI calculations become unfeasible for larger molecules, the EMS measurements are also likely to provide a useful testbed for new quantum mechanical theoretical methods such as Density Functional Theory[43,44] that are now rapidly evolving.

ACKNOWLEDGMENTS

Financial support for this work was provided by NSERC (Canada) and the Canadian National Networks of Centres of Excellence (CEMAID). I would like to acknowledge the extensive contributions made to this work by my co-workers at the University of British Columbia and by Professor E.R. Davidson (Indiana University).

REFERENCES

1. C.E. Brion, *Inst. Phys. Conf. Series* **122**, 171 (1992) and refs. therein.
2. I.E. McCarthy & E. Weigold, *Rep. Prog. Phys.* **54**, 789 (1991) and refs. therein.
3. K.T. Leung, *Sci. Progress Oxford* **75**, 157 (1990).
4 C.E. Brion, *Int. J. Quantum Chem.* **29**, 1397 (1986).
5. A.O. Bawagan, C.E. Brion, E.R. Davidson & D. Feller, *Chem. Phys.* **113**, 9 (1987).

6. D. Feller, C.H. Boyle and E.R. Davidson, *J. Chem. Phys.* **86**, 3424 (1987).
7. K.T. Leung and C.E. Brion, *J. Electron Spectrosc.* **35**, 327 (1986).
8. K.T. Leung. In *Theoretical Models of Chemical Bonding, Part 3;* Z.B. Maksic, Ed.; Springer Verlag: Berlin, 1991; pp 339-386.
9. K. McMillan, M.A. Coplan, J.H. Moore and J.A. Tossell, *J. Phys. Chem.* **94** 8648 (1990).
10. J.A. Tossell, J.H. Moore, K. McMillan and M.A. Coplan, *J. Am. Chem. Soc.* **113**, 1031 (1991).
11. K. McMillan, M.A. Coplan, J.A. Tossell and J.H. Moore, *Molec. Phys.* **73** 619 (1991).
12. J.A. Tossell, J.H. Moore, K. McMillan, C.K. Subramanian & M.A. Coplan, *J. Am. Chem. Soc.* **114**, 1114 (1992).
13. T.A. Daniels, H. Zhu, M.P. Banjavcic & K.T. Leung, *Chem. Phys.* **159** 289 (1992).
14. S.W. Braidwood, M.J. Brunger, W. von Niessen, E. Weigold and V.G. Zakrzewski, to be published.
15. M.P. Banjavcic, T.A. Daniels and K.T. Leung, *Chem. Phys.* **155**, 309 (1991).
16. S.M. Bharathi, S.K. Datta, A.M. Grisogono, R. Pascual, E. Weigold, and W. von Niessen, *J. Electron Spectrosc.* **53**, 51 (1990).
17. S.A.C. Clark, A.O. Bawagan and C.E. Brion, *Chem. Phys.* **137**, 407 (1989).
18. Y. Zheng, E. Weigold, C.E. Brion and W. von Niessen, *J. Electron Spectrosc.* **53**, 153 (1990).
19. B.P. Hollebone, P. Duffy, C.E. Brion, Y. Wang and E.R. Davidson, *Chem. Phys.*, submitted (1993).
20. J. Rolke, B.P. Hollebone and C.E. Brion, to be published.
21. A.O. Bawagan and C.E. Brion, *Chem. Phys.* **144**, 167 (1990).
22. P. Duffy, C.E. Brion, M. Casida and D.P. Chong, *Chem. Phys.* **159**, 347 (1992).
23. L.A. Graham and A.D.O. Bawagan, *Chem. Phys. Lett.* **178**, 441 (1991).
24. L.A. Graham, S.J. Desjardins and A.D.O. Bawagan, *Can. J. Chem.* **71**, 216 (1993).
25. B.R. Todd, N. Lermer and C.E. Brion, *Rev. Sci. Instrum.*, submitted (1993).
26. A.O. Bawagan, C.E. Brion and E.R. Davidson, *Chem. Phys.* **128**, 439 (1988).
27. C.L. French, C.E. Brion and E.R. Davidson, *Chem. Phys.* **122**, 247 (1988).
28. P. Duffy, S.A.C. Clark, C.E. Brion, M.E. Casida, D.P. Chong, E.R. Davidson and C. Maxwell, *Chem. Phys.* **165**, 183 (1992).
29. B.P. Hollebone, Y. Zheng, C.E. Brion, E.R. Davidson and D. Feller, *Chem. Phys.* **171**, 303 (1993).
30. B.P. Hollebone, C.E. Brion, E.R. Davidson and C. Boyle, *Chem. Phys.* **173**, 193 (1993).
31. S.A.C. Clark, T.J. Reddish, C.E. Brion, E.R. Davidson and R. Frey, *Chem. Phys.* **143**, 1 (1990).
32. E.R. Davidson and D. Feller, *Chem. Phys. Lett.* **104**, 54 (1984).
33. I.N. Levine in *Quantum Chemistry*: Prentice Hall, NJ, 1991; pp 474 & 510.
34. C.J. Maxwell, F.B.C. Machado and E.R. Davidson, *J. Am. Chem. Soc.* **114**, 6496 (1992).
35. A.O. Bawagan and C.E. Brion, *Chem. Phys.* **123**, 51 (1988).
36. K. Fukui, T. Yonezawa and H. Shingu, *J. Chem. Phys.* **20**, 722 (1952).
37. A.O. Bawagan and C.E. Brion, *Chem. Phys. Lett.* **137**, 573 (1987).
38. J.H. Moore, M.A. Coplan, T.L. Skillman and E.D. Brooks, *Rev. Sci. Instrum.* **49**, 463 (1978).
39. J.P.D. Cook, I.E. McCarthy, A.T. Stelbovics and E. Weigold, *J. Phys. B* **17**, 2339 (1984).
40. Y. Zheng, I.E. McCarthy, E. Weigold and D. Zhang, *Phys. Rev. Lett.* **64**, 1358 (1990).
41. L. Avaldi, R. Camilloni, E. Fainelli and G. Stefani, *J. Phys. B* **20**, 4163 (1987).
42. J. Neville, Y. Zheng and C.E. Brion, work in progress.
43. R.G. Parr and W. Yang, *Density Functional Theory of Atoms and Molecules*; Oxford University Press, New York, 1989.
44. C. Lee and W. Yang, *J. Chem. Phys.* **96**, 2408 (1992).

RECENT THEORETICAL RESULTS ON ELECTRON-POLYATOMIC MOLECULE COLLISIONS

C. W. McCurdy
Lawrence Livermore National Laboratory, Livermore, CA 94550

ABSTRACT

Until recently, the principal barrier to the accurate theoretical description of electronic collisions with polyatomic molecules was the computational problem of scattering by a nonlocal, arbitrarily asymmetric potential. Effective numerical techniques capable of solving this variety of potential scattering problem for electronic collisions have now matured, and the first applications of methods for treating many-body aspects of collisions of electrons with polyatomic molecules have begun to appear in the literature. The past two years have seen the appearance of a large collection of calculations on electron-polyatomic collisions which compare favorably with experimental determinations. In addition to the dramatic developments in methods which explicitly exploit the methods of quantum chemistry to treat the effects of electron correlation, polarization, etc., parameter-free model potential methods for electronically elastic collisions have also evolved markedly in recent years. Progress in both electronically elastic and inelastic processes is reviewed briefly.

I. INTRODUCTION

The most remarkable fact about the published literature on the accurate theoretical treatment of electron-polyatomic molecule scattering is that there is so little of it. The discipline has lagged ten or fifteen years behind theoretical developments in electron-atom scattering.[1,2] Surprisingly, this disparity in theoretical progress has occurred even for fixed-nuclei calculations and is *not* due to the inherent difficulty of treating phenomena associated with nuclear motion in molecules. The first barriers to the development of successful treatments of electron-molecule scattering were associated with the difficulties of treating collisions with arbitrarily nonspherical targets with nonlocal (exchange) potentials. Even calculations at the static-exchange level on polyatomic targets began to appear only in the late 1980s.[3,4] In this review I will maintain that the recent literature on electron-polyatomic molecule collisions gives ample evidence that the problem of scattering electrons from nonspherical, nonlocal potentials has been solved effectively by several computational approaches. More serious problems in the physics of electron-polyatomic collisions now dominate the field.

To understand recent changes in the capabilities of workers in this area, consider the following comparison. The role of configuration interaction in electron-atom scattering was already the subject of *review* talks by 1976.[5] It is accurate to say that as of 1993 the consequences in electronic collisions of the effects of configuration interaction in target states of polyatomic molecules have hardly begun to be dealt with (although considerable efforts have been made for diatomics). The reason for this disparity is that the technology for treating the bound states of polyatomic molecules, that is, the algorithmic edifice of quantum chemistry, has proved to be difficult to couple to electron scattering calculations. Typical configuration interaction calculations on bound electronic states of polyatomics involve tens or hundreds of thousands (if not millions) of configurations of Gaussian orbitals. The awkwardness of incorporating such target descriptions in old-style scattering calculations involving single-center expansions of the continuum functions is apparent. Less obvious, but even more troublesome, is the problem of consistency in treating the N-electron target and (N+1) electron collisional correlation problems.

However, even these barriers are being surmounted, and that is the real news in this brief review.

There is another story to tell, and that concerns the evolution of model potential calculations on electronically elastic scattering from polyatomic targets. The earliest calculations (by far) on polyatomic molecules which bore comparison with experiment were of this variety, and some such methods have been improved to be parameter-free and have a well understood range of reliability.

II. THEORETICAL METHODS FOR POLYATOMIC COLLISIONS

This section presents a terse catalog of the methods which have been successfully applied in accurate calculations on polyatomic systems.

A. Variational Methods

While variational methods based on the Kato identity[6] or the Schwinger principle[7] have been historically prominent in the development of electron-atom scattering, they were slow to be applied to molecular collisions. Once the problems with resonance-like anomalies in the Kohn variational approach were overcome, that approach found immediate application in the molecular context. The key barrier to be overcome in applying the Schwinger principle was the computation of Green's function matrix elements for molecular wave functions. These two variational methods have an important feature in common, namely that they involve an explicit specification of the trial function in terms of molecular target states. The form of the trial function provides the starting point for incorporating the wave functions and techniques of quantum chemistry.

(1) The Complex Kohn Variational Method

The Kohn variational method, in a version called the "complex Kohn variational method", was first applied to electron-polyatomic molecule scattering[4] shortly after Miller and Jansen op de Haar observed[8] that a simple change in the boundary conditions (from principal-value to traveling-wave) eliminates the numerical pathology of anomalous resonances which plagued applications of the Kohn approach to electron-atom scattering. In the complex Kohn method the Kohn principle is used to determine the T matrix connecting channels Γ and Γ_0 as the stationary value of the functional

$$[T^{\Gamma \Gamma_0}] = T^{\Gamma \Gamma_0} - 2\int \Psi_\Gamma (H-E) \Psi_{\Gamma_0} \tag{1}$$

The trial function is chosen to have the form

$$\Psi_{\Gamma_0} = \sum_{\Gamma} A\,[\chi_{\Gamma}(\vec{r}_1, ..., \vec{r}_N)\, F_{\Gamma\Gamma_0}(\vec{r}_{N+1})] + \sum_{\mu} d_{\mu}^{\Gamma_0} \Theta_{\mu}(\vec{r}_1, ..., \vec{r}_{N+1}) \qquad (2)$$

In Eq.(2) χ_{Γ} denotes an open-channel target state, the Θ_{μ} are square-integrable, antisymmetric (N+1) electron functions, and $F_{\Gamma\Gamma_0}$ is a channel continuum function whose short-range behavior is represented by an expansion in Gaussians and whose long-range behavior is explicitly specified in terms of the Riccati-Bessel functions j_l and $h_l^{(+)}$. The functions χ_{Γ} and Θ_{μ} are constructed, as are typical components of quantum-chemistry calculations on bound electron states, in terms of antisymmetric configurations of molecular spin orbitals which are in turn expanded in Gaussians centered anywhere in the molecule.

An additional twist on this approach,[4,9] which aids in casting as much of the calculation as possible in a form which can be done by standard quantum-chemistry codes, is to use Feshbach partitioning on the trial wave function in Eq.(2) and define the Q-space (closed-channel space) as the space of the Θ_{μ} functions. Then H in Eq.(1) is replaced by $H_{effective}$ which includes the *ab initio* optical potential

$$H_{effective} = H_{PP} + H_{PQ}(E - H_{QQ})^{-1} H_{QP}. \qquad (3)$$

It is this form which has been applied extensively to both electronically elastic and inelastic electron-polyatomic scattering by McCurdy, Rescigno, Lengsfield, Schneider and their coworkers. Only Hamiltonian matrix elements (no Green's function matrix elements) with respect to the functions χ_{Γ} and Θ_{μ} are required so that implementing the connection to quantum chemistry codes is direct and simple.

(2) The Schwinger Multichannel (SMC) Variational Method

In the implementation of this approach to electron-polyatomic molecules collisions[5,10] the (N+1) electron wave functions of the electron-molecule scattering system are expanded in a basis and the coefficients determined by extremizing the Schwinger variational expression for the scattering amplitude

$$f(\vec{k}_m, \vec{k}_n) = -\frac{1}{2\pi} \sum_{i,j} \langle S_m(\vec{k}_m) | V | \chi_i \rangle A_{ij}^{-1} \langle \chi_j | V | S_n(\vec{k}_n) \rangle \qquad (4)$$

In Eq.(5) the χ_i are the expansion basis, $S_m(\vec{k}_m)$ is the unperturbed scattering function

$$S_m(\vec{k}_m) = \phi_{target}^{(m)}(1, 2, ..., N)\, exp(i\vec{k}_m \cdot \vec{r}) \qquad (5)$$

V is the interaction between the scattered electron and the target, and A_{ij}^{-1} are elements of the inverse of the matrix

$$A_{ij} = \langle \chi_i | A^{(+)} | \chi_j \rangle \qquad (6)$$

of the operator $A^{(+)}$. That operator is given by

$$A^{(+)} = \frac{1}{2}(PV + VP) - VG_P^{(+)}V - \frac{1}{N+1}\{\hat{H} - \frac{N+1}{2}(\hat{H}P - P\hat{H})\} \qquad (7)$$

where $\hat{H} = E - H$, and it involves the N-electron projection operator P which projects on open channels and the projected, free-particle Green's function $G_P^{(+)}$

In this procedure both the expansion basis and the target states are objects from the technology of quantum chemistry. Green's function matrix elements are required in addition to matrix elements of the Hamiltonian, but in return for expending the effort to compute them one gains considerable flexibility in the choice of the expansion basis, $\{\chi_i\}$, which can be used to exploit the efficient convergence properties of the Schwinger method. The general approach was formulated by Takatsuka and McKoy[11] and has been applied to polyatomics by McKoy and coworkers and by Huo and coworkers.

B. The R-matrix Method for Polyatomics

While computational applications of R-matrix theory to electronic collisions with atomic[12] and diatomic[13, 14] targets date from the 1970's, only very recently, to my knowledge, has this approach been applied for the first time to a nonlinear polyatomic target in an *ab initio* calculation.[15] From a schematic view of the R-matrix method one can see the computational barriers which have to be overcome in a polyatomic application as well as how an explicit connection with the methods of quantum chemistry can be made.

The essence of the R-matrix idea in this context is to solve the (N+1)-electron problem in a region confined to a finite sphere while enforcing boundary conditions which discretize the spectrum of the (N+1)-electron Hamiltonian. Those discrete states are then used to expand the scattering wave function inside the sphere, and that expansion is matched to the asymptotic form (incorporating explicit N-electron target states) on the surface of the sphere. The point to emphasize is that the (N+1)-electron problem must be solved on a finite region—a task that no quantum chemistry code is constructed to perform.

An implementation which is suitable for polyatomic applications has been developed by Nestmann, Nesbet, Peyerimhoff and Pfingst.[15,16] They solve the key computational problems of the method for polyatomics as follows.[15]

- Cartesian Gaussians are used to form an (N+1)-electron basis.
- The electron-repulsion integrals along with nuclear attraction integrals are confined to the R-matrix sphere by using a multipole expansion to approximate an effective potential outside the sphere. Using this device to subtract off the $r > a$ contribution allows the use of standard codes for the calculation of two-electron molecular integrals.
- Correlation and polarization effects are only taken into account for selected, energetically low-lying (N+1)-electron CI roots (the R-matrix poles), which have nonvanishing amplitudes on the R-matrix sphere. Higher lying roots are approximated by a Static-Exchange calculation.

While complete details of a polyatomic application have not yet appeared in the literature, this approach shows much promise, as the preliminary results of these workers shown in Section III. amply suggest.

C. Parameter-free Model Interactions

While the more general, and potentially exact, methods described in Sections II. A. and II B. appeared a decade and a half after their atomic counterparts, model potential calculations on polyatomics are much older.[17] In very recent years considerable progress has been made on parameter-free versions of such theories by Gianturco and Jain and their coworkers. In a modern implementation[18, 19] of the method the continuum orbital, $\mu_0^\Gamma(\vec{r})$, in a given molecular symmetry (irreducible representation Γ) is the solution of

$$\{-\frac{1}{2}\nabla^2 + V_{st}^{^1A_1}(\vec{r}) + V_{ex}^{^1A_1}(\vec{r};k^2) + V_{CP}^{^1A_1}(\vec{r}) - \frac{1}{2}k^2\}\mu_0^\Gamma(\vec{r}) = \sum_i \lambda_i \mu_i^\Gamma(\vec{r}) \quad (8)$$

where the closed shell target is assumed to be in its $^1A_1$ ground state, and $\mu_i^\Gamma(\vec{r})$ denotes an occupied orbital. The terms in Eq.(10) are the static interaction, $V_{st}(\vec{r})$, an energy-dependent semiclassical exchange interaction, $V_{ex}(\vec{r};k^2)$, a correlation-polarization interaction $V_{CP}(\vec{r})$, and Lagrange multiplier terms on the RHS to enforce the proper nodal behavior in the continuum orbital.

Considerable exploration of the forms of the exchange and correlation-polarization interactions has lead to parameter-free versions based on the Hohenberg-Kohn theorm and perturbation theory considerations. Eq. (8) represents a tremendous simplification of the electronically elastic scattering problem and provides the groundwork, for example, for calculations of rotationally inelastic cross sections for polyatomics.[20]

D. Stabilization Calculations on Resonances

This venerable technique,[21, 22] which involves only the diagonalization of the Hamiltonian for the scattering system in a limited basis, continues to prove useful. It can be applied to largest molecules for which we can perform bound state calculations, provided the resonances are sufficiently narrow. It has been applied to a wide variety of polyatomic systems in explorations of the ways in which shape resonances change as the chemical substituents of a molecule are changed, and has been applied recently to molecules the size of tetracyanoethylene[23]—vastly larger than systems on which full scattering calculations are possible.

III. SOME RECENT COMPUTATIONAL RESULTS

Space restrictions dictate that this section can present only the briefest sampling of the kinds of results from the past two years. I therefore focus on comparisons with experimental results which suggest the level of accuracy now attainable.

The complex Kohn method has been recently applied to CH_4,[24, 25] C_2H_2,[26] SiH_4,[27] NH_3,[28] C_2H_6[29], and H_2O.[30] A principal characteristic of these calculations is their inclusion of correlation effects, either in the form of configurations contributing to target polarization included in the *ab initio* optical potential of Eq.(3) or in the form of open electronic channels. A robust and intimate interface with electronic structure codes is necessary for such calculations. Results for the electron-ethane elastic cross section from a calculation using more than 12,000 correlating configurations[29] are

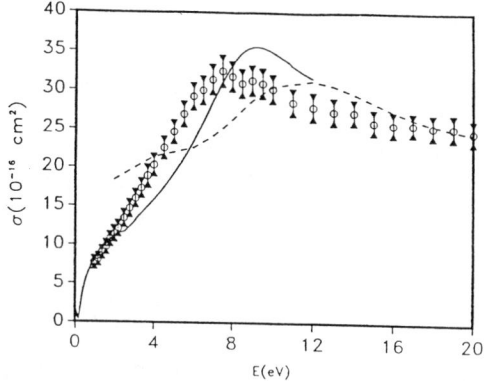

Fig. 1. Elastic cross section for ethane. Solid line, with optical potential; dashed line, static-exchange; points, experiment of Ref. 31.

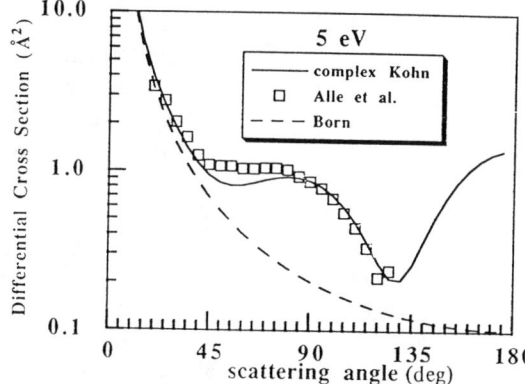

Fig. 2. Differential cross section for e^--NH_3. Solid line, with *ab initio* optical potential; dashed line Born approx.; points, experiment of Ref. 32.

compared with experiment in Fig. 1. The calculations predict a pronounced Ramsauer-Townsend minimum. Polar molecules present their own particular set of additional problems but these methods have also been applied in that case. Results for a differential cross section for electron-ammonia are compared with experiment[32] in Fig. 2. The complex Kohn method has been applied to electronic excitation of CH_2O, CH_4, and H_2O. Very recent results on the excitation of dissociative states of H_2O are shown in Fig. 3.

The Schwinger multichannel method has been applied recently to elastic scattering from CH_4, AsH_3, PH_3,[33] and isomers of C_3H_6,[34] CF_4,[35] and electronic excitation of C_2H_4,[36] CH_2O,[37] CH_4,[38] and C_5H_6.[39] Earlier

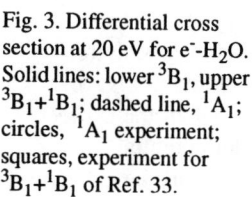

Fig. 3. Differential cross section at 20 eV for e^--H_2O. Solid lines: lower $^3B_1$, upper $^3B_1$+$^1B_1$; dashed line, $^1A_1$; circles, $^1A_1$ experiment; squares, experiment for $^3B_1$+$^1B_1$ of Ref. 33.

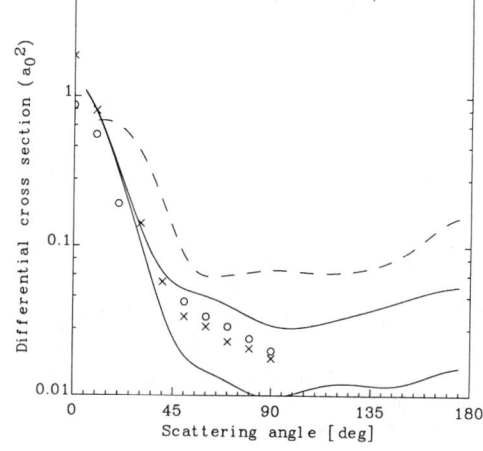

calculations using this method have included extensive polarization effects of closed channels, but these recent efforts have been at the static-exchange level or included some number of single configuration open channels. The static-exchange results from this method, including earlier ones, have made a compelling case for the sufficiency of this approximation for incident energies above 5 or 10 eV. Results for elastic scattering from CF_4,[35] are shown in Fig. 4 together with results by Huo.[40] A calculation of the electron-impact excitation of the $^3A_2''$ state of [1.1.1] propellane[39] (a highly symmetric form of C_5H_6) is shown in Fig. 5.

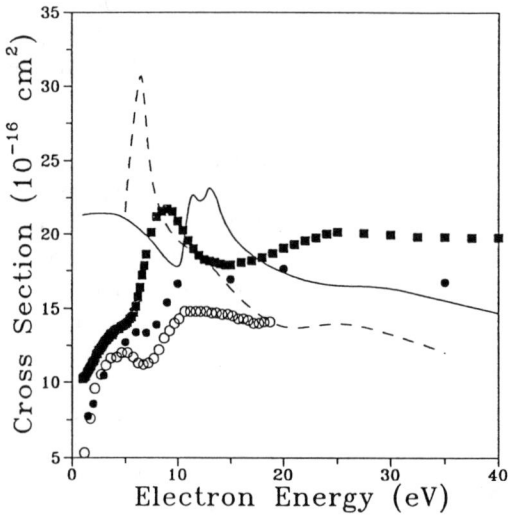

Fig. 4. Elastic cross section for CF_4. Solid line SMC of Ref 35; dashed line SMC of Ref 40; filled circles, exp., Ref. 41; open circles, exp., Ref. 42; filled squares, exp. total cross section, Ref. 43.

The first *ab initio* calculation on a polyatomic target using the R-matrix method has been performed this year. That calculation is on elastic scattering from CH_4 and includes the effects of polarization using CI for the lower R-matrix roots as discussed above. CH_4 has become an important test case because several other calculations have treated the Ramsauer-Townsend minimum region which is critically sensitive to correlation effects, and because good experiments are available. Prelimary results from the R-matrix calculation[15] are shown in Fig. 6.

Parameter-free model potential calculations have been performed on a large array of polyatomic systems, with some recent examples including H_2S,[20] NH_3,[20] H_2O,[48] SiH_4,[49,52]

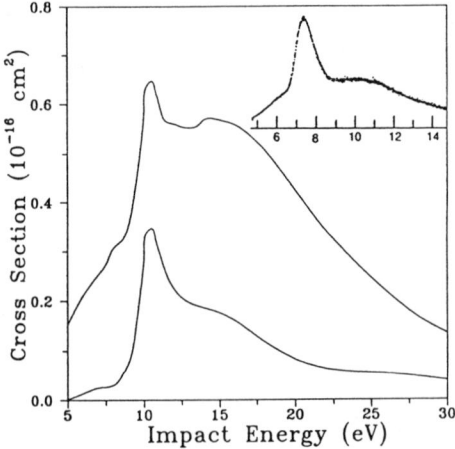

Fig. 5. SMC calculation of electron impact excitation cross section for $^3A_2''$ state of [1.1.1] propellane, upper curve and its $^2A_1'$ component, lower curve; exp. of Ref. 44.

H_2S,[50] GeH_4,[51,52] CF_4.[52] Additional light is shed on the capabilities of the method in

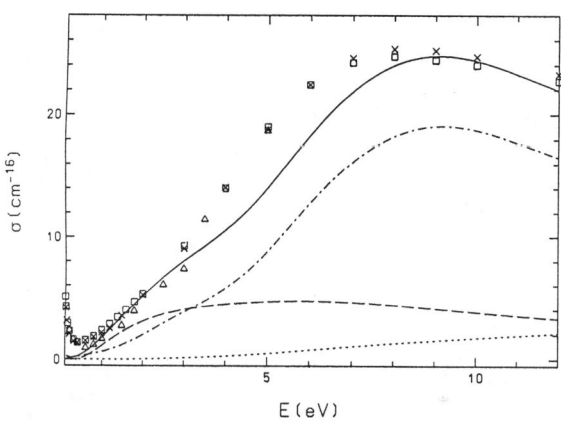

Fig. 6. R-matrix results for elastic scattering cross section for CH_4. Solid, integral cross section; Partial cross sections: $^2A_1$ (dashed), $^2T_2$ (dash-dot) and ^2E (dotted). Experiments: squares, Ref. 45, crosses, Ref. 46, triangles, Ref. 47.

positron scattering from CH_4, SiH_4,[53] and H_2O.[54] The differential cross section for GeH_4 shown in Fig. 7 indicates the level of accuracy attained in many cases for energies greater that about 10 eV. Another recent development in this approach is the incorporation of a complex optical potential to model inelastic effects. Fig. 8 shows the results of such a calculation.[56]

Finally, Fig. 9 shows a stabilization calculation on tetracyanoethylene at the Hartree-Fock level of the sort used routinely now to identify shape resonance features in electron-transmission spectra of large organic molecules.

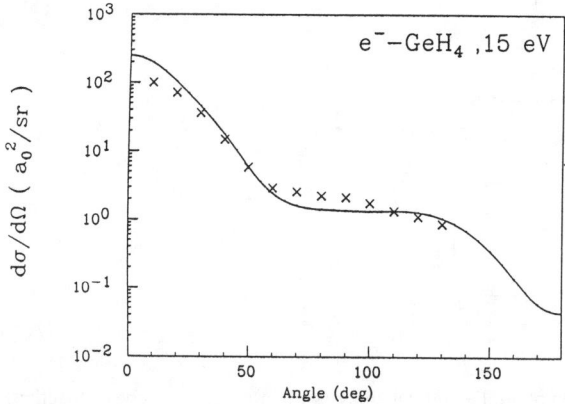

Fig 7. Differential cross section at 15 eV for GeH_4 from a parameter-free model potential calculation. Crosses are experimental points from Ref. 55.

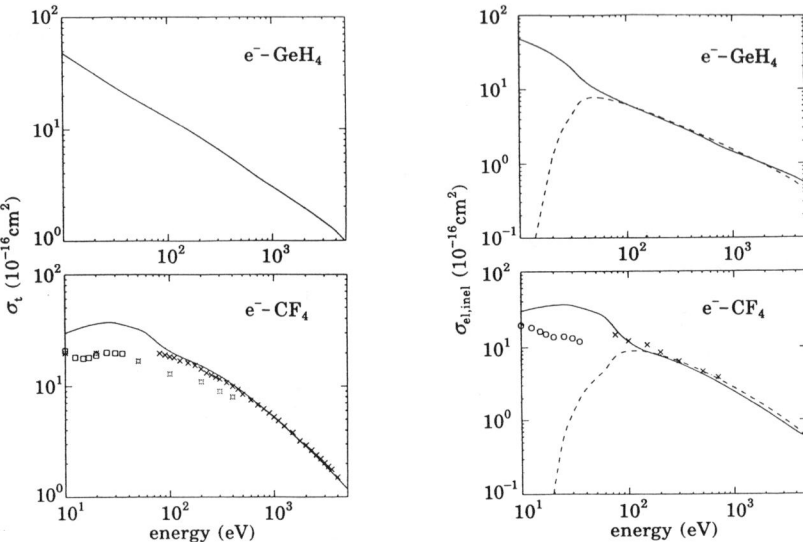

Fig. 8. Left column, Total (elastic plus inelastic) cross sections for CF_4 and GeH_4 compared with experiments summarized in Ref. 56. Right column, elastic (solid) and inelastic (dashed) cross sections from the same calculation compared with results of Huo (Ref. 40) (circles) and experiment of Ref. 57.

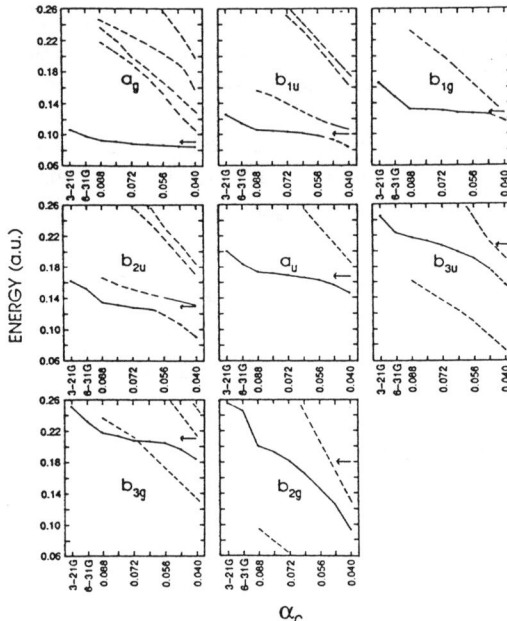

Fig. 9. Stabilization calculation of unfilled orbital energies of tetracyanoethylene as functions of scale factor α.

CONCLUSION AND ACKNOWLEDGEMENTS

Only a minimum impression of the range of new work on electron-polyatomic scattering is given by the few examples included in this review. Sadly, space does not permit inclusion of the entire range of work on vibrational excitation and dissociative attachment of polyatomics which appeared over the last two years.

I would like to thank the authors who made their work available prior to publication for inclusion in this review as well as those who helped by sending re/ preprints. I apologize for the many omissions of outstanding work in this area. This work was performed under the auspices of the U. S. Department of Energy by the Lawrence Livermore National Laboratory under contract No. W-7405-ENG-48.

REFERENCES

1. *Electron and Photon Interactions with Atoms*, edited by H. Kleinpoppen and M.R.C. McDowell (Plenum Press, New York 1976).
2. Robert K. Nesbet, *Variational Methods in Electron Atom Scattering Theory* (Plenum Press, New York, 1980).
3. M. A. P. Lima, T. L. Gibson, W. H. Huo, and V. McKoy, Phys Rev. A **32**, 2696 (1985).
4. C. W. McCurdy and T. N. Rescigno, Phys. Rev. A **39**, 4487 (1989).
5. P. G. Burke in *Electron and Photon Interactions with Atoms*, edited by H. Kleinpoppen and M.R.C. McDowell (Plenum Press, New York 1976) p.1.
6. T. Kato, Prog. Theo. Phys. **6**, 394 (1951).
7. J. Schwinger, Phys. Rev., **78** 135 (1950).
8. W. H. Miller and B. M. D. D. Jansen op de Haar, J. Chem. Phys. **86**, 6213 (1987).
9. T. J. Gil, B. H. Lengsfield and C. W. McCurdy, J. Phys. B: At. Mol. Opt. Phys. **26**, 509 (1993).
10. C. Winstead, P. G. Hipes, M. A. P. Lima, and V. McKoy, J. Chem. Phys. **94**, 5455 (1991).
11. K. Takatsuka and V. McKoy Phys. Rev. A **24**, 2473 (1981); Phys. Rev. A **30**, 1734 (1984).
12. P. G. Burke and W. D. Robb, Adv. Atom. Molec. Phys. **11**, 143 (1975).
13. P. G. Burke, I. Mackey and I. Shimamura, J. Phys. B: At. Molec. Phys. **10**, 2497 (1977).
14. B. I. Schneider and P. J. Hay, Phys. Rev. A **13**, 2049 (1976).
15. B. M. Nestmann and K. J. Pfingst, private communication.
16. B.M. Nestman, R. K. Nesbet and S. D. Peyerimhoff, J. Phys. B: At. Mol. Opt. Phys. **24** 5133 (1991).
17. *See, for example:* N.F. Lane and R. J. W. Henry, Phys. Rev. **173**, 183 (1968).
18. F. A. Gianturco, J. Mol. Struct. (Theochem) **254**, 99 (1992).
19. F. A. Gianturco, Phys. Scr. **23**, 41 (1988).
20. F. A. Gianturco, J. Phys. B: At. Mol. Opt. Phys. **24**, 4627 (1991).
21. A. U. Hazi and H. S. Taylor, Phys. Rev. A **1**, 1109 (1970).
22. H. S. Taylor and A. U. Hazi, Phys. Rev. A **14**, 2071 (1976).
23. P. D. Burrow, A. E. Howard, A. R. Johnston and K. D. Jordan, J. Phys. Chem. **96** 7570 (1992).
24. B. H. Lengsfield, T. N. Rescigno and C. W. McCurdy, Phys. Rev. A **43**, 3514 (1991).

25. T. J. Gil, B. H. Lengsfield. C. W. McCurdy and T. N. Rescigno, *in preparation*.
26. B. I. Schneider, T. N. Rescigno, B. H. Lengsfield, and C. W. McCurdy, Phys. Rev. Letters **66**, 2728 (1991).
27. W. Sun, C. W. McCurdy and B. H. Lengsfield, Phys. Rev. A **45**, 6323 (1992).
28. T. N. Rescigno, B. H. Lengsfield, C. W. McCurdy and S. D. Parker, Phys. Rev. A **45**, 7800 (1992).
29. W. Sun, C. W. McCurdy and B. H. Lengsfield, J. Chem. Phys. **97**, 5480 (1992).
30. T. J. Gil, B. H. Lengsfield. C. W. McCurdy and T. N. Rescigno, *in preparation*.
31. O. Sueoka and S. Mori, J. Phys. B **19**, 4035 (1986).
32. D. T. Alle, R. J. Gulley, S. J. Buckman, and M. J. Brunger, J. Phys. B (in press).
33. C. Winstead, Q. Sun, V. McKoy and M. A. P. Lima, Z. Phys. D - Atoms, Molecules, and Clusters **24**, 141 (1992).
34. C. Winstead, Q. Sun, and V. McKoy, J. Chem. Phys. **96**, 4246 (1992).
35. C. Winstead, Q. Sun, and V. McKoy, J. Chem. Phys. **98**, 1105 (1992).
36. Q. Sun, C. Winstead, V. McKoy, and M. A. P. Lima, J. Chem. Phys. **96**, 3531 (1992).
37. Q. Sun, C. Winstead, V. McKoy, J. S. E. Germano, and M. A. P. Lima, Phys. Rev. A **46**, 2463 (1992).
38. C. Winstead, Q. Sun, V. McKoy, J. L. S. Lino, and M. A. P. Lima, J. Chem. Phys. **98**, 2132 (1993).
39. C. Winstead, Q. Sun, and V. McKoy, J. Chem. Phys. **97**, 9483 (1992).
40. W. M. Huo, Phys. Rev. A **38**, 3303 (1988).
41. L. Boesten, H. Tanaka, A. K. Kobayashi, M. A. Dillon, and M. Kimura, J. Phys B **25**, 1607 (1992).
42. A. Mann and F. Linder, J. Phys. B **25**, 533 (1992).
43. R. K. Jones, J. Chem. Phys. **84**, 813 (1986).
44. O. Schafer, M. Allan, G. Szeimies, and M. Sanktjohansen, J. Am. Chem. Soc. (submitted, 1992).
45. J. Ferch, B. Granitza, and W. Raith, J. Phys. B: At. Mol. Phys. **18**, L445 (1985).
46. B. Lohman and S. J. Buckman, J. Phys. B: At. Mol. Phys. **19**, 2565 (1986).
47. W. Sohn, K. H. Kochem, K-M Scheuerlein, K. Jung, and H. Ehrhardt, J. Phys. B: At. Mol. Phys. **19**, 3625 (1986).
48. F. A. Gianturco, J. Phys. B: At. Mol. Opt. Phys. **24**, 3837 (1991).
49. A. K. Jain, A. N. Tripathi and A. Jain, J. Phys. B **20**, L389 (1987).
50. A. K. Jain, A. N. Tripathi and A. Jain, Phys. Rev. A **42**, 6912 (1990).
51. A. Jain, K. L. Baluja, V. Di Martino and F. A. Gianturco, Chem. Phys. Letts **183**, 34 (1991).
52. F. A. Gianturco, V. Di Martino, and A. Jain, Il Nuovo Cimento **14**, 411 (1992).
53. A. Jain and F. A. Gianturco, J. Phys. B At. Mol. Opt. Phys. **24**, 2387 (1991).
54. F. A. Gianturco, Europhys. Lett. **12**, 689 (1990).
55. L. Boesten and H. Tanaka, private communication in Ref. 51.
56. K. L. Baluja, A. Jain, V. Di Martino and F. A. Gianturco, Europhysics Lett. **17**, 139 (1992).
57. T. Sakae, S. Sumiyoshi, E. Murakami, Y. Matsumoto, K. Ishibashi, and A. Katase, J. Phys. B **22**, 1385 (1989).

RYDBERG ATOMS: A NANO-LABORATORY FOR COLLISION PROCESSES

K. A. Smith and F. B. Dunning

Departments of Physics and Space Physics and Astronomy and
the Rice Quantum Institute, Rice University, Houston, TX 77251 U.S.A.

ABSTRACT

This paper presents a discussion of Rydberg atom techniques applied to the study of electron-molecule collisions. These techniques present new opportunities for study of ultra-low-energy electron collisions and for investigation of chemical dynamics.

INTRODUCTION

The availability of the tunable dye laser permits excitation of atomic species to well-defined Rydberg states having principal quantum numbers n extending to $n > 500$. Studies of these species have provided new information on atomic spectroscopy, on the behavior of atomic systems in the presence of strong electric and magnetic fields, and on the physics of electronic and atomic collisions. This paper focuses on the use of Rydberg atoms as a small laboratory within which one can investigate a remarkable range of collision processes, including both elastic and inelastic electron-molecule collisions, dissociative attachment, ion-ion collisions, and ion-molecule reactions. New details of these processes are revealed; and, in the case of electron-molecule interactions, the range of effective collision energies extends below 100 μeV (1 Kelvin).

Use of Rydberg atoms to study electron collisions is predicated on the "essentially-free electron" model which assumes that the separation between the excited electron and the core ion is large enough that both do not simultaneously interact with a neutral target molecule. This model is reasonable because the mean electron-core separation scales as n^2; and, even at $n = 10$, the atomic diameter is 100 Å. Thus the effective range of the electron-target interaction is usually much less than the radius of the Rydberg atom. The time scale of the electron-molecule interaction is short compared to the orbital period of the Rydberg electron. Thus the Rydberg atom does not interact with the target *as an atom* but rather as a pair of independent particles. Theoretical calculations[1] and comparisons between free-electron data and Rydberg-atom data for electron attachment processes[2] indicate that the free-electron model has quantitative validity. It is only a model, however, and its applicability may depend on the process under study and will clearly diminish at low n.[3] While the electron-target interaction can be viewed as a binary collision, the presence of the ion core *is* important. The electron's velocity distribution is defined by its initial atomic state; and, after the collision, the charged products interact with the core either directly or through its Coulomb field. While the "essentially-free electron model" appears to be satisfactory for describing collisions involving electrons, the same concept is less applicable to collisions of the core ion. Because the times associated with ion-molecule interactions are comparable to the electron orbit period, the excited electron is more likely to play a role in collisions involving the core ion.[4]

Typically, atoms in well-defined high-lying states are produced by photoexcitation. Since the average kinetic energy of the excited electron is equal to its binding energy, the photoexcitation process provides an electron of precisely known average energy. Typical binding energies of the excited electron in Rydberg atoms range between about 200 meV ($n = 8$) and 0.085 meV ($n = 400$). The energy of truly free electrons is affected by stray electric fields in the apparatus. Because the Rydberg atom is itself electrically neutral, small stray fields do not significantly affect its motion, and have a negligible effect on the energy of the bound Rydberg electron. It is, however, only the *average* Rydberg electron energy that is defined, and the collisions between the electron and target species take place with a range of velocities that is determined primarily by the quantum-mechanically-specified momentum distribution for the excited electron. For low l states (which are the only ones that can be directly photoexcited from the ground or low-lying metastable states) these distributions are broad. This is taken into account by noting that, on the basis of the essentially-free electron model, the rate constant k for a Rydberg-atom collision process dominated by the electron-target interaction is given by:

$$k_d = \int_0^\infty v \; \sigma_e(v) \; f(v) \; dv \tag{1}$$

where v is the electron velocity, $\sigma_e(v)$ is the electron impact cross section for the process under study, and f(v) is the electron velocity distribution that corresponds to the initial Rydberg atom state.

The Coulomb potential of the Rydberg core confines the excited electron to a specific volume of space. Because this volume is centered on the relatively slowly-moving core ion, the electron remains in the experimental apparatus for much longer times than a free electron of equivalent energy would, increasing the probability of observing a collision. Charged reaction products interact with the Coulomb field of the core. If the electron is simply scattered elastically or inelastically in a collision, it may remain bound to the core, but in a different quantum state. This state can be determined by selective field ionization (SFI)[5], and the magnitude of any energy transfer in the collision can be assessed. If the energy transfer is large enough, the Rydberg atom is ionized. If a negative ion is formed by electron capture, its interactions with the core can result in conversion of internal energy to translational energy. If the negative ion is short-lived, analysis of its scattering with the core can reveal its lifetime against predissociation or autodetachment, and the energetics of those process. Examples of these are given below.

QUASI-ELASTIC SCATTERING

Among the earliest studies of collisions with Rydberg atoms were measurements of l-changing reactions of the type:

$$K(nd) + X \rightarrow K(nl') + X \qquad (2)$$

where X is a rare gas or a molecule.[6] Such processes result essentially from elastic scattering of the excited electron by the target particle. The cross sections for collisions of this type depend substantially on the energy difference between the initial and final states, favoring small energy transfers. This is not particularly surprising, since the electron kinetic energy *in the rest frame of the core ion* is not substantially changed by an elastic electron-target[7] collision. Theoretical studies of reaction (2) using the essentially-free electron model[8] yield results in excellent agreement with experiment, although so too do models that treat the process in detail as a collision between an excited atom and the target.[9]

INELASTIC SCATTERING

Collisions with polar molecules provide a simple example of how a binary collision between the excited electron and the target can leave its signature in the products. In such collisions, quasi-resonant transfer of molecular rotational energy to the Rydberg electron can result in n-changing reactions of the type:

$$Xe(nf) + NH_3(J) \rightarrow Xe(n'l') + NH_3(J-1) \qquad (3)$$

in which one quantum of rotational energy from the NH_3 is transferred to the Rydberg electron in an inelastic collision, exciting it to a state of higher n. Because states of different n field-ionize at different values of electric field, the n-state distribution of the reaction product can be determined by applying a ramped electric field to the interaction volume and recording the number of electrons resulting from field-ionization as a function of applied electric field.[5] Figure 1 gives such an analysis for products of reaction (3) for initial $n = 31$. As can be seen, the NH_3 *rotational* state distribution is reflected in the *electronic* state distribution of the product Rydberg atoms. Figure 2 depicts the energetics of the reaction. For small initial J values, rotationally inelastic collisions result in formation of bound Rydberg states, while ionization results for large initial J.

Inspection of Figure 2 reveals that if the initial n is increased above about 80, *all* rotational transitions provide sufficient energy to ionize the Rydberg atom. Measurements of the rate constant for collisional ionization as a function of initial n can then reveal the dependence of the cross section for rotationally-inelastic electron-molecule scattering on the energy of the incident electron.

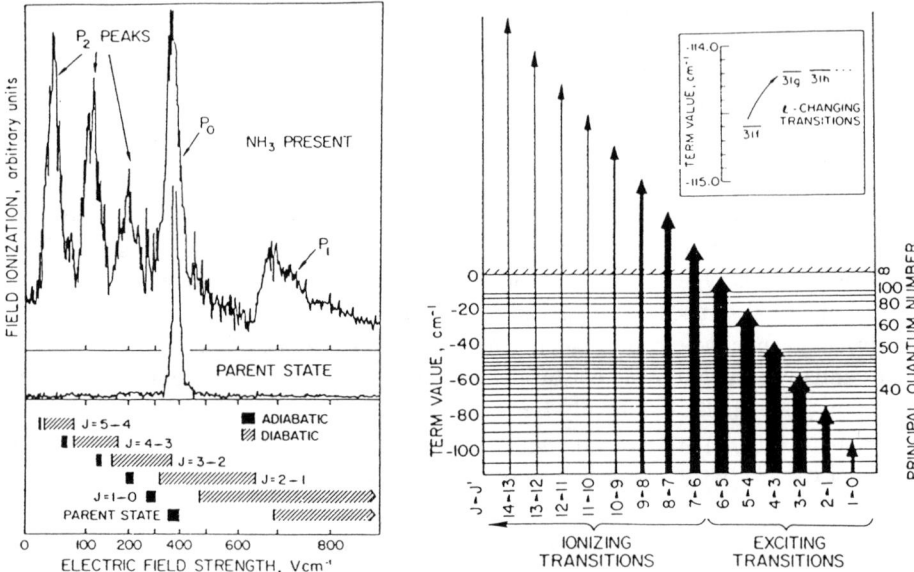

Fig. 1. (left panel) Upper curve is the selective field ionization spectrum observed when the products of reaction (3) for an initial n of 31 are subjected to an ionizing field ramp. The lower curve is the SFI spectrum of the laser-excited $n = 31$ state. The anticipated range of fields required to ionize the products of the J → J' transitions indicated are denoted by the hatched areas in the lower part of the figure.

Fig. 2 (right panel) Term diagram pertinent to reaction (3) for $n = 31$. Arrow lengths correspond to energies given up in the J → J' transitions indicated. These transitions produce product peaks P_2. Arrow widths indicate relative population of J levels of NH_3 at room temperature. Inset shows quasi-elastic process producing product peak P_1.

We have recently measured rate constants Rydberg atom destruction through ionization in collisions between K(np) Rydberg atoms and the polar molecules HF and NH_3, for initial n in the range from 80 to 400[10] which corresponds to average electron kinetic energies between 1.4 meV and 80 μeV, the corresponding de Broglie wavelengths range between 330 and 1300 Å, much less than the mean atomic diameters which vary between about 10,000 and 160,000 Å. The data in Figure 3 show that the rate constant for collisional ionization increases substantially with increasing n. Analysis of the data using Equation (1) shows that the cross section for rotationally-inelastic scattering scales as $1/v^2$, where v is the electron velocity. In some respects, this result is surprizing, because the Born approximation predicts that, in this energy regime, the electron cross section should scale as $1/v$,[11] which would make the rate constant independent of n. However as is evident from Figure 2, the energy imparted to the electron is much greater than its initial energy, and its momentum change is substantial, violating the basic assumption of weak scattering inherent in the Born approximation. At $n = 400$, the electron collision cross section derived for HF is on the order of 100,000 Å^2.

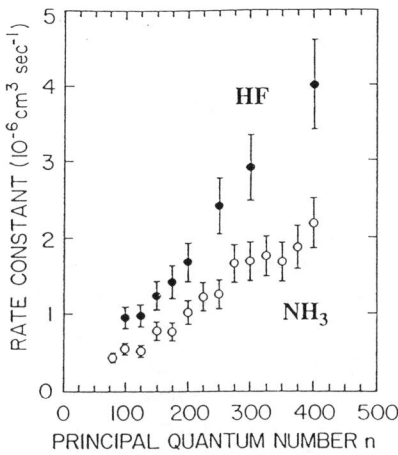
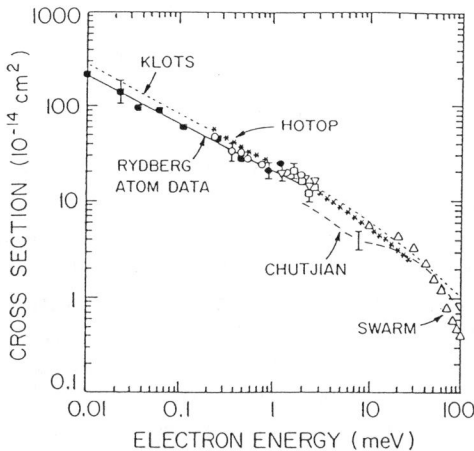

Fig. 3. (left panel) Experimentally-measured rate constants for collisional ionization of HF and NH_3.

Fig. 4 (right panel) Cross section for electron attachment to SF_6 measured by Rydberg-atom and free-electron methods (References 13-16).

These experiments are the first on electron/polar-molecule scattering at submillivolt energies, and there is no detailed theoretical explanation for them at this time. It is worth noting that a simple "hard-sphere" model which assumes that reaction will occur if the electron approaches within about one de Broglie wavelength of the target yields cross sections having the velocity dependence and approximate magnitudes of those observed. The present electron-molecule collisions are unusual in that the change in electron momentum is great; the electron's de Broglie wavelength is large; and the interaction time is comparable to the molecular rotational period. These studies represent a first step into a new energy regime of electron-molecule interactions and we anticipate other developments will follow.

ELECTRON ATTACHMENT

Rydberg atom collisions with many electronegative molecules result in Rydberg atom destruction through electron-transfer reactions of the type:

$$K(nl) + SF_6 \rightarrow K^+ + SF_6^{-*}$$
$$K(nl) + CCl_4 \rightarrow K^+ + CCl_4^{-*} \rightarrow K^+ + CCl_3 + Cl^-$$
(3)

The excited electron is captured by the target, generally into excited states that may undergo dissociation or autodetachment. We report here three recent advances in the use of Rydberg atoms to study low-energy electron attachment: extension of measurements to very high n, investigations at low n that reveal significant energy transfer through post-attachment interactions, and techniques for analyzing the dynamics of dissociative attachment processes.

Measurements of electron-attachment to SF_6 and CCl_4 have recently been extended into the microvolt-electron-energy range by studies of Rydberg atom destruction in collisions with these species. The rate constants were found to be essentially independent of n for $30 < n < 400$, with the values being about 4×10^{-7} cm^3/sec for SF_6 and 7×10^{-7} cm^3/sec for CCl_4.[12] Since k is independent of n, it is independent of the electron velocity v. Equation (1) then requires that the attachment cross section scale as $1/v$, consistent with the threshold behavior expected for s-wave capture. Cross sections derived from Rydberg atom data are shown in Figure 4 along with other data for free electron collisions obtained by swarm,[13] threshold photoelectron spectroscopy,[14,15] and calculations.[16] Note that for the Rydberg atom data it is the median energy of the attached electrons that is plotted, and that the collision energy scale extends down to 10 μeV.

At low values of n, electron transfer produces a negative ion in close proximity to the positive core ion. In order for the products to be observed as free ions, they must have (or attain) enough kinetic energy to overcome their mutual electrostatic attraction and separate. In the case of SF_6, we observe that significantly more ionic products appear at low n than would be expected if the only energy available to the system is the initial kinetic energy of the reagents. This suggests that after attachment occurs, the SF_6^{-*} and K^+ ions collide, and internal energy of the SF_6^{-*} is converted to translational energy of the ion pair, allowing them to separate. In order to investigate this process, we have studied the free ion production as a function of the initial Rydberg atom velocity.[17] This work required making a detailed Monte Carlo model of the collision process that predicts the ion production rate and the angular distribution of the scattered ions.[18] The most compelling evidence for post-attachment inelastic collisions is shown in Figure 5, which includes data for SF_6 and two other species, CH_3I and CF_3I, both of which attach dissociatively, producing I^- ions with diffrent, but known, energy distributions [measured by observing the velocities of the I^- fragments[19] (see below)].

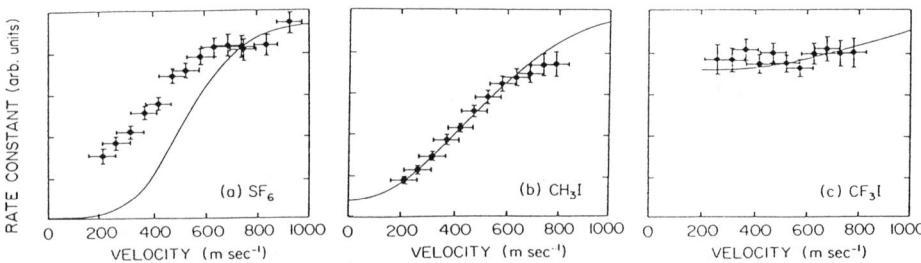

Fig. 5. Free ion production as a function of Rydberg atom velocity for collisions with SF_6, CH_3I, and CF_3I. Points are experimental data, lines are model results assuming no post-attachment energy transfer.

Model calculations that assume no post-attachment energy transfer are in good agreement with experiment for CH_3I and CF_3I, but fail for SF_6. The excess SF_6^- observed at low collision velocities suggests energy transfer. If it is arbitrarily assumed that about 20 meV of energy transfer occurs, the model provides results in good agreement with the data.

In some systems post-attachment collisions between the product ions appear to be responsible for stabilization of otherwise unstable negative ions. Our recent studies of Rydberg atom collisions with O_2, CS_2 and C_2Cl_4[20] show this to be the case, allowing observation of stable O_2^-, CS_2^- and $C_2Cl_4^-$ products. Suggestions of stabilization *via* post-attachment collisions have been made by other workers.[21] It is worth noting that positive-ion/negative-ion collisions can stabilize the negative ion, without mutual neutralization occurring simultaneously. This suggests that Rydberg atom collisions may shed further light on the details of energy transfer in ion-ion collisions.

Use of position-sensitive detector (PSD) techniques, coupled with development of a detailed model for the reaction kinematics provides new insights into the chemical dynamics of post-attachment reactions. The apparatus used is shown in Figure 6.

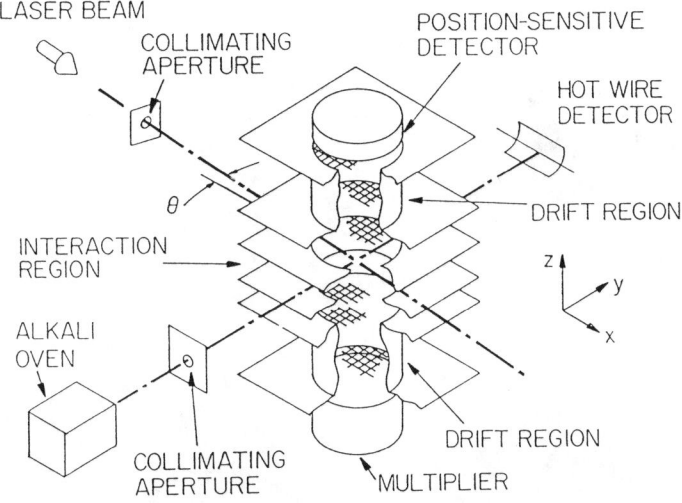

Fig. 6. Apparatus for kinematic studies.

Rydberg atoms are excited by a one-microsecond laser light pulse, and collisions are allowed to take place for one or two microseconds. The positive and negative products are then accelerated to the upper and lower detectors, and reach the detectors after a drift time of about 10 microseconds.

Arrival position distributions at the PSD for I⁻ ions formed in the reaction

$$K(nd) + CF_3I \rightarrow K^+ + CF_3I^{-*} \rightarrow K^+ + CF_3 + I^- \qquad (4)$$

are shown in Figure 7. For $n > 25$, where post-attachment interactions are minimal the I⁻ distribution is symmetric, and the kinetic energies of the fragments can be determined from analysis of the breadth of the ion distributions and knowledge of the ion flight times.[22] At low n, the distributions are clearly peaked in the backward direction. This occurs because the I⁻ products formed traveling in the forward direction have about the same velocity as the K⁺ core ion, and the two species have insufficient relative kinetic energy to overcome their mutual attraction and separate and be detected. Ions formed traveling in opposite directions, however, have enough energy to separate. The strong backward peaking of the I⁻ distribution indicates that the CF_3I^{-*} intermediate dissociates before the K⁺ ion trajectory is significantly altered by interaction with the negative ion. The kinematic model shows that the I⁻ distribution should depend significantly on the CF_3I^{-*} lifetime, and comparisons between model and experiment show us that the lifetime of the CF_3I^{-*} intermediate is below 10 psec. Similar studies for CCl_4 show a lifetime of around 20 psec.[18] Analysis of the scattering that occurs during the short intermediate state lifetime provides a new means for studying chemical dynamics on a picosecond time scale. These processes are now being pursued with improved detector arrangements.

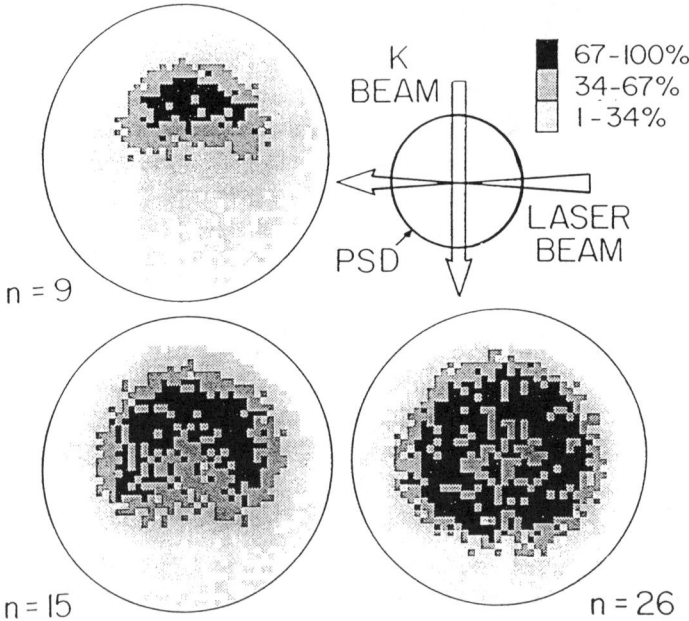

Fig. 7. Arrival position distributions at the position-sensitive detector of I⁻ ions formed in reaction (4). Rydberg atoms are produced at the intersection of the K and laser beams.

In summary, Rydberg atom methods provide significant insights into electron-molecule collisions and into the chemical dynamics of unimolecular reactions. They reveal new aspects of ion-ion collisions, and provide a method for measurement of the lifetimes of transition state complexes. They additionally allow controlled collision experiments with electrons of microvolt energies. We gratefully acknowledge support of this work by both the Physics and Chemistry Sections of the National Science Foundation and by The Robert A. Welch Foundation.

REFERENCES

1. M. Matsuzawa "Theoretical studies of collisions of Rydberg atoms with molecules" in *Rydberg States of Atoms and Molecules* F. B. Dunning and R. F. Stebbings, eds. (Cambridge University Press, 1983), p. 267.
2. F. B. Dunning, J. Phys. Chem. **91**, 2244 (1987).
3. A. P. Hickman, R. E. Olson, and J. Pascale, "Theoretical approaches to low-energy collisions of Rydberg atoms with atoms and ions" in *Rydberg States of Atoms and Molecules*, F. B. Dunning and R. F. Stebbings, eds. (Cambridge University Press, 1983), p. 187.
4. A. Kalamarides, C. W. Walter, B. G. Zollars, K. A. Smith, and F. B. Dunning, J. Chem. Phys. **87**, 4238 (1987); R. A. Popple, K.. A. Smith, and F. B. Dunning, J. Chem. Phys., in press (1993).
5. K. A. Smith, F. G. Kellert, R. D. Rundel, F. B. Dunning, and R. F. Stebbings, Phys. Rev. Lett. **40**, 1362 (1978); J. Chem. Phys. **32**, 3179 (1978).
6. T. F. Gallagher, S. A. Edelstein, and R. M. Hill, Phys. Rev. Lett. **35**, 644 (1975); Phys. Rev. A **15**, 1945 (1977).
7. J. P. McIntire, G. B. McMillian, K. A. Smith, F. B. Dunning, and R. F. Stebbings, Phys. Rev. A **29**, 381 (1984).
8. E. de Prunele and J. Pascale, J. Phys. B **12**, 2511 (1979).
9. A. P. Hickman, Phys. Rev. A **18**, 1339 (1978); Phys. Rev. A **23**, 87 (1981).
10. X. Ling, M. T. Frey, K. A. Smith and F. B. Dunning, Phys. Rev., in press (1993).
11. O. H. Crawford, J. Chem. Phys. **47**, 1100 (1967).
12. X. Ling, B. G. Lindsay, K. A. Smith, and F. B. Dunning, Phys. Rev. A **45**, 242 (1992).
13. R. Y. Pai, L. G. Christophorou, and A. A. Christodoulides, J. Chem. Phys. **70**, 1169 (1971).
14. A. Chutjian and S. H. Alajajian, Phys. Rev. A **31**, 2885 (1985).
15. H. Hotop, private communication.
16. C. E. Klots, Chem. Phys. Lett. **38**, 61 (1976).
17. R. A. Popple, M. A. Durham, R. W. Marawar, B. G. Lindsay, K. A. Smith, and F. B. Dunning, Phys. Rev. A **45**, 247 (1992).
18. X. Ling, M. A. Durham, A. Kalamarides, R. W. Marawar, B. G. Lindsay, K. A. Smith, and F. B. Dunning, J. Chem. Phys. **93**, 8669 (1990).

19. A. Kalamarides, C. W. Walter, B. G. Lindsay, K. A. Smith, and F. B. Dunning, J. Chem. Phys. **91**, 4411 (1989); C. W. Walter, B. G. Lindsay, K. A. Smith, and F. B. Dunning, Chem. Phys. Lett **54**, 409 (1989).
20. R. A. Popple, K. A. Smith, and F. B. Dunning, "Energy Transfer in Rydberg Atom Collision Processes at Intermediate n", Division of Atomic, Molecular, and Optical Physics Meeting of the APS, May 1993, Reno, Nevada, USA.
21. G. Brincourt, S. Rajab Pacha, R. Catella, Y. Zerega and J. Andre, Chem. Phys. Lett. **156**, 573 (1989); A. Pesnelle, C. Ronge, M. Perdrix, and G. Watell, Phys. Rev. A **42**, 273 (1990); H. Carman, C. Klots, and R. Compton, private communication.
22. C. W. Walter, B. G. Lindsay, K. A. Smith, and F. B. Dunning, Chem. Phys. Lett. **154**, 409 (1989).

FORMATION AND DECAY OF TRANSIENT ANIONS PRODUCED BY ELECTRON IMPACT ON SURFACE MOLECULES

Léon Sanche

Group of the Medical Research Council of Canada in the Radiation Sciences
Canadian Center of Excellence in Molecular and Interfacial Dynamics
Faculté de médecine, Université de Sherbrooke
Sherbrooke, Qc, Canada J1H 5N4

ABSTRACT

Various aspects of resonance electron scattering and attachment to molecules condensed on dielectric and metal surfaces are discussed in this article. Results are presented to illustrate the effect of distance from a metal surface on electron scattering via $^2\Pi_g$ state of N_2^- and on the behavior of dissociative attachment to molecular oxygen. Taking C_2D_6 on a rare gas substrate as an example, it is further shown that core-excited resonances can decay by transferring both charge and energy to a surface molecule.

I. INTRODUCTION

The formation of transient negative ions (i.e., electron resonances) is a fundamental phenomenon of wide spread occurrence in low-energy (0-30 eV) electron scattering from gas-phase atoms and molecules.[1] During the last decade, it has been shown that the interaction of low-energy electrons with condensed molecules[2,3] also leads to the formation of transient negative ions. As in the gas-phase both single-particle (e.g., shape resonances) and two-particle one-hole (e.g., core-excited resonances) states have been observed to occur at surfaces[2] and in the bulk[3] of dielectric materials where they can decay by leaving the target molecule in a vibrationally and/or electronically excited state, by dissociative attachment and by stabilization. Other decay channels occurring exclusively in surface experiments include intermolecular[3] (e.g. phonon) and molecule-surface[4] vibrational excitation and electron exchange between dissimilar species[5].

The purpose of the present article is to point out major differences between the gas- and condensed-phase formation and decay of transient anions. Many factors contribute to these differences including the polarization of the media by the transient anion and/or its image charge induced in a metal substrate, the changes in the symmetry of the scattering problem and the possibility of charge and energy exchange with neighboring molecules or the substrate. Compared to their gas-phase counterparts transient anions formed at surfaces[2] have energies usually lowered by about 0.5 to 2.0 eV depending on the polarizability of the surface and the lowering of the symmetry; they usually have a lower number of intramolecular decay channels due to their lower energies but other decay channels appear from excitation of intermolecular energy modes of the solid and electron emission into bulk and surface states; their lifetimes are either shorter or longer depending on the change in the number of decay channels and symmetry; finally, the probabilities for initial electron capture, decay into inter- and intra-molecular excitations and decay into dissociative attachment (DA) may vary *by orders of magnitude* depending on energy, symmetry and molecular orientation. It may not be easy to predict the condensed phase behavior from the gas-phase information because of the interrelationship between these various parameters. In the case of DA, the lifetime is often a crucial parameter which may affect exponentially the magnitude of the anion yields, but breakdown of molecular symmetry may also have drastic effects. A few results are presented in this article to illustrate major differences between gas and condensed transient anion.

II. EXPERIMENTS

The results reported in the following sections were recorded with a high-resolution electron-energy-loss (HREEL) spectrometer and an electron stimulated desorption (ESD) apparatus. In both systems, a monochromatic electron beam was incident on a target composed of rare gas solid (RGS) multilayer films covered by submonolayer amounts of the molecule under investigation. The RGS films were condensed onto a cooled platinum substrate (15-20 K) having a strong azimuthal disorder characterized by microfacets with the (lll) plane parallel to surface.[6] The RGS films which served as a spacer between the metal substrate and the molecular adsorbate were also found to grow according to the fcc (lll) orientation. Varying the thickness of the spacer layers made it possible to study systematically the effect of the metal surface.

With the HREEL spectrometer electrons being scattered by the adsorbate can be energy analyzed and the incident electron energy dependence of a given energy loss (i.e., the excitation function) recorded.[7] With the ESD apparatus, a portion of the anion flux emanating from the surface is measured by a mass spectrometer placed near the sample surface. This arrangement allows the measurement of anion yields as a function of incident electron energy.[5,8] All experiments were performed under ultra-high-vacuum conditions (10^{-10}-10^{-11} torr).

III. THE $^2\Pi_g$ STATE OF N_2^-

In this section the behavior of the well-known $^2\Pi_g$ transient state of N_2^- is reported as a function of its distance from the Pt substrate for N_2 deposited on a RGS film surface and embedded in the bulk of RGS multilayer films. The $^2\Pi_g$ shape resonance of N_2^-, located at 2.3 eV in the gas phase,[1] involves the temporary localization of an electron in the next vacant antibonding orbital of N_2 with the ground state configuration $(\sigma_g 1s)^2(\sigma_u^* 1s)^2(\sigma_g 2s)^2(\sigma_u^* 2s)^2(\pi_u 2s)^4(\sigma_g 2p)^2(\pi_g^* 2p)$. The coupling of this transitory state with intramolecular vibrations leads to the observation of a large enhancement and of an oscillatory structure in the isolated-molecule elastic and vibration cross sections.[1] This latter structure is also observed in the excitation function for vibrations of ground state N_2 physisorbed on a metal, in a pure solid film, as well as for a N_2 in rare gas matrices.[2,3] This condition allows an accurate determination of the resonance energy in various media.

The excitation function for the v=1 vibrational loss of ground state N_2 recorded with a HREEL spectrometer between 0 and 4 eV[9] is shown in Fig. 1. The numbers on the right of each curve indicate the number of monolayers (ML) of the Ar spacer on which 0.1 ML of N_2 was physisorbed. These excitation functions for N_2 isolated on Ar were recorded with an incident beam positioned at 14° from the surface normal; electrons emerging at 45° were energy analyzed. The oscillatory structure, due to vibrational motion of N_2^-, is similar to that recorded at 90° in the gas-phase differential cross section[10] shown in the bottom of Fig. 1. Once appropriately translated by -0.62 eV all peak positions of the gas-phase measurement agree within ±0.02 eV with those of the 32-ML result. The larger overlap in the condensed-phase oscillatory structure may be attributed either to a reduction in the lifetime of the resonance or to an inhomogeneous broadening due to molecules occupying several different adsorption sites. For thicknesses smaller than 32 ML, the overall intensity of the resonance feature decreases by a factor of up to 5, while the oscillatory structure shifts to lower energy and becomes even broader. Even though, the relative separation between the peaks appears almost unperturbed, a substantial energy shift varying from 0.62 to 1.6 eV is obtained from

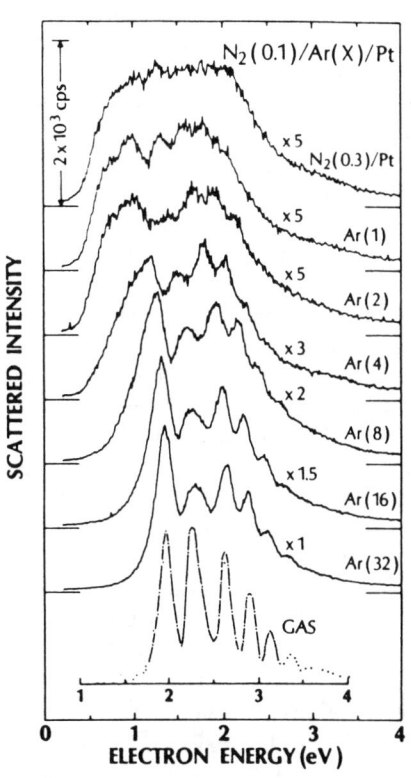

Fig. 1. Excitation function for the v=1 level of N_2 separated from the Pt substrate by an Ar film of thicknesses ranging from 0 to 32 monolayers (ML). At the bottom is shown aligned with the 32-ML result the corresponding differential cross section measured in the gas phase. The N_2 coverage is 0.1 ML except at 0 ML of Ar where it it 0.3 ML.

alignment of the condensed phase spectra with the gas-phase results. Calculations[9] of the electronic polarization energy of N_2^- as a function of the thickness of the Ar film can reproduce the energy shift to an accuracy of a few percent down to a thickness of 2 layers; indicating that the lowering of the N_2^- state with decreasing distance from the metal is essentially due to the increase in induced electronic polarization. Close to the metal surface, the classical picture of a point charge near a semi-infinite dielectric breaks down and gives an overestimate of the polarization energy. Near the surface details of the perturbed charge distribution of the anion as well as that of induced charge at the metal surface must also be taken into account. Furthermore, the symmetry of the scattering problem and molecular orientation may differ. In this case, it is preferable to compare these results to more elaborate calculations which include multiple scattering of the electron wave[11] within the substrate and between the substrate and the molecule or calculations which include, at least, both the image potential and the surface barrier such as the coupled-angular mode method of Teilly-Billy and Gauyacq.[12] Using this latter method, these authors and Rous[13] were able to provide a description of the variation of the resonance energy and width of the N_2^- $^2\Pi_g$ state as a function of its distance from a metal surface.

As seen in Fig. 1, the relative magnitude of each peak of the oscillatory structure is different for the isolated and condensed molecule. This is probably due to the nature of the incoming and outgoing electron wave functions. Rous et al[11] have clearly shown for O_2 on graphite that the free electron wavefunctions are appreciably modified by the presence of the substrate. The electron waves incident on N_2 physisorbed on Ar arise principally from two sources: those which arrive directly from vacuum and those which are reflected from the solid surface. Similarly, the scattered waves can be directly emitted in vacuum or first be reflected from the surface. The multiple scattering of electron waves in the Ar solid and between the molecule and the solid changes the partial wave content of the isolated N_2^- resonance and consequently the parameters of the state (i.e., lifetime, energy, electron capture cross section and symmetry). As the N_2 molecule is placed closer to the substrate (i.e., as the number of Ar ML is reduced from 32 to 0) these parameters are further affected by the metal surface; the lineshape of the N_2^- $^2\Pi_g$ excitation function is consequently modified exhibiting appreciable changes in the magnitude and width of oscillatory structure. At large Ar thick-

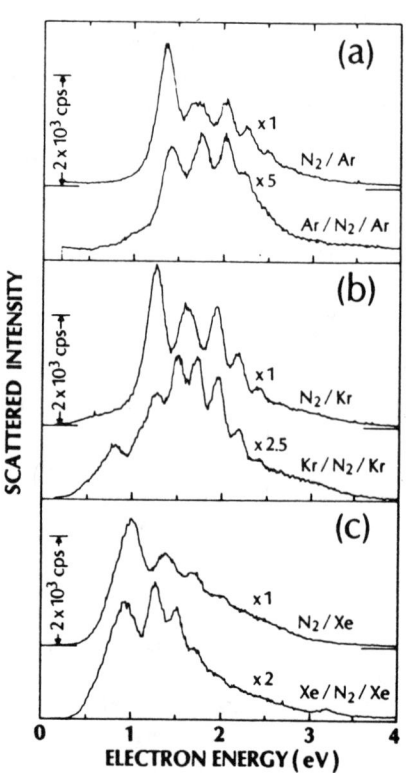

Fig. 2. Excitation function for the v=1 level of N_2 deposited on the surface of a 32-ML rare gas film and within the same rare gas film is shown for Ar, Kr and Xe in (a), (b) and (c), respectively.

nesses these modifications are probably due to changes in multiple scattering conditions in the RGS film substrate and the increased contribution from multiple scattering in the metal; however, when N_2 is very close to the metal surface the appearance of new decay channels in this latter can considerably shorten the lifetime. The orientation of the N_2 molecule may be modified near the metal surface, thus also causing changes in the resonance parameters.

The effect of multiple scattering of the incoming and outgoing electron waves is even more obvious from Fig. 2 where the excitation function of the v=1 level of N_2 deposited on and embedded inside[14] a 32-ML film of Ar, Kr and Xe are shown in (a), (b) and (c), respectively.[9] Here, it may be seen that both the energy and line shape of the $^2\Pi_g$ N_2^- state are dependent on the nature of the substrate and coverage by additional RGS layers, whereas the energy-integrated magnitude of the resonant phenomena is practically the same for all cases. However, the relative magnitude of each peak in the oscillatory structure may differ considerably. One notices that the lowest energy peak in the surface spectra of Ar and Kr has considerably diminished once recorded into the bulk; it is also shifted down by 0.27 and 0.43 eV, respectively, due to the increased electronic polarization in the overlayer. For Xe the lowest energy peak is no longer visible in the overlayer excitation function [Fig. 2(c)].

IV. DISSOCIATIVE ATTACHMENT: THE $^2\Pi_u$ STATE OF O_2^-

The effect of the image charge induced in the metal substrate on the dissociative attachment (DA) process can also be investigated by inserting RGS layers between the condensed molecule and the metal surface. In this section, we compare DA to the O_2 molecule [e + $O_2(^3\Sigma_g^-)$ → $O_2^-(^2\Pi_u)$ → $O^-(^2P)$ + $O(^3P)$] to the dipolar dissociation (DD) process [e + O_2 → O_2^* + e → O^+ + O^- + e]. The effect on the O^- yield of isolating the O_2 molecule from the polycrystalline Pt substrate by 1.3, 2.2 and 4.3 ML of argon,[15] respectively, is shown in the lowest three curves of Fig. 3. For all curves but the upper one, which represents the gas-phase cross section for O^- production,[16] the O_2 coverage is 0.1 ML. The curves labeled 0.37 ML and 0.7 ML are the O^- ESD yields when these amounts of Ar are coadsorbed with 0.1 ML of O_2. In these examples, the O_2 molecules are physisorbed on Pt. The magnitude of the O^- DA peak around 6 eV increases by two orders of magnitude as the distance of the

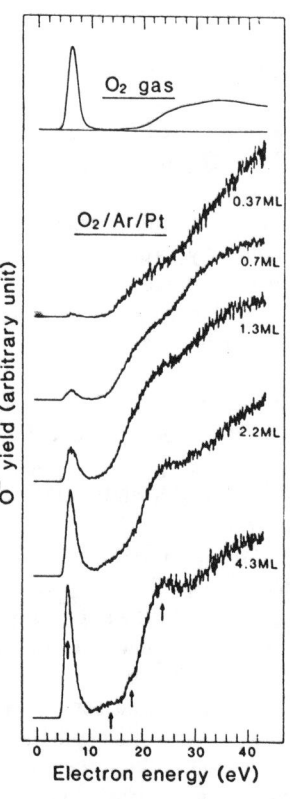

Fig. 3. Comparison of O^- yield curves produced by electron impact on O_2 gas with those from O_2/Ar/Pt samples with a constant (0.1 ML) O_2 coverage and variable (0.37-4.3 ML) Ar thicknesses. The $^2\Pi_u$ state of O_2^- is indicated by the arrow at 6 eV; the others indicate features involving multiple electron scattering.

transient O_2^- anion from the metal is increased by spacer Ar layers up to 32 ML.[17] The DD contribution to the O^- yield is the smooth background signal above ~16 eV, over which multiple electron scattering features (indicated by arrows) are superimposed. These features arise from electrons having lost energy to electronic excitation of O_2 or Ar prior to formation of O_2^- temporary states. Notice that the DD signal is much less affected by the proximity of the metal surface.

The effect of the image charge can be explained by considering the perturbation it induces on potential energy curves of the intermediate states O_2^- and O_2^* involved in the DA and DD processes, respectively. The solid curves in Fig. 4 correspond to the gas-phase O_2^- and O_2^* potential energy curves; the curves under the influence of an image charge are represented by broken lines. In these curves, species on the surface are identified by the symbol /M where M designates the metal surface (e.g., if O^- remains on the surface it is represented by O^-/M). The image charge due to the Pt lowers the energies of the ionic species O^+/M, O^-/M, and O_2^-/M, while the energies of the neutrals O/M, O_2^*/M, and O_2/M remain essentially the same. The magnitude of this lowering depends on the distance between the ion and the metal surface, and hence on the thickness of the rare-gas spacer layer. The image charge also splits the $O + O^-$ dissociation limit (i.e., O_2^-/M has two major dissociation limits O^- + O/M and O^-/M + O depending on

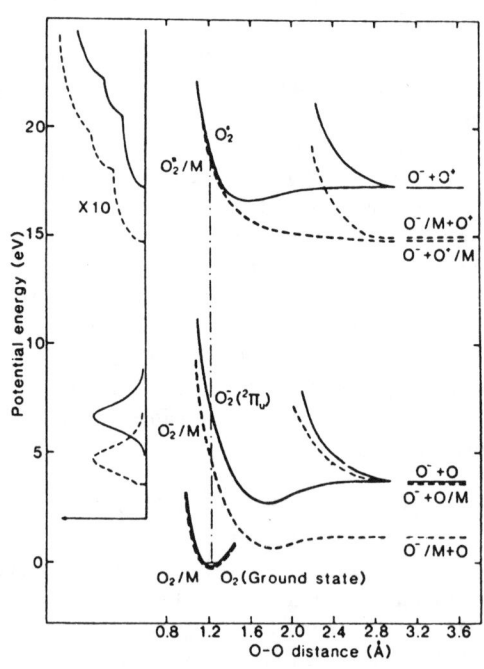

Fig. 4. Potential-energy curve of O_2 gas (solid curves) and the corresponding potential-energy curves under the influence of an image charge (dashed curves). Spectra on the left side show schematically the relative number of vibrational and/or electronic states which have sufficient energy to yield O^- ions, for O_2 gas (solid lines) and for O_2 on surface (dashed lines).

whether or not the O^- ion escapes in vacuum). *Thus, the branching ratio between the trapped O^-/M and desorbed (i.e., free O^-) states can vary considerably depending on the energy difference between the two dissociation limits.* The free O^- yield seen in Fig. 3 is expected to increase as the energy of the O_2^- $^2\Pi_u$ state and the energy of the O^-/M+O limit rise with Ar thickness. In contrast, the separation between the O^-/M + O^+ and O^- + O^+/M DD dissociation limits is negligibly small since the polarization energies of O^+/M and O^-/M are similar. In this case, the competition between the O^+ and O^- channels is expected to remain essentially the same with Ar thickness.

Another effect of the image charge involves the kinetic energy of the escaping O^- ions. In comparison with the corresponding O^- kinetic energy from O_2 gas, the kinetic energy of the O^- ions projected in vacuum via O_2^-/M is decreased while that via O_2^*/M is increased. As can be seen in Fig. 4, the dissociation limit being lower on the surface than in vacuum for DD, the kinetic energy of either O^+ or O^- escaping from the surface is larger. However, the O^- + O/M limit *is essentially the same* for surface or vacuum DA. Thus, the O_2^-/M state produces ESD of O^- ions having a greater chance to be recaptured by the metal. This decreases the DA O^- yield, while the O_2^*/M produces faster O^- ions having a greater chance for escape, thus increasing the relative DD O^- yield. The fourth effect of the image charge involves the quenching rate of the intermediate state. Near a metal surface, some of the O_2^- or O_2^* states may be neutralized or deexcited before dissociation. Since the O_2^- state can be destroyed by either charge or energy transfer to the metal, intermediate-state quenching has a stronger effect on the DA than on the DD process.

Finally, the lowering of the intermediate anion energy by the polarization force induced in the metal, as well as in the RGS layers, changes the probability for decay of the O_2^- state with respect to the isolated anion. For O_2 condensed on a RGS, it has been shown theoretically and experimentally[18] that a reduction in the autoionization of the intermediate $^2\Pi_u$ O_2^- state increase its lifetime considerably, causing the DA cross section to increase by at least an order of magnitude in comparison with the gas-phase value.

V. TRANSIENT ANIONS AS CHARGE AND ENERGY TRANSFER STATES

When an anion is formed at a surface or inside a solid composed of similar atoms or molecules, there exists a finite probability for the additional electron to hop from one site to another. If electron trapping occurs via a core-excited resonance, then simultaneous energy and charge transfer is expected (i.e., the additional electron could move with the electronic excitation energy from site to site). Such a simultaneous energy and charge transfer from one site to another is difficult to demonstrate for interaction between similar species, but recently it has been shown[5,19,20] that such an "electron-exciton complex" formed in a RGS multilayer film can migrate to the film's surface and under certain conditions transfer its energy and charge to a molecular adsorbate. To illustrate this phenomenon, the D^- ESD yield functions[5] are shown in Fig. 5 for (a) a pure C_2D_6 8-ML film, (b) 0.2 ML C_2D_6 deposited on a 80 ML Kr spacer film and, (c) 0.05 ML C_2D_6 deposited on a 50 ML Xe spacer film. The D^- yield distribution measured from pure multilayer C_2D_6 shown in (a) has a single broad peak at 9.9 eV which arises from the DA process e + C_2D_6 → $C_2D_6^-$ → C_2D_5 + D^-. When deposited on Kr and Xe substrates, strong and narrow enhancements are observed at energies slightly below those needed to create an exciton in the rare gas (Fig. 5(b) and (c)). The direct

ESD yield from the C_2D_6 is significantly reduced at incident electron energies well above and below that of the enhancement. For fixed quantities of C_2D_6, the absolute enhancement intensity increases as the rare-gas substrate is made thicker; the intensity of the sharp peak relative to the broad D^- distribution also increases as the C_2D_6 adsorbate coverage is decreased for a fixed substrate thickness. These trends suggest that the initial process occurs in the bulk rather than at the surface, and that it has significant mobility. Enhanced anion yields have also been obtained with an Ar substrate and with other molecular adsorbates.[5,19] In each of these cases, the enhanced ESD yields are related to electronic excitations of the rare-gas; however, the energies of the observed enhancements are *below* the energy of the lowest optically accessible exciton states in the bulk of the respective RGS films. The sharp enhancements, whose energies were found to be independent of the incidence angle of the impinging electrons, have been explained[19] by invoking the formation of a core-excited resonance below the energy of the lowest exciton in the rare gas substrates (i.e., a transient anion formed by trapping an electron in the positive electron affinity well of the lowest excited state of the RGS. Such states are well known for isolated rare-gas atoms (e.g., $Kr^-[4p^55s^2]$ $^2P_{3/2}$ at 9.52 eV and $Xe^-[5p^56s^2]$ $^2P_{3/2}$ at 7.90 eV),[21] where they are observed ~0.5 eV below the lowest excited neutral state of the atoms. Briefly stated, the core excited anion state or "electron-exciton complex" moves to the surface of the rare gas, where it exchanges its charge and energy by forming a core-excited anion of the molecular adsorbate (e.g., $C_2D_6^-$) which afterwards dissociates, thus producing the sharp enhancement in the ESD yield function (e.g. in the D^- signal of Fig. 5(b) and (c)) precisely at the energy of anion formation.

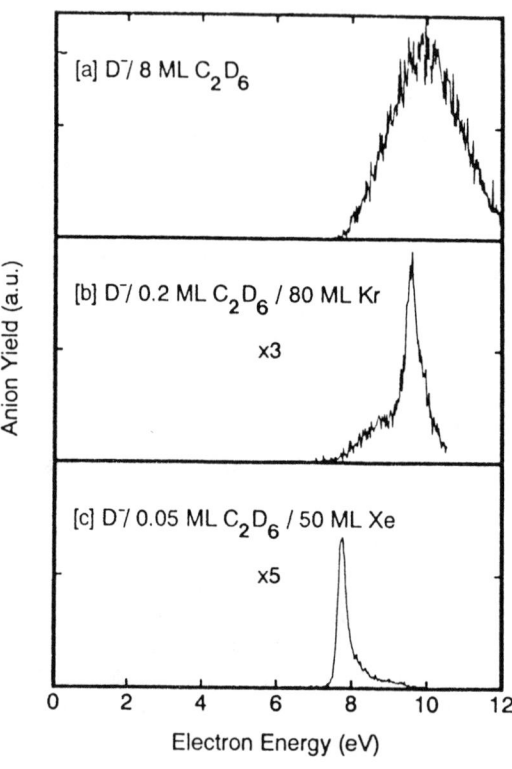

Fig. 5. D^- desorption yield for (a) 8 ML of C_2D_6 condensed on a Pt substrate and (b) and (c) submonolayer C_2D_6 supported on Kr and Xe substrates of the indicated thicknesses. Above and below the energy of the sharp peak due to a core-excited resonance in the substrate, the anion desorption yield is drastically reduced in comparison to (a).

ACKNOWLEDGMENTS

The author would like to thank Marc Michaud and Luc Parenteau for useful suggestions and corrections. This work was supported by the Medical Research Council of Canada and the Canadian Center of Excellence in Molecular and Interfacial Dynamics.

REFERENCES

1. For a recent review see M. Allan, J. Electron. Spectrosc. Relat. Phenom. 48, 219 (1989).
2. For a review of electron interaction with surface molecules see L. Sanche, J. Phys. B23, 1597 (1990); and R.E. Palmer and P.J. Rous, Rev. Mod. Phys. 64, 383 (1992).
3. For a review of electron interaction within dielectric solids see L. Sanche, Chapt 1 "Excess electrons in dielectric media" C. Ferradini and J.-P. Jay-Gerin eds, (CRC Press, Boca Raton, 1991).
4. K. Jacobi and M. Bertolo, Phys. Rev. B42, 3733 (1990).
5. P. Rowntree, H. Sambe, L. Parenteau, and L. Sanche, Phys. Rev. B47, 4537 (1993).
6. M. Michaud, L. Sanche, C. Gaubert, and R. Baudoing, Surf. Sci. 205, 447 (1988).
7. L. Sanche and M. Michaud, Phys. Rev. B30, 6078 (1984).
8. R. Azria, L. Parenteau, and L. Sanche, J. Chem. Phys. 87, 2292 (1987).
9. M. Michaud and L. Sanche, J. Electr. Spectr. Rel. Phenom. 51, 237 (1990).
10. S.F. Wong, J.A. Michejda, and A. Stamatovic, personal communication.
11. P.J. Rous, R.E. Palmer, and R.F. Willis, Phys. Rev. B39, 7552 (1989).
12. D. Teilly-Billy and J.-P. Gauyack, Nucl. Instrum. Methods B58, 393 (1991).
13. P.J. Rous, Surf. Sci. 260, 361 (1991).
14. When small molecules are sandwiched between rare gas layers some diffusion may occur; L. Sanche, A.D. Bass, L. Parenteau, and Z. Gortel (submitted).
15. H. Sambe, D.E. Ramaker, L. Parenteau, and L. Sanche, Phys. Rev. Lett. 59, 236 (1987).
16. D. Rapp and D.D. Briglia, J. Chem. Phys. 43, 1480 (1965).
17. H. Sambe, D.E. Ramaker, L. Parenteau, and L. Sanche, Phys. Rev. Lett. 59, 505 (1987).
18. H. Sambe, D.E. Ramaker, M. Deschênes, A.D. Bass, and L. Sanche, Phys. Rev. Lett. 64, 523 (1990).
19. P. Rowntree, L. Parenteau, and L. Sanche, Chem. Phys. Lett. 183, 379 (1991).
20. M. Michaud and L. Sanche, Phys. Rev. B47, 4131 (1993).
21. L. Sanche and G.J. Schulz, Phys. Rev. A5, 1672 (1972).

THRESHOLD AND RESONANT FEATURES IN THE VIBRATIONAL EXCITATION FUNCTIONS OF HYDROGEN HALIDES

Slobodan Cvejanović*
Institut za fiziku, P.O.Box 57, 11001 Beograd, Yugoslavia

ABSTRACT

The present situation in the understanding of the scattering mechanisms responsible for the unique features of the vibrational excitation cross sections in hydrogen halides is reviewed. A particular attention is applied to the critical evaluation of the available experimental data concerning the location and shape of the threshold peaks. New structure found in the region of higher vibrational thresholds in all hydrogen halides is discussed in terms of resonances and channel interference.

INTRODUCTION

The discovery of large threshold structure in electron–impact vibrational excitation cross sections of hydrogen halides came in 1975 from two different experiments on HCl - a dissociative attachment (DA) measurement[1] showing large, step–like decreases[2] in the Cl$^-$ yield at the openings of higher vibrational excitation (VE) channels, and a differential VE function measurement[3] which indeed showed strong near – threshold peaks. The angle integrated cross sections in the peaks were in the range of 10^{-15} cm^2, much larger than the Born approximation results, indicating that an enhancement mechanism such as resonant scattering should be considered. Further measurements indicated that the phenomenon of large threshold peaks in the VE cross section is a feature common to many molecules, but the hydrogen halides, featuring a permanent dipole moment and appreciable polarisability, remained the test ground on which new theories aiming at describing the experimentally observed features were tried.

Although all main features of the measured hydrogen halide cross sections have been reproduced by a number of different theories, the agreement between different calculations and between theory and experiment is still only qualitative. In this work I will concentrate on the remaining discrepancies, with an emphasis on the critical evaluation of available experimental data, which include recent rotationally resolved measurements, increased detection sensitivity for near–zero energy electrons and better control over the analyzer transmission function in the near threshold region.

ORIGINAL MEASUREMENTS OF ROHR AND LINDER

The differential VEF measurements obtained by Rohr and Linder[4] and Rohr[5] for e–HF(HCl, HBr) scattering were a remarkable achievement, the first of their kind. The authors had to solve the problems of electromagnetic shielding in the electrostatically focused electron spectrometer of high resolution to enable work with very low energy electrons, and to achieve stability over many hours of data accumulation time in the presence of highly corrosive gases such

---

* Present address: Physics Department, University of Manchester, Manchester M13 9PL, UK

as hydrogen halides. Additionally, a precise control was needed of the high-resolution incident beam down to 0.2 eV and of the analyzer transmission for scattered electron energies from a few meV to several eV.

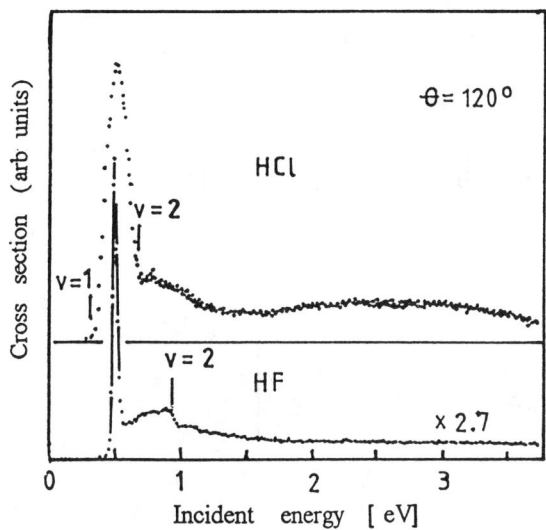

Fig. 1. Vibrational excitation functions for the transition v=0→1 in HF and HCl. Adapted from Rohr and Linder.[4]

A sample of their results is shown in Fig. 1. The main characteristics of all measured cross sections were:

1. *Peaks in the near-threshold region.* For the v=1 states the peak widths and displacements from thresholds were different in every gas. The smallest values have been found in HF: $\leq$30 meV and 19 meV, and the largest in HCl: 180 meV and 150 meV for the peak widths and displacements from threshold respectively. The values of the angle–integrated cross sections increase with the polarisability of the molecule, reaching $4 \cdot 10^{-15}$ cm^2 for HBr, for which the dipole moment is the smallest (0.82 D).

2. *Higher energy peaks* of larger widths laying above respective DA thresholds were found in HCl (at $\approx$2.5 eV, see Fig. 1) and, much more intense, in HBr.

3. *Near isotropic angular distributions* were found in the region of both peaks, in contrast to the elastic cross section which is strongly forward–peaked.

4. *Large threshold cusps* have been found at the opening of the next higher vibrational threshold in HF, but a much smaller one in the v=1 function of HCl (see Fig. 1) and none in HBr.

5. *A similarity in the main features of the cross sections* for the excitation of several vibrational channels was found, most notably in HBr[5].

THEORY

As extensive reviews of related theoretical work prior to 1991 have been written by Domcke[6], Herzenberg[7] and Morisson[8], only an overview of the basic concepts and a discussion of the main results will be given here.

A convenient division of the theoretical approaches into resonant versus non–resonant can be made. The non–resonant theories were stimulated by the apparent paradox of cross section enhancement while the measured angular distributions strongly indicated Σ symmetry for the scattering complex. Predominantly s-wave scattering was inconsistent with the usual shape resonance mechanism, due to the lack of a centrifugal barrier to trap the scattered electrons. The effective range theory of Teillet–Billy and Gauyacq[9] has shown that the HCl⁻ curve, which could be calculated at higher internuclear distances, merges with the ground state curve at a specific (model dependent) $R>R_e$ value, as shown in Fig. 2(a). For smaller internuclear distances the electron is essentially free, suggesting description in terms of a virtual state. The virtual state mechanism has already been shown[10] to yield large and relatively narrow threshold peaks.

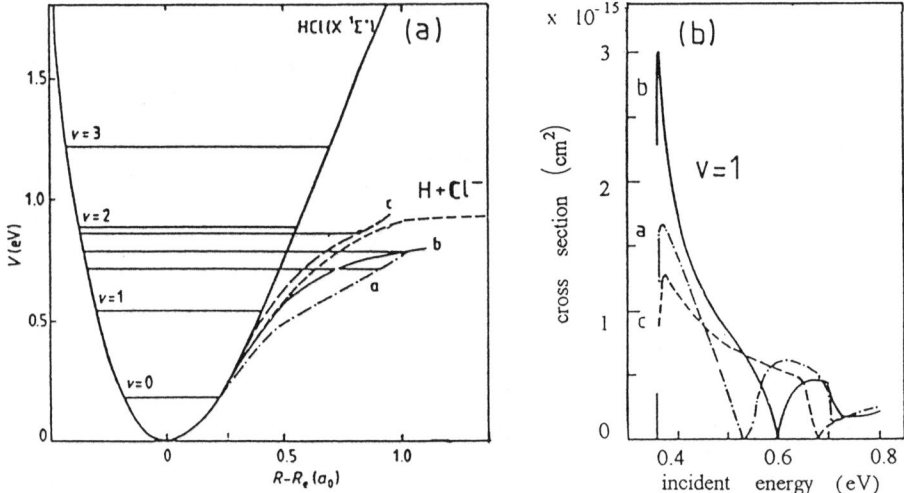

Fig. 2. Effective range theory; (a) Potential energy curves; (b) Integrated cross sections for the excitation of the v=1 state, for the three sets of HCl curve parameters. Adapted from Teillet–Billy and Gauyacq.[9]

Following the diabatic capture concept introduced earlier[11] to describe the DA process without a need to define the intermediate resonant state, strong non-adiabatic interaction between the slow scattered electron and nuclear motion was found responsible for extensive vibrational excitation as well. Parameter-dependent VE functions are shown in Fig. 2(b). A neat explanation was offered for threshold peaks in higher vibrational states - they were threshold enhancements due to nuclear- excited Feshbach (NEF) resonances formed below respective vibrational thresholds, as the binding of the scattered electron increases with internuclear separation. Such a resonance causes the sharp minimum in the v=1 function of Fig. 2(b).

Although this latest development somewhat bridges the gap between resonant and non-resonant approaches, an important difference is that the second broad peak measured in HCl and HBr could not be described by non-adiabatic effects only.

Subsequent ab-initio calculations for HF[12] and for HCl[13] have essentially confirmed the NEF resonance concept whilst stressing the importance of proper representation of the polarisation interaction for the threshold region of the VE cross sections.

In the class of resonant calculations one or more short lived negative ion states of $^2\Sigma^+$ symmetry have been modeled inside the Franck–Condon region of the ground state. Good overall agreement with experiment has been ensured by a suitable choice of resonant state parameters and by using the appropriate non-local treatment of nuclear dynamics in the resonance. The most sophisticated calculation of this kind has been conducted by Domcke and Mündel[14]. Their non-local and energy dependent complex potential resonant treatment, with model parameters which have been adjusted to reproduce eigenphase sums of the ab-initio close coupling calculation of Padial and Norcross[15], is illustrated on Fig. 3. The discrete $^2\Sigma^+$ state curve of HCl$^-$ undergoes an unexpected change due to interaction with the scattering continuum. This is represented by backbending of the adiabatic potential energy curve before the crossing with the neutral state, as shown in Fig. 3(a). The calculated width and level shift for the resonant state are strongly energy dependent. The resulting VE cross section for the v=1 transition in HCl is shown in Fig. 3(b). Apart from the inprecise description of threshold peak, the curve fits the data[4] quite well over the entire energy range including the second broad resonant peak. However, calculated VE functions appear to be very sensitive to the model parameters, as we shall see later. Results similar to that in Fig. 3(b) have been obtained for HCl and HF by other model calculations[16] and by ab-initio R-matrix[17] and close coupling[12] calculations. Good agreement between all theories and experiment was also found for the DA cross sections behavior. However, a serious discrepancy with experiment concerning the precise location of the threshold peaks emerged as all theories agreed that the peaks should be within meV's from thresholds. This raised questions - was there an error in the determination of large peak displacement from thresholds in HBr and especially HCl? If the peaks were shifted from thresholds, would that indicate a completely new interaction mechanism involving higher partial waves, at the same time questioning the isotropic angular distributions found experimentally?

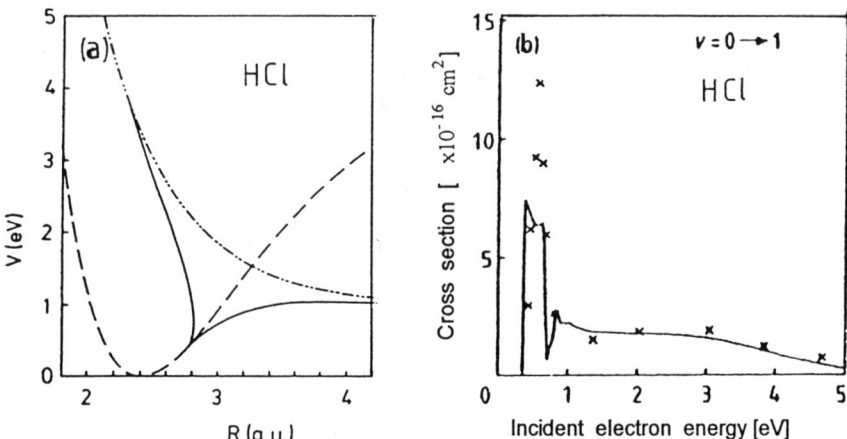

Fig. 3. Non-local resonance model calculation of Domcke and Mündel[14] (a), Potential energy curves: - - - ground state; - ·· - ·· - discrete HCl$^-$ state; ——— adiabatic HCl$^-$ curve; (b) v=1 excitation function: ——— calculation; × experiment[4]

The situation regarding the calculations of the resonant state parameters is rather complicated. Even when the low-lying $^2\Sigma^+$ state was found[18,19] its parameters were hardly distinguishable from the neutral state, suggesting a virtual state description to be more appropriate. However, in several cases[19,20,21] the repulsive part of the molecular negative ion potential curve in the Franck–Condon region was found for HCl, but not for HF. More recently in an ab–initio PO calculation[22] it was shown that the repulsive part of the HCl⁻ curve could be constructed inside the neutral state potential well, using a charge–increase stabilisation method, but the details of the level–shifted curve were found to be very different from the model–calculations[14]. Furthermore, in an analysis of the resonance effects in HCl McCartney et al[23] have found that the bound HCl⁻ state at large R and the 3 eV resonance at R_e are two unrelated poles of the S–matrix.

The role of the permanent dipole potential in the formation of threshold peaks and other characteristic features of the VE cross sections is another controversial issue. Within the effective range theory[9] it appeared that the permanent dipole moment was not needed to reproduce threshold peaks. In several calculations[9,17] it was demonstrated that the proper inclusion of polarisation forces was necessary to bring the calculations and measurement into semi–quantitative agreement, leaving the permanent dipole term almost redundant. However, non–local resonant model calculations[14] stress the importance of the dipole field for both the threshold peaks and the resonant oscillations below the DA threshold in HCl. In recent ro–vibrational ab–initio calculations[12,24] for HF, which has a dipole moment above the critical value necessary to bind an electron, the dipole interaction was found responsible for the sharp onset and cusp–like appearance of the threshold peaks.

To conclude this section let us sumarise the main questions raised by theory which called for experimental verification.

- The precise location of the threshold peaks - the measure of the angular anisotropy.
- The role of the permanent dipole moment - is there a feature of the cross sections related to it?
- The resonance versus non–adiabatic interaction picture - is there a clear signature of a resonant state on VE cross sections? Can NEF resonance levels below VE thresholds be detected experimentally?

ROTATIONALLY RESOLVED CROSS SECTIONS

A new extensive study of the excitation of HCl and HF has been conducted in Kaiserslautern, using an electron spectrometer[25] incorporating two double cylindrical analysers, capable of achieving a very high resolution of 20 meV in the energy loss mode.

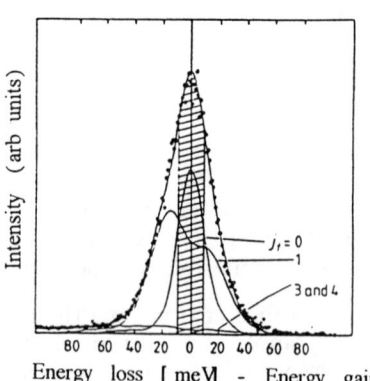

Fig. 4. Rotational analysis of the energy loss peak for the v=1 transition at $\theta=15°$ in HF, from Knoth et al[25].

Vibrational energy loss measurements[25,26] in the near–threshold region have shown substantial peak broadening due to rotationally inelastic transitions, as shown in Fig. 4. Individual rovibrational components up to $\Delta j=\pm 4$ have been extracted by a deconvolution procedure based on the adiabatic nuclei rotational approximation. Rather surprisingly it was found that the most intense component was $\Delta j=1$, which is associated with p–wave scattering. The angular distribution of the total, rotationally integrated cross section was found to possess a definite anisotropy at all energies, and especially in the region of threshold peaks.

Ab initio calculations,[12,17,24] which followed, were in excellent agreement with measured rotationally integrated distributions at higher incident energies, but failed to reveal any large anisotropy in the near–threshold region. Furthermore, the theory could not confirm the dominance of $\Delta j=1$ channel - instead, rotationally elastic $\Delta j=0$ transitions were calculated to be the strongest. Clearly more work is needed to clarify this new discrepancy, including an independent experimental verification.

VE functions have been measured in the rotationally summed mode, where the signal at each incident energy represents the integral over the energy loss peak, and in the predominantly rotationally elastic mode, in which the signal is represented by the height of the energy loss peak. The general shapes of the two cross sections were similar, although the sharp structures became more diffuse in the rotationally summed cross sections[25]. Despite the high resolution claimed, no new features in the v=1 cross sections of HF and HCl were found. The HCl threshold peak displacement of $\approx$150 meV was confirmed, and an average width of 170 meV was found. In HF similar values were found, in striking disagreement with the original measurement[4] of a narrow peak at threshold. These findings have widened the gap between experiment and theory, which was not able to explain threshold peak shifts even after allowing for rotational effects and including several partial waves in ab–initio calculations.

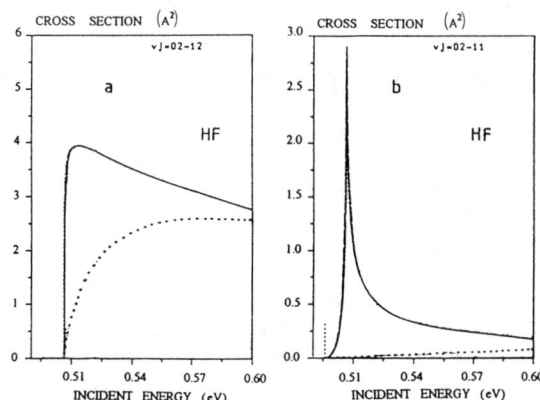

Fig. 5. The dominant partial cross sections for the excitation of the v=1 level of HF: (a) Rotationally elastic component; (b) $\Delta j=-1$ component. The dotted curves correspond to the omission of all long range forces from the calculation. Adapted from Thümmel et al.[24]

The remarkable results of the first calculation[24] of the ro–vibrational excitation functions in the threshold region of the v=1 state of HF are shown in Fig. 5. The rotationally elastic component, curve (a), is dominant over a large region from threshold, but the sharpness of the very narrow peak at the $\Delta j=0$ threshold is provided by a giant cusp structure in the $\Delta j=-1$ component, curve (b).

NUCLEAR EXCITED FESHBACH RESONANCES IN HF

Perhaps the most important result[27] of the Kaiserslautern group concerns the study of the cusp region in the vibrational excitation functions of HF (no cusp or resonant structure in HCl was found). Higher VE functions represented in Fig. 6 exhibit a clear dip below the opening of the next vibrational channel, whose separation from the cusp increases with v. This was interpreted as the first measurement of NEF resonance structures predicted by several theories[9,17].

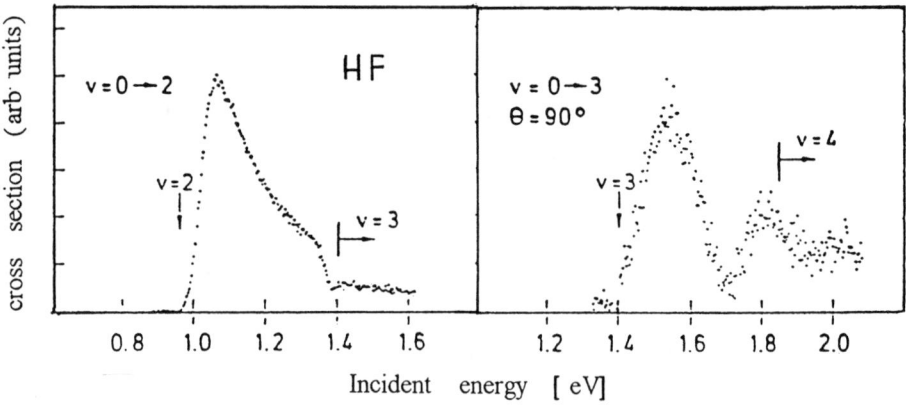

Fig. 6. Vibrational excitation of the v=2 and v=3 levels of HF, measured[27] at $\theta=90°$. The NEF resonance minima below the higher VE thresholds are clearly visible.

THRESHOLD–SENSITIVE STUDIES OF HCl AND HBr

At about the same time another experiment on HCl and HBr was conducted in Belgrade on an apparatus schematically presented in Fig. 7. This electron spectrometer has evolved from a threshold electron spectrometer[28] of the field–penetration type,[29] to which a versatile analyser input lens was added. The monochromator is based on a hemispherical selector of 2 cm mean radius and a combination of aperture zoom lenses. The scattering region is surrounded by a highly transparent mesh, which can rotate with the analyser assembly. Target gas is introduced through a long tube of 0.5 mm bore. The electrons scattered at an angle 30° < Θ <120° pass through the hole in the mesh adjacent to the extracting electrode, the first element of a three–cylinder lens coupling the target region and the analyser entrance. The scattered electron analyser is a double pass cylindrical mirror analyser (DCMA), of a moderate resolution (≈50 meV), followed by an electron–ion separator and a channeltron. The separator consists of a parallel plate deflector in which an alternating high frequency electric field of variable amplitude can be used to disperse the electrons while transmitting negative ions.

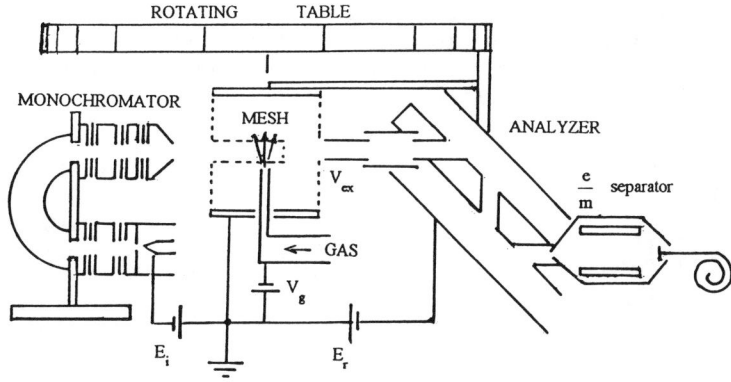

Fig. 7. Schematic diagram of the threshold–sensitive electron spectrometer.

The main functional difference with respect to the standard electron-spectrometer design is a possibility of controling electrical fields in the scattering region. In other words, a field due to contact potential difference and surface charging of the gas orifice can be estimated and compensated for. If the resulting surface charges are not compensated a loss of signal at small scattered electron energies is inevitable, accompanied by the loss of energy resolution in more severe cases.

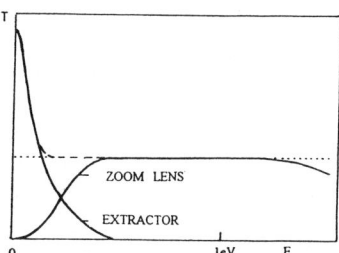

Fig. 8. Schematic representation of the analyser optics transmission function. Small extracting field compensates for the defficiency of the zoom lens for the near zero–energy electrons.

Another important characteristic of this experimental method is that the region of uniformity of the analyser transmission efficiency can be extended further towards lower energies (typically ≈ 0.1 eV) if slight field penetration into the shielded enclosure is allowed, as shown schematically on Fig. 8. This, however, results in the increased sensitivity for smaller energy electrons, peaking at threshold. In the threshold spectroscopy mode the extraction and transmission are optimized for zero energy electrons and only the incident energy is varied. This produces sharp peaks at thresholds for inelastic processes.

The threshold spectrum of HCl, shown in Fig. 9, was obtained[30] over the large incident energy range of 10 eV starting from zero. Compared to the prominent[31] Rydberg excitation peaks above 9 eV, the yield in the v=1 peak is approximately 600 times stronger. The threshold cross section drops by a factor of about 18 between successive vibrational states until v=3, but much more slowly between higher levels. This can be considered as a proof that the cross section enhancement found in the above–threshold region of v=1,2 cross sections[4] extends right down to less than 10 meV, which is the width of the analyser transmission function in the threshold mode. The threshold spectrum of HBr shows a similar behavior.

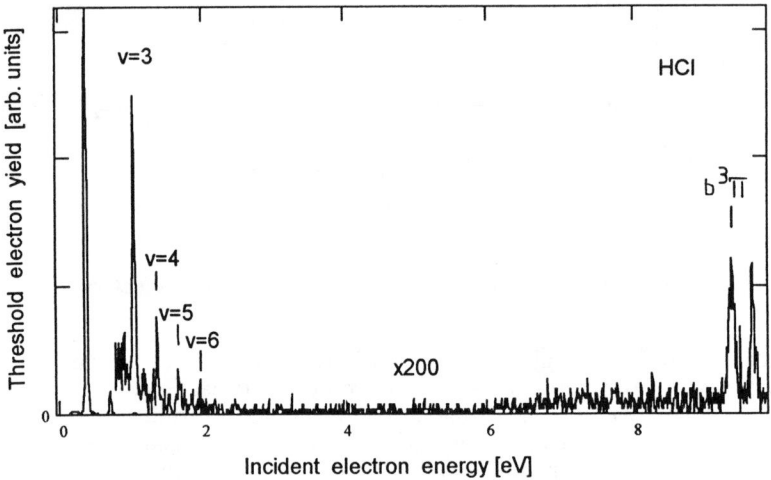

Fig.9. Threshold electron spectrum of HCl from 0 to 9.8 eV incident electron energy.

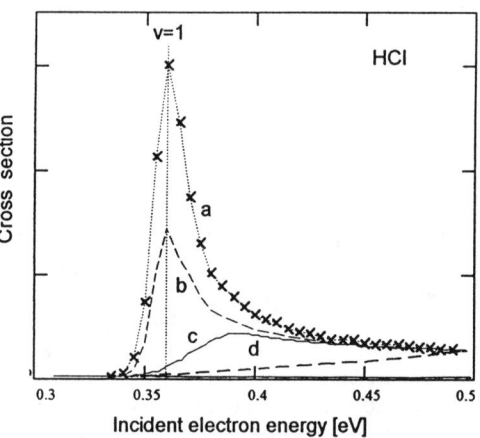

Fig. 10. Measurements of the v=1 excitation function in HCl in the threshold region, normalised at 0.5 eV. a,b – raw and transmission corrected $\theta=90°$ data of Cvejanović et al[30]; c – $\theta=0°$ and $\theta=90°$ data of Schafer and Allan; d – total cross section measurements[4,25].

A comparison of available experimental data for excitation of the v=1 level in HCl very close to threshold is shown in Fig. 10. Between our transmission–corrected[30] result (b), which features a peak at threshold, and the curve (d) which represents the data from Kaiserslautern,[4,25] there are the data of Schafer and Allan, curve (c). Their instrument[32] uses magnetic confinement to control the beam geometry and a double trochoidal monochromator to prepare a monochromatic beam and to analyse the electrons scattered in the forward and backward direction. The detection efficiency function of such an instrument is non–uniform, but of a regular shape whose analytic representation was determined and used to normalize the excitation functions. Curves (b) and (c) are in excellent agreement down to 30 meV above threshold, the energy beyond which the sensitivity of the magnetic instrument drops sharply[32].

A feature related to the v=1 threshold peak is a corresponding cusp in the elastic cross section, which was observed in HCl in the form of a large downward step of 20% height by Knoth et al.[25] The higher resolution experiment[30] confirms the size of the cusp, revealing the asymmetry just below threshold, in agreement with the shapes calculated for HCl[13] and HF[24]. Knoth et al[25] found that the cusp structure in HF is much weaker than in HCl indicating a smaller threshold peak – elastic cross section ratio for HF than for HCl.

OSCILLATIONS BELOW DISSOCIATIVE ATTACHMENT THRESHOLD

New experimental data and the two representative theoretical curves for the excitation of the v=1 level in HCl are shown in Fig. 11. The overall shape of the cross section seems well established. It peaks within 10 meV from threshold, in excellent agreement with all the theories. Due to the proximity to threshold, the exact height of the differential cross section peak is not measurable.

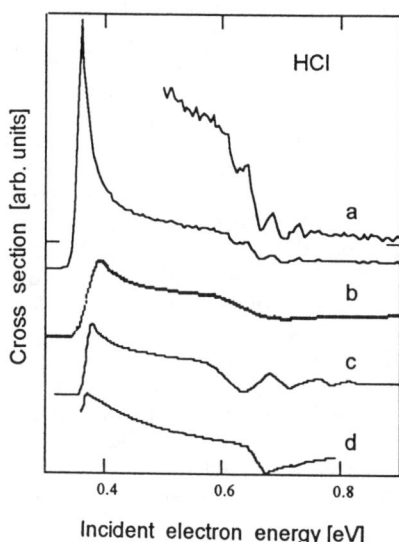
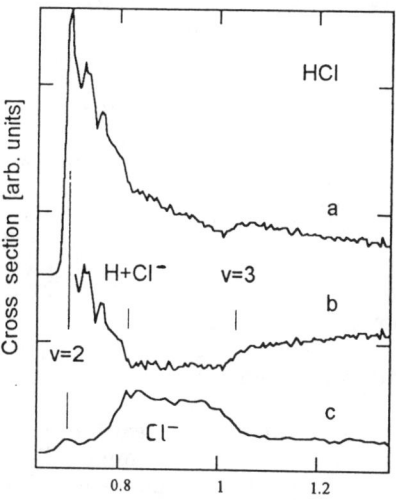

Fig. 11. Comparison between the v=1 excitation functions of HCl. Experiment: a – Cvejanović et al[30]; b – Schafer and Allan[32]. Theory: c – Domcke and Mündel[14]; d – Teillet–Billy and Gauyacq.[9]

Fig. 12. a,b – excitation functions of the v=2 state (see text); c – threshold spectrum obtained with greatly reduced detection efficiency for electrons compared to Cl$^-$ ions.

The region around the v=2 threshold energy is very interesting. Our experiment[30,33] shows well characterized oscillations, with average spacing of 40 meV, which have been confirmed by the lower resolution data of Schafer and Allan.[32] They are superimposed on the rather steep decline of the v=1 cross section towards the v=2 threshold, which could correspond to a NEF resonance unresolved from the v=2 threshold, in general agreement with the effective range theory, curve (d). The threshold cusp is completely obscured by these structures. Oscillations in this region have been actually generated by the resonant model calculations of Domcke and Mündel, curve (c), using somewhat different parameters[34] from their standard set[14]. At present the agreement between theory and experiment, recently extended[35] to all accessible decay channels of the resonant complex, is only qualitative. The failure to detect these threshold and resonant structures in HCl in the experiment by Knoth et al[25] is difficult to explain, unless surface charging effects, discussed earlier, were responsible.

STRUCTURES AT HIGHER THRESHOLDS IN HCl AND HBr

Very interesting details have been found in the v=2 cross section[30] in HCl, shown in Fig. 12. The cross section is again sharply peaked at threshold, but two more peaks at 30 meV intervals can be seen on a gradual decline. Their origin is probably the same as for the v=1 oscillations, which are of the same frequency but of opposite phase[33]. Further structure is seen at the DA threshold and at the v=3 threshold, in the form of sudden changes of the smoothly varying "background" scattering. This is particularly clear after a quadratic function is subtracted from the original spectra, curve (b), leaving a flat depression in the v=2 cross section which matches almost exactly the shape of the yield of slow Cl$^-$ ions from DA, curve (c). The apparent close link between the $\Delta v=2$ and DA cross section over an energy range greater than vibrational interval suggests that the two decay routes of the scattering complex have almost the same probability, and that the structure in one of them directly influences the other. Furthermore, at the threshold of the vibrational excitation process the scattering yield switches from the DA channel not only into the new channel, but into the first lower vibrational channel as well. There is evidence[14,30,32] that this is a general phenomenon affecting higher VE functions in HCl and all measured VE functions in HBr. This would make the observation of Wigner cusps at the opening of higher vibrational thresholds highly improbable, and the interpretation of minima below thresholds in terms of NEF resonances less obvious.

In conclusion, we have seen that the new, highly sophisticated experiments and calculations have helped in removing a long-standing discrepancy between theory and observation concerning the location of the threshold peaks. They have also provided a wealth of new information, which should stimulate further work to resolve the remaining differences and to gain more detailed and quantitative knowledge of the nuclear dynamic and configuration interactions in these unique collisional systems.

Acknowledgment: The author is grateful to his colleagues J. Jureta, D. Čubrić and D. Cvejanović for invaluable assistance in obtaining some of the experimental results shown here for the first time and in preparation of this manuscript, and to the University of Manchester for the hospitality and excellent working conditions provided during the past year.

REFERENCES

1. J. P. Ziesel, I. Nenner and G. J. Schulz, J. Chem. Phys. 63, 1943 (1975)
2. R. Abouaf and D. Teillet-Billy, J. Phys. B 10, 2261 (1977)
3. K. Rohr and F. Linder, J. Phys. B 8, L200 (1975)

4. K. Rohr and F. Linder, J. Phys. B 9, 2521 (1976)
5. K. Rohr and F. Linder, J. Phys. B 11, 1849 (1978)
6. W. Domcke, Phys. Rep. 208, 97 (1991)
7. A. Herzenberg, in Electron – Molecule Scattering and Photoionization, ed. P. G. Burke and J. B. West (New York: Plenum), p.187 (1988)
8. M. A. Morrison, Adv. At. Mol. Phys. 24, 51 (1988)
9. D. Teillet-Billy and J. P. Gauyacq, J. Phys. B 17, 4041 (1984)
10. L. Dube and A. Herzenberg, Phys. Rev. Lett., 38, 820 (1977)
11. O. H. Crawford and B. J. D. Koch, J. Chem. Phys. 60, 4512 (1974)
12. G. Snitcher, D. Norcross, A. Jain and S. Alston, Phys. Rev. A 42, 671 (1990)
13. L. A. Morgan, P. G. Burke and C. J. Gillan, J. Phys. B 23, 99 (1990)
14. W. Domcke and C. Mündel, J. Phys. B 18, 4491 (1985)
15. N. T. Padial and D. W. Norcross, unpublished, see ref. 14
16. I. I. Fabrikant, S. A. Kalin and A. K. Kazansky, J. Chem. Phys. 95, 4966 (1991)
17. L. A. Morgan and P. G. Burke, J. Phys. B 21, 2091 (1988)
18. H. S. Taylor, E. Goldstein and G. A. Segal, J. Phys. B 10, 2253 (1977)
19. M. Bettendorff, R. J. Buenker and S. D. Peyerimhoff, Mol. Phys. 50, 1363 (1983)
20. F. A. Gianturco and D. G. Thompson, J. Phys. B 10, L21 (1977)
21. M. Rajzman, F. Spiegelmann and J. P. Malrieu, J. Chem. Phys. 89, 433 (1988)
22. V. Pless, B. M. Nestmann, V. Krumbach and S. D. Peyerimhoff, J. Phys. B 25, 2089 (1992)
23. M. McCartney, P. G. Burke, L. A. Morgan and C. J. Gillan, J. Phys. B 23, L415 (1990)
24. H. T. Thümmel, R. K. Nesbet and S. D. Peyerimhoff, J. Phys. B 26, 1233 (1993)
25. G. Knoth, M. Rädle, M. Gote, H. Ehrhardt and K. Jung, J. Phys. B 22, 299 (1989)
26. G. Knoth, M. Gote, M. Rädle, F. Leber, K. Jung and H. Ehrhardt, J. Phys. B 22, 2797 (1989)
27. G. Knoth, M. Gote, M. Rädle, K. Jung and H. Ehrhardt, Phys. Rev. Lett. 62, 1735 (1989)
28. S. Čvejanović, J. Jureta and D. Čvejanović, J. Phys. B 18, 2541 (1985)
29. S. Cvejanović and F. H. Read, J. Phys. B 7, 1180 (1974)
30. S. Čvejanović, J. Jureta, D. Čubrić and D. Cvejanović, to be published
31. J. Jureta, S. Cvejanović, D. Cvejanović, M. Kurepa and D. Čubrić, J. Phys. B 22, 2623 (1989)
32. O. Schafer and M. Allan, J. Phys. B 24, 3069 (1991)
33. S. Cvejanović and J. Jureta, 3rd Eur. Conf. on At. Mol. Phys. (Bordeaux 1989), Book of Abstr., p. 638
34. W. Domcke, in Aspects of Electron–Molecule Scattering and Photoionization, AIP Conf. Proc., 204, ed. A. Herzenberg (AIP, New York 1989) p.169
35. S. Cvejanović, J. Jureta, D. Čubrić and D. Cvejanović, 18th Int. Conf. Phys. Electr. At. Coll. (Aarhus 1993), Abstr. Contr. P., p.279

ELECTRON-ION COLLISIONS

ELECTRON-IMPACT EXCITATION OF IONS: EXPERIMENTAL STATUS AND PERSPECTIVES

Ronald A. Phaneuf
University of Nevada, Reno NV 89557-0058, USA

ABSTRACT

The development of new experimental techniques based on electron energy-loss spectroscopy and of powerful new ion sources and devices has facilitated more detailed quantitative studies of electron-impact excitation of positive ions than have heretofore been possible. Data for angular differential as well as total cross sections are providing more stringent tests of the theoretical approximations used for such collisions, and some new insights into the excitation process. Extremely precise measurements of electron-impact ionization cross sections have also revealed rich structures due to inner-shell excitation and resonant excitation processes. A overview is presented of the current status and outlook of experimental studies of electron-impact excitation of positive ions.

INTRODUCTION

More than 25 years have elapsed since Dance, Harrison and Smith[1] first studied electron-impact excitation of positive ions using the colliding-beams technique. Their pioneering experiment on the simplest ion, He^+, established a high standard for research in the field, and set the stage for absolute total cross-section measurements on primarily resonance transitions of a number of positive ions (Li^+, Be^+, C^+, C^{2+}, C^{3+}, N^{4+}, Mg^+, Al^+, Ca^+, Zn^+, Ga^+, Cd^+, Ba^+, Hg^+)[2]. These measurements were accomplished by colliding beams of electrons and ions and detecting a known fraction of the photons radiated as the excited ionic level decays. The absolute cross-section measurement for 4s-4p excitation of Ca^+ by Taylor and Dunn[3] must rank as one of the most accurate collision experiments ever performed. However, the excitation signal count rate in such experiments is severely limited by small ion beam densities, solid angles and efficiencies for photon detection. Photon dispersion and absolute detector calibration become prohibitively difficult for multiply charged ions, where most of the resonance transitions of interest lie in the vacuum ultraviolet or soft x-ray regions.

As important applications in fusion energy, x-ray laser, and other high-temperature plasma research intensified the interest in excitation of multiply charged ions, the need for innovation gained urgency. Only limited experimental data were available to benchmark the theoretical methods[4] that were being applied to modelling and diagnostics of astrophysical and fusion plasmas. In the mid-1980's, several laboratories initiated development of new experimental techniques to study outer-shell electron excitation via electron energy-loss spectroscopy.

Experimental studies of electron-impact ionization of ions have also provided important cross-section data for the excitation of inner-shell electrons. The importance of this mechanism was first demonstrated 25 years ago by Peart and Dolder in crossed-beams measurements of electron-impact ionization of Ba^+. For many ions, and particularly multiply charged ions, such excitations produce distinct signatures in the total ionization cross section via the autoionization process. To deduce accurate excitation cross sections from such features in the ionization cross section, knowledge of the excitation energies and branching ratios for autoionization versus radiative decay of the core-excited levels is required. For ions in the lower ionization stages, these branching ratios strongly favor autoionization, and are near unity. Since ionized ion detection efficiencies in colliding-beams experiments also approach unity, this technique has a high sensitivity. This subject has been comprehensively reviewed by Dunn[5], and more recently by Müller[6].

OUTER-SHELL EXCITATION

<u>Differential Cross Sections</u>

The first application of electron energy-loss spectroscopy to the study of electron-impact excitation was by Chutjian[7] at Jet Propulsion Laboratory (JPL). Using a crossed-beams geometry, he was successful in determining the shape of the differential cross section for 3s-3p excitation of Mg^+ and 4s-4p excitation of Zn^+ at electron scattering angles between 4 and 16 degrees, and electron energies approximately ten times the excitation threshold. These measurements provided the first experimental tests of theory for the relative differential cross section over a limited range of scattering angles. The method, shown in Fig. 1, was subsequently extended[8] to dipole-forbidden as well as resonance transitions in Mg^+, Zn^+ and Cd^+.

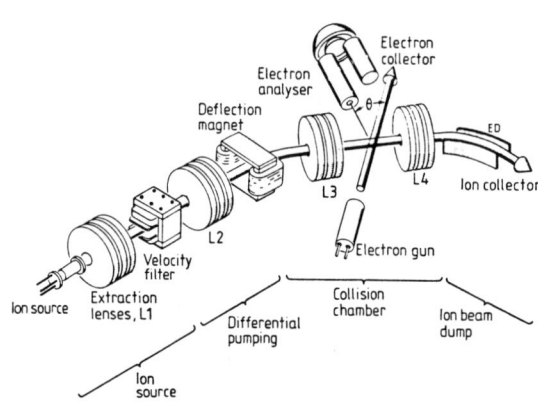

Figure 1. Schematic diagram of the apparatus used by Chutjian et al for differential cross-section measurements, from Ref. 7.

A similar experimental development was undertaken by Huber and collaborators at the Centre d'Etudes Nucléaires de Grenoble. Their focus was excitation of multiply charged ions produced by the electron-cyclotron-resonance (ECR) ion source at the LAGRIPPA facility. They were successful in measuring the differential cross section for 3s-3p excitation of Ar^{7+} at an electron energy of 100 eV and scattering angles between 13 and 29 degrees[9]. Their

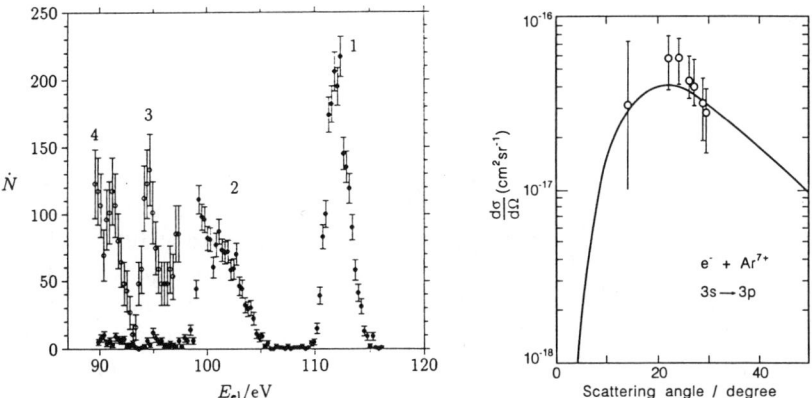

Figure 2. Measured electron energy distribution at a scattering angle of 24° and differential cross section at an electron energy of 100 eV, from Huber et al (Ref. 9). The solid curve is a distorted-wave theoretical calculation.

scattered electron energy spectrum measured at 24° and differential cross section at 100 eV are shown in Fig. 2. The relatively strong elastic scattering feature (1) is clearly resolved from the inelastic signal (3) in the electron energy spectrum. Peaks (2) and (4) are due to electron scattering from residual gas. The differential excitation cross section was placed on an absolute scale by normalizing the measured inelastic scattering signal to the elastic signal, and applying the Rutherford formula. These data at roughly 6 times the threshold energy agree well with semi-classical, Coulomb-Born and distorted-wave calculations of the differential cross section, and suggest the importance of large-angle scattering for multiply charged ions. This apparatus has been modified to extend the angular coverage to 15°- 60°, and new measurements have been made for N^{4+} and Ar^{6+}.

Total Cross Sections

The application of the electron energy-loss technique to absolute total cross-section measurements for multiply charged ions near the excitation threshold was first undertaken by the group of Dunn at the Joint Institute for Laboratory Astrophysics (JILA) in the mid-1980's. A similar development for singly charged ions was initiated by Chutjian and co-workers at JPL. Both approaches employ a merged-beams geometry, in which an ion beam accelerated to several to tens of keV is merged with a 10-100 eV electron beam in a uniform axial magnetic field. Trochoidal (**E x B**) analyzers merge and de-merge the electron beam from the axis of the ion beam, and inelastically scattered electrons are counted on a position-sensitive detector.

The merged-beams apparatus developed by the JILA group[10] is depicted schematically in Fig. 3. The entire apparatus is enclosed by a large solenoid which creates a highly uniform magnetic field of typically 0.05 T parallel to the axis of the ion beam, and facilitates complete collection of inelastically scattered electrons onto the position-sensitive detector. The intensity distributions and spatial overlap of the electron and ion beams are measured directly by a moveable beam probe, which can be inserted into and translated through the collision region. This probe consists of microchannel plate, phosphor screen, fiber-optic bundle and CCD camera. The apparatus developed at JPL[11] utilizes a series of apertures on four vanes that can be rotated sequentially into the beams to measure their spatial intensity distributions. In both experiments, two-beam modulation and gating schemes are used to separate signal counts due to beam-beam collisions from counts produced by interaction of either beam with residual gas. The latter dominate the detector count rates, even with operating pressures in the collision region of 10^{-10} torr.

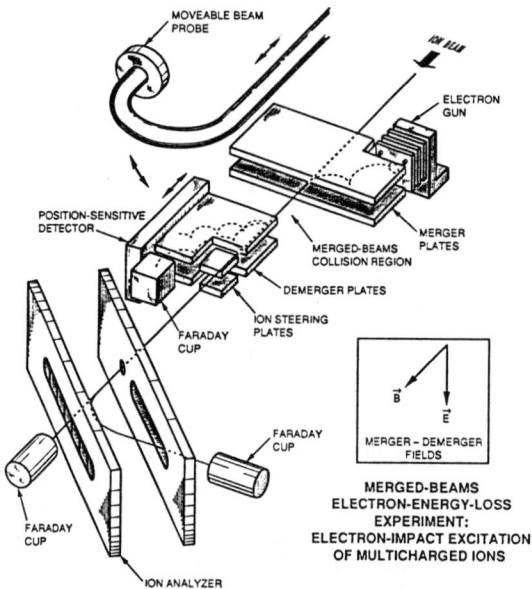

Figure 3. Schematic of merged-beams electron-energy-loss apparatus developed by Dunn and co-workers ar JILA for measurements of absolute total electron-impact excitation cross sections for multiply charged ions.

Careful computer modelling of electron trajectories in the demerger/analyzer is necessary to predict the locations of both elastically and inelastically scattered electrons on the position-sensitive detector. At the same center-of-mass energy, the strength of the electron-ion elastic scattering signal relative to that for inelastic scattering is larger for multiply charged ions by the square of the ionic charge q. The ability to resolve elastically and inelastically scattered electrons on the detector places a limitation on the range of electron energies above the excitation threshold that is accessible with this technique. However, this is not a serious limitation since the near-threshold energy region is most important for most plasma applications (e.g. in magnetic fusion), and also the most critical for tests of theoretical methods and approximations.

In Fig. 4, experimental energy-loss results of Smith et al[11] for $4s^2S - 4p^2P$ excitation of Zn^+ (solid points) are compared with crossed-beams photon-emission measurements of Rogers et al.[12] (open points), and with 31-state close-coupling calculations (curve) of Pindzola et al.[13] Within the experimental uncertainties, the energy-loss measurements are consistent with the photon-emission measurements and theory near threshold, but fall consistently lower at higher energies. Recent energy-loss measurements[14] for 3s-3p excitation of Mg^+ show a similar trend.

For excitation of multiply charged ions, experimental results based on the merged-beams electron-energy-loss method have been obtained by the JILA-ORNL group for Na-like Si^{3+} and Ar^{7+}, and most recently for 2s-2p excitation of Li-like O^{5+}. Ion beams for these measurements are obtained at the ECR Multicharged Ion Research Facility at Oak Ridge National Laboratory. The results of Guo et al[15] for 3s-3p excitation of Na-like Ar^{7+} are compared in Fig. 5 to the 7-state close-coupling calculations of Badnell et al,[16] which predict a large resonance near 21.5 eV. While the experimental measurements are in quite reasonable agreement with the theoretical prediction near the 17.45 eV threshold, attempts to extend them to higher energies to test for the presence of this resonance were unsuccessful. It was suspected that electron backscattering was responsible, and a systematic series of measurements was carried out at different ion energies. Since higher ion energies result in higher center-of-mass velocities, a larger fraction of electrons which have been scattered into the backward direction in the center-of-mass frame will have forward velocities in the lab frame, and therefore be detected by the experiment. These results, presented in Fig. 6, lend additional credence to this hypothesis, which was subsequently confirmed by semiclassical, distorted-wave and close-coupling calculations. Curves labelled "a", "b" and "c" in Fig. 6 refer to computer models of the angular acceptance of the apparatus under different assumptions about the behavior of the inelastic

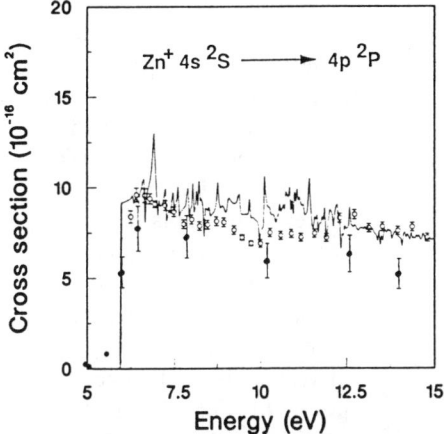

Figure 4. Comparison of merged-beams energy-loss measurements of Smith et al for Zn^+ with photon-emission measurements of Rogers et al and 31-state close-coupling theory of Pindzola et al. [from Ref. 13].

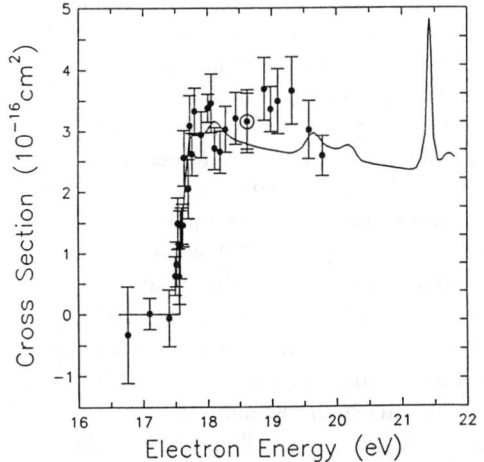

Figure 5. Merged-beams electron-energy-loss measurements of Guo et al for 3s-3p excitation of Ar^{7+}, compared to 7-state close-coupling calculations of Badnell et al. [from Ref. 15].

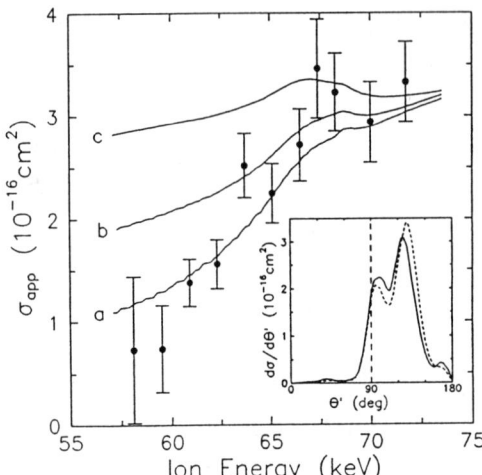

Figure 6. Apparent cross sections for 3s-3p excitation of Ar^{7+} measured as a function of laboratory ion energy. Inset shows calculated differential cross section. [from Ref. 15].

differential cross section. Curve "a" corresponds to either the close-coupling[17] or distorted-wave[18] calculations of the differential cross section shown in the inset. Curves "b" and "c" correspond to assumed isotropic and forward-peaked angular distributions, respectively. This loss of scattered-electron signal in the merged-beams experiment becomes more significant as the relative collision energy increases above the excitation threshold, since the scattered electron retains a larger fraction of its energy after the collision, which in turn results in a significant backscattering in the laboratory frame.

This somewhat surprising result was quite contrary to the "conventional wisdom" which held that cross sections for electron-impact excitation of ions were forward-peaked. Indeed, the importance of large-angle scattering at higher energies was noted[9] in the earlier differential cross-section measurements of Huber et al for Ar^{7+} at 100 eV (see Fig. 2). In hindsight, electron backscattering at threshold is obvious and to be expected! Merged-beams measurements for 2s-2p excitation of Li-like O^{5+} show a similar effect,[19] which likely characterizes near-threshold excitation of dipole-allowed transitions in all positive ions.

The development of the electron beam ion trap[20] (EBIT) at Lawrence Livermore National Laboratory made possible the study of inelastic collisions of electrons with very highly ionized ions (q>80). In EBIT, ions are successively ionized and trapped by the space charge of an intense, high-energy electron beam, and the processes of excitation, ionization and recombination are studied by high-resolution x-ray spectroscopy. To assure the ultra-high vacuum required for creation and survival of ions in such high ionization stages, the trap is made extremely compact and maintained at liquid helium temperature. Operationally, the electron beam can be alternately switched between two energies, the first to create and sustain the population of ions to be trapped, and the second to collisionally interact with these trapped ions. The second energy may be stepped to determine the energy dependence of the cross section for the chosen interaction. Figure 7 compares total cross sections measured by Chantrenne et. al.[21] in EBIT for electron-impact excitation of n=1 → n=2 transitions in He-like Ti^{20+} with theoretical results. The measurements in EBIT are placed on an absolute scale by normalizing the measured photon emission due to radiative recombination into the n=2 levels of Ti^{20+} to theoretical values of the cross section. Structure due to dielectronic resonances is evident in the measurements, which correlate quite well with the theoretical predictions. The latter are based on the relativistic distorted-wave method of Zhang et al[22] with resonance contributions calculated by Chen[23]. Cascade contributions from excitation to n=3 and n=4 levels are also evident in the measurements.

Emission cross sections for He-like Ar^{16+} have also been measured at zero degrees by measuring K x-rays with a Si[Li] detector as a function of electron-beam energy in the electron-beam ion source (EBIS) at Kansas State University.[24] The EBIS is similar in principle and design to the EBIT, but not as compact, and designed primarily to extract ion beams for acceleration. In addition to the n=1 to n=2 direct excitation, 1s 2l nl' dielectronic resonances were resolved, and their well-established cross sections were used to determine the absolute scale for the emission measurements.

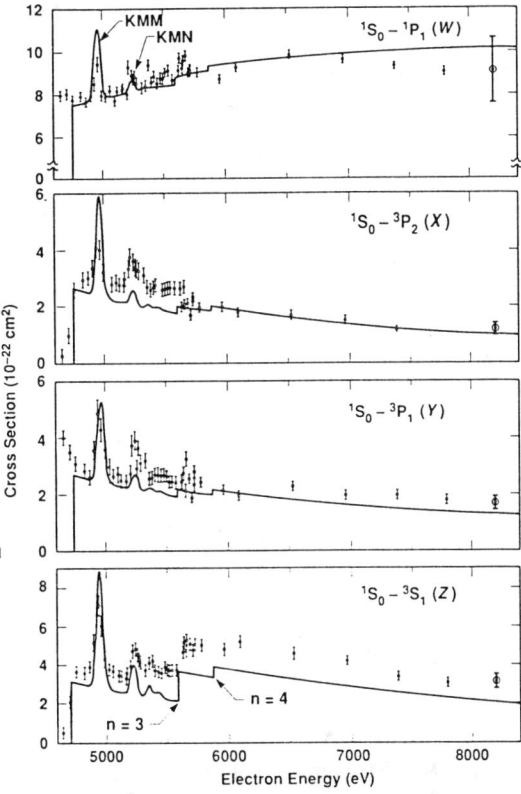

Figure 7. Comparison of cross sections measured in EBIT for n=1 to n=2 excitation of Ti^{20+} with theoretical predictions. [from Ref. 21].

INNER-SHELL EXCITATION

<u>Total Cross Sections</u>

As noted in the Introduction, precision measurements of electron-impact ionization of multiply charged ions often exhibit well-defined structures due to inner-shell excitation followed by autoionization (EA). The sharpness of such EA features is a consequence of the threshold characteristic of the cross section for electron-impact excitation of an ion, which rises abruptly at the threshold energy (with an infinite slope), frequently to its maximum value. Identification and interpretation of such features requires knowledge of the energies of the relevant doubly-excited levels, as well as branching ratios for radiative decay versus autoionization. The relative EA contribution is generally more important for multiply charged ions because the binding energy of the outermost electron increases with ionic charge, causing the direct ionization cross section to decrease (approximately as the inverse square of the binding energy),[2,5,6] whereas the inner-shell excitation cross sections are only minimally affected. Figure 8 presents an illustrative example of the EA process for Li-like O^{5+}, which possesses the

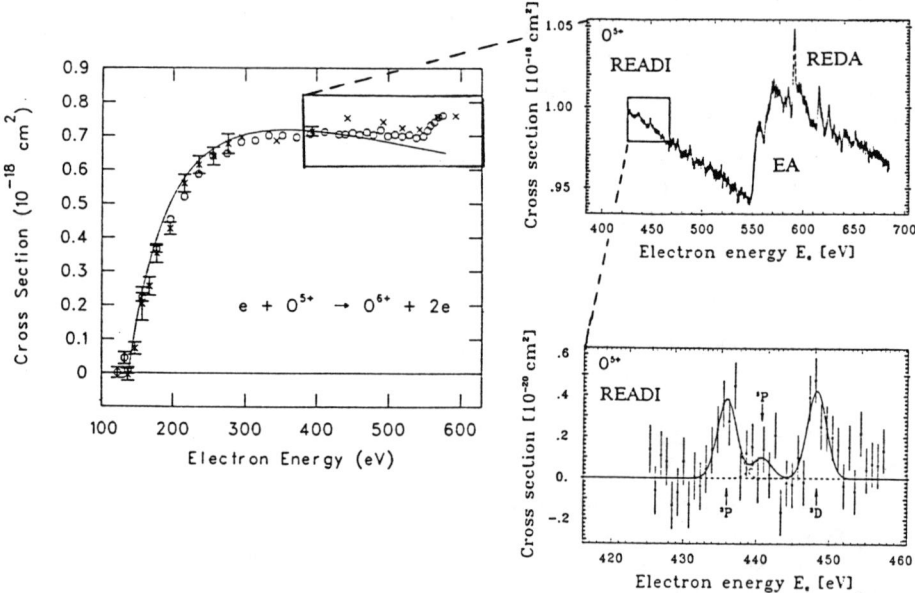

Figure 8. Measured total cross sections for electron-impact ionization of Li-like O^{5+}, showing structures due to the EA, REDA and READI processes [from Ref. 25 (left) and Ref. 26 (right)].

simplest electronic structure for which there exists an inner shell to be excited. The total ionization cross-section curve on the left[25] was measured at ORNL using a crossed-beams apparatus, and shows a distinct rise beginning at 550 eV due to the EA process:

$$e + O^{5+}(1s^2\ 2s) \rightarrow e + O^{5+}(1s\ 2s\ nl) \rightarrow e + O^{6+}(1s^2) + e.$$

The curve on the right is a similar measurement made several years later in Giessen[26] using a high-intensity, space-charge-neutralized electron beam. In addition to the initial step due to 1s-2s excitation, these high-precision measurements indicate the 1s-2p, 1s-3s, ... 1s-nl EA processes, as well as rich structures due to the resonant excitation double autoionization (REDA) process:

$$e + O^{5+}(1s^2 2s) \rightarrow O^{4+}(1s\ 2s\ nl\ n'l') \rightarrow O^{6+}(1s^2) + 2e.$$

Achieving a statistical precision of 0.1% in their cross-section measurements, the

Giessen group were also able to clearly resolve the so-called resonant-excitation auto-double ionization (READI) resonances in the 165-175 eV energy region. These are indicated in the inset box in Fig. 8. In this case, nl and n'l' in the above equation refer to 2s and 2p, and the triply excited resonant states must decay by a simultaneous rather than a sequential double autoionization process in order to manifest themselves in the ionization channel.

While such effects constitute a relatively small fraction (10%) of the total ionization cross section for O^{5+}, they provide important data for benchmarking theoretical cross sections for inner-shell excitation and resonant processes. For ions with more complex electronic structures, on the other hand, the EA process may dominate direct outer-shell ionization by an order of magnitude[2,5,6], and must be taken into account in plasma modelling.

SUMMARY AND OUTLOOK

Technological advances have facilitated dramatic improvements in our ability to study the excitation of ions by electron-impact. These include advanced sources of ions and electrons, position-sensitive particle detectors, and innovative devices for particle beam interaction, analysis and characterization. Heavy ion storage rings in Europe have revolutionized the study of electron-ion recombination processes,[27] and the first ionization experiments under development should provide data of unprecedented precision for inner-shell excitation of very highly charged ions. Steady improvements are also being realized in EBIT, EBIS and ECR ion source technology. These developments have been paralleled by comparable developments in the theory of electron-ion interactions, made possible in part by advances in supercomputer technology. The continued synergistic application of these resources promises a deeper understanding of the excitation process.

The author is most grateful to Gordon Dunn for introducing him to this research area, for nearly two decades of fruitful collaboration in electron-ion collisions, and for insightful comments on this manuscript.

REFERENCES

1. D.F. Dance, M.F.A. Harrison and A.C.H. Smith, Proc. Roy. Soc. A (London) 290, 74 (1966).
2. R. A. Phaneuf, "Experiments on Electron-Impact Excitation and Ionization of Ions," pp. 117-156 in *Atomic Processes in Electron-Ion and Ion-Ion Collisions*, F. Brouillard ed., NATO Advanced Science Institute Series B, Vol. 145 (Plenum, New York and London, 1986).
3. P.O. Taylor and G.H. Dunn, Phys. Rev A 8, 2304 (1973).
4. R.J.W. Henry, Phys. Rep. 68, 1 (1981).
5. G. H. Dunn, "Electron-Ion Ionization", pp. 277-319 in *Electron Impact Ionization*, T.D. Mark and G.H. Dunn, eds. (Springer, New York and Vienna, 1985).

6. A. Müller, "Ion Formation Processes: Ionization in Electron-Ion Collisions," pp. 13-90 in *Physics of Ion Impact Phenomena*, D. Mathur, ed., (Springer-Verlag, Berlin, Heidelberg, 1991).
7. A. Chutjian, Phys. Rev A 29, 64 (1984); A. Chutjian, A.Z. Msezane and R.J.W. Henry, Phys. Rev. Lett. 50, 1357 (1983).
8. I.D. Williams, A. Chutjian and R.J. Mawhorter, J. Phys. B 19, 2189 (1986).
9. B.A. Huber, Ch. Ristori, P.A. Hervieux, M. Maurel, C. Guet and H.J. Andrä, Phys. Rev. Lett. 67, 1407 (1991); Z. Phys. D Suppl. 21, S203 (1991); Ch. Ristori et al., Z. Phys D 22, 425 (1991).
10. E. K. Wåhlin, J.S. Thompson, G.H. Dunn, R.A. Phaneuf, D.C. Gregory, and A.C.H. Smith, Phys. Rev. Lett. 66, 157 (1991).
11. S. J. Smith, K-F. Man, R. J. Mawhorter, I.D. Williams and A. Chutjian, Phys. Rev. Lett. 67, 30 (1991).
12. W.T. Rogers, G.H. Dunn, J.O. Olsen, M. Reading and G. Stefani, Phys. Rev. A 25, 681 (1982).
13. M.S. Pindzola, N.R. Badnell, R.J.W. Henry, D.C. Griffin and W.L. van Wijngaarden, Phys. Rev. A 44, 5628 (1991).
14. S.J. Smith, A. Chutjian, J. Mitroy, S.S. Tayal, R.J.W. Henry, K-F. Man, R.J. Mawhorter and I.D. Williams, Phys. Rev. A (1993, in press).
15. X.Q. Guo, E.W. Bell, J.S. Thompson, G.H. Dunn, M.E. Bannister, R.A. Phaneuf and A.C.H. Smith, Phys. Rev. A 47, R9 (1993).
16. N.R. Badnell, M.S. Pindzola and D.C. Griffin, Phys. Rev. A 43, 2250 (1991).
17. D.C. Griffin, N.R. Badnell and M.S. Pindzola (data reported in Ref. 15).
18. R.E.H. Clark (data reported in Ref. 15).
19. E.W. Bell, X.Q. Guo, G.H. Dunn, M.E. Bannister, A.C.H. Smith, B.D. DePaola, C.A. Timmer and E.K. Wåhlin, Bull. Am. Phys. Soc. 38, 1157 (1993).
20. M.A. Levine, R.E. Marrs, J.R. Henderson, D.A. Knapp and M.B. Schneider, Physica Scripta T72, 157 (1988).
21. S. Chantrenne, P. Beiersdorfer, R. Cauble and M.B. Schneider, Phys. Rev. Lett. 69, 265 (1992).
22. H.L. Zhang, D.H. Sampson and R.E.H. Clark, Phys. Rev. A 41, 198 (1990).
23. M.H. Chen, reported in ref. 20, based on method described by M. H. Chen, Phys. Rev. A 33, 994 (1986).
24. R. Ali, C.L. Cocke, M. Schulz and M. Stöckli, Z. Phys. D. Suppl. 21, S207 (1991).
25. K. Rinn, D.C. Gregory, L.J. Wang, R.A. Phaneuf and A. Müller, Phys. Rev. A 36, 595 (1987).
26. G. Hoffman, A. Müller, K. Tinschert and E. Salzborn, Z. Phys. D 16, 113 (1990).
27. A. Müller, "Electron-Ion Recombination Phenomena: Formation and Decay of Intermediate Resonance States," pp. 155-179 in *Recombination of Atomic Ions*, eds. W.G. Graham, W. Fritsch, Y. Hahn and J.A. Tanis, NATO ASI Series B: Physics Vol 296 (Plenum, New York, 1992).

ELECTRON SCATTERING FROM ATOMIC IONS

M. S. Pindzola and T. W. Gorczyca

Department of Physics, Auburn University, Auburn, AL 36849

D. C. Griffin

Department of Physics, Rollins College, Winter Park, FL 32789

N. R. Badnell

Department of Physics, University of Strathclyde, Glasgow, UK G40NG

Abstract

We review the theory of electron scattering from atomic ions. An independent-processes approximation is used as a starting point for the study of electron-ion recombination, excitation, and ionization. Where possible we compare theory with recent electron-ion crossed and merged beams experiments.

1 Introduction

The scattering of electrons from atomic ions serves as a benchmark for our understanding of quantum mechanical scattering theory. The projectile electron together with the electrons of the target ion form a many-body system dominated by the known laws of quantum electrodynamics. At low incident electron energies and for low charged ions, non-central field electron correlation effects are often important in obtaining accurate collision cross sections and rates. At high incident electron energies and for high charged ions, relativistic effects on electron motion in the strong central field of the nucleus also become important. In this paper we review the scattering theory associated with electron-ion recombination, excitation, and ionization. In each case, the starting point for our theoretical development is an independent-processes approximation in which various scattering pathways are tested in a perturbative distorted-wave approach. Target correlation and relativistic effects are then examined. Finally, the non-perturbative close-coupling method is used to study interference between processes, continuum-state correlation, and interaction between resonance structures.

The layout of the paper is as follows. In Sec. II we review the theory of electron-ion recombination and compare theory and experiment for the Li-like ions O^{5+}, Cu^{26+}, and Au^{76+}. In Sec. III we review the theory of electron-impact excitation of ions and compare theory and experiment for the Na-like ions Si^{3+} and Ar^{7+}. In Sec. IV we review the theory of electron-impact ionization of ions and compare theory and experiment for Na-like Fe^{15+} and K-like Sc^{2+}. In Sec. V we provide a short summary.

2 Electron-Ion Recombination

Electron-ion recombination into a particular final recombined state, in the presence of one projectile continuum, may be schematically represented as

$$e^- + A_i^{q+} \rightarrow A_f^{(q-1)+} + \omega , \qquad (1)$$

and
$$e^- + A_i^{q+} \to \left[A_j^{(q-1)+}\right] \to A_f^{(q-1)+} + \omega , \qquad (2)$$

where q is the charge on the atomic ion A, ω is the frequency of the emitted light, and the brackets in Eq.(2) indicate a doubly-excited resonance state. The first process is called radiative recombination(RR), while the second is called dielectronic recombination (DR).

In the independent-processes approximation the two paths for recombination are summed incoherently. The radiative recombination cross section for Eq.(1), in lowest-order of perturbation theory, is given by

$$\sigma_{RR}(i \to f) = \frac{8\pi^2}{k_i^3} \frac{1}{2g_i} \sum_{\ell_i} |\langle \alpha_f J_f | D | \alpha_i \ell_i J_i \rangle|^2 , \qquad (3)$$

where k_i is the linear momentum and ℓ_i is the angular momentum of the continuum electron, J_i and J_f are total angular momenta, α_i is used to designate the initial level, α_f is used to designate the final bound recombined level, and g_i is the statistical weight of the initial level. The dipole radiation field operator is given by

$$D = \sqrt{\frac{2\omega^3}{3\pi c^3}} \sum_{s=1}^{N+1} \vec{r}_s , \qquad (4)$$

continuum normalization is chosen as one times a sine function, and atomic units ($e=\hbar=m=1$) are used. In the isolated-resonance approximation, the dielectronic recombination cross section for Eq.(2), in lowest-order perturbation theory, is given by

$$\sigma_{DR}(i \to f) = \frac{8\pi^2}{k_i^3} \sum_j \frac{g_j}{2g_i} \sum_{\ell_i} \left| \frac{\langle \alpha_f J_f | D | \alpha_j J_j \rangle \langle \alpha_j J_j | V | \alpha_i \ell_i J_j \rangle}{E_i - E_j + i\Gamma_j/2} \right|^2 , \qquad (5)$$

where α_j designates a resonance level with energy E_j and total width Γ_j, and the electrostatic interaction between electrons is given by

$$V = \sum_{s=1}^{N} |\vec{r}_s - \vec{r}_{N+1}|^{-1} . \qquad (6)$$

By the principle of detailed balance, the radiative recombination cross section of Eq.(3) is proportional to the photoionization cross section from the bound state, while the energy-averaged dielectronic recombination cross section may be written as:

$$\langle \sigma_{DR}(i \to f) \rangle = \frac{2\pi^2}{\Delta \epsilon k_i^2} \sum_j \frac{g_j}{2g_i} \frac{A_a(j \to i) A_r(j \to f)}{\Gamma_j} , \qquad (7)$$

where the autoionization decay rate is given by

$$A_a(j \to i) = \frac{4}{k_i} \sum_{\ell_i} |\langle \alpha_i \ell_i J_i | V | \alpha_j J_j \rangle|^2 , \qquad (8)$$

the radiative decay rate is given by

$$A_r(j \to f) = 2\pi |\langle \alpha_f J_f | D | \alpha_j J_j \rangle|^2 , \qquad (9)$$

and $\Delta\epsilon$ is the energy bin width.

In recent years atomic physics experiments carried out using heavy-ion traps, accelerators, and storage-cooler rings have produced high-resolution mappings of the resonance structures associated with electron-ion recombination. For the most part, the agreement between the high-resolution measurements and theoretical calculations based on the independent-processes and isolated-resonance approximations has been quite good[1-17]. We illustrate the type of agreement with experiment obtained by calculations employing the independent-processes and isolated-resonance approximations with three examples from the Li isoelectronic sequence. The pathways for dielectronic capture, in terms of specific levels, are given by

$$e^- + A^{q+}(2s_{1/2}) \rightarrow \left[A^{(q-1)+}(2p_{1/2}nlj)\right]$$
$$\searrow \left[A^{(q-1)+}(2p_{3/2}nlj)\right], \quad (10)$$

where a $1s^2$ core is assumed to be present. Both Rydberg series autoionize by the reverse of the paths in Eq.(10), while for sufficiently high n the $2p_{3/2}nlj$ levels may autoionize to the $2p_{1/2}$ continuum. Both series radiatively stabilize by either a $2p \rightarrow 2s$ core orbital transition, or by a $nlj \rightarrow n'l'j'$ valence orbital transition, where $2pjn'l'j'$ is a bound level.

Dielectronic recombination cross section calculations[9] in the independent processes and isolated-resonance approximations for O^{5+} are compared with experiment in Fig. 1. Fine-structure splitting of the two series of Eq.(10) is minimal for this light ion, so

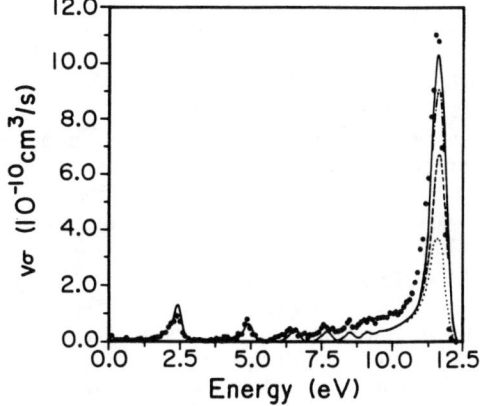

Figure 1: Dielectronic Recombination for O^{5+}

there appears only one Rydberg series. The $2p6\ell$ resonances are located at 2.5 eV, the $2p7\ell$ at 5.0 eV, and so on; accumulating at the series limit around 11.3 eV. Electric field effects on the high n resonances are strong in O^{5+}, so that calculations were done for fields of 0(dot), 3(dash), 5(chain), and 7(solid) V/cm. Since the precise electric field strength in the experiment is not known, the accuracy of an electric field dependent theory has yet to be determined.

Dielectronic recombination cross section calculations[12] in the independent-processes and isolated-resonance approximations for Cu^{26+} are compared with experiment in Fig.

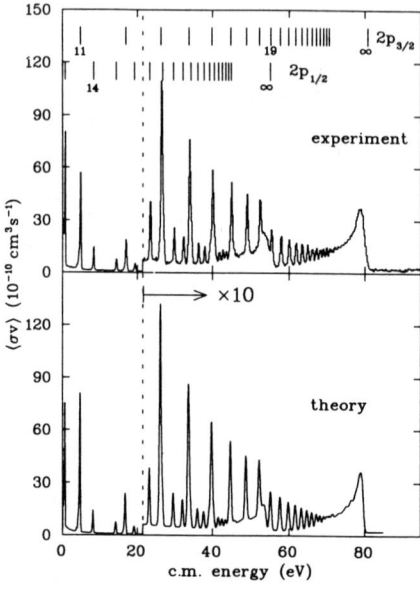

Figure 2: Dielectronic Recombination for Cu^{26+}

2. The two fine-structure Rydberg series are now clearly resolved, the fine-structure splitting is about 27 eV for Cu^{26+}. The $2p_{1/2}13\ell$ resonances are just above threshold, while the $2p_{3/2}11\ell$ resonances are found around 5.0 eV. Electric fields in the range 0-50 V/cm will have little effect on the Cu^{26+} spectrum. Overall the agreement between theory and experiment is excellent.

Electron-ion recombination cross section calculations[13] in the independent-processes and isolated-resonance approximations for low-lying resonances in Au^{76+} are compared with experiment in Fig. 3. The $2p_{1/2}n l j$ series limit is at 217 eV, while the $2p_{3/2}n l j$ series limit is at 2.24 keV; yielding a fine structure splitting of 2.03 keV. QED effects alone shift the $2p_{3/2}n l j$ series limit by 22.0 eV. Thus accurate atomic structure calculations must be made to locate the $2p_{3/2}6 l j$ resonances in the 0-50 eV energy range of the experiment. The perturbative-relativistic (a), semirelativistic (b), and fully-relativistic (c) calculations for the dielectronic recombination cross section ride on top of a strong radiative recombination background. In principle, the fully-relativistic theory contains the most physics and thus it is comforting to see that on the whole it is in good agreement with the experimental results. It is instructive, however, to see how well the computationally simpler perturbative-relativistic and semirelativistic theories do for such a highly charged ion.

There has been a great deal of effort in recent years to develop a more general theory of electron-ion recombination which would go beyond the independent-processes and isolated-resonance approximations to include radiative-dielectronic recombination interference and overlapping (and interacting) resonance structures[18-25]. In almost all cases[25], the interference between a dielectronic-recombination resonance and the radiative-recombination background has been found to be quite small and will be dif-

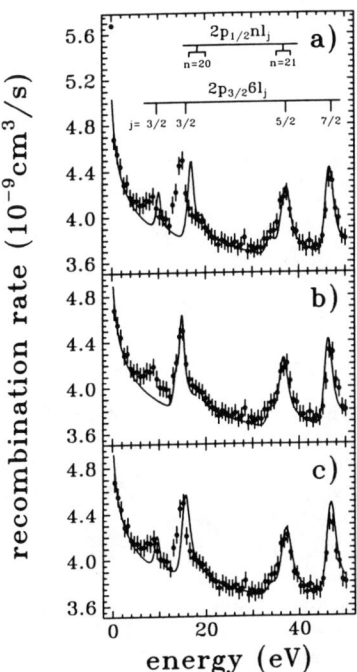

Figure 3: Dielectronic Recombination for Au[76+]

ficult to observe in current generation experiments. For overlapping (and interacting) doubly-excited resonances, the dielectronic recombination cross section for Eq.(2) may be written as:

$$\sigma_{\rm DR}(i \to f) = \frac{8\pi^2}{k_i^3} \frac{1}{2g_i} \left| \sum_j \sum_k \frac{\sqrt{g_j}\,\langle \alpha_f J_f|D|\alpha_j J_j\rangle\,(\Omega^{-1})_{jk}\,\langle \alpha_k J_k|V|\alpha_i \ell_i J_i\rangle}{E_i - E_k + i\Gamma_k/2} \right|^2 , \quad (11)$$

where LaGattuta's dimensionless coupling matrix is given by

$$\Omega_{jk} = \delta_{jk} - (1 - \delta_{jk})\frac{\Lambda_{jk}}{E_i - E_j + i\Gamma_j/2} , \quad (12)$$

and δ_{jk} is the Kronecker delta. In the pole approximation,

$$\Lambda_{jk} = -\frac{i}{2}\left[\tilde{A}_a(j \to i \to k) + \tilde{A}_r(j \to f \to k)\right] , \quad (13)$$

where the generalized autoionization coupling rate is given by

$$\tilde{A}_a(j \to i \to k) = \frac{4}{k_i}\sum_{\ell_i} \langle \alpha_k J_k|V|\alpha_i \ell_i J_i\rangle \langle \alpha_i \ell_i J_i|V|\alpha_j J_j\rangle , \quad (14)$$

and the generalized radiative coupling rate is given by

$$\tilde{A}_r(j \to f \to k) = 2\pi \langle \alpha_k J_k|D|\alpha_f J_f\rangle \langle \alpha_f J_f|D|\alpha_j J_j\rangle . \quad (15)$$

In the cases studied to date, however, the combination of electron and photon continuum coupling selection rules and the requirement of near energy degeneracy make the overlapping (and interacting) resonance effects small and difficult to observe. Heavy ions in relatively low stages of ionization are the best place to look, since there are resonance series attached to the large numbers of LS terms or fine-structure levels.

Finally, the reason the distorted-wave approximation has been so successful in describing dielectronic recombination cross sections for any atomic ion is that for low charged ions the DR cross section is proportional to the radiative rate, while for highly charged ions the DR cross section is proportional to the autoionization rate. Thus the weakness of the distorted-wave method in calculating accurate autoionization rates for low charged ions is masked by a DR cross section that is highly dependent on radiative atomic structure. As one moves to more highly charged ions, the DR cross section becomes more sensitive to the autoionization rates, but at the same time, the distorted-wave method becomes increasingly more accurate.

3 Electron-Ion Excitation

Electron-ion excitation from an initial state i to a final state f, may be schematically represented as

$$e^- + A_i^{q+} \to A_f^{q+} + e^- , \quad (16)$$

and

$$e^- + A_i^{q+} \to \left[A_j^{(q-1)+}\right] \to A_f^{q+} + e^- . \quad (17)$$

The first process is called direct excitation (DE), while the second process involving dielectronic-capture followed by autoionization to an excited-state is called resonant excitation (RE).

In the independent-processes approximation the two paths for excitation are summed incoherently. The direct excitation cross section for Eq.(16), in lowest-order of perturbation theory, is given by

$$\sigma_{DE}(i \to f) = \frac{16\pi}{k_i^3 k_f} \frac{1}{2g_i} \sum_{\ell_i} \sum_{\ell_f} \sum_J (2J+1) |\langle \alpha_f J_f \ell_f J | V | \alpha_i J_i \ell_i J \rangle|^2 \;, \quad (18)$$

corresponding to transitions from states $\alpha_i J_i$ to $\alpha_f J_f$ in the target ion. In the isolated-resonance approximation, the energy-averaged resonant excitation cross section may be written as:

$$\langle \sigma_{RE}(i \to f) \rangle = \frac{2\pi^2}{\Delta \epsilon k_i^2} \sum_j \frac{g_j}{2g_i} \frac{A_a(j \to i) A_a(j \to f)}{\Gamma_j} \;, \quad (19)$$

The expressions found in Eq.(7) for the dielectronic recombination cross section and in Eq.(19) for the resonant excitation cross section are quite similar. In fact, the lure of the independent-processes and isolated-resonance approximations is that the same computer codes written to analyze the dielectronic-recombination experiments could be used for electron-ion excitation. Radiation damping, which becomes important for highly charged ions, is also included in Eq.(19) for the resonant excitation cross section.

For low charged ions, where radiation damping is small, we may check the general validity of the perturbative approach embodied in Eqs.(18-19) for electron-ion excitation by comparing results with the non-perturbative close-coupling method. The close-coupling method includes direct-resonant excitation interference, overlapping (and interacting) resonances structures, and continuum channel coupling effects. Previous comparisons[26-28] between the close-coupling method and the independent-processes approximation using distorted-waves have been made between rate coefficients for excitation, i.e. integrated over all resonance structures. Some detailed comparisons[29,30] have been made between close-coupling and distorted-wave resonances, but mostly within the framework of multichannel quantum-defect theory. We illustrate the type of agreement between the two methods and experiment with two examples from the Na isoelectronic sequence. For the strong dipole-allowed 3s→3p excitation, the pathways for dielectronic capture, in terms of specific configurations, are given by

$$\begin{array}{c} e^- + A^{q+}(3s) \to \left[A^{(q-1)+}(3dnl) \right] \\ \searrow \\ \left[A^{(q-1)+}(4\ell'nl) \right] \;, \end{array} \quad (20)$$

where a $1s^2 2s^2 2p^6$ core is assumed to be present. The multiple Rydberg series of Eq.(20) may autoionize to energetically allowed 3s, 3p, 3d, and $4\ell'$ continua. However, only autoionization to the 3p continuum will contribute to the 3s→3p excitation cross section.

Electron-impact excitation cross section calculations[31] for the 3s→3p transition in Si^{3+} in both the close-coupling(top) and distorted-wave(bottom) methods are compared in Fig. 4. The most significant difference between the results is a large window resonance at 11 eV due to the $^3G^e$ partial wave which is repeated at high energies (14 eV, 16 eV, ...). The independent-processes calculation places the 3d4d $^3G^e$ resonance at about 11 eV, but fails, of course, to include interference with the background direct excitation. The close-coupling calculations convoluted with 0.20 eV FWHM electron energy distribution, are compared with experiment[32] in Fig. 5. The agreement of theory and experiment is seen to be very good in both magnitude and shape. Unfortunately the experiment was not able to reach the interesting window resonance feature at 11 eV.

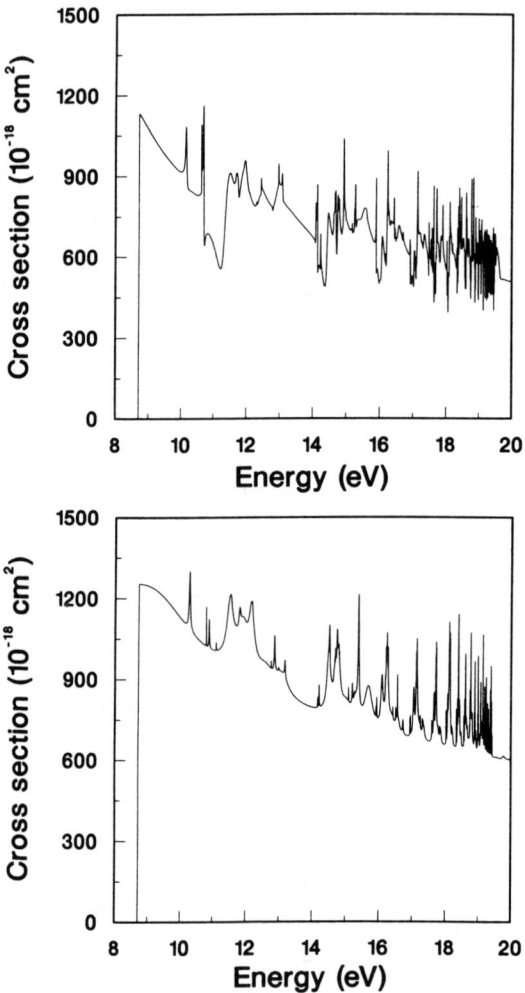

Figure 4: Electron-impact Excitation of Si^{3+}

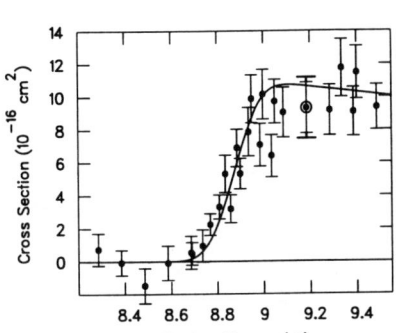

Figure 5: Electron-impact Excitation of Si^{3+}

Electron-impact excitation cross sections calculations[31] for the 3s→3p transition in Ar^{7+} in both the close-coupling(top) and distorted-wave(bottom) methods are compared in Fig. 6. The direct-resonant excitation interference effects are greatly reduced and there is an overall good agreement between the two resonance spectrums. The close-coupling calculations convoluted with a 0.2 eV FWHM Gaussian are compared with experiment[33] in Fig. 7. The agreement between theory and experiment is good over the energy range investigated, but again experimental resonance features are not obtained. The electron energy-loss detection scheme employed in the merged-beams experiments for Si^{3+} and Ar^{7+} is sensitive to the angular distribution of scattered electrons and has sparked a theoretical investigation of the angular differential cross section for the 3s→3p excitation in Na-like ions[34]. Differential cross sections for the 3s→3p excitation in Ar^{7+}, calculated using the distorted-wave method, agree well with experiment[35] at 100 eV.

In support of future experiments, electron-impact excitation cross sections for the $3s^2\,{}^1S \rightarrow 3s3p\,{}^{3,1}P$ transitions in Si^{2+} and Ar^{6+} were calculated in the close-coupling method[36]. Surprisingly, when distorted-wave calculations for the resonance structures were compared to the close-coupling results, significant differences were found due to not only direct-resonant excitation interference effects but overlapping (and interacting) resonance series. Moreover, the overlapping (and interacting) resonance effects did not become smaller for the higher-charged members of the Mg-like isoelectronic sequence. This has important implications for the description of resonant processes which are strongly radiation damped, for which the isolated-resonance approximation is widely used and the standard formulation of the close-coupling approximation is not applicable. Finally, recent fully-relativistic calculations[37, 38] for electron-impact excitation of highly charged ions have shown the importance of extending the electrostatic interaction V of Eq.(18) to include magnetic and retardation effects.

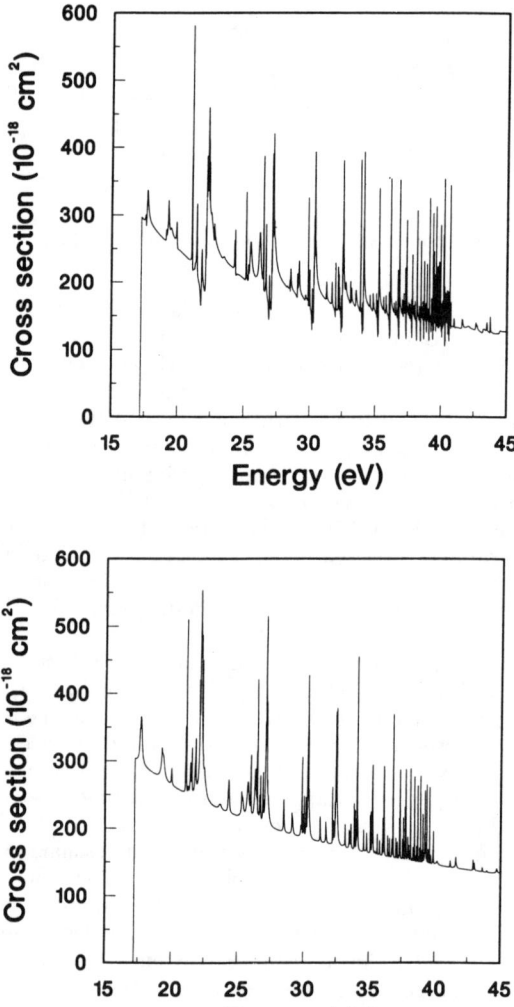

Figure 6: Electron-impact Excitation of Ar^{7+}

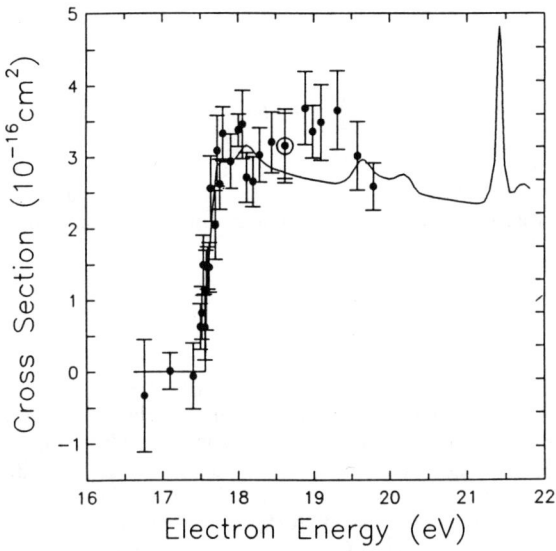

Figure 7: Electron-impact Excitation of Ar^{7+}

4 Electron-Ion Ionization

Electron-impact ionization of an atomic ion from an initial state i to a final state f, may be represented as

$$e^- + A_i^{q+} \rightarrow A_f^{(q+1)+} + e^- + e^- , \qquad (21)$$

$$e^- + A_i^{q+} \rightarrow \left[A_j^{q+}\right] + e^- \rightarrow A_f^{(q+1)+} + e^- + e^- , \qquad (22)$$

and

$$e^- + A_i^{q+} \rightarrow \left[A_j^{(q-1)+}\right] \rightarrow \left[A_{j'}^{q+}\right] + e^- \rightarrow A_f^{(q+1)+} + e^- + e^- . \qquad (23)$$

The first process is called direct ionization (DI), the second process is called excitation-autoionization (EA), while the third process involving dielectronic-capture followed by sequential double autoionization is called resonant-excitation double-autoionization (REDA).

In the independent-processes approximation the three paths for ionization are summed incoherently. The direct ionization cross section for Eq.(21), in lowest-order of perturbation theory, is given by

$$\sigma_{DI}(i \rightarrow f) = \frac{32}{k_i^3 k_f k_e} \frac{1}{2g_i} \sum_{\ell_i} \sum_{\ell_f} \sum_{\ell_e} \sum_{J} \sum_{J'} (2J+1) \int_0^{E_{max}} d\epsilon_e \, |\langle \alpha_f J_f \ell_e J' \ell_f J | V | \alpha_i J_i \ell_i J \rangle|^2 , \qquad (24)$$

corresponding to a transition from state $\alpha_i J_i$ of the N-electron initial ion to the state $\alpha_f J_f$ of the (N-1)-electron final ion. The energy $E_{max} = \epsilon_f + \epsilon_e = \epsilon_i - I_p$, where I_p

is the ionization potential of the target ion. The excitation-autoionization cross section for Eq.(22) is given by

$$\sigma_{EA}(i \to f) = \sum_j \sigma_{DE}(i \to j) \frac{A_a(j \to f)}{\Gamma_j} , \qquad (25)$$

where σ_{DE} may be calculated using Eq.(18). In the isolated-resonance approximation, the energy-averaged resonant-excitation double-autoionization cross section may be written as

$$\langle \sigma_{REDA}(i \to f) \rangle = \sum_{j'} \langle \sigma_{RE}(i \to j') \rangle \frac{A_a(j' \to f)}{\Gamma_{j'}} , \qquad (26)$$

where $\langle \sigma_{RE}(i \to j') \rangle$ may be calculated using Eq.(19).

Evidence exists for selected ions[39-42] that the interference between the direct ionization and excitation-autoionization pathways has little effect on the total ionization cross section. For low charged ions, where radiation damping is small, the independent-processes approximation for the indirect ionization contributions may be checked by again comparing close-coupling and distorted-wave calculations. Extensive calculations [43, 44] on Li-like C^{3+}, N^{4+}, O^{5+}, and F^{6+} ions have shown reasonable agreement between close-coupling and distorted-wave results for the magnitude and shape of dielectronic-capture resonance features. Both theoretical approaches also agree well with crossed-beams experimental measurements [45]. For more highly charged-ions, independent-processes distorted-wave calculations[46, 47] have shown that radiation damping must be included in the contributions from both excitation-autoionization and resonant-excitation double-autoionization.

We illustrate the type of agreement that may be found between theory and experiment by considering the ionization of Na-like Fe^{15+} and K-like Sc^{2+}. The main excitation-autoionization contributions for Na-like ions are given by

$$e^- + A^{q+}(2p^6 3s) \to \left[A^{q+}(2p^5 3s 3\ell) \right] + e^- , \qquad (27)$$

where $\ell = 0, 1, 2$ and a $1s^2 2s^2$ core is assumed to be present. The main pathways for dielectronic capture are given by

$$e^- + A^{q+}(2p^6 3s) \to \left[A^{(q-1)+}(2p^5 3s n\ell n'\ell') \right] , \qquad (28)$$

where $n = 3, 4$. Electron-impact ionization calculations[48] in the independent-processes and isolated-resonance approximations for Fe^{15+} are compared with experiment[49] in Fig. 8. Resonant-excitation double-autoionization contributes about 30% to the total collisional ionization rate for temperatures near the ionization threshold. The inclusion of radiation damping reduces the excitation-autoionization contributions by 40%, while the resonant-excitation double-autoionization contribution is reduced by 50%. The experimental error bars are too large, however, to say much about the accuracy of the detailed resonance spectrum. Planned storage ring experiments should remedy the situation in the near future. Dielectronic-capture resonance structures have also been recently calculated for Na-like Kr^{25+} and Xe^{43+}[50].

The main excitation-autoionization contributions for Sc^{2+} are given by

$$e^- + Sc^{2+}(3p^6 3d) \to \left[Sc^{2+}(3p^5 3d^2) \right] + e^- , \qquad (29)$$

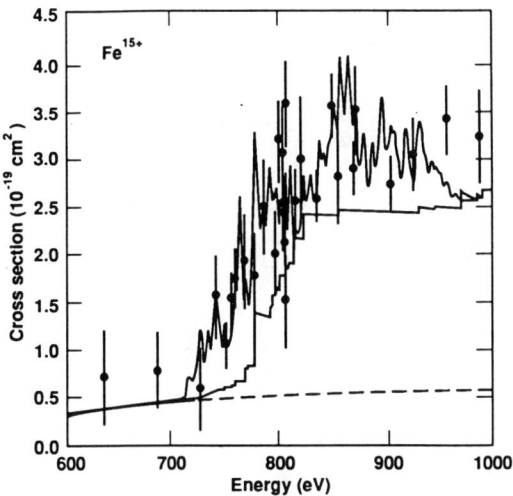

Figure 8: Electron-impact Ionization for Fe^{15+}

where a $1s^2 2s^2 2p^6 3s^2$ core is assumed to be present. The main pathways for dielectronic capture are given by

$$e^- + Sc^{2+}(3p^6 3d) \rightarrow \left[Sc^+(3p^5 3d^2 n\ell)\right] . \tag{30}$$

Electron-impact ionization calculations[51] for Sc^{2+} are compared with experiment in Fig. 9. The distorted-wave(dot) calculations include only excitation-autoionization contributions, while the close-coupling(solid) calculations include both excitation-autoionization and resonant-excitation double-autoionization contributions, as well as interference effects. For doubly-charged ions like Sc^{2+} it is safe to ignore radiation damping. Overall the agreement between theory and experiment is quite good, although the strength of the resonance structures around 32 eV are overestimated by theory.

5 Summary

An independent-processes approximation may be used as a starting point for the development of the theory of electron-ion scattering. Perturbative methods, based on the distorted-wave approximation, may then be employed to calculate the various direct and resonance processes found in electron-ion recombination, excitation, and ionization. What remains are the interesting effects due to quantum mechanical interference between the various processes, some of which can be handled in the non-perturbative

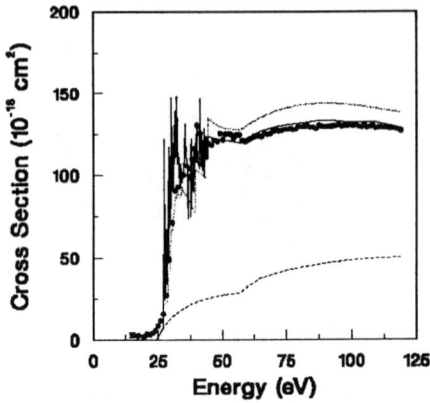

Figure 9: Electron-impact Ionization for Sc^{2+}

close-coupling approximation. Challenges for the future include the development of perturbative methods which include interference effects and the development of nonperturbative methods which include the radiation continuum.

6 Acknowledgements

Our theoretical study of electron-ion collisions has benefited greatly from a long standing collaboration with our now departed friend and colleague, Chris Bottcher. We will miss his keen insight into, and encyclopedic knowledge of, atomic collision physics in the coming years. This work was supported by the U.S. Department of Energy under contract No. DE-FG05-86ER53217 with Auburn University and under contract No. DE-AC05-84OR21400 with Martin Marietta Energy Systems, Inc., contract manager of Oak Ridge National Laboratory.

References

[1] G. Kilgus, J. Berger, P. Blatt, M. Grieser, D. Habs, B. Hochadel, E. Jaeschke, D. Kramer, R. Neumann, G. Neureither, W. Ott, D. Schwalm, M. Steck, R. Stokstad, E. Szmola, A. Wolf, R. Schuch, A. Müller and M. Wagner, "Dielectronic recombination of hydrogenlike oxygen in a heavy-ion storage ring," Phys. Rev. Lett **64**, 737 (1990).

[2] M. S. Pindzola, N. R. Badnell, and D. C. Griffin, "Dielectronic recombination cross sections for H-like ions," Phys. Rev. A **42**, 282 (1990).

[3] D. R. DeWitt, D. Schneider, M. W. Clark, M. H. Chen, and D. Church, "Dielectronic-recombination cross sections of hydrogenlike argon," Phys. Rev. A **44**, 7185 (1991).

[4] L. H. Andersen, P. Hvelplund, H. Knudsen, and P. Kvistgaard, "State-selective dielectronic-recombination measurements for the He-like oxygen ions in an electron-cooler," Phys. Rev. Lett **62**, 2656 (1989).

[5] N. R. Badnell, M. S. Pindzola, and D. C. Griffin, "Dielectronic recombination from the ground and excited states of C^{4+} and O^{6+}," Phys. Rev. A **41**, 2422(1990).

[6] L. H. Andersen, G. Y. Pan, H. T. Schmidt, N. R. Badnell, and M. S. Pindzola, "Absolute measurements and calculations of dielectronic recombination with metastable He-like N, F, and Si ions," Phys. Rev. A **45**, 7868 (1992).

[7] D. A. Knapp, R. E. Marrs, M. B. Schneider, M. H. Chen, M. A. Levine, and P. Lee, "Dielectronic recombination of heliumlike ions," Phys. Rev. A **47**, 2039 (1993).

[8] R. Ali, C. P. Bhalla, C. L. Cocke, and M. Stockli, "Dielectronic recombination on heliumlike argon," Phys. Rev. Lett **64**, 633 (1990).

[9] D. C. Griffin, M. S. Pindzola, and P. Krylstedt, "Distorted-wave calculations of dielectronic recombination for C^{3+} and O^{5+} in small electric fields," Phys. Rev. A **40**, 6699 (1989).

[10] L. H. Andersen, J. Bolko, and P. Kvistgaard, "State-selective dielectronic-recombination measurements for He-like and Li-like carbon and oxygen ions," Phys. Rev. A **41**, 1293 (1990).

[11] L. H. Andersen, G. Y. Pan, H. T. Schmidt, M. S. Pindzola, and N. R. Badnell, "State-selective measurements and calculations of dielectronic recombination with Li-like N^{4+}, F^{6+}, and Si^{11+} ions," Phys. Rev. A **45**, 6332 (1992).

[12] G. Kilgus, D. Habs, D. Schwalm, A. Wolf, N. R. Badnell, and A. Müller, "High-resolution measurements of dielectronic recombination of lithiumlike Cu^{26+}," Phys. Rev. A **46**, 5730 (1992).

[13] W. Spies, A. Müller, J. Linkemann, A. Frank, M. Wagner, C. Kozhuharov, B. Franzke, K. Beckert, F. Bosch, H. Eickhoff, M. Jung, O. Klepper, W. König, P. H. Mokler, R. Moshammer, F. Nolden, U. Schaaf, P. Spädtke, M. Steck, P. Zimmerer, N. Grün, W. Scheid, M. S. Pindzola, and N. R. Badnell, "Dielectronic and radiative recombination of lithiumlike gold," Phys. Rev. Lett. **69**, 2768 (1992).

[14] N. R. Badnell, M. S. Pindzola, L. H. Andersen, J. Bolko, and H. T. Schmidt, "Dielectronic recombination of light Be-like and B-like ions," J. Phys. B **24**, 4441 (1991).

[15] A. Lampert, D. Habs, G. Kilgus, D. Schwalm, A. Wolf, N. R. Badnell and M. S. Pindzola, "High resolution measurement of dielectronic recombination of fluorine-like Se^{25+}," in the proceedings of the VIth International Conference on the Physics of Highly- Charged Ions, Manhattan, Kansas, 1992, p. 537.

[16] D. R. DeWitt, D. Schneider, M. H. Chen, M. W. Clark, J. W. McDonald, and M. B. Schneider, "Dielectronic recombination cross sections of neonlike xenon," Phys. Rev. Lett **68**, 1694 (1992).

[17] M. B. Schneider, D. A. Knapp, M. H. Chen, J. H. Scofield, P. Beiersdorfer, C. L. Bennett, J. R. Henderson, R. E. Marrs and M. A. Levine, "Measurement of the LMM dielectronic recombination resonances of neonlike gold," Phys. Rev. A **45**, R1291 (1992).

[18] P. C. W. Davies and M. J. Seaton, "Radiation damping in the optical continuum," J. Phys. B **2**, 757 (1969).

[19] E. Trefftz, "Time-dependent theory of overlapping resonances interacting with electron and photon continua: dielectronic recombination," J. Phys. B **3**, 763 (1970).

[20] R. H. Bell and M. J. Seaton, "Dielectronic recombination. I. General theory," J. Phys. B **18**, 1589 (1985).

[21] G. Alber, J. Cooper, and A. R. P. Rau, "Unified treatment of radiative and dielectronic recombination," Phys. Rev. A **30**, 2845 (1984).

[22] K. J. LaGattuta, "Interference effects in electron-ion recombination. III. Excited target states and continuum-continuum coupling," Phys. Rev. A **40**, 558 (1989).

[23] S. L. Haan and V. L. Jacobs, "Projection-operator approach to the unified treatment of radiative and dielectronic recombination," Phys. Rev. A **40**, 80 (1989).

[24] N. R. Badnell and M. S. Pindzola, "Unified dielectronic and radiative-recombination cross sections for U^{90+}," Phys. Rev. A **45**, 2820 (1992).

[25] M. S. Pindzola, N. R. Badnell, and D. C. Griffin, "Validity of the independent-processes and isolated- resonance approximations for electron-ion recombination," Phys. Rev. A **46**, 5725 (1992).

[26] R. D. Cowan, "Resonant-scattering (autoionisation) contributions to excitation rates in O^{4+} and similar ions," J. Phys. B **13**, 1471 (1980).

[27] M. H. Chen and K. J. Reed "Direct and resonance contributions to electron- impact excitation of $n = 2$ to $n = 3$ transitions in neonlike ions," Phys. Rev. A **40**, 2292 (1989).

[28] G. P. Gupta, K. A. Berrington, and A. E. Kingston, "An R-matrix calculation of the electron impact excitation of the fine-structure levels in neonlike selenium (Se XXV)," J. Phys. B **22** 3289 (1989).

[29] M. D. Hershkowitz and M. J. Seaton, "The calculation of resonances in electron-ion scattering using the distorted-wave approximation," J. Phys. B **6**, 1176 (1973).

[30] R. E. H. Clark, A. L. Merts, J. B. Mann, and L. A. Collins, "Quantum-defect and distorted-wave theory applied to the resonance contribution on inelastic electron scattering," Phys. Rev. A **27**, 1812 (1983).

[31] N. R. Badnell, M. S. Pindzola, and D. C. Griffin, "Validity of the independent-processes approximation for resonance structures in electron-ion scattering cross sections," Phys. Rev. A **43**, 2250 (1991).

[32] E. K. Wahlin, J. S. Thompson, G. H. Dunn, R. A. Phaneuf, D. C. Gregory, and A. C. H. Smith, "Electron-impact excitation of Si^{3+} (3s→3p) using a merged-beam electron-energy-loss technique," Phys. Rev. Lett **66**, 157 (1991).

[33] X. Q. Guo, E. W. Bell, J. S. Thompson, G. H. Dunn, M. E. Bannister, R. A. Phaneuf, and A. C. H. Smith, "Evidence for significant backscattering in near threshold electron-impact excitation of Ar^{7+} (3s→3p)," Phys. Rev. A **47**, R9 (1993).

[34] D. C. Griffin, M. S. Pindzola, and N. R. Badnell, "Theoretical calculations of angular distributions on inelastically scattered electrons from multiply charged ions," in the proceedings of the VIth International Conference on the Physics of Highly Charged Ions, Manhattan, Kansas, 1992, p. 473.

[35] B. A. Huber, Ch. Ristori, P. A. Hervieux, M. Maurel, C. Guet and H. J. Andra, "Differential cross sections for the 3s→3p excitation of sodiumlike Ar^{7+} ions by electron impact," Phys. Rev. Lett **67**, 1407 (1991).

[36] D. C. Griffin, M. S. Pindzola, and N. R. Badnell, "Low energy total and differential cross sections for the electron-impact excitation of Si^{2+} and Ar^{6+}," Phys. Rev. A **47**, 2871 (1993).

[37] D. L. Moores and M. S. Pindzola, "Electron-excitation of highly charged hydrogenic ions," J. Phys. B **25**, 4581 (1992).

[38] C. J. Fontes, D. H. Sampson, and H. L. Zhang, "Inclusion of the generalized Breit interaction in excitation of highly-charged ions by electron-impact," Phys. Rev. A **47**, 1009 (1993).

[39] H. Jakubowicz and D. L. Moores, "Electron-impact ionisation of Li-like and Be-like ions," J. Phys. B **14**, 3733 (1981).

[40] M. S. Pindzola, D. C. Griffin, and C. Bottcher, "Electron-impact excitation-autoionization of Ga II," Phys. Rev. A **25**, 211 (1982).

[41] D. L. Moores and K. J. Reed, "Electron-impact ionization of Se^{24+}," Phys. Rev. A **39**, 1747 (1989).

[42] M. S. Pindzola, D. C. Griffin, and C. Bottcher, "Correlation enhancement of the electron-impact ionization cross section for excited-state Ne-like ions," Phys. Rev. A **41**, 1375 (1990).

[43] S. S. Tayal and R. J. W. Henry, "Resonant-excitation-double-autoionization process in the electron-impact ionization of Li-like ions," Phys. Rev. A **44**, 2955 (1991).

[44] K. J. Reed and M. H. Chen, "Distorted-wave calculations of cross sections for electron-impact ionization of lithiumlike ions," Phys. Rev. A **45**, 4519 (1992).

[45] G. Hofmann, A. Müller, K. Tinschert, and E. Salzborn, "Indirect processes in the electron-impact ionization of Li-like ions," Z. Phys. D **16**, 113 (1990).

[46] M. H. Chen and K. J. Reed, "Electron-impact ionization of highly charged lithiumlike ions," Phys. Rev. A **45**, 4525 (1992).

[47] N. R. Badnell and M. S. Pindzola, "Resonance contributions to the electron-impact ionization of few-electron highly charged ions," Phys. Rev. A **47**, 2937 (1993).

[48] M. H. Chen and K. J. Reed, "Contributions of resonant excitation double autoionization to the electron-impact ionization of Fe^{15+}," Phys. Rev. Lett **64**, 1350 (1990).

[49] D. C. Gregory, L. J. Wang, F. W. Meyer, and K. Rinn, "Electron-impact ionization of iron ions: Fe^{11+}, Fe^{13+}, and Fe^{15+}," Phys. Rev. A **35**, 3256 (1987).

[50] M. H. Chen and K. J. Reed, "Indirect contributions to electron-impact ionization of Kr^{24+}, Kr^{25+}, and Xe^{43+}," Phys. Rev. A **47**, 1874 (1993).

[51] M. S. Pindzola, T. W. Gorczyca, N. R. Badnell, D. C. Griffin, G. H. Dunn, A. Müller, G. Hofmann, K. Tinschert, and E. Salzborn, "Dielectronic capture processes in the electron- impact ionization of Sc^{2+}," Phys. Rev. (submitted).

EXPERIMENTAL STUDIES OF DIELECTRONIC RECOMBINATION

Lars H. Andersen
Institute of Physics and Astronomy, University of Aarhus
DK-8000 Aarhus C, Denmark

ABSTRACT

The experimental techniques in the field of electron-ion recombination have advanced substantially over the last years. Some techniques are discussed in the present paper with the main emphasis on interacting-beams techniques. Results that emphasize the advances of the different methods are presented and discussed.

INTRODUCTION

Collisions between free electrons and positive ions take place in many places such as in man-made plasmas, the solar corona, gaseous nebulae, and interstellar space. Much of the early concern about electron-ion recombination was indeed related to astrophysics. Recombination between an electron and a positive ion normally results in the emission of one or more light quanta. These light quanta carry information about distant parts of the Universe. We may thus learn about the composition of our Universe and its history by knowing the *physical* origin of these photons as well as the interaction between matter and light in the media in which the photons travel. The field of electron-ion recombination, however, is not motivated only by astrophysics and plasma physics. Since, as will be discussed below, doubly excited states are often involved as well as high Rydberg states, precise knowledge about the *structure* of atoms and ions is required to accurately predict the rate of recombination between free electrons and positive ions and molecules. Some of the states involved may be very sensitive to external fields and interactions between different electronic configurations. This is reflected in the experimental data, and thus experimental information on the recombination processes is essential to guide us to the right theoretical comprehension of atomic structure.

As an electron approaches a highly charged ion, it is accelerated, and large kinetic energies may be obtained, especially in the vicinity of the nucleus. To become bound to the ion, the electron must exchange energy and momentum with a third 'particle'. The process, known as dielectronic recombination (DR) is a resonant process, in which the incoming electron excites an electron which is present on the ion. It thereby loses energy and becomes bound to the ion provided the initial kinetic energy is lower than the excitation energy. In this way, doubly excited ionic states are created. If these states decay radiatively, the electron may not be able to escape from the ion, and the DR process is completed. The DR process can be represented by

$$X^{Z+} + e^- \rightarrow (X^{(Z-1)+})^{**} \rightarrow (X^{(Z-1)+})^* + h\nu_1$$
$$\rightarrow (X^{(Z-1)+})^* + h\nu_2$$
$$\rightarrow .$$
$$\rightarrow .$$
$$\rightarrow X^{(Z-1)+} + h\nu_i .$$
(1)

If the core-excitation energy is ΔE and the initial kinetic energy of the electron in the ion rest frame is E_e, then for capture to a state γ, conservation of energy yields the resonance condition:

$$E_e + E_\gamma = \Delta E ,$$
(2)

where E_γ is the binding energy of the state designated γ.

The DR reaction has several signatures that one may map out in an experiment. First of all, a new ion is formed ($X^{(q-1)+}$) with a corresponding loss of initial X^{q+} ions. In addition, several characteristic photons are emitted — sometimes as X rays. Finally, the cross section exhibits a resonant character.

Before the early 1980's, the only accessible way to study electron-ion recombination was in various kinds of plasmas. The information was mainly limited to total rate coefficients (i.e., the product of the cross section and the electron-velocity distribution averaged over a wide Maxwellian velocity distribution).[1-3] Unfortunately, the plasma methods seem to have some disadvantages. First of all, due to the large energy spread of the contributing particles, the obtained results were not differential in nature. That is, individual resonances in the dielectronic-recombination cross section were not resolved. Problems also arise when several processes are competing in the plasma. Clearly, collisional-ionization processes take place, and the obtained results often rely on plasma models and calculated rates for ionization and charge-exchange recombination rates.

In 1983 several experiments on dielectronic recombination were reported almost simultaneously.[4-6] They were based on two well established experimental techniques, the crossed-beams and the merged-beams methods. A discussion of these 'first-generation' experiments on DR has been presented in an earlier comment by Belič and Pradhan[7] and by Dunn[8]. There is no doubt that several circumstances played a decisive role for the sudden experimental success. Apart from skillful scientists and advanced technology, it is likely that the appearance of calculated cross sections rather than integrated rate coefficients appealed to the experimentalists from the atomic-physics community. Moreover, the increasing activity in fusion research called for accurate cross sections related to the atomic processes in hot-fusion plasmas.

During the last years, the growth in the experimental activity on electron-ion recombination studies has continued, and progress in the field is being promoted by both experimentalists and theorists. Today, the techniques, which are being used to explore the nature of recombination, are well developed. The main ingredients of a good experiment are an assortment of ions in a well defined electronic configuration (this includes a well defined charge state), free electrons of a known density and with a well controlled and narrow velocity distribution relative to the ions, and an effective detection scheme to monitor the recombination reaction. Basically, two types of experiments are being used. One uses ion traps[9-10] as discussed by Stockli[11], another is with merged beams of electrons and ions which will be the main subject of the present contribution.

Naturally, the ideal target of electrons consists of *free* electrons. However, as an electron source one can also use a gas target. The DR resonances then become broad in energy due to the electron momentum distribution of the target electrons. The resonant process in ion-atom collisions akin to DR is called Resonant Transfer and Excitation (RTE)[12]. The RTE method has been used in a number of experiments, and the method looks perhaps most promising for very fast (~100-200 MeV/amu), highly charged ions. As an example, we mention the work by Graham et al. with U^{90+} on H_2.[13] Another spectacular method to study resonant processes that resemble the DR reaction is with channeled ions in crystals[14]. Ions confined to the open space between atomic rows interact only with the loosely bound outer electrons where the electron density may be of the order of ~10^{23} cm^{-3}. Both this and the RTE method are being used today[15,16] and constitute alternative methods to uncover the physics of DR.

MERGING BEAMS

The merged-beams approach is being utilized with electron coolers on ion-storage rings[17-21], and the experimental situation has indeed taken a new step forward with the development of these devices. It is possible to obtain good energy resolution due to the careful transport of electrons in the cooler device and the relatively low density of electrons (typically 10^7-10^8 cm^{-3}). Yet, the electron density is sufficiently high that merged-beam experiments can be performed because of the good vacuum conditions in the coolers which prevent electron capture from the rest-gas molecules to dominate the true recombination signal. An efficient detection scheme to monitor the recombination events is provided simply by counting the ions that have changed charge after the interaction with the electron target.

One of the first high-resolution experiments for DR studies was developed in Aarhus. An electron-cooler device was constructed for the ASTRID heavy-ion storage ring[22-23]. Before the installation on the ring, the electron cooler, or the electron target, was tested and used at one of the beamlines at the 6 MV EN-tandem accelerator in a merged-beams configuration. This particular experiment gave the first DR results in 1989[24], as also reported at the XVI ICPEAC by Hvelplund[25]. At the same conference, Wolf presented the first experimental results with the electron cooler at the Heidelberg Test Storage Ring (TSR).[26]

The success of the 'electron-cooler' experiments is based on the fact that it has been possible to produce a beam of electrons with a relatively low energy spread in an ultrahigh vacuum environment. Moreover, it has been possible to characterize the velocity distribution so that the measured cross sections averaged over the actual electron-velocity profile could be compared with theoretical predictions. A brief discussion of the characteristics of the electron target will follow below. A thorough description of electron-beam properties may be found in the report by Poth[27].

Generally, the relative motion between ions and electrons is governed by the motion of the electrons since the energy spread of the ion beam is generally small. It is therefore desirable to produce an electron beam that has a narrow velocity distribution. For an ordinary gas of electrons at a given temperature, one would describe the velocity distribution by a three-dimensional Maxwell distribution. However, the symmetry of the 'electron gas' in the beam is broken since the electrons are accelerated along a particular direction. This results in two different temperatures, one associated with the two-dimensional motion perpendicular to the beam direction $T_\perp$ and another, $T_\parallel$, associated with the one-dimensional motion parallel to the beam direction. We therefore describe the relative velocity distribution in the beam as

$$f(v) = \frac{m_e}{2\pi k T_\perp} e^{-m_e v_\perp^2 / 2kT_\perp} \sqrt{\frac{m_e}{2\pi k T_\parallel}} e^{-m_e(v_\parallel - \Delta)^2 / 2kT_\parallel}, \quad (3)$$

where k is Boltzmanns constant, m_e is the mass of the electron, and $v_\perp$ and $v_\parallel$ are the electron-velocity components relative to the mean-drift velocity perpendicular and parallel to the beam direction, respectively. Here Δ defines the detuning energy ($E_d = \frac{1}{2} m_e \Delta^2$).

The electron beam is normally generated from a tungsten cathode with a surface of BaO heated to a temperature $T_0 \approx 1200K$. At the surface of the cathode, the electrons are thus associated with an energy spread of about $3/2kT_0 \sim 0.15$eV. After acceleration to the desired energy, the parallel energy spread $kT_\parallel$ in the moving frame becomes[28]

$$kT_\parallel = \frac{kT_0}{8E_L} \cdot kT_0 + C \cdot (n_e)^{1/3}, \quad (4)$$

where E_L is the mean laboratory kinetic energy, n_e is the electron density in cm^{-3} and C is a numerical constant which is approximately equal to $2 \cdot e^2/(4\pi\varepsilon_0) = 2.8 \cdot 10^{-7}$ eV·cm.[28] The last term is caused by the electrostatic repulsion that increases the longitudinal temperature (longitudinal-longitudinal relaxation) during the acceleration. For typical electron densities in the order of $5 \cdot 10^7$ cm^{-3}, this additional term amounts to about 10^{-4}eV, which is an order of magnitude larger than the first contribution. There is no acceleration in the transverse direction and, consequently, no reduction of the transverse temperature.

The actual beam temperatures are found by a fitting procedure to the measured DR resonances.[29] Normally, the two temperatures are $kT_\perp \approx 0.1$-0.15eV and $kT_\parallel \approx 10^{-3}$eV. One main characteristics of the velocity distribution is that it is flattened: $kT_\parallel \ll kT_\perp$. Figure 1 displays a typical velocity distribution for an electron beam at 1keV laboratory energy. The energy spread in the ion frame is for relative energies $E_r \gg kT_\perp$ given by $\Delta E_r = (2 \cdot kT_\parallel \cdot E_r)^{1/2}$. This is shown in Fig. 2.

Since very small relative energies may be considered, the energy spread in the electron beam cannot in general be neglected. Instead of the cross section, the rate coefficient is often considered. The rate coefficient is defined as:

$$<v\sigma> = \frac{N^{(Z-1)+} - N_0^{(Z-1)+}}{N^{Z+}} \frac{v_i}{GL\rho\varepsilon},$$

where $N^{(Z-1)+}$ and $N_0^{(Z-1)+}$ is the count rate with and without the electrons being on, respectively, N^{Z+} is the number of incoming ions of charge Z, L is the length of the merge section, ε is the ion-detection efficiency, v_i is the ion velocity, and ρ is the electron density. G is the overlap function which has to be determined. In many merged-beams experiments, this may often require some extra effort. However, in electron coolers, the electron beam has a diameter larger than the ion-beam diameter. Moreover, as seen in Fig. 3, the electron density is approximately constant across the electron beam as directly measured in the Aarhus experiment.[29] Thus, G is to a good approximation equal to one, and the merged-beams technique offers a way to determine the DR cross section on an absolute scale without the need to normalize to other cross sections.

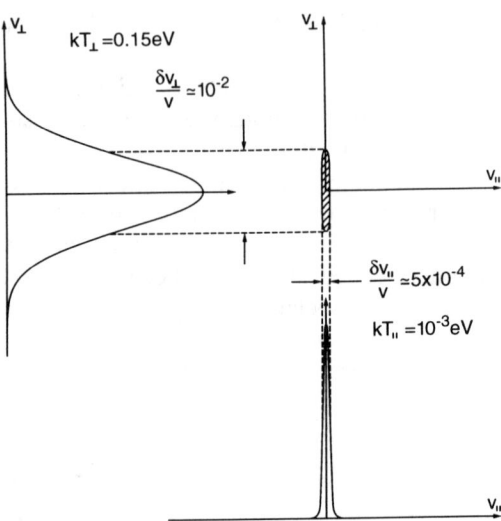

Figur 1 *Transverse and longitudinal velocity profiles of the electron beam at a laboratory energy of 1keV.*

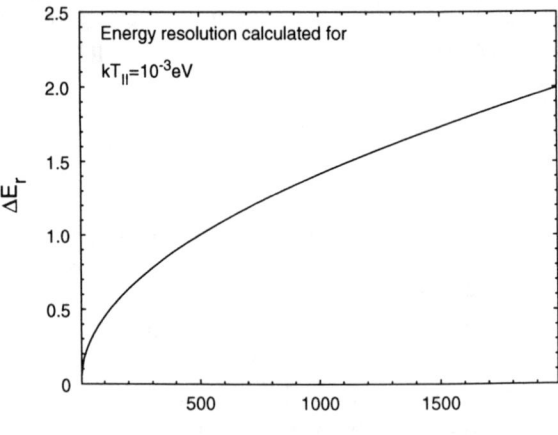

Figur 2 *Energy spread in the ion-rest frame as a function of relative energy in electron coolers.*

RESULTS

In the present section, we shall see some representative results from different experimental techniques that are presently being used in studying DR. At the space available here, it is not possible to discuss all matters that are currently under investigation. The reader may consult the recent NATO conference proceeding[30] on recombination of atomic ions for more details. Here we shall focus on a few different experimental approaches, and as an example, we consider DR with He-like ions.

The first experimental results with He-like ions were obtained with the Electron Beam Ion Trap (EBIT)[10] at the Lawrence Livermore National Laboratory. The trap[31] was used to produce He-like Ni^{26+}. The EBIT clearly demonstrated its excellent ability to produce ions of very high charge states relevant for fusion research. In the measurement, the X rays from the stabilizing transitions were measured as a function of electron-beam energy. The electron-energy resolution obtained at than time was about 55eV (FWHM), which is more that sufficient to resolve many DR resonances. The

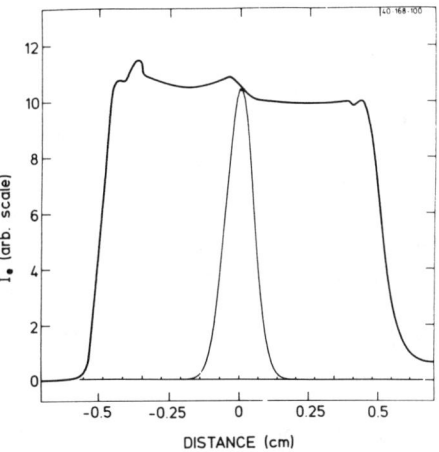

Figur 3 *Measured electron-beam profile of the 1 cm diameter electron beam at the Aarhus electron cooler. A typical ion beam profile is shown inside the electron-beam profile.*

disadvantage of the trap experiment was the lack of knowledge of the beam overlap. Consequently, absolute measurements were impossible. Instead, the DR data were normalized to the non-resonant radiative-recombination (RR) cross section for Ni^{26+} which was calculated. The RR signal was also used to estimate the charge-state distribution which had about 85% He-like Ni. The agreement with a multiconfiguration Dirac-Fock calculation was good.

The same year He-like O^{6+} was considered with the single-pass experiment at Aarhus.[24] For the first time, the interacting-beam technique showed strong signal and at the same time a narrow, sub-eV energy resolution. However, in the measurement only low relative energies were considered (0-~15eV). The DR resonances related to the $1s^2$ ground state were located at a relative energy of several hundreds of eV, and the observed resonances were due to the $1s2s(^1S)$ and $1s2s(^3S)$ metastable states. In the first paper from 1989, the metastable-beam composition was not known, and only relative rate coefficients were reported. Later the metastable fractions were determined by an independent measurement[32], and the data were presented as absolute rate coefficients. The O^{6+} data are shown in Fig. 4 together with a distorted wave calculation both in the LS coupling and the intermediate-coupling scheme. As seen from the figure, the rate coefficient is extremely large. This is due to the large radiative stabilization rate (A_r) for the $1s2p(^1P)$ state and the fact the DR cross section is proportional to the product $A_r \cdot A_a$, where A_a is the Auger rate. Also, A_a is large since it varies as $1/E_e$. There is some disagreement between theory and experiment mainly below the ^3P threshold. This disagreement is probably due to radiative transitions of Rydberg states $(^3P)nl$ into metastable Rydberg states $(^3S)nl$, that are stable on the time scale of the flight time of the ions from the target to the ion-charge-state analyzer as shown by independent measurements with a stronger analyzer field at the Heidelberg storage ring[33].

Dielectronic recombination with ground-state He-like Ar^{16+} ions was measured at the Electron Beam Ion Source (EBIS) at Kansas.[9] Again the trap method proved its ability to produce highly charged ions for DR studies. The energy resolution was about 40eV. As a comparison, the merged-beams method with a longitudinal temperature of 10^{-3}eV would provide a resolution of about 2eV at 2keV. The data from Kansas are displayed in Fig. 5. The cross section was normalized to the electron-impact-ionization cross section for Li-like Ar. The agreement with the calculation by Bhalla and Tunnell[34] is good.

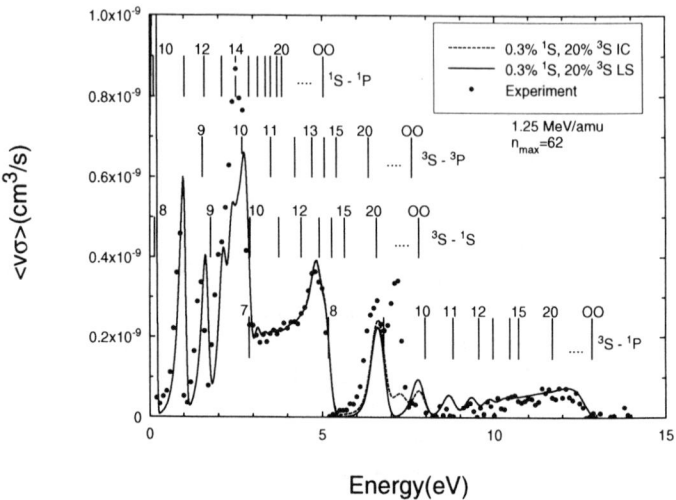

Figur 4 *The DR rate coefficient for O^{6+} as a function of energy. Shown are experimental data together with a distorted wave calculation by Badnell, Griffin and Pindzola.*

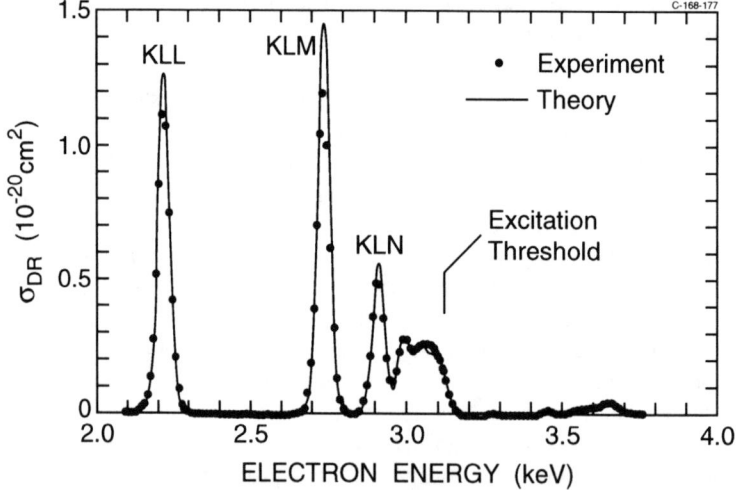

Figur 5 *The DR cross section for C^{4+}. Shown are experimental data together with the distorted wave calculation. From Ref. 36.*

Recently, a high resolution crystal spectrometer was used at the EBIT at Livermore. The resolution was improved to 4.5eV at a study of KLL X rays resulting from DR with He-like Fe^{24+}.[35]

As the last and latest example with He-like ions, we consider DR with ground-state He-like C^{4+} as measured at the Test Storage Ring at Heidelberg.[36] The beam was free from metastable states since the DR measurement starts after several seconds following injection. The energy resolution in the experiment was about 2eV, and several terms of the 1s2l2l' configurations were resolved. The general agreement with a distorted-wave calculation was good. However, as for the initial metastable states discussed above, there was some disagreement for the Rydberg resonances n= 6-8 of the 1s2p(^{1}P) series as shown in Fig. 6. Clearly, the presence of several open continuum channels complicates the situation.

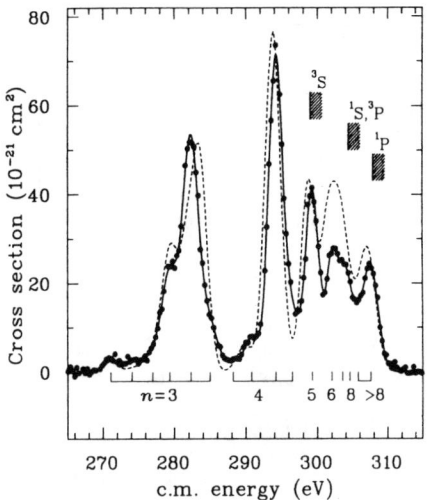

Figur 6 *The DR cross section for C^{4+}. Shown are experimental data together with the distorted wave calculation. From Ref. 36.*

FUTURE EXPERIMENTS

Clearly, there has been a tremendous progress in the experimental investigation of dielectronic recombination. Figure 7 summarizes the past results both for radiative and dielectronic recombination. It is seen that for the heavy elements, Z>20, the effort has mainly been concentrated on few electron systems where theory is available. One goal has been to study QED and relativistic effects. For the low-Z ions, there are still many systems that have not been investigated. In particular, it is striking that only a few singly charged ions have been considered. Moreover, these ions were subject only for the early '1983-experiments' (except for He^+ [37,38]). It therefore seems natural to study DR with singly charged ions with the high energy resolution currently available.

ACKNOWLEDGEMENTS

Discussions and communication of unpublished data with H.T. Schmidt are greatly acknowledged.

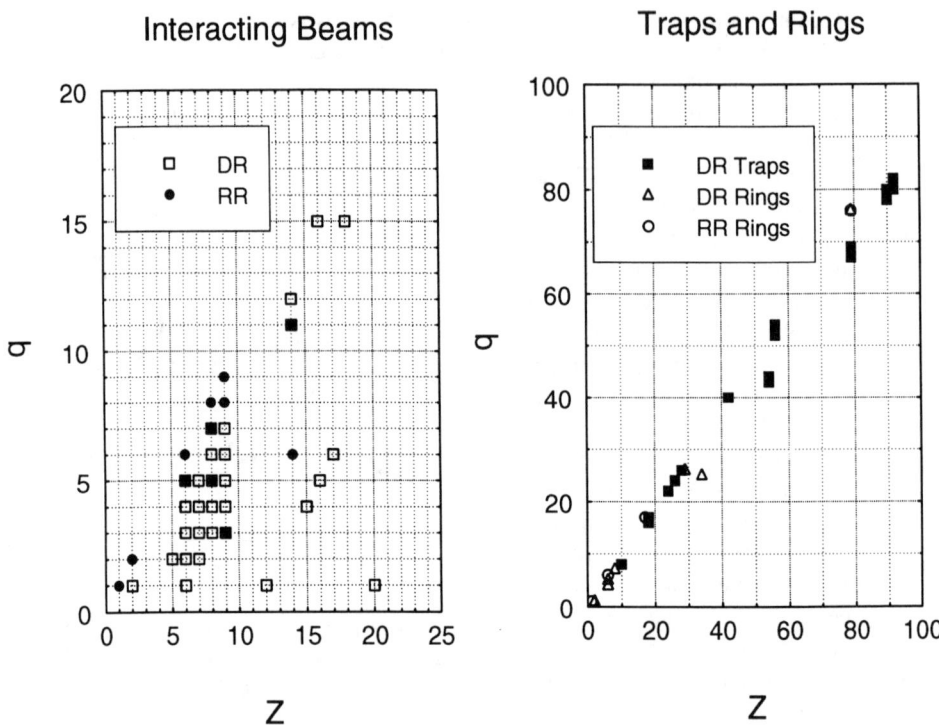

Figur 7 DR and RR studies have been performed for ions with nuclear charge Z and ion charge q. The graph to the left displays results with interacting beams, the other shows results with ion rings and ion traps.

1) R. L. Brooks, R.U. Datla, and H. R. Griem, Phys. Rev. Lett. **41** (1978) 107
2) C. Breton, C. De Michelis, M. Finkenthal, and M. Mattioli, Phys. Rev. Lett. **41** (1978) 110
3) R.C. Isler, E.C. Crume, and D.E. Arnurius, Phys. Rev. A **26** (1982) 2105
4) D. S. Belič, G.H. Dunn, T.J. Morgan, D. W. Mueller, and C. Timmer, Phys.Rev.Lett. **50** (1983) 399
5) P.F. Dittner, S. Datz, P.D. Miller, C.D. Moak, P.H. Stelson, C. Bottcher, W.B. Dress, G.D. Alton, N. Neskovic and C.M. Fou. Phys. Rev.Lett. **51** (1983) 31
6) J.B.A. Mitchell, C.T. Ng, J.L. Forand, D.P. Levac, R.E. Mitchell, A. Sen, D.B. Miko, and J.W. McGowan, Phys.Rev.Lett. **50** (1983) 335
7) D. Belič and A. K. Pradhan, Comments At. Mol. Phys. **20** (1987) 317
8) G.H. Dunn, in: *'Proceedings of a Nato Advanced Research Workshop on Recombination of Atomic Ions'*, eds. W.G. Graham, et al., Plenum Press, New York, 1992, p.115
9) R. Ali, C.P. Bhalla, C.L. Cocke, and M. Stockli, Phys.Rev.Lett. **64** (1990) 633; R. Ali, C.P. Bhalla, C.L. Cocke, M. Schulz and M. Stockli, Phys.Rev. A **44** 223 (1991)

10) D.A. Knapp et al., Phys. Rev. Lett. **62** (1989) 2104; P. Beiersdorfer, M.H. Chen, R.E. Marrs, M.B. Schneider and R.S. Walling, Phys.Rev. A **44** (1991) 396
11) M.P. Stockli, Z. Phys. D **21** (1991) 111
12) D. Brandt, Phys. Rev. A **27** (1983) 1314
13) W.G. Graham et al., Phys. Rev. Lett. **65** (1990) 2773
14) S. Datz et al., Nucl. Instrum. Methods Phys. Res., Sect. B **48** (1990) 114
15) E. M. Bernstein et al., Phys. Rev. A **40** (1992) 4085
16) J.U. Andersen et al., Phys. Rev. Lett. **70** (1993) 750
17) G. Budker et al., IEEE Trans. Nucl. Sci. **NS-22** (1975) 2093
18) G. Budker et al., Part. Accel. **7** (1976) 197
19) C. Habfast et al., Phys. Scr. **T22** (1988) 277
20) H. Poth, Physics Reports, **196** (1990) 135
21) See also proceedings of the workshop on: *'Electron Cooling and New Cooling Techniques'* eds. R. Calabrese and L. Tecchio, World Scientific (1991)
22) R. Stensgaard, Physica Scripta **T22** (1988) 315
23) S.P. Møller, in: *Proc. 14th Biannual Particle Accelerator Conference,* San Fransisco, May 1991
24) L.H. Andersen, P. Hvelplund, H. Knudsen, and P. Kvistgaard, Phys. Rev.Lett. **23** (1989) 2656
25) P. Hvelplund in: *Proc. 16th Int. Cnf. on the Phys. of Electronic and Atomic Collisions (ICPEAC),* eds. A. Dalgarno et al. AIP Conference Proceedings. New York (1990) p.556
26) A. Wolf et al., in: *Proc. 16th Int. Cnf. on the Phys. of Electronic and Atomic Collisions (ICPEAC),* eds. A. Dalgarno et al. AIP Conference Proceedings. New York.(1990) p.378
27) H. Poth, Physics Reports, **196** (1990) 135
28) A. V. Aleksandrov, in: *Proceedings of the Workshop on Electron Cooling and New Cooling Techniques,* World Scientific, Singapore, eds. R. Calabrese and L. Tecchio, (1991) p.279
29) L.H. Andersen, J. Bolko, and P. Kvistgaard, Phys.Rev. A **41** (1990) 1293
30) *'Proceedings of a Nato Advanced Research Workshop on Recombination of Atomic Ions',* eds. W.G. Graham, et al., Plenum Press, New York, 1992, p.115
31) M.A. Levine, R.E. Marrs, J.R. Henderson, D.A. Knapp, and M.B. Schneider, Phys. Scr. **T22** (1988) 157
32) L.H. Andersen, G.-Y. Pan, H.T. Schmidt, N.R. Badnell and M.S. Pindzola, Phys. Rev. A **45** (1992) 7868
33) H.T. Schmidt et al., private communication.
34) C.P. Bhalla and T.W. Tunell, J. Quant. Spectrosc. Radiat. Transfer **32** (1984) 141
35) P.Beiersdorfer et al. Phys. Rev. A **46** (1992) 3812
36) G. Kilgus, et al. Phys. Rev. A **47** (1993) 4859
37) J.A. Tanis et al. Nucl. Instrum. Methods. in Phys. Research. **B56/57** (1991) 337
38) T. Tanabe et al. Phys. Rev. A **45** (1992) 276

BASIC PROBLEMS IN DISSOCIATIVE RECOMBINATION OF DIATOMIC MOLECULES

Hidekazu Takagi

Physics Laboratory, School of Medicine, Kitasato University, Sagamihara, Kanagawa, 228 Japan

ABSTRACT: The configuration interaction (CI) theory and multichannel quantum defect theory for the dissociative recombination is reviewed. Recent progress on the theories and on those applications are reported: rotational effect, higher order effect in the CI theory, and remarks on electronic states.

1. INTRODUCTION

The dissociative recombination (DR) combines molecular dissociation and electron recombination processes:

$$AB^+ + e \to A + B.$$

In order to recombine with the ion, the incident electron must excite the molecular ion electronically or ro-vibrationally. In the case of electronic excitation, the main promoting force is configuration interaction (CI) between the one- and two- electron excited states. The ro-vibrational excitation can occur by the non-adiabatic interaction other than the electronic interaction.

Figure 1 shows three types of DR. Type a is induced by the CI only[1], which is called *direct process* and considered the major process. In type b, the dissociation induced by the CI occurs after the capture process into a ro-vibrationally excited Rydberg state[3] (*indirect process*). The type c is direct dissociation of the molecular ion with electron recombination, which is well investigated in the photodissociation[13].

A formulation for the direct process has been given by Bardsley.[2] About a decade after, the direct and indirect processes are uniformly formulated by Giusti[4] applying the multichannel quantum defect theory (MQDT). In this review, we will discuss the theoretical frameworks of these two theories and the related recent developments.

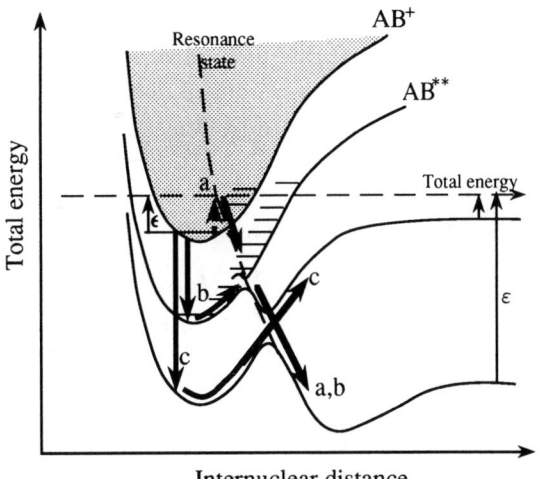

Figure 1. Schematic representation for potential curves and mechanisms of the DR. The solid curve denotes adiabatic state: AB^+ for ion, AB^* for Rydberg state, the lowest one for valence state. The broken curve in the continuum electronic state shows resonance energy, and below the ionization threshold, diabatic dissociative state. The short horizontal lines indicate ro-vibrational levels. The arrows indicate the potentials driving the relative nuclear motion along each path of tree mechanisms: a,b,and c.

2. CI THEORY FOR DR

Table 1 shows the channels and diabatic states related to the DR. Notation

Table 1. Diabatic states for the DR (*1)for a target in the ground state, (*2)including more than two

channel	state		electronic
	electron	nucleus	configuration(*1)
ionizing (initial)	continuum $\phi_\epsilon^{\tilde{\ell}}(r)$	ro-vibration $\chi_v^N(R)$	one-electron excitation
closed	Rydberg $\phi_n^{\tilde{\ell}}(r)$		
dissociative (final)	resonance $\phi_d(r)$	dissociation $F_\varepsilon^J(R)$	two-electron(*2) excitation

for the wave functions is also shown in the table, where ϵ and n indicate the energy and the principal quantum number of slow electrons respectively, v does the vibrational quantum number, and ε does dissociation energy (see figure 1). The coordinates for all electrons are symbolically represented by r, and the relative coordinate of the two nuclei is R. The superscript in the wave functions denotes angular momentum: $\tilde{\ell}$ is for slow electron, N for molecular rotation, and J for dissociating fragments, which is equal to the total angular momentum in the Born-Oppenheimer (BO) approximation.

We adopt $\{\phi_d(r) F_\varepsilon^J(R)\}$ and $\{\phi_\epsilon^{\tilde{\ell}}(r) \chi_v^N(R)\}$ as basis functions in order to represent the total wave function.

$$\Psi_E(r,R) = \sum_d \int a_{\varepsilon d}\, \phi_d(r)\, F_\varepsilon^J(R)\, d\varepsilon + \sum_v \int b_{v\epsilon}\, \phi_\epsilon^{\tilde{\ell}}(r)\, \chi_v^N(R)\, d\epsilon. \quad (1)$$

The coefficients $a_{\epsilon d}$ and $b_{v\epsilon}$ are determined as the $\Psi_E(r, R)$ becomes the eigenfunction of the total Hamiltonian with the total energy E, which is written as the sum of the nuclear kinetic-energy operator and the electronic Hamiltonian.

$$[T_R + H_{el}(r, R) - E]\Psi_E^J(r, R) = 0. \tag{2}$$

In the CI theory introduced by Bardsley[2, 5], the nuclear part of the dissociative channel $\xi_d^J(R)$ is expressed as

$$\xi_d^J(R) = \int a_{\epsilon d} \, F_\epsilon^J(R), \tag{3}$$

where the number of dissociative state is confined to one for simplicity.

Projecting equation (2) on the basis functions except for $\xi_d^J(R)$, we obtain an equation for $\xi_d^J(R)$.

$$\begin{aligned}
\Big[-\frac{1}{2M}\frac{\partial^2}{\partial R^2} &+ \frac{J(J+1) - \Lambda_d}{2MR^2} + E_d(R) - E\Big] \xi_d^J(R) = V_{d,\epsilon\tilde{\ell}}(R)\,\chi_{v_i}^N(R) \\
&+ \sum_{v'} \int dE' \,[\frac{1}{E - E'} + i\pi\delta(E - E')]\chi_{v'}^{N*}(R)\,V_{d,\epsilon'\tilde{\ell}}(R) \\
&\int dR'\,\chi_{v'}^N(R')\,V_{d,\epsilon'\tilde{\ell}}(R')\,\xi_d^J(R'),
\end{aligned} \tag{4}$$

where M is the reduced mass of the nuclei, Λ_d is angular momentum around the molecular axis in d state, and

$$E_d(R) = <\phi_d(r)|H_{el}|\phi_d(r)>, \tag{5}$$

$$V_{d,\epsilon\tilde{\ell}}(R) = <\phi_d(r)|H_{el}|\phi_\epsilon^{\tilde{\ell}}(r)> . \tag{6}$$

Verifying equation (4), we assume $E' = E'_{vN} + \epsilon'$ with E'_{vN} being ro-vibrational energy: we adopt the BO approximation. Because the CI does not change the angular momentum of the nuclei, J is equal to N. The cross section can be obtained from the asymptotic form of $\xi_d^J(R)$.

Two types of approximation is proposed in order to solve equation (4). One is called local approximation[2], where total energy E is assumed to be independent of v. Thus, the last term of equation (4) becomes local complex potential. The other approximation [5] is to neglect the principal part integral $\frac{1}{E-E'}$ assuming this part is included in $E_d(R)$. This assumption is justified if the electronic wavefunction is separated out from the nuclear part.

Another approach is possible starting from equation (2). Projecting the equation on the whole basis functions and imposing a boundary condition of the standing wave for the continuum functions, the equation to be solved becomes the Lippmann-Schwinger equation for the reactance matrix $\mathbf{K}$.[20]

$$K_{\beta\alpha}(E_\beta E_\alpha) = -\pi V_{\beta\alpha}(E_\beta E_\alpha) + \sum_\gamma \wp \int dE_\gamma \frac{V_{\beta\gamma}(E_\beta E_\gamma) K_{\gamma\alpha}(E_\gamma E_\alpha)}{E_\alpha - E_\gamma}, \tag{7}$$

with
$$V^J_{\varepsilon d, v\tilde{e\ell}} = < F^J_\varepsilon |V_{d,\tilde{e\ell}}(R)|\chi^J_v >, \qquad (8)$$

where we denote by α, β, or γ the set of channel indices (εd), or $(v\tilde{e\ell})$ and E is the total energy of the system.

The last term in equation (7) is contribution from the off-the-energy-shell and gives higher order effect in the perturbation theory. The CI theory is incorporated in the MQDT through the reactance matrix **K**, where the first order approximation is most available [4,5,6,14]: $\mathbf{K} = -\pi \mathbf{V}$. Judging from comparison with measured cross sections [7], this simple approximation seems to give fine results.

As for the higher order effect, there are two types of actual calculations. One is according to equation (4) with neglecting the principle part integral, where some configurations of molecular Rydberg states are taken into account.[5,17] Another is based on the second order approximation with neglecting the K_{dd}.[18,11,26] Further systematic investigation is necessary for understanding this effect.

3. MQDT FOR DR

We here consider a Rydberg state or a electron-scattering in low energy, of which asymptotic form is composed of an excited slow electron and an ionic core. In the MQDT, the space is divided into two regions: inner and outer regions. In the inner region near the ionic core, the motion is almost independent of the asymptotic energy because of Coulombic attractive force. At the edge of the inner region, we define the reactance matrix $\mathcal{R}$ from the asymptotic form of the slow electron, which we call smooth reactance matrix (smooth with the incident energy). The smooth matrix contains the full information on the effect of the interaction.

In the outer region, a wave function of the slow electron satisfies the physical boundary condition, which depends on its energy. The ordinary reactance matrix **R** and observable quantities such as cross sections are defined. Matching the wave functions of both regions at the boundary, Seaton has derived the following formula[23]:

$$\mathbf{R} = \mathcal{R}_{oo} - \mathcal{R}_{oc}[\mathcal{R}_{cc} + \tan \pi \nu]^{-1} \mathcal{R}_{co}, \qquad (9)$$

where the suffix o (c) indicates open (closed) channels and ν is the effective principal quantum number representing the energy of a closed channel.

The idea of "frame transformation" introduced by Fano[24] makes MQDT more powerful. The point of this theory is to take a most suitable physical picture for each region by choosing an appropriate approximation or coordinate. For example, Jungen and Atabek[28] have given the Born-Oppenheimer description for the inner region and the close-coupling description for the outer region. Matching the representation between both regions, we can obtain $\mathcal{R}$ itself. Their formulation succeeded to represent strong non-adiabatic interaction between the Rydberg states, to which the perturbative method is inapplicable.[25]

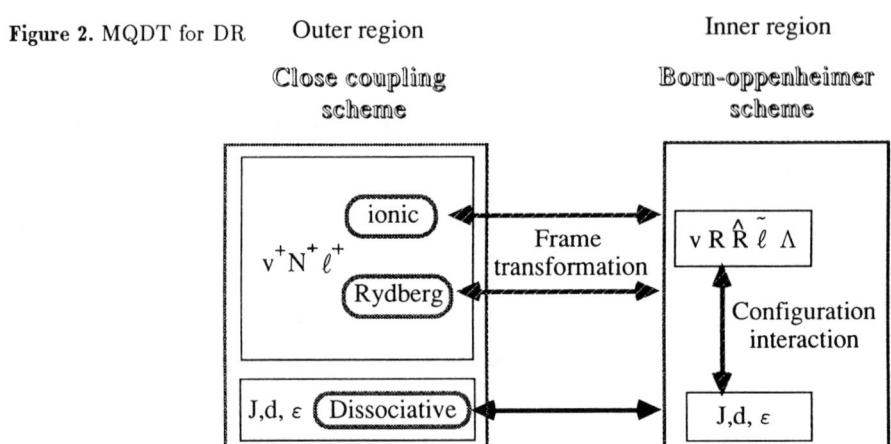

Figure 2. MQDT for DR

The MQDT by Jungen and Atabek just fits for the first capture step of the indirect process. Giusti[4] has incorporated the CI in the MQDT by developing a "two-step" method. First, we generate a basis set in witch matrix elements of the CI are diagonal. Second, using this basis set for the inner region, the MQDT treatment is performed. The scheme of the MQDT for the DR is shown in figure 2.

The MQDT for the DR has been extended to include the angular motion by Takagi at.al.[14] in order to discuss partial wave coupling of the slow electron and Rotational effect.[15] The cross section is given as follows.

$$\sigma^{DR}_{d,v^+N^+} = \frac{\pi}{2\epsilon}\rho \sum_J \frac{2J+1}{2N^++1} \sum_{\ell^+} |\tilde{S}^J_{Jd,v^+N^+\ell^+}|^2, \qquad (10)$$

where ρ is ratio of multiplicity of the two-electron excited state to that of initial electronic state, and the superscript + is given to the quantum number defined in the outer region. The sign $\tilde{S}$ denotes a reduced S matrix, where the word 'reduced' means that the azimuthal angular factor is removed. The reduced S matrix is represented using the reduced reactance matrix $\tilde{R}$, which is given by the reduced smooth reactance matrix $\tilde{\mathcal{R}}$ according to the reduced version of equation (9). The reduced smooth reactance matrix is obtained as follows.

$$(\tilde{\mathcal{R}})^J_{i^{+\prime},i^+} = \sum_\Lambda G(i^{+\prime}J\Lambda)\,(\mathcal{SC}^{-1})^{J\Lambda}_{i^{+\prime},i^+}\,G(i^+ J\Lambda), \qquad (11)$$

where we denote by i^+ set of quantum number $v^+N^+\ell^+$ or $J_d d$.

$G(i^{+\prime}J\Lambda)$ is a matrix transforming the representation from Λ to N^+, which can be presented using Clebsh-Gordan coefficients. The matrix $\mathcal{C}$ is defined as follows.

$$\mathcal{C}^{JN^+\Lambda}_{v^+\ell^+,\alpha} = \sum_{v\tilde{\ell}} \langle \chi^{N^+\Lambda^+}_{v^+} \mid \cos(\pi\mu_{\tilde{\ell}\Lambda}(R) + \eta^{J\Lambda}_\alpha)\, M_{\ell^+\tilde{\ell}}(R) \mid \chi^{J\Lambda}_v \rangle U^{J\Lambda}_{v\tilde{\ell},\alpha}, \qquad (12)$$

and
$$c_{d,\alpha}^{JJ_d\Lambda} = \cos(\eta_\alpha^{J\Lambda}) \, U_{d,\alpha}^{J\Lambda}. \tag{13}$$

The quantities $\mathcal{S}$ are the same as $\mathcal{C}$ except for the cosine function replaced by the sine function. $\tilde{\ell}$ in equation (7) is the quantum number of the eigenstate where the partial wave coupling is diagonalized, the matrix $\mathsf{M}(R)$ diagonalizes this coupling, and $\pi\mu_{\tilde{\ell}\Lambda}(R)$ is its eigen phaseshift; $\mu_{\tilde{\ell}\Lambda}(R)$ is called adiabatic quantum defect. $\eta_\alpha^{J\Lambda}$ is eigen phaseshift caused by the CI, where α indicates those eigenstates and U is the matrix diagonalizing the CI.

4. DIABATIC STATE AND ELECTRONIC PARAMETERS

4.1. Diabatic state

The non-adiabaticity in the adiabatic state is strong around avoided crossings, because the wave functions drastically change with the internuclear distance. Since the change is induced by the CI, we have regarded a diabatic state as a eigenstate of the Hamiltonian without the CI.

It is noteworthy that the serious non-adiabatic effect is brought by the localized CI around an avoided crossing. In the other regions, the effect of the CI never increases the non-adiabaticity but is necessary for accurate description of the electronic states. The delocalized CI must be included in the diabatic states. An example of the delocalized CI is seen in $^2\Pi$ state of CH molecule.[14] The two-electron excited dissociative state contains considerable Rydberg configuration at smaller internuclear distance than R=2.3 a.u.

The DR is sensitive to the crossing between a dissociative potential curve and the ionic one. Thus it is important to calculate those two energies with the same accuracy, especially for many-electron molecules. Although there are no rigorous methods to preserve the same accuracy, it is necessary, at least, to adopt common basis sets and electron configurations for the ionic state equivalent to the neutral excited one.

4.2. Electronic parameters

The electronic parameters required for the MQDT are the potential curve, the adiabatic quantum defect $\mu_{\tilde{\ell}\Lambda}(R)$, the strength of the CI $V_{d,\epsilon\tilde{\ell}}(R)$ (equation (6)), and the coupling magnitude in electronic partial waves $\mathsf{M}(R)$. Those parameters are obtained from

- scattering calculations of a electron by a molecular ion using the R-matrix method[10] or Kohn's variational method,[9] *etc*,
- calculation using Feshbach'spartition technique,[22, 26]
- analysis of adiabatic potentials,[14]
- analysis of spectroscopic data.[12]

Some problems are left even for the system of $H_2^+ + e$, which has been investigated most accurately. Figure 3 (a) shows discrepancy in the quantum defect among various analysis. The discrepancy mainly comes from how to remove the effect of the CI, which is separately taken into account. Figure 3 (b) shows various

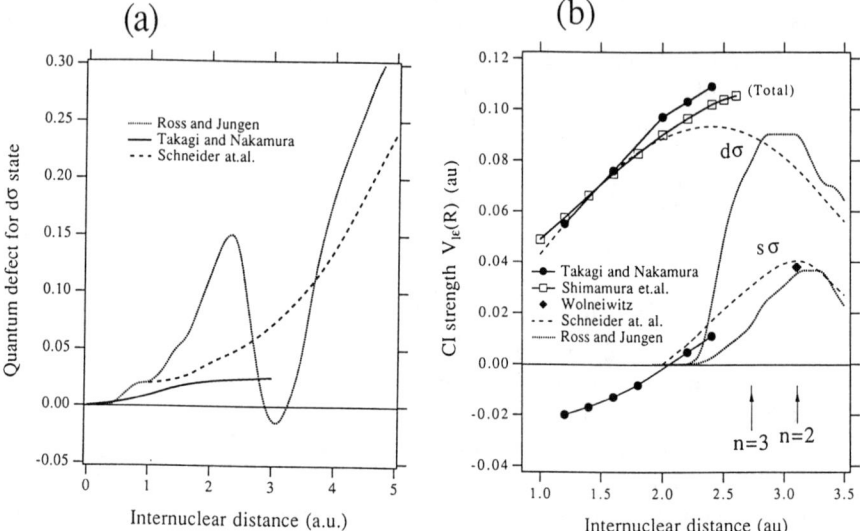

Figure 3. Parameters for H_2. (a) $\mu_{\tilde{\ell}\Lambda}(R)$ with $\tilde{\ell} = 2$. Takagi and Nakamura: Kohn's variational calculation,[9], Ross and Jungen: analysis of photo-absorption experiment by MQDT[12], schneider at.al.[11]. (b) Magnitude of CI: $V_{\tilde{\ell}e}(R)$ for $\tilde{\ell} = 0, 2$. o: amount of avoided crossing in EF and GK states,[8] □: R matrix calculation[10] (total magnitude).
Arrows indicate a crossing point of the two electron excited state with a Rydberg state of principal quantum number n.

analysis of the magnitude of the CI. The DR depends on relative sign between the $V_{d,e\tilde{\ell}}(R)$ of $s\sigma$ and that of $d\sigma$, although the sign has no meaning for the resonance scattering in clamped nuclei. In figure 3(b), the sign are arbitrarily chosen.

In both the quantum defect and the CI strength, we should remember that the electronic energy is different between the large and the small internuclear distances. The energy dependence should be investigated.

5. ROTATIONAL EFFECT

An actual calculation identifying the rotational states of the molecular ion was done for $H_2^+ + e$ according to the MQDT introduced in section 3. The condition of the calculation is actually the same as that in Nakashima et.al.[6] except that the rotational states were included; The first order approximation is employed for the CI (see section 2), and we assume that only the $d\sigma$ partial wave contribute to the DR. In this case, at most, three rotational states couple with one another for each J because $J = N^+ \pm 2, N^+$.

The present calculation shows that the rotational motion induces a number of clear resonance structures in the DR cross section as shown in figure 4. The rotational transition with a change of electronic state occurs easily because the rotational motion cooperatively couples with the CI through the anisotropy of the dissociative electronic state. Moreover, the centrifugal distortion of vibrational states induces a stronger nonadiabatic transition than the vibronic interaction without the rotational transition. On the other hand, the rotational structure is often narrow because the change of the energy is limited to be small by the conservation low of angular momentum.

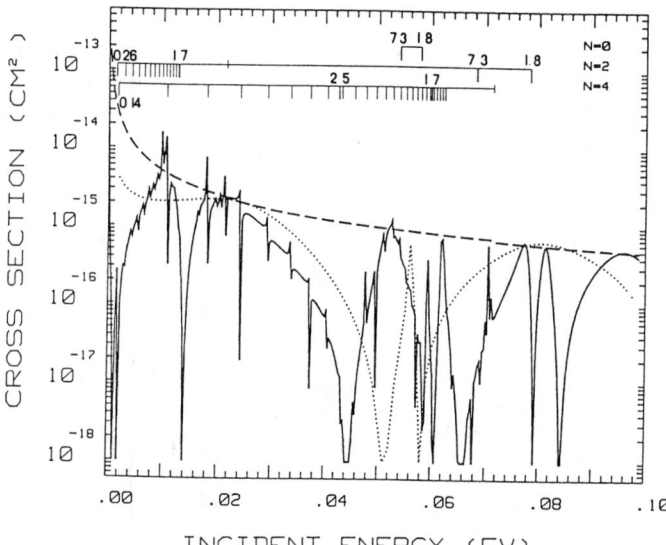

Figure 4. DR cross section for $v^+ = 0, N^+ = 0$. solid line: present full calculation, dashed line: without Rydberg contribution, dotted line: rotation free calculation. The upper scale indicates resonance position; bold vertical mark: principal quantum number $n \leq 8$, thin vertical mark : $9 \leq n \leq 39$, crossed vertical mark: ionization threshold, N: rotational state. The number on the vertical mark; first digit: vibrational state, other digit(s): n.

The magnitude of the CI changes with the total energy of the system mainly through the dissociative wave function. Thus, the rotational motion affects the direct process through the change in the total energy. This is outstanding at around the energy of the zero of $V^J_{ed,vel}$ defined in equation (8). The rate constant also clearly depends on the rotational state. As the result of rotational dependency, the DR is affected by the nuclear spin state: ortho- or para- Hydrogen.

6. HIGHER ORDER EFFECT IN CI

We here introduce the results obtained by solving equation (7) numerically for $H_2^+ + e$. In this calculation, we need to estimate the off-the-energy-shell elements of **V**: elements for all ϵ and ε. On the other hand, it has the following advantages:

- it represents the influence of the electronic distortion by the dynamical CI effect: non-local effect in K_{dd} can be confirmed.
- it is possible to obtain the solution beyond the perturbation theory.

The matrix elements $V_{\bar{\ell}\epsilon}(R)$ were calculated by using the static exchange approximation for the continuum state and a variational calculation for the resonance state.[19] As the method of numerical analysis for equation (7), we tried two method: iterative method and algebraic method.

The iteration started from $\mathbf{K}^{(1)} = -\pi \mathbf{V}$ forms the perturbation sequence. The value of the element of $\mathbf{K}$ strongly oscillates with each iteration and diverges. On the other hand, the algebraic calculation gives a stable result although the reactance matrix slightly fluctuates with energy, which seems to come from numerical error near a pole with the energy. It turns out that the reactance matrix becomes singular at certain energy (the S matrix is regular). This is the reason the usual iterative method does not work. Using the obtained reactance matrix, we did the MQDT calculation as in the previous section. The result is shown in figure 5. Compared with figure 4, the resonance structure is stressed because the

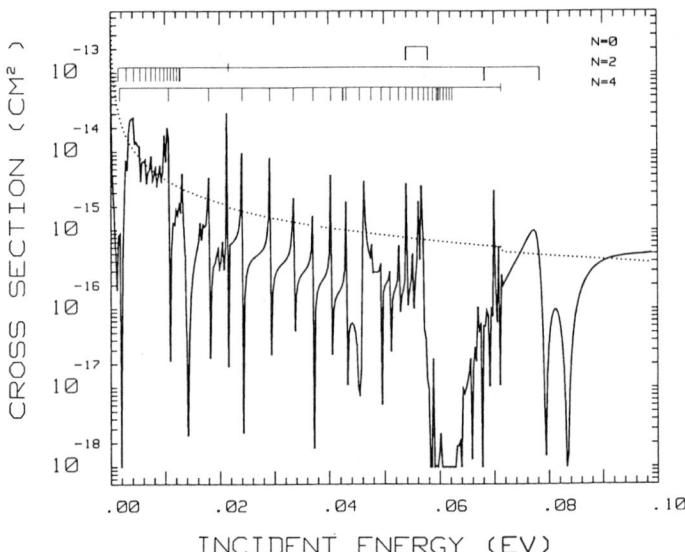

Figure 5. DR cross section for $H_2^+ + e$: higher order effect (see figure 4)

ro-vibronic transition is enhanced. The direct process is affected by about a factor of two typically. All the zeros of the matrix elements $K_{\epsilon d, v \epsilon \bar{\ell}}(E, E)$ in the first order approximation disappear due to the higher order effect. The calculated result clearly represent the large resonance structure at 60 meV in an experiment.[16]

7. CONCLUSION

We tried to present the basic theoretical aspects for the DR, especially for the low energy case. The mechanism of the rotational effect on the DR is not peculiar to H_2 but possible in other molecules. On the higher order effects of the CI, the iterative treatment for the reactance matrix is invalid, and the effect is not negligible within the accuracy of factor two except for happy coincidence.

Acknowledgements. The author would like to thank prof. S. Hara and prof. H. Sato for their collaborations on the study shown in the section 6. The present research is partly supported by Grant-in-Aid for Scientific Research on Priority Area "Theory of Chemical Reactions" from the Ministry of Education, Science and Culture.

References

[1] D. R. Bates, Phys. Rev., **78**, 492 (1950).
[2] J. N. Bardsley, J. Phys. B, **1**, 349 (1968).
[3] J. N. Bardsley, J. Phys. B, **1**, 365 (1968).
[4] A. Giusti, J. Phys. B, **13**, 3867 (1980).
[5] A. Giusti-Suzor, J.N. Bardsley and C. Derkits, Phys. Rev. A, **28**, 682 (1983).
[6] K. Nakashima, H. Takagi, and H. Nakamura, J. Chem. Phys., **86**, 726 (1987).
[7] J. B. A. Mitchell, Phys. Rep., **186**, 215 (1990).
[8] L. Wolniewcz and K. Dressler, J. Chem. Phys., **82**, 3292 (1985).
[9] H. Takagi and H. Nakamura, Phys. Rev. A, **27**, 691 (1983).
[10] I .Shimamura, C. J. Noble, P.G. Burke, Phys. Rev. A, **41**, 3545 (1990).
[11] I. F. Schneider, O. Dulieu, and A. Giusti-Suzor, J. Phys. B, **24**, L289 (1991).
[12] S. Ross and Ch. Jungen, Phys. Rev. Lett., **59**, 1297 (1987).
[13] Ch. Jungen, Phys. Rev. Lett., , (1984).
[14] H. Takagi, N. Kosugi, and M. LeDourneuf, J. Phys. B, **24**, 711 (1991).
[15] H. Takagi, J. Phys. B, submitted.; Dissociative recombination, ed. B.R.Rowe, J.B.A.Mitchell, and A. Canosa, 75(1993) Plenum.
[16] P. Van der Donk, F.B. Yousif, J. B. A. Mitchell and A.P. Hickman, Phys. Rev. Lett., **67**, 42 (1991); **68**, 2252 (1992).
[17] A. P. Hickman, J. Phys. B, **20**, 2091 (1987).
[18] A. P. Hickman, Dissociative recombination, ed. J. B. A. Mitchell and S. L. Guberman, 35 (1989). Wold Scientific Publishing (Singapore)
[19] H. Sato and S. Hara, J. Phys. B, **19**, 2611 (1986).
[20] R. G. Newton, "Scattering Theory of Waves and Particles", 188 (1982) Springer-Verlag, New York.
[21] H. Takagi and H. Nakamura, J. Chem. Phys., **84**, 2431 (1986).
[22] H. Feshbach, Ann. Phys. (NY), **19**, 287 (1962).
[23] M. J. Seaton, Proc. Phys. Soc., **88**, 801 (1966).
[24] U. Fano, Phys. Rev. A, **2**, 353, (1970).
[25] H. Takagi and H. Nakamura, J. Chem. Phys., **74**, 5808 (1981).
[26] S. L. Guberman and A. Giusti-Suzor, J. Chem. Phys. **95**, 2602 (1991).
[27] S. L. Guberman, Dissociative recombination, ed. J. B. A. Mitchell and S. L. Guberman, 45 (1989). Wold Scientific Publishing (Singapore)
[28] Ch. Jungen and O. Atabek, J. Chem. Phys., **66**, 5584 (1977).

POSITRON COLLISIONS AND EXOTIC SYSTEMS

PROGRESS IN POSITRON SCATTERING FROM ATOMS AND MOLECULES

M. Charlton and G. Laricchia
Department of Physics and Astronomy
University College London, Gower Street
London WC1E 6BT, UK

ABSTRACT

Recent advances in positron scattering are reviewed. Topics covered include new measurements of total scattering cross sections for atomic hydrogen and alkali metals, elastic scattering (integrated cross sections near the threshold for positronium formation and angle resolved), positronium formation (atomic hydrogen, near threshold phenomena and cross section behaviour at intermediate energies) and ionization (atomic hydrogen, molecules and double differential cross sections).

TOTAL CROSS SECTIONS

Total cross sections, σ_T, provide a baseline of information on scattering processes and have been extremely important in the development of e^+ physics[1,2]. The noble gases and various molecules have been extensively investigated in the energy range from $\approx$ 1-1000eV and, on the whole, there is reasonable agreement between the various experiments. A nice compendium of the e^+-noble gas σ_Ts has been given by Stein and Kauppila[1] and is shown in figure 1 where the importance of the Ps formation channel is evident. The threshold energy for this process is given by $E_{Ps} = E_I - B$ where E_I is the target ionisation energy and B the Ps binding energy. In the ground state B = 6.8eV thus for all the noble gases $E_{Ps} > 0$. It is also notable here that at low energy the e^+ σ_Ts tends to be lower than those for electrons (e^-). This has been discussed by many authors[1-3] in terms of the static and polarisation interactions which tend to add for e^- but cancel for e^+.

The Detroit group[4] have recently presented preliminary σ_Ts for $e^+(e^-)$-H scattering. This was done using an attenuation method by passing the projectiles through a thick target of H (and H_2) formed by filling a cooled aluminium chamber[5] using a radio frequency discharge tube. The absolute σ_Ts were obtained by normalisation to $e^{\pm}$-H_2 results[6] by taking the dissociation fraction in the H cell to be 0.5 (measured outside) or 1. The data are shown in figure 2 where they compare favourably with semi-empirical e^- results[7] and e^+ calculations [8,9]. Using these latter e^+ results and those of Morgan[10] together with experimental data for Ps formation[11] and ionisation[12], Stein *et al*[13] have constructed an empirical sum σ_T which they find to be higher than those shown in figure 2. This implies some contradiction between the various cross sections (see below for a discussion of Ps formation and ionisation).

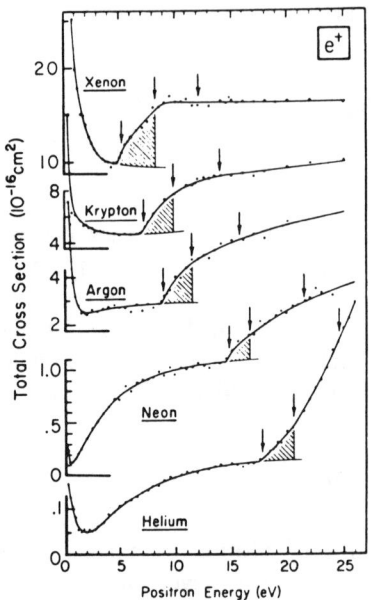

Figure 1. Compendium of e^+ total cross sections[1].

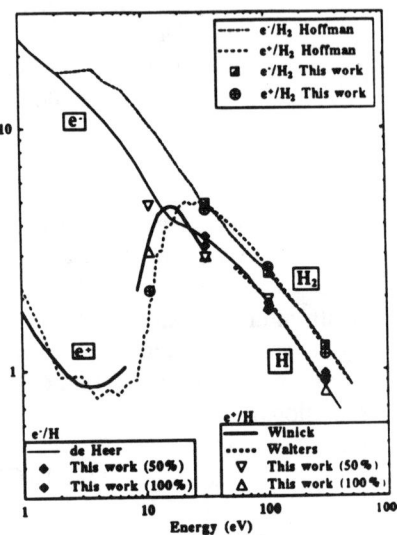

Figure 2. e^+-H total cross sections[4]. See text. Other works mentioned can be found in refs 6-9.

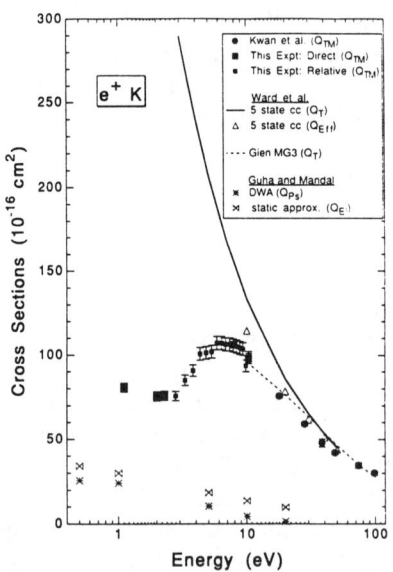

Figure 3. e^+ - K σ_T[15]. See also refs 14 and 16 and references therein.

The Detroit group has also pursued their program of σ_T measurements for the alkali metals. Their apparatus and methods have been described in detail elsewhere[14]. In the case of $e^{\pm}$-Na it has been found[14] that there is reasonable accord between theory and experiment when account is taken of the neglect of small angle forward scattering in the latter. The cross sections are notable for their large size ($\sim 10^{-18}$m^2), their tendency to rise rapidly at low energies and for the merging of the e^+ and e^- results at energies around 40eV below which the e^+ data exceeds that for e^-. This is the opposite of what is found in room temperature gases (see eg figure 1) and may be an indication of the greater relative importance of inelastic effects in e$^+$ scattering.

More curious results appear in the e$^+$ - K and e$^+$ - Rb systems where, as shown in figure 3, a peak is found in σ_T around 6eV[15] whereas theory[16] continues to rise steeply. Though some of this discrepancy may be due to the neglect of forward scattering[15], this could not be fully accounted for by a recent 5 state close-coupling calculation[16]. Further experiment and theory, perhaps concerning the role of Ps formation, are needed to shed more light on this problem.

ELASTIC SCATTERING

The single integrated elastic scattering cross section, σ_{el}, has recently come under scrutiny near E_{Ps} in helium gas[17]. This was an attempt to measure directly the cusp in σ_{el} inferred[18] from separate studies of the Ps formation cross section[19], σ_{Ps}, and σ_T[20]. The cusp was thought to arise due to competition for flux between the newly opening Ps channel and elastic scattering. The principle of the experiment was, adapting a technique pioneered by the Arlington group[21,22], to measure σ_T and σ_{Ps} sequentially using the same apparatus to determine σ_{el}. (The latter is practically equal to σ_T below E_{Ps} and to ($\sigma_T - \sigma_{Ps}$) up to the first excitation threshold). As shown in figure 4 however σ_{el} was found to vary smoothly across E_{Ps} with no hint of an appreciable cusp

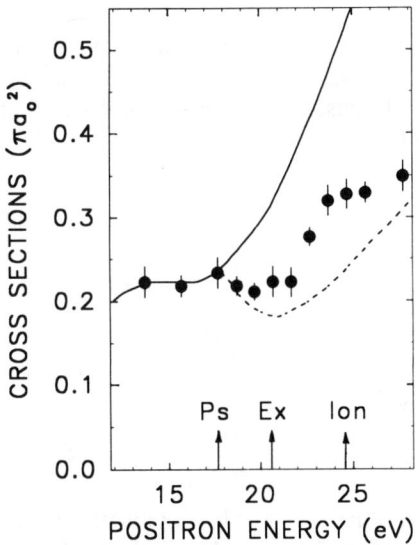
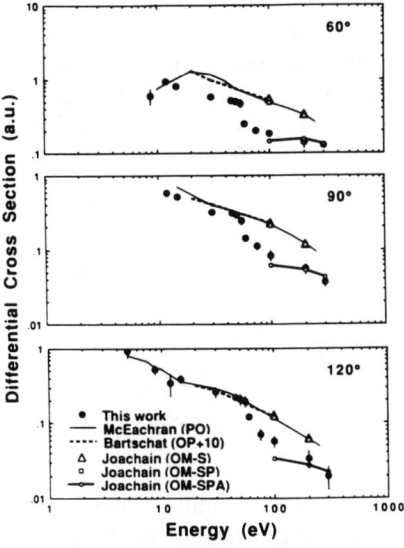

Figure 4. e$^+$-He collisions. The solid line is σ_T[20], the points are $(\sigma_{el}+\sigma_{ex})$[17] (see text) and the dashed curve was by deduced by Campeanu et al[18].

Figure 5. e$^+$-Ar dσ_{el}/dΩ for Ar collisions at three angles and variable energy[23]. Details of the theories can be found in ref 23.

structure. This topic is discussed further in the next section in the light of new near threshold measurements of σ_{Ps}.

Differential elastic scattering cross sections, $d\sigma_{el}/d\Omega$, continue to be measured and shed new light on the importance of absorption effects. Perhaps the most interesting discovery in this field has been that reported by Dou et al[23] who found structure in $d\sigma_{el}/d\Omega$ at intermediate energies in e^+ - Ar scattering. Similar effects have now been observed for other targets[24,25]. The e^+ - Ar $d\sigma_{el}/d\Omega$ are given in figure 5 at three fixed angles but as a function of impact energy. It is notable, particularly at 60° that the cross section shows a sharp drop between 55 and 60eV. The origin of this structure is not clear, though, given the appearance of similar features in other targets at different energies, is presumably quite general in nature. It may[23-25] be related to the "coupled channel shape resonance" found by Higgins and Burke[26] and others[27] in Ps formation in e^+ - H collisions. These appear to be examples of a new class of broad resonance in atomic physics which occur at intermediate energies due to complex coupling effects between the various open channels. It is notable that the broad features observed in $d\sigma_{el}/d\Omega$ appear in the energy range in which the e^+ ionisation cross sections exceed those for electrons and the size of the features may be related to the ratio of these cross sections. Much more experimental and theoretical activity is anticipated.

POSITRONIUM FORMATION

There have been a number of interesting new studies of Ps formation. Here we concentrate on measurements of σ_{Ps} for e^+ - H scattering, the behaviour of σ_{Ps} near threshold in the noble gases and selected molecules and σ_{Ps} at intermediate energies for the e^+ - He system.

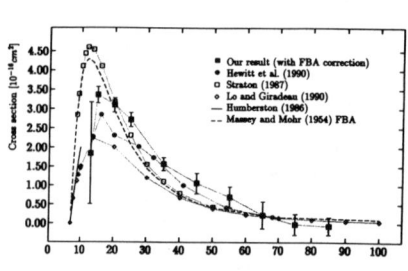
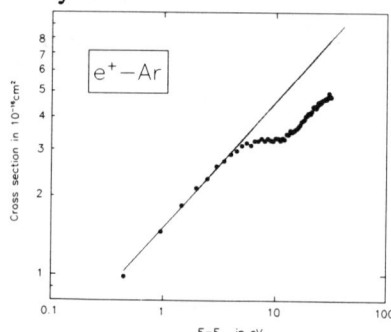

Figure 6. Positronium formation from H[11]. The theoretical work can be found in references 29-33.

Figure 7. Threshold plot of σ_{Ps} in argon. The line has a slope of 0.51 ± 0.01.

Firstly the atomic hydrogen experiment of Sperber et al[11]. In this work e^+ produced at the strong beam facility at Brookhaven National Laboratory, USA[28], were crossed with an atomic hydrogen beam and protons produced in the reaction $e^+ + H \rightarrow p + Ps$ detected. These were distinguished from protons produced by direct ionisation by virtue of the scattered e^+ present in the latter which enabled σ_{Ps} to be deduced, after appropriate normalisation, essentially by a subtraction technique. Their results are shown along with those for several theories[29-33] in figure 6, where σ_{Ps} is seen to rise steeply from threshold (6.8eV), reach a maximum at around 15eV and then fall to a very small value beyond ~ 70eV. The data shown in figure 6 have been subject to a FBA correction in an attempt to allow for large-angle ionisation in which the positron cannot be detected. The corrected data are in reasonable accord with a number of theories[30,32,33] but lower than others[29,31]. Further theory and experiment is anticipated.

The near threshold region of σ_{Ps} has been examined in detail recently for a number of gases in experiments[34-37] in which the total ionisation cross section (ie the sum of Ps formation and ionisation) was measured. Note, from above, that in the 6.8eV range below E_I this cross section is equal to σ_{Ps}. We present two examples of these measurements in figures 7 and 8 for Ar^{37} and $O_2{}^{35}$ respectively.

Turning first to Ar, the power law fit shown in figure 7 is of the form $\sigma_{Ps} = AE^n$, where A and n are constants and E is the Ps kinetic energy, and yields $n = 0.51 \pm 0.01$ for $E \leq 4eV$. From threshold theory which was developed in nuclear physics[38-41], but which is also appropriate to Ps formation[37], the expected law if only one partial wave in the Ps channel of orbital angular momentum number, l, is important is of the form $\sigma_{Ps} \propto E^{l + 1/2}$. The derived value of n points to $l = 0$ which, using parity and angular momentum conservation considerations[37], implies that the outgoing Ps is predominantly s-wave in character. Furthermore, by using σ_{Ps} together with

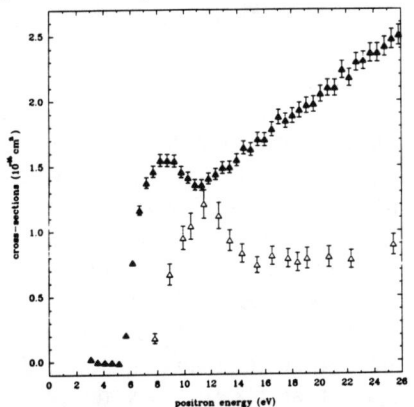

Figure 8. e^+-O_2 collisions. The filled triangles are the total ionisation cross sections[35] whilst the open ones are those for excitation[43].

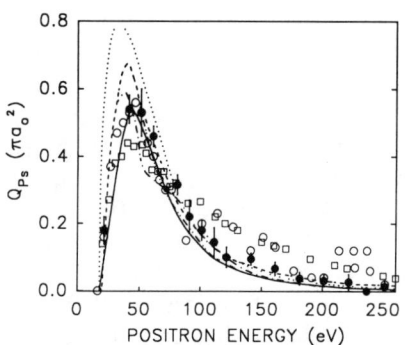

Figure 9. Positronium formation in e^+-He collisions. Experiment: filled circles (48), open squares (19), open circles (21,22). Theory see (48).

phase shift input from theory for σ_{el} calculated with the effect of Ps formation not included[42], Moxom et al[37] have predicted the magnitude and shape of the expected threshold behaviour in σ_{el}. They find for He and Ar that these are small, due to the weak e^+-atom coupling and in line with the smooth behaviour found in σ_{el} across E_{Ps} shown in figure 4[17]. Thus, here the threshold effect is almost entirely confined to σ_T which, as shown in figure 1, often displays an abrupt rise at E_{Ps}. Detailed measurements of σ_T near E_{Ps} may yet reveal the subtle threshold signatures.

The near threshold behaviour of the total e^+ impact ionisation cross section in O_2 is shown in figure 8[35] in which σ_{Ps} rises abruptly from threshold at around 5.3eV, reaches a maximum at ~8.5eV and then falls to a minimum at 11.5eV. This minimum coincides with a peak in the measured excitation cross section[43] which was attributed to a resonance in the excitation to the Schumann-Runge u-v continuum. Whatever the origin of this peak, the relative behaviour of the two cross sections suggests that the two channels are strongly coupled.

Before leaving this section we turn finally to the behaviour of σ_{Ps} at intermediate energies in He gas. Measurements of this cross section[19,22] found the fall-off as a function of impact energy to be E^{-1} to $E^{-1.5}$, in marked discrepancy with most theories which predicted a power in the range 3-5[44-47]. Schultz and Olson[47] attributed this to a failure of the experiments to account for positrons which were scattered at large angles in ionising collisions. This may have resulted in e^+ being lost from the beam by means other than Ps formation and led to an overestimate of σ_{Ps}. We present, in figure 9, some new data taken by Overton et al[48] together with the previous measurements and a selection of the theories. These new data are in much better accord with theory at intermediate energies.

IONISATION

As in previous sections, there is new data to report for atomic hydrogen. Ionisation in this fundamental system has now been studied by two groups who have reported cross sections, σ_i, for the reaction $e^+ + H \rightarrow e^+ + e^- + p$. The data are presented in figures 10(a) and (b) which show the two e^+ experiments[12,49], data for e^- impact[50] and the results of various theories[51-56]. Most of the latter agree reasonably with the data of Jones et al[49] however the recent work of Acacia et al[56] is in better accord with that of Spicher et al[12]. Their results are higher than those for e^- at almost all energies investigated and, most notably, do not merge with the latter even at 600eV. As reviewed elsewhere[57], this is not typical of most other targets which have been studied. At present the cause of the discrepancy between the two experiments is unknown and further experimental and theoretical work is necessary. The preliminary analysis of the partition of σ_T for e^+-H collisions[13] may also be pertinent.

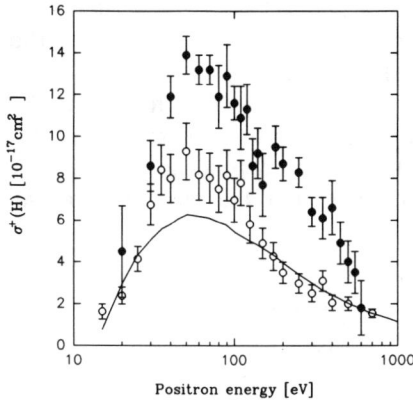

Figure 10a. σ_i for e^+-H collisions. The open circles are for ref 49, the filled circles ref 12 and the solid line is the e^- data.

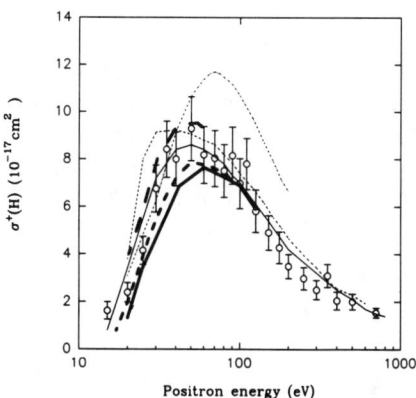

Figure 10b. The e^+-H σ_i data of Jones et al[49] compared to theory. Thick solid (σ_i), solid (52) thick long dash (53) thick dash (54), thin dash (55) and dot dash (the upper curve) (56).

The atomic physics group at the University of Aarhus have made major contributions to the study of ionisation by antiparticle impact[57]. Recently they have undertaken a detailed study of e^+-molecule ionisation, including higher order processes and fragmentation etc. This data has yet to be published[58] however a sample of the complexities involved is shown in figure 11. Here an ionic time-of-flight spectrum for 100eV e^+-CH_4 collisions is displayed - the overlapping features can, apparently, be deconvoluted to yield estimates of the relevant cross

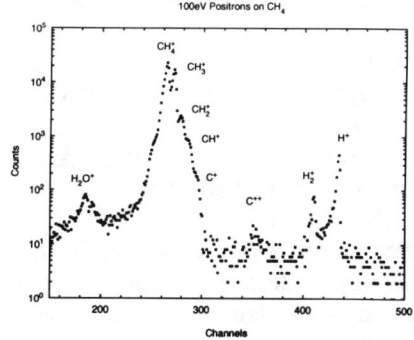

Figure 11. Ionic time-of-flight spectrum resulting from 100eV e^+CH_4 collisions[58].

sections. Comparisons with e^- can also be made. The e^+ - CH_4 system was also used in the first experiment[59] to detect the bound state of positronium hydride, PsH, via the (low cross section) reaction $e^+ + CH_4 \to PsH + CH_3^+$. The threshold for this process lies below that for free Ps and H production by ~1.06eV, the binding energy of PsH[60].

Progress is such that experiment is now beginning to go beyond the determination of integral ionisation cross sections and two groups have reported doubly differential (in angle and energy) cross sections, $d^2\sigma_i/dEd\Omega$[61-63]. (Note that the E and Ω can refer to either the energy and angle of the scattered or ejected particles). Some of these data are shown in figures 12 and 13. The former shows[61] $d^2\sigma_i/dEd\Omega$ at various ejected electron angles but at a fixed energy of 15eV for 100eV $e^{\pm}$-Ar collisions. The most notable feature is the order of magnitude enhancement (though with large errors), at small ejection angles of the e^+ results over those for e^-. The angular dependence of the data compare well with recent theory for H[64] but not with the other (hydrogen) measurements shown[65].

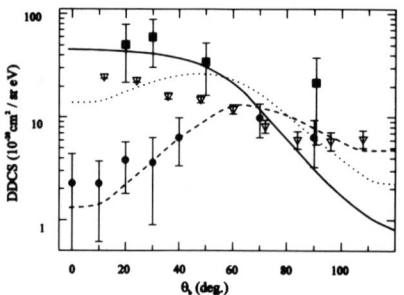

Figure 12. $d^2\sigma_i/dEd\Omega$ for e^+ (squares) and e^- (circles) for 15eV e^- emitted at various angles[61]. Other data for $e^{\pm}$-H: triangles (65) all other lines (64).

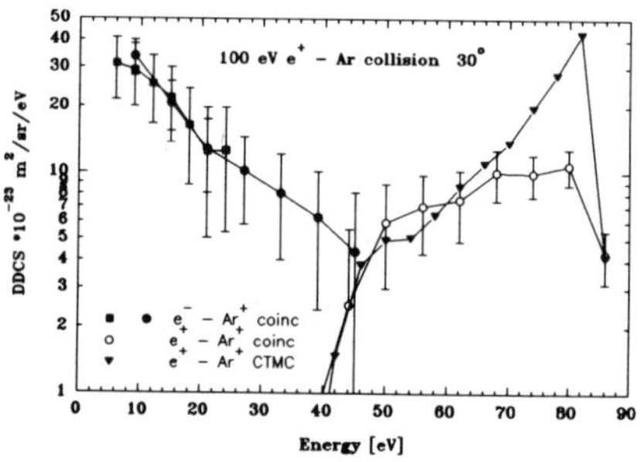

Figure 13. $d^2\sigma_i/dEd\Omega$ for 100eV e^+ on Ar at 30° for both ejected e^- and scattered e^+[63]. The CTMC calculation is from ref 66.

Figure 13 shows data, again for 100eV e^+-Ar collisions but now in which $d^2\sigma_i/dEd\Omega$ is plotted at a fixed angle (30°) versus the energy of the ejected e^- (5-45eV) and scattered e^+ (85-44eV)[63]. Also shown are some recent calculations[66] which are not in accord with experiment at high scattered e^+ energies. Electron measurements were also taken[63] and found to be in good accord with those of DuBois and Rudd[67] to which they were normalised. Comparing e^+ and e^-, a factor of ~2 enhancement of the e^+ over the e^- was found at 15eV ejection energy, rather than the order of magnitude deduced by Schmitt et al[61]. Clearly much work of interest remains to be tackled in this new area of e^+ scattering research.

CONCLUDING REMARKS

In this brief report we have surveyed some of the new things in atomic physics with e^+. In so doing we have had to be very selective. Some of the other notable highlights which, unfortunately, we have had to omit are e^+-molecule bound states[68], progress with the production of excited state Ps[69-71] and positronium beams[72]. These areas are just as important as the covered topics to the field of e^+ physics, which we hope to have shown in a flourishing branch of atomic physics.

ACKNOWLEDGEMENTS

We thank P.G. Coleman, L.M. Diana, N.P. Frandsen, T.T. Gien, W.E. Kauppila, G.R. Massoumi, R.P. McEachran, J.H. McGuire, R.E. Olson, W. Raith, P.J. Schultz, H. Schneider, D.M. Schrader, T.S. Stein, O. Sueoka, C.M. Surko, H.R.J. Walters and M. Weber for sending us preprints and reprints of their latest work. We also wish to thank all of our colleagues and collaborators particularly D.J. Day, G.O. Jones, A. Kover, H. Knudsen, W.E. Meyerhof, J. Moxom and N. Zafar.

REFERENCES

1. T.S. Stein and W.E. Kauppila, Adv. At. Mol. Phys. $\underline{18}$ 53 (1982).
2. M. Charlton, Rep. Prog. Phys. $\underline{48}$ 737 (1985).
3. H.S.W. Massey, Physics Today $\underline{29}$ (3) 42 (1976).
4. S. Zhou, W.E. Kauppila, C.K. Kwan and T.S. Stein, Abstract submitted to ICPEAC 18 (1993).
5. R.G.H. Robertson, private communication to Zhou et al[4].
6. K.R. Hoffman et al, Phys. Rev. A$\underline{25}$ 1393 (1982).
7. F.J. de Heer et al, J. Phys. B$\underline{10}$ 1945 (1977).
8. J.R. Winick and W.P. Reinhardt, Phys. Rev. A$\underline{18}$ 925 (1978).
9. H.R.J. Walters, J. Phys. B$\underline{21}$ 1893 (1988).
10. L.A. Morgan, J. Phys. B$\underline{15}$ L25 (1982).
11. W. Sperber et al, Phys. Rev. Lett. $\underline{68}$ 3690 (1992).
12. G. Spicher et al, Phys. Rev. Lett. $\underline{64}$ 1019 (1990).
13. T.S. Stein, W.E. Kauppila, C.K. Kwan and S. Zhou, Abstract submitted to ICPEAC 18 (1993).
14. C.K. Kwan et al, Phys. Rev. A$\underline{44}$ 1620 (1991).
15. S.P. Parikh et al, Phys. Rev. A$\underline{47}$ 1535 (1993).
16. S.J. Ward et al, J. Phys. B$\underline{22}$ 1845 (1989).
17. P.G. Coleman et al, J. Phys. B$\underline{25}$ L585 (1992).
18. R.I. Campeanu et al, J. Phys. B$\underline{20}$ 3557 (1987).
19. D. Fromme et al, Phys. Rev. Lett. $\underline{57}$ 3031 (1986).
20. T.S. Stein et al, Phys. Rev. A$\underline{17}$ 1600 (1978).
21. L.S. Fornari et al, Phys. Rev. Letts $\underline{51}$ 2276 (1983).
22. L.M. Diana et al, Phys. Rev. A$\underline{34}$ 2731 (1986).
23. L. Dou et al, Phys. Rev. Letts $\underline{68}$ 2913 (1992).
24. L. Dou et al, Phys. Rev. A$\underline{46}$ R5327 (1992).
25. L. Dou, W.E. Kauppila, C.K. Kwan and T.S. Stein, Abstract submitted to ICPEAC 18 (1993).
26. K. Higgins and P.G. Burke, J. Phys. B$\underline{24}$ L343 (1991).
27. R.N. Hewitt et al, J. Phys. B$\underline{24}$ L635 (1991).
28. M. Weber et al, Hyp. Int. $\underline{73}$ 147 (1992).
29. H.S.W. Massey and C.B.O. Mohr, Proc. R. Soc. London A$\underline{67}$ 695 (1954).

30. J.W. Humberston, Adv. At. Mol. Phys. 22 1 (1986).
31. J.C. Straton, Phys. Rev. A35 3725 (1987).
32. C. Lo and M.D. Girardeau, Phys. Rev. A41 158 (1990).
33. R.N. Hewitt et al, J. Phys. B23 4185 (1990).
34. J. Moxom et al, J. Phys. B26 (1993) in press.
35. G. Laricchia et al, Phys. Rev. Letts. 70 3229 (1993).
36. G. Laricchia and J. Moxom, Phys. Letts A174 255 (1993).
37. J. Moxom, G. Laricchia, M. Charlton, A. Kover and W.E. Meyerhof, submitted to Phys. Rev. Letts.
38. E.P. Wigner, Phys. Rev. 73 1002 (1948).
39. A.I. Baz, Soviet Phys. JETP 6 709 (1958).
40. R.G. Newton, Phys. Rev. 114 1611 (1959).
41. W.E. Meyerhof, Phys. Rev. 123 2312 (1962); 129 692 (1963).
42. R.P. McEachran et al, J. Phys. B12 1031 (1979).
43. Y. Katayama et al, J. Phys. B20 1645 (1987).
44. P. Mandal et al, J. Phys. B12 2913 (1979).
45. P. Khan and A.S. Ghosh, Phys. Rev. A28 2181 (1983).
46. N.C. Deb et al, Phys. Rev. A36 1082 (1987).
47. D.R. Schultz and R.E. Olson, Phys. Rev. A38 1861 (1988).
48. N. Overton, R.J. Mills and P.G. Coleman, submitted to J. Phys. B. (1993).
49. G.O. Jones et al, J. Phys. B, in press (1993).
50. M.B. Shah et al, J. Phys. B20 3501 (1987).
51. J.E. Golden and J.H. McGuire, Phys. Rev. Lett. 32 1218 (1974).
52. G. Peach, private communication to Shah et al (ref 50) (1986).
53. K.K. Mukherjee et al, J. Phys. B22 99 (1989).
54. A. Ohsaki et al, Phys. Rev. A32 2640 (1985).
55. K. Ratnavelu, Aust. J. Phys. 44 265 (1991).
56. P. Acacia, R.I. Campeanu, M. Horbatsch, R.P. McEachran and A.D. Stauffer, submitted to Phys. Lett. A (1993).
57. H. Knudsen and J.F. Reading, Phys. Rep. 212 107 (1992).
58. N-P Frandsen, Master thesis, Univ. of Aarhus (unpublished) (1993).
59. D.M. Schrader et al, Phys. Rev. Letts. 69 57 (1992).
60. Y.K. Ho, Phys. Rev. A34 609 (1986).
61. A. Schmitt, U. Cerny, H. Muller, W. Raith and M. Weber, submitted to Phys. Rev. Letts. (1993).
62. A. Kover et al, J. Phys. B, in press (1993).
63. A. Kover, G. Laricchia and M. Charlton, in preparation.
64. H. Klar and J. Berakdar (unpublished).
65. T.W. Shyn, Phys. Rev. A45 2951 (1992).
66. R.A. Sparrow and R.E. Olson, private communication and to be submitted to J. Phys. B.
67. R.D. DuBois and M.E. Rudd, Phys. Rev. A17 843 (1978).
68. S. Tang et al, Phys. Rev. Letts. 68 3793 (1992).

69. T.D. Steiger and R.S. Conti, Phys. Rev. A$\underline{45}$ 2744 (1992).
70. D.C. Schoepf *et al*, Phys. Rev. A$\underline{45}$ 1407 (1992).
71. D.J. Day, M. Charlton and G. Laricchia, to be published.
72. N. Zafar *et al*, Phys. B$\underline{24}$ 4661 (1991).

RECENT DEVELOPMENT IN THE THEORY OF POSITRON-HYDROGEN COLLISIONS

S. J. Ward
Department of Physics, University of North Texas, Denton, TX 76203-5368

ABSTRACT

A review is given of the theory of positron-hydrogen collisions. Recently, within the close-coupling framework resonances have been theoretically predicted to occur at intermediate energies in positron-hydrogen collisions. The mechanism for these resonances is unclear. Experimental measurements of positronium formation in positron-hydrogen collisions and positron-impact ionization of atomic hydrogen reveal that the positron-hydrogen system is not yet fully understood. In this review we (1) present the theoretical calculations which predict the intermediate energy resonances, (2) present the theory of positronium formation, both at low and high energy, and (3) discuss some theoretical work performed for ionization. The question of how significant electron capture to the continuum is for positron-hydrogen ionization is discussed. Comparison is made between theory and experiment for positronium formation and ionization.

I. INTRODUCTION

The recent development in the theoretical treatment of positron collisions has been promoted by the experimental advances made possible by the development of slow positron beams of increasingly high intensity. At the first positron satellite meeting held in 1981 the status of experimental work was primarily at measuring total angle-integrated cross sections for positron collisions from the inert gases.[1] Since that meeting, partial cross sections have been made for various inelastic processes, namely, excitation, positronium (Ps) formation and ionization.[2] Experimental measurements were reported for the first time in 1985 for positron collisions from non-room temperature gases, namely the alkakis.[3] Recently, experimental measurements have been reported for positron impact ionization of atomic hydrogen[4,5] and Ps-formation in positron-hydrogen collisions.[6,7] In this review paper we will focus primarily on theoretical work performed on positron-hydrogen collisions. The reason is twofold: Firstly, these experimental measurements reveal that the positron-hydrogen system is not fully understood. Secondly, resonances have been theoretically predicted to occur for positron-hydrogen collisions at intermediate energies.[8]

Positron-hydrogen collisions are of theoretical interest for several reasons. Firstly, the positron-hydrogen system represents a fundamental system interacting via Coulomb forces. Secondly, positron-hydrogen collisions are of astrophysical significance. Gamma rays with a well defined energy of 511keV arising from positron annihilation have been observed from solar flares and from the galactic center.[9] Measuring the width of the line provides information on the radiating medium which presumably consists of hydrogen atoms.

In section II we present the new kind of resonance theoretically predicted to occur at intermediate energies in positron-hydrogen collisions. In section III we present theory for Ps-formation at low and high energies, and make a comparison with experiment. In section IV we discuss how the theoretical calculations of ionization compare with experiment; in particular we discuss the process of Ps-formation into the continuum. Finally, in section V we give a summary.

II. INTERMEDIATE ENERGY RESONANCES

A new kind of resonance has been theoretically predicted to occur for S-wave positron-hydrogen collisions at an energy above the ionization threshold of hydrogen.[8] This resonance Higgins and Burke[8] predict with the framework of the coupled static approximation (CSA). The CSA retains only the ground-state target and positronium wave functions in the close coupling expansion. The close coupling wave function takes the form

$$\Psi = \sum_\nu \varphi_\nu^H(\mathbf{r}_1) F_\nu(\mathbf{r}_2) + \sum_{\nu'} \varphi_{\nu'}^{Ps}(\mathbf{r}_2) G_{\nu'}(\mathbf{R}) \qquad (1)$$

where an expansion has been made about both the hydrogen target $\varphi_\nu^H(\mathbf{r}_1)$ and the positronium $\varphi_{\nu'}^{Ps}(\mathbf{r}_2)$ states. The functions $F_\nu(\mathbf{r}_2)$ and $G_{\nu'}(\mathbf{R})$ to be numerically determined correspond to the motions of the positron with respect to the hydrogen atom and the positronium with respect to the proton, respectively. The coordinates used in Eq. (1) are as follows: $\mathbf{r}_1(\mathbf{r}_2)$ refers to the position vector of the electron (positron) with respect to the proton, $\mathbf{r}_3$ refers to the position vector of the electron with respect to the positron, and $\mathbf{R}$ refers to the position vector of the C.O.M of the positronium atom with respect to the proton. The position and width of this resonance predicted in the CSA are 35.6eV and 4.2eV, respectively. This resonance occurs totally in the Ps scattering channel.

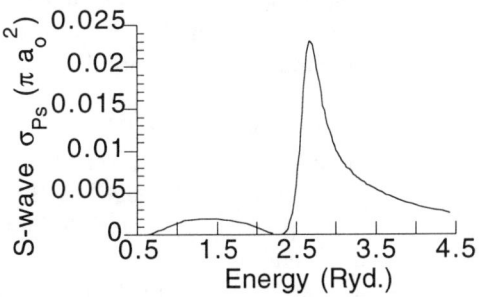

Fig. 1 Higgins and Burke S-wave resonance.[8]

Although it is very pronounced in the S-wave Ps-formation cross section, it is not very prominent in the summed S, P, D and F partial Ps-formation cross section. The mechanism for this resonance is unclear since it occurs at an energy much above the ionization threshold. As Higgins and Burke[8] demonstrate this new resonance differs from the usual shape and Feshbach resonance which both occur as resonances or bound states even when the coupling potential is switched off. In the Higgins and Burke case there are no resonances or bound states in the uncoupled Ps scattering channel.

A number of authors confirm this new coupled channel shape resonance within the close coupling approximation (CCA) in which up to the $n = 2$ states of both hydrogen and positronium are explicitly included.[10,11,12,13] However, with the inclusion of each target and positronium state the position of the resonance shifts to a higher energy and becomes narrower.[12] In the H($1s, 2s, 2p$), Ps($1s, 2s, 2p$) close coupling calculation the Higgins and Burke resonance shifts to 51.5eV.[11] Furthermore, in this close coupling calculation the Higgins and Burke resonance is also present in all other channels.[11] Additional structure is also seen in the vicinity of 10–20eV for all channels.[11] Very recently, Higgins[13] has predicted these coupled channel shape resonances in the S, P, D and F partial waves. Higgins[13] employed the CCA in which the states H($1s, 2\bar{s}, 2\bar{p}$), Ps($1s, 2\bar{s}, 2\bar{p}$) were included. (As is conventional the bar indicates a pseudostate). However, resonance structure at intermediate energies in the S, P, D and F partial waves is not present in the hyperspherical approximation as employed by Igarashi and Toshima.[14] Therefore, to determine whether the resonances predicted by CSA and CCA are true resonances, experimental verification as well as a theoretical calculation which explicitly includes the break-up channel would be valuable.

III. Ps-FORMATION IN POSITRON-HYDROGEN COLLISIONS

A. LOW ENERGY

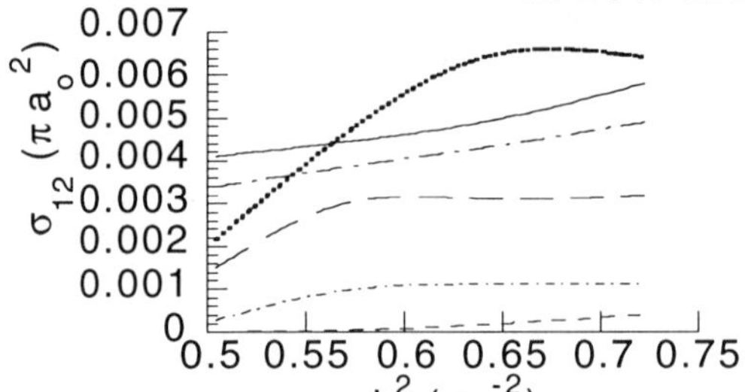

Fig. 2 S-wave Ps-formation cross section The solid line (———), the long-dashed (— —), the short-long dashed line(— -), the dotted line ($\cdots$), the dot-dashed line (-· -), and the dashed line (- - - -) represent the following theoretical calculations: Kohn variational,[15–17] CSA with 26 correlation terms,[21], Hyperspherical,[20], CCA,[23] the first Born (10^{-3}) and the CSA.[22]

The most accurate result for the S-, P-, and D-wave contibutions to the Ps-formation cross section in the ore gap are presumably those obtained by Humberston and co-workers using the two channel Kohn variaional method with very elaborate trial functions.[15–19] The ore-gap is the energy region between the ground-state Ps-formation threshold (6.8eV) and the first excitation threshold of the target (hydrogen) atom (10.2eV). Archer et. al.[20] have employed the hyperspherical coordinate approach to calculate S-wave elastic, excitation, ground and excited state Ps-formation cross sections. Forty-three adiabatic basic functions were used to solve the coupled channel equations for energies below the H(n=4) threshold. Close coupling calculations have been performed by a number of authors to calculate the Ps-formation cross sections. A large number of these calculations have been performed only for energies below the first excitation threshold of hydrogen H(n=2). For example, for S-wave scattering and for energies below the n=2 threshold of hydrogen, Chan and Fraser have performed a CSA with the addition of 26 correlation terms.[21] Abdel-Raouf et. al.[22] have calculated in the CSA all partial waves up to $l = 6$ for energies in the ore gap. Basu et. al. [23] have performed five state H($1s, 2s, 2\bar{p}, 3d$), Ps($1s$) close coupling calculation. Fig. (2) compares the S-wave Ps-formation cross sections obtained by various theoretical calculations: CSA, CCA, hyperspherical, Kohn variational and first Born.[24,25] This figure shows that the S-wave Ps-formation cross section for energies in the ore gap is very sensitive to the number of target states retained in the close coupling expansion. Recently, a close coupling calculation has been performed using by McAlinden et. al.[11] the following expansion schemes:

(a) H($1s, 2s, 2p$), Ps($1s, 2s, 2p$),
(b) H($1s, 2s, 3\bar{s}, 4\bar{s}, 2p, 3\bar{p}, 4\bar{p}, 3\bar{d}, 4\bar{d}$), Ps($1s, 2s, 3\bar{s}, 4\bar{s}, 2p, 3\bar{p}, 4\bar{p}, 3\bar{d}, 4\bar{d}$).

The results obtained with the coupled scheme (b) are in fairly good agreement with the accurate Kohn variational results for the S, P and D partial waves.

Several close coupling calculations have been performed which extend to energies above the H(n=2) threshold.[11,12,26] The first close coupling calculation to include the n=2 states of positronium explicitly was performed by Hewitt et. al.[26] It is important to include the n=2 states because positronium is highly polarizable, having dipole polarizability of 36 a_0^3, and the Ps($2p$) state will account for two thirds of this polarizability. The two coupling schemes employed by Hewitt et. al.[26] are (a) and (c), where (c) is given by

(c) H($1s, 2s, 2\bar{p}$), Ps($1s, 2s, 2p$).

B. HIGH ENERGY

It is fairly well understood that for proton collisions the dominant mechanism for electron capture at asymptotically high velocities is the Thomas double collision mechanism. The double collision process which was first described classically by Thomas[27] in 1927 has subsequently been identified with second Born terms by Drisko.[28] The classical description proposed by Thomas is as follows: The incident high-speed projectile strikes the electron, which is considered free, deflects it toward the target where it undergoes a further collision. The electron emerges from the second collision with almost the same velocity as the scattered projectile and therefore capture can easily occur. The angle of the outgoing particle with respect to the incident projectile is known as the critical angle. This angle depends purely on the kinematics. For proton scattering the critical angle for this Thomas double collision mechanism is $0.47 mrad$ whereas for positron scattering it is 45°.

Another double scattering mechanism can occur which also results in electron capture. The high velocity projectile strikes the electron, which is considered free, knocks it out of the atom, and is itself deflected toward the target where it undergoes a further collision. The scattered projectile and emerged electron have the same velocity so that capture can easily occur. The critical angle for this double scattering mechanism is 60° for proton collisions whereas for positron collisions it is again 45°. Shakeshaft and Wadehra pointed out that since for positron collisions the two double collision mechanisms occur at the same critical angle (45°), interference occurs between the two second order Born amplitudes which describe these mechanisms. The interference is destructive for capture into states of even angular momentum, e.g. 1s, but constructive for capture into states of odd angular momentum, e.g. 2p. At high energy, perturbation and distorted wave methods are often employed because these methods can incorporate in a reasonable way the continuum intermediate states which are necessary to describe double scattering.

Recently, Igarashi and Toshima[30,31] have performed an exact second Born calculation to obtain differential and angle-integrated cross sections for Ps-formation into $1s, 2s$ and $2p$ states. This work represents the first exact calculation of second Born amplitudes for Ps-formation to excited states. Because the interference causes delicate cancellation of second order terms, it is important that each amplitude be evaluated with very high precision.

For electron capture by protons Briggs et. al.[32] demonstrate that the use of Coulomb intermediate states rather than plane wave intermediate states provides a better description of the angular distribution near the Thomas peak. The Strong Potential Born (SPB) and the Impulse Approximation (IA), which use Coulomb Green operators, are in closer agreement with experimental measurements of the angular distribution for proton-hydrogen than are the second Born and higher Born terms. In view of this McGuire et. al.[33] and Deb et. al.[34] apply the SPB approximation, and Ward and Macek[35] apply the IA approximation to calculate Ps-formation cross sections for positron collisions from hydrogenic ions.

McGuire et. al.[33] and Deb et. al.[34] start with the post form of the exact T-matrix, namely,

$$T = T_1 + T_2 - T_3, \qquad (2)$$

where

$$\begin{aligned} T_1 &= \langle \Phi_f | [V_{eT} + V_{PT}] | [1 + G^+ V_{PT}] | \Phi_i \rangle \\ T_2 &= \langle \Phi_f | [V_{eT} + V_{PT}] | [1 + G^+ V_{Pe}] | \Phi_i \rangle \\ T_3 &= \langle \Phi_f | [V_{eT} + V_{PT}] | \Phi_i \rangle. \end{aligned} \qquad (3)$$

In Eq. (3) G^+ is the exact Green operator corresponding to the total Hamiltonian H of the system, and V_{eT}, V_{Pe} and V_{PT} are the electron-target, projectile-electron and projectile-target interactions, repectively. The initial and final states of the system are denoted by Φ_i and Φ_f, repectively. McGuire et. al.[33] and Deb et. al.[34] obtained the SPB approximations to the T-matrix by employing the following approximations:

(i) In the term T_1, the exact Green operator is replaced by the Green operator corresponding to the interaction V_{Pe}—i. e. the interactions $V_{eT} + V_{PT}$ are neglected. The Coulomb intermediate states therefore used in T_1 are the Ps intermediate states.

(ii) In the term T_2 the exact Green operator is replaced by the Green operator corresponding to the interactions $V_{eT} + V_{PT}$—i. e. the interaction V_{Pe} is neglected. Hydrogen-atom intermediate states are therefore used in T_2.

(iii) The off-energy-shell Coulomb wave function is approximated using the near shell approximation.

McGuire et. al.[33] retain only two leading-order terms in a (Z/v) expansion of T_2, whereas Deb et. al.[34] reduce the total transition matrix element to a one-dimensional integral which they evaluate numerically. Both McGuire et. al.[33] and Deb et. al.[34] calculate differential and total cross sections for ground state Ps-formation from atomic hydrogen.

Ward and Macek[35] formulate the IA to study Ps-formation in positron collisions with hydrogenic ions. The IA is of the type developed by Briggs[36] for electron capture in asymmetric ion-atom collisions. A consistent expansion is made of the scattering wave function in terms of the ratio of the weak interaction to the strong, and the leading order term is retained. Ward and Macek[35] consider a nuclear target Z_T much larger than unity and apply the following approximations:

(i) In the exact final state scattering wave function $|\Psi_f^-\rangle$ the exact Green oprator is replaced by the Green operator corresponding to the final state interaction $V_f(= V_{eT} + V_{PT})$—i. e. the interaction V_{Pe} is neglected.

(ii) the Fourier transform of the final state wave function $\varphi_{Ps}(\mathbf{r}_3)$ is made and the off-energy-shell Coulomb wave function is approximated by the on-energy-shell Coulomb wave function.

This impulse approximation to the wave function is substituted into the prior form of the exact T-matrix. Ward and Macek[35] evaluate this T-matrix for the particular problem of ground-state Ps-formation in positron-hydrogen collisions by employing the full peaking approximation. In Fig. (3) these calculations are labelled as I.A. A significant difference between the calculation of Ward and Macek[35] and the SPB approximation of McGuire et. al.[33] and Deb et. al.[34] is that Ward and Macek[35] approximate the exact Green operator by the *same* Coulomb Green operator for the entire T-matrix whereas McGuire et. al.[33] and Deb et. al.[34] employed different Coulomb Green operators. The Thomas peak for $1s - 1s$ electron capture by atomic hydrogen is more pronounced in the fully peaked impulse approximation[35] than in the SPB calculation of McGuire et. al.[33] and it is seen at a lower energy. In the fully peaking approximation it is visible at 500eV but is very apparent at 2000eV, which corresponds to a velocity of 12 a.u. Presently there are no experimental measurements available for single differential cross section for ground-state Ps-formation in atomic hydrogen to compare with these theoretical predictions.

Ward and Macek[35] also employ an alternative impulse approximation based upon a newly developed perturbation expansion method.[37,38] In this method the entire T-matrix rather than the scattering wave function is expanded in powers of the ratio of the weak interaction to the strong, and the lowest order term in the expansion is retained. Ward and Macek[35] evaluate the T-matrix for the opposite limits for ground-state Ps-formation in atomic hydrogen by employing the full peaking approximation. In Fig. (3), the full peaking approximation corresponding to the limit $Z_T \to 0$ ($Z_T \to \infty$) is referred to as I.A. $Z_T \to 0$ ($Z_T \to \infty$).

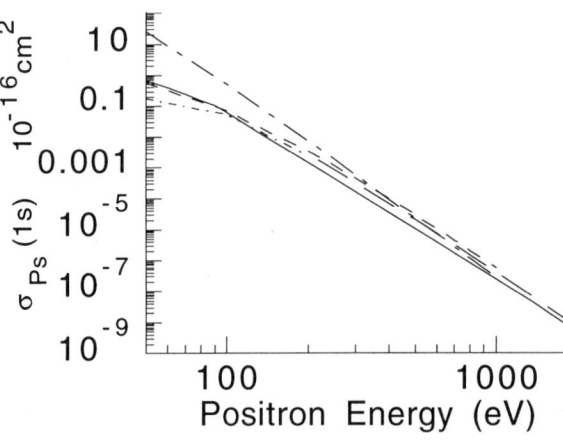

Fig. 3 Ps(1s)-formation cross section obtained from the following second-order calculations: second Born (——), SPB[33] (— —), I.A. (- - -), I.A. $Z_T \to 0$ (— - —), and $Z_T \to \infty$ (- · · -).

Fig. (3) compares the various fully peaked impulse approximations, the SPB approximation and the second Born for the angle-integrated cross sections for Ps(1s) formation from atomic hydrogen. At an incident positron energy of about 300eV, the various fully peaked impulse approximations are in fairly good agreement with each other and with the SPB calculation. The I.A. $Z_T \to 0$ calculation differs significantly from all other calculations for energies less than 200eV, and the disagreement increases with decreasing energy. The impulse approximation corresponding to an expansion of the scattering wave function is in close agreement with the exact second Born calculation for all energies considered. This is interesting since it shows that for angle-integrated cross sections and for positron collisions it makes little difference whether free or Coulomb Green operators are employed.

C. COMPARISON WITH EXPERIMENT

Recently, Sperber et. al.[6] have measured the total angle-integrated Ps-formation cross section for positron collisions with hydrogen atoms in the energy range 13 – 205 eV. These experimental data have been somewhat revised to include, for instance, the fraction at low energy of positrons that ionize hydrogen, and escape the detection system to large angle, and are falsely registered among positrons that form positronium.[7] The fraction is estimated using the first Born approximation. Fig. (4) shows the revised experimental measurements[7] and compares them with the following theoretical calculations: first Born,[31] second Born,[31] Kohn variational[17] and CCA.[26] In the figure, the first Born, second Born and the close coupling results are the sum for formation into the $1s, 2s$ and $2p$. No theory obtains the precise shape of the measurements for the entire energy range. However, by 45eV the second Born[30,31] is in agreement with experiment[7] within the experimental error.

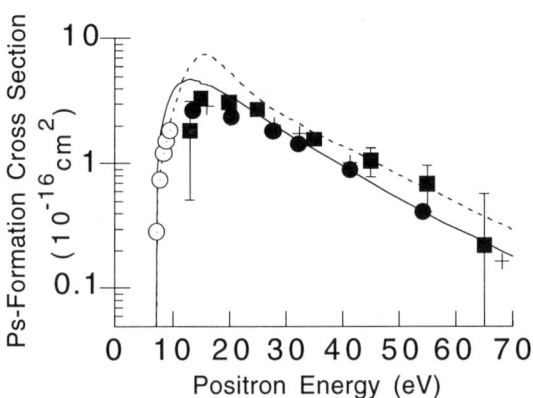

Fig. 4 Ps-formation cross section. Theory: first Born (——), second Born (- - -), Kohn variational (○), H($1s, 2s, 2p$), Ps($1s, 2s, 2p$) CCA, (+), H($1s, 2s, 2p$), Ps($1s, 2s, 2p$) CCA (●). Experimental measurements are denoted by ■.

IV. POSITRON IMPACT IONIZATION OF ATOMIC HYDROGEN

Recent experimental measurements reveal that positron-impact ionization of hydrogen—the simplest of all atoms—is not understood.[4] These experimental measurements are in sharp disagreement with the theoretical calculations (present at the time of the experiment) over the entire energy range. In fact, at the maximum of the cross section the theoretical predictions of Ghosh et. al.[39] using the distorted-wave polarized orbital approximation are too low by 20%. The predictions of Mukherjee et. al.[40] using the distorted-wave approximation are even lower. This is somewhat surprising since Mukherjee et. al.[40] employ the same distorted-wave approximation as employed earlier by Campeanu et. al.[41] for positron-helium collisions, and that this method for positron-helium collisions gives excellent agreement with experiment. It should be noted that recently Acacia et. al.[42] reported an extension to the distorted-wave model of Campaneau et. al.[41] in which a detailed description is provided of the interaction between the two outgoing particles and the residual ion. Their model gave good agreement with the experimental measurements of Spicher et. al.[4] for positron-hydrogen ionization. It should also be noted that the positron-impact ionization of atomic hydrogen has been independently measured by the University College London (UCL) group.[5] The measurements of UCL group[5] are considerably lower than the measurements of Spicher et. al.[4] Thus, presently it is uncertain which experimental measurements and which theoretical predictions are the most reliable. Further work is therefore required both experimentally and theoretically to determine the *correct* shape and absolute angle integrated cross sections for positron-hydrogen ionization.

Spicher et. al.[4] measure electron-hydrogen ionization in the same apparatus they use for positron collisions. For energies less than 450eV, the positron ionization cross section is considerably larger than the electron one; near the peak of the cross section it twice as large. Since the first Born approximation gives identical results for electron and positron impact ionization, the difference between the positron and electron cross sections has to come from the second and higher order terms in the Born series. It has been suggested that the dominant feature of positron-impact ionization near the threshold may be Electron Capture to the Continuum (ECC).[43] If this is the case, this may explain the considerable difference between the positron and electron ionization cross sections. ECC describes the process in which the relative velocity of the outgoing ionized electron with respect to the positively charged projectile is small compared with the velocity of their C.O.M. with respect to the residual ion. For positron collisions, ECC is known as Ps-formation to the continuum. It is therefore crucial to determine how significant ECC is for positron collisions in order to interpret the experimental measurements.

Capture into continuum states as well as capture into bound states can occur via the double collision mechanisms. Shakeshaft and Spruch[44] demonstrate that the double collision mechanisms have a pronounced effect on the distribution of ECC electrons. This implies that if ECC is a dominant feature of positron impact ionization then any first order theory will fail to describe adequately ionization and that at least a second order theory is required.

Presently, ECC has not been experimentally observed for positron impact ionization. Moxom et. al.[45] measure the ejected electron spectra of positron-argon ionization for incident energies of 50, 100 and 150eV. The cusp which is the signature of ECC is not seen in the energy spectrum for any of the incident energies considered, and within the experimental sensitivity. However, the analogous process of anti-electron capture to the continuum recently has been experimentally verified and theoretically confirmed for electron-helium ionization.[46]

To study ECC and the double collision mechanisms in positron-hydrogen scattering, Brauner and Briggs[47] calculate the TDCS using the BBK wave function[48] for the final state scattering wave function. The interference which occurs between the two double collision mechanisms for electron capture in positron collisions will persist even above the ionization threshold. Since the interference is destructive for capture into even parity states there will occur a suppression of even parity contributions to the final state wave function when both particles emerge at the critical angle of 45°. The essential point of the paper by Brauner and Briggs[47] is to show that this interference has a strong influence on the shape of the ECC cross section in the vicinity of the critical angle. Brauner and Briggs[47] compute the TDCS at a fixed scattering and ejected angle of 45° and plot the TDCS as a function of the energy of the ejected electron. For an incident positron energy of 250eV a cusp is seen in the cross section when the energy of the ejected electron matches that of the scattered positron. No significant additional structure corresponding to the interference of the two double collision mechanisms is apparent at this energy. However, for a much higher incident energy of 10keV a very deep minimum is present in the differential cross section. The dip in the TDCS does not coincide with the cusp but occurs at a lower ejected electron energy. Brauner and Briggs[47] show that the dip is only evident for scattering angles close to the critical angle. By an incident positron energy of 100keV, the dip and cusp coincide indicating that the high velocity limit has been reached.

V. SUMMARY

A new kind of coupled channel shape resonance has been predicted in the close-coupling framework for positron-hydrogen collisions at intermediate energies.[8] The resonance parameters (position and width) of the resonances change significantly with the inclusion of more target and Ps states.[11,12] It is an open question whether these are true resonances, and if so, what the mechanism for the resonance is. It would be valuable to perform an experiment to search for these resonances and to perform a theoretical calculation explicitly including the break-up channel.

The status of the theory of Ps-formation in positron-hydrogen collisions is as follows: At low energies, the close-coupling calculations have enable Ps-formation into the ground and first few excited states to be fairly well understood. In the ore gap the Ps-formation and elastic cross sections have been accurately determined with the Kohn and inverse Kohn variational methods.[17] At high energies reasonably good agreement has been obtained for Ps(1s)-formation—at least for the angle-integrated cross sections—between the second Born[30,31], strong potential Born[33,34] and the various fully peaked impulse approximations.[35] This indicates that the angle-integrated Ps(1s)-formation is fairly well determined at high energies. However, accurate determination of the Ps-formation cross section at intermediate energies is still required and the discrepancy with experiment needs to be resolved.

Presently, it is unclear how significant ECC is in positron-atom ionization. The experimental measurements of positron-hydrogen ionization by Spicher et. al.[4] suggest that ECC is a dominant feature of ionization since this may explain the huge difference between the positron and electron results. However, in the experiment performed by Moxom et. al.[45] for positron-argon ionization, ECC is not seen in the energy spectrum for the energies considered and within the experimental sensitivities. If ECC is a dominant feature of positron-ionization then, at least a second order theory must be employed to describe adequately ionization at high energies.

ACKNOWLEDGMENTS

Discussions with Drs. M. Charlton, J. W. Humberston, J. H. Macek, H. R. J. Walters and Mr. M. S. T. Watts were helpful. S. J. W. is thankful to Dr. J. H. Macek for the opportunity to spend the summer at the University of Tennessee where this review article was written. Support from the National Science Foundation under grant No. PHY 9213900 and from ICPEAC is gratefully acknowledged.

REFERENCES

1. H. S. W. Massey, Can. J. Phys. **60**, 461 (1982).
2. M. Charlton and G. Laricchia, J. Phys. B**23**, 1045 (1990).
3. T. S. Stein, R. D. Gomez, Y.-F. Hsieh, W. E. Kauppila, C. K. Kwan and Y. J. Wan, Phys. Rev. Lett. **55**, 488 (1955).
4. G. Spicher, B. Olsson, W. Raith, G. Sinapius and W, Sperber, Phys. Rev. Lett. **64**, 1019 (1990).
5. M. Charlton, Invited Talk, XVIII ICPEAC, 21-27 July, 1993, Aarhus, Denmark.
6. W. Sperber, D. Becke, K. G. Lynn, W. Raith, A. Sinapius, G. Spicher and M. Weber, Phys. Rev. Lett. **68**, 3690 (1992).
7. W. Sperber, Ph.D Thesis Universität Bielefeld, 1955 (unpublished) (M. Weber, private communication (1993)).
8. K. Higgins and P. G. Burke, J. Phys. B**24**, L343 (1991).
9. M. Leventhal and B. L. Brown, in *Proceedings of the Third International Workshop on Positron (Electron)- Gas Scattering*, W. E. Kauppila, T. S. Stein and J. Wadehra, eds. (World Scientific, Sinapore, 1986), p. 140.
10. R. N. Hewitt, C. J. Noble and B. H. Bransden, J. Phys. B**24**, L635 (1991).
11. M. T. McAlinden, A. A. Kernoghan, H. R. J. Walters, private communication (1993) (H. R. J. Walters, Invited Talk at the Positron Interactions with Atoms, Molecules and Clusters Satellite Meeting to XVIII ICPEAC).
12. N. K. Sarkar, M. Basu, M. Mukherjee, A. S. Ghosh, to be published in J.Phys. B (1993). (A. S. Ghosh, Invited Talk at the Positron Interactions with Atoms, Molecules and Clusters Satellite Meeting to XVIII ICPEAC).
13. K. Higgins, Contributed Paper, XVIII ICPEAC, 21-27 July, 1993, Aarhus, Denmark.
14. A. Igarashi and N. Toshima, Contributed Paper, XVIII ICPEAC, 21-27 July, 1993, Aarhus, Denmark.
15. J. W. Humberston, Can. J. Phys. **60**, 591 (1982).
16. J. W. Humberston, J. Phys. B**17**, 2353 (1983).
17. J. W. Humberston, Adv. At. Mol. Phys. **22**,1 (1986).
18. C. J. Brown and J. W. Humberston, J. Phys. B**17**, L423 (1984).
19. C. J. Brown and J. W. Humberston, J. Phys. B**18**, L401 (1985).
20. R. J. Archer, G. A. Parker and R. T. Pack, Phys. Rev. A**41**, 1303 (1990).

21. Y. F. Chan and P. A. Fraser, J. Phys. B6 2504 (1973).
22. M. A. Abdel-Raouf, J. W. Darewych, R. P. McEachran and A. D. Stauffer, Phys. Lett. **100A**, 353 (1984).
23. M. Basu, M. Mukherjee and A. S. Ghosh, J. Phys. B**23**, 2641 (1990).
24. H. S. W. Massey and C. B. O. Mohr, Proc. R. Soc. London A**67**, 695 (1954).
25. P. Khan and A. S. Ghosh, Phys. Rev. A**27**, 1904 (1983).
26. R. N. Hewitt, C. J. Noble and B. H. Bransden, J. Phys. B**23**, 4185 (1990).
27. L. H. Thomas, Proc. R. Soc. London, Ser. A**114**, 561 (1927).
28. R. M. Drisko, Ph. D. Thesis, Carnegie Institute of Technology, 1955 (unpublished).
29. R. Shakeshaft and J. H. Wadehra, Phys. Rev. A**22**, 968 (1976).
30. A. Igarashi and N. Toshima, Phys. Rev. A**46**, R1159 (1992).
31. A. Igarashi and N. Toshima, Phys. Rev. A**47**, 2386 (1993).
32. J. S. Briggs, P. T. Greenland and L. Kocbach, J. Phys. B**15** 3085 (1983).
33. J. H. McGuire, N. C. Sil and N. C. Deb, Phys. Rev. A**34**, 685 (1986).
34. N. C. Deb, J. H. McGuire and N. C. Sil, Phys. Rev. A**36**, 3707 (1987).
35. S. J. Ward and J. H. Macek, Contributed Paper, XVIII ICPEAC, 21-27 July, 1993, Aarhus, Denmark (in preparation for Phys. Rev.A).
36. J. S. Briggs, J. Phys. B**10**, 3075 (1977).
37. J. H. Macek and J. Botero, Phys. Rev. A**45**, R8 (1992).
38. J. Botero and J. H. Macek, Phys. Rev. A**45**, 154 (1992).
39. A. S. Ghosh, P. S. Mazumdar and M. Basu, Can. J. Phys. **63**, 621 (1985).
40. K. K. Mukherjee, N. R. Singh and P. S. Mazumdar, J. Phys. B**22**, 99 (1987).
41. R. I. Campeanu, R. P. McEachran and A. D. Stauffer, J. Phys. B**20**, 1635 (1987).
42. P. Acacia, R. I. Campeanu, M. Horbatsch, R. P. McEachran and A. D. Stauffer, accepted for publication in J. Phys. B.
43. B. H. Bransden, P. G. Coleman and W. Raith, Can. J. Phys. **60**, 565 (1982).
44. R. Shakeshaft and L. Spruch, Phys. Rev. Lett. **41**, 1037 (1978).
45. J. Moxom, G. Laricchia, M. Charlton, G. O. Jones and A. Kover, J. Phys. B**25**, L613 (1992).
46. P. Guang-yan, P. Hvelplund, H. Knudsen, Y. Yamazaki, M. Brauner and J. S. Briggs, Phys. Rev. A**47**, 1531 (1993).
47. M. Brauner and J. S. Briggs, J. Phys. B**24**, 2227 (1991).
48. M. Brauner, J. S. Briggs and H. Klar, J. Phys. B**22**, 2265 (1989).

THE THEORY OF MUON CATALYSED FUSION

E A G Armour and M R Harston
Mathematics Department, University of Nottingham,
Nottingham NG7 2RD, England
M P Faifman
Russian Science Center, "Kurchatov Institute",
Moscow 123182, Russia

ABSTRACT

It has been known since about 1980 that a negatively charged muon in a deuterium-tritium mixture at room temperature can catalyse more than 100 nuclear fusion reactions. The muon first of all binds a deuteron and a triton to form a muonic molecular ion, in which the two nuclei are so close together that fusion occurs very rapidly. Thereafter, the muon is usually released and is thus free to bring about further fusion reactions like a chemical catalyst.

Since 1980 muon catalysed fusion has been the subject of extensive research both by experimentalists and theoreticians. In this article, we describe the theory of muon catalysed fusion, highlighting recent developments and indicating where further work is necessary. This involves a fascinating mixture of atomic, molecular, nuclear and particle physics. At the end of the article we comment on the possibility of using muon catalysed fusion as an energy and neutron source.

INTRODUCTION

It has been known since the 1930's that the sun generates its energy by a fusion reaction in which hydrogen is converted into helium (see, for example, Bethe[1]). However, this reaction requires a temperature of several million degrees. Extensive research into using such a reaction as a source of energy here on Earth has been going on for many years (see, for example, Rebut[2]). Progress has been slow as the high temperatures involved make energy generation by this reaction a very difficult process.

Suggestions[3] in 1989 that fusion can be brought about in an electrolysis process at room temperature are now discredited[4]. However, it has been known for many years that a negatively charged muon, μ^-, can bind two isotopes of hydrogen to form a muonic molecular ion, in which the two nuclei are close enough together that fusion occurs very rapidly. Furthermore, this reaction takes place at room temperature.

The muon is a lepton and is thus a fermion with spin $\frac{1}{2}$. Its mass, m_μ, is 206.8 m_e, where m_e is the mass of the electron. It is formed as a result of a decay process brought about by the weak interaction. A negative pion (pi meson) is first formed in a high-energy nuclear collision. It has a mean lifetime of 2.6×10^{-8} secs and decays into a negatively charged muon and a neutrino. The muon itself has a mean lifetime of 2.2×10^{-6} secs before it also decays through the weak interaction into an electron, a neutrino and an antineutrino (see, for example, Semat and Albright[5]).

In 1947, only ten years after the discovery of the muon and before it was known for certain that the muon and pion were different particles, Frank[6] suggested that a negatively charged muon might catalyse fusion of two hydrogen isotopes i.e. bring about fusion of the isotopes while remaining unchanged at the end of the reaction. Shortly afterwards Sakharov[7] discussed the possibility of energy production by such a process and later it was considered further by Zeldovich[8,9].

Frank was led to his suggestion by scaling considerations. In a muonic molecular ion, $xy\mu$, where x and y are protons, deuterons or tritons, the dimensions of the molecular ion scale as $m_e/m_\mu \simeq \frac{1}{207}$ relative to xye, where e is an electron. At the same time the binding energy scales as $m_\mu/m_e \simeq 207$. Thus the nuclei x and y are brought much closer together than in xye and with sufficient kinetic energy to penetrate the Coulomb barrier and allow fusion to occur. Provided $xy\mu$ is in a state of zero angular momentum and x and y are not both protons, fusion occurs at a rate between $\sim 10^5 \sec^{-1}$ and $\sim 10^{12} \sec^{-1}$, depending on the nuclei involved (see, for example, Ponomarev[10]). The proton-proton fusion process is very much slower as it involves the weak interaction.

These theoretical speculations were unknown to Alvarez et al[11,12] when in 1956 they correctly interpreted particle tracks in their bubble chamber at Berkeley as representing p-d fusion catalysed by a muon. Upon reading about Alvarez's exciting discovery in the New York Times, Jackson[13,14] proceeded to make an analysis of energy production possibilities and concluded that very few p-d fusions could be catalysed by a single muon.

However, interest in the subject revived when experiments with dd fusion catalysed by a muon carried out by Dzhelepov et al[15] revealed a strong and unexpected temperature dependence in the rate of formation of ddμ. They suspected that this was because the formation process is a resonant process.

A form for this resonant process was suggested by Vesman[16] in 1967. He proposed that the formation of ddμ could occur by the reaction

$$d\mu + D_2 \rightarrow [(dd\mu)^* dee] \qquad (1)$$

where the species on the right-hand side is a muonic molecular complex in which ddμ, in a weakly bound excited state, forms one of the nuclei. The energy lost by the system through the formation of the weakly bound state goes into exciting the vibrational and rotational states of the muonic molecular complex. Due to the quantised structure of these states, this will be a resonant process and the formation will be very sensitive to the kinetic energy of the dμ. This accounted qualitatively for the temperature dependence observed by Dzhelepov et al[15].

Such a mechanism depends crucially on the existence of a weakly bound state with binding energy less than the 4.5 eV dissociation energy of D_2. A decade later, Gershtein and Ponomarev[17] and Vinitsky et al.[18] were able to confirm the existence of just such a weakly bound state of ddμ with rotational and "vibrational" quantum numbers (J, v) = (1, 1) and binding energy ~ 2 eV, using the adiabatic representation method[19] in which the kinetic energy of the d nuclei is taken into account by solving close-coupling type equations with the Born–Oppenheimer states of ddμ as basis. They were also able to show by this method that a corresponding weakly bound (J, v) = (1, 1) state exists for dtμ with binding energy ~ 1 eV. In addition, Gershtein and Ponomarev[17] showed that rapid formation of dtμ by the Vesman mechanism would enable a muon to catalyse ~ 100 d-t fusions. The rapid formation rate of dtμ was confirmed experimentally by Bystritsky et al[20] in 1980.

The existence of these weakly bound states of ddμ and dtμ which makes possible the resonant formation process is remarkable. The binding energy of the ground state of ddμ and dtμ is of the order of 300 eV and 2 eV is just 0.08% of the muonic Rydberg unit.

The pioneering work of the above researchers in discovering the theoretical basis for the resonant formation of ddμ and dtμ set the stage for all the subsequent work in the field of muon catalysed fusion. However, as we shall see, the d-t muon catalysed fusion cycle has proved to be far more intricate than could ever have been imagined by these early researchers[21,22,14].

Muon catalysed fusion has actually been observed for all pairs of hydrogen isotopes except two protons. However, ddμ and dtμ are the only two muonic molecular ions for which resonant processes exist which increase their formation rates. Of all the hydrogen isotope reactions catalysed by a muon which have been studied in the past decade, the d-t reaction has attracted by far the most theoretical attention. This is because dtμ is formed much more rapidly than any other muonic molecular ion, undergoes fusion much more rapidly and has one of the highest energy yields per fusion[10].

As a consequence of this high energy yield, the α particle produced in the fusion reaction has a high recoil velocity which reduces the possibility of what is called "sticking". This is the process whereby the muon forms a bound state with the α particle and remains bound until it decays. It is thus unable to catalyse further fusion reactions.

In view of all of these features which have focussed attention on the d-t reaction, we shall concentrate on this reaction in what follows. The d-d fusion reaction and other aspects of muon catalysed fusion are discussed, for example, by Ponomarev[10] and Breunlich et al.[22]

In recent years muon catalysed fusion has been the subject of many review articles. In addition to those by Bracci and Fiorentini[12], Breunlich et al[22], Cohen[14] and Ponomarev[10], which we have already referred to, there are review articles by Leon[23], Bhatia and Drachman[24], Rafelski et al[25], Froelich[26] and Petitjean[27].

THE d-t MUON CATALYSED FUSION CYCLE

1. Muonic Atom Formation

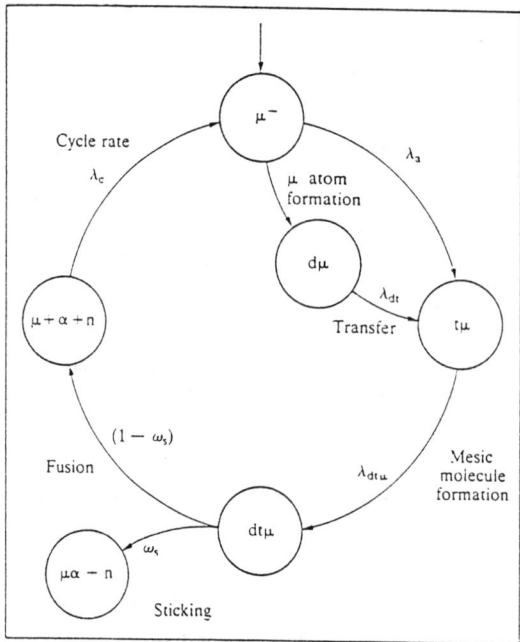

Figure 1 is from Ponomarev[10]. Side chains involving d-d and t-t fusion are not shown.

λ_a = muonic atom formation rate

$\approx 4 \times 10^{12}\ \text{sec}^{-1}$

λ_{dt} = muon transfer rate from $(d\mu)_{1s}$ to $(t\mu)_{1s}$

$\approx 3 \times 10^{8}\ \text{sec}^{-1}$

$\lambda_{dt\mu}$ = resonant formation rate of $dt\mu$

$\approx 4 \times 10^{8}\ \text{sec}^{-1}$

w_s = effective sticking probability

$\approx 0.43\%$

λ_c = cycle rate

Values[22] are for $T = 300K$ and normalised to $\phi = 1$.

Figure 1. The principal muon catalysis fusion cycle in a deuterium and tritium mixture.

The various steps in the d-t muon catalysed fusion cycle (see, for example, Ponomarev[10]) are set out schematically in figure 1. The process starts with the injection of a negative muon into a mixture, either liquid or gaseous, of deuterium and tritium i.e. D_2, DT and T_2. $\phi = 1$ corresponds to the density of liquid hydrogen i.e. 4.25×10^{22} atoms/cm^3. The temperature can range from near zero Kelvin up to 10^4K.

After slowing down, the muon forms a bound state with d or t:

$$\mu + D_2 \to d\mu(n) + d + 2e \qquad (2)$$

$$\mu + T_2 \to t\mu(n) + t + 2e \qquad (3)$$

where n is the principal quantum number. Detailed theory only exists for muon capture by D or T atoms rather than D_2 or T_2 molecules. According to the Classical Trajectory Monte Carlo calculation by Cohen[28], the capture probability density is appreciable only in the range 0–13 eV and shows a broad maximum ~ 7 eV. The probability of Auger capture depends on the overlap between the wave functions of the muon and the electron[29]. This has its largest value for D or T atoms in their ground states when $n = 14$.

When the targets are the molecules D_2 and T_2 as in (2) and (3), it is thought that the first stage in the capture process is the formation of large muonic molecular[10,26] complexes such as $D_2^+\mu^-$ and $T_2^+\mu^-$. According to Korenman and Popov[30], these complexes then dissociate forming highly excited muonic atoms with $n = 10-12$ and kinetic energy of the order of 10 eV.

If the muon is first captured by d, it has to be transferred to t if it is to participate in the resonant formation process for dtμ. $d\mu(n)$ is deexcited in a cascade involving external Auger deexcitations[10] such as

$$d\mu(n) + D_2 \rightarrow d\mu(\bar{n}) + D_2^+ + e \qquad (4)$$

where $\bar{n} < n$. At any stage in this cascade, $d\mu(m)$, $m \leqslant n$, may be transferred to t by reactions such as

$$d\mu(m) + DT \rightarrow 2D + t\mu(\bar{m}) \quad \bar{m} \leqslant m \qquad (5)$$

$t\mu(m)$ is more tightly bound than $d\mu(m)$ by $\dfrac{48}{m^2}$ eV and $t\mu(\bar{m})$ more tightly bound than $t\mu(m)$ by $2711\left(\dfrac{1}{\bar{m}^2} - \dfrac{1}{m^2}\right)$ eV. The dissociation energy of D_2 is about 4.5 eV. Thus there are many possible values of m and $\bar{m}$ for which this reaction is energetically favoured and it proceeds rapidly.

Muon transfer is brought about by the interaction of the t with dμ at distances of the order of one muonic atomic unit i.e. two orders of magnitude smaller than the dimension of the electronic shells. Thus the electrons do not play an important role in the exchange process. The excess of binding energy due to the transfer to the heavier isotope is released into the translational motion of the colliding nuclei rather than into Auger electrons. Thus to a good approximation, ground state transfer can be treated as a binary rearrangement collision. Such a rearrangement can be treated quite adequately by the adiabatic expansion method[19] (see, for example, Melezhik[31] and Adamczak et al.[32]). Detailed calculations have also been carried out by Kamimura, initially using a Kohn-type variational method[33] then using a precise finite difference method to solve the coupled channel equations using Jacobi coordinates[34]. Only semi-classical methods have so far been used for transfer involving excited states (see, for example, Menshikov and Ponomarev[35,36]).

It is believed that the transfer rate from excited states of dμ is comparable to that of the deexcitation rate ($\sim 10^{11}\,\text{sec}^{-1}$). Thus a significant fraction of transfers should occur extremely fast from $(d\mu)_{n>1}$ states before the dμ reaches its ground state. Once in the ground state the rate of transfer is relatively slow, $\sim 3\times 10^8\,\text{sec}^{-1}$. The probability of the dμ reaching its ground state is denoted by q_{1s}. It can also be interpreted as the population of the dμ ground state. The larger q_{1s}, the more muons are held up in the dμ ground state awaiting transfer to t. Unless $\lambda_{dt} \gg \lambda_{dt\mu}$, which is not usually the case (see Figure 1), the population q_{1s} plays an important role in muon cycle dynamics and a significant fraction of the average d-t cycle time is taken up by the transfer of the d to t. The calculation of the dependence of q_{1s} on ϕ and the tritium concentration, c_t, is at present one of the most complicated and urgent problems in the description of this part of the muon catalysed fusion cycle[10]. If tμ is formed in an excited state it cascades rapidly to its ground state with rate $\sim 10^{11}\,\text{sec}^{-1}$ by processes similar to (4).

2. Muonic Molecule Formation

The next step, muonic molecule formation, is a key step in the muon catalysed fusion cycle and of great interest to atomic and molecular physicists. $dt\mu$ can be formed by the non-resonant Auger process

$$t\mu + D_2 \rightarrow [(dt\mu)de]^+ + e \qquad (6)$$

The formation rate is given by[9,37]

$$\lambda_f = \frac{2\pi}{\hbar} \int \rho_f(E) |T_{if}|^2 \, d\omega \qquad (7)$$

where ρ_f is the density of final states and

$$T_{if} = \langle \Psi_i | V | \Psi_f \rangle \qquad (8)$$

Ψ_i is the initial state and Ψ_f is an appropriately normalised final state. The integral is over all solid angles as the Auger electron may be ejected in any direction. V is the interaction operator. The formation rate is slow, having a maximum value of $\sim 5 \times 10^6 \, \text{sec}^{-1}$ for the transition from the initial $J_i = 1$ state into the $(J, v) = (0, 1)$ state of $dt\mu$[38].

However, as pointed out in the introduction, $dt\mu$ forms more rapidly by the resonance process

$$t\mu + D_2 \rightarrow [(dt\mu)_{11} dee]_{Kv} \qquad (9)$$

in which a $t\mu$ atom in its ground state collides with one of the nuclei of a D_2 molecule and raises it to a higher rovibrational state while simultaneously attaching as the $(1, 1)$ state of $dt\mu$.

The muonic molecular complex, $[(dt\mu)_{11} dee]_{Kv}$, is in a meta-stable state (see, for example, Petrov[39]). The formation process is resonant as for appropriate K and v values, $K = 0-4$, $v = 2$, the energy of the muonic molecular complex is very close to the energy of the initial state of the system.

The determination of the exact details of the energy balance requires a precise knowledge of the energies of the various states involved. Starting in 1973 Ponomarev and coworkers overcame theoretical difficulties and developed an appropriate computational technique for calculating binding energies of muonic molecular ions using the adiabatic representation method[10,19].

The first stage of this method is as in the application of the Born–Oppenheimer approximation in the calculation of molecular energies in quantum chemistry: the calculation of muonic wave functions, $\{\Phi_i(r; R)\}$, as a function of the parameter R. r is the position vector of the muon with respect to the centre of mass of the nuclei as origin and R is the internuclear distance vector.

However, the Born–Oppenheimer approximation is not a good approximation for $dt\mu$ as the ratio of the mass of the muon to the mass of the d (or t) is greater than 0.01. Thus the overall $dt\mu$ wave function is expressed in the form

$$\Psi(r, R) = \sum_i \Phi_i(r; R) \frac{1}{R} \chi_i(R) \qquad (10)$$

and the $\{\chi_i(R)\}$ are obtained by solving close-coupling type integro-differential equations[19].

More recently, the Rayleigh–Ritz variational method has been widely used to calculate the energies of the bound states of $dt\mu$ and, in particular, the energy of the weakly bound $(1, 1)$ state involved in the resonance (see, for example, Bhatia and Drachman[40], Frolov and Efros[41], Hu[42], Szalewicz et al[43], Alexander and Monkhorst[44], Vinitsky et al[45], Hara et al[46] and Kamimura[33]).

The origin for the coordinates which describe the internal motion of $dt\mu$ can be located at any of the three nuclei. If it is located at the triton, then the Hamiltonian for this motion is of the form

$$H = -\frac{1}{2m_{13}}\nabla_1^2 - \frac{1}{2m_{23}}\nabla_2^2 - \frac{1}{m_3}\nabla_1\cdot\nabla_2 + \frac{1}{r_1} - \frac{1}{r_2} - \frac{1}{r_{12}} \quad (11)$$

where particle 1 is the deuteron, particle 2, the muon and particle 3, the triton. m_{ij} is the reduced mass of particles i and j. r_i is the distance of particle i from the triton. r_{12} is the distance between the deuteron and the muon. The units are Hartree atomic units in terms of which $e = \hbar = m_e = 1$, where e is the charge on the deuteron and triton and m_e is the mass of the electron.

Several authors[40,42,43] used generalised Hylleraas basis functions. For example, for $J = 0$ states they use basis functions of the form

$$\Psi_i = r_1^{a_i} e^{-\alpha r_1} r_2^{b_i} e^{-\beta r_2} r_{12}^{c_i} e^{-\gamma r_{12}} \quad (12)$$

where a_i, b_i and c_i are non-negative integers. The non-linear parameters α, β and γ, are chosen by optimisation.

Frolov and Efros[41] and Alexander and Monkhorst[44] use Slater geminals which have the same form as the generalised Hylleraas basis (12) except that all powers of the interparticle distances are the same e.g. 0 if $J = 0$ and 1 if $J = 1$. The exponents are chosen by a procedure known as random tempering. Such a basis set has the disadvantage that even when using a relatively small number of basis functions (<300), problems can arise because the set is nearly linearly dependent. However, this type of basis is more suitable for representing weakly bound states such as the (1, 1) state of dtμ.

Vinitsky et al[45] and Hara et al[46,47] carried out their calculations using prolate spheroidal coordinates[48] with origin at the midpoint of the line joining the nuclei and with z axis in the direction of this line. For states for which $J = 0$, they use basis functions of the form

$$\Psi_i = R^{a_i}\xi^{b_i}\eta^{c_i}\exp(-\alpha_i R - \beta_i R\xi) \quad (13)$$

where a_i, b_i and c_i are non-negative integers, R is the internuclear distance and ξ and η are the prolate spheroidal coordinates,

$$\xi = \frac{r_{t\mu} + r_{d\mu}}{R}, \qquad \eta = \frac{r_{t\mu} - r_{d\mu}}{R} \quad (14)$$

The third prolate spheroidal coordinate is the usual azimuthal angle, ϕ. The basis functions depend on ϕ when $J \neq 0$. These coordinates are also used in most calculations involving the adiabatic expansion method. However, the most recent calculations by this method have used hyperspherical coordinates.[49]

The Hughes–Eckart or mass polarisation term, $-\frac{1}{m_3}\nabla_1\cdot\nabla_2$, can be removed from the Hamiltonian (11) by using Jacobi coordinates. In this system the vectors in the wave function are the position vector of the second particle with respect to the first and the position vector of the third particle with respect to the centre of mass of the first two. Calculations of binding energies for muonic molecular ions using these coordinates and Gaussian basis sets have been carried out by Kamimura[33]. This method is also suitable for representing weakly bound states[26].

The values of the binding energy of $(dt\mu)_{11}$ obtained by these different methods using a non-relativistic spinless Hamiltonian of the kind indicated in equation (11) now converge on a value of 660 meV. Since the calculated formation rate of $[(dt\mu)_{11}\,dee]$ is very sensitive to the precise value of this binding energy, an accuracy of approximately 1 meV or better needs to be achieved for satisfactory comparison between theory and experiment. Thus this energy needs to be corrected for relativistic, QED, hyperfine and other effects (see, for example, Bakalov[50], Aissing and Monkhorst[51] and Swe Myint et al[52]). These corrections together give an energy shift which reduces the binding energy by approximately 64 meV in the lowest hyperfine level.[22]

Resonant states above the $d\mu+t$ threshold have been detected by Froelich et al.[53], Hu and Bhatia[54], using the complex coordinate rotation method[55], and Hara and Ishihara[56] using prolate spheroidal coordinates and a basis set proposed by Halpern[57]. These are of Feshbach type, clustering, for example, below the $(t\mu)_{2s}$ threshold[58,26].

All the calculations we have referred to so far deal with the muonic molecular ion, $dt\mu$, and not with the meta-stable muonic molecular complex $[(dt\mu)_{11}\, dee]_{Kv}$ whose existence makes possible the resonant formation process. The size of $(dt\mu)_{11}$ is of the order[59] of $0.05\,a_0$, which is small compared to the mean electron-nucleus distance in D_2. Thus, to a good approximation, the energy levels of $[(dt\mu)_{11}\, dee]$ are equal to the sum of the internal energy of $(dt\mu)_{11}$ and the energies of the states of a D_2-like molecule, N dee, in which N is a fictitious point particle of unit charge and with mass equal to the sum of the masses of d, t and μ. A further correction due to the finite size of $(dt\mu)_{11}$ can then be calculated using perturbation theory[60-63]. If the $dt\mu$ is treated as a fictitious particle, the electronic wave function of the quasi-molecule is identical to the electronic ground state wave function of H_2. Thus the rovibrational levels of the quasi-molecule can be calculated within the Born–Oppenheimer approximation[64] using, for example, the very accurate H_2 electronic ground-state potential of Kołos and Wolniewicz[65]. They have also been calculated by a non-adiabatic method by Scrinzi et al[66].

All the calculations which we have described above make it possible to determine the energy levels of the quasi-molecule very accurately. Hence the degree to which the resonance condition is satisfied can be determined for a given value of the kinetic energy due to the relative motion of $t\mu$ and D_2.

The resonant formation rate, $\lambda_{dt\mu}$, is calculated using the relation[67-74,25]

$$\lambda_{dt\mu} = \pi\rho_d \sum_{K_i K_f} P_{K_i}(T) \int_0^\infty |\langle \Psi_f|V|\Psi_i\rangle|^2 f(E;T)\delta(E-E^{res}_{K_i K_f})\, dE \qquad (15)$$

where Ψ_i is the wave function of the initial state of the system, an incoming $t\mu$ in its ground state and a D_2 molecule in its ground electronic and vibrational state and Ψ_f is a suitably normalised final state wave function. E is the energy of the initial state and $E^{res}_{K_i K_f}$ is the position of the resonance under consideration, both relative to the $d+(t\mu)_{1s}$ threshold. $E^{res}_{K_i K_f}$ depends on (K_i, v_i) and (K_f, v_f), the rotational and vibrational quantum numbers of the initial and final states. v_i and v_f are 0 and 2, respectively. ρ_d is the deuteron density. P_{K_i} is the (temperature-dependent) probability that the target D_2 molecule is in its K_i rotational state and f is the CM distribution of the target D_2 and $t\mu$ kinetic energy. f is usually, but not always[75,76], assumed to be Maxwellian.

Ψ_f is taken to be of the form

$$\Psi_f = \Phi_{df}\Psi_{Mf} \qquad (16)$$

where Φ_{df} is the internal wave function for $dt\mu$ in its (1, 1) state. Ψ_{Mf} is the wave function for a D_2-like molecule with one of its nuclei a deuteron and the other the fictitious particle described earlier. Ψ_f in (16) is a good approximation to the exact wave function for the quasi-molecule. V can be taken to be either the interaction between $t\mu$ and D_2 in the entrance channel or the perturbation which generates the exact final state wave function from the approximate wave function (16)[67,73].

Details of nuclear and muon spin have been suppressed. These can easily be taken into account as V does not involve spin.

The above theory gives very good agreement with experiment when used to calculate the temperature dependence of the formation of rate of $dd\mu$[77-80]. However, the agreement is not as good for the formation rate of $dt\mu$[10,22,26]. In contrast to theoretical predictions, the experimental formation rate remains large and nearly constant for $T < \sim 300K$ and exhibits a strong density dependence indicating three or more body collision effects[22]. These features of the observed formation rate can be accounted for qualitatively by assuming that density dependent collisions broaden the energy levels of the quasi-molecule, replacing the delta function in (15) by a Lorentzian[39,81] and thus allowing states close to $E^{res}_{K_i K_f}$, on either side, to contribute to the resonance[81-83]. However, there are still important theoretical problems which need to be solved if good agreement is to be obtained between theory and experiment.

Finally, Faifman and Ponomarev[84] have recently carried out calculations of resonant formation rates of dtμ in mixtures containing H_2 and HD in addition to D_2, DT and T_2. These show that the formation rate is fastest by the reaction

$$t\mu + HD \rightarrow \left[(dt\mu)_{11} \, pee\right]_{Kv} \tag{17}$$

3. Auger Deexcitation of $(dt\mu)_{11}$

As has been emphasised by Ostrovsky and Ustinov[70] and Lane[71], the existence of a resonant process such as (9) or (17) is not enough to guarantee that the muon will catalyse fusion of d and t. The resonant process might simply reverse before fusion occurred; this is all the more likely as the excited state produced in the reaction is known to have $J = 1$ and thus a low probability of contact between the nuclei.

The dominant process by which stabilisation takes place is an Auger process (or internal conversion) in which the excitation energy of the muonic molecular ion is carried away by an orbital electron[10] e.g.

$$\left[(dt\mu)_{11} \, dee\right]_{Kv} \rightarrow \left[(dt\mu)_{01} \, de\right]^+_{K'v'} + e \tag{18}$$

Several theoretical calculations have been carried out of this deexcitation rate[85-88]. As pointed out in the last section, even in the weakly bound $(J, v) = (1, 1)$ excited state, dtμ is sufficiently small compared with the electronic part of the muonic molecular complex that, to a good approximation, it can be treated as a fictitious particle. The remaining terms due to the non-zero separation between the d, t and μ are treated as a perturbation, V_p, using Fermi's golden rule[89,90]. According to the golden rule, the deexcitation rate, λ_{dex}, is given by

$$\lambda_{dex} = \frac{2\pi}{\hbar} \sum_f |\langle \Psi_{Mf} \Phi_{df} | V_p | \Psi_{Mi} \Phi_{di} \rangle|^2 \tag{19}$$

where Ψ_{Mi} is the initial wave function for the quasi-molecule, Φ_{di} is the wave function for the $(J, v) = (1, 1)$ state of dtμ. Ψ_{Mf} is a suitably normalised wave function for an ionised state of the quasi-molecule, of the appropriate energy, which is coupled to Ψ_{Mi} by the perturbation. Φ_{df} is the wave function for the $(J, v) = (0, 1)$ state of dtμ. The right-hand side of (19) is averaged over all possible initial states and summed over all possible final states. As the dimensions of the dtμ are much smaller than those of the electronic part of the quasi-molecule, it is appropriate to expand V_p as a multipole expansion.[85,86] Since an E1 transition for which $|\nabla J| = 1$ is being considered, only the dipole term in this expansion contributes to the matrix element in (19). This then takes the form of a product of a matrix element involving only the dtμ wave functions and a matrix element involving only the molecular wavefunctions.

The calculations[85-88] take into account the molecular nature of the quasi-molecule in varying degrees ranging from the approximation of $\left[(dt\mu)_{11} \, dee\right]_{Kv}$ by $(dt\mu)_{11}e$ by Bhatia et al[86] to a full treatment by Armour et al[88]. However, all the calculations predict that λ_{dex} is $\sim 10^{12} \, sec^{-1}$, which is amply fast enough to prevent reversal of the $\left[(dt\mu)_{11} \, dee\right]_{Kv}$ formation process. This conclusion also applies to the deexcitation rates of $\left[(dt\mu)_{11} \, pee\right]_{Kv}$ and $\left[(dt\mu)_{11} \, tee\right]_{Kv}$.

4. Fusion, Sticking and Stripping

Once dtμ is in the strongly bound (0, 1) (or (0, 0)) state, there is a high probability of d and t tunnelling through the Coulomb barrier. When they come within a distance of $\sim 3 \times 10^{-5}$ bohr, they experience the strong interaction which brings about fusion through the reaction[9,10,26]

$$\text{dt}\mu \rightarrow {}^5\text{He} + \mu \rightarrow \alpha + n + \mu + 17.58 \text{ MeV} \tag{20}$$

Bogdonova et al[91] have calculated the rate of this process using the simple formula for the fusion rate,[9,23]

$$\lambda_f = \underset{v \rightarrow 0}{\text{Limit}} \frac{\sigma v}{C^2} \int |\Psi_{\text{dt}\mu}(r, r_{\text{td}} = 0)|^2 \, dr \tag{21}$$

where σ is the cross section for the reaction

$$t + d \rightarrow \alpha + n \tag{22}$$

v is the relative velocity of t and d. C is the Coulomb penetration or Gamow factor. $\Psi_{\text{dt}\mu}(r, r_{\text{td}})$ is the wave function for the internal motion of the dtμ and r is the muon position vector relative to the dt centre of mass as origin. They obtain fusion rates of 1.2×10^{12} sec^{-1} for the (0, 0) ground state and 1.0×10^{12} sec^{-1} for the (0, 1) excited state. Similar results have been obtained by, for example, Struensee et al[92] and Kamimura[93], using the R-matrix method[94].

At the end of the fusion reaction (20) it might be thought that the muon would be free to repeat the muon catalysed fusion cycle. Unfortunately, this is not always the case. The α particle and the muon produced in the reaction (20) may combine to form a bound state. This process is referred to as "sticking"; if the muon remains bound in this way, it is not free to continue the catalytic cycle.

The initial sticking probability, w_s^0, can be calculated using the relations[9,10]

$$w_s^0 = \sum_{nl} w_{nl}^0 \tag{23}$$

$$w_{nl}^0 = \left| \int e^{-i\mathbf{k} \cdot \mathbf{r}} \phi_{nl}(r) \Psi(r, r_{\text{td}} = 0) \, dr \right|^2 \tag{24}$$

where $\phi_{nl}(r)$ is the hydrogen-like wave function of $\alpha\mu$ in the (n, l) state, k is the wave vector of a muon moving with the velocity of the $\alpha\mu$ atom. $\Psi(r, r_{\text{td}})$ is assumed to be normalised to unity at $r_{\text{td}} = 0$.

Jackson[13] estimated that $w_s^0 = 1\%$ while Bracci and Fiorentini[95] and Gershtein et al[96] obtained a similar value, $w_s^0 = 1.2\%$. More recent calculations using more accurate methods e.g. adiabatic[97] and adiabatic hyperspherical[98], Monte Carlo[99] and variational[93,42,100] give slightly lower values, for example[42], $w_s^0 = 0.858\%$.

w_s^0 is the initial but not the final sticking probability. $\alpha\mu$ is emitted with a kinetic energy of about 3.5 MeV[9,23,24]. As it slows down in the medium it undergoes many collisions which may result in the stripping of the muon from the α particle. If this process is fast enough, the muon becomes free to resume the catalytic cycle. The degree to which stripping occurs is density dependent. In addition, the muon can become available through transfer reactions[24]

$$\alpha\mu + t \rightarrow t\mu + \alpha$$

$$\alpha\mu + d \rightarrow d\mu + \alpha \tag{25}$$

The effective sticking probability w_s, is given by

$$w_s = w_s^0 (1 - R) \tag{26}$$

where R is the muon reactivation coefficient. It is a measure of the extent to which stripping occurs. Detailed calculations of R have been carried out by Cohen[101], Markushin[102] and

Rafelski et al[103]. These authors carried out a complete treatment of the $\alpha\mu$ kinetics using accurate cross sections for, in particular, stopping, excitation, ionisation and charge transfer. For $\phi = 1.2$, for example, Cohen obtained a value of $R = 0.36\pm0.05$, which implies that $w_s \approx 0.55\%$, considerably less than w_s^0. However, this w_s value is significantly larger than most of the experimental values[26].

ENERGY RELEASE

As can be seen from figure 1, the muonic atom formation rate, λ_a, is much larger than either, λ_{dt}, the rate of transfer of a muon from $(d\mu)_{1s}$ to $t\mu$ or $\lambda_{dt\mu}$ the rate of muonic molecular ion formation. Consequently, the rate of the muon catalysis cycle, λ_c, is practically independent of λ_a. For the same reason, it is practically independent of the deexcitation and fusion rates.

If we take $\phi = 1$ for convenience, then the average muon catalysis cycle time[10,22,104]

$$\tau_c = \frac{1}{\lambda_c} \approx \frac{c_d q_{1s}}{\lambda_{dt} c_t} + \frac{1}{\lambda_{dt\mu} c_d} \qquad (27)$$

where c_d and c_t are the concentrations of deuterium and tritium, respectively, in the mixture $(c_d+c_t = 1)$. As stated earlier, q_{1s} is the probability of the $d\mu$ reaching its ground state before the muon is passed on to tritium. The first term in (27) is the average time the muon spends waiting to transfer to tritium. The second term is the average time the muon spends in the ground state of the $t\mu$ atom.

If sticking did not occur, X_c, the average number of catalysis cycles catalysed by one muon with average lifetime $\lambda_0^{-1} = 2.2\times10^{-6}$ secs, would be equal to λ_c/λ_0. Sticking modifies it to

$$X_c = \frac{1}{\frac{\lambda_0}{\lambda_c} + w_s} \qquad (28)$$

So far a maximum of X_c value of about 150 has been observed with energy release ~ 2.5 GeV/muon[10,24]. There are some indications that $X_c \approx 300$ might be attainable[26]. With the *estimated* cost of ~ 8 GeV to produce a muon[101], breakeven would be achieved for $X_c \approx 480$. Thus methods of increasing X_c must be found before a reactor could produce energy directly using the muon catalysed fusion cycle[14].

However, 14.1 MeV of the 17.58 MeV generated by the $dt\mu$ fusion reaction (20) is carried away by the neutron. The present X_c value of ~ 150 would be adequate for the production of energy by using neutrons obtained in this way to convert ^{238}U into plutonium fuel for conventional fission reactors[10,105,24]. Unfortunately, this method of energy production would not be as "clean" as using the muon catalysed fusion cycle directly.

CONCLUSION

The muon catalysed fusion cycle is a very interesting physical process involving not only atomic and molecular physics but also nuclear and particle physics. The central role of the muon adds novelty to the atomic and molecular processes involved e.g. the initial charge exchange reactions, the resonant $dt\mu$ formation process and the process which leads to sticking.

These three processes play a crucial role, through q_{1s}, $\lambda_{dt\mu}$ and w_s, in determining the average number of cycles catalysed by a muon. Very interesting theoretical problems remain to be solved in connection with all three. We consider this to be particularly the case for the resonant formation process. In our opinion, understanding this process sufficiently well to be able to calculate $\lambda_{dt\mu}$ accurately is among the most interesting and challenging problems in present-day atomic and molecular physics.

Research on muon catalysed fusion has already had a very stimulating effect on atomic and molecular physics. This state of affairs is likely to continue for many years.

ACKNOWLEDGEMENTS

We wish to thank SERC (UK) for a Visiting Fellowship for M.P.F. and a Research Assistantship for M.R.H. We also wish to thank NATO for the award of a Linkage Grant.

REFERENCES

1. H.A. Bethe, Phys. Rev. **55**, 434 (1939).
2. P.H. Rebut, Nuc. Fusion **32**, 187 (1992).
3. M. Fleischmann, S. Pons, M. Hawkins and R.J. Hoffman, Nature **339**, 667 (1989).
4. F. Close, Too Hot to Handle: the Race for Cold Fusion (Princeton University Press, Princeton, 1990).
5. H. Semat and J.R. Albright, Introduction to Atomic and Nuclear Physics, Fifth Edition (Chapman and Hall, London, 1973) p. 589.
6. F.C. Frank, Nature **160**, 525 (1947).
7. A.D. Sakharov, Report of the Physics Institute, Academy of Sciences, USSR (1948).
8. Ya. B. Zeldovich, Dokl. Akad. Nauk. SSR **95**, 493 (1954).
9. Ya. B. Zeldovich and S.S. Gershtein, Sov. Phys. Uspecki **3**, 593 (1961).
10. L.I. Ponomarev, Contemporary Physics **31**, 219 (1990).
11. L.W. Alvarez *et al*, Phys. Rev. **105**, 1127 (1957).
12. L. Bracci and G. Fiorentini, Phys. Reports **86**, 169 (1982).
13. J.D. Jackson, Phys. Rev. **106**, 330 (1957).
14. J.S. Cohen, Proceedings of an International Symposium on Muon Catalysed Fusion - 89, ed. J.D. Davies (Rutherford Appleton Laboratory Report RAL-90-022, 1990) p. 2.
15. V.P. Dzhelepov, P. Ermolov, V.I. Moscalev, V.V. Filchenkov, Sov. Phys. JETP **23**, 820 (1966).
16. E.A. Vesman, Sov. Phys. JETP Lett. **5**, 91 (1967).
17. S.S. Gershtein and L.I. Ponomarev, Physics Letters **72 B**, 80 (1977).
18. S.I. Vinitsky, L.I. Ponomarev, I.V. Puzynin, T.P. Puzynina, L.N. Somov and M.P. Faifmain, Sov. Phys. JETP **47**, 444 (1978).
19. S.I. Vinitsky and L.I. Ponomarev, Sov. J. Part. Nucl. **13** 557 (1982).
20. V.M. Bystritsky *et al*, Sov. Phys. JETP **53**, 877 (1981).
21. L.I. Ponomarev, Muon Catal. Fusion **3**, 629 (1988).
22. W.H. Breunlich, P. Kammel, J.S. Cohen and M. Leon, Annu. Rev. Nucl. Part. Sci. **39**, 311 (1989).
23. M. Leon, AIP Conference Proceedings **181**, Muon Catalysed Fusion, eds. S.E. Jones, J. Rafelski and H.J. Monkhorst (American Institute of Physics, New York, 1989) p. 94.
24. A.K. Bhatia and R.J. Drachman, Comments At. Mol. Phys. **22**, 281 (1989).
25. H.E. Rafelski, D. Harley, G.R. Shin and J. Rafelski, J. Phys. B **24**, 1469 (1991).
26. P. Froelich, Advances in Physics **41**, 405 (1992).
27. C. Petitjean, Progress in Muon Catalysed Fusion, (Paul Scherrer Institut, Switzerland, 1992) PR-92-05.
28. J.S. Cohen, Phys. Rev. A **27**, 167 (1983).
29. J.D. Garcia and N.H. Kwong, Phys. Rev. **35**, 4068 (1987).
30. G. Ya. Korenman and N.P. Popov, AIP Conference Proceedings **181**, Muon Catalysed Fusion, eds. S.E. Jones, J. Rafelski and H.J. Monkhorst (American Institute of Physics, New York, 1989) p. 145.
31. V.S. Melezhik, Muon Catal. Fusion **1**, 205 (1987).

32. A. Adamczak, C. Chiccoli, V.I. Korobov, V.S. Melezhik, P. Pasini, L.I. Ponomarev and J. Wozniak, Physics Letters B **285**, 319 (1992).
33. M. Kamimura, Muon Catal. Fusion **3**, 335 (1988).
34. Y. Kino and M. Kamimura, International Workshop on Muon Catalysed Fusion, μCF-92, Uppsala, Sweden, 1992. To be published.
35. L.I. Menshikov and L.I. Ponomarev, Soviet Physics JETP Lett. **39**, 542 (1984).
36. L.I. Menshikov and L.I. Ponomarev, Z. Phys. D. **2**, 1 (1986).
37. L.I. Ponomarev and M.P. Faifman, Sov. Phys. JETP **44**, 886 (1976).
38. M.P. Faifman, Muon Catalysed Fusion **4**, 341 (1989).
39. V. Yu. Petrov, Phys. Lett. B **163**, 28 (1985).
40. A.K. Bhatia and R.J. Drachman, Phys. Rev. A **30**, 2138 (1984).
41. A.M. Frolov and V.D. Efros, J. Phys. B **18** L265 (1985).
42. C.Y. Hu, Phys. Rev. A **34**, 2536 (1986); A **36**, 4135 (1987).
43. K. Szalewicz, W. Kołos, H.J. Monkhorst and A. Scrinzi, Phys. Rev. A **36**, 5494 (1987).
44. S.A. Alexander and H.J. Monkhorst, Phys. Rev. A **38**, 26 (1988).
45. S.I. Vinitsky, V.I. Korobov and I.V. Puzynin, Sov. Phys. JETP **64**, 417 (1986).
46. S. Hara, T. Ishihara and N. Toshima, Muon Catal. Fusion **1**, 277 (1987).
47. S. Hara and T. Ishihara, Phys. Rev. A **40**, 4232 (1989).
48. C. Flammer, Spheroidal Wave Functions (Stanford University Press, Stanford, 1957).
49. V.V. Gusev et al., Few-Body Systems **9**, 137 (1990).
50. D. Bakalov, Muon Catal. Fusion **3**, 321 (1988).
51. G. Aissing and H.J. Monkhorst, Phys. Rev. A **42**, 3789 (1990).
52. K. Swe Myint, Y. Akaishi, H. Tanaka, M. Kamimura and H. Narumi, Z. Phys. A **334**, 423 (1989).
53. P. Froelich, A. Flores-Riveros and S.A. Alexander, Phys. Rev. A **46**, 2330 (1992).
54. C.Y. Hu and A.K. Bhatia, Phys. Rev. A **43**, 1229 (1991).
55. Y.K. Ho, Phys. Reports **99**, 1 (1983).
56. S. Hara and T. Ishihara, Phys. Rev. A **40**, 4232 (1989).
57. A. Halpern, Phys. Rev. Lett. **13**, 660 (1964).
58. I. Shimamura, Phys. Rev. A, **40**, 4863 (1989).
59. A. Scrinzi, H.J. Monkhorst and S.A. Alexander, Phys. Rev. A **38**, 4859 (1988).
60. L.I. Menshikov, Sov. J. Nucl. Phys. **42**, 918 (1985).
61. A. Scrinzi and K. Szalewicz, Phys. Rev. A **39** 4983 (1989).
62. M.R. Harston, I. Shimamura and M. Kamimura, Z. Phys. D **22**, 635 (1992).
63. M.R. Harston, I. Shimamura and M. Kamimura, Phys. Rev. A **45**, 94 (1992).
64. M.P. Faifman, L.I. Menshikov, L.I. Ponomarev, I.V. Puzynin, T.P. Puzynina and T.A. Strizh, Z. Phys. D **2**, 79 (1986).
65. W. Kołos and L. Wolniewicz, J. Chem. Phys. **43**, 2429 (1965).
66. A. Scrinzi, K. Szalewicz and H. Monkhorst, Phys. Rev. A **37**, 2270 (1988).
67. L.I. Menshikov and M.P. Faifman, Sov. J. Nucl. Phys. **43**, 414 (1986).
68. U. Fano, Phys. Rev. **124**, 1866 (1961).
69. J.N. Bardsley, J. Phys. B **1**, 349 (1968).
70. V.N. Ostrovsky and V.I. Ustinov, Sov. Phys. JETP **52**, 620 (1980).
71. A.M. Lane, Phys. Lett. **98** A, 337 (1983).
72. A.M. Lane, J. Phys. B **20**, 2911 (1987).
73. A.M. Lane, J. Phys. B **21**, 2159 (1988).
74. M.P. Faifman, L.I. Menshikov and T.A. Strizh, Muon Catal. Fusion **4**, 1 (1989).
75. J.S. Cohen and M. Leon, Phys. Rev. Lett. **55**, 52 (1985).
76. P. Kammel, Nuovo Cimento Lett. **43**, 349 (1985).
77. M.P. Faifman, Muon Catal. Fusion, **2**, 247 (1988).
78. M.P. Faifman, L.I. Menshikov and L.I. Ponomarev, Muon Catal. Fusion **2**, 285 (1988).
79. L.I. Menshikov, L.I. Ponomarev, T.A. Strizh and M.P. Faifman, Sov. Phys. JETP **65**, 656 (1987).

80. A. Scrinzi, P. Kammel, J. Zmeskal, W.H. Breunlich, J. Marton, M.P. Faifman, L.I. Ponomarev and T.A. Strizh, Phys. Rev. A **47**, 4691 (1993).
81. L.I. Menshikov, Sov. J. Part. Nucl. **19**, 583 (1988).
82. V. Yu Petrov and Yu. V. Petrov, Muon Catal. Fusion **4**, 73 (1989).
83. J.S. Cohen and M. Leon, Phys. Rev. A **39**, 946 (1989).
84. M.P. Faifman and L.I. Ponomarev, Phys. Lett. B **265**, 201 (1991).
85. S.I. Vinitsky, L.I. Ponomarev and M.P. Faifman, Sov. Phys. JETP **55**, 578 (1982).
86. A.K. Bhatia, R.J. Drachman and L. Chatterjee, Phys. Rev. A **38**, 3400 (1988).
87. A. Scrinzi and K. Szalewicz, Phys. Rev. A **39**, 2855 (1989).
88. E.A.G. Armour, D.M. Lewis and S. Hara, Phys. Rev. A **46**, 6888 (1992).
89. N.F. Mott and H.S.W. Massey, The Theory of Atomic Collisions 3rd edn. (Oxford University Press, Oxford, 1965) p. 802.
90. A. Messiah, Quantum Mechanics (Wiley, New York, 1962) Vol II, p. 732.
91. L.N. Bogdanova, V.E. Markushin, V.S. Melezhik and L.I. Ponomarev, Sov. Phys. JETP **56**, 931 (1982).
92. M.C. Struensee, G.M. Hale, R.T. Pack and J.S. Cohen, Phys. Rev. A **37**, 340 (1988).
93. M. Kamimura, AIP Conference Proceedings **181**: Muon Catalysed Fusion, eds. S.E. Jones, J. Rafelski and H.J. Monkhorst (American Institute of Physics, New York, 1989) p. 330.
94. A.M. Lane and R.G. Thomas, Rev. Mod. Phys. **30**, 257 (1958).
95. L. Bracci and G. Fiorentini, Nucl. Phys. A **364**, 383 (1981).
96. S.S. Gershtein *et al*, Sov. Phys. JETP **53**, 872 (1981).
97. L.N. Bogdanova *et al*, Nucl. Phys. **454**, 653 (1986).
98. V.V. Gusev *et al.* Muon Catal. Fusion **4**, 1467 (1989).
99. D. Cepperley and B.J. Alder, Phys. Rev. A **31**, 1999 (1985).
100. S.E. Haywood, H.J. Monkhorst and K. Szalewicz, Phys. Rev. **37**, 3393 (1988); **39**, 1634 (1989).
101. J.S. Cohen, Phys. Rev. Lett. **58**, 1407 (1987); Muon Catal. Fusion **1**, 179 (1987).
102. V.E. Markushin, Muon Catal. Fusion **3**, 395 (1988).
103. H.E. Rafelski, B. Müller, J. Rafelski, D. Trautmann and R.D. Viollier, Progr. Part. and Nucl. Phys. **22**, 279 (1989).
104. P. Kammel, Muon Catal. Fusion **5**, 111 (1990).
105. Yu. V. Petrov, Nature **285**, 466 (1980); Muon Catal. Fusion **3**, 525 (1988).

ELECTRON-POSITRON PAIR PRODUCTION IN COULOMB COLLISIONS AT ULTRARELATIVISTIC ENERGIES

C. R. Vane, S. Datz, P. F. Dittner, H. F. Krause, C. Bottcher,[+] and M. Strayer
Oak Ridge National Laboratory,[*] Oak Ridge, TN 38731 USA

R. Schuch and H. Gao
Manne Siegbahn Institute, Stockholm, Sweden

R. Hutton
University of Lund, Lund, Sweden

ABSTRACT

We have measured angular and momentum distributions for electrons and positrons created as pairs in peripheral collisions of 6.4 TeV bare sulfur ions with fixed targets of Al, Pd, and Au. Singly- and doubly-differential cross sections have been determined for 1–17 MeV/c electrons and positrons detected independently and in coincidence as pairs. Integrated yields for pair production are found to vary as the square of the target nuclear charge. Relative angular and momentum differential cross sections are effectively target independent. Probability distributions for the pair total momentum, the positron fraction of the pair momentum, and the pair transverse momentum have been derived from the coincident electron-positron data.

I. INTRODUCTION

The study of direct electron-positron pair production in Coulomb collisions of high-energy charged particles has a long history, dating from early calculations of Bhabha,[1] Racah,[2] and Landau and Lifshitz[3] in the 1930's and continuing with modern quantum electrodynamic (QED) calculations through the 1970's.[4-8] Extensive measurements of electron pairs produced by beams of singly-charged electron, pion, and proton projectiles, mainly in nuclear track emulsions, have also been performed.[9-17] At ultrarelativistic energies, peripheral collisions between heavy, highly-charged atoms produce extremely intense, rapidly varying electromagnetic fields which are expected to give rise to copious lepton-pair formation. Electron-positron pairs should be formed at impact parameters $\sim \lambda_c$, the electron Compton wavelength. These collisions are fundamentally different from those involving singly-charged projectiles because the coupling constant $Z\alpha$, where Z is the charge and α is the fine structure constant, can be large (≥ 0.5) in heavy systems. Perturbative calculations typically order scattering amplitudes in ascending powers of $Z\alpha$, neglecting terms beyond some order above which complexity of the calculations become unmanageable. Lepton pair production in these systems is especially interesting because the collision strength can be varied continuously, from regions of low charge and energy, where past applications of low-order perturbative methods are suitable, to higher energy and charge regimes where first-order perturbative calculations are known to occasionally give unphysical results.[18]

[+]Deceased.
[*]Research sponsored by the U.S. Department of Energy, Office of Basic Energy Sciences, Division of Chemical Sciences, under Contract No. DE-AC05-84OR21400 with Martin Marietta Energy Systems, Inc.

Recent progress toward realization of energetic ion-ion colliders such as the Relativistic Heavy-Ion Collider (RHIC) at Brookhaven National Laboratory and the Large Hadron Collider (LHC) at the European Organization for Nuclear Research (CERN) has sparked renewed interest in electromagnetic phenomena at very high energies. As a consequence, a number of theoretical treatments have recently appeared concentrating on lepton-pair production in ultrarelativistic heavy-ion collisions.[18-23] Results of these varied calculations differ substantially in quantitative detail, but generally agree that electron-pair production cross sections should be large, should steadily increase with collision energy, and should lead to single- and possibly multiple-pair formation at rates sufficient to overwhelm contributions from all nuclear processes.[21-22] In particular, direct Coulomb pair production may obviate the suggested[23] use of lepton signals as a penetrating probe for studying formation and decay of the quark-gluon plasma in ultrarelativistic heavy-ion nuclear collisions.[24-26]

In this paper, we present measurements and comparisons where possible with perturbative theory of angular and momentum distributions and correlations of 1–17 MeV/c electrons and positrons from electron pairs generated by 6.4-TeV sulfur ions in Coulomb collisions with targets of Al, Pd, and Au. For these collision systems perturbative calculations, though complex, are expected to adequately describe the pair production phenomena.

II. EXPERIMENTAL METHODS

Fully-stripped 6.4 TeV sulfur ions from the CERN Super Proton Synchrotron (SPS) accelerator were passed through thin foil targets in a high-vacuum chamber centered in the 14 cm gap of a 1 meter long, CERN standard dipole bending magnet located upstream of a large nuclear physics collaboration experiment WA93.[27] Measurements of electrons and positrons generated in the targets were performed parasitically with operation of WA93. A schematic of the apparatus is given in Fig. 1, with the magnetic field directed perpendicular to the drawing. Electrons and positrons generated by ions in the targets were separated in the nearly uniform (± 3%) vertical magnetic field and transported 180° along circular arcs to one of two identical arrays of discrete detectors mounted on either side of the ion beam in the plane containing the chosen target. Two large scintillator detectors sensitive to minimum ionizing particles, each with a 6 cm diameter central hole were mounted symmetrically about the beam axis 2 m upstream and downstream of the spectrometer. Signals from these detectors were used to veto detector noise counts generated by any high energy particles accompanying the primary ions, but lying in an extended halo >3 cm beyond the main beam axis. A second veto signal was generated for similar halo and scattered high energy particles lying >2 mm and <4 cm from the beam axis, by a quartz Cerenkov active collimator located 5 m downstream, in the WA93 experiment.

For forward emitted electrons or positrons, horizontal displacement of the intercept point of a particle's trajectory with the detector plane is closely proportional to its momentum and is only weakly dependent on launch angle, due to first-order focusing in the plane of dispersion. Particle motion in the vertical direction is unaffected by the magnetic field. Hence, the displacement of the detection position in the horizontal plane from the beam axis corresponds to positron or electron momentum while vertical displacement corresponds to the vertical component of the transverse momentum (or the vertical component of the polar launch angle for a given total momentum.) The apparatus thus constitutes a 180°, homogeneous field, pair spectrometer. Each detector array was composed of 41 separate circular (2 cm dia.) silicon surface barrier detectors with depletion depths of 300 μm. Detectors were arranged in five horizontal rows covering 52% of the available area.

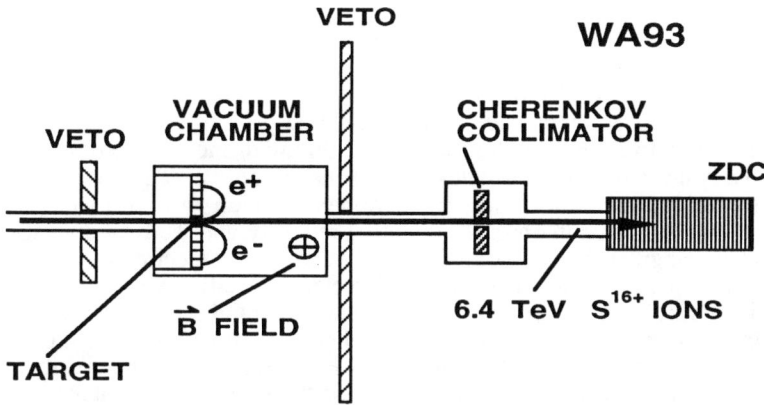

Fig. 1. Schematic of experimental apparatus showing relationships of veto detectors, magnetic analysis region (B is perpendicular to page), and WA93 Zero Degree Calorimeter (ZDC).

Electron and positron counts were stored as functions of detector position for two settings of magnetic field (0.18 and 0.45 T) corresponding to detected momenta of 1.0 – 7.0 MeV/c and 2.5 – 17 MeV/c, respectively. For data presented here, electron and/or positron detector multiplicities for a valid event were set to unity. Data were collected for targets of: 75 µg/cm^2 polypropylene $(CH_2)_x$; 180 µg/cm^2 Al; 4.9 mg/cm^2 Pd; and 0.6, 1.5, and 4.7 mg/cm^2 Au. Less than one sulfur ion in ten thousand underwent nuclear charge-changing collisions in these targets (electromagnetic spallation being the most probable at 2 x 10^{-5}/ion in the thickest Au target used).[28] Signals from the WA93 Zero Degree Calorimeter (ZDC)[29] were used in our experiment to identify, count, and provide timing for full energy projectile sulfur ions.

Sulfur ions from the SPS were delivered in roughly uniform spills of 5.1 sec length with 10^5 to 10^6 ions per spill. For the 1.5 mg/cm^2 Au target, typical counting rates for 1 – 7-MeV/c singles positrons and electrons (primarily Knock On or KO electrons) detected in coincidence with projectile ions were (1 – 10)/sec and (100 – 1000)/sec, respectively. Variations in data acquisition efficiencies as functions of counting rates for the various modes of operation were studied by comparing electron and/or positron yields per ion at different ion spill rates and for a number of targets and thicknesses. Positron and electron yields/ion were found to vary linearly with Au target thickness from 0.6 to 4.7 mg/cm^2.

Light targets of polypropylene and aluminum were used to investigate the KO electrons and their effects as backgrounds, and to study efficiencies for detection and acquisition under beam-on conditions at electron energies characteristic of both field settings. Efficiencies were determined for each detector in an array by comparing of the measured KO yields with results of Monte Carlo simulations based on relativistic First Born calculations for differential KO cross sections.[30] The average detection efficiency for 1 – 17 MeV/c electrons found in this way was 46%. We assumed that the efficiency for positron detection by these thin silicon detectors was the same as for electrons of the same energy.[31]

III. RESULTS AND DISCUSSION

A. Singles Measurements

We first present results of measurements of electrons and positrons detected independently, but in coincidence with unscattered projectile ions. These 'singles' results are used to directly extract singly differential and total cross sections, to quantitatively investigate backgrounds and to solidify confidence in our interpretations of positron-electron coincidence measurements which will follow. As noted previously, the variation in observed counting intensity with vertical displacement is related to the angular distribution of ejected positrons or electrons. The measured distributions represent projections of the source angular differential cross sections on the limited vertical aperture determined by the height of the detector arrays. Electrons and positrons from pairs formed in these peripheral collisions are emitted primarily in the forward direction. Instead of attempting to deconvolute measured distributions, we have mapped selected angular differential cross sections through the magnetic analyzing field to the detector plane and compared the resulting calculated distributions with the data. Using a computer simulation which incorporates effects of multiple Coulomb scattering, and assuming a simple exponential form for the angular cross section ($d\sigma/d\Omega_+ \sim \exp(-\theta/w)$, where θ is the polar angle measured from the beam axis), we have determined 1/e angular widths (w) which most closely reproduce the data, as a function of horizontal position (i.e., positron momentum). The exponential form for $d\sigma/d\Omega_+$ was chosen after reviewing several theoretical predictions for expected angular distributions.[18,19] Results of these simulation fits to the data are displayed in Fig. 2 for positrons from a 1.5 mg/cm^2 Au target. Within error all other targets give the same widths. Figure 2 also shows results for similar angular widths calculated in an exact Monte Carlo evaluation of the two-photon terms in lowest-order QED calculations for point nuclei described in Ref. 18, and hereafter referred to as MCQED.

Since the angular distribution for positron emission is peaked so narrowly in the forward direction, the horizontal intensity distribution on the detector array directly represents the momentum distribution. We have summed counts in nine sets of detectors grouped according to horizontal position and plot overlapping histograms of the resulting momentum distributions in Fig. 3, at the two field settings. The yields have been converted to cross sections and corrected for the fractions of positrons lying outside the vertical acceptance of each detector group. These corrections were made assuming theoretical angular distributions with widths as in Fig. 2. Theoretical cross sections are displayed in Fig. 3 as smooth curves. Results of First Born approximation calculations, which employ Sommerfeld-Maue wave functions for the free electron and positron, are displayed by the dashed line. To construct this curve, cross sections for 200 GeV protons and uranium targets have been interpolated from Fig. 4 of Ref. 19 and scaled according to Z_P^2 and Z_T^2 where Z_P and Z_T are the projectile and target nuclear charges, as directed in Ref. 19. We note that the First Born results underestimate the cross sections below ~3 MeV/c and overestimate them slightly above ~8 MeV/c. Results of MCQED calculations are indicated by the solid curve. Agreement is remarkably good up to ~4 MeV/c. At higher momenta MCQED cross sections are as much as 75% higher than the measurements. Linear extrapolation of the First Born calculation yields a cross section of about 83 barns integrated over 1 – 17 MeV/c. The corresponding MCQED result is 98 barns. The measured cross section is 85 ± 22 barns, where the error includes statistical and background correction errors and estimated uncertainties in detection and acquisition efficiencies. It does not include any

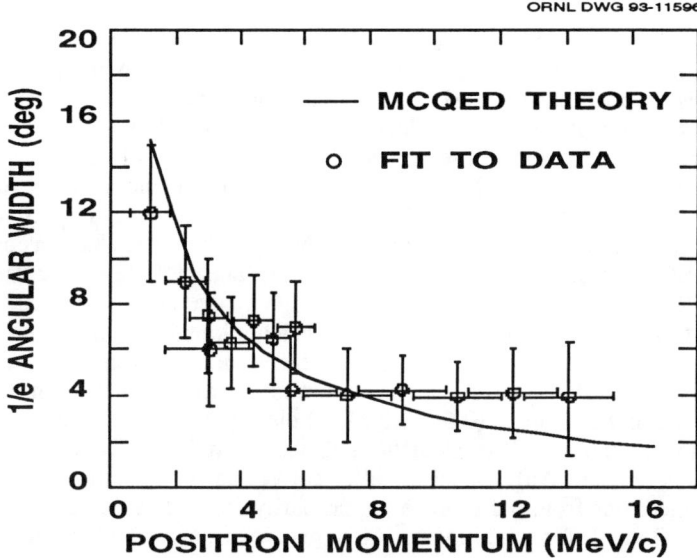

Fig. 2. Angular 1/e widths for $d\sigma/d\Omega_+$ plotted as a function of positron momentum for 6.4 TeV S + Au. Error bars indicate observed fitting uncertainty dominated by experimental angular resolution set by detector size.

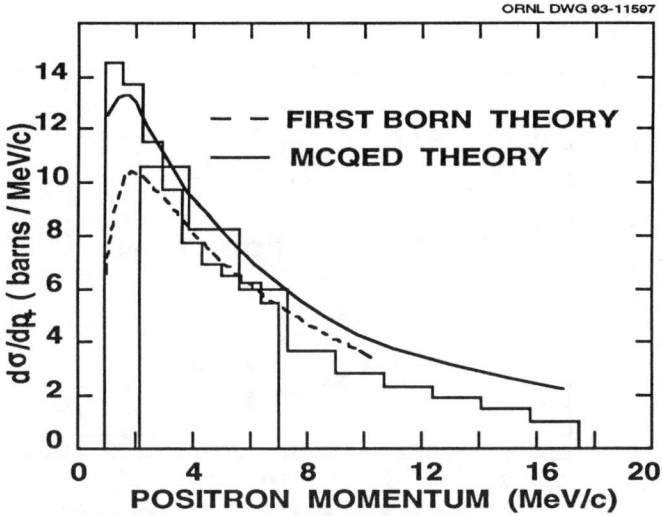

Fig. 3 Measured positron differential cross sections compared with theoretical results. Overlapping solid line histograms are measured cross sections for field settings of 0.18 and 0.45 Tesla.

estimate of errors due to corrections made as noted above for incomplete angular acceptance of the pair spectrometer.

Figure 4 shows a comparison of measured momentum-binned yields per sulfur ion for **singles** electrons and positrons from a gold target. No corrections for angular acceptance by the detectors have been made for these data. Momentum and angular distributions for positrons from Al and Pd are the same (±20% relative error) as found for Au. Assuming pair produced electrons and positrons have similar momentum distributions, only 2% of the observed singles electrons come from pair production. The other 98% arise from primary KO production and from secondary scattering processes. By using thin, light (polypropylene and aluminum) targets where pair production is minimized, we have experimentally isolated the KO electron contribution. Yields for KO electrons integrated over all detectors at each field setting vary linearly (±10%) with electron thickness for all targets. We have compared these measured absolute KO yields with results of trajectory calculations based on relativistic Born approximation differential cross sections for binary ion-electron collisions.[30] Our calculations for absolute yields agree (within ±25% error) with the light target data, except for the lowest energy (0.6 – 1.0 MeV) electrons where secondary scattering effects become important. The calculations also agree well with KO measurements for heavy targets (Pd and Au), especially after correction for the small but significant fraction of electrons from pairs which are primarily emitted in the forward direction. Calculated KO yields for a 1.5 mg/cm^2 Au target are displayed by the smooth curve in Fig. 4.

Calculations and measurements for KO electrons both exhibit nearly uniform counting distributions along the vertical direction within the detector array. The observed uniformity arises from the narrow angular cuts imposed by the vertical height

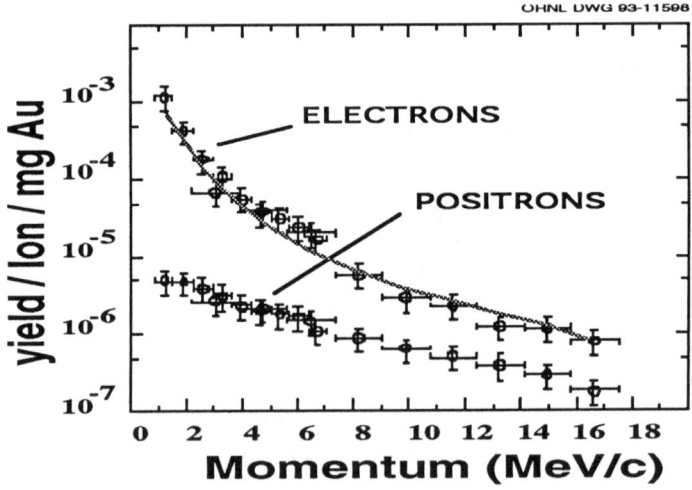

Fig. 4. Singles yields for positrons and electrons from gold detected independently, but in coincidence with sulfur ions. Data from low and high field settings are shown overlapped. Smooth curve (——) represents results of Born Approximation calculations of KO electrons. Error bars for yields represent combined statistical uncertainties and estimates of detection efficiency and acquisition deadtime errors.

accepted by the detector array when compared with the relatively larger KO launch angles. Our calculations indicate that only ~20% of all 1 – 17 MeV/c KO electrons emitted follow paths which intercept the detectors. Most of the low energy KOs spiral into the aluminum walls of the vacuum chamber downstream of the detectors and lose too much energy in penetrating the metal to continue on trajectories leading back to the detector arrays. Higher energy KOs, which are emitted very forward, proceed further downstream and suffer the same fate. Backgrounds in the positron detectors from scattered KOs are small. We have investigated these by measuring 'positron' singles for the low Z_T targets where **real** positron yields are weak and KO yields are still relatively strong.

To examine the dependence of pair production on target nuclear charge Z_T, we have integrated singles positron yields over 2.5-17 MeV/c for targets of Al, Pd, and Au and fitted the resulting relative partial cross sections against Z_T^n. The fit gives n = 1.99 ± 0.02. Cross sections are expected to scale as Z_T^2 by all known lowest-order theories. Significant contributions by KO electron backgrounds would lead to a positron yield component linear in Z_T.

B. Electron-Positron Coincidence Measurements

Having established from the singles data some confidence in our understanding of the angular and momentum distributions for pairs and of the major background, KO electrons, we now proceed to results deduced from measurements of electrons and positrons detected in time coincidence, i.e., 'coincidence' data. Doubly differential yields have been obtained by sorting event data into coincidence counts as functions of various parameters calculated from the (41 x 41) possible pairs of detector positions. To obtain cross sections, corrections have been made for variations in detector efficiencies and for estimated fractions of detector intercepted electrons and positrons according to assumed extended angular distributions as indicated for singles positrons (see Fig. 2).

Cross sections $d^2\sigma/dp_+dp_-$ for positrons and for electrons are compared in Fig. 5, where the coincident particle (electrons or positrons, respectively) momentum has been limited in the sort of the data from 2.5 to 5.5 MeV/c, a range covered by detectors at both magnetic field settings. The only substantial difference between the spectra displayed is a slight shift in the electron distribution toward low momentum. For momenta ≤4 MeV/c, the yield of electrons exceeds that of positrons by 23%. The excess could be explained by coincidences (random and correlated) with low momentum KO electrons, where singles rates (see Fig. 4) are 5 to 100 times larger than for positrons and hence, for pair-generated electrons. However, the random positron-KO electron coincidence fraction is estimated to be very small. A typical randoms/reals fraction calculated from measured singles rates and from the time window defining a coincidence is 10^{-2}, or ~1%. Correlated coincidences between positrons and KO electrons can also occur because of secondary KO production by the pair-producing ion, either in ion-electron collisions within the same atom participating in pair production event or in another collision in traversing target foil. For typical thickness gold targets, the latter is just the measured (1 – 17 MeV/c) KO yield/ion which is $\leq 10^{-2}$. We estimate the former process to occur ~2% of the time, with less than half of these being detected. The sum then of random and correlated KO positron-electron coincidences should amount to less than 3% of the real pair coincidences.

Since the KO electrons are expected to account for only a small fraction of the low-energy electron excess observed in the pairs, we have considered alternative explanations. A small horizontal shift ~6 mm in the beam position at the target could

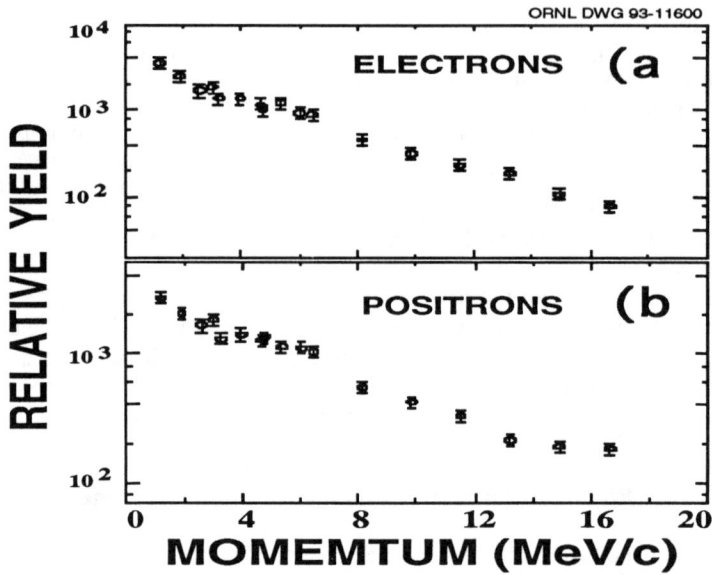

Fig. 5. Momentum distributions of pairs of (a) electrons and (b) positrons from gold detected in coincidence and in coincidence with sulfur ions. Momentum range of the coincident pair partner has been set to 2.5 – 5.5 MeV/c in each case.

also lead to the observed asymmetry. However, by reversing the orientation of the magnetic field, we have exchanged electron and positron detectors and obtained the same result. The difference between the positron and electron distributions may be attributed instead to a real charge-dependent effect due to the Coulomb potential at the origin of the pair. The $d^2\sigma/dp_+ dp_-$ weighted average kinetic energies taken from the data in Fig. 5 are 4.21 and 3.87 MeV, respectively, for 1 – 17 MeV/c positrons and electrons. If we interpret in a purely classical picture this (0.34 MeV) shift in the average kinetic energy as arising from a potential energy difference experienced by the oppositely charged particles at the pair origin, then a rough approximation of the mean distance $<b>$ to the gold target nucleus (neglecting screening and the sulfur ion's field) gives $<b> \approx 700$ fm or $\sim 2\lambda_c$. (The 1 – 17 MeV/c electrons and positrons are localized to $<<\lambda_c$ according to the uncertainty principle.)

As noted for the singles data, the vertical displacement of a detected electron is approximately proportional to the vertical component of its transverse momentum (p_y). We have sorted event data to determine the intensity distributions as a function of pair ($p_{y+} + p_{y-}$) transverse momenta and present results in Fig. 6. As expected, electrons and positrons forming pairs tend to be emitted with opposite transverse momenta, so that total transverse momentum is small. The observed probability distribution peaks at ($p_{y+} + p_{y-}$) ≈ 0 and falls rapidly to ~1 MeV/c where it becomes essentially constant. Both ranges of detected momenta give approximately the same distribution.

We have also investigated the distribution of total pair momentum by adding separate momenta of electrons and positrons forming detected pairs and then binning the corresponding corrected counts to give intensity distributions, with the ranges of either component limited by the momenta covered at each field setting. Results are

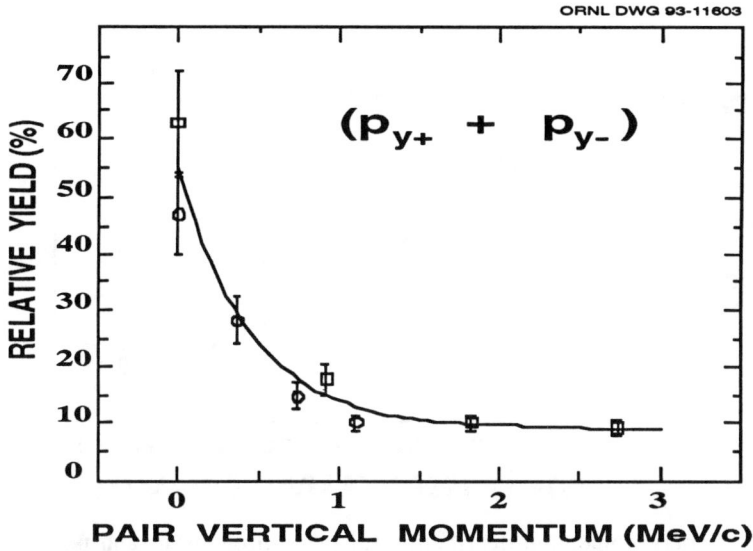

Fig. 6. Transverse vertical (py+ + py-) momentum distribution for detected pairs. (1 – 7 MeV/c) – O; (2.5 – 17 MeV/c) – □. Curve added to guide the eye.

shown in Fig. 7. Remarkably, overlap of the two distributions is nearly continuous, implying a general uncorrelated nature for the momenta of coincident pair electrons and positrons. These same data have been analyzed to determine the fraction of total pair momentum carried away by the positron. Figure 8 shows the probability distribution for momentum sharing measured for pairs covered in the two field settings. The error bars represent statistical and efficiency uncertainties for the ratios of binned counts. It appears that all modes of partitioning of the momentum available are nearly equally probable. We note that for photoproduction, i.e., the conversion of real γ-rays into electron-positron pairs near nuclei, very similar results are obtained.[32] We also note that the positron carries away on average slightly more than 50% of the pair momentum. This is consistent with our previous observation (Fig. 5) of a small energy shift between electrons and positrons, which is not surprising as the two figures are merely different ways of representing the same data.

CONCLUSION

Electrons and positrons formed as pairs in peripheral collisions of 6.4 TeV sulfur ions with fixed targets have been studied using a uniform field 180° magnetic analyzer. The positron and electron momentum distributions agree reasonably well with modern QED calculations. The absolute total cross sections are large (>100 b) and also agree with theory. Momentum is shared between the pair components in a nearly uncorrelated fashion, so that either may carry away any fraction with almost uniform probability. This result is consistent with photoproduction. The pair electrons and positrons are emitted forward (θ < 15°), along the ion's axis, and the transverse component of the pair momentum peaks at zero.

Perturbative QED calculations, as expected, do a reasonably good job of predicting these results. However, as Zα is increased at these energies, a regime will be entered in which perturbation theory must fail. Multiple pair formation is expected in single collisions in the non-perturbative regime. Future (1994) measurements at the CERN

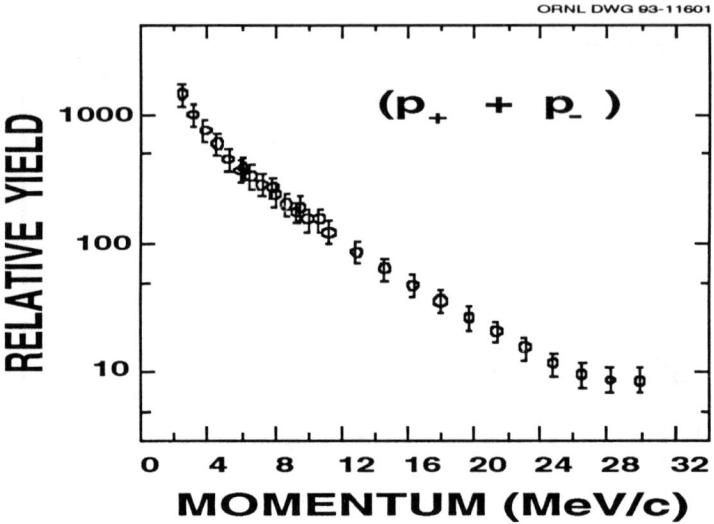

Fig. 7. Total momentum distribution for pairs of electrons and positrons from gold.

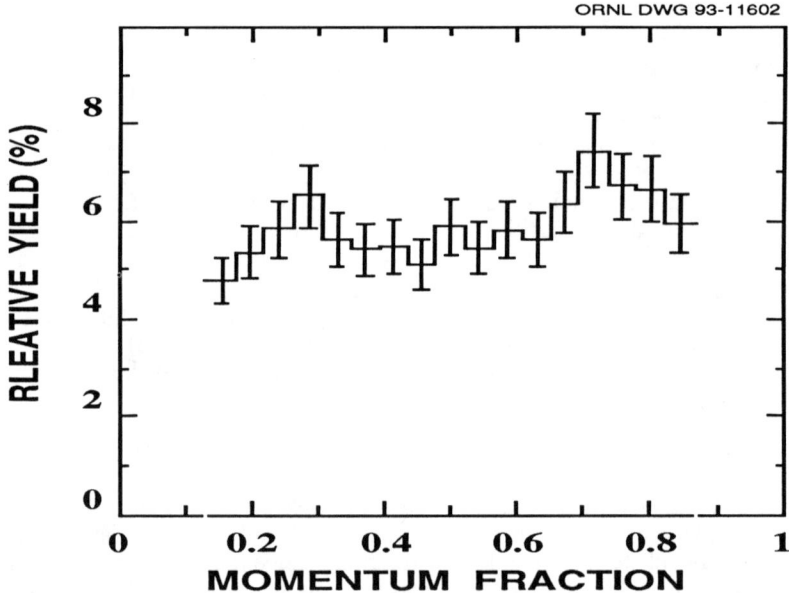

Fig. 8. Probability distribution for momentum sharing for all pairs covered by the two field settings

SPS using 33 TeV lead ions have been approved to test the ability of a variety of new non-perturbative calculations to predict the distributions of pairs occurring in this new regime.

REFERENCES

1. H. J. Bhabha, Proc. R. Soc. London, Ser. A **152**,559 (1935).
2. G. Racah, Nuovo Cimento **14**, 93 (1937).
3. L. D. Landau and E. M. Lifshitz, Phys. Z. Sowjetunion **6**, 244 (1934).
4. T. Murota, A. Ueda, and H. Tanaka, Prog. Theor. Phys. **16**, 482 (1956).
5. F. F. Ternovskii, Zh. Eksp. Teor. Fiz. **37**, 793 (1960) [Sov. Phys. JETP **10**, 565 (1960)]; **37**, 1010 (1960) [**10**, 718 (1960)].
6. S. R. Kelner, Yad. Fiz. **5**, 1092 (1966) [Sov. J. Nucl. Phys. **5**, 778 (1967)].
7. S. R. Kelner and Y. D. Kotov, Yad. Fiz. **7**, 324 (1968) [Sov. J. Nucl. Phys. **7**, 237 (1968)].
8. A. G. Wright, J. Phys. A **6**, 79 (1973); critical review of calculations.
9. M. M. Block, D. T. King, and W. W. Wada, Phys. Rev. **96**, 1627 (1954).
10. M. M. Block and D. T. King, Phys. Rev. **95**, 171 (1954).
11. M. Koshiba and M. F. Kaplan, Phys. Rev. **97**, 193 (1955).
12. J. E. Naugle and P. S. Freier, Phys. Rev. **104**, 804 (1956).
13. A. S. Cary, W. H. Barkas, and E. L. Hart, Phys. Rev. D **4**, 27 (1971).
14. P. L. Jain, M. Kazuno, and B. Girard, Phys. Rev. Lett. **32**, 1460 (1974).
15. P. L. Jain, M. Kazuno, B. Girard, and Z. Ahmad, Phys. Rev. Lett. **32**, 797 (1974).
16. J. E. Butt and D. T. King, Phys. Rev Lett. **31**, 904 (1973).
17. L. R. Fortney et al., Phys. Rev. Lett. **34**, 907 (1975).
18. C. Bottcher and M. R. Strayer, Phys. Rev. D **39**, 1330 (1989); M. J. Rhoades-Brown and J. Weneser, Rev. Mod. Phys., in press.
19. F. Decker, Phys. Rev. A **44**, 2883 (1991); F. Decker and J. Eichler, Phys. Rev. A **45**, 3343 (1992).
20. K. Rumrich et al., Phys. Rev. Lett. **66**, 2613 (1991).
21. G. Baur, Phys. Rev. A **42**, 5736 (1990); J. Eichler, Phys. Rep. **193**, 167 (1990) – review of theoretical progress to 1990.
22. P. B. Eby, Phys. Rev. A **43**, 2258 (1991) and Phys. Rev. A **39**, 2374 (1989).
23. K. Kajantie et al., Phys. Rev. D **34**, 811 (1986); K. Kajantie et al., Phys. Rev. D **34**, 2746 (1986).
24. H. Gould, in "Atomic Theory Workshop on Relativistic and QED Effects in Heavy Atoms", Hugh P. Kelly and Yong-ki Kim, eds. (AIP Conf. Proc. 136), (AIP, New York,1985).
25. C. Bottcher and M. R. Strayer, in "Physics of Strong Fields", W. Greiner, ed. (NATO ASI Ser. B Vol 153) (Plenum, New York, 1987), pg. 629.
26. E. Teller, Nucl. Instrum. Meth. Phys. Res. **B24/25**, 3 (1987).
27. H. Gutbrod et al., CERN Report /SPSLC 91-17 SPSLC/P260, May 1991.
28. B. Price, Ren Guoxiao, and W. T. Williams, Phys. Rev. Lett. **61**, 2193 (1988).
29. G. R. Young et al., Nucl. Instr. Methods **A279**, 503 (1989).
30. C. R. Vane, S. Datz, P. F. Dittner, H. F. Krause, C. Bottcher, M. Strayer, R. Schuch, H. Gao, and R. Hutton, Nucl. Instrum. Meth. Phys. Res. B (1993), in press.
31. Th. Frommhold et al., Nucl. Instr. Meth. A **310**, 657 (1991).
32. C. F. Powell, P. H. Fowler, and D. H. Perkins, "The Study of Elementary Particles by the Photographic Method," Pergamon Press, New York, 959, p. 181.

HEAVY PARTICLE COLLISIONS

COLLISION DYNAMICS OF ORIENTED STATES

Nils Andersen
Niels Bohr Institute, Ørsted Laboratory, Universitetsparken 5
DK-2100 Copenhagen, Denmark

ABSTRACT

This chapter reviews the present understanding of the collision dynamics of oriented states, i.e., states with a preferred sense of circulation of the active electron around the atomic core. The focus is on excitation and charge transfer processes in energetic collisions between two atoms, i.e., collisions for which the collision velocity is comparable to the velocity of the active electron.

INTRODUCTION

The study of alignment and orientation in atomic collisions has a long history. The seminal paper by Fano and Macek [1] starting the modern era is now 20 years old. Many of the key ideas still in use today for interpretation of alignment and orientation effects in heavy particle collisions originated, as we shall see, from the Nebraska group [2, 3, 4] around this time, and served as inspiration for later generalisation of the framework [5].

The schematic diagram of Fig. 1 of the basic collision event may serve as introduction to the problem. In Fig. 1(a), two initially separated particles A and B (e.g. atoms, molecules, electrons, positrons, or clusters) are allowed to interact for some time, and we would like to understand what goes on in this region. When the particles separate, they generally not only have directions different from the initial ones, but they may also be in different quantum states.

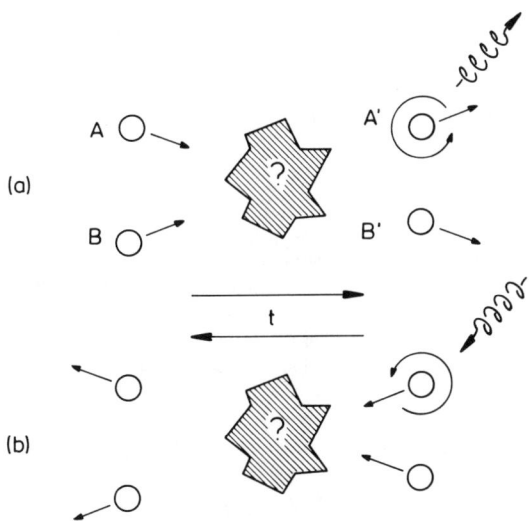

Figure 1: (a) A schematic diagram of the basic collision event. The particles A and B are initially separated, then interact for some time, and finally separate again. (b) shows the same event, but with the arrow of time reversed. The schemes (a) and (b) yield essentially the same information.

506 Collision Dynamics of Oriented States

If in this state the charge distribution of the electron cloud is nonspherical, we call it *aligned*. If the electron has a preferred sense of circulation around the atomic core, i.e. the expectation value of the orbital angular momentum is nonzero, we call it *oriented*. For an excited state, the alignment and orientation are revealed through the polarization properties of the photon emitted when the state decays to a lower state (Stokes parameter analysis).

Essentially equivalent information may be obtained from the scheme shown in Fig. 1(b), obtained from (a) by letting time run backwards. Thus, a particle may be prepared with an aligned or oriented state *before* the interaction by polarized laser light, and one studies how the outcome of the collision depends on the polarization properties of the incident light. Which of the two schemes (a) or (b) is preferable depends mainly on considerations of experimental convenience.

In this paper we shall focus the discussion on *orientation*. Also, to avoid overlap with other contributions to the ICPEAC program, the focus is limited even further. Electron impact excitation is left out, since this is reviewed by A. Crowe [6]. Low energy, or thermal, collisions between heavy particles, for which the collision velocity is much smaller than the velocity of the active electron, are covered in the paper by S.R. Leone [7]. Collisions with oriented Rydberg states are discussed in the Hot Topic session by S.B. Hansen [8]. The present paper thus reviews the current understanding of orientation phenomena in the energy range, where the collision velocity is of the samer order of magnitude as the velocity of the active electron.

Substantial contributions to the present understanding of alignment and orientation phenomena in this area originate from the groups in Aarhus, Argonne, Bergen, Bielefeld, Boulder, Chicago, Copenhagen, Freiburg, Groningen, Lincoln, Madrid, Manhattan/KSU, Orsay, Paris, St. Petersburg, and Vienna. A comprehensive overview of these studies may be found in the forthcoming Vol. II of the JILA Review Series [9].

Here, due to space limitations, we cannot cover all experimental and theoretical results accumulated until now. Rather, the approach selected is to present a few case studies, for which comprehensive experimental and theoretical data exist, and extract general conclusions on the underlying collision dynamics responsible for orientation propensities. The presentation is preceded by a brief introduction to the necessary framework, and at the end some general conclusions are presented.

FRAMEWORK

This section briefly summarizes how orientation is defined and observed. For a complete account, the reader is referred to [5] or a chapter in an earlier volume of the ICPEAC proceedings [10].

A general conservation law greatly facilitates the discussion : Referring to Fig. 1, the reflection symmetry with respect to the collision plane of the total wavefunction describing the (A+B) complex is conserved during the collision. In case the state of B is not changed, this law implies that the reflection symmetry of the A wavefunction alone is unchanged. For the simple example of S $\rightarrow$ P excitation this means that, since the wavefunction of the spherically symmetric S state has positive reflection symmetry, only the p_{-1} and p_{+1} substates of the P state may become excited, while transitions to the p_0 state are not allowed. Here, the magnetic quantum number m of the p_m state refers to a quantisation axis perpendicular to the collision plane (See Fig. 1 of [10]). Consequently, the orbital angular momentum of the P state is in the direction perpendicular to the collision plane. Its expectation value $< L_\perp > = L_\perp$ is, in atomic units, a number between -1 and +1, the actual size depending on the relative population of the p_{-1} and p_{+1} substates.

Detection of this quantity may conveniently be done in an experimental setup such as the one shown in Fig. 2. Here, particles scattered at a selected angle θ are detected in coincidence with polarization-analysed photons emitted perpendicular to the scattering plane. In this way, the set of three Stokes parameters (P_1, P_2, P_3) fully characterizing the light polarization in this direction [11] may be determined. This vector quantity, corrected for fine- and hyperfine structure effects

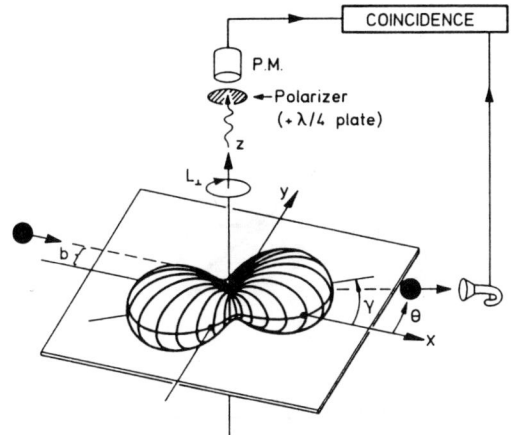

Figure 2: In this experimental setup, particles scattered at a selected angle θ are measured in coincidence with polarization-analysed photons emitted perpendicular to the scattering plane.

if necessary [5], also fully determines the wavefunction of the P state, characterized by the two quantities $L_\perp$, which measures the *orientation* of the state, and the angle γ, which gives the *alignment* of the state with respect to the incident beam direction. They are related to the components of the Stokes vector in the following way

$$P_1 + i P_2 = P_l\, e^{2i\gamma} \tag{1}$$

$$P_3 = -L_\perp = \frac{N(RHC) - N(LHC)}{N(RHC) + N(LHC)} \tag{2}$$

Here, $L_\perp^2 + P_l^2 = 1$ if the excitation is coherent. The minus sign in the equation for circular polarization follows the convention of classical optics [11], i.e. right hand circularly (RHC) polarized light has negative helicity. These relations are simple consequences of the characteristics of electric dipole radiation and conservation of angular momentum, cf. e.g. [5] and Fig. 2 of [10].

We shall now proceed to some specific examples of results obtained with an experimental setup of this type.

CASE STUDIES

Figure 3 presents the three classes of processes to be discussed in detail. In all three cases comprehensive sets of experimental and theoretical studies are available which, as we shall see, allow conclusions of a more general nature to be drawn.

The first case to be discussed is direct excitation, or a transition between two almost parallel potential curves with an energy defect ΔE, Fig. 3(a). The second example is a transition between two crossing curves, i.e., two molecular potential curves which at a specific internuclear distance become energetically degenerate, Fig. 3(b). The final situation is (near-)resonant charge exchange, $\Delta E \simeq 0$; the essential feature in this case is that the active electron is bound to a different center before and after the interaction, Fig. 3(c).

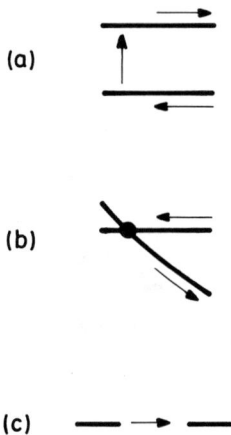

Figure 3: The three case studies discussed in the text : (a) Direct excitation between two nearly parallel energy curves, (b) Excitation at a curve crossing, and (c) Near-resonant charge transfer.

Case (i) : Direct excitation

Excitation of the resonance transition in quasi-one electron systems, such as Mg^+ (3s → 3p), in $Mg^+ - rare\ gas$ collisions has been studied extensively [12]. If the impact parameter is sufficiently large, so that molecular curve crossings due to promotion of rare gas electrons are avoided, the excitation takes place due to direct transitions between the two, nearly parallel, potential curves involved. The transition probability is large in the region of collision velocities v where the so-called "Maximum Rule"

$$\frac{\Delta\epsilon\, a}{\hbar v} \simeq \pi \tag{3}$$

is fulfilled. Here, a is the effective interaction length of the collision, and $\Delta\epsilon$ the effective energy change, in most cases close to its asymptotic value.

Figure 4 illustrates the outcome of an experiment of the type shown in Fig. 2 for the case of Mg^+ (3p → 3s) photons produced in collisions with He with an impact energy of 35 keV and impact parameter $b \simeq 1.0$ a.u.[13]. The corresponding collision geometry is shown in the lower left corner. A strong dominance of RHC photons over LHC photons is observed. The same dominance has been observed for direct excitation in several other quasi-one electron systems in the velocity region around the excitation maximum [14, 15].

However, if the collision velocity is changed, and in particular lowered, so that the collision partners are given more time to interact, the excitation probability is lowered, and the orientation may change, and even exhibits several sign reversals [13], as illustrated for the present case in Fig. 5.

The detailed collision dynamics responsible for this *propensity rule* [16] for orientation for direct transitions has now been understood for some time and is well documented in the literature [17, 18, 19]. Here we summarize the way of reasoning :

From the close-coupling equations [17] one may derive the following first-order expressions for the S - P excitation amplitudes a_m for the two p_m states of opposite helicity, using state amplitudes $(a_s, a_{-1}, a_{+1}) = (1, 0, 0)$ as initial conditions :

$$\begin{cases} a_{-1} = \frac{1}{v}\int_{-\infty}^{\infty} F_{sp}\, e^{i(\frac{\Delta E x}{v} - \phi)}\, dx \\ a_{+1} = -\frac{1}{v}\int_{-\infty}^{\infty} F_{sp}\, e^{i(\frac{\Delta E x}{v} + \phi)}\, dx \end{cases} \quad (4)$$

with $F_{sp}(R) = <s|V|p_{-1}> = - <s|V|p_{+1}>$. V is the interaction between the active electron and particle B, and $\Delta E = \Delta E(R(x))$ is the S - P energy level difference [17, 18] at an internuclear distance R at the point x along the trajectory. The angle $\phi = \phi(x)$ is the corresponding rotation angle of the internuclear axis.

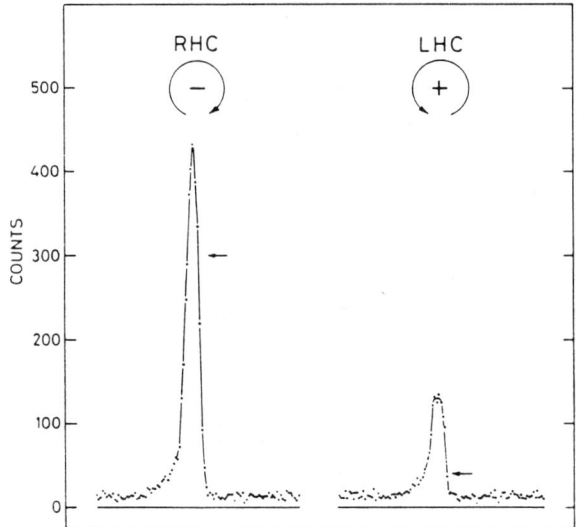

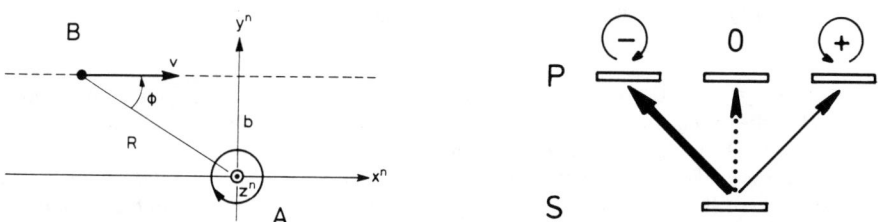

Figure 4: The upper part shows the relative number of RHC and LHC polarized photons detected for $Mg^+(3s \rightarrow 3p)$ excitation in collisions with He in the geometry shown in the lower left corner. The impact energy is 35 keV and the impact parameter is $b \simeq 1.0$ a.u. In this situation there is thus a strong *propensity* for transitions from the 3s ground state to the $3p_{-1}$ state, as shown in the lower right corner.

We now explore under which conditions these amplitudes may become large. The function F_{sp} in the present case of direct excitation is generally a smooth, bell-shaped function with its maximum at $x = 0$. Thus, the size of a_{-1} is mainly determined by the behavior of the phase of the exponential function. This phase consists of two parts. The latter one, the rotation angle of the internuclear axis ϕ, changes by about π during a full collision. The variation of the first one depends on the size of the velocity v. If v is small, this phase term may accumulate many multiples of π during the collision, leading to a lot of cancellation in the integral, and a correspondingly small a_{-1} value. On the other hand, if v is very large, this phase term stays small. However, if v is adjusted so that this term also increases by π during the collision, the two terms may effectively cancel, leading to a stationary phase, and thereby yielding the maximum value of a_{-1}. Furthermore, at this velocity, the phase term of the a_{+1} integral will accordingly vary by about 2π during the collision, causing this amplitude to become very small.

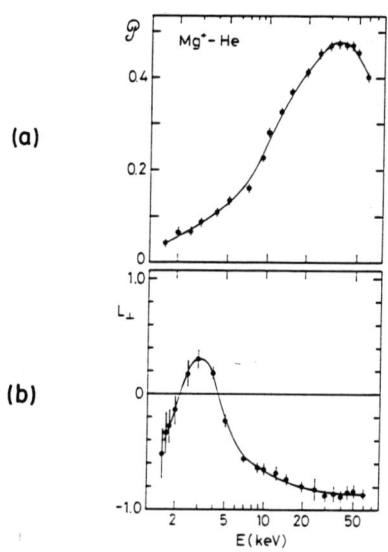

Figure 5: The excitation probability $\mathcal{P}$ (a) and angular momentum $L_\perp$ (b) for $Mg^+(3s \rightarrow 3p)$ excitation in $Mg^+ - He$ collisions with impact parameter $b \simeq 1.0$ a.u. as function of collision energy.

In summary, when the velocity v is selected so that the Maximum Rule stated above is fulfilled, the S $\rightarrow$ P excitation probability has its maximum, and the overwhelming part of the probability goes to the p_{-1} state. Thus, *a propensity for orientation,* $< L_\perp > \simeq -1.0$, *is seen when the phase build-up due to the S-P energy difference matches the rotation angle of the internuclear axis.* It is also clear how *oscillations in the orientation,* as seen in Fig. 5, *may occur* by doubling or tripling the collision time, since this doubles or triples the accumulated phase difference.

Note also that reversal of the sign of ΔE, i.e., *going from excitation to deexcitation* keeping all other parameters unchanged, *will change the sign of the orientation.*

The arguments above, derived for S $\rightarrow$ P transitions may easily be generalized to arbitrary transitions [19] :

$$\frac{\Delta \epsilon\, a}{\hbar v} + \Delta m \pi \simeq 0 \qquad (5)$$

Here, $\Delta m = m_{final} - m_{initial}$ is the change in the magnetic quantum number m along z.

Several other important consequences were pointed out already in an earlier contribution to the ICPEAC Proceedings Series [10].

Case (ii) : Curve crossing

Following the discussion in the previous section, it appears natural to proceed to a case where the excitation takes place at a well-defined curve-crossing, i.e. $\Delta E(R) = 0$, and with as little interference from additional mechanisms as possible. An early discussion of such a geometry was presented by Russek [20]. A possible choice here is the $B^{3+} - He$ system for which the molecular curves of importance have a very simple structure, shown in Fig. 6.

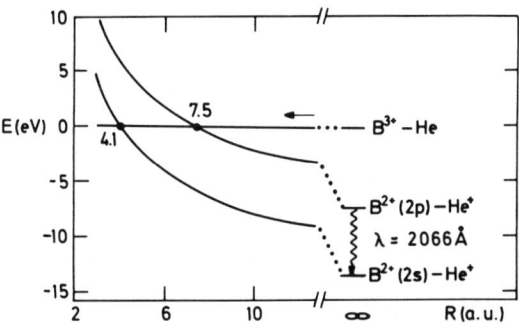

Figure 6: Molecular potential energy curves for the $B^{3+} - He$ collision. Only two curves cross the incoming channel, the Coulomb curves leading to charge exchange into $B^{2+}(2s)$ and $B^{2+}(2p)$, respectively.

The essentially flat incoming $B^{3+} - He$ channel crosses the two Coulomb curves leading to charge exchange into $B^{2+}(2s) - He^+$ and $B^{2+}(2p) - He^+$ at internuclear distances of approximately 4.1 and 7.5 a.u., respectively. No other channels interfere. The $B^{2+}(2p)$ level decays to the ground state by emission of a 2066 Å photon, i.e. in a wavelength region where linear and circular polarization analysis is readily performed with standard techniques.

The first report on alignment and orientation for this process was a theoretical analysis published in 1990 predicting strong orientation effects with a sense of orientation following the rotation direction of the internuclear axis for the main, forward peak of the charge exchange cross section [21]. The paper also illustrated the evolution of the electron charge cloud along the trajectory, and in particular the buildup of the $B^{2+}(2p)$ component. A subsequent experimental investigation mapped the doubly-differential cross section in scattering angle and energy loss, as illustrated in Fig. 7 [22]. The data confirm the expectation from Fig. 6 of electron tranfer into the $B^{2+}(2s)$ and $B^{2+}(2p)$ levels, with capture into the ground state being the main channel.

A photon polarization study of the $B^{2+}(2p)$ channel yielded an orientation with a sign confirming the theoretical prediction, as shown in Fig. 8(a) together with some recent theoretical predictions [23]. We also note a sign reversal when going to larger scattering angles, corresponding to the tail of the differential cross section. For the $B^{3+} - Ne$ collision the $B^{2+}(2p)$ channel dominates $B^{2+}(2s)$, and it is possible to obtain better counting statistics. Fig. 8(b) shows the results in this case, revealing the same sign of the orientation for forward scattering, and a clearly pronounced, regular oscillatory structure in the tails.

The main features of these observations may be understood from the following simple analysis. Figure 9 shows in (a) how the incident Σ potential curve crosses the two degenerate Σ and Π curves of the outgoing channel. Excitation takes place by Σ – Σ radial coupling on the way in, (1), or on the way out, (2). After excitation, efficient rotational Σ – Π coupling will cause the excited orbital to remain space-fixed. Thus, the collision geometry will be as outlined in (b), in which two σ-orbitals are drawn, (1) corresponding to excitation on the way in, (2) to

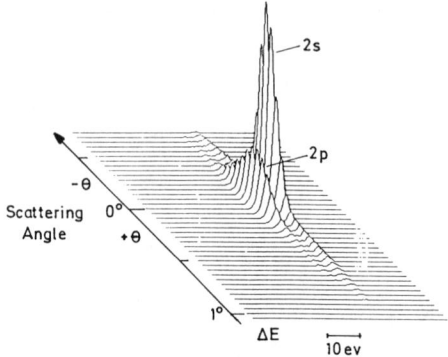

Figure 7: Doubly-differential cross sections in energy loss ΔE and scattering angle Θ for the charge exchange channels, separating clearly electron tranfer into the $B^{2+}(2s)$ and $B^{2+}(2p)$ levels, respectively. The impact energy is 1.8 keV.

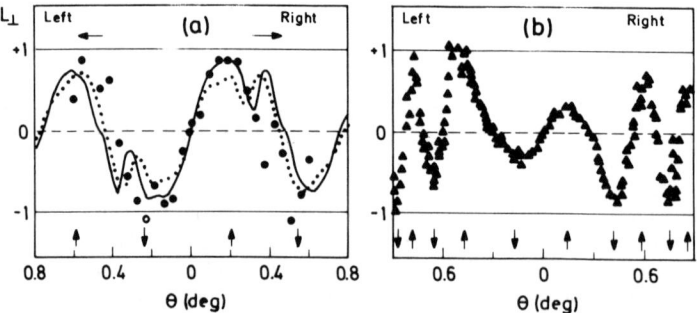

Figure 8: Orientation parameter $L_\perp$ as function of scattering angle for the $B^{2+}(2p)$ state produced in 1.5 keV (a) $B^{3+} - He$ and (b) $B^{3+} - Ne$ collisions, respectively.

excitation at the symmetric point on the way out. The rotation angle of the internuclear axis from (1) to (2) is ϕ. The total amplitude for excitation is obtained by coherently adding the two corresponding dipoles, which differ in two ways : (2) is rotated by an angle ϕ with respect to (1); (1) has accumulated an additional phase $\Phi = A/\hbar v$, where A is the area shown in (c) corresponding to the energy difference between the two alternative potential curve paths that can be followed along the trajectory. Assuming that the transition probability at the crossings is small, superposition of the two contributions gives the following expression for the excitation amplitudes

$$a_\pm = \cos\frac{1}{2}(\frac{A}{\hbar v} \pm \phi) = \cos\frac{1}{2}(\Phi \pm \phi) \qquad (6)$$

This formula was derived recently by Ostrovsky [24], but the line of thinking outlined here can be traced much further back in the literature [3, 4, 27]. The corresponding orientation is easily derived

$$L_\perp = -\frac{\sin\frac{A}{\hbar v}\sin\phi}{1 + \cos\frac{A}{\hbar v}\cos\phi} \qquad (7)$$

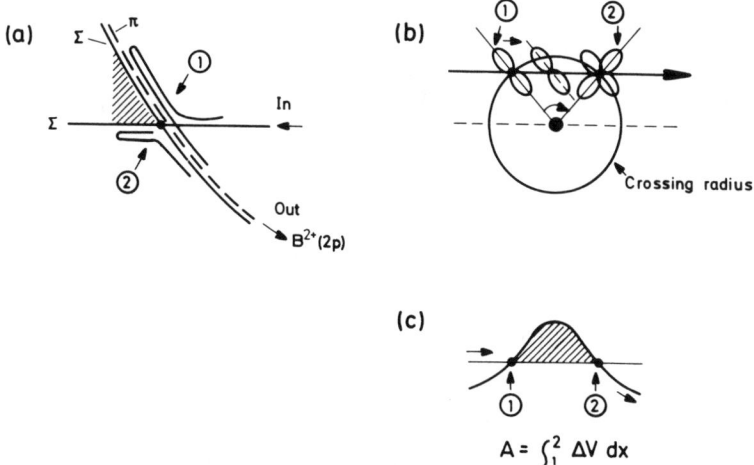

Figure 9: (a) The incoming Σ potential curve crosses the outgoing Σ and Π curves. (b) The corresponding orbital geometry along the trajectory. (c) The molecular potential energy curves along the trajectory.

It may also be concluded that in this simple model the alignment angle γ is equal to $0°$ or $90°$, depending on whether the major axis of the polarization ellipse is parallel or perpendicular to the incident beam direction. An equivalent statement is $P_2 = 0$ for the second component of the Stokes vector, or in the "(λ, χ)-language", $\chi = 90°$, as first noted in [3]. From the equation above it is clear that $L_\perp$ starts out being *negative* at small scattering angles, i.e. it follows the rotation direction of the internuclear axis, and that sign reversals in $L_\perp$ will develop with increasing scattering angle, due to increase of the area A as progressively shorter internuclear distances are reached. It can also be seen that if the geometry is changed so that the potential curve difference changes sign, the corresponding orientation will change sign also due to the sign change of A.

It is appropriate to end this section by comparing the two cases (i) and (ii). Though the geometrical details of the two cases are quite different, the underlying physics responsible for the orientation effects are, by closer inspection, very similar. In both cases, the orientation propensity, which for the main scattering component follows the rotation direction of the internuclear axis, arises from an interplay between a phase Φ, proportional to the energy difference ΔE between the incoming and outgoing potential curves, and the rotation angle ϕ of the internuclear axis. In both cases, an oscillatory pattern in $L_\perp$ may develop due to an increase of the first phase term, Φ, in case (i) by decreasing the collision velocity keeping the impact parameter fixed, in case (ii) by decreasing the impact parameter keeping the collision velocity fixed. In both cases, the orientation propensity reverses by reversal of the ΔE sign. Indeed, the pair of equations (6) may be very simply derived from the pair of equations (4), namely by using for the interaction term $F_{sp}(R)$ a sum of two δ-functions, one centered at the point (1) on the way in, one centered at the point (2) on the way out.

This alternative derivation of the Ostrovsky formulas (6) highlights the fact that the underlying collision dynamical effects responsible for the observed orientation phenomena in the two cases have the same physical origin.

Case (iii) : Near-resonant charge transfer

The results discussed in the previous two sections naturally rise the question of what will be observed in terms of orientation effects in the case of zero (or near-zero) energy defect, in particular for charge transfer, for which the cross section in this situation typically may become very large.

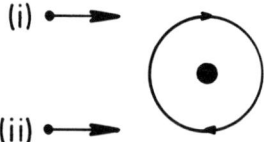

Figure 10: At colllision velocities comparable to the orbital velocity of the electron, a strong left-right scattering asymmetry may be expected for electron capture from a circular state.

Figure 10 compares schematically the case of electron capture from a circular state for the two trajectories where the velocity vector of the incoming charged particle is (i) parallel and (ii) antiparallel, respectively, with the velocity vector of the orbiting electron. Classically, one would expect capture to be more likely when the two velocity vectors are parallel than when they are antiparallel. This "velocity matching" argument may be extracted from the discussion of capture for Rydberg states in the CTMC (Classical Trajectory Monte Carlo) calculations by Kohring et al [25], and confirmed experimentally by Hansen and coworkers [8]. See also [26].

This problem has been studied recently theoretically and experimentally in considerable detail for the case of proton impact on laserexcited sodium [28, 29] :

$$H^+ + Na(3p) \rightarrow H(n = 2) + Na^+ \tag{8}$$

whith an energy defect of 0.36 eV. In the energy region of interest here, this channel dominates the transfer process.

The theoretical treatment of this case must properly take into account that the active electron changes center during the collision by inclusion of so-called electron-translation factors (ETF) to describe the corresponding kinetic energy and momentum transferred to the electron [31, 32].

Another important aspect to pay attention to is the concept of a trajectory. For the present charge transfer processes, the electron transfer may take place at very large distances, and scattering angles typically become so small that both repulsive and attractive forces influence the outcome, and one may enter a region where the concept of a classical trajectory loses its meaning; instead, a large range of impact parameters and azimuthal angles contribute to the scattering at a specific (small) scattering angle, and the scattering should be evaluated using a formalism, the eikonal method, closely analogous to that of Fraunhofer diffraction encountered in classical optics [33].

A theoretical analysis allows these aspects to be displayed in detail [29]. As illustrated in the center of Figure 11, this graph shows what happens for proton impact on a $Na(3p_{-1})$ state, with the electron orbit located in the collision plane. The proton velocity is 0.2 a.u., somewhat below the matching velocity of 0.46 a.u. The lower panel shows the impact parameter dependence of the capture probabilities to the three symmetry-allowed, energetically degenerate $H(n = 2)$ states, $2s$, $2p_{-1}$, and $2p_{+1}$.

Firstly, we see that $H(2s)$ plays a minor role in the transfer channel, compared to $H(2p)$.

Secondly, we notice that electron transfer is more likely on the left side of the atom than on the right side, in agreement with the picture outlined in Figure 10. Also, the left side probability maximum is located considerably *outside* the mean radius of the $Na(3p)$ electron, drawn to scale at the center, while the probability maximum is located near the $3p$ mean radius for impact parameters on the right side of the $Na(3p_{-1})$ state.

Thirdly, we observe that, hitting on the left side, transfer with conservation of the orientation of the electron orbit is by far the most likely; orientation reversal is strongly suppressed.

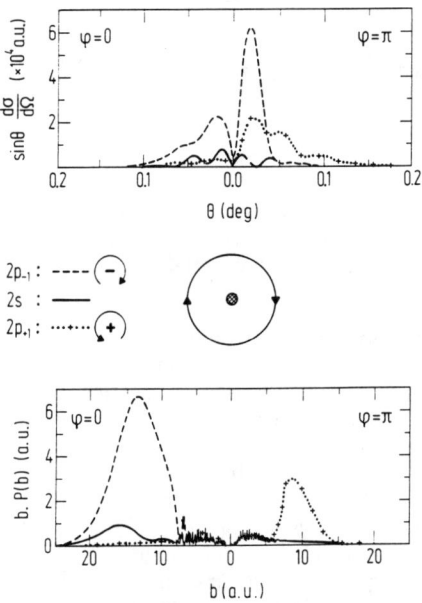

Figure 11: Dependence on impact parameter (lower panel) and scattering angle (upper panel) of electron capture into the three symmetry-allowed H(n=2) states for planar $H^+ - Na(3p)$ collisions. The $Na(3p)$ atom with mean radius drawn to the scale on the lower panel is shown at center. The scattering geometry is shown in Figure 12.

The picture emerging from the lower panel may thus be summarized in the following way :
(i) For proton impact parameters on the *left* side of the p_{-1} atom, the capture probability has a large maximum well outside the electron mean radius; on this side, orientation *conservation* dominates the transfer, the final orientation following the rotation direction of the internuclear axis.
(ii) For impact parameters on the *right* side of the p_{-1} atom, the capture probability is smaller, with a maximum for "head on" collisions with the electron; on this side, orientation *reversal* dominates the transfer.

When converting the impact parameter dependence of the lower panel to scattering angle dependent results, as shown on the upper panel, interesting observations can be made :

Starting with the dominating process, orientation conserving transfer, we notice that the scattering angle maximum is located to the *right*, while the maximum as function of impact parameter is located to the *left*. An interpretation of this may be obtained by assuming that for the corresponding trajectories, at large impact parameters of 10 - 15 a.u., attractive forces dominate the scattering.

Turning to the other process, orientation reversal, we note that even for this channel the scattering angle maximum is located to the *right*, while the impact parameter dependent prob-

516   Collision Dynamics of Oriented States

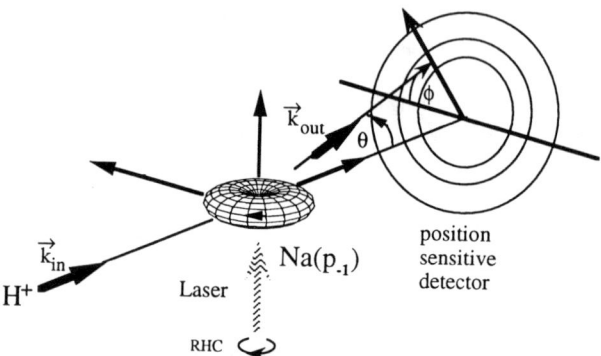

Figure 12: An experiment to study electron capture by protons colliding with sodium atoms excited by circularly polarized laser light to the oriented $Na(3p_{-1})$ state. Neutral atoms scattered in a direction specified by polar angles (θ, ϕ) are detected by a position sensitive detector.

ability also has its maximum to the *right*. An interpretation of this observation leads to the conclusion that in this case the trajectories are dominated by repulsive forces; this statement is not incompatible with the conclusion above for orientation conservation, since the impact parameters that dominantly contribute to orientation reversal are significantly smaller, about 8 a.u., and at this shorter distance repulsion may dominate attraction.

However, in both cases the concept of a trajectory should be used with caution, since it ultimately loses its meaning in the limit $\theta \to 0$. A more detailed discussion has recently been given by Hansen et al [34], but this interesting aspect still deserves further analysis. Calculations at other velocities show a picture similar to the one outlined in Figure 11 [29].

We now briefly discuss an experimental setup designed to address this problem. Figure 12 shows very schematically a triple-beam experiment constructed by Dowek and collaborators in Orsay [28, 35, 36]. A beam of sodium atoms is crossed at right angles by a laser beam, the polarization of which is controlled by a $\frac{\lambda}{2}$-plate and a Pockels cell. The neutralisation of a proton beam, perpendicular to the two other beams, is studied as function of scattering angles (θ, ϕ) using a position sensitive detector, a channel plate located a long distance (~ 7.5 m) away from the point of beam intersection in order to resolve the very small scattering angles. In this way the electron capture process may be studied as function of the alignment and orientation of the $Na(3p)$ state. Capture into the various $H(n)$ levels are separated by the time-of-flight.

Figure 13 shows differential cross sections in the scattering plane ($\phi = 0, \pi$) for capture into the $H(n = 2)$ level for the $3p_{-1}$ target state of Figures 11 and 12. The dashed curve shows the sum of the three convoluted theoretical cross sections of Figure 11, which nicely reproduces the observed, strong left-right scattering asymmetries. The lower part of the figure shows that strong left-right scattering asymmetry may also be observed for a *tilted* p-orbital, the $3p_{45°}$ state, again in agreement with the theoretical predictions [29], but a detailed theoretical analysis similar to the one above for $3p_{-1}$ is not yet available.

Obviously it would be desirable experimentally to resolve the $H(n = 2)$ channel into its sublevel components. However, though some steps have been taken in this direction [37], this will imply polarization analysis of L_α photons, a very difficult task. A more promising alternative might be to study the process

$$Li^+ + Na(3p) \to Li(2p) + Na^+ \qquad (9)$$

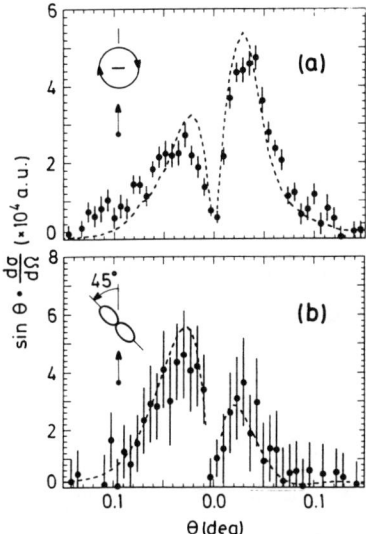

Figure 13: Differential scattering cross sections for capture into the $H(n=2)$ level for (a) the $3p_{-1}$ state shown on Figure 12, and a tilted orbital, the $3p_{45°}$ state. Strong left-right scattering asymmetries depending on scattering angle θ are evident.

This system has the advantages compared to $H^+ - Na(3p)$ that the 2s - 2p degeneracy in the exit channel is lifted, and that the Li(2p) level decays by emission of a red photon for which polarization analysis is readily performed. Preliminary results for this system are encouraging [38], and may form a basis for a complete study of P → P transitions, as discussed by Fano and Pan [39]. Interesting new experimental results on orientation propensities have also recently been obtained by the Lincoln group [40].

CONCLUSIONS

This chapter has reviewed the present understanding of orientation effects for some simple and fundamental systems involving excitation and charge transfer at energies for which the collision velocity is comparable to the orbital velocity of the active electron. Detailed analysis reveals that several different cases can be traced back to a common physical origin. Considerable progress has been made also for the study of orientation effects in charge transfer processes, but several aspects still deserve further attention, such as the case of P → P transfer for which no complete analysis has yet been performed.

ACKNOWLEDGEMENTS

The research program on which this paper is based has been supported over the years by The Danish Natural Science Research Council, The Carlsberg Foundation, The Novo Foundation, the European Community through the CODEST and the Human Capital and Mobility programs, the Nordic Research Academy (NorFA), Institut Français in Copenhagen, and the NIST Office of Standard Reference Data. Svend Erik Nielsen and Jens Olaf Pedersen are thanked for careful reading of the manuscript.

REFERENCES

[1] U. Fano and J. Macek, Rev. Mod. Phys. **45** (1973) 553.

[2] D.H. Jaecks and J. Macek, Phys. Rev. A **4** (1971) 2288.

[3] D.H. Jaecks, F.J. Eriksen, W. de Rijk, and J. Macek, Phys. Rev. Lett. **35** (1975) 723.

[4] F.J. Eriksen, D.H. Jaecks, W. de Rijk, and J. Macek, Phys. Rev. A **14** (1976) 119.

[5] N. Andersen, J.W. Gallagher and I.V. Hertel, Phys. Rep. **165** (1988) 1.

[6] A. Crowe, Invited Paper, XVIII ICPEAC, see this volume.

[7] S R. Leone, Invited Paper, XVIII ICPEAC, see this volume.

[8] S. B. Hansen, Hot Topic, XVIII ICPEAC, see this volume.

[9] N. Andersen, E.E.B. Campbell, J.T. Broad, J W. Gallagher and I.V. Hertel, Physics Reports, in preparation.

[10] N. Andersen, J.W. Gallagher and I.V. Hertel, in : *Electronic and Atomic Collisions*, (D.C. Lorents, W.E. Meyerhof, J.R. Peterson, Eds.), Proceedings of the 14th Int. Conf. on the Physics of Electronic and Atomic Collisions, Palo Alto, California, July 1985 (North-Holland, 1986) 57.

[11] M. Born and E. Wolf, *Principles of Optics, 5. Ed.* (Pergamon Press, 1975)

[12] N. Andersen and S.E. Nielsen, Adv. At. Mol. Phys. **18** (1982) 265.

[13] G.S. Panev, N. Andersen, T. Andersen, P. Dalby and S.E. Nielsen, Europhys. Lett. **3** (1987) 565.

[14] G.S. Panev, N. Andersen, T. Andersen, and P. Dalby, Z. Phys. D **5** (1987) 331.

[15] N. Andersen, T. Andersen, P. Dalby, and T. Royer, Z. Phys. D **9** (1988) 315.

[16] U. Fano, Phys. Rev. **A32** (1985) 617.

[17] N. Andersen and S.E. Nielsen, Europhys. Lett. **1** (1986) 15.

[18] N. Andersen and S E. Nielsen, Z. Phys. D **5** (1987) 309.

[19] S.E. Nielsen and N. Andersen, Z. Phys. D **5** (1987) 321.

[20] A. Russek, D.B. Kimball and M.J. Cavagnero, Phys Rev. **A23** (1981) 139.

[21] J.P. Hansen, L. Kocbach, A. Dubois, and S E. Nielsen, Phys. Rev. Lett. **64** (1990) 2491.

[22] P. Roncin, C. Adjouri, M.N. Gaboriaud, L. Guillemot, M. Barat and N. Andersen, Phys. Rev. Lett. **65** (1990) 3261.

[23] C. Adjouri, Thesis, Université de Paris-Sud (1993).

[24] V. Ostrovsky, J. Phys. B **24** (1991) L507.

[25] G.A. Kohring, A.E. Wetmore and R.E. Olson, Phys. Rev. **A28** (1982) 2526.

[26] E. Lewartowski and C. Courbin, J. Phys. B **25** (1992) L63.

[27] J.P. Gauyacq, J. Phys. B **11** (1978) 85.

[28] C. Richter, N. Andersen, J.C. Brenot, D. Dowek, J.C. Houver, J. Salgado and J W. Thomsen, J. Phys. B **26** (1993) 772.

[29] A. Dubois, S.E. Nielsen and J.P. Hansen, J. Phys. B **26** (1993) 705.

[30] A. Dubois, J.P. Hansen and S.E. Nielsen, J. Phys. B **22** (1989) L279.

[31] D.R. Bates and R. McCarroll, Proc. Phys. Soc. **A245** (1958) 175.

[32] E E.B. Campbell, I V. Hertel and S.E. Nielsen, J. Phys. B **24** (1991) 3825.

[33] E. Hecht, *Optics, 2. Ed.* (Addison-Wesley, 1982)

[34] J.P. Hansen, S.E. Nielsen and A. Dubois, Phys. Rev. **A 46** (1992) R5331.

[35] T. Royer, D. Dowek, J.C. Houver, J. Pommier and N. Andersen, Z. Phys. D **10** (1988) 45.

[36] D. Dowek, J.C. Houver and C. Richter, in : *Electronic and Atomic Collisions*, (W.R. MacGillivray, I E. McCarthy, M.C. Standage, Eds.), Proceedings of the 17th Int. Conf. on the Physics of Electronic and Atomic Collisions, Brisbane, Australia, July 1991 (Adam Hilger, 1992) 537.

[37] Z. Roller-Lutz, Y. Wang, K. Finck and H.O. Lutz, J. Phys. B **26** (1993) 2697.

[38] D. Dowek, private communication.

[39] U. Fano and X.C. Pan, Comments At. Mol. Phys. **26** (1991) 203.

[40] B.W. Moudry, O. Yenen and D.H. Jaecks, this conference, Book of Abstracts, p.558

RECENT ADVANCES AND CHALLENGES OF HEAVY-PARTICLE COLLISION THEORIES

R. E. Olson and J. Wang
Physics Department, University of Missouri-Rolla, Rolla, MO 65401 U.S.A.
J. Ullrich
Gesellschaft für Schwerionenforschung, D-6100 Darmstadt, Germany

ABSTRACT

Within the last decade, considerable progress has been made toward the detailed understanding of the collision dynamics of both simple and complicated systems. At intermediate energies, 10 keV/u ≤ E ≤ 10 MeV/u, the effects of electron correlation have been revealed in p/$\bar{p}$ + He single and double ionization collisions. In general, multiple ionization is poorly understood; however, progress is being made in the description of ejected electron spectra and recoil ion momenta. Alignment collisions continue to challenge theoretical methods, with progress being realized on the prediction of integral parameters.

INTRODUCTION

The past decade has been a challenging time for the development of atomic scattering theories. Increasingly sophisticated experimental programs are providing data whose information content far surpasses the traditional studies of total cross sections. The advent of antimatter sources has allowed investigation of cross section dependencies as a function of only the sign of the projectile charge or of only the mass of the projectile as found in proton/antiproton and electron/positron scattering. Differential cross sections are now commonplace. For bound-bound transitions, electron capture to specific *nl*-levels are supplemented with the preparation of oriented target atoms and the observation of the alignment of the electron cloud after the collision. For ionizing collisions, electron correlation manifests itself in here-to-fore unexpected cross section dependencies. Multiple ionization is complicated by the necessity to include an n-body description of the collision partners. Major differences are found that contrast sharply with the predictions of the independent-electron-model. Observation of the ejected electron spectra and the transverse and longitudinal momenta of the recoil ion in coincidence with the final charge states of the projectile and recoil ions now provide severe tests of theoretical methods.

New technology has not only been a blessing to the experimentalist, but also to the collision theorist. Modern vector- and parallel-processors allow relaxation of many of the restrictions on basis set dimensions and the number of computed trajectories. Presently, relatively inexpensive desktop workstations exceed the computational power of the largest supercomputers of the mid-1980's.

In this review, we will concentrate on the advances and challenges of modern heavy-particle collision theories, and how they are responding to, and driving, new experimental efforts. As a timeframe, we will mainly discuss work accomplished in the last five years. Also we will restrict ourselves to intermediate velocity collisions which normally correspond to a few tens of keV/u to about 10 MeV/u. Both "simple", fundamental collision systems such as H$^+$ + He will be discussed along with multiple ionization and capture from many-electron atomic targets by heavy, highly-charged projectiles.

THEORETICAL METHODS

Within the period of the last two ICPEAC conferences, many of the theoretical methods applicable to intermediate velocity heavy particle collisions have been discussed. These include basis set expansion methods based on molecular orbital[1-3] (MO) and atomic orbital[4,5] (AO) close-coupling techniques. These methods enjoy considerable success at low to intermediate collision velocities. Likewise, distorted wave methods (CDW) have a large region of validity for ionizing collisions that extends from about 100 keV/u to high velocities where relativistic effects become important.[6-8] Excitation is well described by the eikonal impulse approximation and has a demonstrated large region of validity.[9]

The most rigorous theoretical treatment for two-electron targets is that conceived by Ford and Reading[10] and is termed the forced impulse method (FIM). This method applies a large basis set expansion with the inclusion of electron correlation. At intermediate steps during the solution of the scattering coupled-equations, the electron-electron $1/r_{12}$ term is reintroduced to incorporate the dynamical effects of the electron-electron interaction. Thus, electron shake-off is directly included along with the important direct interactions between electrons. The success of this method was highlighted by the description of p/p̄ + He single and double ionization collisions,[11] a case in which no method based on the commonly applied independent electron model is valid.

The classical trajectory Monte Carlo (CTMC) method has a demonstrated region of applicability in the intermediate velocity regime, particularly in the elucidation of both heavy-particle and electron collision dynamics. The method is best termed a semiclassical method in that the initial conditions for the electron orbits are determined by quantum mechanically determined interaction potentials with the parent nucleus. The trajectories are then evolved with the inclusion of all two-body coulomb interactions except the direct electron-electron interactions on individual centers. Post-collision interactions such as electron "cusp" formation are inherently included in a method that simultaneously models excitation, electron capture and ionization. Since the method is most applicable to strongly-coupled systems, it has been applied successfully to a variety of intermediate energy multiply-charged ion collisions.[12,13]

In Fig. 1 we attempt to pictorially describe the regions of validity of commonly applied theoretical models. Both the AO and MO basis set expansion methods work exceedingly well until ionization strongly mediates the collision; the theoretical description of the ionization continuum is not well-founded and relies on pseudo-states to span all ejected electron energies and angles. We have arbitrarily limited these methods to Z_p/Z_t of approximately 8, since above this value the number of terms in the basis set becomes prohibitively large. The CTMC method does not include molecular effects, thus it is restricted from low velocities except in the case of high charge-state projectiles that capture into high-lying Rydberg states which are well-described classically. Likewise, at high velocities the method is inapplicable in the perturbation regime where quantum tunnelling is important and thus is restricted to only strongly-coupled systems. The CDW methods greatly extend the region of applicability of first-order perturbation methods and have demonstrated validity on high-charge state ionizing collisions.

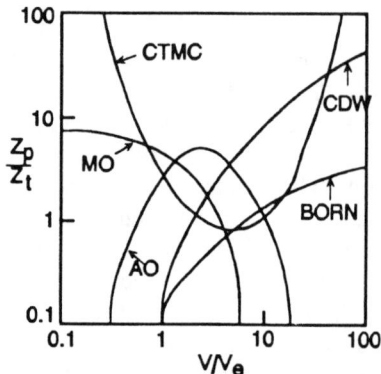

Fig.1. Qualitative regions of validity of various theoretical methods. The ratio of Z_p, the projectile's charge to that of the target, Z_t, is plotted versus the collision velocity divided by the orbital velocity of the active target electron v_e. Theoretical methods: molecular orbital (MO), atomic orbital (AO), classical trajectory Monte Carlo (CTMC), continuum distorted wave (CDW), and first-order perturbation theory (BORN).

In the next sections we will concisely overview some of the recent challenges to theoretical methods within the last couple of years. We will try not to overlap with recent reviews. Also, where possible, we will present systems where more than one theoretical method has been applied so that the continuity of theoretical results can be observed.

ELECTRON CAPTURE

For total single electron capture cross sections the AO method at relatively low energies[5] is complimented by CDW results at high energies.[6,14] For double electron capture from He targets, reasonable results are only obtained for theoretical methods that extend beyond the application of the widely used independent electron model (IEM).[15] An independent event model[16] has been used to extend the CDW method to this difficult class of reactions. Likewise, the use of two-electron wavefunctions within the boundary corrected CDW method has been successful.[17]

nl - Partial Cross Sections

For fully-stripped ions and hydrogenic targets with collision energies above approximately 40 keV/u (depending on the system) the CDW method is accurate. For convenience Belkić et al have made available their computer code and documented several test cases.[18,19] However, much of the data available at intermediate energies are for non-hydrogenic systems, with He and the alkali atoms being prevalent targets. The experimental group in Groningen, Netherlands has provided a considerable resource of data using the photon emission spectroscopy technique.[20,21] Even for fully-stripped projectiles, the differing lifetimes of the individual nl-levels make it possible to determine cross section for degenerate levels.

As a comparison between theory and experiment, we show in Fig. 2 the C^{4+} + Li case. Both AO and CTMC results are in good agreement with each other and the data. Moreover, the individual nl-cross sections for the $n=5$ level are in close agreement.

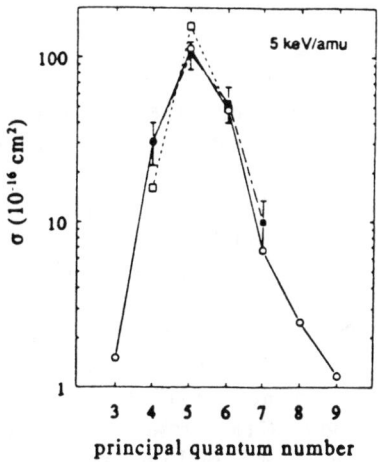

Fig. 2. The dependence of the n-level cross sections for 5 keV/u C^{4+} + Li collisions. Open circles: CTMC results,[22] open squares: AO values,[23] and closed circles: experimental data of Dijkkamp et al.[24]

One of the merits of the CTMC method is that it can provide cross sections for large n-values which are difficult to realize with basis set expansion techniques. Given a quantum mechanically defined potential for the active electron, it has been possible to calculate accurate cross sections for the O^{8+} + Li system where n-values up to 20 were included in order to account for all cascades in the line emission cross section determinations.[25] Likewise, studies of electron capture from highly-excited targets has been very successful.[26]

Alignment and Orientation

Alignment and orientation in atomic collisions is the topic of the review in this book by N. O. Andersen. Thus, we will present only a few illustrative cases that highlight recent theoretical comparisons and developments. In general, alignment and orientation effects in atomic collisions provide a severe test of theory. Alignment parameters after a collision require an accurate description of the nl-cross sections and provide sensitive information about the final orientation of the electron cloud. Comparisons with experimental data severely test a theory's ability to correctly account for multiple-scattering corrections used in perturbation expansions of the electron capture scattering amplitudes.[27] A comprehensive review of alignment collisions has been given by Hippler.[28]

A nearly ideal target for electron capture studies is laser excited Na*(3p). It is a quasi-one electron atom with a loosely bound outer electron which allows experimental manipulation of the orbital alignment with respect to the collision velocity. The effects of initial Na*(3p) alignment on the capture cross sections for 0.5 to 2.0 keV H^+ collisions have been observed by Dowek et al.[29] Very large alignment for the preference of perpendicular orbital alignment over parallel alignment with an anisotropy parameter of ~75% has been observed by Aumayr et al[30] for He^{2+} collisions. Most recently, measurements similar to those of Aumayr et al have been made over a broad energy range, which includes the "velocity-matching" energy of 5.7 keV/u. It has now been shown that the anisotropy parameter for capture to the dominant $n=4$ state of He^+ changes sign from -75% to +75% near the velocity-matching region.[31] The Σ and Π experimental cross sections for this case are shown along with MO and CTMC results in Fig. 3.

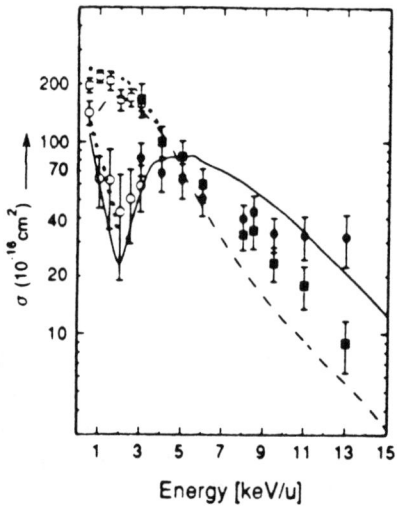

Fig. 3. Σ and Π cross sections for electron capture to $He^+(n=4)$ in He^{2+} + Na*(3p) aligned collisions. The open symbols are data from Aumayr et al[30] and the closed ones from Schlatmann et al [31] with the circles being the Σ results and the squares the Π values. The corresponding CTMC results are given by the dashed and solid line, respectively. MO values[30] are denoted by the short dashed lines.

Orientation of the excited states produced after collisions is being actively pursued. Alignment parameters provide sensitive information about the orientation of the electron cloud after the collision and thus yield detailed insight into the collision dynamics. A common testbed for theory has been H^+ + He electron capture to $H(n=2,3)$. Here, several sets of experimental data exist for the z-component of the collision-induced electric dipole moment **D**, the Runge-Lenz vector **A**, and the first-order moment of the current distribution $<L \times A>$. Simple classical pictures may be used to illustrate the relation of these parameters to the qualitative orientations of the electron orbit after capture. Positive values of $<D_z>$ indicate the extent the electron trails behind the proton after electron capture, the Runge-Lenz vector is a measure of the eccentricity of the electron orbit, and $<L \times A>$ represents a vector which points along the electron's orbital velocity at the perihelion of its orbit.

To illustrate the richness of the data and compare existing theories, we show in Fig. 4 $\langle D_z \rangle$ for electron capture to the H($n=2$) state in H$^+$ + He collisions. At all energies above 30 keV, the captured electron trails behind the H$^+$ projectile. First-order methods yield zero for the dipole moment, thus, the necessity of including higher-order terms in theoretical formulations is emphasized, especially the post-collision interaction effect.[27] Other alignment parameters, such as L_y and D_x, help to specify the orbit and the direction of electron rotation.

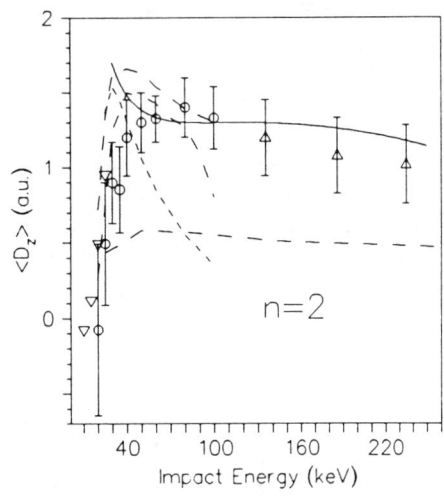

Fig. 4. Comparison of theoretical and experimental values for the z-component of the dipole moment for electron capture into H($n=2$) for H$^+$ + He collisions. Theoretical curves: CTMC, solid line;[32] AO, dashed line[33] and dotted line;[34] MO+AO, dot-dot-dash;[35] semi-classical impact parameter, dot-dash;[36] and CDW with post collision interaction, long dash.[27] Experimental data: circle,[37] square,[38] triangle[39] and inverted triangle.[40]

An even more complete description of the scattering process is realized if the alignment parameters are given differential in the scattering of the projectile. Although the integral values vary slowly with energy, the differential values have been found to vary rapidly in scattering angle, showing regions where the projectile mainly interacts with the active electron at small angles, to large angle collisions where the projectile's deflection is determined by the interaction with the target nucleus.[32]

IONIZING COLLISIONS

Bound to continuum state transitions have been a serious challenge to collision theorists. Basis set expansion methods have met with limited success here because of the difficulty in describing the continuum states. Furthermore, since these wavefunctions are approximate, projecting out multiply-ionizing events in an n-electron quantum mechanical calculation is non-trivial. Thus most of the work in this area rests on applying the independent electron model,[15] within the context of CDW and CTMC methods.

Electron Correlation

The term "electron correlation" means different things to different people. The differentiation between static and dynamic correlation has been reviewed.[41] For heavy-particle scattering at intermediate energies, one of the nicest examples of the effects of electron correlation was illustrated by Reading and Ford[11] in their development of the forced impulse method for application to p/$\bar{p}$ + He single and double ionization collisions. In essence, this method stops the time-evolution of the scattering equations at intermediate steps during the collision and imposes a configuration interaction calculation to include the important electron-electron $1/r_{12}$ correlation term. Most recently, the remaining 35% difference between theory and experiment has been eliminated by expanding the basis set to include s-, p- and now d-waves.[42]

The p/$\bar{p}$ + He cross section difference is sometimes referred to as an "interference" effect. This is because in a Born-type series expansion of the scattering amplitudes, the Z^3 term (Barkas effect) arises as a cross-term between the first and second scattering amplitudes.[43] This is not, however, a true interference but a model-dependent interference effect. Such a conclusion is confirmed by simple 4-body CTMC calculations using the Bohr He atom that includes the $1/r_{12}$ electron repulsion, which show the enhancement of double ionization for $\bar{p}$ over that for p collisions.[44] This same paper predicted the reverse enhancement for the single ionization that was latter confirmed by experiment.[45]

Heavy Particle Dynamics

Experimentalists have increasingly made measurements differential in the scattering of the projectile and recoil ion. Such observations strive to completely define both the transverse and longitudinal momenta of all collision products. One of the challenges to theorists were measurements on the ratio of double to single ionization for p + He collisions that were differential in the scattering angle of the projectile.[46] A peak in this ratio was realized at ~0.9 mrad. The sophisticated FIM method failed to reproduce the peak position or magnitude.[47] Subsequent 4-body CTMC results[48] indicated that the peak was due to uncorrelated scattering of the protons by both He-electrons in hard collisions that deflected the protons to angles larger than that for a single scattering event which has a maximum angle of 0.54 mrad. The basis set used by the FIM approximation was restricted to ejected electrons with energies less than 350 eV, thus, the physical effect could not be displayed.

The complete transverse momentum spectra for all outgoing particles were observed for p + He single ionization.[49,50] The interplay between the projectile scattering from the target electron in large impact parameter, small angle collisions, to large angle collisions, showing the interaction of the projectile and He nucleus were completely elucidated. Non-central scattering for the projectile and recoil ion were later analyzed and are in general agreement with 4-body CTMC calculations.[51]

Although one would have to admit that not everything is understood about intermediate energy collisions of simple systems such as p + He, at least there has been considerable progress in understanding their scattering dynamics. In contrast, multiple ionization and electron capture events for multiply-charged ion, many-electron target atom collisions are poorly described. The only theoretical methods being actively pursued for these collisions are those based on the solution of Hamilton's equations of motion.[52]

For highly-charged ion, many-electron atom collisions, detailed information about the transverse momentum distributions of the recoil ions, in coincidence with their charge state and the outgoing charge of the projectile, is available for selected systems.[52] In principle, recoil ion differential cross sections are similar to those differential in the angular scattering of the projectile. For heavy ions, the deflections are on the order of 10^{-5} rad, so experimental limitations restrict one to the use of recoil ion momentum spectroscopy.[53] Correspondingly, available theories are responding to these new data.

This year, the first measurements of the longitudinal momentum distributions of the recoil ion became available;[54] the system was 1 MeV/u F^{9+} + Ne. From conservation of momentum and energy, the longitudinal momentum of the recoil ion is determined by

$$p_z = - Q/v - nm_e v/2 - p_e ,$$

where m_e is the electron mass, p_e is the net z-momentum of the ionized electrons, Q is the exoergicity of the reaction, and n is the number of electrons captured to the projectile. The average longitudinal recoil ion momenta for the above system are shown in Fig. 5. It is striking that the recoil ions are found in the backward direction, even though reasonable energy losses should force the recoils into the forward direction. Thus, the z-component of the vector sum of the ejected electrons is compensating for the energy loss. Of interest is that both the experiment and theory display the decrease in p_z as the number of electrons captured by the projectile increases. For each electron captured, the recoil rebounds backward by an additional $m_e v/2$. The observation of this quantity is the direct signature of the electron translation factors (ETF's) required in the formulation of MO theories.

Fig. 5. Mean values of p_z plotted versus n, the number of electrons captured by the projectile in 1 MeV/u F^{9+} + Ne collisions[54]. The upper half of the figure shows the experimental results; the lower half, nCTMC calculations.

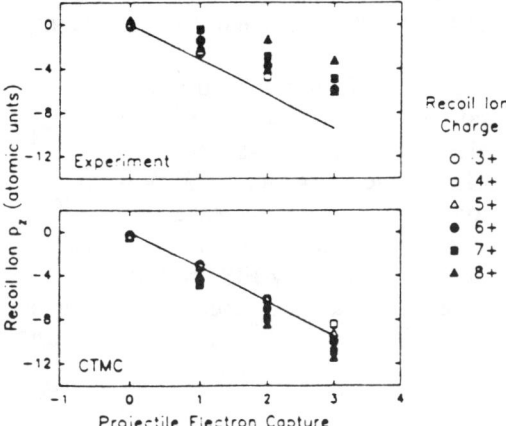

For heavy particle dynamics, much of our theoretical information arises from classical calculations. Although these calculations simulate a quantum mechanical distribution of initial states, they do not require quantization of the final bound states after the collision nor do they account for quantal tunnelling. Their main asset is that they include all the target electrons in the calculations along with all coupling between electronic and nuclear motions. However, the neglect of electron-electron interactions precludes the account of Auger and radiative decay after multiple excitation of the projectile or recoil, which is known to be very important in many cases. Thus, we have a true n-body problem which will require sophisticated theories and experiments to provide a basic understanding of the collision mechanisms.

Electron Ejection

Recent developments in fast ionizing collisions involving fully or partially stripped projectile ions have been marked by various experimental measurements of the ejected electron spectra as the primary source of information on the collision dynamics. Combined experimental and theoretical studies[55,56] of ionization by the impact of highly charged ions have for the most part focused only on a subset of the electron emission spectrum usually referred to as the binary encounter (BE) electrons created through hard collisions between the heavy projectile and the target electron(s). Since the fast BE electrons are well separated from the slow electrons in momentum space, they are relatively easy to measure experimentally. Theoretical description is thought to be facilitated by the underlying nature of strong two-body interactions in the production of BE electrons.

We shall restrict discussion to recent progress in the production of BE electrons by the impact of partially stripped, highly charged ions. Although a number of unique features, such as the bifurcation and the enhancement of binary peaks, have been discovered and explained qualitatively,[57,58] the overall collision dynamics remain poorly understood. Systematic studies using a variety of projectile/target combinations have revealed large discrepancies between experiments and theories, most notably in the position and width of the binary peak, the relative height and the angular range of the double peaks, and the observed target dependence.

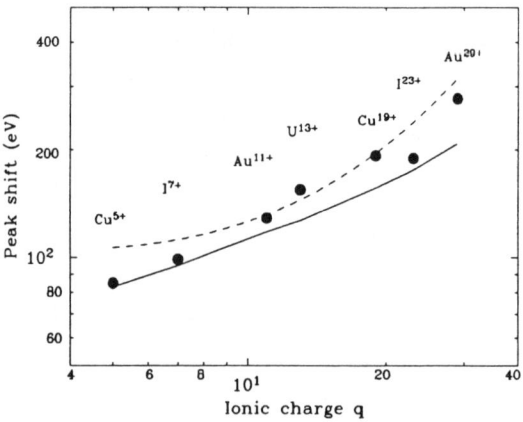

Fig. 6. Shift of the binary peak position with respect to the free electron value $2Vp^2$ as a function of the ionic charge for 0.6 MeV/u ions impact on H_2. Filled circle: Experimental data of Wolf et al;[59] Solid line: CDW-EIS results; Dashed line: CTMC results.

To give a flavor of the current theoretical status, we show in Fig. 6 a comparison of the observed BE peak energy shifts with CDW-EIS and CTMC calculations[60-62] for a variety of projectile ions at zero degree ejection angle. The energy shift is defined as the difference between the free electron energy (2 Vp^2) and the actual BE peak energy. As can be seen from Fig. 6, the shift of the binary peak increases smoothly with the ionic charge q, despite the fact that we are comparing clothed ions of strongly varying degrees of screening. We note that the CDW-EIS calculations could only be performed assuming a bare ion with a pure Coulomb potential of charge q. Yet, CDW-EIS reproduces well the energy shifts, especially for lower charge states, despite its neglect of the screened core of the projectile. The CTMC results, also assuming pure Coulomb ions, become increasingly in agreement with experimental data as the ionic charge increases. The reason that one may neglect the projectile screened core in making these comparisons is that the elastic scattering cross sections for a pure Coulomb projectile potential (Rutherford cross section) and for a screened potential are both smooth and very similar in shape at 180 degrees. This approach is justifiable only if one is interested in the zero degree peak positions. Interestingly, the impulse approximation,[56] the only model that has been able to explain qualitatively the binary peak oscillations, gives the poorest results of the peak shifts at this angle (not shown in Fig. 6).

For many electron targets, multiple ionization presents yet another challenging task. Among many complicating factors, the charge state of the recoil ion must be considered, which can post-distort the ejection spectrum as well. To that end, we show in Fig. 7 the dependence of binary encounter electron production on the charge

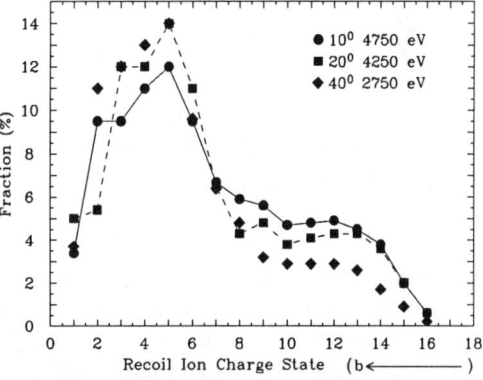

Fig. 7. Fractional distribution of the binary electron cross section as a function of the charge state of the recoil ion for 2.4 MeV/u Xe^{21+} on Ar at three ejection angles. Circle: 10 degrees; Square: 20 degrees; Diamond: 40 degrees. The energies indicated in the figure are the binary peak positions at the three angles, respectively, where the fractions are evaluated.

state of the recoil ion for 2.4 MeV/u Xe^{21+} on Ar as obtained from nCTMC calculations.[63] The contributions to the production cross section are expressed as fractional percentages for each recoil ion charge state evaluated right at the binary peak of the electron spectra at three ejection angles 10, 20, and 40 degrees. Each curve in Fig. 7 peaks at 5, indicating that the largest contribution to the BE electron cross section occurs at a recoil ion charge state of +5. This value is so remarkably low that it is even less than 1/3 of the largest possible recoil charge state of 18. However, Fig. 7 also indicates that when a binary electron is created, the target is mostly multiply ionized, as the fraction for singly charged recoil ions is rather small. The peak at 5+

and the broad shoulder structure near 12+ are indications of the projectile penetrating the L and M shells of Ar, respectively. The recoil ion charge states correspond directly to the impact parameter range of the collision. By measuring the binary electron in coincidence with the recoil ion, one gains direct information on the impact parameter being probed. It affords a much more sensitive test on the current understanding of ionization involving highly charge ions. Unfortunately, no experimental data on the coincidence measurements are currently available for a direct comparison between theory and experiment.

Looking ahead, we expect many exciting and interesting surprises in the study of the electron ejection spectra by partially-stripped ions. The relative interplay between the asymptotic region and the interaction region, which has been a question raised in electron atom collisions,[64] will have a significant impact. Although the two regions cannot be completely decoupled, it is reasonable to assume in our case that the interaction region will affect mostly the magnitude and the bifurcation of the binary peak, while the asymptotic region will polarize the initial state and distort the ejection spectrum. Though the present overriding factor in any theoretical description remains the proper treatment of the projectile screened potential as a whole, explicit electron-electron interactions[65] and exchange effects[66,67] will also play important roles. In addition, it is essential to realize that a great deal of work has been done in the area of electron loss by heavy targets (projectile ionization),[68] especially as shown in recent coincidence measurements.[69] For various reasons, the two fields have developed nearly in parallel. By sharing the wealth of information in both electron loss and target ionization using a simple kinematic frame transformation,[70] our understanding of the subject can be greatly enhanced.

CONCLUSIONS

In this report, we have tried to give a flavor of some of the challenges and successes of heavy-particle collision theories. Clearly, there is much to be done, but an outside observer would have to admit that considerable progress has been made in just the last five years. More work still has to be done, especially in our understanding of n-body collision dynamics. However, advances in computing techniques will eliminate many pragmatic roadblocks. On the fundamental side, rigorous descriptions of continuum (ionization) reactions are needed, along with models that directly include the electron-electron interactions for many-electron systems. In order to successfully describe collisions between structured particles, theoretical models must be able to take into account simultaneous excitation or ionization processes caused by nuclear core and electron-electron interactions. It is an extremely challenging task to treat the latter interaction as it involves at least a four-body problem. In the non-perturbative regime, a combined classical and quantal approach may prove to be necessary and fruitful.

ACKNOWLEDGEMENTS

The authors would like to thank Drs. Jean Pascale, Horst Schmidt-Böcking,

Ronnie Hoekstra and Hans Wolf for many helpful discussions and collaborations. Work supported by the Office of Fusion Energy - Department of Energy.

REFERENCES

1. A. Bárány, *The Physics of Electronic and Atomic Collisions*, ed. by A. Dalgarno, R. S. Freund, P. M. Koch, M. S. Lubell and T. B. Lucatorto (AIP Conf. Proc. **205**, 1990) pp. 246-57
2. K. Taulbjerg, *Atomic Physics of Highly Charged Ions*, ed. by E. Salzborn, P. H. Mokler and A. Müller (Springer-Verlag, Berlin, 1991) pp. 77-80
3. M. Kimura and N. F. Lane, *Advances in Atomic Molecular and Optical Physics*, ed. by B. Bederson and D. R. Bates (Academic Press, New York) **26**, 27 (1989)
4. W. Fritsch, *Physics of Highly Charged Ions*, ed. by P. Richard, M. Stöckli, C. L. Cocke and C. D. Lin (AIP Conf. Proc. **274**, 1993) pp. 24-33
5. W. Fritsch and C. D. Lin, Phys. Rep. **202**, 1 (1991)
6. D. S. F. Crothers, *Electronic and Atomic Collisions*, ed. by W. R. MacGillivray, I. E. McCarthy and M. C. Standage (Adam Hilger, Bristol, 1992) pp. 413-22
7. R. D. Rivarola, P. D. Fainstein and H. Ponce, *The Physics of Electronic and Atomic Collisions*, ed. by A. Dalgarno, R. S. Freund, P. M. Koch, M. S. Lubell and T. B. Lucatorto (AIP Conf. Proc. **205**, 1990) pp. 264-72
8. R. D. Rivarola, *Physics of Highly Charged Ions*, ed. by P. Richard, M. Stöckli, C. L. Cocke and C. D. Lin (AIP Conf. Proc. **274**, 1993) pp. 237-50
9. J. E. Miragalia and V. D. Rodríquez, *Electronic and Atomic Collisions*, ed. by W. R. MacGillivray, I. E. McCarthy and M. C. Standage (Adam Hilger, Bristol, 1992) pp. 423-34
10. J. F. Reading and A. L. Ford, *Electronic and Atomic Collisions*, ed. by H. B. Gilbody, W. R. Newell, F. H. Read and A. C. H. Smith (Elsevier Science Publishers B. V. 1988) pp. 693-8
11. J. F. Reading and A. L. Ford, Phys. Rev. Lett. **58**, 543 (1987)
12. R. E. Olson, *Electronic and Atomic Collisions*, ed. by H. B. Gilbody, W. R. Newell, F. H. Read and A. C. H. Smith (Elsevier Science Publishers B. V. 1988) pp. 271-85
13. R. E. Olson and J. Wang, *Physics of Highly Charged Ions*, ed. by P. Richard, M. Stöckli, C. L. Cocke and C. D. Lin (AIP Conf. Proc. **274**, 1993) pp. 229-36
14. R. Gayet, R. D. Rivarola and A. Salin, J. Phys. B **14**, 2421 (1981)
15. J. H. McGuire and L. Weaver, Phys. Rev. A **16**, 41 (1977)
16. D. S. F. Crothers and R. McCarroll, J. Phys. B **20**, 2835 (1987)
17. D. Belkić, J. Phys. B **26**, 497 (1993)
18. D. Belkić, R. Gayet and A. Salin, At. Data Nucl. Data Tables **51**, 59 (1992)
19. D. Belkić, R. Gayet and A. Salin, Comput. Phys. Commun. **31**, 385 (1984)
20. D. Ćirić, R. Hoekstra, F. J. de Heer and R. Morgenstern, *Electronic and Atomic Collisions*, ed. by H. B. Gilbody, W. R. Newell, F. H. Read and A. C. H. Smith (Elsevier Science Publishers B. V. 1988) pp. 655-60
21. R. Hoekstra, F. J. de Heer and R. Morgenstern, *The Physics of Electronic and*

Atomic Collisions, ed. by A. Dalgarno, R. S. Freund, P. M. Koch, M. S. Lubell and T. B. Lucatorto (AIP Conf. Proc. **205**,1990) pp. 390-95
22. R. Hoekstra, R. E. Olson, H. O. Folkerts, E. Wolfrum, J. Pascale, F. J. de Heer, R. Morgenstern and H. Winter, J. Phys. B (in press)
23. W. Fritsch and C. D. Lin, J. Phys. B **17**, 3271 (1984)
24. D. Dijkkamp, D. Ćirić, E. Vlieg, A. de Boer and F. J. de Heer, J. Phys. B **18**, 4763 (1985)
25. R. E. Olson, J. Pascale and R. Hoekstra, J. Phys. B **25**, 4241 (1992)
26. J. Pascale, *Electronic and Atomic Collisions*, ed. by W. R. MacGillivray, I. E. McCarthy and M. C. Standage (Adam Hilger, Bristol, 1992) pp. 401-11
27. J. Burgdörfer and L. J. Dubé, Phys. Rev. Lett. **52**, 2225 (1984)
28. R. Hippler, J. Phys. B **26**, 1 (1993)
29. D. Dowek, J. C. Houver, J. Pommier, C. Richter and T. Royer, Phys. Rev. Lett. **64**, 1713 (1990)
30. F. Aumayr, M. Gieler, J. Schweinzer, H. Winter and J. P. Hansen, Phys. Rev. Lett. **68**, 3277 (1992)
31. A. R. Schlatmann, R. Hoekstra, R. Morgenstern, R. E. Olson and J. Pascale, Phys. Rev. Lett. (in press)
32. C. Lundy, R. E. Olson, D. R. Schultz and J. P. Pascale, J. Phys. B (in press)
33. A. Jain, C. D. Lin and W. Fritsch, Phys. Rev. A 36, 2041 (1987)
34. R. Shingal and C. D. Lin, J. Phys. B **24**, 963 (1991)
35. M. Kimura, Phys. Rev. A **44**, R5339 (1991)
36. H. A. Slim, E. L. Heck, B. H. Bransden and D. R. Flower, J. Phys. B **24**, 1683 (1991)
37. R. A. Cline, W. B. Westerveld and J. S. Risley, *Proc. 17th Int. Conf. on Phys. of Elect. At. Coll.*, ed. by I. E. McCarthy, W. R. MacGillivray and M. C. Standage (Griffith University, 1991) p. 431
38. J. R. Asburn, R. A. Cline, P. J. M. van der Burgt, W. B. Westerveld and J. S. Risley, Phys. Rev A **41**, 2407 (1990)
39. R. DeSerio, C. Gonzalez-Lepera, J. P. Gibbons, J. Burgdörfer and I. A. Sellin, Phys. Rev A **37**, 4111 (1988)
40. R. Hippler, O. Plotzke, W. Harbich, H. Madeheim, H. Kleinpoppen and H. O. Lutz, Phys. Rev. A **43**, 2587 (1991)
41. N. Stolterfoht, K. Sommer, F. Frémont, X. Husson, D. Lecler, J. K. Swenson, C. C. Havener and F. W. Meyer, *Physics of Highly Charged Ions*, ed. by P. Richard, M. Stöckli, C. L. Cocke and C. D. Lin (AIP Conf. Proc. **274**, 1993) pp. 53-62
42. A. L. Ford and J. F. Reading, Bul. Amer. Phys. Soc. **38**, 1132 (1993)
43. J. H. McGuire and J. C. Straton, *The Physics of Electronic and Atomic Collisions*, ed. by A. Dalgarno, R. S. Freund, P. M. Koch, M. S. Lubell and T. B. Lucatorto (AIP Conf. Proc. **205**,1990) pp. 280-89
44. R. E. Olson, Phys. Rev. A 36, 1519 (1987)
45. D. R. Schultz, R. E. Olson and C. O. Reinhold, J. Phys. B **24**, 521 (1991)
46. J. P. Giese and E. Horsdal, Phys. Rev. Lett. **60**, 2018 (1988)

47. J. F. Reading, A. L. Ford and X. Fang, Phys. Rev. Lett. **62**, 245 (1989)
48. R. E. Olson, J. Ullrich, R. Dörner and H. Schmidt-Böcking, Phys. Rev. A **40**, 2843 (1989)
49. R. Dörner, J. Ullrich, H. Schmidt-Böcking and R. E. Olson, Phys. Rev. Lett. **63**, 147 (1989)
50. R. Dörner, J. Ullrich, O. Jagutzki, S. Lencinas, A. Gensmantel and H. Schmidt-Böcking, *Electronic and Atomic Collisions*, ed. by W. R. MacGillivray, I. E. McCarthy and M. C. Standage (Adam Hilger, Bristol, 1992) pp. 351-60
51. A. Gensmantel, J. Ullrich, R. Dörner, R. E. Olson, K. Ullmann, E. Forberich, S. Lencinas and H. Schmidt-Böcking, Phys. Rev. A **45**, 4572 (1992)
52. C. L. Cocke and R. E. Olson, Phys. Rep. **205**, 155 (1991)
53. J. Ullrich, R. Dörner, H. Berg, C. L. Cocke, J. Euler, S. Lencinas, K. Froschauer, L. Spielberger, K. Ullmann, H. Schmidt-Böcking, R. E. Olson and M. Schulz, *Physics of Highly Charged Ions*, ed. by P. Richard, M. Stöckli, C. L. Cocke and C. D. Lin (AIP Conf. Proc. **274**, 1993) pp. 251-61
54. V. Frohne, S. Cheng, R. Ali, M. Raphaelian, C. L. Cocke and R. E. Olson, Phys. Rev. Lett. (in press)
55. P. Richard, D. Lee, T. Zouros, J. Sanders and J. Shinpaugh, J. Phys. B **23**, L213 (1990)
56. W. Wolff, J. Shinpaugh, H. Wolf, R. Olson, J. Wang, S. Lecinas, D Piscevic, R. Herrmann and H. Schmidt-Böcking J. Phys. B **25**, 3863 (1992)
57. C. Reinhold, D. Schultz, R. Olson, C. Kelbch, R. Koch and H. Schmidt-Böcking, Phys. Rev. Lett. **66**, 1842, (1991)
58. C. Cocke, *Electronic and Atomic Collisions*, ed. by W. R. MacGillivray, I. E. McCarthy and M. C. Standage (Adam Hilger, Bristol, 1992) pp. 49-68
59. H. Wolf *et al*, J. Phys. B (1993, to be published)
60. D. Crothers and J. McCann, J. Phys. B **16**, 3229 (1983)
61. P. Fainstein, V. Ponce and R. Rivarola, J. Phys. B **21**, 287 (1988)
62. R. Olson and A. Salop, Phys. Rev. A **14**, 579 (1976)
63. R. Olson, J. Ullrich and H. Schmidt-Böcking, Phys. Rev. A **39**, 5572 (1989)
64. A. Stelbovics, *Electronic and Atomic Collisions*, ed. by W. R. MacGillivray, I. E. McCarthy and M. C. Standage (Adam Hilger, Bristol, 1992) pp. 21-34, and references therein
65. J. Wang, C. Reinhold and J. Burgdörfer, Phys. Rev. A **45**, 4507 (1992)
66. K. Taulbjerg, J. Phys. B **23**, L761 (1990)
67. Z. Chen, D. Madison and C. D. Lin, J. Phys. B **24**, 3203 (1991)
68. A. Kövér, Gy Szabo, L. Gulyás, K. Tökési, D. Berényi, O. Heil and K. O. Groeneveld, J. Phys. B **21**, 3231 (1988)
69. O. Heil, R. DuBois, R. Maier, M. Kuzel and K. Groeneveld, Phys. Rev. A **45**, 2850 (1992)
70. J. Wang, R. Olson, K. Tökési, J. Burgdöfer and C. Reinhold, *Physics of Highly Charged Ions*, ed. by P. Richard, M. Stöckli, C. Cocke and C. D. Lin (AIP Conf. Proc. **274**, 1992) pp. 271-280

COLLISIONS WITH MULTIPLY CHARGED IONS

Mechanisms for radiative decay of multiply excited states populated in collision of highly charged ions with rare gases

P.Roncin, M.N.Gaboriaud, Z Szilagyi and M.Barat
Laboratoire des Collisions Atomiques et Moléculaires
(Unité de recherche associée au CNRS 281)
Université Paris Sud, Bâtiment 351, 91405 Orsay Cedex, France

ABSTRACT

The present paper summarizes some recent developments in the understanding of double capture mechanisms by multiply charged ions with charges from 6 to 11 colliding with various rare gas targets. The basic electron capture mechanisms are briefly recalled with help of simple models, then various post collisional effects are examined which strongly influence the decay of the doubly excited states formed. In particular, the influence of the highly asymmetrical Rydberg series is outlined. In the second part of the paper, experimental data are examined to stress the consistency and limitation of the available models.

INTRODUCTION

The classical Coulombic Barrier Model (CBM) provides the most simple framework to understand the large cross sections for single and multiple electron capture by highly charged ions. This model considers the electrons as bound to the target with energies given by their successive ionization potential, the ith electron being captured as soon as its motion from the target toward the projectile is classically allowed (over barrier). Since the height of the Coulombic barrier between the two nuclei decreases with internuclear distances, the electrons are captured successively at distance R_i.

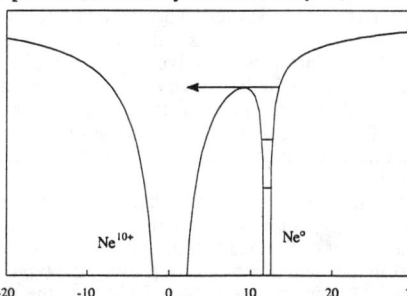

Figure 1: electrostatic potential for the first captured electron in the coulombic barrier model (CBM)

With no other assumptions the model already predicts the order of magnitude of cross section for removal of r electrons from the target.The model also indicates the range of binding energy where the multiply excited states formed are to be found on the projectile (reaction window). Further improvement requires the knowledge of the ionic structure where the electron capture may actually take place. Fortunately, the very large ionic coulomb forces involved in these systems greatly simplify the energy level diagram which may be estimated with help of hydrogenic formula and Slater screening factors.For low projectile velocities (v≈0.1 a.u.) investigated here, a quasi molecular approach is well suited.

In its simplest form the diabatic potential energy curves simply account for the strong internuclear repulsion r(Z-r) between the target ionized r times and the projectile ion which captured r electrons. An example of the resulting curve crossing scheme is seen in fig.2. Following the ideas of the CBM, the dominant interaction at the crossing between two levels (if they differ by only one electron orbital) is the one electron exchange interaction. Even within these simple assumptions which only involve electron-nucleus interaction, double electron capture can take place in one or two steps [1]. This "terminology" is better understood if related to the potential energy diagram. Figure 2 represent a rather general topology with a flat entrance channel, two single and two double capture channels. For instance C^{6+}_Xe, N^{7+}_Ar or_Kr, O^{8+}_Ar, F^{9+}_Ne and Ne^{10+}_He display a curve crossing scheme having this topology and electron capture takes place in the same shells. At keV collision energies, single capture takes place almost exclusively into the n=5 shell at the A crossing. Depending wether the n=5 electron is captured at A, the system will reach the B crossing to produce double capture in *two* steps into the *(4,5)* series or, if the first electron is not captured, the system will reach the C crossing to produce *one* step double capture into the *(4,4)* series.

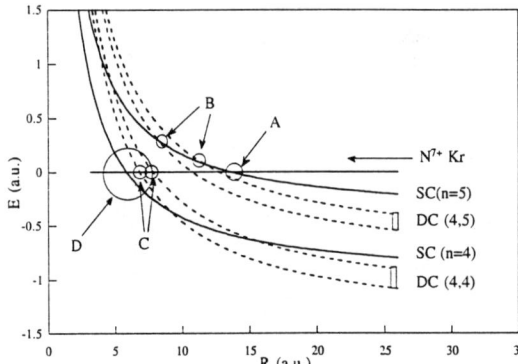

Figure 2: schematic diabatic potential energy curves for the N^{7+}_Kr system. Crossings A and D are responsible for single electron capture and crossings C and B control the double electron capture in one and two steps respectively. The large circle around the D crossing indicates that this crossing behaves adiabatically and strongly "influences" the C crossings.

This second scheme is made possible only because the C crossing for "direct" two electron transition is in fact located in the region of influence of the D crossing with the n=4 single capture channel. This so called "region of influence" of the D crossing is described in the Landau-Zener formalism as the region ΔR in which the transitions may take place. The more adiabatic the transition, the broader the ΔR range and since the one electron exchange interaction varies exponentially with R, the inner crossing associated with the capture into the *(n-1)* shell, generally behaves adiabatically. Indeed, in all the systems mentioned above, the *n=4* channel is hardly populated (transitions both on the way in and on the way out of the collision). The paradox of a second electron being captured before the first one is then removed when realizing that, inside the range ΔR, single electron capture has already started. Though this one step double capture is slightly more tricky than the two step process, it is still explained in an independent particle model and complements the CBM. Finally, the relative importance of the one or two step process (population of the *(4,4)* or *(4,5)* series in the above examples) depends on the transition probability at the A crossing which directs the incident flux toward B or C crossings, lower collision velocities will favor the B crossing. Note that the same discussion holds for the N^{7+}_He and O^{8+}_He systems which have a comparable curve crossing scheme (just change the n values 4→3 and 5→4)

At keV energies, double capture is well understood into (quasi) symmetrical states. The systems investigated so far by energy gain spectroscopy, show autoionizing double capture (ADC) into such $(n,n'{\approx}n)$ series (With the exception [2,3] of q=6+ ions colliding on He for which no (quasi) symmetrical double capture levels exist in the "reaction window"). Since the typical autoionization rates measured [4,5] and calculated [5,6,7] are on the order of $10^{15}s^{-1}$, the net number of electrons kept by the projectile is reduced to one: autoionizing double capture (ADC). In this context the large true double capture (TDC) cross sections where the two electrons removed from the target remain on the projectile are surprising. Two possible interpretation have been put forward

Direct Stabilization

Direct photon decay from a $(n,n'{\approx}n)$ doubly excited state is also possible. Radiative rates show little variations, from state to state and are on the order of magnitude of the hydrogenic value scaled by Z^4. Though this already brings the rates close to 10^{12} s^{-1} for the N^{6+}(3p), it is still too slow to compete efficiently with typical autoionization rates. On the other hand, the autoionization rates of He-like doubly excited states may vary by two orders of magnitude inside a given manifold [6,7,8]. Significant fluorescence yield could be possible for those specific states with particularly slow autoionization rates. These specific states were identified as the one with "unnatural" parity [9] or those with A<1 in $Ar^{16+}(5,6)$ [8] in the Herrick and Sinanoglu [10] classification (not confirmed in $O^{6+}(4,5)$ [11]). There is however a complete lack of data to support the hypothesis of a selective population of theses states by the collision. High resolution spectroscopy in the X-UV range could directly identify the relevant states if populated.

Indirect stabilization

It was then realized that important True Double Capture (TDC) only occurs when the doubly excited (quasi) symmetrical series are degenerate with Rydberg series [12]. This energy overlap occurs first between the *(4,4)* and *(3,n)* series, then between the *(5,6)* and the *(4,n)* series.... During the receding part of the collision, the quasi symmetrical *(n,n')* series have time to interact by e⁻-e⁻ interaction with the Rydberg series known to have reduced autoionization rates. The problem consists now in a quantitative evaluation of the couplings between the *(4,4)* and *(3,n)* . A pure atomic approximation would consider that the collision is so fast that the *(4,4)* states are populated in the isolated atom. Two different approaches have been performed, extensive [11] configuration interaction (C.I.) among the stationary states in $O^{6+}(4,4)$ and *(3l,nl')* or a simple time dependant close coupling [13] between unperturbed series in $N^{7+}(4,4)$) and *(3l,nl')* (fig. 4 and 5).In both approaches, the *(3,n)* series strongly interact and the *(4,4)* fluorescence yield average value reaches around 20% for singlet states. These studies show that the top of the Rydberg series behave as a continuum and "absorbs" the *(4,4)* population in a time scale τ comparable to the autoionization lifetime of the *(4,4)* states located above the ionization threshold. It is clear, in this case that the corresponding width $\Gamma=1/\tau$ can not be interpreted as an autoionization width.

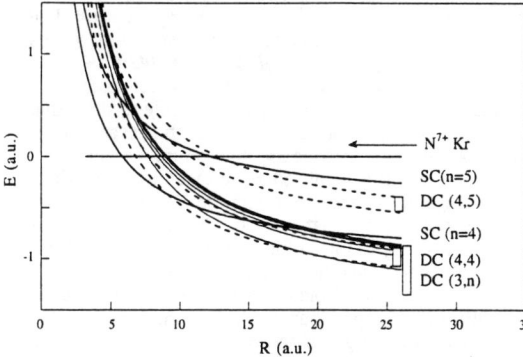

Figure 3: same potential energy curves as in figure 2 but the *(3l,nl')* Rydberg series converging to the n=3 ionization limit are also shown. On the "way out" of the collision, the *(4,4)* states have time to interact by configuration interaction with the straddling *(3l,nl')* Rydberg series. At R=∞, the energy gain of the two series are comparable.

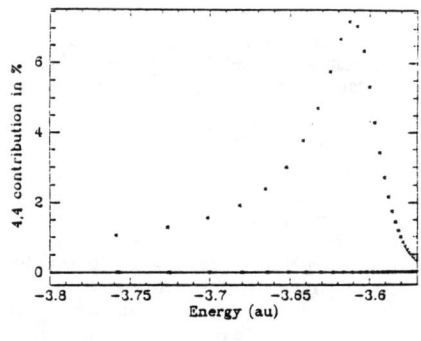

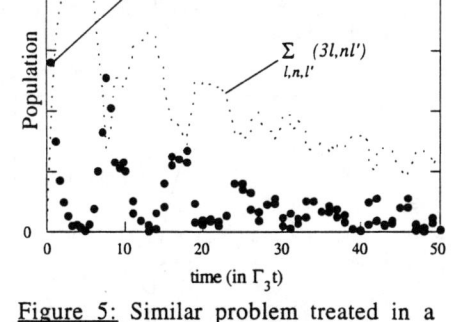

Figure 4: van den Hart et al [11] plot the initial $^1I^e$ $O^{6+}(4f,4f)$ state fraction in the adiabatic states resultings from an extensive configuration interaction with the *(3l,nl')* states. When the *(4,4)* state is located close to the top of the *(3l,nl')* Rydberg series, the width of this distribution amounts to ≈ 5 10⁻² a.u.

Figure 5: Similar problem treated in a simplified time dependant approach [13]. The initial $^1P^o$ $N^{5+}(4,4)$ state population o is transferred to the Rydberg population (dashed line) in a time τ of the order of $1/\Gamma_3$ ≈10⁻¹⁵ sec. The overall population slowly decays into e⁻ and photon continua.

At low velocity, this implies that population transfer occurs at typical distances R~vτ on the order of few ten to few hundred a.u.. These internuclear distances are short enough to allow strong perturbations by the target ionic core. At such distances, both *(4,4)* electrons screen the projectile ionic charge whereas for large enough n values, the outer *(3,n)* Rydberg electron may not. This produces a 2/R potential energy dependance between the *(4,4)* and *(3,n)* series. Above the ionization threshold, this effect is responsible for the Barker and Berry [14] post collisional profile when a diexcited state autoionizes in the Coulombic field of an ionic target. As a result, a given *(4,4)* state does not remain below a given *(3,n)* states but rather spans a large fraction of the whole Rydberg series (fig.6), from the bottom where the successive crossings may be well separated to the top of the Rydberg series which may be seen as quasi continuum.

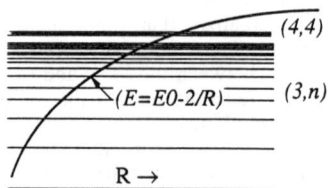

Figure 6: To emphasis the 2/R post collisional energy variation of a *(4,4)* state relative to the ionization threshold. The overall 12/R repulsion of the n=3 ionization limit in fig. 3 have been subtracted.

We have investigated [13] these different regimes altogether with the help of close coupling calculations where the quasi molecular Rydberg and *(4,4)* states are considered atomic-like. The energy of a *(4,4)* state is $E(t) = E_\infty - 2/R(t)$ with $R(t)=vt$ and E_∞ the atomic energy at $R=\infty$. The dielectronic interactions between this *(4,4)* state and the *(3l,nl')* associated states $V_i = <(4,4)|\ 1/r_{12}\ |(3l,nl')>$ are estimated from $<(4,4)|\ 1/r_{12}\ |(3l,\varepsilon l')> \sim (\Gamma_{3l,l'}/2\pi)^{1/2}$, with $\Gamma_{3l,l'}$ the partial autoionization width that this *(4,4)* state would have if located above the ionization limit. Within these simplifying "atomic" approximations, it was possible to show that most of the initial *(4,4)* population is transferred to the Rydberg series (and continua) at a rate Γ close to $\Gamma=\Gamma_1+\Gamma_2+\Gamma_3$ with Γ_i the autoionization rates towards the n=i continuum. Of course, at the bottom of the Rydberg series, the population transfer is localised at well identified crossings but the average "loss" rate is approximately constant in agreement with the Demkov-Osherov model [15]. Since the *(4,4)* state is moving, the population transferred to a given Rydberg state has little chance to return to the *(4,4)* state as it was possible in the "static" pure atomic limit discussed above (fig. 5). The new result shows that Γ_3 has to be included even though the *(4,4)* states may be located few eV below the n=3 ionization limit. Note that since the closest available continuum is generally favored, Γ_3 is usually much larger than $\Gamma_1+\Gamma_2$. In other words, the coupling terms Vi may be interpreted as a width implying that the population transfer to the Rydberg states is an irreversible loss. In this context, the name $\Gamma_{ATR}=\Gamma_3$ (ATR standing for Auto Transfer to Rydberg states) was introduced [16].

From these results, we have developed a simple post collisional decay model [13] in which the *(3l,nl')* and *(3l,εl')* Rydberg and continuum states are treated equally. One considers the *(4,4)* states as coupled to the n=1, n=2 and n=3 "autoionization channels" with partial width Γ_1, Γ_2 and Γ_3 respectively. The probability that a *(4,4)* ^{S}L state "decays" at an energy E is given by the Barker and Berry profile [14] (fig. 7).

$$P(E) \propto \frac{e^{-\Gamma/v(E-E_\infty)}}{(E-E_\infty)^2}$$ with E_∞ the energy at $R=\infty$

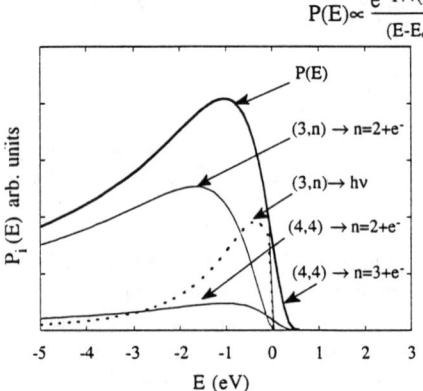

Figure 7: For a given *(4,4)* ^{S}L state, the post collisional Barker and Berry profile is used to derive the *(3l,nl')* Rydberg population. The decay of these Rydberg states is evaluated by n^{-3} scaling law for autoionization and with hydrogenic approximation for the radiative part. The thin curve indicates the Rydberg population which autoionizes to the n=2 continuum whereas the dashed line represents the Rydberg states which have decayed by photon emission and peaks to the top of the Rydberg series. Energies are refered to the n=3 ionization limit

The relative probabilities among the channels i are given by the branching ratio Γ_i/Γ. In fact the n=3 channel (in the sense of MQDT [17]) is made of continuum and bound states, which have different decay schemes. The decay rate of a $(3l,nl')$ ^{5}L Rydberg states is evaluated by scaling the autoionization rates γ_a from those of the corresponding [7] $(3,5)$ ^{5}L series with a n^{-3} power law whereas the radiative decay rates γ_v are supposed constant and governed by that of the inner $3l$ electron taken as hydrogen like (direct radiative decay from the $(4,4)$ state is neglected). Finally the stabilization ratio of the $(4,4)$ ^{5}L is derived by integrating the fluorescence yield $\gamma_v/(\gamma_v+\gamma_a)$ of the $(3l,nl')$ Rydberg states along the Barker and Berry population profile (figure 7).

EXPERIMENTAL DATA

Photon data

According to the barrier model, the radiative decay of the states populated by single capture from a rare gas target was expected to lie in the V.U.V range. Indeed most of the photon spectroscopy focussed to these wavelengths. For alkali target, however the emitted light may appear in the visible. However, it was soon realized by the Lyon group [18], that very large emission cross section in the visible were associated with multiple electron capture *from rare gases*. They built an efficient spectrometer (fig.8) to record coincidences [19] between the analyzed V.U.V. photon and charge state analyzed projectiles and recoil ions.

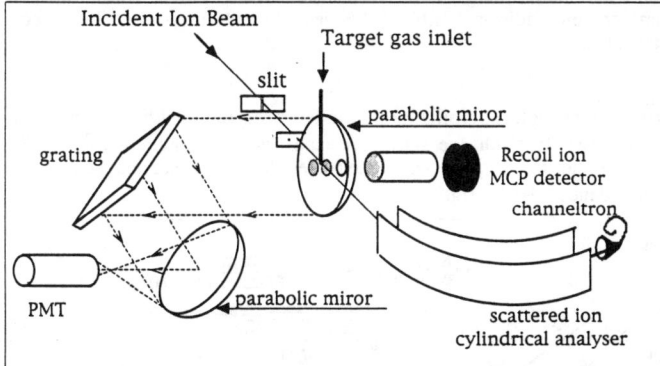

Figure 8: schematic view of the experimental set-up of the Lyon group used to record triple coincidences between visible photon spectra and recoil ion and projectile charge states

The results unambiguously show that, for rare gas target the V.U.V. photons are only correlated with multiple capture events, and in particular with stabilized double capture. They could also record Rydberg transitions from the ionized target. "Unfortunately" a careful cascade analysis shows that most of the Rydberg transitions come from cascade decay of upper states not emitting directly in the visible range. For Kr^{18+} Kr collisions, they could resolve the Rydberg angular momentum of the transition observed. As a striking result, most of the intensity was confined along the "Yrast" chain (between nl Rydberg states with $l=n-1$). This seems in contradiction with the decay model described above which couples the initial (quasi) symmetric $(n,n'\approx n)$ states only with Rydberg states of limited angular momentum (L is supposed conserved). These findings have drawn our attention on the high sensitivity of Rydberg states to external perturbations namely here the target ion coulombic field. So far, our model only accounts for the population and decay of the Rydberg states but not for their evolution, they are treated as independant whereas they should be coupled with the target electric field. Proper account of the very complex curve crossings between the molecular Rydberg states seems a formidable task. From classical mechanics, Kazansky [20,21] derived a first order approximation to evaluate the influence of the ionic target on the projectile Rydberg angular momentum. This post collisional effect is easily understood as a curvature of the Rydberg electron trajectory by the target coulombic field. This effect might have a strong influence on the final stabilization ratio. Indeed the decay model [13] shows that most of the initial $(4,4)$ population is transferred to the Rydberg series but that subsequent autoionization is the dominant decay channel. If the angular momentum increases fast enough (as suggested by Kazansky classical trajectories), the stabilization might be favored since the autoionization rates of such $(3l,nl')$ Rydberg states has been shown to drop exponentially [22,23] with l'.

Electron data

Electron spectroscopy is the most efficient tool to study the autoionizing states formed by the charge exchange collisions. Due to the post collisional profile, both energy and autoionization width of the $O^{6+}(3,3)$ manifold are accurately known. Early indication of population of autoionizing Rydberg states were reported [2]. In case of multiple electron capture processes, the interpretation is more complex [24] and sophisticated coincidence experiments are needed to separate satelite lines [25]. Since the non autoionizing states do not appear in electron spectra, the evidences of large stabilization ratio are only indirect. The two majors effects could be stressed with the example of the *(4,4)* series.

i) If the *(4,4)* population is actually transferred to the Rydberg series at a rate $\Gamma = \Gamma_1 + \Gamma_2 + \Gamma_3$, then the post collisional energy shift should be of the order of $2(\Gamma_1+\Gamma_2+\Gamma_3)/v >> 2(\Gamma_1+\Gamma_2)/v$. The *(4,4)* manifold is not yet as well resolved as the *(3,3)*, however it was underlined by Moretto-Capelle [26] that the overall *(4,4)* appears systematically shifted by few eV toward low electron energy in : O^{8+}_He, N^{7+}_He and N^{7+}_Ar electron spectra. This behavior may be explained either by a surprising selective population of the lower members of the *(4,4)* structure or by taking into account [27] the coupling width Γ_3 with the straddling *(3l,nl')* Rydberg series as actually implicitly done in the total width Γ calculated with the complex scaling method [6,13].

ii) In the decay model described above, the fluorescence mainly originates from those Rydberg states located on top of the Rydberg series. This causes a small dip on the order of 1 to 2 eV which results in an apparent additional shift of the ejected electron profile. This effect remains to be clearly established.

Ion spectroscopy

Evidences of large double capture stabilization ratio has been reported by experiments where both projectile and target final charge states are measured in coincidence [28-30,19]. However, the information is easier to interprete when the projectile energy gain and scattering angle are measured in coincidence [12,31,32] (fig.9).

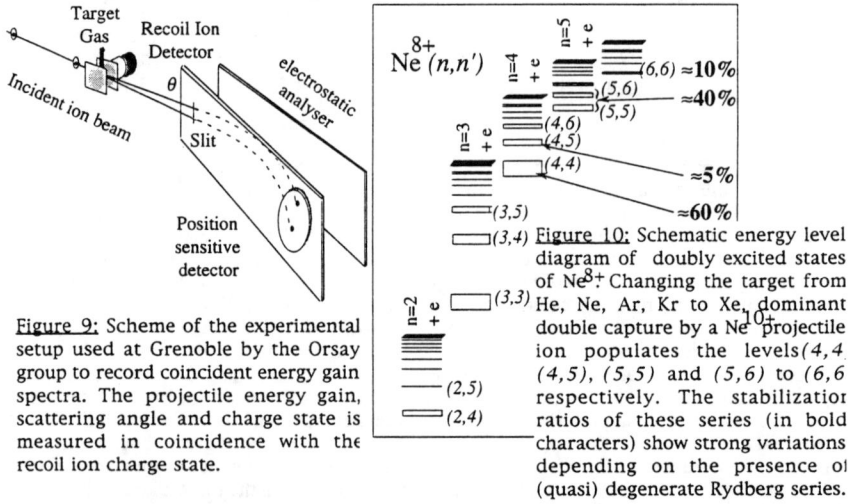

Figure 9: Scheme of the experimental setup used at Grenoble by the Orsay group to record coincident energy gain spectra. The projectile energy gain, scattering angle and charge state is measured in coincidence with the recoil ion charge state.

Figure 10: Schematic energy level diagram of doubly excited states of Ne^{8+}. Changing the target from He, Ne, Ar, Kr to Xe, dominant double capture by a Ne^{10+} projectile ion populates the levels *(4,4)*, *(4,5)*, *(5,5)* and *(5,6)* to *(6,6)* respectively. The stabilization ratios of these series (in bold characters) show strong variations depending on the presence of (quasi) degenerate Rydberg series.

In such experiments, the average stabilization ratio of an individual *(n,n')* manifold can be measured. Indeed large variations are observed from one manyfold to another. For instance, in Ne^{10+} collisions, the stabilization ratios vary from 60% down to 5% depending on the series populated. On the other hand, since double capture often populates two different series simultaneously, the overall double stabilization ratios show less pronounced variations. In Ne^{10+}_Ne collisions (fig. 11), double capture dominantly populates the *(4,5)* series (80%) and the *(4,4)* series (20%) however an overall 15% stabilization ratio is observed due mainly to the high stabilization ratio of the *(4,4)* (>50%) series whereas the *(4,5)* series has a fluorescence yield of only 3%. Figure 11 displays the two doubly differential energy gain spectra associated with double electron capture namely the ADC and TDC spectra.

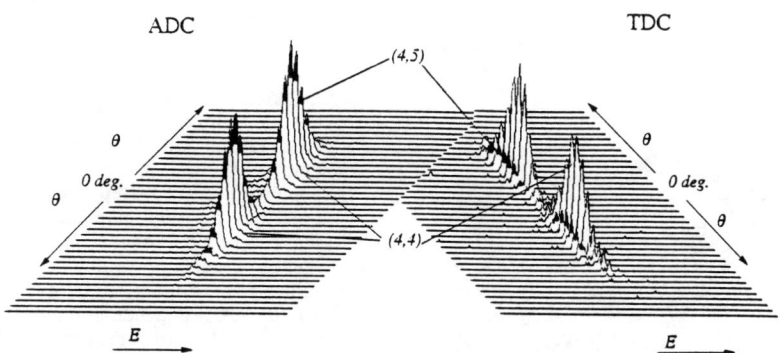

Figure 11 shows the two doubly differential cross sections for autoionizing and non autoionizing double electron capture in the Ne^{10+}_Ne collisions. In ADC, double capture into the (4,5) series (filled in black) is the major channel whereas the (4,4) series largely dominate in TDC.

The peaks in ADC energy gain spectra are broadened by $\Delta E \propto \sqrt{E_e}$ due the ejection in the forward or backward direction of an electron with energy E_e. As a result two series which have different ionization continua have different widths. For instance, the (4,4) and (4,5) are broadened by 20 and 9 eV respectively in Ne^{10+} collisions at 10 keV. In this case the the (4,4) and (4,5) peaks slightly overlap giving rise to a 10-30% uncertainty when an unfolding is needed. The TDC spectra are obviously not affected

Actually, in our post-collisional model, the dominant TDC process is not attributed to the (4,4) states (though initially populated) but rather to the fraction of the (3,n) Rydberg states which have decayed by photon emission. For bare incident ions, the fact that the three $3l$ ionization limits are degenerate allows an interpretation of the TDC energy width. In the present case, the 13 eV energy resolution FWHM is subtracted from the 17eV TDC width to produce a 10 ± 2 eV intrinsic width which corresponds to the binding energy of a $n \approx 10$ Rydberg electron. In N^{7+}_Ar collisions, the 5 ± 1 eV intrinsic spread also corresponds to $n \approx 10$.

Recapture of the Rydberg electron

Recapture of the Rydberg electron by the target (on a Rydberg level) is seen both in N^{7+}_Ar and in O^{7+}_Ar collisions [12]: N^{5+} (3,n) + $Ar^{2+} \rightarrow N^{6+}(n=3) + Ar^+$ (n>>2). This process also seen in coincident photon spectra [19] clearly shows that the population transfer *does not occur* in the separated atoms but when the two collision partners are within reasonable distances $R \approx v.\tau$. This process is also reproduced by classical trajectory calculation [21] and may be seen as another aspect of the strong target influence on the electron trajectory. This recapture process is probably most velocity dependant since the potential wells of the target and projectile rapidly rise up again to separate the two ionic cores.

Direct stabilization

If the population of a given (quasi) symmetrical doubly excited state is transferred to a Rydberg state, the final energy gain will be that of this Rydberg state. Conversely, if in TDC a doubly excited state is found at an energy gain which *does not* correspond to that of Rydberg series, we can conclude that it did not transfer to Rydberg series. The (4,5) structures seen in TDC (fig11) decay by "direct" photon emission. The average fluorescence yield of these series is of the order of 5% with a maximum value of 11% found for the O^{8+}_Ar collisional system [32] at 10.5 keV. This last value is surprising since the "same"(4,5) series populated in collisions with F^{9+} and Ne^{10+} projectiles only amount from 3 to 8 ± 3% and do not show the Z^4 scaling expected from a hydrogen like approximation. Of course, the populations inside the (4,5) series are not necessarily identical but these values are not yet understood.

Differential cross sections

To remove **r** electrons from the target, the projectile needs to reach smaller internuclear distances (more tightly bound ionic orbitals) and the average Coulombic repulsion $\approx$ **r**.(Z-r)/R increases rapidly with **r**. The angular profiles associated with multiple electron capture are indeed primarily characteristic of the number **r** of electrons removed from the target. On the other hand, two different processes corresponding to the same number of captured electrons, produce more subtle variation of the scattering profile.

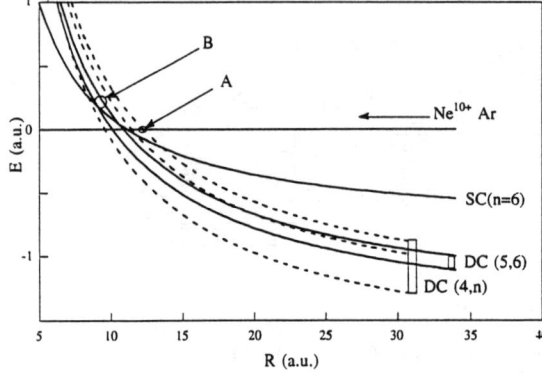

Figure 12: schematic potential energy curves for the Ne^{10+} Ar system. In this case, the doubly excited states $(5,6)$ are populated in two steps at the B crossings. Direct population of the $(4l,nl')$ Rydberg series at the A crossing would produce scattering angles smaller by 10 to 30% (depending on the n value). The differential cross sections in fig. 13 rather suggest a common population mechanism at the B crossing.

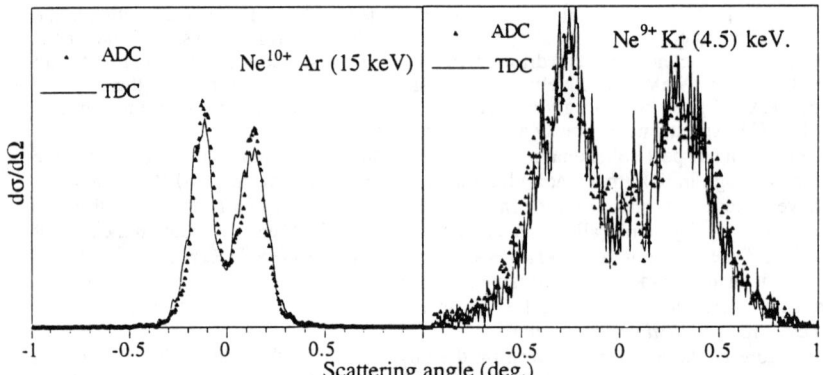

Figure 13: differential cross sections $d\sigma/d\Omega$ for the two system Ne^{10+}_Ar and Ne^{9+}_Kr. In both cases autoinizing double capture (ADC) populates the $(5,6)$ series straddling with the $(4l,nl')$ Rydberg series. TDC is then interpreted as the fraction of the Rydberg states that have not autoionized. The scattering profiles are shown as measured with both positive and negative scattering angles, the ADC and TDC profiles are normalized to each other.

In the Ne^{10+}_Ar and Ne^{9+}_Kr systems, the $(5,6)$ channels are populated (in two steps) at the B crossings (fig. 12) and then the $(5,6)$ population is "slowly" transferred to the energy straddling Rydberg states during the receding part of the collision. Since this population transfer occurs at large distances, the angular profiles of ADC and TDC are expected and actually found identical (fig. 13). In the Correlated Double Capture (CDC) suggested by N. Stolterfoht et al [2], the two electrons should be captured into asymmetrical doubly excited Rydberg states directly at the crossing labelled A (fig. 12), the angular profiles of ADC and TDC are not necessarily connected. For instance classical deflection function would predict 10 to 30% difference between the threshold angles of the $(5,6)$ ADC profile as compared with that of a TDC profile arising from direct CDC. The experimental differential cross sections (fig 13) for ADC and TDC are very similar for all investigated systems, clearly speaking in favor of the indirect process. Of course such similar angular profile could also be explained if the observed ADC "$(4,4)$ structure" was due dominantly to CDC into the Rydberg series with cross section$\approx 10^{-14}$ cm^{-2}.

Influence of the ion core structure

The ion core structure should not modify too much the population transfer to the Rydberg series. On the other hand the radiative and autoionization cascades from the *(3l,nl)* or *(4l,nl)* series may be strongly perturbed. First, the three 3s, 3p and 3d series are not degenerate any more and Coster-Kronig transitions may occur. The radiative rates may also be modified. For instance the fast 3p→1s transition is not allowed any more if the n=1 shell is already filled. Finally as noted by R.Ali et al [30] or Martin et al [19], the inner electron may decay to a core metastable state. For instance, in Ne^{9+} Ne collisions, the *(1s,3l,nl')* Rydberg states may be formed of a *(1s,3l)* triplet core which rapidly forms a *(1s,2l)* ^3L metastable core. If this metastable state survives long enough, the decay of the outer Rydberg electron may form autoinizing *(1s,2l,2l')* states. Our data in table 1 suggest that this last mechanism is more important than the possible Coster-Kronig transition. Indeed the stabilization ratio averaged among the *(4,4)* structure is maximum for bare F^{9+} projectile ions, then strongly decreases for H-like Ne^{9+} ions and increases again for Be-like Al^{9+} projectiles.

	collision system	overall ratio	(4,4) ratio indirect	(4,5) ratio direct
bare ion	F^{9+}-Ne (13.5keV)	35%	**52%**	2%
H-Like (1s)	Ne^{9+}-Ne (9 keV)	15%	**20%**	1%
Be-Like ($1s^2 2s^2$)	Al^{9+}-Ne (9 keV)	20%	**36%**	9%

Table I

Double capture stabilization ratio for 9+ ions colliding on Ne target. Assuming a common primary double capture process, the results (with an estimated 20% error) show a strong core effect which may be explained by the formation of triplet metastable core $((1s\ 2l)^3L\ nl')$ which could eventually autoionize if the outer Rydberg electron decay is fast enough to form *(1s 2l2l')* or *(1s 2l3l')* states.

Comparison with the decay model

With all the partial widths Γ_i which couple a given *(4,4)* ^SL state to all the available autoionisation and "auto transfer" channels the decay model calculates the stabilization ratio for each initial state. Since no initial population is known yet, we only compare the averaged value. For N^{7+} collisions at v=0.2 a.u. the model [13] predicts 21% stabilization of the double capture into the singlet states of the *(4,4)* structure. This value is about twice less than the experimental value for 10 keV N^{7+}_Ar collisions. The same calculation performed with the $Ne^{8+}(4,4)$ atomic data yields a stabilisation ratio of 40% and 64% among the singlet and triplet states respectively. These values directly compare with the 40-50% and 50-70% found in collision with He and Ne targets respectively. The range indicated corresponds to collision energies between 7 and 30 keV. These results show a better agreement for higher projectile charge state suggesting, for intance, a reduction of the target influence on the Rydberg electron trajectory.

The authors would like to acknowledge the numerous collaborators associated with this work.
- The Bordeaux group; H Bachau and C Harel
- The Toulouse group; Moretto-Capelle P, Bordenave-Montesquieu A, Benoit-Cattin P, Gleizes
- The Lyon group; Martin S, Denis A, Delon A, Carre M, Désesquelles J and Ouerdane Y
- The staff who runs the ECR ion source facility at Grenoble where all the experiments have been performed; T Lamy, A Brenac, G Lambollé, HJ Andrä and C Guet .
It is a pleasure to acknowledge the challenging interactions with the Amterdamm group in atomic structure: Van den Hart H.W., Vaeck N. and Hansen J.E. (fig. 4 is reproduced by courtesy of these authors).

REFERENCES

1. Barat M and Roncin P (1992) J.Phys.B: At. Mol. Opt.Phys.**25** 2205 (Topical Review)
2. Stolterfoht N, Havener CC, Phaneuf RA, Swenson JK, Shafroth SM and Meyer F Phys.Rev.Lett. **57** (1986) 74
3. Roncin P, Barat M, Gaboriaud MN, Guillemot L and Laurent H. (1989) J.Phys.B: At. Mol. Opt.Phys.**22** 509
4. Mack M, Nijland J.H. Straten P vd, Niehaus A and Morgenstern R (1989) Phys Rev A **39** 3846
5. Boudjema M, Moretto-Capelle P, Bordenave-Montesquieu A, Benoit-Cattin P, Gleizes A, Bachau H, Galant P, Martin F, Riera A and Yañez J.Phys.B: At. Mol. Opt.Phys. **22** (1989) L121
6. Ho Y K Phys. Rev. A. **35** (1987) p2035-43
7. Bachau H, Galan P, Martin F, Riera A et Yañez M (1990) Atomic and Nuclear Data Tables **44** 305
8. Chen Z and Lin C.D. (1992) J.Phys.B : Atom.Molec.Phys.**26** (1993) 957-963
9. Vaeck N, Van den Hart H.W. and Hansen J.E. (1993) VIthInt. H.C.I. Conf., Kansas, USA Eds P Richard, M Stöckli, CL Cocke and CD Lin AIP #274 p 414
10. Herrick D.R. and Sinanoglu O. Phys. Rev. A. **11**, 97 (1975)
11. Van den Hart H.W, Vaeck N. and Hansen J.E. submitted
12. Roncin P, Gaboriaud MN and Barat M Europhys.Lett.**16** 551 (1991)
13. Roncin et al submitted
14. Barker RB and Berry HW, (1966) Phys.Rev. **151** 14-9
15. Demkov Yu.N and Osherov V.I. (1967) Zh. Eksp. Teor. Phys. **53** 1589 translation JETP (1968) **26** 916
16. Bachau H, Roncin P and Harel C (1992) J.Phys.B : Atom.Molec.Phys. **25** L 109
17. Seaton M J Rep. Prog. Phys. (1983) **46** 167-257
18. Martin S, Denis A, Ouerdane Y and Carre M Phys. Lett A **165**.(1992) 441
19. Martin S, Denis A, Delon A, Désesquelles J. and Ouerdane Y (1993) submitted
20. Kazansky A J.Phys.B: Atom.Molec.Phys..**25** (1992) L381
21. Kazansky A J.Phys.B: Atom.Molec.Phys(1993) *submitted*
22. Poirier M (1988) Phys. Rev. A **38** 3484
23. Nikitin SI and Ostrovsky VN. J.Phys.B: At. Mol.Phys.(1980) **13** 1961-84
24. Benoit-Cattin, Bordenave-Montesquieu A, P, Boudjema M, Gleizes A, Dousson S and Hitz (1988) D J.Phys.B: At. Mol. Opt.Phys. **21** 3387
25. Posthumus J.H. and Morgenstern R (1992) Phys. Rev. Lett. **68** 1315
26. Moretto-Capelle P, Oza D.H,Benoit-Cattin P, Bordenave-Montesquieu A, Boudjema M, Gleizes A, Dousson S and Hitz D J.Phys.B: At. Mol. Opt.Phys. **22** (1989) 271
27. Bachau H, Harel C, Barat M, Roncin P, Moretto-Capelle P, Bordenave-Montesquieu A, Benoit-Cattin P, Gleizes A and Benhenni M (1993) VIthInt. H.C.I. Conf., Kansas, USA Eds P Richard, M Stöckli, CL Cocke and CD Lin AIP #274 p 175
28. Müller A, Groh W., Salzborn E, Phys.Rev.Lett. **51** (1983) 107
29. Cederquist H, Andersson H, Beebe E, Biedermann C, Broström L, Engström Å, Gao H, Hutton R, Levin J.C, Liljeby L, Pajek M,Quinteros T, Selberg N and Sigray P. Phys. Rev. A. **46** (1992) 2592
30. Ali R., Cocke C.L., Raphaelian M.L.A. and Stockli J.Phys.B: At. Mol. Opt.Phys. **26** L177-84 (1993)
31. Guillemot L, Roncin P, Gaboriaud MN, Laurent H. and Barat M J.Phys.B: At. Mol. Opt.Phys.**23** (1990) 4293
32. Gaboriaud MN, Roncin P and Barat M J.Phys.B. Lett. in press

THEORETICAL LIFETIME AND DECAY MODE OF MULTIPLY EXCITED IONS PRODUCED DURING ION-ATOM COLLISIONS

Henri Bachau

Laboratoire des Collisions Atomiques[1]
Université de Bordeaux I
351, Cours de la Libération
33405 Talence, France

ABSTRACT

Electrons captured by multiply charged ions on rare gas targets are known to populate highly multiply-excited states of the projectile with large cross sections. Despite spectacular experimental and theoretical progress have been achieved for the comprehension of double charge exchange, some aspects of the decay process remain to be fully understood. For example, there is some experimental evidence that populating multiply excited states could lead to radiative stabilization of the system on the way out of the collision. This is in contrast with the usually assumed assumption that multiply excited states decay mainly through electron emission (at least for lower series). We will review the mechanisms of de-excitation of autoionizing states and focus on series of resonances interacting with Rydberg states. We will discuss the possibility for two-electron excited states to decay to Rydberg series under post-collisional effects and stabilize . The relative importance of the other competing channels (direct stabilization and autoionization) will be discussed on the basis of new results obtained for Ne^{8+}. The case of triply excited states will be also evocated.

---

[1] Unité Propre du CNRS 260

INTRODUCTION

The development of sources for highly charged ions during the last decade has made possible studies of ion-atom collisions where a relatively large number of electrons are transferred to the projectile. At intermediate velocities multiple electron capture populates multi-excited states which de-excite mainly in the way out of the collision. As a consequence, atomic-structure calculations have been revealed to be a crucial complement to collision studies in order to make a comprehensive evaluation of the collision outcome. Indeed, the charge and radiative balance of this type of collision is governed by, i) the cross sections associated to single and multiple electron exchange and, ii) the partial widths associated to the different open channels. In particular, it is of crucial importance to evaluate the fraction of ions which de-excite radiatively and through electron emission since these two channels lead to differently charged products. The question of properties of multi-excited ions has recently become even more crucial since such systems are populated during ion-surface and ion-complex-atom collisions with a very large number of electrons transferred. The aim of this contribution is to study collisions of the type:

$$N^{7+} + He \rightarrow N^{5+}(n1, n2) + He^{++} \rightarrow N^{6+}(nl) + He^{++} + e$$

at intermediate velocities. For low series of autoionizing states (n1$\simeq$ n2$\simeq$ 3), this type of collision has been experimentally and theoretically widely studied with an overall agreement between theory and calculations (see Harel and Jouin[1] and Barat and Roncin[2] for references and review). Collisions on targets where the electrons have less binding energy populate higher series, Bordenave-Montesquieu et al[3] have observed (using electron spectroscopy techniques) that (4,4) series are populated during $N^{7+} + H_2$ collision at 4.9 KeV amu^{-1}. Two unusual characteristics were observed in the electron spectrum :the spectrum is peaked around the lowest (4,4) state and it vanishes before reaching the N^{6+}(n=3) threshold. Note that the series (4,4) lies in the threshold region and some of the states are in interaction with N^{5+}(3,n) Rydberg states. This pattern, as we will see in the following sections, is not an "accident" proper to (4,4) states but a common property of high lying series of autoionizing states of atoms and ions. Some insight has been recently brought into the problem by Roncin et al[4] who measured energy and angle doubly differential cross sections in coincidence with both projectile and target final state charge. The energy gain spectra for charge exchange reactions following O^{7+}+Ar collisions at 3.5 KeV laboratory energy revealed similarities for differential cross sections associated with autoionization and photon stabilized double capture. It was suggested that autoionization and radiative stabilization proceeded from the same series of autoionizing states initially populated. More precisely, the initial (n1,n2) quasi- symmetric states could "feed" adjacent Rydberg series as well as autoionize directly to open adjacent continua. We will see that the process of "feeding" Rydberg series is the continuation of autoionization below threshold and we will describe it as auto transfer to Rydberg states (ATR). It is worth noting that the production of Rydberg states induced by electron-electron interaction has been extensively investigated by Stolterfoht[5] who analyzed *correlation process associated to autoexcitation* in collisions of the type O^{6+}+He. Basically, the incident channel crosses the non-equivalent 2pnl configurations *at small internuclear distance* then, at the positions of resonances, both electrons are transferred to 2pnl Rydberg states (*correlated double capture*). On the other hand ATR is essentially a post collisional effect governed by post-collisional interactions and multi-excited ion properties. The observations of Roncin et al[4] have been supported by VUV photon experiment (see Martin et al[6]). More recently, Ali et al[7] found some qualitative agreement between the velocity dependence of stabilization (in coincidence experiment) and our model predictions. Also we note the works of Cedersquist et al[8] who observed the effects of the charge of the projectile on radiative stabilization and Danared et al[9] who suggested that the multi-excited ions could not fully autoionize in reactions of the type $Ar^{q+} + Ar \to Ar^{r+} + Ar^{s+} + (s+r-q)e$ with q=6,7,8,9 and 11 and r=4-12.

We have proposed a model[10] which put the ideas of Roncin et al[4] in a more quantitative basis when realizing that the transfer to Rydberg states is the "natural" continuation of autoionization below threshold. Due to long range Coulomb effects, quasi-symmetrical (n1,n2) autoionizing states "sweep" the Rydberg series to which they (partially) decay. We were able to evaluate quantitatively the decay rate to the dense part of the Rydberg series and we found that there is no discontinuity of the physical properties of the system across the threshold. This is in agreement with the ideas sustaining the Quantum Defect Theory and we have shown recently[11] that such a decay persists even in the low density region on the basis of a Landau-Zener analysis. Also it is important to note that all quasi-symmetric states lying in the region of the thresholds (above or below threshold) are concerned by ATR. It is worth recalling that Rydberg series as (3l,nl') are also autoionizing states and only those having a large value of n decay radiatively (and for a given n value, the autoionization rate drops very fast for large l' values). This point was raised recently by Vaeck et al[12] who argue that some quasi-symmetrical states could decay radiatively (direct stabilization) and, on the basis of a very selective process, ATR could not be necessary to explain the post-collisional stabilization. Such considerations are supported by recent calculations of Chen and Lin et al[13] who found that a class of doubly excited states of Ar^{16+} de-excites radiatively. At this point it must be noted that such heavy multicharged ions favor the radiative de-excitation since it scales as Z^4 while the autoionization rate scales as Z^0 (Z being the nuclear charge).

We recall the main theoretical results in the following section and the case of 2 and 3 electrons systems will be discussed in the last section.

RECAPITULATION

The principal aspects of the problem are recalled in this section. We consider the case of the collision of the type $I^{q+}+T \to I^{\,(q-2)+}(n1,n2)+T^{2+}$ and we assume that the internuclear distance R is large enough to retain only the Coulomb effect $\frac{2}{R}$ of the receeding target on each projectile electron. Under this approximation the doubly excited $I^{\,(q-2)+}$ ion may be treated as an isolated ion with shifted levels. Considering that the series of symmetrical or quasi-symmetrical states $(n1,n2)$ lie in the region of (n_s, nl') Rydberg series it is of crucial importance to note that there will be a relative shifting of $\frac{2}{R}$ between the two series due to the Coulomb effect, *as far as the Rydberg electron does not screen the T^{2+} charge*. This defines two regions in the curve crossing topology delimited by the radial value $\rho_n = \frac{3}{2Z}\frac{n^2-l(l+1))}{}$ where Z is the ion charge "seen" by the Rydberg electron. Then, in the inner region, there is a relative shifting between the (n1,n2) and Rydberg series while both series are equally shifted in the outer region (where both electron screens the nuclear charge in both series). In Figure I we show an energy diagram for the case of the $N^{5+}(4,4)$ series, the regions I (where ATR to (n_s, nl') Rydberg states dominates, see dashed area) and region II (where autoionisation to (n_s, kl') or lower threshods dominates) are represented. In the lower part of the spectrum (where the crossings are isolated) we were able[11] to evaluate the probability of populating a particular Rydberg level from a (n1,n2) state on the basis of a Landau-Zener treatment (for isolated crossings), this probability is given by:

$$P(n) = \frac{R_c^2\, Z^2}{2\, v_r\, n^3}\, \Gamma_{k\to 0} \qquad (1)$$

R_c is the internuclear distance where the crossing occurs, v_r the radial collision velocity and $\Gamma_{k\to 0}$ is a rate given by:

$$\Gamma_{k\to 0} = 2\pi\, |<n1,n2\,|\,\frac{1}{r_{12}}\,|\,n_s, k\to 0>|^2 \qquad (2)$$

The normalization of the continuum function $|\,n_s, k \to 0 >$ is assumed on the energy scale. *Note that this rate corresponds to the autoionization rate to the $(n_s, k \to 0\, l')$ continuum the level would have if situated just above threshold.* This important result has been obtained in the Coulomb approximation where the inner electron n_s fully screens the nuclear charge. We have shown[10] that the above formula are also valid in the region of dense Rydberg states (where their energy separation is smaller than $\Gamma_{k\to 0}$). In that region Equation 1 is put in the more appropriate form of density of probability $P(E)$:

$$P(E) = \frac{2}{(E_0 - E)^2\, v_r}\, \Gamma_{k\to 0} \qquad (3)$$

where E_0 and E are the energies of level (n1,n2) at R=∞ and R=R_c respectively. From Equation 3 it is possible to derive a simple formula (Barker and Berry formula) for the energy distribution resulting from the de-excitation in the region I defined above:

$$\mathcal{P}(E) = \frac{N}{(E_0 - E)^2} \exp(\frac{-2}{v_r (E_0 - E)} \Gamma_{k \to 0}) \qquad (4)$$

where N is a normalization factor. We have focused on a particular channel leaving the I^{q-1} in a residual n_s state and it must be kept in mind that autoionizing channels leaving the ion in lower excited states may also be present. For example, the lowest $N^{5+}(4,4)$ states autoionize leaving the residual N^{6+} ion in 1s, 2s and 2p states. Nevertheless, $\Gamma_{k \to 0}$ dominates generally the other partial widths and ATR and autoionization to the n_s threshold (when energetically allowed) will dominate the process.

The above results have been obtained in the context of the Feshbach formalism. This approach is well suited for our problem since the different open channels are treated separately and the associated partial widths are easily calculated. Also the P and Q projection operators (which define the open and closed channels respectively) may be defined for the atomic system and conserved for the study of the quasi-molecular system as far as *only* the long range Coulomb effect is considered. For small internuclear distances these operators must be defined on the basis of molecular orbitals.

DISCUSSION AND CONCLUSION

Results for nitrogen have been presented elsewhere [11] and we will focus here on the case of helium-like neon ions. The evaluation of the stabilization amount necessitates a dynamical treatment, a crude evaluation may be easily calculated using the Eq.4 and a more elaborated approach is possible on the basis of a close-coupling type approach (see the contribution of Ph. Roncin). In order to perform the above analysis, the atomic-structure parameters (decay rates, energies ...) must be calculated. We will focus here on such calculations and on the analysis of energy diagrams. In Table I, we show the energies and widths of the (4,4) and (4,5) $^1D^e$ series of Ne^{8+} (we have performed calculations for $^1S^e$ to $^1I^{e,o}$ symmetries). We have calculated the rates for direct radiative stabilization (Γ_{RAD}), autoionization (Γ_{AIS}) and auto-transfer to Rydberg states (Γ_{ATR}). All (4,4) levels are situated below the $Ne^{9+}(3)$ threshold while the (4,5) series lie above. Figure II presents the energy diagram of the ($Ne^{8+}+T^{2+}$) system, the regions I and II specified in the above sections are also represented. Since the (4,5) states lie above the n=3 threshold during most of the collision, they mainly autoionize leaving the Ne^{9+} ion in 3l states (note that their radiative decay is small compared to autoionization, therefore only few percents of the (4,5) levels should directly stabilize, this is in agreement with the recent findings of Gaboriaud et al.[14]). As seen on Figure II, the (4,4) states overlap the region I and are likely to produce Rydberg (3l,nl') states and, to a lesser extend, to autoionize to the adjacent (2l,kl') continuum. Note that the direct radiative decay is negligible and that stabilisation through ATR is certainly under-estimated on the basis of atomic structure considerations since Kasansky[15] has shown that the dynamical rotation-Stark mixing effect results in the production of high L Rydberg states. Also the question of the contribution of ATR in transfer-excitation (where the Rydberg electron is finally left on the target) remains to be investigated. There is not a unique trend for the behavior of the probability $P(n)$ of populating Rydberg states versus n, it depends on the parameters involved in the problem ($\Gamma_{k\to 0}$, E_0, v_r ...), generally $P(n)$ increases or decreases regularly in the low density region and drops in the dense region of Rydberg states. Depending on the atomic and collisional parameters the total amount of decay to Rydberg series may vary between few percents and close to 100%, with a large amount of stabilization[11]. Ne^{8+} ion is a good candidate for ATR and already under extensive experimental investigations.

We now consider the atomic-structure properties of the isolated doubly excited ion. It must be emphasised that some upper quasi-symmetrical states are located just below threshold (in the dense region, where Γ_{ATR} is larger than the energy separation of adjacent Rydberg states). This raise the question of the definition of the widths and we have already noted that strong discrepancies exist between theoretical calculations[11] for N^{5+}(4,4) states situated below the N^{6+}(n=3) threshold. Such discrepancies exist for the lowest Ne^{8+}(4,4) states as well (compare our earlier calculations[16] for the autoionization widths and those performed with a complex coordinate rotation (CCR) method[17]). Nevertheless we have checked that the CCR calculations compare well with the sum of our ATR and AIS rates in all cases. This result can be understood considering that CCR calculation includes the interactions with Rydberg series, therefore the rates obtained with this method should not be interpreted as autoionization rates since the Rydberg states may decay radiatively. Nevertheless the above analysis should be valid only in the dense region of Rydberg states and the discrepancies remain to be understood for the lower (4,4) states.

In conclusion, we have shown that ATR has important consequences in the domain of ion-atom collisions and atomic structure calculations. Although we have focused on two-electron atoms for which calculations are easier, ATR will have important consequences on multi-electron species produced in ion atom collisions. In figure 4 we present some of our recent calculation for N^{4+}(4,4,4) states for a large number of symmetries. The dashed lines represent the limit the energy spectrum of N^{5+}(3,4) series, at a first glance we see the that the two series strongly overlap and we expect that ATR will have dramatic effects in that case (ATR produces here N^{4+}(3,4,n) Rydberg states). Investigations are in progress in this domain.

REFERENCES

1. C. Harel and H. Jouin, J. Phys. B, 25, p.221 (1992).
2. M. Barat and Ph. Roncin, J. Phys. B, 25, p.2205 (1992).
3. A. Bordenave-Montesquieu, P. Benoit-Cattin, A. Gleizes, A.I. Marrakchi, S. Dousson and D. Hitz, J. Phys. B, 24, p.L127 (1984).
4. Ph. Roncin, M.N. Gaboriaud and M. Barat, Europ. Lett., 16, p.551 (1991).
5. N. Stolterfoht, Phys. Scripta, 42, p.192 (1990).
6. S. Martin, A. Denis, Y. Ouerdanne and M. Carre, Phys. Lett. A, 165, p.441 (1992).
7. R. Ali, C.L. Cocke, M.L.A. Raphaelian and M. Stockli, J. Phys. B, 26, p.L177 (1993).
8. H. Cederquist, H. Andersson, E. Beebe, C. Biedermann, L. Bröstom, Å. Engström, H. Gao, R. Hutton, J.C. Levin, L. Liljeby, M. Pajek, T. Quinteros, N. Selberg and P. Sigray, Phys. Rev. A, 46, p.2592 (1992).
9. H. Danared, H. Andersson, G. Astner, P. Defrance and S. Rachafi, Phys. Scripta, 36, p.756 (1987).
10. H. Bachau, Ph. Roncin and C. Harel, J. Phys. B, 25, p.L109 (1992).
11. H. Bachau, C. Harel, M. Barat, Ph. Roncin, A. Bordenave-Montesquieu, P. Moretto-Capelle, P. Benoit-Cattin, A. Gleizes and M. Benhenni, VI Int. Conf. Highly Charged Ions (Kansas USA, 1992), USA AIP Publishers and paper submitted to J. Phys. B. Also H. Bachau, Habilitation à Diriger des Recherches, november 1992, (not published).
12. N. Vaeck, H.W. van der Hart and J.E. Hansen, same than above (VI I.C.H.C.I, Kansas USA, 1992) and Comment to appear in J. Phys. B.
13. Z. Chen and C.D. Lin, J. Phys. B, 26, p.957 (1993).
14. M.N. Gaboriaud, P. Roncin and M. Barat, to be published in J. Phys. B.
15. A.K. Kasansky, J. Phys. B, 25, p.L381 (1992).
16. H. Bachau, J. Phys. B, 17, p.1771 (1984).
17. Y.K. Ho, Phys. Lett. A, 79, p.44 (1980).

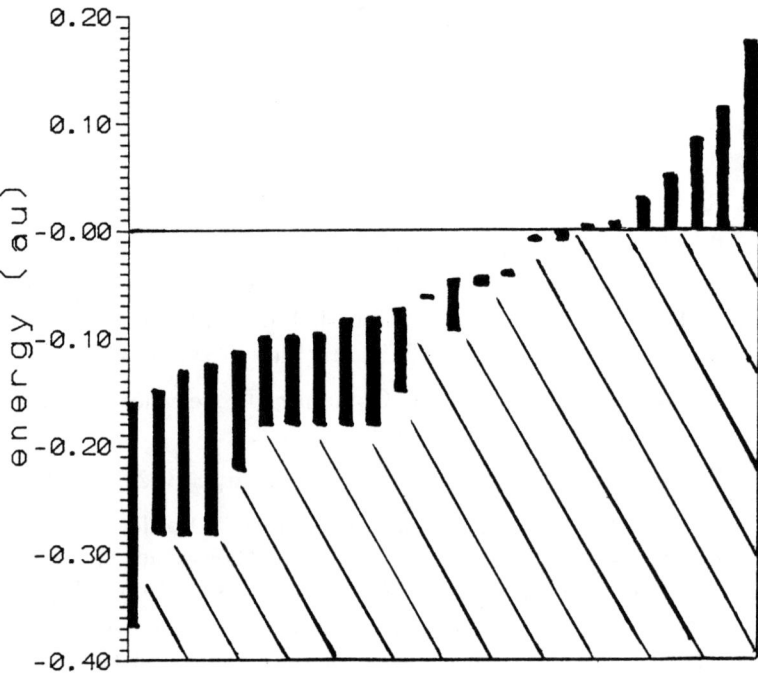

FIGURE I Energy diagram (y axis, $E = E_0 - \frac{2}{R}$) for the 24 singlet (4,4) states (x axis) of N^{5+} in the presence of a receding T^{2+} ion. The zero energy is put at the $N^{6+}(n = 3)$ threshold (horizontal line). The dashed area (region I) represents the region where ATR is allowed. The zone colored in black is the region II where autoionization dominates, below threshold autoionization leaves an $N^{6+}(n = 2)$ ion, above theshold it is autoionization to $(n_s = 3, kl')$ continua. E is limited upward by the energy E_0 of the level of the isolated $N^{5+}(4,4)$ ion (at infinite internuclear distance), downward by a minimum internuclear distance R

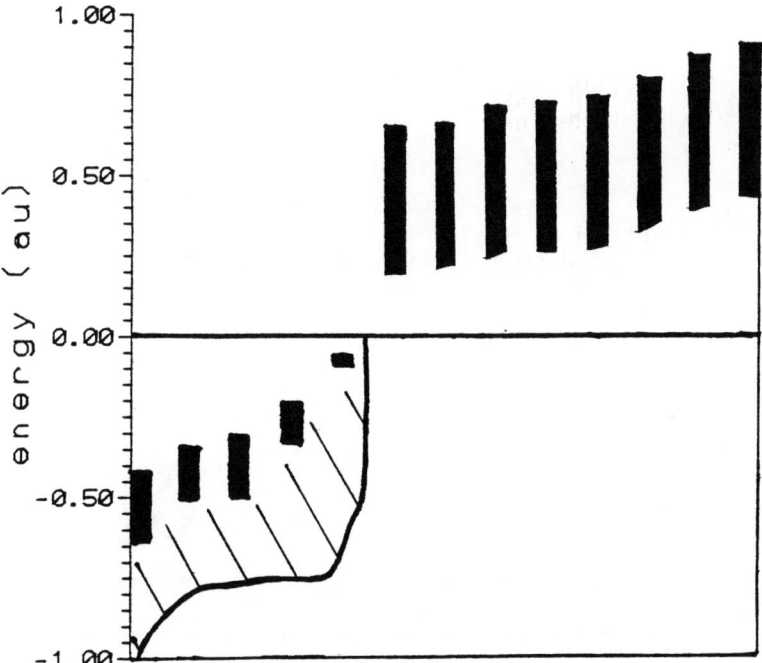

FIGURE II Same than Figure I but for the 13 $Ne^{3+}(4,4)$ and $(4,5)$ levels with symmetry $^1D^e$. The heavy line limitates downward the dashed area.

556 Theoretical Lifetime and Decay Mode

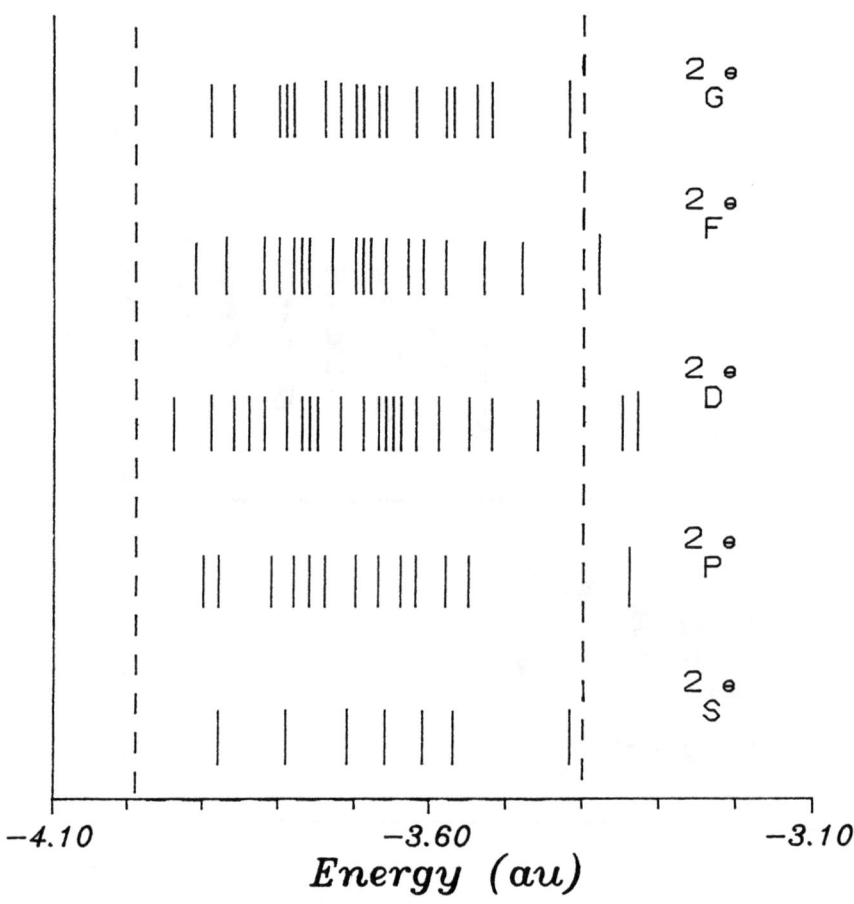

FIGURE III Energies of $N^{4+}(4,4,4)$ series, the two dashed vertical lines represent the limit of the $N^{5+}(3,4)$ series. The zero energy is put at the N^{7+} threshold.

symmetry	$-E_s(au)$	$\Gamma_{AIS}(au)$	$\Gamma_{ATR}(au)$	$\Gamma_{RAD}(au)$
$^1D^e(1)$	5.974	5.66(−4)	2.96(−3)	1.52(−5)
$^1D^e(2)$	5.900	1.03(−3)	8.18(−3)	1.69(−6)
$^1D^e(3)$	5.863	1.97(−4)	9.67(−3)	1.59(−5)
$^1D^e(4)$	5.764	6.47(−5)	1.17(−2)	2.74(−6)
$^1D^e(5)$	5.616	1.69(−4)	4.21(−3)	7.79(−6)
	5.555 (*Ionization threshold* $Ne^{9+}(n=3)$)			
$^1D^e(6)$	4.907	1.87(−4)		4.29(−6)
$^1D^e(7)$	4.900	3.08(−3)		1.18(−5)
$^1D^e(8)$	4.844	6.61(−3)		6.73(−7)
$^1D^e(9)$	4.834	2.79(−4)		4.23(−6)
$^1D^e(10)$	4.819	8.85(−3)		1.27(−5)
$^1D^e(11)$	4.757	5.88(−3)		1.71(−6)
$^1D^e(12)$	4.686	2.55(−3)		4.60(−6)
$^1D^e(13)$	4.652	9.33(−4)		6.63(−6)

TABLE I Energies and widths (associated to autoionization, ATR and direct radiative stabilization) of $Ne^{8+}(4,4)$ and $(4,5)$ $^1D^e$ states. All $(4,5)$ are situated above threshold, they are autoionizing levels.

EXCITATION AND IONIZATION CHARACTERISTICS OF MULTICHARGED IONS

V.P.Shevelko

P.N.Lebedev Physical Institute
Russian Academy of Sciences
Leninsky prospect 53, 117924 Moscow
Russia

ABSTRACT

This short paper is devoted to analytical description of some radiative and collisional characteristics of multicharged ions: photoionization, excitation and ionization by electrons and ions. The corresponding formulas were obtained using accurate calculations and model potentials. The data presented can be used for solving plasma kinetic problems.

INTRODUCTION

Physics of highly charged ions plays an essential role in the investigation of the properties of hot laboratory and astrophysical plasmas. An understanding of elementary processes taking place in high-temperature plasmas is very important for many fundamental aspects as well as the physical applications: fusion research, X-ray spectroscopy, astrophysics, kinetics of non-LTE plasmas etc.

The growing interest in plasma kinetic problems such as line intensities, impurity transport, radiative losses, various diagnostic purposes is stimulated by the needs of astrophysics as well as the development of the new laboratory ion sources and, especially, by challenge of creating the short-wavelength lasers using inversion population in highly charged ions.

The use of numerical methods for solving plasma kinetic problems faces severe obstacles connected with the necessity of including rate coefficients for a large number of atomic states considered. This task is significantly simplified if the rate coefficients required are given in a close analytical form with a minimal number of fitting parameters. Such formulas can be obtained from experimental data or sophisticated calculations of atomic characteristics. Here some analytical formulas for the cross sections and rate coefficients are considered.

PHOTOIONIZATION

One-electron photoionization or, simply, photoionization

$$X_z + \hbar\omega \rightarrow X_{z+1} + e \qquad (1)$$

is investigated in detail theoretically[1] and experimentally. Here z is the spectroscopic symbol of an ion (for neutrals z = 1). Cross sections of n-electron photoionization

$$X_z + \hbar\omega \rightarrow X_{z+n} + ne \qquad (2)$$

are measurable quantities[2,3] as well as the total (or photoabsorption) cross sections[4]

$$\sigma_t(\omega) = \sigma^{abs}(\omega) = \sum_{n=1}^{N} \sigma_n(\omega), \qquad (3)$$

where N is the total number of the target electrons. The quantities $\sigma^{abs}(\omega)$ and $\sigma_n(\omega)$ are very important radiative characteristics which are also required for determination of collisional properties, e.g. multiple electron ionization cross sections by electron and ion impact (see below).

Photoabsorption cross section σ^{abs} is related to oscillator strengths by the well known sum rule[5]

$$\sum_{n'} f_{nn'} + \frac{c}{2\pi^2} \int_I^\infty \sigma^{abs}(\omega) d\omega = N, \qquad (4)$$

where I is the first ionization potential of the target, and with dynamic polarizability $\alpha(\omega)$:

$$\sigma^{abs}(\omega) = \frac{4\pi\omega}{c} \operatorname{Im}\alpha(\omega) \qquad (5)$$

Here c is the speed of light. Dynamic polarizabilities $\alpha(\omega)$ define the van der Waals constant of two interacting atoms in their ground states[6]:

$$C_6 = \frac{3}{\pi} \int_0^\infty \alpha_1(i\omega) \cdot \alpha_2(i\omega) d\omega \qquad (6)$$

Photoionization cross section $\sigma(\omega)$ can be expressed as

$$\sigma(nl) = \frac{Q}{2l+1} n^2 G_{nl} \sigma^{Kr}(n), \qquad (7)$$

where Q is the angular coefficient, σ^{Kr} is the Kramers cross section, G is the Gaunt factor. At high energies the nl-dependence of photoionization cross section is given by[7]:

$$G(nl_0 - \varepsilon l) = (1+x^2)^{-l-1} \exp\left[\frac{4}{\sqrt{\varepsilon_0}}(1-\mathrm{arctg}x/x)\right] \cdot$$

$$\cdot \left[1-e^{-2\pi/\sqrt{\varepsilon_0} x}\right]^{-1} \prod_{s=1}^{l} \frac{1+(xs/n)^2}{1+x^2} \overline{G}, \quad x = (\varepsilon/\varepsilon_0)^{1/2}, \quad \varepsilon \gg \varepsilon_0, \qquad (8)$$

where ε_0 is the binding energy of the initial state, ε is the energy of photoelectron, $\overline{G}$ is the modified Gaunt factor which is constant at high ε (Fig.1). Here $\prod=1$ if $l=0$, and $G=\overline{G}$ if $x=0$.

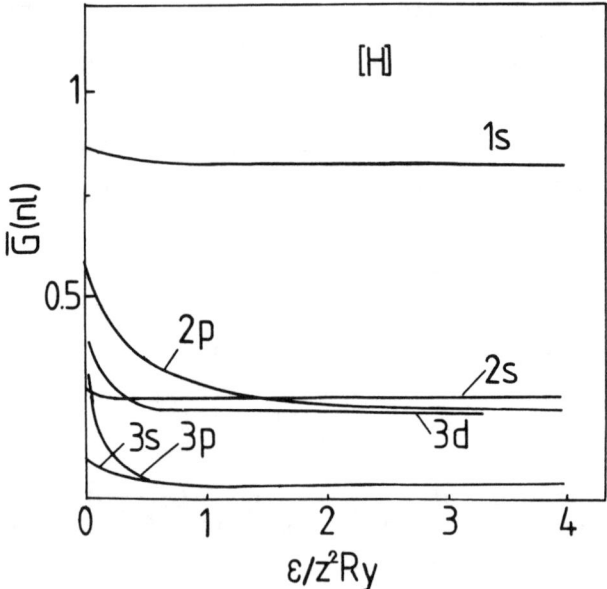

Fig.1. The modified Gaunt factor $\overline{G}(nl)$ for H-like ions as a function of the scaled photoelectron energy ε.

The behaviour of the inverse process - photorecombination or radiative recombination - for highly charged ions was theoretically investigated in[8]. The photorecombination cross section $\sigma^r(nl)$ at low photoelectron energies ε has the following nl-dependence:

$$\sigma^r(nl) \approx \frac{\varepsilon_0}{\varepsilon} \frac{\pi\alpha^3}{3} \left[(l+1) C^2_{l+1}(n,l) + l\, C^2_{l-1}(n,l) \right] [\pi a_0^2] \quad (9)$$

$$\varepsilon/\varepsilon_0 \ll 1, \quad n \leqslant z$$

where α is the fine structure constant, $C(n,l)$ are reduced dipole matrix elements expressed in terms of the confluent hypergeometric functions $F(1-n \pm 1, 2l; 4n)$. According to (9), a free electron is captured predominantly into the states with $l \approx n/2$, if $n \leqslant 10$ and into the states $l \approx n/3$, if $n > 10$.

Calculations of photoionization cross sections for complex ions showed the significant importance of photoexcitation of the core electrons (PEC). PEC processes lead to appearance of large resonances followed by autoionization, and can strongly modify photoionization cross sections over a wide range of frequencies[9].

IONIZATION BY ELECTRON IMPACT

Ionization of ions by electron impact

$$X_z(nl^q) + e \to X_{z+1}(nl^{q-1}) + 2e \quad (10)$$

is described by three types of processes: direct ionization (DI), excitation-autoionization (EA) and resonant ionization (RI).

DI cross section is a smooth function of the incident electron energy E with asymptotics

$$\sigma \sim \begin{cases} u^{3/2}, & u \to 0, \quad z = 1 \\ u, & u \to 0, \quad z > 1 \\ \dfrac{A}{u} + B \dfrac{\ln u}{u}, & u \to \infty, \quad B = \dfrac{1}{\pi\alpha} \int_I^\infty \dfrac{\sigma(\omega)}{\omega} d\omega \end{cases} \quad (11)$$

$$\sigma_{max} \sim I^{-2} \sim z^{-4}, \quad u_{max} \approx 1.5,$$

where $u = E/I - 1$ is a scaled incident electron energy, I is a binding energy of the outer or inner shell of the target

nl^q, q is the number of equivalent electrons, $\sigma(\omega)$ is a photoionization cross section of the target.

DI cross sections σ and Maxwellian rates $<v\sigma>$ have the following scaling laws:

$$I^2\sigma = f(u), \quad u = (E-I)/I \tag{12}$$

$$I^{3/2}e^{\beta}<v\sigma>/q = g(\beta), \quad \beta = I/T, \tag{13}$$

where T is electron temperature. The functions $f(u)$ and $g(\beta)$ have to be universal for ions with a charge $z \leqslant 50$ along a given isoelectronic sequence.

For ions with $z \leqslant 50$ DI cross sections and rates can be estimated using approximation formulas[10]

$$\sigma = q(Ry/I)^2 \frac{Cu}{(u+1)(u+\varphi)}[\pi a_0^2], u = E/I - 1 \tag{14}$$

$$<v\sigma> = 10^{-8} cm^3 s^{-1} q(Ry/I)^{3/2} e^{-\beta}\frac{A\sqrt{\beta}}{\beta + \chi}, \quad \beta = I/T \tag{15}$$

$$0 \leqslant u \leqslant 16, \quad 1/8 \leqslant \beta \leqslant 8 \tag{16}$$

Fitting parameters C, φ, A, χ for nl-states (nl $\leqslant$ 6h) obtained recently[11] on the basis of the accurate CBE calculations[12] are given in Table I. The fitting parameters for the l-averaged cross sections and rate coefficients are given in Table II. Fig.2 shows a contribution of DI estimated with the help of Table I in the case of ionization of Ni^{14+} ions from excited 3p3d state.

For ions with $z > 50$ it is necessary to take into account relativistic effects such as electromagnetic and retardation interactions[15] using sophisticated numerical methods (e.g. MCDF). The similar situation takes place also for ionization of inner-shell electrons in neutral atoms: for heavy atoms with nuclear charge $Z_n > 30$ relativistic effects were found to be very important.

For target ion having more than one electron shell the total ionization cross section can contain a structure on top of DI cross section which is due to indirect multistep) ionization mechanisms. One of such mechanism is excitation of the inner-shell electrons into autoionizing

Table I

Fitting parameters C, φ and A, χ (eqs.(14-15)) for ionization of multicharged ions from nl-states by electron impact.

nl-state	C	φ	A	χ
1s	7.96	2.70	5.65	0.40
2s	6.69	2.03	6.23	0.52
2p	6.93	1.47	9.05	0.73
3s	6.00	1.59	7.37	0.70
3p	6.24	1.31	9.11	0.82
3d	6.57	1.08	11.7	1.00
4s	5.77	1.43	7.76	0.76
4p	6.00	1.26	9.11	0.86
4d	6.23	1.11	10.8	0.97
4f	7.06	1.00	13.5	1.07
5s	5.66	1.36	7.96	0.79
5p	5.88	1.23	9.13	0.87
5d	6.08	1.12	10.4	0.96
5f	6.26	1.08	11.1	1.00
5g	6.47	1.04	11.9	1.03
6s	5.60	1.32	8.07	0.80
6p	5.82	1.22	9.13	0.88
6d	6.00	1.13	10.2	0.95
6f	6.24	1.07	11.2	1.00
6g	6.33	1.04	11.7	1.03
6h	6.44	1.01	12.2	1.06

Table II

The same as in Table I, for the l-averaged cross sections σ and rates $\langle v\sigma \rangle$.

n-state	C	φ	A	χ
n=1	7.96	2.70	5.65	0.40
2	6.82	1.55	8.33	0.68
3	6.44	1.23	10.2	0.90
4	6.30	1.11	10.9	0.97
5	6.24	1.06	11.2	1.01
6	6.21	1.03	11.4	1.03

states followed by autoionization decay:

$$X_z + e \rightarrow X_z^{**} + e \rightarrow X_{z+1} + 2e \qquad (17)$$

which is termed excitation-autoionization (EA). For example, for Li-like ions EA processes are

$$X_z(1s^22s) + e \rightarrow [X_z(1s2snl)]^{**} + e \rightarrow X_{z+1}(1s^2) + 2e, \quad n \geq 2 \quad (18)$$

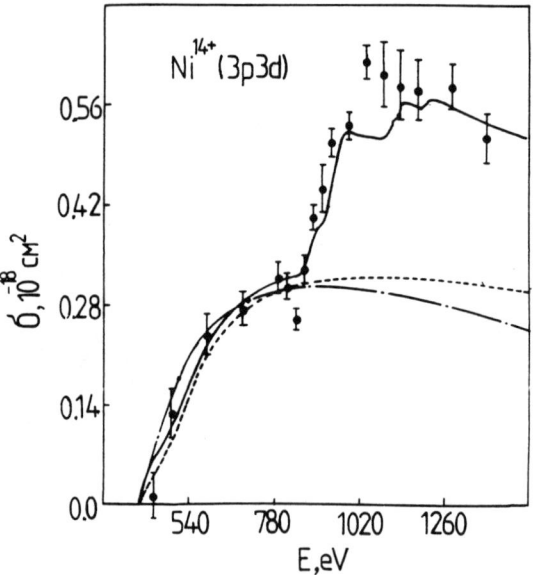

Fig 2. Ionization cross section for Ni^{14+} ions from excited 3p3d state: dashed curve - calculated DI contribution[13], solid curve - the total cross section[13], dot-dashed curve - eq.(14), points - experiment[14].

Besides DI and EA there are another important indirect processes leading to ionization but having resonant nature, i.e. which take place only at definite (resonance) energy of the incident electron. Such resonant processes are also multi-step processes and lead to appearance of narrow resonances on the total ionization cross section. In the case of Li-like ions one has

$$X_z(1s^22s) + e \rightarrow [X_{z-1}(1s2s2p2l)]^{**} \rightarrow X_{z+1}(1s^2) + 2e \quad (19)$$

$$X_z(1s^22s) + e \rightarrow [X_{z-1}(1s2s3p3l)]^{**} \rightarrow$$

$$\rightarrow [X_z(1s2s2l')]^* + e \rightarrow X_{z+1}(1s^2) + 2e \quad (20)$$

The process (19) is termed resonant-excitation-auto-double-ionization (READI), the process (20) is termed resonant-excitation-double-autoionization (REDA). The RI processes were predicted theoretically[16,17] and confirmed experimentally[18]. To perform the experimental measurements was possible only using advanced technique, which, in particular, permits to study narrow features like resonances and excitation thresholds with extremely high precision, i.e. also to obtain spectroscopic information on multiply excited states[19].

The contributions from DI, EA and RI processes depend on ion charge and its electronic configuration. The largest contribution from REDA was measured for ionization of FeXVI ions, where REDA gives up to 30% contribution to the total ionization cross section.

EXCITATION

Excitation of ions by electrons

$$X_z + e \rightarrow X_z^* + e \qquad (21)$$

is described by three main processes: excitation of outer-shell electrons, excitation of inner-shell electrons and resonant excitation.

For dipole (optically allowed) transitions ($\Delta l = \pm 1$, $\Delta S = 0$) the excitation cross sections are finite at threshold and, as a rule, have their maximum there, which can be estimated from the Van Regemorter formula

$$\sigma_{th}^{dip} \approx 2.9 f / (\Delta E/Ry)^2 \quad [\pi a_0^2] \qquad (22)$$

where ΔE is a transition energy, f is an oscillator strength. At high electron energy $E \gg \Delta E$ the dipole cross section falls off according to the Bethe formula:

$$\sigma^{dip} \approx \frac{A}{E} + B \frac{\ln E}{E}, \quad E \gg \Delta E \qquad (23)$$

where A and B are constant: B is related to oscillator strength (B ~ f) and A is obtained from numerical calculations.

At present, absolute experimental excitation cross sections σ are known only for a few ions with charge greater than one: Al^{2+}, C^{3+}, N^{4+}, Hg^{2+} and Si^{3+}.

For ions with z > 50 the basic methods have to include close-coupling, relativistic effects and even QED corrections. QED effects lead to enormous increasing of the excitation cross sections from threshold up to high energy region due to electromagnetic and retardation interactions in highly charged ions[20].

Excitation of inner-shell electrons has properties similar to excitation of outer-shell electrons and is very important for ionization due to excitation-autoionization (EA) processes. EA involves direct excitation of inner-shell electrons and a production of ionized ion

$$X_z + e \to X_z^{**} + e \to X_{z+1} + 2e \qquad (24)$$

Because EA cross sections are non-zero at threshold they give distinct steps on top of the ionization cross sections. EA cross sections can be obtained from the total experimental ionization cross sections and theoretical direct ionization ones (Table III).

To estimate the dipole excitation cross sections and rate coefficients for transitions from outer and inner shells the analytical formulas were suggested[21]. The formulas are based on the model interaction potential

$$V(R) = -\frac{\lambda_\varkappa R^\varkappa}{(R^2 + R_0^2)^{\varkappa + 1/2}} \qquad \varkappa \neq 0 \qquad (25)$$

where λ is the interaction constant, R_0 is the effective radius. For dipole transitions ($\varkappa = 1$) the suggested expression for R_0 has the form:

$$R_0 = \frac{n_0^* n_1^*}{z(n_1^* - n_0^* + 0.5)} \; [a_0], \qquad n^* = \frac{z}{\sqrt{E_{nl}/Ry}} \qquad (26)$$

for transitions from outer shell, and

$$R_0 = \frac{n_1^*}{2\sqrt{I_0/Ry}} \; [a_0], \qquad (27)$$

for transitions from inner shells, where n^* are effective quantum numbers, I_0 is the binding energy of the inner-shell electron.

Table III

The sum of inner-shell excitation cross sections at threshold $\sigma_{th}(1s^2 2s-1s2s^2) + \sigma_{th}(1s^2 2s-1s2s2p)$ for Li-like ions (in units of $10^{-19} cm^2$) (for references see[21]).

Ion	Close-coupling of six states	CBE	Experiment
BeII	9.3	11.5	20.0± 8.0
BIII	4.1	6.96	4.0± 1.0
CIV	2.24	3.77	2.3± 0.7
NV	1.27	1.98	1.6± 0.4
OVI	0.74	1.07	0.8± 0.3

The dipole cross section for ions can be described in the whole energy range E by the formula:

$$\frac{\sigma(E)}{\pi a_0^2} = \frac{8f}{E \cdot \Delta E} \Phi(x), \quad \Phi(x) = \frac{x^2}{2}\left[K_0(x)K_1(x) - K_1^2(x)\right], \quad x = \frac{\Delta E \cdot R_0}{2\sqrt{E}} \quad (28)$$

where f is a oscillator strength, K(x) is the modified Bessel function; energies are in Ry units. At threshold

$$\sigma_{th}(E=\Delta E) = const = \frac{8f}{(\Delta E)^2} \Phi(x_0)[\pi a_0^2], \quad x_0 = \frac{\sqrt{\Delta E} \cdot R_0}{2} \quad (29)$$

At high electron energies $E \gg \Delta E$ eq.(28) gives

$$\sigma(E) = \frac{8f}{E \cdot \Delta E} \ln\frac{1.36\sqrt{E}}{\Delta E \cdot R_0}[\pi a_0^2], \quad E \gg \Delta E \quad (30)$$

The rate coefficient, corresponding to (28) is given by

$$<v\sigma> = 1.74 \cdot 10^{-7} \text{cm}^3/\text{s} \frac{f v \bar{\beta}}{z \Delta E} e^{-\Delta E/T} \ln \frac{1.26 z}{\Delta E \cdot R_0 \sqrt{\bar{\beta}}}, \quad \beta = z^2 \text{Ry}/T \quad (31)$$

where T is an electron temperature.

Due to the Coulomb force, a multicharged ion can capture a free electron and then decay by autoionization channel:

$$X_z(\alpha_0) + e \to X_{z-1}^{**}(\gamma) \to X_z^*(\alpha_1) + e, \quad \gamma = \alpha n l \quad (32)$$

This process is termed resonant excitation, because it has a resonance nature and is possible only if a free electron energy E is equal to the quantity:

$$E_{res} \approx E_{\alpha \alpha_0} - (z-1)^2/n^2 \quad (33)$$

Experimental data on RE cross sections are rather scarce, therefore most of the results are obtained by theoretical calculations. There are three main theoretical methods which lead to equivalent results: R-matrix[22,23], the generalized quantum defect method and the asymptotic expansion approach.

RE leads to appearance of resonance structure on top of the 'usual' potential cross sections. The contribution of resonances is rather small for optically allowed transitions (e.g. $1^2S - 2^2P$ in H-like ions), other strong transitions and transitions between highly excited states. The resonance contribution to the total excitation cross section can be very important for weak transitions (e.g. intercombination transitions with $\Delta S=1$), especially at low temperatures (Fig.3).

The total contribution of RE to the rate coefficient for transition $\alpha_0 - \alpha_1$ in a plasma with a Maxwellian velocity distribution can be estimated by

$$<v\sigma_{res}> = \frac{g(\gamma) 4\pi^{3/2} a_0^3 W(\gamma, \alpha_0) W(\gamma, \alpha_1)}{g(\alpha_0)(\Delta E/\text{Ry})^2 (A(\gamma) + W(\gamma))} \beta^{3/2} e^{-E_{res}/T}, \quad (34)$$

$\Delta E = E_{\alpha_0} - E_{\alpha_1}$, $\beta = \Delta E/T$, $\gamma = \alpha n l$,

where T is a plasma temperature, g is the statistical

weight, A and W are radiative and autoionization probabilities.

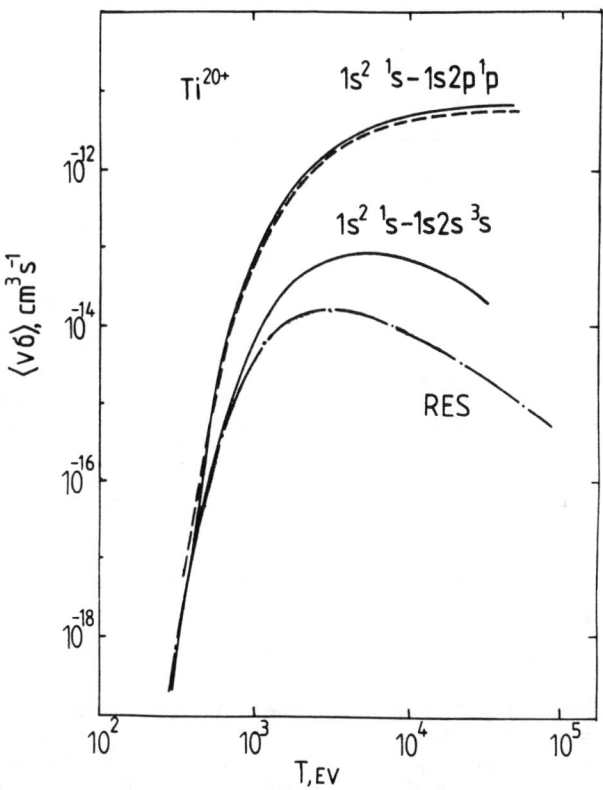

Fig.3. Excitation rates for transitions in Ti^{+20} ions: solid curves - CBE calculations, dot-dashed curve - resonant excitation rate for transition $1s^2$-1s2s, dashed curve - eq. (34) (the code ATOM[10])

The values $<v\sigma_{res}>$ can be fitted by the formula[7]

$$<v\sigma_{res}> = 10^{-13} cm^3 s^{-1} A\beta^{3/2} e^{-\beta\chi}, \quad \beta=z^2 Ry/T, \quad (35)$$

where A and χ are fitting parameters. Parameters A and χ for some He-like ions are given in Table IV.

Table IV

Fitting parameters A and χ (35) for transitions in He-like ions[7] : $0.1 \leqslant \beta \leqslant 10$.

Transition	MgXI		FeXXV	
	A	χ	A	χ
$1s^2\ {}^1S_0 - 1s2s\ {}^3S_1$	1.82	0.85	0.16	0.85
$1s2s\ {}^1S_0$	1.82	0.85	0.16	0.85
$1s2p\ {}^3P_0$	0.94	0.85	0.10	0.85
$1s2p\ {}^3P_1$	2.82	0.85	0.29	0.85
$1s2p\ {}^3P_2$	4.71	0.85	0.48	0.85
$1s2p\ {}^1P_1$	3.53	0.85	0.15	0.85

MULTIPLE IONIZATION OF HEAVY ATOMS BY FAST IONS

Multiple ionization (MI) processes arising in fast ion-atom collisions

$$X^{q+} + A \rightarrow X^{q+} + A^{n+} + ne \qquad (36)$$

are of interest both for fundamental reasons and practical applications. MI is a very effective mechanism for producing highly charged, very slow recoil ions with a large cross section σ in a single encounter. For example, $\sigma \approx 10^{-17} cm^2$ for producing recoil ions Xe^{+25} in U^{+75} + Xe collisions[24].

Available experimental data[24-28] provide the global information on the basic parameters of MI: the recoil ion charge state n production (n=0-35) correlated with the charge q(q=1-92) of the projectile and its energy E (1 keV/u <E < 1.5 GeV/u). The first experimental measurements and theoretical consideration, devoted to this subject, have been performed in the pioneering work by Cocke[25].

The basic mechanisms of MI are not yet fully understood. There are two main approaches in theory of MI: Classical Trajectory Monte-Carlo (CTMC) method[26] and semiclassical independent model[29]. Both theories consider the target electrons as classical particles disregarding the electron-electron correlations and do not comprise a Bethe logarithmic term $\ln E/E$ in the asymptotic behaviour of MI processes. However, CTMC calculations made it possible to evaluate a scaling law for the total ionization cross section σ_t in the form

$$\sigma_t/q = f(E/q), \qquad \sigma_t = \sum_n \sigma_n \qquad (37)$$

Another treatment was suggested in[30] as a modification of the classical impulse approximation in the impact parameter representation where the n-electron ionization cross section σ_n is presented as a sum of contributions from distant (dis) and close (cl) collisions:

$$\sigma_n = \sigma_n^{dis} + \sigma_n^{cl}, \qquad \sigma_n^{dis} = 2\pi \int_0^\infty P_n^{dis}(\rho)\rho d\rho \qquad (38)$$

$$P_n^{dis}(\rho) = \frac{c}{\pi^2} \frac{q^2}{(\pi\rho)^2} \int_I^\infty \sigma_n(\omega) S(\omega v/\rho) \frac{d\omega}{\omega}, \qquad \rho \gg q/v \qquad (39)$$

$$S(x) = x^2 \left[K_0^2(x) + K_1^2(x) \right] \qquad (40)$$

Here v is a relative velocity of colliding particles, $\sigma_n(\omega)$ is a n-electron photoionization cross section of an atom A at a frequency ω, $S(x)$ is a dimensionless function describing a Fourier component of electric field induced by the projectile, I is the first ionization potential.

Each of the terms in (38) has asymptotic

$$\sigma_n^{dis} \approx (q/v)^2 \left(A_n \ln v + B_n^{dis} \right), \qquad v/q \gg 1 \qquad (41)$$

$$\sigma_n^{cl} \approx (q/v)^{2n} B_n^{cl}, \qquad v/q \gg 1 \qquad (42)$$

while the total cross section has the asymptotic behaviour

$$\sigma_t = \sum_n \sigma_n \approx \frac{A}{v^2} + \frac{B \ln v}{v^2}, \qquad B = \frac{1}{\pi\alpha} \int_I^\infty \frac{\sigma(\omega)}{\omega} d\omega, \qquad v/q \gg 1 \qquad (43)$$

The treatment[30] gives the same scaling law (37) for the total ionization cross section. The use of the Thomas-Fermi model for electron density of the target atom leads to the scaling law both on the projectile charge q and the nuclear charge of the target Z_n:

$$\frac{Z_n^2}{q} \sigma_t = F\left(\frac{v}{q^{1/2} Z_n^{3/4}}\right) \quad (44)$$

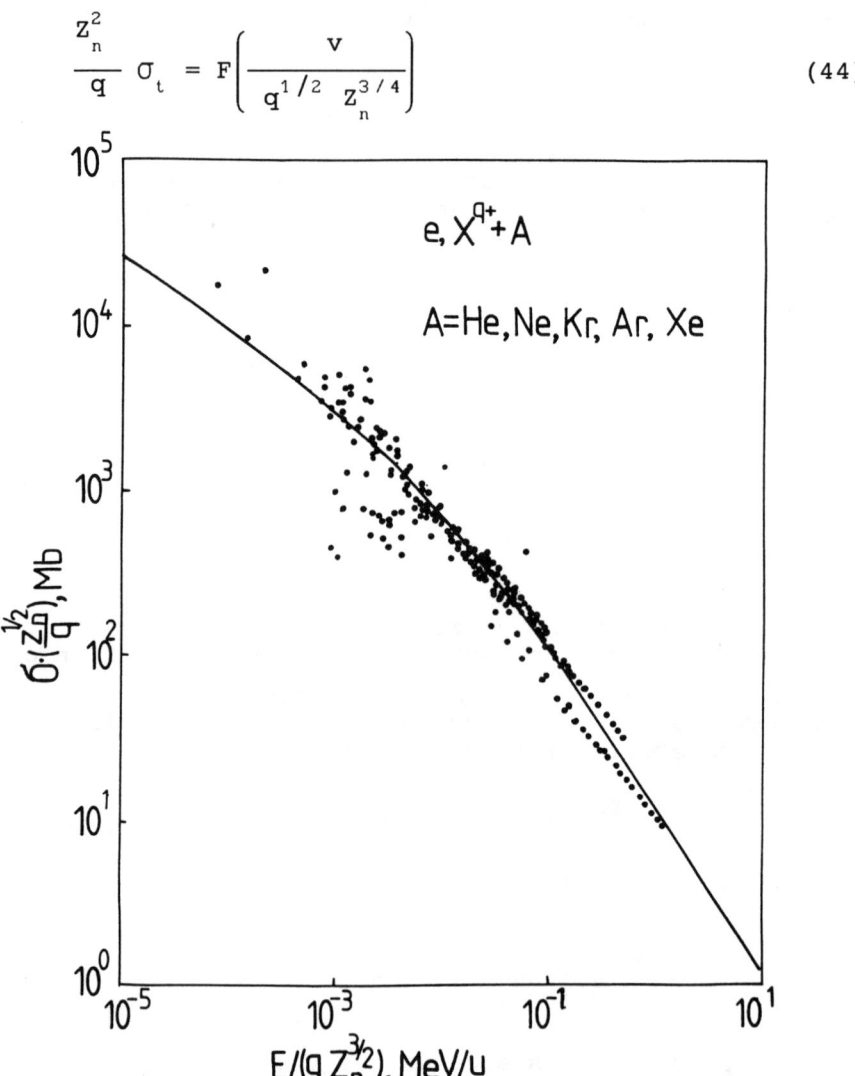

Fig.4. The total universal ionization cross section σ_t (44) for collisions of electrons and highly charged ions with the inert-gaseous targets.

REFERENCES

1. M.Ya.Amusia, Atomic Photoeffect (Pergamon,Oxford,1990).
2. D.M.P.Holland, K.Codling, J.B.West, G.V.Marr, J.Phys. B12, 2465 (1979).
3. M.J.van der Wiel and G.Wiebes, Physica,53, 225 (1971).
4. E.B.Solomon and J.B.Hubell, ADNDT, 38, 1 (1988).
5. I.I.Sobelman, Atomic Spectra and Radiative Transitions (Springer, Berlin, 1992), (2nd edition).
6. H.B.G.Casimir and P.Polder, Phys.Rev.73, 360 (1948).
7. Proc. Lebedev Institute, Moscow, Vol. 218 (ed. I.I.Sobelman), (Nova Science Publ., N.Y., 1993).
8. M.Pajek and R.Schuch, Phys.Rev.A45, 7894 (1992).
9. J.A.Tully, M.J.Seaton and K.A.Berrington, J. de Physique, IV, C1-169 (1991).
10. V.P.Shevelko and L.A.Vainshtein, Atomic Physics for Hot Plasmas (Adam Hilger, Bristol, 1993).
11. A.Barany, F.Nurhussen, I.L.Beigman and V.P.Shevelko, Physica Scripta (1993, to be submitted).
12. R.E.H.Clarke and J.Abdallah, Jr, Phys.Scr.T37, 28 (1991).
13. M.S.Pindzola, D.C.Griffin, C.Bottcher et al., ORNL-TM-11202 (Oak Ridge Natl. Lab.,Tennessee, 1990).
14. L.J.Wang, K.Rinn, and D.C.Gregory, J.Phys. B21, 2117 (1988).
15. D.L.Moores, M.S.Pindzola, Phys.Rev.A41, 1375 (1990).
16. K.J.LaGattuta and Y.Hahn, Phys.Rev.A24, 2273 (1981).
17. R.J.W.Henry, A.Z.Msezane, Phys. Rev.A26,2545 (1982).
18. A.Muller, K.Tinschert, G.Hofmann and E.Salzborn, Phys. Rev. Lett.61, 70 (1988); ibid.p. 1352.
19. A.Muller, Ionization in Ion-Electron Collisions: In: Physics of Ion Impact Phenomena (ed. D. Mathur) (Springer, Berlin 1991).
20. M.S.Pindzola, N.R.Badnell, D.L.Moores and D.C.Griffin, Z.Phys. D21, S23 (1991).
21. V.P.Shevelko, Phys. Scr.46, 531 (1992).
22. P.G.Burke and W.D.Robb, Adv.Atom.Mol.Phys.11,143 (1975).
23. K.A.Berrington, Phys.Scr.T37,19 (1991).
24. S.Kelbch, J.Ullrich, R.Mann, et al., J.Phys. B18, 323 (1985).
25. C.L.Cocke, Phys.Rev.A20, 749 (1979).
26. A.S.Schlachter, K.H.Berkner, H.F.Beyer et al., Phys.Scr.T3, 153 (1983).
27. A.Muller, B.Schuch, W.Groh and E.Salzborn, Z.Phys. D7, 251 (1987).
28. R.E.Olson, J.Ullrich and H.Schmidt-Bocking, Phys.Rev. A39, 5572 (1989).
29. M.Horbatsch, Phys.B25, 3797 (1992).
30. O.I.Tolstikhin, D.B.Uskov and V.P.Shevelko, Proc. VIth Int. Conf. Phys. Highly Charged Ions, p. B3 (KSU Manhattan, 1992).

CHARGE CHANGING PROCESSES IN ION-ION COLLISIONS

Frank Melchert
Institut für Kernphysik, Universität Giessen
D–35392 Giessen, Germany

ABSTRACT

Crossed-beam studies deliver detailed information about collision processes between ions. This report is concerned with collisions between negative ions, between negative and positive ions, and between positive ions at keV impact energies. For the latter case, very recent results of angular differential cross section measurements are reported.

INTRODUCTION

Collisions between projectile ion beams and target atoms or molecules have been studied extensively in the past decades. The neutrality of these targets allows high target densities ranging from relatively thin gaseous targets up to solid state densities. The use of well defined ionic targets reduces the achievable target densities drastically due to Coulomb repulsion between the ions. Nevertheless, ion-ion collisions are of vital interest for the understanding and realistic description of plasma dynamics. The impetus for accurate cross section data has increased in recent years due to research on controlled thermonuclear fusion using either magnetically or inertially confined plasmas.

In magnetic fusion, there is need for auxilliary heating of the fusion plasma by neutral beam injection. Next generation of fusion devices require injection of multi-megawatt neutral $H^o(D^o)$ beams at MeV energies. Neutralization of $H^-(D^-)$ beams in plasma neutralizers[1,2] offers considerably higher neutralization efficiency than gas neutralizers (85% vs 60%). However, design and modelling studies for plasma neutralizers have suffered from the lack of cross section data for single- and double-electron removal from H^- in energetic collisions with multiply-charged positive ions.

In inertial confinement fusion, intense heavy ion beams of Bi^+ or Cs^+ are proposed[3] drivers for igniting a DT-pellet. In the storage rings, severe intensity losses due to collisions between ions within the beam pulses may occur[4]. In order to obtain an accurate assessment of the expected loss rates, the cross sections both for electron capture and for ionization in collisions between identical heavy ions have to be known.

The theoretical description of ion-ion collisions is not as developed as for collisions between ions and atoms. For interactions between negative ions, not a single quantum mechanical calculation exists. Charge exchange between protons and He^+ ions, the simplest ion-ion collision system, is fairly well investigated, both theoretically and experimentally[5], with respect to absolute cross sections. Up to now, not a single experiment delivered differential cross sections for this fundamental collision system. Theoretical approaches which predict reliable results for absolute cross sections have to undergo a stringent test by describing angular differential cross sections.

In this paper, first systematic experimental investigations are reported for the position sensitive detection of reaction products produced in $Cs^+ + Cs^+$ and $He^+ + He^+$ collisions. Advanced experimental technique allows to observe scattering distributions from which angular differential cross sections can be extracted in many cases. On the other hand, this technique resolves speculations about the absolute or partial character of previously measured cross sections[23].

For detailed discussions the reader is referred to several review reports on ion-ion collisions[6–8].

EXPERIMENTAL TECHNIQUE

The experiments discussed here make use of the crossed-beams technique. The basic principles are simple but their realization is invariably complicated. Two well collimated beams are made to intersect and each beam is usually selected by a magnetic analyzer so that it contains only one type of ion with a well defined energy. Reactions occur when the beams collide, but as the energy transferred between reactants is much smaller than the beam energy, the reaction products remain within the beam until they can conveniently be separated and measured.

Low particle density in ion beams gives rise to most of the experimental difficulties. Beam densities are limited by space charge effects, by the availability of high intensity, high brightness ion sources and will rarely exceed $10^7 \frac{ions}{cm^3}$. Consequently the fraction of ions which react can be only 10^{-13} and the rate at which the required collisions are produced is typically between 0.01 and 100 $\frac{reactions}{sec}$. Especially the position sensitive detection of these reaction products is very time consuming and requires stability of all experimental components. Single particle detection and ultra-high-vacuum techniques ($p \approx 10^{-11} mbar$) are absolutely necessary, but even so, the particle density in the beams is often less than in the vacuum through which the beams pass. Therefore the ion beams have to be cleaned shortly before interaction and have to be analyzed as soon as possible afterwards. Nevertheless, beam modulation techniques[9,10] or coincidence methodes[5] have to be applied in order to seperate signal from background events.

In spite of the need for complex apparatus, intersecting-beam techniques are attractive and highly developed. The type of reactant and the collision geometry can usually be sharply defined. Accurate, reliable data can be obtained and the interaction energy spans several orders of magnitude (10^{-1} to 10^6 eV). The cm-energy (center of mass frame) and the energy resolution depend greatly upon the angle ϑ at which the beams are made to intersect. The energy in the cm-frame is given by

$$E_{cm} = \frac{m_1 \cdot m_2}{m_1 + m_2} \cdot \left[\frac{E_1}{m_1} + \frac{E_2}{m_2} - 2\sqrt{\frac{E_1 E_2}{m_1 m_2}} \cdot \cos\vartheta \right] \qquad (1)$$

where E_i and m_i represent the energy and the mass of the particles. The attainable energy resolution δE_{cm} depends upon spreads in the laboratory energies δE_1 and δE_2 for ϑ=const

$$\delta E_{cm} = \frac{m_1 \cdot m_2}{m_1 + m_2} \cdot \left[\frac{1}{m_1} - \sqrt{\frac{E_2}{m_1 m_2 E_1}} \cdot \cos\vartheta \right] \cdot \delta E_1$$

$$+ \frac{m_1 \cdot m_2}{m_1 + m_2} \cdot \left[\frac{1}{m_2} - \sqrt{\frac{E_1}{m_1 m_2 E_2}} \cdot \cos\vartheta \right] \cdot \delta E_2 \qquad (2)$$

For small interaction angles ϑ, the effect of imperfect collimation $\delta\vartheta \neq 0$ yields an energy shift

$$E_\vartheta = E_0 + \frac{m_1 \cdot m_2}{m_1 + m_2} \cdot \sqrt{\frac{E_1 E_2}{m_1 m_2}} \cdot \vartheta^2 \qquad (3)$$

where E_ϑ and E_0 are values of E_{cm} for intersecting angles of ϑ and 0.

Merged-beam experiments give access to low collision energies and to extremely good energy resolution (e.g. $E_{cm} = 0.25 eV$ and $\delta E_{cm} = 0.10 eV$ for identical masses $m_1 = m_2$; $E_1 = 5 keV$, $E_2 = 5.1 keV$; $\delta\vartheta = \delta E_1 = 0$; $\delta E_2 \approx 20 eV$). The absolute cm-energy spread is a factor of 50 lower than the lab frame beam energy spread. In addition to these advantages, merged beams permit particles to interact over a path length much longer than the beam diameter so that the number of collisions may be large. However, the effective interaction length is shorter than the geometrical merged beam path resulting in a vanishing signal in the extreme case of identical velocities where no collision will occur. Presently, ion-ion merged-beam experiments of Prof.Brouillard's group[11] and Peart et al. [12] are operational.

Inclined-beam experiments bypass the problem of beam overlap determination which is a sensitive point in merged-beam experiments. The price is a reduced signal due to the shorter interaction length in the order of the beam diameter. The first inclined beams experiment was set up in 1969 by Rundel et al.[13] to study mutual neutralization of H^+ and H^- ions.

This report concentrates on the arrangement ($\vartheta = 45°$) used in Giessen[5] (fig.1). Two momentum-analyzed and well collimated ion beams are made to intersect in an ultra-high vacuum region of order 10^{-11}mbar. Both ion beams are cleaned, shortly before intersection, by electrostatic deflectors E1, E2 from particles in other charge states resulting from charge-changing collisions in the residual gas on the path downstream from the analyzing magnets. This precaution is imperative in order to reduce background events. Immediately downstream from the interaction region the reaction products formed in both beams are separated from their parent ion beams.

In the slow beam line (energies up to 20 kV · q) a two stage electrostatic analyzer system E4, E5 is used to separate A^{2+} collision products from the A^+ parent beam. The product A^0 atoms pass undeflected through the ion-optical mirror E4. In this beam line also multiply-charged ions A^{q+} produced in a 5 GHz ECR ion source are available.

In the fast beam line (energies up to 400kV · q) an electrostatic deflector separates B^0 atoms or B^{2+} ions from the parent B^+ ion beam. In case of incident negative ions X^- both reaction products X^0 and X^+ can be identified. The parent ion beams are recorded by biased Faraday cups F1, F2 whilst the reaction products are counted individually either by channeltron-based single particle detectors D1, D3, D4 or by a channelplates-detector D2 that uses a position sensitive resistive anode. In some experiments, the position sensitive detector D2 replaced detector D3 to measure the scattering distribution of A^0 atoms. The region of beam collimation and beam dumps is pumped separately in order to avoid degradation of the low residual gas pressure in the interaction region.

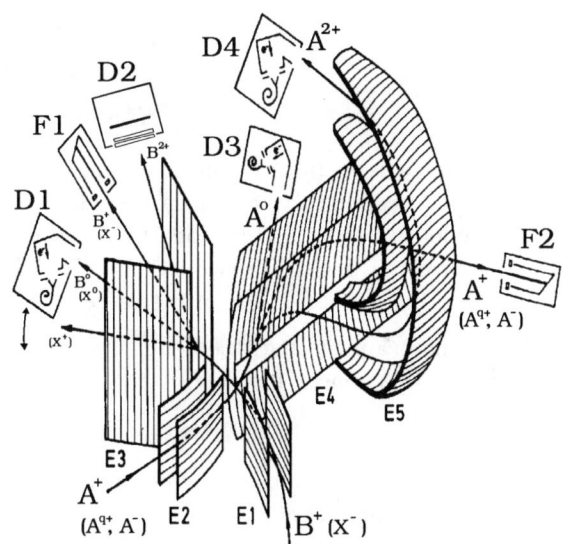

Figure 1:
Crossed-beams arrangement used in Giessen for the collision systems $A^+ + B^+$, $A^{q+} + X^-$, and $A^- + X^-$.

In order to discriminate the ion-ion signal from background counts which are normally some orders of magnitude more frequent, a coincidence technique and a beam pulsing technique are used.

A carefull charge state analysis is required to separate particles which have changed their charge state in ion-ion collisions or in ion-residual gas interactions from their parent ion beams. This intensity ratio can easily exceed 10^{-10}. Since ion-ion collisions take place at the well defined interaction point, fixed flight times to the corresponding detectors yield a constant time delay between both detector signals, in contrast to randomly produced background signals.

If one reactant remains in its charge state, total cross sections for the formation of the other reactant can be determined using a beam pulsing technique for signal recovery. The adequate count rates are recorded as a function of the beam pulsing sequence to distinguish between signal and background events (Refer to Rinn et al[10] for details).

RESULTS

1. Collisions between negative ions

The simplest case in which two negative ions are made to interact is the $H^- + H^-$ collision system. Here, the following reaction channels are possible:

$$H^- + H^- \longrightarrow H^0 + H^- + e^- \quad (\sigma_{o-}) \quad (4)$$
$$\longrightarrow H^0 + H^0 + 2e^- \quad (\sigma_{oo}) \quad (5)$$
$$\longrightarrow H^0 + H^+ + 3e^- \quad (\sigma_{o+}) \quad (6)$$
$$\longrightarrow H^+ + H^+ + 4e^- \quad (\sigma_{++}) \quad (7)$$

Absolute cross sections for mutual ionization (σ_{oo}, reaction 5) and triple ionization (σ_{o+}, reaction 6) have been measured using the coincidence technique. Quadruple ionization (σ_{++}, reaction 7) will be smaller than triple ionization since its endothermicity is nearly twice as big. CTMC calculations[14] predict for quadruple ionization maximal 30% of the values for triple ionization.

Using the beam pulsing technique, absolute cross sections for the total electron loss (σ_{tot}) have been measured

$$H^- + H^- \longrightarrow H^0 + ... \quad (\sigma_{tot}) \quad (8)$$

Following reaction 8, absolute cross sections for the detachment (σ_{o-}, reaction 4) can be determined from

$$\sigma_{o-} = \sigma_{tot} - \sigma_{oo} - \sigma_{o+} \quad (9)$$

The experimental results for reactions 4-6 are shown in figure 2. As expected, the cross sections drop with the number of electrons released. The general behavior can be reproduced by recent CTMC-calculations[15] using the model potentials V_{el} (full curve) and V_{eff} (broken curve):

$$V_{el} = -\frac{0.405}{r}[(1+\alpha r)e^{-2\alpha r} + (1+r)e^{-2r} + 0.94(3+\alpha)e^{-(\alpha+1)r}] \quad (10)$$

$$V_{eff} = -\frac{\alpha}{r} \quad ; \quad \alpha = 0.2354 \quad (11)$$

Theory underestimates the experimental results for the detachment (σ_{o-}) channel, but reproduces the energy dependence. For the mutual ionization channel (σ_{oo}), the solid line lies on top of the data except for collision energies lower than 2 keV where theory and experiment seem to follow different trends. Up to now, the reason for this is not clear. For triple ionization (σ_{o+}) the experimental uncertainties are quite big due to the small cross sections, but the CTMC calculations seem to overestimate the experimental results.

Experimental investigations were performed to study mutual ionization (σ_{oo}) in collision of H^- ions with various other negative ions:

$$H^- + X^- \longrightarrow H^o + X^o \quad (12)$$

$X^- = B^-, Cu^-, O^-, Au^-, F^-$

Compared to the $H^- + H^-$ collision system, preliminary experimental results show a similar dependence on collision energy. The observed cross sections scale roughly with the endothermicity of the pertinent reaction. However, deviations occur for the O^- ion and the metallic ions.

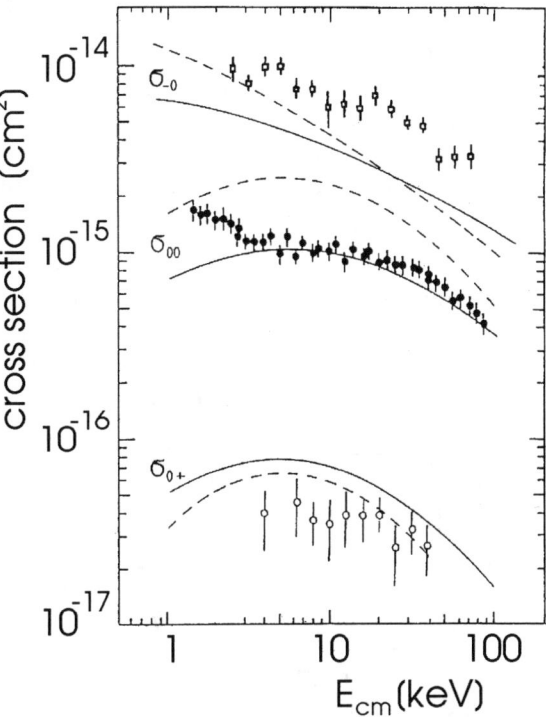

Figure 2:
Cross sections σ_{-o}, σ_{oo} and σ_{o+} for collisions between H^- ions. The lines represent CTMC-calculations[15] using the model potentials V_{el} (full curve) and V_{eff} (broken curve).

2. Collisions between H^- and multiply charged ions

In order to inject intense neutral beams into magnetically confined fusion plasmas, neutralization of the accelerated charged particles is required. Modelling and design studies for plasma neutralizers[16] have suffered from the lack of experimental cross sections for the single-electron removal

$$H^- + X^{q+} \longrightarrow H^o + ... \quad (\sigma_{-o}) \quad (13)$$

and the double-electron removal reactions

$$H^- + X^{q+} \longrightarrow H^+ + ... \quad (\sigma_{-+}) \quad (14)$$

The first reaction feeds the desired neutralization channel, the latter reaction directly degrades the efficiency of a plasma neutralizer. Experimental results for reaction 13 and 14 are shown in figure 3. As the charge state of the ion increases, the q^2 scaling predicted by the first-order Born approximation (Kim et al.[17]) for the single-electron removal becomes increasingly invalid even though the collision energy of 50 keV/u is well above the projectile-electron velocity-matching region of ≈ 1.4 keV/u. The reason for the lack of q^2 scaling is the saturation of transition probabilities for removing the weakly bound electron from H^- for all charge states. Furthermore, the range of single-electron removal probability is quite large and on the order of several tenth of a_0 for even the low-charge state ions. Therefore, the single-electron removal cross section is independent of the species of the multiply-charged ion.

The transition probability for double-electron removal is concentrated at small impact parameters where the outer electron is ionized with nearly 100% probability[18]. This leads to a double-electron removal cross section that is similar to single-electron removal from neutral H^0, indicated by the dash-dotted line in figure 3.

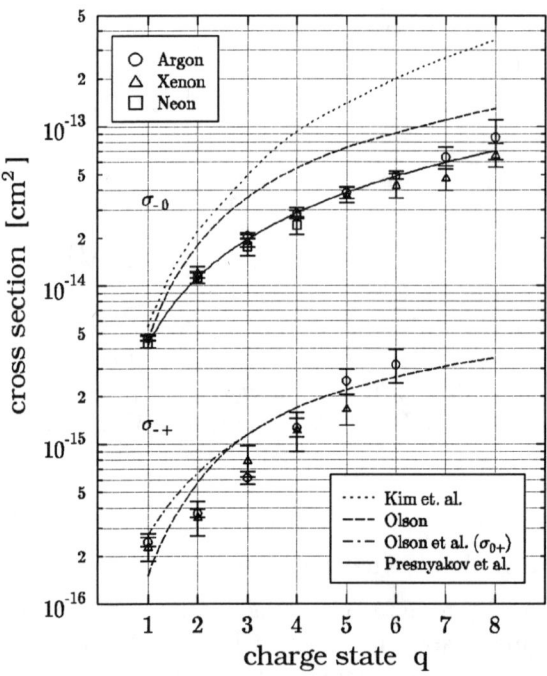

Figure 3:
Cross sections σ_{-o} and σ_{-+} for collisions between H^- and X^{q+} ions (X=Ar,Xe,Ne) as a function of the ionic charge state q at $E_{cm} = 50$ keV. The lines represent different theoretical approaches: Bethe-Born approximation (Kim et al.[17], dotted line). CTMC calculations (Olson[18], dashed line) for the ion-ion channels and for X^{q+} impact on neutral hydrogen (Olson et al.[19], σ_{o+}, dash-dotted line). Quantum calculation using Keldysh theory (Presnyakov et al.[20], σ_{-o}, solid line).

3. Angular differential cross section measurements in ion-ion collisions

Up to now, experimental investigations of charge changing processes in ion-ion collisions delivered absolute cross sections for the processes of interest. In our group, for the first time, angular differential cross sections were measured for collisions between Cs^+ ions. Furthermore, preliminary scattering distributions were observed for collisions between He^+ ions.

Experimental techniques have been developed to detect particles produced in ion-ion collisions with respect to their arrival position both for the coincidence and the beam pulsing method. A paper describing experimental details is in progress.

In short, a position sensitive, channel-plates based single particle detector employing a resistive anode resolves a position matrix of 256·256 channels corresponding to an area of 40·40 mm^2. In addition, a timing signal is generated.

Every single event creates a pair of position–time information. The time signal is related to a corresponding detector event in the other beam line (coincidence method) or to the beam pulsing sequence (pulsing method). The position–time pairs are recorded by a hostcomputer in list mode. After the measurement, time windows can be set (and varied) to allow proper allocation of the events to two position matrices. Figure 4 illustrates the procedure.

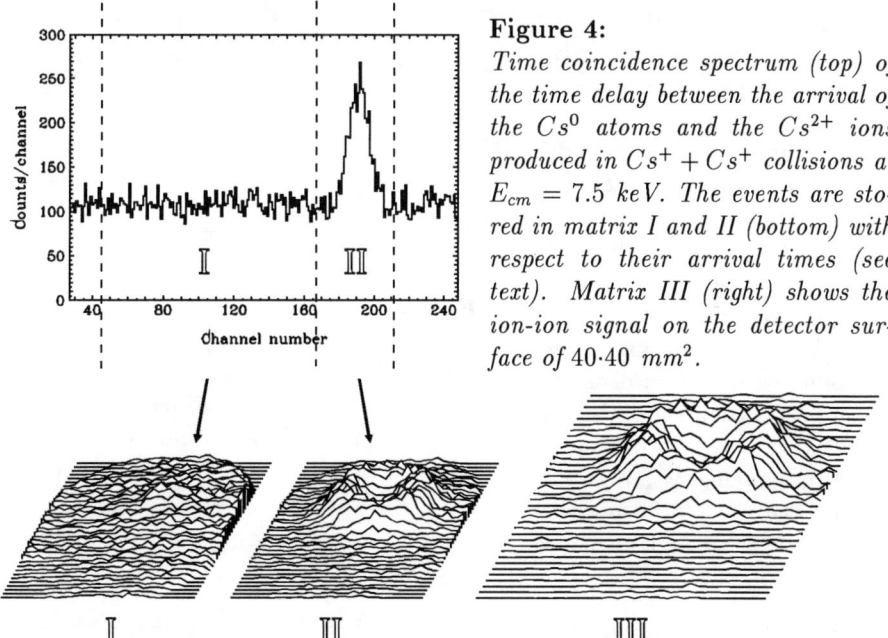

Figure 4:
Time coincidence spectrum (top) of the time delay between the arrival of the Cs^0 atoms and the Cs^{2+} ions produced in $Cs^+ + Cs^+$ collisions at $E_{cm} = 7.5$ keV. The events are stored in matrix I and II (bottom) with respect to their arrival times (see text). Matrix III (right) shows the ion-ion signal on the detector surface of 40·40 mm^2.

After the measurement, the time spectrum is subdivided into sections, e.g. I and II. All events registered in time section I (background) are collected in position matrix I and all events registered in II (background and signal) in matrix II, respectively. The signal matrix III is generated by subtracting the normalized matrix I from II.

As figure 4 clearly shows, the charge exchange process in $Cs^+ + Cs^+$ collisions

$$Cs^+ + Cs^+ \longrightarrow Cs^{2+} + Cs^0 \quad (charge\ exchange) \quad (15)$$

does not peak in forward direction but at lab frame scattering angles around 1°. Integration of matrix III yields the differential cross section for reaction 15 after transformation to the cm-system. The charge exchange process peaks at cm-scattering angles between 2° and 3°.

By means of a beam pulsing technique, the differential cross section for the total Cs^{2+}-production, consisting of charge exchange and ionization processes, was measured.

$$Cs^+ + Cs^+ \longrightarrow Cs^{2+} + Cs^+ + e^- \quad (ionization) \tag{16}$$

$$Cs^+ + Cs^+ \longrightarrow Cs^{2+} + ... \quad (total \; Cs^{2+} - production) \tag{17}$$

The major contribution to the total Cs^{2+}-production (17) is generated via the ionization channel (16). The differential cross section for ionization peaks at cm- scattering angles around 6° and exceeds the differential charge exchange cross section by a factor of five.

The present measurements of differential cross sections for collisions between Cs^+-ions clearify an early discrepancy between measurements of reaction 17 by Dunn et al.[21,22] and Peart et al.[23]. Angular scattering hindered complete detection of all reaction products in the latter experiment. The results of our absolute measurements are confirmed by the use of the position sensitive detector and displayed in figure 5.

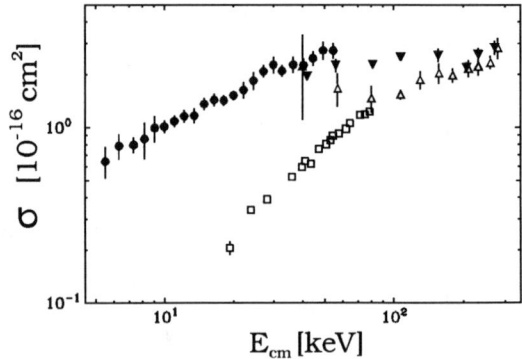

Figure 5:
Absolute cross sections for Cs^{2+}-production (17) in $Cs^+ + Cs^+$ collisions.
▼ : *Neill et al.*[22]   □ : *Peart et al.*[23]
△ : *Dunn et al.*[21]   • : *Giessen results*

Absolute cross sections for charge exchange in collisions between He^+ ions

$$He^+ + He^+ \longrightarrow He^{2+} + He^0 \quad (charge \; exchange) \tag{18}$$

have been investigated both experimentally[24,25] and theoretically[26-29]. Finally, good agreement between experiment and theory was achieved. To deepen the insight into the collision process, theory should be tested on a more sophisticated level. Angular distributions that became measurable by advanced experimental technique should be reproduced by reliable theoretical approaches.

In contrast to heavy collision systems, cross sections for reaction (18) at $E_{cm} = 12.5$ keV are lower ($\approx 3 \cdot 10^{-18} cm^2$) and cm-scattering angles occuring are smaller ($\vartheta_{cm} \ll 1°$). Figure 6 shows a preliminary result of a scattering distribution for reaction (18). The intense peak in the center is produced by small-angle scattering originating form large impact parameter transitions and images the primary beam shape. Small impact parameter transitions result in larger scattering angles generating the elliptical distribution due to 45° crossed-beam scattering kinematics. In order to extract differential cross sections from the scattering distribution shown in figure 7, the primary ion beam profile must be taken into account as well as the transformation between lab- and cm-system.

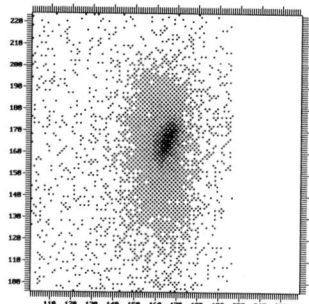

Figure 6:
Scattering distribution of He^0 atoms produced in $He^+ + He^+$ collisions at $E_{cm} = 12.5$ keV measured at the position of detector D3 (figure 1).

CONCLUSIONS

In recent years, significant experimental progress extended the knowledge of ion-ion collisions. For the first time, collisions between negative ions have been studied systematically. ECR ion sources allowed to investigate collisions involving multiply-charged ions. By means of position sensitive single particle detection angular differential cross sections have been measured recently.

Powerful ion sources, advanced detector technology, increased computer power and attainability of ultra high vacuum opens a wide field of experimental investigations becoming feasible for ion-ion collisions, a real challenge for experimentalists. Theory has to come up with angular differential and final state selective cross sections for the various processes to ensure the applicability of theoretical approaches.

ACKNOWLEDGEMENTS

The author would like to thank Prof.E.Salzborn for his support during many years which made the work presented here possible. Special thanks are due to Dr.S.Krüdener and Dr.R.Schulze for fruitful collaboration.
Financial support by the German Federal Minister for Research and Technology (BMFT) under contract 06 GI 303 is gratefully acknowledged.

REFERENCES

1. Dimov G.I., Ivanov A.A. and Roslyakov R.G., Sov. Phys. Techn. Phys. **22** (1976) 1091
2. Ivanov A.A. and Roslyakov G.V., Sov. Phys. Techn. Phys. **25** (1980) 1346
3. HIBALL-II: An Improved Conceptual Heavy Ion Driven Fusion Reactor Study, Kernforschungszentrum Karlsruhe, Report KfK 3840 (1985)
4. Budcin D., Hoffmann I., Conte M., Schulze R., Melchert F. and Salzborn E., Nuovo Cimento, in print
5. Rinn K., Melchert F. and Salzborn E., J.Phys.B**18** (1985) 3783
6. Dolder K. and Peart B., Rep. Prog. Phys. **48** (1985) 1282
7. Gilbody H.B. in: PHYSICS OF ELECTRONIC AND ATOMIC COLLISIONS, ed. S.Datz, Amsterdam, North Holland (1982) p.223
8. Salzborn E. in: THE PHYSICS OF ELECTRONIC AND ATOMIC COLLISIONS, eds. Dalgarno A., Freund R.S., Koch P.M., Lubell M.S. and Lucatoro T.B., AIP Conference Proceedings No. 205 (1990) p.290
9. Dolder K. in: CASE STUDIES IN ATOMIC COLLISION PHYSICS Vol.1, eds. McDaniel E.W. and McDowell M.R.C., Amsterdam, North Holland (1969) p.249
10. Rinn K., Melchert F., Rink K. and Salzborn E., J.Phys.B**19** (1986) 3717
11. Olamba K., Szücs S., Chenu J.P., Naji El Arbi and Melchert F., Proceedings of the XVIIIth International Conference on the Physics of Electronic and Atomic Collisions, Aarhus, Denmark, 21.-27.July 1993, Abstracts of Contributed Papers, p.650
12. Peart B. and Hayton D.A., J.Phys.B**25** (1992) 5109
13. Rundell R.D., Aitken K.L. and Harrison M.F.A., J.Phys.B**2** (1969) 954
14. Olson R.E., private communication
15. Schulze R., Melchert F., Hagmann M., Krüdener S., Krüger J., Salzborn E., Reinhold C.O. and Olson R.E., J.Phys.B**24** (1991) L 7
16. Schlachter A.S., Leung K.N., Stearns J.W. and Olson R.E., in: PRODUCTION AND NEUTRALTZATION OF NEGATIVE IONS AND BEAMS, ed. Alessi J.G., AIP Conf. Proc. No. 158 (1987) p.631
17. Kim Y.K. and Inokuti M., Phys. Rev. A**3** (1971) 665
18. Melchert F., Debus W., Liehr M., Olson R.E. and Salzborn E., Europhys. Lett. **9**(5) (1989) 433
19. Olson R.E., Berkner K.H., Graham W.G., Pyle R.V., Schlachter A.S. and Stearns J.W., Phys. Rev. Lett. **41** (1978) 163
20. Presnyakov L.P. and Uskov D.B., paper in progress
21. Dunn K.F., Angel G.C. and Gilbody H.B., J.Phys.B**12** (1979) L 623
22. Neill P.A., Angel G.C., Dunn K.F. and Gilbody H.B., J.Phys.B**15** (1982) 4219
23. Peart B., Forrest R.A. and Dolder K., J.Phys.B**14** (1981) 1655
24. Peart B., Rinn K. and Dolder K.T., J.Phys.B**16** (1983) 2831
25. Melchert F., Rinn K., Rink K., Salzborn E. and Grün N., J.Phys.B**20** (1987) L 223
26. Henne A., Toepfer A., Luedde H.J. and Dreizler R.M., J.Phys.B**19** (1986) L 361
27. Allan J.R., J.Phys.B**19** (1986) L 683
28. Fritsch W. and Lin C.D., Phys. Lett. **123A** (1987) 128
29. Kimura M., J.Phys.B**21** (1988) L 19

ONE AND TWO ELECTRON PROCESSES IN COLLISIONS OF HIGHLY CHARGED IONS WITH He AT VELOCITIES AROUND 1 a.u.

J.P. Giese, W. Wu, I. Ben-Itzhak, C.L.Cocke, R.Ali, P.Richard, M.Stöckli, and H. Schöne

J.R.Macdonald Laboratory, Kansas State University, Manhattan, Kansas 66506, USA

ABSTRACT

We have studied one and two electron processes in collisions of Ar^{16+} and O^{8+} ions with He atoms at velocities between 0.23 and 1.67 a.u. These processes were identified by measuring the final charges of the projectile and recoil ions in coincidence. Single electron capture (SC) is the dominant process at all our velocities. Transfer ionization (TI), where the He loses both electrons but the projectile captures only one, is next largest. Double capture (DC) and single ionization (SI) are about 10 times smaller, with DC larger for Ar and SI negligible for O. None of these processes show any dramatic velocity dependence. The energy independence of SI is somewhat surprising as our measurements span the velocity region where ionization would be expected to increase. Our analysis suggests that the ionization process is being suppressed by SC and TI processes. The average Q-values for capture channels were determined by measuring the longitudinal momenta of the recoiling target ions. The Q-values for both SC and TI decrease with increasing projectile energy, which indicates that electrons are captured into less tightly bound states as the collision velocity is increased.

INTRODUCTION

A fundamental goal of atomic physics is to understand the dynamics of many-electron systems. Such systems are produced in collisions between highly charged ions and many-electron target atoms. The large potential energy carried into these collisions by the projectile ions strongly perturbs the target. This can result in many target electrons being excited, ionized, or transferred to the projectile. The transfer of many electrons to highly charged projectiles produces multiply excited states not normally found in nature. The study of these collisions therefore offers opportunities to systematically explore the properties of strongly perturbed and unusual many electron systems. We have chosen to study the collisions of highly charged O^{8+} and Ar^{16+} ions with He atoms. Helium was chosen to limit our investigations to one and two electron processes. The $1s^2$ ground state electrons of Ar^{16+} are inactive spectators at our collision velocities, so these collision systems are effectively both two-electron systems.

Collisions between highly charged ions and atoms at velocities near 1 a.u. are interesting because within this velocity region our understanding of capture and ionization processes must shift from low energy molecular orbital (MO) models to higher energy perturbative models. Charge capture is the dominant process at low velocities. Single electron capture at low velocities is well understood in terms of the formation of molecular orbitals during the collision. Multiple electron capture is qualitatively understood using either molecular orbital models or classical overbarrier models. Target ionization by highly charged ions is almost negligible at low velocities and is not particularly well understood. In contrast, ionization is the dominant process at high velocities, where ionization and capture are both understood in terms of perturbative models.

While collisions of highly charged ions and atoms have been extensively studied at velocities both low[1,2] and high[3,4] compared to those of the target electrons, the intermediate velocity region ($v_p \sim v_e$) is not as well explored either experimentally or theoretically. These velocities are interesting because capture and ionization processes must be comparable somewhere in this region. Furthermore, neither low energy molecular orbital models nor high energy perturbation models should necessarily be expected to work well at these velocities. Measurements in this region therefore provide a stringent test of our models of both ionization and capture.

© 1993 American Institute of Physics

We have investigated the following processes:
- Single Ionization(SI): $X^{q+} + He \rightarrow X^{q+} + He^+ + e^-$;
- Single Capture(SC): $X^{q+} + He \rightarrow X^{(q-1)+} + He^+$;
- Double Capture(DC): $X^{q+} + He \rightarrow X^{(q-2)+} + He^{2+}$;
- Transfer Ionization(TI): $X^{q+} + He \rightarrow X^{(q-1)+} + He^{2+} + e^-$.

where $X^{q+} = O^{8+}$ and Ar^{16+}. We present cross sections for all these processes and Q-values for the capture processes. Our velocity range (0.23 to 1.67$a.u.$) covers the region around the average velocity of the He electrons (1.34$a.u.$), and probes the region of intermediate velocity where the limits of the models of capture and ionization can be tested.

EXPERIMENT and DATA ANALYSIS

We have measured cross sections for these processes by observing the final charges of the projectile and recoiling He target ions in coincidence. The experimental apparatus is shown in Fig. 1 and has been described in detail elsewhere.[5,6] An O^{8+} or Ar^{16+} ion beam was extracted from the KSU CRYEBIS, magnetically analyzed, and accelerated to the collision energy. Charge state impurities were minimized by cleaning the beam with a magnet immediately prior to the collision region. After passing through the collision chamber, the projectile ions were charge analyzed by a parallel-plate electrostatic deflector and then detected by a position-sensitive backgammon anode detector (PSD-1). All three final projectile charge states could be simultaneously observed on this detector. This reduced the uncertainty in the relative cross sections and speeded data collection. The He gas target was supplied by a multi-channel array molecular jet. The gas flow was adjusted to minimize double collisions. No more than about 3% of the projectiles that changed their charge state were observed to have undergone double encounters.

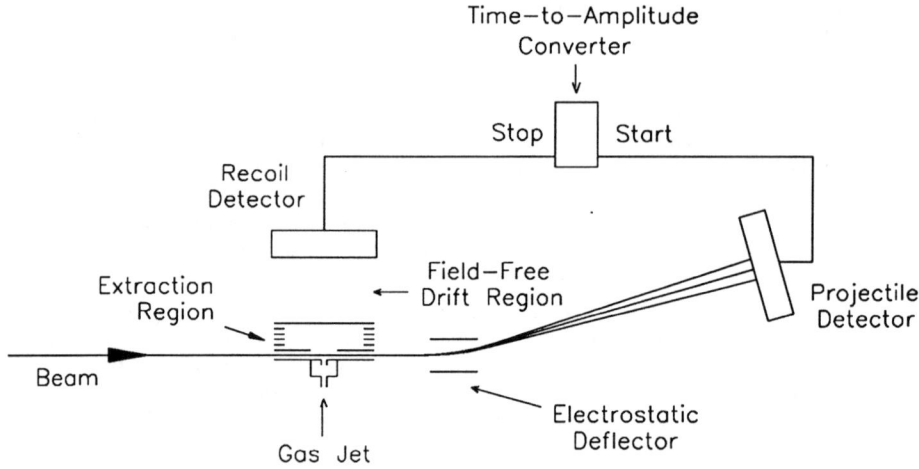

Fig. 1. Schematic of the experimental apparatus.

The He^+ and He^{2+} ions produced in the collision region were extracted transverse to the beam by a uniform electric field and detected by a two-dimensional position-sensitive resistive anode detector (PSD-2). The strength of the recoil extraction field was set high enough (about 60 V/cm) to ensure that all recoil ions were extracted by the field. The recoil charge states were determined by the time-of-flight technique using coincidences between the recoil ions and the projectile ions. The longitudinal recoil momentum was determined from the position of the recoil ions on the detector.

The cross sections for the individual processes (SC, TI, etc) were determined from the corresponding yields of two dimensional coincidence spectra of the projectile position versus the recoil time of flight. It is necessary to correct the coincidence yields for both random coincidences and events due to multiple collisions.[5,6] The magnitude of the correction is less than 3% for SI, DI, SC, and TI and about 30% for DC. The relative cross sections were calculated as the ratio of the corrected coincidence yields to the total number of incident projectiles. We normalized our total single projectile charge change cross sections ($\sigma^{0,1}_{16,15}+\sigma^{0,2}_{16,15}$) for argon to the 2.3keV/q Ar^{16+} on He data of Vancura et al.[7] and for oxygen to the 10 keV/q O^{8+} data of Bliman et al..[8] The measured cross sections for single capture (SC), double capture(DC), single ionization(SI) and transfer ionization(TI) are shown in Fig. 2 as a function of the projectile energy. The error bars shown in the figure account for both statistical errors and reproducibility errors. There is a further uncertainty of about 15% in the absolute normalization.

Q-values for the capture processes were measured by recoil longitudinal momentum spectroscopy.[5] Conservation of energy and momentum can be used to calculate Q-values from the recoil ion momenta as long as no electrons are directly ionized to the continuum. The Q-values for collision processes where i electrons are captured are given by $Q \approx -vP_r^{\parallel} - iv^2/2$, where v is the projectile velocity, i is the number of electrons being captured, and $P_r^{\parallel}$ is the measured recoil longitudinal momentum. The average Q-values for SC and TI are plotted in Fig. 3. The Q-values for TI were calculated by assuming that TI is primarily autoionizing DC. The Q-value error bars are due mainly to uncertainty in locating both the centroid of the recoil ion position spectra and the zero of the recoil position.

RESULTS AND DISCUSSION

A. Single Capture

The single electron capture is the most likely process at all of our collision energies for both Ar and O projectiles (Fig. 2). The SC cross sections are relatively large and fairly independent

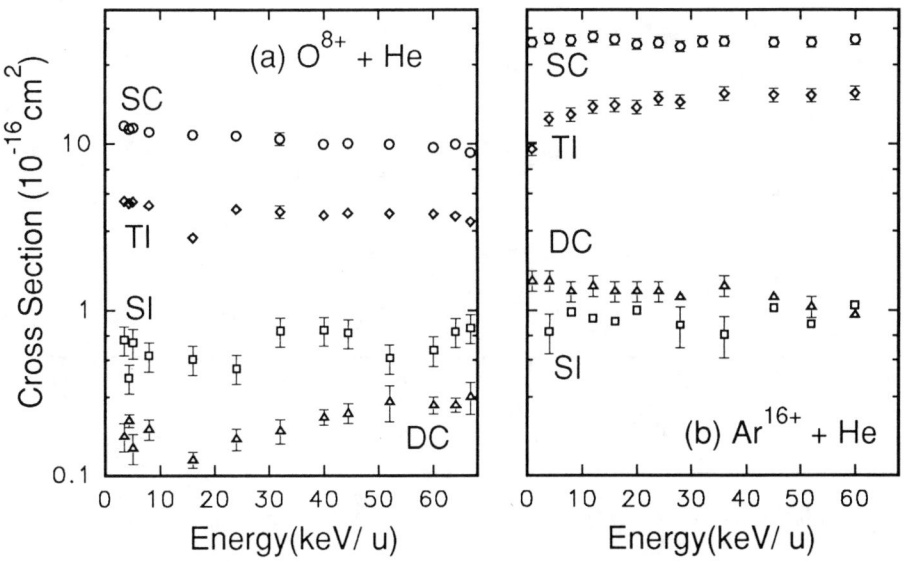

Fig. 2 Measured cross sections for SC (circles), TI (diamonds), DC (triangles), and SI (squares).

of energy. This energy independence can be qualitatively understood using a simple molecular-orbital (MO) model, where electron capture occurs at localized crossings of the initial and final electronic states. When the projectile is highly charged, many of these crossings can be active, i.e. there is a range of states to which the capture reaction can occur. The cross section depends in part on the number of possible final states within this "reaction window". As the collision velocity is increased, the reaction window moves toward smaller internuclear radii and capture proceeds to more tightly bound states. The density of final states around $n = 7$ for Ar^{15+} and $n = 4$ for O^{7+} is fairly uniform. Therefore, as the velocity increases and the reaction window moves, there are always about the same number of crossings in the window. As a result, the total cross section is insensitive to the collision velocity. Our SC cross sections suggest that this simple model of SC works up to velocities over $1.55 a.u.$. It should be noted that the velocity independence of our SC cross sections also qualitatively agrees with detailed calculations for total single capture in the one electron systems of bare projectiles on atomic hydrogen.[9,10,11]

The puzzle is that the general MO model predicts that the Q-value for this collision system would be either relatively energy independent or would slowly *increase* as the projectile velocity increased. Note that the MO prediction is consistent with measurements and calculations using very fast projectiles, which produce capture to more tightly bound states (i.e. larger Q-values) as the velocity is increased. Both the low velocity MO model and the high velocity perturbative models directly contradict our data shown in Fig. 3. The data clearly indicate that capture proceeds to less tightly bound states as the projectile velocity is increased. This trend is most dramatic for single capture by Ar where the average final projectile n-state seems to change from about $n = 8$ at lower energy and to about $n = 11$ at higher energy.

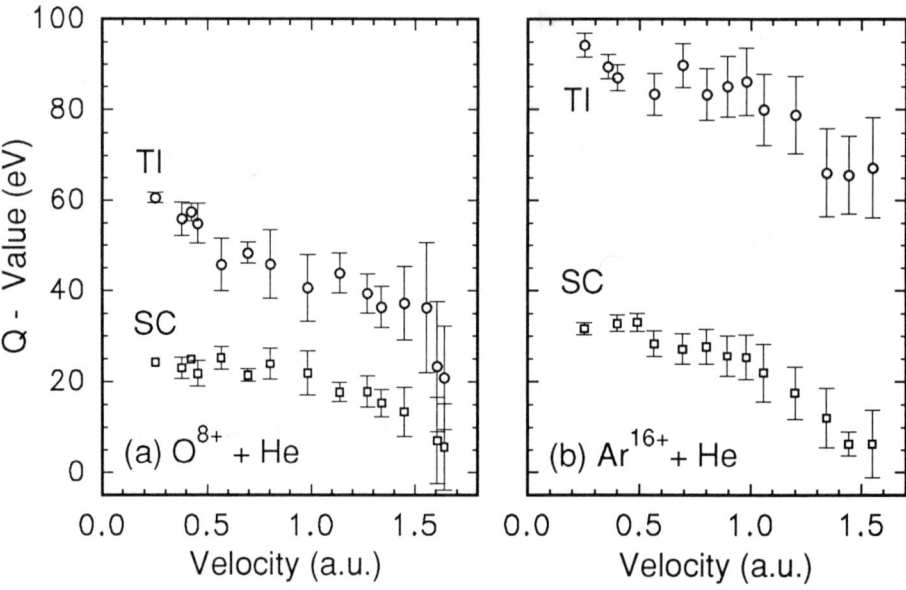

Fig. 3. The measured Q-values for SC (circles) and TI(squares).

Other measurements and calculations show no consistent dependence of the final n-state on velocity. Some authors[12,13] reporting a general tendency of capture to smaller n with increasing velocity, while others[14-17] reported a tendency to larger n. These reports combined with our results seem to suggest that the velocity dependence of the projectile state population may be sensitive to the properties of the individual collision systems. However, the other investigations were at somewhat lower velocities ($v < 0.6 a.u.$) and used projectiles of much lower charge states ($q < 8$) than the ones used in our experiment.

Several new calculations are now available which give the n-state populations for capture by multiply charged projectiles in this velocity region. The Classical Trajectory Monte Carlo (CTMC) model [6] was used to calculate both the recoil longitudinal momenta ($P_r^{\parallel}$) and cross sections as a function of the final n-state of the captured electron for Ar^{16+} on He. The CTMC total single capture cross section agrees with the data in both its energy independence and magnitude. The calculated $P_r^{\parallel}$, shown in Fig. 4, also reproduce both the magnitude and the energy dependence of the data fairly well. The coupled channels method was also used to calculate the cross sections as a function of the final n-state for bare projectiles ($Z = 2,3,5,6$) on atomic hydrogen.[18] The results for C^{6+} on H are shown in Fig. 5, These results seem to indicate that the trend toward smaller q-values is general, i.e. any highly charged ion will show this feature in the velocity region above 1 a.u..

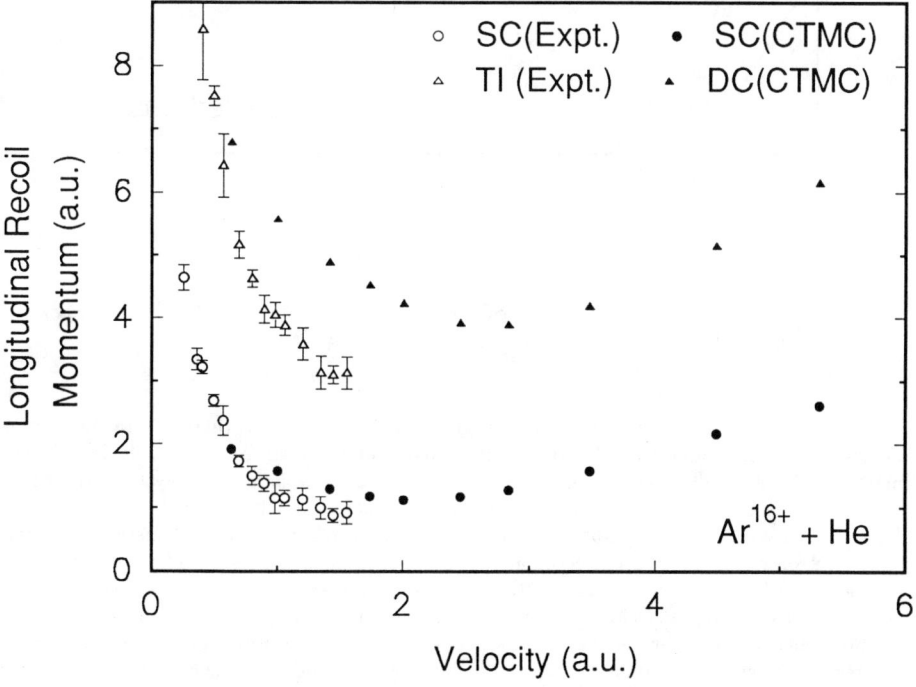

Fig. 4. A comparision of recoil longitudinal momenta from the present data (open symbols) and from the CTMC calculations[6] (closed symbols). The circles represent SC, while the triangles represent TI for the data and DC for the calculation. Note that the TI observed at these energies is thought to result from DC followed by autoionization, hence the comparison of our TI data with the CTMC calculation.

Two interesting features of both the CTMC and coupled channels calculations are worth noting. First, both models indicate that while the n-value of the peak of the reaction window for capture is relatively insensitive to velocity, the width of the reaction window is increasing with velocity. This broadening seems to be mainly towards higher n and therefore leads to smaller average Q-values. Second, both suggest that the Q-value begins to increase with energy at very high velocities. This indicates capture into more tightly bound states as would be expected by high velocity, perturbative models.

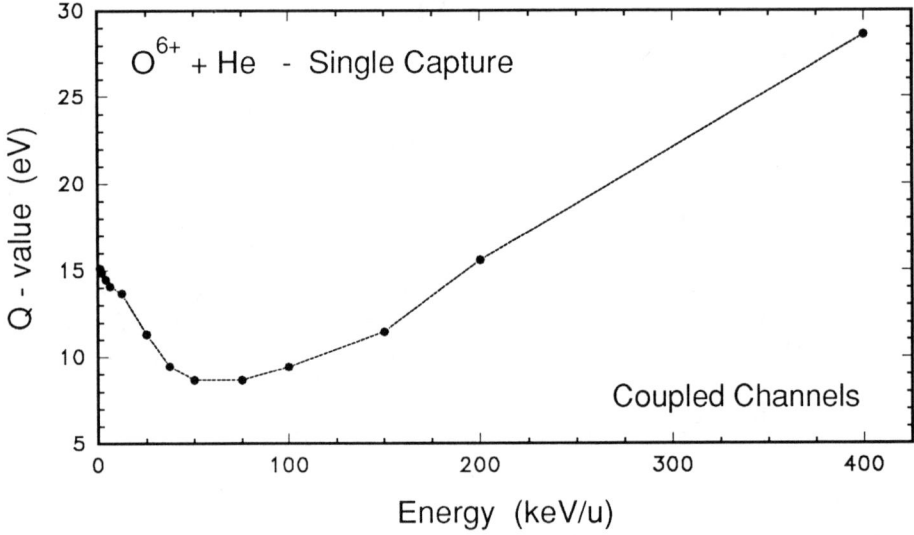

Fig. 5. Calculated Q-values for SC by O^{6+} + H as a function of collision energy by Lin and Toshima.[18]

Our data, CTMC and coupled channels calculations suggest that there is a range of intermediate collision velocities where capture proceeds to higher n-states as the collision velocity increases. We believe the fundamental cause of this trend is the kinetic energy carried into the collision by the target electron. Our SC cross sections are relatively constant, suggesting that the average distance at which capture occurs is fairly constant. The potential energy of the captured electron relative to the projectile at the moment of capture must also be fairly constant. The electron's kinetic energy relative to the projectile, however, increases with the square of the collision velocity. Only for velocites above 0.5 a.u. does this energy become comparable to the potential energy gained by the electron as it is captured by the projectile. This may explain why results at low velocity do not show any definite trend in the dependence of the final n-state on velocity, while a definite trend to higher n is observed at intermediate velocities.

B. Transfer Ionization and Double Capture

Transfer ionization (TI) is the second most likely process at all our collision energies as is seen in Fig. 2. Transfer ionization (TI) by slow, highly charged ions colliding with He is generally thought to occur when the projectile captures both electrons into doubly excited states[2]. These states can decay by either autoionization, leading to TI, or by radiative stabilization, leading to "true" double capture (DC). Note from Fig. 2 that DC is about an order of magnitude smaller than TI in this collision system. Our measured Q-values for TI and DC are the same, suggesting both that our TI cross section is dominated by autoionizing double capture and that the same range of doubly excited states is populated in each process. The measured recoil longitudinal momenta for TI by Ar projectiles are compared to CTMC calculations for DC in Fig. 4 and have about the same magnitude and energy dependence. This further suggests that the measured TI is a product of autoionizing DC.

Using the average Q-values for TI, the final states of the captured electrons for Ar were estimated to be $(n, n') = (6, 7)$ and $(7, 7)$ at the lower energy. The decreasing Q-value suggests that the relative population of $(7, 7)$ slowly increases with increasing projectile energy. A similar calculation for O projectiles indicates capture into $(n, n') = (3, 4)$ and $(4, 4)$ at low velocities, with the relative population of $(4, 4)$ states increasing as the velocity is increased. Our data therefore suggest that the two-electron DC and TI processes, like the one-electron SC process, populate more highly excited states as the projectile velocity is increased.

C. Single Ionization

Single ionization (SI) is about one order of magnitude smaller than TI or SC for both Ar and O projectiles at all our collision energies. Double ionization was negligibly small at all our collision energies. The mere observation of SI at these energies is not surprising because it has been previously observed even at much lower energies[19] for highly charged projectiles. The relative energy independence of the single ionization, however, was surprising to us because we expected the ionization to begin increasing in the region around matching velocity(45keV/u). Our expectations were based in part on simple models of ionization which often do not consider the competing process of capture. However, even coupled channels calculations of bare projectiles on H, which did attempt to account for both ionization and capture, show a rapidly increasing ionization cross section in the velocity range covered by our data.[20,21] We have developed a model which includes the interplay between capture and ionization around the matching velocity and which does help explain the energy independence of single ionization.

One should in principle treat the two-electron target as a whole and, using one model, compute the probabilities of the competing processes of the electrons being captured, ionized, or remaining on the target. Our simple model including both capture and ionization, however, was developed within the independent electron approximation.[22,23] The probability functions for the different processes are given by,

$$P_{SC}(b) = 2\, P_c(b)\, R\ ;$$
$$P_{DC}(b) = P_c^2(b)\ ;$$
$$P_{SI}(b) = 2\, P_i(b)\, R\ ;$$
$$P_{TI}(b) = 2\, P_i(b)\, P_c(b)\ ,$$

where $P_c(b)$ and $P_i(b)$ are the capture and ionization probabilities per active electron, and $R(b) = 1 - P_c(b) - P_i(b)$ is the probability for the spectator electron to remain on the He target. We have used the classical over-barrier model (CBM) to describe capture and the semi-classical Coulomb approximation (SCA) to describe ionization. The two target electrons are treated as equivalent electrons, i.e. in two-electron processes we have neglected effects due to the increase in binding energy of the second electron. Such treatment is not rigorously correct and will tend to overestimate the two electron processes. However, this treatment is computationally easier and does provide qualitative insight into the collision mechanism.

The CBM capture probability is a constant for impact parameters smaller than the capture radius, which is about $R_c = 10 a.u.$ for $Ar^{16+} + He$ collisions, and zero otherwise. The value of this constant, $P_c(b) = 0.4$, was evaluated using the measured SC cross section. This $P_c(b)$ has a sharp cutoff at R_c which is not physical. We remove this sharp cutoff using a functional form suggested by Brandt[24] which takes into account the time the electron spends within the capture radius. The capture probability is then given by $P_c(b) = P_0\sqrt{1 - b^2/R_c^2}$, where $P_0 = 0.5$ and $R_c = 12$ were evaluated using the measured values of the SC and DC cross sections. None of the calculated cross sections were sensitive to the smoothing of this cutoff.

The ionization probabilities were obtained from the SCA tables of Hansteen, Johansen, and Kocbach.[25] This approximation is probably good for this collision system at large impact parameters where $P_i(b)$ is small. However, it overestimates $P_i(b)$ at small impact parameters where perturbation theory is not valid for the large q^2 scaling factor needed for the Ar^{16+}. As a result, $P_c^{CBM}(b) < P_i^{SCA}(b)$ and $P_c^{CBM}(b) + P_i^{SCA}(b) > 1$ for small impact parameters. We know that SC is dominant over SI in our systems, so we have defined the ionization probability as the complement of the capture probability at small impact parameters, i.e. $P_i(b) = 1 - P_c(b)$ and $R(b) = 0$, wherever $P_c^{CBM}(b) + P_i^{SCA}(b) \geq 1$.

The calculated and measured cross sections are in reasonable agreement over the entire energy range. The calculated SI cross section increases only very slowly with increasing collision velocity. The large increase expected around matching velocity seems to be suppressed by the competing capture process. This can be seen by examining the impact parameter dependence of these processes. The probability functions for each multiplied by b are plotted versus b in Fig. 6. SC is the dominant process at large impact parameters. As mentioned above, the measured TI is probably due to autoionizing double capture. The TI and DC channels were therefore added in this figure. These two-electron processes dominate for close collisions where it is unlikely that either electron will remain on the He target. Interestingly, single ionization peaks at intermediate impact parameters of $\sim 4 a.u.$. This exclusive process is suppressed at small b by the requirement that one electron has to remain on the target and at large b by the capture process which dominates due to the large electron-projectile attraction.

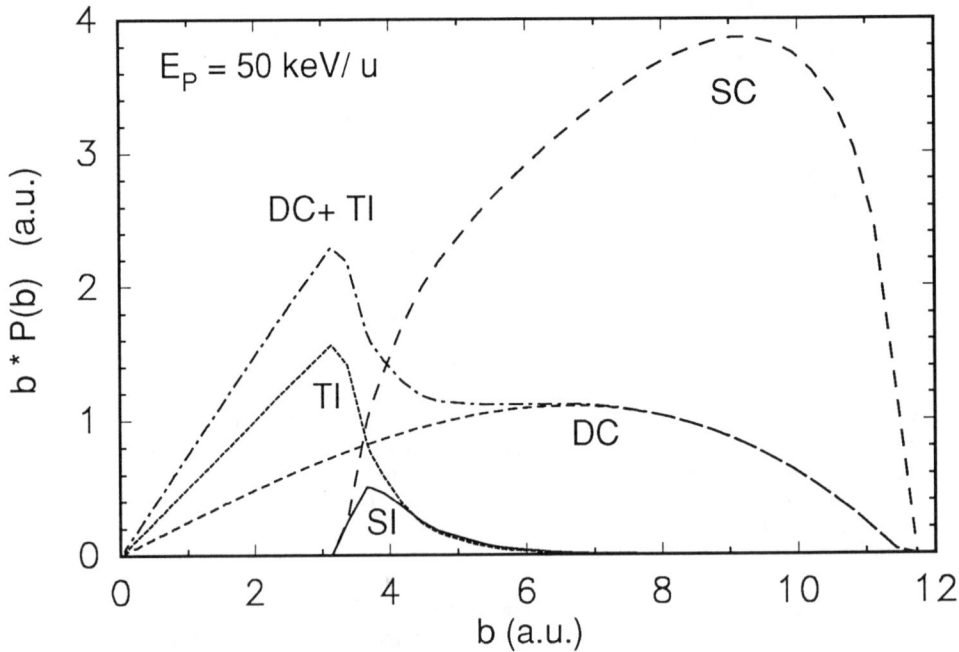

Fig. 6. Probability functions multiplied by b for single capture(SC), single ionization(SI), double capture(DC), and transfer ionization (TI), as a function of the impact parameter. Note that in this calculation TI is 'real' TI, i.e. independent capture and ionization events, while in the data TI is thought to be autoionizing DC.

The maximum in the ionization probability function at intermediate impact parameters disagrees with the known behavior in fast collisions for which ionization is dominant at large impact parameters (soft collisions) while capture is dominant for small impact parameters. This behavior in fast collisions is easily explained by our model as the capture probability is negligible except at very small impact parameters and never suppresses the ionization probability. This analysis lead us to expect that SI will increase rapidly with velocity only at higher collision energies where the SC probability is no longer dominant at large impact parameters.

The prediction of our simple model that SI might only occur at intermediate impact parameters can be understood in terms of MO models of ionization. This model requires that one target electron is first captured by the projectile at a MO crossing at relatively large internuclear distances. As the projectile and target move closer together, the other electron is captured via a MO crossing between the single capture orbital and a double capture orbital. Finally as the projectile and target continue to move closer together, one electron is transferred back to the target and the other into the continuum via a MO crossing between the double capture orbital and a single ionization orbital. This model suggests that SI can occur only via crossings with these capture processes at internuclear separations smaller than those required for double capture, or in this case about $3 a.u.$. These radii are quite similar to the radii predicted for SI in Fig. 6 by our independent-electron model. Note that this model also suggests that SI of He by these slow, highly charged ions might also be thought of as a two electron process. Although this picture seems to have somewhat different physical content from our simple independent-electron model, it does also point to ionization at intermediate impact parameters and the two pictures might not be incompatible.

We should note that ionization by very highly charged Xe ions ($q > 30$) incident on Xe targets has been recently observed at very low velocities ($v = 0.034\sqrt{q}$ $a.u.$).[22] These results are explained in terms of the formation of core-excited, autoionizing, neutral target states. It is proposed that a target electron is first captured by the projectile at a MO crossing at relatively large internuclear distances. As the projectile and target move closer together, the electron is then transferred back to the target via a MO crossing between the single capture orbital and the core-excited target state. Although this model depends on the $3p^6$ sturcture of the target Xe, it is in principle very similar to the MO model of ionization discussed here. The study of ionization by highly charged ions at low and intermediate velocities offers the opportunity for detailed study of this process.

CONCLUSIONS

Our studies of one and two electron processes in collisions of O^{8+} and Ar^{16+} with He have produced two interesting and somewhat surprising results. First, our measurements of the Q-values for SC and TI suggest that both one and two electron capture proceed to less tightly bound states as the projectile energy is increased. This result suggests that the collisional kinetic energy of the captured electron strongly affects the final state populations. Second, our SI cross sections were small and relatively independent of collision energy. This is surprising as our energy range spans the region where SI is expected to rapidly increase. We developed a model which suggests this behavior results from suppression of SI by SC at large impact parameters and by two-electron processes at small impact parameters.

Acknowledgments: We would like to thank C.D. Lin and N. Toshima for the use of their calculations prior to publication. This work was supported by the Division of Chemical Sciences, Office of Basic Energy Sciences, Office of Energy Research, U.S. Department of Energy.

1. R.K. Janev and L.P. Presnyakov, Phys. Rep. **70**, 1(1981).
2. M. Barat and P. Roncin, J. Phys. B**25**, 2205(1992).

3. H.Knudsen, H.K.Haugen, and P.Hevlplund, Phys. Rev. A**23**, 597(1981).
4. J.L.Shinpaugh, J.M.Sanders, J.M.Hall, D.H.Lee, H.Schimt-Böcking, T.N.Tipping, T.J.M.Zouros, and P.Richard, Phys. Rev. A**45**, 2922(1992).
5. R.Ali, V. Frohne, C.L. Cocke, M. Stockli, S. Cheng, and M.L.A. Raphaelian, Phys. Rev. Lett. **69**, 2491(1992).
6. W. Wu, J.P. Giese, I. Ben-Itzhak, C.L. Cocke, P. Richard, M. Stockli, R. Ali, and H. Schöne, accepted Phys. Rev. A (1993).
7. J. Vancura, V.J. Marchetti, J.J. Perotti, and V.O. Kostroun, Phys. Rev. A**47**, 3758(1993)
8. S. Bliman, D. Hitz, B. Jacquot, C. Harel, and A. Salin, J. Phys. **16**, 2849 (1983).
9. R.E. Olson and A. Salop, Phys. Rev. A**16**, 531(1977).
10. H. Ryufuku, Phys. Rev. A**25**, 720(1982).
11. R.K. Janev and L.P. Presnyakov, J. Phys. B**13**, 4233(1980).
12. M.Barat, M.N.Gaboriand, L.Guillement, P.Roncin, H.Laurent, and S.Andriamonje, J. Phys. B**20**, 577(1987).
13. C. Harel and H. Jouin, J. Phys. B**25**, 221(1992).
14. P.Moretto-Capelle, P.Benoit-Cattin, A.Bordenave-Montesquieu, and A.Gleizes, Z. Phys. D**21**, S283(1991).
15. D.Dijkkamp, D.Ćirić, E.Vlieg, A.de Boer, and F.J.de Heer, J. Phys. B**18**, 4763(1985).
16. C. Harel and H. Jouin, J. Phys. B**21**, 859(1988).
17. W. Fritsch and C.D. Lin, J. Phys. B**19**, 2683(1986).
18. C.D. Lin and N. Toshima, private comminication, 1993.
19. H. Cederquist, C. Biedermann, N. Selberg, E. Beebe, M. Pajek, and A. Barany, Phys. Rev. A**47**, R4551(1993).
20. H. Ryufuku, Phys. Rev. A**25**, 720(1982).
21. R.K. Janev and L.P. Presnyakov, J. Phys. B**13**, 4233(1980).
22. J.H. McGuire and O.L. Weaver, Phys. Rev. A**16**, 41 (1977).
23. I. Ben-Itzhak, T.J. Gray, J.C. Legg, J.H. McGuire, Phys. Rev. A**37**, 3685 (1988).
24. D. Brandt, Nucl. Instr. Meth. **214**, 93 (1983).
25. J.H. Hansteen, O.M. Johansen, and L. Kocbach, At. and Nucl. Data Tables **15**, 305 (1975).

RADIATIVE PROCESSES IN ATOMIC COLLISIONS AT STORAGE RINGS

L.R. Andersson* and J. Burgdörfer
Department of Physics and Astronomy, 401 A.H. Nielsen Bldg.,
University of Tennessee, Knoxville, TN 37996-1200, USA
and Oak Ridge National Laboratory, Oak Ridge, TN 37931-6377, USA

ABSTRACT

We discuss radiative electron capture into excited states of fast, highly charged ions from neutral targets. Capture to high n shells and to low-lying projectile continuum states is studied. We show that capture into excited states can account for up to 50% of the radiative capture cross section at the "Bremsstrahlung" threshold. The feasibility of state-resolved measurements at storage rings is discussed.

INTRODUCTION

In collisions involving fast ions, emission of X-rays due to various processes can occur. Besides the emission of characteristic X-ray lines originating in the relaxation of excited projectile or target states or γ-rays from nuclear transitions, continuous emission from transitions between transient molecular states (MO X-rays) and radiative electron capture (REC) will be present. The continuous REC and MO X-ray spectra are, apart from their intrinsic interest, important for analyzing the non-constant background of line-emission spectra. Since MO X-rays are most important in the intermediate and low velocity regime, we are concerned with REC, which represents the dominant contribution to continuous X-rays at high collision velocities.

REC is most conveniently visualized as a free-bound transition of a quasi-free target electron as seen in the projectile rest frame. The electron has a velocity distribution given by the Doppler-broadened momentum distribution of the initially bound target state centered around a mean velocity equal to the collision velocity. In addition to REC, processes where the final state of the electron is in the continuum of both projectile and target (usually termed radiative ionization (RI), radiative capture to the continuum (RCC)) are possible. The latter is analogous to the free-free transition for atomic bremsstrahlung when one approximates the binding to the initial target state by the Galilei-shifted momentum distribution. We will in the following denote this process by RCC in order to stress the continuity across the ionization limit when the final state is a low-lying continuum state in the frame of the projectile. One objective of

---

*Present address: University of Stockholm, Atomic Physics, S-10405 Stockholm, Sweden

the present work is to explore the REC and RCC processes in the vicinity of the "bremsstrahlung threshold" corresponding to transitions to projectile states close to the continuum limit.

REC was experimentally first observed by Schnopper et al.[1] in collisions between heavy ions and solid targets. In addition to the broad X-ray distribution due to RCC and characteristic X-ray lines, a peak deriving from capture to the K shell of the projectile (K-REC) was observed. Considerable experimental[2-5] and theoretical[6-8] work followed, mainly concentrating on K-REC since this is the dominant capture channel at high velocities and its spectral signature can be relatively easily separated from other features. Only very recently, studies of REC into excited states have become available.[9-12] REC into excited shells of the projectile is difficult to measure unless the parameters of the collision system are chosen carefully.

We present in the following REC and RCC spectra in Ne^{10+}–H collisions at a relatively moderate collision velocity calculated in the impulse approximation, in order to analyze how the REC spectrum for high-n capture and the low-continuum RCC spectrum build up the total X-ray spectrum. We, furthermore, illustrate the utility of storage rings for shell-resolved REC experiments. We also stress the interrelation to radiative recombination (RR).

Experimental studies of radiative recombination (RR) at low collision energies have been carried out in single-pass experiments in Århus[13] and in the storage rings in Heidelberg[14] and Stockholm[15] using the electron cooler beam as the target. It is thus interesting to note that the two related atomic processes can be explored in detail in completely different velocity regimes with these new experimental facilities.

CRITERION FOR SHELL RESOLVED REC SPECTRA

We want to establish a simple criterion under which REC peaks resulting from capture into the nth and $n+1^{th}$ shell of a hydrogenic projectile can be resolved. The key is that the width of the REC peaks, originating in the width of the target electron velocity distribution and the Doppler broadening when transformed to the projectile rest frame, is intrinsic and limits the ability to resolve adjacent peaks. Denoting the width of the initial target-electron velocity distribution by Γ (on a velocity scale in a.u.), the corresponding width of the energy distribution associated with the Doppler shifted velocity distribution is $\approx v\Gamma$. Consequently, for bare projectiles, a criterion for resolving the adjacent peaks resulting from capture into the n and $n+1$ shells can be written as

$$\frac{Z_P^2}{2}\frac{2n+1}{n^2(n+1)^2} > v\Gamma, \tag{1}$$

where Z_P is the projectile charge. For an atomic target, Γ is of the order of the (effective) charge Z_T. For conduction band electrons of a solid target, Γ is of the order of the Fermi velocity v_F. Weakly bound electrons with small Γ (or Z_T) are therefore preferable.

A second intrinsic difficulty associated with state-resolved REC spectra is the simultaneous presence of characteristic target and projectile X-ray lines. In order to separate the line spectrum from the REC spectrum, the projectile and target Sommerfeld parameters should not exceed unity. (This is also the parameter regime where perturbative models are expected to give the most reliable results.) Since we consider systems with $Z_P > Z_T$, the condition for a REC spectrum unperturbed by X-ray lines is simply

$$\frac{Z_P}{v} < 1. \qquad (2)$$

Using scaled variables Z_p/Γ and v/Γ, the criteria (Eqs. (1) and (2)) can be represented in one diagram (Fig. 1). Eq. (1) results in a family of curves each signifying the resolution of the nth and $n+1^{th}$ peak ($n = 1, 2 \ldots$) while Eq. (2) represents the 45° line indicating that only the parameter space corresponding to the lower half of the diagram admits X-ray spectra unperturbed from characteristic lines. We show in Fig. 1 also the parameter range of recent REC experiments using gas targets (open symbols) and solid targets (solid symbols). As can be seen, most experiments which satisfied (1) for $n \geq 2$ did not satisfy Eq. (2). The regime with high Z_P and high v, but small Z_P/v is, to a large extent, unexplored, the obvious reason being the small cross section. In Fig. 2 is shown schematically the contours of constant cross section plotted against the reduced parameters Z_P/Γ and v/Γ. For simplicity we have used the recombination cross section for free electrons which is a function of Z_P/v only.

The main reason for the frequent use of a solid target is, of course, the high density of target electrons which improves the count rates counteracting the small cross section. Solid targets introduce, however, other difficulties when making detailed comparison with theory. Secondary processes producing X-rays, such as secondary bremsstrahlung,[10] will contribute an additional background which can, in unfavorable cases, completely overtake the REC spectrum. Further, unless the projectile ions are channeled, the charge state of the projectile is ill-defined. Under channeling conditions, the equilibrium charge state is close to the initial charge state, and, in addition, Γ can be reduced so that REC into excited states can be measured.[11]

In the regime of small Z_P/v heavy-ion storage rings appear to be the most promising candidates for detailed studies of REC in fast collisions.[12] The high luminosity of the circulating beam compensates for the low target density and small cross sections. A caveat is, however, that a much larger mechanical capture cross section must not lead to an alteration of the initial charge state of the beam. It should be noted that Γ is greatly reduced if a free-electron target is used. Fig. 1 therefore indicates that radiative recombination measurements[13-15]

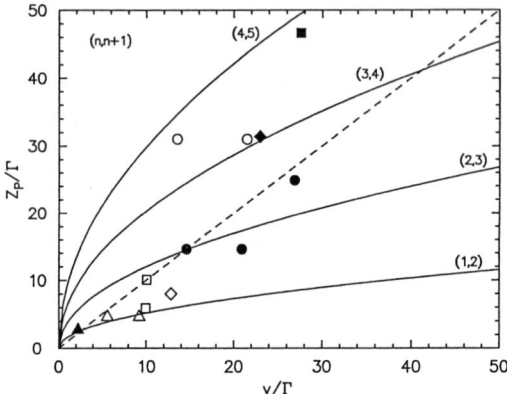

Figure 1: Criteria for well-resolved peaks for REC into the n and $n+1$ shells of bare projectiles with charge Z_P. Γ denotes the width of the velocity distribution of the target electron. Symbols represent the region marked by experiments. □: Kienle et al.[2]; ◇: Schnopper et al.[3]; △: Tawara et al.[4]; ○: Stöhlker et al.[12]; ●: Anholt et al.[10]; ■: Andriamonje et al.[11]; ▲: Jakubaßa et al.[16]; ◆: Datz et al.[17] Open and solid symbols denote gas and solid targets, respectively.

using the electron cooler beam as a target should permit the resolution of high n states. In this case $Z_p/\Gamma \gg 1$ and $v/\Gamma \approx 1$ corresponding to the upper left corner of the parameter space of Fig. 1.

THEORY

The simple picture for REC as a radiative-recombination event in the rest frame of the projectile with the velocity distribution of the incident electron given by the properly transformed momentum distribution of the target bound state leads to the *impulse approximation* (IA).[6,7] The cross section differential in the photon energy and emission solid angle for capture to the final projectile state f is given by

$$\frac{d\sigma_f^{REC}}{d\omega\, d\Omega} = \sum_\lambda \frac{4\pi^2 \omega}{c^3 v} \int d^3q \left|\phi_i^T(\mathbf{q}+\mathbf{v})\right|^2 \left|\langle \psi_f^P | \mathbf{e}_\lambda \cdot \mathbf{p} | \psi_\mathbf{q}^P \rangle\right|^2$$
$$\times\ \delta(E_f^P - E_i^T + \omega + v^2/2 + \mathbf{q}\cdot\mathbf{v}), \qquad (3)$$

where ω is the photon energy, c is the speed of light, $\mathbf{e}_\lambda$ are the polarization vectors of the photon, $\mathbf{v}$ is the collision velocity, E_f^P and E_i^P are the (negative) binding energies of the final and initial states of the electron, respectively, ϕ_i^T is the momentum distribution of the initial target state, ψ_f^P is the final bound state on the projectile, and $\psi_\mathbf{q}^P$ is a projectile continuum state of momentum $\mathbf{q}$.

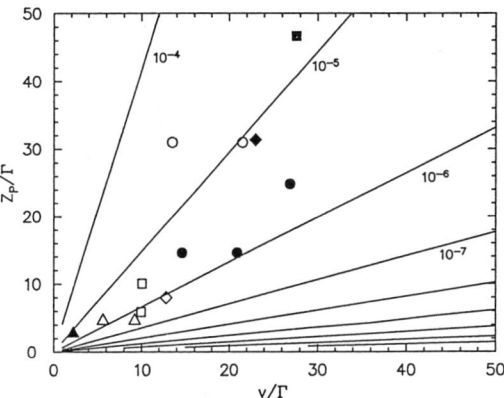

Figure 2: Contour lines of the K-shell radiative recombination cross section in the Z_P/Γ–v/Γ plane. The increment between adjacent contours is one order of magnitude. The cross section in a.u. is indicated at a few contours. Symbols as in Fig. 1.

Correspondingly, the cross section for RCC is obtained by replacing ψ_f^P by a final state in the projectile continuum $\psi_{\mathbf{k}}^P$ [8,18]

$$\frac{d\sigma^{RCC}}{d\omega\, d\Omega\, d^3k} = \sum_\lambda \frac{4\pi^2\omega}{c^3 v}\int d^3q\, \left|\phi_i^T(\mathbf{q}+\mathbf{v})\right|^2 \left|\langle\psi_{\mathbf{k}}^P|\mathbf{e}_\lambda\cdot\mathbf{p}|\psi_{\mathbf{q}}^P\rangle\right|^2$$
$$\times\ \delta(k^2/2 - E_i^T + \omega + v^2/2 + \mathbf{q}\cdot\mathbf{v}), \qquad (4)$$

where k is the final momentum of the electron in the projectile rest frame. The above expressions for the REC and RCC cross sections are on-shell approximations to the *distorted strong potential Born* (DSPB) expressions,[19] in which the state $\psi_{\mathbf{q}}^P$ is off the energy shell since, strictly, the target electron does not propagate with energy $q^2/2$ in the projectile rest frame. For radiative electron capture, differences between the IA and the DSPB are expected to be small.[18,19]

In (3) and (4) the emission angles and energies of the photon are given in the projectile rest frame. The Lorentz transformation to laboratory quantities is given by

$$\cos\theta_{lab} = \frac{\cos\theta + \beta}{1 + \beta\cos\theta}$$
$$\omega_{lab} = \omega\gamma^{-1}(1 - \beta\cos\theta_{lab})^{-1} \qquad (5)$$
$$\frac{d\sigma}{d\omega_{lab}\, d\Omega_{lab}} = \gamma^{-1}(1 - \beta\cos\theta_{lab})^{-1}\frac{d\sigma}{d\omega\, d\Omega},$$

where $\beta = v/c$, $\gamma^{-2} = 1 - \beta^2$, and the angles are taken with respect to the direction of the projectile velocity.

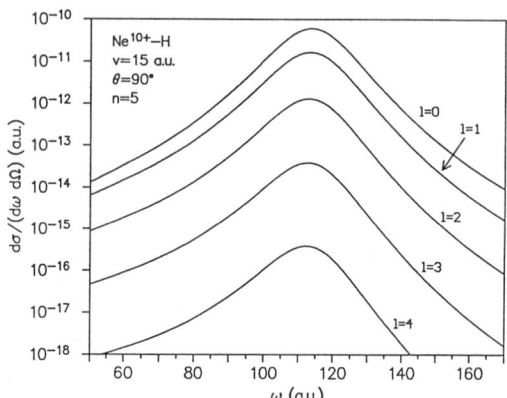

Figure 3: REC cross section for Ne^{10+}–H collisions at $v = 15$ a.u. The photon emission angle is 90° in the projectile frame. The curves represent capture into various l-substates of the $n = 5$ shell of Ne^{9+}.

We restrict ourselves to fully stripped projectiles, for which dipole matrix elements are available in closed form. For free-bound transitions, the dipole matrix elements have been evaluated by Stobbe[20] and for the free-free transitions by Sommerfeld.[21] Assuming further that the target momentum distribution is spherically symmetric and that the magnetic quantum number m of the final state is not observed, one-dimensional quadratures are required in order to evaluate (3) and (4). Notice that in order to obtain $d\sigma^{RCC}/(d\omega\, d\Omega)$ two additional numerical quadratures are necessary. Sommerfeld's ingenious method to integrate out both the photon and electron emission angles[21] can only be used for calculating the singly differential cross section $d\sigma^{RCC}/d\omega$.

RESULTS AND DISCUSSION

We present here results for REC and RCC for the collision system Ne^{10+}–H at $v = 15$ a.u. ($E = 5.6$ MeV/amu). In Fig. 3 is shown the cross section for capture into $n = 5$ broken down into the various l-states. The low angular momenta dominate completely, with $l = 0$ and $l = 1$ accounting for 99% of the cross section. This is in sharp contrast to RR at low energies where the l-distribution peaks at $l/n \approx 0.3$ for high n.[22]

In Fig. 4 is shown the REC-spectrum for $n \leq 10$ summed over all l states for the same collision system. The dashed line indicates the 'RCC-threshold', i.e. the position ω_C of the centroid of the REC peak corresponding to capture into states with $n \to \infty$. Its value is given by

$$\omega_C = v^2/2 + E_i^T. \tag{6}$$

Since $Z_P/v = 2/3$ the Ne X-ray lines are well below ω_C and thus separated from the REC spectrum.

The height of the REC-peaks rapidly approach an n^{-3} dependence with increasing n. However, since the density of bound projectile states close to the RCC-threshold scale as n^3, the high-n peaks will provide a finite contribution to the total X-ray spectrum. In order to study the behavior of the spectrum close

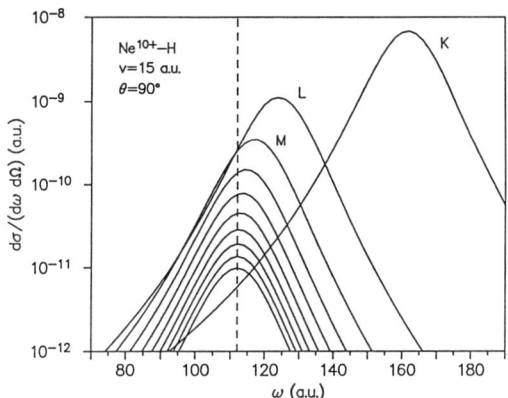

Figure 4: REC cross sections for capture into the 10 lowest lying principal shells of Ne^{9+} in Ne^{10+}–H collisions at $v = 15$ a.u. The photon emission angle is 90° in the projectile frame. The dashed line indicates the position of ω_C (see Eq. (6)).

to $\omega = \omega_C$, the RCC cross section must be included as well. The free-electron bremsstrahlung spectrum has a strict upper cut-off at ω_C, corresponding to the slowing down of the electron to rest with respect to the projectile. With the finite width of the velocity distribution of the initial state the RCC spectrum extends to frequencies above ω_C. In Fig. 5 is shown the REC and RCC spectra and their sum. Shells up to $n = 50$ were included which was necessary in order to obtain convergence. In Fig. 6 is illustrated the convergence of the spectrum as a function of the maximum n included. One remarkable observation is that the shells with $2 < n \leq 50$ contribute about 50% to the total cross section in the vicinity of ω_C.

All the above calculations were performed assuming that the dipole approximation is valid. The wave length of a photon equals the hydrogenic Bohr radius at $\omega \approx 4$ keV, which is close to the K-REC peak energy for the example studied above. For higher projectile charges and/or velocities the photon energies will exceed this value and retardation effects will become increasingly important. The main effect of retardation is to shift the photon angular distribution to backward angles in the projectile frame. For the case of free incident elec-

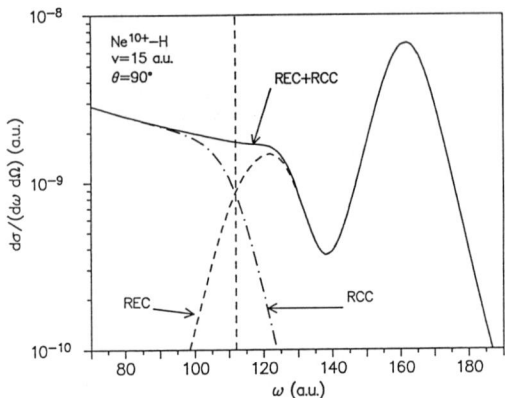

Figure 5: REC and RCC cross sections for Ne^{10+}–H collisions at $v = 15$ a.u. The photon emission angle is $90°$ in the projectile frame.

---

trons captured into the $1s$ state the cross section in the projectile frame (when integrated over photon energies) has the angular dependence

$$\frac{d\sigma^{REC}}{d\Omega} \propto \frac{\sin^2 \theta}{(1 + \beta \cos\theta)^4}. \tag{7}$$

After the Lorentz transformation to the laboratory system the angular distribution is proportional to $\sin^2 \theta_{lab}$, which is equivalent to the angular distribution in the projectile frame without retardation. This cancellation between retardation and Lorentz transformation was experimentally observed by Spindler et al.[4] who explained it based on the free-electron angular distributions. Three-body calculations by Pacher et al.[23] show that the K-REC cross sections are, indeed, symmetric with respect to the perpendicular photon-emission direction to a good degree of approximation when retardation is included and the projectile velocity is not approaching the relativistic region ($v < c/2$). We find the same result for K-REC using the IA. However, for L-REC the cancellation effect is not complete for individual l states. Nevertheless, the *sum* of capture into $2s$ and $2p$ is, indeed, symmetric with respect to $\theta = 90°$. This is a direct consequence of the Fock sum rule for the momentum space wavefunctions of hydrogenic ions,

$$\sum_{lm} |\phi_{nlm}(\mathbf{q})|^2 = n^2 |\phi_{1s}(\mathbf{q})|^2_{Z=Z_P/n}, \tag{8}$$

i.e., the sum over the momentum distributions over all l for a fixed n is equivalent to a $1s$ momentum distribution with an effective charge. This sum rule appears to be applicable to a good degree of approximation even when the full IA is used.

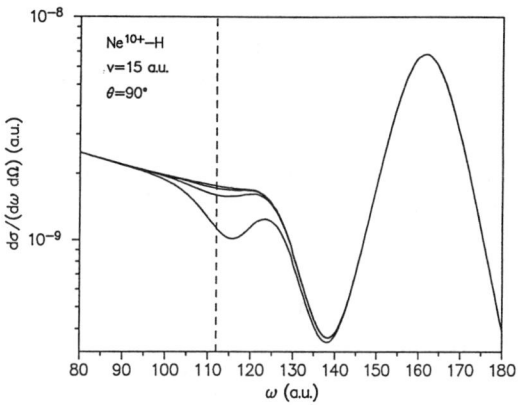

Figure 6: Total REC and RCC spectrum for Ne^{10+}–H collisions at $v = 15$ a.u. The different curves represent the spectrum obtained by summing the REC final states up to various maximum n. From below, the curves correspond to $n_{max} = 2, 5, 10$, and 50, respectively.

The X-ray spectrum (i.e. $d\sigma/(d\omega\, d\Omega)$) as a function of the photon emission angle can be used for indirectly determining capture into individual l states. In our example, the $2p$ content of the L-REC peak is 38 %, 18 %, and 42 % at the laboratory emission angles 30, 90, and 150 degrees, respectively. The backward-forward asymmetry of the ratio of the K- and L-REC peaks is therefore more pronounced than for a pure $2s$ final state.

CONCLUSIONS

We have shown how radiative electron capture into bound and continuum projectile states builds up the total X-ray spectrum. In order to properly represent the spectrum close to the continuum limit of the projectile spectrum, REC into a relatively large number of bound projectile shells must be included. We have discussed criteria for selecting the parameters of the collision system in order to make detailed comparisons between experiments and theory. Heavy-ion storage rings seem to be well suited for this purpose.

REFERENCES

[1] H.W. Schnopper, H.D. Betz, J.P. Delvaille, K. Kalata, A.R. Sohlval, K.W. Jones, and H.E. Wegner, Phys. Rev. Lett. **29**, 898 (1972).
[2] P. Kienle, M. Kleber, B. Povh, R.M. Diamond, F.S. Stephens, E. Grosse, M.R. Maier, and D. Proetel, Phys. Rev. Lett. **31**, 1099 (1973).

3 H.W. Schnopper, J.P. Delvaille, K. Kalata, A.R. Sohlval, M. Abdulwahab, K.W. Jones, and H.E. Wegner, Phys. Lett. **47A**, 61 (1974).
4 E. Spindler, H.-D. Betz, and F. Bell, Phys. Rev. Lett. **42**, 832 (1979).
5 H. Tawara, P. Richard, and K. Kawatsura, Phys. Rev. A **26**, 154 (1982).
6 J.S. Briggs and K. Dettmann, Phys. Rev. Lett. **33**, 1123 (1974).
7 M. Kleber and D.H. Jakubaßa, Nucl. Phys. **A252**, 152 (1975).
8 D.H. Jakubaßa and M. Kleber, Z. Phys. A **273**, 29 (1975).
9 J.E. Miraglia, C.R. Garibotti, and A.D. González, Phys. Rev. A **32**, 250 (1985).
10 R. Anholt, Ch. Stoller, J.D. Molitoris, D.W. Spooner, E. Morenzoni, S.A. Andriamonje, W.E. Meyerhof, H. Bowman, J.-S. Xu, Z.-Z. Xu, J.O. Rasmussen, and D.H.H. Hoffmann, Phys. Rev. A **33**, 2270 (1986).
11 S. Andriamonje, M. Chevalier, C. Cohen, J. Dural, M.J. Gaillard, R. Genre, M. Hage-Ali, R. Kirsch, A. L'Hoir, B. Mazuy, J. Mory, J. Moulin, J.C. Poizat, J. Remillieux, D. Schmaus, and M. Toulemonde, Phys. Rev. Lett. **59**, 2271 (1987).
12 Th. Stöhlker, C. Kozhuharov, A.E. Livingston, P.H. Mokler, Z. Stachura, and A. Warczak, Z. Phys. D **23**, 121 (1992).
13 L.H. Andersen, G.-Y. Pan, and H.T. Schmidt, J. Phys. B: At. Mol. Opt. Phys. **25**, 277 (1992).
14 A. Wolf, J. Berger, M. Bock, D. Habs, B. Hochadel, G. Kilgus, G. Neureither, U. Schramm, D. Schwalm, E. Szmola, A. Müller, M. Wagner, and R. Schuch, Z. Phys. D **21**, S69 (1991).
15 R. Schuch, T. Quinteros, M. Pajek, Y. Haruyama, H. Danared, G. Hui, G. Andler, D. Schneider, and J. Starker, (in press) (1993).
16 D.H. Jakubaßa-Amundsen, R. Höppler, and H.-D. Betz, J. Phys. B: At. Mol. Phys. **17**, 3943 (1984).
17 S. Datz, P.F. Dittner, J. Gomez del Campo, H.F. Krause, T.M. Rosseel, and C.R. Vane, Z. Phys. D **21**, 545 (1991).
18 D.H. Jakubaßa-Amundsen, J. Phys. B: At. Mol. Phys. **20**, 325 (1987).
19 K. Taulbjerg, R.O. Barrachina, and J.H. Macek, Phys. Rev. A **41**, 207 (1990).
20 M. Stobbe, Ann. Phys. (Leipzig) **7**, 661 (1930).
21 A. Sommerfeld, *Atombau und Spektrallinien* 2. Aufl., Bd. 2 (Braunschweig: Vieweg & Sohn) (1939).
22 M. Pajek and R. Schuch, Phys. Rev. A **45**, 7894 (1992).
23 M.C. Pacher, A.D. González, and J.E. Miraglia, Phys. Rev. A **35**, 4108 (1987).

LOW ENERGY HEAVY PARTICLE COLLISIONS

GUIDED ION BEAMS, RF ION TRAPS, AND MERGED BEAMS: STATE SPECIFIC ION – MOLECULE REACTIONS AT meV ENERGIES

D. Gerlich

Fakultät für Physik der Universität, D–79104 Freiburg, Germany

ABSTRACT

This review is concerned with recent progress in the study of ion molecule collisions at very low energies. Special emphasis is put on experiments which make use of ion confinement by inhomogeneous radio frequency (rf) fields. Three areas of development are discussed: 1. combination of the guided ion beam method with lasers for preparation of ions or analysis of reaction products, 2. superposition of a slow ion beam with a neutral beam in a crossed or merged arrangement for obtaining high kinetic energy resolution and very low relative velocities, and 3. variable temperature ion traps for measuring rate coefficients especially at low temperatures and with high sensitivity. Experimental results will be presented for state specific ion preparation, bimolecular reactions, radiative association and cluster growth.

INTRODUCTION

During the last years remarkable progress has been achieved in the development of experimental methods which allow one to study processes with ions at temperatures way below 80 K or at collision energies of a few meV. One of the stimulants for these experimental activities originates from astrophysical applications since in cold interstellar clouds the gas phase chemistry is dominated by ion–molecule reactions. There are, however, also many effects of fundamental interest which occur in collisions with such small relative velocities, that (i) the total available energy is already significantly changed by excitation of the lowest rotational or fine structure states of the reactants or by slight differences in zero point energies in the case of isotope exchange, (ii) only a small number of total orbital angular momenta are involved, (iii) strongly bound intermediates lead to long lifetimes and eventually to unlikely processes such as tunneling or radiative association, and (iv) the angular dependence of the electrostatic interaction between the charge and the neutral can cause orientation effects and rotational state dependent capture probabilities.

A detailed comparative discussion of the various techniques which are available today for the study of ion–molecule collisions at energies of a few K or at very low temperatures can be found in a recent article by M. Smith.[1] In the majority of these experiments the reactant species are surrounded by a dense buffer gas which is cooled by cryogenic methods or by supersonic expansions. This review summarizes recent progress in the study of low energy and state–specific ion–molecule reactions with instruments which make use of different rf based devices in combination with lasers, supersonic beams and cryocoolers. From the wealth of experimental results a few typical examples were selected. These include multiphoton ionization and fragmentation of H_2, state specific cross sections for $Ar^+(^2P)+O_2$ charge transfer, and a discussion of structures in the energy dependence of the $Ar^++H_2 \rightarrow ArH^++H$ cross section. Low temperature studies are reported for $C_2H_2^++H_2 \rightarrow C_2H_3^++H$, for association of C^+ with H_2, and for the growth of H_n^+ clusters with $n-H_2$ and $p-H_2$.

RF FIELDS

The principle of reflecting, focusing, guiding or trapping particles by time–dependent forces has already been treated theoretically by many approaches and applied in quite different experiments. Examples include not only the well–known Paul trap, quadrupole mass spectrometer and the octopole ion beam guide, but also developments in accelerator and plasma physics, the interaction of microwave and laser fields with electrons or the storage of droplets and other macroscopic charged particles in electrical fields alternating with a few Hz. Many aspects of the motion of a charged particle in a fast oscillating electric or electromagnetic field have been discussed recently in a comprehensive review.[2] The central point for understanding the principle of operation of ion guides, traps etc. is the treatment of the equation of motion within an adiabatic approximation, which leads to the introduction of an *effective potential* (also called *quasi–, pseudo– or ponderomotive potential*). In the case of a particle with charge q and mass m, moving in an oscillatory electrical field $E_0 \cos(\Omega t)$, the effective potential is given by

$$V^* = q^2 E_0^2 / 4m\Omega^2 . \qquad (1)$$

In order to remain within the range of validity of the adiabatic approximation one has to operate at sufficiently high frequencies. A more quantitative condition is based on the *adiabaticity parameter* η defined by

$$\eta = 2q |\nabla E_0| / m\Omega^2 . \qquad (2)$$

The semiempirical rule for safe operating conditions, $\eta < 0.3$, and many aspects such as adiabatic conservation of energy, the influence of "defocusing" collisions, the required precision of field geometries and the necessary stability of the amplitude E_0 and the frequency Ω are discussed in the previously mentioned review.[2]

The most prominent example of an rf–based ion trap which has been used for many fundamental experiments in atomic and molecular physics, is the quadrupole or Paul–trap. As illustrated in the left part of Fig. 1 the electrostatic potential of a quadrupole is characterized by a saddle point, a consequence of the Laplace equation $\Delta\Phi=0$. With such a device confinement of a particle in three dimensions only becomes possible if one uses alternating voltages. A vivid description of the influence of the time dependent forces has been given by W. Paul in his Nobel Lecture using a rotating mechanical model as indicated in Fig. 1.[3] Evaluation of Eq. 1 for a quadrupole reveals that not only the electrostatic but also the effective potential depends on the square of the radius. This makes the Paul trap the ideal device for laser sideband cooling and precise localization of a single laser cooled ion.

For guiding or storing a large number of ions in a quadrupole field, rf heating of the ion motion via ion–ion (or space charge) coupling becomes a serious problem. The adiabatic conservation of energy is also impaired if the ions collide with neutral gas. In this case energy can be transferred from the rf field into translational energy, especially if the collision occurs in a region of high field strength. In order to minimize these effects, suitable electrode structures have to be used where the influence of the rf field is reduced. One of the first corresponding extensions was the use of a linear ocotopole[4] which became the central part in more than a dozen Guided Ion Beam (GIB) instruments.[2] Rf ion traps consisting of stacks of ring electrodes have been used successfully in recent years for collision experiments with slow stored ions.[5,6]

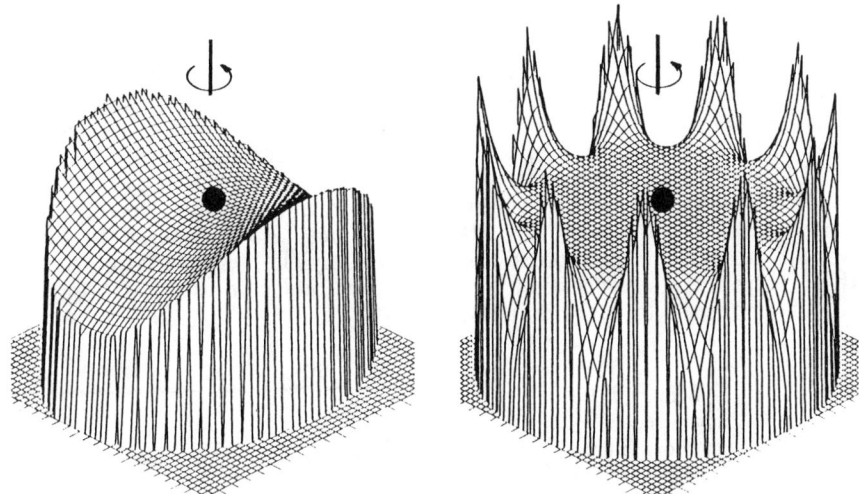

Fig. 1. Mechanical model used by W. Paul for the demonstration of the dynamical stabilization of the indicated particle in a time dependent potential.[3] The surfaces are given by the function $\Phi(r,\varphi)=\Phi_0\,(r/r_0)^n \cos n\varphi$, representing in polar coordinates (r,φ) for $n=2$ and $n=11$ the electrostatic potential of a quadrupole and a 22–pole, respectively. There is no stable region under static conditions whereas rotation of the structures with a sufficiently high frequency Ω leads to confinement of the particle in the vicinity of the saddle point.

Another more recent approach was the construction of a linear 22–pole trap, the electrostatic potential of which is depicted in the right half of Fig. 1. It has a wide plateau in the middle and it is clear that the rf field (or the rotation of the mechanical model) only has an effect if the particle tends to escape. Calculation of the effective potential using Eq. (1) leads to a rotationally symmetric, extremely steep wall which increases proportional to $(r/r_0)^{20}$. Concerning the energy distributions of ions stored in such fields in the presence of collisions the reader is referred again to reference.[2]

GUIDED ION BEAMS

Using ion beam guides in rather simple instruments several experiments have been performed in combination with photoionization methods.[2,7,8] A typical example is a recently developed multiphoton ionization (MPI) source which has unique features concerning collection efficiency, energy and/or mass resolution.[9,10] In this device, neutrals are ionized in a guiding field which is created by 4 × 15 thin rods ($d=1$ mm) arranged along the hyperbolic equipotential lines of a linear quadrupole ($r_0=20$ mm). A pulsed nozzle creates a coaxial molecular beam of neutral precursors. Using a skimmer and a nozzle–laser focus distance of more than 8 cm, ion–neutral collisions are negligible (density$<10^{12}$ cm^{-3}). The mass and energy selectivity of the device is based on the fact that the effective potential of an rf quadrupole is harmonic leading to phase space conserving transmission properties. The position of the image of the small ionization volume (defined by the laser beam and the neutral beam) depends on the frequency of the harmonic secular motion in transverse direction and the flight time. These two parameters can be changed with the rf amplitude, a superimposed DC field, or a gradient between the potential of the ionization volume and that of the quadrupole axis.

Fig. 2 illustrates that an energy resolution of 4 meV can be attained for MPI of H_2 focused via a 33 cm long ion guide onto a 5 mm diameter exit hole. In this example the originally created H_2^+ beam has an energy half width of 33 meV due to space charge effects; however, only certain velocity groups are focused leading to a transmission of 5 well separated peaks. Using in addition at the exit TOF discrimination, an ion beam with a well defined energy can be prepared. In the same way this MPI source has also been used to suppress unwanted masses. For example in the case of acetylene, fragmentation competes with state selective preparation of the mother molecule and C_2H^+ fragments could be suppressed by more than a factor 10.

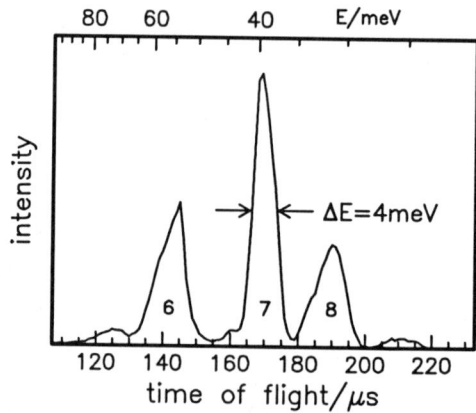

Fig. 2. Time of flight (TOF) spectra of H_2^+ ions, created by REMPI inside of a quadrupole beam guide operated with an extremely good stability parameter of $\eta=0.008$ ($\Omega/2\pi=7.14$ MHz, $V_0 = 34$ V). The focusing properties lead to a transmission which depends critically on the axial ion energy (upper scale). The three main maxima correspond to 6–8 half cycles of the sinusoidal ion trajectories.

In similar arrangements but with shorter nozzle–laser focus distances the effect of rotational energy on collision processes has been studied. In some of these experiments the quadrupole beam guide has been used as a time of flight tube by operating it in the full transmission mode; sometimes a low rf frequency has been chosen to tune to the mass selective mode. For example, the j–dependence of the three–body association process $CO^+(j)+2CO \rightarrow (CO)_2^+ + CO$ has been studied at a density of almost 10^{16} cm^{-3}; the results have been reported in Ref. 7. Another typical measurement using an hydrogen beam with a density of about 5×10^{13} cm^{-3} in the laser focus, is shown in the left panel of Fig. 3. The dominant peak is due to H_2^+ ions, created with resonance enhanced multiphoton ionization (REMPI). A fraction of the ions reacts with the neutral hydrogen and forms H_3^+. Due to the free jet environment the mean collision energy is less than a few meV in the absence of accelerating fields. Tuning the laser to different B(v'=3,j') intermediate states allows preparation of ground–vibrational–state H_2^+ ions with variable rotational energy. A detailed analysis of the H_3^+ production rate has shown, that both vibration and rotation inhibits the reaction, but that the effect of rotation is roughly three times larger.[8,10]

In the same experimental setup, multiphoton fragmentation processes of hydrogen have also been observed. The left panel of Fig. 3 shows at short flight times a small number of ions which are related to 3+1 and 3+2 multiphoton fragmentation of neutral H_2 into $H+H(n\geq2)$. The excited H–atoms are detected by subsequent photoionization within the same laser pulse. The right panel shows the H^+ time–of–flight distributions with higher resolution. The dominant

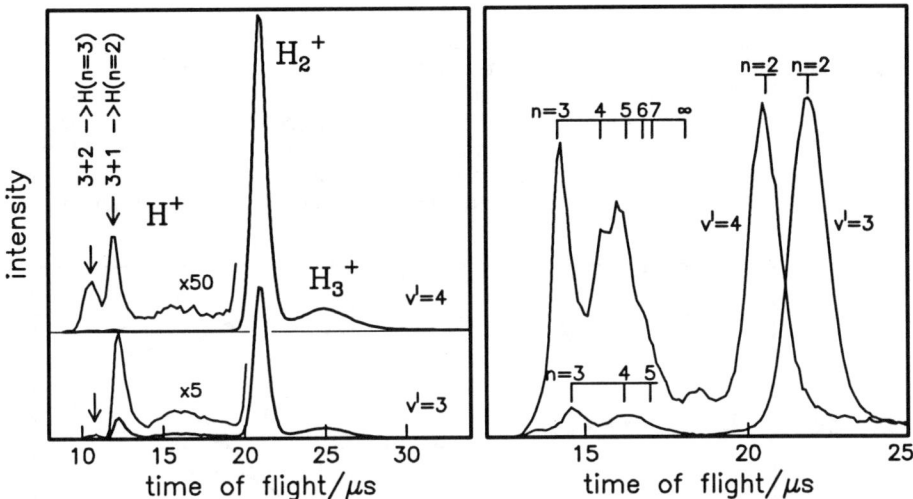

Fig. 3. TOF spectra created by resonant 3+1 multiphoton ionization of H_2 via the B–state (v'=3: 319 nm, v'=4: 315 nm). The main peak of the distribution in the left panel corresponds to H_2^+ ions; the somewhat slower ion peak is due to H_3^+ products from the reaction $H_2^+ + H_2$. The structure at shorter times which is displayed with better resolution in the right panel, is due to multiphoton photofragmentation of the neutral H_2 into $H(n)+H$ with subsequent ionization of the excited hydrogen atom. The marks indicate the expected flight times, calculated for $(3+1)h\nu$ and n=2 or $(3+2)h\nu$ and n≥3 transitions.

peak is due to 3+1 fragmentation where, for reasons of energy, only n=2 can be formed, whereas absorption of 5 photons (before fragmentation) leads to a variety of states with n>2. More information on the dynamics of this fundamental and interesting fragmentation process which proceeds via doubly excited electronic states of H_2 can be found in Ref. 11.

In most applications of the *guided ion beam technique* (GIB) ions are created in a separate part of the machine and, as in the first GIB apparatus[4] they are then injected into the octopole which is usually surrounded by a scattering cell. Such arrangements have outstanding features concerning accessible energy range, 4π collection efficiency, determination of product velocity distributions, and product state analysis by secondary reactions. Today they are routinely used in combination with VUV photoionization sources, ionic clusters, multiply charged ions etc. (for references see Ref. 2).

A selection of results, measured with different GIB instruments for the $Ar^+ + O_2 \rightarrow O_2^+ + Ar$ charge transfer reaction, is plotted in Fig. 4 as effective rate coefficients $<g>\cdot\sigma_{eff}$. Data from a GIB apparatus[12] are represented by the dashed line. They show in qualitative accordance with earlier DRIFT results[13] a pronounced minimum at a few tenths of an eV and a reincrease above the threshold for formation of the a–state. A detailed study of the product angular and kinetic energy distributions and a state selective analysis of the metastable $a^4\Pi$–state by photofragmentation technique has been performed with the same GIB apparatus,[12] emphasizing the universality of this machine. In contrast to what is generally expected from charge transfer processes, efficient excitation of high rotational states has been observed.

In several experimental approaches (for a comparison see Ref. 14) the influence of the Ar$^+$ spin–orbit state on the formation of O_2^+ in the $a^4\Pi$–state has been studied too, especially in the vicinity of the two thresholds indicated by arrows in Fig. 4. Some very recent results, depicted as open symbols, show that the rise of the cross section is significantly shifted towards lower energies if the Ar$^+$ ions are excited. These measurements have been performed using REMPI for state selective preparation of Ar$^+$.[10] From the slopes of the two cross sections it can be concluded that fine structure energy is more efficient than translational energy to produce $O_2^+(a)$.

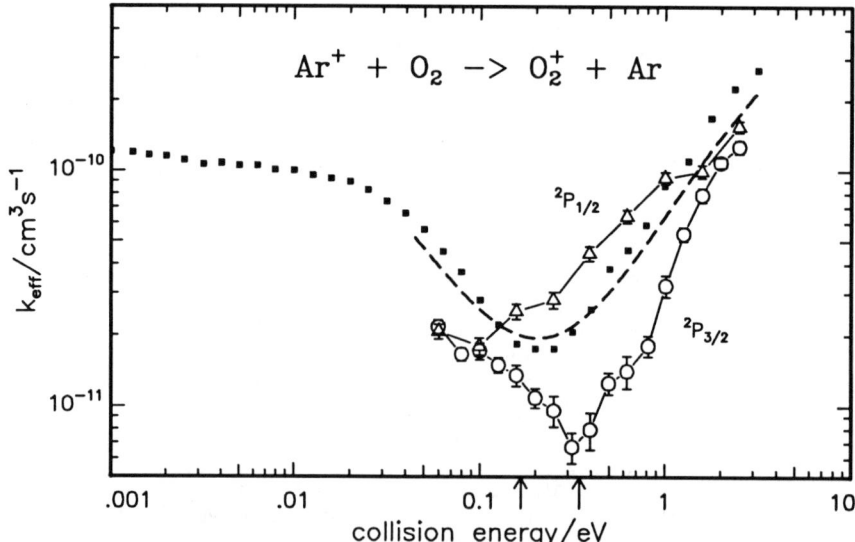

Fig. 4. Effective integral rate coefficients for the charge transfer process Ar$^+$+O$_2$ → O$_2^+$+Ar, measured with different guided ion beam machines (dashed line: standard GIB machine,[12] filled squares: merged beam, open symbols: state selective preparation of the two Ar$^+(^2P)$ using a REMPI source). For the two Ar$^+(^2P)$ states, the arrows mark the thresholds for forming O$_2^+$ in the metastable $a^4\Pi_u$ state.

There are two shortcomings limiting the GIB / scattering cell combination. One is that the lowest attainable energy is often determined by the thermal motion of the target gas, since ion beams with laboratory energies as low as 4 meV have been realized already by using guiding fields. Another disadvantage is that the thermal motion of the target gas causes a significant dispersion of the collision energy. Generally, any experiment provides only information on the so–called effective cross section σ_{eff} which is related to the real function $\sigma(g)$ by the integral equation

$$\sigma_{\text{eff}}(<g>) = \int_0^\infty \frac{g}{<g>} \sigma(g) f(g) \, dg . \qquad (3)$$

The distribution of the relative velocity, $f(g)$, and the mean value $<g>$ is determined by the velocity distributions of the two reactants, $f_1(v_1)$ and $f_2(v_2)$. For thermal target, $f(g)$ is a rather broad distribution,[15] however, significant improvements can be made by using beam–beam techniques as shown in the following section.

CROSSED AND MERGED BEAMS

The first high resolution results obtained by crossing a guided ion beam with a pulsed supersonic beam have been reported for the $C^+ + H_2(j) \rightarrow CH^+ + H$ reaction and a discovered rotational state dependent structure has been explained by a statistical theory which included nuclear spin conservation.[16] Much progress has been accomplished in the last years in Trento with a newly constructed apparatus where a continuous supersonic beam traverses an octopole at right angle.[17] A selection of interesting high–resolution measurements from this group was reviewed recently.[18] Another instrument has been developed in Freiburg by superimposing a slow guided ion beam and a supersonic beam in parallel.[2,19] With this merged beam arrangement, not only is high kinetic energy resolution achievable but processes can also be studied at collision energies as low as one meV. Recent improvements of this apparatus include the integration of a variable temperature ion trap (10 K–300 K) for thermalizing the primary ions and the possibility to operate the neutral beam, in many cases hydrogen, with nozzle temperatures between 35 K and 300 K.[20] A typical experimental study, illustrating not only the features and diversity of the apparatus but also revealing interesting information on low energy reaction dynamics, is the H–D exchange in $D_3^+ + H_2$ collisions, which is endothermic due to differences in zero–point energies. The cross sections, which have been measured at collision energies between 1 meV and 100 meV, show a strong dependence on the internal ion temperature which was varied from 450 K to 20 K.[21]

A well–studied reaction system which has attracted anew a lot of interest in the last years is $Ar^+ + H_2$. Investigation of the hydrogen transfer process with the high resolution guided ion beam apparatus in Trento lead to the discovery of a structure in the energy dependence of the integral cross section.[17] This structure was attributed to v–dependent crossings of the $Ar^+ + H_2$ $^2\Sigma$ potential surface with the $Ar + H_2^+(v)$ vibronic surfaces, which become successively accessible with increasing energy.[22] This qualitative explanation was corroborated recently by theoretical studies performed on empirical potential energy surfaces, both for H_2 and D_2 target.[23] With the dynamical calculations several of the experimental findings could be reproduced, especially the location of the structure in the H_2 case.

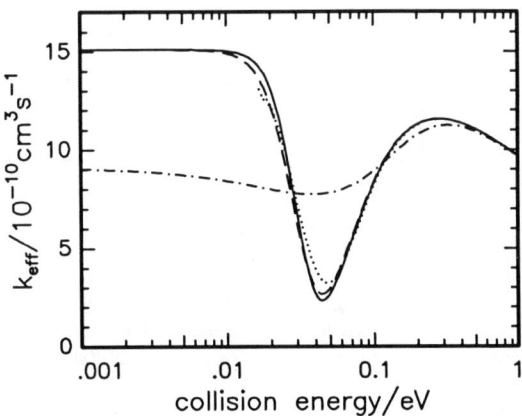

Fig. 5. Illustration of the collision energy resolution of a crossed beam (dotted line), a merged beam (dashed line) and a 300 K scattering cell (dash dotted line) apparatus. The effective rate coefficients have been calculated for $Ar^+ + H_2$ using in Eq. 3 for $\sigma(g)$ the model cross section (solid line) published by Tosi et al.[17] The experimental conditions are given in the caption of Table I.

A serious experimental problem which has not yet been solved is that the structure in the cross section could not be reproduced with the merged beam apparatus.[21] This is surprising since, concerning overall energy resolution and other experimental conditions, the method to superimpose two beams in parallel is in principle superior to crossing them at right angle as exemplified by a few numbers in Table I. The energy resolution, attainable with a scattering cell, a crossed beam, and a merged beam instrument, is also illustrated in Fig. 5 which shows the trial function $\sigma(g)$ published in Ref. 17 and the effective cross sections σ_{eff} calculated with Eq. 3.

Table I. Dependence of E_{Lab} and ΔE_T on E_T for $Ar^+ + H_2$.

E_T/eV	Merged Beam*		Crossed Beam*		Scatt. cell
	E_{Lab}/eV	ΔE_T/meV	E_{Lab}/eV	ΔE_T/meV	ΔE_T/meV
.001	0.52	2.0			
.005	0.81	3.5			
.010	1.07	4.4			
.020	1.50	5.2	0.09	10.0	
.050	2.56	6.4	0.72	12.5	117
.100	4.10	7.3	1.77	15.9	165

*) Experimental conditions Ar^+ ion beam: energy half-width 200 meV, angular spread ±10°, supersonic beam: mach number M=40, nozzle temperature T=77 K.

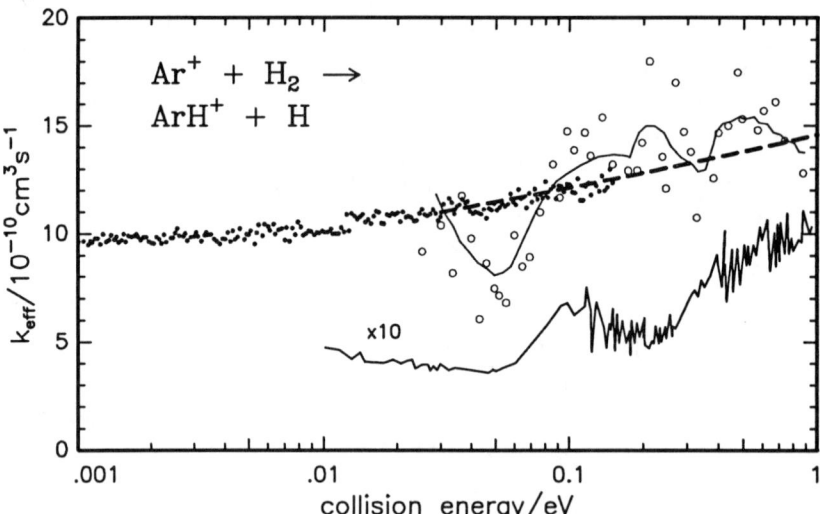

Fig. 6. Effective rate coefficient for the reaction $Ar^+ + H_2 \rightarrow ArH^+ + H$ measured with three GIB instruments. The dashed line represents low resolution scattering cell results.[24] The open circles are results from the Trento high resolution crossed beam experiment,[22] which have been fitted with a structured trial function (upper line). The discrepancy between the smooth merged beam results (solid dots) and the structured crossed beam results is discussed in the text. The fine structured line represents the calculated absolute cross section from Ref. 23 augmented by a factor 10 for better comparison.

Several measured cross sections and the mentioned theoretical result are depicted in Fig. 6. Concerning the representation of the data, two remarks must be made: 1. In order to emphasize deviations from the polarization (or Langevin) cross section, $\sigma(g) \sim 1/g$, and to avoid logarithmic cross section scales, the results are depicted as effective rate coefficients, $k_{\text{eff}} = <g> \cdot \sigma_{\text{eff}}(<g>)$ 2. There are some inconsistencies concerning the determination of the *effective cross section* from experimental data measured with a beam–beam arrangement. In accordance with Eq. 3, σ_{eff} must be calculated from[2]

$$\sigma_{\text{eff}}(<g>) = C\, I_p/I_1 <v_1>/<g>, \qquad (4)$$

where I_p and I_1 are the intensities of the product and the primary ions, respectively, $<v_1> = |<\mathbf{v}_1>|$ is the mean velocity of the ion beam, and $<g> = |<\mathbf{v}_1> - <\mathbf{v}_2>|$ is the mean relative velocity. The vector $<\mathbf{v}_2>$ is the mean velocity of the neutral target. For obtaining absolute values, the proportionality factor $C = (<n_2> L)^{-1}$ has to be determined from the target density distribution $n_2(\mathbf{r})$ and the interaction length L or by using a suitable calibration reaction. σ_{eff} or $<g> \cdot \sigma_{\text{eff}}$ is usually plotted as a function of the mean collision energy, given by $\mu/2 \cdot <g>^2$.

The experimental cross sections, reported in Refs. 22 and 23 are relative values, derived from $\sigma_{\text{exp}} \sim I_p/I_1$ without the velocity weighting factor. They are shown in Fig. 6 as open circles using, in accordance with the above definition, the conversion $k_{\text{eff}} \sim I_p/I_1 \cdot <v_1>$. The thin line, following these data points, is the parameterized function from Ref. 22 which has been used to explain the observed features by endothermic pathways opening at 0.055, 0.19, and 0.37 eV. The dashed line is representative for the data points, measured with a GIB/scattering cell instrument.[24] As can be seen from Fig. 5, such an instrument cannot resolve the minimum which is indicated in the crossed beam data at 0.05 eV. The small solid dots, finally, are results reported from the merged beam instrument.[21] They show a very smooth monotonous increase in good accordance with the scattering cell results but, unfortunately, there is no indication of a structure despite an overall resolution of 6 meV in the region of the pronounced minimum.

At present the discrepancies between the crossed and the merged beam measurements are unexplained. Some possible experimental artifacts have been briefly mentioned in Ref. 18 and also discussed in Ref. 21. In principle, the merged beam apparatus should be superior to the crossed beam apparatus for the following reasons: 1. The energy range where the minimum should appear (20–100 meV) lies in a more convenient laboratory energy range as can be seen from Table I. 2. Problems with product discrimination are less likely due to a more favorable center-of-mass motion of the ArH^+ ions towards the detector. 3. The resolving power is better partly due to the kinematic compression and partly due to the fact, that the angular divergence affects the energy resolution only in second order. 4. The signal rates are larger due to a bigger interaction volume. 5. Ar^+ ions in the excited finestructure state which do not contribute to the structure according to the theory,[23] are at least partly quenched due to the use of the storage ion source.

For all these reasons it is difficult to identify an experimental effect which could have removed a possible structure in the cross section determined with the merged beam apparatus. Moreover, the smooth unstructured cross section was also reproduced in several additional measurements performed with D_2 and $p-H_2$ and using different nozzle temperatures. In order to solve this problem, further measurements are required, e.g. state selective cross sections below 0.1 eV, thermal rate coefficients between 300 and 10 K, cross sections for the charge transfer channel, product velocity and angular distributions, or ArH⁺ product state distributions. Also the results of the theoretical prediction[23] have to be checked in more detail. It is disquieting that the absolute values of the calculated cross sections are too small by more than a factor 20 (see Fig. 6). Also the $\sigma(1/2)/\sigma(3/2)$ ratio deviates very significantly from experimental 300 K results.[10]

To balance the above mentioned advantages of the merged beam method it must be noted that also the crossed beam arrangement is superior in certain respects, e.g. due to the well localized interaction zone or since it is technically less complicated. For a better judgment of both rather new GIB–supersonic beam combinations more tests and more comparative studies are required. A step into this direction is a recent joint experimental study of the $N^+ + D_2 \rightarrow ND^+ + D$ reaction[25] where good agreement between the data from the two instruments corroborates that they both are indeed able to resolve structures at low energies.

There are, concerning beam–beam experiments with slow ions, still many developments and improvements necessary. Especially the overall sensitivity of these machines is for some applications (e.g. threshold studies) still very low. The sensitivity could be augmented on one side by increasing the beam intensities. This, however, is limited by several fundamental and practical boundary conditions. On the other side one could increase the interaction time by extending the overlapping region or by using much slower beams. This would indeed be possible by combining laser cooling of neutrals with very slow guided or stored ion beams. The ion traps, described in the next section, could be regarded as a step into this direction.

ION TRAP EXPERIMENTS

At present, the interaction time in the ion beam experiments described above is below 100 μs whereas in an ion trap, a prepared ion ensemble can be allowed to interact with neutrals for seconds or much longer. This leads to a significant increase of sensitivity and enables one to observe very slow reactions, e.g. the astrophysically important process of radiative association. This has been demonstrated first in the pioneering work of Dunn and coworkers[26] who utilized a Penning trap. Significant progress has been made by the development of variable temperature rf ion traps and a selection of results for radiative association processes[6] and for other reactions of astrophysical interest[20] has been published recently.

The construction of a variable temperature ion trap using a ring electrode geometry or a multipole structure is rather simple. The main complications arise from the need to cool not only all surrounding walls to the desired temperature but also the trap electrodes. Since they always absorb a small amount of rf power, a good thermal contact to the cooling device is required, with the additional condition of electrical insulation. A l–N₂ cooled version has been described in Ref. 5, a ring electrode trap mounted onto a cryocooler in Ref. 6. One of the newest versions, a linear 22–pole ion trap, is shown schematically in Fig. 7. The whole structure is thermally connected to the cold head of a closed cycle refrigeration system. The nominal temperature can be varied between $T_n = 10$ K and 300 K; the actual ion temperature has been determined using different temperature dependent rate coefficients and at present the coldest ion cloud is characterized by 15±5 K. Since these "ion thermometers" are not very accurate, work is in progress to use laser methods.

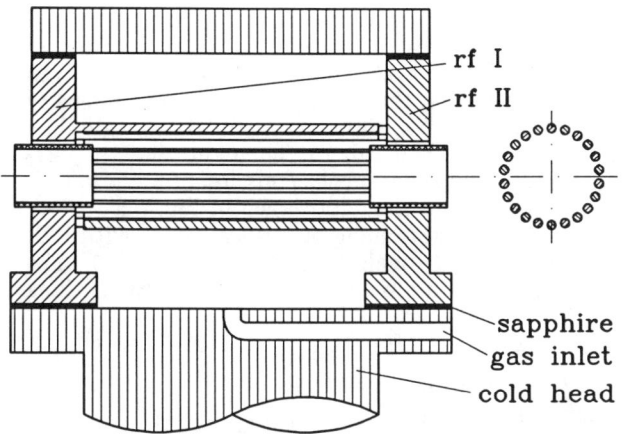

Fig. 7. Variable temperature 22—pole ion trap (rod diameter 1 mm, rod length 36 mm, $r_0=5$ mm). The rf electrodes I and II are mounted onto the cold head using sapphire for electrical insulation. In axial direction the ions are confined by electrostatic voltages, applied to the two cylindrical tubes. These electrodes are operated in a pulsed mode for injecting externally prepared ions and for extracting the trap content for mass analysis.

Some velocity distributions, probed via the Doppler profile, have been presented recently.[27] In the case of molecular ions the rotational population will be used as an independent measure for the ion–buffer gas interaction.

The ion trap is the central part of an apparatus which has been described elsewhere.[2,6,20] Ions are prepared in a separate ion source, mass selected and then injected via the pulsed entrance electrode (see Fig. 7). For thermalization and also for additional capture of fast ions which otherwise would escape, a short (5 ms FWHM, decay time 3.5 ms) and very intense pulse of buffer gas (e.g. [He]$>10^{14}$ cm^{-3}) is injected using a piezo electric valve. The mean storage time of the ions is usually many minutes and mainly limited by reactions with the background gas, the pressure of which is estimated to be below 10^{-11} mbar at $T_n=10$K.

Rate coefficients for reactions with added target gas are determined from the temporal change of the ion composition which is measured by periodically filling the trap, extracting its content after various times, and analyzing its

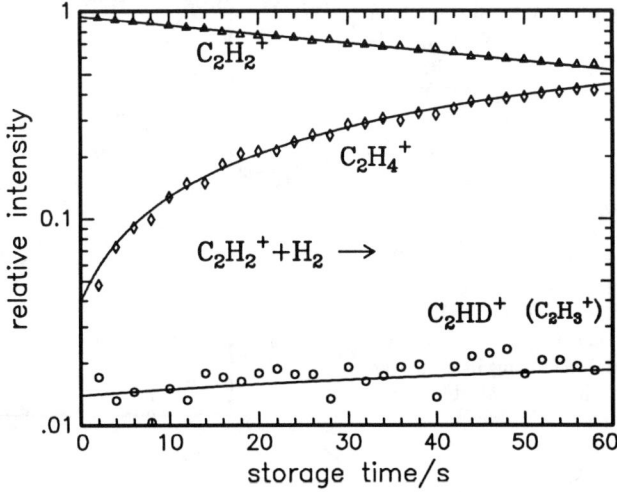

Fig. 8. Time evolution of $C_2H_2^+$ ions stored between 2 s and 60 s in the 22—pole ion trap at 10 K and at a p—H$_2$ density as low as 1.6×10^9 cm^{-3}. The dominant channel is formation of $C_2H_4^+$ via radiative association, the few ions, produced with mass 27, are predominantly C_2HD^+. The lines are solutions of a simplified rate equation approach.

mass composition with a quadrupole mass spectrometer. As an example, Fig. 8 shows reactions of $C_2H_2^+$ with $p-H_2$ at $T_n=10$ K, observed at storage times up to 1 min. At this low temperature almost all acetylene ions react with hydrogen through the association channel. Note that in this example the hydrogen density was so low (1.6×10^9 cm^{-3}) that three body collisions are completely negligible: the rate for tree—body association is lower than 1 day^{-1}! Therefore the $C_2H_2^+\cdot H_2$ complexes can only stabilize themselves by spontaneous emission of a photon.

For a detailed quantitative analysis of the data shown in Fig. 8 a variety of possible reaction paths must be considered (among others: collisions with HD, perturbations from ^{13}C containing ions, a small amount of warm H_2). The curves plotted in Fig. 8 are solutions of a simplified rate equation system using the parameters:

$$k_{26\to 27}=7\times 10^{-14} \text{ cm}^3\text{s}^{-1},\ k_{26\to 28}=6\times 10^{-12} \text{ cm}^3\text{s}^{-1},\ k_{27\to 29}=2.5\times 10^{-12} \text{ cm}^3\text{s}^{-1}$$
$$I_{26}=0.982,\ I_{27}=0.014,\ I_{28}=0.004$$

The indices refer to the masses of the ions; the indicated initial conditions account in an approximate way for processes occurring during the injection and thermalization of the primary ions. A large source of error ($\pm 50\%$) in the determination of absolute rate coefficients at such a low density is its actual value and perturbations by gas impurities.

Despite these uncertainties it can be concluded from Fig. 8 without any doubt that the bimolecular channel to form $C_2H_3^+$ is very slow if present at all. The question whether the reaction $C_2H_2^+ + H_2 \to C_2H_3^+ + H$ is endo— or exothermic has been discussed quite often this year.[20,28,29] It has been known from early ICR and SIFT studies that this reaction is slow at room temperature and becomes immeasurable at lower temperatures.[30] This temperature dependence has been measured with the 22–pole ion trap and thermal rate coefficients are shown in Fig. 9 in an Arrhenius plot. Evaluation of several measurements has lead to an endothermicity of (50 ± 20) meV in good accordance with the earlier conclusions.

The question whether this endothermicity is in reality a barrier was raised by measurements performed with a low temperature flow reactor.[31] In

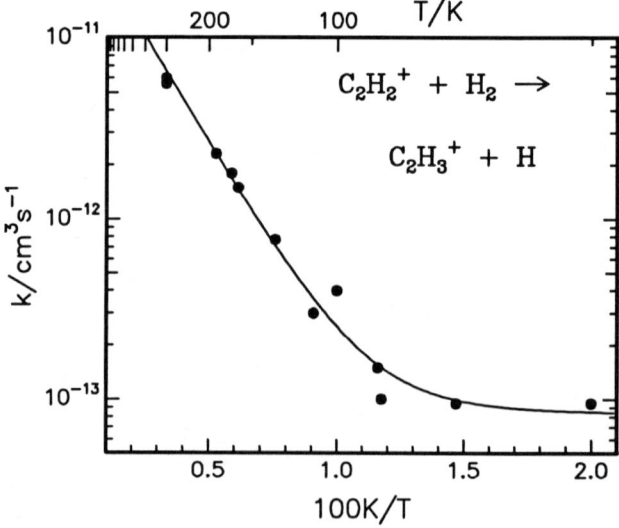

Fig 9. Arrhenius plot of the temperature dependence of rate coefficients for formation of $C_2H_3^+$ in collisions of $C_2H_2^+$ with $n-H_2$. The indicated fit leads to an endothermicity of $\Delta H= 48$ meV. The flattening of the curve at temperatures below 100 K is predominantly due to formation of C_2HD^+ in collisions of $C_2H_2^+$ with HD occurring in hydrogen with a natural abundance of 3×10^{-4}.

this instrument a rate coefficient of 10^{-12} cm^3/s was measured at $T_n=10$ K and an increasing tendency towards 10^{-11} cm^3/s was observed when the temperature was lowered to 1 K. This increase was interpreted with a tunneling mechanism.[31,32] None of the ion trap studies could corroborate these conclusions. In all measurements products with mass 27 have been found to be formed very slowly as in Fig. 8, and, below 60 K, with an almost temperature independent rate coefficient of less than 10^{-13} cm^{3}s^{-1}. The main fraction of these mass 27 products could be attributed to C_2HD^+ instead of $C_2H_3^+$ which are formed in collisions with the natural abundant HD. Experiments with HD depleted hydrogen supported this conclusion qualitatively. Another important experimental observation is that, with p–H$_2$, the lifetime of the $C_2H_2^+ \cdot H_2$ complex increases by a factor 4 relative to n–H$_2$. This has been concluded both from the corresponding increase of the radiative and the ternary rate coefficient. In contrast formation of mass 27 was seen to be slightly smaller with p–H$_2$. For a better judgment of the controversial results which are discussed in more detail in Refs. 20 and 33, two hints must be added. 1. The free jet experiment operates in a non–uniform flow with densities up to 10^{15} cm^{-3} whereas densities in the the ion trap are typically below 10^{11} cm^{-3}. 2. Rf heating of the ions can be neglected since the collision temperature is predominantly determined by the light H$_2$ target.

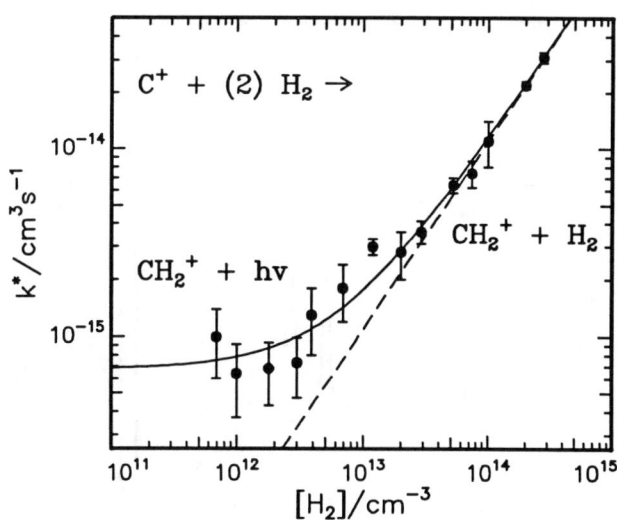

Fig. 10. Ternary and radiative association of C^+ with n–H$_2$ at a nominal temperature of 10 K, measured over a wide range of hydrogen densities. The error bars indicate exclusively the statistical errors. The solid line represents a fit of Eq. 5 to the measured effective rate coefficients with the parameters
$k_3 = 1.1 \times 10^{-28}$ cm^6/s,
$k_r = 6.7 \times 10^{-16}$ cm^3/s.
The dashed line shows the function $k^* = [H_2] k_3$.

A second example for applications of the 22–pole ion trap is concerned with the radiative association process $C^+ + H_2 \rightarrow CH_2^+ + h\nu$. This very fundamental reaction is the first step in the chain of formation of small hydrocarbons in interstellar clouds. Since this process is rather slow, experiments are best performed at densities where stabilization of the collision complex by a third body cannot be neglected as in the example in Fig.8. One measures therefore a density dependent effective rate coefficient which can be approximated by

$$k^* = k_3 [H_2] + k_r , \quad (5)$$

provided the density is not too high. Here k_3 is the termolecular and k_r the bimolecular radiative rate coefficient. Such a dependence is shown in Fig. 10 for

$C^+ + H_2$ association at $T_n = 10$ K. The results cover a density range of three orders of magnitude and can be nicely fitted with the function given by Eq. (5). Contributions from the additive term k_r become measurable below 10^{13} cm^3/s as indicated by the deviations of the experimental points from the dashed line. More details about this system can be found in Ref. 33.

The third and last example, illustrating the versatility of the variable temperature 22–pole ion trap, is concerned with the growth of clusters. In contrast to cluster formation in free jet expansions, where one has only some rather global control of the parameters, ion trapping allows one to perform a systematic study of the dynamics of the association and cooling processes under well defined thermal conditions. Especially important is that the choice of an adequate density enables one to stretch the time scale for the growth processes such that individual steps can be traced in great detail. Another advantage is that externally created clusters of a given size can be injected.

Ternary rate coefficients for reaction of H_3^+ towards H_5^+, H_7^+, and H_9^+ have been measured at various temperatures and densities.[34] Fig. 11 shows a few selected mass spectra, recorded at different storage times for p–H_2 and n–H_2 under otherwise identical conditions ($T_n = 10$ K, [H_2]=3.5×10^{14} cm^{-3}). A comparison of the two middle spectra, recorded at different storage times, reveals that cluster formation is much faster with p–H_2 target than with n–H_2.

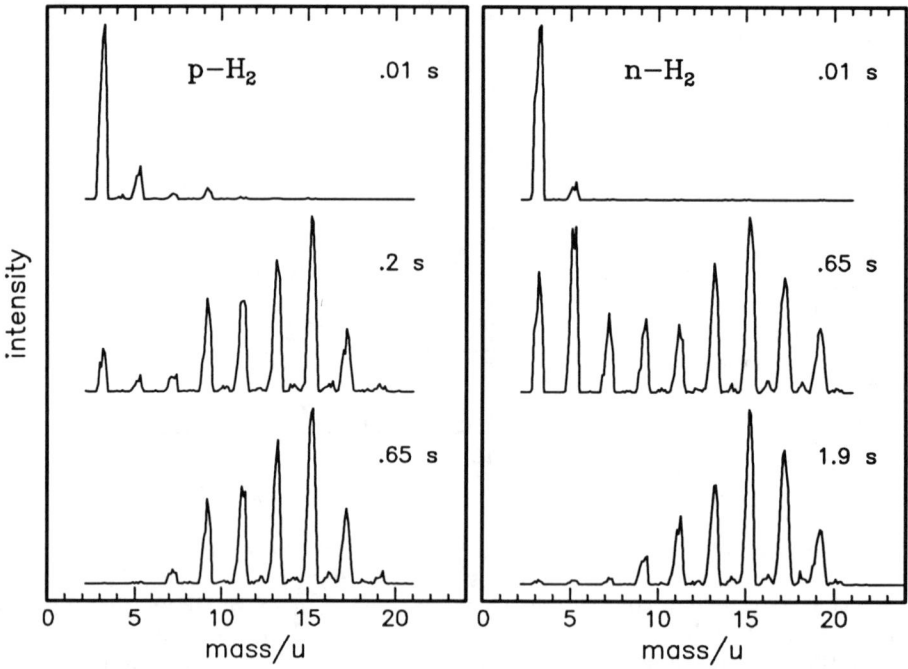

Fig. 11. Time dependence of the ion trap content after injection of H_3^+ (T=10 K, [H_2] = 3.5×10^{14} cm^{-3}). Comparison of the mass spectra recorded at the indicated storage times reveals, that hydrogen clusters grow significantly faster in p–H_2 than in n–H_2. The two lowest spectra represent in both cases already the thermodynamic equilibrium distributions. Their similarity is completely unexpected due to the rotational energy which can be provided by the n–H_2.

This is not too surprising since in n–H_2, 75% of the molecules are in their first rotational state corresponding to a large amount of energy in the 10 K environment. If one takes in addition into account that proton exchange is fast in $H_3^+ + H_2$ collisions, diminishing nuclear spin restrictions, this energy can really affect the reaction $H_3^+ + 2H_2 \rightarrow H_5^+ + H_2$. A quantitative analysis of this reaction has lead to the conclusion that the rotational energy slows down the cluster formation process by a factor 6.

The dominant peaks in Fig. 11 are all at odd masses and belong to clusters with a "structure" $H_3^+ \cdot (H_2)_x$. There are some small peaks at even masses which belong predominantly to deuterated clusters formed in collisions with HD. A qualitative test has been performed by adding small amounts of D_2. It was found that already minor traces ($<.5\%$) lead to a significant shift of the mass distribution towards higher masses. This effect of isotopic fractionation at low temperatures is due to differences in zero point energies and well–known in interstellar chemistry, where certain molecules have been observed to be enriched in heavy isotopes.

A very surprising observation refers to the two lowest mass spectra where, due to the high density and the low temperature, the storage time is already long enough that the cluster distributions are in thermal equilibrium. Further growth is counterbalanced by collision induced fragmentation. Completely unexpected, the two distributions are very similar, and a closer look reveals that for n–H_2 the mean cluster size is shifted to even larger clusters! This is in full contradiction to any interpretation which is based exclusively on energy arguments. What are the explanations? Can they be found in restrictions imposed by the permutation symmetry? Or can the additional angular momentum provided by the rotating hydrogen molecules, lead to a higher centrifugal barrier and therefore to additional stabilization of the weakly bound H_{17}^+ clusters?

CONCLUSIONS

In 1974, the first publication on the guided ion beam technique appeared,[4] in 1985, a progress report was presented on the ICPEAC.[35] In the following years the interplay between technical developments of rf based devices and the study of reactivity and structure of ions has lead to a lot of new discoveries. Since this has been described in a recent very comprehensive review[2], it has been tried in the present paper to combine review material with a few problems of current interest and to conclude with partly unexplained results. Hopefully, this selection has demonstrated that combinations of guided ion beams with laser methods and supersonic beams have lead to new outstanding instruments, that rf ion traps are versatile tools, and that these instruments have provided significant contributions to the understanding of low energy ion molecule reactions.

ACKNOWLEDGMENTS

Significant contributions to this work are due to the former PhD. students S. Scherbarth, G. Kaefer, M. Schweizer, the present PhD. students A. Sorgenfrei, O. Wick, W. Paul, S. Mark and several Diploma students. The author would like to thank especially Prof. Ch. Schlier for stimulating discussions and for providing the scientific environment over many years. Support of the Deutsche Forschungsgemeinschaft (SFB 276) is acknowledged.

REFERENCES

1. M. A. Smith in: *Current Topics in Ion Chemistry and Physics*, Vol. 2, C. Y. Ng, T. Baer, and I. Powis, eds. (Wiley, New York, 1993).
2. D. Gerlich, *Adv. in Chem. Phys.* **LXXXII**, 1 (1992).
3. W. Paul, *Nobel-Lecture,* Stockholm (1989) and *Phys. Bl.* **46**, 227 (1990).
4. E. Teloy, D. Gerlich, *Chem. Phys.* **4**, 417 (1974).
5. D. Gerlich and G. Kaefer, *Ap. J.* **347**, 849 (1989).
6. D. Gerlich and S. Horning, *Chem. Rev.* **92**, 1509, (1992).
7. D. Gerlich and T. Rox, *Z. Phys. D* **13**, 259 (1989).
8. S. Anderson, *Adv. in Chem. Phys.* **LXXXII**, 177 (1992).
9. M. Schweizer, S. Mark, and D. Gerlich in: *ICPEAC 1993,* Book of contributed papers, eds. T. Anderson et al. 40, (1993).
10. M. Schweizer, Dissertation, Freiburg (1993).
11. J. W. J. Verschuur and H. B. van Linden, *Chem. Phys.* **129**, 1 (1989)
12. S. Scherbarth and D. Gerlich, *J. Chem. Phys.* **90**, 1610 (1989).
13. I. Dotan, W. Lindinger, *J. Chem. Phys.* **76**, 4972 (1982).
14. O. Dutuit et al. in: *Symposium on Atomic and Surface Physics*, eds. D. Bassi, M. Scotoni, and P. Tosi, Pampeago, p. 1.26 (1992).
15. P. J. Chantry, *J. Chem. Phys.* **55**, 2746 (1971).
16. D. Gerlich, R. Disch, and S. Scherbarth, *J. Chem. Phys.* **87**, 350, (1987).
17. P. Tosi, F. Boldo, F. Eccher, M. Filippi and D. Bassi, *Chem. Phys. Lett.* **164**, 471 (1989).
18. P. Tosi, *Chem. Rev.* **92**, 1667 (1992).
19. D. Gerlich in: XII *International Symposium on Molecular Beams,* edited by V. Aquilanti, Perugia, 37 (1989).
20. D. Gerlich, *J. Chem. Soc., Faraday Trans.* **89**, 2199 (1993).
21. D. Gerlich, H.J. Jitschin, O. Wick in: *Symposium on Atomic and Surface Physics*, eds. D. Bassi, M. Scotoni, and P. Tosi, Pampeago, 2.32 (1992).
22. P. Tosi, F. Eccher, D. Bassi, F. Pirani, D. Cappelletti and V. Aquilanti, *Phys. Rev. Letters* **67**, 1254 (1991).
23. P. Tosi, O. Dmitrijev. Y. Soldo, and D. Bassi, D. Cappelletti, F. Pirani, and V. Aquilanti, *J. Chem. Phys.* **99**, 985, (1993).
24. K. M. Ervin and P.B. Armentrout, *J. Chem. Phys.* **86**, 2659, (1987).
25. P. Tosi, O. Dmitrijev, D. Bassi, O. Wick, D. Gerlich, in preparation
26. S. E. Barlow, G. Dunn, and M. Schauer, *Phys. Rev. Lett.* **52**, 902 (1984).
27. W. Paul and D. Gerlich in: *ICPEAC 1993,* Book of contributed papers, eds. T. Anderson et al., 807 (1993).
28. E. Herbst, K. Yamashita *J. Chem. Soc. Faraday Trans.* **89**, 2175 (1993).
29. D. Smith, J. Glosik, V. Skalsky, P. Spanel, and W. Lindinger, *Int. J. of Mass Spectr. and Ion Proc.*, in print (1993).
30. D. Smith, N. G. Adams, and E. E. Ferguson, *Int. J. of Mass Spectrom. Ion Proc.* **61**, 15 (1984).
31. M. Hawley and M. A. Smith, *J. Chem. Phys.*, **96**, 1121 (1992).
32. E. Herbst and K. Yamashita, *J. Chem. Soc. Faraday Trans.* **89**, 2175 (1993).
33. D. Gerlich in: *50th Int. Meeting on Physical Chemistry of Molecules and Grains in Space*, Book of invited papers, (1993).
34. D. Gerlich, G. Kaefer, W. Paul, in *Symposium on Atomic and Suface Physics*, edited by T.D. Märk and F.Howorka, Obertraun, 332 (1990).
35. D. Gerlich in: *ICPEAC 1985, Invited paper* eds. D.C. Lorents et al. Elsevier, 541 (1986).

LOW-ENERGY COLLISIONAL AUTOIONIZATION PROCESSES

B.Brunetti, S.Falcinelli, and F.Vecchiocattivi
Dipartimento di Chimica, Università di Perugia
06100 Perugia, Italy

ABSTRACT

The general features of collisional autoionization processes involving metastable rare gas atoms studied in recent years are outlined. For atom-atom system the role of the spin-orbit state of the metastable atom in determining the dynamics of the ionization of rare gas atoms is reported as it results from recent experimental studies. For atom-molecule systems, the cases where a weak interaction between the two partners is present both before and after the ionization event are discussed with reference to the Ne^*-N_2 example. The other case is then considered of atom-molecule systems where a chemical interaction affects the ionization dynamics either before or after the ionization event.

INTRODUCTION

Autoionization processes in slow molecular collisions can occur because the autoionization time is usually shorter than the characteristic molecular collision time at thermal energies ($\sim 10^{-12}$ s). The basic requirement is that the two partners should have enough internal energy to produce a collision complex degenerate with the ionization continuum. Collisional autoionization must present several analogies with photoionization. For example, in both cases when the electron is ejected, the measurement of its energy and momentum provides a detailed spectroscopy of the ionic product, which can then be correlated with its subsequent dynamical evolution. However, while in photoionization the conditions of the autoionized molecules are determined by the energy and polarization of the photon, in the collisional autoionization process these conditions are determined by the collision characteristics of the system under consideration, such as relative velocity, internal states of the reactants, relative orientation, etc.

A collisional autoionization process can be schematically written as follows:
$$X + Y \rightarrow [X...Y]^*$$
where X and Y are atoms or molecules and $[X...Y]^*$ the collision complex in an autoionizing state,
$$[X...Y]^* \rightarrow [X...Y]^+ + e^-.$$
After an $[X...Y]^+$ ionic complex is formed, the collision continues towards the final ionic products
$$[X...Y]^+ \rightarrow \text{ion products}$$
In the literature this collisional autoionization process is often called

"Penning" ionization after the early observation in 1927 by F.M.Penning.[1] This process has long attracted the attention of the scientific community, as shown by the large number of papers and review articles on this topic.[2,3] Specific features make these processes very interesting from a fundamental point of view. However many applications of collisional autoionization to important fields like radiation chemistry, plasma physics and chemistry, combustion processes, and the development of laser sources, are also possible.

Several experimental techniques are used to study the microscopic dynamics of these collisional autoionization processes. It is well established that the most valuable information about the dynamics of a collisional process is provided by molecular beam scattering experiments. In such cases one can study the process by detecting the metastable atoms, the electrons or the product ions. In Fig.1 some possible experiments are schematically shown.

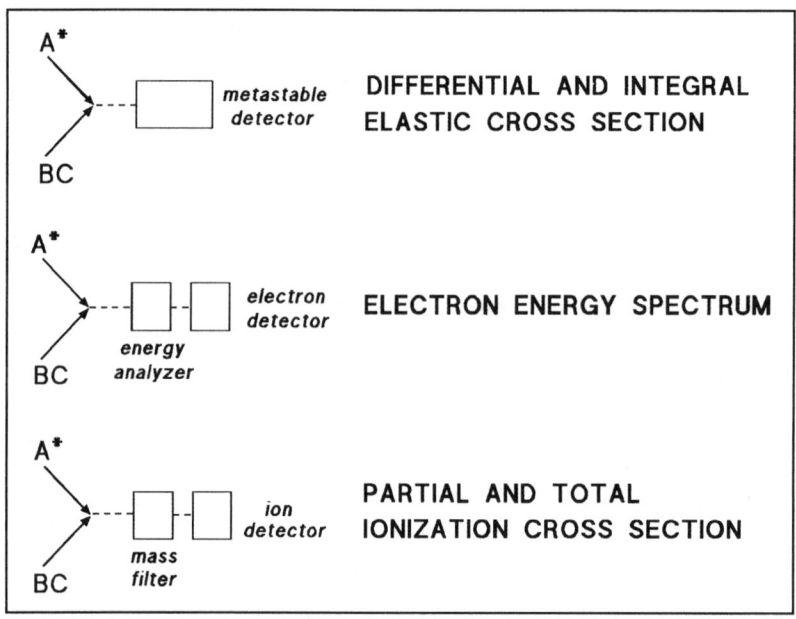

Figure 1
Possible collisional autoionization experiments.

In general elastic scattering experiments provide information mainly about the pre-ionization dynamics, electron energy spectra provide information about the dynamics of the autoionization event, and ion mass spectrometric experiments mainly about the post-ionization dynamics.

The rare gas atoms, when excited to their first metastable levels, are very suitable for these experimental studies because of their high energy content and relatively long life-time, which allows them to survive along beam path lengths typical of the laboratory molecular beam experiments. The electronic energy and life-time values of metastable rare gas atoms are listed in Table I. The energy values are large enough for an autoionizing complex to be formed in most cases. Metastable helium atoms have enough energy to ionize all known atomic and molecular species, except ground state helium and neon atoms, while in the case of metastable neon atoms the exceptions also include fluorine atoms.

Table I
Some characteristics of metastable rare gas atoms

Atoms	Excitation energy (eV)	Lifetime (s)
$He^*(2^3S)$	19.820	4.2×10^3
$He^*(2^1S)$	20.616	3.8×10^{-2}
$Ne^*(^3P_2)$	16.619	24.4
$Ne^*(^3P_0)$	16.716	430
$Ar^*(^3P_2)$	11.548	55.9
$Ar^*(^3P_0)$	11.723	44.9
$Kr^*(^3P_2)$	9.915	85.1
$Kr^*(^3P_0)$	10.563	4.9×10^{-1}
$Xe^*(^3P_2)$	8.315	150
$Xe^*(^3P_0)$	9.447	7.8×10^{-2}

In the present paper we review some general features of autoionization processes, involving rare gas metastable atoms, as have been obtained from recent experimental studies.

ATOM-ATOM SYSTEMS

The $He^*(2^3S)$, $He^*(2^1S)$-H, D systems constitute the natural meeting ground between experiment and theory: the small number of electrons and the simple nature of the interactions allow a direct and rigorous comparison between experiments and theoretical treatments of the collisional autoionization dynamics. These systems have been studied by several more or less sophisticated techniques, up to having presently deep understanding of these processes. Recently Siska and coworkers[4] reported a study of the angular and energy dependence of the product ions leading to a detailed characterization of the dynamics of such processes. Very recently Hotop and coworkers[5] have reported the measurement of the angular and energy distribution of ions and electrons, making it possible a further advancement toward a complete characterization of the dynamics of this system.

The ionization of heavier rare gas atoms by metastable He^* and Ne^* atoms has been also intensively studied in the past recent years. In the case of the ionization of a rare gas atom, Rg, by He^*, the relevant potential energy curves involved are schematically shown in Fig.2. While for the entrance He^*-Rg channel

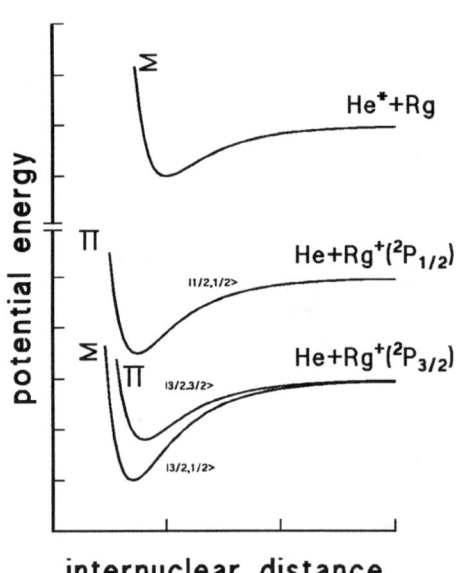

Figure 2

Potential energy curve scheme of He*-heavier rare gas.

only a Σ symmetry potential energy curve describes the interaction, the He-Rg$^+$ exit channel interaction is represented by three $|j,\Omega\rangle$ potential energy curves. These correlate at short range with the Σ and Π states and at long range with the two spin orbit states of Rg$^+$. Hoffmann and Morgner[6] have shown that in these systems during the ionization the He-Rg$^+$ product ion is formed predominantly in the Σ symmetry. Therefore the three $|j,\Omega\rangle$ curves are populated accordingly to their Σ character which varies differently on the internuclear distances for the $|3/2,1/2\rangle$ and $|1/2,1/2\rangle$ states and is constantly zero for the $|3/2,3/2\rangle$ one. This allows to describe the state population of final ions which has been measured in these systems.[7]

In the case of Ne*-Rg the situation is different. In fact in this case also the entrance channel is described by a manifold of states.[8] However the Ne*-Rg curves are practically parallel at long range. Nevertheless this leads to an interesting phenomenon. Morgner has shown that also in this case a Σ symmetry in the entrance and exit channel must be the one which is the most responsible for the ionization.[9] The potential energy curves for Ne*-Rg are schematically shown in Fig.3. The $|j,\Omega\rangle$ curves for the entrance channel give rise to the ionization accordingly their Σ character; the final ion curves are also populated in the same way. This can satisfactorily explain the qualitative behavior of the two j states of the metastable Ne* atoms and the final ion population in these systems.[2]

ATOM-MOLECULE SYSTEMS

In the case of atom-molecule systems the situation can be more complicated than in the atom-atom case. This is not only due to the higher dimensionality, but also to the bond characteristics of the system.

The simplest case is the one where the interaction between the excited atom and the molecule is weak, i.e. of van der Waals type, and the same papplies also for the interaction between the final product, i.e. the molecular ion and the rare gas atom. A typical case of this type is Ne*-N_2. In this case the reaction proceeds as follows:[10] in a first step the collisional complex autoionizes

Ne*-N_2 → [Ne...N_2]$^+$ + e$^-$,

then the ionic complex continues the collision dynamics leading to N_2^+ and NeN_2^+

[Ne...N_2]$^+$ → Ne + N_2^+
→ NeN_2^+.

However some of the NeN_2^+ ions are formed with enough energy to predissociate after a characteristic life time:[2]

NeN_2^+ → Ne + N_2^+.

These observations are conveniently discussed introducing the concept of an optical potential which describes the interaction between the two neutral partners before the ionization event. For this

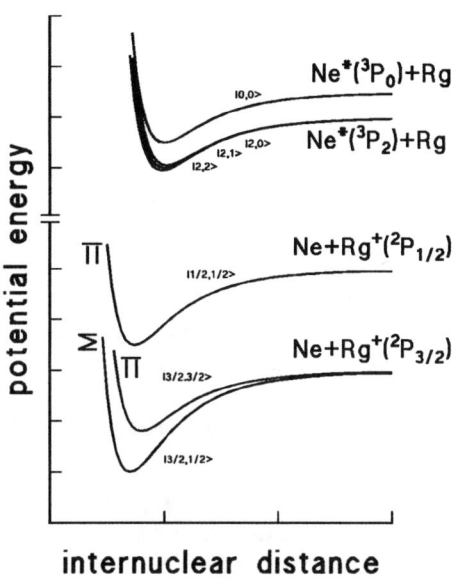

Figure 3

Potential energy curve scheme of Ne*-heavier rare gas.

case it has been accurately determined by Baudon et al.[11] by analyzing the elastic differential and integral cross sections[12] and the total ionization cross sections.[13] The electron energy spectra measured by Hotop and coworkers[14] show the nascent [Ne...N_2]$^+$ energy population, and all the complexes where N_2^+ is with v≥1 have enough energy to predissociate. Sonnenfroh and Leone[15] have shown that the internal state distribution of the final N_2^+ ion is bimodal, confirming a double path for the production of this ion. It should be interesting to analyze such state distribution to understand the dissociation dynamics of the excited NeN_2^+ ion. Such study is presently in progress.[16]

Collisions of metastable rare gas atoms with halogen molecules are examples of systems where a chemical rection can occur before the autoionization event. In such cases the low ionization potential of the metastable atom and the high electron affinity of the molecular partner allow occurring an electron transfer even at long distances, leading to an ion-pair complex. The characteristics of such a complex strongly affect the autoionization dynamics.

Collisional autoionization of this type, involving metastable helium atoms,

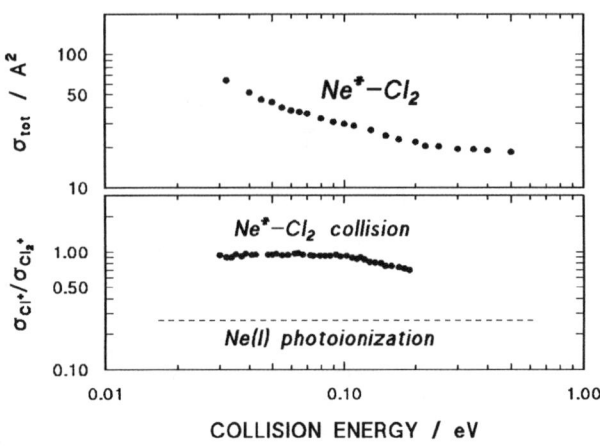

Figure 4
Total ionization cross sections and Cl^+/Cl_2^+ branching ratio for Ne^*-Cl_2 collisions.

has been studied by Morgner and coworkers[17] and the dynamics features have been explained through the assumption of an ion pair formation. Brunetti et al.[18] have studied the collisional autoionization of Cl_2 by Ne^* and have compared the Cl^+/Cl_2^+ product ion branching ratio with the one obtained by $Ne(I)$ photoionization of chlorine molecules. In Fig.4 the total ionization cross section and the Cl^+/Cl_2^+ branching ratio are plotted as a function of the collision energy; the branching ratio is compared with the one obtained by photoionization. It is well evident that the Cl^+ production in the collisional autoionization is higher than the one in photoionization with photons of comparable energy. This and other experimental results give evidence for the formation of a Ne^+-Cl_2^- ion pair before the autoionization event. This reaction also depends on the mutual orientation of the two collision partners. In fact, as schematically shown in Fig.5, the 3s electron of Ne^* moves into the $3p_z\sigma_u^*$ orbital of the diatom. This is obviously possible only for a collinear orientation of the two partners. Once Cl_2^- is formed the lower bond energy leads to a dissociation of the diatom and therefore the autoionization largely occurs when the two chlorine atoms are already separated, with consequent production of the atomic Cl^+ ion.

Another possible type of atom-molecule autoionization system is the one where a chemical reaction occurs after the ionization event. An example of this type is the collisional autoionization of H_2 by He^* or Ne^*. After the electron ejection, the H_2^+ ion can react with the rare gas partner leading to HeH^+ or NeH^+. In the case of metastable neon the possible reactions are[19]

$$Ne^* + H_2 \rightarrow [Ne...H_2]^+ + e$$
$$[Ne...H_2]^+ \rightarrow Ne + H_2^+$$
$$\rightarrow NeH^+ + H$$
$$\rightarrow NeH_2^+.$$

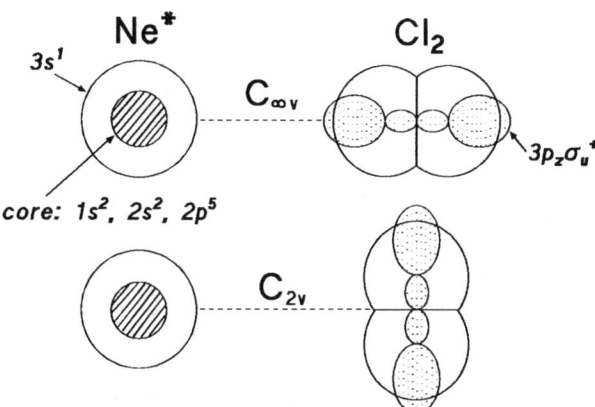

Figure 5
Schematic representation of some atomic and molecular orbitals involved in Ne*-Cl₂ autoionizing collision.

In the case of He* also the H⁺ ion product is possible.[19-21]

Siska and coworkers[20,21] have studied the He*-H$_2$ system in a crossed beam experiment and succeeded in describing the process by using a theoretical scheme where the autoionization event is described by an optical model and the following ion-molecule reaction is described by the Phase Space Theory.[21]

Another system where a chemical reaction can occur after the autoionization event is Ne*-HCl. In this case the autoionization gives rise to two ionic complexes: one in the ground state and the other one in the first excited electronic state,

$$\text{Ne}^* + \text{HCl} \rightarrow [\text{Ne...HCl}^+(X)] + e^-$$
$$\rightarrow [\text{Ne...HCl}^+(A)] + e^-.$$

THey show different reactive behavior:

$$[\text{Ne...HCl}^+(X)] \rightarrow \text{Ne} + \text{HCl}^+(X)$$
$$\rightarrow \text{NeHCl}^+(X)$$
$$[\text{Ne...HCl}^+(A)] \rightarrow \text{Ne} + \text{HCl}^+(A)$$
$$\rightarrow \text{NeH}^+ + \text{Cl}.$$

Therefore in this case only the excited ionic complex can lead to a rearrangement through a proton transfer reaction.

The collisional autoionization of HCl by Ne* has been studied with several techniques in different laboratory. In Kaiserslautern the electron energy spectrum has been measured for selected spin-orbit states of Ne*.[22] The optical spectrum produced by the radiative decay of HCl⁺(A) has been studied in Santa Barbara[23] and in Kaiserslautern.[24] In Perugia the energy dependence for each one of the possible products has been measured also for the isotopic DCl variant.[25,26] A theoretical study recently performed in Perugia has shown that the potential energy profile, along the minimum energy path, for the excited [Ne...HCl⁺(A)] complex, in collinear configuration, has the shape reported in Fig.6. From these results it appears that this intermediate complex is weakly bound when the Ne-H-

Cl angle is 180° and becomes repulsive when the angle is outside the range ~180°±10°.

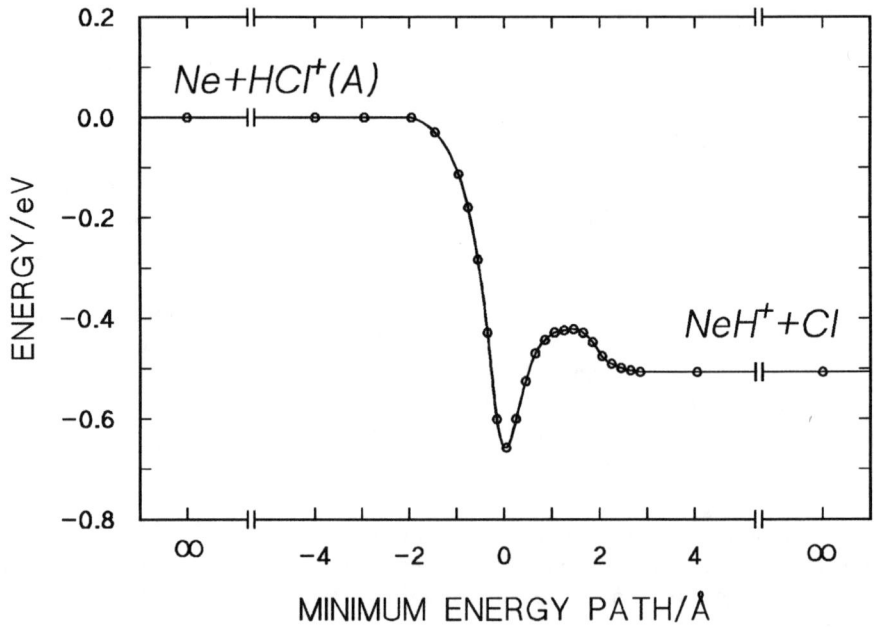

Figure 6
Potential energy profile of the colinear Ne...HCl$^+$(A) ionic complex

As has been done for the He*-H$_2$ system by Siska and coworkers,[21] also for this system the Phase-Space-Theory has been used to describe the post-ionization proton transfer reaction. In Fig.7 the ratio between the cross sections for NeH$^+$ production and for total ionization is reported as a function of the collision energy and compared with the theoretical calculation.

The experimental results for the isotopic DCl variant have shown that the post ionization reaction does not present isotopic effects. This is also in agreement with the theoretical calculations.

CONCLUDING REMARKS AND FUTURE DIRECTIONS

In the last decade the studies of collisional autoionization by metastable rare gas atoms have increased considerably both qualitatively and quantitatively. New experiments are giving further insight into the microscopic dynamics of these processes, and allow a systematic study of a large number of different systems employing both atoms and molecules as collision partners.

For atom-atom systems the studies with j-selected Rg* atoms have clarified several aspects of the microscopic ionization mechanism, revealing the role played during the reaction by the symmetry of the atomic orbitals involved. This has stimulated theoretical effort in the direction of clarifying the potential energy functions describing the interactions between the collision partners, as well as in the direction of a rigorous quantitative description of the ionization dynamics.

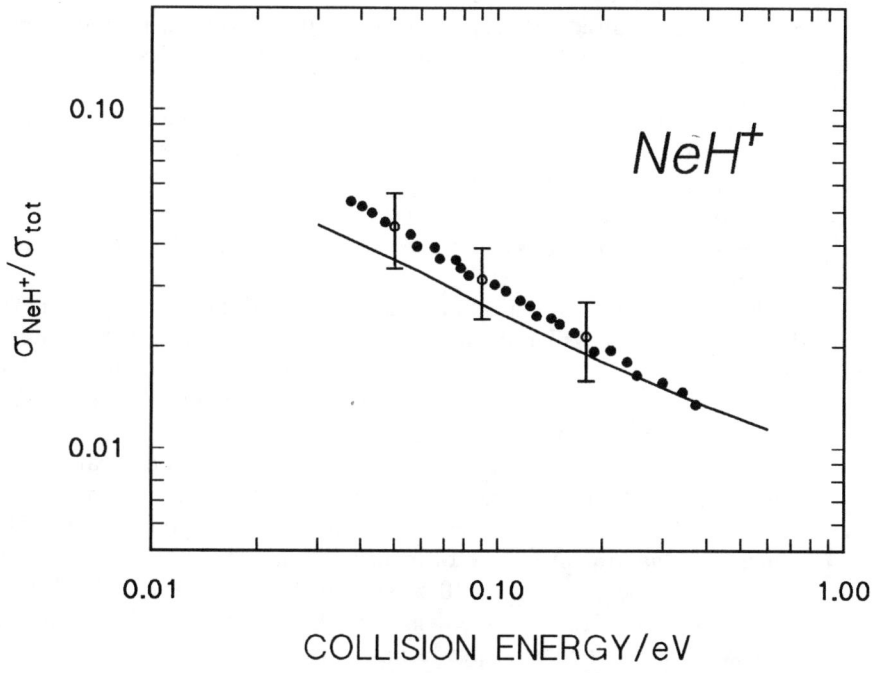

Figure 7
The NeH$^+$ to total cross section ratio in Ne*-HCl collisions. The dots are the experimental cross sections while the line is the Phase Space Theory calculation.

The study of atom-molecule collisional autoionization systems appears to be more complex, but the last few years have seen considerable advances due to the application of powerful experimental techniques: ion-electron coincidence experiments have allowed the correlation among different energy states produced during the ionization and the final ionic products. also, state resolved ionization cross sections and PIES spectra as a function of the collision energy have allowed the understanding of the role played by molecular orbital symmetry in determining the ionization probability.

For collisional autoionization in atom-molecule systems, the theory is currently unable to provide a unified dynamical treatment, as developed in the atom-atom case. However recent progress in chemical reaction dynamics theory can be applied here, with appropriate adjustment, for the ionization occurring during the collision. The "Surface Trajectory Leaking" treatment used for the He*-H$_2$ system[27] could be a promising development for the studies of the dynamics of collisional autoionization involving molecules. The statistical treatment used by Siska and coworkers[21] for the He-H$_2^+$ dynamics, which follows the He*-H$_2$ collisional autoionization, and that used by Brunetti et al.[26] in Ne*-HCl system, also provide interesting examples of a dynamical study of these post-ionization reactions.

The accumulation of a substantial body of quantitative information on the rates of these processes is especially important in view of the possible application to the understanding of the mechanism of processes which take place in plasmas, flames, atmospheres. As an example of these applications we recently reported a study of the influence of the reactions between Ne* and HCl in the series of reactions occurring in excimer lasers.[28]

All the collisional autoionization cases discussed in this paper involve metastable rare gas atoms. As has been mentioned, the reason for such an intense application of these metastable atoms in current studies lies mainly in their long life-time and in their electronic energy, which is large enough to ionize many possible collision partners. The simplest of these reactants is the He* atom which has only two electrons and is therefore accessible to theoretical calculation. However its energy content is so large that, in many cases, the ionization process is complicated by the possibility of populating numerous electronic states of the ionic products. The natural prototype of metastable atoms should be the H*(2s) atom, which has enough energy, 10.2 eV, to lead to several collisional autoionization processes. However the dynamics of such processes have not been studied because of the difficulty in producing a stable H*(2s) beam. In fact this atom can be easily quenched by a weak electrical field, which produces a mixing with the degenerate 2p state causing a decay to the ground state with the emission of the corresponding Lyman-α photon. However Vassilev et al.[29] recently succeeded in producing a stable molecular beam of H*(2s) atoms stable enough to be used for scattering experiments in H*+H$_2$ collisions. These authors have obtained metastable hydrogen atoms by electron bombardment of a H atom beam produced by a radio-frequency discharge source. This experiment indicates that the study of collisional autoionization processes in a crossed beam experiment induced by these simple basic atoms should be possible in the near future.

REFERENCES

1. F.M.Penning, Z.Phys. 46, 335 (1928); 57, 723 (1929); 72, 338 (1931).

2. B.Brunetti and F.Vecchiocattivi, in "Cluster Ions", C.Y.Ng ed., Wiley & Sons (1993), p.359-445 and references therein.

3. P.E.Siska, Rev.Mod.Phys. 65, 337 (1993), and references therein.

4. A.Khan, H.R.Siddiqui, and P.E.Siska, J.Chem.Phys. 92, 2588 (1991).

5. A.Merz, M.W.Ruf, and H.Hotop Phys.Rev.Lett. 69, 3467 (1993).

6. V.Hoffmann and H.Morgner, J.Phys. B12, 2857 (1979).

7. H.Hotop and A.Niehaus, Z.Phys. 228, 68 (1969); 238, 432 (1970).

8. A.Aguilar, B.Brunetti, S.Rosi, F.Vecchiocattivi, and G.G.Volpi, J.Chem.Phys. 82, 773 (1985).

9. H.Morgner, Comm.At.Mol.Phys. 11, 271 (1982); J.Phys. B18, 251 (1985).

10. L.Appolloni, B.Brunetti, F.Vecchiocattivi, and G.G.Volpi, J.Phys.Chem. 92, 918 (1988).

11. J.Baudon, P.Feron, C.Miniatura, F.Perales, J.Reinhardt, J.Robert, H.Haberland, B.Brunetti, and F.Vecchiocattivi, J.Chem.Phys. 95, 1801 (1991).

12. E.R.Th.Kerstel, M.F.M.Janssens, K.A.H.van Leuwen, and A.C.H.Beijerinck, Chem.Phys. 145, 211 (1990).

13. A.Aguilar, B.Brunetti, M.Gonzalez, and F.Vecchiocattivi, Chem.Phys. 145, 211 (1990).

14. H.Hotop, private communication.

15. D.M.Sonnenfroh and S.R.Leone, Int.J.Mass Spectr.Ion Processes 80, 63 (1987).

16. B.Brunetti, S.Falcinelli, and F.Vecchiocattivi, to be published.

17. O.Leisin, H.Morgner, and H.Seiberle, Mol.Phys. 56, 349 (1985); W.Kischlat and H.Morgner, Z.Phys. A312, 305 (1983); A.Benz and H.Morgner, Mol.Phys. 57, 319 (1986).

18. B.Brunetti, S.Falcinelli, S.Paul, F.Vecchiocattivi, and G.G.Volpi, J.Chem.Soc. Faraday Trans. 89, 1505 (1993).

19. P.W.West, T.B.Cook, F.B.Dunning, R.D.Rundel, and R.F.Stebbings, J.Chem.Phys. 63, 1237 (1975).

20. D.W.Martin, D.Bernfeld, and P.E.Siska, Chem.Phys.Lett. 110, 298 (1984).

21. D.W.Martin, C.Weiser, R.F.Sperlein, D.Bernfeld, and P.E.Siska, J.Chem.Phys. 90, 1564 (1989).

22. A.J.Yencha, J.Ganz, M.W.Ruf, and H.Hotop, Z.Phys. D14, 57 (1983).

23. H.L.Sneider, B.T.Smith, and R.M.Martin, Chem.Phys.Lett. 94, 90 (1983).

24. W.Simon, A.J.Yencha, M.W.Ruf, and H.Hotop, Z.Phys. D8, 71 (1988).

25. A.Aguilar, B.Brunetti, S.Falcinelli, M.Gonzalez, and F.Vecchiocattivi, J.Chem.Phys. 96, 433 (1992).

26. B.Brunetti, R.Cambi, S.Falcinelli, J.M.Farrar, and F.Vecchiocattivi, to be published.

27. J.Vojtik and I.Paidarova, in "The Dynamics of Systems with Chemical Reaction", J.Popieleski ed., World Scientific, Singapore 1989, p.241.

28. V.Aquilanti, B.Brunetti, F.Vecchiocattivi, T.Letardi, F.Fang, and S.Fu, Chem.Phys.Lett. 205, 229 (1993).

29. G.Vassilev, F.Perales, Ch.Miniatura, J.Robert, J.Reinhardt, F.Vecchiocattivi and J.Baudon, Z.Phys. D 17, 101 (1990).

ENERGY SHARING IN THERMAL ENERGY COLLISION SPECTROSCOPY

M. Allegrini and S. Milosevic*
Dipartimento di Fisica, Università di Pisa, Piazza Torricelli 2, I-56126 Pisa, Italy

ABSTRACT

The progress recently made on collisions involving laser excited atoms and/or molecules at thermal energies are here illustrated with two examples of experiments: reactive collisions between one molecule and one atom leading to the formation of the IA-IIB intermetallic excimers and energy pooling collisions between two Cs atoms excited to the first resonant level.

INTRODUCTION

The laser is a powerful tool for the preparation of large populations of atoms or molecules into specific electronic states, and it gives the possibility to investigate collisions involving excited species. The measurement of the collision rate constants provide information on the potential curves. In certain ranges of interatomic distances it is still one of the few experimental ways to determine atom-atom and atom-molecule interactions. The knowledge of the potential curves is important also for the collisions involving ultracold atoms, which can be prepared with the recent laser cooling and trapping techniques. Although the dynamics of ultracold collisions is fundamentally different[1] from that at thermal energies regime, the opening or closing of collision channels is determined from the potential curves. In this report, we summarize some recent advances made in our laboratory on reactive collisions (RC) involving intermetallic species and energy pooling collisions (EP) in pure Cs vapor. Both processes are based on energy sharing of the electronically excited species.

Laser excitation of atoms and molecules in intermetallic vapor mixtures at thermal energies produces a complex system of different energy sharing processes, reactive and non-reactive. We have studied mixtures of Li or Na combined with Zn, Cd or Hg, having in mind future investigations on more complex systems with K, Rb, and Cs. The reactive collisions we have investigated are used to prepare excited intermetallic molecules (AB):

$$A_2^* + B \rightarrow (AB)^* + A \qquad (1)$$
$$A_2 + B^* \rightarrow (AB)^* + A \qquad (2)$$

where A and A_2 are an alkali atom and dimer, respectively, and B is an atom of column IIB. A_2^* is usually in the $C^1\Pi_u$ or in the $B^1\Pi_u$ state, whereas B^* is in the $^3P_1$

---

*Permanent address : Institute of Physics, University of Zagreb, P.O.Box 304, 41000 Zagreb, Croatia

metastable state. The electronic excitation of the dimers A_2^* or of the atom B^* is transferred by the reactive collision to the intermetallic excimer $(AB)^*$, the energy balance being assured, as usually, by the kinetic energy of the colliding partners. The entrance and exit channels of the reactions are monitored by observing the fluorescence emission of the A_2^* and $(AB)^*$ molecules. The comparison of the observed $(AB)^*$ emission with spectral simulations yields the information about nascent rovibrational distributions from reactive collisions. Simultaneously, other competing collision energy transfer processes are present in the vapor mixture, such as

$$A_2(\Lambda)^* + A \rightarrow A_2(\Lambda')^* + A \qquad (3)$$
$$B^* + A \rightarrow A^* + B \qquad (4)$$

These collisions produce excited atoms and molecules in different electronic states, all contributing to the fluorescence spectrum. Radiatively trapped resonant excited alkali atoms and metastable B^* atoms could also play an important role in the secondary collisions.

The second experiment we report concerns energy pooling collisions between Cs atoms excited to the upper fine structure component of the first resonance level. Three different energy pooling combinations are expected: $6P_{3/2} + 6P_{3/2}$, $6P_{3/2} + 6P_{1/2}$ and $6P_{1/2} + 6P_{1/2}$. We have measured the cross section for the process

$$Cs(6P_{3/2}) + Cs(6P_{3/2}) \rightarrow Cs(7P_{3/2,\ 1/2}) + Cs(6S) \qquad (5)$$

Such an elementary process might be of interest in the determination of loss mechanisms in current cooling and trapping experiments. Moreover, it is also a prototype of experiments to be performed on alkali-IIB mixtures.

Since all of the above mentioned processes are related to the potential energy surfaces of A_2+B complexes or potential energy curves of A_2 and AB diatomic molecules, our experimental work is accompanied with *ab initio* calculations carried out by a few quantum-chemistry groups (Pisa, Bonn, Leiden, Göttingen), the results of which are used in the interpretation of the observed spectra and are necessary in further steps of studying the dynamics of these processes.

EXPERIMENTAL METHODS

The main problem in studying the alkali-IIB dimers is how to prepare the vapor mixture which gives enough collisions between the alkali dimers and IIB atom to observe the photochemical reactions (1) and (2). The vapor pressures of mixture constituents are quite different and care should be taken in order to prepare the desired mixture. A common approach is to fill a standard heat-pipe oven with the amalgam of a certain mass ratio of alkali and IIB atoms and to rely on the empirical Raoult law which gives the partial vapor pressures. This works well, for example, in LiZn, LiCd, NaHg, NaCd for which the IIB atom vapor pressure is higher than the alkali vapor pressure at a given temperature[2,3,4]. However, if vapor pressures differ very much, as in LiHg, NaZn, KZn mixtures, or if the alkali vapor pressure is higher, the mixture could be prepared by introducing a controlled stream of vapor of the element with higher vapor pressure into the vapor of the element with lower vapor pressure[5,6]. The mixture prepared in these ways contains atoms and dimers in different ratios giving the desired collision frequencies.

In our RC experiments, the excitation is typically made by using a pulsed laser, with 16 ns time duration and a repetition rate of 10 Hz, which excites the $C^1\Pi_u$ state or some nearby electronic states of the alkali dimer A_2. During its lifetime (about 20 ns), the excited A_2^* suffer many collisions with A or B atoms, depending on the amalgam prepared. A sketch of the set-up is shown in Fig 1.

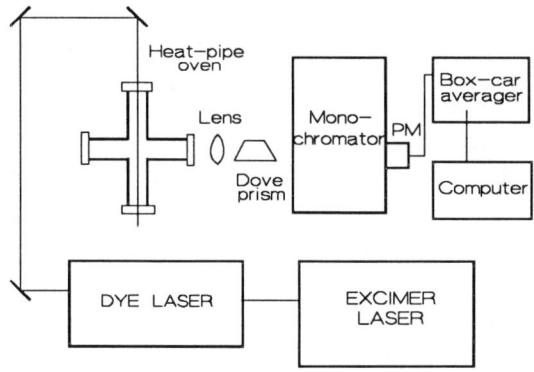

Fig. 1. Experimental setup for the photochemical preparation of alkali-IIB dimers in a heat-pipe oven.

To measure energy transfer, and specifically energy pooling, cross sections, the main experimental requirements are different. For Cs, we have used a capillary cell where the laser-vapor interaction is well defined and we have kept the vapor pressure as low as possible to minimize the radiation trapping phenomenon. The excitation is made by a cw laser diode at 852.1 nm and in saturation condition the concentration of the $6P_{3/2}$ excited state is precisely determined. The emission intensities of $6P_{3/2}$, $6P_{1/2}$, $7P_{3/2}$ and $7P_{1/2}$ are observed and their ratio to the $6P_{3/2} \rightarrow 6S_{1/2}$ is measured.

ALKALI DIMER- IIB ATOM REACTIVE COLLISIONS AND ENERGY TRANSFER

a) Photochemical reactions

In Fig. 2 we show a simplified potential energy diagram for Li-IIB atom combinations. As a product of reaction (1), with the alkali dimer excited to the $C^1\Pi_u$ state, the intermetallic molecules (AB) are formed in the $2^2\Pi$ state, and the emission to the $X^2\Sigma^+$ ground state is observed. The positions of the main emission peaks are shown in Table I. Two main peaks are observed in LiCd, LiHg and NaCd

Table I: Wavelength of the main peaks (in nm) of the blue-green emission in Li,Na-IIB dimers, which is due to the $2^2\Pi \rightarrow X^2\Sigma^+$ transitions.

	Zn	Cd	Hg
Li	477	482	445
		489	467
Na	478	479	450
		484	470

and NaHg cases. *Ab initio* calculations show that the wavefunction of the $2^2\Pi$ state is influenced by the (nsnp) ^3P configuration of the IIB atom[2,3,5]. Going from Zn towards Hg the spin-orbit splitting of the ^3P state increases and consequently the $2^2\Pi$ state has to be described in Hund's coupling case c, e.g as $2^2\Pi_{1/2}$ and $2^2\Pi_{3/2}$. In the case of Zn this splitting is not sufficient to be observed in the present type of experiments.

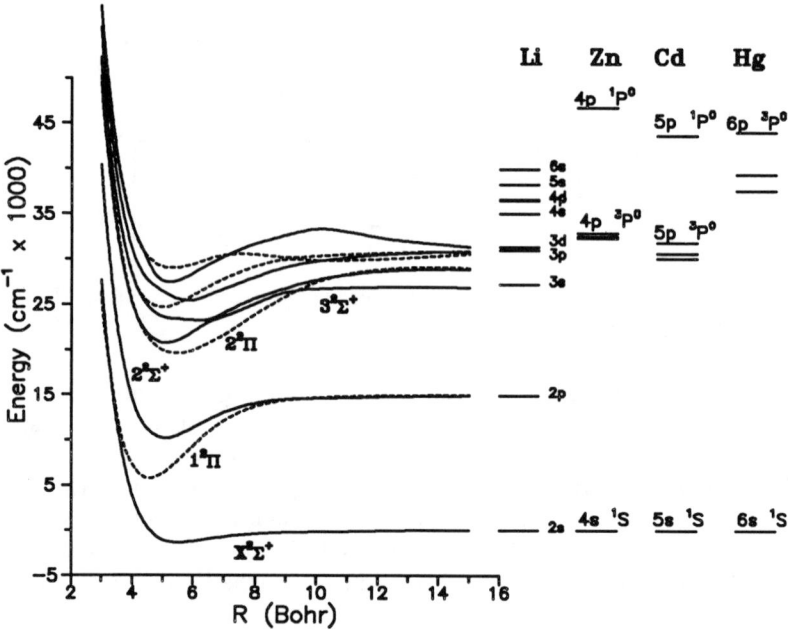

Fig. 2. Energy term diagram of Li - IIB atom combinations. In the left panel the non-relativistic *ab initio* potentials of the LiZn are shown[2].

Fig. 3 gives an example of the observed blue-green emission for LiZn, LiCd and LiHg. In all cases shown, an excimer laser at 308 nm is used for preparing $Li_2(C^1\Pi_u)$ state in a high rovibrational level. The nascent distribution of the Li-IIB($2^2\Pi$) molecular state is found to be different from Boltzmann. However, due to many (AB)* + B and/or (AB)* + A collisions, the observed spectra are termalized.

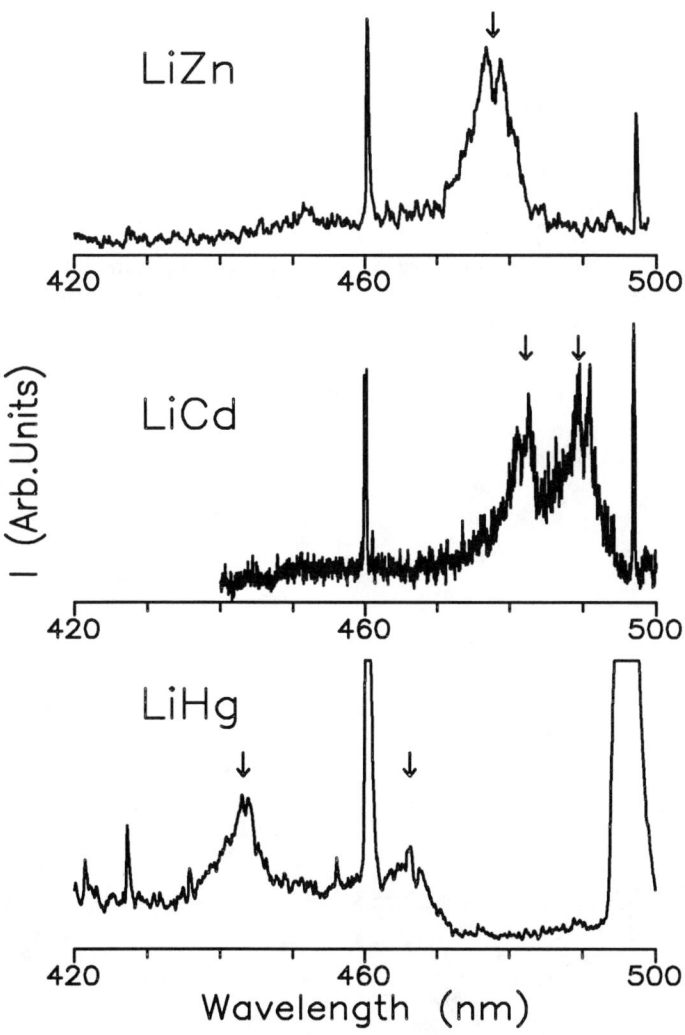

Fig. 3. Observed blue-green bands of LiZn, LiCd and LiHg upon excitation of Li$_2$ with the 308 nm excimer laser line. The lithium atomic lines at 460.3 nm (4^2D-2^2P) and 497.2 nm (4^2S-2^2P) are observed in addition to molecular features.

A detailed spectral simulation of the observed emission shows that both bound-bound and bound-free type of transitions contribute to the spectrum[7]. The bound-bound contributions dominates the spectrum if the population of the upper $2^2\Pi$ state is termalized and this gives additional sharp structures on the top of each emission peak. The interplay between dominance of either bound-bound or bound-free transitions explain the changes of the spectral shapes observed for the blue-green bands. Therefore, by observing (AB)* emission, it is possible to study details of the reactive collisions (1) and how the entrance parameters influence the formation of the AB molecules, provided termalization effects are avoided.

For the mixtures that have been investigated in cw experiments, no differences were observed and the spectroscopic peak emissions were found at the same wavelengths. The advantage of the pulsed laser excitation is the possibility to follow the time behaviour of the fluorescence and to observe termalization processes. Because of the nanosecond range of the laser in our experiments, the time increasing portion of the (AB)* fluorescence is masked by the laser pulse signal. However, with shorter laser pulse, it should be possible to determine directly the cross section for relevant reaction processes[8]. Actually, we plan to perform experiments using a laser with a pulse duration of hundred picoseconds for the excitation of Na_2. This will give us the possibility to measure the nascent distribution of the (AB)* levels formed by reactive collisions (1) and to determine the cross sections of process (1).

b) Energy transfer

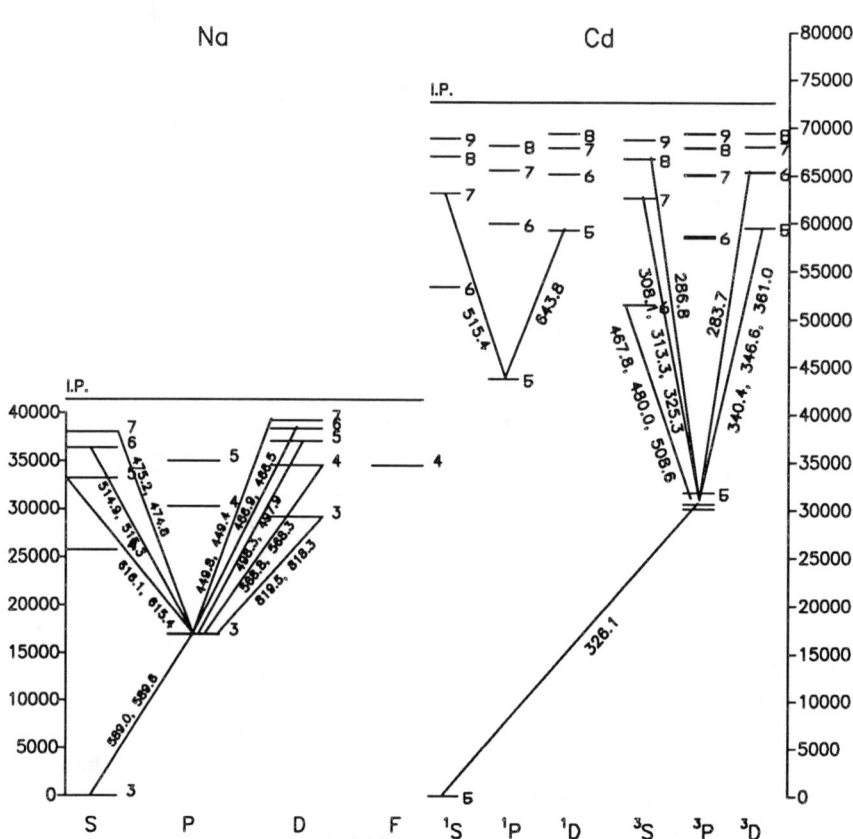

Fig. 4. Energy term diagram relevant for energy transfer Cd*/Na and Cd*/Cd* for the case of irradiation with 326.1 nm laser line.

The collisions between metastable IIB and alkali atoms become important when RC process (2) is used to prepare (AB)* excimers[9]. Indeed, even if the excitation is weak, being a singlet-triplet intercombination transition, the excited IIB atoms are in a metastable state and available for collisions during their long lifetime. Fig. 4 shows a simplified energy diagram and the lines of atomic Na and Cd observed in emission when exciting the Cd $5^1S_0 \rightarrow 5\ ^3P_1$ transition. Since the 5^3P_1 level of Cd is very close to sodium 3D and 4P, efficient energy transfer (4) is possible. This processes can be also used as a probe to study (AB) potential energy curves. Energy pooling processes in Cd take place, as well:

$$Cd(5^3P_1) + Cd(5^3P_1) \rightarrow Cd^* + Cd(5^1S_0) \tag{6}$$

Fig. 5 shows a typical emission spectrum from NaCd vapor mixture where both molecular and atomic features are present.

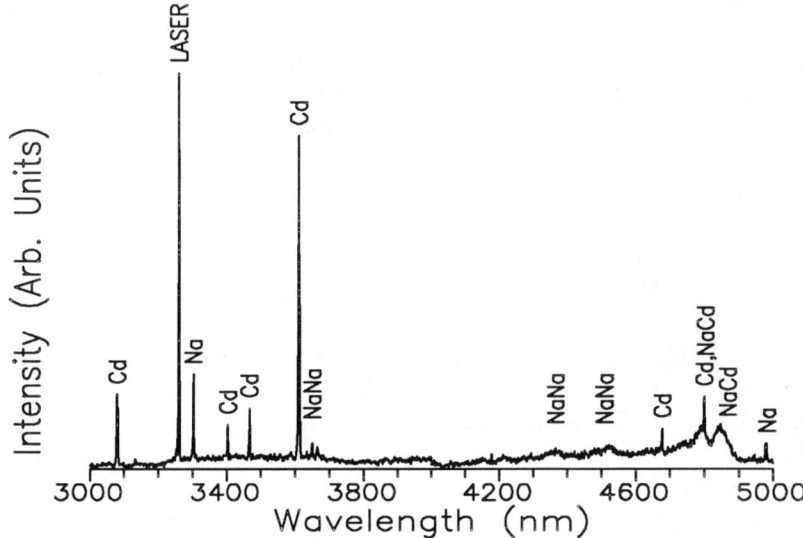

Fig 5. Example of the fluorescence spectrum of Na-Cd mixture upon excitation of the Cd metastable state 5^3P_1.

ENERGY POOLING

For the energy pooling experiment we have used cw diode laser excitation of the $6P_{3/2}$ level and measured the intensities of the $6P_{3/2}$, $7P_{1/2}$, $7P_{3/2}$ fluorescence emissions, I_{6P}, I_{7P_J}. The measurements have been done in the range of cesium atom concentrations $5 \cdot 10^{12}$ to $3 \cdot 10^{13}$ cm^{-3}. The experimental data have been analyzed by using a simple rate equation for the $7P_J$ levels concentration. The steady state solution gives the rate coefficient $k_{7P} = c\ \gamma\ I_{7P}/I_{6P}\ 1/N_{6P}\ \tau^*$, where $c = (\alpha\ \nu)_{6P}/(\alpha\ \nu)_{7P}$

with α a parameter taking into account the instrumental response and ν the transition frequency. γ is the appropriate branching ratio. The excited atom density N_{6P} was determined as a fraction of the total atom concentration because the experiment was performed in saturation conditions of the resonance fluorescence. The effective lifetime τ* was calculated by using Milne theory, valid in the optical thin vapor regime. At T=370 K, having taken the transition probabilities from Warner[10], we obtain for the collisional process (5) the following rate constants and cross sections

$$k_{7P_{1/2}}=(0.95 \pm 0.2) \times 10^{-11} \text{ cm}^{-3}\text{s}^{-1} \; ; \; \sigma_{7P_{1/2}} = (2.8 \pm 0.5) \times 10^{-16} \text{ cm}^{-2}$$
$$k_{7P_{3/2}}=(1.10 \pm 0.2) \times 10^{-11} \text{ cm}^{-3}\text{s}^{-1} \; ; \; \sigma_{7P_{3/2}} = (3.2 \pm 0.6) \times 10^{-16} \text{ cm}^{-2}.$$

These rate constants should be taken as effective rate constants since a certain amount of the $6P_{1/2}+6P_{3/2}$ collisions could contribute, as well. The total atom concentration in our experiment is less than $2 \cdot 10^{13}$ cm^{-3}, therefore the relative population of the $6P_{1/2}$ and $6P_{3/2}$ levels is mainly determined by optical excitation and radiation processes[11]. We estimate from intensity measurements that concentration in the $6P_{1/2}$ level is less than 10% of the concentration in the $6P_{3/2}$ level.

At present it is not simple to compare our values of rate constants with other alkali cases since most of previous measurements were not distinguishing between 3/2+3/2, 1/2+1/2 or 1/2+3/2 collisions, as apparently it is the case in Cs, which has the larger fine structure splitting ($\Delta E \approx$ 2-3 kT). It is interesting, however, to see how our Cs EP cross section values fit into the general dependence of rate constants on energy defect ΔE[12]. In our case ΔE is 1699 cm^{-1} and 1518 cm^{-1} for $7P_{1/2}$ and $7P_{3/2}$, respectively. This gives, at T=370 K, $\Delta E/kT \approx 5.5$ which is much larger than for the other alkalis.

It is also interesting that for the range of vapor temperatures 350-385 K we do not observe emission from the 6D level for which the energy defect is smaller than for the 7P level, but we do observe emission from the 7P level. The emission from the 6D level becomes visible at Cs atom densities $N>10^{14}$ cm^{-3}, having considered the sensitivity of our detection apparatus. At that atom density we have also observed emission from other Cs upper levels, namely 7D, 8P, 8S, 9S and 5F. This fact indicate that besides energy pooling, also associative ionization or secondary collisional processes take place. Therefore, it is not proved that the 6D emission is a consequence of the $6P_{3/2} + 6P_{3/2}$ collisions alone. Previously, Klyucharev and Lazarenko[11] used cesium gas-discharge lamp and populated both $6P_{1/2}$ and $6P_{3/2}$ levels with relative concentrations in range from 0.9 to 1.4. They observed the 6D emission fluorescence and gave an upper limit of the effective cross section $\sigma \leq 10^{-13}$ cm^2 at 528 K. Later, Borodin and Komarov[13] have calculated that the cross section exceeds $1.5 \cdot 10^{-15}$ cm^2 at about 500 K, and that $6P_{1/2}+6P_{3/2}$ collisions give the dominant contribution to the cross section.

Our experiment is the first in which only one fine structure 6P component was preferentially excited and we plan in the near future to excite with a laser diode separately and simultaneously both $6P_{3/2}$ and $6P_{1/2}$ levels. This, together with the *ab-*

initio calculations of relevant potential energy curves, will give more insight into the complicated energy pooling process Cs(6P)+Cs(6P).

Table II gives the present status of cross sections measurements at thermal energies for cesium atom-atom collisions (fine-structure mixing, energy pooling and associative ionization). We note that the ratio between the cross section for the fine-structure mixing and the energy pooling, presently measured, compare well with other alkali cases.

Table II: Measured cross sections for different collision processes in cesium.

process	$\sigma\,[\mathrm{cm}^2]\times 10^{-16}$	T [K]	reference
$6P_{1/2}+6S_{1/2} \to 6P_{3/2}+6S_{1/2}$	6.4	300-320	14
$6P_{3/2}+6S_{1/2} \to 6P_{1/2}+6S_{1/2}$	31	300-320	14
$7P_{1/2}+6S_{1/2} \to 7P_{3/2}+6S_{1/2}$	121 ± 25	443	15
$7P_{3/2}+6S_{1/2} \to 7P_{1/2}+6S_{1/2}$	107 ± 22	443	15
$8P_{1/2}+6S_{1/2} \to 8P_{3/2}+6S_{1/2}$	190 ± 57	420	16
$8P_{1/2}+6S_{1/2} \to 7D_{5/2}+6S_{1/2}$	20 ± 5	420	16
$8P_{1/2}+6S_{1/2} \to 7D_{3/2}+6S_{1/2}$	21 ± 5	420	16
$5D_{5/2}+6S_{1/2} \to 5D_{3/2}+6S_{1/2}$	36 ± 8	601	8
$(6P+6P)_{1/2,3/2} \to 6D+6S$	≤ 1000	528	11
$6P_{1/2}+6P_{3/2} \to 6D_{3/2,5/2}+6S$	15 ≤ σ ≤ 200	500	13, theory
$6D_{3/2}+6S_{1/2} \to 6P+6P$	150	< 500	17
$6P_{3/2}+6P_{3/2} \to 7P_{1/2}+6S_{1/2}$	2.8 ± 0.5	355-385	this work
$6P_{3/2}+6P_{3/2} \to 7P_{3/2}+6S_{1/2}$	3.2 ± 0.6	355-385	this work
$(6P+6P)_{1/2,3/2} \to Cs_2^+ + e$	11 ± 3	450	18
$8P_{1/2}+6S \to Cs_2^+ + e$	0.33 ± 0.09	470	19
$7D+6S \to Cs_2^+ + e$	30	425	20

ACKNOWLEDGMENTS

In addition to the work carried out in Pisa, the reactive collision experiments have currently been performed in Graz and in Zagreb. We wish to thank F.Fuso, F.deTomasi, G.DeFilippo, (Pisa), L.Windholz, M.Musso, D.Gruber, (Graz) and G.Pichler, X.Li, D.Azinovic and I.Vezmar, (Zagreb) for the wonderful cooperation on the experiments which is enlightened by the *ab initio* work of B.Hess (Bonn), M.Persico (Pisa), M.C. van Hemert (Leiden), R.Düren and B.Heumann (Göttingen). We acknowledge the Austria-Italy Cultural Exchange Program for financial support. One of the authors (S.M.) undertook the work on energy pooling with the support of the "ICTP Program for Training and Research in Italian Laboratories, Trieste, Italy".

REFERENCES

1. Y.B. Band and P.S. Julienne, Phys.Rev. A46, 330 (1992).
2. S.Milosevic, X.Li, D.Azinovic, G.Pichler, M.C.van Hemert, A.Stehouwer, and R.Düren, J.Chem.Phys. 96, 7364 (1992).
3. M.C.van Hemert, D.Azinovic, X.Li, S.Milosevic, G.Pichler and R.Düren, Chem.Phys.Lett. 200, 97 (1992).
4. M.Musso, L.Windholz, F.Fuso and M.Allegrini, J.Chem.Phys. 97, 7017, (1992).
5. D.Gruber, M.Gleichman, M.Musso, F.Fuso, B.Hess, L.Windholz and M.Allegrini, to be published.
6. D.Azinovic, X.Li, S.Milosevic, G.Pichler, M.C.van Hemert and R.Düren, J.Chem.Phys. 98 4672 (1993).
7. X.Li, S.Milosevic, D.Azinovic, G.Pichler, R.Düren and M.C.van Hemert, to be published.
8. A.Sasso, W.Demtröder, T. Colbert, C. Wang, E. Ehrlacher and J. Huennekens, Phys.Rev. A45, 1670 (1992).
9. M.Musso, F.Fuso, L.Windholz and M.Allegrini, Proceedings of the Symposium on Atomic and Surface Physics, SASP 92, D. Bassi, M. Scotoni and P. Tosi Eds., p. A24.
10. B.Warner, Mon.Not.R.Astr.Soc. 139, 115 (1968).
11. A.N.Klyucharev and A.V.Lazarenko, Opt.Spectrosc. 32, 576 (1972).
12. C.Gabbanini, S.Gozzini, G.Squadrito, M.Allegrini and L.Moi, Phys.Rev. A39, 6148 (1989).
13. V.M.Borodin and I.V.Komarov, Opt.Spectrosc. 36, 145 (1974).
14. M.Czajkowski and L.Krause, Can.J.Phys. 43, 1259 (1965).
15. P.W.Pace and J.B.Atkinson, Can.J.Phys. 52, 1641 (1974).
16. M. Pimbert, J.Phys. 33, 331 (1972).
17. T.Yabuzaki, A.C.Tam, M.Hou, W.Happer and S.M.Curry, Opt.Commun., 24, 305 (1978).
18. A.N.Klyucharev and N.S.Ryazanov, Opt.Spectrosc. 33, 230 (1972).
19. B.V.Dobrolezh, A.N.Klyucharev and V.Yu Sepman, Opt.Spectrosc. 38, 630 (1975).
20. A.N.Klyucharev and N.S.Ryazanov, Opt.Spectrosc. 31, 187 (1971).

ASSOCIATIVE AND PENNING IONIZATION IN H^*+H^*, Na^*+Na^* AND K^*+K^* COLLISIONS

X. Urbain

Buys Ballotlaboratorium, Rijksuniversiteit Utrecht, PO Box 80000,
NL-3508 TA Utrecht, The Netherlands
permanent address : Département de Physique, Université Catholique de Louvain,
Chemin du Cyclotron 2, B-1348 Louvain-la-Neuve, Belgium

ABSTRACT

Progress made in the knowledge and understanding of associative ionization among the alkali series is presented. New measurements of the total cross section for K(4p)+K(4p) collisions illustrate the smooth connection between associative and Penning ionization across threshold. Possible methods for characterizing the molecular products are reviewed. It is shown that in-flight dissociation of the molecular ions, when performed over a broad wavelength range, offers a direct probe of their nascent vibrational excitation.

INTRODUCTION

The associative ionization is understood as the autoionization of the molecular complex during the close approach due to the promotion of a valence state into the continuum. This implies a di-electronic transition, in order to be efficient within the short collision time. An extensive review of the work devoted to this process has been made by Weiner et al[1]. Doubly excited asymptotic configurations are most likely leading to such valence states of doubly excited character. The purpose of our study is to follow the reaction between excited atoms throughout the first column of the elements table, each atom bringing one active electron to the process, making the calculations somewhat tractable. The systematics of the study doesn't mean that all processes are expected to be alike, since the doubly excited spectrum happens to move dramatically with respect to the singly excited asymptotes when one moves from hydrogen to potassium (fig. 1). Indeed, the first doubly excited limit in hydrogen, namely H(2s)+H(2s), lies well above the first ionization limit H(1s)+H$^+$, while in K_2, the K(4p)+K(4p) asymptote

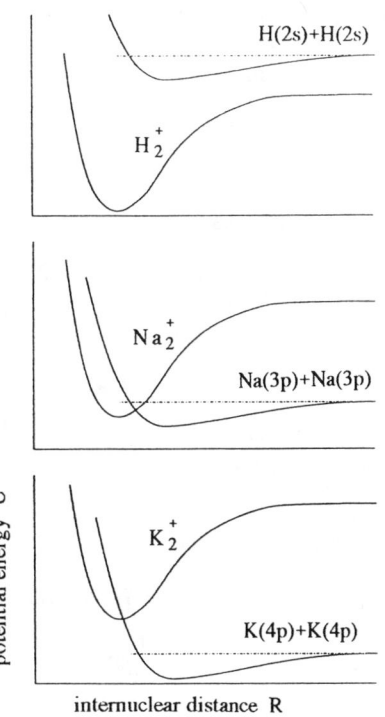

Fig.1 : Schematic potential energy curves showing the change in energy balance from H_2 to K_2.

lies even below the molecular ionization threshold, i.e. the z.p.e. of the K_2^+ ground state. The famous case of Na(3p)+Na(3p) happens to be a singular one, with an asymptotic energy just exceeding the z.p.e. of the Na_2^+ ground state. This special feature leads to an intricate dynamics over the whole range of internuclear distances, that has been and remains a puzzle for theoreticians. Finally, the hyperfine structure makes heavy alkalis much less tractable than hydrogen.

TOTAL CROSS SECTION : K(4p)+K(4p) REACTIONS

These distinctions seem to ruin the hope of finding any systematics in the process among the alkali series. However, the gross dynamics of associative ionization is expected to manifest itself according to the energy balance of the reaction, whatever the nature of the initial state. This first assumption has been tested by considering the K(4p)+K(4p) → K_2^+ + e⁻ reaction, endothermic by 0.3 eV, to be compared with the system we studied previously[2], i.e. H(1s)+H(2s) → H_2^+ + e⁻. The cross section for the latter reaction exhibited the following features (fig. 2) :

- a marked increase of the cross section above the A.I. threshold located at 0.75 eV, originating in the increasing number of accessible rovibrational levels of H_2^+.
- a rapid decrease of the cross section above the Penning ionization threshold, i.e. the dissociation threshold of H_2^+ into $H+H^+$
- an interference pattern resulting from the simultaneous population of degenerate channels interacting at short range.

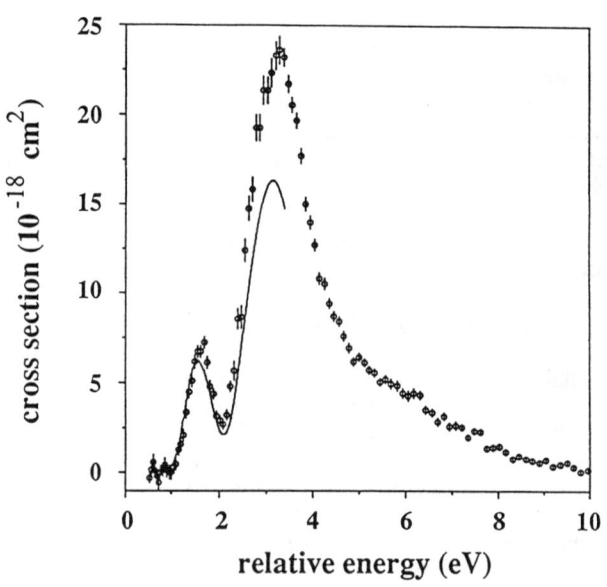

Fig.2 : total cross section for the reaction H(1s)+H(2s) → H_2^+ + e⁻. Experiments and calculation (full line) by Urbain et al[2].

Apart from the last effect, the above mentioned characteristics have been observed for K(4p)+K(4p) collisions by Brencher et al[3].

Their apparatus allowed the observation of K^+, K^- and K_2^+ ions issued from the following processes:

$K^* + K^* \rightarrow K^+ + K^-$, ion pair formation (IPF)

$K^* + K^* \rightarrow K_2^+ + e^-$, associative ionization (AI)

$K^* + K^* \rightarrow K + K^+ + e^-$, Penning ionization (PI)

The ions were collected and sent through a quadrupole mass filter, with the drawback of inconsistency of the yields, the number of detected K^+ ions being quite different from that of K^- ions simultaneously produced in the IPF process. The experiment has been redesigned and measurements have been repeated after some improvement of the ion optics.

As in the previous experiment, an H_2- or He- seeded potassium oven produces a supersonic beam, the distribution of which is narrowed by a Fizeau-type velocity selector before it crosses at right angle an effusive beam of high density (10^{10}-10^{11}cm^{-3}). Both beams are excited perpendicularly (fig. 3) by the beam coming out of a CW, frequency stabilized dye laser pumping the $4s_{1/2}$,F=2 to $4p_{3/2}$,F=3 transition.

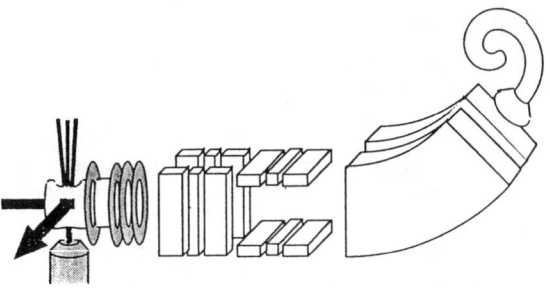

Fig.3 : ion optics for the K(4p)+K(4p) measurements. The thick arrow is the CW laser exciting both supersonic and effusive atomic beams at right angle.

This happens inside an open electrode floated at the ion accelerating potential. Extraction and focusing are performed by a set of ring electrodes, and a pair of planar lenses guides the ion trajectories to the gap of a magnetic sector, that directs the K^+ or K_2^+ ions onto a channeltron, depending on the applied magnetic field. This magnet can be exchanged with an electrostatic analyser, to collect all ions simultaneously. This helps in checking the possible mass-dependent transmission of the optics. Negative ions are detected by reversing voltages and current, and floating the open end of the channeltron to a positive voltage. The reaction volume is surrounded by a set of coils to correct for the residual magnetic field produced by the electro-magnet. The atomic beam intensity is monitored with a Langmuir-Taylor probe, while the laser power is kept constant and chosen high enough in the saturation regime to avoid fluctuations of the excited fraction.

The measured yields appear as follows (fig. 4) :

- The K_2^+ signal, proportional to the AI cross section, rises continuously from threshold (0.3 eV) up to the PI threshold (1 eV) and then drops rapidly, as observed in H(1s)+H(2s) AI

- The K^+ signal, once the contribution of IPF rising at low velocity has been extracted by comparison with the K^- yield, produces the PI cross section, with a sharp onset above that threshold. These contributions to the positive ion yield, when added, perfectly match the total yield measured with electrostatic deflection, and illustrate their common origin in the molecular autoionization that produces either K_2^+ or $K+K^+$ depending on the energy available for the nuclear motion.

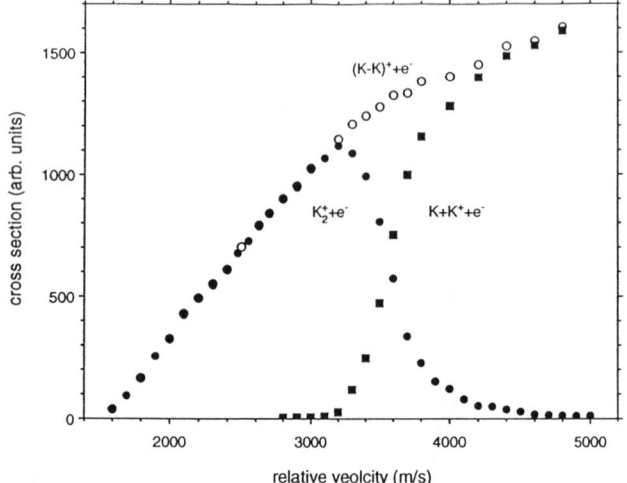

Fig.4 : Relative total cross sections for associative (dots) and Penning (squares) ionization in K(4p)+K(4p) collisions. Open circles are the independently measured total ionization yield.

The IPF is out of the scope of the present paper, but could be used as an indirect calibration of all cross sections. Similarly, the polarization dependence has been addressed with details by Nawracala *et al* and will not be discussed here since no *ab initio* calculations are available to motivate a symmetry analysis. This case study indicates that the Penning ionization is the natural prolongation of AI and should not be forgotten in a comprehensive study of the process.

PARTIAL CROSS SECTIONS : Na(3p)+Na(3p) REACTIONS

In most of the experiments performed so far on associative ionization, total cross sections only have been obtained, in the sense that no characterization of the products has been achieved. Two major exceptions exist however, namely the work of Keller *et al* [4] on one hand, that of Meijer *et al* [5] on the other hand, both devoted to the reaction Na(3p)+Na(3p) → Na_2^+ +e⁻. The latter work rests on the energy conservation that makes the ejected electron energy distribution be a mirror image of the internal energy distribution of the ions.

In their set-up (fig. 5), a CMA is placed on top of the excitation volume defined by the crossing of a CW laser beam with a single effusive sodium beam. Special care was taken of the contact potential that arises from the sodium deposition on the surfaces and affects the transmission and energy shift of the electrons.

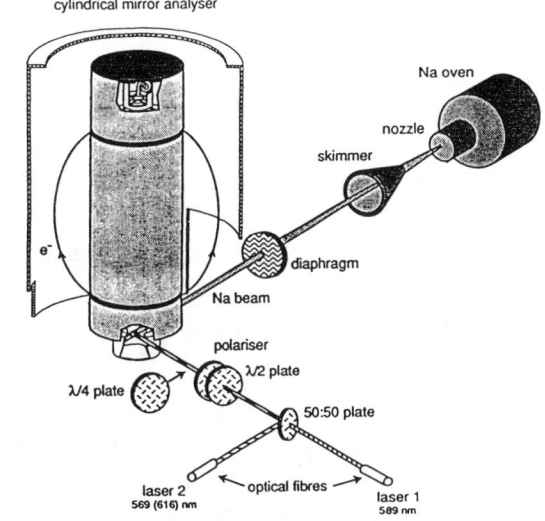

Fig. 5 : experimental set-up of Meijer et al[5]

The obtained spectrum (fig. 6) peaks at 30 meV and is insensitive to the polarization of the exciting light. The authors drew two main conclusions:
- the reaction proceeds through a single autoionizing channel, or symmetry
- the most populated vibrational level is v=3

Unfortunately, the latter conclusion does not hold when the rotational shift of the levels is taken into account.

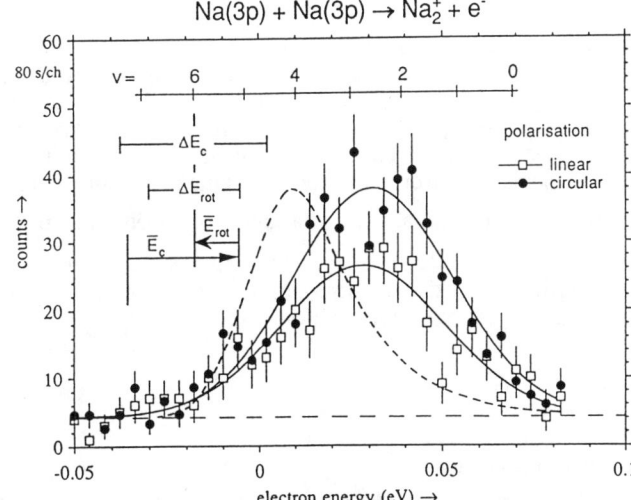

Fig.6 : electron spectrum as measured by Meijer et al. Full lines are gaussian fits, dashed line is a simulation based on the calculations of Dulieu et al[6].

Furthermore, the poor resolution of the analyser, some 25 meV, makes useless any interpretation of the shape of the peak. The authors claimed however to be in good agreement with the theoretical calculations of Dulieu et al[6]. By constructing the electron spectrum corresponding to the computed partial cross sections and folding it with an apparatus function (dashed line on fig. 6), one gets a better comparison with the experimental results, and the agreement is seen to be only qualitative, the peaks being shifted with respect to one another. Such measurements should be repeated with better resolution and better confidence in the energy shift before any definitive conclusion can be drawn.

A second approach was chosen by Keller et al, that consists in photodissociating the Na_2^+ in flight and detecting the photofragments (fig. 7), the difference in arrival time reflecting the c.m. recoil energy. If colinear dissociations are selected, some resolution can be gained at the expense of the signal. The limiting factor is, as in the case of electron spectrometry, the extremely small vibrational quantum of Na_2^+, 14 meV, to be compared with the electronvolt shared by the atomic fragments. A rather involved analysis of the time of flight spectrum recorded at a single wavelength (570 nm) led Wang and Weiner[7] to locate the transition region at the left turning point of the potential well and to question the intervention of a doubly excited state in the associative process, in total disagreement with the recent analysis by Dulieu et al.

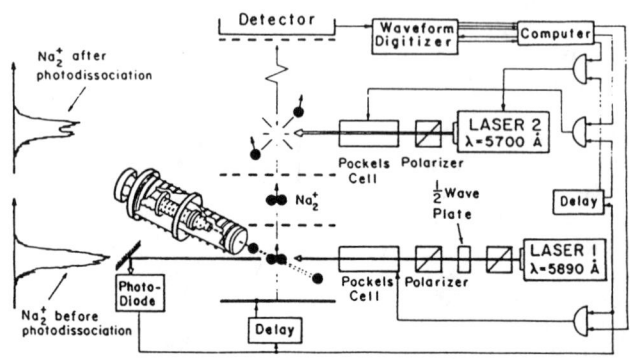

Fig.7 : Schematics of the apparatus of Keller et al[4]. Ions are extracted and accelerated by grids into a drift tube where photodissocation takes place. Both lasers are pulsed.

The idea of photodissociating the molecular ions to analyse their rovibrational excitation should not be discarded, since it adds one degree of freedom to the analysis of the reaction. In the experiments of Keller and Weiner, the wavelength of the pulsed dissociating laser was fixed at a value producing a significant dissociation rate whatever the vibrational level, through excitation from the $^2\Sigma_g$ ground state to the dissociating $^2\Sigma_u$ state of Na_2^+. Our approach is, at the opposite, to vary the wavelength of the dissociating laser to explore the outer turning points of the successive vibrational wavefunctions.

Indeed, the steep $^2\Sigma_u$ potential curve makes the continuum wavefunction peak as a delta function, shifting radially as the photon energy is varied, and the Franck-Condon factor entering the cross section ressembles much the square of the bound wavefunction, projected in wavelength space.

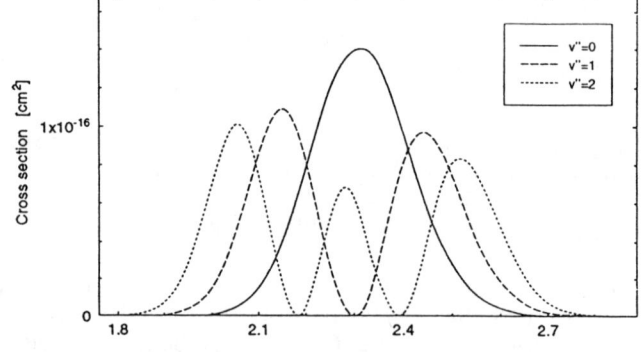

Fig.8 : Photodissociation cross section for the first vibrational levels of Na_2^+. Note the shift of the peaks from level to level.

Such cross sections are shown on fig. 8, where it appears that the outer turning point produces a peak at long wavelength that shifts to the red by some 25nm from level to level. This huge shift is to be compared with the comparatively small vibrational quantum. Moreover, the large cross sections (~10^{-16} cm^2) ensure a measurable dissociation rate, even with a commercial dye laser running multimode.

The set-up makes use of the two ovens apparatus developed in Utrecht by Meijer et al[8] for A.I. and Nijland et al[9] for energy pooling experiments. The counterpropagating beams of sodium are intersected by a laser beam of adjustable polarization, reflected on itself to ensure the selection of a velocity class in both atomic beams, where it pumps the $3s_{1/2}$ F=2 to $3p_{3/2}$ F=3 transition. The interaction volume (fig. 9) defined by this intersection is surrounded by a pair of grids to extract the ions, and these are accelerated further by a third grid up to 100 eV. A first set of plates focuses the ions that emerge from a slab, 6 wide, 8 mm high and 0.6 mm thick to form a narrow beam 10 cm above the interaction plane. At that point, it is crossed by a second laser beam, inclined under an angle of 7° and retroreflected to increase the irradiance in the dissociation region. The atomic ions that emerge from this region are collected by an electrostatic lens prior to detection. The channeltron is placed off axis behind a deflector in order to separate the 50 eV Na$^+$ ions from the 100 eV Na$_2^+$ ions, with a typical Na$^+$/Na$_2^+$ ratio of 10^{-3} for 500 mW. This ratio is found to vary linearly with the power, and all measurements are normalized to that parameter. The overlap of the laser beam with the ions trajectory is optimized by scanning one across the other with a computer driven mirror. The measurements are performed by counting alternatively the parent and product ions.

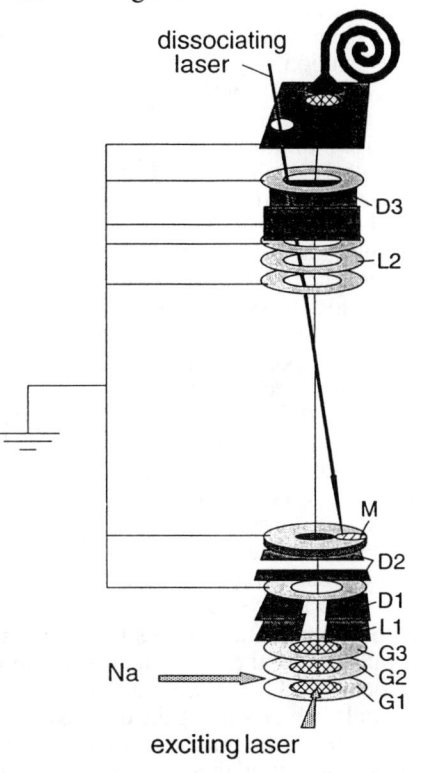

Fig.9 : Schematics of the phodissociation set-up. G1-3 : accelerating grids ; L1-2 : electrostatic lenses ; D1-2 : steering deflectors ; D3 : analysing deflector ; M :retroreflecting mirror. The channel electron multiplier is placed off-axis

By tuning the laser on the whole gain curve of the rhodamine 6G and part of the DCM, the photodissociation rate has been measured between 565 and 670 nm, for a single sodium beam excited at right angle with circularly polarized light. A rich structure has been found in the wavelength dependence (fig. 10), that peaks at 630 nm. When compared with the photodissociation (PD) cross section of the first 6 levels, it indicates a dominant population of the v=3 and 4 levels, in qualitative agreement

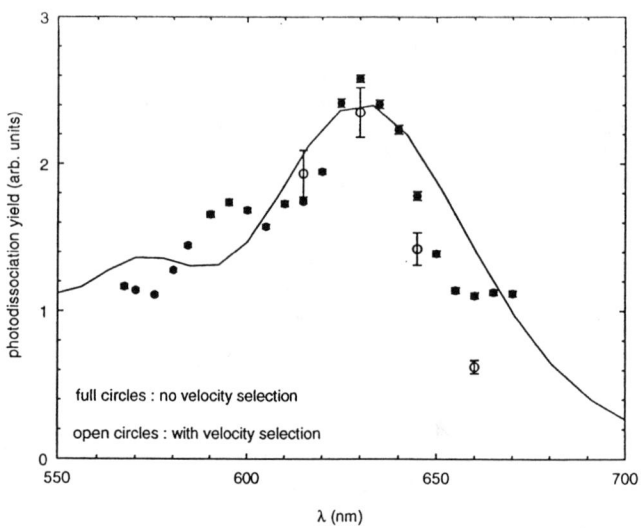

Fig.10 : photodissociation yield versus wavelength. Circles : experiment with a single beam excited by circularly polarized light. Full curve : simulation based on the calculations of Dulieu et al[6].

with the calculations of Dulieu *et al*. Further comparison requires the knowledge of photodissociation cross sections for any (v,J) level, in order to predict the global PD rate from the AI partial cross sections. These cross sections have been computed with the potential curves and dipolar momentum of Magnier *et al*[10]. A full simulation of the spectrum (full line on fig. 10) based on the AI calculations of Dulieu *et al* and the present PD calculations reproduces the position of the maximum but fails in reproducing the detailed structure.

In our preliminary measurements, we did vary the polarization of the exciting light from circular to linear, and the latter from parallel to perpendicular to the collision axis, with no observable change in the photodissociation response. This fact, if confirmed by more accurate measurements, bears out the conclusions of Meijer *et al*, that a single autoionizing state is active in the reaction. Another stringent test is to vary the average collision energy. This is performed by inclining the laser beam to select a velocity class in the beam, reducing the average collision energy from 15 to 5 meV. In this case, some change is observed in the tail at long wavelength (open circles on fig. 10), that we attribute to the closing of the high vibrational levels (v≥5) at low energy. Such measurements have to be repeated with various polarizations and velocities, and extended to higher collision energies by operating the two ovens.

CONCLUDING REMARKS

The main interest of this new method is its straightforward application to the study of cold and ultracold collisions. Indeed, the ions always have to be extracted out of the magneto-optical trap or the optical molasses prior to detection[11]. Such an analysis of the excitation of the products would tell if other channels are populated when the reaction proceeds, as suggested, via two successive excitations of the Na_2 collisional complex, instead of atomic excitation prior to collision.

The detection of both photofragments would bring additional information about the rotational excitation, completing the picture of the dynamics. Kinetic Energy Release (KER) measurements, as achieved by van der Zande and coworkers[12], could address this question.

Finally, internal energy of the product ion can be almost directly measured by KER analysis of the resonant dissociative charge transfer, as it occurs for H_2^+ colliding with cesium[13]. Total cross sections have been measured for the reactions[14,15] $H(2s)+H(2s) \rightarrow H_2^+ +e^-$ and $H^+ + H^- \rightarrow H_2^+ +e^-$, that are good candidates for such a study, presently in preparation in Louvain-la-Neuve.

ACKNOWLEDGEMENT

The author thanks S. Magnier, O. Dulieu and F. Masnou-Seeuws for having made available their numerical data before publication. The author was visiting scientist at the MPI für Strömungsforschung with a stipendium of the Max-Planck-Gesellschaft and is now research associate of the 'Fonds National de la Recherche Scientifique' (F.N.R.S.) of Belgium. This work was furthermore partly supported by the 'Stichting voor Fundamenteel Onderzoek der Materie' (FOM), which is financially supported by the 'Nederlandse organisatie voor Wetenschappelijk Onderzoek' (NWO).

REFERENCES

1. J. Weiner, F. Masnou-Seeuws and A. Giusti-Suzor, Adv. At. Mol. Opt. Phys. 26, 209 (1990).
2. X. Urbain, A. Cornet, F. Brouillard and A. Giusti-Suzor, Phys. Rev. Lett. 66, 1685 (1991).
3. L. Brencher, B. Nawracala and H. Pauly, Z. Phys. D10, 211 (1988).
4. J. Keller, R. Bonnano, M-X. Wang, M.S. de Vries and J. Weiner, Phys. Rev. A33, 1612 (1986).
5. H.A.J. Meijer, S. Schol, H. Dengel, M.W. Müller, M.W. Ruf and H. Hotop, J. Phys. B24, 3621 (1991).
6. O. Dulieu, A. Giusti-Suzor and F. Masnou-Seeuws, J. Phys. B24, 4391 (1991).
7. Y. Wang and J. Weiner, Phys. Rev. A41, 2873 (1990).
8. H.A.J. Meijer, Th. Zeegers, T.J.C. Pelgrim, H.G.M. Heideman and R. Morgenstern, J. Chem. Phys. 90, 729 (1989) ; H.A.J. Meijer, Z. Phys. D17, 257 (1990).
9. J.H. Nijland, J.A. de Gouw, H.A. Dijkerman and H.G.M. Heideman, J. Phys. B25, 2841 (1992).
10. S. Magnier, O. Dulieu and F. Masnou-Seeuws, to be published.
11 P. Gould, P. Lett, P. Julienne, W. Phillips, H. Thorsheim and J. Weiner, Phys. Rev. Lett. 60, 788 (1988) ; M.E. Wagshul, K. Helmerson, P.D. Lett, S.L. Rolston, W.D. Phillips, R. Heather and P.S. Julienne, Phys. Rev. Lett. 70, 2074 (1993).
12 J.M. Schins, R.B. Vrijen, W.J. van der Zande and J. Los, Surf. Sci. 280, 145 (1993) and references therein.
13. D.P. de Bruijn, J. Neuteboom and J. Los, J. Chem. Phys. 85, 233 (1984).
14. X. Urbain, A. Cornet and J. Jureta, J. Phys. B25, L189 (1992).
15. G. Poulaert, F. Brouillard, W. Claeys, J.W. Mc Gowan and G. Van Wassenhove, J. Phys. B11, L671 (1978).

REACTIVE SCATTERING AND ELECTRON DETACHMENT IN H⁻+H$_2$/D$_2$ COLLISIONS AT LOW eV ENERGIES

M. Zimmer and F. Linder
Department of Physics, University of Kaiserslautern
D–67653 Kaiserslautern, Germany

ABSTRACT

We report on a crossed beam study of various H⁻+H$_2$/D$_2$ collision processes in the collision energy range E_{rel} = 0.3–3 eV. Special emphasis has been placed on the rearrangement reaction H⁻+D$_2$(v=0) → HD(v')+D⁻. From the vibrationally state–resolved energy and angular spectra of the D⁻ product ions, we derive state–specific differential cross sections in the CM system. Absolute units are obtained by using elastic H⁻+He scattering as a reference system. The measurements allow detailed conclusions concerning the reaction mechanism. By integrating over angles and summing over partial cross sections (for individual v' states of the HD product molecule), we determine the absolute total cross section of the reaction. The reaction has a threshold at E_{rel} = 0.42 ± 0.12 eV. The cross section rises to a maximum of 2 x 10⁻¹⁶ cm² at 1.5 eV and then rapidly decreases again. For energies above the detachment threshold (about 1.45 eV), a strong competition between reactive scattering and electron detachment is expected which will be an interesting subject for future studies.

INTRODUCTION

Reactive scattering is one of the fundamental processes in atomic and molecular collision physics. Reactions of hydrogen systems are of particular interest in this respect, because they are amenable to the most rigorous theoretical treatment and thus represent ideal prototype cases for a detailed comparison between theory and experiment. This is best exemplified by the neutral hydrogen system H+H$_2$ (we choose the isotopic variant H+D$_2$ to distinguish between elastic/inelastic and reactive scattering):

$$H+D_2(v,j) \rightarrow H+D_2(v',j')$$
$$\rightarrow HD(v',j')+D$$

Quantum chemistry has provided a very accurate potential energy surface (PES) for this system. The collision dynamics are treated by quasi–classical trajectory (QCT) calculations or, more recently, by rigorous, fully converged quantum calculations. Furthermore, there is a considerable number of very detailed experimental results on which this theory can be tested. The H+H$_2$ reaction has a long history, and there is an enormous amount of work on this system. In recent years, there has been a booming interest due to some spectacular advances in both theory and experiment. For a review of these developments, we refer to the article by Miller [1] which gives an impressive picture of the state of the art in this field.

In a similar way, the (positively charged) ionic hydrogen system H^++H_2 is often considered as a prototype system for the study of ion–molecule reaction dynamics. The scheme of competing processes can be illustrated as follows:

$$\begin{aligned}H^++D_2 &\rightarrow H^++D_2\\ &\rightarrow HD+D^+\\ &\rightarrow H+D_2^+\\ &\rightarrow HD^++D\end{aligned}$$

Numerous studies, both experimental and theoretical, have been carried out for this system. Again, one needs contributions from three sides: PES, dynamics calculations, and experiment. On the whole, the main features of this system are quite well understood, although the state of the art is not quite as advanced as in the case of the neutral hydrogen system, both regarding theory and experiment. A recent survey of various particle interchange reactions for this system can be found in the paper by Schlier et al. [2].

One should note, of course, that the reaction dynamics of the neutral hydrogen system $H+H_2$ and the ionic hydrogen system H^++H_2 are completely different. The $H+H_2$ system represents a reaction with a barrier. In contrast, the PES of the H^++H_2 system is strongly attractive and has a deep potential minimum which influences the dynamics of the H^++H_2 rearrangement reaction in a completely different way than in the $H+H_2$ case. Furthermore, there is a strong coupling of the ground state PES with the next higher PES of H_3^+ which correlates to $H+H_2^+$. This coupling gives rise to charge transfer processes with and without rearrangement. All reactions given above have cross sections of comparable magnitude at low eV energies.

Compared to the systems $H+H_2$ and H^++H_2, relatively little work has been done on the negative ion hydrogen system H^-+H_2, both experimentally and theoretically. In this case, the scheme of competing processes at low eV energies is written in the following way:

$$\begin{aligned}H^-+D_2 &\rightarrow H^-+D_2\\ &\rightarrow HD+D^-\\ &\rightarrow H+D_2+e^-\\ &\rightarrow HD+D+e^-\end{aligned}$$

Electron detachment has been studied in several papers [3–5]. However, no distinction is made between processes (3) and (4) of the above scheme. State–resolved measurements of vibrational excitation have been reported only for higher collision energies where this process becomes more efficient [6]. For the rearrangement reaction $H^-+D_2 \rightarrow HD+D^-$, there are two experiments which give information on the total cross section as a function of collision energy [3,7]. The results of these measurements deviate strongly from each other for energies below 4 eV. Concerning the PES for this system, there are some preliminary calculations [7]. However, these calculations are rather incomplete and probably not very accurate, while other calculations are mainly concerned with the stability of the H_3^- van der Waals complex [8,9]. To our knowledge, there are no dynamics calculations and no detailed scattering experiments on the H^-+D_2 reaction.

In this paper, we report on a detailed experimental study of the rearrangement reaction $H^-+D_2(v=0) \rightarrow HD(v')+D^-$ in the collision energy range $E_{rel} = 0.3$–3 eV. Using the crossed beam technique, we measure vibrationally state–resolved energy and angular spectra of the D^- product ions. From these data, we obtain state–specific differential cross sections in the CM system. Detailed conclusions can be drawn concerning the reaction mechanism. A striking similarity to the neutral $H+H_2$ system is observed. By integrating over angles and summing over partial cross sections (for individual v'), we determine the absolute total cross section of the reaction. This cross section is compared with the above–mentioned previous measurements for this reaction. In addition to the rearrangement reaction, we have also studied rotationally inelastic scattering of H^- ions from H_2 and D_2. In the case of the H^-+H_2 system, state–resolved measurements could be performed. A first report on the present work has been given recently [10].

EXPERIMENTAL

The crossed beam apparatus is similar to the one described by Bischof and Linder [11]. A mass and energy selected beam of H^- ions is crossed at right angles with a supersonic nozzle beam of H_2 or D_2 molecules. The scattered product ions are detected in the plane of the crossed beams by a rotatable detector system which allows mass and energy analysis of the product ions. The angular range covered in the present measurements is from $\Theta_{lab} = 0°$ to $\Theta_{lab} = 90°$, the angle being defined with respect to the primary H^- beam.

The experimental conditions are briefly characterized as follows. The measured peak width for elastic H^-+He scattering is 160 meV (FWHM). This value can be taken as a measure of the overall energy resolution in the present experiment. The effective angular resolution is $\Delta\Theta_{lab} \lesssim \pm 5°$, which corresponds to $\Delta\theta_{cm} \lesssim \pm 8°$ in the CM system of the H^-+D_2 reaction. The uncertainties in the calibration of the detector energy scale and the collision energy scale are $\Delta E_{det} = \pm 40$ meV and $\Delta E_{rel} = \pm 60$ meV, respectively. The rotational temperature of the H_2/D_2 target beam is estimated to $T_{rot} = 180$ K. This means that the D_2 molecules are mostly in the rotational states j = 0,1, and 2. It is clear that all target molecules can be assumed to be in the vibrational ground state. Some more details of the experiment are given in the above–mentioned papers [10,11].

The transmission properties of the detector as a function of the energy of the detected ions are determined by using elastic H^-+He scattering as a reference system. The measured intensities are compared with calculated differential cross sections which are based on an accurately known H^-+He interaction potential [12,13]. This allows us to construct a correction function which is then applied to the energy spectra of the H^-+D_2 measurements. In a similar way, namely by comparing the measured intensities of the H^-+D_2 reaction with those of elastic H^-+He scattering, it is possible to give all measured cross sections of the present work directly in absolute units.

RESULTS AND DISCUSSION

Fig. 1 shows some examples of kinetic energy spectra of the D^- product ions, measured at a detection angle of $\Theta_{lab} = 30°$ and different collision energies. One observes clearly resolved peaks which can unambiguously be assigned to vibrational states of the HD product molecule. At low collision energies, the HD product molecule is only formed in the vibrational ground state v'=0. With increasing collision energy, more and more vibrational states of the HD molecule are excited.

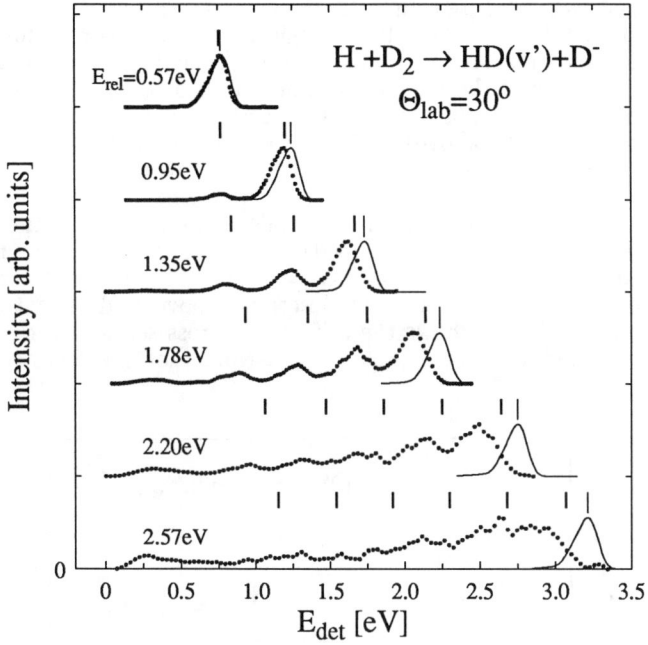

Fig. 1. Kinetic energy spectra of D^- product ions (dotted curves) from the reaction $H^-+D_2 \rightarrow HD(v')+D^-$ at $\Theta_{lab} = 30°$ and different collision energies. The elastic H^-+He peak (full line) is shown for comparison. Marks above each spectrum: see text.

Rotational excitation is not resolved. However, some conclusions can be drawn from the measured positions and widths of the peaks. In Fig. 1, the peak of elastic H^-+He scattering is shown directly in comparison with the measured spectra of the D^- product ions. The H^-+He peak can be used to calibrate the energy scale and to demonstrate the actual instrumental energy resolution for each spectrum. The long thin bar indicates the position of the H^-+He peak, whereas the short thick bars above each spectrum indicate the kinematically expected positions of the D^-(v') peaks assuming $\Delta j=0$ transitions from reactants to products. The results reveal the following. At the lowest collision energies, the reaction occurs with no or only little rotational energy transfer. With the increase of the collision energy, the amount of rotational excitation is found to increase as is indicated by the shift and the broadening of the D^- peaks. At collision energies higher than 2.5 eV, the vibrational structure of the D^- spectra gets completely smeared out. A qualitative interpretation of this behaviour in terms of the properties of the H_3^- PES will be given below.

We note that for collision energies above 2 eV the cross section decreases rapidly and the D^- signal becomes very weak. Because of the weak signal and the increasingly diffuse character of the spectra, the present measurements could not be extended beyond 3 eV.

Special attention was given to the determination of the reaction threshold. The collision energy was varied in small steps over the range $E_{rel} = 0.3$–0.5 eV and the D^- signal was measured. In order to ensure the proper functioning of the ion gun and the detector at these low energies, each H^-+D_2 measurement was accompanied by a corresponding measurement of elastic H^-+He scattering. In this way, the threshold of the reaction was determined to $E_{rel,th} = 0.42 \pm 0.06$ eV. The quoted error does not include the uncertainty in the calibration of the collision energy scale (see above), so that the maximum total error must be given as $\Delta E_{rel,th} = 0.12$ eV.

Fig. 2 gives an example of the differential cross sections obtained from the present measurements. The raw data are D^- energy spectra taken at fixed laboratory scattering angles in the range $\Theta_{lab} = 0$–$90°$. The data are corrected for transmission effects of the detector as described above and then transformed into the CM system. Fig. 2 shows the differential cross sections (angular distributions) for v'=0 and v'=1 at $E_{rel} = 1.1$ eV. Note that the cross sections are given in absolute units.

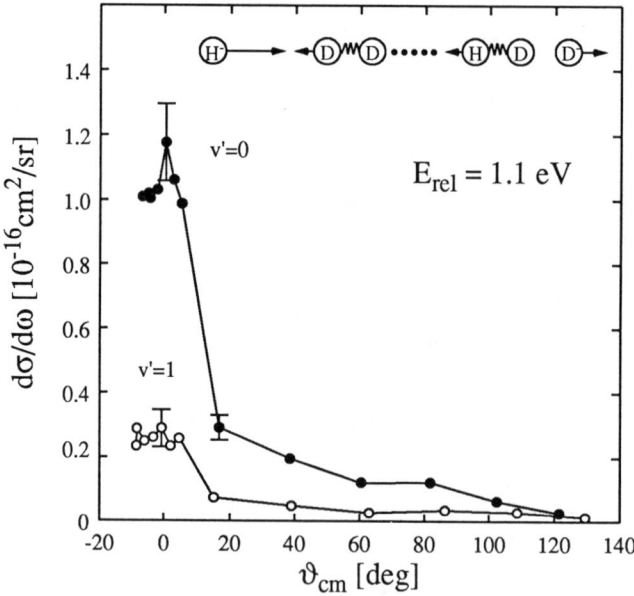

Fig. 2. Differential cross sections in the CM system for the channels v'=0 and v'=1 of the reaction $H^-+D_2 \rightarrow HD(v')+D^-$ at $E_{rel} = 1.1$ eV. The angle θ_{cm} describes the scattering angle of the D^- product ion in the CM system with respect to the velocity vector of the incoming H^- ion.

As is clearly evident from Fig. 2, the angular distribution of the D^- product ions is strongly forward peaked. Correspondingly, the HD product molecule is scattered in the backward direction. Together with the observation that rotational excitation is nearly missing at near–threshold energies, this leads to the conclusion that the reaction occurs preferentially for a collinear configuration of the reactants and that it is restricted to a relatively narrow range of small impact parameters. The collinear reaction mechanism is schematically indicated in the insert of Fig. 2.

Some further results for additional collision energies are shown in Fig. 3. The observations can be summarized as follows: (i) with increasing collision energy, more and more vibrational states of the HD product molecule get excited; (ii) the D$^-$ product distribution is always peaked in forward direction, but the angular distribution gets broader as the collision energy increases; (iii) as discussed above, the amount of rotational excitation is small near threshold and increases with the collision energy.

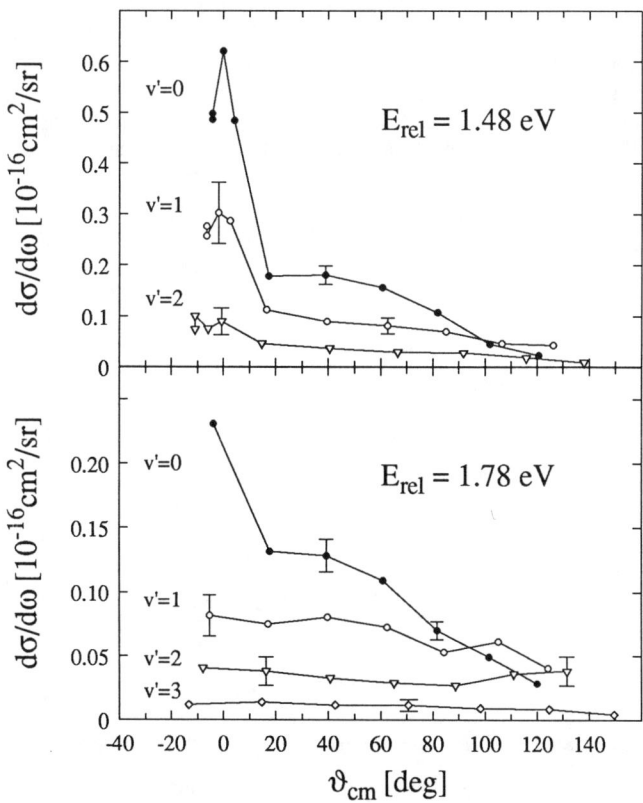

Fig. 3. Similar results as in figure 2 for somewhat higher collision energies.

This behaviour can be understood from the properties of the PES. As already noted, there is a striking similarity between the present observations for the H$^-$+D$_2$ reaction and the corresponding results for the neutral H+H$_2$ system. This is not too surprising, though, since the above—mentioned calculations of the H$_3^-$ PES indicate a qualitative similarity to the neutral H$_3$ system. In both cases, we have a reaction with a barrier. The saddle point energy is found to be lowest for the collinear configuration, while for the bent configuration it is significantly higher. The barrier increases monotonically with the angle γ. From these properties, we can qualitatively explain the results for the H^-+D_2 reaction. Near the reaction threshold, only collinear geometries contribute to the reaction. This gives a sharply forward–peaked angular distribution of the D^- product ions and allows only little rotational energy transfer in the reaction. As the collision energy increases, more and more contributions come from bent configurations of the reactants. The angular distribution of the product ions gets broader and the amount of rotational excitation increases. Quantitative calculations based on ab initio theory are needed now in order to provide a more thorough understanding of the H^-+D_2 reaction.

When the differential cross sections are integrated over angles, we obtain the partial cross sections $\sigma_{v'}(E_{rel})$ and the branching ratios $\Gamma_{v'}(E_{rel}) = \sigma_{v'}/\Sigma\sigma_{v'}$, of the reaction, each specified by the vibrational quantum number v' of the HD product molecule. The results are shown in Fig. 4 and Fig. 5, respectively. These data show in quantitative terms what was already reflected in the D^- product spectra. With increasing collision energy, the formation of the HD molecule in vibrationally excited states becomes of growing importance. However, v'=0 remains the dominant channel for collision energies up to 2 eV.

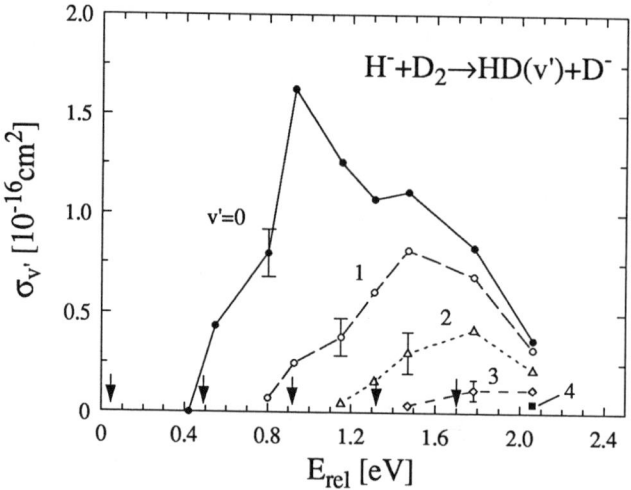

Fig. 4. Angle–integrated partial cross sections $\sigma_{v'}(E_{rel})$ for the reaction $H^-+D_2 \rightarrow HD(v')+D^-$. The arrows above the energy scale indicate the thermodynamical thresholds for the reaction with v'=0–4 (see text).

Two remarks may be added. There appears to be a structure emerging at around $E_{rel} = 0.9$ eV in the v'=0 cross section. However, it seems too early to speculate about the significance of this structure. If real, this observation requires further study. Another feature observed in the partial cross sections of Fig. 4, however, is definitely significant and worth pointing out. Besides the v'=0 cross section which has a threshold at 0.42 eV, also the other cross sections apparently show a delayed onset compared to the value at which the reaction is energetically possible. The latter values for v'=0–4 are indicated by arrows above the energy scale in Fig. 4. In all cases, there is clearly a delayed onset of the measured cross section. It would be interesting to see this feature verified in theoretical studies of the H^-+D_2 reaction.

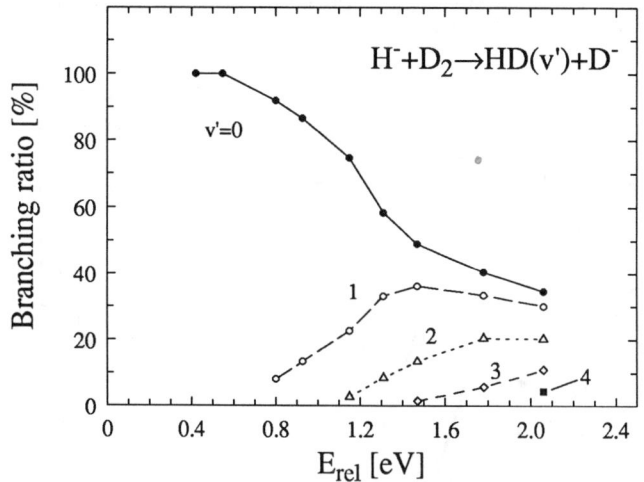

Fig. 5. Branching ratios $\Gamma_{v'}(E_{rel}) = \sigma_{v'}/\Sigma\sigma_{v'}$ of the partial cross sections for the reaction $H^-+D_2 \rightarrow HD(v')+D^-$.

By summing up the partial cross sections of Fig. 4, we obtain the absolute total cross section of the reaction. This cross section is shown in Fig. 6. The full dots and the solid line represent the results of the present work, the line being a smoothed spline fit to the experimental points. The estimated error is about 20 % in the maximum of the curve and increases to about 30 % at the higher energies.

Also shown in Fig. 6 are the results of previous measurements of the total cross section. The results reported by Michels and Paulson [7] have been obtained using the tandem mass spectrometer (TMS) technique. These authors find a threshold at about 1.3 eV and a cross section maximum of 0.65×10^{-16} cm^2 near 3 eV, which is clearly at variance with the present measurements. We believe that the discrepancy observed for energies below 3 eV must be attributed to transmission and/or collection efficiency problems in the TMS measurements at low energies. On the other hand, one may suppose that the TMS data represent an approximately correct continuation of the reaction cross section for energies above 3 eV.

These conclusions are supported by measurements of Huq et al. [3] which are also included in Fig. 6 for comparison. The curve σ_e gives the total electron detachment cross section, whereas σ_i represents the total cross section for slow ion production in H^-+D_2 collisions. It should be mentioned that Huq et al. could not distinguish between H^- and D^- products, since no mass analysis was provided in their experiment. An unambiguous interpretation of σ_i is therefore difficult. Besides D^- products resulting from the reaction $H^-+D_2 \rightarrow HD+D^-$, the ion signal probably also contains a considerable portion of H^- ions resulting from large angle elastic and inelastic scattering. Thus, σ_i must probably be regarded as an upper limit for the reaction cross section. In any case, the general shape of σ_i for the energy range $E_{rel} = 2-8$ eV is consistent with the interpretation given above, namely that the total cross section for the H^-+D_2 rearrangement reaction is correctly represented by the low energy data of the present work and the TMS results for energies above 3 eV.

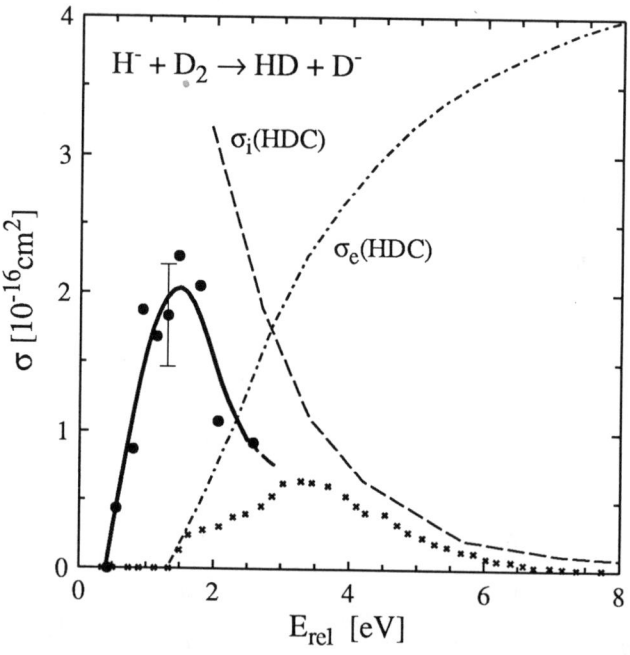

Fig. 6. Total cross section for the reaction $H^-+D_2 \rightarrow HD+D^-$. Full dots and solid line: results of the present work. Crosses: results of Michels and Paulson [7]. The curves σ_e(HDC) and σ_i(HDC) represent the total cross sections for electron detachment and slow ion production, respectively, measured by Huq et al. [3].

It is interesting to note that the energy, at which the reaction cross section reaches its maximum and begins to decrease again, practically coincides with the onset of electron detachment which has been determined to 1.45 eV by Huq et al. [3]. This is certainly not fortuitous, but one has to assume that the decrease of the $HD+D^-$ cross section is directly connected with the opening of the electron detachment channel. Since the rearrangement reaction is confined to small impact parameters, it is not surprising that electron detachment rapidly becomes the dominant process once that this channel is open. This does not preclude, however, that rearrangement still occurs, i.e. that the reaction $H^-+D_2 \rightarrow HD+D+e^-$ takes place. The measurements of the electron detachment cross section σ_e shown in Fig. 6 do not distinguish between simple direct detachment $H+D_2+e^-$ and reactive detachment $HD+D+e^-$. It would certainly be of high interest, if the competition between the various processes for energies above the detachment threshold could be studied in more detail.

SUMMARY AND OUTLOOK

We have presented a detailed study of the rearrangement reaction $H^-+D_2 \rightarrow HD(v')+D^-$ at low eV energies. The results include state–specific differential cross sections, angle–integrated partial cross sections, and finally the total reaction cross section summed over all product states. All cross sections are given in absolute units. The measurements thus provide a rather complete picture of the reaction. Detailed conclusions can be drawn concerning the reaction mechanism. A discrepancy between two previous measurements bearing on the determination of the total cross section could be resolved.

A striking similarity is observed between the present findings for the H^-+D_2 reaction and corresponding results for the neutral $H+H_2$ system. This similarity is based on the fact that the potential energy surfaces of the two systems are similar. Theoretical calculations, which explore the parallel between the two systems in more detail, are highly desirable. The theoretical methodology of reactive scattering, which has been greatly advanced in recent years [1], is available. The theoretical methods have been successfully applied to the neutral $H+H_2$ reactions. From this point of view, an extension to the H^-+H_2/D_2 system seems straightforward. On the other hand, however, one has to realize that the present system poses new problems which are due to the opening of the electron detachment channel at 1.45 eV collision energy. A complete and quantitative description of the various competing processes in this energy range requires not only a detailed knowledge of the H_3^- and H_3 potential energy surfaces, but also a proper treatment of the detachment process at the crossing seam of the two surfaces. The H^-+H_2/D_2 system may therefore represent a challenging new problem for theory.

Very recently, accurate ab initio calculations of the H_3^- potential energy surface have been carried out by Stärck and Meyer at the University of Kaiserslautern [14]. Using MR–CI and CEPA methods, the PES has been calculated for more than 400 nuclear configurations. It is cast in analytical form by a fit with 23 parameters. The fit to the ab initio points is better than 7 meV for the whole surface and better than 2 meV for all points below 1.4 eV. The position and the height of the barrier are found to be $r_b = 1.997\ a_o$ and $E_b = 0.457$ eV, respectively. Taking appropriate account of zero–point energies, a threshold energy of 0.49 eV is obtained for the reaction $H^-+D_2 \rightarrow HD+D^-$, in good agreement with the experimental value of 0.42 ± 0.12 eV. The crossing seam of the H_3^- and H_3 potential energy surfaces has been determined as a function of all three internuclear coordinates. Its lowest point is found for the perpendicular geometry ($\gamma = 90°$) with an energy of 1.2 eV which also agrees quite well with experimental values for the detachment threshold [3].

With the H_3^- and H_3 surfaces being available now, the way is open for quantitative dynamical calculations. Such calculations are expected for the near future; they will make possible a detailed comparison between theory and experiment. This, in turn, will stimulate further experimental work. With the combined efforts of theory and experiment, the goal can be envisaged to establish the H^-+H_2/D_2 system as a benchmark system for the present class of processes, i.e. reactive processes of negative ions including detachment channels. Further work towards this end is in progress.

ACKNOWLEDGEMENT

This work has been supported by the Deutsche Forschungsgemeinschaft under SFB 91.

REFERENCES

1. W.H. Miller, Annu. Rev. Phys. Chem. 41, 245 (1990)
2. Ch. Schlier, U. Nowotny, and E. Teloy, Chem. Phys. 111, 401 (1987)
3. M.S. Huq, L.D. Doverspike, and R.L. Champion, Phys. Rev. A 27, 2831 (1983)
4. V.A. Esaulov, J.P. Grouard, R.I. Hall, M. Landau, J.L. Montmagnon, F. Pichou, and C. Schermann, J. Phys. B: At. Mol. Phys. 17, 1855 (1984)
5. V.N. Tuan, V.A. Esaulov, J.P. Gauyacq, and A. Herzenberg, J. Phys. B: At. Mol. Phys. 18, 721 (1985)
6. U. Hege and F. Linder, Z. Phys. A 320, 95 (1985)
7. H.H. Michels and J.F. Paulson, in "Potential energy surfaces and dynamics calculations", ed. D.G. Truhlar, Plenum Press, New York 1981 (p. 535)
8. H.H. Michels and J.A. Montgomery, Chem. Phys. Lett. 139, 535 (1987)
9. G. Chalasinski, R.A. Kendall, and J. Simons, J. Phys. Chem. 91, 6151 (1987)
10. M. Zimmer and F. Linder, Chem. Phys. Lett. 195, 153 (1992)
11. G. Bischof and F. Linder, Z. Phys. D 1, 303 (1986)
12. R.E. Olson and B. Liu, Phys. Rev. A 22, 1389 (1980)
13. J. Stärck and W. Meyer, to be published
14. J. Stärck and W. Meyer, Chem. Phys., in press (1993)

PHOTOABSORPTION STUDIES OF QUASIMOLECULES IN THERMAL COLLISIONS OF ALKALINE-EARTH ATOMS

Yukinori SATO
Res. Inst. Sci. Measur., Tohoku University, Sendai 980, Japan

ABSTRACT

Results of our recent observation for continuum (free-free) absorption bands in binary gas-mixtures of some metal atoms (Sr, Ba, Yb, and Hg) with rare-gas atoms and N_2 are briefly presented. Observations are made of (a) far-wing spectra of the metal resonance lines exhibiting the resonance broadening, the oscillatory satellite-structures, and collision-induced-dipole absorption bands, (b) the pair absorption in the Ba-Ba collisions, and (c) radiative collisions in the $Hg+N_2+\hbar\omega$ system. By means of a high precision dispersion and absorption method using a Mach-Zendar interferometer, the absorption coefficients of the metal–metal and the metal–rare-gas quasi-molecular bands are separated, and the absorption coefficients are determined in the absolute scale. The data are analyzed to discuss the relevant interaction potential energies at the internuclear distances from 8 to $40a_0$. The oscillatory satellite structures are discussed in terms of the semiclassical collision theory.

INTRODUCTION

Optical absorption and emission of collisional quasimolecules are often discussed in terms of two different classes. The process of the first class is called *optical collisions*[1], and can be expressed as

$$A + B + \hbar\omega \rightleftharpoons A^* + B. \tag{1}$$

The process (1) is observed in far-wings of the pressure-broadened lines of A or in the collision-induced-dipole (CID) bands[2] associated with the forbidden transitions of A. The process of the second class, called *radiative collisions*[3], can be expressed as

$$A^* + B + \hbar\omega \rightleftharpoons A + B^*. \tag{2}$$

Process (2) changes states of both atoms whereas process (1) changes state of only one atom. From this viewpoint, *pair absorption*[4], which is expresses as

$$A + B + \hbar\omega \rightleftharpoons A^* + B^*, \tag{3}$$

may be included in the second class. All of the above processes can be regarded[1,5,6] as the photon-assisted collisions in the weak field limit where a single photon is absorbed or emitted by a collisional quasimolecules. We have started a series of spectroscopic measurements for studing these types of photon-assisted collisions in the systems including some metal atoms (Ca, Sr, Ba, Yb, and Hg) as a collision partner and the rare-gas atoms or some diatomic molecules (H_2, N_2) as the other collision partner.

© 1993 American Institute of Physics

EXPERIMENTAL

Our experimental approach is based on the photon-beam–gas-cell method. A heat-pipe oven is used as a gas cell which contains a binary mixture of metal vapor and rare gas. In the absorption studies, we have made efforts to separate the spectral component of the metal–rare-gas collisions from that of the metal–metal collisions. We have made efforts also to determine absolute values of the reduced absorption coefficients. For these purposes, we adopt a conventional double-beam dispersion and absorption method.

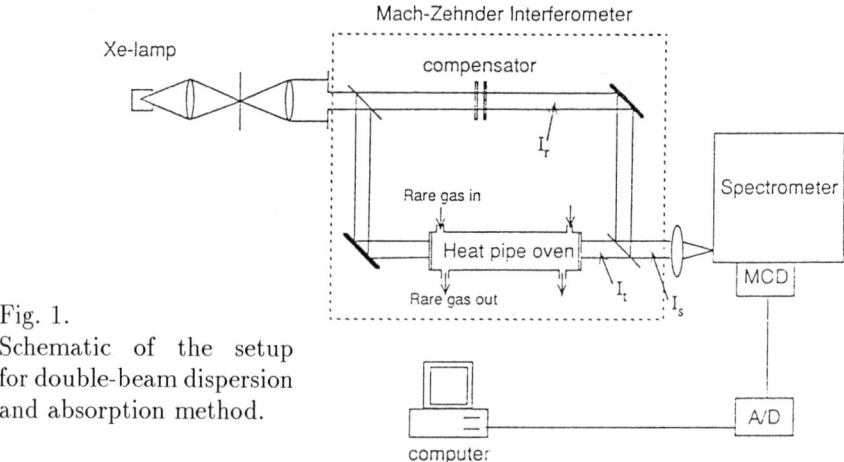

Fig. 1.
Schematic of the setup for double-beam dispersion and absorption method.

Figure 1 illustrates the method. Metal vapor is generated in a heat-pipe-oven absorption cell heated to 500–1100°C. One of the rare gases (RG: He, Ne, Ar, Kr. and Xe) is admixed in the cell at a pressure in the range of 50-600 Torr. The absorption cell is placed in a test-beam section of a Mach-Zender interferometer. A light beam from a xenon lamp is divided into two: one passes through the absorption cell, while the other serves as a reference for the absorption. These two light beams and the interferece beam which is the superposition of these two beams are alternately dispersed by a Czerny-Turner spectrometer and are detected by a self-scanning 1024-channel photodiode array. From the interference spectrum in the vicinity of the metal resonance line, the number density of the metal atoms integrated over the line-of-sight (or column density) $N_M l = \int_{vapor-length} N_M dl$ is determined with the statistical uncertainty of less than 1%.[7,8,9] Absorption spectrum, in terms of the optical depth, is given by

$$k(\lambda)l = -\ln[I_T(\lambda)/I_R(\lambda)], \qquad (4)$$

where $I_T(\lambda)$ and $I_R(\lambda)$ are the relative spectral intensities of the test beam and the reference beam, respectively.

In Fig. 2, examples of the absorption spectra of the Sr/He mixtures are shown in the vicinity of Sr: $5s^2\,{}^1S_0 - 5s4d\,{}^1D_2$ electric-quadrupole transition. It is clearly seen that the absorption depends on the He pressure. Assuming that the far-wing spectrum consists of a superposition of the Sr-He and Sr-Sr components, we have

$$k(\lambda) = \gamma_{\text{SrHe}}(\lambda) N_{\text{Sr}} N_{\text{He}} + \gamma_{\text{SrSr}}(\lambda) N_{\text{Sr}}^2, \tag{5a}$$

and, therefore,

$$\frac{k(\lambda)l}{(N_{\text{Sr}}l)^2} = (\gamma_{\text{SrHe}}) \frac{N_{\text{He}}}{N_{\text{Sr}}l} + \frac{\gamma_{\text{SrSr}}}{l}, \tag{5b}$$

where $\gamma_{\text{SrHe}}(\lambda)$ and $\gamma_{\text{SrSr}}(\lambda)$ are the reduced absorption coefficients (RAC's) for the Sr-He and the Sr-Sr systems, respectively. Thus, we can obtain $\gamma_{\text{SrHe}}(\lambda)$ and $\gamma_{\text{SrSr}}(\lambda)/l$ through the linear fit on the plot of $k(\lambda)l/(N_{\text{Sr}}l)^2$ against $N_{\text{He}}/(N_{\text{Sr}}l)$ at every element of wavelength λ dispersed onto the 1024ch array detector. Aiming to obtain the RAC γ_{MM} for the metal-metal collisions, we have proposed a method to estimate the effective absorption length l_{eff} by observing the resonance broadening of the metal resonance line. [10,11]

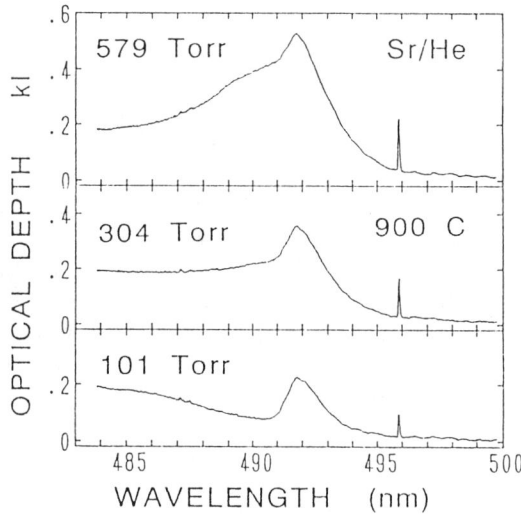

Fig. 2. Absorption spectra of the Sr/He mixture in the vicinity of the Sr $5s^2\ ^1S_0 - 5s4d\ ^1D_2$ E2 line, which is seen as a sharp line at 496nm. He pressure is indicated for each spectrum. The cell temperature is 900°C.

RESULTS AND DISCUSSIONS

A. Resonance broadening in homonuclear metal-metal collisions

Figure 3 shows the measured γ_{SrSr} for the wings of the Sr $^1P_1 \leftarrow\ ^1S_0$ resonance line broadened by Sr-Sr collisions. The spectrum of Fig. 3 shows the following features;
  (a) resonance broadening which extends symmetrically to both wings,
  (b) satellite structures of the blue wing for $1000 \leq \Delta\nu \leq 1600\text{cm}^{-1}$, and
  (c) a small peak at $\Delta\nu = -1200\text{cm}^{-1}$, which we have assigned to the CID band[9] associated with the Sr $5s^2\ ^1S_0 - 5s4d\ ^1D_2$ electric-quadrupole transition at $\lambda = 496$nm.

The wings in region (a) of Fig. 3 can be assigned to the Sr$_2$ transition $^1\Pi_u \leftarrow\ ^1\Sigma_g$ for the blue wing and $^1\Sigma_u \leftarrow\ ^1\Sigma_g$ for the red wing. Assuming the R^{-3} dependence of the upper state potentials due to the first-order dipole-dipole (d-d) interaction and using the quasistatic theory (QST) of line broadening, the RAC for the resonance broadening can be approximated as

$$\gamma_{\text{SrSr}}(\Delta\nu) = \frac{e^4 f^2}{9m^2 c\nu_0} \frac{1}{(\Delta\nu)^2}, \quad (6)$$

where ν_0 is the wave number of the line center and f is the absorption oscillator strength of the resonance line.

Adjusting the vapor-length l to an appropriate value, the measured $\gamma_{\text{SrSr}}(\Delta\nu)$ of Fig 3 is well explained by Eq. (6) in the range of $\Delta\nu$ from -50 to $-1150\,\text{cm}^{-1}$ for the red wing and from 50 to 900 cm^{-1} for the blue wing. These ranges of detuning correspond to the range of internuclear distances from 15 to $44 a_0$ for the red wing and from 14 to $34 a_0$ for the blue wing. The potential curves thus obtained are shown in Fig. 4. Quite similar features of the symmetric $\Delta\nu^{-2}$-dependence are found for $\gamma_{MM}(\Delta\nu)$ of the self-broadened first resonance lines:

Ca $4s^2\,{}^1S_0$–$4s4p\,{}^1P_1$ at 423nm,
Sr $5s^2\,{}^1S_0$–$5s5p\,{}^1P_1$ at 461nm,
Ba $6s^2\,{}^1S_0$–$6s6p\,{}^1P_1$ at 554nm, and
Yb $6s^2\,{}^1S_0$–$6s6p\,{}^1P_1$ at 399nm.[12]

B. Blue-wing satellite of the Sr$_2$ band

The satellite structure in Fig. 3 is due to the existence of a maximum on the difference potential $\Delta V(R)$ between the excited B $^1\Pi_u$ and the ground X $^1\Sigma_g$ states of the Sr$_2$ molecule.

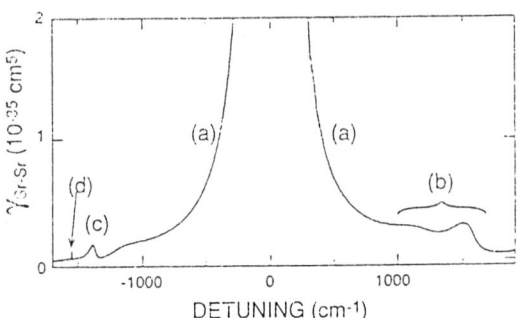

Fig. 3. Sr-Sr absorption bands in the vicinity of the Sr $5s^2\,{}^1S_0$-$5s5p\,{}^1P_1$ line. γ_{SrSr} is plotted as a function of detuning $\Delta\nu$ (cm^{-1}) from the line center. (a) regions of the resonance broadening, (b) blue-wing satellite structures. (c) s-d type CID band, (e) Sr $5s^2\,{}^1S_0 - 5s4d\,{}^1D_2$ electric-quadrupole transition.

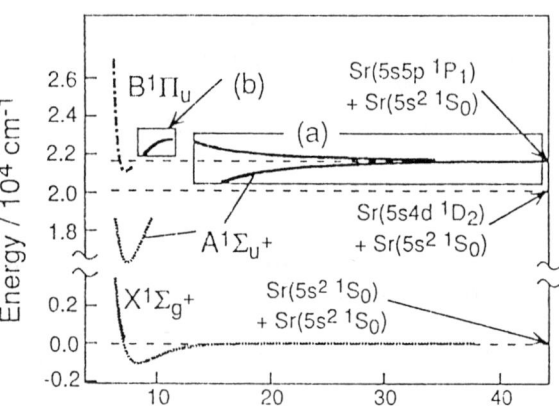

Fig. 4. Sr$_2$ potential curves. Region (a) corresponds to the resonance interaction (C_3/R^3) potential, region (b) comes from the satellite analysis. The curves of $X^1\Sigma_g^+$ and $A^1\Sigma_u^+$ at $R < 8a_0$ are from ref.14. The curve of $B^1\Pi_u$ at $R < 7a_0$ is from ref. 15.

The maximum on $\Delta V(R)$ may arise from either a van der Waals (vdW) well of the the lower X $^1\Sigma_g$ state or a hump of the upper B $^1\Pi_u$ state potential. Applying the unified Franck-Condon treatment[13] to the primary satellite peak to estimate $\Delta V(R)$ around the maximum and adding it to the X $^1\Sigma_g$ state potential of Gerber et al.[14], we estimate the B $^1\Pi_u$ potential as shown in Fig. 4. Our potential curve is smoothly connected to the curve given by Bordas et al.[15] We see, therefore, that the satellite peak at $\Delta\nu \sim 1500 \text{cm}^{-1}$ reflects a hump of the B $^1\Pi_u$ state potential located at $R \sim 12a_0$ as shown in Fig. 4.

C. Blue-wing satellite of the Yb and the Ba first resonance lines broadened by rare-gas atoms.

Both of the Ba and Yb first resonance lines are assigned to the $6s^2$ $^1S_0$ - $6s6p$ $^1P_1$ transition. The far-wing spectra of them show a good similarity. Figure 5 gives plots of the γ_{BaRG}[16] and γ_{YbRG}[12] measured for blue-wing absorption of the first resonance line broadened by rare-gas collisions. A feature common to both of the Ba-RG and the Yb-RG collisions is the oscillatory satellite structures. A blue-wing satellite appears when a maximum exists on the difference potential $\Delta V(R)$ between upper and lower states of the transition. When $\Delta V(R)$ has an extremum, photoabsorption can occur at two Condon points for a given photon energy. In Fig. 6, two Condon points are illustrated as the crossing points, R_1 and R_2, between the

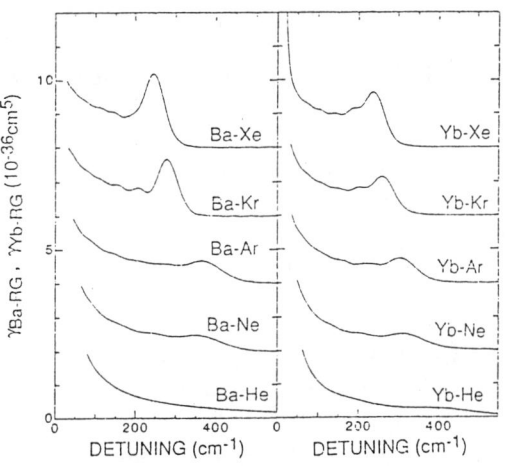

Fig. 5. Blue wing spectra of Ba and Yb $6s^2$ $^1S_0$-$6s6p$ $^1P_1$ lines broadened by rare-gas collisions. Reduced absorption coefficient is plotted as a function of detuning $\Delta\nu$ (cm^{-1}) from the corresponding line center.

potentials of the excited state and the photo-dressed ground state. The photoabsorption process can be regarded as the collision process accompanied with a transition from V_i to V_f at one of the crossing points. For a given photon energy, there are four possible scattering pathways in the limit of weak field as shown in Fig. 6. The total cross section is then semiclassically approximated by

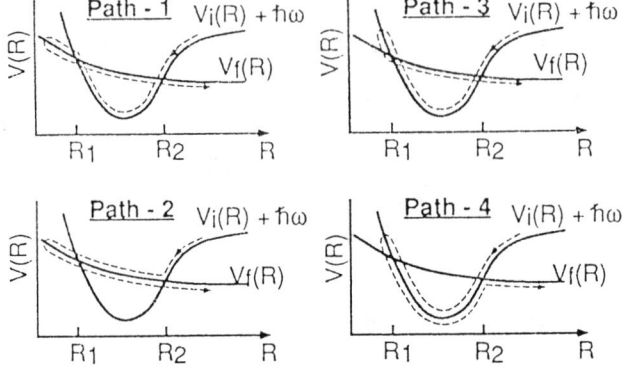

Fig. 6. Scattering pathways for a collision during which photoabsorption may occur at either of two crossing points.

$$Q = 4\pi \int_0^{b_1} p_1(1-p_1)(1+\cos 2\phi_1)b\,db + 4\pi \int_0^{b_2} p_2(1-p_2)(1+\cos 2\phi_2)b\,db$$

$$+8\pi \int_0^{b_1} [p_1(1-p_1)(p_2(1-p_2)]^{1/2}[\cos(\phi_{12}+\sigma_{12})+\cos(\phi_1-\phi_{12})]b\,db, \quad (7)$$

where b is the impact parameter, b_n the maximum of b for a transition to occur at the crossing point R_n, p_n the probability of photoabsorption at R_n, and ϕ's are the semiclassical phase differences that can occur due to the branching of the scattering path. The three ϕ's in Eq. 7 are the phase differences between the pathways shown in Fig. 7. The phase difference ϕ_n changes rapidly with the change of b. The corresponding interference term oscillates, therefore, rapidly to give a negligible contribution to the total cross section. The phase difference ϕ_{12}, on the other hand, depends weakly on b. Then we have

$$Q \simeq Q_1 + Q_2 + Q_{12}, \quad (8a)$$

with

$$Q_n = 4\pi \int_0^{b_n} p_n(1-p_n)b\,db, \quad (8b)$$

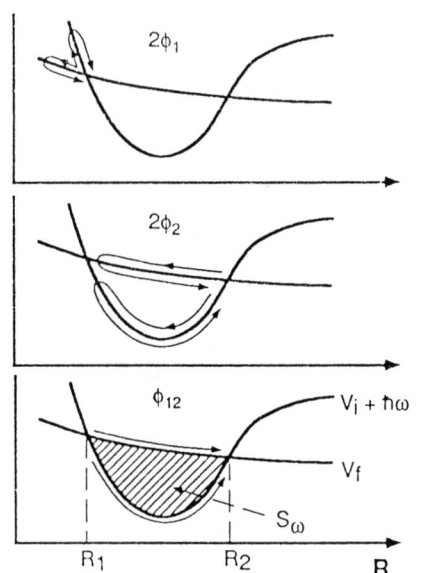

Fig. 7. Phase differences between two branches of scattering paths, which are shown by arrows along the potential curves.

$$Q_{12} = 8\pi \int_0^{b_n} [p_1(1-p_1)p_2(1-p_2)]^{1/2} [\cos(\phi_{12} + \sigma_{12}) b db. \tag{8c}$$

Q_n is the absorption cross-section at the Condon point R_n and Q_{12} is the interference term which gives an oscillatory structure to the absorption cross-section as a function of photon energy. If one assumes the Landau-Zener (LZ) form for the probability p_n and expands $p_n(1-p_n)$ in powers of the photon flux (I_p) to keep only the leading term, then $<vQ_n>_T/I_p$, where v is the collision velocity and $<>_T$ means the thermal average at a temperature T, coincides with the QST expression. The phase difference ϕ_{12} is estimated roughly as

$$\phi_{12} \sim \frac{1}{v\hbar} \int_{R_1}^{R_2} [V_f(R) - V_i(R) - \hbar\omega] dR, \tag{9}$$

and, therefore, is roughly proportional to the area surrounded by $V_i(R) + \hbar\omega$ and $V_f(R)$ between the inner and the outer crossing points, (the shaded area in Fig. 7). The above mechanism which gives the oscillatory structure to the absorption coefficient bears a resemblance to Rosenthal oscillation[17] in inelastic atomic collisions.

In the weak field limit, the collisional approach based on the semiclassical expression for the S-matrix is consistent with the Franck-Condon-factor (FCF) approach based on the WKB wavefunctions. An analytic expression of the probability p_n, such as the LZ-formula, is highly desirable to simulate the observed profile with an iterative manner for seeking a potential energy curves. The LZ-formula, however, fails to explain the absorption profiles of the primary satellite peak. When $\hbar\omega$ is close to the maximum of ΔV, the two curves, $V_i(R) + \hbar\omega$ and $V_f(R)$, cross with very small intersection angles. The LZ-probability is not accurate in this case. Consider the following two models of the difference potential:

(i) *Parabolic model*

$$\Delta V(R) = -\frac{1}{2}\Delta V''(R_0)(R-R_0)^2 + \epsilon, \tag{10}$$

(ii) *Morse-function model*

$$\Delta V(R) = -\epsilon[e^{-2\alpha(R-R_0)} - 2e^{-\alpha(R-R_0)}] + \hbar\omega_0, \tag{11}$$

The crossing points R_n's are given by the equation $\Delta V(R_n) = \hbar\omega$. Using the semiclassical formula[18], the transition probability at the crossing points may be given by

$$p = \exp(-\delta), \tag{12a}$$

$$\sigma_0 + i\delta \simeq \frac{1}{\hbar v_c} \int_{Z_c}^{Z_c^*} \{[\Delta V(Z) - \hbar\omega]^2 + 4V_{12}^2\}^{1/2} dZ. \tag{12b}$$

Inserting Eq.(10) or Eq.(11) into Eq.(12b) and assuming that V_{12} does not depend on R, we have

$$p = \exp[-(\frac{\epsilon}{\hbar v \alpha}) S_M], \tag{13a}$$

$$S_M = \sqrt{2}\rho^{3/2} \int_0^\phi (\cos\theta - \frac{\xi}{\rho})^{\frac{1}{2}} \frac{\cos\theta + M\sqrt{\rho}\cos(\theta/2)}{1 + |M|\rho + 2M\sqrt{\rho}\cos(\theta/2)} d\theta, \quad (13b)$$

where

$$\rho = \sqrt{\xi^2 + \eta^2}, \quad \cos\phi = \frac{\xi}{\rho}, \quad \xi = 1 - \frac{\hbar(\omega - \omega_0)}{\epsilon}, \quad \eta = \frac{2V_{12}}{\epsilon}.$$

Equation (13) expresses the probability for the parabolic model when $M = 0$ and for the Morse-function model when $M = \pm 1$. For $0.2 \leq \xi \leq 0.8$, the above Morse-function model (Eqs. (13) with $M = \pm 1$) is in good agreement with the LZ-model. As ξ decreases from 0.2, the LZ-model deviates from the Morse-function model. When $\xi = 0$, then $\Delta V'(R_n)=0$, and δ_{LZ} from the LZ-formula goes to infinity, while the Morse-function model gives a definite value of δ.

D. Asymmetric self-broadening of the Ba second resonance line and pair-absorption in Ba-Ba system

The far-wing absorption bands of the second resonance line of Ba, $6s^2\ ^1S_0$ – $5d6p\ ^1P_1$ at 350nm, broadened by Ba-Ba collisions show an asymmetric feature[17] with its red wing being more extended than the blue wing. The red-wing band corresponds to the process

$$\text{Ba}(6s^2\ ^1S_0) + \text{Ba}(6s^2\ ^1S_0) + \hbar\omega \rightarrow \text{Ba}^*(5d6p\ ^1P_1) + \text{Ba}(6s^2\ ^1S_0). \quad (14)$$

The $\gamma_{\text{BaBa}}(\Delta\nu)/l$ of the red wing behaves as $(\Delta\nu)^{-3/2}$ as a function of detuning $\Delta\nu$ from the line center. This implies that self-broadening of the second resonance line is not *resonance-like* but *van der Waals-like*. The process (14) is coupled with the pair-absorption process

$$\text{Ba}(6s^2\ ^1_0) + \text{Ba}(6s^2\ ^1S_0) + \hbar\omega \rightarrow \text{Ba}^*(6s6p\ ^1P_1) + \text{Ba}^*(6s5d\ ^1D_2), \quad (15)$$

which was first observed by White et al.[19]. The γ_{BaBa}/l of the process (15) is plotted in Fig. 8 as a function of detuning from the sum of the energies of the $6s6p\ ^1P_1$ and $6s5d\ ^1D_2$ levels. The potential energy curves associate with the final excited states of processes (14) and (15) can be calculated for the asymptotic region based on the first and second order d-d interaction.

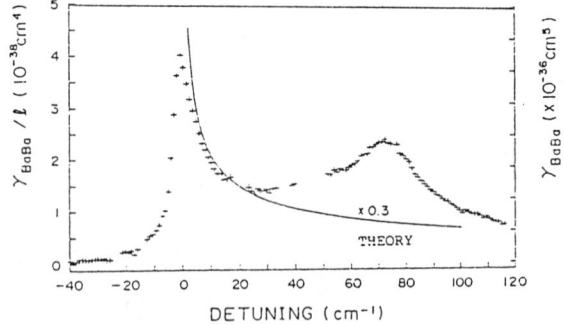

Fig. 8. Reduced absorption coefficients of the Ba-Ba pair absorption plotted as a function of detuning from the sum of Ba $6s5d\ ^1D_2$ and $6s6p\ ^1P_1$. +; measurements, solid line;calculatons. See text for the detail.

The resonance (1st order d-d) interaction gives energy splitting between a group of states correlating to the final state of process (14) as well as between the other group of states correlating to that of process (15). The van der Waals (2nd order d-d) interaction causes further splitting between the two groups of states, which makes the final states of the process (14) attractive and the final states of the process (15) repulsive. Combination of the QST with the calculated potentials explains well the observed $\gamma_{BaBa}(\Delta\nu)$ for the self-broadening in the range of $\Delta\nu$ from -25 to 200 cm^{-1}. Similar calculation explains qualitatively the $\gamma_{BaBa}(\Delta\nu)$ for the pair absorption at the blue-shaded part of the peak around $\Delta\nu = 0$. The observed broad maximum around $\Delta\nu \sim 70$cm^{-1} is left unexplained.

E. Photon-assisted atom-diatom collisions

Photon-assited collisions in atom-diatom systems such as

$$Hg(6s^2\ {}^1S_0) + N_2 + \hbar\omega$$
$$\to Hg^*(6s6p\ {}^3P_0) + N_2. \quad (16)$$

are studied using a laser-pump/probe method. A pulsed pump-laser is directed into the Hg/N$_2$ mixture to excite the above process in far-wings of the Hg resonance line: $6s6p\ {}^3P_1 \leftarrow 6s^2\ {}^1S_0$ at 254nm. After a short time-delay (2ns), a pulsed probe-laser is triggered to detect the product Hg$^*(6s6p\ {}^3P_0)$ with a laser-excited fluorescence (LIF) method. Figure 9 shows yield spectra of the Hg$^*(6s6p\ {}^3P_0)$ excited around 254nm. The wavenumber of the pump-laser is given as the detuning from the resonance line center. The upper trace, taken under the single collision condition, stands for the action spectrum of the Hg$^*(6s6p\ {}^3P_0)$ in the vicinity of the Hg resonance line.

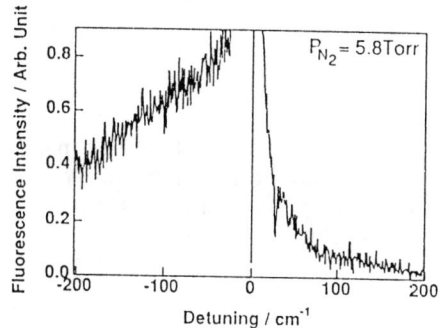
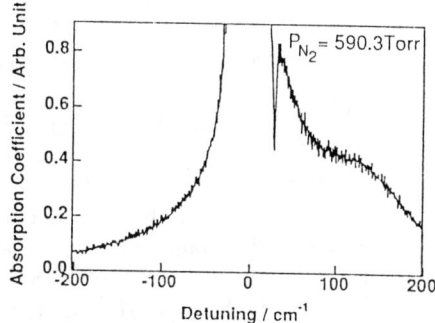

Fig. 9. Yield spectra of the Hg$^*(6s6p\ {}^3P_0)$ excited around 254 nm for the admixtures of Hg with low- and high-pressure N$_2$.

The high-pressure data (lower trace), on the other hand, are considered to represent the absorption spectrum of the Hg-N$_2$ complex. It is seen that the red-wing region is much more effective in inducing the process (16) than the blue-wing region. A similar pump/probe method is also applied to study the chemical radiative collisions such as

$$Hg(6s^2\ {}^1S_0) + H_2 + \hbar\omega \to HgH(X^1\Sigma, v, j) + H, \quad (17)$$

aiming to investigate the transition state of the reaction.

In summary, we have observed several types of photon-assisted atom-atom and atom-diatom collisions in weak photon fields. The observation has been made by means of the dispersion/absorption and the laser-pump/probe methods. A convensional photoabsorption and emission spectroscopy is still of value to approach directly the quasimolecules formed during two-body collisions. The method can provide us information about the interatomic potential energies over a wide range of distances. The oscillatory structures are commonly observed on the blue wing spectrum of the pressure-broadened first-resonance lines of the metal atoms (Sr, Ba, and Yb). The structures are discussed in terms of the phase interference due to branching of the collisional pathways.

ACKNOWLEDGEMENT

I am grateful to all of the members of our laboratory (laboratory of VUV spectroscopy at RISM), especially to Dr. K. Ueda, Dr. K. Ohmori, Dr. M. Okunishi, Mr. H. Chiba, Mr. H. Ito, Mr. S. Moriyama, and Mr. T. Maeyama.

REFERENCES

1. V. S. Lisitsa and S. I. Yakovlenko, Sov. Phys. JETP **39**, 759 (1974).
2. P. D. Kleiber, A. K. Fletcher and K. M. Sando, Phys. Rev. A **37**, 3584 (1988), and references therein.
3. L. I. Gudzenko and S. I. Yakovlenkov, Sov. Phys. JETP **35**, 877 (1972).
4. L. I. Gudzenko and S. I. Yakovlenkov, Phys. Lett. **46A**, 475 (1974).
5. A. Gallagher and T. Holstein, Phys. Rev. A **16**, 2413 (1977).
6. P. S. Juline, Phys. Rev. A **26**, 3299 (1982), and references therein.
7. W. T. Hill III, Appl. Opt. **25**, 4476 (1986).
8. K. Ueda and Y. Sato, J. Spectrosc. Soc. Jpn. **38**, 28 (1989) (in Japanese).
9. K. Ueda, T. Komatsu, and Y. Sato, J. Chem. Phys. **91**, 4495 (1989).
10. K. Ueda, H. Sotome, and Y. Sato, J. Chem. Phys. **94**, 1907 (1991).
11. K. Ueda, O. Sonobe, H. Chiba, Y. Sato, and T. Namioka, Rev. Sci. Instrum. **63**, 1690 (1992).
12. K. Ueda, O. Sonobe, H. Chiba, and Y. Sato, J. Chem. Phys. **95**, 8083 (1991).
13. J. Szudy and W. E. Baylis, J. Quant. Spectrosc. Radiat. Trans. **15**, 641 (1975).
14. G. Gerber, R. Moller, and H. Schneider, J. Chem. Phys. **81**, 1538 (1984).
15. C. Bordas, M. Broyer, J. Chevaleyer, and P. Dugourd, Chem. Phys. Lett. **197**, 562 (1992).
16. T. Maeyama, H. Ito, H. Chiba, K. Ohmori, K. Ueda, and Y. Sato, J. Chem. Phys. **97**, 9492 (1992).
17. H. Rosenthal and H. M. Foley, Phys. Rev. Lett. **23**, 1480 (1969).
18. H. Nakamura, Int. Rev. Phys. Chem. **10**, 123 (1991).
19. J. C. White, Z. A. Zasiuk, J. F. Young, and S. E. Harris, Opt. Lett. **4**, 137 (1979).

ORBITAL ALIGNMENT AND VECTOR CORRELATIONS IN INELASTIC ATOMIC COLLISIONS

Eileen M. Spain,[a] Chris J. Smith, Mark J. Dalberth, and Stephen R. Leone[b]

Joint Institute for Laboratory Astrophysics, National Institute of Standards and Technology and University of Colorado and Department of Chemistry and Biochemistry, University of Colorado
Boulder, Colorado 80303-0440

ABSTRACT

We present a brief commentary on two-, three-, and four-vector correlations in the inelastic scattering of orbitally aligned calcium atoms.

INTRODUCTORY REMARKS

To better understand the dynamics of fundamental atomic processes, experiments are devised which define and control various vector quantities. These vectors may correspond to the rotational alignment, relative velocity, and asymptotic molecular states to name just a few. With straightforward transformations, these vectors may be correlated in a single reference frame, usually the collision frame. Mechanistic interpretation and quantitative information about the dynamical event may then be extracted. Many two-[1] and three-vector[2,3] correlations of atomic processes have been reported. Even some four-vector correlations have appeared in the literature,[4,5] although quantitative results are sparse. We refer the reader to several excellent feature and review articles that provide in-depth information.[6-8]

These proceedings briefly consider n-vector correlations of near-resonant electronic energy transfer, where one of the controlled vectors is a laser-aligned atomic orbital of the Ca atom. The energy transfer process occurs by crossings between the asymptotically prepared molecular van der Waals potential curves during the collision. We focus on the stereospecificity exhibited by the state-changing collisions of calcium with rare gases. These studies along with various theoretical treatments have matured to provide robust quantitative interpretations. We *emphasize* the importance of coherence in describing dynamical events with an n-vector correlation of n≥3. Next, we explore the kinds of information that may be gleaned from a four-vector correlation. We close our commentary with a discussion of the qualitative differences between the scattering of an aligned valence orbital vs. a Rydberg orbital.

[a] National Science Foundation Postdoctoral Fellow, 1992-94
[b] Staff Member, Quantum Physics Division, National Institute of Standards and Technology.

TWO-VECTOR CORRELATION

The dynamics of near-resonant electronic energy transfer are first examined by a two-vector correlation. The two controlled vectors are the initial atomic orbital alignment, determined by the laser polarization E_i (linear) or laser propagation k_i (circular), and the initial relative velocity vector, v_i, which is well defined by a crossed beam experiment. Figure 1(a) provides a diagram for a two-vector correlation. The collision dynamics is most easily viewed in the collision frame, where the z axis corresponds to v_i, and the aligned orbital makes an angle β with respect to this axis. The fundamental question asked in this kind of study is how the rate of energy transfer is dependent on the initial orbital alignment. For example, we ask if the energy transfer process is more efficient when the prepared atomic orbital (e.g. a p orbital) is parallel or perpendicular to the approach of the collision partner.

A Ca p, d, or f orbital can be excited and aligned using polarized laser light. The direction of the aligned atomic orbital may be rotated with respect to the initial relative velocity vector, v_i. The rate for energy transfer is then monitored as a function of the angle the orbital makes with respect to v_i. Dramatic signal variations in the rate are observed, and different types of signals are observed for the p, d, and f orbitals. It has been shown that the energy transfer rate varies smoothly as a sum of cosine functions, in general expressed as

$$a_0 + \sum_{n=1}^{n_{max}} a_{2n}\cos(2n\beta + \phi),$$

where ϕ is a phase and where $n_{max}=1,2,3$ for a p, d, or f orbital. The most significant and beautiful result of this approach is that the variation of the cross section with orbital alignment is directly related to the rotational properties of the initially prepared orbital wave function.

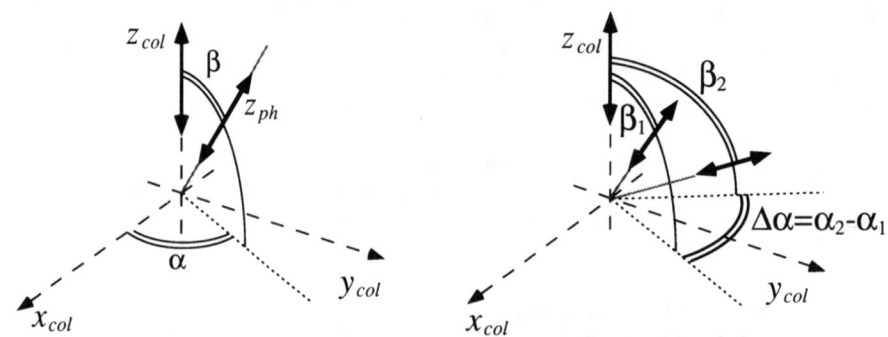

a) two vector correlation   b) three vector correlation

Fig. 1. Relevant angles in two- and three-vector correlations.

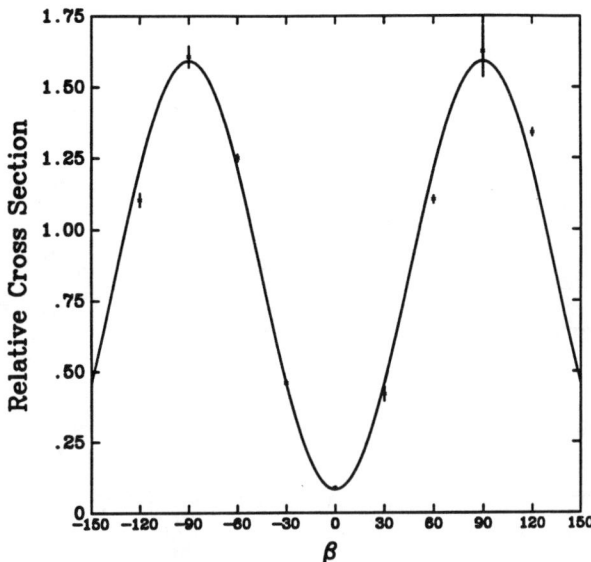

Fig. 2. Alignment curve of the two-vector correlation study of Ca(4s4p $^3P_1$) → Ca(4s4p $^3P_0$) collisional energy transfer.

For example, our group recently investigated the intramultiplet mixing of the Ca $4s4p$ $^3P_1$ state with He.[3] Consider the two-vector correlation of the $^3P_1 \to {}^3P_0$ energy transfer process. The energy transfer rate as a function of the initial p orbital alignment is measured. The experimental alignment curve is shown in Fig. 2. One can see from the data that a perpendicular (m=±1) reagent approach (β=90°) is strongly preferred over a parallel (m=0) approach (β=0°). The relative m-sublevel integral cross sections for the energy transfer $^3P_1 \to {}^3P_0$ are measured to be $\sigma^{|m|=1} = 0.96 \pm 0.02$ and $\sigma^{|m|=0} = 0.04 \pm 0.01$. In addition, a satisfying qualitative mechanistic interpretation follows the quantitative values for the integral cross sections. The schematic Born-Oppenheimer molecular states of the Ca $4s4p$ ^3P_J + He system are given in Fig. 3. The two curves that derive from the $^3P_1$ + He interaction are $^3\Pi_1$ and $^3\Pi_{0+}$. Parallel preparation of the $^3P_1$ (m=0) state produces flux on the $^3\Pi_{0+}$ curve, while perpendicular alignment (m= ± 1) prepares flux on the $^3\Pi_1$ curve. Only the $^3\Pi_{0-}$ curve derives from the $^3P_0$ + He interaction. When the laser prepares Ca on the $^3\Pi_{0+}$ curve, no simple mechanism can couple an $\Omega=0^+$ state to an $\Omega=0^-$ state. Therefore, no transfer of flux can occur for an atomic state initially aligned parallel to the approach of He. For perpendicular preparation, flux is produced on the $^3\Pi_1$ state which can transfer flux to the $^3\Pi_{0-}$ state through a rotational coupling mechanism. The total angular momen-

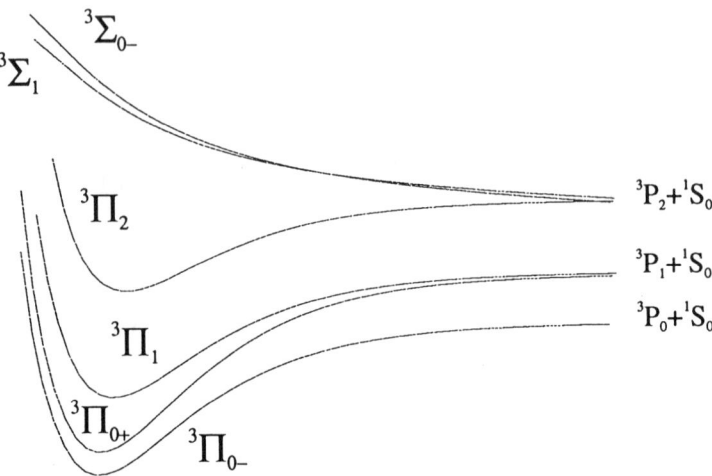

Fig. 3. Schematic potential energy curves for the Ca 4s4p ^3P_J + He interaction.

tum Ω is a good quantum number throughout the flux redistribution and the experimental results demonstrate the strict symmetry rules. The large alignment effect of $\sigma^{|m|=1}/\sigma^{|m|=0} = 24 \pm 1$ follows from this satisfying physical picture and indicates good orbital locking.

THREE-VECTOR CORRELATION

In the two-vector correlation studies, the initial m-sublevel state dependence of the rate of energy transfer was determined. Now the question is whether the final state is aligned *after* the collision. That is, we ask what the final state alignment will be and what are the m_i-to-m_f (initial → final) state resolved cross sections. In order to address these questions, a third vector is added to the dynamical correlation, the third vector representing the polarized laser (E_f or k_f) that probes the final m distribution or alignment. The three correlated vectors are diagrammed in Fig. 1(b).

As documented by Driessen and Leone,[8] when three or more vectors are correlated in a scattering event, coherence (interference) effects are introduced and required to describe the process fully. In addition to collisional population transfer, the transfer of phase or coherence must be included and cannot be ignored in the data analysis. A theoretical prescription for extracting both population (conventional) cross sections and phase (coherence) cross sections from laboratory frame data is provided by Driessen and Eno.[9]

The general expression equating the experimental data to collision frame cross sections presented in a density matrix formalism is

$$\sigma_{j_2m_f \leftarrow j_1m_i}(\Delta\alpha,\beta_1,\beta_2) = \sum_m \rho_{m_1m'_1}(\alpha_1,\beta_1)\rho^*_{m_2m'_2}(\alpha_2,\beta_2)\sigma_{j_2m_2m'_2 \leftarrow j_1m_1m'_1}$$

with

$$\rho_{m_1m'_1} = D^{j_1}_{m_1m_i}D^{j_1*}_{m'_1m_i} \; , \; \rho^*_{m_2m'_2} = D^{j_2*}_{m_2m_f}D^{j_2}_{m'_2m_f}$$

where m_i and m_f are the magnetic quantum numbers defined in the pump and probe laser frames, respectively, m_1 and m_2 are the magnetic quantum numbers defined in the collision frame, and $\rho_{mm'}(\alpha,\beta)$ are the elements of the density matrix which describe the $|jm_i\rangle$ or $|jm_f\rangle$ state in the laser frames as a coherent mixture of $|jm\rangle$ states in the collision frame. The cross sections in the collision frame are the fundamental cross sections that we determine. The diagonal elements $\rho_{mm'}(\alpha,\beta)$ for $m=m'$ give population information and are termed conventional cross sections. The off-diagonal elements $\rho_{mm'}(\alpha,\beta)$ for $m \neq m'$ give coherence or phase information between two initially prepared m states and two final m states, and are termed coherence cross sections.

Let us return to the recent intramultiplet mixing investigation from our group. A three-vector correlation was applied to the Ca $4s4p\ ^3P_1 \rightarrow 4s4p\ ^3P_2$ energy transfer process, where relative state-to-state m resolved integral cross sections were obtained. In the three-vector study, eight conventional (real valued) and six coherence (complex valued) cross sections can be measured. Due to symmetry, the 20 cross sections can be reduced to 15 unique cross section parameters. To determine these, both linear and circular polarized laser light are used to prepare and probe the initial and final alignment, respectively. In our study, 13 of the 15 parameters were measured and the values are displayed in Figs. 4 and 5.

From Fig. 4, it is seen that for initial parallel orbital alignment in the $^3P_1$ state, the $^3P_2$ final state is aligned. The lack of population transferred into the final $m = 0$ sublevel reflects a symmetry restriction in the collision process. For initial perpendicular alignment all the m_i-to-m_f collisional transitions are possible. A preference for the final $m = -1$ transition is observed. We do not resolve the collisional transitions into the final $m = \pm 2$ sublevels; the average of their values is measured and reported in the figure. The magnitude of the real and imaginary parts of the coherence cross sections, displayed in Fig. 5, are not easily interpreted. In principle, they describe phase information and how this is transferred from an initial superposition state to the final superposition state. The transfer of phase contains information about the rotation of the orbital in the plane perpendicular to v_i. Our group is actively pursuing intuitive and quantitative understandings of the coherence terms.

A similar type of study of the Ba($6s6p\ ^3P_1 \rightarrow\ ^3P_2$) transition with N_2, O_2 and H_2 has been reported by Mestdagh and co-workers.[10] In their study, a magnetic field is used to remove the degeneracy of the m_j states and

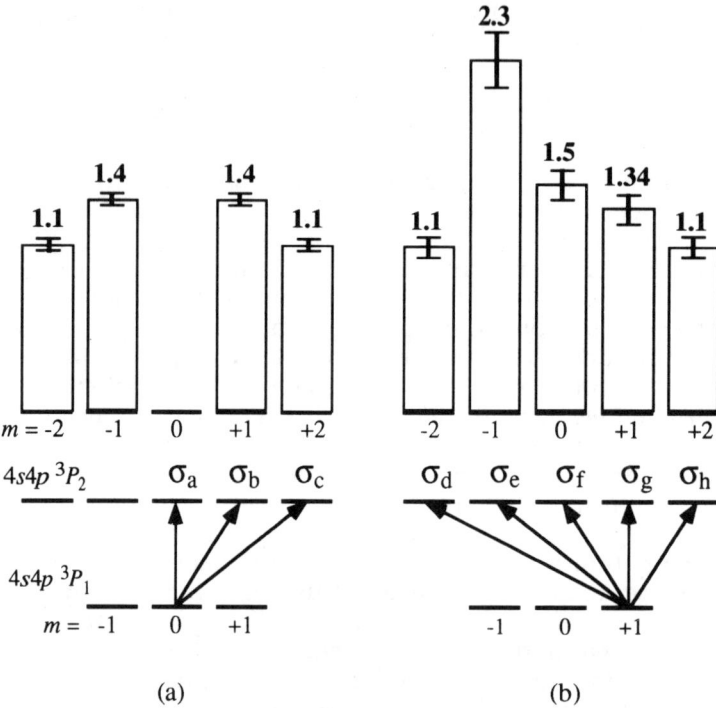

Fig. 4. Relative cross sections for (a) parallel preparation, $\beta_1 = 0°$, (b) perpendicular preparation, $\beta_1 = 90°$. The individual cross sections σ_d and σ_h are not separated, and only the average of the two is shown.

narrow-band lasers excite individual Zeeman transitions of the final $^3P_2$. While the experimental technique is quite elegant, their analysis is unable to extract the complete details of the cross section terms because the relation between the magnetic field and the collision frame is not included. A full specification of the collision dynamics can be obtained when the angle between the magnetic field and the relative velocity vector (this angle introduces coherence terms) is taken into account.

Coherence terms have been extracted from other vector correlation investigations. A number of research groups have addressed coherence terms in atom/atom and atom/ion collisions, and a recent paper has discussed coherence when using electric quadrupole and magnetic dipole transitions to obtain orientation and alignment information.[11] Richter et al. have investigated the $H^+ + Na(3p)$ charge transfer process.[12] In this study, the Na atom is initially aligned or oriented, the initial and final relative velocities are specified, and a complete determination of the density matrix is achieved.

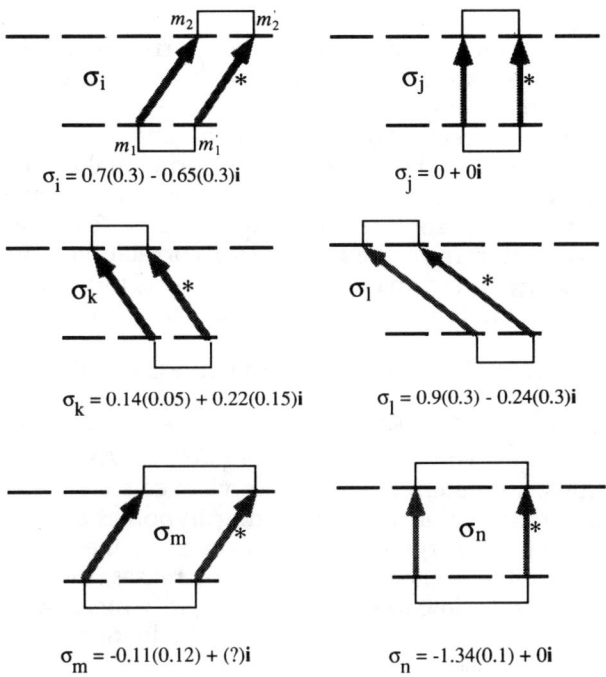

Fig. 5. Depiction of the coherence cross sections with experimentally determined values for the real and imaginary parts lists below.

FOUR-VECTOR CORRELATION

We briefly evaluate two four-vector correlation investigations that have recently appeared in the literature. Mestdagh and coworkers[4] report on the spin-changing collisions of Ba with H_2 using Doppler spectroscopy. The four controlled vectors in this investigation are the initial and final relative velocity and the initial and final polarization of the Ba $6s6p\ ^1P_1 \to 6s6p\ ^3P_2$ energy transfer process. In their work, the four vectors are not correlated with respect to one quantization axis; the objective was to observe qualitative differences in the scattering into the final Ba m sublevels. Their experimental findings show a different line shape for scattering into the final $m = \pm 2$ levels than into the $m = \pm 1, 0$ levels. The investigators attribute this difference to a true dynamical event.

The second study by Collins et al.[5] presents a two-color sub-Doppler circular dichroism technique to study the rotationally elastic and inelastic collisions of Li_2 with rare gases. A narrow circularly polarized pump laser prepares the Li_2 in a selected relative velocity and j orientation. Following collision, a single frequency circularly polarized probe laser measures the line shape shift and width. Thus, their study correlates the initial and final relative velocity and the initial and final j orientation. It is anticipated that more detailed results and analyses from this experimental method will be forthcoming.

To aspire to four-vector correlations experiments is laudable, and it is important to appreciate what Herculean experimental efforts are required. However much work remains to obtain quantitative results and interpretations from these sophisticated experiments. The paucity of quantitative information from the four-vector correlation is in large part due to the complexity of the analysis. In the future, these four-vector analyses will be quantitative and must include coherence effects.

ALIGNED VALENCE AND RYDBERG ORBITALS

We are currently engaged in extending our studies of alignment effects to state-changing collisions of Rydberg atoms. As discussed for two-vector correlations, alignment effects in the valence states of calcium are amenable to qualitative interpretations using hypothetical Born-Oppenheimer potential energy curves. However, as the principal quantum number of the excited atomic state increases, the number of states energetically accessible to the collision rapidly increases. As a result, potential energy diagrams for the collision of a Rydberg state and a rare gas atom contain many crossings and polarization effects into the final state may be washed out, since flux prepared on a single curve will be quickly redistributed.

However, the collisional scattering of a Rydberg atom and a rare gas atom may be viewed as a three body problem. Interactions will exist among the Rydberg electron, the structureless rare gas atom, and the ionic atomic core. Fortunately, previous studies of Rydberg atom collisions with rare gases have shown that the process may be described theoretically by taking into account only the scattering of the Rydberg electron with the rare gas collision partner.[13,14] This leads us to expect that alignment effects for Rydberg states might actually be observable due to the large spatial extent of the Rydberg electron cloud. Recent calculations using a quasi-free electron model with the impulse approximation show alignment effects for the 13d → 12f collisional transfer of Ca induced by a rare gas.[15] One of our goals is to understand the contrast between the collision mechanism in the valence and Rydberg regimes.

CLOSING REMARKS

Two-vector correlations with orbitally aligned atoms are now beginning to be understood, and more is to be learned from other systems such as collisions of orbitally aligned Rydberg atoms. Three-vector correlations are much more difficult but still quantitative results can be obtained. Interpretation of interference terms for three or more correlated vectors is now an important goal. Four-vector correlations are exceedingly difficult experimentally and theoretically and provide a challenge to future research in atomic and molecular dynamics.

ACKNOWLEDGMENT

This work was supported by the National Science Foundation.

REFERENCES

1. See for example: R.L. Robinson, L.J. Kovalenko, C.J. Smith, and S.R. Leone, J. Chem. Phys. **92**, 5260 (1990); J.P.J. Driessen, C.J. Smith, and S.R. Leone, *J. Phys. Chem.* **95**, 8163 (1991); A.G. Suits, P. de Pujo, O. Sublemontier, J.P. Visticot, J. Berlande, J. Cuvellier, T. Gustavsson, J.M. Mestdagh, P. Meynadier, and Y.T. Lee, *J. Chem. Phys.* **97**, 4094 (1992).
2. See for example: C.J. Smith, J.P.J. Driessen, L. Eno, and S.R. Leone, *J. Chem. Phys.* **96**, 8212 (1992).
3. C.J. Smith, E.M. Spain, M.J. Dalberth, S.R. Leone, and J.P.J. Driessen, *J. Chem. Soc. Faraday Trans.* **89**, 1401 (1993).
4. J.M. Mestdagh, J.P. Visticot, P. Meynadier, O. Sublemontier, and A.G. Suits, *J. Chem. Soc. Faraday Trans.* **89**, 1413 (1993)
5. T.L.D. Collins, A.J. McCaffery, and M.J. Wynn, *Faraday Discuss. Chem. Soc.* **91**, 91 (1991).
6. E.E.B. Campbell, H. Schmidt, and I.V. Hertel, *Adv. Chem. Phys.* **72**, 37 (1988).
7. S.R. Leone, *Acct. Chem. Res.* **25**, 71 (1992).
8. J.P.J. Driessen and S.R. Leone, *J. Phys. Chem.* **96**, 6136 (1992).
9. J.P.J. Driessen and L. Eno, *J. Chem. Phys.* **97**, 5532 (1992).
10. J.M. Mestdagh, P. Meynadier, P. de Pujo, O. Sublemontier, J.P. Visticot, J. Berlande, J. Cuvellier, T. Gustavsson, A.G. Suits, and Y.T. Lee, *Phys. Rev. A* **47**, 241 (1993).
11. C.A. Taatjes and S. Stolte, *J. Chem. Soc. Faraday Trans.* **89**, 1551 (1993).
12. C. Richter, N. Andersen, J.C. Brenot, D. Dowek, J.C. Houver, J. Salgado, and J. W. Thomsen, *J. Phys. B: At. Mol. Opt. Phys.* **26**, 723 (1993).
13. T.F. Gallagher, W.E. Cooke, and S.A. Edelstein, *Phys. Rev. A* **17**, 904 (1978).
14. R.F. Stebbings, F.B. Dunning, eds., *Rydberg States of Atoms and Molecules*, (Cambridge University Press, London, 1983).
15. W.A. Isaacs and M.A. Morrison, preliminary calculations, private communication.

ELECTRON CAPTURE BY LOW-ENERGY PROTON IMPACT ON ATMOSPHERIC GASES

Bert Van Zyl
Department of Physics, University of Denver
Denver, CO 80208 USA

ABSTRACT

The available cross-section data on electron capture by H^+ incident on N_2 and O_2 molecules and on O atoms are reviewed. The contributions to such total cross sections made by electron capture into selected excited states of the resulting H atoms are discussed, and some information about the product-ion states populated during the interactions is presented. The applications of such data to understanding the proton aurora are summarized.

INTRODUCTION

The discovery of Doppler-shifted line emissions from atomic hydrogen (H) in the auroral spectrum by Vegard[1] in 1948 provided the first direct evidence that fast protons (H^+) could "precipitate" into the Earth's high atmosphere during auroral activity. Electron capture by these H^+ to form fast excited H atoms could directly yield such emissions, or they could result from excitation of electron-capture-produced H atoms in subsequent collisions.

During such aurorae, the repetitive conversion of H^+ into H via electron-capture reactions and of H back to H^+ via ionization-stripping reactions eventually results in a mixed flux of H^+/H moving through the atmosphere having "charge-state fractions" given approximately by

$$F(H) = \sigma_{10}/(\sigma_{10} + \sigma_{01}), \quad F(H^+) = \sigma_{01}/(\sigma_{10} + \sigma_{01}).$$

Here σ_{10} and σ_{01} are the total cross sections for electron capture and ionization stripping, respectively. Values of these fractions are needed as input variables for many of the simpler proton-auroral models in use today.[2]

In addition, electron-capture reactions of the type $H^+ + N_2 \rightarrow H + N_2^+(B, {}^2\Sigma_u^+)$ can contribute to such aurorally bright band emissions as the first-negative system of N_2^+. This reaction, at the lower H^+ energies, also results in "unusual vibrational excitation" of the $N_2^+(B)$ product-ion state[3] (to be described later) and has therefore received attention as a diagnostic tool for auroral analyses.[2]

The primary auroral-H^+ energies above the atmosphere span the range from about 1 keV (magnetic-cleft aurora), to between 5 and 50 keV (typical nighttime aurora), up to above 1000 keV (energetic polar-cap events). Low-energy

H^+ can only penetrate the atmosphere down to near 200 km, above which atomic oxygen (O) is the primary constituent. In contrast, nighttime auroral emissions usually maximize well below 150 km, where N_2 is the principal atmospheric species, and the O/O_2 ratio is rapidly decreasing. Accurate electron-capture cross sections for a variety of resultant states for H^+ impact on N_2, O_2, and O targets are therefore all needed over a large range of H^+ energies to meet various modeling needs. I will attempt to summarize here our progress in satisfying these needs.

TOTAL ELECTRON-CAPTURE CROSS SECTIONS

A sampling of the available total-electron-capture cross sections for $H^+ + N_2$ and $H^+ + O_2$ collisions is shown in Fig. 1. For clarity, not all the available data could be presented, and not all the individual measurements of the workers cited are plotted. (Inferred data points are shown for those workers giving their results as graphical line curves.) However, the data included are representative, both in magnitude and in claimed uncertainties.

The overall dependences of these cross sections on H^+ energy are well established by these data, but Fig. 1 shows considerable spread among the absolute values plotted. At low H^+ energies, I favor the results of Koopman[11] because of the excellent agreement between his data for $H^+ + H_2$ collisions,[15] our own relative cross sections for that reaction,[16] and the highly-definitive measurements of McClure.[17] For H^+ energies above 4 keV, many workers have used the pioneering work of Stier and Barnett[4] (extended up to 1000 keV by Barnett and Reynolds[5]) as the standard. For H_2 targets, however, these Stier and Barnett[4] data are 10% below the minimally-uncertain values of McClure.[17]

Because of their importance for O-atom targets to be discussed below, the relative data of Stebbings et al.[18] and of Williams et al.[19] have been included in the O_2 data in Fig. 1b. These results were normalized to the data of Stier and Barnett.[4] In contrast, the relative results of Van Zyl and Stephen[20] were referenced to Koopman's[11] data.

The recent data of Gao et al.[14] shown in Fig. 1 have been increased by 10% over the "integral cross sections" given in their paper. These workers measured the differential cross sections for H^+ electron capture for angles from 0.02° to 1.0°, and integrated these $d\sigma_{10}/d\Omega$ to obtain their results. For $H^+ + O_2$ collisions, however, the angular range covered by these studies was recently extended to 3.0° by Sieglaff et al.[21] At 1.5 keV H^+ energy, the integrated $d\sigma_{10}/d\Omega$ increased by 7.4%. A rough estimate suggests that another few % increase (I used 2.6%) should be made to include still larger-angle-scattered H atoms. A 10% increase was thus made to all the Gao et al.[14] data shown here, and 2.6% to all Sieglaff et al.[21] values.

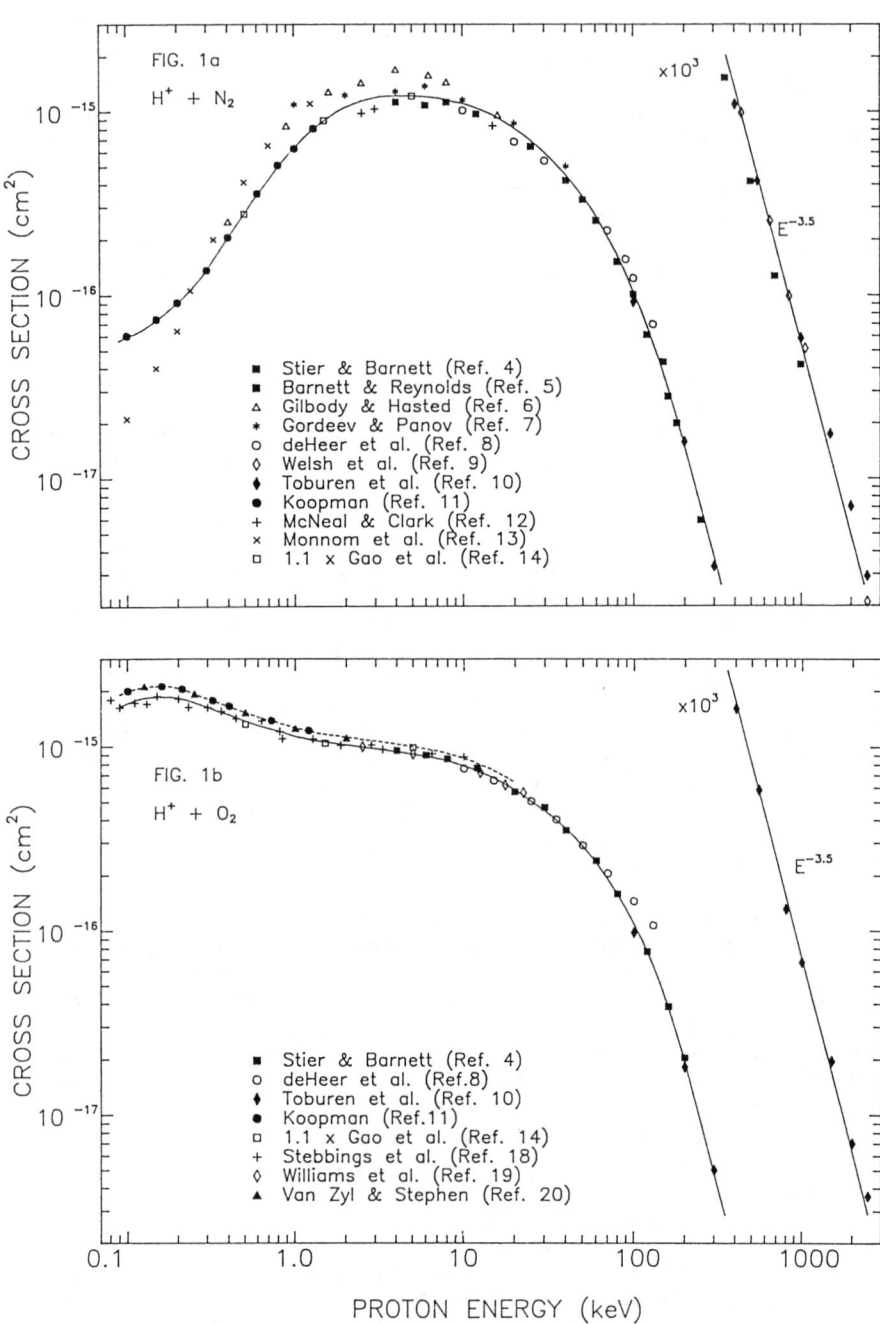

Fig. 1. Total electron-capture cross sections for H^+ impact on N_2 and O_2 targets.

For N_2 targets, these increased Gao et al.[14] values at 0.5 and 1.5 keV H^+ energy are in excellent agreement with Koopman's[11] data, and lie about 10% above Stier and Barnett's[4] results at 5 keV H^+ energy. The similarity of this situation to that for H_2 noted above has led me to draw the line curve shown in Fig. 1a as a reasonable estimate of this aurorally-important cross section.

In contrast, Fig. 1b shows that this 10% increase of the data of Gao et al.[14] for O_2 targets still leaves them about 10% below the data of Koopman[11] at 0.5 and 1.5 keV H^+ energy. Their 5 keV data point, however, is still approximately 10% above Stier and Barnett's[4] results as was found for N_2, and is on a reasonable (dashed-line) extrapolation of the value of Van Zyl and Stephen[20] at 2 keV H^+ energy. Thus the "best normalization" for these various results for H^+ impact on O_2 must remain uncertain for now.

This uncertainty also influences the available data for O targets. Stebbings et al.[18] and Williams et al.[19] both measured an O/O_2 cross-section ratio, basically normalizing their O results to Stier and Barnett's[4] O_2 data. However, Van Zyl and Stephen[20] used Koopman's[11] larger O_2 values for their O/O_2-ratio reference. These results are all shown here in Fig. 2, along with those of Rutherford and Vroom[22] (extending down to 1 eV), and the preliminary values of Sieglaff et al.[21] presented at this conference (and increased by 2.6%, as noted above).

Considering the difficulty of making suitable O-atom targets for these studies (using high-temperature O_2 dissociation furnaces,[19,20,22] or radio-frequency[18] or microwave[21] discharges), the agreement among the low-H^+-energy results in Fig. 2 is reasonable. The preliminary value of Sieglaff et al.[21] at 5 keV H^+ energy is again on a line with Van Zyl and Stephen's[20] point at 2 keV, and both sets

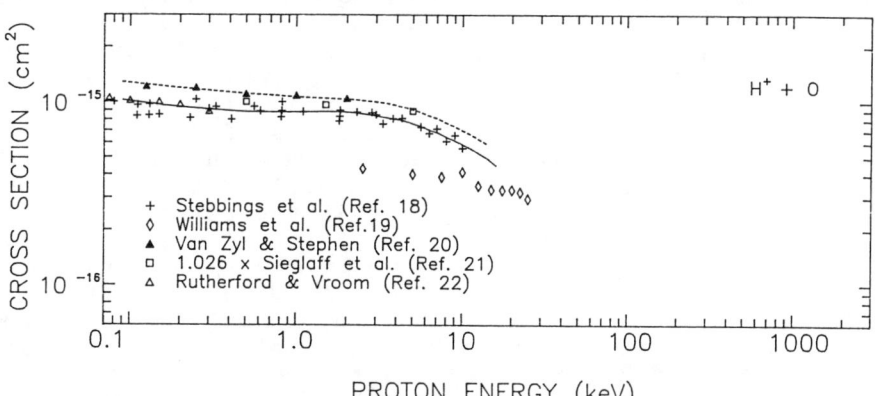

Fig. 2. Total electron-capture cross sections for H^+ impact on O-atom targets.

of data are above the older results of Stebbings et al.[21] and of Rutherford and Vroom.[22] Clearly, however, more data are needed for H⁺ energies above 10 keV.

ELECTRON CAPTURE INTO EXCITED H-ATOM STATES

As noted above, hydrogen-line emissions have played and continue to play a fundamental role in proton-auroral analyses. Historically, the Balmer-alpha (H_α at 656 nm) and Balmer-beta (H_β at 486 nm) emissions have served as proton-auroral sensors,[2] even though H_α is often obscured by nearby N_2 band emissions. More recently, Lyman-alpha and Lyman-beta emissions (122 and 103 nm) have been monitored from satellite-based instruments. Cross sections for reactions involving electron capture by H⁺ into $H^*(n\ell)$ states are needed to interpret all such observations.

It would not be possible here to review all the progress which has been made towards this need. However, as an example of the available information, Fig. 3 shows a selected set of cross sections for electron capture into low-lying $H^*(ns)$ states for H⁺ + N_2 collisions. The decay of the 3s and 4s states to the 2p state yield the H_α and H_β emissions of interest. The (metastable) 2s state, of course, does not normally decay via Lyman-alpha emission, but can be a source of this radiation during aurorae (as will be discussed later).

According to Born-approximation calculations[34] for H⁺ + H reactions, most electron capture occurs into H(ns) states at higher H⁺ energies, where the state populations should follow the n^{-3} scaling law. Thus, 83.3% of the H atoms produced should be in the 1s state, 10.4% in the 2s state, etc. The line curves in Fig. 3 assume this to be valid for H⁺ + N_2 collisions as well, the cross sections for the 2s, 3s ,and 4s states being simply the product of such fractions and the total σ_{10} curve from Fig. 1a.

Note that these predicted cross sections agree reasonably well with the data for H⁺ energies near and above 100 keV, particularly with the high-energy 3s results of Ford and Thomas.[29] At lower H⁺ energies, I favor the 3s data of Lenormand[31] (his absolute H_α-detector calibration shown by Williams et al.[35] to agree well with that used by Van Zyl and Neumann[32]). Note that if the relative results of Loyd and Dawson[30] shown in Fig. 3 for H⁺ energies near 5 keV were about doubled, they would nicely join the low-energy values of Van Zyl and Neumann[32] with the relative data of Yousif et al.[33] who, in turn, normalized their results to those of Lenormand.[31] This implies that the pioneering data of Hughes et al.[26] for 3s-state decay (used for normalization by Loyd and Dawson[30]) may underestimate the H_α emission at lower H⁺ energies, even though it may be accurate near 100 keV. A similar situation likely occurs for H_β emission from the 4s state.

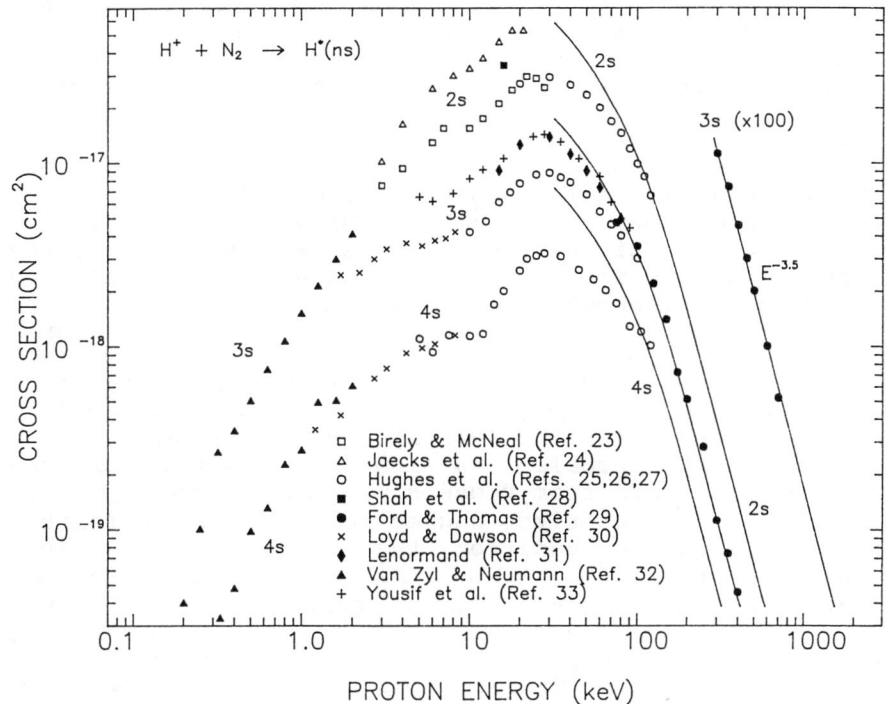

Fig. 3. Cross sections for electron capture into excited $H^*(ns)$ states for H^+ impact on N_2 targets.

An even more tortuous series of arguments is needed to trace the absolute magnitudes of the 2s cross sections shown in Fig. 3. Such results are generally obtained by observing the Lyman-alpha emission from electric-field-induced mixing of the 2s and 2p states of H, usually depending on the observation of Lyman-alpha emission from some previously-measured reaction (or often from several successive reactions) for a photon-detector calibration. I believe the data of Hughes et al.[25] for H^+ energies near 100 keV are again reasonably good, and the 16 keV value of Shah et al.[28] would be my choice at lower H^+ energies.

Many of the references cited in Fig. 3 also contain data for O_2 targets. However, only $H^*(2s)$ production has been measured[36] for O-atom targets.

Van Zyl and Neumann[37] have recently deduced values of the Lyman-alpha emission and 2p-state capture cross sections for H^+ impact on N_2 and O_2 molecules for H^+ energies less than 10 keV. At larger energies, the available data were judged too inconclusive to analyze. As an example, most workers have observed the Lyman alpha from such collisions a few cm into their "target cells." At higher H^+ velocities, therefore, most relatively long-lived $H^*(ns)$

atoms produced in these few cm will decay well past this observation region, so that Lyman alpha from the cascade processes ns → 2p → 1s for n ≥ 3 will escape detection. Because electron capture into ns states is the dominant process at high H^+ energies, only a small fraction of the total Lyman alpha produced is experimentally observed.

However, electric fields of only a few V/cm can alter the radiative lifetimes and branching ratios[38] for decay of $H^*(n\ell)$ atoms for n ≥ 4, changing these Lyman-alpha cascade contributions. For example, in such fields, the 4s and 4d states mix with the 4p state, which then decay primarily by Lyman-gamma emission instead of feeding the 2p state via H_β emission. Stray electric fields of this size can easily be present in ion-beam apparatus,[38] and stray magnetic fields (or the Earth's magnetic field) can induce comparable electric fields in a moving H-atom's reference frame from the (**v** x **B**) Lorentz force.

This latter problem can also affect H_β emission during proton aurorae from those $H^*(4\ell)$ moving through the atmosphere at large pitch angles relative to the Earth's magnetic field. In addition, the induced electric fields experienced by fast $H^*(2s)$ can reduce their lifetime for Lyman-alpha decay to ~ 10^{-4} sec,[39] representing a (high-altitude) source of auroral Lyman alpha not generally recognized by aeronomers. Thus much remains to be learned about how $H^*(n\ell)$ are formed in collisions, and how to apply the information to auroral analysis.

EXCITED PRODUCT-ION PRODUCTION

Probably the most widely-studied example of excited target-ion formation during H^+ reactions with atmospheric gases is that for formation of $N_2^+(B, {}^2\Sigma_u^+)$. About one-third of the available results are shown in Fig. 4, in the form of N_2^+ first-negative (1N) emission cross sections (the 0,0 band of the $B^2\Sigma_u^+ \to X^2\Sigma_g^+$ transition, at 391.4 nm).

The solid-line curve is my best judgment of the true magnitude of this important cross section. For their results at low H^+ energy, Van Zyl and Neumann[3] directly compared (1N) emissions produced by H^+ and e^- impact on N_2, using the e^- cross-section data of Borst and Zipf[47] as a calibration standard. The data point labeled "$e^- + N_2$" on this curve in Fig. 4 is also from Borst and Zipf's[47] work for 1 keV e^- impact, but plotted where the e^- and H^+ velocity are both 1.9 x 10^8 cm/sec. At this high velocity, the 1N emissions from both e^- and H^+ impact on N_2 should come almost entirely from the simple ionization of N_2 to $N_2^+(B)$, as estimated by the short-dashed curve in Fig. 4. Furthermore, the (Born approximation) cross sections for this process should be the same for e^- and H^+ projectiles (at the same velocity), so the solid-line curve in Fig. 4 is similarly normalized at both high and low H^+ energies.

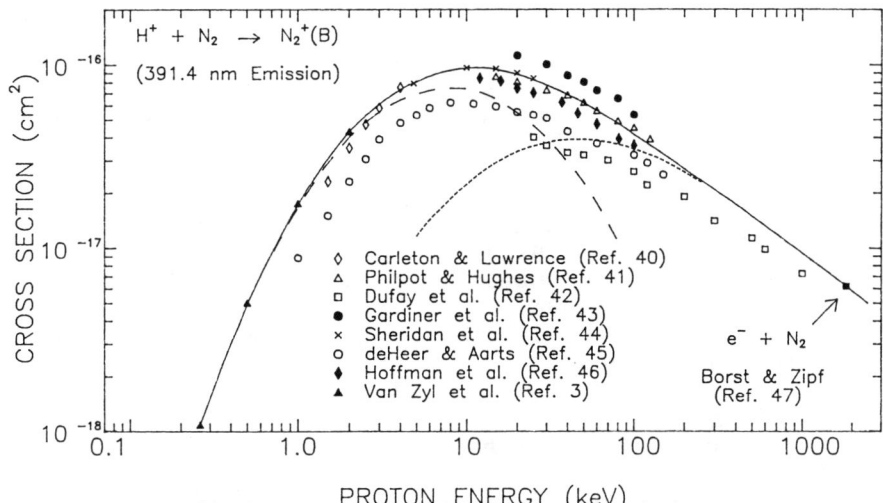

Fig. 4. Cross sections for N_2^+ first-negative (0,0) band emission for H^+ impact on N_2 targets.

At lower H^+ energies, $N_2^+(B)$ production during $H^+ + N_2$ collisions should be dominated by electron-capture reactions, as estimated by the long-dashed curve in Fig. 4. Among the many reasons to study this process at such energies is the fact that the resultant populations of the $N_2^+(B,v)$ vibrational levels are different from the Frank-Condon-factor predictions for $N_2(X) \rightarrow N_2^+(B)$ transitions. The definitive early work of Moore and Doering[48] for various ionic projectiles, and its extension to lower H^+ energies by Van Zyl and Neumann,[3] show significant enhancement of the $v \geq 1$ populations (normally small) relative to that for $v = 0$ (normally dominant). While this "unusual vibrational excitation" has found application as a (1N emission) monitor of auroral H^+ energies,[2] no satisfactory explanation of the phenomena exists at this time.

However, in conclusion, I note that about 40 years ago Lockwood and Everhart[49] found that electron capture by low-energy H^+ on specific-impact-parameter trajectories led to a classical picture of the interaction where an e^- could "jump back and forth" between projectile and target during a collisional encounter. When averaged over all impact parameters, of course, these e^- oscillations lost visibility, but nevertheless were still occurring. For $H^+ + N_2$ reactions, then, might not the final $N_2^+(B,v)$ populations be likely to reflect such multiple transitions as $N_2 \rightarrow N_2^+ \rightarrow N_2 \rightarrow N_2^+ \rightarrow N_2 \rightarrow N_2^+$, for example? We are now using a simple model to compute final $N_2^+(B,v)$ populations using this assumption, but this fascinating research area is clearly in need of more attention in future years.

ACKNOWLEDGMENTS

The author thanks T. Stephen and J. Gibas for help with this paper. Supported in part by the Division of Atmospheric Sciences, U. S. National Science Foundation.

REFERENCES

1. L. Vegard, *Reports of the Gassiot Committee*, Physical Society, London (1948).
2. B. Van Zyl, M. W. Gealy, and H. Neumann, J. Geophys. Res. **89**, 1701 (1984).
3. B. Van Zyl, M. W. Gealy, and H. Neumann, Phys. Rev. A **28**, 2141 (1983).
4. P. M. Stier and C. F. Barnett, Phys. Rev. **103**, 896 (1956).
5. C. F. Barnett and H. K. Reynolds, Phys. Rev. **109**, 355 (1958).
6. H. B. Gilbody and J. B. Hasted, Proc. Roy. Soc. (London) A **238**, 334 (1956).
7. Y. S. Gordeev and M. N. Panov, Sov. Phys. Tech. Phys. **9**, 656 (1964).
8. F. J. deHeer, J. Schutten, and H. Moustafa, Physica **32**, 1766 (1966).
9. L. M. Welsh, K. H. Berkner, S. N. Kaplan, and R. V. Pyle, Phys. Rev. **158**, 85 (1967).
10. L. H. Toburen, M. Y. Nakai, and R. C. Langley, Phys. Rev. **171**, 114 (1968).
11. D. W. Koopman, Phys. Rev. **166**, 57 (1968).
12. R. J. McNeal and D. C. Clark, J. Geophys. Res. **74**, 5065 (1969).
13. G. Monnom, M. Eliot, J. Guidini, and F. P. G. Valckx, C. R. Acad. Sc. Paris **281**, 425 (1975).
14. R. S. Gao, L. K. Johnson, C. L. Hakes, K. A. Smith, and R. F. Stebbings, Phys. Rev. A **41**, 5929 (1990).
15. D. W. Koopman, Phys. Rev. **154**, 79 (1967).
16. M. W. Gealy and B. Van Zyl, Phys. Rev. A **36**, 3091 (1987).
17. G. W. McClure, Phys. Rev. **148**, 47 (1966).
18. R. F. Stebbings, A. C. H. Smith, and H. Ehrhardt, J. Geophys. Res. **69**, 2349 (1964).
19. I. D. Williams, J. Geddes, and H. B. Gilbody, J. Phys. B **17**, 1547 (1984).
20. B. Van Zyl and T. M. Stephen, Abstracts of Contributed Papers, XVII ICPEAC (Brisbane, 1991) p. 437.
21. D. R. Sieglaff, B. G. Lindsay, K. A. Smith, and R. F. Stebbings, Abstracts of Contributed Papers, XVIII ICPEAC (Aarhus, 1993) p. 573.
22. J. A. Rutherford and D. A. Vroom, J. Chem. Phys. **61**, 2514 (1974).
23. J. H. Birely and R. J. McNeal, J. Geophys. Res. **76**, 3700 (1971).

24. D. H. Jaecks, D. H. Crandall, and R. H. McKnight, *Proceedings of the International Conference on Mass Spectroscopy*, edited by K. Ogata and T. Hayakawa (Kyoto, 1970) p. 881.
25. R. H. Hughes, E. D. Stokes, S. S. Choe, and T. J. King, Phys. Rev. A **4**, 1453 (1971).
26. R. H. Hughes, C. A. Stigers, B. M. Doughty, and E. D. Stokes, Phys. Rev. A **1**, 1424 (1970).
27. R. H. Hughes, H. R. Dawson, and B. M. Doughty, Phys. Rev. **164**, 166 (1967).
28. M. B. Shah, J. Geddes, and H. B. Gilbody, J. Phys. B **13**, 4049 (1980).
29. J. C. Ford and E. W. Thomas, Phys. Rev. A **5**, 1701 (1972).
30. D. H. Loyd and H. R. Dawson, Phys. Rev. A **11**, 140 (1975).
31. J. Lenormand, J. Physique **37**, 699 (1976).
32. B. Van Zyl and H. Neumann, J. Geophys. Res. **85**, 6006 (1980).
33. F. B. Yousif, J. Geddes, and H. B. Gilbody, J. Phys. B **19**, 217 (1986).
34. D. R. Bates and A. Dalgarno, Proc. Roy. Soc. (London) A **66**, 972 (1953).
35. I. D. Williams, J. Geddes, and H. B. Gilbody, J. Phys. B **15**, 1377 (1982).
36. I. D. Williams, J. Geddes, and H. B. Gilbody, J. Phys. B **17**, 1547 (1984).
37. B. Van Zyl and H. Neumann, J. Geophys. Res. **93**, 1023 (1988).
38. B. Van Zyl, B. K. Van Zyl , and W. K. Westerveld, Phys. Rev. A **37**, 4201 (1988).
39. G. Lüders, Z. Naturforsch **5a**, 608 (1950).
40. N. P. Carleton and T. R. Lawrence, Phys. Rev. **109**, 1159 (1958).
41. J. L. Philpot and R. H. Hughes, Phys. Rev. **133**, A107 (1964).
42. M. Dufay, J. Desesquelles, M. Druetta, and M. Eidelsberg, Ann. Geophys. **22**, 614 (1966).
43. H. A. B. Gardiner, W. R. Pendleton, Jr., J. J. Merrill, and D. J. Baker, Phys. Rev. **188**, 257 (1969).
44. W. F. Sheridan, O. Oldenberg, and N. P. Carleton, J. Geophys. Res. **76**, 2429 (1969).
45. F. J. deHeer and J. F. M. Aarts, Physica **48**, 620 (1970).
46. J. M. Hoffman, G. J. Lockwood, and G. H. Miller, Phys. Rev. A **11**, 841 (1975).
47. W. L. Borst and E. C. Zipf, Phys. Rev. A **1**, 834 (1970).
48. J. H. Moore and J. P. Doering, Phys. Rev. **177**, 218 (1969).
49. G. J. Lockwood and E. Everhart, Phys. Rev. **125**, 567 (1962).

CLUSTERS

COLLISION EXPERIMENTS WITH FULLERENES

E.E.B. Campbell, R. Ehlich, M. Westerburg, I.V. Hertel

Max Born Institut für Nichtlineare Optik und Kurzzeitspektroskopie,
Rudower Chaussee 6, D-12474 Berlin-Adlershof, Federal Republic of Germany

ABSTRACT

Relative fragmentation cross sections for fullerene ion collisions with rare gas atoms and SF_6 are presented over a range of collision energies. Structure in the cross sections and threshold energy determinations can shed some light on the fragmentation dynamics. Cluster cluster collisions with fullerenes are also described which show evidence of fusion reactions.

1. INTRODUCTION

Collision experiments with fullerenes have become an important method for investigating the collisional dynamics of large clusters since macroscopic amounts of C_{60} and C_{70} became available about three years ago [1]. Since then a large range of types of collisions with fullerenes have been carried out from photon collisions [e.g. 2-5], electron collisions [e.g. 6,7] collisions with atoms and molecules over a large energy range from a few eV to hundreds of keV [e.g. 8-12], cluster cluster collisions [13] and surface collisions [e.g. 14, 15]. There are still many puzzles that have to be solved concerning the dynamics of the various processes observed. The bimodal fragmentation pattern of C_{60} first observed in photo-fragmentation experiments carried out by the Smalley group in 1988 [2] and since seen in collisional fragmentation at laboratory energies of a few keV (corresponding to collision energies in the centre of mass frame on the order of 50-100 eV)[9,10] is one of these. Can the fragmentation be described by a statistical evaporation model or does "fission" of the C_{60} have to be invoked to explain the presence of the small fragments? We have recently carried out a series of experiments to try to shed some light on the fragmentation dynamics. These experiments are described in section 1. In section 2 the first cluster cluster collisions with fullerenes[13] are described which show evidence for the occurrence of molecular fusion and are compared to other experiments in which the production of larger fullerenes is observed.

2. FRAGMENTATION IN COLLISIONS BETWEEN FULLERENE IONS AND ATOMS OR MOLECULES

A schematic diagram of the apparatus used in these studies is shown in Fig. 1. A C_{60}^+ or C_{70}^+ beam is produced by low fluence uv laser desorption of commercially available fullerite deposited on a rotating metal plate. Positively charged fullerene ions are produced by the laser with very little fragmentation occurring [16] and are accelerated away from the surface by a constant electric field. The fullerene ion required is mass selected by using a pulsed electric field to remove any other ions present from the flight path and enters the collision cell containing the target gas. After the collision the parent and product ions are accelerated by a pulsed electric field and mass selected in a linear time-of-flight mass spectrometer.

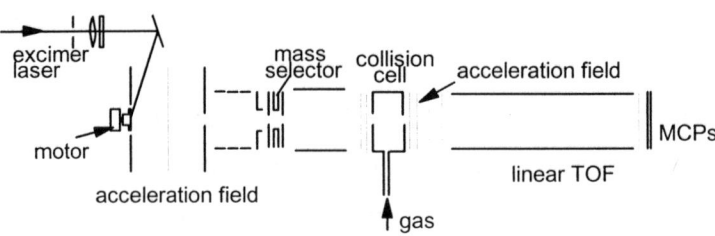

Fig. 1. Schematic diagram of Apparatus

The laser desorption fullerene ion source is ideal for such collision experiments since the ions are produced with a low energy spread ($\Delta E/E \sim 0.01$) thus enabling us to obtain good energy resolution. The experiments are carried out under single collision conditions and the collision energy in the laboratory frame of reference is varied from 100 eV to 5 keV.

Figure 2 shows three fragment mass spectra obtained at very different laboratory collision energies but with the same centre of mass collision energy. The patterns are very similar in all three cases and show that the amount of energy which can be transferred during the collision is the important parameter. They also show that the fragmentation from C_{70}^+ is very similar to that from C_{60}^+.

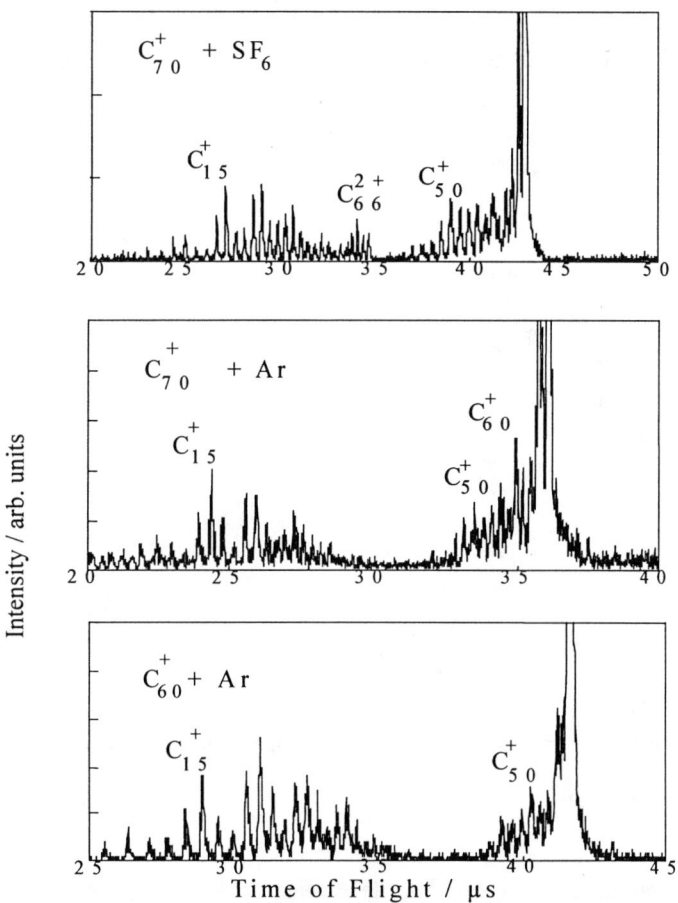

Fig.2. Time of flight mass spectra obtained for different collision partners at a centre of mass collision energy of 100 eV. Note the similar structure and relative fragment intensities in each case.

One obvious difference can be seen in the mass spectra of Fig. 2. In collisions with SF_6 relatively large intensities of doubly charged fullerene ions can be seen which is a consequence of the large electron affinity of SF_6 compared with Ar. In the collision energy range probed in our experiments we see no evidence for stable doubly charged fullerenes produced in collisions with rare gases.

In order to obtain more information on the collision processes the relative fragment intensities were recorded over a large energy range. Fig. 3 shows the collision energy dependence of the relative intensities of three typical fragment ions from collisions between C_{70}^+ and SF_6. The singly charged ions typically show a narrower distribution than the doubly charged ions with a well-defined maximum.

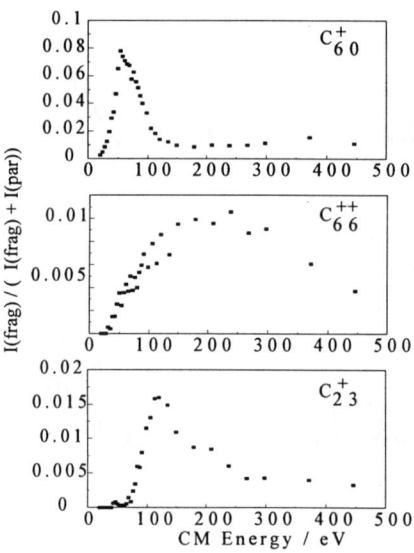

Fig. 3 Relative ion intensities as a function of centre of mass collision energy for $C_{70}^+ + SF_6$ collisions.

For some fragment ions a second, clear peak, seen as a "shoulder" in the bottom spectrum of Fig. 3, could be identified. The position of the fragmentation cross section maxima are plotted in Fig. 4a as a function of the centre of mass collision energy for the $C_{70}^+ + SF_6$ collisions. The position of the first maximum increases more or less linearly with decreasing fragment mass from C_{68}^+ to about C_{50}^+ and then remains approximately constant until C_{20}^+ after which it rises steeply. The doubly charged ions with $30 < n/q \leq 35$ have their maxima at considerably higher collision energies, corresponding to the second maxima of the singly charged ions, which is again relatively independent of fragment mass. This strongly suggests that the second peak observed in the cross sections for production of the singly charged fragments (circles in Fig. 4a) is due to fission of doubly charged fullerenes leading to two singly charged fragments. Such processes have been observed in collisions with doubly charged parent ions[12]. In Fig. 4(b) the first maxima in the SF$_6$ data are compared with the maxima found in collisions between C_{70}^+ and Ar and C_{60}^+ and Ar. The results from C_{70}^+ + Ar collisions follow the SF$_6$ results closely down to n ~ 40. Unfortunately the maximum centre of mass collision energy in this series of experiments was ~ 100 eV so that it was not possible to follow the development of the position of the maxima for the smaller fragments. The maxima for the large fragments are shifted slightly to lower collision energies in the Ar collisions. This shift can be accounted for by considering that the centre of mass collision energy transferred to internal energy in the collision is equally partitioned among the vibrational degrees of freedom. This would mean that only 93% of the energy is transferred to C_{70}^+ in the SF$_6$ collisions compared to 100% for the rare gas case. The C_{60}^+ + Ar results show exactly the same dependence but shifted by n=10 to

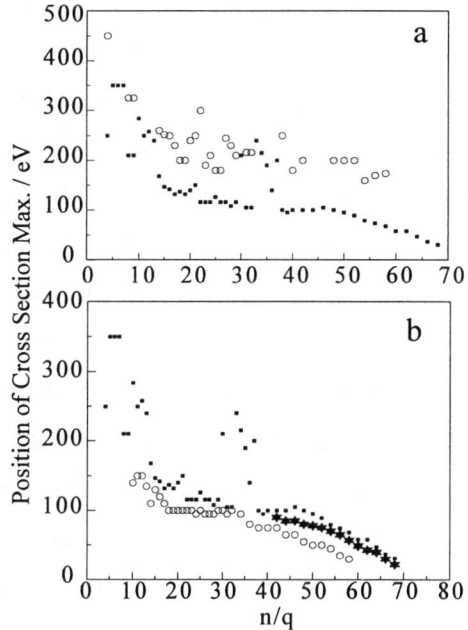

Fig. 4 a) Position of cross section maxima vs. fragment size for collisions between C_{70} ions and SF$_6$. (b) Comparison between C_{70} + SF$_6$ first maxima (rectangles), Ar (stars) and C_{60} + Ar (circles) collisions.

smaller masses. In this data the plateau region for fragments in the middle of the mass range can again be clearly seen.

The energetic thresholds for the production of the different fragments in C_{70}^+ + SF_6 and C_{70}^+ + Ne collisions are shown in Fig. 5. In the fragment range down to about n = 50 the two sets of data again follow one another very closely with just a slight shift of the Ne data to lower collision energies as discussed above for the fragmentation maxima. In the range 20 < n/q < 50 where doubly charged fullerenes and their fission products contribute to the SF_6 data but probably not to the Ne data the threshold is considerably lower for SF_6 (as seen in Fig. 3 the energy distributions for the doubly charged species are much broader). For n/q < 20 the SF_6 thresholds rise steeply but the Ne values remain constant at about 40 eV. This could indicate that passage of the Ne atom through the fullerene cage with collisions on entrance and exit is a particularly efficient way of producing the small fragments.

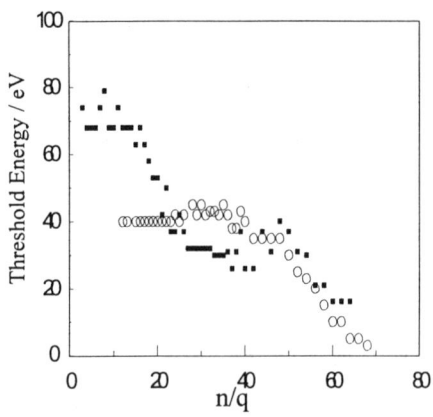

Fig. 5 Energetic thresholds as a function of fragment size for C_{70}^+ + SF_6 (rectangles) and Ne (circles)

The collision energy dependence of the large fragment intensities shown in Fig. 6 for C_{60}^+ + Ar collisions is very reminiscent of data obtained from electron impact ionisation work[6] in which the experimentally obtained breakdown graphs were compared with two statistical models: the finite heat bath model of Klots[17] and the Rice-Ramsperger-Kassel-Marcus (RRKM) theory. We have also applied the RRKM theory using the same assumptions as in Ref. 6 and calculating the breakdown graphs for the sequential evaporation of C_2 from internally highly excited C_{60}. Very good agreement with the experimantally obtained intensity maxima is found if we assume that all the centre of mass collision energy is transferred to the fullerene (from molecular dynamics simulations[18] we know that fullerene collisions are exteremely inelastic) and if we assume an initial temperature in our fullerene beam of 2250 K.

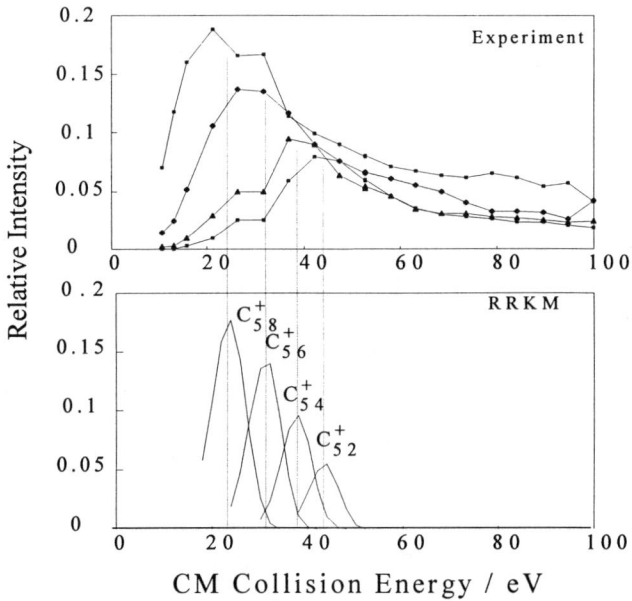

Fig. 6 Comparison of experimental data for C_{60}^+ + Ar collisions with RRKM theory. The calculations assume that all the centre of mass collision energy is converted to internal energy and that the initial temperature of the fullerene beam is 2250K.

The larger width of the experimental data can be accounted for by considering that not all collisions will be completely inelastic (shift to higher collision energy) and that there will be a temperature distribution in the source. The temperature of the fullerene ions produced by the laser desorption source is not known but from other considerations [e.g.3] it would seem that a temperature on the order of a few thousand Kelvin is reasonable. These preliminary investigations show that the production of the large fragments (n/q > 50, 60 for C_{70}^+, C_{60}^+ collisions respectively) can be satisfactorily explained by invoking a statistical model of successive unimolecular fragmentation of C_2. This simple picture can, however, not explain the formation of the fragments with n/q < 50, 40 respectively. This conclusion is borne out by considering the dependence of the fragment intensities on the fluence of the laser used to desorb the fullerene ions in the source. The results are displayed in Fig. 7.

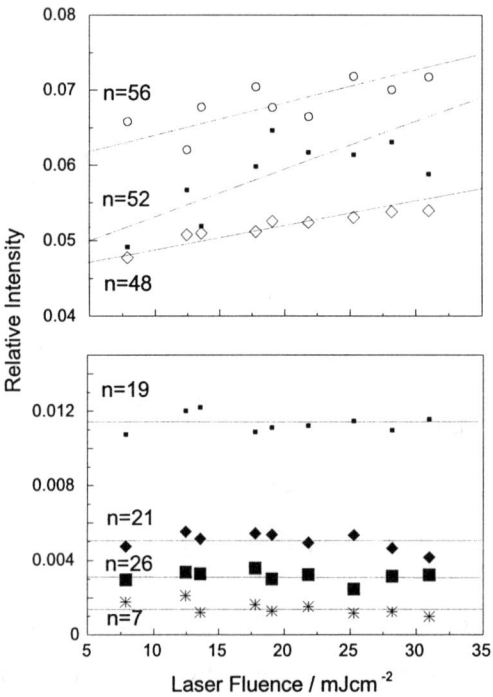

Fig. 7 Relative fragment intensity as a function of the desorbing laser fluence for C_{60}^+ + Ar at a collision energy of 58 eV (cm). The large fragments show a marked dependence on laser fluence, the small fragments are laser fluence independent.

As the laser fluence is increased, the temperature of the fullerene beam is expected to rise. The intensity of the large fragments shows a significant dependence on laser fluence as would be expected from a statistical evaporation mechanism. The intensity of the small fragments, on the other hand, is independent of laser fluence within the range investigated which is another strong indication that a different fragmentation mechanism is responsible for the production of these species.

Photo-fragmentation measurements carried out in our group[5] have shown that the kinetic energy released in the photo-fragmentation is independent of fragment mass which is indicative of a binary fission mechanism, perhaps as a consequence of the excitation of collective oscillation modes. Work is in progress in our laboratory to investigate this in more detail. A similar process may be responsible for the collision results. Another possibility could be the "unzipping" of the carbon cage suggested by deMuro et al.[19]. The activation energy for removal of carbon chains from the fullerene cage may be more or less independent of chain size in the range leading to charged fragments with $20 < n < 50$ (C_{70}^+ collisions). This has to be looked into in more detail but if it is the case it may again be possible to explain the results by unimolecular fragmentation from transition states leading to chain removal.

3. CLUSTER CLUSTER COLLISIONS

The availability of large amounts of neutral mass-selected atomic clusters with well-defined structures in the form of the fullerenes has opened up the possibility of new kinds of cluster experiments. Cluster cluster collisions are an exciting new area of collision physics[20] which has benefited from the availability of fullerenes. We have recently carried out fullerene-fullerene collisions[13], $C_{60}^+ + C_{60}$, C_{70} at collision energies of a few hundred eV which show direct experimental evidence for the occurence of molecular fusion. A mass spectrum produced by colliding C_{60}^+ with a C_{60}/C_{70} target under single collision conditions at a collision energy of 200 eV is shown in Fig. 8(a). The mass spectrum was obtained with a reflectron which allowed the "fused" fullerenes which survived passage through the mass spectrometer to be separated from ions produced by fragmentation of the highly excited "fused" species in the field free region between the collision cell and the reflectron. The lines and numbers in Fig. 8(a) indicate the expected flight times for the unfragmented species with n=120, 130 and fragments produced from them. A similar fusion process has also been observed in our group in collisions between C_{60}^+ and a fullerite surface at a collision energy of 275 eV. The mass spectra of ions scattered from the surface after

C_{60}^+ impact is shown in Fig. 8(b). The fragmentation mechanism of the excited C_{120}^+ in the surface experiments is a statistical evaporation of C_2 in contrast to the gas phase experiments where the results suggest that fragmentation into two or more large fragments is the predominant process[13], as has also been seen in molecular dynamics simulations of the gas phase collisions using phenomenological two- and three-body forces[21]. The "fusion" observed in these controlled collisions is a different process from the coalescence of very hot fullerenes that can be seen under certain conditions during laser desorption of fullerite[16,22].

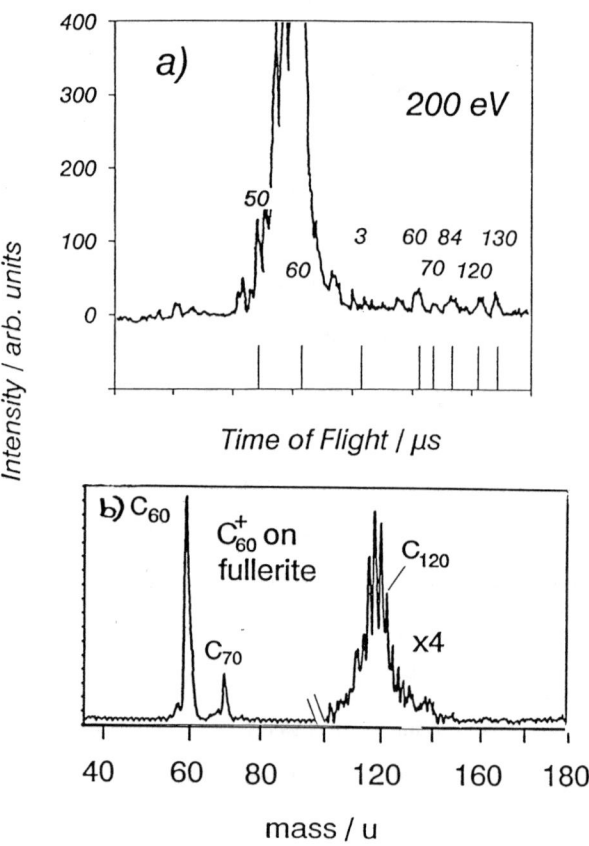

Fig.8 a) Mass spectrum obtained in collisions between C_{60}^+ and C_{60}/C_{70} at a collision energy of 200 eV in the centre of mass reference frame. Evidence of molecular fusion can be seen. (b) Mass spectrum of fullerene ions after surface impact of 275 eV C_{60}^+ on a crystalline fullerite film (from Ref. 15).

ACKNOWLEDGEMENTS

We would like to thank Dr. M. Foltin for sending us his RRKM program. Financial support from the Deutsche Forschungsgemeinschaft through SFB 276 (TP C7) is gratefully acknowledged.

REFERENCES

1. W. Krätschmer, L.D. Lamb, K. Fostiropoulos, D.R. Huffman, Nature 347, 354 (1990)
2. S.C. O'Brien, J.R. Heath, R.F. Curl, R.E. Smalley, J. Chem. Phys. 88, 220 (1988)
3. E.E.B. Campbell, G. ULmer, I.V. Hertel, Z. Phys. D 24, 81 (1992)
4. I.V. Hertel, H. Steger, J. de Vries, B.Weisser, C. Menzel, B. Kamke, W. Kamke, Phys. Rev. Lett. 68 784 (1992)
5. H. Gaber, R. Hiss, H.-G. Busmann, I.V. Hertel, Z. Phys. D 24, 307 (1992)
6. M. Foltin, M. Lezius, P. Scheier, T.D. Märk, J. Chem. Phys. 98, 9624 (1993)
7. D.R. Luffer, K.H. Schram, Rapid Comm. Mass Spec. 4, 552 (1990)
8. T. Weiske, J.Hrusak, D.K. Bohme, H. Schwarz, Helv. Chim. Acta 75, 79 (1992)
9. M.M. Ross, J.H. Callahan,J. Phys. Chem. 95, 5720 (1991)
10. E.E.B. Campbell, A. Hielscher, R. Ehlich, V. Schyja, I.V. Hertel in Nuclear Physics Concepts in Atomic Cluster Physics, ed. R. Schmidt, H.O. Lutz, R.M. Dreizler, Lect. Notes in Physics 404, 185 (Springer, Berlin 1992); E.E.B. Campbell, R. Ehlich, A. Hielscher, J.M.A. Frazao, I.V. Hertel, Z. Phys. D 23, 1 (1992)
11. Z. Wan, J.F. Christian, S.L. Anderson, J. Chem. Phys. 96, 3344 (1992)
12. P. Hvelplund, L.H. Andersen, H.K. Haugen, J. Lindhard, D.C. Lorents, R. Malhotra, R. Ruoff, Phys. Rev. Lett. 69, 1915 (1992)

13. E.E.B. Campbell, V. Schyja, R. Ehlich, I.V. Hertel, Phys. Rev. Lett. 70, 263 (1993)
14. R.D. Beck, P.St. John, M.M. Alvarez, F. Diederich, R.L. Whetten, J. Chem. Phys. 95, 8402 (1991)
15. Th. Lill, H.-G. Busmann, B. Reif, I.V. Hertel, Appl. Phys. A 55, 461 (1992); Th. Lill, F. Lacher, H.-G. Busmann, I.V. Hertel, Phys. Rev. Lett., submitted
16. G. Ulmer, E.E.B. Campbell, R. Kühnle, H.-G. Busmann, I.V. Hertel, Chem. Phys. Lett. 182, 114 (1991)
17. C.E. Klots, Z. Phys. D 21, 335 (1991)
18. R. Ehlich, E.E.B. Campbell, O. Knospe, R. Schmidt, Z. Phys. D, in press
19. R.L. DeMuro, D.A. Jelski, T.F. George, J. Phys. Chem. 96, 10603 (1992)
20. R. Schmidt, G. Seifert, H.O. Lutz, Phys. Lett. A 158, 231 (1991); G. Seifert, R. Schmidt, H.O. Lutz, Phys. Lett. A 158, 237 (1991)
21. O. Knospe, R. Schmidt, private communication
22. C. Yeretzian, K. Hansen, F. Diederich, R.L. Whetten, Nature 359, 44 (1992)

ENERGETIC FULLERENE ION-ATOM COLLISIONS. FRAGMENTATION, CHARGE EXCHANGE, AND LIFETIMES

Preben Hvelplund
Institute of Physics and Astronomy, Aarhus University
DK-8000 Aarhus C, Denmark

ABSTRACT

Recent experiments involving energetic fullerene ions are discussed. In collisions between fullerene ions and atoms, fragmentation is normally the dominant reaction channel, but also charge stripping and electron capture play a role. Lifetime measurements both from single-pass experiments and multiple-turn storage-ring experiments are also described.

INTRODUCTION

The discovery[1] of the exceptional stability of C_{60} (Buckminsterfullerene) and its ions has prompted a large number of experimental and theoretical investigations concerning this soccerball-shaped molecule. After the discovery of a method[2], by which macroscopic quantities of C_{60} are available, this molecule has become an obvious candidate for structural and dynamic investigations by atomic-collision experiments.

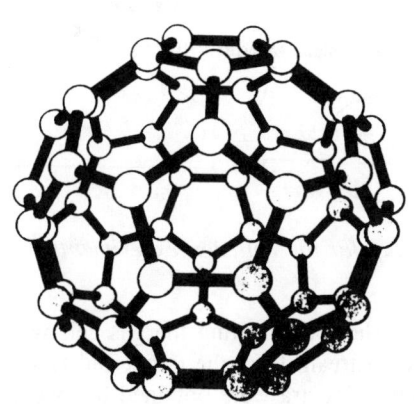

Fig. 1. Illustration of the structure of Buckminster fullerene (C_{60}).

A C_{60} cluster has the structure of the cover of a soccerball (see Fig. 1). The carbon atoms occupy the vertices of 20 regular hexagons and 12 pentagons which cover the surface of a sphere in such a way that no pentagons have common boundaries. The C_{60} molecule has a cage diameter of around 7.1 Å and has been produced with charge states ranging from -2 to +5. C_{60} vapourizes at temperatures from 300 to 600° C. The C_{60}/C_{70} mix which we have used has been produced at the Molecular Physics Laboratory of SRI International. Rods of graphite are butted together, and a high current is passed through the rods. Carbon vapourizes near the contact, producing a carbon plasma that condenses on the walls of the surrounding chamber. It is found to be absolutely crucial for the production of C_{60} by this method that the chamber is filled with helium at a pressure of around 100 torr. The He gas cools the carbon clusters and allows them to form C_{60} and C_{70}. The C_{60} and C_{70} fullerenes are separated from the soot by dissolving them in toluene. The solution containing fullerenes is then dried, and a C_{60}/C_{70} powder is obtained.

© 1993 American Institute of Physics

Several collision studies utilizing mass spectrometry have been reported[3-5]. Typical laboratory collision energies of C_{60}^+ have been less than 10 keV. Collisionally induced dissociation studies are reported by Young, Cousins, and Harrison[3]. They found, in agreement with photodissociation studies, that fragmentation occurs by the loss of an even number of carbon atoms. In collisions involving C_{60}^{2+} and C_{60}^{3+}, they also found that fragmentation occurs by loss of an even number of neutral carbon atoms. Singly charged fragment ions resulting from collisions between C_{60}^{2+} and O_2 were later reported by Doyle and Ross[4] but were attributed to the formation of excited C_{60}^+ resulting from electron capture and subsequent dissociation. Both electron capture and loss without fragmentation have been observed[4], but no absolute values of cross sections are reported. Campbell et al.[5] have measured mass spectra resulting from collisions between 0.2–6-keV C_{60}^+ and various target gases. They conclude that the distribution with peaks around C_{15}^+ and C_{50}^+ originates from collisionally induced fission of C_{60}^+. Another interesting collision aspect of C_{60} is its ability to form endohedral cluster compounds. Inclusion of rare gases and D_2 has been reported[6,7].

EXPERIMENTAL

The experimental setup is shown in Fig. 2. C_{60} ions were generated in a plasma ion source[8] and electrostatically accelerated to energies between 20 and 300 keV. After acceleration, the ions were momentum-selected by a 90° analyzing magnet before entering the 3-cm long gas cell. The ions exciting the gas cell were deflected by an electrostatic analyzer 35 cm downstream from the target and finally counted by a ceratron electron-multiplier detector. Spectra were obtained by sweeping the analyzer voltage and counting charged C_{60} ions or fragments.

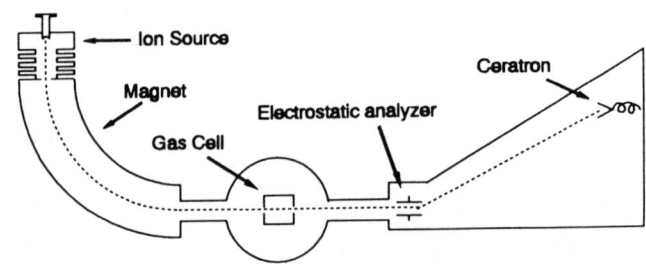

Fig. 2. Experimental setup for the collision experiments.

Lifetime studies were performed at the Aarhus STorage RIng Denmark (ASTRID). The ring is 40 m in circumference with 90° bending magnets in each of the four corners. Ions were injected in the ring with energies from 25 to 50 keV. After injection of a number of ions in the ring, their lifetimes were recorded by a neutral-particle detector installed in a chamber behind one of the bending magnets (see Fig. 3). the average pressure in the ring was ~5×10^{-11} mbar, and the rest gas was dominated by H_2.

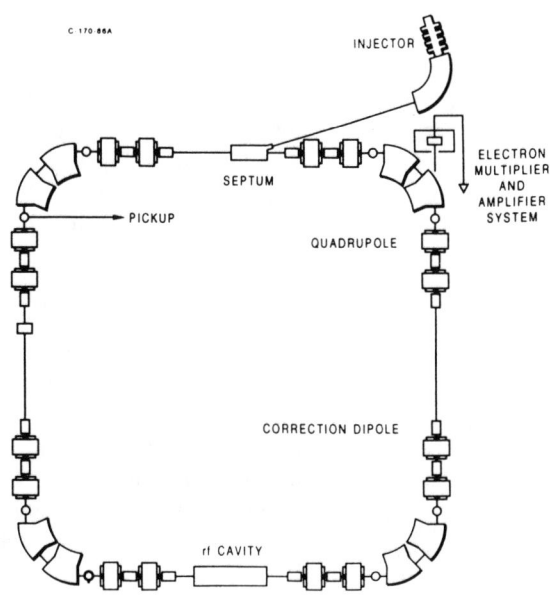

Fig. 3. Schematics of ASTRID showing the neutral-particle detector.

RESULTS

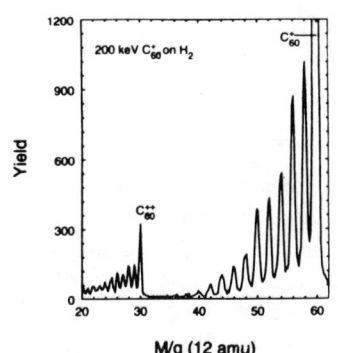

Fig. 4. M/q (mass divided by charge) spectrum.

Collisionally induced fragmentation

Figure 4 shows a mass spectrum obtained by colliding 200-keV C_{60}^+ with H_2 (Ref. 9). The dominant process is the loss of an even number of carbon atoms, resulting in even-numbered peaks ranging from C_{36}^+ to C_{58}^+. The intensity of odd-numbered peaks relative to the even-numbered peaks in this mass range is less than 0.01. It should also be noted that electron loss — and electron loss accompanied by fragmentation — are quite probable. Also, in this process, only even-numbered fragments are observed. The absolute cross sections for attenuation of the primary beams are found to be around $(3-3.5) \times 10^{-15}$ cm^2 which is almost of geometrical size ($\sim 4 \times 10^{-15}$ cm^2) and in agreement with qualitative calculations of atom-atom scattering based on a screened Coulomb potential. At the velocities in question, 'elastic' collisions between gas atoms and a carbon atom have large cross sections, $\sim \pi a_0^2$, where a_0 is the Bohr radius, for substantial energy transfer. Accordingly, a gas atom which encounters

C_{60} will suffer a number of collisions with individual carbon atoms. It seems reasonable to suppose that a large rift in C_{60} will then be produced.

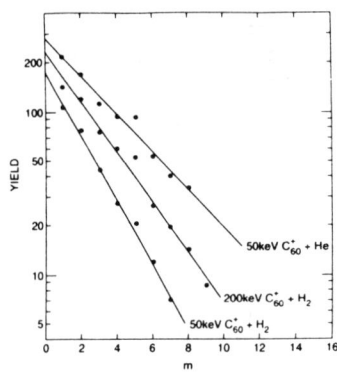

Fig. 5. *Fragment-peak intensities as a function of lost number of C_2's.*

In Fig. 5, the fragment intensities versus lost C pairs are plotted on a semilogarithmic plot for various energies and target gases. The intensities are seen to follow an exponential behaviour rather well, with C_{50}^+ and sometimes C_{58}^+ as exceptions. The extra stability of C_{50}^+ compared to its neighbours has been observed also in several other fragmentation experiments. The general shape of the fragmentation patterns can now be given in the simple form $I(m) = \text{const} \times p^m$, where m is the number of lost pairs of C atoms and p is an experimentally determined number. In Table I, we list the experimentally determined p values for different collision energies in collisions of C_{60}^+ with H_2 and He.

Table I. $I = cp^m$

E (keV)	Target	Proj.	Prod.	p	t_1/t_2
50	H_2	C_{60}^+	C_n^+	0.63	0.59
100	H_2	C_{60}^+	C_n^+	0.66	0.52
200	H_2	C_{60}^+	C_n^+	0.70	0.43
50	He	C_{60}^+	C_n^+	0.76	0.32
100	He	C_{60}^+	C_n^+	0.74	0.35
200	He	C_{60}^+	C_n^+	0.76	0.32

The spectrum shown in Fig. 4 suggests that the C_{60} ions break up either by successive losses of pairs of C or by single-step loss of a larger even number of neutral carbon atoms. Note that nothing can be concluded about the actual structure of the lost carbon-atom groups, i.e, whether they actually stay together either in pairs or in larger groups of an even number of atoms. In the collision experiments reported by Young, Cousins, and Harrison[3], the available center-of-mass collision energy is as low as 28 eV, and it is argued that sequential loss of C_2 is impossible when a larger number of carbon atoms are missing, simply from energy-conservation considerations.

In our experiments, where the available center-of-mass collision energy is between 150 and 1000 eV, such energy-conservation arguments are not essential. In fact, the fragment distributions shown in Fig. 5 suggest a mechanism where fragmentation occurs by sequential loss of 'pairs' of carbon atoms.

If we accept the exponential behaviour of ions evaporated as a basic general phenomenon for a wide interval of energy and for light background gases, there is still only little choice as to the mechanism of decay. We have been able to account for it only in terms of the following competition between an evaporation lifetime t_1 and a closure lifetime t_2. Suppose that the C_{60}^+ ion has got a large 'rift' as a result of a collision with the target atom, and that C pairs can evaporate with a characteristic time t_1 while the system is open. Since the available energy is large, we can assume that the losses of C pairs are independent events, all with the same characteristic time. The probability for m evaporations in the time interval $(0,t)$ is then

$$P_t(m) = \frac{1}{m!}\left(\frac{t}{t_1}\right)^m e^{-t/t_1}. \qquad (1)$$

Suppose further that the remaining fullerene closes in $(t, t+\delta t)$ with a probability $e^{-t/t_2}\delta t/t_2$, where t_2 is a characteristic lifetime of the rift. Then the total probability for m evaporations is

$$P(m) = \frac{t_1}{t_2}\left(\frac{t_2}{t_1+t_2}\right)^{m+1}. \qquad (2)$$

The distribution described by this formula is equivalent to the experimental exponential behaviour if we make the following substitution:

$$p = t_2/(t_1+t_2). \qquad (3)$$

In Table I, values for t_1/t_2 are also listed, and it is found that the times for which the molecule is open in collisions with H_2 are approximately twice the characteristic evaporation time, whereas the ratio is around three for He.

It should be noted from (3) that the constancy for p for successive evaporations requires that the ratio t_1/t_2 remains constant during the evaporation. The result that C_{50} lies slightly higher will then correspond to a slightly shorter closure time. As to the magnitudes of t_1 and t_2, they must of course be long compared with the collision time $\sim(1-3)\times 10^{-15}$ sec. They must further be short compared to the passage time from the target to the analyzer, $\sim 3\times 10^{-6}$ sec. Finally, it is implied that the evaporation time of the system after closure of the rift should be longer than the passage time. On the basis of the above discussion, we propose that the dominant fragmentation mechanism in collisions between fast C_{60} ions and light targets is independent sequential loss of neutral pairs of the C atom.

Charge transfer

The measured charge-transfer cross sections for 120-keV C_{60}^{3+} in various target gases are shown in Fig. 6 as a function of target-ionization potential. The line at 15.6 eV indicates the third ionization potential of C_{60}, as reported by Petrie et al.[10]. It is readily observed that the cross sections decrease with increasing target-ioni-

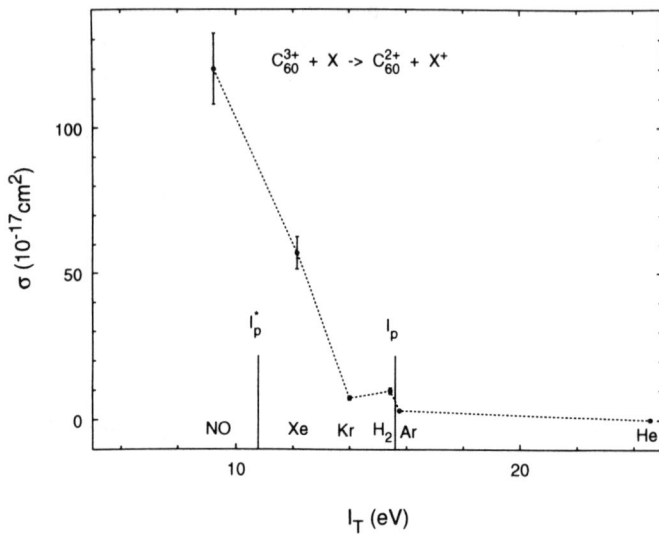

Fig. 6. Charge-transfer cross section as a function of target-ionization potential for 120-keV C_{60}^{3+}. The lines indicate the third ionization potential I_p and I_p^* for C_{60}.

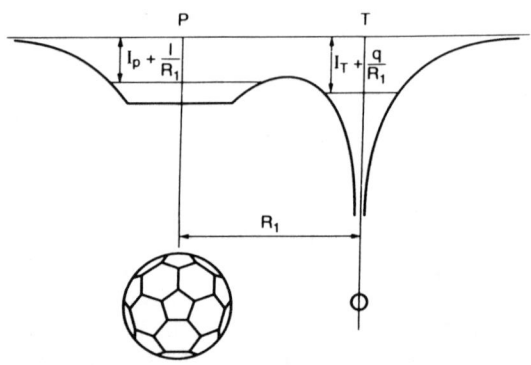

Fig. 7. The classical over-barrier transition model. The projectile P is a q times charged C_{60} molecule with ionization potential, I_p and the target T is an atom with ionization potential I_T.

zation potential. It should further be noted that the absolute value of the cross section is at most 10% of the geometrical cross section for a C_{60} molecule.

It is found that the capture cross section is small also for targets with ionization potential smaller than 15.6 eV. To understand this cross-section dependence on target-ionization potential, we have applied the over-barrier model which has proven to be quite successful for multiply charged ions[11,12]. This model states that capture may take place at an internuclear separation where the Stark-shifted initial-energy level of the electron rises above the barrier separating the two ionic cores (see Fig. 7). This situation is met for an internuclear distance (atomic units),

$$R = \frac{1}{I_T}[2\sqrt{q} + 1], \tag{4}$$

where I_T is the target-ionization energy and q is the projectile-charge state. However, if at this distance, the Stark-shifted initial-energy level is below the Stark-shifted projectile ground state, capture is excluded. The internuclear distance R_1, at which the energy-resonance condition is fulfilled, is found from

$$I_T + \frac{q}{R_1} = I_p + \frac{1}{R_1}, \tag{5}$$

where I_p is the binding energy of an extra electron on the projectile. The condition for capture to take place can now be formulated as

$$R_1 \leq R \tag{6}$$

or

$$\frac{I_p}{I_T} \geq \frac{q-1}{2\sqrt{q}+1} + 1. \tag{7}$$

The values of the right-hand side of Eq. (7) are listed in Table II for various q values.

Equation (7) readily indicates that the 'bracketing' method can be used only to determine the ionization potential for neutral species which are determined in collisions involving $q=1$ ions. The 'bracketing' method is built on the assumption that charge transfer at low energies can take place only when $I_T \leq I_p$. In order for a doubly charged ion to capture an electron from a neutral particle, $I_p(=I_{II})$ has to be 26% larger than the target-ionization potential. For $I_p(=I_{III})$, this value is 45%, etc. In general, charge transfer can only take place at low energies when $I_T \leq I_p^*$ where

Table II

q	$(q-1)/(2\sqrt{q}+1)+1$
1	1
2	1.26
3	1.45
4	1.60

$$I_p^* = I_p/[\frac{q-1}{2\sqrt{q}+1} + 1]. \tag{8}$$

With the values $I_{II}=11.8$ eV and $I_{III}=15.6$ eV, we find that low-energy capture will take place in gases with $I_T \leq 9.37$ for C_{60}^{2+} and with $I_T \leq 10.75$ eV for C_{60}^{3+}.

The cross-section behaviour on target-ionization potential shown in Fig. 6 can readily be understood on the basis of this simple model.

Lifetimes

Fullerene ions are known to decay by fragmentation or by electron emission after formation in excited states. The internal energy is stored either as vibrational energy or in the form of electronic excitation[13,14]. We have measured lifetime for unimolecular dissociation. In this process, an excited fullerene ion emits a C_2 unit, i.e.,

$$C_n^+ \rightarrow C_{n-2}^+ + C_2. \qquad (9)$$

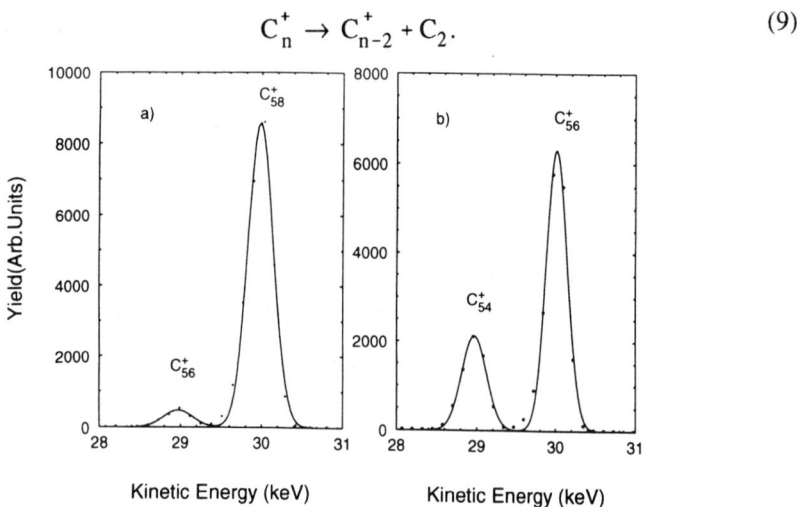

Fig. 8. Unimolecular dissociation spectra for (a) C_{58}^+ and (b) C_{56}^+ parent ions.

Two spectra showing this process are displayed in Fig. 8. The low-mass peak originates from ions which have emitted a C_2 unit between the analyzer magnet and the electrostatic analyzer (see Fig. 2).

From such spectra, the lifetime τ of the parent ions can be estimated if we assume that they decay according to the formula

$$I(t) = I_0 \exp(-t/\tau). \qquad (10)$$

The lifetimes for unimolecular decay of a few ions are listed in Table III.

The lifetimes for C_{58}^+ and C_{56}^+ are in agreement with earlier measurements by Radii et al.[15] and Campbell et al.[16], where the ions were produced directly by desorption from graphite and polyimide, respectively.

The long lifetime of C_{60}^+ in the present experiments as compared with a lifetime of ~1 ms in the measurements by Radii et al. and Campbell et al. is probably caused by the different ways of production. In our measurements, a vapour of C_{60} is present in the ion source, and soft ionization leaves the C_{60}^+ ions in a low-energy state (low temperature).

Table III

C_n^+	τ (µs)
C_{60}^+	~∞
C_{58}^+	337
C_{56}^+	66

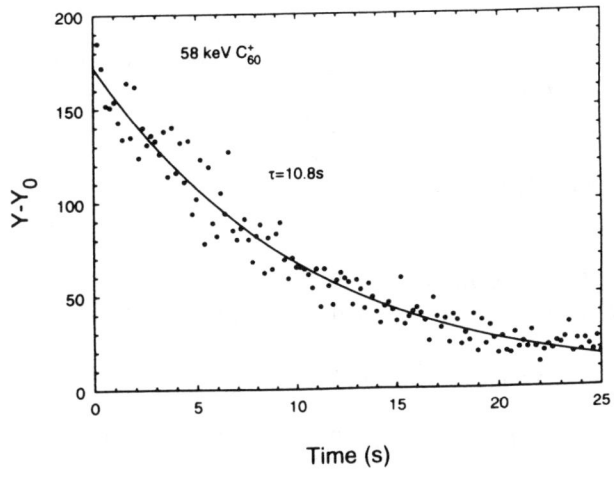

Fig. 9. *The yield of neutral particles at the detector as a function of time after injection for storage of 58-keV C_{60}^+.*

Inspired by the long lifetimes (≥ 1 ms) reported for C_{60}^{2-} by Hettich *et al.*[17], we have performed lifetime studies of fullerene ions in the heavy-ion storage ring ASTRID.

In Fig. 9 is shown a typical exponential decay for 58-keV C_{60}^+ ions. The lifetime (~10 s) is caused by collisions with rest gas. It is thus possible to store these ions for seconds and eventually interact with them in a controlled way. For several fullerene ions, we have measured lifetime components in the 10-ms range and thus proved the feasibility of storage rings for lifetime studies of these ions. Our lifetime measurements will be published soon after this Conference.

ACKNOWLEDGEMENT

I would like to express my gratitude to my co-workers, L.H. Andersen, C. Brink, H.K. Haugen, J. Lindhard, D.C. Lorents, R. Malhotra, R. Ruoff, and D.H. Yu, who participated in this work.

REFERENCES

[1]H.W. Kroto, J.R. Heath, S.C. O'Brien, R.F. Curl, and R.E. Smalley, Nature (London) **318**, 162 (1985)
[2]W. Krätschmer, L.D. Lamb, K. Fostiropoulos, and D.R. Huffman, Nature (London) **347**, 354 (1990)
[3]A.B. Young, L.M. Cousins, and A.G. Harrison, Rapid Commun. Mass Spectrom. **5**, 226 (1991)
[4]R.J. Doyle and M.M. Ross, J.Phys.Chem. **95**, 4954 (1991)
[5]E.E.B. Campbell, A. Hielscher, R. Elich, V. Schyja, and I.V. Hertel, in: *Nuclear Physics Concepts in Atomic Cluster Physics*, eds. R. Schmidt *et al.*, Lecture Notes in Physics, Vol. 404 (Springer, Berlin, 1992) p. 185
[6]M.M. Ross and J.H. Callahan, J.Phys.Chem. **95**, 5720 (1991)
[7]T. Weiske, D.K. Böhme, J. Hruzák, W. Krätchmer, and H. Schwartz, Agnew Chem. Int.Ed.Engl. **30**, 884 (1991)

[8] O. Almén and K.O. Nielsen, Nucl.Instrum.Methods **1**, 302 (1957)
[9] P. Hvelplund, L.H. Andersen, H.K. Haugen, J. Lindhard, D.C. Lorents, R. Malhotra, and R. Rouff, Phys.Rev.Lett. **69**, 1915 (1992)
[10] S. Petrie, J. Wang, and D.K. Böhme, Chem.Phys.Lett. **37**, 2631 (1993)
[11] A. Bárány et al., Nucl.Instrum.Methods **B9**, 397 (1985)
[12] P. Hvelplund et al., Nucl.Instrum.Methods **B9**, 421 (1985)
[13] C.W. Walter, Y.K. Bae, D.C. Lorents, and J.R. Peterson, Chem.Phys.Lett. **195**, 543 (1992)
[14] Y. Zhang and M. Stuke, Phys.Rev.Lett. **70**, 3231 (1993)
[15] P. Radii, M.T. Hsu, M.E. Rincon, P.R. Kemper, and M.T. Bowers, Chem.Phys.Lett. **174**, 223 (1990)
[16] E.E.B. Campbell, G. Ulmer, H.G. Bugmann, and I.V. Hertel, Chem.Phys.Lett. **175**, 505 (1990)
[17] R.L. Hettich, R.N. Compton, and R.H. Ritchie, Phys.Rev.Lett. **67**, 1242 (1990)

DIFFRACTION AND VIBRATIONAL DYNAMICS OF LARGE CLUSTERS FROM HE-ATOM SCATTERING

U.Buck
Max-Planck-Institut für Strömungsforschung, D 37073 Göttingen, Germany

ABSTRACT

Large Ar_n clusters which are generated by adiabatic expansion through conical nozzles are probed by He atom scattering. The diffraction oscillations in the elastically scattered angular distributions are used to derive information on the size distribution of these clusters by comparison with model calculations. The inelastically scattered He atoms which are detected by time-of-flight analysis contain detailed information on the vibrational dynamics and thus on the phonon spectra of the clusters. The peak of the energy tranfer increases with increasing scattering angle for one average cluster size. At constant and comparable scattering angle, the position of the peak slightly decreases as a function of average cluster size $\bar{n}$ and converges to the position of surface phonons of solid Ar.

INTRODUCTION

The knowledge of the frequency spectrum of a cluster plays an important role both for the interpretation of static and structural properties as well as for the dynamical behavior. This includes such quantities as the thermal free energy and the specific heat, the melting and condensation processes, and the excitation or deexcitation of the vibrational modes.[1] The latter process is of special interest, since it is very sensitive to the transition of the cluster from the discrete spectrum of vibrational modes of the molecular system to the lattice vibrations and the continuous phonon dispersion curves of the solid. The theoretical methods to study these topics range from the harmonic normal mode analysis over classical molecular dynamics simulations to complete quantum calculations of the lattice dynamics. The experimental methods include IR- and Raman- spectroscopy for the molecular systems and neutron, electron or He atom scattering for probing bulk and surface phonons, respectively. From all these methods the scattering of He atoms appears to be the most general process to study the vibrational spectra of clusters, since this method is mainly sensitive to surface properties and obeys nearly no selection rules. In the last 10 years it has been developed into a very successful and reliable method for measuring surface phonons of the solid and for deriving a series of very interesting surface properties.[2]

Here we report measurements of He atom scattering from large Ar_n clusters in the range from n = 11 to 4600. There are mainly two different types of observables: (1) The angular distributions, which contain among other things information on the geometry and the size of the investigated objects through the diffraction oscillations. (2) The inelastic energy transfer, which is related to the excitation of the vibrational modes and which is measured by time-of-flight analysis of the scattered He atom. A preliminary account of some of the results appeared in two conference proceedings.[3,4]

EXPERIMENTAL

The experiments have been carried out in a crossed molecular beam machine which is described in detail elsewhere.[5] Essentially it consists of a He supersonic nozzle beam, a cluster beam for the target which intersect at an angle of 90°, and a detector with an electron bombardment ionizer and a quadrupole mass filter operating under ultrahigh vacuum conditions. The angular dependence is measured by rotating the source assembly relative to the fixed detector position. The velocity of the scattered He atoms is measured by time-of-flight analysis using the pseudorandom chopping technique with a flight path of 450 mm.

The helium atom beam is produced by the expansion of the gas under high stagnation pressure (typically 30 bar) through a small orifice (diameter 30 μm) into the vacuum. Most of the measurements were carried out at a temperature of $T_0 = 77$ K. With a speed ratio of $S = 90$ an internal temperature of lower than 0.1 K and a corresponding relative width of the velocity distribution of $\Delta v/v = 0.018$ is obtained. This leads at collision energies of about 25 meV to a resolution of better than 1 meV.

The cluster beam is generated by expansion from stagnation pressures of 1.2 to 5.0 bar through different nozzles of conical shape [6] which vary in diameter (60 μm to 130 μm), length (1 mm, 10 mm), and opening angle (20° to 30°). It is interesting to note that the velocity analysis of the Ar-beam which contains a fraction of large clusters gave two contributions: a faster one around 615 m/s, which is attributed to the remaining Ar monomers and a slower one around 500 m/s, which is attributed to the clusters and their fragments including the monomer mass. The contribution of the remaining monomers gradually decreases with the production of larger clusters.

The beam contains only a distribution of cluster sizes the maximum of which can be shifted to nearly any desired position by varying the different shape parameters of the nozzle and the stagnation pressure. The key parameter which relates the corresponding average cluster sizes $\bar{n}$ to the stagnation pressure p_0, the source temperature T_0, and the nozzle diameter d is given by [7]

$$\Gamma^* = p_0 d^{0.85} K_{ch} T_0^{-2.2875} \qquad (1)$$

where K_{ch} is a constant which is characteristic for the material. The numbers refer to Ar with $K_{ch} = 1646$, if the units are mbar, μm, and K, respectively.

DIFFRACTIVE SCATTERING

Angular distributions or, in the language of scattering experiments, total differential cross sections, have been measured for series of cluster sizes ranging

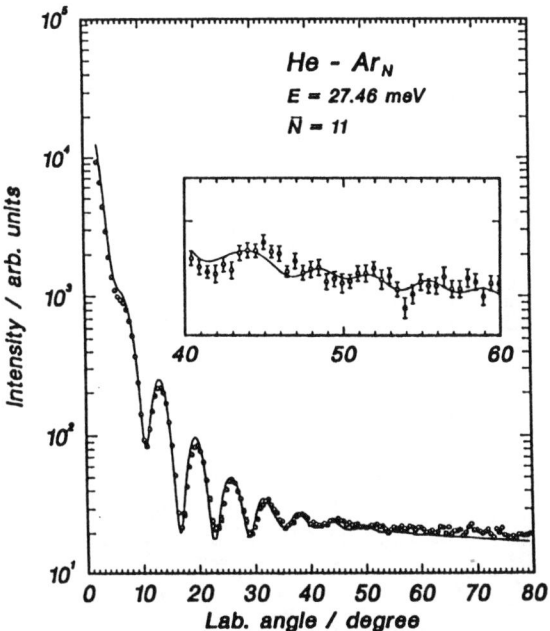

Figure 1: Measured total differential cross sections of He scattered from the averaged cluster size Ar_{11}. The solid lines are calculations based on the model described in the text.

from $\bar{n} = 11$ to $\bar{n} = 4600$ for deflection angles between 3° and 80° and collision energes of about 27 meV. They can be characterized by three different regions. (1) At small angles up to 30° they are all dominated by an oscillation of the same large angular separation. (2) Then an angular range (30°- 50°) follows with oscillations of smaller angular separation which vary with cluster size. (3) Finally, an unregular type of intensity fluctuations is observed between 50° and 80° which also depends on cluster size. A typical example is shown in Fig. 1 for $\bar{n} = 11$ which clearly shows the oscillations of type (1) and (2), but not the intensity fluctuations of type (3). In order to better visualize the oscillations of type (2), part of the cross section for $\bar{n} = 11$ is reproduced enlarged in the upper part of the figure.

The oscillations can be attributed to diffraction which gives for the angular separation [8]

$$\Delta\Theta = \pi/(kR_0) \qquad (2)$$

in which k is the wavenumber, and R_0 the range of the repulsive potential which is closely related to the diameter of the particle. Simple calculations of the elastic differential cross sections show that the oscillations of type (1) can be reproduced by He-Ar scattering, that is the scattering from the monomers in the beam. The detailed comparison with the measured data, however, reveals

722 Diffraction and Vibrational Dynamics

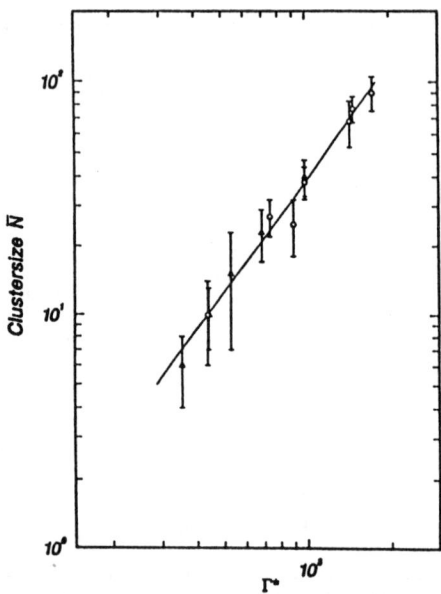

Figure 2: Averaged cluster size as a function of the scaling parameter Γ^* determined from diffractive scattering for sharp edged($\triangle$) and conical($\bigcirc$) nozzles.

disagreement in the amplitudes. In order to reproduce the oscillations of type (2), similar calculations have been carried out for different cluster sizes using the model of a sphere of homogeneously distributed Lennard-Jones (12-6) potentials.[9] Again there is reasonable agreement in the positions, but a complete disagreement in the amplitudes. In order to test the potential model, a more precise evaluation of the interaction potential of He-Ar_{55} based on realistic interactions and the icosahedral structure of this cluster was carried out. The result gives about the same value $R_0 = 8.0$ Å for the most probable configuration of the approach of the He-atom to the face of the $n = 55$ icosahedron.

To improve the comparison between measured and calculated data, the distribution of cluster sizes in the beam is accounted for. The measured cross sections are fitted to a weighted sum of cluster cross section σ_n and the monomer cross section σ_1 according to

$$\sigma_{meas} = \sum_n g_n \sigma_n + \sigma_1 \tag{3}$$

The results are shown in Fig. 1 by the solid lines. Now the agreement is very satisfatorily both in the monomer and in the cluster part. The average cluster size $\bar{n}$ is then given by

$$\bar{n} = \sum_n g_n n / \sum_n g_n \qquad (4)$$

The width in the distribution is roughly given by $\Delta n = \bar{n}/2$. The averaged cluster sizes determined in this way are now plotted as function of the parameter Γ^* in Fig. 2. The plot contains both data from sharp edged and conical nozzles. The data are reproduced by the simple formula

$$\bar{n} = 38.4(\Gamma^*/1000)^{1.64} \qquad (5)$$

It is noted that these new results are in good agreement with values of $\bar{n}$ taken from measured mass spectra.[10,11] They deviate, however, from the results obtained by electron diffraction [12], which give larger values. The reason for this discrepancy is not clear. Perhaps the measurements of electron diffraction are biased against smaller clusters.

INELASTIC SCATTERING

The information on the vibrational excitation is obtained from time-of-flight (TOF) spectra taken at different laboratory deflection angles. Measurements were carried out at a collision energy of about 27.0 meV for a range of averaged cluster sizes from $\bar{n} = 25$ to 4600 and different scattering angles from $\Theta = 5°$ to $80°$. A typical example for $\bar{n} = 77$ at $5°$, $12°$, $18.75°$, and $25°$ is shown in the left hand part of Fig. 3. From the measured beam velocities, the positions of elastically scattered monomers and clusters are known. They are used in a Monte-Carlo fitting procedure in which also the experimental resolution of the apparatus based on the measured velocity distributions and angular divergencies are properly taken into account.[13] The remaining intensities are attributed to inelastic transitions and the TOF-spectra are marked by the corresponding energy transfer ΔE in meV starting from the elastic scattering of the cluster. The monomer scattering, marked by $n = 1$, can be easily distinguished from the cluster scattering because of the different velocities. It is noted that for small cluster sizes, the TOF spectra give directly the vibrational modes which are excited in the collisions with He. The additional constraint for the momentum of the phonon which leads to a different procedure for obtaining bulk and surface phonon spectra of the solid [2] is probably only valid for very large clusters. The results for $\Theta = 5°$ and $10°$ indicate that the energy transfer is small with a most probable value of 1 to 2 meV. The results for $\Theta = 18.75°$ and $25°$ are quite different. Now the monomer and the elastic scattering is clearly separated from the inelastic part with a most probable energy transfer of about $\Delta E = 4$ to 5 meV and the largest detectable energy transfer which does not exceed $\Delta E = 8$ meV. The results for the other cluster sizes and larger angles differ in details but agree as far as the general shape with a maximum and a cut off of the intensity is concerned.

An example for the dependence of the spectra from the cluster size is shown in the right hand side of Fig. 3. Here the spectra taken at $18.75°$ for the clusters $\bar{n}$

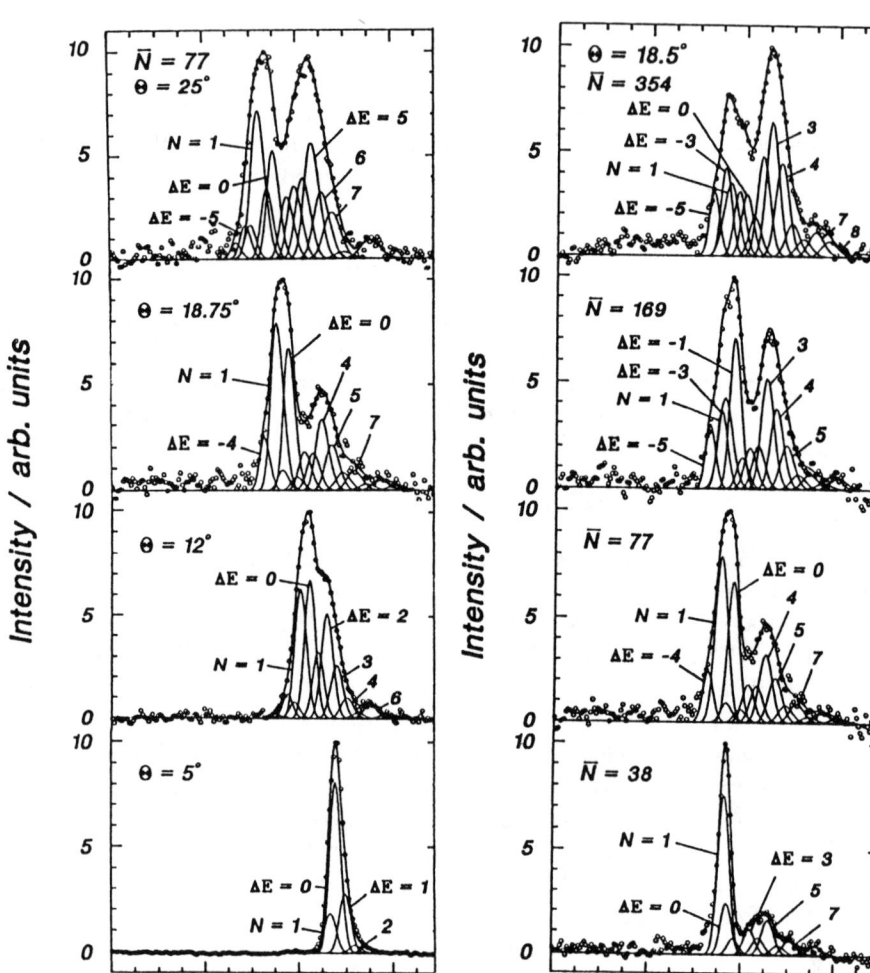

Figure 3: Measured time-of-flight spectra for the averaged cluster size $\bar{n} = 77$ and different deflection angles (left panel) and for a fixed deflection angle $\Theta = 18.75°$ and different cluster sizes (right panel). The numbers refer to the energy transfer in meV.

= 38, 77, 169, and 354 are displayed. Again the monomer and the elastic cluster scattering are clearly separated from the inelastic part which increases in size relative to the former one. The maximum in the distribution decreases from ΔE = 5 meV at $\bar{n}$ =38 to ΔE = 3 meV at $\bar{n}$ = 354. This general trend is also observed for many other cluster sizes.

VIBRATIONAL DYNAMICS

How do these experimental findings compare with what is known about the density of vibrational states from experiments or calculations? The measured bulk phonon density of solid Ar at 10 K shows a large peak at 8 meV which is caused by the longitudinal motion and a second peak around 5 meV from the transverse branch with the characteristic van Hove singularities.[14] The spectra of surface phonons exhibit maxima between 2 and 3 meV which can be traced back to the well known Rayleigh mode.[2,15]

For clusters or particles of finite size only calculations are available. The methods include Molecular Dynamics (MD) simulations from which the frequency spectrum is obtained by the Fourier transform of the velocity autocorrelation function.[16-19] Other calculations are based on the collective motion of elastic vibrations of solid [20,21] or density fluctuations of liquid material [22] which manifests itself in the excitation of spheroidal or torsional modes [20,21] and shape or compression oscillations [22], respectively. The results of all calculations treating large clusters n = 55, 135, 419 and 14000 [17,20] or finite layers and nearly spherical blobs [16] can be summarized as follows. In addition to the peaks which already appear for the bulk material at 5 meV and 8 meV two further peaks are recognized at smaller frequencies at about 2-3 meV and at about 1.2 meV. They are attributed to surface modes with amplitudes normal to the surface and edge atoms, respectively. The overall size of the particles is reflected only in the relative amount of the intensities of the different modes to each other. Bulklike modes are less probable than surface modes in particles with a large surface-to-volume ratio and vice versa.

We demonstrate this in calculations of the vibrational modes of Ar_{55} which has icosahedral structure with one central atom, a complete inner and a complete outer shell. Figure 4 shows the spectrum obtained in MD-simulations from the Fourier transform of the velocity autocorrelation function for T = 10 K to 37 K. At the lower temperature the different contributions can be identified. The single peak at 10 meV is mainly attributed to the motion of the central atom, the group of lines between 5 and 8 meV is caused by the inner shell, while the frequencies around 2-3 meV are resulting from the outer shell. This picture is in good agreement which the simulations of smaller systems [18,19] as well as with the results discussed earlier.[21] With increasing temperature the structure gradually disappears, but still the same general shape remains. This is an indication of the solid-like behaviour of the cluster. At the largest temperature studied which is already liquid-like we find a completely different behaviour. Here there are a

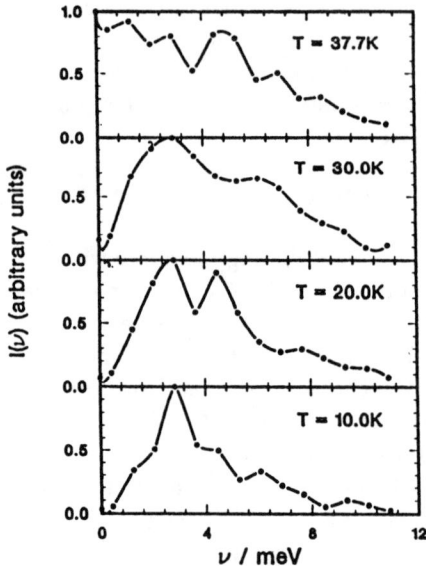

Figure 4: Calculated frequency spectra of Ar_{55} as a function of temperature using Molecular Dynamics simulations.

large number of low frequency modes which are characteristic for the diffusive behaviour. Hopefully we will get information on these temperatures by direct comparison of measured and calculated vibrational spectra.

The measured trend that the maximum of the surface vibrations slightly decreases with increasing cluster size is also reproduced in the model calculations of Ref.21. It converges to the value of the lowest surface mode of the bulk phonons.

SUMMARY

These first results on the measurement of the vibrational frequency spectrum of free clusters by He atom scattering look very promising and open a range of new experimental possibilities. The great advantage of this technique is the general applicability to any kind system and any kind of vibrational spectrum. The disadvantage is, of course, the low intensity of the scattered signal and the preparation of the cluster target which consists of a distribution of cluster sizes.

ACKNOWLEDGEMENTS

I acknowledge with gratitude the contributions of my coworkers R. Krohne and J. Siebers who did the experimental and theoretical work presented here. The financial support by the Max-Planck Research Award also is gratefully acknowledged.

REFERENCES

1. M.R. Hoare and P. Pal, Adv.Phys.**24**, 645 (1975)
2. J.P. Toennies, Physica Scripta **T19**, 39 (1987)
3. U. Buck, R. Krohne, and J. Siebers, Nuclear Physics Concepts in Atomic Cluster Physics, edited by R. Schmidt, H.O. Lutz, R. Dreizler (Springer, Berlin,1992) p.178
4. U. Buck, R. Krohne, and J. Siebers, Z. Phys. D **26**, 169 (1993)
5. U. Buck, F. Huisken, J. Schleusener, and J. Schaefer, J.Chem.Phys. **72**, 1512 (1980)
6. W. Obert, Rarefied Gas Dynamics, ed. R. Campargue, CEA Paris, **11**, 1181 (1979)
7. O.F. Hagena, Z.Phys.D **4**, 291 (1987)
8. U.Buck, in Atomic and Molecular Beam Methods, ed. G. Scoles, (Oxford, New York, 1988) Chap. 20, p. 499
9. J. Gspann and H. Vollmer, in Rarefied Gas Dynamics, ed. R. Campargue, CEA, Paris, **11**, 1193 (1979)
10. O.F. Hagena and W. Obert, J.Chem.Phys. **56**, 1793 (1972)
11. R. Müller, Diplomarbeit, Universität Hamburg, 1990; see also J. Stapelfeldt, J. Wörmer, and T. Möller, Phys.Rev.Letters **62**, 98 (1989)
12. J. Farges, M.F. de Feraudy. B. Raoult and G. Torchet, J.Chem.Phys. **84**, 3491 (1986)
13. U.Buck, in Atomic and Molecular Beam Methods, ed. G. Scoles, (Oxford, New York, 1988) Chap. 22, p. 525
14. Z. Fujii, N.A. Lurie, R. Pynn, and G. Shirane, Phys.Rev.B **10**, 3647 (1974)
15. R.E. Allen, G.P. Alldredge, and F.W. de Wette, Phys. Rev. B **4**, 1661 (1971)
16. J.M. Dickey and A. Paskin, Phys.Rev.B **1**, 851 (1970)
17. W.D. Kristensen, E.J. Jensen and R.M.J. Cotterill, J.Chem.Phys. **60**, 4161 (1974)
18. T.L. Beck and T.L. Marchioro II, J.Chem.Phys. **93**, 1347 (1990)
19. J.A. Adams and R.M. Stratt, J.Chem.Phys. **93**, 1358 (1990)
20. A. Tamura and T. Ichinokawa, J.Phys.C **16**, 4779 (1983)
21. Y. Ozaki, M. Ichihashi and T. Kondow, Chem.Phys.Lett. **182**, 57 (1991)
22. A. Tamura and T. Ichinokawa, Surf.Science **136**, 437 (1984)

SURFACES

HOLLOW ATOMS BELOW, ABOVE, AND AT SURFACE

Jean Pierre Briand
Laboratoire de Physique Atomique et Nucléaire-Institut du Radium
Université Pierre et Marie Curie, 4 place Jussieu 75252 Paris Cedex 05, France
Unité associée au CNRS n°771

ABSTRACT

It is now quite obvious that, in most cases, when a highly charged ion approaches, or penetrates a surface, many electrons are captured in excited states of the projectile. The nature of the hollow atoms formed depends on the velocity of the ion, and whether or not capture has occured above, below, or at the surface. I would like in this talk to discuss the nature, namely the electronic configuration of the hollow atoms formed in various circumstances. In the first two sections I shall summarize recent results, some of them are already published, or have been presented in other conferences.[1] Section I will be devoted to the study of hollow atoms formed inside the surface, section II to those formed far from the surface. In a third section I will present some new results, obtained at very low velocities, on hollow atoms **at surface**.

These results have been obtained through a large international collaboration: J.P. DESCLAUX, CEN, Grenoble; B. d'ETAT, G. GIARDINO, L. de BILLY, S. BARDIN, LPAN-Université P&M Curie, Paris; D. SCHNEIDER, M. BRIERE, M. CLARK, D. KNAPP, V. DECAUX, LLNL, Livermore-Californie; R. ALI, N. RENARD, M. STOCKLI, P. RICHARD, KSU, Manhattan-Kansas; A. BRENAC, G. LAMBOLLEY, AIM, Grenoble; J. FAURE, Laboratoire National Saturne, Saclay.

I. HOLLOW ATOMS INSIDE A SOLID

First I would like to briefly summarize what is known or still unknown about the hollow atoms formed below the surface, i.e., in the most frequently encountered experimental situations but yet difficult to understand. In the case of ions of atomic numbers higher than 18, it has been proved[2] that the M and N shells are very quickly filled up, while the innermost ones are empty, or quasi empty, leading to the formation of a first kind of hollow atoms. First I shall summarize what we know about the mechanism leading to the formation of these species and their decay model.

The quick filling of the M (N) shell of the ion was first explained[2] by using the resonant neutralization model. In the case of argon ions on carbon targets (or any surface contaminated by carbon) there is an approximate energy matching between the K shell of carbon and the M shell of argon, which might explain this quick filling. We soon realized,[1] in studying on the same targets much heavier ions like krypton or iron, where such an energy matching would populate more highly excited states (n>5) (in contradiction with experimental results), that some other, or additional, processes have to be taken into account to explain the formation of these hollow atoms. We have then suggested that the Auger neutralization process, which should lead to a quite selective population of the n=3, 4, 5 shells of these ions, could be responsible of such a formation below the surface. These assumptions will be discussed in more detail by B. d'Etat at the ISIAC meeting.

A much more puzzling question is to know which processes lead to the filling of the innermost K and L shells, and how can compete the direct collisional filling of these shells [the so-called side feeding][3] and the Auger decay from outermost electrons.

I would like to discuss some experimental results that may provide information on

© 1993 American Institute of Physics

the neutralization time of these ions, and some insights about the competing processes filling the K and L innershells. What is observed is the transition filling of the K hole (or the two K holes), the Kα x-rays,[2] or the KLL Auger.[3,4,5] In some cases (e.g. Ar^{16+}) the Auger transitions from the M shell to the L shell has also been observed.[6] These transitions, occuring in presence of any number of L, M (N) spectator electrons, are usually made of a complex array of satellite lines providing some snapshots of the hollow atoms. By studying the electronic configurations of these ions at the time of the filling of the K shell, one may infer what could have been the different steps of the cascade filling the innermost shells. Figure 1 shows these snapshots in the case of argon.

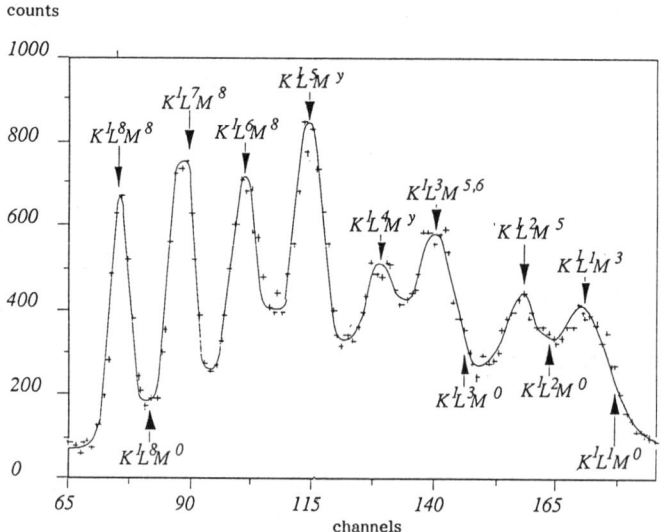

Fig. 1. *Ar Kα satellites observed[2] during the collision of Ar^{17+} ions with a surface at 20kV/q and normal incidence.*

One can observe the (8) satellites lines corresponding to the (8) occupation states of the L shell (KL^X satellites). The exact energy of these lines giving the mean number of M electrons, one can easily see on this figure that the M shell is always 2 or 3 times more populated than the L one (the M shell is then in most cases closed). In the frame of a stepwise (time dependent) picture neutralization of the ion this would mean that the M shell is 2-3 times faster filled than the L one. The M shell being most of the time closed up, the L shell can also be directly filled up through LMM Auger transitions, as we first assumed,[2] and as it has been recently observed.[6] It was then supposed that if this process was the dominant one (the L shell side feeding being at least, from the usual scaling laws, 5 times slower than the M shell filling), the filling process of the innershells would have occured through a cascade of 8 Auger transitions.

Now I would like to discuss briefly some recent experiments which demonstrated the time dependent picture of the neutralization process and gave some approximate values of the neutralization and decay times. The idea[7] is to make use of fully stripped ions whose two K vacancies are sequentially filled through the hypersatellite-satellite cascade. By looking at the electronic configuration of the ion at the time of the decay (KL^X satellite structure of the Kα hypersatellite and satellite) it would be possible to study separately two steps (snapshots) of the time evolution of the electronic configuration of the hollow atoms, the two lines being emitted one after the other. We observed[7] such photographs of the electronic configurations at two different times of the decay process, which clearly demonstrated the stepwise filling of the L shell (a completely different KL^X distribution and more L electrons were observed on the satellite spectrum than on the hypersatellite one). We recently carried out at the Livermore EBIT, in collaboration with D. Schneider et al, a coincidence experiment between the hypersatellite and the satellite on Fe^{26+} ions, which gave a clear proof of this time evolution of the neutralization process. Figure 2 shows in a biparametric set-up the peaks corresponding to the hypersatellite-satellite coincidences. Moreover, from this experiment, one can study the progressive filling of the innershells of the ions from the outermost ones. One can see, for instance, a larger number of L electrons in the ion in the second step (satellite) of the K shell filling than in the first (hypersatellite) one. From this kind of experiment one can extract, if one knowns the lifetime of the satellite, the mean time for the filling of one L hole.

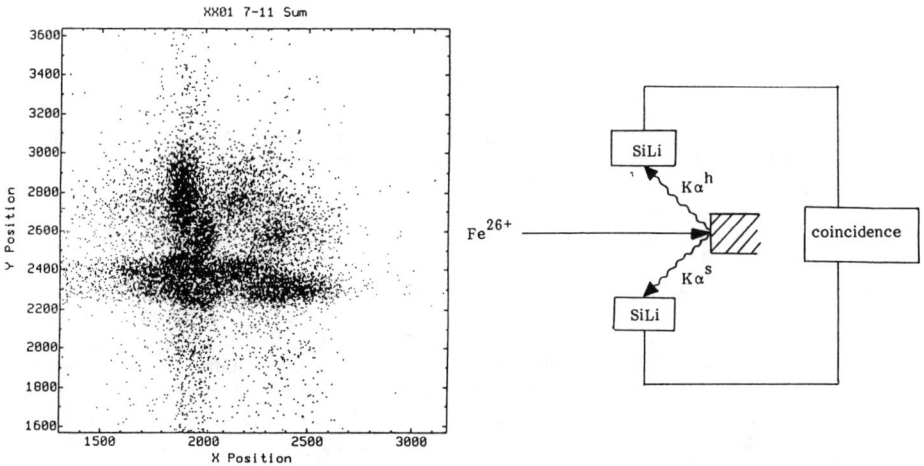

Fig. 2. *Hypersatellite-satellite biparametric coincidences for Fe^{26+} ions on metal target.*

Another interesting feature of this biparametric spectrum is the progressive moving of the M and N electrons to the L and K shells (The hypersatellite spectrum in coincidence with the satellite -the lowest horizontal line- clearly shows strong Kβ and Kγ peaks whereas the satellite spectrum -just above the previous one- has a weaker β line and no γ ray. This clearly demonstrates the descent of the N, M electrons to the L shell).

We have measured the meantimes for the filling of one L hole for argon, iron and krypton ions which have been found to be of the order of $3-5 \times 10^{-16}$ s per L hole. These findings may provide some insights on (i) the mechanisms for the filling of the L shell, and (ii) the neutralization time.

(i) The meantime for the filling of one L hole must be compared with the LMM Auger lifetimes which were first evaluated with the Larkins statistical procedure.[2,8] More accurate Dirac Fock calculations[9] led to some mean lifetimes significantly longer than first assumed, and of the order of 10^{-15} s (within $\neq 20\%$) for argon, iron and krypton ions (the Auger decay not scaling at first approximation with Z). These meantimes being roughly two times shorter than the lifetime for a LMM Auger decay, one can thus assess that both LMM Auger decay and side feeding processes may fill the L shell. The exact mechanism of the side feeding process remains however an open question (even if it was supposed[1] to be another Auger, the Auger neutralization, process). Although the time evolution of the neutralization and decay processes is now well established, it is however not yet possible to say precisely in how many steps the L shell is filled up.

(ii) The M shell being roughly two or three times faster filled up than the L one, it is now possible to give an estimate of the meantime for the filling of the whole M shell which has to be 8/3 time the meantime for the filling of a L hole, i.e., of the order of 1×10^{-15} s. At velocities of the order of 10^6 m.s^{-1} (20 kV/q) such a filling process would occur along a mean range of 10 Å, i.e., 3 atomic monolayers. If we take into account the fast depopulation of the M shell via LMM Auger cascades this means, roughly speaking, that a mean number of about 2 or 3 electrons are captured in the M shell through an atomic row, plus an unknown number of N electrons (may be of the same order of magnitude as demonstrated for krypton).

II. "OUTSIDE" HOLLOW ATOMS

Far from the surface highly charged ions can capture through the resonance neutralization process many electrons in highly excited states of the ions. In the case of Ar^{17+} for instance, one can assume that up to 17 electrons can be captured in n~20 states at 15-20 Å from the surface. These very excited states have only a short time to reach the surface. Generally the cascade of Auger decay to the ground state lasts much longer than the transit time and no signal coming from the last step of the decay can be observed outside the surface. These highly excited ions are then peeled off while touching the surface and new hollow atoms are prepared below the surface as described in section I. However by slowing down the ions, or by working at grazing incidence, it is possible to observe a very small amount of the "outside" decay (the slowing down of the ions being naturally limited by their acceleration, close to the surface, by their own image). Many groups through the world have successfully attempted to observe this outside decay which has been found to represent no more than few percents of the decay.[10,11] The observed spectra are then mainly due to the "inside" decay described in section I on which a small, usually very characteristic signal of the outside decay is superimposed. In the case of Ar^{17+}, we have studied with a crystal spectrometer this "outside" decay by looking, at grazing incidence, at the Kα lines emitted by the ion.[9] We have also studied in Ref.(9) the theoretical decay of these very excited states produced far from the surface. I shall only describe in this section some of the gross features of the electronic configuration, at different steps of the decay, of an ideal hollow atom in which up to 17 electrons are captured in e.g. the n=10 state of argon. This will then allow to make some comparison with the electronic configuration of the hollow atoms produced below the surface (section I) or at surface (section III).

The capture of many electrons in high n states is a very complex process involving at least few n states of unknown angular momentum. Capture occuring in presence of many "flying" electrons and their image, as well as the image of the nucleus itself, the study of an unperturbed (n,l)[17] state must then be considered as an "academic" case, but that certainly will help understanding the main characteristics of the Auger cascade. The number of substates in such a configuration is enormous and few tens of thousands of specific Auger rates have been calculated by J.P. Desclaux[9] in order to determine the mean values of the lifetimes and the scaling laws.

The most characteristic features of the Auger transitions from this high n states is that the decay proceeds stepwise through the smaller energetically allowed Δn states (the lower the energy of the Auger electron, the faster the transition). Another main characteristic of the decay of these very excited states is that the total rate scales with the number x of electrons not like x(x-1)/2 as supposed in the statistical model but is very roughly proportional to x. As a consequence of these two properties, at the first stage of the decay, many transitions from n=10→n=8 have to occur before there is enough electrons in this n=8 shell to allow a significant Auger rate from this state. One can thus immediately assess that even if the lifetime of the first n=10 transition is rather fast (10^{-14} s, i.e., comparable to the lifetime of the decay in the n=2 shell), the full cascade process may take quite a long time (~10^{-12} s), usually much too long compared to the transit time to the surface. The full "outside" decay is therefore very unlikely (the large number of steps slows down the whole process).

The second interesting feature of the decay is that the electronic population of these species has, at any stage of these successive waves of Auger decay, more electrons in a state of n quantum number than in the next available one, leading to the so-called "triangular" population (Figure 3). This kind of electronic configuration characteristic of the "outside" decay is then dramatically different from that observed for the outside decay. This is what we observed[9] by looking at the Ar Kα lines, at grazing incidence with a crystal spectrometer (e.g. the observation of the decay of the KL^0M^2 states, through LMM Auger transitions, followed by the emission of the $K\alpha L^1 M^0$ satellite, which appeared superimposed to the $K\alpha L^1 M^3$ line, characteristic of the "inside" decay).

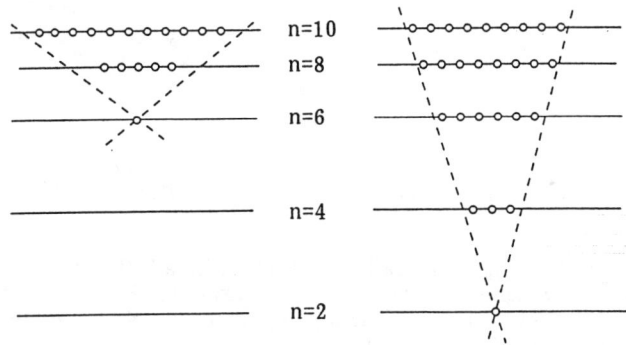

Fig. 3. *Typical electronic configurations of hollow atoms outside the surface (triangular population).*

III. AT SURFACE HOLLOW ATOMS

As quoted above, at relatively high velocities, most of the observed decay of the hollow atoms take place below the surface, but as discussed in section II the slowing of the ions, down to e.g. 10^5 m.s^{-1}, does not allow anyway the observation of substantial "outside" decay. In this section we shall deal with the study of the hollow atoms at the lowest allowed velocities.

We have studied the K x-ray spectrum emitted by Ar^{17+} ions on metal surfaces in a very large range of energies (from 0.5 V/q up to 470,000 V/q), in various places such as Grenoble, KSU, Livermore and Saturne at Saclay.

In Figure 4 is shown the Ar K lines observed at 20 kV/q at normal incidence with a crystal and a Si(Li) detector.

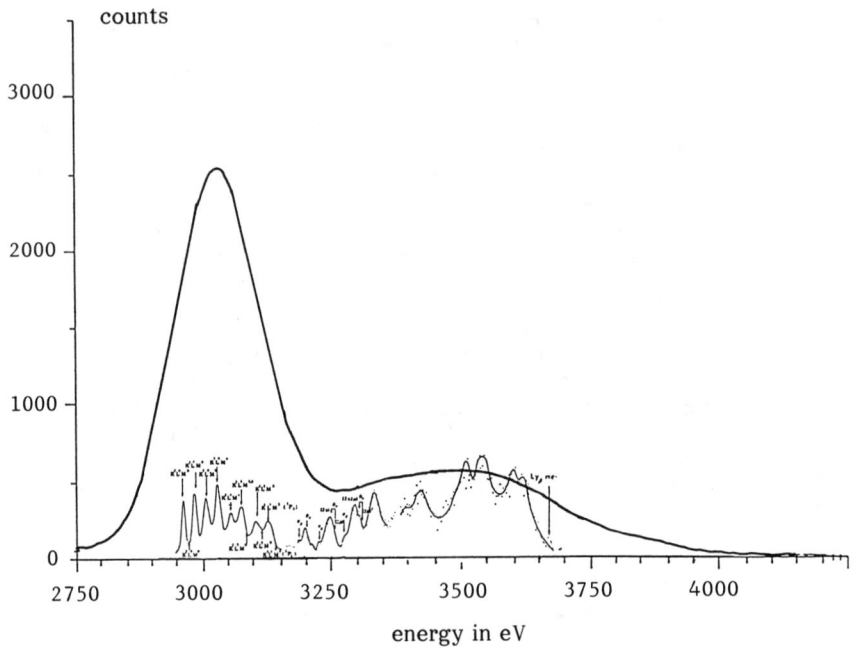

Fig. 4. *Ar K x-ray spectrum observed with a SiLi and a crystal spectrometers during the interaction of Ar^{17+} ions with a metal surface at 20 V/q.*

The comparison of these spectra clearly shows that with a Si(Li) detector one can easily determine, through the measurement of the main energy of the $K\alpha$ line, the mean number of L electrons at the time of the filling of the K hole, and that the KL^X satellite distribution can be deduced from the shape of the $K\beta$ line.

We have studied the high energies at KSU (150→3kV/q) and at Saturne (470 kV/q), and the very low energies at Grenoble from 3 kV/q→0.5 V/q. The x-ray spectra observed in this energy range dramatically change when the velocity decreases.

When the energy decreases from 20 kV/q down to 200 eV/q (i.e., the velocity by a factor of 10), the $K\alpha$ line is slightly shifted to the highest energies, i.e., to states owning less L spectator electrons and we observed roughly the same spectrum that the one observed at grazing incidence at $\theta=6°$,[9,11] i.e., still decreasing the normal velocity by a factor of 10. The mean number of L electrons in these cases is roughly decreased by one unit as shown by measuring the $K\alpha$ line. More interesting are the shape and the intensity of the $K\beta$ line: the shape of the "β" structure does not change with the ion energy while the $K\beta$ total intensity decreases from a certain asymptotic limit. This decrease coming from the variation of the radiative electron capture cross section inside the solid is not in the scope of this talk (cf poster Tu 133) so I shall not comment on this result.

By using the AIM facility in Grenoble we further decreased, by an extra factor of 10, the velocity of the ions, at roughly normal incidence. The design of the experiment is shown in Figure 5.

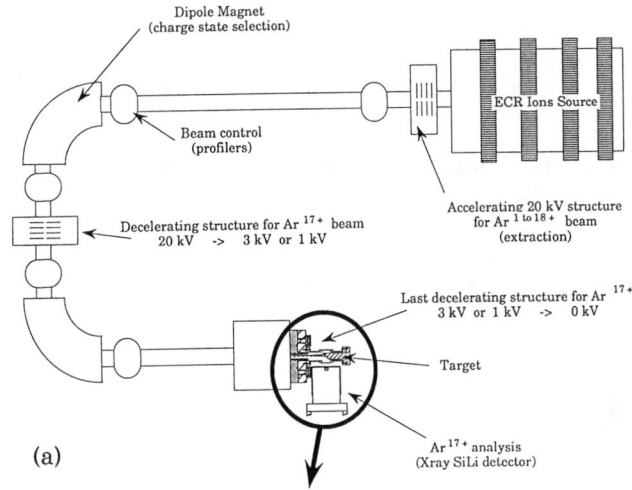

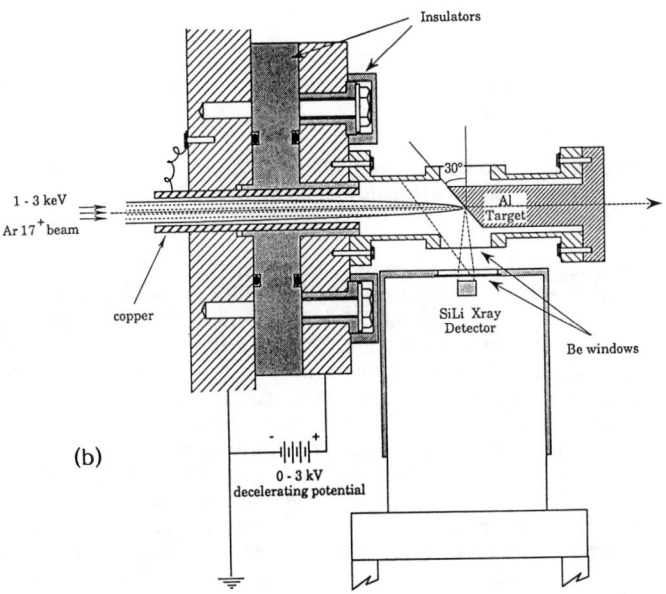

Fig. 5. *(a) Schematics of the deceleration of the Ar^{17+} ions, (b) experimental set-up.*

The Ar^{17+} ions extracted at 20 kV/q from the old LAGRIPPA ECR source were first decelerated down to 3 or 1 kV/q between the two dipoles of the achromatic beam line. While arriving on our target this beam was further decelerated from 3 or 1 kV/q down to 0.5 V/q, or below, by polarizing a purposely designed solid target. It iwas then possible, theoretically, to decelerate the beam down to zero without loosing too many ions, and still having large incidence angles. The spectrum observed when slowing down the ions by this extra factor of ten, presented Figure 6, clearly shows dramatic changes with respect to those obtained at higher velocities (Fig. 4).

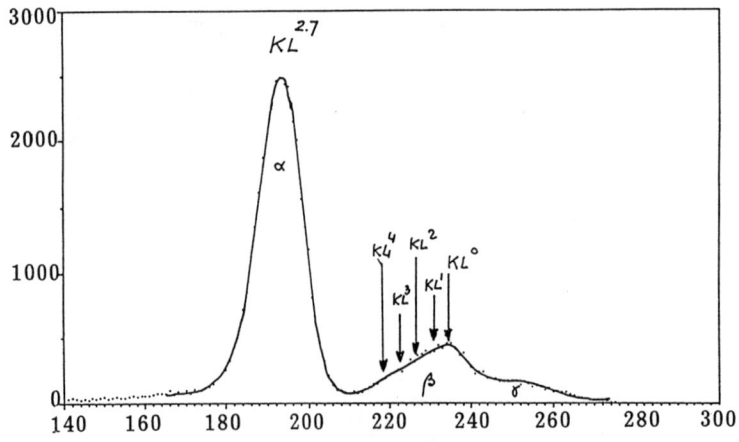

Fig. 6. *Ar K x-ray spectra observed with a SiLi detector during the interaction of Ar^{17+} ions with a surface at 300 eV/q and 1.5 V/q.*

The mean energy of the Kα line increases from 3,046 eV up to 3,090 eV, i.e., from a mean number of L electrons of 4 down to 2.7 while the width of this line decreases, clearly showing a narrowing of the the KL^x distribution. The Kβ line also dramatically changes and a deep valley between the α and β line is drilled while decreasing the energy [as compared to the spectrum of Fig.(4)].At the lowest energies, for instance when the K x-ray intensity reaches zero (the ion cannot approach the surface) one can clearly see that only the KβL^x lines between $0 \geq x \geq 4$ are observed (there is no room for any KβL^5 line for instance). The mean number of L electrons, as deduced from the mean energy of the Kα line, which corresponds to n_L=2.7, clearly indicates that the Kα satellites are strongly peaked on KL^3 because there is no KL^5 line and only few KL^4 ones as deduced from the study of the Kβ's. One can then assess that while the KαL satellite distribution is peaked on the last KL^3 line, the intensity of the KL^2 and KL^1 lines decreases with x. This behavior is obviously the opposite to that of the KβL^x lines which decrease in intensity when x increases. The intensity of the β line being roughly proportional to the number of M electrons, this clearly indicates that when there is only one L electron we have a certain number (about 3 from a more detailed analysis) of M electrons, while there is no, or extremely few, M electrons for the KL^3 satellite. These results are completely different from those obtained in all the other experiments where one observes inside the solid a quasi closed M shell whatever is the number of L electrons, or always a larger number of M electrons than that of the L ones above the surface (triangular distribution). These findings are also consistent with the experiments where one observes inside the solid a quasi closed M shell whatever is the number of L electrons, or always a larger number of M electrons than that of the L ones above the surface (triangular distribution). These findings are also consistent with the intensity of the overall β line intensity, which clearly shows that the mean number of M electrons is much smaller than in any of the other cases we studied.

The main characteristic of the hollow atoms produced at these very low velocities could be explained, in this case, by considering some of the orders of magnitude of the capture and decay lengths as compared to an atomic size. One must first determine what is the minimum asymptotic normal velocity allowed when decelerating the ions, taking into account the acceleration of an ion by its own image.[12] The energy gain due to this image has been discussed by various authors (see H. Winter's contribution at this conference) and would range for Ar^{17+} ions between 60 to 120 eV - total energy, i.e., ~ 4 to 7 V/q. The minimum asymptotic velocity is then of the order of 2×10^4 m.s^{-1}, which means that an atomic size is crossed by an ion in about 10^{-14} s, a time much longer than the capture and decay characteristic times.

As previously described the Ar M shell is usually quickly filled up in the bulk in about 2 or 3 atomic rows. In our case, the penetration depth of the ion might not exceed one atomic monolayer (it is about 1,500 Å at an energy 20,000 larger...) and one cannot expect, using the same models as those considered at 20 kV/q, to get more than 3 or 4 M electrons at surface, in agreement to what is observed (the KL^0 or KL^1 line ownes 3 to 4 M electrons). The first interesting number to be considered in this low energy range of ion-surface interaction is the distance from the surface at which the M shell of the ion becomes populated through the resonant neutralization model. The ion interacting simultaneously with few atoms of the surface, it is difficult to calculate precisely this minimum distance of capture. In ion-atom collisions, taking account of the screening effect, a reasonable estimate of this length could be the barrier radius, which in our case is of the order of 0.5 Å. This number has to be compared to the mean range of a hollow atom owning 4 M electrons decaying via a LMM Auger transition, which is at the considered velocities of the order of 0.3 Å. The ion then decays "in situ" or has enough time, while crossing an atomic diameter, to experience few LMM Auger transitions. The ion having rapidly captured few M electrons, at 0.5 Å from the first atoms of the surface, may then loose two M electrons two or three times along an atomic diameter, which may explain why the last KL^3 line does not owne any (or extremely few) M electron. The very slow velocity of the ion thus explains this original behavior of the ion as compared to what is observed in all the other cases, namely at 20 kV/q.

At higher energies, e.g. 20 kV/q, the mean free path for a LMM Auger decay is of the order of 4 atomic monolayers inside the solid. The ion which looses two electrons in the M shell along such a range may then be quickly refed in the M shell through 4 atomic monolayers, as observed (3 M electrons are captured in one monolayer; the M shell is then in most cases closed up). At these low velocities the ion, which is fed in the M shell while crossing one monolayer, may also decay via 2 or 3 LMM Auger cascades in one monolayer, and cannot be refed.

We then have to consider at very low velocities another kind of hollow atom population. At relatively high velocities the hollow atoms, which are formed below the surface, have few, or no, K and L electrons while the M (N) shell is quasi filled up. At lower velocities or at grazing incidence hollow atoms may be filled in high n shells and decay to the ground state, leading, at the latest stage of the cascade, to the so-called triangular population. As previously quoted only very few of these atoms may emit their characteristic Kα lines prior to touching the surface where they are peeled off. At the lowest authorized velocities the ions do not penetrate the solid by more than an atomic monolayer. While the outermost shells are peeled off when touching the surface some capture processes in the innershells may independently occur forming another kind of hollow atom electronic distribution **at surface**.

REFERENCES

1. J.P. Briand, Invited talk at the X-Ray and Innershell Processes Conference (X'93), Debrecen, July 1993, to be published.
2. J.P. Briand, L. de Billy, P. Charles, S. Essabaa, P. Briand, R. Geller, J.P. Desclaux, C. Ristori, Phys. Rev. Lett. **65**, 159 (1990).
3. L. Folkerts and R. Morgenstern, Europhys. Lett. **13**, 377 (1990).
4. F.W. Meyer, C.C. Havener, K.J. Snowdon, S.H. Overbury, D.M. Zehner, Phys. Rev. A**35**, 3176 (1987), and F.W. Meyer, Invited talk at this conference.
5. J. Das, Invited talk at this conference.
6. R. Köhrbrück, K. Sommer, J.P. Biersaak, J. Bleck-Neuhaus, S. Schippers, P. Roncin, D. Lecler, F. Fremont, N. Stolterfoht, Phys. Rev. A**45**, 4653 (1992).
7. J.P. Briand, L. de Billy, P. Charles, S. Essabaa, P. Briand, R. Geller, J.P. Desclaux, S. Bliman, C. Ristori, Phys. Rev. A**43**, 565 (1991).
8. F.P. Larkins, J. Phys. B**9**, L29 (1971).
9. B. d'Etat, J.P. Briand, G. Ban, L. de Billy, J.P. Desclaux, P. Briand, to appear in Phys. Rev. A**48** (1993).
10. J. Burgdörfer, P. Lerner, F. Meyer, Phys. Rev. A**44**, 5674 (1991).
11. M. Schulz, C.L. Cocke, S. Hagmann, M. Stöckli, H. Schmidt-Böcking, Phys. Rev. A**44**, 1653 (1991).
12. H. Winter, Europhys. Lett. **18**, 207 (1992), and Invited talk at this conference.

ELECTRON SPECTROSCOPIC STUDIES OF THE NEUTRALIZATION OF SLOW MULTICHARGED IONS DURING INTERACTIONS WITH A METAL SURFACE

F.W. Meyer, L. Folkerts[*], C.C. Havener, I.G. Hughes[*], S.H. Overbury, D.M. Zehner, and P.A. Zeijlmans van Emmichoven[†]
Oak Ridge National Laboratory, Oak Ridge, Tennessee USA 37831-6362

ABSTRACT

Recent experimental results of electron spectroscopic studies of the neutralization and relaxation of slow multicharged ions interacting with metal surfaces obtained at Oak Ridge National Laboratory are summarized. Discussed are measurements of projectile K-Auger electron emission during interactions of N^{6+} ions with clean and cesiated Au, as well as clean Cu single crystal targets. While the dominant component in the measured K-Auger spectra is shown to be due to sub-surface electron emission, at incident perpendicular velocities below about 10^{-2} au, small features become evident in the measured K-Auger spectra that are consistent with above-surface projectile electron emission. The experimental results are compared with results of modelling studies of above- and sub-surface electron emission. Also presented are measurements of low energy electron emission during multicharged ion-surface interactions, together with a discussion and analysis of the various contributing low energy emission mechanisms.

INTRODUCTION

The neutralization and relaxation of slow multicharged ions during interactions with surfaces has been the focus of numerous recent studies. A variety of experimental approaches have been used, ranging from total electron yield[1] and electron spectroscopic[2-6] measurements, to analysis of scattered ions,[7] and low and high resolution x-ray spectroscopy[8-10]. Although still evolving, a consistent picture is beginning to emerge of the overall scenario by which the neutralization and relaxation is thought to occur. It is by now generally accepted that neutralization of multicharged ions can already start well above the surface by classical overbarrier transitions of target valence band electrons to high lying Rydberg levels of the approaching projectile. The critical distance above the surface at which such transitions first become possible is given (in atomic units) by[11]

$$R_c(W) \approx \sqrt{8q + 2}/2W \quad (1)$$

where W is the surface workfunction of the target material and q is the initial charge state of the incident projectile. The principal quantum number, n_c, into which the valence electrons are initially captured scales[11] as $W^{1/2}$. For N^{6+} ions incident on clean Au, R_c and n_c are approximately 20 a.u. and 7, respectively. Due to the long time scales required for at least two electrons to cascade down to the projectile L-shell, contributions of above-surface KLL-Auger transitions to the total observed projectile K-Auger electron emission are seen only at very low incident perpendicular velocities.

© 1993 American Institute of Physics

Projectile image-force-acceleration imposes a limit[1,5,11] on the lowest perpendicular velocity achievable, with the consequence that the observed above-surface K-Auger yield is limited to less than about 20% of the total K-Auger emission[15]. After penetration of the surface, screening effects facilitate direct capture into lower projectile n-levels (for N^{6+} projectiles, n=2-3), leading to a very fast filling of the surviving K-shell vacancies. The total K-Auger electron emission thus arises predominantly as a result of sub-surface processes.

In this progress report, recent experimental results of studies on low energy multicharged ion surface interactions obtained at Oak Ridge National Laboratory are summarized. Comparisons of the measurement results and results of modelling studies of above- and sub-surface electron emission are presented and discussed within the framework of the above-outlined scenario.

EXPERIMENTAL APPROACH

Most of the details of the experimental apparatus and procedure have been previously described[4]. Briefly, an ion beam of selected energy extracted from the ORNL ECR ion source is magnetically analyzed and then collimated by passage through two small apertures to give a 1-mm-diam spot size on a clean single-crystal metal target at normal incidence. Prior to multicharged-ion-induced electron-emission measurements the metal targets were sputter cleaned using 1 keV Ar^+ ions; surface cleanliness was verified using electron-induced Auger spectroscopy. The angle of incidence of the ion beam, which determines the perpendicular approach velocity, could be varied in the entire range between normal and extreme grazing conditions. The perpendicular ion approach velocity could, of course, also be varied by changing the total ion kinetic energy at fixed incidence angle. However, in this case, it is more difficult to compare spectral features at different perpendicular velocities, since strongly energy-dependent collisional broadening effects and different sub-surface neutralization dynamics alter the sub-surface projectile K-Auger peak shape.

Electron spectra were measured at selected angles relative to the beam direction using a small spherical sector energy analyzer with a nominal energy resolution of 2.8%. As part of the normalization procedure used to place the spectra on an absolute scale, a

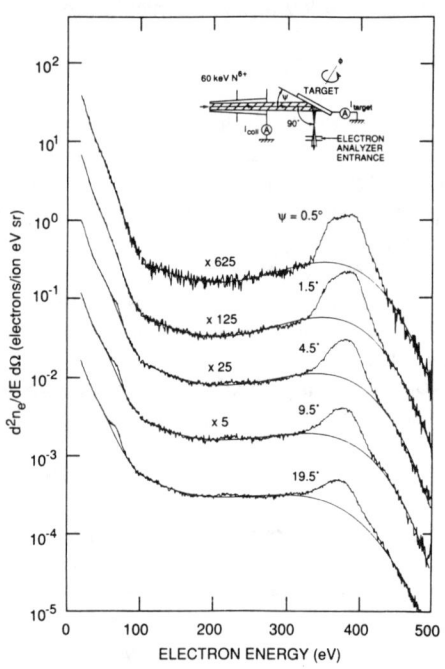

Fig. 1. Electron spectra for 60 keV N^{6+} incident on Au(001) for various incidence angles.

geometric correction factor was applied for angles of incidence less than about 15°, when the target region illuminated by the incident beam was larger than the experimentally determined 3.3-mm-diam viewing region of the electron spectrometer. For incidence angles below 5° the effective beam divergence was less than ±0.2°. Figure 1 shows typical electron spectra acquired for 60 keV N^{6+} ions incident on clean Au(001) for incidence angles in the range 0.5 - 19.5°; the inset of the figure gives details of the experimental configuration used.

For reliable quantitative measurements at extreme grazing angles, an accurate calibration[12] of the incidence angle of the ion beam on the metal target is required. This was accomplished through an auxiliary measurement in which the K-Auger electron signal observed perpendicular to the incident beam direction was measured as a function of target manipulator rotation angle. For incidence angles sufficiently small (e.g. for Cu, <3.4°) that the total <u>normalized</u> K-Auger yield has reached its maximum value[4] (corresponding to one K-Auger decay per incident ion), the decrease in <u>raw</u> K-Auger signal with further decreasing incidence angle is solely due to the decreasing fraction of the incident beam illuminating the target region viewed by the spectrometer, which vanishes at zero incidence angle. As shown in Ref. 12, extrapolation of the raw K-Auger signal to zero provides an accurate zero incidence angle calibration, while the width of the transition region around "zero" provides a measure of the divergence of the incident beam. It is noted that this technique can also provide a guide for reproducing a given grazing incidence condition subsequent to target sputter cleaning, which requires target rotation away from the orientation used for the grazing incidence measurements.

PROJECTILE K-AUGER EMISSION: CLEAN CU(001) RESULTS

Figure 2a shows normalized K-Auger electron spectra measured for incidence angles in the range 0.2 - 2.4°, after background subtraction using a procedure out-

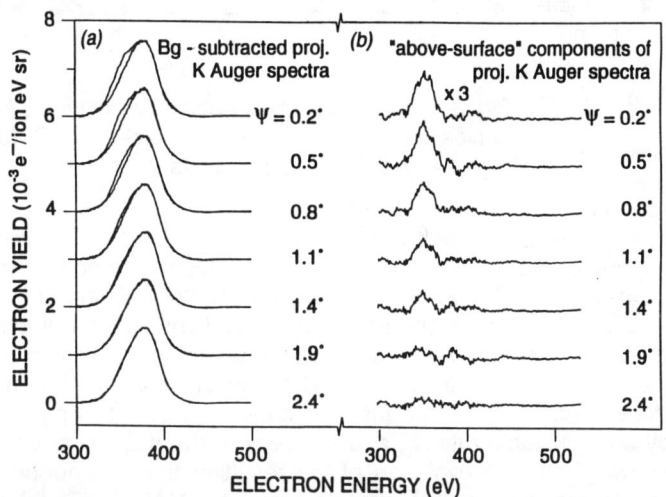

Fig. 2. Background subtracted projectile K-Auger spectra at extreme grazing incidence angles for 60 keV N^{6+} ions on Cu(011), with superimposed scaled 7.4° reference spectra (a), and (b) after subtraction of same [see text].

lined in detail in Ref. 4. In this range of incidence angles, a noticeable evolution of the K-Auger peak shape occurs, as a progressively more intense component peaked at about 350 eV is seen to appear in addition to the main sub-surface[4] K-Auger lineshape peaked at about 380 eV. Prior to the appearance of this additional component, i.e. at incidence angles larger than 2.4°, the shapes of the background-subtracted spectra were found to remain remarkably constant, the only variation being in the overall intensity[4]. The peak energy of the additional component is consistent with a minimum L-shell population at the time of the K-Auger decay, as is expected from the slow above-surface autoionization cascade. This "above-surface" component is shown in Fig. 2b as a difference between a particular grazing angle spectrum and a representative large incidence angle ($\Psi=7.4°$) spectrum that has been superimposed on it in Fig. 2a, after scaling to have the same intensity at 380 eV (where only sub-surface emission contributes) as the grazing angle spectrum.

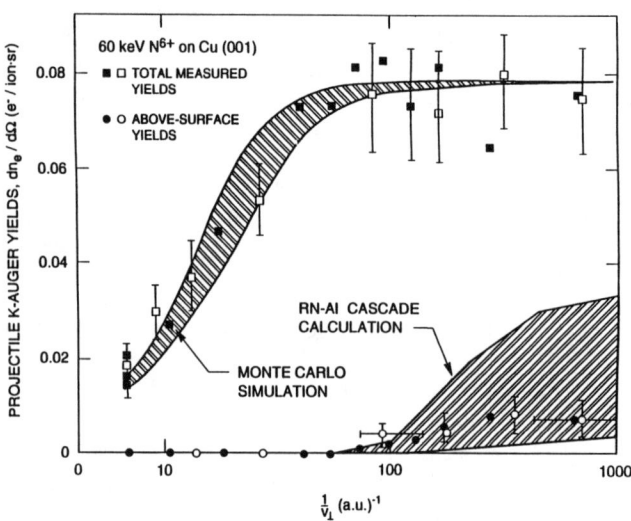

The inverse perpendicular velocity dependence of the intensities of both K-Auger components is summarized in Fig. 3, where the integrated K-Auger peak areas are shown for incidence angles up to 20° (corresponding to the minimum investigated inverse perpendicular velocity of 7 a.u.$^{-1}$). The open and solid symbols refer to measurements first published in Refs. 4 and 12, respectively. Similar results have also been obtained[4] for a Au(011) single crystal target. The identification of the

Fig. 3. Above- and sub-surface K-Auger yields (squares and circles, resp.) for 60 keV N^{6+} incident on Cu(011) vs. $v_\perp^{-1}$, together with simulation results for same.

main component of the projectile K-Auger peak as sub-surface emission is supported by the observation of significant target azimuth dependence found for the intensity of this component, as illustrated in Fig. 4. The experimentally observed target azimuth dependence as well as the $v_\perp^{-1}$ dependence is in excellent accord with a Monte Carlo simulation of the sub-surface K-Auger emission, also shown in Figs. 3 and 4. The simulation, described in detail in Refs. 4 and 13, assumes a two step mechanism in the sub-surface K-Auger emission, the first consisting of the filling of the L-shell, described by an adjustable rate R_L, and the second the KLL transition itself whose rate R_K is known. The time evolution of the resulting three component system is solved along actual ion trajectories calculated using the MARLOWE code[14]. Transport of the emitted Auger electron to the target surface is treated using tabulated electron inelastic mean free paths. The band delimiting the sub-surface simulation results in

Fig. 3 represents the variation in simulated K-Auger yields obtained within the 5° uncertainty range about the nominal [110] mechanical azimuth setting.

Also shown in Fig. 3 are the above-surface K-Auger peak areas. As can be seen, there is a clear saturation[4,12] of the above-surface K-Auger electron emission at small perpendicular velocities, as also observed by Das and Morgenstern[5], which is ascribed to projectile image-potential acceleration. In addition, the results show a threshold of above-surface K-Auger electron emission in the inverse perpendicular velocity interval 57 - 73 a.u.$^{-1}$ (incidence angle range 2.4° > Ψ > 1.9°), in excellent agreement with classical over-the-barrier simulation[11] results for the above-surface resonant-neutralization/autoionization (RN-AI) cascade, also shown in Fig. 3. This calculation features a self-consistent and dynamical treatment of the resonance neutralization process and the projectile effective charge, that allows for multiple electron capture as well as loss. The calculation explicitly includes the effect of the dynamical image potential both on the projectile energy levels as well as on the projectile trajectories. Image-potential-acceleration of the projectile is thus accounted for. The indicated band of results shown in Fig. 3 represents the variation of results obtained under different assumptions for the forms of the image potentials and screening functions, as well as different choices of the Auger rates (see Ref. 11 for additional details).

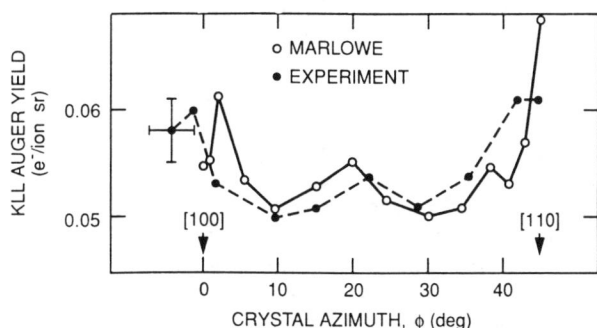

Fig. 4. Comparison of simulated and measured Cu target azimuth dependence of K-Auger electron emission for 60 keV N^{6+} incident on Cu(001) at 5.°

PROJECTILE K-AUGER EMISSION: CESIATED AU(110) SURFACE RESULTS

To determine the effect of surface-workfunction-reduction on the above-surface neutralization dynamics, measurements of projectile K-Auger electron emission were performed using a cesiated Au surface[15]. To facilitate these measurements, a Cs dispenser with shutter was mounted in our experimental chamber approximately 6 cm from the single crystal Au target. By resistive heating of the dispenser above a threshold temperature monitored by a chromel-alumel thermocouple, a reproducible flux of Cs could be obtained to which the Au crystal could be exposed. Using a retarding field method with normally incident 200 eV electrons, the surface work function was found to decrease in proportion to Cs dosing time up to about 0.5 Cs monolayer coverage, corresponding to a work function decrease of approximately 3 eV.

Using the above procedure, projectile K-Auger spectra were measured at incidence angles in the range 0.3-5° for clean and cesiated Au. The results are illustrated in Fig. 5. For the 1.4° and 0.7° incidence measurements (top two panels of Fig. 5), the surface was first prepared by Cs dosing and the change in work function (relative to that for clean Au) measured as described above. *In situ* thermal desorption was then employed to remove the Cs overlayer, after which another electron spectrum was acquired (right half of each panel). Subsequent to this measurement the loss of Cs was confirmed by a second work function determination. The only difference in experimental conditions between the two spectra was thus the amount of Cs coverage. While significantly reducing the Cs coverage in each case (see measured work function changes noted in each panel), the thermal desorption cycles were unsuccessful in completely removing the Cs overlayer. For this reason, the spectra for the smallest incidence angle were measured in reverse order, starting with a freshly sputter cleaned sample.

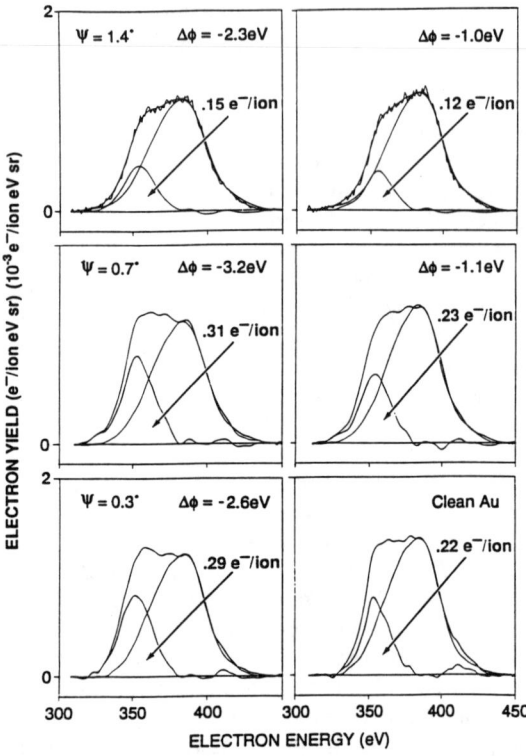

As has been previously verified both for Au and Cu targets[4,12], at an incidence angle of 5° the K-Auger peak arises solely from sub-surface neutralization and relaxation processes.

Fig. 5. Background subtracted projectile K-Auger peaks for 60 keV N^{6+} grazingly incident on cesiated and clean Au(110), together with 5° reference spectra used to estimate the sub-surface components.

In the present measurements, the shapes of the 5° K-Auger peaks, also shown in Fig. 5, were found to be independent of Cs coverage in the 0 - 0.7 monolayer range investigated, and were used as estimates of the sub-surface components in the more grazing incidence angle spectra in a manner already described in the previous section. As can be seen from the spectra shown on the right in Fig. 5, pronounced above-surface components are evident in all the grazing incidence spectra even in the case of minimal Cs coverage. Furthermore, significant enhancements of these above-surface K-Auger components were observed for the reduced work function surfaces produced by Cs deposition (spectra on left in Fig. 5).

From Eq.(1) it can be seen that, for the work function reductions indicated in Fig. 5, more than a factor of two increase in the above-surface critical distance should have been realized. With neutralization starting at larger distances above the surface, the image potential acceleration of the projectile will likely be significantly reduced, which should increase the time available for the above-surface cascade and subsequent K-Auger emission, and therefore increase the observed above-surface K-Auger yield. On the other hand, the increase in the critical distance for first electron capture by the projectile is accompanied by an increase in the principal quantum number initially populated, due to the reduced binding energy of the neutralizing target (i.e., Cs) electrons, and the reduced upward projectile energy level shifts due to the image potential interaction at the larger above-surface distances. It is therefore to be expected that the subsequent deexcitation cascade populating the projectile inner shells will require more time. In addition, certainly for a thick Cs layer, the surface electron density of states at the Fermi level may be significantly reduced relative to the clean Au case, thereby reducing the number of metal electrons available for projectile neutralization. The latter two effects would tend to decrease the above-surface K-Auger electron emission for a particular interaction time. In light of these competing factors, it is interesting to note that an overall enhancement of the above-surface K-Auger yield, approaching 35% at some of the incidence angles, was observed.

The experimentally determined above-surface K-Auger electron emission is shown in Fig. 6, together with simulation results from the previously described over-the-barrier above-surface neutralization code, modified[15] to permit treatment of a cesiated surface. The figure shows the inverse perpendicular velocity dependence of the simulated K-Auger yields for a number of different surface workfunctions in the range 0.1-0.187 a.u. The other simulation parameters were kept fixed to reproduce approximately the "upper bound" simulation results for clean Au (see Ref. 4). The dominant effect of the work function reduction on the simu-

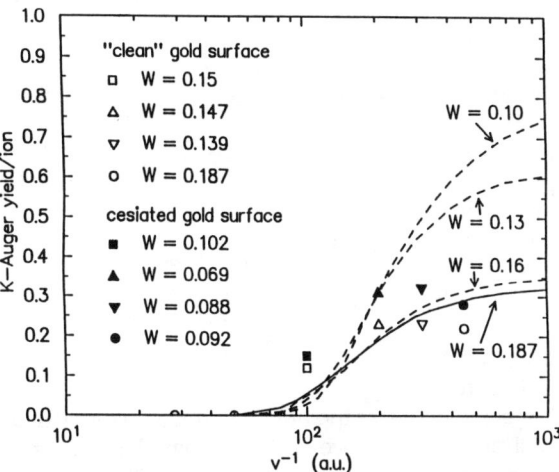

Fig. 6. $v_\perp^{-1}$ dependence of the simulated above-surface K-Auger yields for N^{6+} incident on clean and cesiated Au, for range of surface work function values compared with experimental results.

lated above-surface yields is found[15] to be the suppression of the image acceleration for the ion in front of the cesiated surface at large distances due to the screening by the charge cloud of electrons already captured in outer shells. The increased interaction time available for the Auger cascade resulting from the increased critical distance is found to be largely offset by slower Auger rates in the higher n-shells. While reasonable agreement is found between the simulation results and the experimentally deduced above-surface K-Auger yields at perpendicular velocities above 3×10^{-2} a.u., at lower velocities the simulation appears to significantly overestimate the observed above-surface K-Auger yield enhancement. The reasons for this discrepancy are at present not completely understood.

LOW ENERGY ELECTRON EMISSION

The results shown in the preceding sections have suggested that, even at the lowest incident perpendicular velocities, only a small fraction of incident ions carrying deep lying inner shell vacancies will be completely neutralized and relaxed prior to impact on the target surface, the majority still being in a variety of multiexcited states. Upon surface penetration the sudden screening of the projectile core charge by target valence band electrons will result in any electrons bound to the projectile in shells $n \geq 4$ becoming unbound and "peeled" from the projectile. Those peel-off electrons that escape into the vacuum are expected to have energies less than 50 eV. Low energy electrons are also expected to be produced during the early stages of the above-surface autoionization cascade involving shells $n \geq 3$, as well as during above-surface promotion into the continuum of projectile Rydberg electrons due to their interaction with the projectile image potential or due to screening effects. At our investigated projectile energies, there is the additional large contribution of low energy kinetically emitted electrons which complicates the evaluation of the relative contributions of the above potential emission processes to the total electron yield.

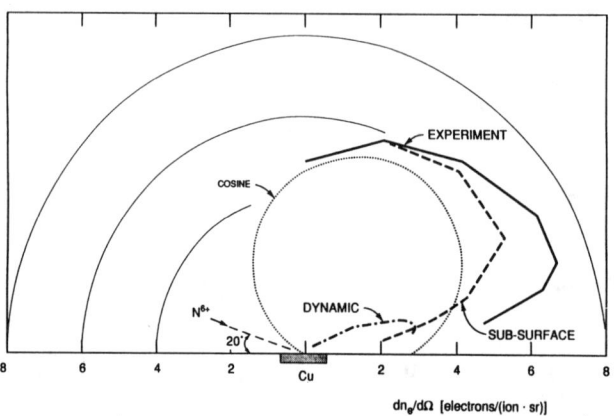

Fig. 7. Polar plot of the angular dependence of the energy-integrated electron emission for 100 keV N^{6+}-Cu(001) collisions, showing "sub-surface" and "dynamic" components.

We have made measurements of energy and angle resolved low energy electron emission[16] as well as total electron yields[17] during multicharged ion interactions with Au and Cu surfaces. As is illustrated in Fig. 7, the low energy electron emission was found to consist of a dominant "sub-surface" component characterized by a $\cos(\theta)$-like angular distribution skewed in the forward direction, and a much smaller "dynamic" component, possibly created during binary encounters with metal electrons at the surface, which is strongly peaked at extreme forward angles.

It was found in addition that both the emission at a particular analyzed electron energy as well as the total electron yields exhibit characteristic variations in magnitude with target azimuthal orientation relative to the incident multicharged ion beam. This is illustrated in Fig. 8 for electron emission at 20 eV for 30 keV N^{2+}, N^{5+}, and N^{6+} ions incident at 20° on Au(011). As can be seen in Fig. 9, we find similar azimuthal variations in the total electron yields. In the velocity range 0.25 - 0.55 a.u., the amplitude of the azimuth variations was observed to increase with increasing velocity.

In the same velocity range, the electron yield differences between different charge states were remarkably target-azimuth, as well as projectile-velocity <u>independent</u>. Calculations of projectile energy loss during traversal of the top 40 Å of the target, made using the MARLOWE code, exhibit the same azimuthal variations, as demonstrated in Fig. 9. We therefore attribute the azimuthally varying component of the electron yield to kinetic electron emission. The additional velocity-independent emission for the higher charge states (e.g. N^{6+} compared with N^{5+}) is ascribed to potential emission processes.

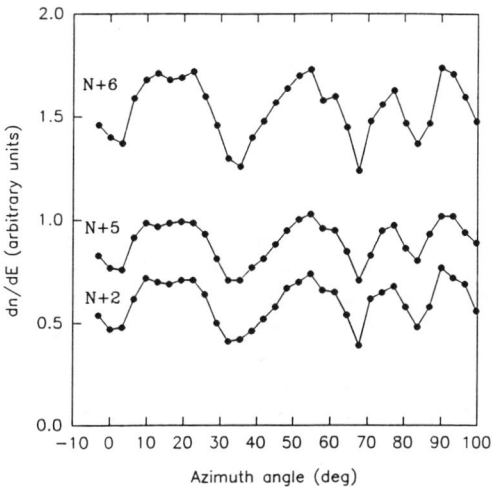

Fig. 8. Target azimuth dependence of 20 eV electron emission for 30 keV N^{q+} incident on Au(011).

Simulation results for the above-surface electron emission processes enumerated at the beginning of this section suggest that, in the investigated velocity regime, the above-surface Auger cascade and promotion to the continuum make only minor contributions to the electron yield. In addition, the maximum possible contribution from peel-off electrons to the yield difference between N^{5+} and N^{6+}, shown in Fig. 9, is on the order of one electron. The observed yield difference of 3.5 e⁻/ion can thus not be accounted for by above-surface processes alone. It is suggested that a major contribution to the

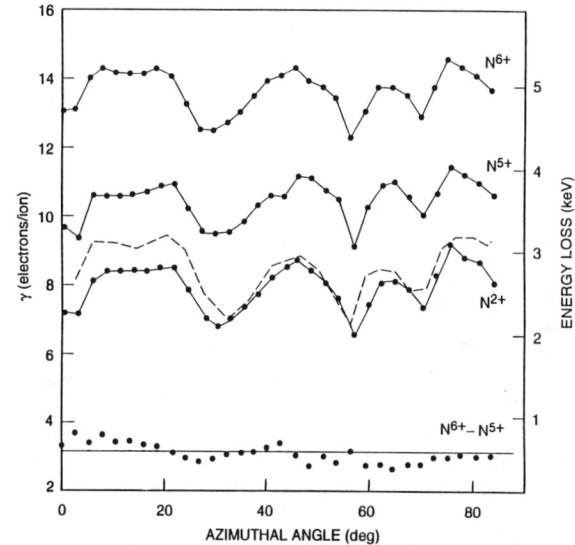

Fig. 9. Target azimuth dependence of γ for 30 keV N^{q+} incident on Au(011) - solid points; dashed line - MARLOWE results for projectile energy loss.

observed yield difference must be due to the presence of the initial K-shell vacancy on the incident N^{6+} ion. The sub-surface filling of this inner shell vacancy in the first 30 Å of the target gives rise to isotropically emitted 380 eV electrons, which can in turn create low energy secondary electrons. Using published data[18] of electron induced secondary electron emission, a rough estimate of 1.5-2e⁻/ion is obtained. "Indirect" potential electron emission[17] due to secondary cascade process within the target bulk is thus believed to be a dominant contributor to potential emission in the investigated energy regime.

ACKNOWLEDGEMENTS

We are indebted to J. Burgdörfer and M.T. Robinson for stimulating discussions and important insights, and to G. Ownby and J. Hale for skilled technical assistance. This work was supported by the Office of Basic Energy Sciences, Division of Chemical Sciences, and the Division of Applied Plasma Physics, Office of Fusion Energy, of the U.S. Department of Energy, under Contract No. DE-AC05-84OR21400 with Martin Marietta Energy Systems, Inc.
*Appointed through the ORAU postdoctoral research program.
†Present address: Buys Ballotlaboratorium Utrecht, University of Utrecht, NL-3584 CC Utrecht, The Netherlands.

REFERENCES

[1] H. Kurz, K. Töglhofer, HP. Winter, F. Aumayr, and R. Mann, Phys. Rev. Lett. **69**, 1140 (1992).
[2] L. Folkerts and R. Morgenstern, Europhys. Lett. **13**, 377 (1990).
[3] H.J. Andrä, A. Simionovici, T. Lamy, A. Brenac, G. Lamboley, J.J. Bonet, A. Fleury, M. Bonnefoy, M. Chassevent, S. Andriamonje, and A. Pesnelle, Z. Phys. D **21**, S15.
[4] F.W. Meyer, S.H. Overbury, C.C. Havener, P.A. Zeijlmans van Emmichoven, J. Burgdörfer, and D.M. Zehner, Phys. Rev. A **44**, 7214 (1991); F.W. Meyer, S.H. Overbury, C.C. Havener, P.A. Zeijlmans van Emmichoven, and D.M. Zehner, Phys. Rev. Lett. **67**, 723 (1991).
[5] J. Das and R. Morgenstern, Phys. Rev. A **47**, R758 (1993).
[6] L. Folkerts, I.G. Hughes, and F.W. Meyer, Abstracts of Contributed Papers of the XVIII ICPEAC, Aarhus, Denmark.
[7] H. Winter, Europhys. Lett. **18**, 207 (1992).
[8] M. Schulz, C. Cocke, S. Hagmann, M. Stöckli, and H. Schmidt-Böcking, Phys. Rev. A **44**, 1653 (1991).
[9] J.J. Bonnet, A. Fleury, M. Bonnefoy, M. Chassevent, T. Lamy, A. Brenac, A. Simionovici, and H.J. Andrä, Z. Phys. D Suppl. **21**, 343 (1991).
[10] B. d'Etat, J.P. Briand, G. Ban, L. de Billy, J.P. Desclaux, and P. Briand, Phys. Rev. A (to be published).
[11] J. Burgdörfer, P. Lerner, and F.W. Meyer, Phys. Rev. A **44**, 5674 (1991); J. Burgdörfer and F.W. Meyer. Phys. Rev. A **47**, R20 (1993).
[12] F.W. Meyer, C.C. Havener, and P.A. Zeijlmans van Emmichoven (submitted to Phys. Rev. A).
[13] S.H. Overbury, F.W. Meyer, and M.T. Robinson, Nucl. Instrum. Methods Phys. Res. B **67**, 126 (1992).
[14] M.T. Robinson, Phys. Rev. B**27**, 5347 (1983); O.S. Oen and M.T. Robinson, Nucl. Instrum. Methods Phys. Res. **132**, 647 (1976).
[15] F.W. Meyer, L. Folkerts, I.G. Hughes, S.H. Overbury, and J. Burgdörfer (submitted to Phys. Rev. A).
[16] P.A. Zeijlmans van Emmichoven, C.C. Havener, I.G. Hughes, D.M. Zehner, and F.W. Meyer, Phys. Rev. A**47**, 3998(1993).
[17] I.G. Hughes, J. Burgdörfer, L. Folkerts, C.C. Havener, S.H. Overbury, M.T. Robinson, D.M. Zehner, P.A. Zeijlmans van Emmichoven, and F.W. Meyer, Phys. Rev. Lett **71**, 291 (1993).
[18] S. Thomas and E.B. Pattinson, J. Phys. D **3**, 349 (1969).

ION-SURFACE COLLISIONS AT GRAZING INCIDENCE : ATOMIC COLLISIONS UNDER SPECIAL CONDITIONS

H. Winter
Institut für Kernphysik der Westfälischen Wilhelms-Universität Münster
Wilhelm-Klemm-Str.9, D-48149 Münster, Germany

ABSTRACT

A brief review on the scattering of fast atoms and ions from metal surfaces under a grazing angle of incidence is presented. We discuss similarities and specific differences between fast atomic collisions with atoms in the gas-phase and with a metal surface. Charge exchange phenomena will be main subject of this presentation.

INTRODUCTION

The interaction of atoms with condensed matter is a field that has been paid increasing attention in the series of ICPEAC-meetings. This holds in particular for processes which are localized at the surface of a solid. We will discuss here the interaction of fast atoms with a surface in a scattering geometry, where the projectiles are scattered from the target surface under a grazing angle of incidence. This type of collision provides a number of attractive features that allows for detailed studies of the interactions of atoms with surfaces. This holds especially for processes which take place *in front* of the surface plane, i.e. above the topmost layer of surface atoms.

In this brief review we will outline basic features of grazing collisions of fast atoms/ions with the surface of a metal and will point out relations to collisions with atoms in the gas phase. Thereafter some representative experiments and their results are presented. We will focus our discussions on electron transfer phenomena between projectile-atom and surface and show that studies of scattering in the grazing incidence geometry lead to detailed informations on the interaction mechanisms. Many items have to be outlined here in a rather concise way, and we refer with respect to detailed discussions to recent reviews that cover aside from a general treatment of atom-surface collisions also the scattering under grazing incidence[1-3].

TRAJECTORIES

In Fig.1 we sketch in a simple classical picture an atomic collision, where a fast projectile atom or ion collides with a target under an impact parameter b. Due to the repulsive interatomic potential the projectile is deflected by an angle Θ. The trajectory shown is a special case for the trajectories of an ensemble of projectiles as indicated in a more complete version of the sketch. In general, this variety in the motion of the projectiles contributes to the outcome of a scattering experiment; by corresponding experimental procedures, however, specific trajectories can be isolated in atomic collisions. Inspection of Fig.1 reveals an interesting feature: the formation

© 1993 American Institute of Physics

of a "shadow cone", i.e. a region behind the target atom that is classically forbidden for the projectiles during the collision.

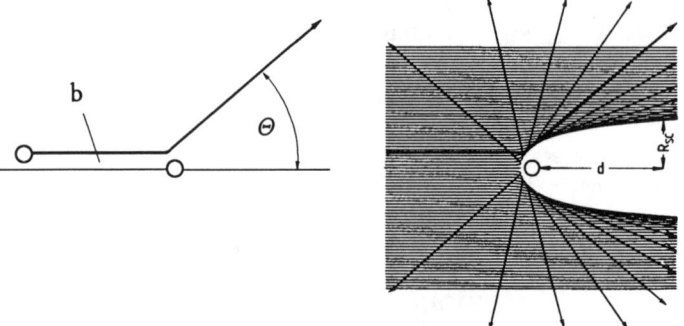

Figure 1.

Trajectories in an atomic collision forming a shadow cone structure. On the left side a selected trajectory is shown.

This cone structure of atomic dimensions cannot be observed in single collision experiments. However, for collisions with arrangements of atoms on a microscopic scale the concept of shadow cones provides informations on the general properties of the trajectories of the scattered projectiles. This is illustrated for the scattering of 1 keV Ar-projectiles with a string of Al-atoms separated by a distance d = 6 a.u., a typical interatomic spacing in a solid; the string of atoms may represent here also a section of the topmost layer of atoms at a surface. In Fig.2 we display for a constant spatial flux of projectiles some representative trajectories for decreasing angles between incident beam and atomic string (surface plane). The shadow cone structures are evident, and the overlay of cones shown in the right part of Fig.2 prevents the projectiles from penetrating the atomic string. Then a pronounced forward scattering dominated by binary type of collisions is apparent.

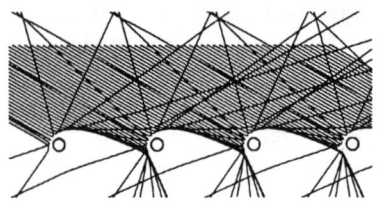
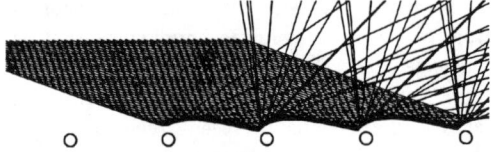

Figure 2. Trajectories of 1 keV Ar-projeciles scattered from a string of Al-atoms

For a further reduction of the angle the overlay of shadow cones affects a successive number of target atoms, and the scattering proceeds by a steering of the projectiles in terms of a major number of small angle deflection events. In the limit of scattering at *grazing incidence* the trajectories can be considered to result from the reflection due to constant repulsive potential planes parallel to the surface. This regime of "channeling" is sketched in Fig.3.

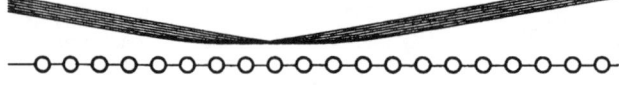

Figure 3.
Trajectories for a grazing scattering of 1 keV Ar from a string of Al-atoms.

Then the motions of projectiles normal and parallel with respect to the surface plane can be considered to be independent. An important aspect are the very different regimes for these motions: For a grazing angle Φ the parallel motion proceeds practically with the projectile energy E; i.e. $E_{\parallel} = E \cos^2\Phi \approx E$, whereas the normal motion scales according to $E_y = E \sin^2\Phi \approx E \Phi^2$. For typical angles $\Phi \approx 1°$ the normal motion for keV-beams proceeds in the eV-regime so that the projectiles cannot overcome the potential formed by surface atoms and will not penetrate into the bulk (see for illustration Fig.3). Furthermore, the normal motion determines the time scale for the interaction with the surface, where the small energies allow for the application of quasi-adiabatic concepts for the description of charge exchange phenomena. This statement, however, holds only, if the (fast) parallel motion can be treated independently to a wide extent. Then a simple incorporation of parallel velocity effects into established concepts concerning electron transfer is possible.

In summary, the trajectories in grazing collisions of atoms with surfaces show clearly different features from collisions with atoms in the gas phase. Grazing surface collisions proceed under "channeling" conditions, where a "fast" *and* a "slow" component for the impact with the target can be clearly separated. All projectiles are scattered from the surface with practically identical trajectories under a common "impact parameter", a distance of closest approach to the surface plane ($\sim 1 - 2$ a.u.). Such a built-in selection of impact parameters and trajectories simplifies the interpretation of scattering experiments. Similar features can be obtained for gas phase collisions only by application of selection procedures, e.g. coincidence techniques.

CHARGE EXCHANGE

The surface of a metal defines a plane of a constant electrostatic potential. When an electric charge is brought into the vicinity of a conductor, the effect of the dielectric response of the metal can asymptotically be approximated by the concept of an image charge. This image charge interaction plays an important role in modeling charge exchange processes in atom-surface scattering and is one of the profound differences to collisions with free atoms, where those collective effects are absent. In Fig.4 we present a simple sketch of the effective charges giving rise to Coulomb-interactions in atom-atom as well as in atom-surface collisions.

Figure 4. Coulomb interactions in atom-atom and atom-surface collisions

Whereas in atom-atom (ion-atom) collisions we have interactions between the ion-cores and electrons of two independent atomic systems, the interactions of an atom in front of a surface can be represented by the presence of a "mirror-atom" inside the bulk of the solid. This subsurface system with opposite charges follows the projectile atom with respect to any motion; i.e. also parallel to the surface (retardation effects can be neglected here). Of specific relevance is the image charge interaction of the active atomic electron with its counterpart, since this interaction - described outside the solid by the asymptotic limit of the classical concept of image charges - merges close to the metal into the electronic surface potential. The electronic potentials for "active" electrons in planes containing the atoms are sketched in Fig.5.

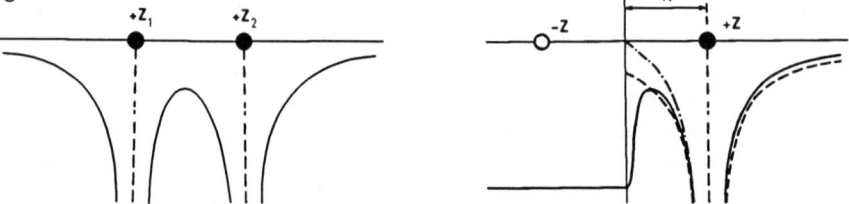

Figure 5. Interaction potentials in atom-atom (left) and atom-surface collisions

When the collision partners come closer, the potential barrier is reduced, and the probabilities for electron tunneling phenomena increase. In a simple classical picture the one-electron transfer can be described by an "over-barrier model", where the pathway for an electron of a given electronic energy is opened up, when the potential barrier is lowered to this energy. This concept developed for the charge exchange between atoms has been applied also in recent years to atom-surface interactions, in particular for charge exchange into Rydberg levels of multicharged ions in front of a surface[4]. With the over-barrier concept we derive for the atom-surface potential - given by the Coulomb potentials due to charge (dashed line in Fig.5) and image charge (dash-dotted line) of the ion core and the electronic surface potential - a rule of thumb for a critical distance $R_c = (2Z)^{1/2}/E_a$ as an upper bound for the resonant transfer of electrons with binding energy E_a (Z = charge of ion core; atomic units)[3].

The Coulomb interactions with the target atom result in a modification of the binding energies of the active electron. In slow atomic collisions (relative motion of projectile and target atom slower than the electronic motion) this interaction can be often viewed as the formation of a quasi-molecule. As an example we sketch in Fig.6 the molecular-orbital (MO) picture for the simple case of a H^+-H collision the variation of the energies for different electronic states with the internuclear distance. The shifts from the energies of the unperturbed atom are essential for the description of excitation or electron transfer phenomena.

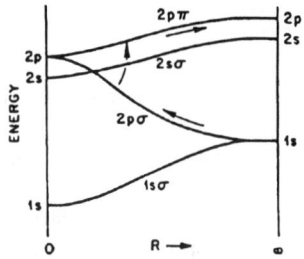 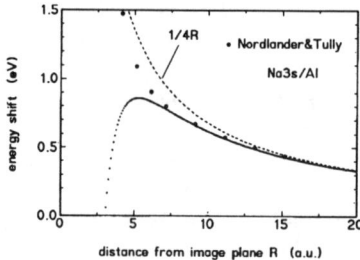

Figure 6. H^+-H-correlation diagram   Figure 7. Level shift of NaI 3s at an Al-surf.

For the atom-surface interaction the energy shift of atomic levels can be considered to stem from the Coulomb-interaction with the "mirror-atom" as sketched in Fig.4. From the Hamiltonian $H = H_A + Z/D - 1/4d$ (H_A denotes the Hamiltonian of the unperturbed atom) we find in a first order perturbational treatment for the energy shift ΔE of the level of an atom at a distance R from the surface (more precise: distance from the electronic reference plane)[5]

$$\Delta E = (2Z-1)/4R - \langle r^2 \rangle / 12R^3 + O(1/R^5) \qquad (1)$$

This level shift can play an important role for the electron transfer processes at surfaces, since we have to consider here effective binding energies which differ from the energies of unperturbed atoms. The shift ΔE is dominated at distances relevant for charge exchange by the 1/4R term in eq.1, whereas the $1/R^3$ (van der Waals) term leads to small corrections. We will show below that we can obtain detailed informations on the level shift of an atom in front of the surface by making use of specific features of grazing ion-surface scattering. In Fig.7 we show ΔE computed from eq.1 for the NaI 3s groundterm (solid/dotted line). Comparison of the results with non-perturbational treatments[6] (dots) shows good agreement for distances where a perturbational approach is applicable; these distances are relevant for the final formation of the 3s-level of Na atoms in the scattering event.

The features discussed so far with respect to the atom-surface interaction are related to the slow motion of the projectile normal to the surface plane. In grazing incidence geometry we have to consider also the fast parallel motion. This proceeds with the velocity/energy of the projectiles so that we are simultaneously to the adiabatic domain in the regime of fast atomic collisions; i.e. the relative motion between the two collision partners has to be taken into account in the description of the electronic states. The corresponding Galilei-transformation is achieved by introducing "electron-translation factors" (ETF) which represent the effective momenta and kinetic energies of active electrons[7]. Then an electronic state Φ_a transforms in the moving frame according to $\psi_a = G(v) \Phi_a$ where G(v) is the operator for the Galilei-transformation.

When we describe the electronic states of the metal in the framework of the free-electron model ("jellium"), the Galilei transformation due to the parallel motion in grazing scattering is particularly simple to perform, since the components of the jellium-wavefunctions parallel to the surface plane are plane waves. As a consequence electronic metal states Φ_k transform into the frame of the projectile with the parallel velocity $v_\parallel = v$ according to $\psi_k = \Phi_{k-v}$. This relation allows for a simple visualization of the effect of the parallel velocity on the electronic structure of the target. Whereas for a target-atom one generally has the one-electron states forming for a hydrogenic atom density of states $D(E_a) = Z / (2E_a)^{-3/2}$, we have for a free electron metal $D(\varepsilon) = 2^{1/2} V \varepsilon^{1/2} / \pi^2$ and a macroscopic number of electrons. V is the macro-volume of the target so that $D(\varepsilon)$ is extremely large; i.e. we can consider a continuum of electronic metal states (Note that the electronic energies ε for metal electrons are referred to the bottom of the conduction band). The occupation of electronic metal states is described by the "Fermi-Dirac-distribution" so that a continuum of states up to the Fermi energy ε_F is filled completely at temperature T = 0; for energies $\varepsilon > \varepsilon_F$ we have a continuum of empty electronic levels. It is well established that these properties for the occupation of electronic levels can be characterized by a constant phase space in momentum representation (dispersion relation $\varepsilon = k^2/2$) where occupied metal states can be found inside of the "Fermi-sphere" with radius $k_F = (2\varepsilon_F)^{1/2}$; ε_F = Fermi-energy.

Based on the features of the Galilei-transformation outlined above, van Wunnik et al.[8] pointed out that the effect of the parallel velocity component on the population of metal states can be visualized by a shift of the "Fermi-sphere" along parallel momenta by $k_\parallel$ = -v. From this procedure it is straightforward to obtain "Doppler-Fermi-Dirac-distributions" (DFDD)[9] which play an essential role for the understanding of neutralization phenomena in grazing surface scattering. In Fig.8 we display the shift of the "Fermi-sphere" in momentum space and DFDD for $v < v_F$ and $v > v_F$, respectively. Main result of the parallel velocity is the elimination of the clear cut separation of occupied and unoccupied states at the Fermi-energy ε_F. Since the DFDD is a result of the dynamics in the scattering, the effective occupation of electronic levels in the restframe of the projectile is related to kinematic thresholds and specific velocities. The consequent use of the parallel velocity as an additional parameter is the basis for new type of studies in atom-surface interactions.

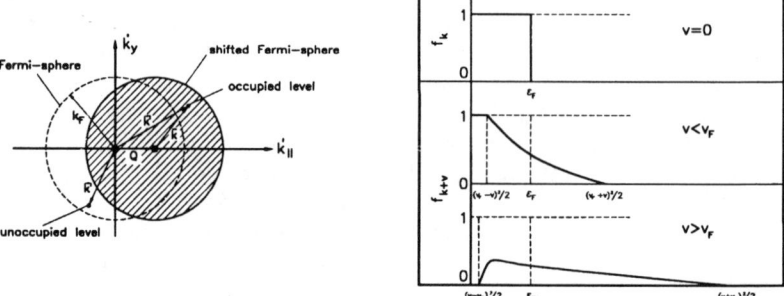

Figure 8. Density of occupied and unoccupied electronic states of the metal in the restframe of a projectile with velocity v ("Doppler-Fermi-Dirac distribution").

In a quantum-mechanical treatment the electron transfer between solid and atom is a tunneling process. Due to the long interaction times with the metal in grazing scattering charge exchange is dominated by mechanisms of relative long range; in particular: resonant one-electron tunneling and Auger processes. The transition rates for those processes show an about exponential dependence on the distance R $\Delta(R) = \Delta_o \exp(-R/R_d)$ as indicated in Fig.9 by results from nonperturbative calculations obtained by Nordlander and Tully[6] with the "complex scaling method" for resonant tunneling between Na I 3s and an Al-surface. At finite parallel velocities the distribution of electronic metal states for the atom is affected by the Doppler-effect, so that atomic levels can be found in resonance with occupied *and* unoccupied metal states. Then tunneling of electrons from the solid to the surface ("resonant neutralization", RN) as well as vice versa ("resonant ionization", RI) can take place, and charge transfer proceeds in a sequence of electron capture and loss events. The final formation of atomic terms is then determined by the survival against loss on the outgoing trajectory. Due to the exponential dependence of the transition rates this survival is localized to a well defined interval of distances from the surface. In Fig.9 we show capture/survival probabilities and the distribution function F(R) for the final formation of Na-atoms receding from an Al-surface with normal velocity v_y = 0.01 a.u.. The function F(R) is well localized around a maximum R_s, the so called "freezing distance". From a rate equation approach one finds from the transition rate parameters $R_s = R_d \ln(\Delta_o R_d/v_y)$. The final formation of NaI 3s is then effective at distances around $R_s \approx 9$ a.u.; i.e. a relatively large distance from the surface.

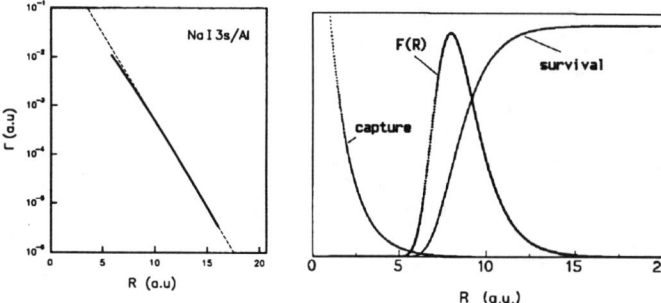

Figure 9. Level width (inverse of total transition rate) of NaI 3s as function of distance R calculated by Nordlander and Tully (ref.6) and corresponding probabilites for capture and subsequent survival. F(R) is the well localized distribution function for the final formation of the atomic levels.

RESULTS OF SOME REPRESENTATIVE EXPERIMENTS

Angular distributions

Since in the grazing incidence geometry large lateral extensions of the target are probed (typically several 100 a.u.), an essential problem in performing scattering experiments is the preparation of very clean targets with low concentrations of imperfections of the surface structure. Adsorbed atoms or defects (e.g. steps) may serve as targets for binary type of collisions and result in dechanneling phenomena. Then the conditions for the theoretical concepts outlined above are no longer met. As a consequence considerable efforts have to be invested in the preparation of the surface of monocrystalline targets, which is performed in our laboratory primarily by cycles of cleaning of the samples via sputtering with 25 keV Ar^+-ions under grazing incidence and subsequent annealing. During the preparation cycles the target surface is inspected by Auger-spectroscopy and a SPA-LEED system. Typical time spans for obtaining clean targets with large widths of terraces formed by topmost surface atoms (generally larger than 500 a.u.) are several weeks up to some months.

A simple method to obtain an estimate on the contributions of surface imperfections during the scattering is the recording of angular distributions of the scattered projectiles. Furthermore one gets from those studies information on the actual angular settings of the target. In Fig.10 we show the experimental concept and two angular distributions obtained for the scattering of 25 keV Na$^+$-ions from the (111)-face of two different Al-targets. For the target of poorer quality we find a pronounced peak indicating a predominant forward scattering of the projectiles, however, the extended tail clearly indicates contributions of binary type collisions leading to larger angles of scattering. For a better target and increased efforts in the preparation we achieve distributions with angular FWHM of only about 0.2° and practically no signal at larger angles of scattering. Then most of the projectiles are scattered within narrow bounds, so that the criteria for trajectories in grazing scattering are fulfilled. We note that angular distributions thus bear information on the "quality of the target surface", and the FWHM is a good measure for the state of preparation of the target.

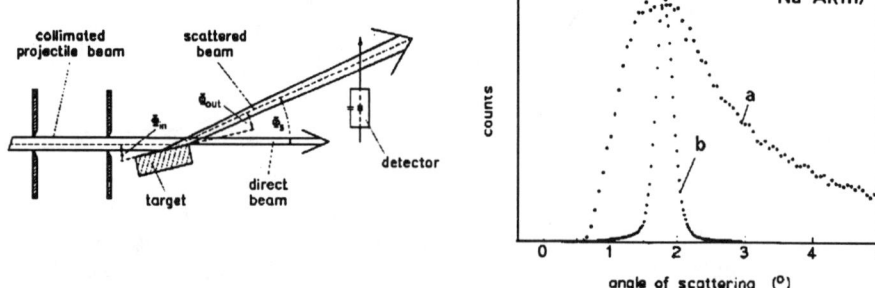

Figure 10. Sketch of the experimental concept and data obtained with 25 keV Na$^+$-ions scattered from two different Al(111)-targets of "good" (a) and "high" (b) quality. The angular FWHM for case b is only 0.2°.

Neutralization of alkali ions

The binding energies of alkali atoms are of the order of typical workfunctions for metal surfaces, so that charge exchange will proceed predominantly into the atomic groundterms via resonant one-electron tunneling. The workfunction (or the position of the Fermi edge) plays an important role in the neutralization process as indicated in Fig.11 for the energies of NaI 3s and KI 4s in front of an Al-surface. The final formation of the atomic terms is expected to take place around distances R_s, where the atomic binding energies are reduced by ΔE (see eq.1 and discussion below). Then the shifted NaI 3s level lies still below the Fermi edge, the KI 4s, however, above of it. In the static case (v ≈ 0) NaI 3s is in resonance exclusively with occupied metal states, and via resonant neutralization the projectiles will leave the surface completely neutralized. The opposite holds for KI 4s, where resonant ionzation is effective ("surface ionization"). By a consequent variation of the (parallel) velocity we can tune the DFDD of electronic states as seen from the restframe of the moving projectile, so that for the NaI 3s unoccupied levels come into resonance and an electron loss channel via resonant ionization is opened up. For KI 3s occupied metal states are tuned into resonance, and a fraction of projectiles will be neutralized.

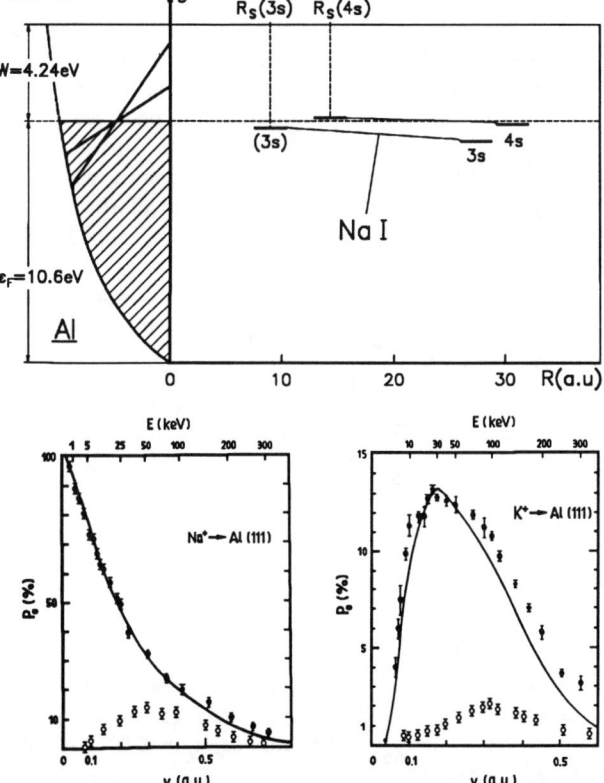

Figure 11.

Sketch of NaI 3s and KI 4s in front of an Al(111)-surface and experimental neutral fractions for Al(111) (dots) and Ni(111) (open circles). The solid lines represent model calculations from ref.10. The slight deviation from the experiment at larger velocities can be ascribed to contributions due to the population of excited terms (Borisov, unpublished).

In Fig.11 we display the velocity dependence of the neutral fractions obtained in experiments with an Al(111)-surface (Al is prototype for a free electron metal) for an angle of incidence and emergence of $\Phi_{in} \approx 0.5°$. We find very different dependences of the neutral fractions on the velocity: A monotonic decrease for Na and a resonance structure for K. These dependences are direct consequences of the DFDD, and the type of dependence of velocity is based on the position of the shifted atomic level relative to the Fermi edge. It is evident from these results that the neutralization depends extremely sensitive on the actual workfunction of the target. This has been demonstrated by us in quite a few experiments, and we show in Fig.11 data obtained with a Ni(111)-surface at otherwise unchanged conditions. Due to the larger workfunction of Ni(111) (W = 5.35 eV) in comparison to Al(111) the NaI 3s level lies then above the Fermi edge, and we find the expected resonance structure for the dependence of the neutral fraction on the velocity. Note the extreme sensitivity of the neutral fraction at lower velocities on the workfunction of the target. This feature has been made use of to deduce workfunctions with good accuracy[10].

These studies indicate that the grazing incidence geometry provides with the parallel velocity an important additional parameter in the study of atom-surface interactions. As an example of the detailed information one can obtain with the help of this parameter, we make use of the fact that the variation of the neutral fractions is characterized by a number of specific velocities, which are related directly to properties of the DFDD; e.g. kinematic thresholds for electron loss (Na-Al) or capture (K-Al), etc.. From a velocity dependence of the neutral fraction as observed for Na one deduces for a fraction of P_o = 2/3 = 66.6 % at a velocity v_{66} the simple relation v_{66}^2 = 2 (W - $|E_a|$ - ΔE). At this velocity the atomic level is in resonance with equal portions of occupied and unoccupied metal states and incorporation of the spin degree of freedom enhances the electron capture channel by a factor of two relative to electron loss.

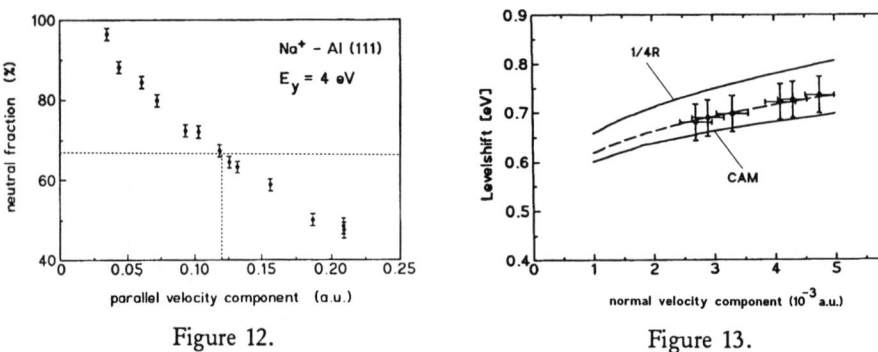

Figure 12.   Figure 13.

In Fig.12 we display a study based on this feature, where the energy for the normal motion is kept constant at E_y = 4 eV by proper settings of the grazing angles. For Na-Al(111) we find in this case v_{66} = 0.118 a.u. or E_{66} = 8.00 keV. With help of the workfunction measured by photoemission here W = 4.27 eV and the binding energy of the unperturbed NaI 3s level E_a = -5.14 eV we deduce for the level shift of the atom at the instant of final formation ΔE = 0.68(3) eV. In Fig.13 we show results for the study of the dependence of ΔE on the normal velocity component. Comparison with theory - F(y) is obtained from electron transition rates as indicated in Fig.9 - reveals that our data lie between the simple 1/4R approach and the inclusion of the van der Waals term in eq.1 or results from nonperturbative methods as "complex scaling"[6] or "coupled angular mode" (CAM)[11] (see also Fig.7) . Zimny and Borisov[12] find good agreement with the data, when dynamic corrections with respect to the van der Waals term are incorporated (dashed line).

Effect of charge exchange and image charge on the outgoing trajectory

Since in grazing incidence geometry the energy for the normal motion E_y shows a strong dependence on the angle, this energy can be obtained with good precision from angular distributions. The feature is of particular interest to investigate an eventual modification of E_y during the scattering. We demonstrate for the neutralization of Na-atoms, how one can obtain in this way important information on the interaction mechanisms. Based on the concepts outlined above final formation of atomic terms proceeds on the outgoing trajectory in an interval of distances around R_s. Then scattered projectiles can be found in defined charge states at distances from the surface larger than about R_s. Projectiles emerging from the target as ions will be attracted via image charge forces towards the surface plane, whereas neutral projectiles are not affected in this way. This leads to an angular shift between the distributions for atoms and ions given by $\Delta\Phi = \Phi_{out}(1 - (1 - V_{im}/E_y)^{1/2})$, where V_{im} is the amount of normal energy lost on the outgoing trajectory by an ionized projectile due to image interaction. Since the image interaction is "switched on" at the instant of final formation of the atomic terms, we have $V_{im} = 1/4 R_s$ for singly charged ions; i.e. V_{im} and thus also $\Delta\Phi$ bear information on the distance of formation.

As an example we present in Fig.14 data for the scattering of 25 keV Na^+-projectiles from an Al(111)-surface. For a distance of formation for NaI 3s of $R_s = 9$ a.u. we have $V_{im} = 0.7$ eV and obtain for $\Phi_{out} = 0.9°$ from the formula in the preceding paragraph $\Delta\Phi \approx 0.05°$. The angular distributions for neutral Na-atoms (full circles) and for Na^+-ions (open circles) indicate that such a small angular shift can be resolved in an experiment, where a "difference method"[13] is applied in these studies. The maxima of the distributions in Fig.14 are shifted by $\Delta\Phi \approx 0.04°$, which is in view of the accuracy of the data very consistent with the more implicit information obtained from our studies discussed above.

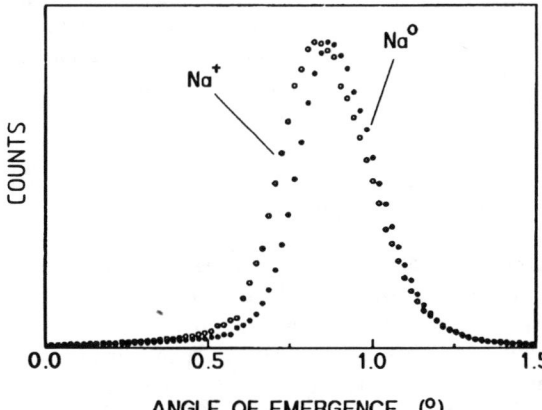

Figure 14.
Angular distributions selected with respect to the charge state of the scattered projectiles (neutral atoms = dots, ions = open circles) for 25 keV Na^+-ions scattered from Al(111). The data are obtained with a "difference method" and the maxima of the two distributions are normalized to the same height. Note the slight angular shift of the receding ions towards smaller angles of emergence by about 0.04°.

Controlled variation of the workfunction of the target

In the discussion on charge exchange outlined above we showed that the position of the Fermi-edge relative to the energy of the atomic levels plays an important role; i.e. resonant neutralization phenomena depend strongly on the workfunction W of the surface. Instead of using various targets with different W as described above, one can also vary W of a target by the controlled adsorption of alkali-atoms.

A corresponding study in this respect has been reported by Geerling and Los[14] for the scattering of 1 keV Li$^+$-ions from a cesiated W(110)-surface, where the projectiles leaving the surface under an angle of exit of Φ_{out} = 10° are analyzed with respect to their ion-fractions. The data for the ion-fractions in Fig.15 show the expected trend: complete ionization for a clean surface with W = 5.25 eV and a drastic decrease of the ion yields for lower W. The solid line is the result of model calculations by the authors, where the energy interval for the transition from complete ionization to neutralization is basically due to the width of the LiI 2s level at the distance of formation. A clear deviation from the data is apparent. A straightforward explanation for this discrepancy has been pointed out recently by Zimny[15], who showed that already at the relative low projectile velocity parallel velocity effects cannot be neglected. In fact, the DFDD for this case corresponds to a Fermi-Dirac distribution for a temperature of about 10^4 K! The dashed line indicates the result of an analysis of data based on the incorporation of these kinematic effects.

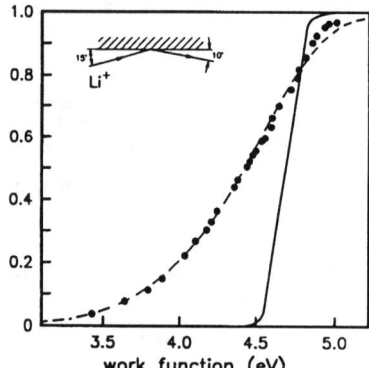

Figure 15.
Ion-fractions as function of the workfunction of the target observed for the scattering of 1 keV Li$^+$-ions from a cesiated W(110)-surface with an angle of emergence of 10°. The data and their analysis (solid line) have been reported by Geerlings et al[14]. The dashed line represents a recent reanalysis of data by Zimny[15] based on the incorporation of parallel velocity effects; i.e the DFDD.

Formation of negative hydrogen atoms

Detailed studies on the neutralization of protons in grazing collisions with a clean metal surface show that the final formation of the HI 1s level takes place at a distance R ≈ 3 a.u.; for details we refer to refs. 13 and 16. Due to the much higher transition rates a formation of negative ions takes place at clearly larger distances R. Since the binding energy of the unperturbed H$^-$-ion is only E_a = - 0.75 eV, the energy gap with respect to the Fermi level is relatively large. Therefore, similar as for the formation of KI 4s sketched in Fig.11, H$^-$ can be formed in collisions with a clean metal only by a kinematic resonance, where occupied metal states are brought into resonance with the (shifted) affinity level of the ion. The larger energy gap in comparison to the alkali atoms results in poor fractions of H$^-$-ions emerging from the surface. In Fig.16 we show a study on the yields of negative ions as function of the projectile velocity performed by Wyputta et al.[17], where the expected kinematic resonant structure of the data is evident. The solid line represents calculations by Borisov et al.[18] performed with the nonperturbative CAM-method. Note that in the experiments with a clean Al(111)-surface indeed very small fractions of typically 10^{-3} are observed. Based on the theoretical description of data the final formation of the negative ions takes place at distances $R_s \approx 8$ a.u. from the image plane.

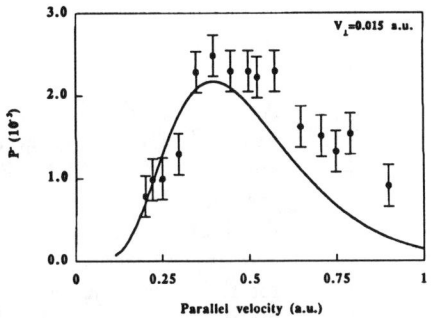 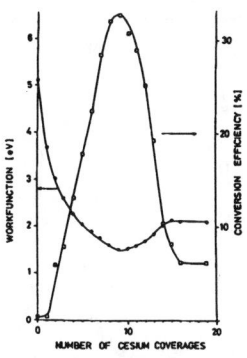

Figure 16. H⁻-fractions for grazing scattering of protons from an Al(111)-surface

Figure 17.

A reduction of the workfunction of the target reduces the energy gap and thus leads to an enhanced formation of negative ions. In Fig.17 we display results from van Wunnik et al.[19] for H⁻-fractions observed as function of the coverage of a W(110)-surface with cesium atoms for the grazing scattering of 400 eV protons. Note the clear correspondence between workfunction and ion yield; for the lowest workfunction of about 1.5 eV a conversion efficiency for negative ions of about 30 % is observed.

Effect of image charge for (multicharged) ions on the incident trajectory

In a similar way as for the experiments on image charge effects on the outgoing trajectory (see Fig.14) one can perform those studies also for the incident trajectory, where an image force attraction of projectile ions is effective before the screening of the projectile charge via neutralization. This feature has been made use by us in experiments with e.g. He⁺-ions to study in detail the Auger-neutralization in front of the surface[20]. Of particular interest is the application of this method to the study of the interaction of (slow) multicharged ions with surfaces. Due to the high projectile charge the image interaction is more pronounced and results in corresponding larger shifts in the angular distributions for the scattered projectiles. In Fig.18 we show a typical result of such studies obtained with 25 keV Ar^{5+}-ions scattered from an Al(111)-surface; for comparison we also show the angular distribution obtained with neutral projectiles of the same energy. In both cases the projectiles leave the surface almost completely neutralized, so that an image charge interaction on the outgoing trajectory can be neglected. The attraction of the ions on the incident path results in a larger angle of incidence, and the angular distributions for incident ions are shifted towards larger angles of scattering in comparison to neutral projectiles (this is opposite to the effect on the outgoing trajectory). From the position of the maxima of the angular distributions it is straightforward to find the energies for the normal motion of the projectiles and to deduce the image interaction energies gained by the projectile on the incident trajectory, until the image charge interaction is ceased by the neutralization sequence[21].

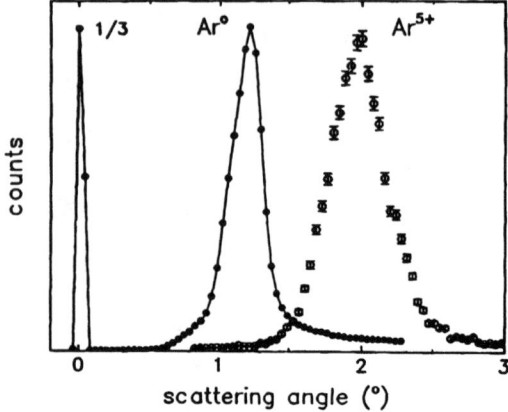

Figure 18.

Angular distributions of 25 keV Ar^0- and Ar^{5+}-projectiles scattered from Al(111). From the angular shift of the distribution for the ions we deduce an energy of 12 eV for the normal motion gained by the ions on the incident part of the trajectory.

In Fig.19 we show the image energies for Xe^{q+}-ions gained in front of an Al(111)-surface for projectile charges up to q = 33, where energies for the normal motion of about 150 eV are observed[22]. As we have said, these energies are related closely to the neutralization of the multicharged ion in front of the surface and important information in this respect can be obtained from the data. The solid line in the figure represents a $q^{3/2}$-dependence predicted by the classical "over-barrier-model"[4] for the population of highly excited and inverted Rydberg-levels in front of the surface plane. Grazing incidence geometry is particularly suited to study those processes which are located "above" the surface. More details in this respect can be found in the contributions by Meyer et al. and Das et al. in these proceedings.

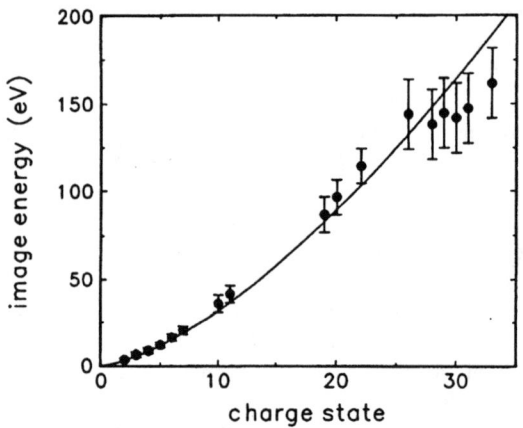

Figure 19.

Image charge interaction energies gained by Xe^{q+}- projectiles in front of an Al(111)-surface[22]. The solid line represents the $q^{3/2}$ dependence predicted by the classical "over-barrier" model.[4]

CONCLUSIONS

In this brief review we discussed some similarities and specific differences between atomic collisions with free atoms and with a metal surface. In our discussions we focused on charge phenomena based on resonant electron transfer processes which are affected in characteristic ways by specific features of surface scattering under a grazing angle of incidence. In recent years, grazing collisions with surfaces have become a technique for a wide field of activities comprising studies of charge exchange phenomena, electron emission (see e.g. contribution of A. Niehaus on previous ICPEAC), "skipping motion", Coulomb-explosion of molecular ions, polarized electron capture from magnetized targets, highly charged ions, energy loss, atomic spectroscopy, nuclear spin polarized beams, etc., which clearly indicate the attractive features of these atomic collisions under special conditions.

REFERENCES

[1] J. Los and J.J.C. Geerlings, Phys. Reports 190, 133 (1990).
[2] J. Burgdörfer, in "Progress in Atomic and Molecular Physics", ed. C.D. Lin, World Scientific Publ., 1993, in press.
[3] H. Winter and R. Zimny, in "Coherence in atomic collision physics", eds. H.J. Beyer, K. Blum, and R. Hippler, Plenum, New York 1988, p.283.
[4] J. Burgdörfer, P. Lerner, and F.W. Meyer, Phys. Rev. A44, 5674 (1991).
[5] J.W. Gadzuk, Phys. Rev. B1, 133 (1967).
[6] P. Nordlander and J.C. Tully, Surf. Sci. 211/212, 207 (1989).
[7] J.B. Delos, Rev. Mod. Phys. 53, 287 (1981)
[8] J.N.M. van Wunnik, R. Brako, K. Makoshi, and D.M. Newns, Surf. Sci. 261, 618 (1983).
[9] D.M. Newns, Comments Cond. Mat. Phys. 14, 295 (1989).
[10] R. Zimny, H. Nienhaus, and H. Winter, Rad. Effects and Defects in Solids 109, 9 (1989).
[11] D. Teillet-Billy and J.P. Gauyacq, Surf. Sci. 239, 343 (1990).
[12] R. Zimny and A. Borisov, to be published.
[13] H. Winter, Phys. Rev. A46, R13 (1992).
[14] J.J.C. Geerlings, L.F.Tz. Kwakan, and J. Los, Surf. Sci. 184, 305 (1987).
[15] R. Zimny, Surf. Sci. 223, 333 (1990).
[16] R. Zimny, Z. Miskovic, N.N. Nedljkovic, and Lj.D. Nedeljkovic, Surf. Sci. 255 (1991) 135.
[17] F. Wyputta, R. Zimny, and H. Winter, Nucl. Instr. Meth. B58, 379 (1991).
[18] A. Borisov, D. Teillet-Billet, and J.P. Gauyacq, Phys. Rev Lett. 68, 2842 (1992).
[19] J.N.M. van Wunnik, J.J.C. Geerlings, E.H.A. Granneman, and J. Los, Surf. Sci. 131, 17 (1983).
[20] H. Winter, J. Physics E, (1993), in press.
[21] H. Winter, Europhys. Lett. 18, 207 (1992).
[22] H. Winter, C. Auth, R. Schuch, and E. Beebe, Phys. Rev. Lett., (1993), in press.

K-AUGER EMISSION DURING HIGHLY CHARGED ION-SURFACE INTERACTION

J. Das, H. Limburg and R. Morgenstern.
Kernfysisch Versneller Instituut, University of Groningen
Zernikelaan 25, 9747 AA Groningen, The Netherlands

ABSTRACT

We present K-Auger spectra arising from collisions of H-like C, N, O, F, and Ne ions with Si(100), Ni(110), and polycrystalline W for different kinetic energies and observation angles. We find that only a small fraction of the K-vacancies is filled prior to collision because the average L-shell filling rate is slow compared to the K-Auger decay time and the time available in front of the surface. We also find that the measured electron spectra are target specific indicating that the properties of the target are of importance during the electron emission process.

INTRODUCTION

Collisions of highly charged ions with surfaces result in the emission of large numbers of electrons. These electrons are characteristic of the ion-surface interaction mechanism and detailed studies of the electron spectra can provide information about the various processes occurring during the collision. A multiply charged ion approaching a solid target is confronted with a large reservoir of electrons and responds by capturing electrons from the target. The process of electron capture from the surface, the charge state evolution of the ion as it approaches the surface and the fate of the ion after it has collided with the surface are the major issues that can be studied by means of electron spectroscopy. The general picture of the processes taking place during ion–surface collisions is well established although the details still remain unclear. It is known that in the early stages of the collision process multiply excited neutral atoms with empty or sparsely populated inner shells are formed. This is due to electron capture from the target occurring mainly via resonant neutralization into high-lying states of the projectile. It is also known that these atoms start to deexcite by means of autoionization processes(AI)[1] but the slow rates of autoionization do not allow a complete relaxation of the projectile prior to collision with the surface. A significant fraction of inner shell vacancies are therefore carried into the surface and filled only after the projectile has suffered close encounters with the target atoms[2,3]. One of the important questions that still remain unanswered is the role of the target in the collision process. Does it simply act as a reservoir of electrons or does it play a more subtle role? To answer this question among others we have measured the electron spectra emitted during collisions of C^{5+}, N^{6+}, O^{7+}, F^{8+}, and Ne^{9+} with different targets viz., Ni(110), Si(100), and polycrystalline W. We find that the electron spectra are target specific indicating that the emitted electrons are indeed influenced by the properties of the target.

EXPERIMENTAL METHOD

The electron spectra were measured in an ultrahigh vacuum μ metal collision chamber with a base pressure below 2×10^{-8} Pa. C^{5+}, N^{6+}, O^{7+}, F^{8+}, and Ne^{9+} beams were extracted from the KVI-ECR ion source[4] and then transported to the experimental setup. Floating the complete apparatus on a positive voltage allowed us to decelerate the ion beam from the ECR source to the desired collision energy. The targets used for the experiment included crystals of Ni(110), doped Si(100) and polycrystalline W. The targets were sputter cleaned with Ar ions prior to measurement. The electron detector was a 180° spherical electrostatic analyzer equipped with a channeltron and rotatable over a large range of detection angles. The energy resolution of the analyser is $5 \times 10^{-3} E$ FWHM and its acceptance at the center of the target is $11.2 \times 10^{-8} E$(sR eV), with E being the energy of the detected electrons in eV. A more detailed description of the apparatus is presented elsewhere[5].

ABOVE-SURFACE ELECTRON EMISSION

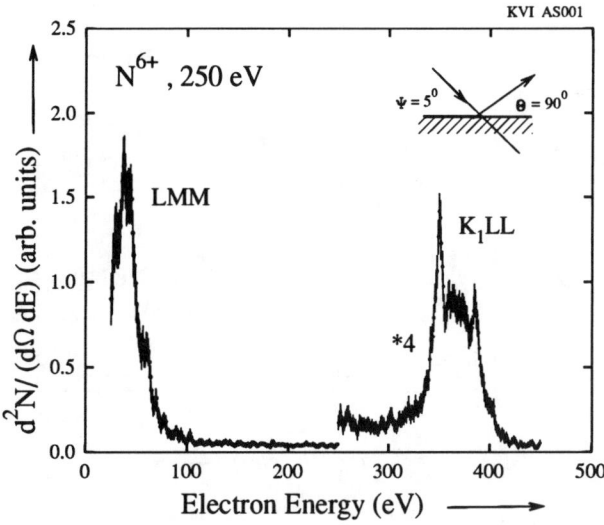

Figure 1: Electron spectrum measured during collisions of 250 eV N^{6+} on a Ni(110) surface. The high-energy part of the spectrum has been magnified by the indicated scale factor.

Figure 1 shows a typical electron energy spectrum measured during collisions of N^{6+} on a Ni(110) surface. The peaks above 300 eV are ascribed to KLL processes which lead to the filling of the K-shell vacancy in N^{6+} while those around 60 eV arise from

LMM transitions. The low energy part of the spectrum also contains contributions from processes other than Auger emission and has been investigated in detail by the group of Winter[6,7]. In the following we focus our attention on the high–energy electrons and use the K–Auger spectra to extract information about neutralization processes occurring during ion–surface collisions. The KLL peak in figure 1 is composed of an above- and a below-surface part and it has been shown previously that the K-Auger peak consists predominantly of electrons that are emitted within the bulk of the material[2,3]. However, a small percentage of inner shell vacancies is filled above the surface and the narrow structure on the low-energy side of the K-Auger peak is ascribed to processes that take place prior to collision[2,8,9].

The intensity of the above-surface component is critically dependent on the interaction time available to the projectile before colliding with the target surface. Lowering the collision velocity increases the fraction of K-vacancies that are filled above the surface thereby increasing the intensity of the above-surface peak. This enhancement does not however continue indefinitely but the intensity of the peak saturates at a certain fraction of the total K-Auger intensity. This saturation effect is a direct consequence of the acceleration of the projectile by its image charge[10] which imposes an intrinsic lower bound to the collision velocity. Figure 2 shows the change in the intensity of the above-surface component as the initial velocity perpendicular to the surface is varied.

The velocity dependence of the intensity can be used to directly determine the time

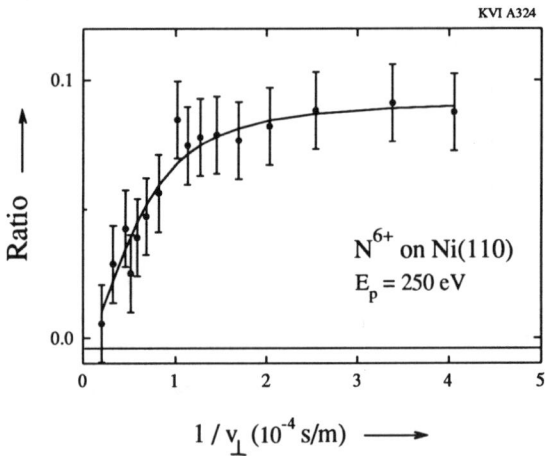

Figure 2: The ratio R of the intensity of the above-surface Auger peak to the total K-Auger intensity as a function of the vertical velocity $v_\perp$ for the collision system $N^{6+} \rightarrow$ Ni(110). The solid line is a fit to the data using eqn. 1

scales involved in electron emission above the surface. We have assumed for the sake of simplicity that the filling of the K-vacancy above the surface is described in terms of three states–the initial state where the projectile is a hollow atom with an empty L-shell, the intermediate state with at least two electrons in the L-shell and finally the

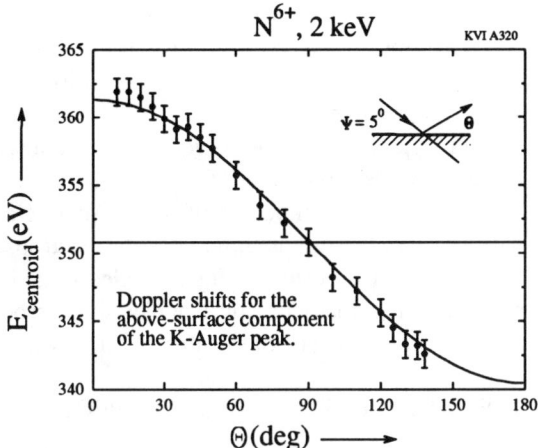

Figure 3: Position of the centroid of the above surface component of the K_1LL peak, for 2keV $N^{6+} \rightarrow$ Ni(110), as a function of the detection angle θ. The solid curve is a calculation of the Doppler shift assuming the decay occurs while the projectile is still in its initial trajectory.

state where the K-vacancy has been filled[8]. We assume that, above the surface, there is no direct electron transfer into the inner shells. The ratio, R, of the intensity, I_{as}, of the above-surface component to the the total K-Auger intensity, I_{tot}, can then be written as

$$R = \frac{I_{as}}{I_{tot}}$$
$$= 1 + \frac{\tau_{av}}{\tau_k - \tau_{av}} e^{-\frac{d}{\tau_{av} v_\perp}} - \frac{\tau_k}{\tau_k - \tau_{av}} e^{-\frac{d}{\tau_k v_\perp}} \qquad (1)$$

where $v_\perp$ is the effective perpendicular velocity after taking account of image acceleration, d the distance from the surface where the Auger cascade is initiated and τ_k is the lifetime of the inner-shell vacancy before it is filled via the KLL process. τ_{av} is a measure of the time required to populate the L-shell to the minimum level required for a KLL process. The solid curve in figure 2 is a fit to data points using eqn. 1. It is clear from figure 2 that this simple picture reproduces the general trend of the data points rather well. From the fit we have an average L-shell filling time, τ_{av}, of 200 fs which is considerably longer than the K-Auger decay time, τ_k, of only 70 fs. The result is in agreement with theoretical estimates[11] that predict time scales of the order of 100 – 1000 fs for the completion of the AI cascade. However one has to remember that τ_{av} represents the average decay time for a large variety of processes that eventually lead to a transfer of the initial population in the high n-shells(n=7 for N^{6+} on Ni) to the L-shell. Besides the Auger cascades there is also a dynamic exchange of electrons between the projectile and the metal: as the ions approach the surface their energy levels are - due to the image charge - shifted above the Fermi level of the metal such that reionization can take place, which is followed again by neutralization in lower n levels. Furthermore capture processes leading to transfer of electrons from the core levels of the target to the projectile L-shell also have to be considered. A direct comparison of calculated Auger rates with τ_{av} is therefore not possible.

Measurement of the Doppler shift of the Auger electrons allows one to probe the emitter velocity at the time of electron emission. Figure 3 shows the Doppler shift measurements for the above-surface component of the K-Auger electrons. The solid curve is a calculation of the Doppler shift assuming that the electrons are emitted while the projectile is still in its initial trajectory. Figure 3 clearly demonstrates that the above-surface electrons show the characteristic Doppler shift associated with 2 keV N^{6+} projectiles providing further evidence of the above surface origin of the electrons.

TARGET EFFECTS IN K-AUGER SPECTRA

In order to investigate the effect of the target on the electron emission spectra we have measured the Auger electrons emitted for different target and projectile combinations. Figure 4 is a plot of the K-Auger electrons emitted during collisions of H-like ions on Si(100). The increase in the energy of the emitted Auger electrons with increasing Z of the projectile is clearly evident. The high resolution of our spectrometer allows us to resolve the sub-structures within each K-Auger peak. These sharp structures are characteristic of the initial configuration of the projectile prior to K-Auger decay and provide information about the electron distribution within the various shells. shows the K-Auger spectra measured during collisions of O^{7+} with Si(100), Ni(110), and polycrystalline W. The collision geometry and the projectile energy was identical in all cases. It is clearly evident that the properties of the target have a large influence on K-Auger emission during ion–surface collisions. The spectrum recorded for a Si target is significantly different from those recorded for W and Ni. The two sharp structures at the low energy side of the K-Auger spectrum have a much larger intensity for collisions of O^{7+} on Si. In fact the peak at 477 eV is hardly distinguishable in case of the Ni target. The major difference between the targets is in their free electron densities–the Si target being a p-doped semiconductor while W and Ni are metals. If it is assumed that these peaks are due to above-surface emission processes then the enhanced intensity could perhaps be due to a reduced image charge acceleration in Si, providing more time for the autoionization cascade prior to collision. The intensity of these peaks should then increase with decreasing velocity of the projectile. Figure 6 shows the ratio of the intensity of the two peaks to the total K-Auger intensity plotted as a function of the vertical velocity inverse. We find that intensities of both peaks show a saturation effect, similar to the case of N^{6+} colliding on Ni (see figure 2). We have also measured the velocity dependence of the K-Auger spectrum, for collisions of N^{6+} on Si(100), where the peak at 350 eV has been previously identified as arising from above-surface processes[2,8,9]. As shown in figure 6 the saturation of the intensity at lower projectile velocities is also evident in this case implying the presence of an image charge. The fraction of the total K-Auger intensity contained in the above-surface peaks is higher

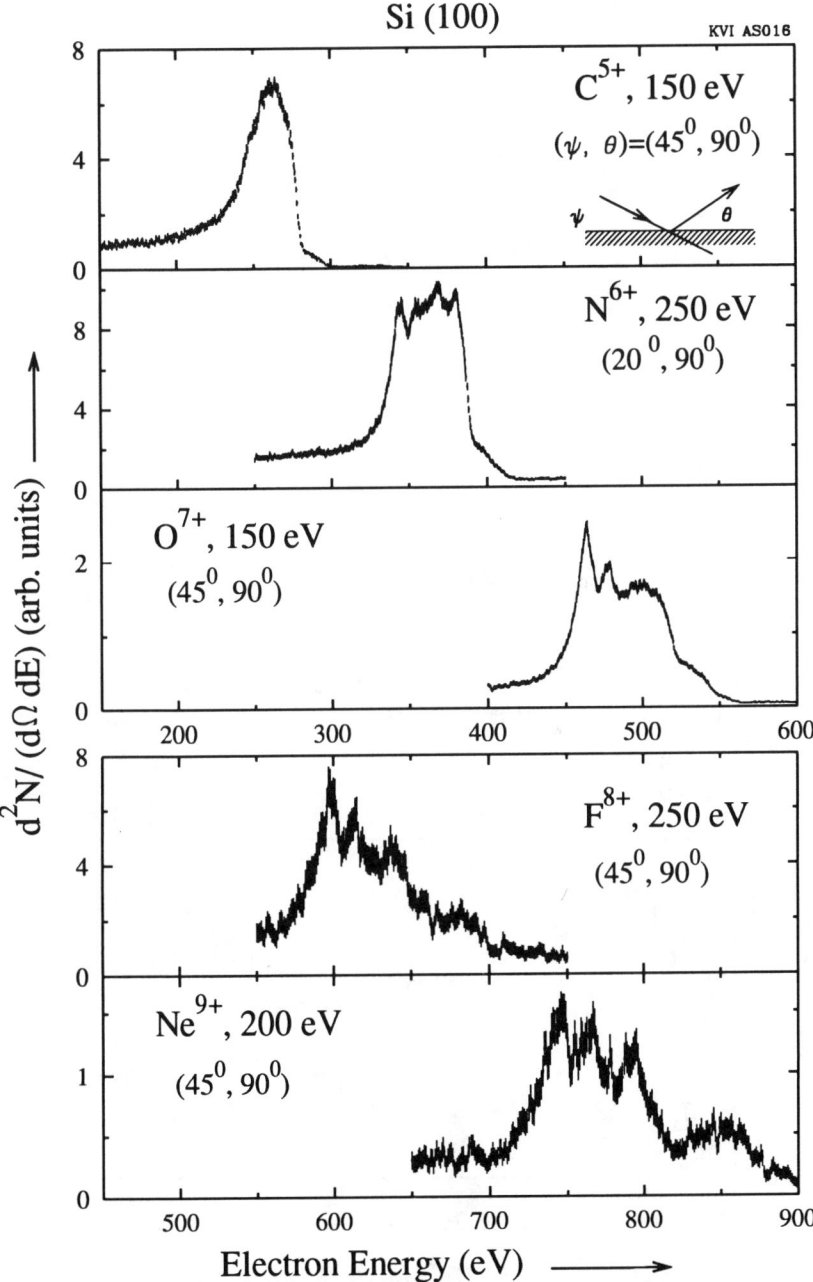

Figure 4: K-Auger spectra measured during collisions of C^{5+}, N^{6+}, O^{7+}, F^{8+}, and Ne^{9+} with a Si(100) target.

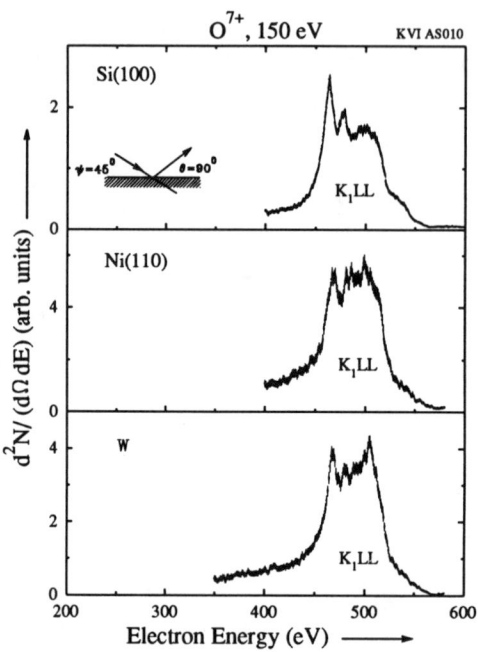

Figure 5: K-Auger spectra measured during collisions of 150 eV O^{7+} with Si(100), Ni(110), and polycrystalline W targets. The collision geometry was identical in all cases.

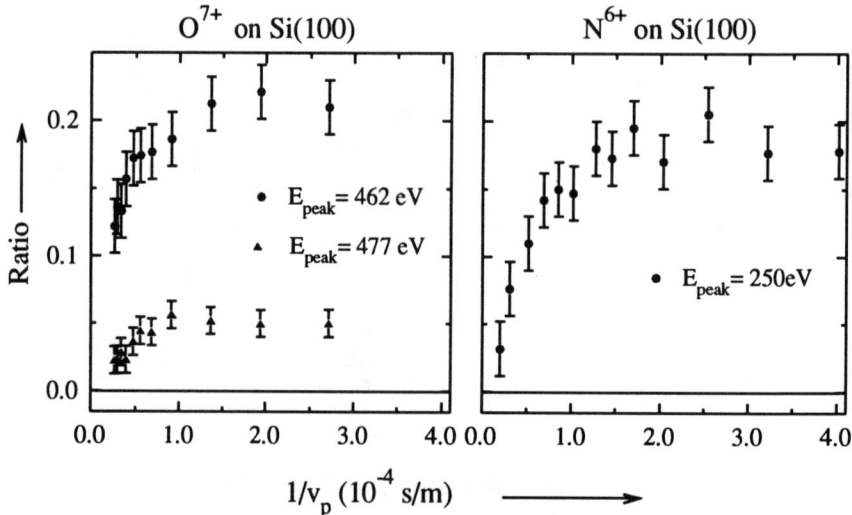

Figure 6: The ratio R of the intensity of the above-surface Auger peak to the total K-Auger intensity as a function of the vertical velocity $v_\perp$ for different collision systems.

for Si ($\approx$ 20% as compared to $\approx$ 10% in the case of Ni) and may partly be due to a weaker image-charge acceleration in Si targets. Although the concept of an image charge is strictly valid only for metals it is nevertheless possible that the presence of the highly charged projectile leads to a distortion of the electronic environment within the Si target which then mimics the effects of an image charge.

The narrow width of the above-mentioned peaks is an indication that only a small number of initial states are involved in the emission process. The energy positions of these peaks are in agreement with those for KLL processes involving only two electrons in the L-shell and the rest in outer shells. In order to identify the precise configurations we have measured the K-Auger spectra for metastable O^{6+} where the ions start out with one electron in the 2s level. Figure 7 compares the K-Auger spectra for O^{7+}

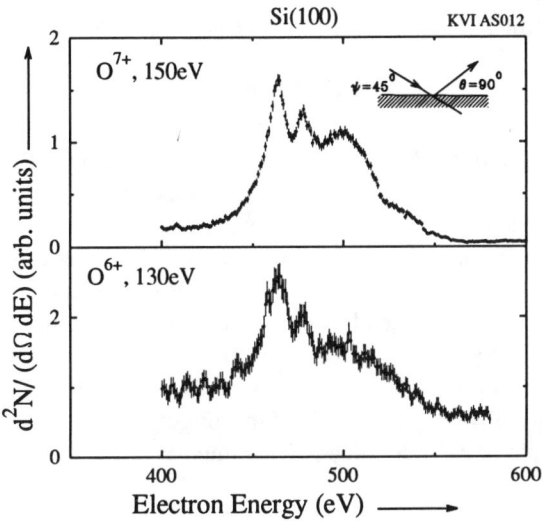

Figure 7: K-Auger spectra for collisions of O^{7+} and metastable O^{6+} with a Si(100) target.

and metastable O^{6+} colliding on Si and it is apparent that the spectra are similar in both cases. The two sharp peaks at 462 eV and 477 eV are present in both spectra implying that they arise from configurations where at least one of the 2 L-shell electrons is in the 2s level. Hartree-Fock calculations using the COWAN code can be used to further narrow the range of possible configurations[12,13]. We preliminarily assign the peaks at 462 eV and 477 eV to have initial configurations arising from $1s2s^2(3\ell)^5$ and $1s2s2p(3\ell^5)$ respectively. If this assignment is correct then the filling rate of the 2s level in the projectile determines the intensity of the peak at 462 eV. The higher intensity of this peak in the case of a Si target as compared to W and Ni could be due to an extra mechanism that preferentially populates the 2s level. In fact for an oxygen projectile in the $1s2s(3\ell)^5$ configuration the 2s level is energetically close to the 2p core-level of the Si target. Quasi-resonant charge-exchange processes can therefore populate the 2s level in O very efficiently. This would however mean that the projectile comes very close to the Si atom before the K-Auger process takes place implying that these electrons are emitted very close to the surface, perhaps even from the first monolayers under it. The peaks under discussion then no longer originate strictly above the surface. A more detailed analysis will be presented in a forthcoming publication[14]. It should also be noted that assigning initial state configurations to the Auger peaks is much more difficult if the process takes place inside the solid. Recent calculations by Arnau et al. have shown that the presence of highly charged ions within a metal can induce a significant charge density in the valence band and this in turn effects the energy levels of the ion[15]. They find that for all charge states of a N ion the 3p electrons are merged into the continuum. However these electrons still retain their p-like character and it is difficult to distinguish whether the p electron is in a bound or a scattering state. In view of this it may be impossible to assign specific initial atomic states to the peaks absorved in the K-Auger spectra.

CONCLUSION

We have measured the K-Auger electrons emitted during collisions of C^{5+}, N^{6+}, O^{7+}, F^{8+}, and Ne^{9+} on Si(100), Ni(110), and polycrstalline W. For the collision system $N^{6+} \rightarrow$ Ni(110) the average L-shell filling time is 200 fs, considerably longer than the K-Auger decay time of only 70 fs. As a result only 10% of the K-vacancies are filled in the pre-collision phase while the rest survives until close encounters with the bulk particles occur. It should be noted however that the exact amount is target dependent. We find that the exact shape of the electron spectra are target specific implying that electron emission is not independent of the properties of the target. Measurements on the $O^{7+} \rightarrow$ Si(100) collision system indicate that direct charge-exchange processes with the core-levels of the target atom may play an important role in populating the L-shell of the projectile.

ACKNOWLEDGEMENTS

We would like to thank R. Kremers and J. Sijbring for their excellent technical support and S. Schippers for help in calculations using the COWAN code . This work is part of the research program of the 'Stichting voor Fundamenteel Onderzoek der Materie' (FOM) with financial support from the 'Nederlandse Organisatie voor Wetenschappelijk Onderzoek' (NWO).

REFERENCES

1. U. A. Arifov, E. S. Mukhamadiev, and E. S. Parilis, Sov. Phys. Tech. Phys. **18**, 118, (1973).
2. F. W. Meyer, S. H. Overbury, C. C. Havener, P. A. Zeijlmans van Emmichoven, and D. M. Zehner, Phys. Rev. Lett. **67**, 723 (1991).
3. J. Das, L. Folkerts, and R. Morgenstern, Phys. Rev. A **45**, 4669, (1992).

4. A. G. Drentje and J. Sijbring, in Oak Ridge National Laboratory Report No. Conf. 9011136 (unpublished), p.17.
5. S. T. De Zwart, A. G. Drentje, A. L. Boers, and R. Morgenstern, Surf. Sci. **217**, 298 (1989).
6. H. Kurz, K. Töglhofer, HP. Winter, and F. Aumayr, Phys. Rev. Lett. **69**, 1140, (1992).
7. F. Aumayr and HP. Winter, Comm. At. Mol. Phys., to be published.
8. J. Das and R. Morgenstern, Phys. Rev. A **47**, R755 (1993).
9. H. J. Andrä, A. Simionovici, T. Lamy, A. Brenac, G. Lamboley, J. J. Bonet, A. Fleury, M. Bonnefoy, M. Chassevent, S. Andriamonje, and A. Pesnelle, Suppl. Z. Phys. D **21**, S135, (1991).
10. H. Winter, Europhys. Lett. **18**, 207, (1992)
11. J. Burgdörfer, P. Lerner, and F. W. Meyer, Phys. Rev. A **44**, 5674, (1991).
12. R. D. Cowan, The Theory of Atomic Structure and Spectra, University of California Press, Berkeley (1981).
13. S. Schippers, S. Hustedt, W. Heiland, R. Köhrbrück, J. Bleck-Neuhaus, J. Kemmler, D. Lecler, and N. Stolterfoht, Nucl. Inst. Meth. Phys. Res. B **78**, 106, (1991).
14. H. Limburg et.al, in preparation
15. A. Arnau, P. A. Zeijlmans van Emmichoven, J. I. Jurasti, and E. Zaremba, Proceedings of the International Seminar on Surface Science, Kaprun, May 1993.

HOT TOPICS

EXPERIMENTAL VISUALIZATION OF PHOTOELECTRON ANGULAR DISTRIBUTIONS

H. Helm, N. Bjerre, M. J. Dyer, M. Saeed, and D. L. Huestis
Molecular Physics Laboratory, SRI International, Menlo Park, CA 94025

ABSTRACT

A novel experimental technique is presented that permits the visualization of energy and angular distributions of photoelectrons. The experimental setup is exceedingly simple, but it is at the same time highly sensitive in that all electrons are monitored simultaneously, irrespective of their energy and angle of ejection. An added advantage is that the detection efficiency of the instrument is independent of electron energy. We describe the technique and give several examples of its application to multiphoton ionization of atoms and molecules.

INTRODUCTION

Photoelectrons that are generated at a point source with constant energy travel on the surface of a sphere that expands with time. For example, electrons produced at time t=0 with an energy of 1 eV are to be found 20 ns later on the surface of a sphere of 25 mm diameter. This sphere can be projected onto a flat screen using an external electric field. A circular image results with a diameter that is related to the electron energy and a filling pattern that reveals the spatial distribution of the electrons on the surface of the sphere. In this way, the squared angular wavefunction of the free electrons is accessible to direct observation.[1] Obviously, many electrons have to be projected in order to represent a statistically meaningful image.

A schematic representation of this projection technique is shown in Figure 1. In this figure a multiphoton process is simulated which leads to constant energy photoelectrons that are emitted with an angular distribution represented by a spherical harmonic function Y_{30}^2. The quantization axis is parallel to the screen, along the vertical direction.

EXPERIMENTAL TECHNIQUE

The primary requirement for realizing the experiment in Figure 1 is the existence of a well defined spatial origin of the photoelectrons. The point source properties can be realized by using a focused laser when a multiphoton process is investigated. In this case the spatial distribution of electron formation is confined to the focal volume due to the nonlinearity of the excitation process. Alternatively one could use two intersecting laser beams or else a crossed laser - molecular beam arrangement.

In our actual experiment we use a DC electric field applied perpendicular to the screen to accelerate electrons formed at the point source on to the screen on the right. Examining the equation of motion of the electron trajectories in a DC field we find that under the condition that the energy gained by the electron from the origin to the screen is very much greater than the nascent electron energy (factor 10 to 100), this DC field projection comes very close to the ideal projection technique in which electrons are first

© 1993 American Institute of Physics

allowed to expand away from the point source for a certain time, and are then instantly projected to the screen with a pulsed electric field. The simulated impact pattern shown on the screen on the right of Figure 1 is that calculated for 1 eV electrons accelerated by a field of 100 V/cm over a distance of 5 cm. It is evident that the projected image represents a rather faithful projection of the three-dimensional electron cloud.

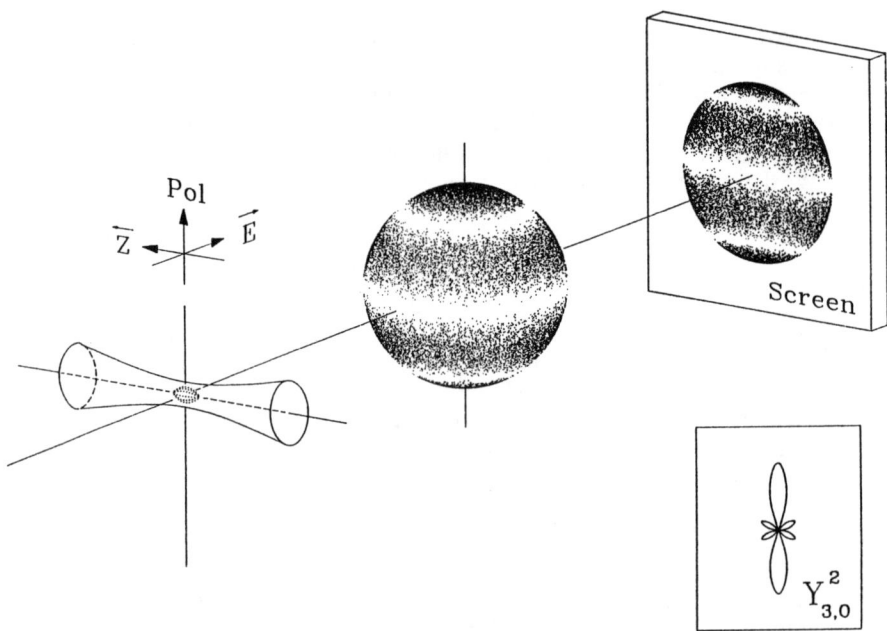

Fig. 1. Schematic view of expanding electron cloud of constant energy and its projection onto a two dimensional screen.

The actual experimental setup is shown schematically in Figure 2. A cylindrical drift space (10 cm long, 10 cm diameter) is formed by ring electrodes which establish a homogeneous electric field along its axial direction, between two grids. A pulsed laser, focused in the center of the drift space, is used to produce the photoelectrons, and a pair of channel-plates at one end of the drift region amplifies the electron signal. Electrons are subsequently accelerated onto a phosphor screen, and the screen image is recorded with a video camera for each laser shot, digitized and stored. Typically 10^4 images are added in computer memory, each containing the signature of 10 to 100 electrons.

RESULTS

We present here several typical images obtained in multiphoton ionization of xenon and helium. These images serve to show the principal operation of the technique. More detailed descriptions of the actual ionization phenomena can be found in the papers quoted.

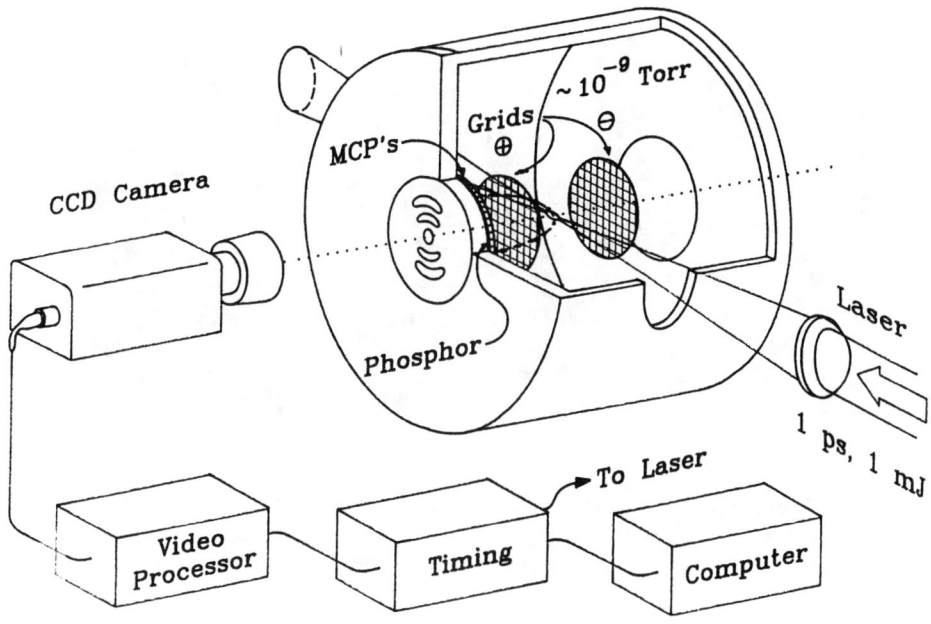

Fig. 2. Schematic of experimental setup.

The images in Figures 3 and 4 result from 6+1 photon resonant ionization of xenon via the Xe(4f) states.[2] The pulse duration of the linearly polarized laser is 1 ps and its peak intensity is $1.5 \cdot 10^{13}$ W/cm^2. The laser wavelength is 640 nm. The 4f state is aligned in the 6-photon excitation process and the resultant photoelectron is predominantly of g (l=4) character owing to the propensity of $\Delta l = +1$ in excitation into the continuum.

In Figure 3, left, the experimental image is shown as obtained for the laser polarization along the vertical axis, parallel to the screen. Four regions of minimal intensity, perpendicular to the polarization axis and perpendicular to the figure plane are evident, showing the predominantly l=4, m= 0 character of the outgoing photoelectron. As a comparison, we give on the right hand side of Figure 3 a simulated image, that is expected under our experimental conditions in the case that the electron angular distribution is given by a pure Y_{40}^2 spherical harmonic function.

Figure 4 shows an image recorded when the laser polarization is perpendicular to the screen, under otherwise identical conditions to those in Figure 3. When the quantization axis is set perpendicular to the screen the four regions of minimum intensity collapse into two dark rings, with most intensity located at the center, along the direction of the quantization axis. On the right hand side of Figure 4 we give the expected projection of a Y_{40}^2 function for this polarization direction. We see that the cylindrical symmetry of the free electron distribution, that is expected from electric dipole transitions,[3] is authentically reproduced in the experimental image.

In Figure 5 we show an image that results from 3+1 photon resonant ionization of xenon via the 5d' resonance. This resonance shares core character from the

782 Visualization of Photoelectron Angular Distributions

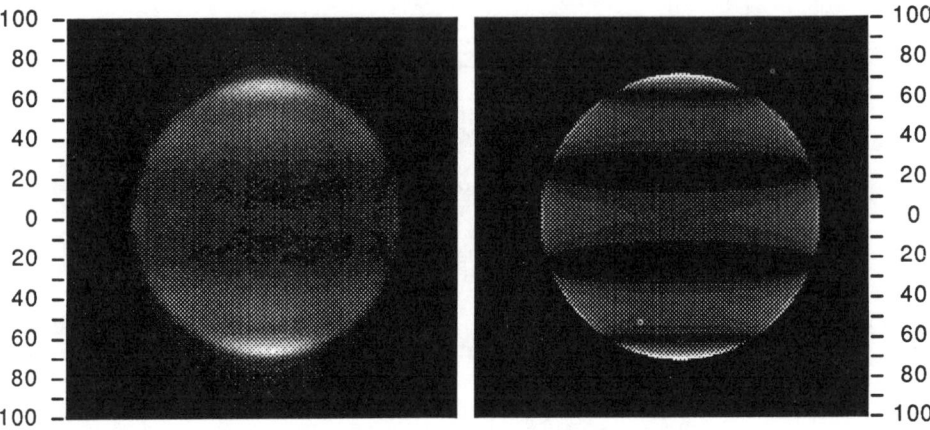

Fig. 3. Left: experimental image from 6+1 photon ionization via the Xe(4f) states, Right: simulation of image expected for a pure Y_{40}^2 function. The laser polarization (quantization axis) is parallel to the screen, along the vertical direction.

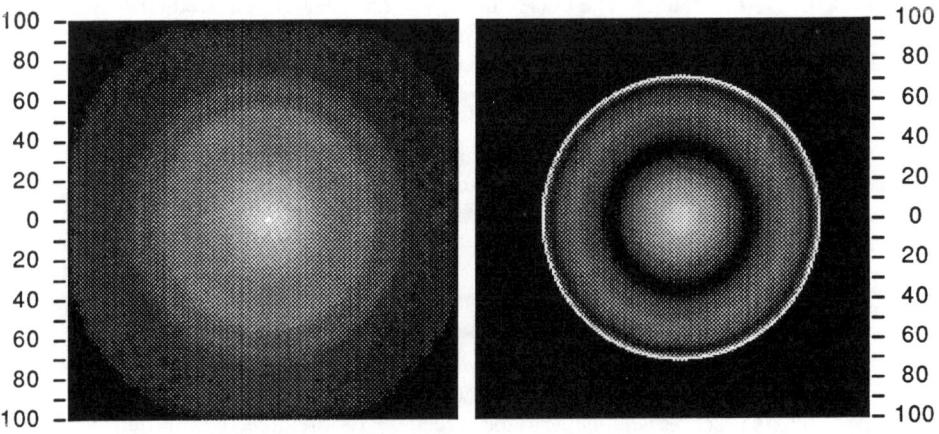

Fig. 4. As in Figure 3 but for a laser polarization perpendicular to the screen.

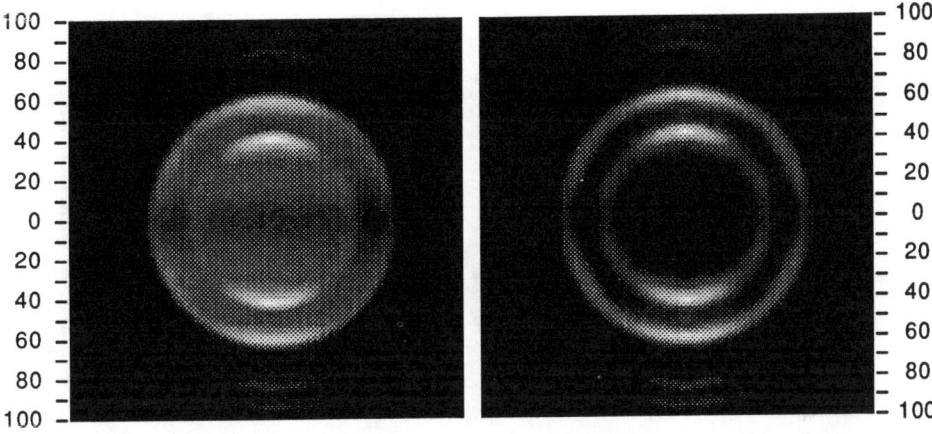

Fig. 5. Image resulting from 3+1 photon ionization of Xenon at 332 nm. The raw image is shown on the left. On the right the Abel-inverted image is given.

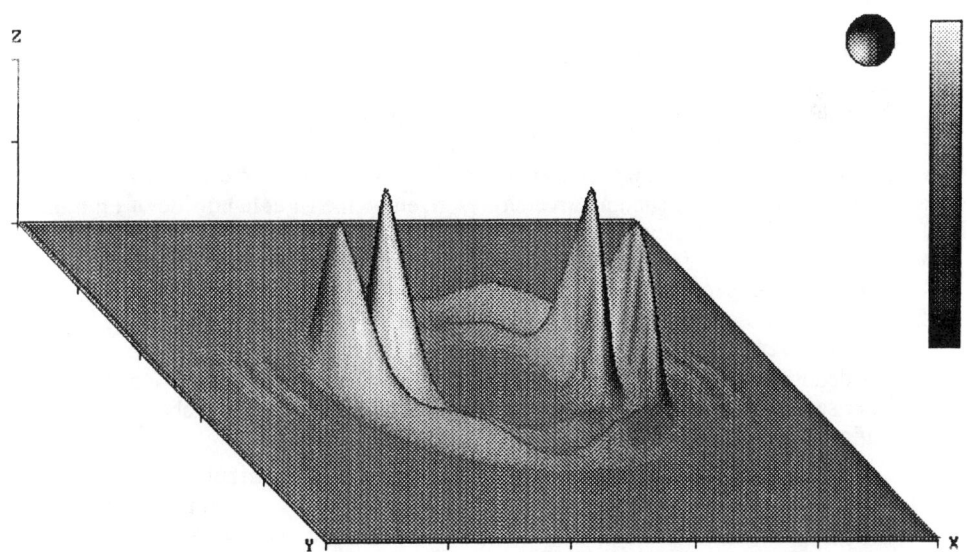

Fig. 6. Three-dimensional representation of energy and angular distributions of photoelectrons in Figure 5.

784 Visualization of Photoelectron Angular Distributions

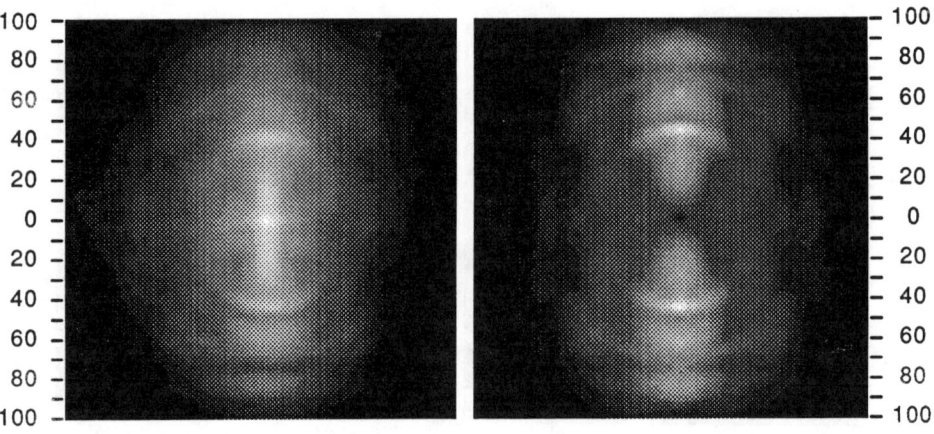

Fig. 7. Image obtained from 7-photon nonresonant ionization of helium at 320 nm. The raw image is shown on the left, the Abel-inverted image is shown on the right.

$Xe^+(^2P_{3/2})$ and from the $Xe^+(^2P_{1/2})$ final ion states and the electron energy distribution results in two groups of electrons being emitted, the slower producing $Xe^+(^2P_{1/2})$, the faster producing $Xe^+(^2P_{3/2})$. When expanding from the ionization region, the two electron groups appear on two concentric spheres and in the projection two concentric circular images result. The distance between the structures is related to the velocity difference of the two groups, as appropriate for the 1.3 eV fine structure splitting of the xenon ion.

Since the observed projections are from distributions that are rotationally symmetric around the quantization axis, one may use one-dimensional tomography, an Abel-inversion, to reduce the projected images into distributions of electron velocity and angle. The result of such an inversion is given on the right hand side of Figure 5. Here the distance from the center of the image represents the magnitude of the electron velocity, and the intensity variation along a ring of constant radius represents the angular distribution of ejected photoelectrons, of velocity specified by the radial distance from the center. A three dimensional representation of the Abel-inverted image in Figure 5 is given in Figure 6. It is evident that energy and angular distributions are readily deduced from the experimental images. In Figure 5 and 6 fainter rings appear at larger separation from the image center. These electrons arise from above threshold ionization[4] (3+2 photon resonant excitation).

When photoelectrons are produced with a continuous distribution of electron energies, the emitted electrons fill a spherical shell of thickness related to the range of electron velocities present. Such distributions are also amenable to investigation by this technique as we show in Figure 7. The process responsible for photoelectron formation here is nonresonant seven-photon ionization[5] of helium at a peak intensity of $4 \cdot 10^{14}$ W/cm^2. The distribution of intensities over the focal volume, combined with the effect of ponderomotive shifting of the ionization limit, and the lack of recovery of the ponderomotive energy from the laser pulse owing to its short time duration (700 fs) lead to a distribution of electron energies. The Abel inverted image on the right of Figure 7 shows the peculiar angular and continuous energy distributions that result from this process. The electron energies span a range from about 20 meV to above 1eV. The faint ring at larger diameter is due to 6+1 photon resonant ionization via the He 1s3d state, which is ac-Stark shifted into resonance at very low intensity.

CONCLUSION

The photoelectron imaging technique described here gathers simultaneously all of the information that can be obtained by conventional techniques. Actually seeing the energy and angular distributions, for even very low energy electrons, helps in the assignment of specific ionization paths and should provide a useful tool for characterizing photoionization processes. Numerous other applications in studies of processes where electrons are formed can be envisioned.

ACKNOWLEDGMENTS

This research was made possible through support by the National Science Foundation under grant PHY-9249199.

REFERENCES

1. H. Helm, N. Bjerre, M. J. Dyer, D. L. Huestis, and M. Saeed, Phys. Rev. Lett. **70**, 3221 (1993).
2. M. Saeed, M. J. Dyer, and H. Helm, Phys. Rev. A, submitted for publication.
3. S. J. Smith and G. Leuchs, Advances in Atomic and Molecular Physics **24**, 157 (1988).
4. P. Agostini et al., Phys. Rev. A **36**, 4111 (1987).
5. M. J. Dyer, and H. Helm, Phys. Rev. A submitted for publication.

CLASSICAL - QUANTUM CORRESPONDENCE FOR IONIZATION OF ATOMS BY ION IMPACT AND ELECTROMAGNETIC PULSES

Carlos O. Reinhold and Joachim Burgdörfer
Dept. of Physics and Astronomy, Univ. of Tennessee, Knoxville, TN 37996–1200, USA
and Oak Ridge National Laboratory, Oak Ridge, TN 37931–6377, USA

ABSTRACT

We present a study of the ionization of atoms by pulses and collisions with ions whose main objective is to disentangle its classical and quantum mechanical aspects and to assess the validity of classical trajectory Monte Carlo methods. The existence of a classical limit is investigated as a function of the momentum transferred to the electron during the collision, the impact parameter, the energy and angle of the emitted electron, and the initial quantum level of the target. Applications related to recent data for electron ejection at large angles are presented. The connection between collisional ionization by charged particles and ionization by radiation fields is discussed with emphasis on ionization of atoms by half cycle pulses.

INTRODUCTION

The theoretical understanding of ionization caused by strong and non-periodic pulses is of fundamental interest. Such pulses can be delivered either by the time-dependent Coulomb field of fast ions passing by an atom or by ultrashort laser pulses with durations corresponding to a fraction of an optical cycle. For fast ion-atom collisions, well established approximations exist at high impact velocities (v_p) or small projectile charges (Z_p) for which the interaction of the target with the impinging ion can be treated as a small perturbation. For collision velocities comparable to the orbital velocity of target electrons and projectile charges of the order of or greater than the target nuclear charge (Z_t), more elaborate approaches are necessary which can treat the target and projectile nuclear fields on equal footing and can account for the strong coupling among all reaction channels. It is under these circumstances that classical trajectory Monte Carlo (CTMC) methods[1-3] have become widely and successfully utilized. The principal assumption underlying these models is that during and after the collision all interacting particles evolve classically in time. Quantum mechanical features enter solely via the initial boundary conditions of the evolution, i.e. the atomic initial state.

In this work we present an analysis of the classical character and limit of the quantum mechanics of ionization. We analyze the "existence of a classical limit" as a function of different parameters and identify regions in parameter space in which classical results mimic their quantum counterparts. We apply our analysis to two problems: the ejection of electrons into large angles in ion-atom collisions and the thresholds for ionization by ultrashort half-cycle electromagnetic pulses. Atomic units will be used throughout.

CLASSICAL SCALING INVARIANCE

In elucidating the classical - quantal correspondence, the exploitation of classical scaling invariances is very useful. Scaling invariances are based on the fact that the form of the classical equations of motion for Coulomb interactions is invariant under the scaling transformation $\vec{r}' = \vec{r}/\theta$, $\vec{p}' = \vec{p}\,\theta^{1/2}$, $t' = t/\theta^{3/2}$ (see e. g. Refs. 1,2). Quantum corrections break the scaling invariance. Deviation of results from classical scaling rules are therefore a clear signature of non-classical contributions to the expectation values of dynamical observables. Choosing $\theta = n'^2/n^2$, the scaling invariance implies, for example, that the ionization probabilities for a given impact parameter $\vec{b}$ from different n levels represented by microcanonical ensembles, P_n, are interrelated to each other; i.e.

$$P_{n'}(\vec{v}_p, \vec{b}) = P_n(\vec{v}_p \theta^{1/2}, \vec{b}/\theta) = P_1(\vec{v}_p\, n', \vec{b}/n'^2) \qquad (1)$$

In other words, the ionization probabilities from any initial state can be reduced to a single universal function.

MOMENTUM TRANSFER ANALYSIS OF IONIZATION

We first analyze a simplified problem which we name the "kicked hydrogen atom". The underlying idea is to study the classical and quantal responses of an electron in a hydrogenic atom to a δ-shaped ultrashort pulse or "kick" $\Delta \vec{p}$ which occurs at a time t_0. The Hamiltonian for this problem is

$$H = \frac{p^2}{2} - \frac{Z_t}{r} - \vec{r}\cdot\Delta\vec{p}\;\delta(t - t_0) \qquad (2)$$

where $\vec{p}$ and $\vec{r}$ are the momentum and position vectors of the electron. It turns out that the ionization by short electromagnetic pulses and by fast ion impact can be analyzed in terms of generalizations of this simplified model. The beauty of this model is the availability of exact solutions in both classical and quantum mechanics.

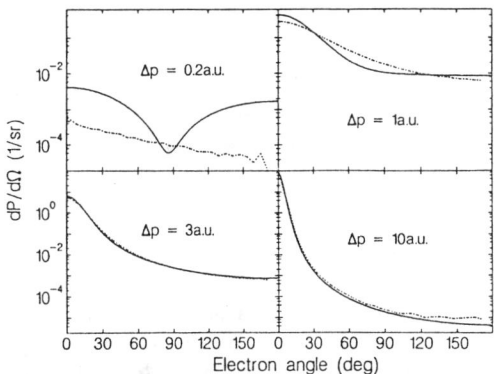

Figure 1: Quantal (solid line) and classical (dashed line) angular distribution of ejected electrons as a function of the emission angle with respect to the z-axis for a hydrogen atom subject to momentum transfers of different magnitude directed towards the positive z-axis.

The classical angular distribution of ejected electrons is compared in Fig. 1 with its quantal counterpart for a hydrogen atom initially in its ground state. For Δp =0.2a.u. the classical calculations clearly underestimate the angular distribution of ionized electrons. On the other hand, for increasing values of Δp the classical results describe the spectrum increasingly accurately, in particular around the region of high intensity. The discrepancies for small Δp values can be easily traced to non-classical transitions. The shape of the quantal angular distribution results from the fact that the angular momentum of the electron is quantized and can only change in one atomic unit. Obviously, dipole transitions are missing in the classical distributions.

The break-down of classical scaling invariances is shown in Fig. 2 for the total ionization probability for different initial energy levels $n = 1, 3$, and 10 as a function of the scaled momentum transfer $n\Delta p$. Note that if classical scaling invariances were to hold, results for all n-levels would lie on the same universal curve as classical results do. The larger n, the larger the range of momentum transfers for which the quantal curve agrees with the scaling invariant classical result. Fig. 2 may be viewed as an illustration of the correspondence principle at work. In the limit $n \to \infty$ the quantum results converge to the same universal curve which, moreover, coincides with the classical result. The convergence is, however, non-uniform in Δp which is the cause of the break-down of the classical approximation for any finite n level for sufficiently small Δp values.

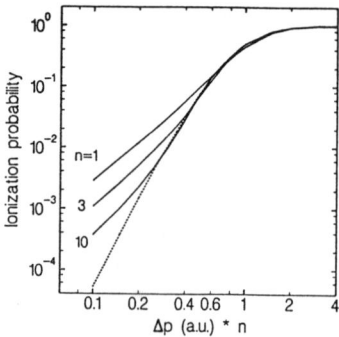

Figure 2: Quantal (solid lines) and classical (dashed line) total ionization probability of H(n) as a function of the scaled momentum transfer $n\,\Delta p$.

The critical momentum transfer Δp_{cr} below which the classical ionization probability begins to separate from the quantal result is approximately given by

$$\frac{\Delta p_{cr}^2}{2} \simeq \frac{1}{2}\left(\frac{Z_t^2}{2n^2} - \frac{Z_t^2}{2(n+1)^2}\right) \simeq \frac{\omega_{Bohr}}{2} \qquad (3)$$

The meaning of this equation is that ionization becomes "classically suppressed" when the mean energy transferred to the electron is smaller than half of the quantum energy gap between the initial state and the closest neighboring level. As soon as the energy gap can be overcome, many excited states are populated whose superposition can be increasingly successfully mimicked by classical dynamics. Note that classical ionization is possible for any value of Δp. However, since classical ionization in the limit of small Δp relies on the high-momentum components of the atomic momentum distribution, the classical ionization probability becomes proportional to $(\Delta p)^5$ whereas the quantum mechanical one becomes proportional to $(\Delta p)^2$ in the dipole limit.

IMPACT PARAMETER ANALYSIS OF IONIZATION

An alternative separation of the ionization process into classical and quantal transitions is often discussed in the literature in terms of the magnitude of the impact parameter.[2,4] If the impact parameter is large enough, the perturbation produced by the impinging projectile is so small that ionization becomes classically suppressed. The quantum mechanical origin of ionization at large impact parameters can be clearly illustrated within the first Born (B1) approximation for which the transition amplitude for a given impact parameter is given by

$$A^{B1}(i \to f, \vec{b}) = -i <\varphi_f| \int_{-\infty}^{+\infty} dt\, U_p e^{i(\epsilon_f - \epsilon_i)t} |\varphi_i> \quad (4)$$

where φ_i (φ_f) is the initial (final) state of the electron with binding energy ϵ_i (ϵ_f), and U_p denotes the field of the projectile. For large impact parameters, a dipole approximation can be used for U_p: $U_p(\vec{r},t) = -Z_p/|\vec{r} - \vec{R}(t)| \simeq -Z_p/R(t) - \vec{r} \cdot \vec{E}(t)$, where $\vec{E}(t) = Z_p \vec{R}(t)/R^3(t)$ is the average perturbing force experienced by the electron which resembles a time-dependent uniform electric field. Using $\omega = (\epsilon_f - \epsilon_i)$,

$$A^{B1}(i \to f, \vec{b}) \simeq i <\vec{r}>_{f,i} \cdot \vec{\tilde{E}}(\omega) \quad (5)$$

where $<\vec{r}>_{f,i} = <\varphi_f|\vec{r}|\varphi_i>$ and $\vec{\tilde{E}}(\omega) = \int_{-\infty}^{+\infty} dt\, e^{i\omega t} \vec{E}(t)$. We note that Eq. 5 is equivalent to the Weizsäcker-Williams method of virtual quanta.[4,5] That is, the time-dependent field of the projectile for large impact parameters can be replaced by that arising from an equivalent flux photons and, therefore, the collisional ionization process becomes very similar to the photoionization process. A given transition is quantum mechanically favored if the frequency associated with the jump in the binding energy of the electron has a relatively important weight in the spectrum of frequencies of the projectile field. This contrasts with classical ionization by photons or charged particles which is mainly governed by the strength of the field or, equivalently, the magnitude of Z_p. Ionization becomes "classically suppressed" when the projectile charge is small and the contribution from large impact parameters becomes appreciable relative to the contribution associated with small impact parameters. Analysis of $\tilde{E}(\omega)$ shows that the region of impact parameters that contribute appreciably to the ionization cross section is given by $b < b_{max} = v_p/\omega$ which is often referred to as the adiabatic cut-off distance. The larger the collision velocity, the larger the region of impact parameters which contribute appreciably to the ionization cross section. This is a direct consequence of the long range nature of the Coulomb field of the projectile. In contrast, the range of impact parameters that contribute to classical ionization at high impact energies for finite Z_p values is approximately given by the radius of the atom.

Conversely, ionization at small impact parameters has been traditionally associated with "hard" collisions. The existence of a well defined classical limit of the quantal ionization probability for large momentum transfers discussed in the previous section would, in turn, imply the validity of classical dynamics for small impact parameter collisions. That this view is not correct is demonstrated in Fig. 3 which displays the ionization probability for 5MeV p + H(1s) collisions as a function of the impact parameter integrated over all energies and angles of the ejected electron. The quantum results in first Born approximation which is expected to be valid at this energy is in sharp disagreement with the CTMC result over the whole range of impact parameters. Of particular importance is here the large deviation at small impact parameters. To verify whether this disagreement is due to the improper radial distribution associated with an initial microcanonical ensemble, we have also performed CTMC calculations with the initial ensemble proposed by Cohen[6] which reproduces exactly the radial distribution of H(1s). The resulting probabilities are seen to extend beyond the cutoff of the microcanonical probabilities at 2 a.u. . However, the small impact parameter disagreement with the B1 calculations still persists.

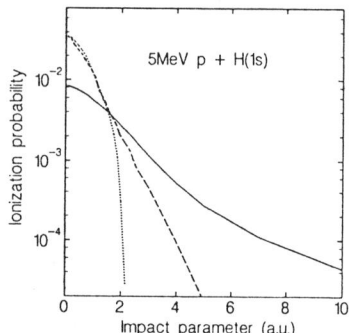

Figure 3: Total ionization probability as a function of the impact parameter in 5MeV p + H(1s) collisions: quantum mechanical B1 result (solid line); CTMC results using an initial microcanonical distribution (dotted line) and the initial distribution of Ref. 6 (dashed line).

To gain further insight into this discrepancy we have analyzed[7] the dependence of the ionization probability at small impact parameters ($b \simeq 0$) on the collision energy and the initial quantum level of the target n. At high impact energies and small impact parameters, the projectile passes so rapidly through the target atom that its main effect consists of transferring an impulsive momentum to the electron which is approximately given by $\Delta \vec{p}(\vec{r}) \simeq -2Z_p \hat{\eta}/(\eta v_p)$, where $\vec{\eta}$ is the initial coordinate of the electron with respect to the projectile in the plane perpendicular to $\vec{v}_p$. Hence, the ionization problem becomes equivalent to the ionization problem of a kicked hydrogen atom with a Hamiltonian

$$H = \frac{p^2}{2} - \frac{Z_t}{r} - \Delta S(\vec{r}) \; \delta(t) \qquad (6)$$

where $\Delta S(\vec{r}) = -2Z_p ln(\eta)/v_p$ represents the change of the phase of the wavefunction associated with the local momentum transfer $\Delta \vec{p}(\vec{r}) = \nabla_{\vec{r}} \Delta S(\vec{r})$ at $t = 0$. Note that there exists an essential difference between the kicked hydrogen atoms described by the Hamiltonians of Eqs. 2 and 6: the atom is subject to a uniform momentum transfer in the former case, whereas the electron suffers a momentum transfer that depends on its coordinate in the latter case.

This difference has immediate consequences for the $b \simeq 0$ behavior of classical ionization. In the case of ionization at high collision energies and small impact parameters, only electrons in the neighborhood of the path of the projectile can be ionized. The reason why the classical ionization probability at $b \simeq 0$ is larger than the quantum mechanical one is related to the fact that the momentum distribution of these electrons contains very high momentum components. That is, in the neighborhood of $r \simeq 0$ the momentum of the electron is very large since $p^2/2 = E_i + Z_t/r$. While the average value $< \Delta p >$ is below the critical value in Eq. 3, it is the high-momentum tail of the Compton profile that leads to an enhancement of classical ionization in comparison with quantum mechanical ionization. In other words, the break down of the classical treatment is due to the existence of a strong correlation between r and p in violation of the lack of correlation between conjugate variables in quantum mechanics.

APPLICATIONS

1 - ANGULAR DISTRIBUTIONS OF EJECTED ELECTRONS

The results of the previous sections indicate that ionization becomes classically suppressed for small momentum transfers. Thus, an improved description of the spectrum of ejected electrons in ion-atom collisions can be obtained by correcting the classical spectrum for the quantal portion which is associated with classically suppressed ionization due to small momentum transfers (i.e. for momentum transfers smaller than Δp_{cr} in Eq. 3). The conceptual advantage of this combined model is that the quantum correction comes from a regime where perturbation theory is expected to be valid and classical mechanics fails while the non-perturbative regime can be treated exactly, though classically.

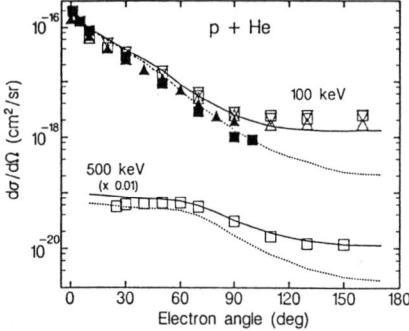

Figure 4: Angular distribution of ejected electrons in 100 and 500keV p + He collisions. Experiments: Ref. 9 (open symbols), Ref. 10 (full squares), and Ref. 11 (full triangles). Theory: CTMC (dotted line), CTMC plus small momentum transfer contribution in first Born (solid line).

It is noteworthy that there exists evidence of classical suppression of ionization even at intermediate impact energies for which CTMC methods are generally employed. This is shown in Fig. 4 in which we analyze the contribution of small momentum transfers in first Born approximation to the angular distribution of ejected electrons in 100 and 500keV p-He collisions. Corrections to the CTMC calculation for small momentum transfers enhance the cross sections at backward emission angles up to a magnitude consistent with the measurements of Rudd et al.[9] This enhancement originates from non-classical dipole emission channels and explains the large disagreement previously observed between classical and quantal calculations at backward angles.

It is interesting to compare the angular distributions of Fig. 4 with the ones arising from the simplified problem of the kicked hydrogen atom (Fig. 1). Clearly, such a comparison can refer only to qualitative features since comparatively slow collisions cannot be reduced to an impulsive momentum transfer from the projectile to the electron and the collision amounts to an average over a distribution of momentum transfers. Nevertheless, the shape of the angular distribution at an impact energy of 100keV is

sum over all momentum transfers at this energy, large momentum transfers dominate at forward angles but small momentum transfers have a non-negligible effect at backward angles.

2 - IONIZATION OF RYDBERG ATOMS BY HALF-CYCLE PULSES

Very recently, Jones et al[12] have studied experimentally the ionization of Rydberg states by subpicosecond "half-cycle" electromagnetic pulses. In this case the atom is subject to a strong unidirectional electrical field confined to a very short time interval. The authors have presented results on the ionization of Rydberg Na atoms in the region where the duration of the pulse T_p is shorter than or of the order of the classical orbital period T_n, n being the effective quantum level of the atom. This regime is of considerable conceptual interest as it provides a bridge between ionization by radiation fields and collisional ionization by charged particles at high energies. A hydrogen atom subject to a pulsed electric field is described by the Hamiltonian

$$H = \frac{p^2}{2} - \frac{1}{r} - F(t)\,z \qquad (7)$$

where $F(t)$ denotes an external electric pulse which is directed towards the positive z-axis. Clearly, this problem becomes equivalent to the simplified problem of the kicked atom in the limit of ultrashort pulses. The corresponding "kick" or momentum transfer is given by $\Delta p = \int_{-\infty}^{\infty} F(t)\,dt$.

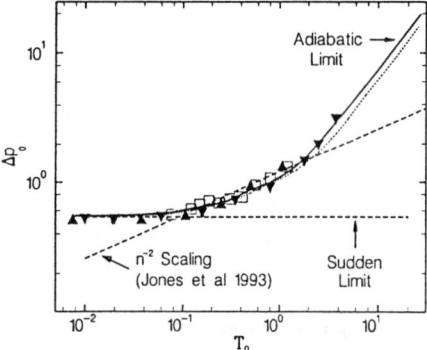

Figure 5: Scaled momentum transfer for 10% ionization threshold of H(n,l=2,m=0) atoms as function of the scaled pulse duration: classical results for inverse parabolic (solid line) and rectangular (dotted line) pulses, quantum mechanical result for ionization of 10d (solid triangles) and 5d (inverted solid triangles) states by rectangular pulses, and experimental data of Ref. 11 for Na(nd) atoms (equivalent to H(n,l=2,m=0)) multiplied by a factor of 2.5 (open squares).

In Fig. 1 we display the data for the 10% ionization thresholds of Ref. 11 for different nd (m=0) states in terms of the scaled variables $T_0 = T_p/T_n$ and $\Delta p_0 = n\Delta p$ together with our classical and quantum calculations. These calculations were performed for hydrogen using the fact that the quantum defect in d states of Na is very small ($\delta = 0.01$). Classical calculations were also performed for Na atoms using realistic core potentials to verify that core effects are negligible in the region $T_0 < 1$. In order to show that the results are insensitive to the shape of the pulse, we present two classical calculations for pulses of different shapes: a rectangular pulse and an inverse parabolic pulse. Classical calculations have also been performed for different quantum levels and pulses of different duration to verify that classical invariances hold for a fixed initial angular momentum of the electron. The quantum mechanical calculations have been performed for $n = 5$ and 10 and we display results for the regions of T_p for which convergence has been reached with the basis sizes used.[13]

All classical and quantum calculations for 10% ionization thresholds as well as the data lie on the same universal curve in the $(\Delta p_0, T_0)$ diagram, verifying the classical-quantal correspondence for ionization by short pulses. Two limiting cases of this curve are of particular interest. In the adiabatic limit of long pulses we find $\Delta p_0 \simeq Const.T_0$ which is equivalent to the $F \propto n^{-4}$ dependence for static electric fields. The coefficient depends, however, on the shape of the pulse and the substate populated. In the opposite limit of ultrashort pulses ($T_0 \ll 1$) the curve approaches the asymptote $\Delta p_0 \simeq 0.54$ giving rise to a dependence $\Delta p \simeq 0.54 n^{-1}$ a.u. . It is important to note that this asymptote is exact and can be obtained independently from inelastic transition form factors.

The complete classical-quantum correspondence for the 10% ionization threshold by half-cycle pulses can be easily understood in terms of the criterion in Eq. 3. From Fig. 2 it is obvious that 10% ionization (i.e. an ionization probability of $P_n = 0.1$) by ultrashort pulses requires a scaled momentum transfer of $n\Delta p \simeq 0.5$. Except for very small n, the latter corresponds to a large mean energy transfer compared to $\omega_{Bohr}/2$. According to the criterion in Eq. 3, we therefore expect that the ionization threshold can be described by classical dynamics. While this argument is rigorously valid only in the limit of ultrashort pulses, it approximately applies to pulses of duration $T_0 \leq T_n$ employed in the experiment.

ACKNOWLEDGEMENTS

Work supported by NSF and the US DOE, Office of Basic Energy Sciences under contract No. DE-AC05-84OR21400 with Martin Marietta Energy Systems Inc.

REFERENCES

[1] R. Abrines, I. C. Percival, Proc. Phys. Soc. London **88** (1966) 861.
[2] I. C. Percival and D. Richards Adv. At. Mol. Phys. **11** (1975) 1.
[3] R. E. Olson and A. Salop, Phys. Rev. A **16** (1977) 531.
[4] M. R. Flannery 1983 *Rydberg States of Atoms and Molecules*, edited by R. F. Stebbings and F. B. Dunning (Cambridge University Press, New York), p. 393.
[5] J. D. Jackson 1962 *Classical Electrodynamics* (John Wiley and Sons, New York), Ch. 13.
[6] J. S. Cohen, J. Phys. B **18** (1985) 1759.
[7] C. O. Reinhold and J. Burgdörfer, to be published in J. Phys. B (1993).
[8] W. Fritsch and C. D. Lin, 1991 Phys. Rep. **202** (1991) 1.
[9] M. E. Rudd et al At Dat Nuc Dat Tables **18** (1976) 413.
[10] D. K. Gibson and I. Reid J. Phys. B **19** (1986) 3265.
[11] G. C. Bernardi et al Phys. Rev. A **40** (1989) 6863.
[12] R. R. Jones et al, Phys. Rev. Lett. **70**, 1236 (1993).
[13] C. O. Reinhold, M. Melles, and J. Burgdörfer, Phys. Rev. Lett. **70**, 4026 (1993).

RADIATIVE STABILIZATION IN DOUBLE–ELECTRON CAPTURE

Nathalie Vaeck[1], Hugo W. van der Hart and Jørgen E. Hansen
Van der Waals-Zeeman Laboratory, University of Amsterdam,
Valckenierstraat 65, NL-1018 XE Amsterdam, The Netherlands

ABSTRACT

Calculations of the degree of radiative stabilization that can be expected for states belonging to the $4\ell4\ell'$ and $4\ell5\ell'$ manifolds in the free O^{6+} atom are described. In this atom, the $4\ell4\ell'$ manifold overlaps the $3\ell n\ell'$ Rydberg series and the resulting interaction leads to an increase in stabilization for the $4\ell4\ell'$ terms while the $4\ell5\ell'$ states are located above the $N = 3$ threshold and therefore decay predominantly by autoionization. We obtain 27% stabilization for the $4\ell4\ell'$ and 3% for the $4\ell5\ell'$ states assuming a statistical population distribution. This should be compared with the recent experimental values of 50% and 11%, respectively, for the two configurations. Since a similar discrepancy has been obtained with the dynamic approach due to Bachau and coworkers, which describes the interaction between the $4\ell4\ell'$ states and the $3\ell n\ell'$ Rydberg series as a function of the distance between the two during the collision, we propose that another mechanism is needed to understand the experimental results. One possibility is that the $3\ell n\ell'$ states are populated more directly in analogy with the population of $2\ell n\ell'$ states in collisions between C^{6+} and He, for example, which has been discussed extensively the last few years.

INTRODUCTION

The transfer of electrons between a neutral target and a highly charged ion during slow ion-atom collisions, can to a good approximation be understood using an independent particle model in which the electrons are captured one by one from the target into excited states of the ion. One approach based on the independent particle model is the Extended Classical over Barrier Model, ECBM[1,2] which has been used successfully to predict cross sections and principal quantum numbers for the multiply-excited states populated by the charge transfer process. The model predicts in particular the existence of a "reaction window" a concept which has been very successful in predicting the excitation energy of the projectile after the collision and thus the energy range of the electrons emitted in the ensuing autoionization processes[3]. One of the consequences of the model is that only configurations consisting of electrons with equivalent or quasi-equivalent principal quantum numbers, n, are populated. Theoretical investigations[4,5] of the decay properties of doubly–excited $n\ell n'\ell'$ states with $n \simeq n'$ indicate that, in a single-configuration approach with a statistical population distribution, these

---

[1]Senior Research Assistant of the National Fund for Scientific Research (Belgium)

states decay by autoionization although it was emphasized that the distance to the closest limit does influence the radiative branching ratio markedly[4].

However, for double-electron capture in collisions between O^{6+} and He at 60 keV, in addition to the expected $3\ell n\ell'$ configurations with $n \simeq 3$, Stolterfoht et al[6] observed strong lines in the electron spectrum arising from the emission of Coster-Kronig electrons from $1s^2 2pn\ell$ Rydberg states with n values between 6 and 12. Evidence for the production of similar states was later found[7] for the system C^{6+} + He at 60 keV. In this case the 1s shell of the projectile is empty and therefore the $2pn\ell$ states can decay **radiatively** to the $1sn\ell$ states, while the $3\ell n\ell'$ states primarily autoionize, making it necessary to consider the radiative branching in order to obtain correct cross sections for the production of $2pn\ell$ ($n > 6$) states relative to the production of the lowest members of the $3\ell n\ell'$ series. Thus for this system **"radiative stabilization"** of the two captured electrons was found to play an important role in a double-capture process. In this case there is no nearby limit that can speed up the non-radiative decay while radiative decay, on the other hand, is boosted by the high energy of the photon corresponding to the $2p \rightarrow 1s$ transition.

Stolterfoht et al[6,7] explained the production of $(1s^2)2pn\ell$ configurations by a one-step double-capture process (Autoexcitation of Rydberg states), induced by electron-electron correlation, in which the potential curve corresponding to the incident channel (two electrons in the 1s shell of He) crosses the potential curve of the $(1s^2)2pn\ell$ states at some internuclear distance, where the two electrons interact and one electron is transferred into the $2p$ orbital while another is excited into a Rydberg orbital $n\ell$. Since then several other mechanisms have been proposed (for a recent discussion see Fremont et al[8]), however, the significance in the present context is that a fairly direct method of populating such states has been demonstrated. A characteristic of this process is that the $3\ell n\ell'$ and the $2\ell n\ell'$ configurations have rather different energies in the free atom. Thus it is possible to determine the relative populations of the two fairly precisely although experiment does not allow to determine with a similar certainty how the populations have come about.

Recently stabilization has been observed for another class of states where the $n\ell n\ell'$ states are degenerate with a Rydberg series in the **free** atom. Then it is not possible to determine the relative populations via the electron spectrum and evidence for the character of the intermediate states is rather speculative. To this category belongs the observation of very large degrees of stabilization for several collision systems by Roncin et al[9] and similar results have now been obtained by Gaboriaud et al[10] for some light projectiles. For example for O^{8+} in collision with Ar, 25% stabilization has been obtained. The initially populated configurations were identified by the authors as $4\ell 4\ell'$ and $4\ell 5\ell'$. However, while the $4\ell 5\ell'$ manifold corresponds to the ECBM predictions and therefore to a two-step sequential capture process, the $4\ell 4\ell'$ states lie outside the reaction window and Gaboriaud et al[10] proposed that they are formed in a direct one–step process.

The $4\ell 4\ell'$ states are straddling the $N = 3$ threshold and therefore a large number of them are lying in the same range of energy as the highest members of the $3\ell n\ell'$ Rydberg series. This fact is, as pointed out by Roncin et al[9], a common characteristic of all the states populated in collisional systems for which large degrees of stabilization have been observed until now. Bachau et al[11] proposed a mechanism (Auto Transfer to Rydberg states) based on the idea that those $4\ell 4\ell'$ states that lie above the $N = 3$ limits in the unperturbed ion might lie below the $N = 3$ limits during the collision due to a type of Post Collisional Interaction effect. During this part of the collision they can lose their $4\ell 4\ell'$ character by mixing with the $3\ell n\ell'$ ($n > 18$) states, which Bachau et al[11] assumed to stabilize radiatively, and these authors proposed that this mechanism is responsible for the large degrees of stabilization observed in these atoms.

This year, Roncin et al[12] have presented quantitative estimates of this effect for N^{5+} based on close coupling calculations taking the time dependent interaction between the $4\ell 4\ell'$ states and the $3\ell n\ell'$ Rydberg series into account. However, the results corresponds to a degree of stabilization a factor of 2 smaller than observed. This is one reason that we have found it of interest to calculate the degree of stabilization which can be expected for the free ion when the interaction between the overlapping $4\ell 4\ell'$ and $3\ell n\ell'$ configurations is taken into account in a static approximation. This information has not been available before for all the 44 terms in the $4\ell 4\ell'$ configurations in any of the ions of interest. We present our final results for O^{6+} here, while a detailed discussion will be presented elsewhere[13].

CALCULATIONAL METHOD

As just mentioned the computational method is described by van der Hart et al[13] and we give here only the most salient points. The difficulty in the calculation is to take the interaction with very high Rydberg states into account. We have used a B-spline basis defined in a box[14] and the high Rydberg states makes the use of a much larger box necessary than for more tightly bound states and in fact even for continuum electrons interacting with tightly bound electrons.

Energies and wavefunctions of the $4\ell(4,5)\ell'$ states have been determined via the Truncated Diagonalisation Method (TDM) using a B-spline based basis set. B-splines of order 7 were used. In the TDM all two-electron basis states which includes a $1s, 2\ell$, or 3ℓ orbital are left out of the calculations. The B-splines are defined in a cavity with a radius of 40 au using an exponential knot set. In the determination of the interaction between the $4\ell 4\ell'$ and the $3\ell n\ell'$ states, the $n\ell$ Rydberg states have been obtained using a B-spline basis set of order 9 in a cavity with a much larger radius (500 au), which allows for an accurate description of states with n up to 40. For the $^1G^e$ symmetry, the cavity radius was extended to 1000 au, allowing a description of $n = 50$ states.

The continuum wavefunctions have been obtained in the small cavity using a partially exponential, partially linear knot set, in order to describe the oscillations of the continuum wavefunctions properly. In the determination of interaction integrals, all orthogonality constraints are fulfilled due to small range of the $(1,2,3,4)\ell$

wavefunctions, which makes the dielectronic interaction with these electrons negligible for $r > 40$ au. The dispersion of the $4\ell 4\ell'$ states in the Rydberg series is determined by a CI approach in which all Rydberg series and the $4\ell 4\ell'$ states with the same LS symmetry are included. Using the result of this CI calculation, the autoionization width of all mixed $4\ell 4\ell'$ and $3\ell n\ell'$ states is determined using Fermi's golden rule without the coupling between degenerate continua included. This is expected to be a good approximation for total widths[15].

The radiative decay rate of each state is obtained from the following formula

$$A^r = A^r_{Ryd} + C^2_{44} \times (A^r_{44} - A^r_{Ryd}) \qquad (1)$$

in which C^2_{44} is the total (summed) $4\ell 4\ell'$ character. A^r_{Ryd} and A^r_{44} are the average radiative decay rates of the Rydberg series of a particular LS symmetry, mainly decay of the inner electron, and of the $4\ell 4\ell'$ terms of the same symmetry, respectively. The spread in radiative decay rates among the $3\ell n\ell'$ as well as the $4\ell 4\ell'$ states is much smaller than the spread in the autoionization rates. From these decay rates, a branching ratio for radiative decay, R^r, can be obtained for each state as the contribution from radiative decay to the total decay. The **"stabilization ratio"** of the $4\ell 4\ell'$ states within one LS symmetry can then be obtained from

$$<R^r> = \frac{\sum C^2_{44} R^r}{\sum C^2_{44}} \qquad (2)$$

in which the summation is over all states. For the $4\ell 4\ell'$ terms above the $N = 3$ threshold, R^r is very small because of the importance for the non-radiative decay of the opening of a new decay channel. Finally the total stabilization ratio can be derived by including the proper statistical weights for each symmetry. This method of calculation assumes that only the $4\ell 4\ell'$ component of each state is populated initially.

RESULTS AND DISCUSSION

In collisions with Ar, Gaboriaud et al[10] have interpreted their results to mean that bare O^{8+} ions initially capture two electrons in the $4\ell 4\ell'$ and the $4\ell 5\ell'$ states with probabilities of 29% and 71%, respectively. Using coincidence energy gain spectroscopy, Roncin et al[9] and Gaboriaud et al[10] observed a total stabilization ratio of 25% for these states and determined that this corresponds to 50% radiative stabilization for the $4\ell 4\ell'$ states and 11% for the $4\ell 5\ell'$ states[10]. In Table I we summarize our results for the stabilization ratios (eq. 2) for each LS symmetry in the $4\ell 4\ell'$ and $4\ell 5\ell'$ manifolds.

The total stabilization ratio for the $4\ell 4\ell'$ manifold is 26.8% while it is only 2.6% for the $4\ell 5\ell'$ manifold. This is in disagreement with the $(K, T)^A$ model[16] which predicts larger fluorescence yields for $n\ell n'\ell'$ configurations with non-equivalent than with equivalent n values and the opposite result in O^{6+} is probably due to the fact that the $4\ell 5\ell'$ states can decay to the nearby $N = 3$ limit while most of the $4\ell 4\ell'$ terms autoionize to the much deeper lying $N = 2$ limit.

Table I: Stabilization ratios for the $4\ell 4\ell'$ and $4\ell 5\ell'$ states in O^{6+}.
Ratios are given averaged over all states with the same LS symmetry. The number of states in each LS symmetry is indicated as well and (in parenthesis) the number of states below the $N = 3$ threshold.

$4\ell 4\ell'$			$4\ell 5\ell'$		
LS	States	$<R^r>$ %	LS	States	$<R^r>$ %
$^1S^e$	4(3)	16.7	$^1S^e$	4	2.6
$^1P^o$	3(2)	8.8	$^3S^e$	4	18.
$^3P^o$	3	21.4	$^1P^o$	7	1.6
$^3P^e$	3(2)	39.1	$^3P^o$	7	5.5
$^1D^e$	5(4)	11.1	$^1P^e$	3	18.
$^3D^e$	2	38.0	$^3P^e$	3	4.1
$^1D^o$	2	64.2	$^1D^e$	8	0.7
$^3D^o$	2	34.9	$^3D^e$	8	3.3
$^1F^o$	3(2)	7.1	$^1D^o$	5	19.
$^3F^o$	3	23.3	$^3D^o$	5	2.1
$^1F^e$	1	53.4	$^1F^o$	8	0.9
$^3F^e$	3	22.4	$^3F^o$	8	1.3
$^1G^e$	3	15.2	$^1F^e$	5	17.
$^3G^e$	1	17.5	$^3F^e$	5	1.5
$^1G^o$	1	27.9	$^1G^e$	6	0.3
$^3G^o$	1	31.3	$^3G^e$	6	3.3
$^1H^o$	1	16.6	$^1G^o$	4	1.9
$^3H^o$	1	5.8	$^3G^o$	4	1.3
$^3H^e$	1	72.2	$^1H^o$	4	0.2
$^1I^e$	1	19.0	$^3H^o$	4	0.7
			$^1H^e$	2	11.
			$^3H^e$	2	0.1
			$^1I^e$	2	0.0
			$^3I^e$	2	0.6
			$^1I^o$	1	0.1
			$^3I^o$	1	0.1
			$^1K^o$	1	0.0
			$^3K^o$	1	0.0

We note a significant difference between the stabilization ratios for the singlet and triplet terms. For the singlet $4\ell 4\ell'$ terms the stabilization ratio is 19.7% and for the triplets 29.4%. Another significant difference between the results in Table I is found between the terms with "unnatural parity" ($\pi = (-1)^{L+1}$) for which the stabilization ratio is 38.9% and those with "natural parity" ($\pi = (-1)^L$) for which the ratio is only 18.1%. For the configuration as a whole we obtain, as mentioned, a stabilization ratio of 27% assuming a statistical population distribution. However, the dependence on the population distribution is not very pronounced.

If no weighting is used, the stabilization ratio is 23.9% for the configuration as a whole while weighting by $(2S+1)(2L+1)^2$ leads to a value of 26.8% thus the same value as if a strictly statistical population is assumed. If particular terms are populated, Table I shows that stabilization ratios larger than 50% can be obtained. However, at the moment there is little direct evidence for a non-statistical population distribution.

The 27% stabilization is very high in comparison with the results obtained previously for states with equivalent n values (see for instance van der Hart and Hansen[17]) but nevertheless it is still very far from the 50% observed by Gaboriaud et al[10]. The dynamic calculations of Roncin et al[12] for the $4\ell 4\ell'$ singlet terms in N^{5+} give a stabilization ratio of 20%, very similar to our results for O^{6+} (20% stabilization for the singlet terms). However, also in N^{5+} a much larger stabilization ratio (>52%) is necessary to explain the experimental results[10]. Therefore it seems clear that another mechanism must be invoked.

Both we and Roncin et al[12] have assumed that initially the $4\ell 4\ell'$ states are populated. This is mainly because the ECBM predicts a capture producing states with near-equivalent n values. However, as pointed out above, the ECBM does not, in principle, apply in this case. One can instead imagine a more direct population of the $3\ell n\ell'$ states, similar to the situation for the $2\ell n\ell'$ states produced in O^{6+} + He collision mentioned earlier. If the $3\ell n\ell'$ states are populated directly, it is not difficult to obtain stabilization ratios above 50%. For example, the average stabilization ratios of the $3\ell n\ell'$ $^1F^o$ series as a function of the principal quantum number are shown in Figure 1.

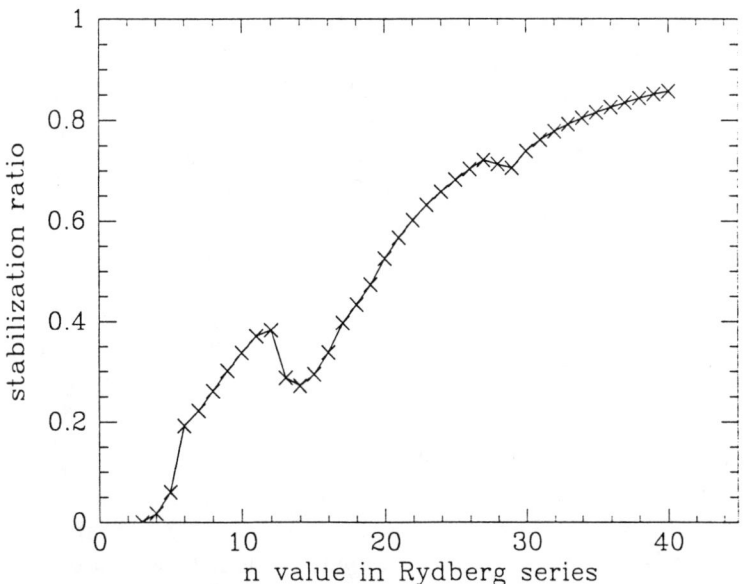

Figure 1: Average stabilization ratios for the six $3\ell n\ell'$ $^1F^o$ Rydberg series as a function of n.

In this figure we assume that initially only the $3\ell n\ell'$ Rydberg series and not the $4\ell 4\ell'$ states are populated and the numbers are obtained by modifying eq. 2 accordingly. The figure shows that for $n = 20$ or higher the stabilization ratio is larger than 50%. This behaviour is the same for the other LS symmetries and if we assume that the population is independent of n we obtain the average stabilization ratios for $3\ell n\ell'$ terms with $n = 10$ to $n = 30$ and $n = 10$ to $n = 18$ given in Table II.

Table II: Stabilization ratios for the $3\ell n\ell'$ Rydberg states in O^{6+}.
The ratios are obtained by averaging over all states with the same LS symmetry assuming that the population is independent of n for the states with n between 10 and 30 (column 2) and between 10 and 18 in the third column. The numbers assume that initially only the Rydberg series are populated, see the text

	$3\ell n\ell'$	
	$n = 10 - 30$	$n = 10 - 18$
LS	$< R^r >$	
	%	
$^1S^e$	34.0	29.5
$^1P^o$	56.2	40.4
$^3P^o$	66.9	56.1
$^3P^e$	62.5	49.9
$^1D^e$	46.8	29.2
$^3D^e$	81.7	69.3
$^1D^o$	89.7	82.2
$^3D^o$	70.8	60.9
$^1F^o$	51.6	34.6
$^3F^o$	83.0	69.5
$^1F^e$	90.8	84.0
$^3F^e$	68.5	56.2
$^1G^e$	49.7	40.7
$^3G^e$	86.6	77.6
$^1G^o$	87.9	78.7
$^3G^o$	81.2	73.2
$^1H^o$	57.4	42.9
$^3H^o$	85.5	77.4
$^3H^e$	92.8	88.6
$^1I^e$	83.8	83.1

Stolterfoht et al[6] found that $2\ell n\ell'$ states with n between 6 and 12 are populated in O^{6+} + He collisions so that a population of a range of n values is reasonable as is also suggested by the width of the observed electron spectra[10]. However, it is of course not realistic to assume that the population is independent of n, even over a limited range of n, and the results in Table II are shown primarily to indicate the large degrees of stabilization which can be reached by directly populating the Rydberg series.

CONCLUSIONS

One of the main uncertainties in the proposed explanations is clearly the first step of the reaction since the $4\ell 4\ell'$ states lie outside the reaction window and this problem is compounded by the fact that the stabilization is observed in collisions with Ar for which potential curves are less reliable. That also triplet terms can be populated in collisions with Ar adds to the complexity of the problem. We have found that the assumption, that the initial population goes into $4\ell 4\ell'$ states, leads to stabilization ratios a factor of two smaller than the experimental values, independent of whether a dynamic or a static approach is used. However, the very large stabilization ratios for some specific $4\ell 4\ell'$ terms show that explanations in terms of non-statistical population distributions are possible although explanations based strictly on a preferential population of high L values, as observed in other cases[18], cannot without further assumptions explain the results. Since such assumptions at the moment cannot be justified, it is worthwhile to explore other possibilities. We have considered the possibility of a more direct population of the Rydberg series (a mechanism which has been established already[6,7]). A population of $3\ell n\ell'$ terms with n between 10 and 20 and L around 3 would be able to explain the experimental results, while if the electrons acquire larger ℓ values following the initial pick-up, as suggested by Kazansky[19], stabilization is likely to be even more probable. In practise, it can be expected that several different effects contribute to the experimental observations.

ACKNOWLEDGEMENTS

We would like to thank Drs P. Roncin and A. Bárány for stimulating electron(ic) exchanges.

REFERENCES

1. A. Bárány, G. Astner, H. Cederquist, H. Danared, S. Huldt, P. Hvelplund, A. Johnson, H. Knudsen, L. Liljeby and K.G. Rensfelt, Nucl. Instr. Meth., B9, 397 (1985).

2. A. Niehaus, J. Phys. B, 19, 2925 (1986).

3. E.M. Mack, Ph.D. thesis, University of Utrecht (1987) unpublished.

4. J.E. Hansen, J. Phys., 50 C1, 603 (1989).

5. E. Luc-Koenig and J. Bauche, J. Phys. B, 23, 1763 (1990).

6. N. Stolterfoht, C.C. Havener, R.A. Phaneuf, J.K. Swenson, S.M. Shafroth and F.W. Meyer, Phys. Rev. Lett., 57, 74 (1986).

7. N. Stolterfoht, K. Sommer, J.K. Swenson, C.C. Havener and F.W. Meyer, Phys. Rev. A, 42, 5396 (1990).

8. F. Fremont, K. Sommer, D. Lecler, S. Hicham, P. Boduch, X. Husson and N. Stolterfoht, Phys. Rev. A, 46, 222 (1992).

9. P. Roncin, M.N. Gaboriaud and M. Barat, Europhys. Lett., 16, 551 (1991).

10. M.N. Gaboriaud, P. Roncin and M. Barat, J. Phys. B, 26, L303 (1993).

11. H. Bachau, P. Roncin and C. Harel, J. Phys. B, 25, L109 (1992).

12. P. Roncin, M.N. Gaboriaud, M. Barat, A. Bordenave–Montesquieu, P. Moretto-Capelle, M. Benhenni, H. Bachau and C. Harel, J. Phys. B, submitted (1993).

13. H.W. van der Hart, N. Vaeck and J.E. Hansen, J. Phys. B, to be published.

14. J.E. Hansen, M. Bentley, H.W. van der Hart, M. Landtman, G.M.S. Lister, Y.-T. Shen and N. Vaeck, Phys. Scr., T47, 7 (1993).

15. F. Martín, I. Sánchez and H. Bachau, Phys. Rev. A, 40, 4245 (1989).

16. Z. Chen and C.D. Lin, J. Phys. B, 26, 957 (1993).

17. H.W. van der Hart and J.E. Hansen, J. Phys. B, 25, 2267 (1992).

18. J.H. Posthumus and R. Morgenstern, J. Phys. B, 23, 2293 (1990).

19. A.K. Kazansky, J. Phys. B, 25, L381 (1992).

DISSOCIATIVE RECOMBINATION OF STORED AND PHASE-SPACED COOLED MOLECULAR IONS IN CRYRING

M. Larsson, G. Sundström, M. Carlson
Physics Department I, The Royal Institute of Technology (KTH), S-100 44
Stockholm, Sweden

H. Danared, A. Källberg, K.G. Rensfelt, M. af Ugglas,
L. Broström, S. Mannervik, P. Sigray
Manne Siegbahn Institute of Physics, S-104 05 Stockholm, Sweden

A. Filevich
Physics Department, CNEA, Tandar, 1429 Buenos Aires, Argentina

S. Datz
Physics Division, Oak Ridge National Laboratory, P.O. Box 2008, Oak Ridge, Tennessee 37831, USA

J.R. Mowat
Department of Physics, North Carolina State University, Raleigh, North Carolina 27695-8202, USA

ABSTRACT

Ion storage rings offer several advantages for studies of recombination of molecular ions by electrons, a process which in general leads to dissociation. During the last year dissociative recombination of H_3^+, HD^+, H_2^+, D_2^+, $^3HeH^+$, $^4HeH^+$ and $^4HeD^+$ have been studied in storage rings in Stockholm, Heidelberg, Tokyo and Aarhus. Only a fraction of the results obtained at these places have been processed to the point where they have been published in the literature. Hence, the discussion of some of the results in this article must be regarded as tentative.

INTRODUCTION

The capture of a free electron by a molecule may be regarded as an elementary step in the general field of electron transfer reactions. If the molecule is neutral, the process leads to the formation of a temporary negative ion the fate of which depends on the electron affinity of the molecule. One possible outcome is unimolecular decomposition of the negative molecular ion into one neutral and one negatively charged fragment; this process is called dissociative attachment. If the free electron is captured by a positively charged molecular ion, the process is called dissociative recombination, the term "dissociative" being added in order to emphasize that the capture normally is followed by a unimolecular decomposition of the neutralized molecule into neutral (or, however less likely, positively and negatively charged) fragments. The process can thus be described as $AB^+ + e \rightarrow A + B$.

There is a fundamental difference between dissociative attachment and dissociative recombination: attachment of a free electron to a neutral molecule is a direct process which only involves a transition between the initial neutral molecule plus a free electron and the final negative ion state. Thus, the cross section for dissociative attachment is in general negligible for very low electron kinetic energies (say << 1 eV) and peaks at 1-10 eV (roughly speaking). Dissociative recombination, on the other hand, may occur via capture into predissociating Rydberg states of the neutral molecule, or through a direct transition to a repulsive doubly excited state. The cross section may have structures, similar to those for dissociative attachment, at several eV, but at low energies the cross section scales as $1/E^n$, where E is the collision energy and n is a number close to unity. This fact complicates the experimental study of dissociative recombination enormously compared to the study of dissociative attachment. Likewise, the competition between different dissociation pathways at low energies complicates a theoretical treatment of the process. A further complication with recombination studies is that it is very difficult to prepare a molecular ion in a well defined quantum state. The approach to an experimental access to low energy dissociative electron-ion collisions has been the merged-beam technique.[1] In this paper we will describe the most recent development of the merged-beam technique for studies of dissociative recombination: the use of beams of molecular ions stored and phase-space manipulated in heavy-ion cooler rings. The first results from dissociative recombination ion storage ring experiments have recently appeared in the literature.[2-4]

BASIC ASPECTS OF DISSOCIATIVE RECOMBINATION STUDIES IN STORAGE RINGS

A good starting point for a description of the basic aspects and main advantages of studies of dissociative recombination in a storage ring is Fig. 1, which schematically shows CRYRING in Stockholm. It should, however, be pointed out that the experiments performed at TSR in Heidelberg, TARN II in Tokyo and ASTRID in Aarhus are in practice very similar to the CRYRING experiments. The ions are produced in an electron-impact ion source (MINIS), extracted at 30 keV, accelerated to 300 keV/u in a four-rod radiofrequency quadrupole (RFQ) and injected into the storage ring. In CRYRING the ions are further accelerated to $96(q/M)^2$, where q is the charge state and M is the molecular mass. The acceleration to full energy takes about 2 s. After reaching full energy the ions are cooled by a collinear beam of velocity-matched electrons (E-COOLER). For H_3^+, Schottky noise spectra showed that the longitudinal velocity spread was reduced from 8×10^{-4} to 6×10^{-5}, which also took about 2 s. Transverse cooling is

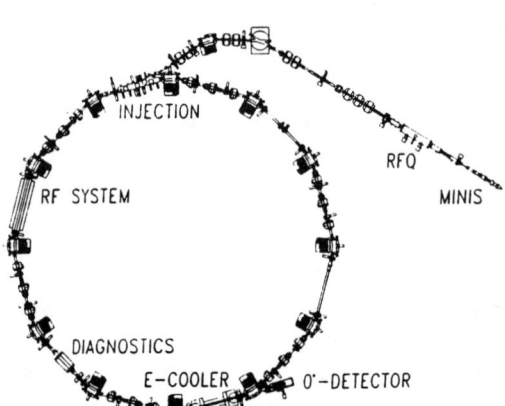

Fig. 1 Schematic diagram of CRYRING.

expected to proceed slower, however, the cooling time has not been measured. After a variable time of a few to tens of seconds, during which cooling is continuously applied, the electron beam energy is switched from E_{cool}, for which the electrons are velocity matched with the ions, to $E_{meas} \sim (\sqrt{E_{cool}} + \sqrt{E_{rel}})^2$, where E_{rel} is the value of the relative collision energy for which the cross section is to be determined and E_{meas} is the energy applied to the electron cooler. In order to avoid electron drag force and heating of the ion beam, the electron cooler is only kept at E_{meas} for 0.3-0.5 s and then switched back to E_{cool}. During the E_{meas} phase, the number of neutral particles produced in the electron-cooler section are counted by an ion implanted silicon detector (0^0-DETECTOR in Fig. 1). The switching between E_{cool} and E_{meas} is repeated a couple of times per injection cycle. Owing to collisions in the rest gas, which is kept at a vacuum in the low 10^{-11} torr region, the stored beam is eventually destroyed and after 10-30 s the storage ring is filled with new ions. In order to obtain absolute cross sections, the current of the circulating beam is measured by a current transformer.

The procedure described above offers the following advantages for measurements of dissociative recombination:

i) The ion beam is circulated in the storage ring so that each ion is passing the electron target of the order of a million times per second. This greatly increases the effective current. Combined with the high electron beam current provided by a cooler designed for high ion beam velocities, ion storage ring technology provides a large enhancement of luminosity compared to single-pass experiments.

ii) Phase-space cooling of the stored molecules reduces the energy spread in the ion beam to the point where it can be neglected when cross sections are extracted. The cooling process also aligns the ion beam with respect to the electron beam, thus minimizing deterioration of the resolution due to beam misalignment.

iii) Ion storage facilitates relaxation of excited vibrational levels in infrared active molecules (i.e. essentially all molecules except homonuclear) so that experiments can be performed with ions populating only their vibrational ground state. This may be the single most important advantage of the ring technique for handling molecular ions.

iv) By performing the experiments at high energies, processes like $AB^+ + Rg \rightarrow A + B$ (Rg represents a rest gas atom) are strongly suppressed. In essence this means that the dissociative recombination data can be recorded against a zero background.

The rate of the reaction $AB^+ + e \rightarrow A + B$ in the storage ring is given by the relation $R = \rho_e v_{rel} \sigma_{DR} l/L$ ion^{-1}s^{-1}, where ρ_e is the electron density in the merged-beam region, v_{rel} is the relative collision velocity, σ_{DR} is the dissociative recombination cross section, l is the length of the interaction region (l=0.9 m for CRYRING) and L is the circumference of the storage ring (L=51.6 m for CRYRING). This relation holds of course only strictly for the ideal case of monoenergetic beams. In practice, at least for very low collision energies, one must consider the deviation from monoenergetic conditions. For a cooled beam, the ions have the same temperature, and hence the same energy spread, as the electrons. Since the ions are several thousands times heavier than the electrons, the ion velocity spread can be neglected. An electron beam accelerated to several keV is characterized by transverse and longitudinal temperatures $T_\perp$ and $T_\parallel$, which differ by orders of magnitude. The term $v_{rel}\sigma_{DR}$ in the expression for R is then replaced by $<v_{rel}\sigma_{DR}> = \int v_{rel}\sigma_{DR}(v)f(v)dv$, i.e. an average of a velocity-weighted cross section over a "flattened" electron velocity distribution,

$$f(v)dv = \left(\frac{m_e}{kT_\perp}\right)\exp(-m_e v_\perp^2/2kT_\perp)v_\perp dv_\perp \left(\frac{m_e}{2\pi kT_\parallel}\right)^{1/2}\exp(-m_e(v_\parallel-\Delta)^2/2kT_\parallel)dv_\parallel \quad (1)$$

where Δ is the detuning velocity corresponding to $E_{rel}=m_e\Delta^2/2$. The transverse electron temperature has for the CRYRING electron cooler been measured to $kT_\perp=0.1$ eV. The longitudinal temperature is much lower owing to the acceleration cooling mechanism. For H_3^+, it was demonstrated that $<v_{rel}\sigma_{DR}>$ could be replaced by the simpler expression $v_{rel}\sigma_{DR}$ for collision energies down to 0.0025 eV under the assumption of $T_\parallel=0$.[4] More recently it has been demonstrated that $<v_{rel}\sigma_{DR}>$ is insensitive to whether one uses the ideal case of $T_\parallel=0$ or a more realistic value of $kT_\parallel=10^{-4}$ eV.[5]

The upshot of the discussion above is that cross sections for collision energies much smaller than $kT_\perp=0.1$ eV can readily be measured in an electron cooler. However, resonances located at energies smaller than 0.1 eV, and with widths much smaller than 0.1 eV, would be difficult to observe.[5]

SOME EXAMPLES: H_3^+ AND HeH^+

The first dissociative recombination storage ring experiments[2-4] all concerned different molecules. In Tokyo (TARN II) recombination of HeH^+ was studied,[2] in Heidelberg (TSR) the HD^+ molecule was chosen for a first experiment,[3] and in Stockholm (CRYRING) H_3^+ was the first molecule on the agenda.[4] All three experiments exploited the infrared radiative cooling mechanism and were thus able to obtain results for vibrationally relaxed ions. Only the CRYRING experiment provided absolute cross sections.

The simplest polyatomic molecular ion H_3^+ is very important in many natural environments. It occupies a central role in interstellar chemistry and it has recently been observed in emission in polar regions of Jupiter and very recently in Uranus and Saturn.[6] The importance of H_3^+ may be illustrated by the fact that it has been the subject of several review articles.[7-9] The dissociative recombination rate of H_3^+ has been a controversial subject for a long time. In the most recent compilation of measured rates one can notice a more than four orders of magnitude spread in the data (few times 10^{-7} - 10^{-11} cm^{3}s^{-1}).[10] These results have been obtained by means of various techniques, such as extraction of rates from measured cross sections with the merged-beam technique,[11] infrared absorption spectroscopy,[12] and different types of flowing afterglow experiments.[10,13,14] An additional problem has been that, despite some efforts,[15,16] theory has been unable to provide any quantitative guidance.

The rate of the reaction $H_3^+ + e \rightarrow H + H + H$ or $H_2 + H$ was measured in CRYRING and converted to cross sections by the relation $R=\rho_e v_{rel}\sigma_{DR}l/L$ ion^{-1}s^{-1}. The measurements were carried out for collision energies E_{rel} in the range 0.0025-30 eV. The results are shown in Fig. 2.

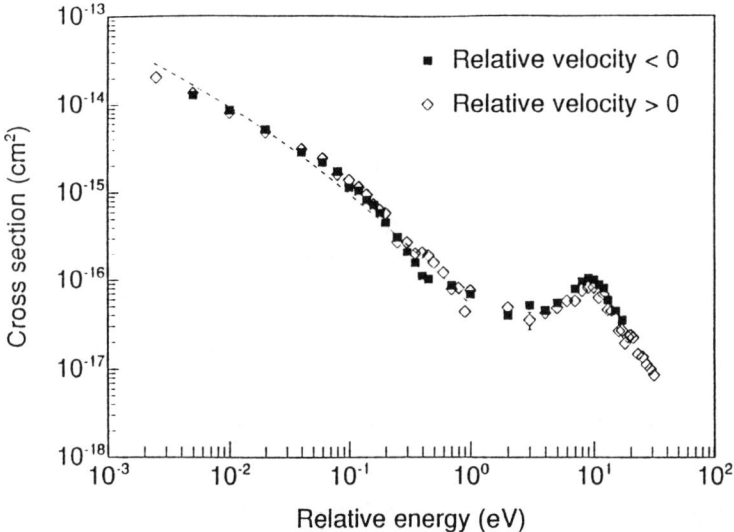

Fig. 2. Total cross sections for forming H+H+H and H+H$_2$. Diamonds illustrate the results obtained by letting the electrons travel *faster* than the ions (denoted "Relative velocity > 0"), and filled squares illustrate the results obtained by letting the electrons travel *slower* than the ions (denoted "Relative velocity < 0").

The low energy part of the cross section behaves as expected, i.e. it displays a $1/E^n$ dependence. At high energy, the cross section increases and peaks at 9.5 eV. This peak represents the direct, resonance channel of the dissociative recombination process in H$_3^+$. Simple model calculations employing potential curves of H$_3^+$ and H$_3$ show that the shape of the 9.5 eV peak indicates that the cross sections were obtained for vibrationally relaxed ions. The data were taken about 10 s after the H$_3^+$ ions were produced in the ion source MINIS. This time scale is sufficient to allow a full vibrational relaxation, as follows from *ab initio* calculations of transition probabilities in H$_3^+$.[17] The observation of the direct channel at 9.5 eV has stimulated theoretical calculations of this process.[18]

The expected cross section behaviour for crossings of neutral dissociative states with a bound ion state has been investigated by model calculations.[19] The quantitative agreement between the measured cross sections in Fig. 2 and the one calculated in ref. 19 (case B crossing) is amusing, but of course purely accidental. In H$_3^+$ the potential crossing occurs at an energy corresponding to vibrational quantum number $v=3$ (as compared to $v=1$ in ref. 19) and hence one would expect a much smaller cross section at low energy than shown in Fig. 2. In fact, the absence of a favourable potential crossing for vibrationally cool H$_3^+$ has been advanced as an argument for a low recombination rate for H$_3^+$ ($v=0$), and for a strongly vibrationally dependent recombination rate.[13,16] In order to facilitate a direct comparison between our measured cross sections and directly measured dissociative recombination rates,[10,12-14] the rate coefficient must be extracted from the measured cross sections via

$$\alpha(T) = \int \frac{8\pi m_e E}{(2\pi m_e kT)^{3/2}} \sigma(E)\exp(-E/kT)dE \qquad (2)$$

The integral is sensitive to the cross sections at very low collision energies, and for this reason we have carried out measurements in the region 0.00005-0.0025 eV. The data indicate a $1/E^{1.2}$ cross section dependence and favour a recombination rate of the order of 10^{-7} cm^{3}s^{-1} at 300 K. These results are preliminary; the final results will be given in a forthcoming publication. It would seem like a 10^{-7} cm^{3}s^{-1} rate is irreconcilable with theory;[16] however, a new, multistep recombination mechanism for H_3^+ has recently been proposed[19] which seems to accommodate a high recombination rate. Finally the claim for a very low rate ($\sim 10^{-11}$ cm^{3}s^{-1}),[13] obtained with the flowing afterglow/Langmuir probe method, has been withdrawn. New measurements give $(1-2) \times 10^{-8}$ cm^{3}s^{-1} at 300 K,[20] which is still low, but not in a drastic disagreement with the storage ring result. The H_3^+ dissociative recombination story has taken a big leap forward during the last year.

The ground state potential curve of HeH$^+$ is not crossed by any neutral state, and for this reason it was for a long time assumed that the recombination cross section for this molecule should be very small. The first merged-beam results[21] for HeH$^+$ therefore came as something of a surprise; the cross section was much larger than one would have expected from a study of the relevant potential curves. Efforts to understand the mechanism were unsuccessful,[22] and it was for some time speculated whether the merged-beam results could have been affected by the presence of excited triplet state population in the ion beam.[21] This possibility was effectively refuted when the final results were published; measurements of dissociative excitation cross sections were used to demonstrate that, under certain conditions, the HeH$^+$ ions did not populate excited states.[23]

HeH$^+$ was chosen as the first molecule for a dissociative recombination experiment at TARN II,[2] possibly because HeH$^+$ is known to cool radiatively on a time scale amenable for storage ring experiments.[24] A previously unobserved peak in the cross section at 20 eV was found and explained as dissociation through two-electron excited states which form Rydberg manifolds converging to excited states of HeH$^+$. Similar high-energy structures were also found for HD$^+$ in the Heidelberg-experiment.[3] Only relative cross sections were obtained in the TARN II-experiment, and the resolution at low energy was insufficient to reveal any structures. Recently absolute cross sections were measured for ^{3}HeH$^+$ in CRYRING. The preliminary results are shown in Fig. 3. The 20 eV peak is observed and there is an additional peak close to 30 eV which is discernible also in the TARN II-data.[2] The cross section at low energy is smaller than the single-pass merged-beam result,[23] however, the difference is within a factor of five. The question is now whether there is a suitable mechanism which could account for a large cross section despite the absence of a potential crossing. A hint is contained in the experiments with charge exchange between HeH$^+$ and Cs.[25,26] These experiments show that Rydberg states of HeH converging to the HeH$^+$ ground state predissociate by coupling to the repulsive ground state of HeH.[27] *Ab initio* calculations of elastic

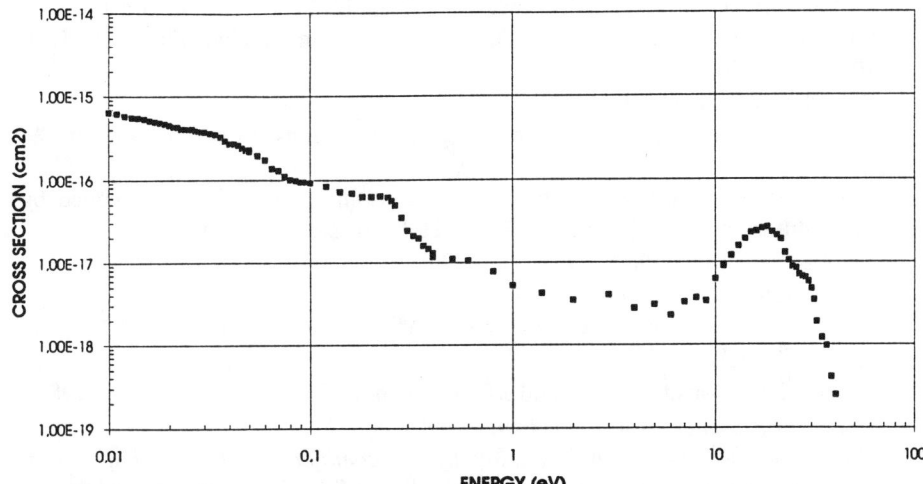

Fig. 3. Preliminary cross sections for the process ^{3}HeH$^+$ + e → ^{3}He + H.

scattering cross sections for electron HeH$^+$ collisions[28] have further indicated a breakdown of the Born-Oppenheimer approximation at low collision energies, a prerequisite for large recombination cross section. Very recently, calculations of dissociative recombination cross sections for HeH$^+$ have provided compelling evidence that a large recombination rate indeed can be obtained also in the non-crossing case.[29,30] In view of this theoretical development, the recent claim that measured recombination at low energy is due solely to the presence of triplet state population in the ion beam appears the more puzzling.

Ion storage rings have proved to be very powerful for studies of dissociative recombination of molecular ions, and it seems that the development of this new experimental tool has been stimulating for theorists. The full potential of the storage rings has not yet been exploited and an exciting future can be anticipated.

ACKNOWLEDGEMENTS

We are indebted to L.H. Andersen, D.R. Bates, A. Dalgarno, S.L. Guberman, D. Habs, K.C. Kulander, D. Smith, T. Tanabe and J. Tennyson for valuable correspondence and discussions, preprints and reprints.

REFERENCES

1. J.B.A. Mitchell, Phys. Rep. **186**, 215 (1990).
2. T. Tanabe, I. Katayama, N. Inoue, K. Chida, Y. Araki, T. Watanabe, M. Yoshizawa, S. Ohtani, and K. Noda, Phys. Rev. Lett. **70**, 422 (1993).

3. P. Forck, M. Grieser, D. Habs, A. Lampert, R. Repnow, D. Schwalm, A. Wolf, D. Zajfman, Phys. Rev. Lett. **70**, 426 (1993).
4. M. Larsson, H. Danared, J.R. Mowat, P. Sigray, G. Sundström, L. Broström, A. Filevich, A. Källberg, S. Mannervik, K.G. Rensfelt, and S. Datz, Phys. Rev. Lett. **70**, 430 (1993).
5. R.J. Mowat (to be published).
6. T. Oka and M.-F. Jagod, J. Chem. Soc. Faraday Trans, to be published in **89**, (1993).
7. T. Oka, in *Molecular Ions: Spectroscopy, Structure and Chemistry,* edited by T.A. Miller and V.E. Bondybey (North-Holland, New York, 1983), p. 73.
8. T. Oka, Rev. Mod. Phys. **64**, 1141 (1992).
9. A. Dalgarno (to be published).
10. A. Canosa, J.C. Gomet, B.R. Rowe, J.B.A. Mitchell, and J.L. Queffelec, J. Chem. Phys. **97**, 1028 (1992).
11. H. Hus, F.B. Yousif, A. Sen, and J.B.A. Mitchell, Phys. Rev. A **38**, 658 (1988).
12. T. Amano, J. Chem. Phys. **92**, 6492 (1990).
13. N.G. Adams and D. Smith, in *Dissociative Recombination: Theory, Experiment and Applications,* edited by J.B.A. Mitchell and S.L. Guberman (World Scientific, Singapore, 1989), p. 124.
14. The reader is referred to ref. 10 for a complete list of dissociative recomination results for H_3^+.
15. K.C. Kulander and M.F. Guest, J. Phys. B **12**, L501 (1979).
16. H.H. Michels and R.H. Hobbs, Astrophys. J. **286**, L27 (1984).
17. B.M. Dinelli, S. Miller, and J. Tennyson, J. Mol. Spectrosc. **153**, 718 (1992).
18. K.C. Kulander and A.E. Orel (private communication).
19. D.R. Bates, M.F. Guest, and R.A. Kendall, Planet. Space Sci. **41**, 9 (1993).
20. D. Smith and P. Španel (preprint, 1993).
21. J.B.A. Mitchell and F.B. Yousidf, in *Dissociative Recombination: Theory, Experiment and Applications,* edited by J.B.A. Mitchell and S.L. Guberman (World Scientific, Singapore, 1989), p. 109.
22. H.H. Michels, in *Dissociative Recombination: Theory, Experiment and Applications,* edited by J.B.A. Mitchell and S.L. Guberman (World Scientific, Singapore, 1989), p. 97.
23. F.B. Yousif and J.B.A. Mitchell, Phys. Rev. A **40**, 4318 (1989).
24. S. Datz and M. Larsson, Phys. Scripta **46**, 343 (1992).
25. W.J. van der Zande, W. Koot, D.P. de Bruijn, and C. Kubach, Phys. Rev. Lett. **57**, 1219 (1986).
26. J.R. Peterson and Y.K. Bae, Phys. Rev. A **34**, 3517 (1986).
27. A.B. Alekseyev, V.S. Ivanov, A.M. Paravilov, and O.D. Shestakov, Chem. Phys. **155**, 173 (1991).
28. B.K. Sarpal, J. Tennyson, and L.M. Morgan, J. Phys. B **24**, 1851 (1991)
29. S.L. Guberman (private communication).
30. J. Tennyson (private communication).
31. J.B.A. Mitchell, M. Chibisov, and F.B. Yousif, DAMOP meeting, TC 2 (1993).

ABSOLUTE CROSS SECTIONS FOR DISSOCIATION OF H_2O BY ELECTRON IMPACT

M. Darrach and J.W. McConkey
University of Windsor, Windsor, Ontario N9B 3P4
Canada

ABSTRACT

Saturated laser induced fluorescence (LIF) techniques are used to probe the production of OH following electron impact on water over an energy range from 20 to 300 eV. Production is dominated at low energy by excitation of triplet parent states of H_2O which lead to low energy OH fragments. Absolute cross-sections for production of OH (X, $v''=0$) are obtained using the cross-section for production of CO_2^+ (X, $v''=0$) as a standard. The cross-section is sharply peaked at low energy and has a value of $5.3 \times 10^{-17} cm^2$ at 100 eV.

INTRODUCTION

Although the dissociation of H_2O has been widely studied in recent years using electron impact most of the work has involved detection of fragments which have been either excited[1-10] or ionized.[11,12] Significant disagreement among published cross-sections is evident. Kawazumi and Ogawa[13] studied the formation of OH ground state, X($^2\Pi$), following electron impact on an effusive beam of H_2O and measured rotational and translational temperatures of this fragment. They did not, however, attempt to measure any cross-sections for production of this ground state species. Their work was extended by Darrach and McConkey[14] who used a supersonic expansion of H_2O and were able to measure relative cross-sections for production of OH(X), again using LIF techniques. In this work we report the first absolute cross-sections for production of OH(X) following electron impact dissociation of water.

EXPERIMENTAL DETAILS

A schematic diagram of the apparatus is shown in Figure 1 of Ref. 15. Briefly it consists of a triple beam arrangement of a pulsed-molecular beam, an electron beam and a probing laser beam. The molecular beam was produced by either an in-house developed molecular beam valve for work above 140°C, or a pulsed valve supplied by General Valve Corp. for work in the lower temperature regime. The unskimmed molecular beam pulse (~1ms duration) was intersected by the laser and electron beams a distance of 8 nozzle diameters (0.5mm orifice) downstream of the nozzle.

The probing laser beam was produced by a Continuum Nd-YAG pumped dye laser combination with an auto-tracking UV generation system and was capable of producing 5mJ pulses in the 306-330 nm wavelength region at 30Hz with a linewidth of about 0.15cm^{-1}. During the course of this investigation the laser power was constrained to the saturated regime by using power levels above 1mJ/pulse and collimating the laser beam to less than a 0.5mm diameter as it passed through the interaction region. Further details of the control and detection systems are given by Darrach and McConkey.[15]

CALIBRATION TECHNIQUES

The LIF signal intensity is directly related to the rotational population in the ground state of the molecule being probed by the laser. More specifically the relationship between the number of molecules $n_{N''}$ in rotational level N'' of the lower state is related to the observed fluorescence intensity $I_{N''}$ by[16]

$$I_{N''} = C(\omega) S_N n_{N''} \eta \tag{1}$$

where $C(\omega)$ is a term which depends only weakly on the transition frequency and includes the transition moment. S_N is the line strength for the transition. η is a factor which takes account of the total efficiency of the detection system and includes various geometrical factors. If the transition between states N'' and N' is saturated, $I_{N''}$ is given by[17]

$$I_{N''} \propto \left(\frac{g_{N'}}{g_{N'} + g_{N''}} \right) n_{N''} \eta = G n_{N''} \eta \tag{2}$$

where $g_N = (2N + 1)$ is the statistical degeneracy of the molecular level, N. We note that the G factors are $\frac{(2N''+3)}{4(N''+1)}$, $\left(\frac{2N''-1}{4N''}\right)$ and ½ for the R, P and Q branches respectively. Clearly the first two G factors also rapidly approach ½ as N'' gets large. Thus for large N'' the G factors can be neglected.

The density, $n_{N''}$, of fragments in the state N'' will be related to the probability of fragmentation of the molecule into such fragments. We may write this as [13,18]

$$n_{N''} = \frac{\sigma(X) L I_e n_z f_{N''} d_{N''}}{k_{out}} \tag{3}$$

where $\sigma(X)$ is the cross-section for production of ground state fragments by electron impact, n_z is the parent molecular density at the interaction region, (distance z downstream from the origin of the jet), I_e is the electron beam current, L is the effective interaction length, k_{out} is the rate of diffusion (or fly out) of the fragments from the detection volume where they may radiate into the detector, $f_{N''}$ is a distribution function relating the number of molecules in state N'' to the total number in all rotational states of a particular manifold and $d_{N''}$

is a factor which takes account of the other rotational manifolds of the same electronic transition.
Thus

$$f_{N''} = \frac{g_{N''} \exp(-\epsilon_{N''}/kT_R)}{\Sigma g_{N''} \exp(-\epsilon_{N''}/kT_R)} \quad (4)$$

where $g_{N''}$ and $\epsilon_{N''}$ are the statistical weighting factor and rotational energy, respectively, of state N" and T_R is the rotational temperature. Combining equations (2) and (3),

$$\sigma(X) \propto \frac{k_{out}}{LI_e n_Z f_{N''} d_{N''}} \cdot \frac{I_{N''}}{G\eta} \quad (5)$$

If the LIF signals from two different species A and B are compared under the same excitation conditions, i.e. same L and I_e, then

$$\frac{\sigma(X,A)}{\sigma(X,B)} = \frac{(k_{out})_A}{(k_{out})_B} \frac{G_B}{G_A} \frac{(I_{N''})_A}{(I_{N''})_B} \frac{(n_z)_B}{(n_z)_A} \frac{(f_{N''})_B}{(f_{N''})_A} \frac{(d_{N''})_B}{(d_{N''})_A} \quad (6)$$

The η factors cancel provided LIF signals at the same wavelength are considered. Equation (6) points the way to how an unknown cross section may be evaluated provided a reference cross-section $\sigma(B)$ is available. In the case of OH(X) the best choice of secondary standard is CO_2^+ production from CO_2. Here we probe the $(X^2\Pi_g - A\ ^2\Pi_u)$ system in the neighbourhood of 308 nm. We note that we assume that (i) the laser power is sufficient to saturate all the transitions from the ground state of both molecules, (ii) the electron and molecular beam characteristics remain the same for measurements of the LIF intensities from the two species, (iii) the sensitivity of the detection system is the same for both species since the fluorescence wavelengths of the observed transitions in OH and CO_2^+ are almost the same.

It is clear from Equation (6) that an accurate number density ratio, $n_Z(CO_2)/n_Z(H_2O)$, is required. Andersen and Fenn[19] give the following expression for n_Z in a supersonic expansion

$$\frac{n_Z}{n_o} = \left(1 + \frac{\gamma-1}{2} M(z)^2\right)^{-\frac{1}{(\gamma-1)}} \quad (7)$$

where γ is the ratio of specific heats, n_o is the reservoir gas density and $M(z)$ is the Mach number of the expansion. In order to check the validity of Equation (7) we carried out an auxiliary experiment. This makes use of the fact that the ionization cross-sections of the rare gases and some molecular gases are known to high accuracy.[12,20,21] Thus an electron beam of energy E and current I_e intersecting the supersonic gas jet a distance, Z, downstream of the expansion, over an effective path length, L, produces an ion current at a suitably placed detector given by

$$I_+ = I_e n_Z L\, \eta_+\, \sigma^+(E) \qquad (8)$$

where $\sigma^+(E)$ is the total ionization cross-section of the target species, and η_+ represents the ion collection efficiency. By taking the ratios of ion currents for two different gases, A and B, whose ionization cross-sections are well known, we obtain (assuming equal η_+ for both)

$$\frac{n_Z(A)}{n_Z(B)} = \frac{I_+(A)}{I_+(B)} \cdot \frac{\sigma^+(B)}{\sigma^+(A)} \qquad (9)$$

This "measured" ratio can then be compared with that calculated using Equation (7).

The apparatus used to measure the accumulated ion yield consisted of an extraction grid of 1mm square copper mesh and a cylindrical ion collector, both clamped below the interaction block. The grid ensured that the low energy (K.E. < 0.25eV) ions created in the electron beam did not impact on the interaction block surface, but were accelerated downwards towards the ion collector. The collection cup consisted of a cylindrical brass sleeve approximately 4cm in diameter with a bare "repeller" wire running down one side and insulated from the ion collector. The "repeller" wire served to establish a field gradient inside the cup and ensured ions travelling down the axis of the ion collector would not escape. Typically the extraction grid was maintained at -3VDC, the collection cup at -30VDC and the "repeller" wire was earthed. Ion trajectories determined using the computer simulation, SIMION, predicted that all ions would be collected inside the cup, even those produced on the edge of the overlap of the gas jet and electron beam.

Measurements were made using He, Ar, N_2 and CO_2 targets. We found that the best agreement between calculated and measured ratios occurred at the higher impact energies and when He and Ar were being used as the test gases. This reflects the accuracy with which these ionization cross-sections are known for these species. Even for these species a considerably poorer level of agreement was obtained when 30eV electron impact was being used. This is almost certainly due to the fact that the calibration of the energy scale was not established to an accuracy greater than ±1eV. Thus at 30eV, which is only 5.5eV above the ionization threshold for He, there is considerable uncertainty in which value to use for $\sigma^+(He)$. A further factor which must be considered when comparing atomic and molecular targets is the fact that fast fragment ions are often produced in considerable quantities when ionization of molecules is being considered.[21-23] This means that our ion collection efficiencies may be not exactly equal for atomic and molecular features. Our conclusion from this experiment was that it is possible to use Equation (7) to predict relative effective densities in the interaction region with an accuracy better than 10%.

LIF SPECTRA

LIF spectra for CO_2^+ were obtained by probing the gas jet $2\mu s$ after termination of the electron beam pulse. There was no significant change in the LIF spectra obtained for delays over the range 1-3μs, demonstrating that collisional effects inside the gas jet were minimal. Assuming the rotational population can be described by a Boltzmann distribution at temperature T, the data should lie on a straight line in a so-called Boltzmann plot[14]. The slope of this line yields the rotational temperature directly. A value of 70 ± 5K was obtained for T_R of CO_2^+ at 300 eV incident energy. A slight dependence of T_R on electron energy similar to that noted by Nakajima et al.[24] was found although our T_R were rather higher overall than recorded by these authors, possibly because our molecular beam valve was operated at an elevated temperature in order that beams of both CO_2 and H_2O could be prepared with identical stagnation pressures.

Attempts to probe the CO_2^+ (X; v'' = 1) and CO_2^+ (X; v'' = 2) states yielded negligible LIF intensities, indicating that the higher vibrational levels of the ground state had relaxed into the v'' = 0 state by the time the laser probed the ions.

The absolute total cross section for the production of CO_2^+ (X) by electron-impact ionization of CO_2 will include contributions due to the formation of excited states of CO_2^+. However both the A and B states of CO_2^+ decay rapidly to the ground state with lifetimes of 124 and 140 ns respectively.[25,26] Thus we can safely assume that all CO_2^+ ions will have relaxed into the ground state by the time they are interrogated by the laser pulse. Thus we take the absolute cross section of the CO_2^+ (X; v'' = 0) ground state at an impact-energy of 100eV to be 2.88×10^{-16} cm^2. In order to check the performance of our electron beam optics and photon detection system, we measured the relative cross-section for CO_2^+ ($A^2\Pi_g$ - $X^2\Pi_u$) emission and found excellent agreement in the shape of the ionization cross-section with that measured earlier.[27] We also carried out a separate experiment in another apparatus to verify that negligible clustering occurred in our CO_2 supersonic beam under the beaming parameters used in the experiment.

LIF spectra of OH(X) obtained using H_2O targets were very similar to those presented earlier.[14] Conditions were chosen where minimal clustering should occur[28] and checks were carried out to demonstrate that the results were independent of laser polarization. Nomenclature and analysis follows that given earlier.[29-31] Rotational temperatures were extracted from Boltzmann plots in a similar fashion to that used for CO_2^+. These were found to be close to 200K (agreeing with our previous measurements[14]) and remained approximately constant over the entire impact energy range. The relative population of the $^2\Pi_{3/2}$ and $^2\Pi_{1/2}$ levels was established as being close to what would be expected on statistical grounds. Interestingly this is opposite to what was found by Andresen et. al.[32] in their study of H_2O photodissociation where dissociation occurred following an optically allowed transition to the repulsive $^1B_1$ parent state and relatively energetic (~0.1 eV) OH fragments resulted. In our case dissociation occurs predominantly via a parent triplet state of H_2O and results in much less energetic fragments. (See Ref. 14 and later discussion).

The relative population of the $\Pi(A')$ and $\Pi(A'')$ levels was measured to be 1.5 ± 0.1 by taking the ratio of LIF intensities from the R_1 and Q_1 transitions, respectively. This is in excellent agreement with Kawazumi and Ogawa[13] and lower that Andresen et.al.[32] who found that the ratio was approximately 2 for $N''=1$ and 2 and then increased rapidly with N''. The fact that this population inversion, between the states of opposite symmetry but same value of N'', is quite different for photon and electron impact dissociation, reflects the different dissociation mechanisms which are occurring.

CROSS-SECTION DETERMINATION

The first-step in determining the absolute cross-section for electron impact dissociation (EID) of H_2O into the OH ground state channel using Equation (6) is to establish how the rotational populations within the ground state are distributed both for OH and the calibration gas, CO_2^+. This is accomplished using the spectroscopic measurements outlined in the previous section and allows us to obtain the ratio $f_{N''}(CO_2)/f_{N''}(OH)$, Equation (6). We now proceed to measure the relative LIF intensities for specific lines within the OH and CO_2^+ spectra. These are chosen to be as close as possible in wavelength to minimize possible errors due to variation of detector sensitivity with wavelength. The relative densities of the two target species in the interaction region are obtained via Equation (7) and measurements of the driving pressures using an accurate pressure transducer. Since the G factors are known the only remaining factor in Equation (6) which remains to be determined is the ratio $k_{out}(OH)/k_{out}(CO_2^+)$. This was determined as follows. Figure 1 shows the LIF intensity from CO_2^+ (X, $v''=0$) and OH (X, $v''=0$) obtained by probing with the laser at different delay times after the termination of the electron beam. Since the ionization process does not change the kinetic energy of the CO_2^+ we expect the predominant factor governing the diffusion of CO_2^+ (X) out of the interaction region will be the flow velocity of the supersonic expansion of CO_2. Similarly we might expect the diffusion out rate of the very slow OH fragments, which we observe, to be dominated by the flow velocity of the parent H_2O. We note that, indeed, the ratio of the diffusion rate (0.69) measured experimentally for $K_{out}(CO_2^+)/K_{out}(OH)$ compares favourably with the value (0.64) which would be predicted assuming the usual expression for the local sonic velocities.[19]

Closer examination of the "fly-out" of OH(X) in Figure 1 suggests the existence of a second faster OH(X) fragment being produced in the EID process with a much larger fly-out rate. This is an important factor for cross section measurements as the faster radical will have left the interaction region if the laser probes the dissociation products more than $2\mu s$ after the electron-beam termination. For these measurements we selectively measure the cross section of the EID process leading to the slow fragments by probing with the laser at least $2\mu s$ after the electron beam. We note that Kawasumi and Ogawa identified various channels for EID of H_2O yielding most probable translational energies of the OH(X) fragments of 0, 0.09 and 0.22eV. The higher energy fragments are discriminated against in our experiment and thus, since the major optically allowed channels[13,32] produce the faster fragments, we automatically discriminate against these.

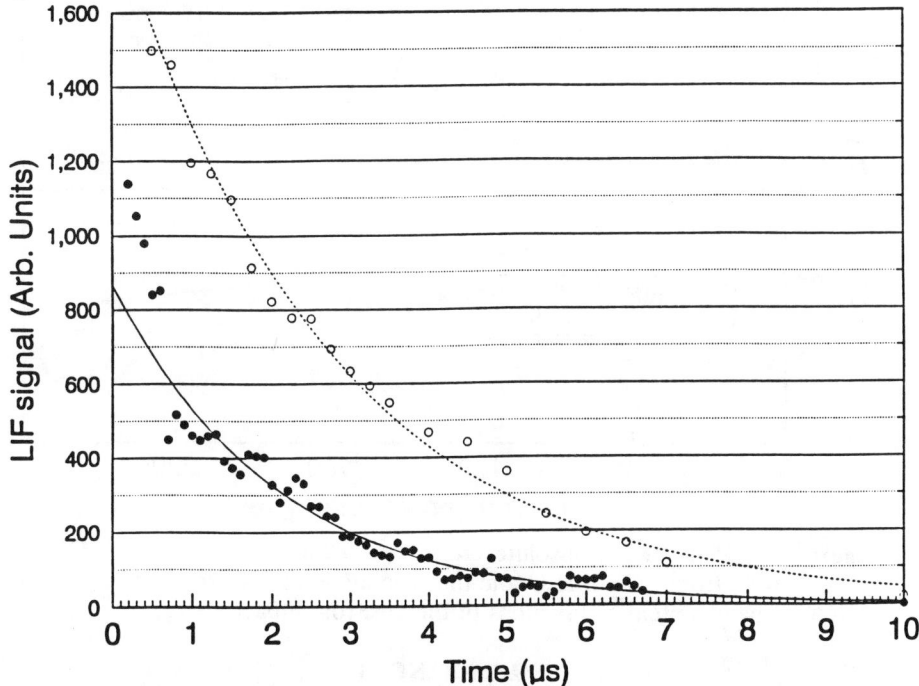

Figure 1. $CO_2^+(X)$ (top) and OH(X) (bottom) LIF intensities as a function of delay between electron and probing laser beams. Characteristic fly-out rates are $0.34 \pm 0.03 \times 10^{-6}$ s^{-1} and $0.49 \pm 0.04 \times 10^{-6}$ s^{-1}, respectively.

This becomes more evident from Figure 2 which shows our data for OH(X) production as a function of incident electron energy. The data are strongly peaked at low energy suggesting that, at low energy, the excitation is predominately via an exchange process. Also shown in Figure 2 is the cross-section for EID of H_2O into the A $^2\Sigma^+$ state of OH[6]. Clearly the shape of the cross-section functions are very similar. Beenakker et.al.[6] found evidence for strong triplet excitation dominating the threshold region but singlet excitation becoming important at higher incident energies. The shape of our curve at higher incident energies is also indicative of singlet-processes playing a role. Kawazumi and Ogawa[13] suggest that very low energy OH(X) fragments also result from an ionization process. This would be optically allowed in nature. An absolute error of some 36% is estimated for the data in Figure 2. A complete analysis will be presented separately.[32]

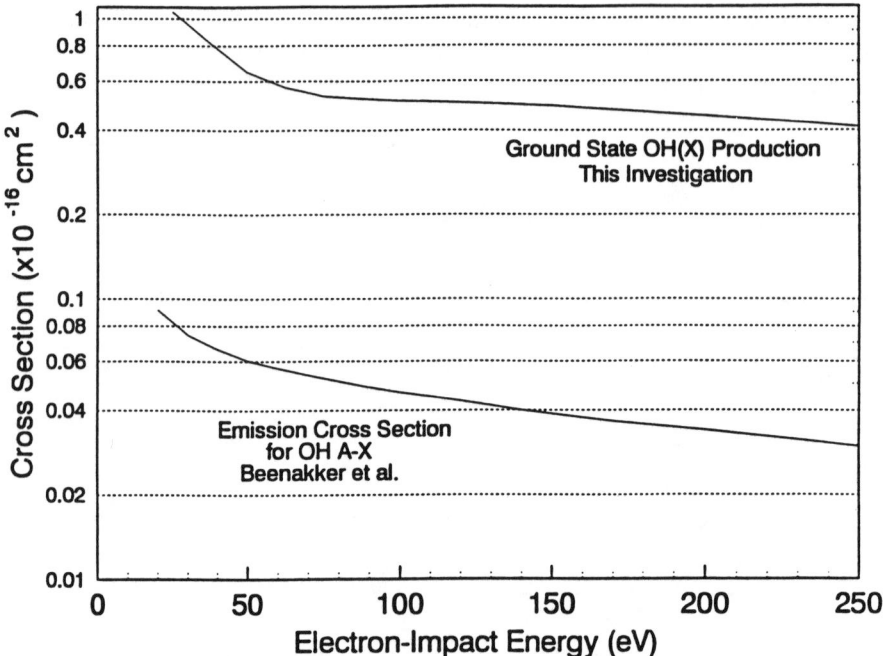

Figure 2. Plots of the absolute cross section for the formation of OH(X) (upper curve) from EID of H_2O and the absolute emission cross sections of the OH(A-X) band system as functions of the electron impact energy.

REFERENCES

1. D.A. Vroom and F.J. de Heer, J. Chem. Phys. <u>50</u>, 1883 (1969).
2. S. Tsurubuchi and S. Isaka, J. Phys. Soc. Japan <u>38</u>, 1224 (1975).
3. K. Becker, B. Stumpf and G. Shultz, Chem. Phys. <u>53</u>, 31 (1980).
4. K. Becker, B. Stumpf and G. Shultz, J. Phys. <u>B14</u>, L517 (1981).
5. G.R. Mohlmann, C.I.M. Beenakker and F.J. de Heer, Chem. Phys. <u>10</u>, 375 (1976).
6. C.I.M. Beenakker, F.J. de Heer, H.B. Krop and G.R. Mohlmann, Chem. Phys. <u>6</u>, 445 (1974).
7. T. Fujita, T. Iwai, K. Ogura, S. Watanabe and Y. Watanabe, J. Phys. Soc. Japan <u>42</u>, 1296 and 1305 (1977).
8. G.M. Lawrence, Phys. Rev. <u>A2</u>, 397 (1970).
9. S. Tsurubuchi, T. Iwai and T. Horie, J. Phys. Soc. Japan <u>36</u>, 537 (1974).
10. J.J. Olivero, R.W. Stagat and A.E.S. Green, J. Geophys. Res. <u>27</u>, 4797 (1972).
11. J. Schutten, F.J. de Heer, H.R. Moustafa, A.J.H. Boerboom and J. Kistemaker, J. Chem. Phys. <u>44</u>, 3924 (1966).
12. O.J. Orient and S.K. Srivastava, J. Phys. <u>B20</u>, 3923 (1987).
13. H. Kawazumi and T. Ogawa, Chem. Phys. <u>114</u>, 149 (1987).

14. M. Darrach and J.W. McConkey, Chem. Phys. Lett. **184**, 141 (1991).
15. M. Darrach and J.W. McConkey, J. Chem. Phys. **95**, 754 (1991).
16. G. Herzberg, Spectra of Diatomic Molecules, Vol. 1 (Van Nostrand, Princeton 1950).
17. W.H. Fisher, T. Carrington, S.V. Filseth, C.M. Sadowski and C.H. Dugan, Chem. Phys. **82**, 443 (1983).
18. J.E. Solomon and D.M. Silva, J. Appl. Phys. **47**, 1519 (1976).
19. J.B. Anderson and J.B. Fenn, Phys. Fluids **8**, 780 (1965).
20. F.J. de Heer and M. Inokuti, "Electron Impact Ionization" Eds T.D. Mark and G.H. Dunn (Springer-Verlag; 1985) p. 232.
21. E. Krishnakumar and S.K. Srivastava, J. Phys. B., **23**, 1893 (1990).
22. A. Crowe and J.W. McConkey, J. Phys. B **8**, 1765 (1975).
23. A. Crowe and J.W. McConkey, Phys. Rev. Lett. **31**, 192 (1973).
24. A. Nakajima, T. Nagata, T. Kondow and K. Kuchitsu, Chem. Phys. Lett. **151**, 511 (1988).
25. F. Bueso-Sanllehi, Phys. Rev. **60**, 556 (1941).
26. J.P. Maier and F. Thommen, Chem. Phys. **51**, 319 (1980).
27. J.W. McConkey, D.J. Burns and J.M. Woolsey, J. Phys. B, **1**, 71 (1968).
28. D. Dreyfuss and H.Y. Wachman, J. Chem. Phys. **76**, 2031 (1982).
29. G.H. Dieke and H.M. Crosswhite, J. Quant. Spectrocs. Radiat. Transfer, **2**, 97 (1962).
30. E.A. Moore and W.G. Richards, Phys. Scr., **3**, 223 (1971).
31. M.H. Alexander et.al., J. Chem. Phys. **89**, 1749 (1988).
32. P. Andresen, G.S. Ondrey, B. Titzer and E.W. Rothe, J. Chem. Phys. **80**, 2548 (1984).
33. M. Darrach and J.W. McConkey (to be published).

ELASTIC ELECTRON ION SCATTERING

B.A. Huber, C. Ristori, C. Guet, D. Jalabert, M. Maurel and J.C. Rocco

CEA / Departement de Recherche Fondamentale sur la Matière Condensée - LI2A
85X, F-38041 Grenoble-Cédex, France

ABSTRACT

Differential cross sections for the elastic scattering of electrons by multiply charged ions have been measured in a crossed-beams device. They are compared with theoretical predictions obtained within the framework of a Hartree Fock calculation. Results are discussed for the ionic systems Ba^{2+} and Xe^{6+} for electron energies below 50 eV and scattering angles between 30° and 80°. Owing to the structure of the ionic core the differential cross sections deviate strongly from the Rutherford cross section but show good agreement with the theoretical predictions.

INTRODUCTION

The elastic scattering of free electrons by neutral atoms has been studied for many years, beginning with the pioneering experiments of Ramsauer and Kollath[1,2] in the 1920s. In the meantime, many characteristic features in the total as well as in the differential cross section have been found and they are nowadays rather well described by theory[3].

In the case where a free electron collides with a bare nucleus, the interaction potential is given by the Coulomb potential and the scattering amplitude and cross section can be obtained easily in analytical form yielding the Rutherford cross section. The situation is very different if we regard the elastic scattering of electrons by partially stripped ions. In this case the interaction potential is given as the superposition of a long-range Coulomb potential and the short-range Hartree-Fock potential of the ionic core. Both experimental as well as theoretical studies are very limited in this case. To our knowledge no experimental studies have been performed with free electrons; neither total nor differential cross sections have been obtained with merged beams or crossed beams devices.

On the other hand, experiments with so-called 'quasi-free' electrons have been performed in heavy particle collisions[4-7]. In this type of experiment energy distributions of electrons, which are emitted in collisions between ions (for example Mg^{6+} or U^{21+}) and targets like He or H_2, are measured at different emission angles with respect to the ion beam direction. The so-called binary encounter peak within this spectrum arises from very close binary collisions between the projectile and target electrons. Thus, from the study of the binary encounter peak information on electron ion collisions can be extracted. In this case a zero-degree emission angle of the electron with respect to the ion beam direction corresponds to a 180-degree scattering angle for the electron in the center-of-mass system. In contrast to an experiment with an electron beam, the energy distribution of the 'quasi-free' electron is rather wide, being described by the Compton profile of the electron in the He atom or H_2 molecule. Furthermore, the collision energy in the center-of-mass system lies normally in the keV range, characterizing the region of high energy electron scattering.

In the literature some calculations of differential cross sections have been published recently, in most cases in relation to the analysis of the binary encounter measurements[8-10]. Very often model potentials have been used to represent the ionic core, and specific aspects - like the scaling of the backward scattering cross section with projectile charge - have been studied.

If one compares the elastic scattering of electrons from neutral atoms with that from the corresponding isoelectronic ions, two aspects should be mentioned:

1. Recently, calculations[11] have been performed regarding the elastic scattering of electrons by neutral Xe atoms and by doubly charged Ba^{2+} ions at low electron energies ($\approx$ 1 a.u.). It has been shown that the effective interaction potential is dramatically different in both cases for partial waves with higher angular momentum. Whereas in the Xe-case the f-wave function stays well outside the ionic core due to the strong repulsive centrifugal barrier, in the Ba^{2+}-case a collapse of this wave function towards smaller nuclear distances occurs, giving rise to a larger phase shift and scattering amplitude. Hence, in both cases the scattering potential and the contributions from various partial waves are quite different.

2. In contrast to the neutral-atom case, the temporary capture of the incoming electron creating a doubly excited ion state - like in the first step of the dielectronic recombination process - is very likely to occur. In this resonance process the angular distribution of the scattered electron might be expected to be rather different from that which results from a direct scattering process as shown recently for the excitation of multiply charged ions by electron impact[12]. Therefore, angular differential measurements may serve as a rigorous test for theoretical predictions including these resonance processes.

It is the aim of these studies to measure angular differential cross sections for the elastic scattering of free electrons by multiply charged ions. In particular, the influence of the ionic core as well as the importance of doubly excited autoionizing states will be studied for heavy ionic systems.

EXPERIMENT

The experimental apparatus has been used before, studying the excitation of multiply charged ions by electron impact, and it is described in more detail elsewhere[13-15]. In brief, a beam of multiply charged ions is crossed with an electron beam under a crossing angle, that can be varied between 45° and 135°. The energy distributions of the scattered electrons are measured with a two-step cylindrical analyser. With the present device, scattering into the angular region between 20° and 120° in the laboratory frame can be studied.

In figure 1a a typical energy distribution is shown for electrons being scattered at a scattering angle of 30°. The wider peak at 38.4 eV is due to the scattering by the residual gas; its position coincides with the primary electron energy. The shifted peak at about 47.4 eV is caused by the scattering from Cl^{7+} ions. Its exact position depends on the ion and electron velocities, the crossing angle between both beams, as well as on the scattering angle. This so-called kinematic shift is an important feature of the experimental method, as it allows a clear separation of the elastic contributions from electron-ion and electron-residual gas scattering. Furthermore, the electron-ion signal is shifted into an energy range where the general background signal due to the primary electron beam is very small. As can be seen in figure 1b, the shift increases with increasing scattering angle (proportional to $\sin(\Theta)$).

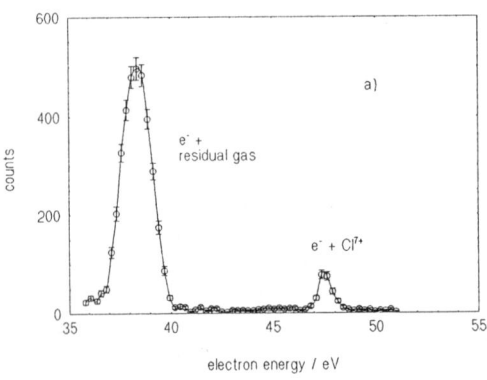

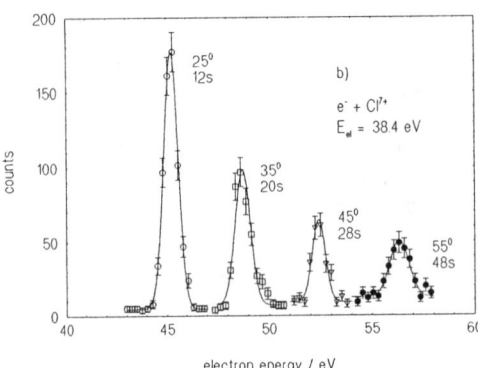

The main difficulties in this type of experiment are due to the low target density represented by the ion beam. The density value of 10^5 ions / cm^3 corresponds to a target pressure of about 10^{-11} mbar. Therefore, the residual gas pressure has to be of the same order of magnitude. The expected signal rate depends on the scattering angle and may vary between 0,01 and 20 counts/s. Some typical values, characterizing the experimental conditions, are given in Table I.

Figure 1: Energy distributions of scattered electrons at a primary energy of 38,4 eV. a) scattering signals from residual gas and Cl^{7+} ions at $\Theta=30°$; b) shift of the elastic electron-ion signal with scattering angle Θ. The indicated time values give the measuring time for an individual point.

Table I: Overview on typical experimental parameters

electron energy	50 eV
electron current	4 µA
ion energy	(20*q) keV
ion current	5 µA
residual gas pressure	10^{-10} mbar
scattering angle	25°-90°
diameter of overlap volume	2-3 mm

From an analysis of the peak areas relative cross sections are obtained. These values are put on an absolute scale by comparing the signals obtained with those being measured for another ionic system, for which the differential cross section is known, for example a bare nucleus system. In the case of Ba^{2+}, which we will discuss below, electron scattering by $^3He^{2+}$ ions has been used for normalizing the Ba^{2+} data. For He^{2+} the differential cross section is given by the Rutherford cross section. This procedure requires that the shape of both ion beams as well as the overlap region is identical in both cases. This was controlled by a beam monitor device, which could be moved directly into the desired interaction region.

DISCUSSION OF RESULTS

Differential cross sections have been calculated by solving the Schrödinger equation for a free particle in the potential

$$V(r) = -q/r + V_{sr}(r), \qquad (1)$$

where the first term represents the Coulomb potential of the ionic system with charge state q. The second one is the short range potential of the target ion, which was taken to be the Hartree Fock potential. The scattering amplitude may be written as a sum of a Coulomb part and a short range part:

$$f(\Theta) = f_c(\Theta) + f_{sr}(\Theta). \qquad (2)$$

If the wave function of the incoming electron is expanded into partial waves, the scattering amplitude is given as sum over different partial waves. In the case of the short range potential the number of partial waves is rather limited and only small angular momenta have to be considered. The angular differential cross section is given by the following expression:

$$(d\sigma/d\Omega)(\Theta) = |f(\Theta)|^2. \qquad (3)$$

In the present calculation electron exchange is included automatically, and a polarisation potential term, describing the polarisation of the ion by the incoming and outgoing electron, has to be included particularly for heavy ions in low charge states and for large scattering angles. Contributions from excited states and autoionizing resonances are not included in the present calculations.

In figure 2 theoretical predictions are shown for the scattering of 50 eV-electrons from Xe^{6+} ions. The Hartree Fock results are compared with Rutherford cross sections for bare ions in the charge states 6 and 54. Due to the structure of the ionic core, strong oscillations are to be expected in the shape of the differential cross section. They are due to the interference of contributions from different partial waves. The strong increase at scattering angles near 180° can be interpreted within the semiclassical scattering theory as a glory scattering effect[5,16]. At small scattering angles, which correspond to large impact parameters, the differential cross section is close to the Rutherford cross section for a bare nucleus of the same charge (in the present case C^{6+}). This shows that at large distances the interaction potential is dominated by the Coulomb field from a point charge of q=6. In the case of backward scattering, where the electron penetrates into the ionic core, the cross section is strongly enhanced compared to the Rutherford cross section showing the interference structure due to the non-Coulombic part of the potential.

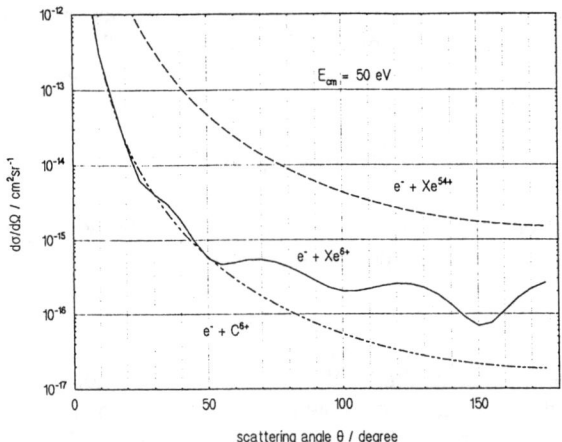

Figure 2: Calculated differential cross sections for the elastic scattering of electrons by multiply charged ions. Full line: HF-result for Xe^{6+}; dashed lines: Rutherford cross sections for bare nuclei C^{6+} and Xe^{54+}. Collision energy in the center-of-mass system: 50 eV.

Figure 3 shows a comparison between experimental and theoretical results for scattering angles below 90° in the center-of-mass system. The error bars of the

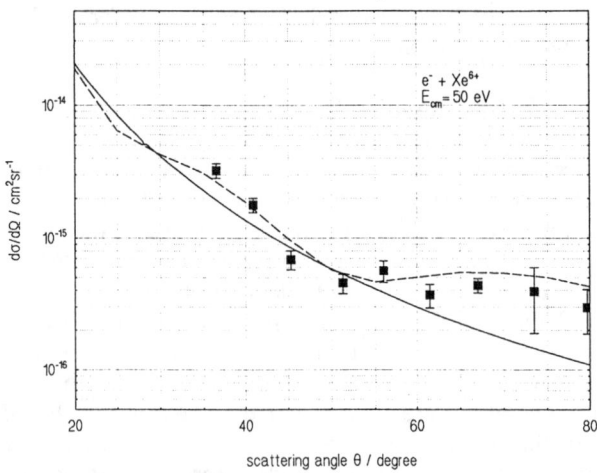

Figure 3: Differential cross sections for the elastic scattering of electrons by multiply charged Xe^{6+} ions. Collision energy: 50 eV. Full squares: present experiment; dashed line: HF-calculation; full line: Rutherford cross section for C^{6+}.

experimental values are statistical ones which include to some extend uncertainties in the determination of the scattering angle and crossing angle. Systematic errors due to the normalization with two different ion beams are not included. In the present case data for electron scattering by Ar^{6+} ions have been used to normalize the Xe^{6+}-data. The cross section for $e^- + Ar^{6+}$ has been calculated in the manner desribed above and has been found to be rather close to the Rutherford cross section in the angular region of interest here. As predicted by theory, the measured cross section shows at scattering angles <90° a strong oscillating behavior and deviates from the pure Coulomb cross section. Taking into account the experimental error bars, the agreement with the theoretical prediction is found to be surprisingly good.

Similar results are obtained for the elastic scattering of electrons by doubly charged Ba^{2+} ions. In this system the Coulombic part of the potential is weaker and hence a larger deviation from a Rutherford behavior might be expected. Furthermore, as mentioned in the introduction, large contributions from the f-waves are to be expected changing the interference structure. The results are shown in figure 4 for a collision energy of 41.4 eV. The experimental values have been normalized to scattering data obtained on He^{2+} ions.

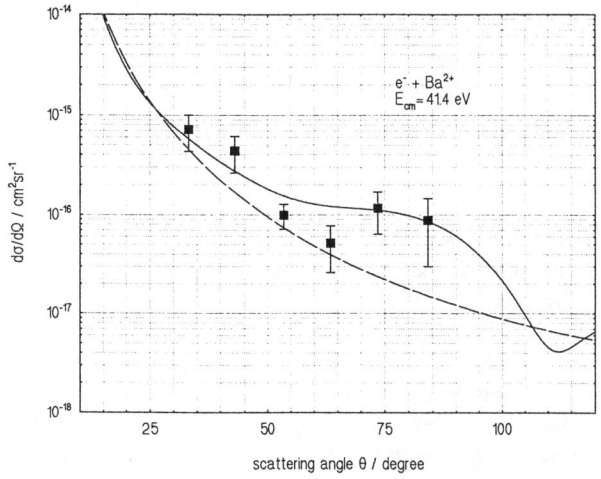

Figure 4: Differential cross sections for the elastic scattering of electrons by doubly charged Ba^{2+} ions. Squares: experimental values; full line: HF-calculation; dashed curve: Rutherford result for He^{2+}. Collision energy: 41.4 eV.

A reasonably good agreement between theoretical and experimental values is found for Ba^{2+}. The experimental values tend to show more pronounced oscillations than theory; however, due to the relatively large error bars, this difference cannot be regarded as significant. The deviation from the Coulomb cross section can be seen more clearly in figure 5, where the ratio of the differential cross section and the Rutherford cross section is shown. At a scattering angle of 50° the cross section deviates from the Rutherford cross section by a factor of 2. It shows a rather complex interference pattern at larger scattering angles and its value exceeds the corresponding Rutherford value at backward scattering (Θ=180°) by almost a factor of 100.

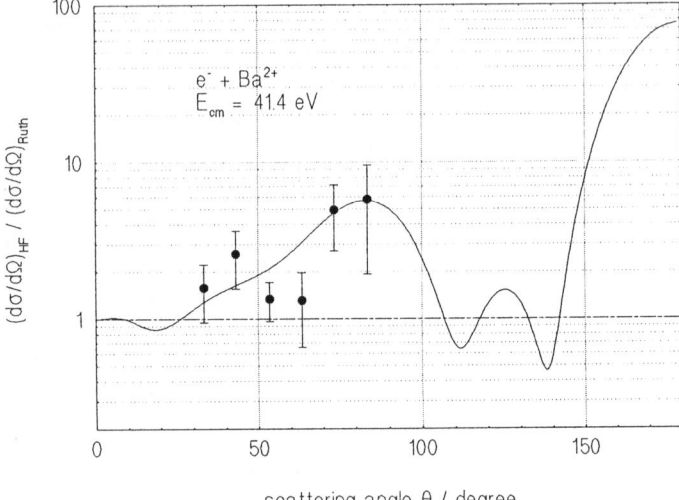

Figure 5: Ratio of the differential cross section for e-+ Ba^{2+} and e-+ He^{2+} (Rutherford) at a collision energy of 41.4 eV. Full line: theoretical result, full dots: experimental values.

CONCLUSION

For the first time, angular differential cross sections for the elastic scattering of free electrons by multiply charged ions have been measured in a crossed-beams device. Results are presented for the systems e-+ Ba^{2+} and e-+ Xe^{6+} at collision energies of 41.4 eV and 50 eV, respectively. For the heavy ions, the influence of the ion core on the scattering process manifests itself at rather small scattering angles. The measured cross section deviates strongly from a Rutherford behavior, showing pronounced interference structures that are due to contributions from different partial waves. In general, good agreement is found with theoretical predictions obtained within the framework of a Hartree-Fock calculation. In order to test the theory more rigorously, measurements at lower collision energies and at larger scattering angles will be performed.

ACKNOWLEDGEMENTS

These experiments have been performed at the AIM accelerator in Grenoble, a common facility of CEA and CNRS. The authors are indebted to A. Brenac, G. Lamboley and T. Lamy for providing a high quality ion beam. Valuable discussions with P. Defrance and C. Belenger are acknowledged. The financial support by a grant from the EC is highly appreciated.

REFERENCES

1. C. Ramsauer, Ann. Phys. (Leipzig) 64, 513 and 66, 546 (1921)
2. C. Ramsauer and R. Kollath, Ann. Phys. (Leipzig) 12, 529, 837 (1932)
3. See for example the article of P.G. Burke in this volume and further contributions on electron-atom and electron-molecule collisions.
4. N. Stolterfoht, D. Schneider, D. Burch, H. Wieman and J.S. Risley, Phys. Rev. Lett. 33, 59 (1974)
5. C.O. Reinold, D.R. Schultz, R.E. Olson, C. Kelbsch, R. Koch and H. Schmidt-Böcking, Phys. Rev. Lett. 66, 1842 (1991)
6. P. Richard, D.H. Lee, T.J.MJ. Zouros, J.M. Sanders and S. Shinpaugh, J. Phys. B: At. Mol. Opt. Phys. 23, L213 (1990)
7. P. Hvelplund, private communication; submitted for publication (1993)
8. G.F. Drukarev and N.B. Berezina, Sov. Phys.-JETP, 42, 423 (1976)
9. R. Shingal, Z. Chen, K.R. Karim, C.D. Lin and C.P. Bhalla, J. Phys. B: At. Mol. Opt. Phys. 23, L637 (1990)
10. Z. Chen, D. Madison and C.D. Lin, J. Phys. B: At. Mol. Opt. Phys. 24, 3202 (1991)
11. W.R. Johnson and C. Guet, J. Phys. B: At. Mol. Opt. Phys.; to be submitted (1993)
12. D.C. Griffin, M.S. Pindzola and N.R. Badnell, Phys. Rev. A 47, 2871 (1993)
13. B.A. Huber, C. Ristori, P. A. Hervieux, M. Maurel, C. Guet and H.J. Andrä, Phys. Rev. Lett. 67, 1407 (1991)
14. C. Ristori, P.A. Hervieux, M. Maurel, H.J.Andrä, A. Brenac, J. Crançon, G. Lamboley, T. Lamy, P. Perrin, J.C. Rocco, F. Zadworny and B.A. Huber, Z. Phys. D 22, 425 (1991)
15. B.A. Huber, C. Ristori and D. Küchler, in AIP Conference Proceedings 274, eds. P Richard, M. Stöckli, C.L. Cocke and C.D. Lin, 455 (1993)
16. Y.N. Demkov, in AIP Conference Proceedings 274, eds. P. Richard, M. Stöckli, C.L. Cocke and C.D. Lin, 512 (1993)

ELECTRON CAPTURE FROM CIRCULAR RYDBERG ATOMS

S.B. Hansen, T. Ehrenreich, and E. Horsdal-Pedersen
Institute of Physics and Astronomy, Aarhus University, DK-8000 Aarhus C, Denmark

K.B. MacAdam
Department of Physics and Astronomy, University of Kentucky, Lexington, KY 40506-0055.

ABSTRACT

Relative cross sections for electron capture by ions from oriented, circular Rydberg atoms have been measured as a function of the impact velocity v and the angle φ between the ion velocity and the angular momentum of the circular orbital. The specific system studied is $^{23}Na^+$ on $Li(1s^2,nlm)$ with $n=25$ and $l=m=n-1$, where m is defined relative to an external magnetic field. The ion energy was varied within the interval 0.6-3keV. Strong dependences on v and φ were found. We expect that studies, such as the present, will lead to an improved understanding of the three-body problem in the quasiclassical limit.

INTRODUCTION

The recently introduced crossed-fields method[1,2] has been used to form a target of oriented, circular Rydberg atoms for the study of charge transfer in ion-atom collisions involving initial states of high angular momentum. The circular state selected is oriented parallel to an external magnetic field **B** (<**L**> ∥ **B**, where **L** is the orbital angular momentum), and it is highly concentrated both in configuration and in momentum space near circular paths of radii $r_0=n^2$ and $p_0=1/n$ (atomic units, a.u.), respectively. This strongly localized and quasiclassical character of the initial state is of particular interest for charge transfer[3]. The energy range studied corresponds to reduced impact velocities $v_r = v/v_0$ in the range 0.81-1.81, where $v_0=p_0$ (a.u.) is the initial orbital velocity. In this contribution we discuss experimental procedures and show the first experimental results of our new initiative.

EXPERIMENT

The experimental arrangement is shown in fig.1. Three laser beams from three dye-lasers pumped by a 14Hz pulsed Nd:YAG laser cross a vertical beam of Li atoms from an oven (400°C). At the crossing point, there is a homogeneous, vertical, electric field **E** defined by a set of plates held at suitable potentials. The laser beams populate the highest state of the $n=25$ Stark manifold of the Li atoms (see fig.2) *via* the intermediate states 2P and 3D. This is a *linear* Stark state with electric dipole moment antiparallel to **E**. It is transformed[1,2] into a *circular* state with magnetic dipole moment antiparallel to an external magnetic field **B** (30gauss), perpendicular to **E**, when it slowly drifts into the region of the horizontal ion beam (see fig.1) where the electric field is nearly zero. The magnetic field **B** is formed by a quadrupole arrangement of electromagnets which makes it possible to set the field direction freely within the horizontal plane and to adjust the magnitude within the range 0-100 gauss.

The Rydberg target is monitored by the Selective Field Ionization[4] method (SFI). The high-voltage (HV) ramp used for this is applied to a pair of field-ionization plates and the resulting ions are counted by an electron multiplier (see fig.1). Representative SFI-spectra are shown in fig.3 for various delays t_d of the HV-ramp relative to the laser shot (duration 8nsec). The main peak seen in each spectrum for $t_d \geq 10\mu sec$ is the so-called diabatic peak[4] corresponding to circular atoms with $n=25$ and magnetic moment parallel to the ionizing field.

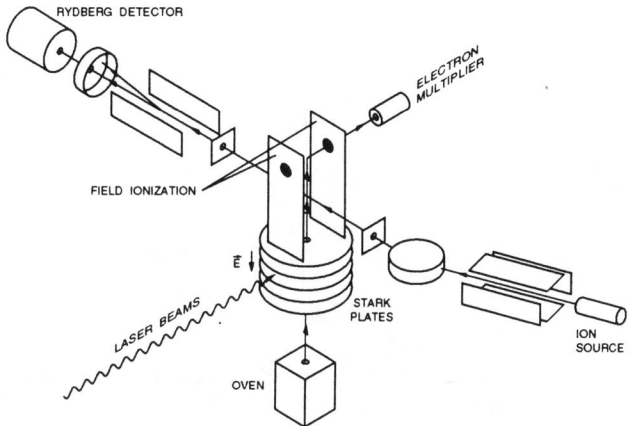

Fig. 1. Experimental arrangement. A horizontal magnetic field **B** is present in the region where the ion beam and the thermal Li beam cross. The field **B** can be rotated 360^0.

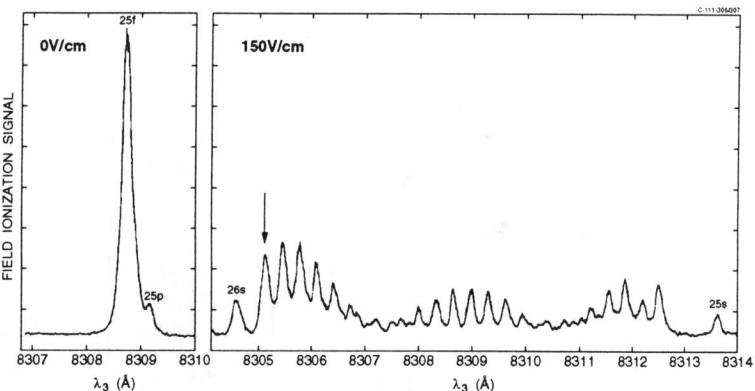

Fig. 2. Excitation function of Rydberg states *vs.* wavelength λ_3 of the third laser near the $n=25$ level of Li for **E**=0 and 150V/cm. The linear Stark state is marked by an arrow.

The side peaks are due to circular atoms with $n=23-27$ formed from the $n=25$ state by interaction with blackbody radiation from the liquid-nitrogen-cooled surroundings[2]. These circular states have geometrical characteristics nearly identical to those of the $n=25$ state, and their symmetrical dispersion of n values yields a first-order compensation of their effect on the measured cross sections, which are attributed to the $n=25$ state alone. The spectrum with $t_d = 5\mu sec$ is completely different from the rest of the spectra. With this short time delay one selects the fastest atoms in the thermal beam and these atoms move into the region with **E**=0 so fast, that the linear atoms do not have enough time to evolve into perfect circular atoms as $|\mathbf{E}|/|\mathbf{B}|$ suddenly approaches zero in the reference frame of the atoms.

The quantum defect δ_s of the Li ns Rydberg states causes quasidegeneracy of the highest-lying $m=0$ Stark state and the highest lying $m=1$ Stark state[5]. The $m=1$ state is transformed

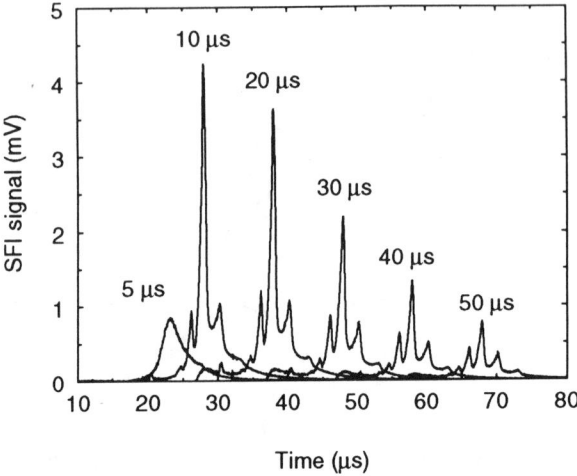

Fig. 3. SFI-spectra. Each spectrum shows a strong diabatic peak from the circular state $(n,l,m) = (25,24,24)$. HV-ramp (160V/μsec, V_{MAX}=3KV) started 5-50μsec after laser shot.

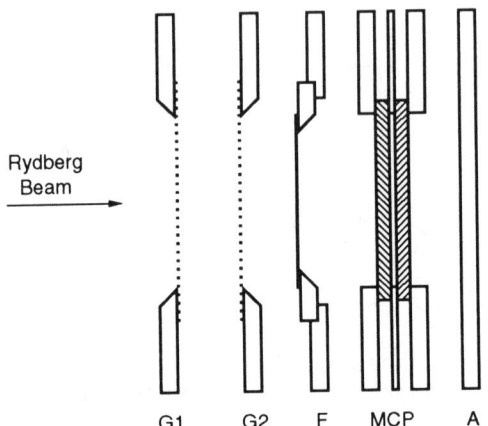

Fig. 4. Schematic diagram of detector for fast Rydberg atoms[6].

adiabatically into a quasicircular state[2]. We do not expect that the capture cross section for the quasicircular state is significantly different from the one for the true circular state.

The current of ^{23}Na$^+$ ions is measured in a Faraday cup, and the fast ^{23}Na(nl) Rydberg atoms formed by charge transfer from the circular atoms are field-ionized and counted by a special detector arrangement designed to respond only to Rydberg atoms and to have a large sensitive area (3cm^2, circular)[6]. The detector is shown in fig.4. It consists of two grids G1 and G2, a 3μg/cm^2 carbon foil F, microchannel plates MCP, and an anode A. Incoming Rydberg atoms with $n\geq16$ are field ionized close to G1 and subsequently accelerated by a strong field

(12kV/cm) formed between G1 at ground potential and G2 at V=12kV. These energetic ions (12.6-15keV, depending on beam energy) penetrate the foil F and are detected by the microchannel plate (MCP) followed by the anode (A), but ground state atoms formed in the rest gas (<3keV) have only a small probability for penetrating F and being detected as a background signal.

The following sequence of measurements (running at the 14Hz pulse repetition rate of the laser) is repeated until sufficient statistics is obtained. After each laser shot (at time t=0), i) Rydberg atoms plus background are counted in the period T_{S+B} (t=12-32µsec) resulting in N_{S+B} counts, ii) at t=32µsec, the HV ramp is started, and the resulting SFI-spectrum obtained from the field-ionization of the Rydberg target is recorded by a digital oscilloscope which also averages successive SFI spectra, and iii) pure background is counted in the period T_B (t=50-60msec) resulting in N_B counts. A cross-section measurement typically requires 50,000 laser shots. In terms of the total area A of the averaged SFI spectrum, the total charge Q collected in the cup, and the quantities mentioned above the charge-transfer cross section is given by

$$\sigma = K \cdot (\sum N_{S+B} / T_{S+B} - \sum N_B / T_B) / (A \cdot Q) \qquad (1)$$

where the summations are over the number of laser shots, and K collects constants of proportionality.

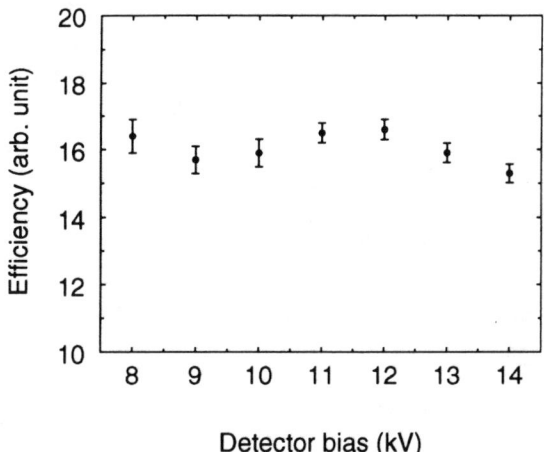

Fig. 5. Specific detection efficiency ε for Rydberg atoms vs. detector bias V.

One such factor is the detection efficiency of the Rydberg detector which depends on beam energy and detector bias. The energy dependence is determined by the product of a foil-transmission factor and an MCP efficiency factor. It was found by directing a small 1keV ion beam onto the detector and measuring the count rate, R_1, as a function of detector bias V (or impact energy at the foil). For Rydberg atoms the efficiency may also depend on the detector bias V through the V-dependent field-ionization rates of the various Rydberg states populated in the beam. The specific efficiency for Rydberg atoms, ε, was determined by measuring the apparent cross section $R_2=\sigma/K$ (eq.(1)) as a function of V and forming the ratio $\varepsilon=R_2/R_1$ which is found to be independent of V as shown in fig.5. We conclude that all Rydberg states in the

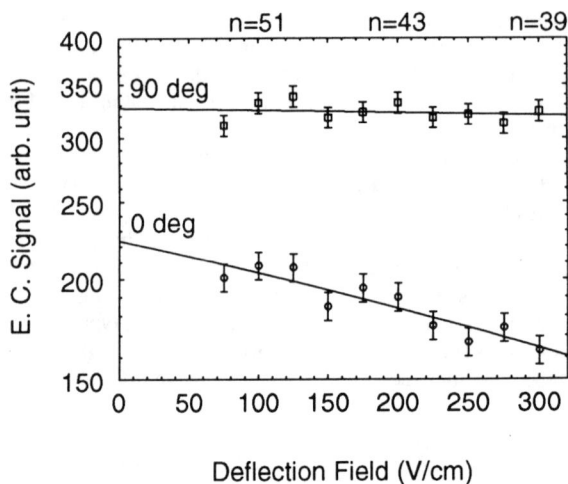

Fig. 6. Electron capture signal EC for 1keV ^{23}Na$^+$ vs deflection field for $\varphi=0°$ and $90°$.

beam have $n>16$ (ε constant) and that the detection efficiency for constant bias V depends only on the impact energy at the foil which is determined mainly by V. The variation of the efficiency is therefore small (~30%) and is given by $R_I(V)$. The cross sections are corrected for this effect.

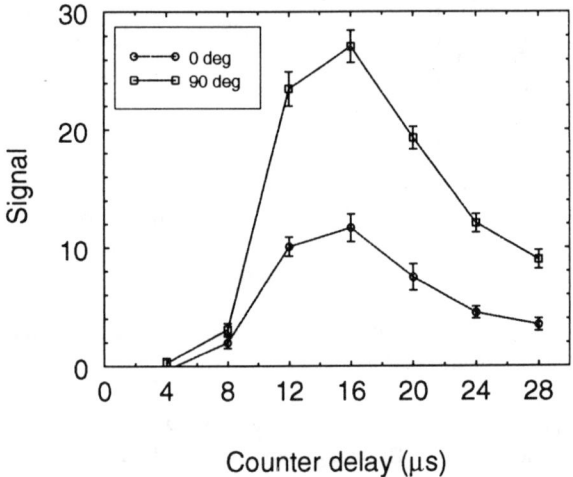

Fig. 7. Capture signal vs. counter delay for 1.5keV ^{23}Na$^+$. $T_{S+B} = 4\mu$sec.

Rydberg atoms with large principal quantum number are field-ionized by the deflection field in front of the Faraday cup. The experiment is therefore sensitive only to final states with n in an interval from 16 to an upper limit which depends on the deflection field. For the present fields the upper limit is within the range $n=39-51$. The dependence of the electron capture signal (eq.(1)) on the upper limit was studied by observing the signal as a function of the deflection field. The result is shown in fig.6. The data for $\varphi=90^0$ are independent of deflection field showing that the final states are dominated by $n<39$ for this orientation, but for $\varphi=0^0$ a significant variation is observed. The final n-distribution therefore depends on φ. All cross sections are corrected to refer to a deflection field of 100V/cm corresponding to an upper limit on n of 51. The present and other[7] experimental evidence suggests that the dominant part of final states produced by capture in the present range of impact velocities will have n-values in this range.

The field **B** causes some deflection of the ion beam at low energies. The deflection depends on the orientation of **B** *i.e.* on the orientation of the circular state measured by the angle φ. This may influence the normalization of the date and thus introduce a spurious dependence on φ. This source of error was controlled by measuring the φ-dependence for Rydberg atoms initially in the 25f-state (see fig.2). Even small stray electric fields effectively couple this state with the nearly degenerate states of higher l-value. The capture cross section for the resulting state is not expected to depend on the direction of **B**, and a small variation observed for this capture signal is used as a correction factor for the circular state.

Fig.7 shows capture signals for $\varphi=0^0$ and 90^0 measured at various counter delays. The variation of the signals is dominated by the near maxwellian velocity distribution of the Li atoms. The ratio $\sigma(90^0)/\sigma(0^0)$ is constant for delays of 12µsec or longer but changes suddenly for smaller delays. This shows that the conditions for adiabatic switching are fulfilled only for delays $>$ 12µsec in agreement with the conclusion reached from the measurements of SFI spectra at various HV ramp delays (fig.3).

RESULTS

The dependence of the measured capture cross sections on the orientation of the circular state is shown in fig.8 and 9 for the reduced impact velocities v_r =1.05 and 1.66, respectively.

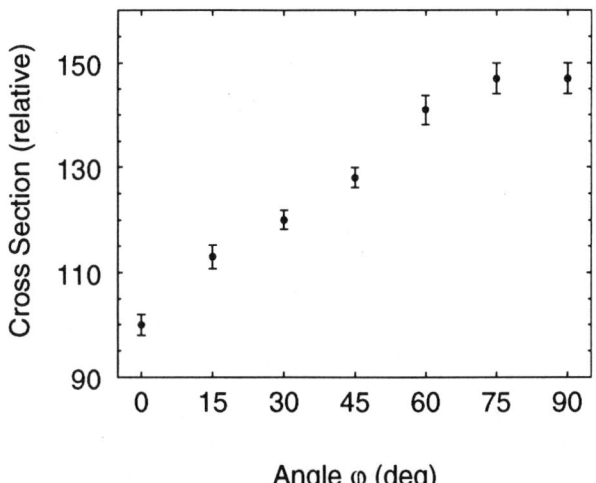

Fig. 8. Dependence of the capture cross section for circular states with $n=25$ on the angle of orientation φ of the circular state relative to the beam direction for 1keV ^{23}Na$^+$.

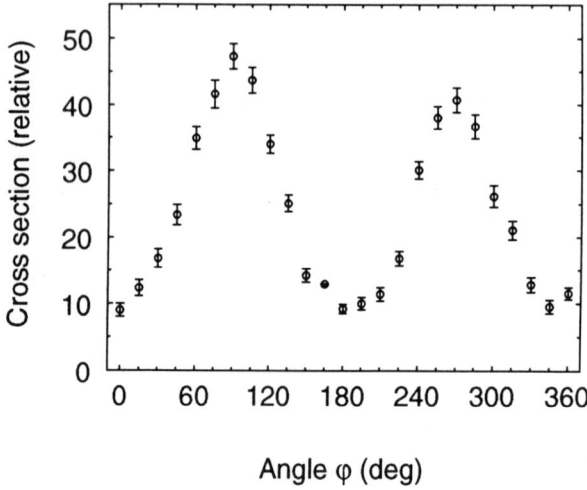

Fig. 9. Same as fig.8 except that impact energy is 2.5keV.

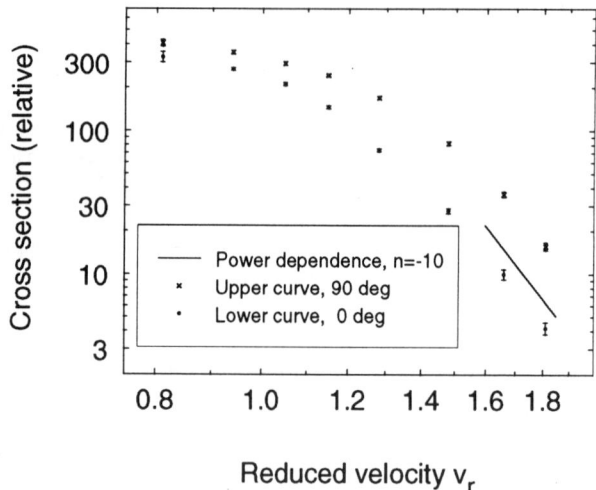

Fig. 10. Velocity dependence of electron capture cross section for circular Rydberg atoms with $n=25$ and angle of orientation $\varphi=0°$ and $90°$.

The most striking result is the strong dependence on φ at both values of v_r but strongest at $v_r =1.66$. For a true circular state the high degree of symmetry in the collision implies that the data should be periodic with two full periods in the angular interval $0°$-$360°$ and with each period being symmetric about the midpoint ($90°$ and $270°$). A small difference in amplitude of the two periods (fig.9) observed in some runs is sensitive to the potential of the uppermost Stark-plate (fig.1). This shows that the state studied is slightly elliptic due to a small, vertical, electric field at the crossing point between the ion beam and the thermal Li beam. It was not possible to reduce the field at this point below ~0.1V/cm. By averaging the angular data according to the fourfold symmetry one obtains a set of cross sections covering the range 0^0-90^0 and representing the case of true circular states. The data on figs. 8 and 9 show that the electron is much more likely to be captured when the ion velocity is parallel to the circular orbit than when it is perpendicular to it, and that the size of this geometrical effect increases with increasing impact velocity. This is seen more directly in fig.10 where cross sections for $\varphi=0^0$ and 90^0 are shown as a function of reduced impact velocity v_r. Note that the cross section is plotted on a logarithmic scale covering more than two orders of magnitude. Near $v_r=1$, the cross sections depend only weakly on velocity, but near $v_r=1.8$ the cross section for impact perpendicular to the plane of the orbit ($\varphi=0^0$) decreases very strongly, $\approx 1/v^{10}$, and the other cross section ($\varphi=90^0$) nearly as fast.

CONCLUSIONS

Relative electron capture cross sections for oriented, circular Rydberg atoms have been measured as a function of the orientation of the circular orbital for intermediate to high ion-impact velocities. The dependence of the cross section on orientation and impact velocity is in overall qualitative agreement with Classical Trajectory Monte Carlo[8] (CTMC) and second Born calculations[9], but a detailed theoretical analysis, directly applicable to the present data, is not available at the moment.

ACKNOWLEDGMENTS

We would like to acknowledge most valuable correspondence and advice by M. Gross in the initial phase of the experiment. Discussions with J.C. Day and L.J. Dubé are also gratefully acknowledged. The work was supported by the Danish Natural Science Research Council (SNF), the Carlsberg Foundation (Denmark), and the National Science Foundation (USA) under grant no. INT-9117374.

REFERENCES

1. D. Delande and J.C. Gay, Europhys. Lett. **5**, 303 (1988).
   A. Bommier, D. Delande, and J.C. Gay, "Atoms in Strong Fields", Ed. C.A. Nicolaides *et al.*, Plenum Press, New York (1990).
2. J Hare, M. Gross, and P. Goy, Phys. Rev. Lett. **61**, 1938 (1988).
3. R. Shakeshaft and L. Spruch, Rev. Mod. Phys., **51**, 369 (1979).
4. R.J. Damburg and V.V. Kolosov, J. Phys. B: At. Mol. Phys., **12**, 2637 (1979).
   T.H. Jeys, G.W. Foltz, K.A. Smith, E.J. Beiting, F.G. Kellert, F.B. Dunning, and R.F. Stebbings, Phys. Rev. Lett. **44**, 390 (1980).
5. C. Fabre, Y. Kaluzny, R. Calabrese, Liang Jun, P. Goy, and S. Haroche, J. Phys. B: At. Mol. Phys **17**, 3217 (1984).
6. S.B. Hansen, L.G. Gray, E. Horsdal-Pedersen, and K.B. MacAdam, J. Phys. B: At. Mol. Phys., **24**, L315 (1991), and corrigendum, J. Phys. B: At. Mol. Phys., **24**, 4475 (1991).
7. K.B. MacAdam, L.G. Gray, and R.G. Rolfes, Phys. Rev. **A42**, 5269 (1990).
8. G.A. Kohring, A.E. Wetmore, and R.E. Olson, Phys. Rev. **A28**, 2526 (1983).
9. L.J. Dubé, private communication (1993).

PHOTOABSORPTION AND COMPTON SCATTERING IN IONIZATION OF HELIUM AT HIGH PHOTON ENERGIES

L.R. Andersson* and J. Burgdörfer
Department of Physics and Astronomy, 401 A.H. Nielsen Bldg.,
University of Tennessee, Knoxville, TN 37996-1200, USA
and Oak Ridge National Laboratory, Oak Ridge, TN 37931-6377, USA

ABSTRACT

Production of singly and doubly charged helium ions by impact of keV photons is studied. The ratio $R_{ph} = \sigma_{ph}^{++}/\sigma_{ph}^{+}$ for photoabsorption is calculated in the photon-energy range 2–18 keV using correlated initial- and final-state wave functions. Extrapolation towards asymptotic photon energies yields $R_{ph}(\omega \to \infty) = 1.66\%$ in agreement with previous predictions. Ionization due to Compton scattering, which becomes comparable to photoabsorption above $\omega \sim 3$ keV, is discussed.

INTRODUCTION

Many-electron transitions in atomic systems induced by photon impact are of considerable interest since the Hamiltonian coupling of the electronic degrees of freedom to the electromagnetic field is built up of one-body operators. A transition involving more than one electron must therefore proceed via the interelectronic interaction (correlation). The simplest systems for studies of these processes are two-electron atoms and ions. Considerable work was carried out in the late 50s and early 60s on the dipole matrix elements for two-electron transitions in helium for the purpose of evaluating the Lamb shift of the ground state.[1-3] In the late 60s, when measurements of the ratio $R_{ph} = \sigma_{ph}^{++}/\sigma_{ph}^{+}$ of the double- to single-photoionization cross sections were reported from threshold up to 625 eV,[4] it was realized that this quantity is very sensitive to the usage of highly accurate wave functions.[5-10] Apart from the theoretical efforts to obtain R_{ph} for photoabsorption in the low-energy regime, predictions of the non-relativistic asymptotic value $R_{ph}(\omega \to \infty)$ also became available.[5-8] The experimental verification of this fundamental quantity has become possible only very recently with the advent of synchrotron-light sources having sufficient intensity. This progress on the experimental side[11-13] has stimulated renewed theoretical interest[14-19] in double ionization of He at high photon energies.

A complication in the interpretation of the experiments arises, however, when the photon energy exceeds approximately 3 keV.[20] The photoionization cross section decays rapidly as $\omega^{-7/2}$ while the Compton scattering cross section is essentially independent of ω in this energy regime. The cross sections are equal at about 6 keV.[21] Based on the energy transfer to the atomic system, the approximate thresholds for single- and double-ionization due to inelastic Compton scattering are 2.5 and 4.5 keV, respectively. Since the present experiments cannot distinguish between the competing photoabsorption and Compton processes, the measured ratio R is expected to be a weighted average of R_{ph} and the corresponding ratio for Compton scattering R_C. Above ~ 8 keV the experimentally measured R is primarily determined by Compton scattering ($R = R_C$).

---

*Present address: University of Stockholm, Atomic Physics, S-10405 Stockholm, Sweden

We present calculations of ionization-excitation and double ionization cross sections for photoabsorption in the 2–18 keV energy range employing correlated initial and final states and sum rules. We discuss the single and double ionization process by Compton scattering and estimate the contribution to the apparent R as measured by the recent experiments.

THEORY

The cross section for ionization of one electron into a continuum state labeled by the momentum k and angular momentum quantum numbers L and M and simultaneous excitation of the other electron to a He$^+(nlm)$ state by photoabsorption is, in the dipole approximation, given by (we use atomic units throughout unless otherwise stated)

$$\sigma_{ph}^{+*}(kLM, nlm) = \frac{2\pi^2}{c} \int dE \, \frac{df(kLM, nlm)}{dE} \delta(E + E_n - \omega + I_1), \tag{1}$$

where c is the speed of light, $df(kLM, nlm)/dE$ is the oscillator strength for the transition from ground state helium to a bound He$^+(nlm)$ state and a continuum state (kLM) with energy $E = k^2/2$, ω is the incident photon energy, I_1 is the first ionization potential of He, and E_n is the excitation energy of the n-manifold of He$^+$ measured from the ground state.

In the *acceleration* gauge the oscillator strength is

$$\frac{df^A(kLM, nlm)}{dE} = \frac{2k}{\omega^3} |\langle kLM, nlm |(\nabla_1 V + \nabla_2 V)_z| i \rangle|^2, \tag{2}$$

where V is the atomic potential energy and the polarization direction is taken along the z axis. Alternatively, the oscillator strength can be expressed in the *length* and *velocity* gauges. With exact initial- and final-state wave functions the various gauges of the oscillator strengths are equivalent while for approximate wave functions this is, in general, not true. The sensitivity of the oscillator strength to the gauge provides in the latter case a measure of the quality of the wave functions.

For the ground-state of He we use a 20-parameter Hylleraas-type wave function[22] and for the final state we use a wave function of the form

$$\Psi_{k,nlm}^{(-)}(\mathbf{r}_1, \mathbf{r}_2) = \frac{1}{\sqrt{2}} \left[\Phi_{nlm}(\mathbf{r}_1) \Phi_{\mathbf{k}}^{(-)}(\mathbf{r}_2) D_{\mathbf{k}_{12}}^{(-)}(\mathbf{r}_{12}) + \mathbf{r}_1 \leftrightarrow \mathbf{r}_2 \right], \tag{3}$$

where Φ_{nlm} and $\Phi_{\mathbf{k}}^{(-)}$ are bound and continuum wave functions defined in the unscreened field of the He^{2+} nucleus and

$$D_{\mathbf{k}_{12}}^{(-)}(\mathbf{r}_{12}) = \exp(-\pi\alpha/2)\Gamma(1 - i\alpha)_1F_1[i\alpha, 1, -i(k_{12}r_{12} + \mathbf{k}_{12} \cdot \mathbf{r}_{12})] \tag{4}$$

is a Coulomb distortion factor which accounts for the electron-electron interaction. The continuum states $\Phi_{\mathbf{k}}^{(-)}$ are normalized to a δ function on the momentum scale. In (4) $\mathbf{k}_{12} = \mathbf{k}/2$ is the interelectronic momentum and $\alpha = 1/(2k_{12})$. The states $|kLM, nlm\rangle$ in (2) are obtained by expanding $\Phi_{\mathbf{k}}^{(-)}$ and $D_{\mathbf{k}_{12}}^{(-)}$ in partial waves and recoupling to (LM) states.

In order to obtain the total cross section for ionizing one electron and leaving the second electron bound to the nucleus, the ionization-excitation cross sections (1) are summed over all bound states and over angular momenta of the continuum electron

$$\sigma_{ph}^{+}(\omega) = \sum_{LM}\sum_{nlm} \sigma_{ph}^{+*}(kLM, nlm). \tag{5}$$

with $L = l \pm 1$ and $M + m = 0$.

The cross section for double ionization with ejection of two electrons having energies $E = k^2/2$ and $E' = k'^2/2$ can be defined in analogy to (1) as

$$\frac{d\sigma_{ph}^{++}(kLM, k'lm)}{dE'} = \frac{2\pi^2}{c} \int dE \, \frac{d^2 f(kLM, k'lm)}{dE \, dE'} \delta(E + E' - \omega + I_2), \tag{6}$$

where I_2 is the total ionization potential of He. The total double-ionization cross section is

$$\sigma_{ph}^{++}(\omega) = \sum_{LM}\sum_{lm} \int_0^{\omega - I_2} dE' \, \frac{d\sigma_{ph}^{++}(kLM, k'lm)}{dE'}. \tag{7}$$

In the acceleration gauge the oscillator strength in (6) is

$$\frac{d^2 f^A(kLM, k'lm)}{dE \, dE'} = \frac{2kk'}{\omega^3} |\langle kLM, k'lm|(\nabla_1 V + \nabla_2 V)_z|i\rangle|^2, \tag{8}$$

and the length and velocity gauges are similarly defined.

The state $|kLM, k'lm\rangle$ in (8) is the analogue to (3) obtained by replacing the bound state Φ_{nlm} by $\Phi_{k'}^{(-)}$, by partial wave expanding the two continuum wave functions and the distortion factor, and by recoupling to angular momenta (LM) and (lm) of each electron.

In the case of two continuum electrons the Sommerfeld parameter α in $D^{(-)}$ depends on the relative angle between the emission directions of the two electrons. This complicates the direct evaluation of σ_{ph}^{++} as compared to ionization-excitation cross sections. However, the problem can be circumvented by employing a closure approximation. At high energies the photo electron carries nearly all the available energy $\omega - I_2 \approx \omega$, and the second electron is 'shaken up' to a low-lying continuum state.[17,19] The error introduced by fixing the energy of the fast electron and extending the upper limit in the integration in (7) to infinity is therefore small at high photon energies. The sum of single- and double-ionization cross sections

$$\sigma_S = \sigma_{ph}^{+} + \sigma_{ph}^{++} = \sum_{LM}\sum_{lm} \left[\sum_n \sigma_{ph}^{+*}(kLM, nlm) + \int_0^{\omega - I_2} dE' \, \frac{d\sigma_{ph}^{++}(kLM, k'lm)}{dE'}\right] \tag{9}$$

can be evaluated with these approximations by using the closure property of the He$^+$ eigenfunctions. The double ionization cross section can now be obtained without reference to the two-electron continuum states, as $\sigma^{++} = \sigma_S - \sigma^+$ In the limit $\omega \to \infty$ this procedure becomes exact.[2,23]

Theoretical investigations of Compton scattering of bound electrons are usually restricted to the coherent and incoherent cross sections.[24] For the problem at hand, this approach cannot be applied since we are concerned with final-state specific processes.

A high-energy approach as described above for photoabsorption is not justified because the dominant energy transfers ΔE from the photon to the electron(s) ranges from zero to an upper limit ΔE_{max} approximately given by the value for Compton scattering off free electrons

$$\Delta E_{max} = \frac{2\omega^2}{mc^2 + 2\omega} \tag{10}$$

and the distribution of energy transfers is essentially independent of ΔE in this range. One other important distinction between Compton scattering and photoabsorption is the distribution of angular momenta in the final state. While for photoabsorption only the final-state P sector is reached from the ground state of He (in the dipole approximation), a large number of final-state angular momenta will contribute to the transition amplitude for Compton scattering.

In order to estimate the influence of Compton scattering on the measured R we make here an impulse (or 'binary-encounter') approximation to obtain the Compton cross section differential in the energy transfer ΔE

$$\frac{d\sigma_C(\omega)}{d\Delta E} = \int d^3q \int_0^{p_{max}} dp \, \frac{p}{c} \, \delta(\mathbf{q}\cdot\mathbf{p} + p^2/2 + \epsilon_B - \Delta E) \, |\phi(\mathbf{q})|^2 \, \frac{d\sigma_{KN}(\omega)}{dp} \tag{11}$$

where p_{max} is the electronic momentum corresponding to maximum energy transfer (see Eq. 10) in a binary encounter between the photon and one electron, ϵ_B is the orbital binding energy of one electron, $\phi(\mathbf{q})$ is the momentum-space wave function of one electron in the ground state, and $d\sigma_{KN}(\omega)/dp$ is the free-electron Compton cross section differential in the momentum transfer to the electron for which we use the Klein-Nishina formula. The single-ionization Compton cross section $\sigma_C^+(\omega)$ is then obtained by integrating (11) between I_1 and ω and multiplying by two to account for the number of electrons. For double ionization we use

$$\sigma_C^{++}(\omega) = 2\int_{I_2}^{\omega} d\Delta E \, \frac{d\sigma_C(\omega)}{d\Delta E} \, R_C(\Delta E), \tag{12}$$

where $R_C(\Delta E)$ specifies the ratio of double to single ionization at a given energy transfer. Of course, the exact knowledge of $R_C(\Delta E)$ would imply that the problem at hand was solved. We make here the following approximation: for final states in the P sector $R_C^{L=1}(\Delta E)$ is assumed to equal the photoabsorption ratio at the photon energy ΔE, for higher angular momenta in the final state the shake-off value $R_C^{L>1}(\Delta E) = 0.73\%$ is used. The justification for this approximation relies on calculations for ionization-excitation by Compton scattering[24] described below.

RESULTS AND DISCUSSION

While it has been established that the ionization-excitation and double ionization cross sections are independent of correlation in the final state as $\omega \to \infty$, provided an accurate wave function of the initial state is used,[5,6,15] this is not the case at finite ω. We illustrate this for $\omega = 2$ keV in Fig. 1 where the difference $\Delta B = B_{corr} - B_{uncorr}$ of the branching ratios $B(n) = \sigma_{ph}^{+*}(n)/\sigma_S$ calculated with and without the final-state distortion $D^{(-)}$ are shown. The effect of final-state correlation is to redistribute probability for ionization without excitation to the ionization-excitation and double-ionization channels.

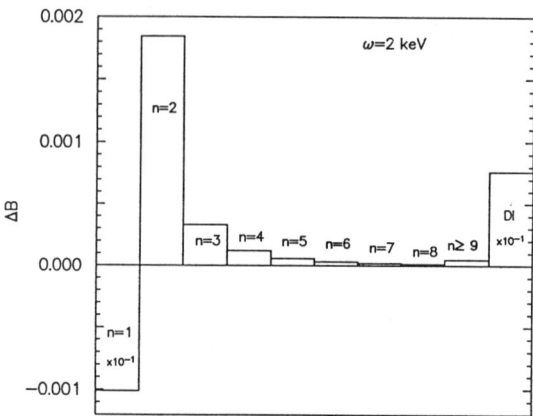

Figure 1: The change ΔB in the branching ratios $B(n) = \sigma_{ph}^{+*}(n)/(\sigma_{ph}^{+} + \sigma_{ph}^{++})$ calculated with and without correlation in the final-state wave function. The bars for $n = 1$ and double ionization (DI) have been multiplied by 10^{-1}.

In Fig. 2 we show our result for R_{ph} as function of ω^{-1} together with the recent calculations by Teng and Shakeshaft,[17] Hino,[18] and the MBPT calculation by Hino et al.[19] In the calculations of Hino and of Teng and Shakeshaft the double-ionization cross section was calculated directly using the two-electron continuum analogue to the final state (3). Teng and Shakeshaft used the velocity form of the dipole operator, while Hino used the acceleration form but took only the monopole contribution from the distortion factor $D^{(-)}$ into account. The MBPT calculation used various forms of the dipole operator. The acceleration form, shown in Fig. 2, the length and velocity forms all give similar results in the high-energy region.[19]

Our present result reaches an ω^{-1} behavior for $\omega > 5$ keV and extrapolation to infinite photon energy yields the correct non-relativistic limit for photoabsorption $R_{ph}(\infty) = 1.66\ \%$, which was obtained much earlier by other authors[5,6] using only a correlated initial state. The value of the coefficient of the leading ω^{-1} term is 0.90 keV. The short-dashed line in Fig. 2 represents an extrapolation of this linear behavior in ω^{-1} to both larger and smaller energies.

The various calculations[17-19] do not converge to the correct high-energy limit, even though they differ by relatively small amounts. We attribute this discrepancy to inaccurate initial-state wave functions used in the calculations. For $\omega^{-1} > 0.2$ the results start to diverge significantly. It is important to realize that the final state (3) and its two-electron continuum analogue constitute high-energy approximations and their use is not justified for lower photon energies. Our present result and the result of Teng and Shakeshaft have very similar slopes from 8 down to about 4 keV. At lower ω our result has a much stronger dependence on ω^{-1}. We have traced this strong dependence to the contributions from the $l \neq 0$ multipoles of $D^{(-)}$. Apart from the validity of the final state (3), the accuracy of the present result relies on the accuracy of the closure

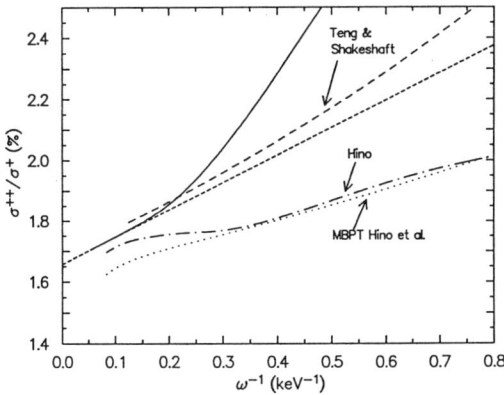

Figure 2: The ratio $R_{ph} = \sigma_{ph}^{++}/\sigma_{ph}^{+}$ for photoabsorption as function of ω^{-1}. Solid curve: present result; long-dashed curve, Teng and Shakeshaft;[17] dash-dotted curve, Hino;[18] dotted curve, Hino et al.[19] Short-dashed curve: the high-energy behavior of the present result accurate to order ω^{-1}.

approximation. This approximation breaks down if the energy sharing between the two electrons is not highly asymmetric. We have verified for an uncorrelated final state that the accuracy is sufficient at least down to $\omega = 2$ keV. However, the validity of the closure method for correlated final states at low energies remains to be verified.

In Fig. 3 is illustrated the excitation-ionization by inelastic Compton scattering. We show the ratios between the cross sections for ionization and excitation to $He^+(5s)$ and ionization without excitation as functions of the energy E of the ionized electron, broken down into the final-state angular-momentum components. We find the angular-momentum decomposition, shown here at $\omega = 10$ keV, to be approximately universal functions of the energy transfer but only weakly dependent on the primary photon energy. The two arrows indicate the asymptotic ratios for photoabsorption and for shake-off,[15] respectively. As can be seen, the $L = 1$ curve is close to the former value over a significant range of E, while the average of the $L > 1$ ratios, weighted by their partial cross sections, is close to the latter value. (The $L = 0$ component is small except for the lowest E and does not significantly contribute to the total cross section.) It is this observation which motivates our choice for $R_C^L(\Delta E)$ discussed in the context of Eq. (12).

In Fig. 4 is shown the ratio R_{ph} for photoabsorption (dash-dotted curve) together with the corresponding ratio for Compton scattering R_C (dotted curve) and the weighted mean of both processes (solid curve) which should be compared to the experimental points. The agreement is, considering the simplicity of the approximation, satisfactory. We also show the linear extrapolation of the photoabsorption ratio (dashed curve) which appears to improve the agreement with the experiments below $\omega = 3$ keV. Further work on two-electron processes by photoabsorption and inelastic scattering of photons is in progress.

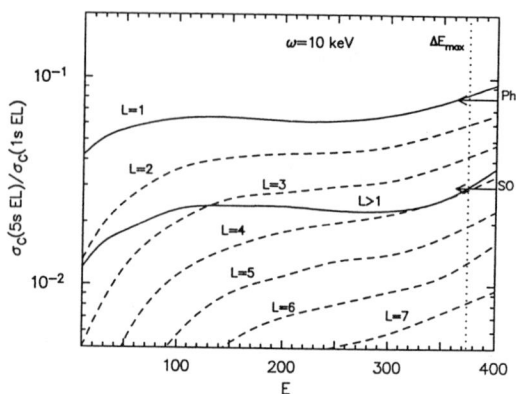

Figure 3: Ratios between the cross sections for ionization and excitation to $5s$ and ionization without excitation by Compton scattering as functions of the energy of the ejected electrons. The solid lines represent the $L = 1$ and $L > 1$ contributions. The dashed lines show the individual contributions from $L = 2$–7. Arrows indicate the asymptotic ratios for photoabsorption (Ph) and shake-off (SO). The dotted line indicates ΔE_{max} given by Eq. (10).

ACKNOWLEDGEMENTS

This work was supported in part by NSF; by U.S. DOE, Office of Basic Energy Sciences under contract No. DE–AC05–84OR21400 with Martin Marietta Energy Systems Inc.; and the Swedish Natural Science Research Council (NFR).

REFERENCES

[1] P.K. Kabir and E.E. Salpeter, Phys. Rev. **108** 1256 (1957).
[2] A. Dalgarno and A.L. Stewart, Proc. Phys. Soc. London **76** 49 (1960).
[3] E.E. Salpeter and M.H. Zaidi, Phys. Rev. **125** 248 (1962).
[4] T. Carlson, Phys. Rev. **156** 142 (1967).
[5] F.W. Byron Jr., and C.J. Joachain, Phys. Rev. **164** 1 (1967).
[6] T. Åberg, Phys. Rev. A **2** 1726 (1970).
[7] R.L. Brown and R.J. Gould, Phys. Rev. D **1** 2252 (1970).
[8] M. Ya. Amusia, E.G. Drukarev, V.G. Gorshkov, and M.P. Kazachkov, J. Phys. B: At. Mol. Phys. **8** 1248 (1975).
[9] R.L. Brown, Phys. Rev. A **1** 341 (1970).
[10] R.L. Brown, Phys. Rev. A **1** 586 (1970).
[11] J.C. Levin, D.W. Lindle, N. Keller, R.D. Miller, Y. Azuma, N. Berrah Mansour, H.G. Berry, and I.A. Sellin, Phys. Rev. Lett. **67** 968 (1991).
[12] J.C. Levin, I.A. Sellin, B.M. Johnson, D.W. Lindle, R.D. Miller, N. Berrah Mansour, Y. Azuma, H.G. Berry, and D.-H. Lee, Phys. Rev. A **47** R16 (1993).
[13] R.J. Bartlett and J.A.R. Samson, (private communication) (1993).
[14] T. Ishihara, K. Hino, and J.H. McGuire, Phys. Rev. A **44** R6980 (1991).

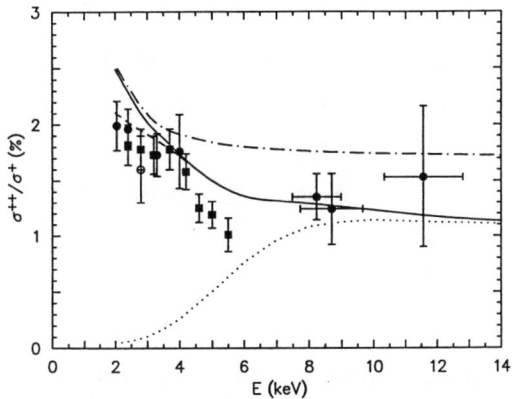

Figure 4: The ratio of double to single ionization of He by photon impact. Dash-dotted curve, photoabsorption; dotted curve, Compton scattering; solid curve, weighted mean of both processes; dashed curve, linear extrapolation of the high-energy behavior of R_{ph} to order ω^{-1}, corrected for Compton scattering. Experimental data: open circle, Levin et al.;[11] solid circles, Levin et al.;[12] solid squares, Bartlett and Samson.[13]

---

[15] A. Dalgarno and H.R. Sadeghpour, Phys. Rev. A **46** R3591 (1992).
[16] L.R. Andersson and J. Burgdörfer, Phys. Rev. Lett., in press (1993).
[17] Z. Teng and R. Shakeshaft, Phys. Rev. A **47** R3487 (1993).
[18] K. Hino, Phys. Rev. A **47** 4845 (1993).
[19] K. Hino, T. Ishihara, F. Shimizu, N. Toshima, and J.H. McGuire, Phys. Rev. A, in press (1993).
[20] J.A.R. Samson, C.H. Greene, and R.J. Bartlett, Phys. Rev. Lett., in press (1993).
[21] D.E. Cullen, M.H. Chen, J.H. Hubbell, S.T. Perkins, E.F. Plechaty, J.A. Rathkopf, and J.H. Scofield, UCRL-504000, Vol. 6 Part A, Rev. 4 (Lawrence Livermore National Laboratory) (1989).
[22] J.F. Hart and G. Herzberg, Phys. Rev. **106** 79 (1957).
[23] A. Dalgarno and R.W. Ewart, Proc. Phys. Soc. London **80** 616 (1962).
[24] J.H. Hubbell, Wm.J. Veigele, E.A. Briggs, R.T. Brown, D.T. Cromer, and R.J. Howerton, J. Phys. Chem. Ref. Data, Vol. 4, No. 3 471 (1975).
[25] L.R. Andersson and J. Burgdörfer, (to be published)

Measurement of Electron Capture From e^+- e^- Pair Production by 0.956 GeV/u U^{92+} on Au, Ag, Cu and Mylar targets

A. Belkacem, Harvey Gould, B. Feinberg, and R. Bossingham

Lawrence Berkeley Laboratory, Berkeley, CA 94720 USA

W. E. Meyerhof

Stanford University, Stanford, CA 94305 USA

ABSTRACT

We describe the first experimental observation of electron capture from electron-positron pair production in relativistic heavy ion collisions. We have used a novel new spectrometer to make the measurement of the cross section for a 0.956 GeV/u U^{92+} beam produced at the BEVALAC facility at LBL on Au, Ag, Cu and Mylar targets. We also measured the energy and angular distribution of the positrons for the Au target. The total cross section for a Au target is measured to be 2.19 (0.25) barns for capture from pair production and 3.30 (0.65) barns for pair production without capture.

INTRODUCTION

At relativistic energies, the capture of electrons by ions (recombination) occurs by the well-understood collisions processes of radiative electron capture (REC) and nonradiative electron capture (NRC) [1]. These processes, which require an electron in the initial state, have cross sections that decrease rapidly with increasing collision energy. This decrease largely accounts for the high ionization states produced in relativistic collisions, as the ionization cross sections are far less energy dependent at this energy [2]. REC is the capture of a target electron by the ion with the simultaneous emission of a photon (to balance momentum and energy). NRC is the capture of an electron initially bound to a target atom.

Until recently, REC and NRC were thought to be the dominant processes for electron capture at all relativistic energies. However, for heavy ions colliding at relativistic energies, the probability for electron positron pair production with the electron from the pair being created directly in the K-shell, was predicted to be significant [3-15]. The cross section for this "capture from pair production" mechanism is expected to increase with energy (as does the cross section for producing free electron-positron pairs), making it the dominant electron capture mechanism at highly relativistic energies. Since capture from pair production requires no electron in the initial state, it can take place between two bare ions, becoming the major source of beam loss for the heaviest ions at Relativistic Heavy Ion Colliders such as RHIC and LHC [16]. Achieving the highest luminosity thus requires an understanding of this capture process. Until now no experimental measurement has existed to check the validity of the different theoretical predictions reported in the literature. We report in this paper the first observation and measurement of electron capture from electron-positron pair production in relativistic heavy ion collisions.

EXPERIMENT

The experiment is performed at Lawrence Berkeley Laboratory's Bevalac accelerator using 0.956 GeV/u U^{92+} (bare uranium ions) incident on thin targets of Au, Ag, Cu. In the absence of a relativistic heavy ion collider, a fixed target and a far lower collision energy must be substituted. Due to the lower energy, the cross section is much smaller, requiring a high collection efficiency, and due to the electrons in the neutral target, there are competing capture processes (probably not present at higher energies) and a large background of knock-on electrons which must be distinguished from the positrons.

The U^{92+} ion passes through the target located inside the Advanced Positron Spectrometer (APS) shown in figure 1 and described below. In the case of capture from pair production, the electron is created directly bound to the uranium (initially bare) projectile, changing its charge state by one unit to U^{91+}. The experimental signature is the detection of the positron emitted at the target during the collision in coincidence with the charge changed U^{91+}. The U^{91+} is magnetically separated from the main beam of U^{92+} and each charge state is detected by a scintillator-photomultiplier tube detector. However, because we use a fixed target, the main fraction of the charge changed beam is due to the capture of a target electron by the projectile by an NRC or an REC process. We used target thicknesses of less than a few mg/cm^2 to keep the fraction of U^{91+} due to these processes below 1% of the beam. This fraction is still three orders of magnitude higher than the charge changing resulting from capture from pair production. Thus, the clean detection of a positron emitted at the target during the collision, in coincidence with the U^{91+}, is necessary to signal the capture from pair production process. Accidental events due to free pair production followed by capture of a target electron were kept to below a few percent and will be discussed at the end of this paper.

The Advanced Positron Spectrometer shown in figure 1 is designed to detect the positrons and measure their energy and angular distribution. APS is 2.6 m long and contains a solenoid magnet with a dipole at each end. The solenoid generates a strong longitudinal field at the center (B=0.8 T max), that adiabatically decreases to reach a value of 0.2 T at each end. This magnetic field transports the electrons and positrons away from the target and, most importantly, converts much of their transverse motion into longitudinal motion. The longitudinal field is then sharply decreased to zero within a few cm using a field clamp consisting of a 12 cm diameter exit hole in a 5 cm thick steel end plate. The positrons and the electrons, which acquire a large longitudinal momentum by the adiabatic field, are deflected in opposite directions by the transverse dipole field into their respective detectors. In addition to the very clean separation of electrons and positrons, this careful shaping of the magnetic field allows a very high detection efficiency. Without the adiabatically decreasing field, the acceptance of the spectrometer would be very low because most of the positrons emitted at relatively large angles, upon reaching the end of the solenoid field would have a large transverse momentum causing them to strike the apparatus rather than be deflected into the spectrometer detectors. Tests of the apparatus using radioactive sources and studying knock on electrons ejected by different ion beams from fixed solid targets have shown a detection

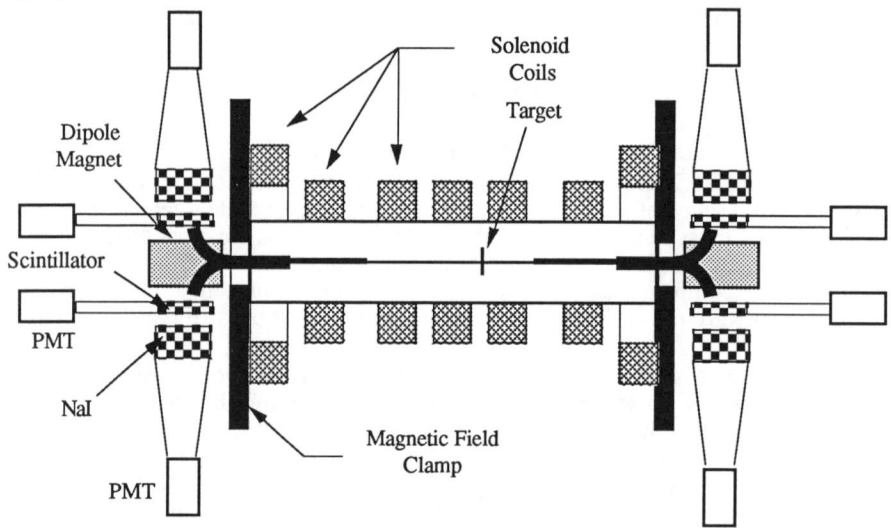

Figure 1 - Schematic diagram of the Advanced Positron Spectrometer (top - sectional view). The solenoidal field decreases adiabatically towards the ends causing the divergence of the particles to decrease, allowing them to be swept by the dipole magnets into the detectors. The target, or source is located near the center of the solenoid and the heavy ion beam travels horizontally through the apparatus.

efficiency close to unity for emission angles of up to 75° forward and backward. This acceptance is independent of the electron or positron energy in the energy range investigated (from 0.1 to 2.5 MeV).

The time of flight of the positrons through the strong field of the solenoid is used to measure their emission angle with respect to the beam direction. Four fast timing scintillator-photomultiplier detectors (two forward and two backward) are used to measure the energy and the time of flight of the positrons and the electrons. The energy resolution is about 17% and the time resolution is about 150-200 ps which translates into an angular resolution of 15° or better. The timing reference for the positron is given by the magnetically separated U^{91+} ion that produced the positron, detected by a fast timing scintillator-photomultiplier detector set at the end of the beamline.

As mentioned above, the initial discrimination between electrons and positrons is made by the dipole magnets that deflect electrons and positrons in opposite directions into their respective scintillator detectors. However, at 1 GeV/u, a large number of knock-on electrons are ejected from the target by the collisions with the uranium ion. Approximately 3 to 4 electrons with an energy above 100 keV are ejected from a 1 mg/cm^2 gold target for every collision, while only a few positrons are expected for every 10^6 collisions. Roughly 0.2% to 0.3% of these knock-on electrons backscatter from the electron scintillator into the positron scintillator, simulating a positron. In order to discriminate against these scattered electrons, we require the detection of one of the two 511 keV photons that are emitted back-to-back when the positron comes to rest and annihilates in the plastic scintillator.

Fig. 2 shows a photon spectrum measured by a 12.5 cm diameter and 15 cm long NaI set behind the forward positron detector. The spectrum displays a narrow 511 keV

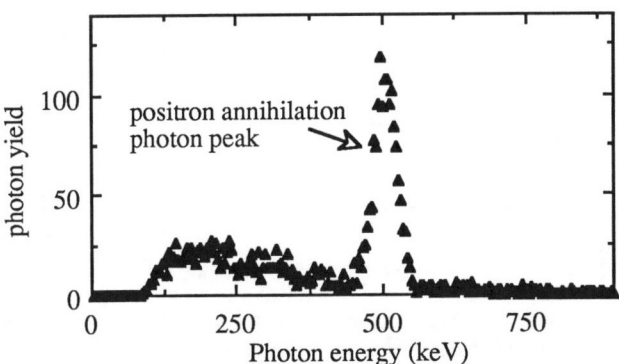

Figure 2 - Photon spectrum measured by the NaI set behind the forward positron detector. The energy resolution of the positron annihilation peak is 8%.

peak corresponding to the annihilation of the positron. The detection efficiency of the 511 keV photons by the NaI detectors is measured to be 42% with approximately 60% of the photons appearing as the narrow single peak with an 8% energy resolution and the remainder as a broad Compton distribution. In our data analysis, we use only the narrow peak to reject efficiently any backscattered electrons.

RESULTS AND DISCUSSION

Figure 3 shows the energy distribution for positrons emitted in coincidence with a charge-changed U^{91+}, for a 0.956 GeV/u U^{92+} beam incident on a 1 mg/cm^2 Au target. The data for the forward and backward directions have been integrated over emission angles between 0° and 75° and between 105° and 180°, respectively. The two spectra are taken simultaneously, and thus are normalized to the same number of incident uranium ions. Both spectra show a relative suppression of low energy positrons due to the repulsion of the gold nucleus, at rest in the laboratory frame. The forward energy spectrum displays a broad maximum around 800 keV while the backward spectrum displays a maximum for positrons energies around 400 keV. The backward and the forward distributions indicate that the positrons with high energy are emitted preferentially in the forward direction.

Figure 4 shows the positron angular spectrum integrated over all positron kinetic energies between 100 keV and 2500 keV. This distribution is the result of an integration by the detector over 360° of the azimuthal emission angle. The two parts of the spectrum (from 0° to 75° for the forward and 105° to 180° for the backward) were recorded simultaneously. A smooth continuation from one part of the spectrum to the other is seen. The positrons are emitted preferably at angles of about 30° and not at zero. This effect agrees qualitatively with predictions based on perturbation theory.

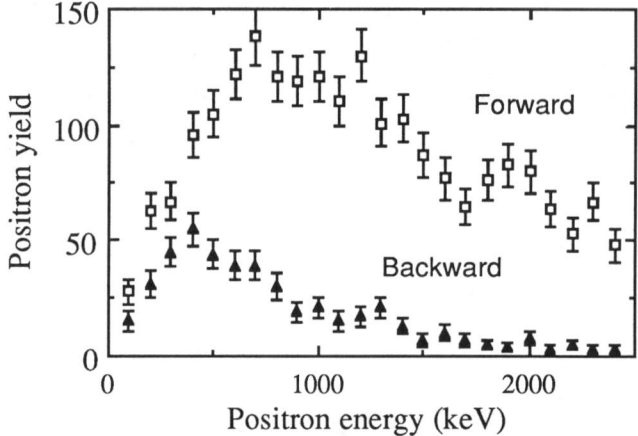

Figure 3 - Yield of positrons versus positron kinetic energy measured for electron capture from pair production by 0.956 GeV/u U^{92+} on a Au target. Each data point is an integration over all angles between 0° and 75° (forward and backward) with respect to the beam and over an energy interval of 100 keV.

To obtain the total cross section for capture from pair production we integrate the spectra over the positron energy and angle. A cut-off of positron energies above 2500

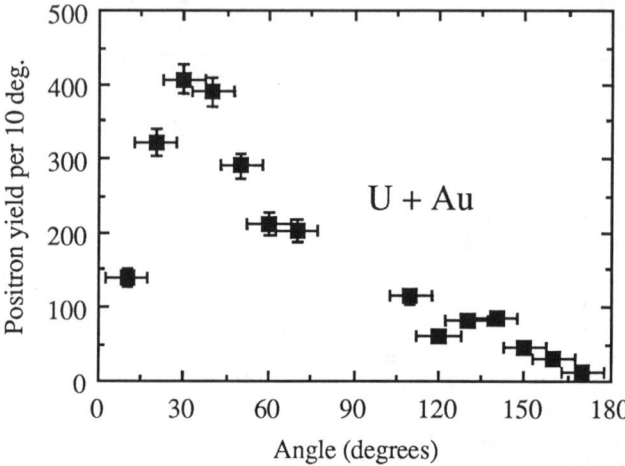

Figure 4 - Angular distribution with respect to the beam direction of the positrons for capture from pair production of a 0.956 GeV/u U^{92+} beam on a Au target. Each data point is a result of an integration over positron kinetic energies between 100 keV and 2500 keV and a 10° angular interval.

keV, seen in figure 3, is due to the energy range set for the analog-to-digital converter during the data acquisition. Higher energies are put in an overflow channel that is counted for the total cross section estimate, and only amounts to about 10%. Also, using the angular distribution displayed in figure 4, we correct for the angular region between 75 and 105 degrees, not accounted for by the spectrometer. We measure the total cross section for capture from pair production by a 0.956 GeV/u U^{92+} on a gold target to be 2.19 (0.25) barns. The major contributions to the uncertainty, shown in parenthesis, are statistics (0.1), target thickness measurement (0.1) and possible systematic effects (0.2).

If we assume that a similar fraction of free pairs are not detected by the spectrometer, we obtain a cross section of 3.3 (0.65) barns for the free pair process (electron - positron pair production *without* capture of the electron). Although the angular distributions for capture from pair production and free pair production are expected to be different, the error we make, in estimating the fraction between 75 and 105 degrees for free pair production, is well within our overall error estimate.

We find it striking that, at 1 GeV/u, the electron of the electron-positron pair is almost as likely to emerge from the collision bound to the uranium ion as it is to emerge free. A calculation based on perturbation theory, and summed over all possible final bound states, yields a value of 1.01 barns, which underestimates the measured total cross section for capture from pair production by U^{92+} in Au by about a factor of 2.2. Recently, a non-perturbative, coupled-channels calculation of capture from pair production has been published[14] for Pb on Pb at 1.2 GeV/u . To the best of our knowledge, a coupled-channels calculation for 1 GeV/u U^{92+} on Au has not yet been performed, making it difficult to compare our measurement to this calculation. (Using scaling from perturbation theory to extrapolate the results in Ref. 14 to U on Au gives a value that overestimates our measured cross section by about a factor of 2.) Our measured cross section for free pair production by U^{92+} in Au disagrees with the value of 5.1 barns of Becker, Grün and Scheid[15] and with the value of 1.25 barns of Decker[17].

Table 1 shows the capture from pair production measured for Au, Ag, Cu and Mylar

Table 1 - Capture from pair production measured for various targets for a 0.956 GeV/u U^{92+} beam. Also given is the measured total cross section of a target electron capture (NRC + REC) by the U^{92+} ion. The calculated values for NRC and REC cross sections are taken from figure 6 of ref. [2]. We estimate a 10% error from reading the theory numbers from the graph.

Process	Au	Ag	Cu	Mylar
Capture from pair production(barn)	2.19 (0.25)	0.485(0.058)	0.129(0.019)	0.0041(0.0025)
Total capture NRC+REC (barn)	3147 (250)	1184(120)	482(50)	110(10)
NRC(theory) (barn)	2300	420	100	15
REC (theory) (barn)	1000	590	400	85

targets. A least-squares fit to these data shows that the cross section varies with the target atomic number Z_t roughly as $Z_t^{2.8\pm0.25}$ which is not in good agreement with a Z_t^2 dependence predicted by perturbation theory for a bare ion on a bare target. Also shown is the target electron capture measured for the various targets. The cross section for the capture of a target electron are in very good agreement with the sum of the predicted values of NRC and REC [2] and are about three orders of magnitude higher than the capture from pair production cross sections.

BACKGROUND DISCUSSION

Two sources of background introduce small corrections to our measurement of the total cross section. First, ionization of the U^{91+} in the target may follow capture from pair production leading to an underestimate of the cross section. In a 1 mg/cm2 Au target an electron that has been captured by the uranium ion has only a 3% chance probability of being ionized in a subsequent collision in the target.

Second, a false signature for the process may be given by the production of a free electron-positron pair, followed by the capture by the U^{92+} of a target electron (by REC or NRC), resulting in an overestimate of the cross section. This can originate from a collision with a single atom or two successive collisions with two separate atoms. In order to minimize the background involving two successive collisions, we used very thin targets so that the U^{92+} has less than 1% probability to capture an electron from the target by all of the capture processes combined. To check for this background, which varies quadratically with target thickness, we verified that the capture from pair production varied linearly with target thickness over a factor of 10 in target thicknesses.

The process in which free pair production occurs in a collision with a target atom and the U^{92+} ion captures an electron by REC or NRC from the target atom in the same collision varies linearly with the target thickness. Its contribution is directly related to the probability of a target electron capture at the impact parameters where pair production occurs. At 1 GeV/u pair production is thought to occur at extremely small impact parameters (of the order of the Compton wavelength or smaller) where the probability for NRC is small and the probability for REC is negligible. The observation of only a small deviation from the Z_t^2 dependence for capture from pair production suggests that the contribution from this latter background is very small; at most a few percent. To support that argument we show in Table 1 the target electron capture cross sections (NRC + REC) measured for the Au, Ag, Cu and Mylar targets. The cross section for NRC+REC decreases by almost an order of magnitude between an Au target and a Cu target. The NRC contribution part (theoretical estimate) is also shown in table 1. It, especially the probability at small impact parameters, which in this case is the main source of the background, decreases even faster with decreasing Z_t [18]. That decrease, which should be on top of the Z_t^2 due to pair production [19-20], is certainly not observed in our data. We estimate that the contribution from these false events to our measurement is not more than 6%.

ACKNOWLEDGMENTS

We thank Steven Abbott for the engineering design of the APS, Richard Leres for the data acquisition software development and Harvey Oakley and the Bevalac technical support staff for helping put together the experiment in a record time in order to meet the deadline set by the shutdown of the Bevalac. We thank Donald Jourdain and Cory Lee for helping us to design and build very efficient scintillating detectors. We thank Charles Munger and Lynette Levy for their assistance during the data taking, and we thank Klaus Momberger for fruitful discussions about the theory of capture from pair production. This work was supported by the Director, Office of Energy Research, Office of Basic Energy Sciences, Division of Chemical Sciences, of the U.S. Department of Energy (DOE) under contract No. DE-AC-03-76SF00098. One of us, (BF) was supported by The Office of High Energy and Nuclear Physics, Division of Nuclear Physics, of the U.S. DOE. One of us, (WM) was partially supported by NSF grant No. PHY8614650.

REFERENCES

1. See, for example, J. Eichler "Theory of Relativistic Ion-Atom Collisions" Phys. Rep. 193, 167 (1990); R. Anholt and U. Becker, Phys. Rev. A36, 4628 (1987).
2. R. Anholt, W.E. Meyerhof, X. -Y. Xu, H. Gould, B. Feinberg, R.J. McDonald, H.E. Wegner, and P. Thieberger, Phys. Rev. A36, 1586 (1987); W. E. Meyerhof, R. Anholt, J. Eichler, H. Gould, Ch. Munger, J. Alonso, P. Thieberger and H.E. Wegner, Phys. Rev. A32, 3291 (1985).
3. H. Gould, "Atomic Physics Aspects of a Relativistic Nuclear Collider" Lawrence Berkeley Laboratory Report, LBL-18593 (1984).
4. C. A. Bertulani and G. Baur, Nucl. Phys. A458, 725 (1986).
5. U.Becker, J. Phys. B20, 6563 (1987); values for U on Au were computed by K. Momberger.
6. U. Becker, N. Grün, and W. Scheid, J. Phys. B20, 2075 (1987).
7. C. A. Bertulani and G. Baur, Phys. Rep. 163, 299 (1988).
8. G.R. Deco and R. D. Rivarola, J. Phys. B21, 1229 (1988).
9. G. Deco and N. Grün, J. Phy. B22, 3709 (1989).
10. M.J. Rhoades-Brown, C. Bottcher and M.R. Strayer, Phys. Rev. A40, 2831 (1989).
11. A.J. Baltz, M.J. Rhoades-Brown and J. Weneser, Phys. Rev. A44, 5569 (1991).
12. M. R. Strayer, C. Bottcher, V. E. Oberacker, and A. S. Umar, Phys. Rev. A41, 1399 (1990).
13. K. Momberger, N. Grün and W. Scheid, Z. Phys. D18, 133 (1991).
14. K. Rumrich, K. Momberger, G. Soff, W. Greiner, N. Grün and W. Scheid, Phys. Rev. Lett. 66, 2613 (1991), and K. Momberger, Private communication.
15. U. Becker, N. Grün and W. Scheid, J. Phys. B19, 1347 (1986)
16. "Conceptual Design of the Relativistic Heavy Ion Collider-RHIC", BNL Report BNL-52195, pp 117-121. See especially Table IV. 3-10.
17. F. Decker, Phys. Rev. A44, 2883 (1991).
18. N. Toshima and J. Eichler, Phys. Rev. Lett. 60, 573 (1988).
19. See, for example, G. Racah, Nuovo Cimento 14, 93 (1937) and references therein.
20. C.R. Vane, S. Datz, P.F. Dittner, H.F. Krause, C. Bottcher, M. Strayer, R. Schuch, H. Gao, and R. Hutton, Phys. Rev. Lett. 69, 1911 (1992).

EXPERIMENTAL AND THEORETICAL INVESTIGATION OF THE AUTOIONIZATION DYNAMICS IN THE EXCITED COLLISION COMPLEX He*(2^3S) + H(1^2S)

A. Merz, M.-W. Ruf, H. Hotop, M. Movre ‡ and W. Meyer ‡

Fachbereich Physik / Chemie ‡, Universität Kaiserslautern, D-67663 Kaiserslautern

ABSTRACT

We present angle-dependent electron energy spectra for the autoionization process He*(2^3S) + H(1^2S) leading to Penning ($\rightarrow$ He + H$^+$ + e$^-$) and associative ionization ($\rightarrow$ HeH$^+$(v$^+$, J$^+$) + e$^-$) obtained from high-resolution experiments and from ab initio theory. Nearly complete agreement between the two sets of spectra is demonstrated. Non-isotropic electron emission is observed, reflecting angular momentum transfer to the electron and a correlation between the heavy-particle dynamics and the internal emission anisotropy. The ab initio calculations include angular momentum dependent resonance-to-continuum couplings and reveal a strong contribution of autoionization into an electronic d-wave.

INTRODUCTION

The autoionization dynamics of collisional complexes is a lively topic of research [1-11] and recent developments in laser cooling and trapping of atoms have created additional interest in ionization processes of excited atoms at low collision energies E_{rel} [3,4,12]. Besides studies of the velocity-dependence of ion production [3,10,11,13,14] and analyses of ion angular distributions [7,8], high resolution electron spectrometry of ionizing collision systems has proven especially fruitful in revealing details of the autoionization dynamics [1,2,4-6,11,15-17] and in yielding direct information on the underlying interaction potentials [4,5,15-18]. Most previous electron spectrometric experiments were carried out at the fixed electron detection angle $\Theta = 90°$ (relative to the collision velocity v_{rel}); this is also true for the basic system

$$\text{He}^*(2^3\text{S}) + \text{H}(1^2\text{S}) \rightarrow \begin{cases} \text{He}(1^1\text{S}) + \text{H}^+ + e^- \, (\epsilon; \Theta) & \text{(PI)} \quad (1a) \\ \text{HeH}^+(v^+, J^+) + e^- \, (\epsilon; \Theta) & \text{(AI)} \quad (1b) \end{cases}$$

which is highly attractive in the excited entrance channel potential V*(R) ($^2\Sigma$) (ab initio well depth $D_e^* = 2.28$ eV at $R_e^* = 3.42$ a_0 [16,30]) as well as in the ionic exit channel V$^+$(R) ($^1\Sigma$) (well depth $D_e^+ = 2.04$ eV at $R_e^+ = 1.46$ a_0 [16]). The electron energy spectrum for reaction (1) shows rich structure due to Airy-type oscillations associated with Penning ionization (1a) at distances around the potential minimum of V*(R) and due to formation of (quasi) bound HeH$^+$(v$^+$,J$^+$) rovibrational levels in associative ionization (1b); the spectrum extends from 3.5 eV to 7.8 eV according to the variation of the classical electron energy function $\epsilon(R) = V*(R) - V^+(R)$ [2].

The good agreement between the high resolution electron energy spectrum of Waibel et al. [16] - partially resolving the rovibronic structure due to HeH$^+$(v$^+$, J$^+$)-formation - and a quantum mechanical, optical potential (fit-)calculation seemed to indicate that the ionization process is well described by the underlying theory [16]. The calculation involved accurate ab initio potentials for the real parts $V*(R)$ and $V^+(R)$ of the He* + H and He + H$^+$ systems, an adjusted imaginary part (autoionization width) $\Gamma(R)$ for the coupling of the entrance channel to the continuum and the assumption that the heavy particle angular momentum is conserved (J* = J$^+$) in the electron emission process, implying isotropic electron emission (s-wave only). It has to be noted, however, that the adjusted width $\Gamma(R)$ exhibits a much stronger decrease towards larger R than the ab initio widths [5,26,30]. Recent work on the rather similar, attractive system He*(2^{3}S) + Li(2^{2}S) revealed a substantial angular variation of the shape of the electron energy spectrum [5,17] which was attributed to angular momentum exchanges $|\Delta J| = |J* - J^+|$ between the heavy particle system and the emitted electron. A classical estimate based on recoil arguments yields in this case $|\Delta J_{max}| \approx 4\,\hbar$, compatible with the significant angle-dependent changes of the spectral shapes [17]. A similar consideration for He* + H results in $|\Delta J_{max}| \lesssim 1\cdot\hbar$, but an alternative mechanism, the Coulomb attraction between the ejected electron and the charge center of the cation, may well lead to significant angular momentum transfer in the case of HeH where the centers of charge and mass are relatively far apart.

The discrepancy between the adjusted and the ab initio width and the angular dependence observed for the He*(2^{3}S) + Li spectrum motivated us to carry out the first experimental study of the angular-dependent electron and ion energy spectra for He*(2^{3}S) + H and - in parallel work - an ab initio calculation of the ℓ-dependent autoionization coupling elements (ℓ = orbital angular momentum of partial electron wave) and the angular-dependent electron spectra for the relevant experimental conditions. The experimental spectra show strong anisotropies. By reference to trajectory calculations, they allow - especially in the Al channel - to extract information on the internal electron angular distribution, i.e. the angle-dependent electron intensity, as emitted in the frame of the autoionizing quasi-molecule for a particular internuclear distance R (here R $\approx$ 2.3 a$_0$). The theoretical spectra, obtained from the ab initio data without any adjustment, are in excellent agreement with the observed spectra and verify a strongly non-isotropic internal electron distribution. They demonstrate that the underlying local complex potential theory in the frame of the Born-Oppenheimer approximation provides an adequate description for the autoionization dynamics in the He*(2^{3}S) + H(1^{2}S) system.

EXPERIMENTAL

Our experimental setup involves crossed atomic beams and two double hemispherical condensers for electron and ion detection (energy resolution $\approx$ 30 meV). The metastable helium atoms are produced in a differentially pumped cold cathode dc discharge [18] (2^{3}S-beam intensity: $1.5 \cdot 10^{15}$ atoms /(s$\cdot$sr)). The 2^{1}S-species, also generated in our metastable source, is removed from the beam via optical pumping [16]. The average metastable velocity is $\overline{v}$(He) = 1750 m/s with a width of $\Delta v/\overline{v}$(He) $\simeq$ 30%. The atomic hydrogen beam is produced in an air cooled microwave discharge outside the vacuum chamber and guided to the reaction zone by a PTFE-covered stainless steel tube, yielding an effective density ratio of n(H)/n(H$_2$) $\approx$ 0.3. Its velocity distribution was not measured, but can be assumed to be a Maxwell-distribution for an effusive quasi beam (T = 300 K, $\overline{v}$(H) = 2500 m/s) [16]. The resulting average relative collision energy is $\overline{E}_{rel}$ = 50 meV.

One of the two spectrometers detects charged particles at 90° with respect to both atomic beams and serves as a monitor to allow the determination of angle-dependent, energy-integrated cross sections; its detection geometry corresponds to the one always used in previous work. The second spectrometer is rotatable to detect electrons (ions) in the collision plane spanned by the two atomic beams. The electron energy spectra are corrected for contributions due to Penning ionization of ground state molecular hydrogen (below 4.5 eV). The energy scale is calibrated accurately (± 8 meV) and the data are also corrected for the energy dependence of the spectrometer transmission [28]. The overall angular resolution of the experiment is limited by the angular and velocity widths of the effusive H atom beam and estimated to be better than ± 20°. (A drawing of the apparatus and further experimental details are included in ref. 32).

THEORY

In the ab initio work we have calculated all the relevant electronic structures and subsequently treated the scattering process without further approximations. Our treatment can be summarized as follows [30]:

1) The resonance state is defined through Feshbach projection - based on orbital occupancy restrictions - in the frame of multi-reference configuration-interaction calculations. The potential is determined with an accuracy of about 10 meV and is very close to our preliminary potential displayed and used in ref. 16.
2) The energy dependent coupling with the continuum is represented in the form of a compact (L^2) "Penning MO" $\varphi = <\Psi^*(N) |H| \Psi^+(N-1)>_{N-1}$ without any phase information being lost.
3) This MO is projected onto the states of the continuum electron in the exit channel, which are calculated within the static-exchange approximation for up to 30 coupled angular momentum channels. This yields for the first time reliable angular dependent coupling elements which have a significant size up to $\ell = 4$.
4) The scattering calculation is based on a complex Numerov algorithm, using a converged set of complex, ℓ-dependent coupling matrix elements.
5) Weighting with experimental collision energy distributions finally gives the angle-dependent as well as the angle-integrated spectra.

RESULTS AND DISCUSSION

Figure 1 shows six electron energy spectra measured at different detection angles. The angle Θ is related to the relative velocity vector, with Θ = 180° labelling the backward direction of

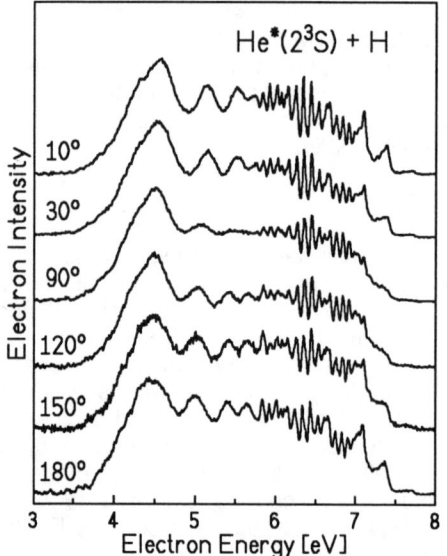

Fig. 1. Energy spectra of electrons from He*(2^3S)+H(1^2S) collisions, as observed at the indicated detection angles Θ. The spectra have a common intensity scale, proportional to the cross section.

electron emission from the He atom in the center-of-mass frame. The geometry allowed us to include almost the total range of possible emission angles, including the for- and backward direction. The spectra show significant angular dependence in the Penning as well as in the associative ionization part. The essential angular variations are:

- In the PI-part of the spectrum the supernumery Airy peaks at 5.2 eV and 5.5 eV (at $\Theta = 10^0$) shift towards lower energies with increasing angle Θ, allowing a third peak to appear in the $\Theta = 180^0$ - spectrum, while the main Airy peak shifts only little.
- The Airy-structure is more strongly modulated in forward and backward direction than at angles close to $\Theta = 90^0$.
- The main Airy-peak of the $\Theta = 10^0$ - spectrum contains a shoulder at $\epsilon \simeq 4.3$ eV which is not present for larger scattering angles.
- The associative ionization structures due to forming rovibronically excited $HeH^+(v^+, J^+)$ ions, as well as the contributions due to "quasi-associative ionization" (ionic states trapped within a rotational barrier in the exit channel potential) are most strongly modulated at 10^0. The modulation structure gets weaker with the angle approaching $\Theta = 90^0$ and stronger again for backward scattering, especially the peaks at 7.1 eV and 7.4 eV.

The shape-variations of the electron energy spectra shown in figure 1 are a consequence of an angle-dependent summation of the different heavy-particle angular momentum contributions [32]. Diagram 2 shows these J-contributions for the associative ionization regime, calculated quantum-mechanically under the assumption of angular momentum conservation ($J^*=J^+$). The data are averaged over the

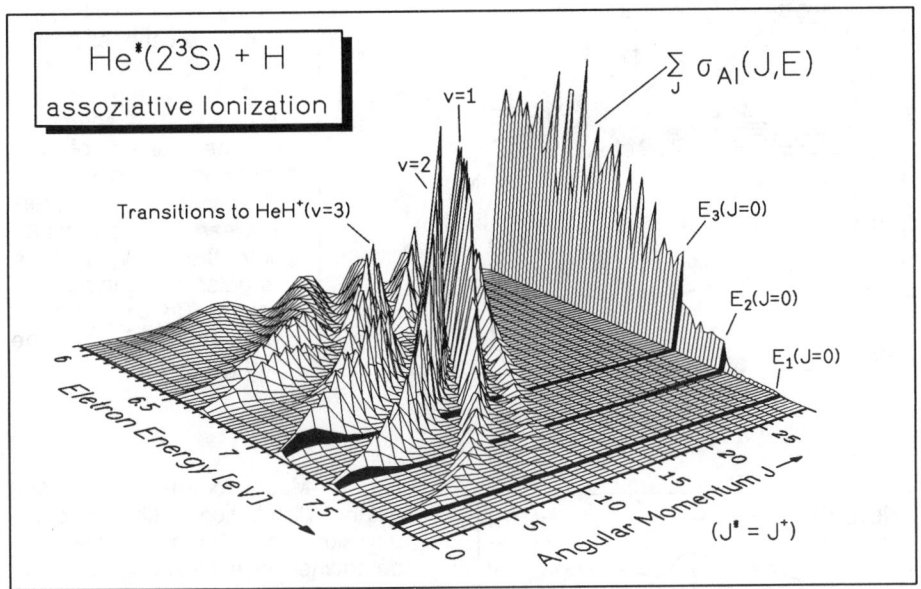

Fig. 2. Quantum mechanical model-calculation, demonstrating the composition of the associative ionization regime through the different heavy-particle partial waves. The partial J-spectra and the summed spectrum in the back are averaged over the experimental collision energy distribution, see text.

experimental collision energy distribution and convoluted with the experimental energy resolution. The spectrum at the back is the sum over the shown partial J-contributions. Note that the shape of the summed spectrum is quite similar to the experimental $\Theta = 90°$ spectrum of figure 1, as already discussed by Waibel et al. [16].

If one follows the lines starting from the peaks labelled $E_v(v=1-3, J=0)$ in the back spectrum and leading from high to low J-values it becomes clear that these peaks are due to transitions to different vibrational levels in the HeH$^+$-potential at very low J-values ($J \leq 5$). Comparing figures 1 and 2 it is obvious that especially the spectra for small J contribute more strongly to the forward and backward direction than to angles around 90°.

The quasi-bound structures, ranging from 5.8 eV to 6.3 eV, show a similar behaviour as the structures due to associative ionization. The peak-modulation is particularly pronounced in the experimental spectra taken at $\Theta = 10°$ and 180°. These features are due to the heavy-particle partial waves around $J \simeq 25$ (see fig. 2) which have the largest contribution and which are significantly more effective in the forward and backward direction than normal to the relative velocity vector.

If the different J-contributions of figure 2, are related to the positions of the internuclear axes at the moment of autoionization, the internal angular distribution for electron emission from the quasimolecule can be extracted from the observed spectra, at least for the R-region where associative ionization occurs, i.e. close to the

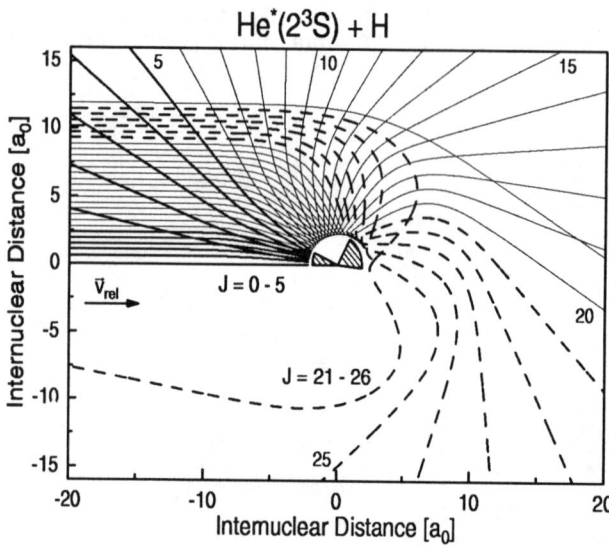

Fig. 3 Classical trajectory calculation - computed with the ab initio entrance channel potential - for the average collision energy E_{rel} = 50 meV. The segments in the center of the diagram indicate the directions of the internuclear axes connected with the heavy particle angular momentum ranges $0 \leq J^* \leq 5$ and $21 \leq J^* \leq 26$ at the turning point.

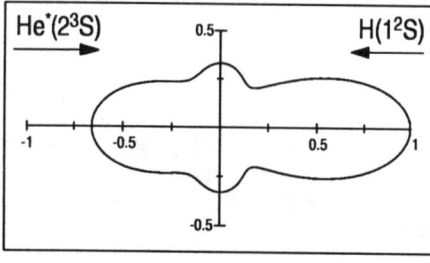

Fig. 4 Model for the internal angular distribution of electron emission from the quasi-molecule, determined in a fit-calculation. It is based on the partial spectra of figure 2 and information taken from the trajectory calculation. Note that the shown intensity distribution is only valid for the energy regime of associative ionisation, as disscused in the text.

turning point. The information concerning the internuclear axes can be read from a classical trajectory calculation.

Inspection of figure 3, which shows such a trajectory calculation for the (average) collision energy E_{rel} = 50 meV, reveals that for small J-values up to five the internuclear axes at the turning points are lying more or less parallel to the relative velocity direction in an angular segment of ± 25°. Although the internuclear axes connected with the highest angular momenta J* are not as directed as those corresponding to the low ones, their average orientation is also parallel to the relative velocity vector. Combining this result with the information read from figures 1 and 2, it is obvious that the internal angular distribution of electron emission should have clear maxima in forward and backward direction of the collision system. A possible model distribution, which is the result of an iterative fit-procedure, is presented in figure 4. It shows that the e^--emission-maximum in forward direction is three times as large as the perpendicular contribution and the backward maximum is supposed to be twice as large as the perpendicular one. In addition, the forward maximum is more strongly focussed than the other two.

Of course the intensity of the internal angular distribution of e^--emission is expected to depend on the internuclear distance, but as mentioned above, we restrict this discussion to AI- and QI-processes which take place in a well-defined R-range between 2.2 a_0 and 2.4 a_0.

If we finally put the results of figures 3 and 4 together, we can generate angle-dependent model spectra, based on the quantum mechanically calculated partial J-spectra (fig. 2). For symmetry-reasons this is only possible in this simple way for the forward and backward direction. In figure 5 we compare the experimental Θ = 10°-spectrum with such a model-spectrum (top) and with the "isotropic" result of figure 2 (bottom). The overall-agreement is substantially improved by the angle-dependent model-spectrum, especially in the energy ranges of the peaks labelled E_1-E_3 and in the quasi-associative ionization regime (5.8 → 6.3 eV). This also holds for the backward direction. The rather satisfying agreement of the model spectrum was achieved even without accounting for the collision energy dependence of the directions of the internuclear axes at the moment of ionization and the limited experimental angular resolution.

The semi-empirical internal distribution of figure 4 can be compared to those which follow from the ab initio l-dependent coupling elements between the resonance and continuum states [30]. The latter are displayed in the complex plane in figure 6 for a set of internuclear separations. The characteristic spiral curves emerge from a strong decrease

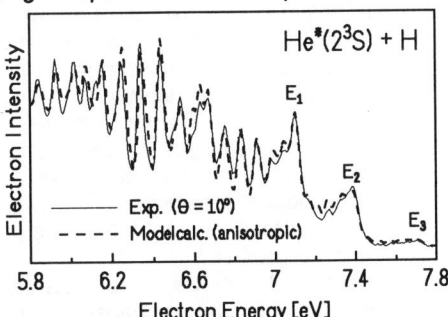

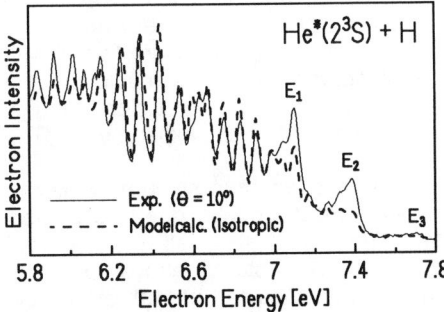

Fig. 5. Comparison of the AI-part of the experimental Θ = 10-spectrum with an "angle-dependent", quantum mechanically calculated model-spectrum (top) and an "isotropic" spectrum (bottom).

of the norms with internuclear separation, as known from the total width, and phase changes which are partly due to the related energy change. For short distances including the turning point region discussed above, the $\ell = 2$ coupling is dominant. Further out the isotropic component $\ell = 0$ is the largest, but significant couplings exist for ℓ up to 4. Note that this angular decomposition refers to the center of mass, as required for the heavy-particle dynamics. There appears to be little resemblance to the only other coupling elements available in the literature [26]. The resulting internal electron distributions are shown in figure 7 for several R. The similarity of that for R = 2.25 a_0 with figure 4 is obvious.

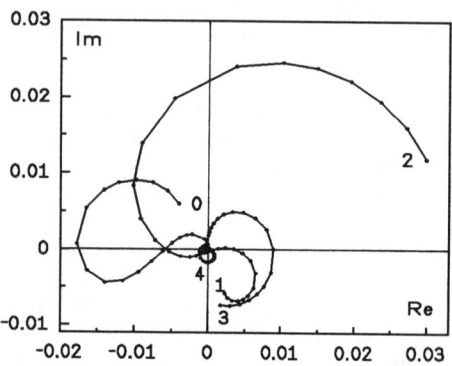

Fig. 6. Complex CM coupling elements for $\ell=0$ to 4; points are at R = 1.6 → 2.1 (ΔR=0.1), 2.25 → 5 (0.25), 6 →10 (1) [au].

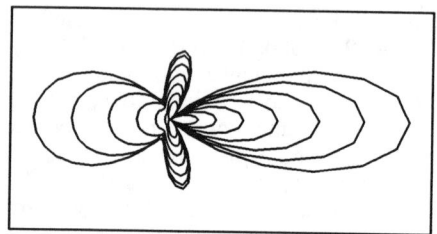

Fig. 7. Ab initio internal angular distribution for R = 2.0 → 3.5 [au], ΔR=0.25 [au].

The ab initio result for the total autoionization width (i.e. the sum over the squares of the different partial amplitudes) is - concerning the shape of the curve - similar to the earlier result of Hickman et al. [26], but the absolute values are lower by a factor of two. The fit-result of ref. 16 deviates significantly in shape, especially in the transition region between small and large internuclear distances and in the long-range behaviour. Our results for the width as well as for the internal angular distribution are in line with recent work of Sarpal and coworkers [33]. Their study of electron scattering by HeH$^+$ and the bound states of HeH, using the molecular R-matrix method, shows that the resonance is anisotropic with considerable "d-wave" character at small internuclear separations, implying a rotational angular momentum transfer during the collision ($|\Delta J| \leq 2$) [33].

Figure 8 shows experimental spectra (full line) in comparison with those from the ab initio calculations (dashed line) at $\Theta = 10^0$, 90^0 and 180^0. The theoretical spectra were folded with the experimental energy resolution and weighted with the collision energy distribution of the experiment. The spectra are found to be in almost perfect agreement, apart from some very narrow structures in the AI regime. These discrepancies are most probably due to a neglect of the divergence of the effusive H-beam. Apart from this, all the different shape-variations in the PI-, QI- and AI-part of spectra, stated above, are well reproduced in the theoretical work. Even the sound shoulder at 4.3 eV in the 10^0-spectrum is confirmed.

The close agreement of the experimental and theoretical spectra, demonstrated in figure 8, proves the adequacy of the local complex potential approach in the frame of the Born- Oppenheimer approximation for this basic autoionization process. Furthermore, it turns out that the angular momentum transfer between the heavy-particle system and the emitted electron plays an important role,

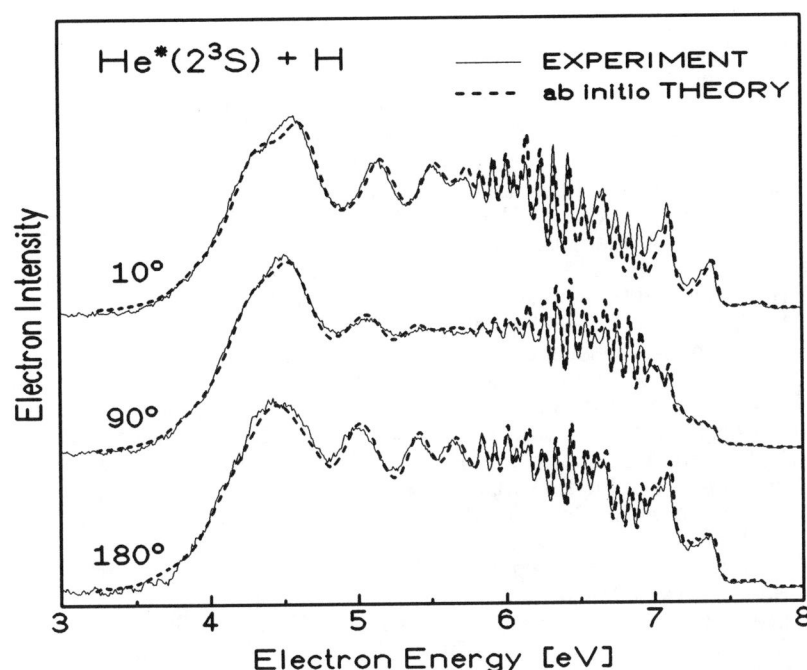

Fig. 8. Experimental spectra for the indicated detection angles compared to the theoretical work completely based on ab initio data.

even in collision systems where (classical) recoil effects are negligible [5,15,16].

REFERENCES

1. H. Morgner, Comments At. Mol. Phys. **21**, 195 (1988)
2. A. Niehaus, Phys. Rep. **186**, 149 (1990)
3. J. Weiner, F. Masnou-Seeuws, A. Giusti-Suzor, Adv. At. Mol. Phys. **26**, 209 (1990)
4. a) M.W. Müller et al., Phys. Rev. Lett. **59**, 2279 (1987)
   b) M.W. Müller et al., Z. Phys. D **21**, 89 (1991)
5. A. Merz et al., Chem. Phys. **145**, 219 (1990)
6. K. Mitsuke, K. Kusafuka, K. Ohno, J. Phys. Chem. **93**, 3062 (1989)
7. A. Khan, H.R.Siddiqui, P.E. Siska, J. Chem. Phys. **94**, 2588 (1991)
8. A. Khan, H.R. Siddiqui, P. E. Siska, J. Chem. Phys. **95**, 3371 (1991)
9. J.P.J. Driessen et al., Phys. Rev. A **42**, 4058 (1990)
10. a) X. Urbain et al., Phys. Rev. Lett. **66**, 1685 (1991)
    b) X. Urbain, A. Cornet, J. Jureta, J. Phys. B **25**, L 189 (1992)
11. B. Brunetti and F. Vecchiocattivi, in "Current Topics in Ion Chemistry and Physics", C.Y. Ng (Ed.), Wiley and Sons (in press)

12. a) P.D. Lett et al., Phys. Rev. Lett. **67**, 2139 (1991)
    b) P.S. Julienne, R. Heather, Phys. Rev. Lett. **67**, 2135 (1991)
13. J. Fort et al., Phys. Rev. A **18**, 2063 (1978)
14. R.H. Neynaber, S.Y. Tang, J. Chem. Phys. **69**, 4851 (1978)
15. H. Morgner, A. Niehaus, J. Phys. B **12**, 1805 (1979)
16. H. Waibel, M.-W. Ruf, H. Hotop, Z. Phys. D **9**, 191 (1988)
17. A. Merz et al., Chem. Phys. Lett. **160**, 377 (1989)
18. M.-W. Ruf, A.J. Yencha, H. Hotop, Z. Phys. D **5**, 9 (1987)
19. T. Ebding and A. Niehaus, Z. Phys. **270**, 43 (1974)
20. D.A. Micha and H. Nakamura, Phys. Rev. A **11**, 1988 (1975)
21. V. Hoffmann and H. Morgner, J. Phys. B **12**, 2857 (1979)
22. A. Niehaus, Adv. Chem. Phys. **45**, 399 (1981)
23. H. Hotop and A. Niehaus, Chem. Phys. Lett. **8**, 497 (1971)
24. A. Le Nadan et al., J. Physique **43**, 1607 (1982)
25. A. Hertzner et al., Abstracts of Contributed Papers, 14th ICPEAC (M.J. Coggiola, D.L. Huestis, R.P. Saxon, Eds.), p. 364, Palo Alto, 1985
26. A.P. Hickman, A.D. Isaacson, W.H. Miller, J. Chem. Phys. **66**, 1483 (1977)
27. T. Bregel et al., Z. Phys. D **13**, 51 (1989)
28. J. Lorenzen et al., Z. Phys. A **310**, 141 (1983)
29. W.H. Miller, J. Chem. Phys. **52**, 3563 (1970)
30. W. Meyer, M. Movre (to be published)
31. A. Hertzner, Dissertation, Univ. Witten (1990, unpublished)
32. A. Merz, M.-W. Ruf, H. Hotop, Phys. Rev. Lett. **69**, 3467 (1992)
33. a) B. K. Sarpal et al., J. Phys. B **24**, 1851 (1991), ibid. 3685 (1991)
    b) B. K. Sarpal, J. Phys. B (1993) submitted

We acknowledge support of this work by the *Deutsche Forschungsgemeinschaft* through *Sonderforschungsbereich 91* and grant *Ho 427/19-1*. We thank B. K. Sarpal for sending a preprint of ref 33 b.

AUTHOR INDEX

A

Ali, R., 585
Allegrini, M., 635
Andersen, N., 505
Andersen, L. H., 432
Andersson, L. R., 595, 836
Armour, E. A. G., 478
Avaldi, L., 61

B

Bachau, H., 547
Badnell, N. R., 415
Barat, M., 537
Bartschat, K., 251
Becker, K., 234
Belkacem, A., 844
Ben-Itzhak, I., 585
Bjerre, N., 779
Bosch, F., 3
Bossingham, R., 844
Bottcher, C., 491
Briand, J. P., 731
Briggs, J. S., 221
Brion, C. E., 350
Broström, L., 803
Brunetti, B., 623
Buck, U., 719
Burgdörfer, J., 595, 786, 836
Burke, P. G., 26

C

Campbell, E. E. B., 697
Carillon, A., 208
Carlson, M., 803
Charlton, M., 455
Charron, E., 127
Cocke, C. L., 585
Crowe, A., 286
Cvejanovic, S., 390

D

Dalberth, M. J., 675
Danared, H., 803
Darrach, M., 811
Das, J., 766
Datz, S., 491, 803
Dawber, G., 61
Dhez, P., 208
Dittner, P. F., 491
Dunning, F. B., 371
Dyer, M. J., 779

E

Ehlich, R., 697
Ehrenreich, T., 828
Ellis, K., 61

F

Falcinelli, S., 623
Faifman, M. P., 478
Feinberg, B., 844
Filevich, A., 803
Folkerts, L., 741
Furst, J. E., 276

G

Gaboriaud, M. N., 537
Gallagher, T. F., 173
Gao, H., 491
Gavrila, M., 115
Gay, T. J., 276
Gerlich, D., 607
Giese, J. P., 585
Giusti-Suzor, A., 127
Goedtkindt, P., 208
Gorczyca, T. W., 415
Gould, H., 844

Griffin, D. C., 415
Guet, C., 820

H

Hall, R. I., 61
Hansen, J. E., 794
Hansen, S. B., 828
Harston, M. R., 478
Haugen, H. K., 105
Havener, C. C., 741
Helm, H., 779
Hertel, I. V., 697
Horsdal-Pedersen, E., 828
Hotop, H., 852
Huber, B. A., 820
Huestis, D. L., 779
Hughes, I. G., 741
Hutton, R., 491
Hvelplund, P., 709

J

Jaeglé, P., 208
Jalabert, D., 820
Jamelot, G., 208

K

Kabachnik, N. M., 73
King, G. C., 61
Kirby, K. P., 48
Klisnick, A., 208
Korsch, H. J., 339
Kosloff, R., 152
Krause, H. F., 491
Källberg, A., 803

L

Laricchia, G., 455
Larsson, M., 803
Lavollée, M., 139
Leone, S. R., 675

Limburg, H., 766
Linder, F., 654

M

MacAdam, K. B., 183, 828
MacDonald, M. A., 61
MacGillivray, W. R., 296
Mannervik, S., 803
Martín, F., 316
Marxer, H., 198
Maurel, M., 820
McConkey, A. G., 61
McConkey, J. W., 811
McCurdy, C. W., 360
Melchert, F., 574
Merz, A., 852
Meyer, F. W., 741
Meyer, M., 139
Meyer, W., 852
Meyerhof, W. E., 844
Mies, F. H., 127
Milosevic, S., 635
Morgenstern, R., 766
Morin, P., 139
Mota-Furtado, F., 198
Movre, M., 852
Mowat, J. R., 803
Muller, H. G., 115

O

O'Mahony, P. F., 198
Olson, R. E., 520
Overbury, S. H., 741

P

Pahler, M., 83
Phaneuf, R. A., 405
Pindzola, M. S., 415

R

Reinhold, C. O., 786
Renn, M. J., 173
Rensfelt, K. G., 803
Richard, P., 585
Ristori, C., 820
Rocco, J. C., 820
Roncin, P., 537
Ruf, M.-W., 852
Rus, B., 208

S

Saeed, M., 779
Sanche, L., 381
Sato, Y., 665
Schuch, R., 491
Schöne, H., 585
Shevelko, V. P., 558
Simon, M., 139
Smith, C. J., 675
Smith, K. A., 371
Spain, E. M., 675
Standage, M. C., 296
Strayer, M., 491
Stöckli, M., 585
Sundström, G., 803
Szilagyi, Z., 537

T

Takagi, H., 442
Takayanagi, T., 326
Tarnovsky, V., 234
Thomson, D. S., 173
Thumm, U., 263

U

Ueda, K., 93
Ugglas, M. af, 803
Ullrich, J., 520
Urbain, X., 645

V

Vaeck, N., 794
van der Hart, H. W., 794
Van Zyl, B., 684
Vane, C. R., 491
Vecchiocattivi, F., 623

W

Wang, J., 520
Ward, S. J., 466
Westerburg, M., 697
Wijayaratna, W. M. K. P., 276
Winter, H., 751
Wu, W., 585

Z

Zehner, D. M., 741
Zeitoun, P., 208
Zeijlmans van Emmichoven, P. A., 741
Zetner, P. W., 306
Zimmer, M., 654
Zubek, M., 61